THE
FISHES OF OHIO

THE
Fishes of Ohio

WITH ILLUSTRATED KEYS

Revised Edition

By

Milton B. Trautman

Emeritus Professor of Zoology
College of Biological Sciences, Ohio State University

The original paintings used as color illustrations
are furnished by courtesy of the Ohio Department of Natural Resources

Ohio State University Press
in Collaboration with the
Ohio Sea Grant Program
Center for Lake Erie Area Research
1981

Library of Congress Catalogue Card Number 80-29521
International Standard Book Number 0-8142-0213-6
Revised Edition, 1981

Library of Congress Cataloguing in Publication Data

Trautman, Milton Bernhard, 1899–
The fishes of Ohio.

Bibliography: p.
Includes index.
1. Fishes—Ohio. 2. Fishes—Identification. 3. Fishes—Ohio —Identification. I. Title.
QL628.03T7 1980 597.092′9771 80-29521
ISBN 0-8142-0213-6

The original 1957 edition was published in collaboration with the Ohio Division of Wildlife and the Ohio State University Development Fund.

To Mary

Contents

Maps

Figures

Outline Drawings for "Key to Families of Ohio Fishes"

Colored Plates

(Following page 739)

Foreword to the Revised Edition

As I begin to write this foreword to the revised edition, my thoughts drift back twelve years to student days at Franz Theodore Stone Laboratory and to the ichthyology class where I first met Professor Milton B. Trautman. Together Milt and Mary Trautman introduced the class to the fascinating study of fish through their marvelous book, *The Fishes of Ohio.* The original 1957 edition had been sold out by that time and the Laboratory preciously guarded the 20 copies it held for class use. Today, the Laboratory is still using the same 20 volumes of the first edition—a tribute to the timelessness of this monumental treatise. This is just one of countless tributes to the excellence of their work.

In 1957, some 35 years of work culminated in the publishing of *The Fishes of Ohio.* The original edition of 683 pages contained descriptions, detailed drawings, distribution maps, and habitat information for 172 forms of Ohio fish. This book was quick to gain the acclaim of ichthyologists worldwide. It provided documentation of the state's fish resources with a degree of precision and accuracy unequalled by any other study on the continent. Authorities called it a classic regional treatise that was the most complete ever published. The limited number of copies (5,000) printed by the Ohio State University Press were sold out within ten years. Demand for this publication has been enormous in the past ten years causing used book prices to rise over 25 times the original new book price of $6.50.

Dr. Trautman did not favor a simple reprinting of *The Fishes of Ohio*, instead, in the late 1960's he and Mary began work on a complete revision of the book. The revised edition contains several additional species and subspecies, revised distribution maps for all fish, more detailed habitat and life history information, and recent ecological findings. The new edition includes 179 forms of Ohio fish (166 species and 13 subspecies). Nothing has been eliminated from the first edition. The primary effort was to bring the book up to date by adding the fish data acquired in the 23 years since the original publication. The revised manuscript has been skillfully molded into this printed form through the editorial efforts of Jane C. Shaw, of Woods Hole, Massachusetts, and Robert S. Demorest, of the Ohio State University Press.

The high standards set by Dr. Trautman in creating this revised edition of *The Fishes of Ohio* can serve as an inspiration to all students of ichthyology, young and old, amateur and professional. After 60 years of work as a field biologist, Dr. Trautman's enthusiasm has not waned, as is demonstrated by his painstaking efforts to gather data from all reliable sources for this revision. His professional experience spans that period of Ohio's agricultural and industrial development in which many aquatic habitats have been radically altered. One of the great values of his habitat descriptions, in addition to characterization of present habitats, is documentation of optimal conditions for each species. This information will prove invaluable to future stream and lake restoration efforts.

Publication and distribution of this book have been made possible through a grant to the Ohio Sea Grant Program from the Office of Sea Grant, National Oceanic and Atmospheric Administration, U.S. Department of Commerce. Resources of the Ohio State University, College of Biological Sciences, Graduate School, Museum of Zoology, University Press, and Center for Lake Erie Area Research and the Ohio Department of Natural Resources, Division of Wildlife have been graciously provided to make the production of *The Fishes of Ohio*, Revised Edition, a reality.

CHARLES E. HERDENDORF
Franz Theodore Stone Laboratory
The Ohio State University
Put-in-Bay, Ohio

Foreword to the First Edition

Over a period of forty years or more I have often heard this naturalist or that—occasionally myself—characterized by some disciple of the rising school of laboratory biologists as "the last of the naturalists." Milt Trautman, as the author of this grand volume is familiarly known to his host of friends, has, I presume, also been so dubbed, for he is indeed a great field naturalist. But the tribe lives on, and will continue to flourish, long after many present-day theories in biology have been abandoned.

That natural history still deserves a prominent place in the affairs of science and education, that natural history itself has expanded and evolved, that natural history commands the interest and concern of the general public as well as the scientist, the author has abundantly demonstrated in this excellent book on the fishes of Ohio.

In this treatise Dr. Trautman presents us with the results of half a lifetime of devoted research—research that has encompassed collecting trips to every stream in the state; critical studies of many thousands of specimens, representing the 172 kinds of fish; appraisal of everything others have written on the subject; studies of the present hydrographic conditions, such as stream gradient and flow, that affect the occurrence and distribution of the various species; a scholarly and detailed interpretation of the vast changes in fish life that have occurred during two full centuries of human use (and all-too-often misuse) of the land and water; and an able review of the evidence on the even more dramatic changes in water and in fish faunas that resulted from the ebb and flow of the continental glaciers during the million-year-long Pleistocene epoch.

I have watched this project grow from humble beginnings through a long-continued and well-supported research program to this finished monograph. I often talked with Mr. E. B. Williamson and Dr. Raymond C. Osburn about their pioneering work, conducted with little help and almost no financial support, at the turn of the century. In the early stages of my own ichthyological career, when science was still largely a matter of individual effort, I cooperated to some extent with Mr. Edward L. Wickliff in his resumption of the Ohio fish survey, under the stimulus of Dr. Osburn. Since 1925, when Dr. Trautman began his active participation, and on through the long years while he painstakingly continued and developed the project (first between 1925–1934 when employed by the Ohio Division of Conservation; then, as my colleague, at the University of Michigan Museum of Zoology between 1934 and 1939; and, finally, as a staff member of the Ohio State University both at the Franz Theodore Stone Laboratory, under the directorship of Dr. T. H. Langlois, and the Department of Zoology and Entomology on the campus), I have had the privilege and pleasure of lending the project moral support and constant encouragement, occasionally a helping hand. So may I be pardoned in assuming a measure of pride in this distinguished contribution to ichthyology, ecology, hydrography, economic zoology, and good, sound natural history in general.

Long live such naturalists as Milton B. Trautman!

Carl L. Hubbs
University of California
La Jolla, California

Jacket Note from the First Edition

Surface water and the fishes that live within it are inseparable natural resources of great economic and aesthetic importance to people everywhere. Use of both resources has greatly accelerated. This is particularly true of industrial states such as Ohio which are placing heavy demands on naturally flowing waters for agricultural, industrial, domestic and recreational purposes. Use of water for recreation, particularly fishing, is sometimes adversely affected by other uses. Under wise management this need not always occur. To manage water wisely for recreational fishing we need to know how and to what extent these uses are impaired. We need also to be able to identify fish resources specifically and to accurately measure their living requirements. *The Fishes of Ohio* attempts to do these things for the sportsman, the commercial fisherman and every citizen with an interest in conservation of our natural resources. In this book, the author traces the history of the state's surface waters from the original clear vegetated natural impoundments and clear, shaded sand or rubble-bottomed streams to the present poor condition of the natural and artificially impounded water.

We are now in an era in which man is attempting to apply the technologies of modern science through new social tools to provide better recreational use of our waters. New laws and techniques for pollution control, the organization of watershed conservancy districts to control floods and store water for future use, liberalized fishing regulations, artificial stocking of fish predators, and use of specific toxicants to control fish populations are part of this movement. There are already signs that these efforts may have a profound effect on the future availability and use of these inseparable resources of fish and water. This book, which interprets the past and gives an insight into the future, is a milestone by which future progress can be measured.

Charles A. Dambach
Director, Natural Resources Institute
Ohio State University

Acknowledgments

First Edition

Since 1925 more than 400 individuals have directly or indirectly contributed some valuable information which has been incorporated into this report. Much as I wish to mention each by name and to thank them individually, such procedure is impractical because of the space required, and impossible because during the many years I have, unwittingly, forgotten many of them or their names have become buried in the huge mass of data; in some cases there were contributors whose names I never even heard. It seems more logical therefore to express to this host of naturalists, biologists, teachers, curators, librarians, conservation workers and officials, typists, fishermen, farmers, and even small boys, my sincere thanks for their contributions, and to restrict personal mention to those who have contributed the most or who have been most instrumental in making possible this report.

From 1922 until his death in 1930, the late Professor James Stewart Hine (Kennedy, 1931: 510–11) gave invaluable assistance. It was he who first urged me to undertake the writing of a fish report, who first showed me how to acquire the necessary grasp of the literature, and who was ever ready with instruction and advice. He spent many hours in blazing sun, snowstorm and midnight rain, helping me to seine and to collect survey data, just as more than a quarter of a century before he had aided Dr. R. C. Osburn to acquire some of the data upon which was based his "Fishes of Ohio."

Since 1925 Edward L. Wickliff has contributed much, by helping to obtain data, directly or indirectly, by obtaining funds for equipment and by much helpful advice.

Since 1925 Dr. Edward S. Thomas has been ever willing to offer encouragement, a helping hand and sound advice.

Between 1925 and 1932 Robert Browning Foster was my frequent companion in the field, spending hundreds of hours with me in the arduous labor of seining Ohio waters.

The thought-provoking suggestions and criticisms of Dr. Charles F. Walker throughout the accumulating of data and writing of the first draft of the manuscript are deeply appreciated.

Since 1925 Dr. Carl L. Hubbs has unstintingly given countless hours from his busy life in guiding, advising and teaching me, and in aiding in the identification of thousands of fishes. His influence appears on every page.

In 1930 the late Mrs. Lydia M. (Hart) Green made for the Ohio Division of Wildlife and under my supervision, the colored paintings of the 21 species of Ohio fishes. Several years previously she had painted many of the fish plates which appeared in "The Fishes of Illinois" (Forbes and Richardson, 1920).

Between 1930 and 1934 Dr. Thomas Huxley Langlois offered much encouragement. In 1939 he secured my appointment to the staff of The Franz Theodore Stone Laboratory and throughout his directorship of this Laboratory offered encouragement and assistance in the prosecution of the project.

Between 1943 and 1946 Mrs. Owen B. Weeks assisted with the first draft of some of the drawings of fishes and in outlining with ink some of the final drafts. She assisted on those drawings which carry the initials "E.R.W."

Between 1950 and 1955 Dr. Charles A. Dambach, as Chief of the Wildlife Division of the Ohio Department of Natural Resources, and his assistant, Lee S. Roach, gave helpful advice and encouragement. Since October, 1955, Dr. Dambach, as Director of the Natural Resources Institute of The Ohio State University, has been responsible for the securing of the funds necessary for the publication of this report and its supervision through the press.

Dr. Frederic W. Heimberger, Vice President of The Ohio State University, has been particularly helpful in giving encouragement and guidance and in securing means for publication.

Since 1940 my wife, Dr. Mary Auten Trautman, has spent countless hours assisting me, mostly in the collecting of fish specimens; in transcribing my field notes, descriptions and measurements; in checking thousands of locality records and in spot mapping them; in the final editing and proof reading of the manuscript and in relieving me of much office routine so that I might have more time to devote to this report.

To these persons I am most grateful, and to them I offer my heartfelt thanks. Without their help this report could not have been completed in its present form.

I wish to thank Dr. Richard P. Goldthwait for reading, and giving advice upon, the section concerning the glacial and early post-glacial history of Ohio; Dr. Earl L. Green for advice upon the section on hybridization; Alvin Staffan, staff artist, Ohio Division of Wildlife, for designing the cover; John H. Melvin for authority to use four maps from Bulletin 44 of the Geological Survey of Ohio (Maps II, III, VII and X); Professor A. K. Lobeck for authority to use the eastern section of his "Physiographic diagram of the United States" (Map IV); Professor Reuel B. Frost for authority to use his "Physiographic map of Ohio" (Map V); Dr. Guy-Harold Smith and the *Geographical Review* for permission to use the map of "Relative relief of Ohio" (Map VI); the Museum of Zoology of the University of Michigan for use of a portion of its stream map of the United States (Map I); the Ohio Historical Society for permission to quote from some of the unprinted manuscripts in its possession.

Additionally, I am most grateful to the Ohio State University and the Ohio Division of Wildlife for the financial support which has made completion of this project possible and to the Ohio State University Development Fund for its contribution to the cost of publication.

Lastly I wish to thank Dr. James E. Pollard, Director of the Ohio State University School of Journalism, who has taken time from his busy schedule to edit this book, and Mrs. Ardath Moore, who has spent many hours outside of her official duties in reading proof to expedite this publication.

M.B.T.

Revised Edition

Since 1955 several hundred individuals have, in some manner, contributed information which has been used in this report. To this host of individuals, who have not been personally mentioned, I wish to express my sincere appreciation.

The late Dr. Charles A. Dambach, as he had done before 1955 and continued until his death in 1969, assisted in innumerable ways in aiding my efforts to obtain information relative to the current changes in the distribution and populations of fishes in Ohio. He was instrumental in making possible the extensive field work by a group of fishery students and myself, which resulted in hundreds of collections of fishes and data obtained primarily from the Scioto River watershed.

Many present or former Ohio State University faculty members and colleagues have assisted in various ways, especially Drs. Tony J. Peterle, Charles E. Herdendorf, III, David H. Stansbery, Ted M. Cavender, N. Wilson Britt, Bernard L. Griswold, Charles C. King, Walter T. Momot, Loren S. Putnam, Carol B. Stein, Roy A. Stein, Ronald L. Stuckey, Stephen H. Taub, Richard A. Tubb, and George W. Wharton, Jr.

Many former or present members of the Ohio Department of Natural Resources have contributed much, including Darrell F. Allison, Eric R. Angle, D. Barry Apgear, Daniel C. Armbruster, George Billy, Denis S. Case, Clarence F. Clark, Clayton H. Lakes, Carl L. Mosley, David F. Ross, and Russel L. Scholl.

It is with much pleasure that I acknowledge the considerable assistance, some of which was voluntary, of former students, several who called themselves "Riffle Rangers," including Roger K. Burnard, Raymond F. Jezerinac, Bruce May, E. Bruce McLean, Donald I. Mount, Richard Moerchen, James M. Norrocky, William L. Pflieger, George J. Phinney, and K. Roger Troutman.

Another group of collaborators was on the faculty of various colleges, and public and private schools; some were fishery personnel and others were in related occupations: all contributed materially. Outstanding in this group were the late Dr. Gerald C. Acker and Dr. Jane L. Forsyth of Bowling Green State University, Ronald R. Preston and Harry D. Van Meter of the U.S. government, and Dr. Andrew M. White of John Carroll University, who with several of his students, Mark S. Caroots, Miles M. Coburn, Eric J. Foell, Michael P. Kelty, and others, investigated the fishes of Lake Erie and its tributaries in northeastern Ohio. Daniel Margulies and Orville Burch, members of Wapora, Inc., supplied locality records of fishes captured near five generating stations along the Ohio River and adjacent to the state of Ohio. Other contributors in various capacities were William M. Clay, Carter R. Gilbert, F. George Griffith, Charles F. Hodgen, Edward F. Hutchins, Carolyn Jenkinson (nee Cooper), Marshall A. Moser, Jr., Arnold J. Stoltz,

Donald Veth, Bernard J. Zahuranic, and Richard A. Zura.

I especially wish to most sincerely thank those primarily responsible for the editing of this publication: Robert S. Demorest, Weldon A. Kefauver, and Jane C. Shaw; and Bill Snyder, who assisted in readying the manuscript for the printer.

Finally I am most deeply indebted to my wife, Mary Auten Trautman, who for many years assisted in the collecting of fishes, recording of data, in handling correspondence, preparing manuscripts, and proofreading. Without her assistance and encouragement the completion of this volume would have been impossible.

Preface

The recording of facts concerning the past and present distribution and abundance of fishes in what is now Ohio started about 1730, when the first English traders began to invade the Ohio Country. As customary with all fact recording, data accumulation has not progressed at a steady rate but has occurred in an irregular fashion. The periods of data gathering have coincided with, and have been largely the result of, the efforts of certain individuals. So long as one or more individuals were interested, data were accumulated and progress was rapid. There were several periods, however, when no one appeared to be interested. Then little progress was made, such as during the periods between 1850–70 and 1902–20.

A number of persons, through their publications, have contributed considerably to our knowledge of Ohio fishes. Following are the more important contributions:

Between 1818–20, C. S. Rafinesque published the first important work upon fishes of the Ohio River drainage. Between 1838–54, Jared P. Kirtland, in a series of excellent publications, first listed the fishes then known to occur in Ohio, giving a definite conception of distribution and abundance prior to 1850.

Between 1874–78, John H. Klippart reported upon many of the important food and game fishes. In 1882, David Starr Jordan produced his "Report on the Fishes of Ohio" which listed the species of fishes known or believed to occur in Ohio. Unfortunately his evidence was largely obtained indirectly, therefore much of it is hearsay, rather than factual evidence.

In 1888–89, James A. Henshall, in his "Contribution to the Ichthyology of Ohio," added several species to the ever growing list of Ohio fishes and contributed some data upon distribution and abundance.

Between 1890–1900 several studies appeared pertaining to sections of Ohio, of which three are outstanding. In 1892, L. M. McCormick published an annotated list of the fishes present in Lorain County. In 1895, Philip H. Kirsch contributed much concerning the fishes of the Maumee River system; this is an outstanding work. In 1898, E. B. Williamson and R. C. Osburn gave us their excellent report upon the fishes occurring in Franklin County. In 1901, "The Fishes of Ohio" appeared. This important contribution by R. C. Osburn summarized much of the data of his own work and that of others.

The present report is an attempt to bring together, as completely as possible, facts obtained prior to 1920, and to incorporate these with the great amount of data, largely unpublished, which has been accumulated later. Since 1920 hundreds of individuals have aided in the accumulation of data, as have many state agencies. About 1920 the State Conservation Department (forerunner of the present Ohio Department of Natural Resources) inaugurated a series of stream surveys in which it was assisted by faculty members and students of The Ohio State University. This cooperative effort by the two agencies has continued until the present time, to the benefit of the public.

Since the preparation of the original edition, during the early 1950s, there has been an ever-increasing interest by researchers, students and the general public in the distribution, abundance, and commercial and sporting importance of the fishes in the waters of Ohio. Interest has markedly increased in the teaching of ichthyology in the universities, colleges and high schools of Ohio. There has been a correspondingly large annual increase in the number of students and others interested in the many phases of ichthyology and related aquatic disciplines. Fish surveys were begun in many streams, ponds, lakes and Lake Erie, which have been partially or entirely reported upon by university and college faculties and students. These individuals have given me their data to incorporate with mine and that of my students for this report. This increased interest in fishes has been in part the result of the aroused awareness of the continued erosion of the natural environment in Ohio and its effects upon the fish populations.

The present edition has several objectives:

(1) To demonstrate the changes in distribution and abundance of Ohio fishes which have occurred between the years 1750–1980, and to present reasons for these changes.

(2) To provide three keys for the identification of the 166 species and 13 additional subspecies known to be present within the state during the 1750–1980 period. Two of the keys were designed principally for students of ichthyology and are of the more conventional type. The third, illustrated with a figure of each species and occasional anatomical drawings, was made as simple as possible in the hope that it might aid the layman in identifying the various species.

MAP I. Geographical position of the State of Ohio and its two major drainages

Introduction

Part I

Location and Size of Ohio

The State of Ohio, situated in the midlands of the United States, is squarish in outline. It consists of approximately 44,803 square miles (116,040km^2), extends approximately 225 miles (362km), in an east-west direction and 210 miles (338km) in a north-south direction, and is situated between 38° 27″ and 41° 57″ N. latitude and 80° 34″ and 84° 49″ W. longitude (map I). Part of its northern border bisects Lake Erie; its southern border is the low water mark along the northern shore of the Ohio River. The northern third of the state lies within the Great Lakes-St. Lawrence watershed; the southern two-thirds is in the Ohio-Mississippi watershed. The eastern part of the state is in that province generally known topographically as the Appalachian or Allegheny Plateau; this section is generally hilly or at least rolling. The western portion is in that province known as the Mississippi Valley Plain; it is quite level or of a gently rolling sag and swell topography (Stout, Ver Steeg and Lamb, 1943:49).

Events prior to 1750

In this report are numerous references to pre-, inter-, and post-glacial conditions and their possible effects upon the recent distribution and abundance of fishes. So as to present these conditions and their effects more fully, it seems feasible to outline them briefly in their chronological sequence.

Bedrocks underlying Ohio

All of the rocks which crop out in the State are sedimentary in nature, having been formed beneath waters during the Paleozoic era (Fenneman, 1938: Fig. 30, p. 122). The older rocks are predominantly limestone, were formed during the Ordovician, Silurian, and Devonian periods which presumably occurred between 305,000,000 and 550,000,000 years ago, and crop out chiefly in the western half of Ohio. Above these older rocks are more recent ones which consist primarily of sandstones and shales; these rocks were formed during the Mississippian, Pennsylvanian, and Permian periods which occurred presumably between 195,000,000 and 305,000,000 years ago. They crop out principally in the eastern half of the State (Carman, 1946: 242–43).

Glacial Invasions

Deep Stage Drainage.—The glacial invasions occurred during the Pleistocene period of the Cenozoic era, between 10,000 and 1,000,000 years ago. Prior to the first known glacial invasions much of the Ohio Country was drained by the huge Teays River system. The Teays entered Ohio from West Virginia, flowing northward through Scioto County, probably continuing northwestward through Ross, Clark and Mercer counties until it entered Indiana. Many headwater streams in unglaciated Ohio, now in existence, were present during Teays time (map II).

Kansan or Pre-Kansan Invasion.—This is the first recognized glacial invasion. The meager evidence now obtainable suggests that this glacier invaded northern Ohio, coming far enough southward to block the mighty Teays in the vicinity of Lake St. Marys. This blocking resulted in bringing about great drainage modifications.

Lake Tight.—Wolfe (1942: fig. 1) has indicated the presence, at this time, of a large lake which flooded the lowland valleys of unglaciated Ohio and portions of neighboring states to an elevation of 900 feet above sea level.

During and after Lake Tight time the Ohio River, known to geologists as the Cincinnati River in part and the Pomeroy River in part, was partially formed, and much of the present upper Muskingum, and most of the Scioto drainages were connected to form the interglacial Newark River system. The Newark River flowed southward through much of the present lower Scioto River valley to its confluence with the major stream (Stout, Ver Steeg and Lamb, 1943: map opp. p. 78).

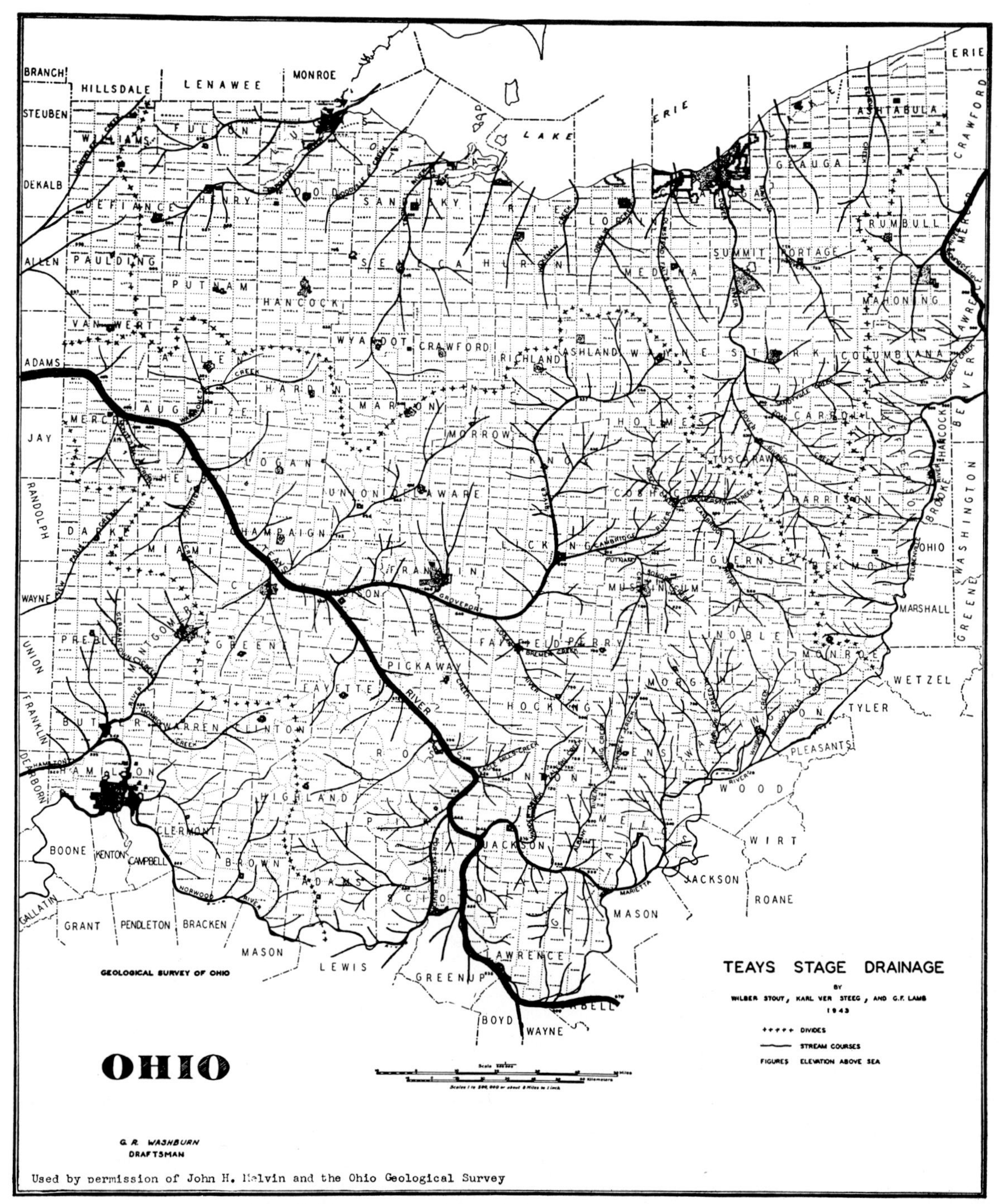

MAP II. Stream systems before the first known glacial invasion

Illinoian Invasions.—This glacier leveled the surface and/or buried in drift all of north-central and western Ohio, an area of approximately 23,000 square miles (59,570km^2).

Post-Illinoian Stage.—Again great modifications occurred in the drainage systems, for only those drainages in the unglaciated portion were not eliminated by glacial action (Stout, Ver Steeg and Lamb, 1943: map opp. p. 86).

Wisconsin Invasions.—This is the last of the glacial epochs, and because it was the last it had a far greater influence upon the recent fish fauna than

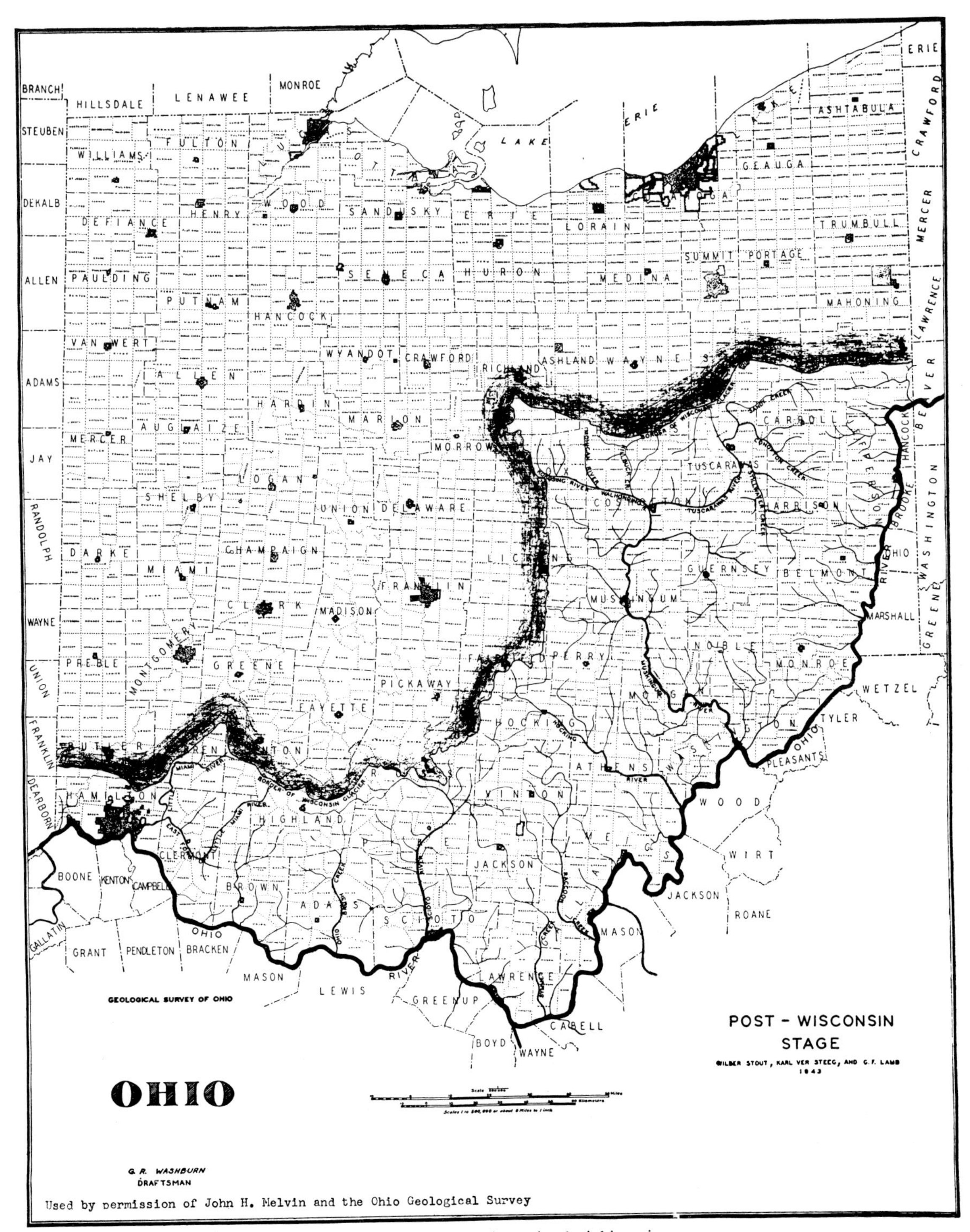

MAP III. Extent of the Wisconsin glacial invasions

had any earlier one. The Wisconsin invasions did not extend as far south in southwestern Ohio as did the Illinoian; the Wisconsin approximated the Illinoian in central Ohio but continued beyond the Ohio-Pennsylvania state line in northeastern Ohio. These invasions covered approximately 26,463 square miles (68,539km^2) in Ohio. During this glaciation the Muskingum River system was essentially the same as it is today and it, the Hocking and lower portions of the Scioto, Little Miami and Great Miami rivers were outlets for the glacial waters (map III, above).

Post-Wisconsin Times.—As the ice disappeared the present Ohio River tributaries were completely established. The Ohio River as the master stream followed rather closely the original course of the Deep Stage Cincinnati and Pomeroy rivers. As the tongues of ice retreated and readvanced periodically northward, a divide became established between the Ohio River drainage and that newly-forming one which eventually became the Lake Erie drainage. During the retreat of the ice and later, several huge glacial lakes existed before the present Lake Erie was established. These lakes left lake beaches, some of which have affected the distribution of Ohio fishes. The more important of these lakes and their beaches are discussed briefly.

Glacial Lake Maumee and the Maumee Beach Ridge.—After the ice had retreated from much or all of inland Ohio and the western third of the present Lake Erie, it was still present in the remainder of the Lake Erie basin. This ice blocked the present eastern flow, causing the impounded waters to rise sufficiently so as to overflow southwestward into the Wabash drainage near the present city of Ft. Wayne, Indiana. This outlet is known as the Maumee or Fort Wayne outlet (Greene, 1935: 16) and the impounded waters as Glacial Lake Maumee. This lake remained a sufficient length of time to form the rather well-defined Maumee Beach Ridge and the extensive lake plains (map XI, below). The other Maumee levels, which formed only minor beaches, apparently were of little significance to the present fish fauna.

Glacial Lake Arkona.—A drastic lowering of Lake Maumee, because of the formation of an outlet from the present Saginaw Bay westward across the state of Michigan to the Glacial Lake Michigan basin, and thence through the Chicago outlet into the Mississippi drainage, resulted in the formation of Lake Arkona. This lake was of comparatively short duration and left no well-defined beach ridge.

Glacial Lake Whittlesey and Lake Whittlesey Beach Ridge.—A readvance of the ice front caused the waters of the Erie basin to rise again, forming a glacial lake which remained sufficiently long to produce the Lake Whittlesey Beach Ridge. This occurred approximately 11,000 years ago.

Glacial Lake Wayne.—Another lowering of the waters occurred when an eastward outlet was formed through the Mohawk and Hudson valleys; it left no prominent beach ridge.

Glacial Lake Warren and Lake Warren Beach Ridge.—Another rise in water levels, caused by the blocking of the Mohawk-Hudson outlet, resulted in a reversal of flow from east to west, again through southern Michigan into the Michigan Basin. This lake remained long enough to form the well-defined Warren Beach Ridge (map XI, below).

Glacial Lake Lundy.—The reopening of the Mohawk outlet, resulting in another lowering of waters, produced this very temporary lake, but before a well-formed beach could be developed the retreating ice opened the present St. Lawrence River outlet. Radiocarbon dating gives the time since the St. Lawrence outlet was opened as 8,500 years ago.

Climatic Condition Previous to 1750

Recently the fossil pollens found in bogs have been studied, resulting in the concept that five major climatic periods have occurred during late Wisconsin times (Deevey, 1949: 1396; Sears, 1941: 225–34; Transeau, 1941: 207–11; Braun, 1934: 252–57). These periods were:

(1) A cool period in which a northern flora was present, during which time such northern species as white* and black spruces were abundant.

(2) This first period was followed by a warm and dry period during which a more southern flora flourished, as indicated by the presence and abundance of pine pollen mixed with that of oaks and hickories.

(3) During this period warm conditions continued but there was considerably more moisture, resulting in the dominance of American beech and eastern hemlock.

(4) A xerothermic period, during which the climate was warm and dry, and there was a preponderance of oak and hickory pollens.

(5) A cooler and more moist period, which has continued until the present, resulting in the possible reinvasion of some northern species of plants.

* Scientific and common names of plants have been taken from Fernald (1950); see index for scientific names of plants, and animals other than fishes.

Possible Presence of a Flora and Fish Fauna in Unglaciated Ohio during Glacial Times

A difference of opinion exists as to whether fishes and some southern plant species were present in unglaciated Ohio during glacial times. Deevey (1949: 1375) is of the opinion that glacial chilling was so severe that warmth-loving species of plants and animals could have survived only as far north in North America as peninsular Florida and Mexico; that they invaded Ohio only after the ice had disappeared and a warm climate prevailed. It is possible that fishes did not invade the state until comparatively recent times, for even as late as 100 years ago during wet periods many species could have crossed the watershed marshes which connected the Ohio River and Lake Erie drainages, and thus invaded the latter drainage.

Others believe that during glacial times conditions within a few miles of the ice front were suitable for many of the plant and animal species which are found in Ohio today. Transeau (1941: 210), Braun (1951: 139–46), and Wolfe (1951: 134–38) believe that a preglacial flora existed throughout the Pleistocene in the Allegheny Plateau within a short distance of the glacial boundaries. Thomas (1951: 153–67) presents evidence supporting the theory that many present day Ohio animals were within a short distance of the glacial boundaries during the Pleistocene.

Goldthwait (1953: 1 and 3), basing his conclusions upon what is happening in Greenland today, states that the average summer temperature is about 50° F (10° C) in the green hills near the Greenland Glacier, and suggests that under similar conditions in Ohio the summers during glacial periods might have been only 10–20 degrees F (-7°— -12° C) cooler than at present. Also, that a temperature of 70°F (21° C) might have been attained "on many summer days."

The present evidence suggests that at least some fish species were present during glacial times in the unglaciated portion of Ohio or immediately to the south of it. Surely the cold-water inhabiting species, such as ciscoes, whitefishes, lake trout, and burbot could have remained close to the ice front during glacial times as they do today in the far north. Greene (1935: 12–18) and many other contemporary workers are of the opinion that fishes followed closely the retreating ice in the Great Lakes region, utilizing the Wabash-Maumee outlet during the Glacial Lake Maumee stage, at a time when the eastern portion of the Erie basin was still blocked by ice.

Effects upon the Fish Fauna by the Five Major Climatic Periods

It seems logical to reason that when warm climatic conditions prevailed during the second and third major climatic periods, that most or all of the fish species known to be present in the Erie basin during historic time were already present, or if not they invaded that basin during one of these warm periods. Also, that during these early periods the paddlefish, bigmouth buffalofish and flathead catfish were more numerous than they have been recently. Hubbs (1940A: 293–97) records the finding of the cranium of a freshwater drum lying upon solid red clay beneath 2′ (0.6 m) of black muck and 4′ (1.2 m) of marl beside Burt Lake, in Cheboygan County of northern Michigan, indicating that this warm-water-inhabiting species must have ranged formerly much farther north than it has during recent historic time. The bones of buffalofishes have been found in a prehistoric village site in Greenville County, Ontario, far north of the present range of this group of fishes, suggesting the possibility that they may have been present farther to the north during an earlier period (Rostlund, 1951: 295).

It seems logical to assume that during xerothermic periods conditions were favorable for the invasion of such western species as the bigmouth shiner, and that with the return of more humid and hence more heavily forested conditions, these plains- and/or prairie-inhabiting forms decreased in abundance again, leaving at present only isolated populations in such areas as Michigan, northern Ohio and New York. During recent years the turbid water-plains type of streams have returned to Ohio because of favorable conditions produced by the white man's agricultural practices; as a result, plains-inhabiting species such as the suckermouth minnow and orangespotted sunfish have been extending their ranges eastward in Ohio during recent years.

Topography

Areas covered by water.—Of the 44,803 square miles (116,040 km^2) which comprise the state of Ohio, 3,450 (8,936 km^2) are covered by the waters of Lake Erie, including Maumee and Sandusky bays.

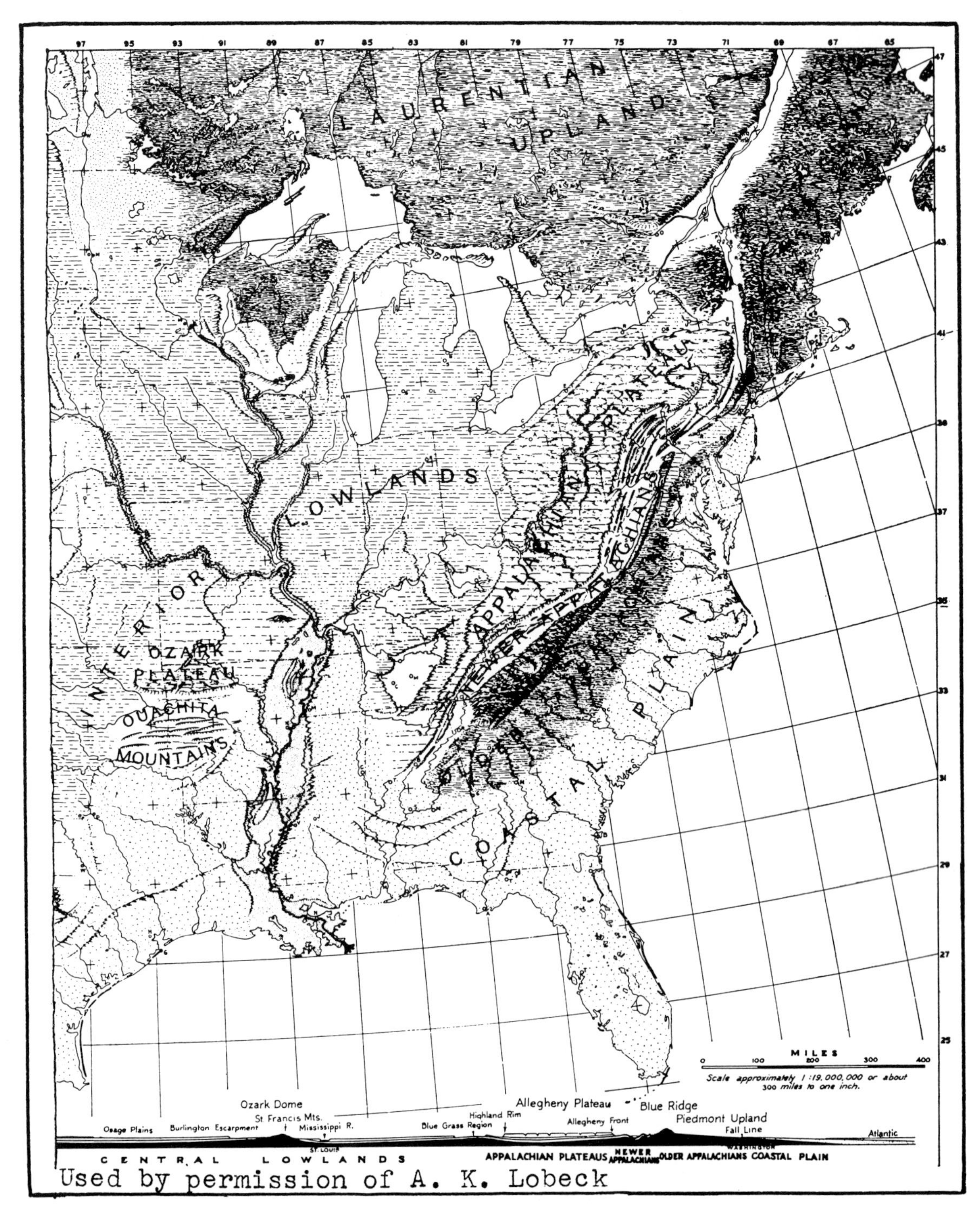

MAP IV. Physiographic diagram of the eastern United States

In 1933, according to Sherman (1933:17), approximately 200 square miles (518 km^2) in inland Ohio were more or less permanently covered by the waters of rivers, ponds, lakes, and other impoundments. By 1950 the number of square miles under water obviously had become much larger, for during the 1933–50 period hundreds of farm ponds, many headwater lakes, and several large industrial

MAP V. Physiographic regions of Ohio

and flood control impoundments were constructed.

Elevations above sea level and relief.—The highest elevation in the state, 1550′ (472 m) above sea level, is at Campbell Hill in Logan County. The lowest elevation, 430′ (131m) above sea level, is at the mouth of the Great Miami River in extreme southwestern Ohio. The mean elevation of Lake Erie is 573′ (175 m).

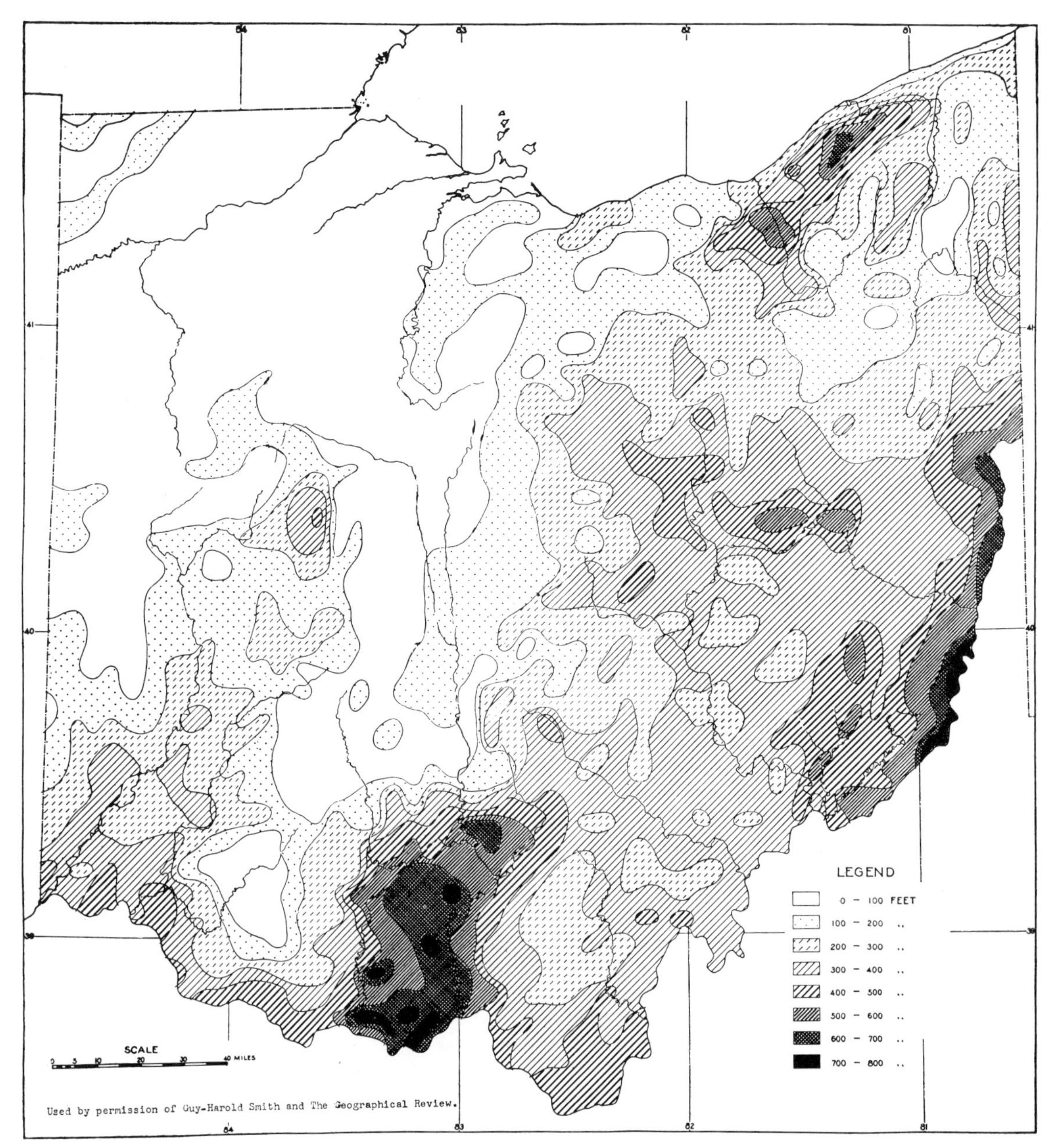

MAP VI. Relative relief of Ohio

The maximum relief in Ohio is relatively slight when compared to the more mountainous states in the Appalachian Plateau (*see* Smith, 1935; 277–84). The northwestern quarter has an especially low relief, for in few areas is the maximum relief greater than 100′ (30 m) (*see* distribution map of southern redbelly dace). The eastern half of the state has a more pronounced relief, for most of it has a maximum relief greater than 200′ (61 m), as have those areas in southern Ohio which are drained by the Little Miami, Mad, and lower portion of the Great Miami rivers (*see* distribution of least brook lamprey). The largest area having a maximum relief greater than 500′ (152 m) is situated in the Blue Grass Region of southcentral Ohio; a smaller area extends in a narrow band along the Ohio River from Washington County to Columbiana County, and very small areas are present in parts of Cuyahoga,

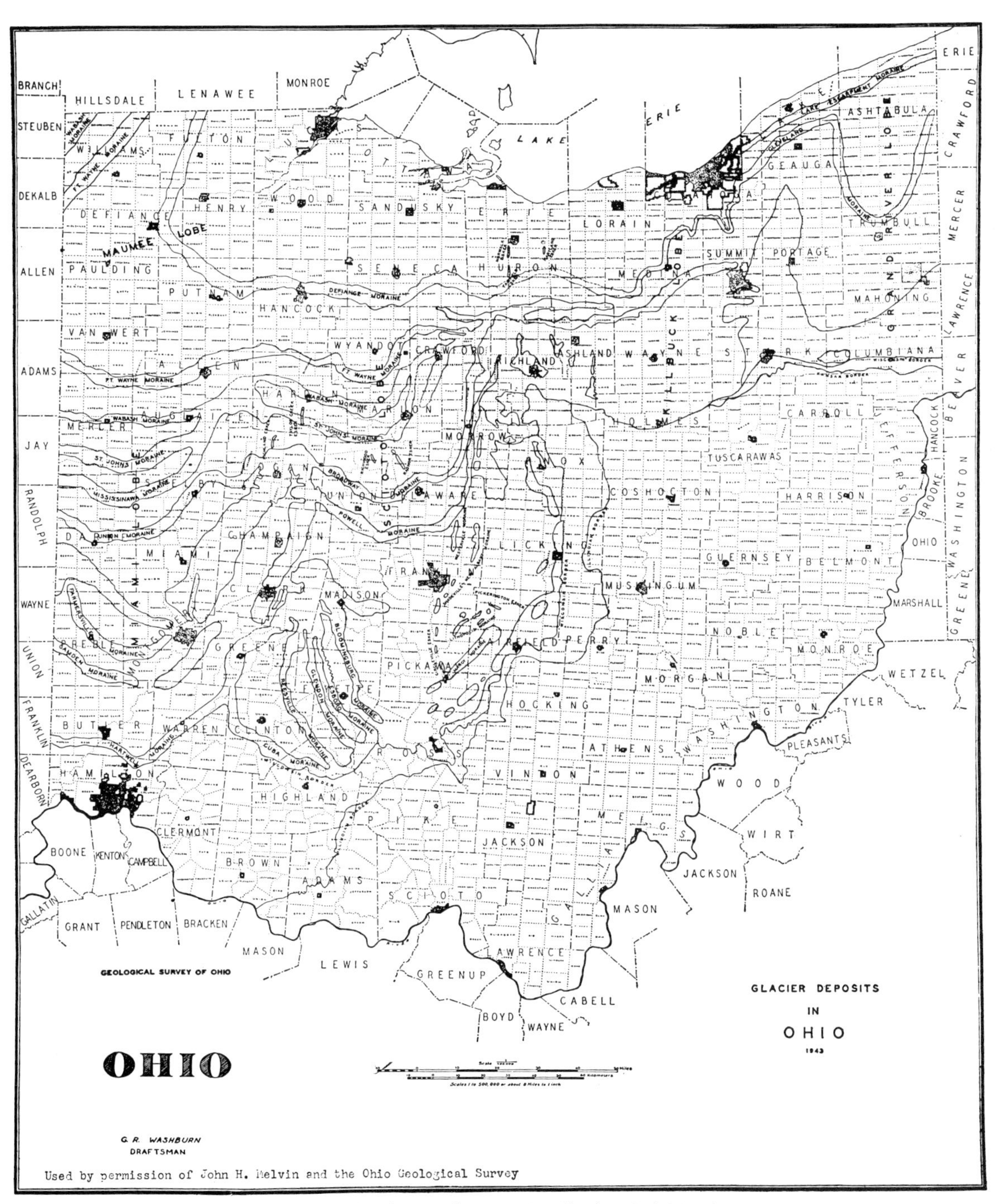

MAP VII. Glacial moraines and the Illinoian glacial border

Summit, Geauga and Lake counties in the northeast, and in Tuscarawas, Washington, Noble, Guernsey and Belmont counties in the east (*see* distribution map of rosyside dace).

The amount of relief appears to be of major importance as a contributing factor in the distribution and abundance of fish species only when it definitely affects the amount of gradient in streams and the speed of eroding the land.

The glaciated portion.—Glacial action has greatly modified the topography in part or all of 62 of the 88 Ohio counties (*see* map V). This glaciated portion readily divides into three easily recognizable types; these are known as the Lake Plains, Till Plains, and the Glaciated Plateau.

Lake Plains.—These plains comprise that area which lies between the southern shore of Lake Erie and the Glacial Lake Maumee Beach Ridge (*see* map V, above). This area was formerly covered by the waters of Lake Maumee, and its soil is comprised chiefly of fine lake silts. In the lake plains are found the largest amount of the most level lands in the state, and only along streams and moraines is much dissection of the topography evident. The streams with few exceptions have very low gradients, a decidedly youthful topography which is the result of intense glacial action. Formerly much of the land contained swamp forests, wet prairies and marshes, and these were flooded during several months of the year. That part of the lake plains which was in the northwest section of the state was known as the "Black Swamp." The area contained a fish fauna dominated by species which required water almost entirely free of clayey silts, and contained much aquatic vegetation; the western banded killifish is an example.

Till Plains.—These plains comprise that part of glaciated Ohio which lies west of the Allegheny Front Escarpment, exclusive of the above-mentioned lake plains (*see* map V). These plains have a sag and swell topography generally, and contain almost all of the large and more prominent moraines found in the state. In the vicinity of these moraines and along some streams the topography is often decidedly hilly, and the maximum relief may be more than 100′ (30m). The streams in these plains range from those having very low gradients such as the St. Marys River in Auglaize, Mercer and Van Wert counties (Trautman, 1942: 216–17) to those of high gradients in the upper Mad River system. The streams of this section have been considerably modified by glacial action and those of moderate or high gradients formerly had their bottoms covered with glacial boulders and sand. Most of the prairies occurred in this section and there was considerable marshy land, especially along the divide between the Lake Erie and Ohio River drainages. This section contained a goodly number of bog ponds and pothole lakes, and springs were formerly abundant, especially in the vicinity of the moraines. The prairies contained a rather characteristic fish fauna of species requiring low or moderate gradients and water free from much turbidity. The creek chubsucker and blackstripe topminnow are excellent examples; note how their distributional patterns outline the original prairies of Ohio. Along the glacial boundary and in the Mad River system the streams contained moderate and high gradients, and originally their bottoms were composed of sand, gravel, bedrock and boulders; clayey silt was virtually absent. These streams contained an abundance of riffle-inhabiting species.

Glaciated Plateau.—This is part of the extensive Allegheny Plateau. In Ohio it is located in the northeastern part north of the glacial boundary and east of the Allegheny Front Escarpment. It is comprised chiefly of rolling, rounded hills. In some sections there is a maximum relief of more than 500′ (152m). The streams show considerable modifications, the result of moderate glacial action. A few streams have low gradients and are swampy; these lie chiefly along the watershed. The redfin shiner is a characteristic species of these low-gradient streams. Most of the streams of the glaciated plateau contain moderate gradients and well developed riffles and are found throughout the area; the sand shiner is representative of such streams. Along the glacial boundary and the Allegheny Front Escarpment are many streams of high gradients. They are characterized by having bottoms composed of boulders, bedrock and gravel, and pools in which there is normally a good current. The rainbow darter and redside dace displayed a decided preference for such streams. In this section were many springs originally and in no other area were bog ponds and pothole lakes so abundant. The mudminnow was a characteristic species of the bogs and ponds.

The unglaciated portion.—All, or the major parts, of 26 southeastern counties are unglaciated. These are characterized by having a well dissected topography in which the hills range between 50–600′ (15–183m) in height. This section contains five definite types of streams.

Glacial outlet streams.—The Muskingum, Hocking, lower Scioto, and several other lesser streams served as outlets for the glacial waters. In so doing they were considerably modified, with the result that they have a rejuvenated appearance and are characterized by having bottoms containing much gravel, sand and many boulders which only recently have begun to be covered with silts. Wherever such streams had moderate or high gradients there were large populations of several species of riffle-inhabiting fishes of which the rainbow darter and gravel chub are examples.

Streams east of the Flushing Escarpment.—The Flushing Escarpment parallels the Ohio River from Monroe County northward to Columbiana County, and is the divide between eastward flowing streams which empty directly into the Ohio River and westward flowing tributaries of the Muskingum River (*see* map XI). Streams east of the escarpment have a youthful or immature appearance, undeveloped courses and high gradients (Stout and Lamb, 1938: 54–55). Such streams elsewhere in Ohio have large populations of riffle-inhabiting species. These streams do not, apparently because "the subsurface and surface rocks and soils contain 70 per cent shales which contribute to the establishment of a rather uniform gradient, producing a very rapid or 'flash' type of run off" and entire streams "are normally reduced in summer to a series of widely separated pools" (Trautman, 1942: 219–20). Also the thin, slate-like pieces of shale and the shale bedrock offer little protection and harbor little food for riffle-inhabiting fishes. In this respect these streams are unlike moderate or high gradient streams elsewhere in Ohio, for the latter have gravel and boulders, around and under which aquatic animals can hide and exist. Since 1900 mine and other industrial and domestic pollutants have been so prevalent in these streams as to affect the fish fauna adversely. The absence of the riffle-inhabiting central mottled sculpin exemplifies this unusual condition; *see* that species, under Distribution.

Low gradient, unglaciated streams.—In that area between the glacial boundary on the west, the Flushing Escarpment on the east, Lawrence and Jackson counties on the south, and Carroll County on the north, are many streams which have remained undisturbed so long that they have succeeded in cutting their courses to almost base level. Wills Creek from its confluence with the Muskingum to its junction upstream with Seneca Fork is an example. This stream is 77 miles (124km) long and has a fall of only 76′(23m), or an approximate gradient of one foot per mile (.2m/km). Such streams have no well-developed riffles; as a result, large populations of riffle-inhabiting species are absent. The only streams having high gradients in this area are the glacial outlet streams mentioned above and the small headwater brooks. The latter usually cease flowing in summer and contain at most only small, isolated populations of small-creek, riffle-inhabiting species.

Lowland streams in the partly-filled preglacial Teays stage valleys.—In that triangular part of unglaciated Ohio, bounded on the west by the glacial boundary, on the east by the Muskingum River, and on the south by the Ohio River, are many preglacial valleys of Teays time which have been covered by the waters of Glacial Lake Tight, and have become partly filled with glacial deposits. These streams have low or moderate gradients, and formerly contained an abundance of aquatic vegetation. The Ohio muskellunge is a characteristic species of the vegetated pools in these streams and the dusky darter of the vegetation and debris of the riffles.

Upland streams of the Bluegrass Region.—The distinctive Bluegrass Region, which is centered in Kentucky, extends across the Ohio River into Ohio, to occupy those sections of Adams, Scioto, Pike and Ross counties which have a maximum relief greater than 500′(152 m); for outline of area *see* distribution map of rosyside dace. The small upland streams which arise on the Lexington Peneplain (Fenneman, 1938: 440–41) in this area have a clear-water fauna, in which the rosyside dace and bigeye shiner are prominent. The peneplain is deeply dissected, and the lowland streams in the partly-filled valleys were once covered by the waters of Lake Tight.

Allegheny Front Escarpment and adjacent streams.—This escarpment enters Ashtabula County from Pennsylvania in northeastern Ohio, and extends diagonally southwestward to Richland County on the Lake Erie-Ohio River watershed, thence continues southward into Adams County where it crosses the Ohio River (*see* map XI). Between the Pennsylvania line and Richland County it is sometimes called the Portage Escarpment; for sake of brevity the entire escarpment in Ohio is referred to in this report as the Allegheny Front Escarpment. South of the Ohio River the escarpment is usually known as the Highland Rim (Stout, Ver Steeg and Lamb, 1943: 48–50).

The Allegheny Front Escarpment is of major importance in the distribution and/or abundance of many fish species, for it divides sharply the more dissected Allegheny or Appalachian Plateau to the east from the more level Lake Plain to the north and Mississippi Valley Plain to the west. In addition, the streams situated in a broad band on each side of the escarpment (but especially on the east) have moderate and high gradients and well developed riffles which contain bottoms of glacial gravels and boulders. Because of these environmental conditions many riffle-inhabiting and swiftwater species reach their maximum abundance in Ohio along this band. The distribution of the rainbow darter is an example.

The Prairies.—Transeau (1935: fig. 1) has demonstrated that the prairie areas of Ohio are an eastward extension of the Prairie Peninsula. The Ohio prairies probably were established during a xeric period or periods, which were sufficiently severe to cause the death by drought of trees in swamp forests and moderately moist habitats, thereby making possible the successful invasion of the prairie. During the height of the xeric period or periods the prairies extended, possibly unbroken, along the southern shores of the Great Lakes as far eastward as New York, and during the height of prairie invasion many species of prairie- and plain- (steppe) inhabiting animals, including some fish species such as the bigmouth shiner (Hubbs and Lagler, 1947:67), invaded as far eastward as Ohio and New York (Schmidt, 1938; Conant, Thomas, and Rausch, 1945; and Thomas, 1951). Later, when moist conditions returned, the forests reinvaded portions of the former prairies, leaving relict populations of western plants and animals in Ohio, Pennsylvania and New York.

Part II

Conditions between the Years 1750 and 1800

There exists an early Ohioana literature sufficient to convey a fairly accurate conception of conditions prevailing in the Ohio Country at the time of arrival of the first white men or shortly thereafter. Some of this literature was written for the purpose of enticing others to buy or do something special; such literature did not always adhere to facts, was usually readily recognizable, and has been discarded from our considerations. Only those statements were considered trustworthy when (1) no reason for embellishment of facts was evident, (2) when these statements agreed with ecological reasoning, (3) when the same statement was made by several authors.

So that some conception of conditions prevailing in Ohio between 1750–1800 may be had, the following description of conditions and statements by authors considered reliable is given:

Overwhelming evidence indicates that between 1750–1800 forests covered most of the Ohio Country, leaving the remainder in marshes, swamps, brush, "bowling greens," sandy beaches, and in wet, dry, tall grass, and wild plum prairies.

The unglaciated section contained a larger percentage of unbroken forest and fewer and smaller treeless areas than did the glaciated section; the land was better drained, and the forests consisted chiefly of a chestnut-oak-mixed mesophytic association (Sears, 1941: fig. 1; Braun, 1951: 236).

The forests of the glaciated section were of three major types. The beech-maple association dominated the hills and intervales, with a chestnut association on the more over-drained portions. This combination was most prevalent in a broad band extending from northeastern to southwestern Ohio. In and about the prairie areas an oak-hickory association prevailed, the oaks often forming "oak islands." The Oak Openings region of Lucas, Wood and Fulton counties was so named because of the oak stands or islands which were surrounded by wet prairie "openings." On lands ill-drained and water-covered throughout much of the year, and especially in spring and early summer, there grew an elm-ash-soft maple association. This association was dominant in the Black Swamp which was that area of northwestern Ohio comprising parts or all of 18 counties of which in early postglacial time were covered by the waters of Glacial Lake Maumee. For map of vegetative cover *see* Sears (1941: fig. 1).

The glaciated section contained a rather large amount of open land which was sufficiently large to modify the fish fauna, as was indicated by the large element of a prairie fish fauna existing in Ohio prior to 1900. The following statements of early authors will give some indication of the amount and types of the original prairies.

Scott (1793–94; ms.) was a soldier under the command of Gen. "Mad Anthony" Wayne. Scott's unpublished and unpaginated journal or diary described some of the prairies which he encountered in western Ohio. On October 21, 1793 while in Darke County he observed "Prieres to the right and left." On October 26, 1793 while returning to Hamilton in Butler County from Darke County he observed many prairies, oak islands, beech flats and ponds. On August 1, 1794, in Mercer County, and presumably near the present town of Rockford, he "struck a large Beautiful Prierie well cloathed with Grass and interspersed with delightful Clumps of Trees" which was about two miles (3 km) wide and was said to be 18 miles (29 km) long. In the near vicinity on August 4, Scott described severe drought conditions and stated that the land was "Low, flat, marshy."*

Capt. Daniel Bradley (Wilson, 1935: 17) was also a soldier under General Wayne's command. While in Butler County he described "a prairia of two [81 hectares] or three hundred acres [121 hectares] where the grass or wild oats is 8 or 10 feet [2–3 m] high and very thick." On October 12, in Darke County he "struck a large prairia in our course—found it impassable," elsewhere the land was hilly and "interspersed with low prairias."

In 1802 Kilbourn (Lee, 1892: 186) describes a "clear meadow" of a thousand acres (404.7 hectares) in which grass grew "higher than a horse's back," and "without a tree or bush in the whole extent."

* From Scott's statements and those of others, severe drought occurred before 1800. During such droughts the larger streams fell far below their normal summer levels.

This prairie was situated at or near Worthington in central Ohio.

Brown (1815: 134–38) describes in detail the northern Ohio prairies as they were before 1815. He mentions the many "open and extensive prairies" in the vicinity of Sandusky Bay, which were "always surrounded by fine oak and chestnut land;" also "a natural meadow independent of the immense prairies," mentioned above, which was "ninety miles [145 km] long and from two to ten [3–16 km] wide, extending from the mouth of Portage" River in Ottawa County, Ohio around the western end of Lake Erie "to Brownstown" south of Detroit, Michigan. Continuing, Brown states that this prairie "cannot contain less than two hundred thousand acres [80,940 hectares]." Brown and several companions explored that part of the prairie "which lies between Miami [Maumee] bay and Portage River" and found "the grass higher than our heads and as thick as a mat, confined together by a species of pea vine." Near the mouth of Toussaint Creek in Ottawa County the prairie grass "was about seven feet [2 m] high and so thick that it would easily sustain one's hat—in some places a cat could have walked on its surface."

Riddell (1837: 13–17) in his authoritative Report 60 to the Governor and Legislature of Ohio, gives excellent accounts of the amount and extent of the lowland and upland prairies of central, western, and northern Ohio; especially of the "Sandusky plains" near Upper Sandusky, Grand Maumee prairie "from the Maumee River to the headwaters of the Portage," Oak Openings prairies, and the "low, level prairies stretching, when seen from some points, as far as the eye could reach" between Sandusky and Port Clinton (Trautman, 1981).

From the few references given above, and from the many others in the literature, it is apparent that (1) the vegetative cover, both living and dead (as mulch or organic debris), was sufficient to keep soil erosion of the land at a minimum, except along stream banks and steep hillsides, (2) there was a sufficient amount of treeless area in the western half of Ohio to affect fish distribution and abundance.

One of the impressive facts, abundantly verified, concerning conditions in primeval Ohio was the great profusion of "durable springs and small brooks," both flowing throughout the year, and the great amount of bog, prairie, and swamp and forest lands which were covered with water during all or much of the year. This abundance of underground and surface water was a prime factor contributing to the great abundance of fishes in Ohio before 1800.

Lee (1892:273) in his history of Columbus (central Ohio) stresses the point that the small brooks before 1800 were "living streams throughout the year." Discussing the "numerous marshes, quagmires and ponds," of the swamp forests from which Columbus later emerged, Lee states that the land was "saturated intermittently from the clouds above and constantly from the springs beneath" and that the land "had the sponge-like quality of retaining much of the water it received." He describes the difficulties encountered in attempting to erect a town where springs and surface waters prevented ditch digging, and where the soft nature of the bogs (as in the vicinity of the present Spring Street) made travel with a horse often impossible, and dangerous for a man on foot.

Brown (1817: 297–308) mentions two springs which with their brooks almost encircled the village of Columbus before 1818. These springs were large enough to be "deemed capable of moving machinery, sufficient for most manufactures of mills, a large part of the year," and later they were so used. Brown also mentions that "durable runs, springs and brooks" were abundant in Champaign, Adams, and Athens counties.

Hildreth (1848: 487 and 492) wrote of the "regular" supply of water throughout the year from the springs of southeastern Ohio before 1800, and that it was this continuous supply of water from springs, and from water draining from the surface, which kept even the small brooks flowing throughout most or all of the year.

Prior to 1800 floods seemed to have crested at heights similar to the flood-crests which occurred after 1900, but apparently with the important difference that before 1800, severe floods which included much of Ohio occurred only during winter and early spring, either when the land was supersaturated and could hold no more water, or when the subsoil was frozen and could not absorb it. Under these conditions the run-off apparently was as rapid as it has been after 1900. Great floods were recorded for the Ohio River in 1762 and 1763 (Moseley, 1939: 224). Dun (1884: 111) describes three Ohio River floods between 1774–92 whose crests were above 60′ (18 m). Lee (1892: 304) describes the flooding of the Scioto River in central Ohio in 1798, the year after the village of Franklinton (now part of Columbus) was laid out.

In 1818 Cramer (1818:35) implied that the best seasons for navigation on the Ohio River were in spring and fall when there was a goodly stage of water, but that "freshes" were not entirely confined to these seasons, for heavy rains frequently raised the river during the summer months sufficiently to make navigation possible for a short period. He added that summer "freshes" were not to be depended upon since the "water subsides rapidly." Hildreth (1848: 184) corroborates Cramer's statements concerning floods and freshets, also the clearness and cleanness of the Ohio River and its tributaries prior to 1800. Hildreth stresses the rapidity with which the river could rise 30′ (9 m), and its shallowness during droughts when it was "not more than 3 feet [1 m] deep the whole way across" in many places, making it impossible for larger boats to cross the shallows. Howe, (1900: I, 52) states that during its lowest stage the Ohio River was fordable in several places between Cincinnati and Pittsburgh.

Before 1800 the streams presented a far different appearance than they did after 1900. In early days the banks were heavily wooded or at least contained a fringe of brush. Trees grew over the banks down to the normal water level; their roots caught and held brush, earth and rubbish, and this had a tendency to keep the streams narrow and deep. The trees on the banks leaned over the water so as to obtain more light, and in doing so made a natural canopy over the stream which partly or entirely shaded it. The leaning trees eventually fell into the water, and one of the characteristics of the early Ohio streams was the great amount of down timber and brush which they contained. This characteristic was frequently commented upon by early writers, especially when it seriously impeded canoe travel or made the dragging of brush nets to capture fishes impossible. The brush and down timber in the streams afforded much protection to the fishes and to their food.

Hildreth (1848: 205 and 320) well describes the above conditions. Writing about conditions prevailing before 1810 he states that "The banks of this stream [Muskingum River at its mouth] were thickly clothed with large sycamore trees whose lofty tops and pendant branches leaned over the shores, obscured the outlet [mouth] so much, that a boat in the middle of the Ohio [River], in a cloudy day, might pass without observing it [the Muskingum] at all," and that the roots of the trees "acted as so many ligatures and fillets of net-work in protecting the earth from the wash of waters."

Despite the protection afforded by tree roots, some bank erosion occurred, but erosion of land other than along streams and steep hillsides was at a minimum. As a result, the small amounts of clayey silts which entered the streams could be carried away during freshets, leaving the stream bottoms clean and containing only sand, gravel, boulders, bedrock and organic debris. Many early writers testify to the siltless bottoms of streams. The testimony of Kilbourn (Lee, 1892: 186) is an example. Describing the streams immediately to the north of Columbus in 1802, Kilbourn states that they were "clear lively streams of pure water as ever flowed from a fountain, with small gravel and in places large pebble bottom."

Cramer (1818) in his detailed description of the Ohio River along the Ohio border gives a great many references to sandbars, sandbanks, clean banks, rock and gravel bars, bedrock and rock ledges, but only once does he mention a mud bank (near the mouth of a creek), all testifying to the clean bottoms and clearness of the waters except during floods or freshets.

The statements of many other early writers also testify to the clearness and purity of the streams, several stating that the early pioneers drank as readily from flowing streams as they did from springs.* It was not until after 1800, after increased soil erosion became evident, that the pioneers stopped drinking stream waters and confined themselves to drinking water from springs and wells. However, bog waters which were stained dark brown or contained much decaying plant and animal remains were considered most unhealthy and were not drunk.

Unshaded clear waters, especially when static, produced an abundance of aquatic vegetation, although shading by overhanging trees and other land vegetation undoubtedly reduced the growth of aquatic vegetation in some waters. Many of the early references to aquatic and land vegetation in the water is indirectly implied with such statements as "the waters were unfit to drink because of the

* The earliest writers seldom used the term "clear" to denote water free from turbidity which was caused by soil or organic matter in suspension. "Beautiful" and "pure" were much used as synonyms of clear, as was "fair" which was used as today we describe a sunny, cloudless day as a "fair day." "Fair water" always appears to have been used to denote clarity, never as it is sometimes used today to denote good or moderate, such as "a fair stage of water." "Yellow" and "turbid" waters were used before 1800, but only when describing severe floods.

profuse vegetation," or that there were "immense numbers of waterfowl feeding upon the vegetation."

Scott (1793: 94) mentions in his unpublished journal, that on July 30, 1794, Beaver Creek in Mercer County contained no current (the waters were low because of drought), it had a mud (muck?) bottom and was "full of grass" (meaning aquatic vegetation). Describing the Maumee River near Grand Rapids he gave its width as "near 600 yards [549 m] wide and near the head of the Rapids it resembals a Meadow flooded over." He described the same river a little farther downstream as having little or no current and looking "like a flooded Meadow with long grass entirely across."

Brown (1815 and 1817) gives excellent descriptions of the great abundance of aquatic vegetation, principally of wild rice, in Sandusky and Maumee bays and in the estuaries and lower courses of the tributaries flowing into western Lake Erie. He (1815: 141–42) states that Miami (Maumee) Bay is like Sandusky Bay, both resembling a little lake; that "Within the bosom of this [Maumee] bay grow several thousand acres of *follie avoine* [wild oats = wild rice]." His many references to waterfowl abundance in Portage, Toussaint, Swan and other creeks and rivers also imply an abundance of aquatic vegetation upon which the waterfowl fed.

As is ecologically apparent, the conditions described above were highly conducive to the production of a huge population consisting of the larger and better food and game fishes (pikes, walleyes, catfishes, buffalofishes, suckers, drums and sturgeons), and to a large population of fish species of smaller size which required normally clear waters and bottoms of clean sand (sand darter), gravel (streamline/gravel chubs), boulders (various riffle darters), or aquatic vegetation, down timber and brush (several species of minnows, topminnows). Such conditions were *not* conducive to production of large populations of species of fishes requiring, or which were tolerant of, soft bottoms of clayey silts or other soils, and/or of great turbidity of the water.

A few of the many references will be used which refer to the abundance of fishes and their importance as food to the Indians and pioneers. Hildreth (1848: 2) writes that the Indians were able to travel long distances by canoe because the waters "afforded them a constant magazine of food in the multitude of fishes which filled its waters."*

Hildreth, Zeisberger (Hulbert and Schwarze, 1910) and other early writers stress the importance of fishes as food for the Indians. Much of the fishing was done by women and children, who could tend to the brush traps and fish with lines, leaving the men free to hunt and travel. Fishes were of particular importance as food during periods of game scarcity which occurred not infrequently, during periods of crop failure, while the Indians were traveling through enemy territory, and when their enemies forced them to remain near their villages.

Hildreth (1848) discusses in detail the abundance of fishes in southeastern Ohio, listing the kinds and their importance to the early settlers, and methods of capturing them. He wrote that the waters "teemed with fish" (1); that the waters "were filled with delicious fish in such abundance, that at certain seasons of the year [spawning?] the smaller tributaries might be said to have been 'alive with them' " (485); that a "multitude of fish" (1) filled the waters; that fishes "furnished no small part of their [settlers] animal food" (499); that in 1792 James Patterson supported his family with the proceeds obtained from the fishes he caught (333); that while journeying by canoe down the Muskingum from Waterford to Marietta, Allen Devoll and David Wilson "often took four to six hundred pounds" (181–272 kg) (467) of fishes by gigging or spearing them; that many fishes were speared through holes in the ice and that many were taken with hook and line and on trotlines. On trotlines "a half barrel or more" were frequently taken in a night (499). Hildreth, in his listing of the more important food fishes includes the pike (Ohio muskellunge), white perch (drum), salmon (walleye and sauger), spotted bass (presumably the spotted blackbass), sturgeon (both species), buffalo (one or more species of buffalofishes, carpsuckers, and possibly other suckers), the yellow cat (flathead and channel) which weighed 5–50 lbs (2–23 kg), and a black cat (flathead) which was caught in 1792 and weighed 96 lbs (44 kg) (498).

Jones (1774: 110–11), while journeying down the Ohio River from Pittsburgh to the mouth of the Scioto, and up that stream into central Ohio, in the years 1772–73, repeatedly refers to the "great abun-

* Hildreth (1848: 498) describes a method which the Indians had of catching large pikes (muskellunges) in the Ohio River and its tributaries, and one which was used by the pioneers. The Indian fastened a fish, weighing a pound (0.5 kg) or more, to a strong hook and line. The hooked fish "was thrown out thirty or forty yards [27–37 m] into the river near a mouth of a creek, where pike lie in waite." The bait was quickly retrieved by the fisherman, in a manner which made it jump along the surface so that the pike mistook it for a living fish.

dance of fine fishes." He describes several of the food fishes, including the "prodigious large pikes" (Ohio muskellunge), the sturgeons (both species) and a catfish so large "that after seven of us had eaten twice of it, part was given to the Indians." Jones concludes that "the riches of the waters are not fully known, the people not having seines made as yet."

Bradley (Wilson, 1935: 53–54) was apparently much impressed with the abundance of fishes in the Ohio, Great Miami and Maumee rivers in 1793–94. He relates that near Hamilton, Butler County, he and other soldiers of the garrison made a "fish dam" . . . "across the Miami with a funnel & basket like for the fish to run in" (presumably the usual type of dam made of brush). On September 3, "we caught 2500 weight of Fish and about as many on the 4th, which makes 5000 weight in two nights." The fishes taken were "Buffelow, Pearch, Catfish, Ells, etc. We have more fish than the whole garrison can make use of." Recent evidence indicates that early September is a period of major downstream migration of our fishes.

Scott (Journal), a natural ecologist and ardent fisherman, in 1793–94 fished several streams in western Ohio, particularly those of the Maumee drainage. He fished with jack light and spear, hook and line, and by "dragging brush nets" (August 2, 1794). He states that pike and/or muskellunge and salmon (walleye and/or sauger) were usually taken by spearing, and that other "fine but smaller fishes" were taken in the nets.

Brown (1815: 144), in discussing the fishes about and below the rapids of the Maumee River, states that their numbers were "almost incredible" and that "So numerous are they at this place, that a spear may be thrown into the water at random, and will rarely miss killing one! I saw several hundred taken in this manner in a few hours. The soldiers of the fort, used to kill them in great quantities, with clubs and stones. Some days there were not less than 1000 taken with the hook within a short distance of the fort, and of an excellent quality."

Summary of 1750–1800 period.—Prior to 1800 the living and dead vegetation covering the Ohio Country was sufficient to prevent erosion of soil except in restricted areas along stream banks, beaches and steep hillsides. The banks of streams were covered with brush or trees, the latter forming a canopy which shaded much of the waters. The streams were narrow and deep, they contained much brush and down timber but where shaded there was little aquatic vegetation. The waters were normally clear, containing little soil in suspension except during some freshets and floods. The bottoms of the waters were free of clayey silts, and were largely composed of sand, gravel, boulders, bedrock, and organic debris. Aquatic vegetation was abundant in quiet, unshaded waters, and it was especially abundant in marshes, bays, glacial bogs and ponds.

Beyond doubt the huge fish population, existing in Ohio waters prior to 1800, was composed largely of the better-flavored and larger food fishes, and of species, both large and small, which required clear waters, clean bottoms of sand, gravel, boulders, bedrock, and organic debris and/or aquatic vegetation.

Conditions between the Years 1801 and 1850

The permanent settlement of the future state of Ohio by the white man started shortly after 1730, when English traders began at an increased rate to trade in the Ohio Country. In 1790 the pioneer population was only 3,000 and because of Indian unrest, the settlement of the country continued to progress slowly until after the Treaty of Greenville was signed in 1795. In this treaty the Indians ceded a considerable part of Ohio to the United States. After 1795, the pioneers came in increasing numbers, and by 1800 the population totaled 45,365. Immediately after Ohio was admitted into the Union in 1803, the annual number of settlers increased sharply so that by 1820 the population had risen to 581,434. The building of the National Road (1820–40) and construction of the Ohio canals (1825–45) opened two great arteries of travel and commerce, and over these arteries came ever increasing numbers of settlers. By 1850 the population had increased to 1,980,329.

This huge increase in human population had profound influence upon the flora and fauna. With the increasing population there was a corresponding increase in amounts of forest removal, land ditched and drained, and virgin prairie plowed, resulting in the conversion of much of the forest and prairies into farms and towns, and a modification of the fauna. During this period much of the big game was extirpated from Ohio, or their numbers were greatly reduced. Several factors became important as modifiers of the fish fauna.

The 1801–50 period was one of water power for manufacturing purposes, and the rapid speed of settlement was made possible because of the abundance of fine mill sites. The first mill erected within the limits of Ohio apparently was the one completed in 1790. It was on Wolfe Creek in Washington County, a short distance from Marietta (Hildreth, 1848: 423–24). After 1790 the number of mills constructed increased annually at a rapid rate, and especially after 1800. By 1808 there were in operation at Worthington, in central Ohio, "three saw-mills, two grist-mills and several other useful water machines, and three other mills are now building" (Lee, 1892: 220). By 1817 there were "about 40" mills in Columbiana County (Brown, 1817: 319). In conversations with Professor Wilber Stout, he told me that at one time 12 mills were in operation along one tributary of the Little Scioto system in southern Ohio; that by 1815 there were hundreds of mills scattered over the state, and by 1850 they numbered between 500–1000.

With few exceptions the finest of Ohio's food fishes were migratory to some degree, these going upstream to near the headwaters where they found suitable spawning habitats, and conditions were favorable for the development of their young. The many dams of the white man were extremely effective in preventing these migrating fishes from reaching their spawning grounds, especially those dams which were near the mouths of the Ohio River and Lake Erie tributaries such as the one which in 1817 dammed the Grand River within a half of a mile (0.8 km) of Lake Erie (Brown, 1817: 274). A decrease in abundance of the muskellunges, pikes and lake sturgeons first became apparent during this period, a decrease apparently caused in part by dams preventing their upstream migrations.

The overdraining of marshes, bogs and ponds, the ditching of streams, timber removal on stream banks, burning of vegetation and other agricultural practices caused many springs to cease flowing and small streams to flow intermittently. In most instances these practices caused a destruction of habitats of some of the more important food fishes, and of some smaller-sized species.

Other factors adverse to fish life first became of considerable importance during this period. The many lumber mills dumped their sawdust into streams, which killed fishes by asphyxiation through compacting their gills with sawdust, and by covering their spawning grounds. Breweries and slaughterhouses dumped their refuse into streams. When only a small amount of such refuse was deposited, it did little harm; in fact, some of it became food for fishes. But when deposited in large amounts and its bacterial action removed all of the oxygen from the water, the fishes were killed.

Commercial fishing in inland waters and in the bays and shallows of Lake Erie developed greatly during this period, and by 1830 had attained considerable economic importance. Seines and nets of twine began to supplant the brush drag nets and brush weirs. The pools below dams were concentration areas for upstream-migrating fishes, and mill owners "rented out" the seining privileges of these pools to commercial fishermen. Dam owners placed a crib on the aprons of their dams, and by diverting the water into the crib, downstream migrating fishes fell into it and were captured. When the dams were first constructed, observers saw that these dams prevented the fishes from migrating upstream, and that shortly after construction there was a sharp decline in the abundance of desired food fishes. These observers therefore condemned the dams as harmful to the fish population. But later observers, not having seen the former great abundance of fishes and the subsequent decrease in their abundance after dam construction, praised the dams because they concentrated fishes in sufficient numbers to make their capture profitable.

Although the amount of land diverted to agriculture was considerable during this period, either it was not sufficient to increase greatly the turbidity of the waters, or enough top soil or humus still remained to prevent the clayey silts from eroding. At any rate, there are few references to "yellow" or "muddy" waters during this period, except when describing the highest floods.

Following are quotations from authors substantiating the statements made above:

Hildreth (1848: 319–20) comments at considerable length upon changes in southeastern Ohio between 1790–1848. Concerning the effects of tree removal along the banks of the Muskingum River at its mouth, at the site of Fort Harmar, he writes "The rivers [Muskingum and Ohio] have made sad inroads on the site of the old fort. At this day [1848] not only the whole space between it and the river is washed away, but more than half of the ground occupied by the walls; so that the stone wall of the well, which was near the center, is now tumbling down the banks of the river. This continual washing of the banks has widened the mouth of the Muskingum so much that during the summer months a sandbar or island occupies the spot that used to afford ten to twelve feet of water. Before any clearings were made, the huge sycamore trees, as they inclined over the water on the opposite shores, narrowed the mouth of the river" (p. 15 of this report).

Concerning the effects of agriculture upon the streams Hildreth (1848: 487–88) comments "Our rivers and creeks already feel the effects of cultivation, and afford a less uniform and steady flow of water in the summer months, than they formerly did. At the first settlement of Marietta, a small creek which passes through the southern half of the town, called 'the Tiber,' rose from springs within two miles of the city. During the few first years, it was a steady stream all the year, and the early settlers thought it would be permanent, and when collected in a reservoir, furnish them with water by means of an aqueduct. But of late years the bed of the stream is often dry in the month of May. The springs which supplied it, while sheltered by the forest, were perennial; but as soon as the trees were cut away, letting in the sun and air, they failed." This failing and diminishing water supply is commented upon by many authors.

Kirtland (1850 A:1) stresses the changes made in northeastern Ohio between 1797–1850: "Fifty-three years have nearly elapsed since the first surveys and settlements were made on the Connecticut Western Reserve. Within that period of time a perfect revolution has been effected in its condition. Its forests have been displaced by farms, villages and cities; canals, railroads and other important thoroughfares are extending in every direction; telegraphs are furnishing increase of facilities for communication; commerce has spread over the Lake [Erie], and the whole face of nature has been changed." He records the great extent of the forest removal, the extirpation or decrease in abundance of large game animals, game birds and waterfowl, the rapidity with which wild fruits were disappearing, and the increase in abundance of such song birds as robin* and bluebird. He continues, "While the tributaries of Lake Erie and the Ohio river were unobstructed by dams and were not swept by seines, they abounded with large and valuable species of fish, which, in their vernal migrations, crowded in immense shoals on the ripples. Sturgeon and Muskallonge often run up the Cuyahoga several miles, and large numbers of Pike, Pickerel [walleye and sauger] and white Perch [white bass] visited the upper waters of the Mahoning during Spring and Summer." Commenting upon conditions as they existed about 1850, Kirtland writes "Still greater changes, if possible, have occurred with the finny tribes. The Sturgeon has nearly forsaken this [south] shore of the Lake [Erie]; the Muskallonge has become scarce, and no longer seeks the mouth of the rivers to deposit its spawn. All the migratory species have been excluded from the Mahoning River by the construction of dams on the Big Beaver. Many smaller species have increased in all our waters since the larger and more voracious have been reduced in numbers. The slaughter houses about the rivers, afford them large supplies of food and contribute to their increase. Artificial slack waters, canals and basins have also in many localities effected similar results."

* Scientific and common names of birds have been taken from Borror (1950).

Despite the rapid increase of adverse conditions during this period there remained through much of Ohio, large populations of fine food fishes and smaller species requiring clear, clean waters. Two of the many illustrations of abundance are as follows:

Capt. Riley was the first permanent white settler within the present Van Wert County, building a cabin at the present site of Willshire. During the rather mild winter and spring of 1821 Capt. Riley, aided by other pioneers, erected a large grist mill, built a dam 8′ (2 m) in height across the St. Marys River and constructed a mill race 1/4 mile (0.4 km) in length. Apparently the mill race was completed during a period of fish migration for as soon as the water entered the race it "seemed to be perfectly filled with pike, pickerel, lake salmon [walleyes], white fish, large muskalonge, black bass and suckers."† Fishes "swarmed" in the race "so thick that, with a dip net they could be thrown out as fast as a man could handle his net" (Howe, 1900, II: 723–25).

Ichthyologists during this early period and well into the next, found a profusion of fish species, and especially of those inhabiting clear waters. Prof. Spencer F. Baird's letter to Dr. P. R. Hoy, dated

† White fish was a name commonly applied before 1860 to many species of white- or silvery-colored fishes, often to the carpsuckers and frequently to many species of suckers, white bass and drum. Before 1850 the whitefish was virtually unknown in inland Ohio.

October 10, 1853, relates Baird's experiences in northern and northeastern Ohio during August of 1853; of "how I went to Elyria alone, or rather with Prof. Andrews and caught prodigious stores of *Etheostomas*: and returning nabbed Dr. Kirtland and posted to Poland where one day we caught 41 species" (Dall, 1915: 308). Baird's collections, preserved in the USNM, testify to the abundance of *Etheostomas* in the Black River at Elyria, Lorain County, a stream which since 1925 has contained relatively few darter species and individuals. Baird's preserved collections from the Mahoning River system, including Yellow Creek, indicate that he and Kirtland collected more than 41 species since their collections contain such unrecognized species as *Hybopsis micropogon*. They also collected such fishes as *Etheostoma maculatum* and *E. camurum*, species which have long ago disappeared from this now heavily-polluted section of the Mahoning system.

Summary of 1801–1850 period.—The populations of many species of fishes showed a definite decrease or increase in abundance. There was a decrease in abundance of several of the large and important food fishes. Habitat changes, however, favored some of the smaller-sized species which were less valued or worthless as human food; these were recorded by Kirtland as having increased in abundance. Principal adverse factors were: (1) construction of many dams which prevented the more migratory species from reaching their headwater spawning grounds, (2) polluting of waters with sawdust, and with brewery and slaughterhouse slops, (3) draining of much marsh land which destroyed valuable spawning grounds, (4) increased amount of ditching of streams, (5) drying up of many springs and small streams for part or all of the year, (6) removal of top soil and humus by burning, (7) great increase in amount of commercial seining, especially in inland waters. The silting of stream bottoms with clays and the great turbidity of water which later became so evident had not become apparent at this time.

Conditions between the Years 1851 and 1900

Throughout this period the large, annual increase in human population continued, so that by 1880 there were 3,198,062 inhabitants; by 1900 there were 4,157,545. As the population increased in numbers the amount of land used for various agricultural purposes likewise increased, so that by 1886 there were 9,705,735 acres (3,927,911 hectares) under cultivation, 6,180,875 (2,501,237 hectares) in pasture, 4,854,473 (1,964,605 hectares) in woodland, and only 640,699 acres (259,291 hectares) remained as "waste land" (Howe, 1900, I: 52).

Except for the white-tailed deer* and those mammals extirpated prior to 1850, all of the large fur bearers such as the river otter, "varmints" such as the bobcat, and the larger game such as the black bear were extirpated or reduced to an occasional stray during this period. Some white-tailed deer still remained in the Black Swamp (Fisher, Klippart, Cummings, 1878:65). On the other hand, those species of small mammals markedly increased which inhabited remnant woodlands, brush and open fields, such as the fox squirrel and cottontail rabbit (Trautman, 1939C: 140–41). The larger game birds as a group suffered decreases in abundance or were extirpated, whereas many small land birds increased greatly in numbers (Trautman, 1940A; for Buckeye Lake but representative of conditions throughout Ohio).

The number of functioning water mills annually increased between 1850–1880; totalling more than a thousand about 1880, according to Professor Wilber Stout. After 1880 their numbers declined because of the revolutionary change from water power as the source of manufacturing power to that of steam. By 1900 the majority of the water mills had fallen into disuse, their dams were being destroyed or were decaying. With the water barriers removed the upstream fish migration could be resumed.

This period saw a steady annual increase in the number of miles of streams which were ditched and dredged, and in the number of acres of marsh land drained. In 1850, the Black Swamp of northwestern Ohio, which was about 120 miles (193 km) long and averaged about 40 miles (64 km) wide, was still undrained except for isolated areas about its periphery. Shortly after 1850, ditching and draining activities greatly increased so that by 1875 much of the swamp had been drained. By 1900, partly through the help given by county commissioners, all except isolated portions of the great swamp had been ditched, drained and tiled, and immense crops grew where a few years previously muskellunges, pikes and sturgeons had spawned and their young had developed.

In this period the first evidences of over-draining of land became apparent. More and more springs

* Scientific and common names of mammals have been taken from Burt and Grossenheider (1952).

ceased to flow, or they flowed only during wet seasons. Many mills became useless because of lack of water to run them, except during freshets. The well was replacing the spring and the more turbid and polluted streams as a source of water supply. As the water table was lowered, these wells had to be dug deeper and deeper. The ditching, tiling and draining increased the rapidity of the run-off, and this together with the lowering water table resulted in many streams becoming intermittent in flow, or reduced to trickles during summer and fall. On the other hand, the increased speed of the run-off caused more and higher floods, especially of the local "flash-flood" type.

The kinds of pollutants and their severity to fish life increased greatly as human population and the amount of manufacturing increased. Pollution was especially harmful during drought. It was during this 1851–1900 period that pollution first became of major importance as a detrimental factor to most fish life. The more destructive pollutants were:

Mining wastes.—The amount of effluent from coal and iron mines, and salt mines and wells, increased so greatly that during the last half of the period many streams had become fishless waters. These effluents were most prevalent in southeastern Ohio, and especially along the Flushing Escarpment and in Athens County.

Oil and gas wells.—Oil wastes, salt and sulphurous waters, and other effluents from wells made many ponds and streams unfit for fish life. The majority of these wells were in northwestern Ohio, where after the great Karg gas well came into production near Findlay in 1886, one of the early, large oil fields soon developed.

Industrial wastes.—Before 1850 Ohio was primarily an agricultural state. During the 1851–1900 period, with the rapid growth of cities, came a corresponding growth in the manufacturing industry. As the manufacturing industry increased in volume it dumped ever increasing amounts of deleterious wastes into the streams. By 1900 many sections of streams, or entire stream systems, were badly polluted and some were devoid of fish life. Principal among the inorganic industrial wastes were those of the rapidly-increasing iron and steel industries, located principally in northeastern Ohio. Slaughterhouses, breweries, dairies, and other food processing plants contributed ever increasing amounts of organic wastes.

Domestic sewage.—Before 1850 comparatively few cities or individuals dumped domestic sewage directly into streams, and in few localities was the amount sufficient to be harmful. During the 1851–1900 period, and especially in the last decade, many municipalities had installed or were installing large sewers. When completed these sewers deposited raw, untreated sewage directly into streams.

Turbid waters and siltation of stream bottoms.—It was during this period that turbidity and siltation first became apparent. Before 1850 there was little mention of turbid waters except during floods, but after 1860 reports of turbid waters became increasingly numerous, and naturalists, agriculturists and fishermen began to be alarmed. After 1890, erosion of land, and the resultant turbidity of the waters and siltation of water bottoms, became almost universal. The covering-over of the stream and lake bottoms with clayey silts and the disappearance of aquatic vegetation was well under way by 1900.

Before 1850 commercial fishing gear consisted principally of spears, shore seines, brush drag seines, brush weirs, brush dams, trot lines, and hook and line. Fishing activity was confined almost entirely to streams, inland lakes, reservoirs, and the bays, harbors and shallows of Lake Erie. Commercial fishing in the deeper waters of Lake Erie was conducted mostly with hook and line or trotline. Around 1850 the gill net, made of twine, came into greater use and poundnets were introduced. These revolutionized commercial fishing. By 1875 commercial fishing in Lake Erie, as well as inland, had become a great industry (Smith and Snell, 1891: 247–80).

It was during this period that commercial fishing reached its peak of production in such large inland reservoirs as Buckeye Lake (Licking), Loramie, Indian and St. Marys (Grand Reservoir). These reservoirs, totaling more than 32,000 acres (12,950 hectares), were established before 1850. They apparently reached their greatest fish producing capacity slightly before 1875, at which time they contained vast amounts of down timber and stumps, and were so choked with aquatic vegetation that holes had to be cut in it in order to set the hoop nets with which the fishes were captured. Thousands of barrels of largemouth blackbasses, bluegills, pumpkinseeds, and bullheads (mostly yellow bullheads) were shipped annually from these reservoirs. After 1875 a decrease in fish abundance became evident in these reservoirs, in the shallows of Lake Erie, and especially in the streams.

After 1870 sport fishing began to assume considerable importance. The decrease in abundance of some species of fine flavored fishes, and the increase in the number of sport fishermen greatly increased an interest in fish protection. In 1857 the Ohio Legislature passed its first law relative to the protection of fishes; on May 3, 1873 an act became law authorizing the appointment of a "Commission of Fisheries" (Klippart and Hussey, 1874: 39–40).

In 1853 Theodatus Garlick (1857: 7) began his classical experiments near Cleveland, Ohio, with artificial fish production. By 1873 this type of propagation had been sufficiently perfected so that one of the first objectives of the newly-appointed Ohio Fish Commission was the installation of fish hatcheries. During the last half of the 1851–1900 period the propagating of fishes played an ever increasing role (Fisher, Cummings, Klippart, 1877: 5–17; Klippart and Hussey, 1874: 14–29).

Another function of the Fish Commission was the introduction of many exotic species of fishes, and introductions from outside Ohio of species which were rare or of local occurrence in this state, such as the eel. Introductions of these species reached a peak between 1880–1895.

A third function was the general protection of fishes. At first this centered largely about fishing restrictions, open seasons and methods of enforcing the law.

As the decrease in abundance of some fish species became increasingly evident, the public became aroused. As so frequently happens a minority group was chosen as the cause of the fish decrease. This group was the commercial fishermen and they were charged with the depletion of the fisheries. Mill owners were blamed to a lesser extent as were owners of slaughterhouses and any individual who deposited offal or refuse in streams. But as usual, the general public failed to blame themselves, and were either blind to or unaware of the havoc created by the destruction of fish habitats through land erosion and municipal use of streams as sewers.

By 1900 or shortly thereafter the decline in the abundance of some species had become so great that their commercial capture and sale were prohibited. Thereafter such species were allowed to be taken only by sporting methods, usually only with hook and line (smallmouth blackbass).

Following are quotations substantiating statements made above:

Howe (1900, II: 245) records the declining rate of flow in the upper Miami valley streams. He writes "On the first establishment of these mills [he mentions several] they would run ten months in a year, and sometimes longer, by heads. The creek would not now [1900] turn one pair of stones two months in a year, and then only on the recurrence of freshets. It is thought this remark is applicable to all streams of the upper Miami valley, showing there is less spring drainage from the country since it has become cleared of its timber and consolidated by cultivation."

Floods and droughts are frequently mentioned as being more severe after 1850 than they were before. Dun (1884: 110) lists 32 periods between 1857–84 when the Ohio River rose more than 40′ (12 m). Of these 28 periods occurred between December 17–April 19; they crested between 41–71′ (12–22 m), and the crests averaged 52.2′ (15.9 m) in height. Only four floods occurred between April 20–December 16; they crested between 40–55′ (12–17 m) and the crests averaged 45.5′ (13.9 m) in height. It is apparent, therefore, that although forest removal and ditching—draining—tiling were very great, the floods involving the Ohio River were still largely confined to the late winter-early spring months. However, summer freshets, involving inland streams, were more numerous and were very destructive to property. Lee (1892: 301–06), describing the floods of the Scioto River at Columbus, records only four major floods between 1798–1847, but mentions 11 between 1852–83. Of the 11, one was the great August 5, 1875 flood which involved not only the Scioto system but the Ohio River as well. Another summer flood was in 1866 when "the greatest September flood took place, which, until that time, had ever been known since the earliest settlement of the country." During this flood the "turbid torrent" bristled with floating trees. A flood, occurring on April 10–11, 1860, contained a "clay-colored current" and when this flood crested the waters were "literally darkened" with floating timbers. This 1860 flood contains one of the earliest, central Ohio references to turbid waters. After 1860, mention of turbid waters was made with ever-increasing frequency; it becomes apparent that considerable eroding of land had begun by that time, and that no longer were stream bottoms composed entirely of sand, pebbles, gravel, rocks, and organic debris; but that the silting over with "yellow" silts had begun. Lee sums up the situation thus, "With the clearing away of the forests, as usually results from that change of conditions, these freshets seem to have increased in suddenness and violence." It must be remembered, however, that floods of a later date which inconvenienced large numbers of individuals were naturally considered to be more sudden and more violent than were floods of an earlier period which involved few persons.

Edward Orton (Howe, 1900, I:89), State Geologist for Ohio during the latter part of this period, aptly discusses the causes and frequency of floods and droughts: "We have been busy for a hundred years in cutting down forests, in draining swamps, in cleaning [of timber and debris] and straightening the channels of minor streams, and finally, in under draining our lands with thousands of miles of tile; in other words, in facilitating by every means in our power the prompt removal of storm water from the land to the nearest water courses. Each and all of these operations tend directly and powerfully to produce such floods as have been described, and it cannot be otherwise than that under their combined operations our rivers will shrink during summer droughts to smaller and smaller volumes, and, under falling rain and melting snow, will swell to more threatening floods than we have hitherto known."

The several types of pollutants received their share of comment during this period. Klippart (1874:7) remarks, "Deleterious substances prevent the increase of fishes. The habit of throwing all of the offal and waste material from factories into the river, not only prevents the increase but actually destroys myriads of fishes annually. The waste discharged into the river from distilleries often destroy millions of fish; the waste discharges from paper mills consist of lime and other alkalies; from woolen mills the waste is mostly refuse dye stuffs, containing acids in various chemical combinations; from tanneries, acids, etc."

It is extremely important to note that by 1881 the streams entering southwestern Lake Erie were carrying heavy loads of silts which had become sufficiently great to influence the shallows of the western end of the lake, and undoubtedly had begun to coat the aquatic vegetation with silt. Howell (1882: 13) reporting as superintendent of Ohio fish hatcheries, states that during the late fall of 1881 he made extensive preparations for the taking of whitefish spawn in western Lake Erie between Locust Point in Ottawa County, Ohio, and the mouth of the Detroit River in Michigan. At this time the now-hidden gravel and bedrock reefs of western Lake Erie still supported a whitefish and cisco spawning population which in turn supported a fishery. After Howell and his associates had obtained only about 15,000,000 eggs they "were visited by a freshet of such magnitude as to swell the various streams, which, emptying into the lake their muddy waters, drove the white fish and herring [cisco] entirely off the shore into deep and clear water, rendering the further taking of spawn on the shore fisheries impossible, this proving that the white fish is most emphatically a clear or blue water fish." These turbid conditions caused Howell to go to the islands and Canadian waters for the remainder of the eggs. Obviously the silting-over of the once clean bottoms of the shallows of western Lake Erie, and the smothering of the once-abundant aquatic vegetation had begun.

Orton (Howe, 1900, I:89) sums up the pollution situation by lamenting the ever-increasing contamination which occurs in the streams during low water stages and "the base use to which we put these streams, great and small, in making them the sole receptacle of all the sewage and manufacturing waste." He continues, "as the amount of water in the river grows less during summer droughts . . . the polluted additions to the water are growing not only relatively but absolutely larger." Since these rivers contain the water supply which "cannot possibly be replaced" from other sources it becomes essential that the waters be kept pure, adding "During the first century of Ohio history not a single town has undertaken to meet this urgent demand of sanitary science, but the signs are multiplying that before the first quarter of the new century goes by the redemption of the rivers of Ohio from the pollution which the civilized occupation of the State has brought upon them and their restoration to their original purity will be at least well begun."

Concerning the decrease in the fish populations Potter (1877: 30) suggests in 1877 that additional reasons other than prevention by dams of upstream migration have caused this decrease, stating that "Many water courses have been strengthened and obstructions to the rapid transit of water removed, destroying the necessary hiding places of the fishes. Drains and ditches have been made all over the State, drying up the swamps and small streams, the natural breeding grounds of most of our river fish. . . . The wonder is that there are so many fishes left as there are. . . . Fishes must have cover as well as game, and when the white man destroys this, he must not expect either to remain with him long."

Commenting upon the "Causes of decrease of the supply of fish," Klippart (1874: 5–6) in 1873 summarizes the prevalent ideas of that period "Since Ohio has become a State [1803], seventeen millions of acres of forest has been removed. This extensive and rapid removal has most assuredly caused the drying up of many swamps and ponds which not only afforded shelter for many kinds of game, but were at the same time sources from which numerous brooks and streamlets derived their supply of water, and which served also as spawning beds for many varieties of fishes. The conversion of these swamps into arable land—as most of them now are—necessarily destroyed the spawning-beds and very seriously reduced the supply of fishes in the larger streams, because the instinct of the fish is to return to its birth-place [?] to spawn. In every part of the State are streams the waters of which forty years since [1834] afforded sufficient motive power to operate a grist or saw-mill, but which at present are absolutely dry during a period of more than four months each ordinary year. Of course such streams produce no fishes now, although all manner of native or indigenous fishes abounded in them at the time the mills were erected. The erection of dams for slack-water navigation have very seriously interfered with the propagation, and have very greatly diminished the supply. The greater number of the mill-dams are impassable for the ordinary fish of the streams, and renders it impossible for the gravid fish to find their way to the shallow streams to deposit their spawn."

Commenting upon the exploitation of fishes by commercial fishermen and the adverse effects of pollution upon fish life Klippart (1874: 9) declares that these "active causes are operating to reduce the quantity and thus diminish the supply of fish in the river, creeks and streams of the State. Even in the reservoirs and the lake [Erie] the supply of fish has very perceptibly decreased."

Lastly Klippart (Fisher, Klippart, Cummings, 1878: 66–68) concludes that in 1878 fishes lacked adequate protection from man, that in the future fishes must be considered more and more as objects of sport, and "as a luxury upon our tables, rather than as a necessary of life," that forest removal had "dried up many springs which formed a body of water of sufficient force to operate a saw mill or fulling mill, but of which, except during freshets, nothing but dry channel remains" and that the driftwood in streams, so abundant 50 years ago [1825] was now largely gone, leaving few hiding places for fishes or for their food. He points out, as did others, that the streams were becoming shallower and wider, and suggests that streambanks might be planted with willows, other trees or shrubs to improve habitat conditions.

Summary of 1851–1900 period.—The human population increased tremendously during this period whereas the large game and fur-bearing mammals, and better food fishes declined in abundance. The water mills reached their maximum number about 1880, then decreased in number rapidly as steam replaced water power. Each year the amount of land increased which had been drained, ditched and/or tiled. The over-draining of land first became evident during this period. Many streams became intermittent which formerly had a rather uniform flow throughout the year. The water table, once at or near the surface of the land, continued to sink lower and lower with the result that wells had to be dug deeper and deeper. Inorganic and organic pollutants increased greatly in number and amount. Clayey silt-bearing waters became conspicuous and widespread, and the silting over of stream and lake bottoms developed rapidly. Droughts were severe, and floods became increasingly disastrous to human life and property. Commercial fishing became a major industry.

Chiefly because of decreases in the abundance of better food fishes, the first law attempting to increase fish abundance was passed in 1857. It was an act of the Legislature and made illegal any device which prevented fishes from ascending streams. In 1873 an act of the Legislature created the Ohio Commission of Fisheries. This commission attempted to increase fish production through (1) propagating fishes in hatcheries, (2) introducing exotic species, (3) obtaining fishes elsewhere which were native to Ohio and liberating them in Ohio waters, (4) protection from exploitation through enforcement of law. Public sentiment against destruction of habitat, especially by pollution, increased during the period.

Conditions between the Years 1901 and 1954

The annual increase in human population was great during this period. The population had risen to 4,767,121 by 1910, to 5,759,394 by 1920, to 6,331,136 by 1930, to 6,907,612 by 1940, and to 7,946,627 by 1950. Population density by 1950 was about one person to every 3.6 acres (1.5 hectares) of water and land in Ohio, as compared with one person (of all races) to more than 20,000 acres (8094 hectares) in 1790.

Destruction of game habitat continued largely unabated between 1901–20. By 1920 all large game and large fur-bearing mammals had disappeared except for an occasional deer; the eastern turkey and greater prairie chicken had been extirpated and the ruffed grouse nearly so, and the bob-white quail had so decreased in abundance that in 1917 it was removed from the game bird list. Waterfowl populations had reached new lows. The sale of gamebirds had been prohibited by 1920. To offset the decreasing game bird supply, exotic species were introduced such as the ringneck pheasant and European partridge.

During the first half of the period it became increasingly evident that destruction of habitat was a primary factor in the decrease of mammals and birds; even so, little was done about it. Instead, stress was placed upon shorter open seasons, smaller bag limits, vermin control, and more stringent law enforcement. But after 1920, habitat requirements for game were given an ever increasing amount of consideration by the public and conservation authorities, and by 1950 efforts to increase the amount of suitable habitats for game species had become a prime objective. After 1930 the ruffed grouse population slowly increased, as worn-out lands were allowed to revert to brush and second-growth forest. Recently the beaver has reappeared in eastern Ohio, and attempts have been made to re-establish the eastern turkey.

The damming of streams entered a new phase during this period. By 1920 all except a few of the many mill dams had disappeared. Between 1925–40 much interest was displayed in low dams, and many were constructed. Their supposed function was to aid in fish production by impounding water which would serve as wintering areas and refuges during drought. But the dams tended to prohibit upstream migration as had the old mill-dams, many were destroyed during floods, or the impoundments behind the dams filled rapidly with silt or gravel.

As the human population increased, destruction of life and property by floods increased greatly. The great flood of 1913 aroused Ohioans, so that immediately afterward flood control measures were undertaken. Notable among these was the construction of huge flood control dams in the upper Great Miami watershed. These dams proved successful in preventing major flooding. Later, similar prevention measures were undertaken. Outstanding among these were the huge impoundments of the Muskingum Conservancy District. Flood control dams usually impounded relatively little water during summer, and were subjected to great fluctuations in water levels. Despite this many retained sufficient water so that on the whole they augmented fish production considerably.

The demand for ever-increasing amounts of water by municipalities and industries continued throughout the period. Many huge impoundments were constructed to meet this demand, such as Pymatuning Lake and Berlin Reservoir. These reservoirs were built for water storage, not fish production, and because of the manner in which the water was used, water levels often fluctuated greatly. These fluctuations produced much submarginal fish habitat, and as a consequence fish production was usually low, or the fishes were of the unwanted kinds such as carpsuckers, or the food and game species were dwarfed in size.

During the latter half of the period many headwater lakes were established. These were situated far enough upstream so that there was little interference with fish migration. They were built to produce fishes for sport fishing, their dams and other structures were built under the supervision of the conservation department, and the building and maintenance were paid for by the sportsmen. Considerable thought was given to the managements of these impoundments; as a result, many of them have become the principal sport fishing areas in their section.

The ponding for navigation purposes of such streams as the Muskingum and Ohio rivers had a most pronounced effect upon their fish populations. The ponding almost or entirely eliminated riffles. It increased the amount of silt settling in the ponded areas and the silt covered the former sand, gravel and muck bottoms, sometimes to a depth of several feet. As a group those species requiring riffles and/or clean, hard bottoms were the most adversely affected; in fact, such species as the crystal darter appear to have been extirpated, whereas others such as the mooneye, river and gravel chubs, smallmouth blackbass, and river darter were greatly reduced in numbers. Conversely, species demanding turbid, quiet waters and/or silted bottoms were benefited and some of these, such as the carp, are now among the most numerous species found in these waters.

Ditching, dredging, draining and tiling continued, and often at an accelerated pace. Much rivalry existed among the county commissioners, each trying to ditch, drain, straighten, or entirely eliminate the greatest number of miles of streams. By the end of the period, over-draining of land had become so evident that reaction against it had occurred. This reaction, if continued, eventually may have considerable beneficial results.

By the latter part of the period the once universally abundant springs had become a rarity in many sections, or they flowed only during the wettest portions of the year. There was a great increase in the use of ground water by industry and municipalities and for air-conditioning of large buildings, resulting in the water table (ground water) dropping ever lower beneath the surface. The average drop during the 1920–40 period was almost 20′ (6 m), and in one locality it was 90′ (27 m) (Ranney, 1940: 8). Many wells repeatedly ran dry and had to be dug deeper; but in a few localities near huge impoundments the water table raised and wells long dry again contained water.

Although human population increase in some cities was great, the amount of water used increased far faster in proportion. In many cities there was an insufficient supply of water during the drier months and, as a result, many municipalities had to curb water use. The streams below water storage reservoirs were often reduced to a small part of their normal flow or became dry because the water in the reservoirs was used by the cities.

Streams, some of moderate size, ceased to flow during droughts even though their waters were not directly utilized. Leading Creek in southeastern Ohio was an example. This stream normally flowed continuously throughout the year but on August 3, 1930 it was completely dry except for a few isolated pools. These pools produced only 23 species of fishes, whereas under normal conditions 30–43 species could be taken at a good seining station. The fish population of Leading Creek continued to be low after 1930 until 1933 after which it appeared to have attained its 1929 abundance.

Floods of a general nature involving all of one or both major drainages occurred with few exceptions only during the first quarter of the year. Localized, flash floods also occurred most frequently during the first quarter of the year but were not uncommon during the remainder. It was the prevalent belief among fishermen that many fishes were swept far downstream during floods and that fishing was poor after floods. Percy Viosca obtained evidence partly substantiating this theory. He told me that immediately after huge floods in the upper Mississippi watershed, he noted an increase in the number of northern smallmouth blackbasses in the lower Missippi River in Louisiana.

Pollutants of many types were highly detrimental to fish life (Trautman, 1933: 69–72). Many stream sections which before 1950 had contained an abundant fish fauna had become almost or entirely devoid of fish life by 1950; "Loveland's Ripple," in the Mahoning River between Youngstown and Poland in Mahoning County is an example. Before 1850 J. P. Kirtland collected from this riffle many species of fishes, including the clear-water inhabiting spotted and variegate darters. Since 1925 this section of the Mahoning has been bordered by steel mills which when working have heated and polluted the stream beyond the point where the waters were inhabitable to fish life.

Studies made since 1925 have proved that since then, if not before, soil suspended in water has been the most universal pollutant in Ohio, and *the* one which has most drastically affected the fish fauna. Clayey soils, suspended in water, prohibited the proper penetration of light, thereby preventing development of the aquatic vegetation, of the food of fishes, of fish eggs and of fry. Some types of water-suspended clays impacted about the gills of fishes, asphyxiating them (*see* greater redhorse under Distribution; also Wallen, 1951: 1–27). Settling over the formerly clean water bottoms, silt destroyed the habitats of those fish species requiring bottoms of sand, gravel, boulders, bedrock or organic debris.

Many streams of low gradient formerly contained much aquatic and land vegetation, especially those streams flowing through prairies, and the till- and lake-plains. These streams greatly overflowed their banks, their flood plains remaining water-covered usually from January until early summer. The flood plains contained much decaying land and/or aquatic vegetation, and it was this water-soaked, dead vegetation which presumably aided greatly in reducing soil turbidity to a minimum. Irwin (1945: 3–16), and Irwin and Stevenson (1951: 1–54) demonstrated recently "that organic matter clarifies turbid waters primarily through the action of liberated hydrogen ions," and that wherever aquatic vegetation is well established clear water is maintained. I have seen this phenomenon abundantly demonstrated in several Ohio waters of which Long Pond of Indian Lake is an illustration. Between 1908–12 Long Pond was choked with aquatic vegetation and its waters were exceedingly clear, even when the open waters of Indian Lake were turbid. As the stumps and logs were removed, the aquatic vegetation became less profuse, and with the decrease of aquatic vegetation came a corresponding increase in turbidity, until by 1950 the vegetationless portions of Long Pond contained as turbid waters as were the waters of the open lake. It was only in those remaining sections of dense vegetation that clear water could still be found.

Prior to 1850 the clear-water, prairie-type streams contained a fish fauna dominated by species requiring the clearest of waters and/or an abundance of aquatic vegetation. Sheet and gully erosion of the fine lake- and till-plain silts and prairie soils became great shortly after the lands were plowed which were drained by these streams, with the result that by 1950 these formerly clear, much-vegetated streams became the most turbid, least-vegetated waters in the state. As these streams became increasingly turbid the habitats of the clear-water inhabiting species were almost or entirely destroyed, and habitats for turbid-water and silt-bottomed species came into existence, or increased greatly in amount. As the formerly clear, prairie streams of Illinois and Indiana also became turbid, conditions in them became favorable for such inhabitants of the turbid, western plains streams as the suckermouth minnow and orangespotted sunfish. As a consequence, such species migrated eastward to become a part of the present, turbid-water fauna of Ohio. The result has been the extirpation or decrease in abundance of those species least tolerant to turbidity and siltation, and the great increase in abundance of those species most tolerant to turbidity and siltation.

In this 1901–50 period soil erosion became of increasing interest to state and federal agencies, and to the public, and during the latter half determined attempts to curb erosion were inaugurated. A notable example is the effort of the U.S. Soil Conservation Service in the Muskingum Conservancy District (Morse, 1939: 1–39).

Soil and other pollutants, and ditching and dredging, affected adversely the aquatic vegetation prior to 1900, and these adverse effects continued during the 1901–50 period. I observed the decrease in amount, or elimination, of aquatic vegetation in flowing waters. Large beds of aquatic vegetation were present until 1930 in sections of such streams as the Maumee and Muskingum rivers, and portions of their tributaries were choked with it. By 1950 all except a few beds or small remnants of the vegetation had disappeared.

The decrease in amount of aquatic vegetation in static waters was likewise great. Before 1940 Middle Harbor in Ottawa County contained a profusion of vegetation consisting of a wild celery-water lily-hornwort-cattail association. It also contained a fish fauna consisting of 51 species which included a large population of fine food and game fishes. Throughout the years silt from the eroding watershed accumulated in the harbor, and when after 1940 some of the harbor bottom was destroyed by dredges building dikes and canals, the disturbed silt was sufficient, when in suspension, to smother completely and kill almost all of the aquatic vegetation. With the vegetation almost gone many fish habitats were eliminated. Consequently, many of the 51 fish species disappeared or were greatly reduced in numbers. By 1948, when the remaining fishes were purposely killed with rotenone, more than 90 per cent of the fish population consisted of carp, goldfish, their hybrids, and dwarfed bullheads.

When the old canal reservoirs were constructed, before 1850, much or all of the timber, brush and other vegetation was left on that land which became water-covered with impoundment. This vegetation, plus the great fertility of the swamp soils, soon produced a profuse amount of aquatic vegetation; in fact, the down timber, stumps and vegetation were so dense that progress in a boat in mid-summer was almost impossible (Trautman, 1940A: 30–34, for Buckeye Lake). The profusion of organic matter was thought to be detrimental to fishes; that it made their flesh taste "woody." As a result, it was suggested as early as 1874 that state agencies remove the timber, stumps, etc. (Klippart and Hussey, 1874: 10–11). Little removal was done, however, before 1911. During the period when down timber and vegetation were abundant, between 1860–1911, fish production in the reservoirs was very great, and those species such as the largemouth blackbass which require logs, snags, stumps, and/or aquatic vegetation, were very abundant, remaining so until about 1914. Between 1907–14 I saw many "strings" each totaling more than 30 large blackbass which were caught in a day at Indian Lake by two men in a boat.* The late Carl E. Balz told me that in August of 1902 or 1903 he and Carl Buchsieb caught 51 largemouths in Indian Lake during a 5-hour period, the fishes weighing between 3/4–3 1/4 lb. (.34–1.5 kg) each. Many old market fishermen have told me that they caught between 60–90 large saleable largemouths in a day. By 1930 most of the timber and stumps had been removed, much of the vegetation was gone, turbidity had increased and strings of 50 bass in a day were impossible, had such strings been legal. With the environmental changes in these lakes there was a corresponding change in the fish population from that of a largemouth blackbass-bluegill-pumpkinseed-yellow bullhead association to that of a white crappie-brown and black bullhead-channel catfish association.

Summary of 1901–50 period.—Human population continued to increase so that by 1950 there were 7,946,627 persons living in Ohio. Large game mammals and game birds disappeared early in this period; later a few species became re-established. Water mills became obsolete. Many huge impoundments were constructed; their primary purpose was to impound water for industrial and domestic uses, or they were built to prevent disastrous floods. Neither of these types of impoundments was suitable for a high rate of fish production. Before 1940 many low dams were built in streams, whose purpose was to increase fish production, but they were largely unsuccessful in this regard. After 1935 headwater reservoirs were constructed for the purpose of increasing fish production and the amount of water acreage available for sport fishing. Many of these reservoirs were highly successful. The ditching, draining, and dredging of streams and lands, and tiling of fields, continued at an increasing rate until about 1940, after which the public became conscious of the evils of over-draining, and such practices were lessened in some counties. Springs became a rarity in many sections. The water table continued to sink lower as the demand for water increased. Many streams, some of moderate—or large—size, ceased flowing during dry periods, temporarily disrupting the fish fauna. Fishing often became temporarily poor after high floods. Industrial and domestic pollutants became more prevalent and pollutants in some stream sections caused a partial or complete elimination of the former fish fauna. It became obvious during this period that turbidity from eroding soils, particularly of fine clays, was the most detrimental and the most universal of all pollutants, and that because of turbidity and resultant siltation of water bottoms the fish fauna of many waters was drastically modified. Aquatic vegetation on the whole decreased in all waters, and after its disappearance these waters often became more turbid. Commercial fishing was almost entirely restricted to Lake Erie. In the first half of the period the trend in conservation was toward greater and more restrictions; during the latter half it was towards more liberalized sport fishing and the emphasis was placed upon habitat improvement.

Summary for entire 1750—1950 period.—In 1750, the land of that portion of the Ohio Country, comprising the future state of Ohio, was covered with forests, except in parts of the western, northwestern, and central northern sections where there was a considerable amount of "prairie." Those streams draining the larger prairies had a typical "prairie" fish fauna.

On the whole there was little eroding of the soils because the mass of living and dead vegetation was sufficient to prevent erosion, except during great floods, and in small areas such as hillsides and stream banks. Stream banks were heavily wooded or covered with brush, which tended to make the streams narrow and deep. The trees overhanging the streams provided much shade, and there was considerable down timber and brush in the streams. Because of the slight amount of erosion, the waters contained virtually no clayey silts in suspension, and were clear except during floods, or when stained by decaying vegetation. The stream bottoms consisted almost entirely of clean sands, gravels, boulders, bedrock, and muck, peat and other organic debris. The amount of clayey silt of stream and lake bottoms was negligible.

There were few substances in the waters which were deleterious to fish life. Occasionally the amount of decaying vegetable matter was sufficient to cause a local fish mortality, and some springs contained enough chemicals to make their waters unfit for fish life. Aquatic vegetation was abundant in many localities, especially in static waters.

The population of fishes was very great, especially of large fishes desired as human food. There were

* These fishes were usually caught by "flipping." A long cane pole had a line attached to its tip, the line being of the same length as the pole. Attached to the free end of the line was a "flip jack," or Joliet Spinner. One man rowed, the other stood in the stern of the boat and "flipped" the lure around the snags, logs, stumps, and beds of vegetation.

great upstream migrations of fishes each spring. Those species were most numerous which required clear waters, and/or clean, firm bottoms, and/or soft bottoms of organic debris, and/or much aquatic vegetation and down timber. Man utilized the fishes as food, especially during periods of game and vegetable food scarcity.

The streams and their fish fauna have been greatly modified since 1750. The removal of the trees and brush along stream banks has been conducive to the widening and shallowing of the streams. Many streams have been ditched or eliminated, and many marshes drained. Many streams and springs have ceased to flow or flow intermittently following the drop in the water table which followed over-draining. Forest removal and agricultural practices have caused greatly increased erosion of land, resulting in greatly increasing the turbidity of the waters and the silting over of the once clay-free stream and lake bottoms. Many pollutants of industrial and domestic origins have adversely affected many species of fishes. Aquatic vegetation has decreased greatly in amount.

These drastic modifications have considerably modified the fish fauna, changing it from a species complex, dominated by fishes requiring clear and /or vegetated waters to one dominated by those species tolerant of much turbidity of water and of bottoms composed of clayey silts. There has been a shift from large fishes of great food value to smaller species unfit as human food, or large fishes of inferior quality as human food.

Before 1800, fishes were captured primarily for food and not as sport. After 1800 commercial and sport fishing rapidly increased in amount. After 1850 commercial fishing in inland Ohio became increasingly restricted by law, until early in the 20th century it was almost entirely prohibited, leaving only Lake Erie open to commercial fishing. After 1850 sport fishing continued to increase until it became a major pastime in Ohio.

Before 1800 there were few inland lakes. During the 19th century some large reservoirs were built, and in the 20th an ever-increasing number of headwater lakes, farm ponds, water storage reservoirs, flood control and other impoundments were constructed. Before 1900 most of the sport fishing was done in streams, and many Ohio streams were noted for their excellent smallmouth blackbass fishing. Since 1900 stream fishing has deteriorated because of increasingly unfavorable habitat conditions in streams. The great increase in acreage of impounded waters has considerably increased the amount of fishing in such waters.

The trend after 1950 will continue towards increased fishing in impounded waters unless successful measures to rehabilitate the streams can be undertaken.

Conditions between the Years 1955 and 1980

This period was characterized by a continuation, and in many instances a great acceleration, of trends begun in previous periods. The annual increase in human population was greater than ever before, rising from 7,946,627 in 1950 to 9,606,397 in 1960, then to an approximate 10,652,017 on April 1, 1970 and to 10,735,000 by 1975. Ohio population density in 1950 was 3.6 acres (1.5 hectare) of land and water per person; in 1960 it was approximately 2.9 acres (1.2 hectare) per person; and by 1970, approximately 2.7 acres (1.1 hectare) per person.

Not only did the human population increase more rapidly than previously, but industrialization, urbanization and the number of highways increased more rapidly in proportion than did the population. Consequently, many thousands of acres were converted from good farmland for use by industry, for home sites and for multilane, interstate and other types of highways. Agriculture became more mechanized and specialized. This resulted in the need of many farmers to enlarge the size of their farms because the increasingly expensive and diverse farm machinery made smaller farms unprofitable. The average field was increased in size to facilitate mechanized cultivation. Hundreds of miles of fence rows were eliminated. Those allowed to remain were reduced in width from several to a few feet, and brush or other cover was largely eliminated. The amount of fall and winter plowing was greatly increased, thereby exposing bare land to wind and water erosion through the winter and early spring. Because of improved efficiency and decreased price, the number of power saws to cut brush and timber increased several times. Although some brush and tree removal was for the purpose of obtaining lumber or for more land for agriculture, some of the cutting appeared to be done because of the ready availability of power saws. Bulldozers increased greatly in size and numbers, some so powerful that several acres of wood- and brush-land could be cleared and leveled in a day. There was an increase in strip mining; such monster machines as the "Big Muskie," which is as tall as a 32-story building and whose bucket could remove 220 cubic yards (168 cubic meters) of earth in a single bite, produced great and fairly permanent scars across the landscape of eastern and southeastern Ohio.

In 1950, 26 low-lift dams spanned the Ohio River between the Pennsylvania and Indiana state lines, providing a continuous channel at least 9′ (3 m) deep. Since 1950 a modernizing program eliminated 15 of these structures because of the construction of 6 high-lift dams, the minimum channel depth remaining at 9′ (3 m). At least one more high-lift dam is under construction. This increased ponding of the Ohio River by the higher dams permanently flooded thousands of acres of adjacent lands and many tributaries from their mouths to varying distances upstream. This increased ponding further altered the environment of this originally free-flowing river (Carl B. Ballengee, personal communication, December 15, 1969).

By the latter part of the 25-year period, a fair proportion of the former natural Ohio landscape disappeared, resulting in the establishment of a more artificial one and a realization by the public of the changes rapidly taking place. Organizations, such as the Nature Conservancy, came into being whose functions were to save the more noteworthy, scenic, vegetative, and faunal areas of the state. State agencies such as the Ohio Division of Wildlife bought submarginal lands with public and hunters' license monies, some of these lands containing hundreds of acres. Unfortunately, instead of allowing some of these areas to continue as swamps, marshes, prairies or woodlands, and allowing the flora and fauna to remain or increase, they were partially or completely destroyed because the land was filled in, leveled off and the vegetation largely eliminated. Sometimes this provided a wholly artificial, recreational complex consisting of hotels, trailers, camps, toilets, golf courses and innumerable picnic tables, thereby changing an interesting remnant of the Ohio landscape, which should have been left intact for future Ohioans to enjoy. In some woodlands, whose primary purpose was game for sportsmen, the "wolf", den, and mature trees, which produced maximum crops of mast, were cut down to allow young, straight, denless trees, producing little mast, to grow for future lumbering, thereby largely destroying these areas for the purpose for which they were purchased. Despite attempts at saving relics of natural Ohio, the overall effect was an increasingly barren and uninteresting landscape.

The increased number and width of highways, increased size of cities and villages with their paved streets and parking complexes, increased amount of winter plowing, vegetation removal and land drainage resulted in a more efficient and rapid runoff of water, which in turn resulted in more and larger flash floods. Streams that heretofore flowed throughout the year and contained a more-or-less stable fish population became raging torrents following heavy rains, later to become intermittent during droughts. Many stream valleys were narrowed and otherwise encroached upon by the building of dikes, urban and industrial complexes and highways, so that the flood crest was considerably increased, thereby increasing the flooding in some cases until floods became of more-or-less annual occurrence. Former stream beds were filled in and highways built in their places, and canal-like dredged ditches, devoid of most water and land-animal habitats, were substituted. Many streams were channelized, especially by the Federal government, resulting in more rapid drainage of water, which was sometimes of value to the farmer but not always.

Channelization usually resulted in drastically changing the composition and abundance of the various species of plants and animals, not infrequently eliminating many species, some of which were prize game and food fishes (Trautman and Gartman, 1974). The Mad River above Springfield, in Clark and Champaign counties, is an example. The composition of the fish fauna and the condition of the Mad River before it was dredged were learned many years ago through conversations with old fishermen and farmers who lived in the area before as well as after dredging. These conversations led to the conclusion that before dredging there was a "normal" fish fauna of suckers, catfishes, crappies, rockbass, blackbasses and other sunfish species compatible with a stream of its size, and that "At that time the stream meandered greatly, containing alternating deep pools and riffles, prime habitats for bass and associated species" (Trautman, 1942:218).

Dredging of this stream to drain swamps and facilitate stream flow began before 1915 (U.S. Geological Survey, 1915: East Liberty Quadrangle), by which time it had been dredged at least once throughout almost its entire length. This dredging completely modified the stream by creating a straight, artificial ditch without riffles or pools. It almost entirely abolished the former medium- or large-sized stream fish fauna, composed of such species as members of the sucker genus *Moxostoma*, catfishes, and blackbasses (only strays of which have been recorded since), and substituted a fauna dominated by such essentially brook species as the blacknose dace, creek chub, redbelly dace, redside dace, and mottled sculpin.

The change of the Mad River from a straightened ditch to a "normal" stream again would take many years or centuries even if it were left unmolested. Some of the high spoil banks, remnants of former dredging are so old that trees 2′ (0.6 m) DBH are growing on them, yet the riffles and pools of the stream remain undeveloped and the fish fauna remains that of a small-brook association.

To partially offset the accelerated and increasingly rapid run-off, Federal and State governments constructed a larger and an ever-increasing number of impoundments, hoping to reduce property damage downstream. In some instances this resulted in the loss of considerable valuable agricultural land, land which in the future may become needed for agriculture to feed an ever-expanding human population.

In other instances, impoundments flooded or destroyed forever areas of great natural and scenic value. The natural vegetation, often scenic, surrounding other impoundments was removed, and huge trailer camps and picnic areas were built. Large numbers of speedboats, with their noise and pollution, plied the waters to the disgust of the die-hard fishermen, contributing to the rapid and continued depletion of the world's supply of fossil fuels. Opportunities for fishing, however, were increased in some impoundments.

As the period progressed, the public became increasingly aware of, and alarmed by, the harmful effects of pollutants and of the increased refuse and garbage littering roads, streams, lakes and the countryside in general; but it was not sufficiently aroused to result in a marked decrease in pollution or littering. Disposal of trash and garbage became an ever-increasing problem. City, State and Federal governments established agencies to find methods of reducing the polluting of air, water and land and to slow down, if not stop, the general deterioration of the environment. The amounts of some kinds of pollutants were increased, decreased or their chemical composition altered.

On the whole, the amount and speed of the silting of stream substrates from erosion of land appeared to decrease slightly. At least, there apparently was a noticeable decrease in the amount of silt deposited in those sections of streams which we have been studying for the past forty years, such as a 2-mile (3.2 -km) section of Big Darby Creek southwest of Fox, Ohio. This decrease possibly may have been the result of two or more principle factors: (1) increased soil conservation practices, (2) the previous removal of the humus and lighter soils, leaving the heavier clays, which did not erode so readily.

The increasing human population and the increased amount of industrialization produced larger quantities of organic and inorganic pollutants than previously. The amounts of water used daily for domestic consumption and industry increased more rapidly in proportion than did the human population, resulting in some areas in shortages of potable water during droughts. More waters than ever before became so polluted that none or only the most resistant forms of fish and other aquatic life remained. The lower Cuyahoga River in Cleveland became so inflammable that it was declared a fire hazard (White, et al; 1975).

Commercial fertilizers had been used in Ohio for many years prior to 1950, but never in such huge amounts as recently and especially in the highly cultivated lands of northwestern Ohio. Leaching through soil and into waters of fertilizers, and especially liquid ones, was comparatively rapid, usually increasing fertility of the waters. This increase was beneficial in those few waters in Ohio which were low in fertility, but in already highly fertile water this added enrichment resulted in oxygen removal through its use by algae, plankton, and so on, creating undesirable or intolerable situations for the public, some fishes and other aquatics.

Hard and soft detergents were little used before 1950. The amount used has increased remarkably since then. Foam from detergents has occasionally become 5′ (1.5 m) or more deep at some disposal plants. Large masses of foam 5′ (1.5 m) high and 10′ (3 m) in diameter drifted down the Scioto River below Columbus. However, toward the end of the period detergents contained fewer foam-producing agents; at least, there was a marked reduction in the amount of foam evident. Some detergents adhered to the substrate of streams, reducing the number or destroying the habitat of many aquatic species.

The use of pesticides, insecticides and herbicides increased enormously, resulting in some fish kills.

At the end of the period the detrimental effects of DDT (dichlorodiphenyltrichloroethane), Dieldrin, and other persistent or "non-degradable" pesticides became so apparent that in 1970 the use of DDT was greatly restricted. Fishes from a few waters contained sufficient pesticides so that their use as human food was discouraged.

The amount of trash (garbage, refuse or rubbish such as containers, plastics, papers, metals, and other domestic and industrial wastes) increased until there was an estimated 5 to 11 lbs (2–5 kg) per person per day! Originally trash was largely deposited to fill natural depressions, but such convenient places of disposal became uncommon. There was a tendency to throw more and more trash into streams or along their banks, anticipating that a rise in water level would carry it downstream. Great quantities of aluminum, glass, tin and other containers so littered the stream bottoms that swimming, or wading while fishing, became hazardous. Following floods, the banks of some streams presented an incongruous sight with cans, bottles and other litter lining the banks, and sheets of various-colored plastics festooning the trees. Toward the end of the period a law was passed prohibiting the dumping of trash in streams or along their banks, and attempted enforcement was begun.

In Ohio during this 20-year period, Lake Erie became a most publicized area regarding the rapid changes and deterioration of an aquatic habitat (White et al.; 1975), especially as concerned the western third of the lake. Between the years 1939 and 1955 I lived on South Bass Island where it was possible for me to observe continuously the changing environment. Many of these changes were recorded in the 1957 publication of this volume, as were the marked changes in abundance and distribution of many species of vertebrates, including several fish species of considerable commercial and/or sport value, past or present, including the lake sturgeon, ciscoes, whitefish, northern pike, Great Lakes muskellunge, and smallmouth blackbass (Van Meter and Trautman; 1970).

Because of the rather slow and gradual rate of change in the environment of the lake prior to 1950, I assumed that the rate of change in the abundance of fishes and other animals would continue at the same speed as it had in the past. In this I was mistaken. Beginning about 1952 there occurred a series of rapid changes. Britt (1955:242–44; 1966:14–15) has demonstrated that prior to 1953, dominant organisms in the soft mud bottom of western Lake Erie were the mayflies of the genus *Hexagenia*; that after 1953 a catastrophic decline in numbers occurred from the previous average population of approximately four hundred larvae per square meter of lake substrate to virtually none by 1965. Conversely, the midge (Chironomidae) larval population, which averaged only seventeen individuals per square meter in 1930 rose to 4,425 larvae per square meter by 1959. A further significant change was a drastic increase in sludge worms (Oligochaetes), which in 22 collections taken in 1930 averaged only five individuals per square meter whereas in six collections taken from the same locality in 1965, averaged 7,280 per square meter, with the largest number for a collection containing 12,570, and by 1977 to more than 50,000 per square meter.

Beeton (1965:2;7–49) points out that many kinds of dissolved solids have "increased significantly" in amount in Lake Erie during the past 50 years; that there was a marked increase or change in abundance of plankton; that the number and sizes of blooms of blue-green algae (p. 175) have increased. Others have substantiated these conclusions.

Beeton (1961:153 and 1969:173–77 and fig. 13), in discussing the changes of abundance of some Lake Erie fish species of commercial importance, shows a decrease from millions of pounds taken annually between 1870 and 1900 to commercial or near commercial extinction by 1965. Hartman (1970, 1972), Van Meter and Trautman (1970:66–78) and White et al., (1975) likewise discuss decreases and a few increases in many fish species during the past hundred years. Some of these changes in fish abundance are discussed in this volume under their respective species. For fish population trends in Lake Erie tributaries, see Clark and Allison, 1966.

From the above, it is obvious that during the past century Lake Erie changed from a slower-aging to a far more rapidly aging, highly enriched body of water, in which amounts of organic and inorganic pollutants continue to increase. During the latter part of the period many diverse agencies began the formulation of plans, some drastic and entailing vast sums of money. It is believed that through the execution of some of these plans more favorable conditions will ensue in the future. The other Great Lakes likewise have been experiencing modifications in fauna and flora (Morman, 1979:2; Crossman and Van Meter, 1979:6).

Not all of the changes during the 1955–80 period

have adversely affected fish life however. The 30± mile-(48±km-) section of the Scioto River from Columbus downstream to Circleville is an example of a recent increase in the number of fish species and individuals.

Pollution adversely affecting aquatic life in this stream section began early in Ohio history. The first important adverse factor, I believe, appears to have been forest removal. Sheet and gully erosion began with the removal of the protective trees, which allowed the raindrops to dislodge, and the water to remove, the accumulated light humus from the forest floor. Sawdust from the many water-powered saw mills contributed to the accumulation of organic matter on stream bottoms. Humus and sawdust were capable, through bacteriological action, of removing sufficient oxygen during low-water periods to eliminate many fish species and other oxygen-needing animals. Naiads, or fresh-water mollusks, were particularly susceptible to silting and other pollutants, largely because of the relative immobility of these bottom-dwelling animals. As early as 1858, Higgens (1858:548–51), writing concerning the great decreases of "The Mollusca of the Vicinity of Columbus," stated that "Gentlemen who collected the shells of this vicinity in early times found many species in great abundance which have at this day either totally disappeared or are represented by occasional straggling specimens, and all species, with but few exceptions, have gradually decreased in numbers." Of the 48 species recorded by Higgens as originally occurring in the Scioto River, by 1858 six of these had become extirpated and four others were on the verge of extirpation. Stansbery (1961:21–22) points out that since 1858 five additional molluscan species have been extirpated and that an additional 13 species are on the verge of extirpation, which, by 1961, leaves only 24 species of the original 48 recorded by Higgens still occurring in goodly numbers.

Apparently a normal fish population remained in the Scioto River between Columbus and Circleville until about 1850. Kirtland (1850L:21) reported at that time of having obtained specimens of the variegate darter from the Scioto River at or near Columbus. It is assumed that if this clear-water-inhabiting species were capable of existing in this stream section before 1850, some of its associates could do likewise. A decrease in species and individuals occurred after 1850, probably at a slower rate than did the mollusks because the fishes were more mobile and migratory. Presumably by 1900, few fish species and individuals remained, largely because of pollution, Orton (Howe, 1900 I:89) considering such streams as the Scioto River below Columbus as "open sewers" and "the sole receptacle for all sewage and manufacturing wastes." Osburn and Williamson (1898:11–20), working Franklin County streams in 1897, recorded 48 species of fishes (no variegate darters) from the Scioto River from both *above* and below Columbus. During many conversations with R. C. Osburn and E. B. Williamson between the years 1925 and 1940, they described in considerable detail the deplorable conditions existing in the Scioto River below Columbus during 1897. They told me that during high waters or immediately following there was a small population of fish species and individuals in this stream section whose numbers were reduced during droughts to only a few highly resistant individuals of such species as the common carp and black bullhead. I found similar conditions between 1922 and 1932 in the same stream section, with the waters blackish from bacteriological action, the substrate heavily coated or buried to a depth of 2′ (0.6 m) with a slimy, black muck, and the few fishes present usually having their fins eroded away.

During the 1930's (Anonymous 1960:8–11) Columbus built a new sewage treatment plant, which in 1950 was enlarged to a daily capacity of 60 million gallons (227,124,707 l). Since then the capacity has been more than doubled. By 1955 the first 10± miles (16± km) of river below Columbus still remained blackish during low stream levels and with hundreds of sludge worms per square meter in that section nearest sewage plant outlets. The middle 10± miles (16± km), below the confluence with Big Walnut Creek and especially during low streamflow, usually was greenish in color, caused largely by an extraordinarily large population of phytoplankton. There were huge floating masses of duckweed (Lemnaceae). Many riffles were choked with Sago pondweed (*Potamogeton pectinatus*). There was sufficient oxygen in the water to support many aquatic invertebrates and vertebrates. The 10± miles (16± km) between the mouth of Little Walnut Creek and Circleville were grayish in color supporting such enormous numbers of zooplankton that visibility through the water was at times reduced to less than 3″ (8 cm).

Although since 1959 all three sections contained fishes at all seasons of the year, the lowest section above Circleville contained by far the greatest number of species and individuals. In this section between 1959 and 1969 there were recorded a total of 76 species, some occurring as strays, others in immense numbers even during drought periods. The capturing through test netting of large walleyes, three species of blackbasses, and other species, when examined proved to be in excellent condition and to have grown rapidly. Smaller individuals of most species likewise had a high coefficient of condition, the yearling smallmouth blackbasses growing to an average total length of 5″ (13 cm) by late fall whereas specimens from adjacent Big Darby Creek averaged only 3″ (8 cm). Fry of such clear-water-inhabitating species as the variegate darter were found.

During some late nights and at daybreak in summer, usually when foggy conditions prevailed and/or when streamflow was low, the water contained no oxygen (0.00 ppm.). At such times apparently the entire and enormous fish population from the smallest in size to those weighing many pounds could be seen sucking air from the water's surface. Despite this oxygenless condition, there occurred little or no fish mortality.

How long this large population of species and individuals remains depends upon the ability of the sewage treatment plants to process the effluent properly and to allow sufficient clean water to escape beyond Columbus. Once the human population grows so large and the amount of water used becomes so great that there is an insufficient supply of clean water flowing downstream from Columbus, then this stream section will again return to 1900–1925 conditions.

As the fishing in streams declined in general throughout the period, there occurred a sharp increase in the amount of fishing in private farm ponds, "pay lakes" and other small artificial impoundments. Prior to 1950 there existed in Ohio several hundred small artificial impoundments; these were largely utilized for the watering of livestock. After 1950 the number of farm ponds constructed annually increased greatly. H. Granville Smith formerly of the U.S. Soil Conservation Service informed me that by June 30, 1969, that agency had given advice upon the building of 22,256 ponds in this state. In addition, hundreds of ponds were built without assistance from the Federal Government, so that by 1970 the total number of farm ponds in existence may have approximated 40,000. These ponds ranged in size from a fraction of an acre to more than 10 acres (4 hectares). All except a small percentage of these artificial impoundments contained fishes. The vast majority were stocked with largemouth blackbasses and bluegills, the remainder with redear, green, or other species of sunfishes, black or white crappies, smallmouth or spotted blackbasses, channel or bullhead catfishes, walleyes, trouts, and so on. In many ponds there occurred an over-production of stunted bluegills, usually the result of over-cropping of the principal predator, the largemouth blackbass, too little cropping of bluegills by man, or too favorable conditions for the production of bluegills, or a combination of these factors. By 1970 the total number of bluegills taken each year in Ohio surely exceeded the total number taken annually between 1920 and 1950.

Fish production in those large, newly constructed impoundments or reservoirs, the result of damming large streams, usually followed a fairly uniform cycle. Shortly after impoundment there occurred a considerable increase in the number of food and game fishes, and these grew rapidly. This condition was later followed by a decline which after several years sometimes resulted in populations of fewer and/or stunted fishes. Primary causes for a rapid increase followed by a decrease in numbers is largely undetermined, but heavy silting over the substrate, which destroyed their habitats, great fluctuations in water levels, and turbidity appear to be in part responsible for this cycle.

Since early in the last century, man has attempted, with the idea of improvement, to introduce exotic species of fishes into Ohio waters. With few exceptions, these introductions have been failures; the two introduced species most successful were the common carp and the goldfish. Throughout this 25-year period the introductions of exotics have continued, and during the latter years highly publicized attempts have been made to introduce the striped bass in inland Ohio waters and the coho and chinook salmons in Lake Erie (Parsons, 1973, 1974).

Prior to 1950 some private individuals disposed of unwanted aquarium fishes by releasing them into the lakes, ponds and streams of Ohio. After 1950 this method of disposal of unwanted aquarium fishes increased because of the increase in numbers of persons having aquaria and aquarium fishes. The result has been that, especially during the last decade, several exotic fish species and other species of vertebrates have been captured while seining for native or established exotic fish species. An example occurred on October 22, 1966, when an Ohio State University class, seining Big Darby Creek in Union County, collected a Mexican tetra, *Astyanax mexicanus* (Filippi) (OSUM 14147). It is expected that in the future there will be an increase in the numbers of exotic fishes dumped into Ohio waters, a few of which may become permanently established, possibly to become part of an unwanted fish fauna. For an excellent picture of the Mexican tetra, *see* Miller and Robinson (1973:51).

Summary of 1955–80 period.—Human population increased more rapidly than ever before, as did industrialization, urbanization and the number of highways. Much agricultural land was converted to other uses. The average field became larger, coincidental with the advent of bigger and more efficient farm machinery. Attempts to retain relict portions of the original flora and fauna were in many instances aborted. The Ohio landscape became increasingly uniform in appearance. The channelization of streams, digging and re-dredging of ditches and more thorough draining of land, originally largely conducted by the owners of the land or by county commissioners, were aided and accelerated by state and federal agencies. Great increases in the use of commercial fertilizers, detergents and pesticides affected the waters and their fish populations, usually adversely. The amount of trash increased to an estimated 5 to 11 lbs (2–5 kg) per day per person, and there was increasing concern as to its disposal. Lake Erie became considerably modified, and some commercial fish species, formerly of great importance, became commercially extinct or largely so. The 30± miles (48± km) of the Scioto River immediately below Columbus, after containing highly polluted waters with small populations of resistant fish species for many years, again became sufficiently "clean" to support a fish population of at least 76 species. Some of these were present in enormous numbers. This return to favorable fish conditions was a result of increased efficiency of the newly constructed sewage disposal plants. Fish populations in large artificial reservoirs usually declined after an early increase. Introductions of exotic fish species continued without outstanding success despite much effort and expenditures of monies.

Major detrimental trends will continue after 1980 if the human population of Ohio continues to increase, if the public continues to litter at its present rate and remains largely unconcerned about the deposition of trash, if the pollution of water, air, and land is not better controlled and if the continued destruction of the natural environment of Ohio is disregarded. *See* Trautman, 1977:1–25.

Part III

The Literature Examined

An extensive perusal of the literature relating to Ohio fishes and Ohioana was undertaken during the 1925–53 period. The majority of this research was done in the private library of Carl L. Hubbs, and in the institutional libraries of the University of Michigan, the Ohio State University and the Ohio State Museum. The Hubbs library was particularly helpful because it contained many ichthyological papers not readily available elsewhere, and the Ohio State Museum library because of its abundance of publications relative to Ohio.

A published record or statement was accepted when verified by a preserved specimen. When not so verified, a record was accepted only if it contained a satisfactory description, figure, plate, or photograph, and if there was no reason to doubt its authenticity. Records unacceptable were ignored whenever possible; otherwise reasons for not accepting them were given.

Species Erroneously Recorded for Ohio

This list includes currently valid species which are removed from the Ohio faunal list because they were recorded upon insufficient evidence; if a name does not appear in this list *see* "Interpretations of scientific names, etc."

For the key to abbreviated references *see* list of authors and their publications, p. 38).

Name used by former author	Current name	Abbreviated reference	Reason for removal of species from Ohio list
Amblyopsis spelaeus	*Amblyopsis spelaeus*	J.'82:901 E.'09:71	No Ohio record given "fish said to have been caught" in *bog*, not a cave, near Hiram, Ohio
Ammocrypta clara	*Ammocrypta pellucida*	T.'46:36	Re-examination of specimens proved them to be *pellucida*
Boleosoma olmstedi	*Etheostoma nigrum olmstedi*	J.'82:967	Range lies east of Ohio
Centrarchus macropterus	*Centrarchus macropterus*	J.'82:926	No Ohio record given
Chologaster agassizi	*Chologaster agassizi*	J.'82:902	No Ohio record given
Coregonus hoyi	*Coregonus hoyi*, until recently *Leucichthys hoyi*	J.'82G881	No Ohio record given
Coregonus nigripinnis	*Coregonus nigripinnis*, until recently *Leucichthys nigripinnis*	J.'82:884	No Ohio record given
Coregonus quadrilateralis	*Prosopium cylindraceum quadrilaterale*	J.'82:878	No Ohio record given
Couesius prosthemius	*Hybopsis plumbea* until recently *Couesius plumbeus*	J.'82:862	No Ohio record given
Etheostoma peltatum	*Percina peltata*	Mc.'92:29	Not present west of crest of the Appalachians
Etheostoma squamiceps	*Etheostoma squamiceps*	J.'82:978	No Ohio record given
Exoglossum maxillilingua	*Exoglossum maxillingua*	J.'82:841	No Ohio record given
Fundulus dispar	*Fundulus dispar dispar*	O.'01:73 O-W-T.'30:174	No exact locality given. *See* O.'01:73 and Jordan and Evermann: 1896–1900:658
Minnilus scabriceps	*Notropis scabriceps*	J.'82:850	Species restricted to upper Kanawha system
Noturus exilis	*Notorus exilis*	J.'82:800	No Ohio record given
Phenacobius teretulus	*Phenacobius teretulus*	J.'82:855	Species restricted to upper Kanawha system
Phoxinus neogaeus	*Phoxinus neogaeus*	J.'82:866	No Ohio record given
Poecilichthys virgatus	*Etheostoma virgatum*	J.'82:979	No Ohio record given

Name used by former author	Current name	Abbreviated reference	Reason for removal of species from Ohio list
Pygosteus pungitius	*Pungitius pungitius*	J.'82:999	No Ohio record given
Roccus interruptus	*Roccus interruptus*	J.'82:956	No Ohio record given
Schilbeodes nocturnus	*Noturus nocturnus*	O-W-T.'30-174 T.'46:30	A hybrid; *see* Trautman, 1948:166–74
Triglopsis thompsonii	*otriglopsis thompsoni*	J.'82:986	No Ohio record given
Typhlichthys subterraneus	*Typhlichthys subterraneous*	J.'82:901	No Ohio record given
Uranidea gracilis	*Cottus cognatus*	J.'82:988	No Ohio record given
Zygonectes dispar	*Fundulus dispar dispar*	J.'82:910	No exact locality given

List of the More Important Fish Species, Exotic and Native, Brought from Elsewhere and Liberated in Ohio Waters[1]

Species	Period of largest liberations	Numbers liberated	Degree of success of liberations
American shad *Alosa sapidissima* (Wilson)	1870–1900	hundreds of thousands	no
Silver salmon *Oncorhynchus kisutch* (Walbaum)	1875 to present	thousands	partial
King salmon *Oncorhynchus tschawytscha*(Walbaum)	1876 to present	thousands	partial
Atlantic salmon *Salmo salar salar* Linnaeus	1876–1885	thousands	no
Sebago salmon *Salmo salar sebago* Girard	1876–1885	thousands	no
Brown trout *Salmo trutta* Linnaeus	1885 to present	thousands	partial
Rainbow trout *Salmo gairdneri* Richardson	1884 to present	thousands	partial
Brook trout *Salvelinus fontinalis* (Mitchill)	1853 to present	thousands	partial
Chain pickerel *Esox niger* Lesueur	1935	hundreds	high
Common Carp *Cyprinus carpio* Linnaeus	1879–1900	hundreds of thousands	very high
Goldfish *Carassius auratus* (Linnaeus)	1888 to present	thousands	moderately high
Tench *Tinca tinca* (Linnaeus)	1898	few	no
Northern redbelly dace *Chrosomus eos* Cope	Oct. 26, 1935 from Michigan	7,000	no
American eel *Anguilla rostrata* (Lesueur)	1882–1900	hundreds of thousands	partial
Mosquitofish *Gambusia affinis affinis* (Baird and Girard)	1947	few	partial
Eastern common gambusia *Gambusia affinis holbrooki* (Girard)	1947	few	apparently no
Striped bass *Morone saxitalis* (Walbaum)	1968 to present	thousands	no
Redear sunfish *Lepomis microlophus* (Günther)	1931–1935	67	high

[1] It is important to note that of the comparatively few species whose introductions were highly successful, only the redear sunfish is today greatly desired; the carp, although of commercial importance in Lake Erie is, like the goldfish, an undesirable species in many waters and one upon which a far larger amount of money was spent in attempts to eliminate it than was spent in introducing it; the chain pickerel is confined to only one lake and because of its small size and voracious appetite might be a liability in many waters. Also it is worthy to note that among such desirable species as the shad, salmons, and trouts, only the trout introductions were successful in a very few localities.

Interpretation of Scientific Names as Used by Some Former Authors

A name as used by an author was included only when data accompanying it were used in this publication in whole or in part *and*,

If the name differed in *any* manner from the name used in this publication for a species; *or*

If the name referred to more than one species (in which case the species thought to be more important appears first in the list); *or*

When the name was identical with the one used by me, but it was important to stress the fact that I agreed with the author's identification; an example is Kirtland's ('51K:157) reference to *Lepisosteus productus* which is *productus* and not *platostomus*.

A name was *not* included when it was identical in spelling and it was not important to stress the fact that I agreed with the author's conclusions; *or*

When the name was of a currently valid species which had been erroneously recorded (*see* list of "Species erroneously recorded for Ohio," p. 36–37).

I based my identification as to species (or complex of species) upon verbal descriptions, photographs, drawings, ecological conditions in the locality at the time, and my personal opinion as to the author's knowledge of the species. In such a list there is a distinct possibility that some of my identifications are incorrect.

To conserve space, references to names of authors and their publications are abbreviated as follows:

Kirtland, 1838:	K.'38:
Kirtland, 1841A:	K.'41A:
Kirtland, 1841B:	K.'41B:
Kirtland, 1844:	K.'44:
Kirtland, 1850:	K.'50:
Kirtland, 1851:	K.'51:
Klippart, 1877:	Kl.'77:
Klippart, 1878:	Kl.'78:
Jordan, 1882:	J.'82:
Henshall, 1888:	H.'88:
Henshall, 1889:	H.'89:
McCormick, 1892:	Mc.'92:
Kirsch, 1895A:	Kirs.'95A:
Williamson and Osborn, 1898:	W-O.'98:
Osburn, 1901:	O.'01:
Eigenmann, 1909:	E.'09:
Osburn, Wickliff and Trautman, 1930:	O-W-T.'30:
Trautman, 1946:	T.'46:
Trautman, 1950:	T.'50:
Trautman, 1957:	T.'57:

A

Name by Former Author	Abbreviated References	Current Name
Abramis crysoleucas	O.'01:49; W-O.'98:28	*Notemigonus crysoleucas*
Accipenser maculosus	K.'38:170	*Acipenser fulvescens*
Accipenser ohioensis	K.'38:170	*Acipenser fulvescens*
Accipenser platorynchus	K.'38:170; K.'51H:233	*Scaphirhynchus platorynchus*
Accipenser rubicundus	K.'38:170; K.'51-I:229	*Acipenser fulvescens*
Acipenser platorynchus	K.'47:25	*Scaphirhynchus platorynchus*
Acipenser rubicundus	K.'44:303; J.'82:766; H.'88:76; O.'01:19; Kirs.'95A:327	*Acipenser fulvescens*
Alburnus nitidus	K.'54B:44	*Notropis atherinoides atherinoides*
Allotis humilis	O-W-T.'30:176	see *Lepomis humilis* of this list
Alosa chrysochloris	K.'51D:117	*Alosa chrysochloris*
Alvordius aspro	J.'82:972	*Percina maculata*
Alvordius evides	J.'82:972	*Percina evides*, this record not acceptable because no Ohio locality was mentioned.
Alvordius macrocephalus	J.'82:972	*Percina macrocephala*, this record not acceptable because no Ohio locality was mentioned.
Alvordius phoxocephalus	J.'82:972	*Percina phoxocephala*
Alvordius variatus	J.'82:973	*Etheostoma variatum*
Ameiurus lacustris	O.'01:23; W-O.'98:15	*Ictalurus punctatus*
Ameiurus melas	Mc.'92:12; Kirs.'95A-327; W-O.'98:11; O.'01:25; O-W-T.'30:174	*Ictalurus melas*
Ameiurus melas melas	T.'46:29; T.'50:29	*Ictalurus melas*
Ameiurus natalis	H.'89:124; Mc.'92:12; Kirs.'95A:327; W-O.'98:11; O.'01:24; O-W-T.'30:174	*Ictalurus natalis*

Name by Former Author	Abbreviated References	Current Name
Ameiurus natalis natalis	T.'50:29	*Ictalurus natalis*
Ameiurus nebulosus	W-O.'98:11 and Pl. 4	Plate 4 is of an *Ictalurus natalis;* description likewise fits *natalis* better than *nebulosus*
Ameiurus nebulosus	H.'89:124; Mc.'92:12; Kirs.'95A:327; W-O.'98:11; O.'01:25; O-W-T.'30:174	*Ictalurus nebulosus*
Ameiurus nebulosus nebulosus	T.'46:29; T.'50:29	*Ictalurus nebulosus*
Ameiurus vulgaris	H.'89:124; O.'01:24; W-O.'98:16; Mc.'92:12	Long-lower-jawed individuals of the genus *Ictalurus*
Amiurus catus	J.'82:793	*Ictalurus nebulosus*
Amiurus marmoratus	J.'82:792	*Ictalurus nebulosus*
Amiurus melas	H.'88:77	*Ictalurus melas*
Amiurus nigricans	J.'82:789; H.'88:77	*Ictalurus punctatus*
Amiurus vulgaris	J.'82:791	Long-lower-jawed individuals of the genus *Ictalurus*
Amiurus xanthocephalus	J.'82:976	*Ictalurus melas*
Ammocetes bicolor	K.'38:170	Unidentifiable species of lamprey
Ammocoetes concolor	K.'41B:473; K.'51R:213	Unidentifiable; possibly *Ichthyomyzon greeleyi* in part and *Lampetra aepyptera* in part
Ammocoetes niger	J.'82:756	Presumably *Lampetra aepyptera*
Ammocrypta clara	T.'46:36	*Ammocrypta pellucida*
Amphiodon alosoides	O-W-T.'30:171; T.'50:3	*Hiodon alosoides*
Anguilla anguilla	H.'89:125	*Anguilla rostrata*
Anguilla bostoniensis	O-W-T.'30:174; T.'50:31	*Anguilla rostrata*
Anguilla chryspa	Kirs.'95A:330; O.'01:65; W-O.'98:36	*Anguilla rostrata*
Anguilla laticauda	K.'38:170	*Anguilla rostrata*
Anguilla lutea	K.'44:234; K.'51-0:189	*Anguilla rostrata*
Anguilla vulgaris	J.'82:781	*Anguilla rostrata*
Anguilla xanthomelas	K.'38:170	*Anguilla rostrata*
Aphredoderus sayanus gibbosus	T.'46:33; T.'50:33	*Aphredoderus sayanus*
Aplites salmoides	O-W-T.'30:176	*Micropterus salmoides salmoides*
Apomotis cyanellus	O.'01:80; O-W-T.'30:176	*Lepomis cyanellus*
Apomotis cyanellus	W-O.'98:42	*Lepomis cyanellus* mostly, *see* Pl.16; but includes some hybrids, *see* Pl. 15 (presumably a *L. cyanellus* by *L. megalotis* hybrid) and Pl. 27 (presumably a *L. cyanellus* by *L. macrochirus* hybrid)
Argyrosomus artedi	O.'01:69	*Coregonus artedii*; both types
Argyrosomus clupeiformis	Kl.'77:63	Probably *Coregonus artedii*
Argyrosomus tullibee	O.'01:69	*Coregonus artedii*, possibly the *albus* type
	B	
Bodianus flavescens	K.'38:168	*Perca flavescens*
Boleichthys fusiformis	O.'01:102	*Etheostoma exile*
Boleosoma nigrum	J.'82:966; O.'01:94	*Etheostoma nigrum*, presumably both subspecies
Boleosoma nigrum	W-O.'98:49	*Etheostoma nigrum nigrum*
Boleosoma nigrum eulepis	T.'50:35	*Etheostoma nigrum eulepis*
Boleosoma nigrum nigrum	O-W-T.'30:175	*Etheostoma nigrum*, presumably both subspecies
Boleosoma nigrum nigrum	T.'50:35	*Etheostoma nigrum nigrum*
Bubalichthys bubalus	Kl.'78:110; J.'82:807	*Ictiobus bubalus*
Bubalichthys urus	Kl.'78:110; J.'82:807	*Ictiobus niger*
	C	
Campostoma anomalum	H.'88:78; W-O'98:27	*Campostoma anomalum anomalum*; possibly some intergrades
Campostoma anomalum	O.'01:43; O-W-T.'30:174	*Campostoma anomalum*; both subspecies and intergrades
Carpiodes bison	J.'82:811	*Carpiodes carpio*; if *bison* is a valid subspecies of *C. carpio*, its range would be west of Ohio.
Carpiodes cutisanserinus	J.'82:812	*Carpiodes carpio carpio*
Carpiodes cyprinus	J.'82:810	*Carpiodes cyprinus hinei*

Name by Former Author	Abbreviated References	Current Name
Carpiodes cyprinus	O-W-T.'30:171	*Carpiodes cyprinus*; both subspecies and intergrades
Carpiodes cyprinus cyprinus	T.'50:7	*Carpiodes cyprinus hinei*
Carpiodes cyprinus thompsoni	T.'50:7	*Carpiodes cyprinus cyprinus*
Carpiodes difformis	Kl.'77:86; J.'82:813; O.'01:32	*Carpiodes velifer*
Carpiodes thompsoni	J.'82:811; Mc.'92:14; O.'01:32	*Carpiodes cyprinus cyprinus*
Carpiodes velifer	Kl.'77:87; Kirs.'95A:328; O.'01:33	*Carpiodes cyprinus hinei*; possibly intergrades
Carpiodes velifer	W-O.'98:20	*Carpiodes cyprinus hinei*
Catonotus flabellaris flabellaris	O-W-T.'30:175; T.'50:43	*Etheostoma flabellare flabellare*
Catostomus anisurus	K.'47:269; K.'51BB:389	*Moxostoma anisurum*
Catostomus aureolus	K.'38:169; K.'41B:349; K.'51G:309	*Moxostoma macrolepidotum*
Catostomus bubalus	K.'38:169	Genus *Ictiobus*, presumably all three species
Catostomus bubalus	K.'47:266; K.'51W:341	Presumably only *Ictiobus bubalus*
Catostomus catostomus	O-W-T.'30:171	*Catostomus catostomus catostomus*
Catostomus commersonii	O.'01:34	*Catostomus commersoni commersoni*
Catostomus commersonnii commersonnii	O-W-T.'30:171; T.'50:8	*Catostomus commersoni commersoni*
Catostomus communis	K.'47:265; K.'51C:317	*Catostomus commersoni commersoni*
Catostomus duquesnie	K.'51Z:365	Includes several species of *Moxostoma*, especially *Moxostoma erythrurum* and *Moxostoma duquesnei*
Catostomus duquesnii	K.'38:169	Includes several species of *Moxostoma*
Catostomus duquesnii	K.'47:268	Describes *Moxostoma duquesnei* as the male, *Moxostoma erythrurum* as the female
Catostomus elongatus	K.'38:169; K.'47:267; K.'51E:349	*Cycleptus elongatus*
Catostomus erythurus	K.'38:169	Includes several species of *Moxostoma*
Catostomus gibbosus	K.'51AA:333	Drawing is of an *Erimyzon*; its high number of dorsal rays and locality suggest *Erimyzon sucetta kennerlyi*
Catostomus gracilis	K.'38:169	*Catostomus commersoni commersoni*
Catostomus longirostris	J.'82:816	*Catostomus catostomus catostomus*
Catostomus melanops	K.'47:271; K.'51F:413	*Minytrema melanops*
Catostomus melanopsis	K.'38:169 and 192	Unidentifiable
Catostomus nigrans	K.'38:169	*Hypentelium nigricans*
Catostomus nigricans	K.'47:273; K.'51Y:397; J.'82:819; Mc.'92:14; Kirs.'95A:328; W-O.'98:20; O.'01:35	*Hypentelium nigricans*
Catostomus oscula	K.'41B:350	*Aplodinotus grunniens*
Catostomus teres	Kl.'77:84; J.'82:817; H.'88:77; Mc.'92:14; Kirs.'95A:328	*Catostomus commersoni commersoni*
Catostomus velifer	K.'38:169	Presumably includes all three species of *Carpiodes*
Centrarchus aeneus	K.'44:239; K.'50BB:77	*Ambloplites rupestris rupestris*
Centrarchus fasciatus	K.'47:28	*Micropterus dolomieui dolomieui*
Centrarchus fasciatus	K.'50CC:93	Presumably includes all three species of *Micropterus* but chiefly *Micropterus dolomieui dolomieui*
Centrarchus hexacanthus	K.'41B:480	*Pomoxis annularis*
Centrarchus hexacanthus	K.'50N:69	Presumably includes both species of *Pomoxis*
Ceratichthys amblyops	J.'82:860	*Hybopsis amblops amblops*
Ceratichthys biguttatus	J.'82:861	*Hybopsis biguttata* and *Hybopsis micropogon*
Ceratichthys perspicuus	T.'50:25	*Pimephales vigilax perspicuus*
Chaenobryttus antistius	J.'82:939	*Lepomis gulosus*
Chaenobryttus coronarius	T.'46:45; T.'50:45	*Lepomis gulosus*
Chatoessus ellipticus	K.'38:169; K.'44:235; K.'50C:1	*Dorosoma cepedianum*
Chiope heterodon	J.'82:846	*Notropis heterodon*
Cichla aenea	K.'38:168	*Ambloplitis rupestris rupestris*
Cichla fasciata	K.'38:168	Genus *Micropterus*, but mostly *Micropterus dolomieui dolomieui*

Name by Former Author	Abbreviated References	Current Name
Cichla minima	K.'38:168	Genus *Micropterus* presumably *Micropterus dolomieui dolomieui*
Cichla ohioensis	K.'38:168	Presumably all three species of *Micropterus*
Cichla storeria	K.'38:168	Probably both species of *Pomoxis*
Cliola vigilax	H.'88:78; W-O.'98:28; O.'01:50	*Pimephales vigilax perspicuus*
Clupea chrysochloris	H.'88:79	*Alosa chrysochloris*
Coregonus albus	K.'38:169; K.'41B:477; Kl.'77:58	*Coregonus clupeaformis*
Coregonus artedi	Mc.'92:22	*Coregonus artedii*, both types
Coregonus artedia	K.'38:169	*Coregenus artedii*, presumably both types
Coregonus clupeaformis latus	T.'46:4; T.'50:4	*Coregonus clupeaformis*
Coregonus clupeiformis	J.'82:879; H.'89:124; Mc.'92:22; Kirs.'95A:330; O.'01:68	*Coregonus clupeaformis*
Coregonus labradoricus	J.'82:881	*Coregonus clupeaformis*
Coregonus tullibee	J.'82:885	Possibly the hybrid *Coregonus artedii* by *Coregonus clupeaformis*
Corvina oscula	K.'51 I:133	*Aplodinotus grunniens*
Cottogaster copelandi	J.'82:969; O.'01:93; O-W-T.'30:175; T.'50:36	*Percina copelandi*
Cottogaster shumardi	O.'01:93; O-W-T.' 30:175	*Percina shumardi*
Cottus bairdii bairdii	O-W-T.'30:176; T.'50:50	*Cottus bairdi bairdi*
Cottus bairdii kumlieni	T.'50:50	*Cottus bairdi kumlieni*
Cottus gobio	K.'47:342	*Cottus bairdi bairdi*
Cottus ictalops	W-O.'98:55; O.'01:104	*Cottus bairdi bairdi*
Cottus richardsoni	Mc.'92:33	*Cottus bairdi bairdi*
Cristivomer namaycush	J.'82:893; O.'01:69; O-W-T.'30:171	*Salvelinus namaycush*
Cristivomer namaycush namaycush	T.'50:5	*Salvelinus namaycush*
Crystallaria asprella	O.'01:95; O-W-T.'30:175	*Ammocrypta asprella*
Crystallaria asprella asprella	T.'46:39; T.'50:39	*Ammocrypta asprella*
Cyprinus carpio coriaceus	Kirs.'95A:328	*Cyprinus carpio*
	D	
Diplesion blennioides	W-O.'98:49; O.'01:94	*Etheostoma blennioides*
Diplesium blennioides	J.'82:967	*Etheostoma blennioides*
	E	
Entosphenus lamottenii	T.'50:2	*Lampetra appendix*
Erimystax dissimilis	J.'82:859; T.'50:14	*Hybopsis dissimilis dissimilis*
Erimystax dissimilis	O-W-T.'30:172	*Hybopsis dissimilis dissimilis* in part and *Hybopsis x-punctata trautmani* in part
Erimyzon sucetta	Kl.'78:107; Mc.'92:14	*Erimyzon sucetta kennerlyi*
Erimyzon sucetta	J.'82:821; Kirs.'95A:328	Both species of *Erimyzon*
Erimyzon sucetta	W-O.'98:21	*Erimyzon oblongus claviformis* entirely or in part
Erimyzon sucetta kennerlii	O-W-T'30:172; T.'50:8	*Erimyzon sucetta kennerlyi*
Erimyzon sucetta oblongus	H.'88:77	*Erimyzon oblongus claviformis?*
Erimyzon sucetta oblongus	O.'01:36	Presumably both species of *Erimyzon*
Esox cypho	Kl.'78:95	*Esox americanus vermiculatus?*
Esox estor	K.'38:169; K.'47:339; K.'51A:61	*Esox masquinongy masquinongy* possibly also *Esox masquinongy ohioensis*
Esox masqualongus	Kl.'77:82	*Esox masquinongy*
Esox masquinongy	Mc.'92:24	*Esox masquinongy masquinongy*
Esox masquinongy ohiensis	O-W-T.'30:174	*Esox masquinongy ohioensis*
Esox niger	K.'38:169	Presumably *Esox americanus vermiculatus*
Esox nobilier	Kl.'78:92	*Esox masquinongy masquinongy*
Esox nobilior	J.'82:917; H.'89:125	*Esox masquinongy*, both subspecies
Esox nobilis	K.'54A:79	*Esox masquinongy masquinongy*
Esox ohiensis	K.'54A:79	*Esox masquinongy ohioensis*
Esox reticulatus	K.'83:169; K.'44:233; K.'51EE:45	*Esox lucius*

Name by Former Author	Abbreviated References	Current Name
Esox salmoneus	Kl.'78:94	*Esox americanus vermiculatus* and possibly young of *Esox lucius*
Esox salmoneus	J.'82:914	*Esox americanus vermiculatus*
Esox umbrosus	K.'54A:79	*Esox americanus vermiculatus*
Esox vermiculatus	O-W-T.'30:174; T.'50:31	*Esox americanus vermiculatus*
Etheostoma asprellus	H.'89:125	*Ammocrypta asprella*
Etheostoma aspro	H.'88:80; Mc.'92:30; Kirs.'95A:331	*Percina maculata*
Etheostoma blennioides	K.'50AA:45	Presumably mostly *Etheostoma blennioides*
Etheostoma blennioides blennioides	T.'50:37	*Etheostoma blennioides*
Etheostoma caprodes	K.'50Z:37; H.'88:80	*Percina caprodes caprodes*
Etheostoma caprodes	Mc.'92:29; Kirs.'95A:331	Both supspecies of *Percina caprodes* and intergrades
Etheostoma coeruleum	H.'88:80	*Etheostoma caeruleum*; presumably also *Etheostoma spectabile spectabile*
Etheostoma coeruleum	Mc.'92:31; Kirs.'95A:331; W-O.'98:52; O.'01:99	*Etheostoma caeruleum*
Etheostoma copelandi	H.'89:125; Mc.'92:29; Kirs.'95A:331	*Percina copelandi*
Etheostoma maculata	K.'38:168; K.'41A:276; K.'50K:29	*Etheostoma maculatum*
Etheostoma nigrum	Mc.'92:28; Kirs.'95A:330	*Etheostoma nigrum nigrum*, possibly also *Etheostoma nigrum eulepis* and intergrades
Etheostoma pellucidum	H.'88:80; Mc.'92:28; Kirs.'95A:331	*Ammocrypta pellucida*
Etheostoma phoxocephalum	H.'88:80	*Percina phoxocephala*
Etheostoma phoxocephalum	Mc.'92:30	Not acceptable because locality is Lake Erie and no specimens
Etheostoma scierum	H.'89:125	*Percina sciera sciera*
Etheostoma shumardi	H.'89:125	*Percina shumardi*
Etheostoma spectabile	Kirs.'95A:331; W-O.'98:52; O.'01:100	*Etheostoma spectabile spectabile*
Etheostoma variata	K.'38:168; K.'41A:274; K.'50L:21	*Etheostoma variatum*
Etheostoma wrighti	Mc.'92:30	Presumably the hybrid *Percina caprodes* × *Percina maculata*
Eupomotis aureus	Kl.'77:80	*Lepomis gibbosus*
Eupomotis euryorus	Mc.'92:27; O.'01:83	Presumably the hybrid *Lepomis cyanellus* × *Lepomis gibbosus*
Eupomotis gibbosus	J.'82:928; W-O.'98:44; O.'01:84; O-W-T.'30:176	*Lepomis gibbosus*
Eupomotis heros	O.'01:83; O-W-T.'30:176	Supposed to be *Lepomis microlophus* but records not acceptable; *see Lepomis notatus*
Eupomotis notatus	J.'82:931	*Leopmis microlophus* but not acceptable because no Ohio locality record given
Exoglossops laurae	O-W-T.'30:173	*Exoglossum laurae hubbsi*
Exoglossum dubium	K.'47:272	*Campostoma anomalum anomalum*, intergrades and possibly *Campostoma anomalum pullum*
Exoglossum lesurianum	K.'38:169	Presumably *Campostoma anomalum*
Exoglossum maxillingua	O.'01:64	*Exoglossum laurae hubbsi*
Extrarius aestivalis hyostomus	T.'50:14	*Hybopsis aestivalis hyostoma*
Extrarius hyostomus	O-W-T.'30:172	*Hybopsis aestivalis hyostoma*
	F	
Fundulus diaphanus	J.'82:903; Mc.'92:23; Kirs.'95A:330	*Fundulus diaphanus menona*
	G	
Gasterosteus inconstans	K.'38:168; K.'41A:273; K.'50FF:125	*Culaea inconstans*

Name by Former Author	Abbreviated References	Current Name
	H	
Hadropterus aspro	W-O.'98:48; O.'01:91	*Percina maculata*
Hadropterus evides	O.'01:92; O-W-T.'30:175; T.'46:38; T.'50:38	*Percina evides*
Hadropterus macrocephalus	T.'46:38; T.'50:38	*Percina macrocephala*
Hadropterus maculatus	O-W-T.'30:175; T.'46:38; T.'50:38	*Percina maculata*
Hadropterus phoxocephalus	W-O.'98:16; O.'01:91; O-W-T.'30:75; T.'46:39; T.'50:39	*Percina phoxocephala*
Hadropterus scierus	O.'01:92; O-W-T.'30:175	*Percina sciera sciera*
Hadropterus scierus scierus	T.'46:39; T.'50:39	*Percina sciera sciera*
Haploidonotus grunniens	Kl.'77:81; J.'82:982	*Aplodinotus grunniens*
Helioperca incisor	O-W-T.'30:176	*Lepomis macrochirus macrochirus*
Helioperca pallida	Kl.'77:80	*Lepomis macrochirus macrochirus*
Hudsonius analostanus	J.'82:845	*Notropis spilopterus*, presumably also *Notropis whipplei*
Hudsonius fretensis	J.'82:844-45	Presumably *Notropis spilopterus*
Hudsonius haematurus	J.'82:845	Presumably *Pimephales notatus*
Hudsonius storerianus	J.'82:843	*Hybopsis storeriana*
Hudsonius stramineus	J.'82:844	*Notropis stramineus*
Hudsonius volucellus	J.'82:843	*Notropis volucellus volucellus*
Huro salmoides	T.'50:45	*Micropterus salmoides salmoides*
Hybopsis dissimilis	H.'88:79	Presumably *Hybopsis x-punctata trautmani* (because of locality), possibly also *Hybopsis dissimilis dissimilis*
Hybopsis dissimilis	W-O.'98:34; O.'01:62	*Hybopsis dissimilis dissimilis* and *Hybopsis x-punctata trautmani*
Hybopsis hyostomus	H.'89:124;O.'01:62	*Hybopsis aestivalis hyostoma*
Hybopsis kentuckiensis	H.'88:78;Mc.'92:20; W-O.'98:34; O.'01:64	Presumably both *Hybopsis biguttata* and *Hybopsis micropogon*
Hybopsis storerianus	H.'88:78; Mc.'92:20; Kirs.'95A:329;O.'01:63; O-W-T.'30:172; T.'50:13	*Hybopsis storeriana*
Hyborhynchus notatus	J.'82:840; O-W-T.'30:174; T.'50:26	*Pimephales notatus*
Hydragira limi	K.'38:169	*Umbra limi*
Hydrargira limi	K.'41A:277	*Umbra limi*
Hydrargyra limi	K.'51FF:37	*Umbra limi*
Hyodon alosoides	J.'82:875; H.'88:79	*Hiodon alosoides*
Hyodon clodalus	K.'38:170	Probably both species of *Hiodon*
Hyodon tergissus	K.'38:170; K.'47:338	Presumably only *Hiodon tergisus*
Hyodon tergisus	K.'51V:133;Kl.'77:65; J.'82:875; H.'88:79	*Hiodon tergisus*; *Hiodon alosoides* probably included
Hyodon vernalis	K.'38:170	Probably both species of *Hiodon*
Hypargyrus velox	O-W-T.'30:174	*Pimephales vigilax perspicuus*
Hypentelium macropterum	K.'38:168	Unidentifiable
	I	
Ichthaelurus furcatus	J.'82:785	*Ictalurus furcatus*
Ichthaelurus punctatus	Kl.'78:115; J.'82:786	*Ictalurus punctatus*
Ichthaelurus robustus	J.'82:786	*Ictalurus furcatus*?
Ichthyobus bubalus	J.'82:805	*Ictiobus cyprinellus*
Ichthyomyzon concolor	O.'01:15	Presumably *Ichthyomyzon unicuspis*, *Ichthyomyzon bdellium*, *Ichthyomyzon greeleyi* and *Lampetra aepyptera*
Ichthyomyzon concolor	O-W-T.'30:170	*Ichthyomyzon unicuspis* and *Ichthyomyzon bdellium*
Ictalurus furcatus furcatus	T.'50:28	*Ictalurus furcatus*
Ictalurus lacustris lacustris	T.'50:28	*Ictalurus punctatus*

Name by Former Author	Abbreviated References	Current Name
Ictalurus lacustris punctatus	T.'50:28	*Ictalurus punctatus*
Icthyobus bubalus	Kl.'78:109	*Ictiobus cyprinellus*
Ictiobus carpio	H.'88:77	*Carpiodes carpio carpio*
Ictiobus cyprinella	H.'88:70; O.'01:30	*Ictiobus cyprinellus*
Ictiobus difformis	H.'88:77	*Carpiodes velifer*
Ictiobus thompsoni	H.'89:124	*Carpiodes cyprinus cyprinus*
Ictiobus urus	H.'89:124; O.'01:31	*Ictiobus niger*
Ictiobus velifer	H.'88:77	*Carpiodes cyprinus hinei*
Imostoma shumardi	J.'82:968; O-W-T.'30:175; T.'50:36	*Percina shumardi*
	L	
Labrax multilineatus	K.'47:21; K.'50J:53	*Morone chrysops*
Lampetra lamotteni	O-W-T.'30:170	*Lampetra aepyptera*
Lampetra wilderi	O.'01:16	*Lampetra aepyptera (*not *Lampetra appendix*)
Lepibema chrysops	O-W-T.'30:175;T.'50:33	*Morone chrysops*
Lepidosteus platystomus	J.'82:773	*Lepisosteus platostomus* and *Lepisosteus productus*
Lepisosteus ferox	K.'38:170; K.'44:18; K.'51J:149	*Lepisosteus spatula*
Lepisosteus oxyurus	K.'38:170; K.'44:16	*Lepisosteus osseus*
Lepisosteus platostomus	O.'01:20	*Lepisosteus platostomus* in part and *Lepisosteus productus* in part
Lepisosteus platystomus	H.'89:123; Kirs.'95A:327	*Lepisosteus productus*
Lepisosteus productus	K.'51K:157	*Lepisosteus productus*
Lepisosteus tristoechus	O.'01:21; O-W-T.'30:171; T.'50:3	*Lepisosteus spatula*
Lepomis euryorus	Mc.'92:27	Presumably the hybrid *Lepomis cyanellus* × *Lepomis gibbosus*
Lepomis humilis	H.'88:79; O.'01:82	Other species of sunfishes but not *Lepomis humilis*
Lepomis megalotis	H'88:79; W-O.'98:42	*Lepomis megalotis megalotis*
Lepomis megalotis	Mc.'92:27; Kirs.'95A:331	*Lepomis megalotis peltastes* and/or intergrades
Lepomis megalotis	O.'01:81	*Lepomis megalotis megalotis* in part, *Lepomis megalotis peltastes* in part, and intergrades
Lepomis notatus	H.'89:125	Presumably not *Lepomis microlophus*
Lepomis pallidus	J.'82:936; H.'88:80; Mc.'92:27; Kirs.'95A:330; W-O.'98:43; O.'01:82	*Lepomis macrochirus macrochirus*
Leptops olivaris	H.'89:124; Mc.'92:12; O.'01:26	*Pilodictis olivaris*
Leucichthys artedi albus	O-W-T.'30:171; T.'50:4	*Coregonus artedii albus*
Leucichthys artedi artedi	O-W-T.'30:171;T.'50:4	*Coregonus artedii artedii*
Leuciscus americanus	K.'50V:221	*Notemigonus crysoleucas*
Leuciscus atromaculatus	K.'50E:213	*Semotilus atromaculatus atromaculatus*
Leuciscus biguttatus	K.'50GG:197	*Hybopsis biguttata*
Leuciscus compressus	K.'50W:229	*Notemigonus crysoleucas*
Leuciscus cornutus	K.'51CC:269	*Notropis cornutus*, presumably both subspecies
Leuciscus crysoleucas	K.'44:305	*Notemigonus crysoleucas*
Leuciscus diplema	K.'47:276	*Notropis cornutus*?
Leuciscus dissimilis	K.'50Q:189	*Hybopsis dissimilis dissimilis*
Leuciscus dorsalis	K.'47:274	*Semotilus atromaculatus atromaculatus*
Leuciscus elongatus	K.'50T:181; O.'01:47	*Clinostomus elongatus*
Leuciscus erythrogaster	K.'50B:213	*Phoxinus erythrogaster*
Leuciscus kentuckiensis	K.'50R:245	*Hybopsis micropogon* in part, *Hybopsis biguttata* in part
Leuciscus longirostris	K.'50V:85	*Clinostomus elongatus* (young)
Leuciscus plagyrus	K.'47:26; K.'50F:237	*Notropis cornutus*
Leuciscus storerianus	K.'47:30; K.'50S:256	*Hybopsis storeriana*
Litholepis spatula	J.'82:774	*Lepisosteus spatula*
Lota lacustris	Kl.'78:83	*Lota lota lacustris*
Lota lota	H.'89:125	*Lota lota lacustris*
Lota lota maculosa	T.'50:51	*Lota lota lacustris*

Name by Former Author	Abbreviated References	Current Name
Lota maculosa	K.'38:170; K.'44:24; K.'51N:181; J.'82:995; Mc.'92:33; O.'01:104; O-W-T.'30:176	*Lota lota lacustris*
Lucioperca americana	K.'38:168; K.'44:237; K.'50P:61	Both species of *Stizostedion*
Lucio-perca canadensis	Kl.'77:70	*Stizostedion canadense*
Lucius lucius	Kirs.'95A:330; O.'01:71	*Esox lucius*
Lucius masquinongy	Kirs.'95A:330	*Esox masquinongy masquinongy*
Lucius masquinongy	O.'01:72	*Esox masquinongy masquinongy* in part and *Esox masquinongy ohioensis* in part
Lucius vermiculatus	Kirs.'95A:330; W-O.'98:36; O.'01:71	*Esox americanus vermiculatus*
Luxilus cornutus	J.'82:853	*Notropis cornutus*, both subspecies
Luxilus dissimilis	K.'41B:341	Presumably only *Hybopsis dissimilis dissimilis*
Luxulus chrysocephalus	K.'38:169	Presumably *Notropis cornutus*
Luxulus elongatus	K.'38:169; K.'41B:339	*Clinostomus elongatus*
Luxulus erythrogaster	K.'38:169; K.'44:23	*Phoxinus erythrogaster*
Luxulus kentuckiensis	K.'38:169; K.'47:27	Presumably *Hybopsis biguttata* in part and *Hybopsis micropogon* in part
Lythrurus diplaemius	J.'82:851	*Notropis umbratilis cyanocephalus* in part, *Notropis ardens lythrurus* in part

M

Name by Former Author	Abbreviated References	Current Name
Megastomatobus cyprinella	O-W-T.'30:171; T.'50:6	*Ictiobus cyprinellus*
Microperca microperca microperca	T.'50:43	*Etheostoma microperca*
Microperca punctulata	J.'82:981; O-W-T.'30:175	*Etheostoma microperca*
Micropterus dolomieu	J.'82:948	*Micropterus dolomieui dolomieui*, presumably also *Micropterus punctulatus punctulatus*
Micropterus dolomieu	H.'88:80	*Micropterus dolomieui dolomieui* and *Micropterus punctulatus punctulatus*; both species found in material collected by Henshall
Micropterus dolomieu	O.'01:85	Mostly *Micropterus dolomieui dolomieui*; presumably in part *Micropterus punctulatus punctulatus*
Micropterus pallidus	Kl.'77:73	*Micropterus salmoides salmoides*
Micropterus pseudaplites	O-W-T.'30:175	*Micropterus punctulatus punctulatus*
Micropterus salmoides	Kl.'77:73	*Micropterus dolomieui dolomieui*
Micropterus salmoides	H.'88:80	*Micropterus salmoides salmoides* in part and *Micropterus punctulatus punctulatus* in part; both species found in material collected by Henshall
Minnilus dinemus	K.'38:168	Unidentifiable
Minnilus dinemus	J.'82:848	Those recorded from Lake Erie presumably were *Notropis atherinoides atherinoides*; those from inland Ohio were *Notropis photogenis*, and some *Notropis rubellus*
Minnilus photogenis	J.'82:849	*Notropis photogenis*?
Minnilus rubrifrons	J.'82:847	*Notropis rubellus*
Moxostoma aureolum	W-O.'98:22	*Moxostoma erythrurum* in part and *Moxostoma duquesnei* in part
Moxostoma aureolum	O.'01:38	*Moxostoma erythrurum*, many; *Moxostoma duquesnei*, *Moxostoma aureolum aureolum*, *Moxostoma breviceps* (determined by locality and preserved specimens)
Moxostoma aureolum	O-W-T.'30:172; T.'50:11	*Moxostoma aureolum aureolum*
Moxostoma breviceps	O.'01:39	*Moxostoma aureolum aureolum* mostly, *Moxostoma breviceps* in part
Moxostoma crassilabre	H.'88:77; H.'89:124	*Moxostoma carinatum*
Moxostoma duquesnii	O-W-T.'30:172	*Moxostoma duquesnei*
Moxostoma duquesnii duquesnii	T.'50:9	*Moxostoma duquesnei*
Moxostoma macrolepidotum	H.'88:77	*Moxostoma erythrurum*, and probably also *Moxostoma duquesnei*

Name by Former Author	Abbreviated References	Current Name
Moxostoma macrolepidotum duquesnei	Mc.'92:15; Kirs.'95A:328	*Moxostoma erythrurum* mostly, possibly also *Moxostoma duquesnei* and *Moxostoma valenciennesi*
Moxostoma rubreques	O-W-T.'30:172; T.'50:10	*Moxostoma valenciennesi*
Myxostoma anisurum	J.'82:827	*Moxostoma anisurum*
Myxostoma aureolum	J.'82:827	*Moxostoma aureolum aureolum* in part and *Moxostoma breviceps* in part
Myxostoma carpio	Kl.'78:106	*Moxostoma anisurum*
Myxostoma carpio	J.'82:830	Possibly *Moxostoma anisurum* and/or *Moxostoma valenciennesi*
Myxostoma macrolepidotum	J.'83:828	Probably includes several species, principally *Moxostoma erythrurum* and *Moxostoma duquesnei*
Myxostoma macrolepidotum duquesni	Kl.'78:105	Presumably includes several species of *Moxostoma*, particularly *Moxostoma erythrurum*, *Moxostoma duquesnei* and possibly *Moxostoma carinatum*
Myxostoma velatum	J.'82:826	*Moxostoma anisurum*
	N	
Nanostoma tessellatum	J.'82:975	*Etheostoma variatum*
Nanostoma zonale	J.'82:975	*Etheostoma zonale zonale*
Nocomis biguttatus	O-W-T.'30:172; T.'50:13	*Hybopsis biguttata*
Nocomis micropogon	O-W-T.'30:172; T.'50:13	*Hybopsis micropogon*
Notemigonus crysoleucas auratus	T.'46:17; T.'50:17	*Notemigonus crysoleucas*
Notemigonus crysoleucas crysoleucas	O-W-T.'30:173	*Notemigonus crysoleucas*
Nothonotus camurus	J.'82:976	*Etheostoma camurum*
Nothonotus maculatus	J.'82:976	*Etheostoma maculatum*
Notropis ardens	H.'88:78	*Notropis ardens*
Notropis ardens	Mc.'92:19; Kirs.'95A:329	*Notropis umbratilis cyanocephalus*
Notropis arge	H.'88:78; Kirs.'95A:329; O.'01:57	*Notropis photogenis*
Notropis arge	Mc.'92:19	*Notropis atherinoides atherinoides*; McCormick considers this species to be the same as Jordan's *Notropis dinemus* (J.'82:848) which is *N. a. atherinoides*
Notropis atherinoides	Mc.'92:19	*Notropis atherinoides atherinoides*
Notropis atherinoides	W-O.'98:31	Apparently all *Notropis photogenis*
Notropis atrocaudalis atrocaudalis	O-W-T.'30:173	*Notropis heterolepis*
Notropis atrocaudalis heterolepis	O-W-T.'30:173	*Notropis heterolepis*
Notropis blennius	W-O.'98:29; O.'01:52	*Notropis volucellus volucellus* in part and *Notropis stramineus* in part (photograph that of *Notropis boops*)
Notropis cayuga	Kirs.'95A:328; W-O.'98:29; O.'01:51	*Notropis heterolepis*
Notropis cornutus	W-O.'98:31	*Notropis cornutus chrysocephalus*
Notropis cornutus	O.'01:55	Both subspecies of *Notropis cornutus* and intergrades
Notropis deliciosus	H.'88:78	*Notropis stramineus* in part, *Notropis volucellus volucellus* in part, possibly also *Notropis volucellus wickliffi* in part
Notropis deliciosus	Mc.'92:18; Kirs.'95A:329	*Notropis stramineus* in part, *Notropis volucellus volucellus* in part
Notropis deliciosus missuriensis	O-W-T.'30:173	*Notropis stramineus*
Notropis dilectus	Mc.'92:19	*Notropis rubellus* inland, at least in part; those "in the lake" were *Notropis atherinoides atherinoides*
Notropis dilectus	Kirs.'95A:392	*Notropis rubellus*
Notropis hudsonius hudsonius	T.'50:20	*Notropis hudsonius*
Notropis jejunus	H.'88:78; O.'01:56	*Notropis blennius*, at least in part
Notropis lythrurus	O-W-T.'30:173	*Notropis ardens*
Notropis megalops	H.'88:78	*Notropis cornutus chrysocephalus*
Notropis megalops	Mc.'92:18; Kirs.'95A:329	Mostly *Notropis cornutus chrysocephalus* and intergrades, remainder *Notropis cornutus frontalis*
Notropis rubrifrons	H.'88:78; W-O.'98:32; O.'01:58	*Notropis rubellus*
Notropis shumardi	W-O.'98:29; O.'01:53	*Notropis boops*
Notropis stramineus	Mc.'92:18	Presumably *Notropis stramineus*
Notropis umbratilis lythrurus	W-O.'98:32	*Notropis ardens*

Name by Former Author	Abbreviated References	Current Name
Notropis umbratilis lythrurus	O.'01:59	*Notropis umbratilis cyanocephalus* in part, *Notropis ardens* in part
Notropis volucellus	O.'01:53	*Notropis volucellus volucellus*
Notropis volucellus buchanani	T.'50:19	*Notropis buchanani*
Notropis whipplei	Mc.'92:18; Kirs.'95A:329	*Notropis spilopterus*
Notropis whipplii	H.'88:78; W-O.'98:30; O.'01:54	*Notropis spilopterus*; possibly *Notropis whipplei* also
Notropis whipplii	T.'50:21	*Notropis whipplei*
Notropis whipplii spilopterus	O-W-T.'30:173	*Notropis spilopterus* in part and *Notropis whipplei* in part.
Noturus miurus	Kirs.'95A:327	*Noturus miurus*; probably also *Noturus furiosus*
Noturus sialis	J.'82:801	*Noturus gyrinus*
	O	
Opsopoeodus megalops	O.'01:48	*Notropis emiliae*
	P	
Parexoglossum hubbsi	T.'50:15	*Exoglossum laurae hubbsi*
Pelodichthys olivaris	J.'82:797	*Pilodictis olivaris*
Pelodictis limosus	O-W-T.'30:174	*Pilodictis olivaris*
Peocilichthys camurus	O-W-T.'30:175	*Etheostoma camurum*
Peocilichthys coeruleus	O-W-T.'30:175	*Etheostoma caeruleum*
Peocilichthys exilis	O-W-T.'30:175	*Etheostoma exile*
Peocilichthys spectabilis	O-W-T.'30:175	*Etheostoma spectabile spectabile*
Peocilichthys tippecanoe	O-W-T.'30:175	*Etheostoma tippecanoe*
Peocilichthys variatus	O-W-T.'30:175	*Etheostoma variatum*
Perca americana	Kl.'77:65; J.'82:958	*Perca flavescens*
Percina caprodes	J.'82:970; W-O.'98:48; O.'01:89	*Percina caprodes caprodes*
Percina caprodes zebra	O.'01:90	*Percina caprodes semifasciata*
Percina manitou	J.'82:971	*Percina caprodes semifasciata*
Percopsis guttatus	J.'82:899; H.'88:79; Mc.'92:23; W-O.'98:37; O.'01:75	*Percopsis omiscomaycus*
Petromyzon argenteus	K.'30:169; K.'41B:342; K.'51Q:205	*Ichthyomyzon bdellium*
Petromyzon concolor	H.'89:123	*Ichthyomyzon bdellium* or *Ichthyomyzon unicuspis*
Petromyzon concolor	Mc.'92:9	*Ichthyomyzon unicuspis*
Petromyzon nigrum	K.'38:169	Presumably several species of lampreys
Phoxinus elongatus	Mc.'92:21	*Clinostomus elongatus*
Pimelodus catus	K.'47:330	All three species of *Ictalurus* having square tails
Pimelodus catus	K.'50G:141	Either *Ictalurus melas* entirely, or that species plus *Ictalurus nebulosus* and/or *Ictalurus natalis*
Pimelodus cerulescens	K.'38:169	*Ictalurus punctatus*
Pimelodus coerulescens	K.'47:332	Presumably *Ictalurus punctatus* in part and *Ictalurus furcatus* in part
Pimelodus coerulescens	K.'50H:173	*Ictalurus punctatus*
Pimelodus cupreus	K.'38:169	Unidentifiable
Pimelodus cupreus	K.'47:333; K.'50X:157	Apparently a composite of *Pilodictis olivaris* and *Ictalurus natalis*
Pimelodus limosus	K.'47:335; K.'50Y:165	*Pilodictis olivaris*
Pimelodus nebulosus	K.'38:169	Presumably *Pilodictis olivaris*
Pimelodus pallidus	K.'38:169	*Ictalurus punctatus*
Pimelodus xanthocephalus	K.'38:169	Presumably all three species of *Ictalurus* having square tails
Pimephales promelas	K.'51DD:285	*Pimephales promelas promelas*
Placopharynx carinatus	J.'82:831; H.'88:78; O-W-T.'30:172; T.'50:11	*Moxostoma carinatum*
Placopharynx carinatus	Mc.'92:16	Considered not to be *Moxostoma carinatum; see* text under that species
Placopharynx duquesnii	W-O.'98:22	Description is of *Moxostoma carinatum*; Pl. 13, labelled *Placopharynx duquesnii* is of *Moxostoma anisurum* whereas Pl. 12, labelled *Moxostoma anisurum* is of *Moxostoma carinatum*

Name by Former Author	Abbreviated References	Current Name
Placopharynx duquesnii	O.'01:39	*Moxostoma carinatum*
Platinostra edentula	K.'38:169; K.'47:22	*Polyodon spathula*
Poecilichthys caeruleus caeruleus	T.'50:41	*Ethoestoma caeruleum*
Poecilichthys camurus	T.'50:41	*Etheostoma camurum*
Poecilichthys coeruleus	J.'82:979	*Etheostoma caeruleum*
Poecilichthys eos	J.'80:980	*Etheostoma exile*
Poecilichthys exilis	T.'50:42	*Etheostoma exile*
Poecilichthys maculatus	T.'50:40	*Etheostoma maculatum*
Poecilichthys spectabilis	J.'82:980; T.'50:42	*Etheostoma spectabile spectabile*
Poecilichthys tippecanoe	T.'50:42	*Etheostoma tippecanoe*
Poecilichthys variatus	T.'50:40	*Etheostoma variatum*
Poecilichthys zonalis zonalis	T.'50:40	*Etheostoma zonale zonale*
Poecilosoma erythrogastrum	K.'54:4-5	*Etheostoma caeruleum*
Polyodon folium	K.'44:21; K.'51V:249; K.'78:111; J.'82:764	*Polyodon spathula*
Pomotis machrochira	K.'50DD:101	*Lepomis macrochirus macrochirus*
Pomotis macrochira	K.'41B:469	*Lepomis macrochirus macrochirus*
Pomotis nitida	K.'41B:472; K.'50M:117	*Lepomis megalotis;* probably mostly *Lepomis megalotis peltastes*
Pomotis vulgaris	K.'38:168	Presumably *Lepomis gibbosus* in part and *Lepomis megalotis* in part
Pomotis vulgaris	K.'41B:470	*Lepomis gibbosus*
Pomotis vulgaris	K.'50EE:109	Plate is figure of *Lepomis gibbosus*; verbal description apparently a composite of *Lepomis gibbosus* and *Lepomis cyanellus*
Pomoxis sparoides	H.'88:79; Mc.'92:26; Kirs.'95A:330; W-O.'98;41; O-W-T.'30:176; O.'01:78	*Pomoxis nigromaculatus*
Pomoxys annularis	Kl.'77:77; J.'82:924	*Pomoxis annularis*
Pomoxys hexacanthus	Kl.'77:77	*Pomoxis nigromaculatus*
Pomoxys spariodes	J.'82:925	*Pomoxis nigromaculatus*

Q

Name by Former Author	Abbreviated References	Current Name
Quassilabia lacera	Kl.'78:104; J.'82:832	*Lagochila lacera*

R

Name by Former Author	Abbreviated References	Current Name
Rhinichthys atratulus obtusus	T.'50:15	*Rhinichthys atratulus meleagris*
Rhinichthys atronasus	J.'82:857; H.'88:78; Mc.'92:20; Kirs.'95A:329; W-O.'98:33; O.'01:61	*Rhinichthys atratulus meleagris*
Rhinichthys atronasus meleagris	O-W-T.'30:172	*Rhinichthys atratulus meleagris*
Rhinichthys atronasus obtusus	O-W-T.'30:172	*Rhinichthys atratulus meleagris*
Rutulus amblops	K.'38:169	*Hybopsis amblops amblops?*
Rutulus compressus	K.'38:169; K.'44:306	*Notemigonus crysoleucas?*
Rutulus crysoleucas	K.'38:169	*Notemigonus crysoleucas*

S

Name by Former Author	Abbreviated References	Current Name
Salmo amethystus	K.'51P:101	*Salvelinus namaycush*
Salmo fario	O-W-T.'30:171	*Salmo trutta*
Salmo fontinalis	K.'44:305; K.'51T:69	*Salvelinus fontinalis*
Salmo gairdnerii irideus	T.'50:5	*Salmo gairdneri*
Salmo irideus	O-W-T.'30:171	*Salmo gairdneri*
Salmo manycash	K.'38:169	*Salvelinus namaycush*
Salmo namycush	K.'44:25	*Salvelinus namaycush*
Salmo trutta fario	T.'50:5	*Salmo trutta*
Scaphirhynchops platyrhynchus	H.'88:77	*Scaphirhynchus platorynchus*

Name by Former Author	Abbreviated References	Current Name
Scaphirrhynchops platyrhynchus	J.'82:768	*Scaphirhynchus platorynchus*
Schilbeodes eleutherus	W-O.'98:12; O.'01:28; O-W-T.'30:174; T.'46:31; T.'50:31	*Noturus eleutherus* in part, *Noturus furiosus* in part
Schilbeodes furiosus	O-W-T.'30:174	*Noturus furiosus*
Schilbeodes gyrinus	W.O.'98:11; O.'01:27	*Noturus gyrinus*
Schilbeodus gyrinus	O-W-T.'30:174	*Noturus gyrinus*
Schilbeodes miurus	W-O.'98:12, O.'01:29; O-W-T.'30:74; T.'46:31; T.'50:31	*Noturus miurus*
Schilbeodes mollis	T.'46:30; T.'50:30	*Noturus gyrinus*
Schilbeodes nocturnus	O-W-T.'30:174; T.'46:30	Hybrid between *Noturus gyrinus* × *N. miurus*
Sciaena grisea	K.'38:168	*Aplodinotus grunniens*
Sciaena oscula	K.'38:168	*Aplodinotus grunniens*
Sclerognathus cyprinus	K.'47:275; K.'51S:373	Presumably all three species of *Carpiodes*
Scolecosoma argentum	J.'82:757	Presumably *Ichthyomyzon unicuspis* in part and *Ichthyomyzon bdellium* in part.
Semotilus biguttatus	K.'41A:344	*Hybopsis biguttata*
Semotilus cephalis	K.'38:169	*Semotilus atromaculatus atromaculatus*
Semotilus cephalus	K.'41A:345	*Semotilus atromaculatus atromaculatus*
Semotilus corporalis	J.'82:863	*Semotilus atromaculatus atromaculatus*
Semotilus diplema	K.'38:169	*Semotilus atromaculatus atromaculatus?*
Semotilus dorsalis	K.'38:169	*Semotilus atromaculatus atromaculatus*
Stizostedion canadense canadense	T.'50:34	*Stizostedion canadense*
Stizostedion canadense griseum	O.'01:88	*Stizostedion canadense*
Stizostedion glaucum	O-W-T.'30:175	*Stizostedion vitreum glaucum*
Stizostedion vitreum	Mc.'92:31; O.'01:88	*Stizostedion vitreum vitreum* in part, and *Stizostedion vitreum glaucum* in part
Stizostedion vitreum	Kirs.'95A:331; O-W-T.'30:175	*Stizostedion vitreum vitreum*
Stizostethium canadense	Kl.'78:86; J.'82:961	*Stizostedion canadense*
Stizostethium vitreum salmoneum	Kl.'78:91; J.'82:963	*Stizostedion vitreum vitreum* in part and *Stizostedion vitreum glaucum* in part
Stizostethium vitreum vitreum	Kl.'78:89; J.'82:962	*Stizostedion vitreum vitreum*
Stizosthethium salmoneum	Kl.'77:67	*Stizostedion vitreum glaucum*
Stizosthethium vitreum	Kl.'77:67	*Stizostedion vitreum*
	T	
Telestes elongatus	J.'82:865	*Clinostomus elongatus*
Teretulus oblongus	Kl.'77:85	Presumably mostly *Erimyzon sucetta kennerlyi* but some *Erimyzon oblongus claviformis*, probably included
	U	
Uranidea franklini	J.'82:988	*Cottus bairdi kumlieni*
Uranidea hoyi	J.'82:987	*Cottus bairdi kumlieni*
Uranidea richardsoni	J.'82:989	*Cottus bairdi bairdi*
Uranidea spilota	J.'82:993	*Cottus ricei*
	V	
Villarius lacustris	O-W-T.'30:174	*Ictalurus punctatus*
	X	
Xernotis megalotis megalotis	O-W-T.'30:176	*Lepomis megalotis megalotis*
Xenotis megalotis peltastes	O-W-T.'30:176	*Lepomis megalotis peltastes*
	Z	
Zygonectes notatus	J.'82:910; H.'88:79; Kirs.'95A:330	*Fundulus notatus*

Changes in Scientific Names since 1957 Edition

Name by Former Author	Abbreviated References	Current Name
Chaenobryttus gulosus	T.'57:496	*Lepomis gulosus*
Chrosomus erythrogaster	T.'57:326	*Phoxinus erythrogaster*
Clinostomus vandoisulus	T.'57:332	*Clinostomus funduloides*
Etheostoma blennioides	T.'57:571	*Etheostoma blennioides blennioides*, (in part)
		Etheostoma blennioides pholidotum, (in part)
Eucalia inconstans	T.'57:617	*Culaea inconstans*
Hybopsis biguttata	T.'57:292	*Nocomis biguttatus*
Hybopsis micropogon	T.'57:295	*Nocomis micropogon*
Lampetra lamottei	T.'57:000	*Lampetra appendix*
Lepisosteus productus	T.'57:163	*Lepisosteus oculatus*
Moxostoma aureolum aureolum	T.'57:253	*Moxostoma macrolepidotum macrolepidotum*
Moxostoma breviceps	T'57:255	*Moxostoma macrolepidotum breviceps*
Notropis cornutus chrysocephalus	T.'57:352	*Notropis chrysocephalus chrysocephalus*
Notropis cornutus frontalis	T.'57:355	*Notropis cornutus cornutus*
Notropis deliciosus stramineus	T.'57:377	*Notropis stramineus*
Noturus furiosus	T.'57:438	*Noturus stigmosus*
Opsopoeodus emilae	T.'57:335	*Notropis emiliae emiliae*
Parexoglossum laurae hubbsi	T.'57:320	*Exoglossum laurae hubbsi*
Pilodictis olivaris	T.'57:430	*Pylodictis olivaris*
Pomolobus chrysochloris	T.'57:177	*Alosa chrysochloris*
Pomolobus pseudoharengus	T.'57:180	*Alosa pseudoharengus*
Roccus americanus	T.'57:474	*Morone americana*
Roccus chrysops	T.'57:471	*Morone chrysops*

Part IV

The Scientific Names of Fishes

Every correctly described species of fish, as well as every other species of animal and plant, has a scientific name which is binomial and consists of the generic and specific names, or trinomial and has a subspecific name following the generic and specific names. Scientific names are of Greek, Latin or barbaric origin; are simple or compound, and usually describe some characteristic such as color or shape, something such as a ship, some geographic feature such as a river, some political subdivision as a state, or are patronymic. (For translations of Greek and Latin terms *see* Woods, 1944 and 1947, Hough, 1953, Brown, 1956). The composition of the scientific names, their endings, and which names may be used are governed by the rules, opinions, and recommendations as outlined in the international rules of zoological nomenclature.

Fishes belong to an animal phylum. This phylum is composed of classes which are divided into orders, the orders into families, the families into genera, the genera usually into more than one species, and some species are divided into subspecies. Each group also may be subdivided, such as orders into suborders, families into subfamilies, and genera into subgenera or into groups of superspecies. Each group, from a class to a genus, presumably contains only those species having the same phylogenetic origin, with those within the genus bearing the closest relationship to one another.

My present concepts of genus, species and subspecies are as follows:

The genus.—"A genus is a systematic unit including one species, or a group of species of presumably common phylogenetic origin, separated by a decided gap from other similar groups (Mayr, 1947:283)." The genus is largely a human conception whose function is to group species in order to stress their relationships to each other. Unfortunately all taxonomists do not place the same value upon the various characters, so they are divided into those who consider any clear-cut character to be of generic value, and those who believe that stressing such characters destroys the purpose of the genus, which is to stress relationship. Such diversity of opinion results in disagreement relative to the number of genera, the "lumpers" stressing relationships, leaving subgenera to stress the minor gaps between groups within a genus, and the "splitters" preferring to give generic rank to what the "lumpers" designate as subgenera.

The species.—In bisexual animals a species is a natural population, or a group of populations, normally isolated by ecological, ethological, and/or mechanical barriers from other populations with which they might breed and which usually exhibit a loss of fertility when hybridizing. The above definition applies to all except a few bisexual animals. No single definition can be applicable to all animals however, because in the large series of imperceptible and intermediate stages in the evolutionary transition from a subspecies to a species there is no well-defined gap separating the two; hence it is sometimes impossible to know whether two populations are specifically or only subspecifically distinct.

The subspecies.—"The subspecies, or geographic race, is a geographically localized subdivision of a species, which differs genetically and taxonomically from other subdivisions of the species (Mayr, 1947:106)." Unlike species, the subspecies comprising a species can freely interbreed without loss of fertility, but mass interbreeding is usually hindered or prevented by some type or degree of geographical, ecological, ethological and/or mechanical barrier or barriers. As indicated under the species, there are many stages in the evolutionary process of species production from an original homogenous group into two homogenous groups or subspecies. Some subspecies are so similar to one another that the two can be separated only through the application of statistical methods, others are so distinct as to be recognizable at a glance. Taxonomists usually consider a subspecies to be valid if 75% or more of one population can be separated statistically from the other population.

The name of the describer or describers.—The name (or names) of the person (or persons) who described and named a species or subspecies may follow the scientific name of the species. His (or their) name indicates that he (or they) described the species if the name is binomial, or the subspecies name if trinomial. If the describer's name is *not* enclosed within parentheses he described the species or subspecies in the genus in which it appears; if his name *is* enclosed in parentheses he described the species or subspecies under a generic name which is different from that in which it appears. The use of parentheses aids in determining the authorship.

The Common or English Names of Fishes

Species or subspecies which are usually recognized by laymen frequently have one or more well-established common or English names. Unfortunately, many of our best-known fishes have many colloquial names and/or two distinct species are considered to be only a single species. An example is the black and white crappies, which as late as 1930 were considered by the majority of Ohio fishermen to be one species, and were known as "shad" in northeastern Ohio, as "croppies" in northwestern Ohio, as "Lake Eries" in central Ohio, as "Campbellites," "tinmouths," or "bank lick bass" in southern Ohio. Recent attempts by various agencies have resulted in the standardization of black crappie, white crappie, and many other names.

There are usually only two common names for species which are not divided into subspecies, as green sunfish—*Lepomis cyanellus.* A third common name is added if the species is divided into two or more subspecies, as central longear sunfish—*Lepomis megalotis megalotis* and northern longear sunfish—*Lepomis megalotis peltastes.*

Reasons for the Changing of Scientific Names

Frequently laymen or fish students are exasperated by what they presume to be the unnecessary changing of well-established scientific or common names, not realizing that name changing is an unavoidable form of progress, even though it causes confusion at the time.

Some reasons for name changing are: It was impossible for the majority of the early describers of fishes such as Rafinesque to review adequately the pertinent literature in order to learn whether their supposedly "new species" had been described, or to compare their specimens with the types of closely related species. As a result, they described many species which had been properly described one or more times previously, and consequently their name for the species, according to the law of priority, had to be synonymized. Some species were thus unknowingly described many times, such as the longnose gar which has been described more than two dozen times (Jordan, Evermann and Clark, 1930: 36–37).

Recent studies of original descriptions of many species have shown that the vague descriptions given by the describers have misled later taxonomists so that they applied scientific names to the wrong species; upon realizing the error another name for these species had to be given. Researchers frequently find a species description in some previously overlooked, obscure publication, which description is of an earlier date than the name applied at present to the species. In that case the law of priority demands that the earliest available name is used, even though the other name has been in usage many years. An example is the recent change of the scientific name of the burbot from *Lota lota maculosa* (Lesueur) to *Lota lota lacustris* (Walbaum); *see* Speirs, 1952: 99–103.

Recent Merging of Genera

It was the custom of the earliest taxonomists to place many species in a single genus, such as the genus *Lepomis.* Later when the various species were compared with one another certain differences became apparent. As a result, the large genus was divided into many genera, until in some instances the majority of the genera had become monotypic, such as the division of *Lepomis* into *Apomotis, Lepomis, Allotis, Xenotis, Helioperca, Eupomotis* and *Chaenobryttus.* Still later, when adequate material had been assembled, a monographic study of the group was undertaken, resulting in the accumulation of abundant evidence indicating that a drastic reduction in the number of genera was necessary, as was the reduction of the above seven genera into two genera, *Lepomis* and *Chaenobryttus.* This swinging from one genus to many, then back to a few when an adequate study of the group was made, is what normally happens.

Between 1944–53 I made a critical study of many Ohio genera. From these studies it became apparent that many genera were based upon a few specific characters, and since many supposed genera were obviously of the same phylogenetic origin (such as *Hiodon* and *Amphiodon*) these genera must be merged. As a single genus they stressed their relationship to each other, and emphasized the gaps between this combined genus and other genera. Recently other workers have merged genera, several such as Bailey (1951) and Legendre (1952 and 1954) publishing their findings. Unfortunately it was impossible, because of the delay it would have caused in completing this report, for me to have published detailed accounts as to why these many genera should be merged. In this report I have merged *only* those genera which I have studied adequately and which were the most obvious, and I have not followed others when I had not studied the group sufficiently. Examples are the genus *Clinostomus* which Bailey (1951: 191) placed in the genus *Richardsonius*, and *Pomolobus* which he (1951: 190) placed in the genus *Alosa*.

Recent Changes in the Scientific and English or Common Names

As indicated above, the developing science of ichthyology sometimes requires the changing of scientific and common or English names. As knowledge accumulates the relationships of groups and species need to be modified. Occasionally the generic or specific name of some fish is changed several times to a new name, then back to the old as additional evidence accumulates.

In 1948, in an attempt to standardize the scientific and common names of fishes, the American Fisheries Society published a list of scientific and common names of fishes, Special Publication No. 1, followed in 1960 by Special Publication No. 2 and in 1970 by Special Publication No. 6 (Robins, et al.) and in 1980 (in press). These publications have aided considerably in the standardization of names.

As relationships between families, genera, species and subspecies have become better known, name changing has become less frequent; however, changes continue (Bailey, et al.; 1960 and 1970). Since the publication in 1957 of *The Fishes of Ohio* there have been at least 42 changes in the scientific or common names of Ohio fishes. Recent investigations have indicated the need of merging such genera as *Pomolobus* with *Alosa*, *Parexoglossum* with *Exoglossum*, *Chrosomus* with *Phoxinus*, *Opsopoeodus* with *Notropis*, *Chaenobryttus* with *Lepomis* and *Eucalia* with *Culaea*. The genus *Hybopsis* has been split and some of the species, formerly in *Hybopsis* have been placed again in the genus *Nocomis*. In the freshwater sea basses, the genus name has continued to be changed, most recently from *Roccus* to *Morone*. It has been demonstrated that the name *aureolum*, long the specific name of the northern shorthead redhorse, was not valid. It has been replaced by the specific name *macrolepidotum*. The northern shorthead and the Ohio redhorses are now considered conspecific.

In the genus *Lepisosteus*, the specific name of the spotted gar has been changed from *productus* to *oculatus*. In the genus *Clinostomus*, the specific name of the rosyside dace has been changed from *vandoisulus* to *funduloides*.

Before 1930 many species of fishes found in Ohio waters had different colloquial names in various parts of the State, causing confusion among sportsmen regarding which species were referred to in the state laws (*see* above, under The Common or English Names of Fishes). As is taking place in other disciplines, ichthyologists have been attempting to standardize common names, designating *only* one common name for each species. Unfortunately, in the second and third editions of *A List of Common and Scientific Names of Fishes* (Bailey et al., 1960 and 1970) the species *Esox americanus americanus* was given the common name redfin pickerel, whereas its subspecies *Esox americanus vermiculatus* was given the name grass pickerel; *Stizostedium vitreum vitreum* was named walleye whereas its subspecies *Stizostedium vitreum glaucum* became the blue pike. The general public has difficulty in grasping the species and subspecies concepts and the retention of two common names for the same species adds to the confusion. It is hoped that eventually the rule of one common name per species will have no exceptions.

In 1841 Kirtland (1841B: 341–42 and Pl. 4) described *Luxilus dissimilis*, giving it the common name of spotted shiner. The highly descriptive name of "spotted" was retained in the literature until 1960 (op. cit., 1960 and 1970) when this long established name was changed to streamline, certainly a less descriptive name for this species. It is hoped that such long established names will be retained in the future, thereby causing less confusion as to the species indicated.

During the past century the combining of common names has followed a definite pattern. As an example: Formerly *Lepomis macrochirus* had the book name of blue gill, which later became blue-gill and now is bluegill. Such names as "perch", now used as a species word, in the families of pirate perches (Aphredoderidae), temperate basses (Percichthyidae), sea basses (Serranidae), sunfishes (Centrarchidae), perches (Percidae), drums (Sciaenidae), sea chubs (Kyphosidae), cichlids (Cichlidae), and surfperches (Embiotocidae) should be combined, at least with a hyphen, except in the case of the true perches (Percidae).

Part V

The Artificial Keys

Individual variations within a species.—Considerable variation in the numbers and/or sizes of such body parts as rays and scales and in body proportions exists among individuals of the same fish species or subspecies. Quantitative, qualitative and meristic variations occur with great frequency between sexes, between young and adults, between well-fed, rapidly-growing individuals and starved, slowly-growing ones, between heavily parasitized and diseased fishes and less parasitized and healthy individuals, and between fishes of different year classes which inhabit the same water and have the same genetic background (Hile, 1937: 105–18). Because of these variations it was necessary to construct keys containing a combination of characters, rather than of a single character. These multi-character keys make possible the identification of all except a few of the normal individuals, the aberrant specimens, and hybrids.

Types of keys.—The keys are of the conventional, dichotomous type, since they offer two alternate choices. They are largely artificial because they do not follow phylogeny. They should offer little difficulty to those persons having had a course in systematic ichthyology, but persons without previous experience or training with such keys will naturally have difficulty at first. These difficulties will eventually disappear, and rapidly if the glossary and procedures for measuring and counting are referred to frequently.

How to use the keys.—Make certain as to which family your fish belongs.

To do this, begin with Number 1 in the "Key to Families of Ohio Fishes," p. 69, and determine carefully which of the two opposable groups of characters fits your fish.

If your specimen fits into the first group it will be, in this case, a member of the family Petromyzontidae.

If it fits into the second group it will lead to a number, in this case Number 2.

Go to 2 continuing until the name of the family is reached in which your specimen belongs.

Essentially the same procedure is followed in using the "Key to the Species and Subspecies of Ohio Fishes," p. 79 provided there is more than one species in the family.

Start with Number 1 and choose the group to which your fish belongs.

If that group leads to the name of a fish you have identified it.

But if it leads to a number, follow that number, and as many other numbers as necessary, until the name of your fish is reached.

Be certain that you understand *every* statement, character referred to, and how to measure and count; if not consult the glossary and methods of counting and measuring.

Great care *must* be taken in counting scales and rays, and in measuring such body parts as eye, mouth, snout and head length; otherwise you will misidentify your fish.

ADDITIONS TO THE 1957 GLOSSARY

The following additions should be made to the glossary of the original edition, which follows: *Crepuscular.*—Pertaining to twilight, dim; *Ecotype.*—A group of freely interbreeding individuals responding to environmental conditions, which through evolutionary processes become comparable to a taxonomic subspecies; *Gill filaments* or *lamellae.*—The pleated folds of skin, richly supplied with small blood vessels, attached to the posterior edge of gill arches; *Jack.*—A precocious male which spawns a year or two earlier than its siblings do and which is notably smaller than the average size of the spawning male; *Hyoid.*—A bone or bones at the base of the tongue, in some species containing teeth.

GLOSSARY OF TECHNICAL TERMS[1]

Abbreviate-heterocercal tail.—Midway between a heterocercal and homocercal condition, in which the posterior end of the vertebral column (backbone) is definitely flexed upward and only partly invades the upper lobe (or half) of the caudal fin.

Abdomen.—That portion of the body which contains the viscera.

Abdominal.—That which pertains to the belly; often used to refer to the ventral portion of the body from the thorax to the anal opening.

Abdominal cavity.—That which contains the viscera.

Abdominal fins.—In the most restricted sense only the pelvic fins, but sometimes includes the pectoral fins in those soft-rayed species in which the origins of the pectorals are situated almost as ventrally as are the pelvics.

Aberrant.—Abnormal; deviating sharply from the average condition and with characteristics not in accordance with the type.

Acanthopterygian.—A spiny-rayed fish; includes almost all of the teleosts.

Acuminate.—Tapering to a slender point.

Acute.—Ending in a sharp point.

Adherent.—Attached or joined together, at least at one point.

Adipose eyelids.—Usually two transparent membranes, one of which is attached to the anterior edge of the eye and extends backward partly across the eye, the other is attached to the posterior edge of the eye and extends forward partly across the eye, the two covering the eye to a greater or less extent; *see* river herring.

Adipose fin.—A fleshy, fin-like, rayless structure situated on the dorsal ridge between the dorsal and tail fins, which in some species is fused to the tail and separated from it only by a slight notch.

Adnate.—Congenitally united; said of adipose fins which are attached to the back throughout their entire length; conjoined.

Adult.—A mature individual, capable of producing offspring at some season of the year.

Air bladder.—A membranous, gas-filled sac lying just beneath the backbone of fishes and adherent or not to the walls of the visceral cavity. Also called the swim bladder.

Alimentary canal or tract.—The tubular passage extending from the mouth to the anus, including the pharynx, esophagus, stomach, small and large intestines, and having a nutritive function.

Allopatric.—Two forms or species which exclude each other geographically; is primarily useful when applied to geographic representatives; *see* Sympatric.

Ammocoetes.—Larval form of any species of lamprey.

Anal.—That which pertains to the vent or anus.

Anal fin.—The fin situated medially and normally immediately behind the vent (except in the pirate perch), and between the posterior end of the abdomen and anterior end of the caudal peduncle.

Anal opening.—*See* Anus.

Analogous.—Similar in function, but not in structure or development.

Anal papilla.—A protuberance, usually bilobed, in front of the genital pore and behind the anus in darters, sculpins, and some other species.

Anomaly.—Any departure from type characteristics or the usual condition.

Anterior.—That which is in the front or fore part, or before or forward to another part; opposite of Posterior.

Anterior fontanel.—A medial, longitudinal opening, covered only by skin, in the dorsal ridge of the skull, immediately before the frontal bones, and in *Carpiodes* situated midway between and slightly behind the nostril openings; *see* Gregory, 1933: 192.

Antrorse.—Directed forward; opposite of Retrorse.

Anus.—The external opening of the intestine, situated, except in the pirate perch, immediately anterior to the origin of the anal fin.

Articulated.—Jointed, as are the component parts of a ray.

Association.—The largest natural plant and/or animal community consisting of a recognizable assemblage of species, and in which a few species dominate; the association is usually named for these dominants such as an elm-ash-soft maple association.

Asymmetrical.—As used in fishes, that condition in which the two halves differ; consequently they cannot be divided into similar parts; having the two sides unlike.

Atrophy.—A shriveling, diminution or wasting away of a part.

Average.—(Abbreviations are Av., Ave., and Avg.)—An arithmetical mean which is obtained by calculating the mean of several.

Axillary process.—An accessory, enlarged scale attached to the upper or anterior base of the pectoral or pelvic fins in certain herring-like fishes.

Barbel.—A fleshy tactile, enlarged flap or icicle-shaped projection, usually situated about the lips, chin or nose; varies considerably in number and size in the various species of fishes.

[1] Although many of these terms are found in standard dictionaries their definitions do not always convey the same shade of meaning as when used ichthyologically. In preparing this glossary, and methods of counting and measuring, the following publications were of great value and many of the definitions are similar—Bailey, 1951: 231–35; Forbes and Richardson, 1920: CXXXI-CXXXVI; Henderson and Henderson, 1953: 1–506; Hubbs and Lagler, 1947: 8–20; Legendre, 1954:55–178.

Bicolored.—Of two colors.

Bicuspid.—A two-pronged tooth on one base.

Bilaterally symmetrical.—Capable of being halved in one, and only one, plane in such a way that the two halves are approximately mirror images of each other.

Binomial.—Consisting of two names; nomenclatorially the first of which is the generic name, the second the species name.

Branched ray.—A soft ray which forks into two or more parts distally.

Branchiae.—Gills or respiratory organs.

Branchial.—Pertaining to the gills.

Branchial arches.—Bony or cartilaginous arches extending dorso-ventrally behind the hyoid arch, and placed one behind the other on each side of the pharynx so as to support the gills.

Branchial basket.—*See* Gill basket.

Branchiostegal membrane.—A membrane or membranes situated on the ventral edges of the gill covers and containing the elongated branchiostegal rays.

Branchiostegals or branchiostegal rays.—Elongated bones arranged fanwise within the branchiostegal membranes.

Breast.—In fishes that part of the body situated ventrally between the isthmus of the gill covers and to a point immediately behind and below the pectoral fins.

Buccal.—Pertaining to the mouth.

Buccal disc.—The oral opening in lampreys including the circular jawless lip with its fimbriae, the tooth bearing roof of the mouth, and the esophageal opening.

Buccal funnel.—The circular cavity within the mouth disc in lampreys.

Caducous.—Pertaining to scales of modified form which are not persistent and may fall off.

Caecum.—A blindly-ending sac connected with the alimentary canal at the posterior end of the stomach in fishes.

Canine teeth.—Strong, conical, sharply pointed teeth on fore part of the jaws which usually are longer than other teeth.

Cardiform teeth.—Sharp teeth arranged in series.

Carnivorous.—Flesh eating.

Cartilage.—A bluish-white tissue, commonly called "gristle" which forms part or all of the skeleton of fishes.

Cartilaginous.—Pertaining to the cartilage.

Caudal or caudal fin.—The fin on the tail of fishes.

Caudal peduncle.—That region of the body between the base of the posterior ray of the anal fin and the base of the tailfin; *see* fig. I.

Character.—As used here, an attribute, quality or property which distinguishes one form or forms from others.

Cheek.—That portion of the lateral surface of the head lying between the eye and the opercle, usually not including the preopercle.

Chin.—In fishes the lower surface between the mandibles.

Chromatophores.—Color cells which under control of the sympathetic nervous system can be altered in shape to produce color changes.

Chute.—A section in a large river in which the water is rather deep and flows swiftly, and in which its surface, although containing eddies or small whirlpools, is not conspicuously broken up by protruding rocks as is the surface of a riffle or rapids.

Circumoral series.—*See* Circumoral teeth.

Circumoral teeth.—The innermost circle of horny teeth surrounding the esophagus in lampreys.

Circumorbitals.—A series of bones about the eye, those bones lying in front of the eye are known as the preorbitals, those below the eye are the suborbitals, and those behind the eye are the postorbitals.

Class.—A comprehensive division of a phylum ranking above an order in the classification of plants and animals.

Classification.—The systematic arrangements of animals and plants in groups which are based upon natural relationship.

Coalesced.—Joined or fused in growth.

Compressed.—Flattened laterally.

Concave.—Arched inward; a *concaved fin* is one in which the central soft-rays or spines are shorter than the anterior and posterior soft-rays or spines, thereby causing an inwardly curved distal edge to the fin; opposite of Convex.

Confluent.—A flowing or merging together or into one.

Convex.—Arched or rounded outward; a *convex fin* is one in which the central soft-rays or spines are longer than the anterior and posterior soft-rays or spines, thereby causing an outwardly curved distal edge to the fin; opposite of Concave.

Ctenoid scales.—Having a comb-like margin of tiny prickles (ctenii) on the exposed or posterior field; the ctenii cause the scales to feel rough to the touch when stroked.

Cusp.—A projection or point on a tooth.

Cycloid scales.—Smooth-edged scales of soft-rayed fishes having an evenly curved posterior border and without a trace of minute spines or ctenii.

Deciduous.—Falling off at some stage of maturity.

Decurved.—Curving downward.

Dentary or dental bone.—The principal bone of the lower jaw in which the lower teeth are set when present.

Dentate.—Having tooth-like notches.

Depressed.—Flattened from above downwards and therefore broader than deep.

Disc or disk.—*See* Buccal disc.

Disc teeth.—All teeth on the roof of the mouth or buccal disc of lampreys, excepting those on the tongue.

Distal.—Farthest from the place of attachment; opposite of Proximal.

Distal edge.—When referring to a fin it is the free margin farthest removed from the basal or proximal portion of the fin.

Dorsal.—The back or upper third of the fish; the opposite of ventral.

Dorsal fin or fins.—A single or double ray and/or spine-bearing fin situated medially on the back and before the adipose fin if one is present.

Dorsal origin.—The apex of an angle which is formed by the dorsal ridge of the body and the anteriormost ray or spine of the dorsal or first dorsal fin.

Dorso-ventral or dorsal-ventrad.—Pertaining to both dorsal and ventral surfaces.

Ecological isolation.—A situation in which the potential parents of two species or subspecies are confined to two different habitats in the same general region, and seldom or never come together during the reproductive season.

Ecologist.—One who studies plants and animals with reference to their environment, and to the factors which control or have controlled their distribution.

Ecology.—Biology dealing with the mutual relations between organisms and their environment.

Emarginate.—Notched but not definitely forked, as is the shallow notch in the tail and between the two dorsal fins of the largemouth blackbass.

Endoskeleton.—The entire bony or cartilagenous framework of the body.

Entire.—Referring to a smooth edge which is not serrated, denticulated or spine-bearing.

Environment.—The sum total of all external influences and conditions affecting the life and development of an organism.

Esophagus or Oesophagus.—That part of the alimentary tract between the pharynx and the stomach or its equivalent.

Ethological barrier.—A difference in the behavior pattern of two animals of different species, or of the same species, which prevents them from successfully mating, or associating with each other in a certain manner.

Exterior.—That which is turned outwards or pertains to the outside; opposite to interior.

Falcate.—Curved like a sickle, a fin is falcate when it is deeply concave, having the middle rays *much* shorter than are the anterior and posterior rays.

Family.—A group of organisms forming a category ranking below an order and above a genus; used in classification to signify a group of related genera.

Fimbria.—A threadlike filament; many fimbriae surround the buccal disc of adult lampreys.

Fimbriate.—Fringed at the margin.

Fin.—A membrane or fold of skin containing in most fishes, soft rays, spines, or other skeletal supports.

Fin rays.—Usually called soft rays and usually, though not always, branched, flexible and bilaterally paired and segmented; may also refer to spiny rays.

Forked.—Connected basally but separated distally; usually applied to the tail whenever there is a pronounced upper and lower lobe.

Free spine.—A spine not connected to another spine by membrane.

Frenum.—A connecting membrane which binds a part or parts together, such as the binding together with skin of the upper jaw to the snout; *see* fig. II.

Ganoid or ganoid scales.—Diamond- or rhombic-shaped scales consisting of bone covered with superficial enamel; also a group of fishes containing the sturgeons, paddlefish, gars, bowfins and many extinct forms.

Genital papilla.—A small blunt, fleshy, projection behind the anal opening; present in darters.

Genus.—*See* p. 51.

Gill.—A respiratory organ in aquatic animals consisting chiefly of filamentous outgrowths for breathing oxygen contained in water.

Gill arch.—The branchial skeleton which contains the gill rakers and the gill lamellae.

Gill basket.—The branchial section or skeleton of lampreys.

Gill cleft.—As used here, the opening between the opercle and side of head; *see* Gill opening.

Gill cover.—*See* Opercle.

Gill membrane.—The thin wall of skin or membrane which closes the gill cavity below and which in many fishes extends as a small flap along the free edge of the opercle.

Gill opening.—The opening between the opercle and side of the head of the higher fishes; lampreys have seven of these openings.

Gill raker.—The blunt knob-like projections on the anterior edge of the first gill arch; when occurring on other gill arches they are not well developed and are usually not used taxonomically.

Gill slit.—The narrow space between the gill arches.

Gonad.—The ovary or the testis.

Gradient.—As used in relation to streams; the rate of descent in feet per mile of streams; *See* pp. 139–40.

Graduated.—Becoming progressively longer or shorter, such as when the first spine is the shortest, the second longer than the first, etc.

Gravid.—Restricted to a female with eggs, especially when she appears to be very turgid; rarely used for males, then usually only those males having greatly enlarged testes in which the sperm are ready to be released.

Gular fold.—A transverse membrane across the throat.

Gular plate.—A large bony plate between the anterior third of the lower jaws of the bowfin, *see* fig. 15A.

Habitat.—The locality or external environment in which a species or subspecies normally lives, and which is controlled by such primary factors as the presence or absence of other organisms, climate, substratum and soil, water and vegetation.

Habitat niche.—One of the many micro-habitats which make up the overall habitat of a species or subspecies.

Haemal spine.—The lowermost spine of a caudal vertebra in fishes.

Herbivorous.—Eating or living upon vegetation.

Heterocercal.—A caudal fin is said to be heterocercal when the posterior end of the vertebral column (backbone) is flexed upward, entering and continuing to near the end of the upper or dorsal lobe of the caudal fin (but does not enter the filament if one is present), that lobe being better developed and often longer than the lower lobe.

Homocercal.—A caudal fin is said to be homocercal when the posterior end of the vertebral column does not flex upward and does not enter either lobe of the caudal fin, but ends in a hypural plate, the two lobes of the tail being nearly equal or equal.

Homogenous.—More or less alike because of descent from the same ancestral stock.

Homologous.—Having the same structure and origin.

Humeral process.—As used here for those species of catfishes in which it is developed, this process is a backward extension of the cleithrum. The process lies immediately beneath the skin, and parallel to, and directly above, the anterior portion of the depressed pectoral spine. By probing one can readily see and feel the outline and length of this process. *See* Bailey and Taylor, 1950: 31 and Pl. 2.

Humeral scale.—A large, often darkly pigmented, scale-like structure in darters near the midline of their bodies and immediately behind the head and above the pectoral origin; apparently it is the postcleithrum.

Hyaline.—Clear, glassy, translucent.

Hybrid.—*See* pp. 119–26.

Hybridize.—To interbreed and to produce hybrids.

Hypural.—The expanded, fused, bony haemal spines of the last few vertebrae which support the caudal fin in certain fishes.

Imbricate.—When the scales overlap like the shingles on a roof.

Inarticulate.—Not segmented or jointed.

Incipient.—In the first stages of becoming something; an incipient species is one which has most of the characteristics of a species and apparently may eventually become a species.

Incisor.—A front or cutting tooth.

Inferior.—Beneath, lower, or on the ventral side.

Inferior mouth.—A mouth is inferior when located near to or on the ventral side of the head, and when the snout more or less overhangs the upper lip; when comparing the positions of the mouths of two species, the mouth most posterior ventrally is the most inferior.

Infraoral cusp or lamina.—The tooth or teeth situated immediately posterior to the esophageal opening in lampreys.

Infraorbital canal.—That portion of the lateral line system which encircles most or all of the eye, except the upper section, and which encroaches onto the snout; *see* fig. II.

Insertion (of pectoral and pelvic fins).—The line along which the fin is attached to the body. Dorsal insertion and anal insertion do not mean dorsal and anal origin; *see* Dorsal origin.

Intergrades.—*See* p. 119.

Interhaemal spines.—The small, bony elements internally supporting the anal fin.

Interior.—That which is inside or pertains to the inside; opposite of exterior.

Intermuscular.—Between or among the muscle segments or fibers; the intermuscular grooves between the myomeres are very apparent in lampreys.

Interneural spines.—The small, bony elements internally supporting the dorsal fins.

Interopercle.—The membrane-bone which lies just below the preopercle bone; *see* fig. II.

Interorbital.—The space between the eyes on top of the head.

Interorbital width.—The distance across the top of the head between the eyes.

Interpelvic, or interpelvic space.—The shortest distance between the insertions of the pelvic fins.

Inter-radial membranes.—Membranes between fin rays or spines.

Interspecific.—Between members of two or more distinct species.

Intraspecific.—Among members of the same species.

Isthmus.—The narrow portion of the breast lying between the gill chambers and separating them.

Jugular.—Pertaining to the throat; the pelvic fins are "jugular" when located anterior to the pectorals.

Keeled.—Having a sharp, median ridge.

Larva, (larvae pl.).—The immature stage of lampreys and fishes before they metamorphose or transform and assume the characteristics of the adults.

Lateral.—Pertaining to the side or toward one side.

Lateral line.—A line formed by a series of sensory tubes and pores, extending backward from the head along the sides of the body. The lateral line is *complete* when all pores are present and the line reaches to the base of the caudal fin; *incomplete* when not extending as far as the base of the caudal fin; *absent* if no tubes or pores are present. The lateral line *system* also extends forward from the body onto the head, there branching into several parts.

Lateral line scales.—Those scales bearing the lateral line organs (if these are present) beginning back of the shoulder girdle and ending at the caudal base opposite the end of the hypural.

Lateral series of scales.—The transverse scale rows along the sides of the body. When part or all of the lateral line is present one member of each transverse scale row contains a pore which is counted in counting the number of scales in the lateral line. When pores are not present one scale of each transverse scale row is counted in order to determine the number of transverse scale rows.

Lentic or Lenitic (water).—Not flowing; static; opposed to Lotic.

Linear scales.—Scales that are higher than deep, like the lateral line scales of the mimic shiners.

Locality.—A particular place; often used to denote the habitats used by one or more species.

Lotic (water).—Water in motion such as the current produced by a riffle, or current and motion produced by wind.

Mandible.—The lower jaw.

Mandibular pores.—Small openings along a tube (usually hidden) on the lower side of each jaw, the tube being part of the cephalic canal system; *see* fig. 33B.

Mandibular symphysis.—The tip of the lower jaw where the two mandibles unite, sometimes called the chin.

Marginal.—At or on the edge or border.

Marginal teeth.—The continuous circle of countable teeth encircling the outermost edge of the buccal funnel of lampreys; *see* fig. 7A.

Maxilla.—The bones on each of the two halves of the upper jaw behind the premaxillae. These bones and the premaxillae form the two sides of the upper jaw.

Maxillary.—Pertaining to or in the region of the upper jaw; sometimes used to denote the upper jaw.

Maxillary symphysis.—The tip of the upper jaw where the two maxillaries unite, sometimes called the tip of the snout, or the nose.

Mechanical barrier.—A difference in the body structure of two animals of different species, or of the same species which prevents them from successfully mating, or associating with each other.

Median.—Lying in the middle or axial plane.

Melanophore.—A black chromatophore or pigment cell.

Membrane.—A thin skin or layer of tissue.

Meristic.—Divided into a number of parts, such as the spines and rays in the dorsal fin; segmented.

Meristic variation.—Differences in such numerical proportions as the number of spines in the dorsal fins in individuals of a given species.

Mid-dorsal streak.—A line lying along the median ridge of the back.

Minute.—Exceedingly small.

Molar teeth.—Flattish-topped teeth used for grinding rather than tearing, present on the pharyngeals of some Ohio fish species.

Monotypic.—Having only one type; a monotypic genus has only one species in the genus; opposite of polytypic in which two or more species occur in a genus.

Myomere.—A muscle-segment divided from others by connective tissue; the ridge-like myomeres in lampreys are plainly discernible, *see* fig. 1.

Nape.—That small area on the back of a fish beginning immediately behind the occipital region of the head and extending posteriorly to the dorsal fin origin in most species of spiny-rayed fishes, and backward from the occiput about the same distance as the length of the occiput in the soft-rayed species; *see* Occiput. The occiput of many species of fishes is naked and when the nape is fully scaled the anterior-most scale row of the nape is the dividing line between occiput and nape.

Nasal.—Pertaining to the region of the nose and nostrils.

Nest.—A spawning place for many species of fish. Most often refers to the globular structure placed above the substrate; for example, the stickleback.

Neural spine.—Uppermost spine of a vertebra in fishes.

Niche.—*See* Habitat niche.

Nomenclature.—The system of names as used in the classification of fishes, other animals and plants, and as distinguished from other technical terms.

Non-protractile.—As used here, the upper jaw is non-protractile when a frenum of membrane binds the anterior portion (premaxillae) of the upper jaw to the snout and there is no groove separating upper jaw from snout along the midline; *see* fig. 151A.

Nostril.—The external opening of the olfactory organ; the nares.

Notochord.—A longitudinal, elastic rod of cells which in the prevertebrates such as lampreys, and in embryos of higher teleosts, forms the supporting axis of the body; it is replaced by the ossified backbone in the higher orders of fishes.

Nuchal.—Pertaining to the nape.

Nuptial.—Pertaining to some courtship behavior or to mating.

Nuptial tubercles (breeding tubercles).—Hardened protuberances which reach their greatest development just before or during the breeding season, especially in adult males, and which normally fall off or disappear after the breeding season, leaving scars on some individuals.

Obsolete.—Scarcely or not evident.

Obtuse.—With blunted or rounded end.

Occipital.—Pertaining to the occiput.

Occiput.—That posterior portion of the dorsal surface of the head, beginning above or immediately behind the eyes of lampreys and fishes and extending backward to the point between the end of the head and the beginning of the nape.

Ocellate.—Containing ocelli.

Ocellus (ocelli pl.).—An eye-like spot, usually roundish and having a lighter or darker border.

Oesophagus.—*See* Esophagus.

Opercle or operculum.—The large, very flat and thin bones on each side of the head of fishes which cover the gills; also called gill cover.

Operculomandicular canal.—This subcutaneous canal has a series of openings, called pores, at irregular intervals and it is these pores which are counted. The first is above or near the upper angle of the preopercle, the remainder continuing downward and forward along, or close to, the posterior edge of the preopercle, thence forward along the ramus of the lower jaw to the symphysis. In some species the anterior-most pore lies immediately to one side of the center of the chin. In others it is medial in which case it is counted. The count includes only one side of the head, from the symphysis to above the upper preopercular angle. The pores in individuals of some species, such as the madtoms, are most difficult to find, necessitating the injection of air into the canal, or the staining or clearing of the specimens in order to see the pores.

Oral valve.—Thin membranes attached near the front of each jaw which prevent water from escaping through the mouth as it is pushed past the gills.

Orbital rim.—Part of the socket in which the orbit (eyeball) rests. In bony fishes the rim is formed by the circumorbital series of bones which include the prefrontal, lachrymal, suborbitals, dermosphenotic, and supraorbitals.

Order.—A category of classification of a group of closely allied organisms, ranking above the family and below the class.

Origin of fins.—The anterior-most point at which the dorsal, anal or other fins are in contact with the body; *see* Dorsal origin.

Oviparous.—Producing eggs which hatch outside the body.

Paired fins.—The pectoral and pelvic fins.

Palatines.—A pair of membrane bones on the roof of the mouth, one on each side and extending from the vomer.

Palatine teeth.—Teeth borne on the paired palatine bones which lie behind the median vomer on the roof of the mouth.

Papilla.—A small fleshy projection.

Papillose.—Covered with papillae.

Parr.—A young trout (or salmon) which still retains the parr marks.

Parr marks.—Refer to the color pattern of young trouts before they have developed the color pattern of adults, and especially to the squarish or oblong blotches along the sides.

Patronymic.—As used here, a scientific name derived from the name of a person or persons by adding "i" when that person is masculine and "ae" when feminine.

Pectoral.—Pertaining to the breast region; *see* Breast.

Pectoral arch.—The shoulder girdle which in most fishes is a framework of bones connected to the skull and to which framework the pectoral fins are attached.

Pectoral fins.—The anterior or uppermost of the paired fins of fishes, one on each side of the breast just behind the head, corresponding to the anterior limbs of the higher vertebrates.

Pellucid.—Translucent, hyaline.

Pelvic arch or girdle.—The aggregate of bones to which the ventral fins are attached.

Pelvin fin or Pelvics.—A paired fin on the ventral surface, abdominal in position when well behind the pectoral fins, or thoracic in position when beneath those fins.

Peritoneum.—The membranous lining of the abdominal cavity.

Pharyngeal bones.—Usually an aggregate of bones, consisting of one pair below and two pairs above, situated behind the gills and immediately before the esophagus in fishes; they may be provided with teeth. These bones represent the fifth gill arch.

Pharyngeal teeth.—Teeth attached to the pharyngeal bones.

Phylogenetic.—Pertaining to phylogeny.

Phylogeny.—History of the development or evolution of a species or group.

Phylum.—A primary division of classification; such as one of the primary divisions of the plant or animal kingdom; the group above a Class.

Physiography.—Physical geography; a study of the features of the earth's surface.

Plankton.—Very small or microscopic plants and animals which swim passively or float weakly in a body of water such as a pond, lake, or stream.

Plicae.—A series of wrinkle-like folds of skin.

Plicate.—Having a group of parallel or transverse folds of skin.

Pollutant, Pollution.—That which pollutes, defiles or makes physically unclean. Used mostly here to denote any water-borne substance which is harmful, directly or indirectly, to fishes.

Polytypic.—*See* Monotypic.

Pool.—A small and rather deep body of water which may be isolated as is a small pond or puddle; a pool in a stream usually is deeper and has less current than the adjacent riffles.

Posterior.—*See* Anterior.

Posterior fontanel.—A medial, interparietal opening, covered only by skin in the dorsal ridge of the skull; in *Carpiodes* and *Ictiobus* situated on the occiput immediately behind the eyes.

Post-larval.—As here used, that transitory stage in the development of a young fish in which many of its adult characters have developed sufficiently far to be evident, although it still retains many of its larval characteristics.

Post-ocular (portion of head).—The distance between the posterior edge of the orbital rim and the posterior edge of the opercle membrane.

Post-orbital.—Region behind the eye; *see* Circumorbitals.

Predorsal scales.—Scales lying along the dorsal ridge between the occiput of the head and the origin of the dorsal fin.

Premaxilla or premaxillary.—The anterior-most bone of each upper jaw which forms part or all of the border of the jaw and may bear teeth.

Preopercle or preoperculum.—The bone lying in front of the opercle and below and behind the eye, and comprising the forepart of the gill cover; *see* fig. II.

Preoperculomandibular canal.—Same as Operculomandibular canal.

Preorbital.—Large bone lying before the eye; *see* Circumorbitals.

Preorbital length.—Same as snout length.

Principal ray.—A branched or unbranched ray which is not rudimentary.

Protractile.—As used here, the upper jaw is protractile when a frenum is absent and a groove separates the upper jaw from the remainder of the snout along the midline, the groove making possible the forward projection or thrusting out of the upper jaw.

Proximal.—Nearest the body, or center or base of attachment; opposite of Distal.

Pseudobranchiae.—Small gill-like structures developed on the under side of the opercle near the junction with the preopercle.

Pyloric caeca.—Finger-like blind tubes or diverticula, usually glandular, which open into the alimentary canal of most fishes at the junction of the stomach and intestine, in the region of the pylorus.

Quadrate.—As used here, squarish or four-sided.

Qualitative analysis.—Separation of anything into component parts to determine their nature and relationships.

Quantitative analysis.—Determination of the amount of a thing usually to ascertain its relation to other things.

Radii.—Grooves on a fish scale which radiate outward from its central part.

Range.—The geographical region throughout which an organism is distributed, or the limits in time of appearance of anything.

Ray.—The term ray applies to all of the soft- and hard-rays of the fins, as well as to all spines.

A *soft-ray* is usually flexible, branched, bilaterally paired and segmented; it may be either a principal or a rudimentary ray; *see* those terms, p. 63.

A *hard-ray* is a hardened soft-ray which may be a simple spine-like ray as in the carp, or the consolidated product of branching as in the catfishes; hard rays may bear serrations as they do in the carp and some catfishes.

A true *spine ray* is an unpaired structure without segmentation, usually stiff, and sharpened apically.

See Branchiostegal ray, Rudimentary ray, and Principal ray.

Recurved.—Curved upward.

Redd.—An excavation or mound (in America often called a nest) made by many species of fish in various ways.

Reticulate. —Containing a network of lines, as in the pattern on the sides of the chain pickerel.

Retrorse.—Curved or turned backward or downward.

Rhomboid or rhombus.—A parallelogram in which the angles are oblique and the adjacent sides are unequal.

Riffle.—A section of stream in which the water is usually shallower than in the connecting pools, and over which the water runs more swiftly than it does in the pools; a riffle is smaller than a rapids and shallower than a chute.

Rudimentary.—Undeveloped.

Rudimentary ray.—A small more or less poorly developed, unbranched ray.

Saddle-bands (or saddle-bars).—The squarish, rectangular or linear bars or bands which cross the dorsal ridge of the backs of fishes, to extend partly or entirely downward (either vertically or obliquely) across the back and sides.

Scale-plate.—Referring to the unique bony plate on the under surface of the lower jaws of the bowfin.

Scapular arch.—*See* Pectoral arch.

Scute.—A horny or bony plate which is often spiny or keeled.

Seasonal or temporal isolation.—Usually used in reference to two or more species of fishes which reach their respective breeding periods at different times of the year, or reach their adult period at different seasons.

Serrated.—Notched or toothed, something like a saw.

Shoulder girdle.—*See* Pectoral arch.

Snout.—That portion of the head projecting forward from the anterior orbital rim of the eye.

Soft dorsal fin.—The entire dorsal fin when consisting of only soft rays, or the posterior part of the dorsal fin if it is composed of soft rays.

Soft ray.—*See* Ray.

Spatulate.—Paddle- or spoon-shaped.

Species.—*See* p. 51.

Spine.—*See* Ray.

Spinelet.—A small spine.

Spinous.—Spine-like; stiff or composed of spines.

Spinous dorsal fin.—That portion of the anterior part of the dorsal fin which consists of unbranched, unarticulated spines.

Spiracle.—An opening above and behind the eye in some fish species.

Squamate.—Scaly.

Squamation.—The arrangement of scales.

Static.—At rest, without visible motion, similar to lentic and opposed to lotic.

Striated.—Marked or streaked by narrow lines, grooves, etc. which usually lie parallel.

Subadult.—An individual similar to an adult and approaching adulthood in age and size, but still incapable of breeding.

Subopercle.—The elongated membrane-bone which lies just below the large opercle bone; *see* fig. II.

Suborbitals.—Those circumorbital bones lying below the eye; *see* Circumorbitals.

Suborbital region.—The area immediately below the eye and the anteriormost part of the cheek.

Subspecies.—*See* p. 51.

Subterminal.—Extending almost, but not entirely to the end.

Subterminal mouth.—*See* Inferior mouth.

Subtriangular.—Almost triangular.

Sucking disc.—Buccal disc.

Supramaxilla.—A wedge-shaped, small movable bone attached to the upper edge of the maxilla near its posterior tip.

Supraoral cusps.—The points, normally two or three in number and usually having as a base a single large horny tooth (sometimes two bases) situated immediately anterior to the opening of the mouth inside the buccal disc.

Supraorbital canal.—Those portions of the lateral-line system which extend along the upper edge of each eye forward onto the snout; *see* fig. II.

Supratemporal canal.—That portion of the lateral-line system which connects the two lateral canals by crossing the top of the head at the occiput; *see* fig. II.

Sympatric.—Two presumably closely related species are sympatric when they have the same or overlapping areas of geographical distribution; the gaps between sympatric species must be absolute, or they could not be good species.

Symphysis.—The point of coalescence of two bones which are usually connected by fibrocartilage; *see* Maxillary symphysis and Mandibular symphysis.

Synonymy.—A collection of properly described names for the same thing, species, or group; also a list of these names.

Tail.—Usually refers to the caudal peduncle and caudal fin, but may refer to the caudal fin only.

Taxonomist.—One who deals with, or is versed in taxonomy.

Taxonomy.—*See* Classification.

Tear-drop.—A more or less vertical, dark, tear- or drop-shaped spot below the eye, *see* fig. 33.

Teleost.—A fish of a group (*teleostei*) including most living bony fishes, as distinguished from ganoids, dipnoans and elasmobranchs.

Terete.—Nearly cylindrical in cross section and usually tapering.

Terminal.—Extending entirely to the end; that part which forms the end of something.

Terminal mouth.—When the upper and lower jaws form the extreme anterior tip of the head.

Tesselated.—Checkered, with the markings or colors arranged in squares or forming a mosaic.

Thoracic.—In the region of the thorax or chest; the pelvic fin may be called the *thoracic fin* when inserted beneath, or in front of, the pectoral fins, instead of considerably posterior to the pectoral fin insertions as is more often the case.

Toothed.—Containing teeth such as the teeth in the jaws, or containing teeth-like serrations as on the posterior edge of the pectoral spine of the brindled madtom; *see* fig. 118.

Transverse.—Crosswise or somewhat oblique.

Transverse scale rows.—The rows of obliquely-vertical scale rows on the body of fishes; *see* Lateral series of scales.

Trenchant.—Very sharp, clear and distinct; also having a sharp point.

Tricuspid.—A three-pronged tooth on one base.

Trinomial.—Consisting of three names, of which nomenclatorially the first is the generic name, the second, or binomial, is the specific name, and the third, or trinomial, is the subspecific name.

Trotline.—A stout line (staging) reaching across a stream or lake for some distance from one bank or from one anchor to another, bearing at frequent intervals single hooks which are attached to the staging by short lines.

Tubercle.—Hardened, usually cone-like protuberances; *see* Nuptial tubercles.

Tuberculate.—Having tubercles.

Turgid.—As used here, a fish appears to be turgid when it is so distended in one or all directions as to give the appearance of being swollen, inflated or very fat.

Type.—The sum of the characteristics common to a large number of individuals, or of some other thing or place. The *type* or *holotype* is the actual specimen upon which the original specific (or generic) description was based.

Type locality.—The locality in which the holotype (*see* above) was found.

Typical.—Conforming to, or exhibiting the essential characteristics of, the type.

Unbranched ray.—A simple, soft ray which is fully developed, and not small and rudimentary; it does not divide into two or more branches distally.

Unicolor.—Of one color.

Unicuspid.—A one-pronged tooth on a single base.

Venom pore.—This opening lies under the depressed pectoral spine of certain catfishes such as madtoms, and connects with a poison gland which is situated under the skin. When one is pricked by the sharp point of a pectoral spine of a madtom the gland discharges part of its poison, resulting in a painful sting which has been likened to the sting of a bee.

Vent.—*See* Anus.

Ventral.—The lower surface; opposite of Dorsal.

Ventral fin.—The pelvic fin.

Vermiculate.—Marked with worm-like lines or bands which tend to form a sinuated pattern.

Vertebra.—One of the bony or cartilaginous segments which make up the backbone.

Vertebral column.—The spinal column, the backbone.

Vertical fins.—The dorsal, anal and caudal fins which are single and on the median line of the body, in contradiction to the pectoral and pelvic fins which are paired.

Villiform teeth.—Teeth so slender and crowded together as to give the appearance of a velvety band.

Viscera.—The internal organs contained in the various cavities of the body.

Viscous.—Sticky, glutinous, slimy.

Viviparous.—Bringing forth living young instead of laying eggs; *see* Oviparous.

Vomer.—An unpaired bone immediately behind the maxillaries in the front part of the roof of the mouth; if tooth bearing, the teeth are called *vomerine teeth.*

Webbing.—The membrane between the spines and rays of the fins.

Equipment and Materials Necessary for The Identification of Fishes

The majority of the larger fishes may be identified satisfactorily without the aid of special instruments, but these are often necessary for the identification of the smaller individuals, or of adults of small size such as the species of darters and minnows. Following are the instruments and materials most frequently used:

Magnification.—The most useful type of instrument for magnification is a binocular microscope having a magnification range of 7 to 40 or 50 times. If a microscope is not available, fair substitutes are a good hand lens, an optician's monocular eye lens, or similar device having a magnification of between 5 and 20 times. Magnification is essential when studying or counting pharyngeal teeth, small scales, spines and rays, and when measuring small body parts.

Dividers.—Dividers should be 6.0″ to 8.0″ (15–20 cm) in overall length. The most satisfactory ones are those having a hinge in each leg, thereby enabling the leg to be bent at an angle and thus giving greater accuracy.

Forceps.—Those 4.0″ to 6.0″ (10–15 cm) in length having straight or curved tips are the ones most frequently used in the lifting of gill covers, opening of mouths, spreading of fins and removing fishes from pans and small bottles. A larger pair 10.0″ to 12.0″ (25.4–30.5 cm) in length is useful in removing the larger fishes from containers.

Scalpel.—One with a blade of between 1.0″ (2.5 cm) and 2.0″ (5.1 cm) in length and a handle between 4.0″ (10 cm) and 6.0″ (15 cm) in length is essential for dissecting purposes.

Scissors.—One having blades between 1.0″ (2.5 cm) and 2.0″ (5.1 cm) long is necessary for making abdominal incisions.

Ruler.—A good quality, accurate ruler, preferably of steel, is essential for the taking of exact measurements. It should contain both the metric and the English linear systems of measurements.

Dissecting needle.—A moderate-sized sewing needle with the "eye-end" mounted in a wooden stick which is about 3/8″ (0.9 cm) in diameter and 3.0″ to 5.0″ (7.6–13 cm) long has many uses and is helpful in lifting up small scales in order to count them.

Preservatives.—Fishes to be preserved for later use should be submerged in a solution of 1 part of commercial formalin to 9 parts of water. If the temperature is above 80° F (27° C) at the time of collecting it is best to strengthen the solution to 1 part of formalin to 7 or 8 parts of water, and especially if the fishes are large or for some unavoidable reason must be closely packed. Fishes should be placed in preservative as quickly after capture as possible in order to best retain their chromatophore patterns. If placed in a too strong formalin solution they die with their mouths widely agape or are otherwise distorted. Preserve fishes carefully and only those which are needed. Even the well-preserved small specimens can be difficult to identify.

Hypodermic syringe.—The abdominal cavities of fishes longer than 6.0″ (15 cm) in standard length should be injected with the formalin solution, or one or more short incisions should be made in the abdominal wall so that the solution can enter the cavity. A hypodermic syringe having a capacity of 1 to 3 ozs (30–89 ml) and a series of needles of various sizes is the most desirable.

Compressed air.—A continuous, or interrupted, flow of air through a glass tube whose opening is about 1 mm in diameter greatly aids in forcing liquid from the orbital canals, thereby making these canals and their pores visible, and in drying out and lifting up tiny scales so that they may be counted and observed. Compressed air whose rate of flow can be regulated is the most satisfactory. If such is not available, a bulb which can be repeatedly squeezed, forcing air through a rubber hose which has a glass tube attached to the other end, is a fair substitute.

Collecting equipment.—Collecting seines, nets and other gear are of many diverse types. For general use a Common Sense or similar minnow seine is excellent. It should be of 1/8″ or 1/4″ (0.3 cm or 0.6 cm) mesh, 4′ to 10′ (1.2 m to 3.0 m) in length and 3′ to 5′ (0.9 m to 1.5 m) in depth. For general survey work use a seine of 1/4″ (0.6 cm) mesh, 15′ to 20′ (4.6 m to 6.1 m) in overall length and 4′ or 5′ (1.2 m or 1.5 m) in depth, having a cone-shaped bag in its center whose mouth is about 4′ or 5′ (1.2 m or 1.5 m) in diameter and about 5′ (1.5 m) in length. Fishes enter this bag and are trapped. Excellent survey work can be done with only these two types of gear.

Collecting permits.—*Before* collecting fishes by methods other than those allowed the possessor of a sport-fishing license, contact your state conservation department relative to the necessary collecting permits.

Note books and filing cards.—Carefully made field notes, written at the time of collecting, are of inestimable value. A loose-leaf note book containing lined paper is needed; also 3″ × 5″ (8 × 13 cm) filing cards of rather stiff cardboard which contain the following or something more appropriate for your particular type of research:

(Front of Card)

No. of Sp.: ....
Species: .... Cat. No.: .....
Name of water: ....
Tributary of: ....
Locality; Co., Twp., Sec., etc: ....
Condition of water: ....
Flood crest: ....
Stability of flow: ....
Pollution: ....
Riffles: ....
Pools: ....
Gradient: ....
Bottom: ....
Aquatic vegetation: ....
Banks: ....

(Back of Card)

Land vegetation: ....
Type of valley: ....
Stream width: ....
Depth of water: ....
Depth seined: ....
Method of capture: ....
Collected by: ....
Date and time: ....
Remarks: ....
....
....
....
....

Methods of Counting[2]

Spines.—These are rays which are either true spines or hardened soft rays (*see* Glossary); they are designated by the use of Roman numerals (fig. I, 9, below).

Principal, branched and unbranched soft-rays.—These include all soft-rays other than rudimentary ones excepting in some instances as explained in the following. They are designated by the use of Arabic numerals (fig. I, 21).

Rudimentary soft-rays.—In such families as the Catostomidae and Cyprinidae rudimentary rays normally are not included in the count of the dorsal and anal fins, but in other families such as the Salmonidae, Esocidae and Ameiuridae the rudimentary rays are included because in these groups the rudimentaries grade gradually into developed rays thereby offering no definite break between the two ray types (fig. I, 36).

Spines and soft rays in separated and joined fins.—When the spines (Roman numerals) are separated from the soft-rays (Arabic numerals) the two sections are separated by a dash thus, VIII to IX—12 to 14; when the two are definitely conjoined a comma separates the counts thus, VIII to IX, 12 to 14.

Last ray of dorsal and anal fins.—Is counted as only one ray even though it is separated to its very base (fig. I; see last ray of dorsal and anal fins).

Caudal rays.—Usually only the principal branched and unbranched rays are counted; if rudimentary rays are counted the fact is generally so stated (fig. I, 37).

Soft-rays and spines in the paired fins.—In these all rays are counted including the smallest at the inner end of the fin base. In the pelvic fins of sculpins a fleshy sheath surrounds the very short spine and the large first ray, making it necessary to remove the sheath and examine under magnification in order to see the tiny splint-like spine (fig. I, 23, 31, 33).

Lateral line scale count.—The count begins behind the head with the scale that touches the pectoral arch (shoulder girdle), continuing backward along the lateral line and/or along the scale rows when pores are absent, until reaching the structural base of the caudal (hypural plate) where the count ends. The caudal base is determined without dissection by bending the caudal fin sidewise, and the scales are counted back to the apex of the angle thus produced. If a scale lies over the apex it is counted only if most of its exposed field is closer to the body than to the tail. Scales posterior to the apex are never counted even though they contain pores. All normal-appearing scales in the lateral line are counted. Small, deformed or obviously supernumerary scales are not counted nor are those not belonging to a lateral series in which most of their exposed field lies above or below the line (fig. I, 20).

Scales above lateral line.—Normally this count begins with the scale, often small, situated at the origin of the dorsal fin (or first dorsal if more than

[2] I have closely followed the methods of counting and measuring as used by Hubbs and Lagler (1947: 8–15) since their methods are very similar to my own and since this aids in establishing greater uniformity in counting and measuring.

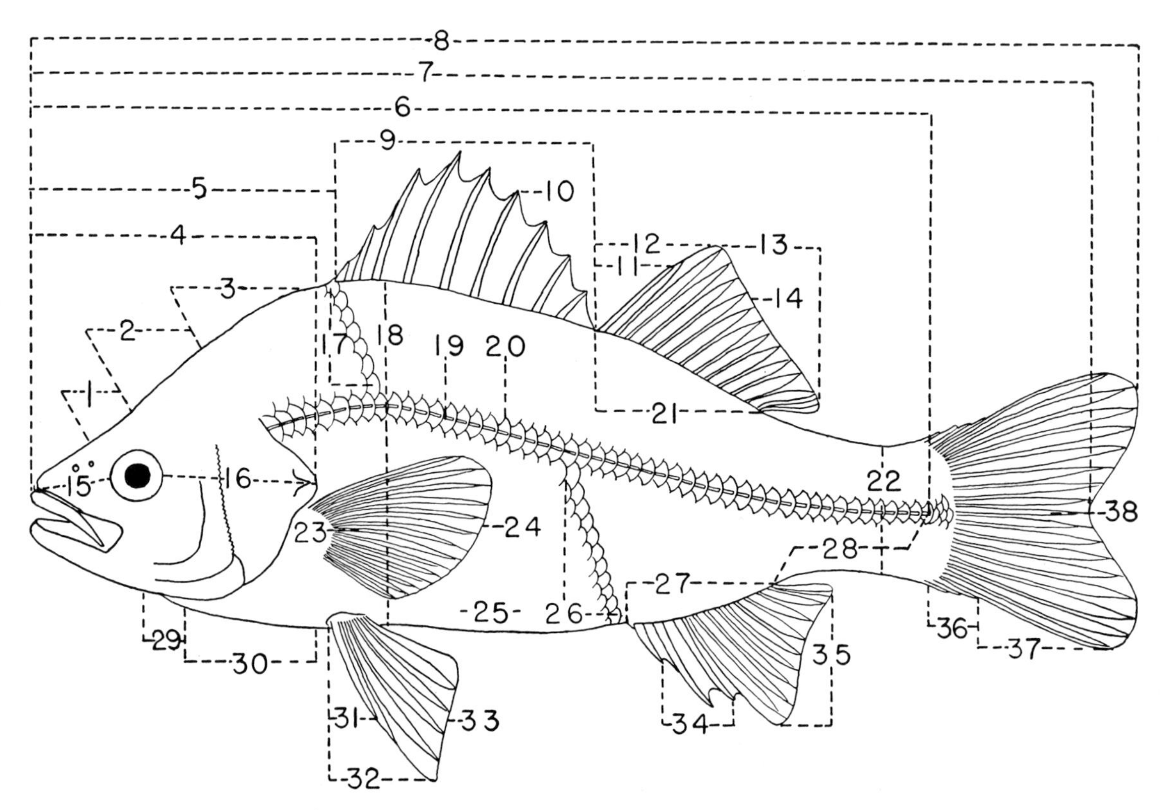

Fig. I. A spiny-rayed fish illustrating parts and methods of counting and measuring: 1—interorbital; 2—occipital; 3—nape; 4—head length; 5—predorsal length; 6—standard length; 7—fork length; 8—total length; 9—length of base of the spinous or first dorsal fin; 10—one of the spines of the dorsal fin; 11—one of the spines of the second or soft dorsal fin; 12—height of second dorsal fin; 13—length of the distal, outer or free edge of second dorsal fin; 14—one of the soft rays of the second dorsal fin; 15—snout length; 16—postorbital head length; 17—scales above the lateral line or lateral series which are counted; 18—body depth; 19—one of the lateral line pores (since in this figure all of the scales in the lateral series are pored the lateral line is complete); 20—one of the lateral scales which with the remainder form the lateral series; 21—length of base of the second or soft dorsal fin; 22—least depth of the caudal peduncle; 23—the pectoral fin; 24—one of the soft rays of the pectoral fin; 25—abdominal region; 26—scales below the lateral line or lateral series which are counted; 27—length of the base of the anal fin; 28—length of the caudal peduncle; 29—the isthmus; 30—the breast; 31— the pelvic spine; 32—height of pelvic fin; 33—one of the soft rays of the pelvic fin; 34—spines of the anal fin; 35—soft rays of the anal fin; 36—the rudimentary rays; 37—one of the principal rays of the caudal fin; 38—the caudal fin.

one is present) extending downward and backward following its scale row and terminating with the scale immediately above the lateral line scale. Do not count the lateral line scale. Occasionally, or in some species, it is more convenient to count from the dorsal origin downward and *forward* following the scale row as described above (fig. I, 17).

Scales below lateral line.—This count begins with the scale at the origin of the anal fin, extending upward and forward following the scale row and ending with the scale immediately below the lateral line scale (fig. I, 26).

Scale rows on cheeks.—This count follows an imaginary line which begins at the eye and extends in a straight line obliquely backward and downward to the preopercle angle. All scale rows crossing the line are counted (fig. IIB, 18).

Circumference scale count.—Begin with the scale immediately before the dorsal fin origin, counting all scale rows around the body until returning to the dorsal origin. Do not recount the first scale at the dorsal origin which was the first one counted (fig. 83 E).

Caudal peduncle scale count.—All scale rows around the narrowest part of the peduncle are counted, beginning with a dorsal- or ventral-ridge scale or a lateral line scale, counting all scale rows around the body until reaching the first counted scale row (fig. I, 22).

Scales before dorsal fin.—Count all scales which

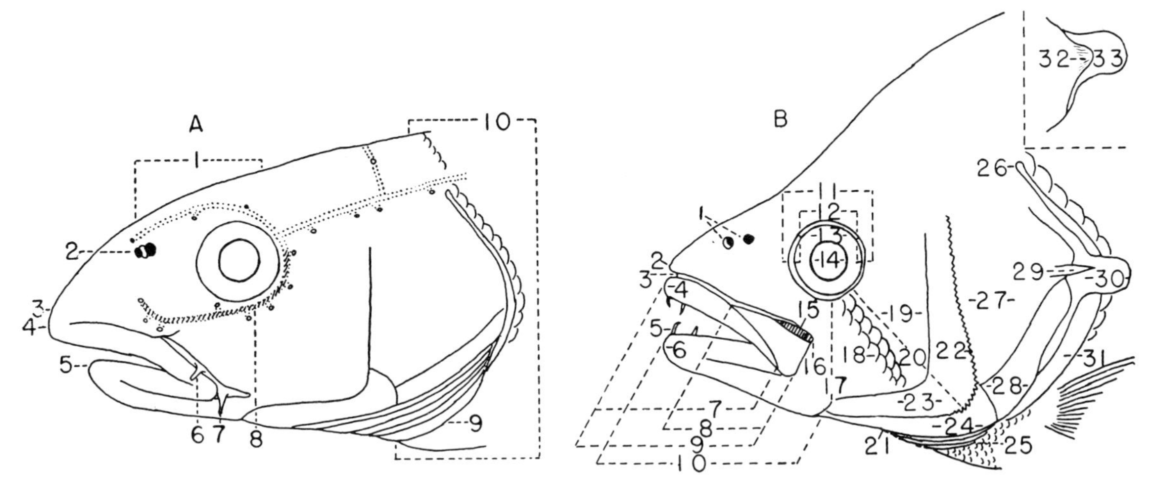

Fig. II. A. Head of a soft-rayed fish showing parts: 1—forehead; 2—nostrils; 3—the frenum, the membrane which binds together the tips of the upper jaw and snout, resulting in a non-protractile upper jaw; 4—tip of the upper jaw, the symphysis of the premaxillaries; 5—tip of the lower jaw, the chin, the symphysis of the dentaries—since the tip of the lower jaw is posterior to the tip of the upper jaw the lower jaw is inferior or included and the mouth is inferior, subterminal and ventral; 6—a barbel which is in advance of the posterior end of the upper jaw and above the maxillary and in the groove between maxillary and pre-orbital bones; 7—barbel at the posterior end of the upper jaw; 8—part of the anterior portion of the lateral canal system the shaded portion of which is the infraorbital canal—if all pores are connected to the canal, the infraorbital canal is said to be complete but if one or more pores are isolated it is incomplete; 9—branchiostegals; 10—head depth.

Fig. II. B. Head of a spiny-rayed fish showing parts: 1—nostrils; 2—tip of the terminal snout; 3—groove between the tips of the snout and symphysis of the upper jaw—since there is no binding frenum the upper jaw may be moved forward, resulting in a protractile upper jaw; 4—tip of upper jaw and symphysis of the premaxillaries; 5—a canine tooth; 6—tip of lower jaw and symphysis of the dentaries—the chin—since the tip of the lower jaw extends as far anteriorly as does the tip of the upper jaw the mouth is terminal; 7—premaxillary length; 8—maxillary length; 9—upper jaw length, sometimes called "maxillary length;" 10—lower jaw or mandibular length; 11—eye or orbital length; 12—eyeball length; 13—iris; 14—pupil; 15—suborbital width; 16—supramaxillary bone (shaded); 17—cheek depth, or height; 18—row of cheek scales which is counted; 19—cheek; 20—length from eye to preopercular angle; 21—isthmus; 22—serrated posterior edge of preopercle; 23—preopercle; 24—interopercle; 25—branchiostegals; 26—upper angle of gill cleft; 27—opercle; 28—subopercle; 29—opercular spine, absent in many species—*see* Insert; 30—membranous opercular flap—the "earflap;" 31—gill cleft, which in most species extends from its upper angle to the isthmus; *see* Insert—32—posterior extension of opercular bone without a spine, *see* spine in 29; 33—membranous opercular flap, the "earflap."

wholly or in part intercept a straight line extending along the predorsal ridge, from the occiput backward to the dorsal fin origin. Normally the count is made on those species in which the occiput is scaleless (fig. 11, 13).

Scale rows before dorsal fin.—Those scale *rows* are counted from the head backward along one side of the back (not on the dorsal ridge) to below the dorsal fin origin. This is a count seldom employed.

Pharyngeal tooth counts.—The two curved bones comprising the pharyngeal arch in the sucker and minnow families are removed and cleaned so that their teeth may be counted and observed.

To remove the pharyngeal arch place the fish upon its right side, its head to your left. While holding the fish firmly, lift its left opercle cover until it is approximately at right angles to the body. To do this it may be necessary to enlarge the gill-slit opening which binds the anterior part of the head to the throat and breast; this is done by using a sharp scalpel and cutting the membrane from the lower gill slit angle forward toward the tip of the lower jaw. The filamentous gills are now exposed. The flesh- or skin-covered left pharyngeal bone (actually several bones fused together) lies immediately behind the last filamentous gill. To expose the bone insert the scalpel point between the pharyngeal bone and the anterior edge of the shoulder girdle, beginning at the upper angle of the gill cleft and cutting downward following the curving shoulder bones; next make a similar cut between the last filamentous gill and the hidden left arch. Carefully pry apart and enlarge these incisions thereby exposing the left arch, then carefully sever the fleshy tendons holding its upper and lower ends in position, after which the arch may be pulled gently until it is removed. The right half of the arch may be removed through the same incision, or the fish may be placed on its left side and the right half of the arch removed in the same manner as described for the left half. Great care must be taken so as not to break the delicate and fragile bones. A low power microscope is essential to aid in removing and cleaning the bones and teeth of small fishes. If a tooth has been broken, the tooth socket is found and counted. After completing your examination of the arch *always* replace it in the throat cavity, pushing the gill covers in place so that the bones do not become lost (fig. 143, Nos. 8–12).

Each of the two pharyngeal bones bears one to three rows of teeth, depending upon the species. Suckers and some minnow species have only one row; in these a tooth formula such as 5–4 is used, indicating that only one row is present on each bone and that the left bone has 5 and the right bone 4 teeth. Other minnow species have two or more rows of teeth; in those having two rows on each bone a formula such as 1, 5–4, 1 is used, indicating that the left bone has one tooth on its outer row and five on the inner, and the right bone has four teeth on its inner row and one on the outer (Fig. 95 B).

The upper or lower pharyngeal bones of sunfishes, drums and other species may be removed in a similar fashion (fig. 143, Nos. 8–12).

Gill-raker count.—This count is made on the anteriormost or first gill arch (either on the right or left side) unless otherwise stated. Include all rakers, even the rudimentaries, in the count. The gill arch is divided by an angle into a lower and upper limb and usually a single count represents the total number on both limbs; however, infrequently a single count has been used to imply only rakers on the lower limb. Usually it is stated as to whether the lower limb only has been counted or both, and if both were counted separately the two counts are separated by a plus sign. If a gill raker is present astride the apex of the angle, it is included in the lower limb count.

Branchiostegal ray count.—The branchiostegal membranes, situated on the ventral edges of the gill covers anteriorly, contain the elongated, slender branchiostegal bones or rays. All rays must be counted including those short, small and almost concealed anteriormost ones (fig. 33 B).

Pyloric caeca.—Each finger-like, diverticular tube or caecum opens into the alimentary canal at the junction of the stomach and intestine, and in the region of the pylorus. These tubes may be attached singly at their bases or in groups of two or more; in each case all tips are counted unless otherwise specified.

Methods of Measuring

Exacting care is essential in the taking of measurements if individuals of some species are to be identified correctly. Lower power magnification is necessary when taking measurements with dividers of small body parts such as snout or eye length. In transferring such lengths from dividers to ruler in order to obtain the numerical value it is best to use the metric system, attempting to measure to the nearest 0.5 millimeter. It is also best to divide the dividend by the divisor to obtain the quotient, rather than to use the dividers to "step off" the divisor into the dividend, such as is done when "stepping off" the head length into the standard length.

Total length.—Is the greatest distance in a straight line (not following body curves) from the anteriormost projecting part of the head to the farthest tip of the caudal fin when its rays are squeezed together (fig. I, 8).

Fork length.—Is the distance in a straight line from the anteriormost projecting part of the head (or from the snout or upper lip if so specified) to the apex of the angle produced by the two caudal lobes. If the tail is emarginated, straight or rounded posteriorly the measurement is taken to the center of the posterior edge of the fin, and then is similar to total length. Forked length is especially valuable when the caudal lobes of a species vary greatly in length, because of sex, or between young and adult, or because they are subject to much wearing, to breaking off, or to individual variation (fig. I, 7).

Standard length.—Is the distance in a straight line from the anteriormost part of the snout or upper lip (not the lower jaw unless so specified) to the caudal base; *see* Lateral line count for discussion of caudal base (fig. I, 6).

Body depth.—Is the greatest vertical distance in a straight line from the midline of the dorsal surface of the body to the midline of the ventral surface exclusive of fleshy or scaly structures connected with the fin bases (fig. I, 18).

Body width.—Is the greatest lateral distance from one side of the body to the other (fig. 40, A).

Head depth.—Is the vertical distance from the midline of the occiput to the ventral midline of the head or breast (fig. IIA, 10).

Head width.—Is the greatest distance between the opercles when they are in their normally closed position.

Head length.—Is the greatest distance from the tip of the snout or upper lip backward to the posteriormost point on the opercular membrane, unless it is specified that the measurement is only to the posteriormost point of the opercular bone, exclusive of the membrane (fig. I, 4).

Eye length.—In its strictest sense, eye length is the distance between the margins of the cartilaginous eyeball, as contrasted with *Orbital length* which is the greatest distance between the free orbital rims. Here, eye length in the bony fishes is considered to be the horizontal distance between the orbital rims when enough pressure has been applied to press the flesh moderately against the rims. This measurement has proven to be the most satisfactory; it avoids excessive pressure against the rims which in fragile specimens might result in the fracturing of one or both rims thereby giving a false value; it eliminates guesswork relative to the location of the eyeball margin; in most species it is a fairly accurate measurement of the greatest diameter of the eyeball (fig. IIB, 11).

Snout length.—Is the distance from the anteriormost point on the midline of the snout or upper lip to the front margin of the orbital rim. Press slightly the posterior leg of the dividers against the bony orbital rim so that the measurement does not include the fleshy portion of the orbital rim (fig. I, 15).

Postorbital head length.—Is the greatest distance in a straight line from a point on the lower posterior edge of the orbital rim obliquely backward and downward to the posterior edge of the opercular membrane (fig. I, 16).

Length from eye to preopercle angle.—Is the greatest distance in a straight line obliquely downward from the orbital rim to the angle of the preopercle, including a spine at that angle if one is present (fig. IIB, 20).

Upper jaw length.—Is the distance from the anteriormost point of the premaxillary (normally the symphysis of the jaw) to the posteriormost point of the maxillary. Supplants the old term "maxillary length" which is incorrect because both the maxillary and premaxillary are measured (fig. IIB, 9).

Lower jaw, or mandible, length.—Is the greatest distance from the anterior point of the symphysis of the lower jaw backward to the posterior edge of the mandibular joint. Press firmly against the flesh-covered mandibular joint so that an accurate measurement may be obtained (fig. IIB, 10).

Cheek length.—Is the distance in a straight line from a point directly below the anterior rim of the eye backward horizontally to the preopercular margin (fig. IIB, 11 to 22).

Cheek height.—Is the least distance from the orbital rim downward to the ventral edge of the preopercle (fig. IIB, 17).

Suborbital width.—Is the least distance from the orbital rim downward to the suborbital or preorbital margin. Do not squeeze the dividers tightly (fig. IIB, 15).

Interorbital widths.—Is the distance between the bony dorsal edges of the orbital rim. To obtain the *least bony width* place one point of the dividers tightly against the bony dorsal rim of one eye, the other point against the bony rim of the opposite eye; this is the measurement usually employed. The *least fleshy width* is obtained by not squeezing the dividers; it is more inaccurate and is seldom used (fig. 148, B).

Width of gape.—Is the greatest distance in a straight line from the posterior edge of one mandibular joint to the posterior edge of the other mandibular joint (fig. 83, A).

Depth of caudal peduncle.—Is the least vertical distance in a straight line from the midline of the dorsal surface to the midline of the ventral surface (fig. I, 22).

Length of caudal peduncle.—Is the oblique distance from the posterior end of the anal fin base to the caudal fin base (fig. I, 28).

Predorsal length.—Is the distance in a straight line from the tip of the snout or upper lip to the dorsal fin origin (fig. I, 5).

Dorsal origin to caudal base.—Is the distance in a straight line from the dorsal fin origin backward obliquely to the caudal fin base.

Length of depressed dorsal or anal fin.—Is the greatest distance from the dorsal or anal fin origin to the farthest point on the distal edge when the fin is flattened down.

Height of dorsal or anal fin.—Is the distance from the dorsal or anal fin origin to the tip of its anterior lobe (fig. I, 12).

Length of a spine or soft-ray.—Is the distance from the apex of the angle formed by the junction of the anterior edge of the ray with the body to the tip of the ray. Be sure to press the divider point firmly against the base of the ray at its junction with the body. Spines are measured only to their tips and never include soft-rayed or filamentous extensions such as occur on the pectoral spines of catfishes. Soft-rays are measured to their extreme tips (fig. I, 11).

Length of pectoral or pelvic fin.—Is the distance, when the fin is asymmetrical, from the base of the outermost or anteriormost ray to its farthest tip (including a filament if present). When the fin is symmetrical or when the longest ray is at or near the middle of the base of the fin the measurement is from the middle of the fin to the tip of the longest ray (fig. I, 32).

Key to Families of Ohio Fishes

Note: The characters in this and the other keys appear to be valid for Ohio specimens, but may not always hold for specimens taken outside of Ohio waters.

1 ADULTS AND AMMOCOETES: Pectoral and pelvic fins absent. Seven pore-like gill openings on each side of head. Nostril single and median in position.
ADULTS or transformed SUBADULTS: Mouth without jaws, oval in shape, and used as a sucking and clasping organ; teeth present.
Ammocoetes (larvae): Mouth without jaws or teeth, oval in shape and partly shielded by an overhanging hood. (Class Monorhina, subclass Cyclostomi.)
LAMPREYS Petromyzontidae Family 1 . . . . (pp. 79–81)

Pectoral fins always present; pelvic fins usually. One slit-like gill opening on each side of head. Nostrils paired. Mouth with true jaws.
(Class Osteichthyes, subclass Teleostomi) . . . . 2.

2 Caudal heterocercal (vertebral column extending into upper lobe of caudal fin almost to its tip) *or* abbreviate—heterocercal (vertebral column turned upward but ending before invading far into upper lobe of caudal fin) . . . . 3.
Caudal homocercal (vertebral column not bent upward and does not enter upper lobe of caudal fin) . . . . 6.

3 Caudal heterocercal, forked, the lower lobe present. Mouth sub-terminal. Endoskeleton mostly cartilaginous . . . . 4.
Caudal abbreviate-heterocercal; the fin rounded and therefore without lobes. Mouth terminal; jaws with many sharp teeth. Endoskeleton mostly bony . . . . 5.

4 Snout extremely long and spatulate (paddle-like), with 2 minute barbels on ventral surface. Skin of body naked, without plates or scales.
PADDLEFISH Polyodontidae Family 2 . . . . (p. 81)

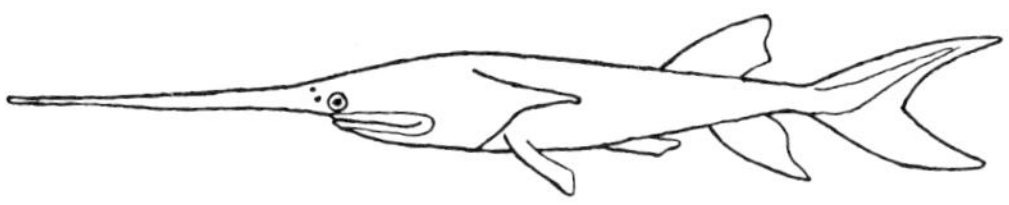

Snout relatively short, and rounded or shovel-shaped, with 4 long barbels on ventral surface. Body with several rows of bony plates.
STURGEONS Acipenseridae Family 3 . . . . (p. 81)

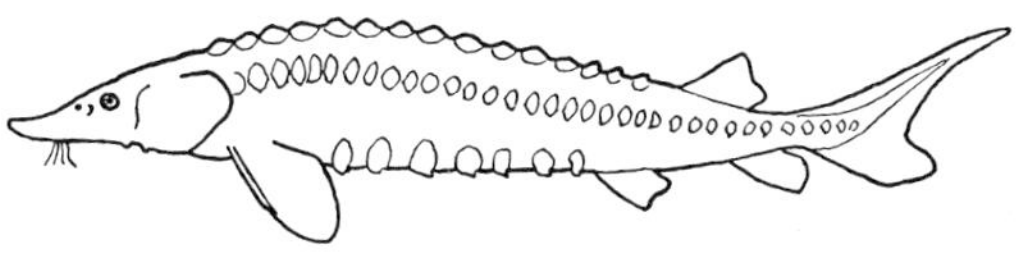

5. Snout produced into an elongated beak. No gular plate. Dorsal fin of fewer than 15 rays, beginning behind anal fin insertion. Ganoid scales, rhombic in shape and not overlapping.
GARS Lepisosteidae Family 4 . . . . (p. 82)

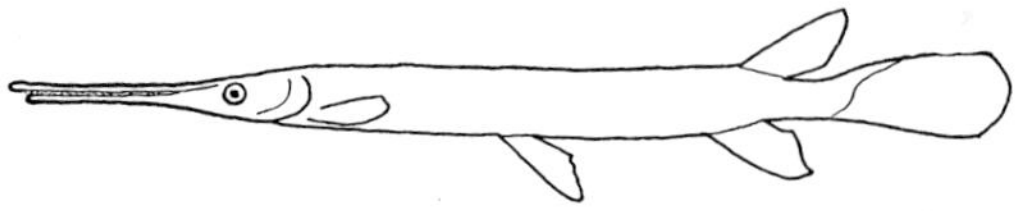

Snout rounded and short. Gular plate present. Dorsal fin of more than 45 rays, beginning far anterior to anal fin insertion. Scales cycloid, overlapping as in ordinary bony fishes.
BOWFIN Amiidae Family 5 . . . . (p. 82)

6. No pelvic fins. Dorsal, caudal, and anal fins continuous. Body lamprey-shaped. Scales so small that skin of body appears to be scaleless.
EEL Anguillidae Family 15 . . . . (p. 103)

Pelvic fins present (except when absent in aberrant individuals). Dorsal, caudal, and anal fins separated . . . . 7.

7. Adipose fin present . . . . 8.
Adipose fin absent . . . . 11.

8. Usually 8, but always more than 4, barbels about mouth. Pectorals and dorsal fins each with a large spiny ray. No scales.
CATFISHES Ictaluridae Family 14 . . . . (pp. 100—103)

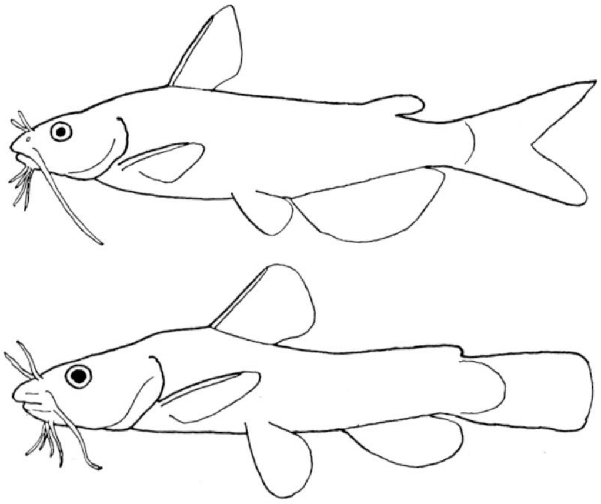

No barbels about mouth. Scales present . . . . 9.

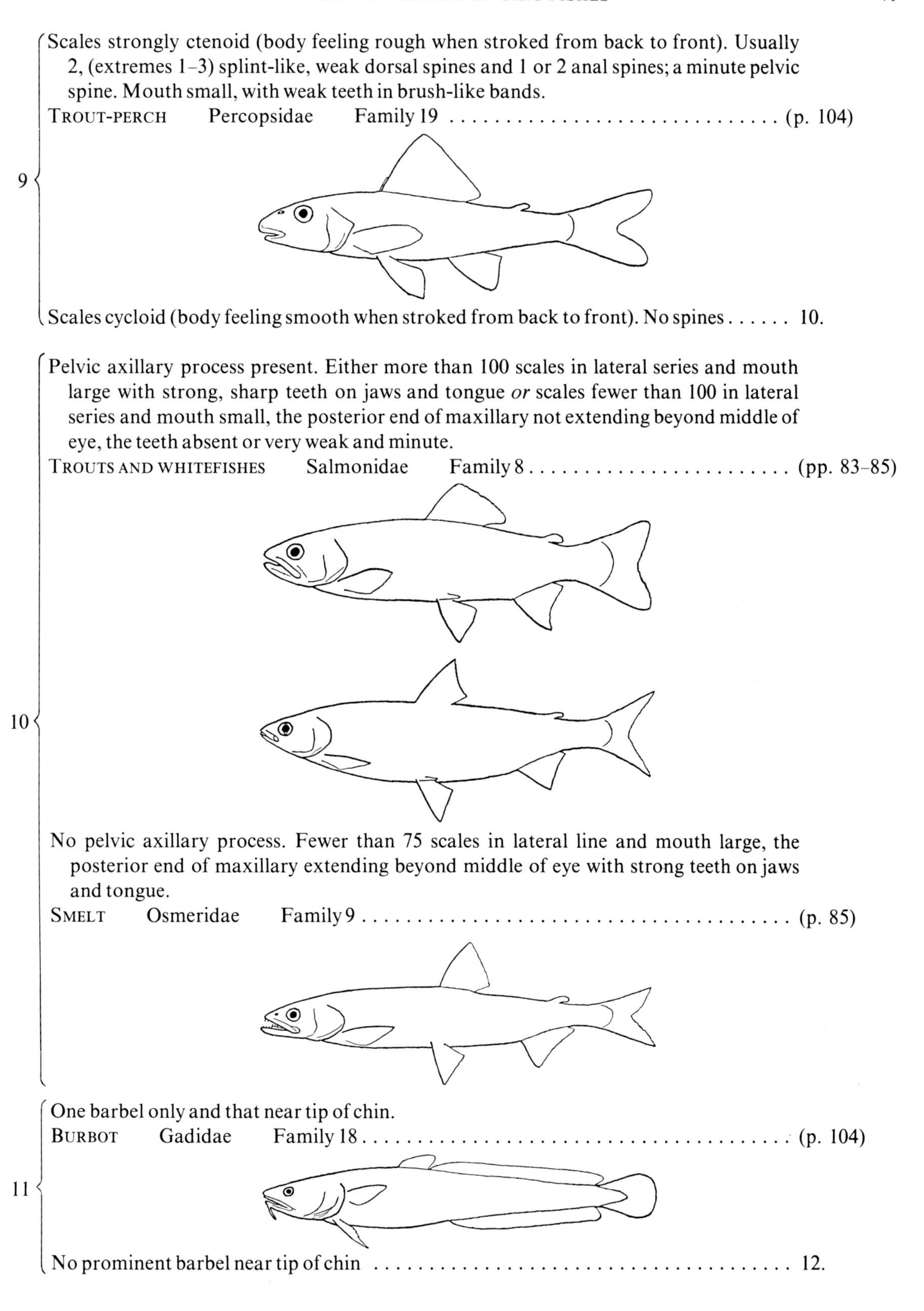

9 Scales strongly ctenoid (body feeling rough when stroked from back to front). Usually 2, (extremes 1–3) splint-like, weak dorsal spines and 1 or 2 anal spines; a minute pelvic spine. Mouth small, with weak teeth in brush-like bands.
TROUT-PERCH Percopsidae Family 19 . . . . (p. 104)

Scales cycloid (body feeling smooth when stroked from back to front). No spines . . . . . . 10.

10 Pelvic axillary process present. Either more than 100 scales in lateral series and mouth large with strong, sharp teeth on jaws and tongue *or* scales fewer than 100 in lateral series and mouth small, the posterior end of maxillary not extending beyond middle of eye, the teeth absent or very weak and minute.
TROUTS AND WHITEFISHES Salmonidae Family 8 . . . . (pp. 83–85)

No pelvic axillary process. Fewer than 75 scales in lateral line and mouth large, the posterior end of maxillary extending beyond middle of eye with strong teeth on jaws and tongue.
SMELT Osmeridae Family 9 . . . . (p. 85)

11 One barbel only and that near tip of chin.
BURBOT Gadidae Family 18 . . . . (p. 104)

No prominent barbel near tip of chin . . . . 12.

12 Anal opening anterior to pelvic fin (except in small young in which it occurs somewhat more posteriorly).
PIRATE PERCH Aphredoderidae Family 20 . . . . (p. 105)

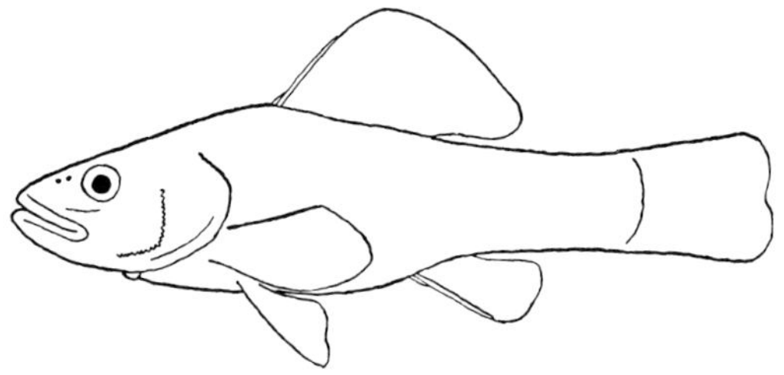

Anal opening in customary position, which is just before anal fin insertion . . . . 13.

13 The 4–6 stout dorsal spines not connected to one another by membrane. A stout spine on each pelvic fin. Body without scales.
STICKLEBACK Gasterosteidae Family 27 . . . . (p. 118)

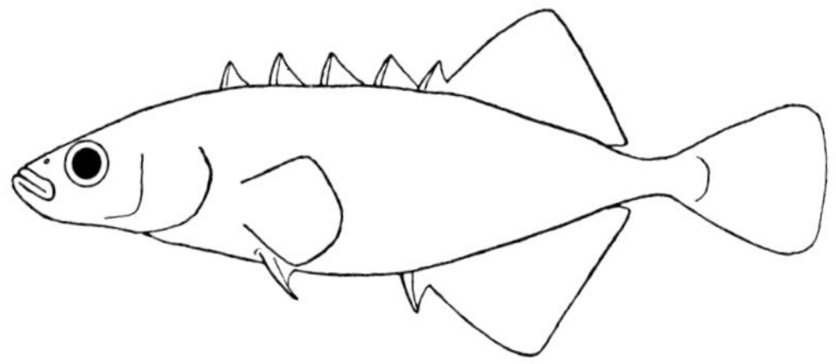

Dorsal fin with or without spines; when spines are present they are connected to one another by membrane . . . . 14.

14 Either one continuous dorsal fin without spines *or* with only one stout spine at the anterior insertion of fin; this spine doubly serrated along its posterior edges. No pelvic spines . . . . 15.
Either two well-separated dorsal fins, *or* if they are conjoined, then the first fin contains stiff spines . . . . 23.

15 Dorsal fin with one stout spine doubly serrated posteriorly. Exotic carp and goldfish only—Cyprinidae in part (*see* 19b) Family 13 . . . . (p. 91)

One continuous soft dorsal fin without spines . . . . 16.

16 { Head scaleless . . . . 17.
Head partly scaled, the cheeks always partly or entirely scaled . . . . 20.

17 { Pelvic axillary process present. Branchiostegal membranes free from isthmus (the gill slit extending forward to below the eye, *see* orangethroat darter, fig. 171 A). Jaws with or without teeth. Eyes partly covered with adipose eyelids . . . . 18.
No pelvic axillary process. Branchiostegal membranes united to isthmus and broadly conjoined (gill slit not extending forward to posterior edge of preopercle; *see* central quillback carpsucker, fig. 90 A). Jaws toothless. Pharyngeal arches with teeth . . . . . . 19.

18 { Dorsal fin base situated partly or entirely over anal fin base. Lateral line complete or well developed. Stout, sharp teeth on tongue. Gill-rakers few, short and knob-like.

MOONEYES Hiodontidae Family 6 . . . . (p. 83)

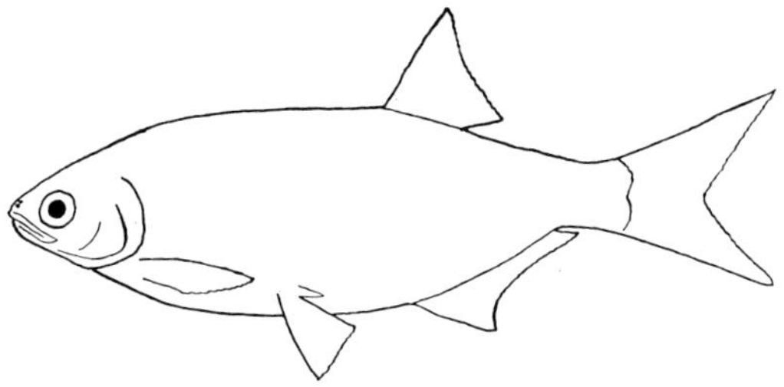

Dorsal & fin base situated over pelvic fin base or slightly behind. No lateral line. Teeth, if present on tongue, are very small and feeble. Gill rakers many, long and slender.

HERRINGS Clupeidae Family 7 . . . . (p. 83)

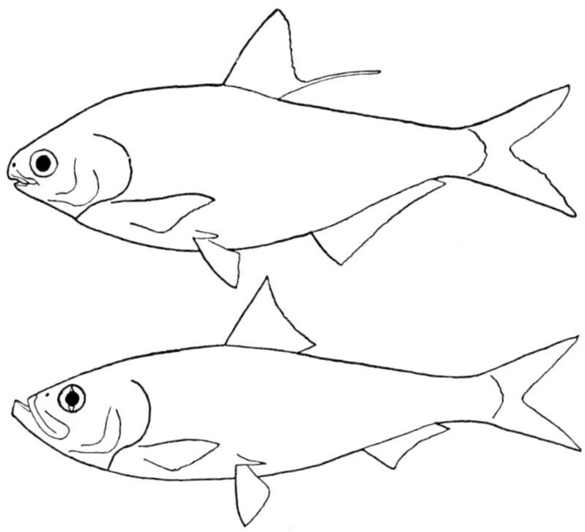

19

Either dorsal rays are 10 or more, or if only 9, then lateral line is absent or reduced to a few pores and mouth inferior, sucker-like, and with striate and papillose lips (9 rays only in some specimens of the genus *Erimyzon*). Anal fin placed far back; distance from its anterior insertion to caudal base usually more than 1.8 times in distance from anal fin insertion to posterior edge of opercle. Principal caudal rays typically 18. Pharyngeal arch with a single row of more than 15 teeth.

SUCKERS Catostomidae Family 12 . . . . (pp. 86–91)

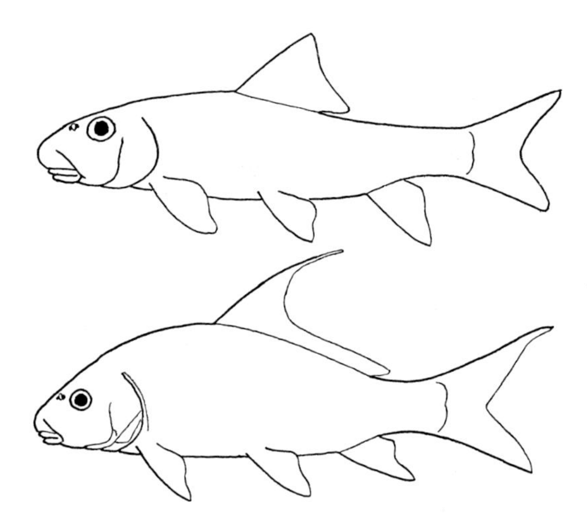

Dorsal rays normally 8, except 9 in pugnose minnow; if 9, then lateral line is present and mouth not sucker-like. Anal fin placed farther forward; distance from its anterior insertion to caudal base usually less than 1.8 times in distance from anal insertion to posterior edge of opercle. Principal caudal rays typically 19. Pharyngeal arch with 1–3 rows of teeth; the principal row having fewer than 7 teeth.

NATIVE MINNOWS Cyprinidae in part (*see* 15a) Family 13 . . . . (pp. 91–100)

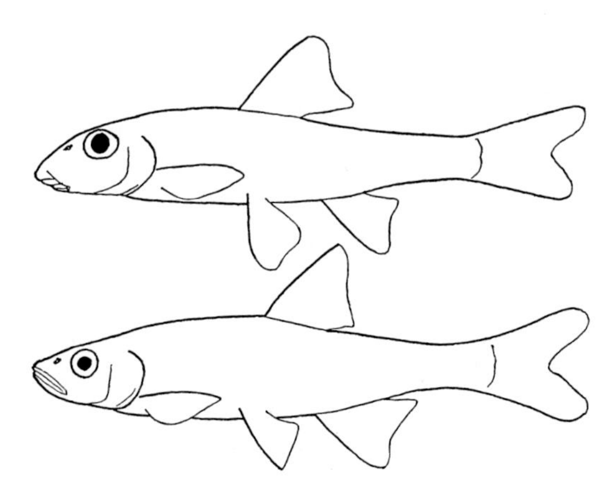

20

Tail deeply forked. Occiput without scales. Jaws produced into a duck-like snout and have large, canine teeth. More than 100 transverse scale rows in lateral series. Branchiostegal rays 11–19.
PIKES Esocidae Family 11 . . . . (p. 85–86)

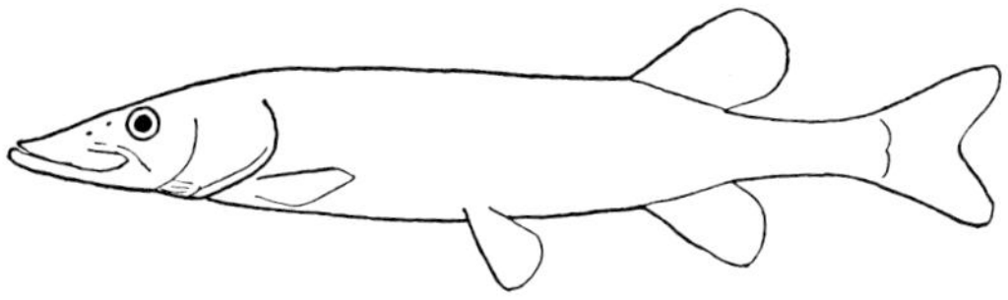

Tail rounded. Occiput with large scales. Jaws short; teeth villiform. Fewer than 50 transverse scale rows in lateral series. Branchiostegal rays fewer than 10 . . . . 21.

21

Pre-maxillaries not protractile (no groove separating tip of upper lip and forehead or snout; *see* blackside darter, fig. 151 A). Anterior origin of pelvic fins closer to caudal base than to tip of snout.
MUDMINNOW Umbridae Family 10 . . . . (p. 85)

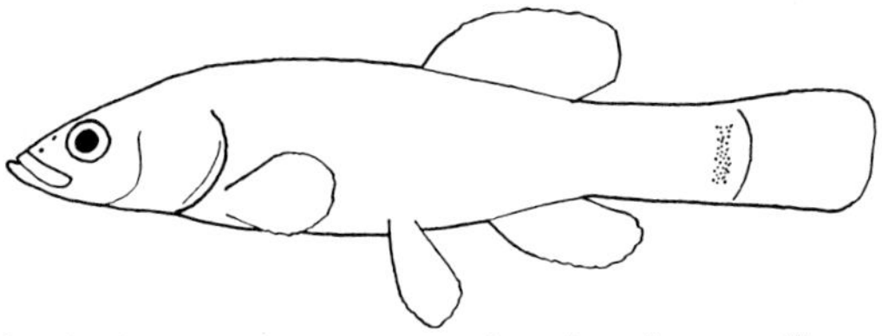

Premaxillaries protractile (a deep groove separating tip of upper lip and snout; *see* scaly johnny darter, fig. 162 B). Anterior origin of pelvic fins midway between caudal base and tip of snout, or nearer to snout . . . . 22.

22

Anal fin of male similar to that of female; third anal ray branched, including rudimentary rays. Female lays eggs.
KILLIFISHES Cyprinodontidae Family 16 . . . . (p. 104)

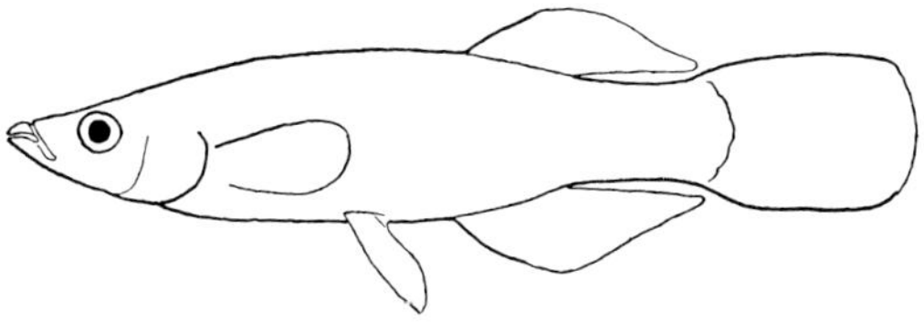

Anal fin of male unlike that of female for it is modified into an intromittent organ; third anal ray unbranched. Brings forth young alive.
LIVEBEARERS Poeciliidae Family 17 . . . . (p. 104)

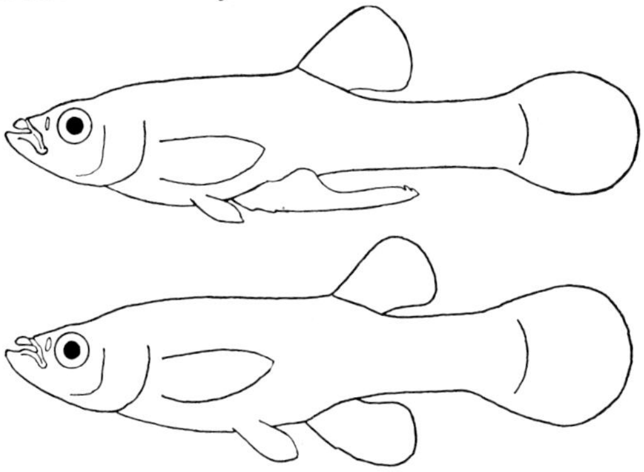

23 Skin of body covered with cycloid scales or body naked. Dorsal fins well separated; spines in first dorsal frail, are readily bent, and have the superficial appearance of simple, flexible rays . . . . 24.
Body partly or entirely covered with ctenoid scales. Dorsal fins separated or conjoined, the spines stiff and sharp . . . . 25.

24 Body with cycloid scales. Soft dorsal base much shorter than anal fin, fewer than 6 spines in first dorsal. Pelvics with one short feeble spine and 5 rays.
SILVERSIDES Atherinidae Family 21 . . . . (p. 105)

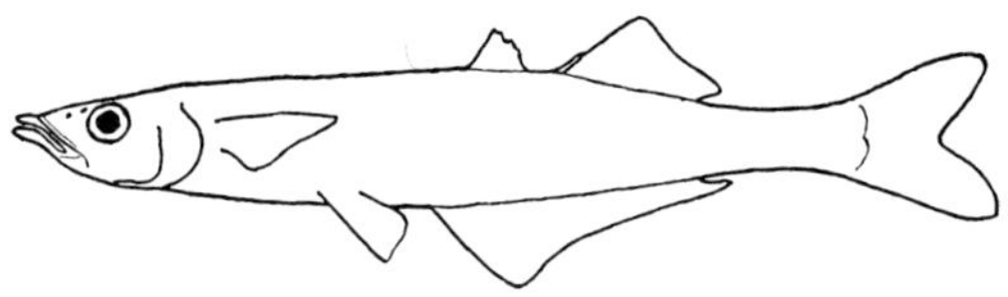

Body naked; sometimes prickly. Soft dorsal base as long or longer than anal fin. More than 6 flexible, ray-like spines in first dorsal. Pelvics with one hidden, small spine and 3 or 4 rays.
SCULPINS Cottidae Family 26 . . . . (p. 118)

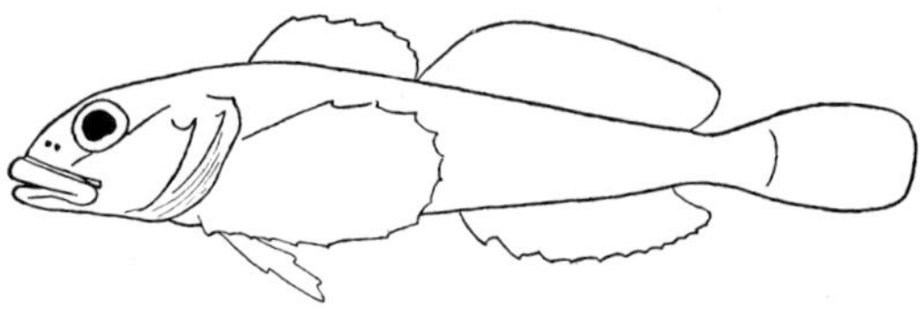

25 Anal spines 3 or more; stiff and sharply pointed . . . . 26.
Anal spines one or 2; in some species weak and flexible . . . . 27.

26

Dorsal fins completely separated or very slightly conjoined at base. A spine on the opercle flap.
BASSES Serranidae Family 22 . . . . (p. 105)

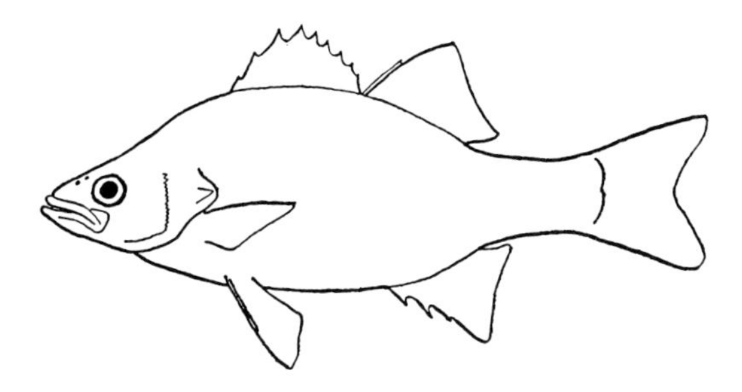

Dorsal fins completely conjoined, or separated at most by a deep notch. No spine on the opercle flap.
SUNFISHES (includes BLACKBASSES) Centrarchidae Family 23 . . . . (pp. 105–10)

27

Lateral line not extending onto caudal fin. Body torpedo-shaped; body depth less than length of head. Second anal spine, if present, neither stout nor unusually long. Lower pharyngeal bones slender and separated; these bones bear sharp teeth.
PERCHES Percidae Family 24 (additional key, *see* Lutterbie, 1975) . . . . (pp. 110–17)

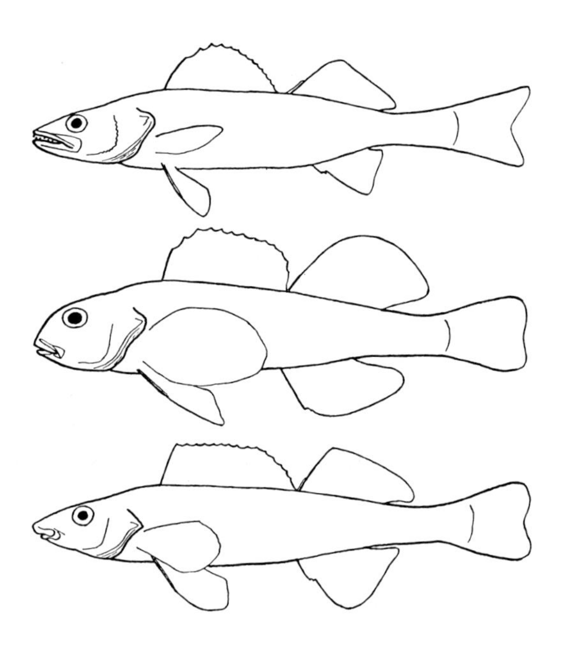

Lateral line extending well back onto caudal fin. Body slab-sided; body depth much greater than length of head. Second anal spine extremely long and heavy. Lower pharyngeal bones broad, heavy and fused; these bones bear flattish molar teeth.
FRESHWATER DRUM Sciaenidae Family 25 . . . . (p. 118)

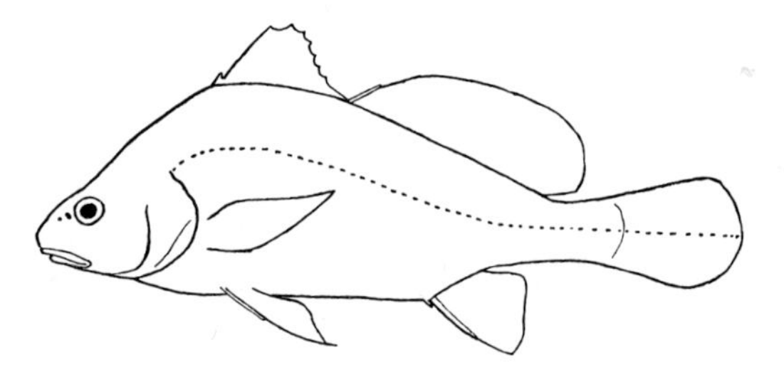

Key to the Species and Subspecies of Ohio Fishes*

Family 1 Petromyzontidae LAMPREYS . . . . (pp. 143–66)

1
- Mouth jawless but with teeth; its outer rim encircled with fimbriae; mouth used as a clasping and sucking organ. Eyes well developed and functional. Newly transformed adults usually more than 5.0″ (13 cm) TL. Free swimming and not habitually living in tunnels, although sometimes hiding under objects.
  Transformed SUB-ADULT and ADULT lampreys . . . . 2.
- Mouth jawless and without teeth; a lip-like hood overhanging the oral opening. Eyes poorly developed and non-functional. Total length up to 7.0″ (18 cm). Living in tunnels in stream bottoms and banks; normally not freely swimming about.
  AMMOCOETES (larvae or untransformed young) of lampreys . . . . 8.

2
- Dorsal fin not divided into 2 separate fins, although the single fin may be deeply notched (fig. 2).
  Genus *Ichthyomyzon* . . . . 3.
- Two distinct, well separated dorsal fins (fig. 5).
  Genera *Petromyzon* and *Lampetra* . . . . 6.

3
- Circumoral teeth (innermost teeth of radiating series including lateral teeth; fig. 1 A) all unicuspid or singly-pointed. Myomeres, 47–56, rarely more than 55 (myomeres counted from last gill opening to anus). . . . . 4.
- Usually 6–10 (extremes, 1–11) circumoral teeth bicuspid. Myomeres 53–62, rarely less than 55 . . . . 5.

4
- Teeth large and sharply pointed. Snout long, the sucking disc (buccal funnel or mouth) may be expanded wider than body width. Parasitic upon fishes. Alimentary tract functional until individual is about to spawn, whereupon it begins to atrophy. Total length of spawning adults, 10.0″–14.0″ (25.4–35.6 cm).
  SILVER LAMPREY *Ichthyomyzon unicuspis* Hubbs and Trautman . . . . (p. 143)
- Teeth small; bluntly pointed, many are partially or entirely hidden by skin. Snout short; sucking disc cannot be expanded wider than body width. Non-parasitic. While transforming from ammocoete to adult, the alimentary tract becomes shriveled into a nonfunctional, thread-like strand. Total length of spawning adults, 5.0″–7.0″ (13–18 cm).
  NORTHERN BROOK LAMPREY *Ichthyomyzon fossor* Reighard and Cummins . . (p. 147)

5
- Snout long; sucking disc may be expanded wider than body width. Parasitic upon fishes. Alimentary tract functional until individual is about to spawn, after which it begins to atrophy. Total length of spawning adults, 10.0″–14.0″ (25.4–35.6 cm).
  OHIO LAMPREY *Ichthyomyzon bdellium* (Jordan) . . . . (p. 151)
- Snout short; sucking disc cannot be expanded wider than body width. Non-parasitic. While transforming from ammocoete to adult, the alimentary tract shrivels into a thread-like strand. Total length of spawning adults, 5.0″–8.0″ (13–20 cm).
  MOUNTAIN BROOK LAMPREY *Ichthyomyzon greeleyi* Hubbs and Trautman . . . . (p. 154)

* *See* p. 51 for explanation of usage of parentheses with names of certain describers of species.

6

Teeth radiating in series from throat outward (as in *Ichthyomyzon*). Teeth large and sharply pointed. Base of supra-oral not as wide as throat opening, the cusps (points) close together. Myomeres usually 65–76, rarely less than 64. Sides of body heavily mottled with darker shades. Parasitic on fishes. Alimentary tract functional until individual is about to spawn, after which it begins to atrophy. Total length of spawning adults, 13.0″–30.0″ (33.0–76.2 cm).

SEA LAMPREY *Petromyzon marinus* Linnaeus . . . . (p. 157)

Teeth in clusters, not in radiating series. Teeth blunt and many partially or entirely hidden by skin. Base of supra-oral as wide as, or wider than, throat opening; the cusps widely separated. Myomeres 53–74. Sides of body bicolored or unicolored; not mottled. Non-parasitic. After transforming, alimentary tract becomes a thread-like strand. Length of spawning adults, 5.0″–8.0″ (13–20 cm).

Genus *Lampetra* . . . . 7.

7

Myomeres less than 62, usually 54–60. No teeth visible on posterior field of sucking disc other than marginals (fig. 6 B). Size of spawning adults averaging smaller, usually less than 6.0″ (15 cm) TL.

LEAST BROOK LAMPREY *Lampetra aepyptera* (Abbott) . . . . (p. 160)

Myomeres more than 62, usually 63–70. A row of teeth, other than marginals, usually visible on posterior field of sucking disc (fig. 7 A). Size of spawning adults averaging larger, usually more than 6.0″ (15 cm) TL.

AMERICAN BROOK LAMPREY *Lampetra appendix* (De Kay) . . . . (p. 164)

Note Concerning Identification of All Ammocoetes

Ammocoetes do not have sufficient morphological characters to make positive identification of specimens possible through the use of morphological characters alone. Fortunately only one or two species occur in any one locality in Ohio; therefore where the species for a locality appears to be well known, it is possible to identify the specimen provisionally through a combination of locality (hence elimination of most species), number of myomeres (elimination of some species), and developing gonads, if present (elimination of parasitic species). As an example; it may be assumed that an ammocoete from the Mad River, Logan County, is an American brook lamprey because it is the only species recorded from that system in Logan County; the high myomere count eliminates all other species except the sea lamprey which does not occur in the Ohio drainage, and if developing gonads are present, it further eliminates the sea lamprey and indicates that the ammocoete belongs to a non-parasitic species.

8
- (*See* above note concerning ammocoetes.) Myomeres less than 55. Probably ammocoetes of: *silver lamprey* when occurring either in the lower portions of streams entering Lake Erie or the Ohio River, or in the upper portions of small brooks which flow directly into the lake or river; or *northern brook lamprey* when found in the Maumee or Grand systems, these ammocoetes may be rather readily found at the place where moderate or large numbers of adults are known to spawn, or immediately below a known spawning area.
- Myomeres between 55–63. Probably ammocoetes of: *Ohio lamprey* when found in streams of southern Ohio near the Ohio River; or *mountain brook lamprey* when found in the Mahoning system where the adults are known to spawn or immediately to several miles below a spawning area; or *least brook lamprey* when found in the smaller brooks and streams of unglaciated Ohio and along the line of glaciation, ammocoetes of this species are often abundant at or below a spawning area, the largest numbers occurring in stream gradients which are lower than the spawning area.
- Myomeres 64 or more. Probably ammocoetes of: *sea lamprey* when found in the lower portions of the larger streams entering Lake Erie; or throughout brooks that flow directly into Lake Erie; or *American brook lamprey* in both drainages when occurring at or below known spawning areas of the adults.

Family 2 Polyodontidae PADDLEFISH . . . . (pp. 174–76)

Snout (even in young only 4.0″ [10 cm] TL) greatly elongated into a thin, horizontal blade like a canoe paddle. Snout as wide as, or wider than, body width, and longer than remainder of head. Skin of body naked. Gill rakers long, and there are hundreds on each gill arch.
PADDLEFISH *Polyodon spathula* (Walbaum) . . . . (p. 174)

Family 3 Acipenseridae STURGEONS . . . . (pp. 167–73)

1
- Snout conical. Caudal peduncle roundish in cross section and short (fig. 8 B). Bony plates on caudal peduncle restricted to a row along lateral band, and a few along dorsal ridge. Barbels not fringed. Lower lip with two, non-papillose lobes. Caudal fin without a filament. Spiracles present.
  LAKE STURGEON *Acipenser fulvescens* Rafinesque . . . . (p. 167)
- Snout flattened dorso-ventrally, resulting in a shovel-shaped snout. Caudal peduncle flattened dorso-ventrally in cross section, long, and completely covered with bony plates. Barbels coarsely fringed. Lower lip with 4 papillose lobes. Upper lobe of caudal fin produced into a long filament which often has been broken off. No spiracle.
  SHOVELNOSE STURGEON *Scaphirhynchus platorynchus* (Rafinesque) . . . . (p. 171)

Family 4 Lepisosteidae GARS . . . . (pp. 177–88)

1
Snout beak-like, narrow and very long. Width of upper jaw at nostrils less than diameter of eye. Distance from posterior edge of eye to posterior edge of opercle membrane contained more than 3.5 times in head length. Scales in transverse row from anal scale-plate to mid-dorsal scale, both included, 17–19. Scales along dorsal ridge from head to dorsal fin origin. 50–52 (fig. 14 A and B).

LONGNOSE GAR *Lepisosteus osseus* (Linnaeus) . . . . (p. 186)

Snout more like bill of duck in shape, shorter and broader. Width of upper jaw at nostrils greater than diameter of eye. Distance from posterior edge of eye to posterior edge of opercle membrane contained less than 3.5 times in length of head . . . . 2.

2
Snout very broad. Diameter of eye contained 1.5 or more times in width of upper jaw at nostrils. Distance from posterior edge of eye to posterior edge of opercle membrane contained 2.5–2.9 times in head length of specimens over 10.0″ (25.4 cm) TL. Scales in transverse row from anal scale-plate to mid-dorsal scale, both included, 23–25. Scales along dorsal ridge, from head to dorsal fin origin, 50–52.

ALLIGATOR GAR *Lepisosteus spatula* Lacepède . . . . (p. 177)

Snout narrower. Diameter of eye contained 1.0–1.5 times in width of upper jaw at nostrils. Distance from posterior edge of eye to posterior edge of opercle membrane contained 2.9–3.5 times in head length of specimens over 10.0″ (25.4 cm) TL. Scales in transverse row from anal scale-plate to mid-dorsal scale, both included, 17–23. Scales along dorsal ridge from head to dorsal fin origin, 46–55 . . . . 3.

3
Scales in lateral series, 59–63 (counted straight backward from first large scale behind head to caudal fin, including the small, ill-formed scales at caudal base). Scales in transverse row from anal scale-plate to mid-dorsal scale, both included, 20–23. Scales along dorsal ridge from head to dorsal fin origin, 50–55. Dark spotting usually restricted to posterior third of body, or spotting absent entirely. No distinct spots on dorsal or ventral surfaces of head.

SHORTNOSE GAR *Lepisosteus platostomus* Rafinesque . . . . (p. 180)

Scales in lateral series, 54–58. Scales in transverse row from anal scale-plate to mid-dorsal scale, both included, 17–20. Scales along dorsal ridge, 46–49. Dark spots usually present on all parts of body including dorsal and ventral surfaces of head.

SPOTTED GAR *Lepisosteus oculatus* (Winchell) . . . . (p. 183)

Family 5 Amiidae BOWFIN . . . . (pp. 189–91)

Only species of fish in Ohio waters with a large, gular plate that occupies the anterior two-thirds of the space between the lower jaws (fig. 15 A). A single dorsal fin, not connected with caudal fin, of more than 45 rays. Scales in lateral line about 65. Nostrils with a short barbel.

BOWFIN *Amia calva* Linnaeus . . . . (p. 189)

Family 6 Hiodontidae MOONEYES .... (pp. 207–12)

1 Dorsal rays 9 or 10. Origin of dorsal fin inserted slightly behind anal fin origin. Dorsal base about a third the length of anal base. A fleshy keel usually extending along mid-ventral line from immediately behind pectorals to vent. The smaller eye has some golden in the iris.
GOLDEYE *Hiodon alosoides* (Rafinesque) .... (p. 207)
Dorsal rays 11 or 12, rarely 10. Origin of dorsal fin inserted slightly before anal fin origin. Dorsal base about half the length of anal base. The fleshy, mid-ventral keel not extending forward in front of pelvic base. The larger eye has a silvery iris.
MOONEYE *Hiodon tergisus* Lesueur................................. (p. 210)

Family 7 Clupeidae HERRINGS.. (pp. 195–206)

1 Mouth terminal, the lower jaw longer or both jaws equal. Upper jaw of the large mouth contained less than 3.2 times in head length. Snout sharp. Last ray of dorsal fin not elongated into a filament.
Genus *Alosa* .... 2.
Mouth subterminal or almost so, the lower jaw included. Upper jaw of the small mouth contained more than 3.2 times in head length. Snout blunt. Last ray of dorsal fin elongated into a prominent filament, which may be broken off in some specimens. Stomach muscular and like the gizzard of a chicken.
Genus *Dorosoma* .... 3.

2 Upper jaw of very large mouth extending past middle of eye. Gill rakers on lower angle of first gill arch less than 30 (Ohio River drainage).
SKIPJACK HERRING *Alosa chrysochloris* (Rafinesque) .... (p. 195)
Upper jaw of smaller mouth not reaching to middle of eye. Gill rakers on lower angle of first gill arch usually more than 30 (both drainages).
ALEWIFE *Alosa pseudoharengus* (Wilson) .... (p. 198)

3 Anterior origin of dorsal insertion distinctly behind pelvic insertion. Dorsal fin rays usually 12, range 11 to 13. Anal fin rays 27–36, usually more than 29; many melanophores along base of anal fin in young less than 30 mm TL. Lateral line scales usually 58–65, range 52–70. Dark postopercular spot as large as eye or larger. Vertebrae 48–51.
EASTERN GIZZARD SHAD *Dorosoma cepedianum* (Lesueur) .... (p. 201)
Anterior origin of dorsal insertion over pelvic insertion. Dorsal fin rays usually 14 or 15. Anal fin rays 17–27, usually 20–25; few or no melanophores along base of anal fin in young less than 30 mm TL. Lateral line scales fewer than 50, usually 40–43. Dark postopercular spot smaller than eye diameter. Vertebrae 40–45.
THREADFIN SHAD *Dorosoma petenense* (Günther) .... (p. 205)

Family 8 Salmonidae SALMONS, TROUTS and WHITEFISHES................... (pp. 213–38)

1 More than 100 extremely small scales in lateral series. Mouth very large with strong, sharp teeth on jaws and tongue. Body color not predominately silvery in trouts.
Genera *Oncorhynchus, Salmo* and *Salvelinus* .... 2.
Fewer than 100 scales in lateral series. Mouth smaller, either without teeth on jaws and tongue, or with the teeth very small and weak. Body always predominantly silvery.
Genus *Coregonus* .... 7.

2
- Anal rays 9–12, rarely 13.
  Genera *Salvelinus* and introduced genus *Salmo* . . . . 3.
- Anal ray 13–20.
  Introduced genus *Oncorhynchus* . . . . 6.

3
- Back of body and dorsal fin with a few to many sharply-defined black and brown spots. No distinct vermiculations (worm-like tracings) on back. The flattened vomer with one or two rows of teeth extending along the shaft (posterior teeth sometimes lost with age). Posterior end of maxillary in mature specimens extending only to posterior edge of eye, or slightly beyond. Scales fewer than 140 along lateral line.
  Introduced genus *Salmo* . . . . 4.
- Dorsal half of body including the dorsal fin with distinct vermiculations and without distinct, small, black or brown spots. The boat-shaped vomer has the shaft toothless, posterior end of maxillary in mature specimens extending far beyond eye. Scales more than 190 along lateral line.
  Genus *Salvelinus* . . . . 5.

4
- Dark spots larger, fewer and more irregular; these spots normally absent or faint on top of head and on caudal fin. Orange and reddish spots surrounded by bluish rings usually present on body; these often numerous and strongly developed. No pink or reddish stripe along sides of adults. Adipose fin of young without a blackish margin or spots. Anal rays (includes the unbranched first ray) typically 9. Dorsal fin usually originating much closer to tip of snout than to caudal base.
  BROWN TROUT *Salmo trutta* Linnaeus . . . . (p. 218)
- Dark spots smaller, very numerous, more regular in outline and distinct, and almost invariably distinct and numerous on top of head and on caudal fin. Sides without orange and reddish spots; sides usually with a pink or reddish stripe (most developed in large young and adults). Adipose fin of young with a blackish margin and/or spots. Anal rays including the unbranched first ray, 10–12 (occasionally 9 in young in which a ray has not yet become branched). Dorsal fin usually originating farther back, about equidistant between base of caudal and tip of snout in sub-adults, closer to snout in adults.
  RAINBOW TROUT *Salmo gairdneri* Richardson . . . . (p. 221)

5
- Caudal fin squarish or only slightly forked. Sides usually with brilliant blue and red spots. Caudal and dorsal fins heavily speckled and vermiculated. First ray of pectoral, pelvic and anal fins milky-white in large young and adults. Usually more than 210 scales in the lateral series.
  BROOK TROUT *Salvelinus fontinalis* (Mitchill) . . . . (p. 224)
- Caudal fin deeply forked. No red or blue spots on body; instead body is spotted and vermiculated with gray. First rays of lower fins not whitish or notably different in color from the other rays. Usually fewer than 210 scales in the lateral series.
  LAKE TROUT *Salvelinus namaycush* (Walbaum) . . . . (p. 228)

6
- Anal rays usually 13–15; black spotting normally confined to back, dorsal fin, usually, and *upper* lobe of caudal fin, although occasional specimens have faint spottings on lower lobe of caudal fin; absence of black pigment (melanophores) along bases of the firmly-set, needle-like teeth.
  COHO SALMON *Onchorhynchus kisutch* (Walbaum) .... (p. 215)
- Anal rays usually 15–19, occasionally fewer; black spotting normally on back, dorsal fin and *both* lobes of caudal fin; black pigment usually along bases of rather loose, conical teeth.
  CHINOOK SALMON *Oncorhynchus tshawytscha* (Walbaum) .... (p. 213)

7
- Snout sharp and mouth terminal. Gill rakers on first branchial arch, 41–53 ....
- Snout rounded and decidedly sub-terminal. Gill rakers on first branchial arch fewer than 31, extremes 25–30. Scales in lateral line 75–86, extremes 73–93.
  LAKE WHITEFISH *Coregonus clupeaformis* (Mitchill) .... (p. 236)

8
- Depth usually more than 3.7 times in standard length. Lateral line scales usually more than 79. Length of longest pelvic ray usually contained more than 1.7 times into distance between origins of pelvic and anal fins.
  GREAT LAKES CISCO *Coregonus artedii artedii* Lesueur .... (p. 231)
- Depth usually less than 3.7 times in standard length. Lateral line scales usually less than 79. Length of longest pelvic ray usually contained less than 1.8 times into distance between origins of pelvic and anal fins.
  LAKE ERIE CISCO *Coregonus artedii albus* Lesueur .... (p. 232)

Family 9 Osmeridae SMELT .... (pp. 239–41)

- Combination of: adipose fin; fewer than 75 scales in the lateral line; more than 12 anal rays; upper jaw extending to below center of eye or beyond; jaws and tongue bearing sharp, canine teeth; coloration silvery.
  RAINBOW SMELT *Osmerus mordax* (Mitchill) .... (p. 239)

Family 10 Umbridae MUDMINNOW .... (pp. 242–44)

- Combination of: caudal fin rounded; dorsal surface of head slightly flattened; head, cheeks and opercles covered with large scales; a dark, vertical bar at caudal base; basic color of head and body brownish-olive mottled with darker; no lateral line; no spinous dorsal fin; upper jaw not protractile.
  CENTRAL MUDMINNOW *Umbra limi* (Kirtland) .... (p. 242)

Family 11 Esocidae PIKES .... (pp. 245–58)

1
- Opercles (also cheeks) fully scaled (fig. 33 A). A prominent dusky, vertical bar extending downward from eye. .... 2.
- Opercles scaleless on their lower halves (fig. 35 A). The dusky, vertical bar below eye normally absent, when present the bar is very faint. .... 3.

2. Young and adults usually barred and mottled conspicuously with dark browns, but never with distinct spots or vermiculations. Branchiostegals usually 11–13 (fig. 33 B). Scales in lateral line usually fewer than 110. Snout usually shorter, its distance from snout tip to center of pupil usually less than distance from center of pupil to edge of opercular margin. Adults small, 5.0″–15.0″ (13–38 cm) TL.
GRASS PICKEREL *Esox americanus vermiculatus* Lesueur .... (p. 245)
Young under 8.0″ (20 cm) TL usually barred. In individuals between 8.0″–12.0″ (20–30.5 cm) TL, the barring changes into the adult pattern of dark, chain-like reticulations. Branchiostegals usually 14–16 (fig. 35 B). Scales in lateral line usually 112–135. Snout usually longer, its distance from snout tip to center of pupil usually greater than distance from center of pupil to edge of opercle margin. Adults larger, 10.0″–25.0″ (25.4–63.5 cm) TL.
CHAIN PICKEREL *Esox niger* Lesueur .... (p. 248)

3. Cheeks fully scaled (fig. 35 A). Body and vertical fins in individuals longer than 9.0″ (23 cm) SL are conspicuously spotted with whitish or yellowish; these tending to form oblique rows. In young less than 9.0″ (23 cm) long the spots are merged and form oblique bars (fig. 35 C). Branchiostegals usually 14–16 (fig. 35 B). Sensory pores on each lower jaw large and usually only 5 in number.
NORTHERN PIKE *Esox lucius* Linnaeus .... (p. 250)
Cheeks (like opercles) scaleless on their lower halves. Body and vertical fins normally with a few to many distinct, dark or dusky (not whitish) spots which tend to form oblique rows in adults; in young less than 9.0″ (23 cm) SL these spots are merged to form oblique bars (fig. 36 C). Branchiostegals usually 17–19. Sensory pores on each lower jaw small and usually 6–9 in number. .... 4.

4. Scales usually confined to upper halves or thirds of cheek and opercle, with little tendency to encroach downward to form a scaly, vertical bar on the anterior edge of the opercle (fig. 36 A). Dusky spots on sides definitely tending to form well-defined, oblique rows; these spots usually very distinct on the posterior half of body, and often evident on the anterior half.
GREAT LAKES MUSKELLUNGE *Esox masquinongy masquinongy* Mitchill ...... (p. 253)
Scales on upper halves or thirds of cheek and opercle often encroaching downward into the lower halves, usually forming a scaly, long, vertical bar on the anterior edge of opercle (fig. 37 A). Dusky spots on sides tending to occur at random, or to form ill-defined, oblique rows; these spots faint (when fish are taken from silty water) or distinct only on posterior third of body, and usually very faint or absent on the anterior two-thirds. In the least spotted specimens, the dusky spot at upper angle of gill cleft becomes the most prominent spot.
OHIO MUSKELLUNGE *Esox masquinongy ohioensis* Kirtland .... (p. 256)

Family 12 Catostomidae SUCKERS .... (pp. 403–71)

1. Dorsal fin longer and of more than 20 principal rays; its base contained less than 4.0 times in standard length. .... 2.
Dorsal fin shorter and of 18 or fewer principal rays; its base contained more than 4.0 times in standard length. .... 9.

2
- Lateral line scales more than 50. Eye closer to posterior edge of opercle membrane than to tip of snout. Depth contained more than 4.0 times in standard length. Lips papillose.
  BLUE SUCKER *Cycleptus elongatus* (Lesueur) . . . . (p. 403)
- Lateral line scales fewer than 50. Eye closer to tip of snout than to posterior edge of opercle membrane. Depth contained less than 4.0 times in standard length. . . . . . 3.

3
- Subopercle broadest at its middle, its posterior (free) edge forming an evenly-curved arc. In specimens more than 6.0″ (15 cm) SL, the distance from eye to lower angle of preopercle is usually contained 1.0 or more times in distance from eye to upper angle of gill cleft. Anterior fontanelle closed or much reduced, posterior fontanelle open. Coloration brownish and darker; not silvery. Has 7 anal rays.
  Genus *Ictiobus* . . . . 4.
- Subopercle broadest below its middle, at the angle on its posterior (free) edge; this angle giving the subopercle a somewhat triangular shape. In specimens more than 6.0″ (15 cm) SL, the distance from eye to lower angle of preopercle is usually contained 1.0 or less times in distance from eye to upper angle of gill cleft. Anterior fontanelle open as is the posterior fontanelle. Basic color silvery. Has 8 anal rays; fin lies more horizontal.
  Genus *Carpiodes* . . . . 6.

4
- Mouth more terminal than in any other Catostomid; very oblique and large. Tip of upper lip about on level with lower margin of eye. Upper jaw about as long as snout. Lips thin, only faintly striated.
  BIGMOUTH BUFFALOFISH *Ictiobus cyprinellus* (Valenciennes) . . . . (p. 406)
- Mouth sub-terminal, almost horizontal, and smaller. Tip of upper lip far below lower margin of eye. Upper jaw definitely shorter than length of snout. Lips fuller and more striated. . . . . 5.

5
- Body more torpedo-like in shape; body depth usually contained 2.9–3.5 (extremes, 2.6–3.5) times in standard length. Eye smaller; in specimens more than 6.0″ (15 cm) SL the eye is contained 5.1–7.4 times in head length and is equal to, or shorter than, length of upper jaw (from tip of upper jaw to posterior end of maxillary); eye contained 2.0–2.5 times in length of snout. Mouth larger and usually less inferior. Thickness of head at opercular bulge (fig. 87 A) contained 4.7–5.4 times in standard length.
  BLACK BUFFALOFISH *Ictiobus niger* (Rafinesque) . . . . (p. 409)
- Body more slab-sided and large specimens more sunfish-like in shape; body depth usually contained 2.4–2.8 (extremes, 2.2–3.0) times in standard length. Eye larger; in specimens more than 6.0″ (15 cm) SL the eye is contained 4.4–5.9 times in head length and is equal to, or longer than, length of upper jaw; eye contained 1.5–2.0 times in length of snout. Mouth smaller and usually more inferior. Thickness of head at opercular bulge contained 5.2–6.1 times in standard length (fig. 88 A).
  SMALLMOUTH BUFFALOFISH *Ictiobus bubalus* (Rafinesque) . . . . (p. 412)

6
- No small knob- or nipple-like projection on tip of lower lip (fig. 90 A). Tip of lower lip definitely in advance of anterior nostril. Distance from tip of snout to anterior nostril equal to length of eye in small specimens, this distance usually much greater (1.2–3.0 times) than length of small eye in adults. Nostrils situated above the posterior third of lower jaw, or behind it. Snout long, in adults usually contained 3.0–3.5 times in head length. Scales in lateral line usually 36–40. Anterior rays of depressed dorsal usually extending over half the length of the dorsal base, unless broken off. 7
- Small knob present on tip of lower lip (fig. 92 A). Tip of lower lip scarcely or not at all in advance of anterior nostril. Distance from tip of snout to anterior nostril less than length of eye in small fishes, and equal in length in large adults. Nostrils above anterior two-thirds of lower jaw or before it. Scales in lateral line usually 33–37. Anterior dorsal rays short or very long. . . . . 8.

7
- Body deeper, the larger specimens sunfish-like in shape; body depth usually 2.3–2.7 (extremes, 2.2–3.4) times in standard length. Eye averaging smaller, usually 5.4–7.0 (extremes, 4.0–8.8) times in head length and more than 4.1 times in fishes less than 4.0″ (10 cm) SL and usually more than 5.5 times in moderate- or large-sized individuals. EASTERN QUILLBACK CARPSUCKER *Carpiodes cyprinus cyprinus* (Lesueur) . . . . . . (p. 416)
- Body moderately slender, its depth usually 2.7–3.0 (extremes, 2.5–3.9) times in standard length. Eye averaging larger, usually 4.4–5.6 (extremes, 3.0–5.8) times in head length; usually less than 4.5 times in specimens less than 3.0″ (7.6 cm) SL; less than 5.0 times in moderately-sized fishes; 5.5 times in large adults. CENTRAL QUILLBACK CARPSUCKER *Carpiodes cyprinus hinei* Trautman . . . . (p. 419)

8
- Anterior dorsal rays short at all ages; usually the longest ray extends less than two-thirds the length of the fin base. Body of large young and adults slender; body depth usually contained 2.7 (young)–3.3 (old adults) times in standard length. Eye of large young and adults smaller; contained 3.0 (young less than 3.0″ [7.6 cm])–6.2 (adults) times in head length. Snout of adults rather long; this appearance of length is heightened by the small eye. Young less than 4.0″ (10 cm) TL difficult, often impossible, to separate from young of highfin carpsucker. NORTHERN RIVER CARPSUCKER *Carpiodes carpio carpio* (Rafinesque) . . . . (p. 422)
- Unbroken anterior dorsal rays extremely long in fishes of more than 5.0″ (13 cm) TL; usually longer than fin base when depressed. Body of large young and adults deeper and more sunfish-like in shape; body usually contained 2.0 (large adults)–3.0 (young) times in standard length. Eye of large young and adults larger and more oblongated horizontally; contained 3.0 (young less than 3.0″ [7.6 cm])–4.8 (adults) times in head length. Snout of adults very blunt and rounded, and because of large eye the snout appears to be very short. Young less than 4.0″ (10 cm) TL difficult, often impossible, to separate from young of river carpsucker. HIGHFIN CARPSUCKER *Carpiodes velifer* (Rafinesque) . . . . (p. 425)

9
- Lateral line complete and well developed. Lips plicate or papillose. Air bladder divided into 2 or 3 parts. . . . . 10.
- Lateral line incomplete or absent. Lips plicate. Air bladder divided into 2 parts. . . 20.

10
- Lateral line scales 50 or fewer. . . . . 11.
- Lateral line with 55 or more scales. Lips heavily papillose. Air bladder divided into 2 parts. Genus *Catostomus* . . . . 19.

11
- Head convex (rounded) between the eyes as with most species of Catostomids. Air bladder divided into 3 parts. . . . . 12.
- Head strongly depressed between the eyes, the depression forming a hollow (fig. 101 A). Body with 4–6 dark, usually prominent, oblique bars. Lips heavily papillose. Air bladder divided into 2 parts.
  HOG SUCKER *Hypentelium nigricans* (Lesueur) . . . . (p. 455)

12
- Premaxillaries protractile (upper lip separated from remainder of snout by a deep groove). Halves of lower lip widely joined together; the lips plicate (fig. 98 A) or papillose (fig. 103 A).
  Genus *Moxostoma* . . . . 13.
- Premaxillaries not protractile (no groove separating tip of upper lip and snout). Halves of lower lip completely separated; the lips smooth or slightly roughened (fig. 100 A) (probably extinct).
  HARELIP SUCKER *Lagochila lacera* Jordan and Brayton . . . . (p. 452)

13
- Body without dark spots at scale bases. Tail always slate-colored in life (broken blood vessels in a frayed tail give the false impression of a reddish tail). . . . . 14.
- Body with distinct, dark spots at scale bases (least noticeable in small young; spots not as pronounced and black as on the slate-tailed spotted sucker, 20a). Tail always pink, red or carmine in life; palest in young; color absent in specimens long preserved. . . .16.

14
- Dorsal rays usually 15–16 (extremes, 14–18). Distal edge of dorsal convex (rounded outward; adults) or straight (small young). Length of longest ray of depressed dorsal, when projected forward, equal to distance from dorsal origin to space between the eyes (sometimes less in specimens over 15.0″ [38.1 cm] TL). Plicae (folds) of lips more or less completely broken up into papillae-like elements. Body depth deeper; usually contained 2.8 (adults)–3.7 (young) times in standard length. Body color pale and silvery.
  SILVER REDHORSE *Moxostoma anisurum* (Rafinesque) . . . . (p. 428)
- Dorsal rays usually 12–13 (extremes, 11–15). Distal edge of dorsal slightly falcate (large young and adults) or straight (small young). Length of longest ray of depressed dorsal, when projected forward, extends from dorsal origin to occiput (region between eyes and nape). Plicae of lips more striated; not greatly broken up into papillae-like elements. Body more torpedo-like in shape; usually contained 3.4–4.7 times in standard length in large young and adults. Body color darker and more golden. . . . . 15.

15
- Scales in lateral line 44–47 (extremes, 42–49). Rays of one or both pelvic fins usually 10 (extremes, 9–11). Body slender. Least depth of caudal peduncle usually contained more than 1.7 times in its length. Eye smaller; usually contained more than 2.2 times in snout of large young and adults, about 1.8 times in snout of young less than 4.0″ (10 cm) TL. No tubercles on snout of breeding males.
  BLACK REDHORSE *Moxostoma duquesnei* (Lesueur) . . . . (p. 431)
- Scales in lateral line 39–42 (extremes, 37–44). Rays of pelvic fins normally 9, rarely 8 or 10. Body deeper. Least depth of caudal peduncle usually contained less than 1.6 times in its length. Eye larger, usually less than 2.2 times in snout of large young and adults, about 1.4 times in snout of young less than 4.0″ (10 cm) TL. Tubercles present on snout of breeding males.
  GOLDEN REDHORSE *Moxostoma erythrurum* (Rafinesque) . . . . (p. 434)

16
- Snout sub-conical; head small and short; head usually contained 4.3–5.4 times in standard length of yearlings and adults, 3.5–4.0 in young of less than 3.0″ (7.6 cm) TL. Mouth small. Posterior edge of lower lip straight; the 2 halves not forming a distinct angle. Distal edge of dorsal fin decidedly falcate. . . . . 17.
- Snout blunt; head bulky and long, usually contained less than 4.3 times in standard length of yearlings and adults, usually 3.0–3.8 times in young of less than 3.0″ (7.6 cm) TL. Mouth large. Lips coarse, the posterior edge of lower lip forming a slight angle. Distal edge of dorsal fin straight or convex in large young and adults, occasionally slightly falcate in small young. . . . . 18.

17
- Pelvic rays normally 9, rarely 10. Dorsal fin less falcate, particularly in small young. Anterior rays of depressed dorsal not extending to end of last ray (fig. 96 A). Mouth averaging larger in size.
  NORTHERN SHORTHEAD REDHORSE *Moxostoma macrolepidotum macrolepidotum* (Lesueur). . . . . (p. 438)
- Pelvic rays usually 10 in one or both fins, rarely 9 in both. Dorsal fin deeply falcate, except in young, where it is shallower. Anterior rays of depressed dorsal extending to, or beyond, end of last ray (fig. 97 B); these anterior rays very long in old males. Mouth averaging smaller in size.
  OHIO SHORTHEAD REDHORSE *Moxostoma macrolepidotum breviceps* (Cope) . . (p. 441)

18
- Eye smaller; usually contained 4.0 times or more in specimens of less than 6.0″ (15 cm) SL; 5.0 times or more in specimens 7.0″–12.0″ (18–30.5 cm); 7.0 times or more in specimens more than 15.0″ (38.1 cm) long. Forehead and occipital region rounded. Pharyngeal arch moderately weak; its breadth usually less than its depth in cross section. All teeth strongly compressed, comb-like, and comparatively weak. No tubercles on snout of breeding males.
  GREATER REDHORSE *Moxostoma valenciennesi* Jordan . . . . (p. 444)
- Eye larger; contained 4.0 times or less in specimens of less than 6.0″ (15 cm) SL; 5.0 times or less in specimens 7.0″–12.0″ (18–30.5 cm); 7.0 times or less in specimens more than 15.0″ (38.1 cm) long. Forehead and occipital region flattened and sloping forward sharply, except in the largest breeding males in which the occiput may be swollen and the snout bulbous. Pharyngeal arch very heavy; its breadth as great as, or greater than, its depth in cross section. Teeth of lower half of arch large and squarish; those functioning are worn flat and molar-like. A dentate, pencil-line of melanophores usually posterior to the scales at base of caudal fin. Tubercles on snout of breeding males.
  RIVER REDHORSE *Moxostoma carinatum* (Cope) . . . . (p. 448)

19
- Between 55–85 scales in lateral line. The rounded snout projects slightly or not at all beyond tip of upper lip. Lower lip thin; posterior end of mouth extending backward only to nostrils. No rosy lateral band on males throughout the spring.
  COMMON WHITE SUCKER *Catostomus commersoni commersoni* (Lacepède) . . . . . . (p. 458)
- More than 85 scales rows in lateral line. The bulbous snout projects considerably beyond the upper lip. Lower lip widely flaring; posterior end of mouth extending backward to beyond nostrils. A rosy lateral band present on breeding males throughout the spring.
  EASTERN LONGNOSE SUCKER *Catostomus catostomus catostomus* (Forster) . . . . (p. 461)

20
- Lateral scale rows usually 43–45 (extremes, 42–47). A distinct blackish spot on each scale base, resulting in a series of longitudinal stripes which are most distinct (except in the smallest young where the faint spots and stripes may be restricted to the region above the anal fin base). Body slender; its depth usually contained more than 4.0 times in standard length. Distal edge of dorsal fin slightly falcate in adults; almost straight in small young. Usually 11–12 rays in dorsal fin (extremes, 10–13). Lateral line somewhat anteriorly developed in adults.
  SPOTTED SUCKER *Minytrema melanops* (Rafinesque) .... (p. 463)
- Lateral scale rows usually 35–41 (extremes, 33–43). No dusky spot on scale bases. Body deeper; usually contained less than 3.5 times in standard length. Distal edge of dorsal fin convex (rounded); almost straight in young. Lateral line lacking at all ages.
  Genus *Erimyzon* .... 21.

21
- Dorsal rays 11–12, (extremes, 10–13). Lateral scale rows 35–37 (extremes, 33–40). Body depth usually contained 3.3 times or less in standard length. Unbroken, blackish lateral band very distinct in young; least distinct in largest adults.
  WESTERN LAKE CHUBSUCKER *Erimyzon sucetta kennerlyi* (Girard) .... (p. 466)
- Dorsal rays 9–10 (extremes, 8–11). Lateral scale rows 39–41 (extremes, 37–43). Body depth usually contained 3.3 times or more in standard length. The dusky lateral band broken into a series of more or less confluent blotches; these blotches sometimes very faint or absent on large adults.
  WESTERN CREEK CHUBSUCKER *Erimyzon oblongus claviformis* (Girard) ...... (p. 469)

Family 13 Cyprinidae CARPS and MINNOWS .... (pp. 259–402)

1
- Dorsal fin with more than 12 soft rays; dorsal and anal fins each with a strong, serrated, spinous-ray (fig. 38 A and B).
  Introduced genera *Cyprinus* and *Carassius* .... 2.
- Dorsal fin with fewer than 10 developed, soft rays. No spinous-rays in dorsal or anal fins.
  Native cyprinids .... 3.

2
- Upper jaw with two fleshy barbels on each side. Lateral line with more than 32 scales, usually 35–38 (except in the almost, or completely, scaleless "leather or mirror" types). A dark spot at each scale base. Pharyngeal teeth in 3 rows 1, 1, 3–3, 1, 1; teeth in the main row short, heavy and molar-like. Gill rakers on anterior (first) arch, 21–27. (Hybrids between this species and the goldfish, 2b, occur wherever both species are present in the same locality; in some localities the number of hybrids is greater than the total number of both parent species.)
  CARP *Cyprinus carpio* Linnaeus .... (p. 259)
- No barbels on upper jaw. Lateral line with fewer than 32 scales, usually 26–30. No dusky spot at each scale base. Pharyngeal teeth in one row, 4–4; not molar-like. Gill rakers on anterior arch, 37–43.
  GOLDFISH *Carassius auratus* (Linnaeus) .... (p. 262)

3
- Abdomen behind pelvic fins with a mid-line of bare skin (a fleshy keel) over which the scales do not pass (fig. 40 B). Anal fin deeply falcate; rays usually 11–13 (extremes, 10–14). Lateral line greatly decurved. Scales in lateral line, 44–54. Body color predominantly yellowish; except in small young which are more silvery. Teeth 5–5, hooked.
  GOLDEN SHINER *Notemigonus crysoleucas* (Mitchill) .... (p. 265)
- Abdomen behind pelvic fins rounded over and scaled in the normal manner ...... 4.

4. A slender, small, barbel present, which is *either* round in cross section and situated at the posterior end of the maxillary (fig. 48 A and B) *or* is flap-like in cross-section and placed in the groove above the maxillary well in advance of its posterior end (fig. 50 A) where it is usually difficult to observe or entirely hidden when mouth is closed. Occasionally a barbel may be missing on one side, especially in the genera *Semotilus* and *Exoglossum* . . . . 5.
   No maxillary barbel (a tubercle, superficially resembling a barbel, is present on the posterior end of maxillary in breeding males of some bluntnose minnows) . . . . 15.

5. Barbel at posterior end of maxillary.
   Genera *Nocomis, Hybopsis* and *Rhinichthys* . . . . 6.
   Barbel in groove above maxillary and well in advance of its posterior end (fig. 50 A).
   Genera *Semotilus* and *Exoglossum* . . . . 14.

6. Upper jaw protractile (separated from snout by a groove). Scales in lateral series 57 or fewer. Scale radii restricted to the posterior (exposed) field. Teeth 1, 4–4, 1, or 4–4, or some combination of these.
   Genera *Nocomis* and *Hybopsis* . . . . 7.
   Upper jaw not protractile (snout not separated from tip of upper jaw by a groove). Scales in lateral series 56 or more. Scale radii present on all fields. Teeth usually 2, 4–4, 2.
   Genus *Rhinichthys* . . . . 13.

7. Eye smaller and mouth larger. Eye contained 1.0 (small young)–2.2 (large young and adults) times in length of upper jaw. Mouth slightly oblique and scarcely overhung by snout, or almost terminal . . . . 8.
   Eye larger and mouth smaller. Eye usually contained 0.4–0.9 (may be 1.0 in small young) times in length of upper jaw. Mouth horizontal and considerably overhung by snout. 9.

8. Snout shorter, the least suborbital width (least distance between edge of eye and edge of mouth, not including maxillary) usually contained 2.0–2.5 times in postorbital length of head (greatest distance between posterior edge of eye and membranous opercular margin). Snout length usually contained 1.2–1.7 times in post-orbital length of head. Base of caudal fin with a dusky spot; this fin orange or red in young. Teeth usually 1, 4–4, 1.
   HORNYHEAD CHUB *Nocomis biguttatus* (Kirtland) . . . . (p. 268)
   Snout longer, the least suborbital width contained 1.3–2.0 times in postorbital length of head. Snout length usually contained 0.8–1.2 times in postorbital length of head. Caudal spot absent or indistinct. Caudal fin slatish, (may be a faint orange in some young). Teeth usually 4–4.
   RIVER CHUB *Nocomis micropogon* (Cope) . . . . (p. 272)

9. Snout shorter; the bony interorbital space contained 0.7–1.4 times in snout length. Body without distinct W- or X-markings, or roundish spots (body may have blackish cysts of parasites, which superficially look like melanophoric spots). Teeth 1, 4–4, 1 . . 10.
   Snout longer; the bony interorbital space usually contained 1.5–2.4 (occasionally 1.3 times in smallest young) times in snout. Body usually with distinct W- or X-shaped markings, or with roundish spots; these markings sometimes very faint in large adults. Teeth 4–4. . . . . 11.

10
- No dark, distinct lateral band; when a faint lateral band is present, it does not extend forward to encircle snout. Dorsal fin more anterior; distance from dorsal fin origin to caudal base much greater than distance from dorsal origin to snout. Size larger; over 4.0″ (10 cm) TL.
  SILVER CHUB *Hybopsis storeriana* (Kirtland) .... (p. 275)
- A conspicuous, unbroken, dusky, lateral band which usually extends forward to encircle snout. Dorsal fin more posterior; distance from dorsal fin origin to caudal base about equal to distance from dorsal origin to snout. Size smaller; seldom reaching a total length of 4.0″ (10 cm).
  NORTHERN BIGEYE CHUB *Hybopsis amblops amblops* (Rafinesque) .... (p. 279)

11
- Mouth smaller; posterior end of maxillary not reaching a point directly below anterior edge of eye. Snout moderately overhangs the mouth. Scales in lateral line, 38–47, rarely less. Spots on dorsal half of body mostly W- or X-shaped, seldom round ...... 12.
- Mouth larger, the posterior end of maxillary reaching, to or extending beyond, the anterior edge of eye. Snout bulbous and greatly overhangs the mouth. Roundish spots on dorsal half of body usually very conspicuous.
  OHIO SPECKLED CHUB *Hybopsis aestivalis hyostoma* (Gilbert) .... (p. 288)

12
- A broken lateral band consisting of 7–11 oblong, blackish spots, including the distinct caudal spot. Caudal spot especially prominent in young. Several sharply-defined alternating light and dark spots along dorsal ridge; these conspicuous only in life. Snout shorter. Preorbital length shorter than, or equal to, postorbital head length. Lateral line scales usually 44–47, rarely 43.
  OHIO STREAMLINE CHUB *Hybopsis dissimilis dissimilis* (Kirtland) .... (p. 282)
- No distinct oblong, blackish spots along lateral line. Scattered W- or X-shaped markings over body and a roundish caudal spot; all markings and spots sometimes obsolete in large adults. No distinct spots along dorsal ridge. Snout longer. Preorbital length of head usually longer, rarely equal to (except in smallest young), postorbital length. Lateral line scales 38–43, usually 42 or less.
  EASTERN GRAVEL CHUB *Hybopsis x-punctata trautmani* Hubbs and Crowe. ... (p. 285)

13
- Snout scarcely projecting beyond the oblique and subterminal mouth. Tip of upper lip about on the level with lower edge of eye. Snout shorter; preorbital length of head contained 1.1 (large adults)–1.7 (small young) times in postorbital length. Eye larger, usually contained less than 5.0 times in head length; extremes, 3.8 (small young)–5.3 (large adults). Dark lateral band usually contrasting sharply, both above and below, with the lighter color of the side. Air bladder fairly well developed, its posterior tip extending well behind pelvic insertion.
  WESTERN BLACKNOSE DACE *Rhinichthys atratulus meleagris* Agassiz .... (p. 291)
- Snout projecting far beyond the almost horizontal mouth. Tip of upper lip well below level of lower edge of eye. Snout longer; preorbital length of head usually contained 0.7 (large adults)–1.2 (small young) times in postorbital length. Eye smaller, usually contained more than 5.0 times in head length; extremes, 4.8 (small young)–6.3 (large adults). Dorsal and ventral edges of lateral band merging into the basic color of the sides. Air bladder of adult rudimentary, its posterior tip well in advance of pelvic insertion.
  LONGNOSE DACE *Rhinichthys cataractae* (Valenciennes) .... (p. 294)

14
- Premaxillaries protractile. Lips of lower jaw similar to those of most species of minnows. Mouth large. The posterior end of maxillary extends well beyond anterior edge of eye. Mouth terminal. Jaws equal. Tip of upper lip extending above anterior edge of eye. A prominent dusky blotch at base of anterior dorsal rays.
  NORTHERN CREEK CHUB *Semotilus atromaculatus atromaculatus* (Mitchill) . . . (p. 297)
- Premaxillaries not protractile. Fleshy lobes of lips of lower jaw covering only the posterior two-thirds of each side of jaw, the anterior third a horny sheath which is somewhat similar to the tips of the lower jaws of the barbelless suckermouth (*see* 16a), and stoneroller (*see* 16b) minnows. Mouth smaller; posterior end of maxillary does not reach to below anterior edge of eye. Mouth subterminal; lower jaw included in upper jaw; tip of upper lips considerably below lower edge of eye. No dusky blotch on anterior dorsal rays.
  WESTERN TONGUETIED MINNOW *Exoglossum laurae hubbsi* (Trautman) . . . . (p. 300)

15
- Lips of lower jaw obviously specialized in appearance, *either* having the fleshy lips restricted to lobes on the posterior halves of each jaw and the anterior halves covered only with skin (*Phenacobius, see* 16a), *or* having a cartilaginous sheath or ridge replacing the fleshy lips (*Campostoma, see* 16b). . . . . 16.
- Lower jaw with fleshy lips and without any obvious specialization. . . . . 18.

16
- Fleshy lips restricted to lobes on the posterior halves of lower jaw. Prominent black lateral band and large black caudal spot. Basic coloration of body silvery. Peritoneum silver. Alimentary tract about as long as head and body. Teeth 4–4.
  SUCKERMOUTH MINNOW *Phenacobius mirabilis* (Girard) . . . . (p. 303)
- A cartilaginous sheath or ridge replacing the fleshy lips which are present in most species of minnows. No distinct lateral band or caudal spot. Basic coloration light brownish with darker scales scattered over body. Peritoneum black. Alimentary tract 1.5 (smallest young)–9.0 (large adults) times as long as head and body; alimentary tract usually coiled about air bladder. Teeth 4–4.
  Genus *Campostoma* . . . . 17.

17
- Scales in lateral line usually 45–50 (extremes, 43–51). Scale row around body immediately before dorsal insertion (fig. 83 E) usually 38–42 (extremes, 35–43). Width of gape usually 3.8–5.0 (extremes, 3.6–5.1, widest in breeding males; fig. 83 C) in head length (fig. 83 A).
  OHIO STONEROLLER MINNOW *Campostoma anomalum anomalum* (Rafinesque) . (p. 397)
- Scales in lateral line usually 50–56 (extremes, 49–64). Scale row around body immediately before dorsal insertion usually 40–50 (extremes, 39–54). Width of gape usually 4.8–5.4 (extremes 4.6–5.6, widest in breeding males) in head length.
  CENTRAL STONEROLLER MINNOW *Campostoma anomalum pullum* (Agassiz) . . . . (p. 401)

18
- Scales in lateral series smaller, 48–95, and rarely less than 51 (then only in the rosyside dace).
  Genera *Phoxinus* and *Clinostomus* . . . . 19.
- Scales in lateral series larger, 28–48, and seldom more than 45. 21.

19
Mouth small; length of upper jaw contained more than 3.2 times in head length. Jaws equal. Scales minute; more than 70 in the lateral series. Lateral line very incomplete. Peritoneum black. Intestine long, with two crosswise coils and a loop. Two dusky lateral bands. Teeth 5–5.
SOUTHERN REDBELLY DACE *Phoxinus erythrogaster* (Rafinesque) . . . . (p. 306)
Mouth very large; length of upper jaw contained less than 2.8 times in head length. Lower jaw extends forward beyond upper jaw. Scales larger; fewer than 70 in the lateral series. Lateral line complete. Peritoneum silvery, speckled with dusky. Intestine short, with only a single loop. One dark, lateral band. Teeth 2, 5–5, 2.
Genus *Clinostomus* . . . . 20.

20
Scales in lateral line more than 58. Body depth contained 4.4 (adults)–5.4 (small young) times in standard length; usually more than 4.5. Caudal peduncle slender, its depth usually contained more than 2.7 times in head length.
REDSIDE DACE *Clinostomus elongatus* (Kirtland) . . . . (p. 309)
Scales in lateral line fewer than 58. Body depth contained 3.9 (adults)–4.7 (small young) times in standard length; usually less than 4.5. Caudal peduncle deeper; its depth usually contained less than 2.7 times in head length.
ROSYSIDE DACE *Clinostomus funduloides* Girard . . . . (p. 312)

21
Dorsal fin with *9* developed soft rays (very rarely 8). Mouth very small and nearly vertical. Peritoneum silvery with darker specklings. (Pugnose shiner, 38, has a similar mouth but has only *8* dorsal rays and a black peritoneum.) Teeth 5–5 or 5–4.
PUGNOSE MINNOW *Notropis emiliae emiliae* (Hay) . . . . (p. 315)
Dorsal fin with *8* developed, soft rays, with or without a thickened and separated anterior half-ray. . . . . 22.

22
Dorsal fin with the anterior half-ray slender and so tightly bound to the first developed ray, as to be scarcely evident.
Genera *Notropis, Ericymba* and *Hybognathus* . . . . 23.
Dorsal fin with a rather stout, blunt-tipped, anterior half-ray that is distinctly separated from the first fully-developed ray, although connected with it by membrane (least noticeable in smallest specimens). Scales on the rather flat pre-dorsal region are much crowded, are much smaller than are the scales on the sides, and are usually in more than 20 rows. Scales in lateral line, 49–49. Anal rays normally 7, rarely 8. A dark blotch on anterior portion of dorsal fin. Teeth usually 4–4.
Genus *Pimephales* . . . . 44.

23
Intestine short; less than twice the length of head and body (excludes *Hybognathus*). No large, squarish, cavernous spaces evident on ventral half of head (excludes *Ericymba*). Peritoneum usually silvery, speckled with darker; definitely brown or black only in common, bigeye and pugnose shiners.
Genus *Notropis* . . . . 24.
Either the ventral surface of head and lower cheeks have large cavernous spaces, the intestine is only about as long as head and body, and the peritoneum silvery (*Ericymba*); *or* the ventral head surface is without large cavernous spaces, the intestine is more than twice as long as head and body and peritoneum black (*Hybognathus*). . . . . 43.

24
Anal fin rays, 9–13; very rarely 8. Teeth almost invariably 2-rowed. . . . . 25.
Anal fin rays 7 or 8; rarely 9. Teeth 2- or 1-rowed. . . . . 32.

25 Anal rays normally 10–13; occasionally 9. When individuals of this group have only 9 rays they *either* have a dusky spot at the dorsal fin origin, or lack this spot and have the anterior lateral line scales less than 2.0 times as high as wide and webbing of dorsal fin without dusky spotting on its posterior half. Teeth 2, 4–4, 2. . . . . 26.
Anal rays normally 9; rarely 8 or 10. Teeth 2, 4–4, 2; 1, 4–4, 1; 1, 4–4, 0. . . . . 30.

26 No dusky spot at base of anterior dorsal fin rays. . . . . 27.
A distinct dusky spot at base of anterior dorsal fin rays (spot faint or undeveloped in smallest young). . . . . 29.

27 Snout blunter and shorter, its length usually contained 1.5 or more times in postorbital head length. Body more compressed and slàb-sided. Mid-dorsal streak indistinct or absent.
COMMON EMERALD SHINER *Notropis atherinoides atherinoides* Rafinesque. . . . . (p. 318)
Snout longer, its length usually contained less than 1.5 times in postorbital length of head. Body rounder and more torpedo-shaped. Mid-dorsal streak usually distinct. . .28.

28 Two crescent-shaped areas of melanophores between the nostrils (fig. 58 A). Eye very large; contained 3.5 times or more in head length. Snout more rounded. Origin of dorsal fin usually over, or slightly behind, the pelvic fin origin. Mid-dorsal streak rather broad and black.
SILVER SHINER *Notropis photogenis* (Cope) . . . . (p. 321)
No crescent-shaped areas of melanophores between the nostrils. Eye smaller, usually contained 3.5 times or less in head length. Snout sharper and more pointed. Origin of dorsal fin considerably behind pelvic fin origin. Mid-dorsal streak narrower, fainter, and more brownish. Head and breast rosy in breeding males.
ROSYFACE SHINER *Notropis rubellus* (Agassiz) . . . . (p. 327)

29 Anal rays usually 11–13, occasionally 10. Body slab-sided. Width of body usually contained more than 1.8 times in its depth. Head length usually less than body depth. Saddle-bands across back and sides usually faint or absent in living fishes, and often absent in preserved specimens. Breeding males with tubercles on cheeks.
NORTHERN REDFIN SHINER *Notropis umbratilis cyanocephalus* (Copeland). . . . (p. 330)
Anal rays 9–11, usually 10, occasionally 11. Body rounder, its width usually contained less than 1.8 times in depth. Head length usually greater than body depth. Saddle-bands across back and sides, 8–11; usually prominent in live and preserved specimens. Breeding males without tubercles on cheeks.
ROSEFIN SHINER *Notropis ardens* (Cope) . . . . (p. 333)

30
- Webbing of dorsal fin unspotted. . . . . 31.
- Dorsal fin with black blotches on membranes between the posterior rays; blotches faint or undeveloped in smallest young. Distance from dorsal origin to caudal base about equal to distance from dorsal origin to snout. Scales in anterior portion of lateral line diamond-shaped; their height less than 2.0 times their exposed width. Peritoneum silvery. Head triangulate. Snout pointed. Eye small; usually contained more than 3.7 times (3.0 times in smallest young) in head length. Teeth normally, 1, 4–4, 1 or 1, 4–4, 0. *Note:* This species differs from the superficially similar spotfin shiner, (*see* 33a); in having normally 9 instead of 8 anal rays; in having an abundance of melanophores present on the webbings between the anterior dorsal rays in all except the smallest young (whereas these melanophores are present in the spotfin shiner only in the oldest adults); in having the dark, lateral band on the caudal peduncle centered over the lateral line scales (seen best in preserved specimens).
  STEELCOLOR SHINER *Notropis whipplei* (Girard) . . . . (p. 355)

31
- Scales in anterior portion of lateral line greatly elevated, their height 2.0–3.0 times greater than their exposed width (fig. 63 C). Distance from dorsal origin to caudal base usually considerably greater than distance from dorsal origin to tip of snout or upper lip. Peritoneum dark brown, head large; snout rounded. Eye usually contained less than 4.0 times (2.5 times in smallest young) in head length. Body definitely slab-sided; body depth usually contained 4.0 or fewer times in standard length (as high as 4.3 times in some small young) . . . . 32.
- Scales in anterior portion of lateral line slightly elevated, their height normally less than 2 times their exposed width. Distance from dorsal origin to caudal base usually slightly less than distance from dorsal origin to tip of snout or upper lip. Peritoneum silvery, uniformly overlain with brown pigment. Eye very large, usually contained 2.5–3.1 times in head length. Body not slab-sided, its depth usually contained more than 4.0 times in standard length. Predorsal scales fewer than 18. Upper jaw length about as long as eye length. Anal rays 9, rarely 8 or 10. Teeth 2, 4–4, 2.
  POPEYE SHINER *Notropis ariommus* (Cope) . . . . (p. 324)

32
- Scales before dorsal origin larger; usually fewer than 22 and rarely more than 25 (all scales counted which cross the midline). Dark, wavy lines on upper sides in adults follow scale rows backward to meet their corresponding fellows on the dorsal ridge, thereby forming narrow V's, which are readily seen when viewed from above (fig. 63 A). Pigment usually present on chin and anterior gular area. Hybridizes or intergrades with common shiner.
  CENTRAL STRIPED SHINER *Notropis chrysocephalus chrysocephalus* (Rafinesque)(p. 336)
- Scales before dorsal origin smaller; usually more than 24. The V-shaped wavy lines are replaced, especially in males, by one or two dark, wavy lines above; these lines are parallel to the lateral band (fig. 64 A). Little or no pigment on chin and anterior gular area.
  COMMON SHINER *Notropis cornutus* (Mitchill) . . . . (p. 340)

33
- Anal fin rays normally 8; rarely 7 or 9. . . . . 34.
- Anal fin rays normally 7; seldom 8 (occasionally pugnose shiner has 7; *see* 37B) . . . . 42.

34
- Dorsal fin with black blotches on webbings between the posterior rays; blotches absent only on smallest young. Similar to steelcolor shiner (*see* above), except that: this species has *8* instead of 9 anal rays; melanophores absent on anterior membranes of dorsal fin, except in large adults; the dark, lateral band on the caudal peduncle is below the center of the lateral line scales and its dorsal edge is distinct and contrasts sharply with the basic color of body instead of merging gradually into it. Teeth normally, 1, 4–4, 1 or 1, 4–4, 0.
  SPOTFIN SHINER *Notropis spilopterus* (Cope) .... (p. 358)
- Dorsal fin without definite spottings on webbings at all ages. .... 35.

35
- A conspicuous black spot at base of caudal fin; spot very pronounced in moderately-sized young, occasionally faint in large adults. Dorsal and anal fins definitely falcate in large young and adults. Lateral band usually obsolete in life. Teeth often 2, 4–4, 2, varying from 0, 4–4, 0, through various combinations to 2, 4–4, 2.
  SPOTTAIL SHINER *Notropis hudsonius* (Clinton) .... (p. 346)
- No conspicuous, black caudal spot. .... 36.

36
- Dark lateral band usually very distinct on sides, cheeks and snout. Scales in anterior portion of lateral line not elevated; their height less than 2.0 times their width. .... 37.
- Lateral band absent or faint. *Either* the ventral surface of head is notably flattened, giving head a triangular appearance in cross-section; *or* the anterior lateral scales are greatly elevated, their height more than 2.0 times their exposed width. .... 39.

37
- Tip of lower jaw (chin) definitely black. Mouth more oblique and terminal. .... 38.
- Tip of lower jaw whitish; not tipped with black. Mouth more horizontal and distinctly subterminal. Lateral line incomplete. Dark borders of lateral line pores expanded to form prominent, black, crescent shaped, vertical crossbars which extend downward from the lateral band. Teeth 4–4.
  BLACKNOSE SHINER *Notropis heterolepis* Eigenmann and Eigenmann .... (p. 376)

38
- Mouth moderately oblique and medium in size; posterior end of upper jaw extending backward to below, or beyond, the posterior half of nostril, and almost to the anterior edge of eye. Teeth usually, 1, 4–4, 1; rarely 4–4. .... 39.
- Mouth nearly vertical and extremely small. Posterior end of upper jaw extending backward only to the anterior half of nostril; far short of the anterior edge of eye. Peritoneum black. Anal rays usually 8, occasionally 7. Teeth 4–4.
  PUGNOSE SHINER *Notropis anogenus* Forbes .... (p. 379)

39
- Lateral line complete. Mouth larger; length of upper jaw equal to the length of the large, prominent eye. Peritoneum black.
  BIGEYE SHINER *Notropis boops* Gilbert .... (p. 352)
- Lateral line incomplete. Mouth smaller; length of upper jaw usually shorter than eye length. Peritoneum silvery.
  BLACKCHIN SHINER *Notropis heterodon* (Cope) .... (p. 349)

40

Ventral surface of head notably flattened; head triangular in cross-section, and head shape similar to that of silverjaw minnow; (*see* 44a). Height of lateral line scales less than 2.0 times their exposed width. Mouth larger and horizontal. Length of upper jaw longer than eye length, except in smallest young. Teeth normally 1, 4–4, 1.
CENTRAL BIGMOUTH SHINER *Notropis dorsalis dorsalis* (Agassiz) . . . . (p. 361)

Ventral surface of head rounded. Lateral line scales greatly elevated, their height more than 2.0 times their exposed width. Mouth smaller and somewhat more oblique. Length of upper jaw about equal to diameter of eye. Teeth 4–4. . . . . 41.

41

Color darker and melanophores more numerous, especially along lateral line. Body depth averaging less than head length; both averaging more than 3.9 times in standard length. Depth of caudal peduncle usually contained less than 2.8 times in head length. Length of caudal peduncle 4.2–5.1 times in standard length. Fins shorter and less falcate. Scales in lateral line, 33–38. Infraorbital canal complete, extending from lateral canal, below the eye, across preorbital to a point in front of nostril.
Notropis volucellus . . . . 42.

Color paler and melanophores few, faint, and small. Body depth and head length about equal; both usually contained less than 3.9 times in standard length. Depth of caudal peduncle contained more than 2.8 times in head length. Length of caudal peduncle 3.8–4.3 times in standard length. Fins higher and more falcate. Body more compressed, its width contained 1.9–2.2 times in depth. Scales in lateral line 30–35; scales greatly elevated, usually about 4.0 times as high as wide. Infraorbital canal undeveloped, or rarely represented by a short section of the tube.
GHOST SHINER *Notropis buchanani* Meek . . . . (p. 373)

42

Melanophores usually many and color pattern prominent. Body more torpedo-like in shape, its depth usually contained more than 4.5 times in standard length. Body width usually more than 1.7 times in depth. Depth of caudal peduncle usually contained more than 2.5 times in head length. Height of depressed dorsal fin averaging more than 2.5 times in predorsal length. Anterior lateral line scales usually about 3.0 (extremes 2.5–3.8) times as wide; (fig. 73 A). Scales in lateral line usually 35–37 (extremes, 34–38).
NORTHERN MIMIC SHINER *Notropis volucellus volucellus* (Cope) . . . . (p. 367)

Melanophores usually fewer and color pattern less prominent. Body usually turgid and robust; its depth usually contained less than 4.5 times in standard length. Body width usually less than 1.7 times in depth. Depth of caudal peduncle usually contained less than 2.5 times in head length. Height of depressed dorsal fin usually averaging less than 2.5 times in predorsal length. Anterior lateral line scales usually about 2.5 (extremes, 2.1–3.2) times as high as wide. Scales in lateral line usually 33–36 (extremes, 33–37).
CHANNEL MIMIC SHINER *Notropis volucellus wickliffi* Trautman . . . . (p. 370)

43

Mid-dorsal stripe distinct, broad, and uniform in width, completely surrounding base of dorsal fin (fig. 65 A); stripe is *not* expanded immediately in front of dorsal origin. Mouth larger; length of upper jaw longer than eye length. Eye smaller; usually contained more than 3.5 times in head length except in smallest young. Anterior dorsal insertion directly over pelvic insertion. Teeth 1, 4–4, 1, or 2, 4–4, 2, or combination of these.
RIVER SHINER *Notropis blennius* (Girard) . . . . (p. 343)

Mid-dorsal stripe variable in width, not surrounding base of dorsal fin (fig. 72 A); stripe is expanded into a wedge-shaped spot directly in front of dorsal origin. Mouth smaller; upper jaw length shorter than eye length. Eye larger; usually contained less than 3.5 times in head length. Anterior dorsal insertion slightly behind pelvic insertion. Teeth 4–4.
SAND SHINER *Notropis stramineus* (Cope) . . . . (p. 364)

44

Cavernous spaces prominent on ventral surface of head and lower cheeks (the superficially similar bigmouth shiner lacks these cavernous spaces; *see* 40 a). Ventral surface of head flattened, giving head a triangular appearance in cross-section. Intestine about as long as head and body. Peritoneum silvery. Scales in lateral line 31–36. Teeth 1, 4–4, 1 or 4–4.
SILVERJAW MINNOW *Ericymba buccata* Cope . . . . (p. 382)

No large cavernous spaces on ventral half of head. Intestine more than 2.0 times the length of head and body. Peritoneum black. Scales in lateral line, 35–38. (The superficially similar common shiner has the intestine only as long as head and body, and anterior lateral line scales greatly elevated; *see* 30.) This species was last taken in Ohio around 1888.
MISSISSIPPI SILVERY MINNOW *Hybognathus nuchalis nuchalis* Agassiz . . . . (p. 385)

45

A small crescent (half moon) of melanophores present on side of snout between anterior third of upper lip and eye, evident even in many small young. Peritoneum silvery. Intestine short and S-shaped, but not coiled or looped. Lateral line complete. Pharyngeal teeth rather strongly hooked.
NORTHERN BULLHEAD MINNOW *Pimephales vigilax perspicuus* (Girard) . . . . . . (p. 388)

No crescent of melanophores on side of snout. Peritoneum black. Intestine long, with one or more coils or loops. Lateral line complete or incomplete. Pharyngeal teeth weakly hooked or not hooked at all. . . . . 45.

46

Lateral line incomplete and very short. Mouth terminal and quite oblique. Body compressed and deeper. Caudal spot and lateral band faint. Cross-hatching absent or indistinct on scales of back.
NORTHERN FATHEAD MINNOW *Pimephales promelas promelas* Rafinesque . . . . (p. 391)

Lateral line complete. Mouth sub-terminal and almost horizontal. Body more torpedo-like in shape. Caudal spot and lateral band dusky and conspicuous. Cross-hatching distinct on scales of back. Breeding male with a tubercle on posterior angle of mouth that superficially resembles a barbel (fig. 82 A).
BLUNTNOSE MINNOW *Pimephales notatus* (Rafinesque) . . . . (p. 394)

Family 14 *Ictaluridae** CATFISHES . . . . (pp. 472–511)

1

Posterior third of adipose fin a free lobe, attached to skin of back only anteriorly (fig. 108) 2.

Entire basal half of adipose fin connected to back, forming a keel-like, fleshy fin, which lacks a distinct, posterior free edge; adipose and caudal fins joined together with only a notch separating them. . . . . 8.

2

Caudal fin deeply forked. Basic body coloration white, silvery or bluish-silver, becoming blue-black in some breeding adults. . . . . 3.

Posterior edge of caudal fin straight, slightly rounded, or slightly notched. Basic body coloration mostly in shades of yellows, browns, and blacks. . . . . 5.

* Formerly Ameiuridae; *see* Taylor, 1954: 43.

3
- Anal fin with 30–36 rays including all rudimentary rays. Anal fin base long; usually contained 2.9–3.3 times in standard length. Distal (outer) edge of anal fin nearly straight. Color milky-white, silvery or bluish, without black spots at all ages. Eyes more ventral, giving the appearance of being situated in the ventral half of head.
  BLUE CATFISH *Ictalurus furcatus* (Lesueur) . . . . (p. 472)
- Anal fin with 30 or fewer rays including all rudimentary rays. Anal fin base shorter; usually contained more than 3.4 times in standard length. Distal edge of anal convex. Eyes more dorsal in position and situated definitely in the dorsal half of head. . . 4.

4
- Anal fin with 24–30 rays including all rudimentaries. Anal base usually contained 3.4–3.7 (extremes, 3.2–4.0) times in standard length. Distal edge of anal moderately convex. Color usually silvery or bluish with blackish spots on body, except in adults which are blue-black and spotless; blackish spots prominent and numerous on body of young less than 12.0″ (30.5 cm) SL; spots becoming faint, but at least a few are usually retained until fishes reach 5–9 lbs (2–4 kg) in weight after which the spots disappear. Many young and small-sized adults have distal edges of anal and caudal fins distinctly bordered with dusky. The large, spotless, dark blue adults, especially the broad-headed males, are the "blue cats" of the fishermen, many of whom consider these large fishes to be a species distinct from "silver, channel, lady, and spotted" catfishes.
  CHANNEL CATFISH *Ictalurus punctatus* (Rafinesque) . . . . (p. 475)
- Anal fin with 19–23 rays (extremes 18–24) including rudimentaries. Anal fin base usually contained 4.3–5.2 (extremes, 4.1–5.4) times in standard length. Distal edge of anal very convex. No distinct dark spots on body. Color usually whitish, bluish or bicolored, except in breeding adults which are blue-black. No distinct dusky border on anal and caudal fins. Introduced.
  WHITE CATFISH *Ictalurus catus* (Linnaeus) . . . . (p. 478)

5
- Anal rays usually 18–27 (extremes, 16–28) including all rudimentary rays. Jaws normally equal in length; the lower rarely longer (these aberrant, long-jawed individuals, formerly considered specifically distinct, were known as *Ameiurus vulgaris*). Head rounded between the eyes. Premaxillary bands of teeth without backward extensions. Adipose fin smaller. Maximum weight under 4.0 lbs (1.8 kg). . . . . 6.
- Anal rays including rudimentaries usually 14–17 (extremes, 13–18). Lower jaw always longer and extending beyond the upper jaw. Head decidedly flattened or even depressed between the eyes. Premaxillary bands of teeth with backward extensions (*see* arrow, fig. 114 A). Adipose fin very large. Young with a pronounced whitish area on dorsal tip of tail, this light area becoming faint or obsolete in adults. Adults of huge size.
  FLATHEAD CATFISH *Pylodictis olivaris* (Rafinesque) . . . . (p. 491)

6
- Chin (mental) barbels whitish without visible melanophores. Anal rays including rudimentaries, usually 25–26 (extremes 23–28). Caudal fin convexly rounded; straighter in small young. Anal fin usually has the basal third light colored; a dusky band, often vague, crossing the middle thirds of the rays and webbings; a dusky border to the margin of the fin. Anal fin neither conspicuously mottled nor with the webbing between rays more or less uniformly dusky.
  YELLOW BULLHEAD *Ictalurus* natalis* (Lesueur) . . . . (p. 481)
- Chin barbels entirely gray or black; or at least sprinkled plentifully with melanophores at their bases. Anal rays including rudimentaries, 16–24. 7.

* Formerly *Ameiurus; see* Taylor, 1954:43.

7

Pectoral spines with sharp, rather stout barbs, teeth, or spinelets on their posterior edges; (fig. 111 A) except in some old adults in which the spinelets have become blunted (to note these barbs, grasp pectoral fins between thumb and forefinger and pull outward; if the spines prick, it is this species). Anal rays usually 22–23 (extremes, 21–24). Black pigment on anal fin usually densest on membranes near their margin; or in spots forming a vague basal bar; or in vague blotching on rays and membranes. In the palest and least mottled specimens the rays and membranes are about equally pigmented. Sides of body usually mottled; in some individuals mottling is intense. No light, vertical bar at caudal base. Adults usually with a whitish or yellowish belly which merges gradually into the mottled sides. Posterior edge of adipose fin often more pointed. Hybridizes with black bullhead.
BROWN BULLHEAD *Ictalurus* nebulosus* (Lesueur) . . . . (p. 484)

Pectoral spines normally without distinct, strong barbs, teeth, or spinelets on their posterior edges (fig. 112 A), although in some young the pectoral is roughened or serrated (when forefinger is drawn outward along posterior edge of spine the smooth or roughened edge does not prick the finger). Anal rays usually 17–21 (extremes, 16–22). Distal two-thirds of anal fin notably darker than basal third, the pigment denser and more uniform on membranes than on rays. Anal fin never mottled, barred, or uniformly pigmented upon both webbing and rays. Adults usually with the belly and lower sides yellow or white, in marked contrast to the darker sides; thus giving the fish a bicolored appearance. The light color of the lower sides extends upward at caudal base to form a light, vertical bar, a most distinctive feature of the adults in life. Posterior edge of adipose fin more rounded.
BLACK BULLHEAD *Ictalurus* melas* (Rafinesque) . . . . (p. 488)

8

No barbs, teeth, or spinelets on posterior edges of pectoral spines (posterior edge of spine smooth to touch when finger is rubbed back and forth across it). Saddle-bands crossing back faint or absent. Basic body colors mostly of yellows, browns, and blue-blacks. . . . . 9.

Strong barbs, teeth, or spinelets on posterior edges of pectoral spines. Saddle-bands across back conspicuous. Basic body colors consist of blacks, grays and whites. Premaxillary bands of teeth without backward extensions. Maximum length 5.0″ (13 cm). 10.

9

Lower jaw much shorter than upper jaw and definitely included. Body terete, especially in young. Light-colored belly usually contrasts sharply with the dark sides resulting in a bicolored appearance, especially in fishes from streams. Adipose fin low. Caudal fin more squarish; its border distinctly lighter in color. Premaxillary bands of teeth with backward extensions (fig. 114 A). Maximum length 12.0″ (30.5 cm).
STONECAT MADTOM *Noturus flavus* Rafinesque. . . . . (p. 494)

Lower and upper jaws equal in length. Body tadpole-shaped. Body and caudal fin brownish, without a bicolored appearance. A conspicuous dark streak along the lateral line; dark lines between myomeres often evident. Adipose fin high and large. Caudal roundish. Premaxillary bands of teeth without backward extensions. Maximum length 5.0″ (13 cm).
TADPOLE MADTOM *Noturus gyrinus*† (Mitchill) . . . . (p. 509)

* Formerly *Ameiurus*; *see* Taylor, 1954:43.

† Formerly *Schilbeodes mollis*; *see* Taylor, 1954:44.

10

Blackish saddle bar on caudal peduncle extending upward to the very edge of adipose fin. Dorsal fin with a roundish, blackish spot on the distal half of the first four rays and their webbings. Body chubby. Caudal fin usually rounded. Eye larger and snout shorter; the eye usually contained 2.0 or fewer times in snout length.
BRINDLED MADTOM *Noturus miurus* Jordan . . . . (p. 506)

Blackish saddle bar on caudal peduncle *not* extending on adipose fin, or if it does, not upward to its distal edge, resulting in a whitish or light border to the adipose fin. When saddle bar extends onto adipose fin it forms a subdistal bar that is sometimes indistinct, especially in the young. Caudal fin squarish. Eye smaller and snout longer; the eye contained 2.0 or more times in snout length, except in smallest young which have very large eyes. . . . . 11.

11

Humeral process (backward extension of the cleithrum) usually absent or very short; posterior edge of process may be somewhat roughened. Posteriormost dark saddle bar *present* on caudal peduncle, but not extending upward onto the yellowish adipose fin. Anal rays 13–16, usually 14. Caudal fin has 17–20 upper simple rays, 14–18 branched rays and 12–16 lower simple rays, totaling 43–54 usually 48–51.
SCIOTO MADTOM *Noturus trautmani* Taylor . . . . (p. 503)

Humeral process well developed, usually longer than width of pectoral spine shaft plus serrae. Posteriormost dark saddle bar extending onto adipose fin, at least half way. 12.

12

Caudal fin without a dark mid-caudal bar or crescent. Humeral process longer than is width of pectoral spine shaft, but never longer than shaft plus serrae (spinelets or barbs). Body and caudal peduncle averaging more slender, depth of caudal peduncle of large young and adults usually contained more than 1.0 times in snout length. Coloration more uniform, the saddle bars and other body markings frequently faint. Caudal rays 39–53, typically fewer than 49. Preoperculomandibular pores usually 10.
MOUNTAIN MADTOM *Noturus* eleutherus* Jordan . . . . (p. 497)

Caudal fin *with* a dark mid-caudal bar or crescent, except in smallest young. Humeral process usually longer than width of pectoral spine shaft plus serrae. Body and caudal peduncle averaging chubbier, the depth of the caudal peduncle of large young and adults usually contained 1.1 times or less in snout length. Coloration bolder, saddle bars and other body markings often contrasting sharply with the ground color. Caudal rays 47–56, usually 49–53. Preoperculomandibular pores typically 11.
NORTHERN MADTOM *Noturus* stigmosus* Taylor . . . . (p. 500)

Family 15 Anguillidae AMERICAN EEL . . . . (pp. 192–94)

Combination of: no pelvic fins but with pectoral and vertical fins present; body lamprey-shaped; jaws present; gill opening reduced to a small slit situated immediately in front of pectoral fin.
AMERICAN EEL *Anguilla rostrata* (Lesueur) . . . . (p. 192)

* Formerly *Schilbeodes*; *see* Taylor, 1954:44.

Family 16 Cyprinodontidae KILLIFISHES and TOPMINNOWS .... (pp. 521–29)

1 Dorsal rays usually 12–14; rarely 11. Origin of dorsal fin distinctly in advance of anal fin. Scales in lateral series, 37–50. Body with dark vertical bars along the sides in both sexes. .... 2.

Dorsal rays usually 7–9, rarely 10. Origin of dorsal fin behind anal fin origin. Scales in lateral series, 28–36. Body with a broad, dark, lateral band which in males tends to break up into a series of short, vertical bars.
BLACKSTRIPE TOPMINNOW *Fundulus notatus* (Rafinesque) .... (p. 527)

2 Vertical bars on sides narrower and more regular in outline; those on caudal peduncle often short but not fused into a median, length-wise band. Scale pockets usually outlined sharply with dark lines, sometimes giving a cross-hatched appearance. Scales in lateral series usually more than 42. Dorsal rays usually 13–14. Anal rays usually 11–12. Bars on sides before dorsal origin in adult males usually 9 or more. Maximum length 4.0″ (10 cm). Introduced.
EASTERN BANDED KILLIFISH *Fundulus diaphanus diaphanus* (Lesueur) .... (p. 521)

Vertical bars on sides broader and less regular in outline; those on caudal peduncle generally fused into a median length-wise band. Cross-hatching of scales poorly defined or absent. Scales in lateral series usually less than 42. Dorsal rays usually 12–13. Anal rays usually 11. Bars on sides before dorsal origin of adult males usually 8 or less. Maximum length, 3.0″ (7.6 cm).
WESTERN BANDED KILLIFISH *Fundulus diaphanus menona* Jordan and Copeland .(p. 524)

Family 17 Peociliidae LIVEBEARERS .... (pp. 530–32)

Combination of: a rounded tail; top of head covered with large scales; premaxillaries protractile; cross-hatchings of scales pronounced; body without vertical bars or lateral band; scales in lateral series, 27–30; anal fin in male modified into an intromittent organ; female does not lay eggs but brings forth its young alive. Introduced.
MOSQUITOFISH *Gambusia affinis* (Baird and Girard) .... (p. 530)

Family 18 Gadidae COD .... (pp. 518–20)

Combination of: one barbel at tip of chin; no dorsal spines although two dorsal fins are present; second dorsal of more than 65 rays; anal fin of more than 60 rays; scales present but so minute as to be almost invisible to the naked eye; somewhat catfish-like in shape.
EASTERN BURBOT *Lota lota lacustris* (Walbaum) .... (p. 518)

Family 19 Percopsidae TROUT-PERCH .... (pp. 515–17)

Combination of: an adipose fin; strongly ctenoid (rough to touch) scales; basic body coloration yellowish; a series of dark spots along lateral line, another series along dorsal ridge, and a third in between these two.
TROUT-PERCH *Percopsis omiscomaycus* (Walbaum) .... (p. 515)

Family 20 Aphredoderidae PIRATE PERCH . . . . (pp. 512–14)

Combination of: an anal opening that in fishes longer than 1.0″ (2.5 cm) SL is in between, or anterior to, the pelvic fins; in smallest young, anus is more posterior but is always farther forward than immediately in front of the anal fin (in all other species of fishes in Ohio the anal opening is situated immediately before the anal fin). Scales strongly ctenoid; no adipose fin.
PIRATE PERCH *Aphredoderus sayanus* (Gilliams) . . . . (p. 512)

Family 21 Atherinidae SILVERSIDE . . . . (pp. 533–35)

Combination of: 2 well-separated dorsal fins, both of which are situated above the long anal base; the first dorsal of 3–7, usually 4, weak, flexible spines; anal fin of one weak spine and more than 20 soft rays; scales in lateral series, 75–79; lateral line very incomplete; body elongate; jaws produced into a short beak; upper jaw protractile.
BROOK SILVERSIDE *Labidesthes sicculus* (Cope) . . . . (p. 533)

Family 22 Serranidae BASSES . . . . (pp. 539–48)

1 The 3 anal spines graduated, the second definitely shorter than the third and not notably stouter. Anal rays 10–12, rarely 13. Coloration bluish-silvery with a series of 6–12 dark, longitudinal stripes along sides and back. . . . . 2.
Second anal spine almost, or as long, as third anal spine and considerably heavier. Anal rays 8 or 9, rarely 10. Coloration more slaty-olivaceous, the longitudinal lines, if present, very faint.
WHITE PERCH *Morone americana* (Gmelin) . . . . (p. 546)

2 Body depth 2.5 (young)–3.2 (large specimens) times in standard length. Anal rays usually 11 or 12, rarely 10 or 13. Second anal spine contained less than 4.0 times in head length. Scales in lateral series normally 52–56 (extremes, 51–58). A series of 6–12 longitudinal lines along sides and back, resulting from dark spotting on scales (absent or faint in smallest young). Hyoid teeth (base of tongue) in a single patch. Adult weight less than 5 lbs (2 kg).
WHITE BASS *Morone chrysops* (Rafinesque) . . . . (p. 542)
Body depth 3.45 (young)–4.2 (large specimens) times in standard length. Anal rays usually 10–12. Second anal spine contained more than 4 times in head length. Scales in lateral series normally 57–67. A series of 7–9 longitudinal lines along the sides and back. Hyoid teeth in two parallel patches. Size large, 60 lbs (27.2 kg) have been recorded. Introduced.
STRIPED BASS *Morone saxatilis* (Walbaum) . . . . (p. 539)

Family 23 Centrarchidae BLACKBASSES, CRAPPIES, SUNFISHES . . . . (pp. 549–602)

1 Anal spines 5–7; usually 6. Dorsal spines 5–9, or 11–13, rarely 10. . . . . 2.
Anal spines 3; very rarely 2 or 4. Dorsal spines usually 10. 4.

2
- Dorsal spines usually 5–8 (extremes, 4–9). Length of anal fin base about equal to length of dorsal base. Branchiostegal rays 7. Preopercle finely serrated. More than 28, long slender gill rakers. Basic colors silvery with bluish and blackish markings.
  Genus *Pomoxis* .... 3.
- Dorsal spines usually 11–12 (extremes, 10–13). Length of anal fin base about one-half (1.6–2.2 times) as long as dorsal fin base. Branchiostegal rays 6. Preopercle smooth or slightly roughened. Fewer than 15 gill rakers of moderate length. Basic body colors chiefly olive-green with brassy reflections. Many of the body scales have a squarish, black, basal spot; these spots form longitudinal rows along the sides which are most conspicuous below the lateral line. Usually 5 rather conspicuous saddle-bands across back.
  NORTHERN ROCKBASS *Ambloplites rupestris rupestris* (Rafinesque) .... (p. 556)

3
- Dorsal spines 5–6 (extremes, 4–7). Length of dorsal fin base much less than distance from dorsal insertion to above eye (fig. 132 A), usually only to above opercle. Chain-like bands on sides usually distinct and regular in shape. Mouth less oblique. Body more slender.
  WHITE CRAPPIE *Pomoxis annularis* Rafinesque .... (p. 549)
- Dorsal spines 7–8 (extremes, 6–9). Length of dorsal fin base equal, or almost equal, to distance from dorsal insertion to above eye (fig. 132 B). Sides of body speckled and mottled, the chain-like bands, if recognizable as such, broken and highly irregular in shape. Mouth quite oblique. Body deeper.
  BLACK CRAPPIE *Pomoxis nigromaculatus* (Lesueur) .... (p. 553)

4
- Body more elongate; body depth usually contained 3.0–4.5 (extremes, 2.5 in largest females–5.0 in smallest young) times in standard length. Scales smaller; more than 55 in lateral line.
  Genus *Micropterus* .... 5.
- Body deeper; body depth usually contained 1.9–2.8 (extremes, 1.7 in largest adults–3.0 in smallest young) times in standard length. Scales larger; fewer than 55 in lateral line.
  Genus *Lepomis* .... 7.
  Note: In some localities as much as 90 per cent of the sunfish population may consist of hybrids of one or more combinations; *see* "List of hybrid combinations."

5. Distal edge of spinous dorsal fin slightly curved. Notch between the 2 dorsal fins shallow; shortest dorsal spine (usually the second-last) at center of notch usually contained 1.2 (largest adults)–1.9 (smallest young) times in length of longest spine (third or fourth). Mouth smaller. In fishes over 6.0″ (15 cm) TL, the posterior end of upper jaw extends only to the posterior edge of eye or slightly beyond; tail is unicolored or bicolored. In fishes less than 6.0″ (15 cm) TL, the mouth is slightly smaller; tail is conspicuously tricolored, with the basal half a whitish-yellow or orange bordered posteriorly, above and below with a dark border, and with a white, terminal bar across the lobes (fig. 135 A). Pyloric caeca typically unbranched. .... 6.

Distal edge of spinious dorsal fin greatly curved (sickle-shaped). Notch between the 2 dorsal fins so deep as almost to divide the fins; shortest dorsal spine (usually the second-last) at center of notch usually contained 2.0 (largest adults and smallest young) –3.3 times in length of longest spine (third or fourth). Mouth larger. In fishes more than 6.0″ (15 cm) TL, the posterior end of upper jaw extends well past the posterior edge of eye and the tail is uniformly slatish or bicolored, with the posterior third a darker slate. In fishes less than 6.0″ (15 cm) TL, the mouth is usually smaller and about the same size as in the spotted blackbass; however, the present species has a rather uniformly colored tail, whereas the tail of the spotted is tricolored. The lateral band is rather uniform in width; band usually wider than eye diameter; is broken up into vertical bars in young; is sometimes absent in oldest adults. Pyloric caeca usually branched at base. Dorsal rays usually 12–13 (extremes, 11–14). Anal rays 11 (extremes, 10–12). Lateral line scales usually 58–69, rarely more.
NORTHERN LARGEMOUTH BLACKBASS *Micropterus salmoides salmoides* (Lacepède)(p. 569)

6. Color pattern extremely variable, may be almost uniformly colored, lightly mottled on sides and back, or with 9 (adults)–16 (small young) vertical bars along the sides. These bars never become confluent to form a lateral band. No longitudinal streaks on sides below lateral line. Caudal spot indistinct or absent in young. Smallest young densely covered with black melanophores; these disappearing when young are about 1.0″ (2.5 cm) TL. Dorsal rays usually 13–15. Anal rays 11 (extremes, 10–12). Lateral-line scales usually 67–81; rarely less. Body deeper, especially in young.
NORTHERN SMALLMOUTH BLACKBASS *Micropterus dolomieui dolomieui* Lacepède(p. 560)

Blotches along lateral line usually forming an irregular, longitudinal black band (somewhat similar to band of largemouth); which is usually solid in young and breaks up into short bars in adults. Scales below lateral line each with a dusky spot at base; these spots forming longitudinal streaks which are most evident above anal fin; streaks least distinct in smallest young and largest adults. A large, black caudal spot in young. Opercle spot in adults usually larger and blacker than in other blackbasses. Tricolored tail of young of less than 6.0″ (15 cm) TL, similar to that of smallmouth. Smallest young never black; always lighter-colored. Dorsal rays usually 12 (extremes, 11–13). Anal rays 10 (extremes, 9–11). Lateral line scales usually 60–68 (extremes, 55–77). Body averaging more slender, especially in young which are near pike-like in shape. Patch of teeth on tongue.
NORTHERN SPOTTED BLACKBASS *Micropterus punctulatus punctulatus* (Rafinesque)(p. 565)

7 Combination of: larger mouth in which the posterior end of upper jaw extends well past the anterior edge of eye (except in smallest young) *and* a short, rounded or bluntly-pointed pectoral fin whose tip when laid forward across the cheek, usually reaches only to, or about to, anterior edge of eye. .... 9.

Combination of: smaller mouth in which the posterior end of upper jaw extends almost to (young), or to (largest adults), the anterior edge of eye, *and* a long, pointed pectoral fin whose tip when laid forward across the cheek, usually reaches almost to tip of snout in adults, and at least to anterior edge of eye in young. .... 12.

8 The stiff opercle bone extends entirely across the opercle flap (but not across the narrow membranous opercle margin); opercle bone so stiff it usually fractures if bent sharply forward. 9

Opercle bone thin, flexible and short, leaving the posterior half of opercle flap largely membranous (opercle flap so flexible it may be doubled forward until its posterior end touches the cheek without fracturing the opercle bone). .... 10.

9 Teeth present on tongue (fig. 138 no. 9). Mouth very large. Posterior end of upper jaw extends to posterior edge of eye in adults; middle of eye in young. Three conspicuous, dark bars radiate backward from eye. Longest gill rakers about 1/3 the eye diameter. The stiff opercle bone extends to the posterior edge of opercle (exclusive of the narrow opercle membrane). Opercle flap black; its margin light olive and white or yellowish; no red. Pectoral fin rounded and short. Soft dorsal mottled with light eye-like spots or ocelli which tend to form oblique rows; breeding males have a brilliant orange spot at base of last three rays. Lower pharyngeal bones narrow, with sharp, conical teeth.
WARMOUTH SUNFISH *Lepomis gulosus* (Cuvier) .... (p. 573)

Rarely teeth on tongue; hybrids between warmouth and this species normally have teeth on tongue (fig. 138 no. 9). Mouth large. Cheeks with emerald mottlings and broken, wavy bars. Gill rakers long and thin, the longest about 1/3 the eye diameter. Opercle flat, stiff and black, its margin bordered with light olive, whitish or yellowish, but not with red. Pectoral fin very short, rounded at tip. Soft dorsal usually plain (sometimes mottled, especially in young), except for a dusky spot at base of last 3 soft rays (spots sometimes faint or absent, especially in small young). Lower pharyngeal bones narrow, with sharp, conical teeth.
GREEN SUNFISH *Lepomis cyanellus* Rafinesque .... (p. 577)

10

Sensory openings along free edges of preopercle unusually large and wide, the greatest width of largest openings usually wider than diameter of anterior nostril opening. Gill rakers long, thin, and straight.
Combination of: large mouth; cheeks and body of male with orange spots, these spots brownish in females and young; gill rakers about 1/3 the eye diameter; opercle flap very flexible, the flexible opercle bone extending only to middle of flap; opercle flap black bordered conspicuously with white, the white having an orange blush in some males; pectoral fin bluntly pointed, its tip extending only to anterior half of eye in adults (shorter in young) when laid forward across cheek; soft dorsal flushed with orange in males, vaguely mottled in females and young; lower pharyngeal bones narrow, with sharply-pointed teeth; maximum size, 4.5″ (11 cm) TL. (Hybridizes freely, particularly with pumpkinseed.)
ORANGESPOTTED SUNFISH *Lepomis humilis* (Girard) .... (p. 584)

Sensory openings along free edges of preopercle small and roundish, their width much less than diameter of anterior nostril opening. Gill rakers short, blunt and often crooked. Combination of: large mouth; cheeks of adults with wavy, emerald-blue bars radiating backward; gill rakers short, almost as wide as long; opercle flap flexible; flap in adults may be doubled forward upon itself; opercle flap black, bordered conspicuously with white; border may contain red spots; pectoral fin short and rounded at tip; soft dorsal flushed with chestnut in males without conspicuous mottlings, vaguely mottled in females and young; lower pharyngeal bones moderately wide, the teeth pointed. (Seldom hybridizes.)
Lepomis megalotis .... 11.

11

Adult males with greatly produced, almost horizontal opercle flaps in which several, small reddish spots may be present in the whitish border. Chain-like bars prominent on sides of smallest young; often obsolete in adults. Maximum total length, 9.0″ (23 cm); largest individuals found in extreme southern Ohio.
CENTRAL LONGEAR SUNFISH *Lepomis megalotis megalotis* (Rafinesque) ...... (p. 588)

Adult males with moderately produced opercle flaps which usually extend upward at a 45° angle, and with a large red spot on the posterior border. Chain-like bars prominent on sides of young; bars often present in adults. Maximum total length 4.0″ (10 cm); largest individuals occurring in Lake Erie tributaries near the crest of the Lake Erie-Ohio River watershed (effects of intergradaton?).
NORTHERN LONGEAR SUNFISH *Lepomis megalotis peltastes* Cope .... (p. 592)

12

Gill rakers long, thin and straight; opercle flap flexible and without red; dark spot on webbings near bases of last 5 soft dorsal rays.
Combination of: small mouth; in larger young and adults, 2 light emerald bars, separated by a dark bar, extending obliquely backward and downward from posterior angle of mouth, giving the "blue-gill" appearance; gill rakers long, about 1/2 eye diameter; opercle flap black and flexible; in adults, flaps may be doubled forward to touch cheek; pectoral fin very long and pointed; black spot on posterior base of soft dorsal conspicuous (except in smallest young), remainder of fin plain; lower pharyngeal bones moderately narrow, the teeth sharply pointed. Young, less than 1.5″ (3.8 cm) TL, have very thin bodies, with 10 or more vertical chain-like bars on sides, and with a few very large melanophores irregularly scattered over the breast. (Hybridizes freely.)
NORTHERN BLUEGILL SUNFISH *Lepomis macrochirus macrochirus* Rafinesque .. (p. 580)

Gill rakers short, thick and often curved; opercle flap has a conspicuous red (young males and adults) or orange (small males and females) spot on the posterior membrane; no dark spot on webbings near bases of last 5 soft dorsal rays. .... 13.

13 Opercle bone rather flexible so that the opercle flap may be bent moderately forward without fracturing the bone. Combination of: small mouth; cheeks with dark mottlings and spots; gill rakers short, blunt and often crooked; opercle flap moderately flexible, its posterior margin flushed with red in adults and orange in young females and small males; pectoral fins pointed and long, reaching almost to tip of snout when laid forward; no distinct spottings on soft dorsal; lower pharyngeal bones wide; teeth bluntly rounded. Introduced. (Hybridizes freely.)
REDEAR SUNFISH *Lepomis microlophus* (Günther) .... (p. 595)

Opercle bone in flap so long and stiff that it fractures if flap is bent sharply forward. Combination of: small mouth; cheeks of adults with wavy, emerald bars; gill rakers short, blunt and somewhat curved; opercle flap stiff, a sharply defined orange (females and young) or turkey-red (males) spot on posterior membrane (may be absent in smallest young); pectoral fin sharply pointed and long; fin reaching almost to tip of snout; distinct spotting on soft dorsal, except in small young; lower pharyngeal bones very wide and heavy; teeth bluntly rounded. (Hybridizes freely.)
PUMPKINSEED SUNFISH *Lepomis gibbosus* (Linnaeus) .... (p. 599)

Family 24 Percidae WALLEYES, PERCH, DARTERS .... (pp. 603–96)

1 Free edges of preopercle strongly serrated (toothed) (fig. 148 C). Tail deeply forked. Branchiostegal rays 7 (rarely 8). No prominent genital papilla. Maximum adult total length usually more than 12.0″ (30.5 cm). .... 2.

Free edges of preopercle smooth (weakly serrated in some individuals of the dusky darter; *see* 12a). Tail rounded or weakly emarginate, never deeply forked (except in large crystal darters; *see* 6a). Branchiostegal rays 6 (rarely 5). Genital papilla large. Darters: maximum length of most species 4.0″ (10 cm) or less; of largest species, 7.0″ (18 cm). 5.

2 Large, sharp, canine teeth on jaws and palatines. Pelvic fins widely separated; distance between pelvic origins almost equal to the width of the base of either fin. Body more terete. When present, the oblique saddle-bands across back and sides are irregular in size and shape. Anal rays 12–13.
Genus *Stizostedion* .... 3.

No canine teeth on jaws. Pelvic fins so close together that the inner edges of their bases almost touch. Body deeper and more compressed. Back and sides crossed with 6–8 black, vertical bands. Anal rays 6–8.
YELLOW PERCH *Perca flavescens* (Mitchill) .... (p. 614)

3 Spinous dorsal fin with round clear-cut, dusky spots that form oblique rows (faint or absent in young under 6.0″ [15 cm]); spinous dorsal without large black, basal blotches on webbings between the last 3 spines. Cheeks usually partly scaled. Rays of soft dorsal 17–21; usually less than 20. Back crossed with 3–4 saddle-bands; these expanded laterally on sides to form 3 oblong blotches, one beneath each dorsal fin and one on the caudal peduncle (saddle-bands very pronounced in Ohio drainage specimens). Pyloric caeca 5–8; the 4 longest much shorter than length of stomach.
SAUGER *Stizostedion canadense* (Smith) .... (p. 603)

Spinous dorsal fin with obscure mottlings on webbings which do not form rows. Prominent, black, basal blotches on webbings between the last 3 spines (faint or absent in young less than 6.0″ [15 cm]). Cheeks usually with few or no scales. Rays of soft dorsal 19–23; usually more than 19. Back and sides usually with 4–14 narrow, dark saddle-bands; bands absent on fishes taken from turbid waters. Pyloric caeca 3; each about as long as stomach.
Stizostedion vitreum .... 4.

4
In life, pelvic fins have a decidedly yellowish cast; body has brassy and yellowish mottlings and reflections. Eyes smaller and farther apart; bony, interorbital width usually contained 1.1–1.4 times in length of eye in fishes less than 10.0″ (25.4 cm) in length, and 0.8–1.2 times in eye length in adults.
WALLEYE (yellow pickerel of the Lake Erie fishermen) *Stizostedion vitreum vitreum* (Mitchill) . . . . (p. 607)
In life, pelvic fins bluish; body grayish-blue without brassy mottlings or reflections. Eyes larger and closer together; bony interorbital width usually contained 1.4–2.0 times in eye length.
BLUE PIKE *Stizostedion vitreum glaucum* Hubbs . . . . (p. 611)
Note: Intergrades between walleye and blue pike are called "gray pike" by the Lake Erie fishermen.

5
One weak, flexible, anal spine. . . . . 6.
Two anal spines which are usually quite stiff; the first often as long or longer than the second. . . . . 9.

6
More than 60 lateral line scales. Coloration pellucid; body occasionally having a slight yellowish cast. Body extremely elongate; depth 7.1–11.0 times in standard length.
Genus *Ammocrypta* . . . . 7.
Between 39–55 lateral line scales. Basic coloration yellowish or straw-colored with dark W-, X-, and V-shaped markings; some of these markings form a series of blotches along the lateral line. Breeding males are flushed with dusky. No reds, greens, or blues. Body deeper; depth less than 7.0 times in standard length. Dorsal spines 7–10. Gill covers slightly joined across isthmus by membrane. Premaxillaries protractile; the upper lip being separated from snout by a groove. Lateral line complete.
JOHNNY DARTERS: genus *Etheostoma* but only in part. . . . . 8.

7
Ventral half of body scaled except for an oblong, naked, ventral strip extending from pelvics to anal fin, its width less than the eye diameter. Broad, oblique, dark, saddle-bands across back; 3–7 in number. More than 80 scales in lateral line. Tail forked in large specimens; notched in young. Anal with one spine and 12–14 soft rays. (Not collected in Ohio since 1900.)
CRYSTAL DARTER *Ammocrypta asprella* (Jordan) . . . . (p. 645)
Ventral half of body almost scaleless, except for 1–3 (rarely 4) scale rows immediately adjacent to the lateral series. Twelve–19 oblong, dusky spots along lateral line; fewer than 80 scales. Anal with one spine; 8–10 soft rays.
EASTERN SAND DARTER *Ammocrypta pellucida* (Putnam) . . . . (p. 647)

8
Nape, cheek, and breast scaleless. Few or no scales on belly immediately behind pelvics. In some adults, especially spring males, the strip from pelvics to anal fin origin is scaleless, except for a bridge of scales before the anus as in the river darter.
CENTRAL JOHNNY DARTER *Etheostoma nigrum nigrum* Rafinesque . . . . (p. 650)
Nape, cheek, and breast scaled. Belly strip from pelvics to anal fin scaled in the usual manner.
SCALY JOHNNY DARTER *Etheostoma nigrum eulepis* (Hubbs and Greene) . . . . . (p. 653)

9
- Combination of: a *complete* lateral line (occasionally a few pores missing) *and* a partly or entirely scaleless strip on midline of belly between pelvic and anal fin insertions; this strip may be entirely scaleless, or with a single row of many-pointed, very large scales, or with a bridge of scales before the anus.
  Genus *Percina* .... 10.
- Combination of: *either* a scaled belly with or without a complete lateral line (9 species), *or* with an incomplete or no lateral line and a scaleless strip between pelvic and anal fin insertions (Tippecanoe and least darters).
  Genus *Etheostoma* (all except johnny darters) .... 18.

10
- Mouth terminal or at most having the lower jaw very slightly included. No conical snout projecting considerably beyond the wholly inferior mouth. Sides of body with no, or fewer than 16, narrow, vertical, dusky bars. Lateral line scales usually fewer than 80. Interorbital space rounded, not depressed. .... 11.
- Mouth decidedly subterminal; the conical snout projecting considerably beyond the wholly inferior mouth. Sides of body with more than 16 narrow, vertical, dusky bars. Lateral line scales usually more than 85 (extremes, 78–103). Interorbital space slightly depressed. .... 17.

11
- Premaxillaries not protractile (fig. 151 A); frenum broad, its width usually 2.0 or more times as wide as the diameter of the anterior nostril (occasionally narrower in the gilt darter). Combination of: usually 12–15 (extremes, 11–16) dorsal spines *and* 55 or more (usually 58–85) lateral line scales.
  Note: Occasional specimens have fewer than 12 dorsal spines and fewer than 58 lateral line scales; such individuals belong to this group *when* the gill covers are broadly connected with membrane (slenderhead darter) or *when* the cheeks are scaleless (gilt darter); if otherwise, go to 11b. .... 12.
- Premaxillaries *either* distinctly protractile (fig. 162 B), or they have a very narrow frenum that is less than 1.5 times the diameter of the anterior nostril, and which is usually crossed by a shallow groove. Combination of: 9–11 (rarely 12) dorsal spines *and* normally 58 or fewer (extremes, 44–60) lateral line scales .... 16.

12
- Gill covers not connected with membrane at the isthmus (fig. 151 B). Distance from junction of gill covers to the moderately sharp snout contained more than 2.0 (usually 2.2–2.8) times in head length, and is usually less than distance from snout to posterior edge of eye. Cheeks and opercles usually partly scaled. Blotches along lateral line usually confluent, never becoming vertical bars and meeting across the back. Young without a tear-drop. Dorsal spines usually 13–15; rarely 12. Lateral line scales usually 65–75 (extremes, 63–81).
  BLACKSIDE DARTER *Percina maculata* (Girard) .... (p. 621)
- Gill covers moderately or broadly connected with membrane across the isthmus (fig. 163 A). Distance from angle of gill-cover membrane to snout usually contained 1.3–2.2 times in head length; usually considerably more than distance from snout to posterior edge of eye. .... 13.

13

- Snout rounded and blunt. Head broadly triangulate; ventral surface of head flat and broad. Body deeper; depth usually contained 1.5 or less times in head length of adults (sometimes more in small young). Blotches along lateral line partly confluent in young, forming, or tending to form, vertical bars in adults. . . . . 14.
- Snout sharply produced. Head narrow laterally, not distinctly triangulate; ventral surface of head rather narrow and gently rounded. Body very slender; body depth usually contained 1.6 or more times in head length of adults. Blotches along lateral line more or less confluent; never completely separated into vertical bars in adults. . . . . 15.

14

- Cheeks partly or entirely scaled. No tear-drop or only a trace. Gill covers rather broadly connected across isthmus with membrane (fig. 163 A); distance from anterior junction or angle of membrane to snout contained less than 2.0 (extremes, 1.3–1.8) times in head length; this distance extending from snout to center of cheek. Caudal base with a broken, vertical bar consisting of 3 dark blotches, the lower 2 often confluent. Preopercle slightly serrated in some individuals. Dorsal spines usually 12–13. Lateral line scales usually 61–65 (extremes, 60–70). No reds or blues.
  NORTHERN DUSKY DARTER *Percina sciera sciera* (Swain) . . . . (p. 618)
- Cheeks usually scaleless; rarely a few scales. A well-defined tear-drop, except in smallest young. Gill covers slightly connected across isthmus with membrane; distance from anterior angle of membrane to snout usually contained 2.0 (extremes, 1.9–2.2) or more times in head length; this distance extending from snout to immediately beyond posterior edge of eye. Caudal base with a single, distinct blotch. Preopercle smooth. Dorsal spines usually 12–13 (extremes, 10–13). Lateral line scales usually 55–60 (extremes, 52–67). Adults have oranges, reds, and blue-blacks; the lateral blotches extending vertically to form bars which meet across the back. (Not collected in Ohio since before 1900.)
  GILT DARTER *Percina evides* (Jordan and Copeland) . . . . (p. 636)

15

- Ventral surface of head with 1–3 dusky spots, the most posterior of which is directly bebelow eye and in adults connects with the suborbital bar to form a sickle-shaped tear-drop. Opercles without scales. Blotches on sides merged into a continuous, dusky lateral band. Gill covers less broadly connected by membrane; distance from anterior angle of membrane to snout usually contained 1.7–2.0 times in head length. Dorsal spines usually 13–14 (extremes, 13–16). Lateral line scales usually 75–80 (extremes, 74–80).
  LONGHEAD DARTER *Percina macrocephala* (Cope) . . . . (p. 624)
- No tear-drop; no prominent markings on ventral surface of head. Opercles with some deeply embedded scales. The 10–16 narrow, lateral blotches only slightly merging together. Gill covers more broadly connected by membrane; distance from anterior angle of membrane to snout usually contained 1.4–1.7 times in head length. Dorsal spines usually 12–13 (extremes, 11–13). Lateral line scales usually 60–70 (extremes, 59–72).
  SLENDERHEAD DARTER *Percina phoxocephala* (Nelson) . . . . (p. 627)

16

Premaxillaries usually bound to snout by a narrow frenum which in occasional specimens is crossed by a shallow groove. Cheeks usually densely scaled. The 10–15 short, vertical, dark bars on sides never meet across back; bars may become somewhat confluent posteriorly along the lateral line. Usually a dusky blotch on webbing between first 2 dorsal spines; 2 more blotches on webbings between last 3 dorsal spines. Midline of belly naked, or crossed before anus with a narrow bridge of scales. Dorsal spines usually 9–11 (rarely 12). Lateral line scales usually 52–58 (extremes, 48–60). Body more compressed.

RIVER DARTER *Percina shumardi* (Girard) . . . . (p. 630)

Premaxillaries usually protractile; when frenum is present it usually is crossed by a shallow groove. Cheeks usually naked, or with an occasional scale. Between 7–14 more or less confluent, horizontal spots along lateral line. Dark, basal blotches on webbings between all dorsal spines. Midline of belly naked, or with a row of enlarged, spiny scales, with such a scale sometimes present between the pelvic fins (fig. 151 B). Dorsal spines usually 10–11 (extremes, 9–12). Lateral line scales usually 48–52 (extremes, 44–56). Body rounder; this species superficially resembling the johnny darter, but has 2 instead of one anal spines.

CHANNEL DARTER *Percina copelandi* (Jordan) . . . . (p. 633)

17

Nape usually closely scaled (fig. 157 A). Vertical bands along sides of body more regular in size and shape, and with little tendency to become confluent along the lateral line.

OHIO LOGPERCH DARTER *Percina caprodes caprodes* (Rafinesque) . . . . (p. 639)

Nape with a completely scaleless triangular area or with a few scattered scales (fig. 158 A). Vertical bands on body more irregular in size and shape; these bands tend to become confluent along the lateral line.

NORTHERN LOGPERCH DARTER *Percina caprodes semifasciata* (De Kay) . . . . . . (p. 642)

18

A deep groove separates tip of upper lip from the rounded, decidedly overhanging and bulbous snout. Groove on side of snout, which separates the maxillary from skin of the preorbital portion of snout, is extremely short, restricted to posterior third of maxillary bone. Snout blunt, rounded and usually slightly overhangs tip of upper lip. Tip of upper lip in large young and adults often produced into a knob or slight projection (fig. 163 A). Mouth almost horizontal. Ventral surface of head flattened, giving head a triangular appearance in cross section. Gill covers broadly joined with membrane across isthmus. Cheeks and opercles usually with a few or many scales, opercles almost invariably with some scales. Belly scaled. Two anal spines. Dorsal spines 12–14, extremes 11–16. Complete lateral line usually with 60–67 scales, extremes 57–70. Color usually light olive green (small young) to green (large young and adults). In small young the lateral line contains a series of blotches; in large young these develop into W-, V-, and Y-shaped markings; in non-breeding adults they merge to form a wavy, confluent lateral band; in breeding adults, especially males, they develop into 5–8 broad, dark green, vertical bars which often meet across the back. Spinous dorsal of breeding adults has a basal band of rufous.

Etheostoma blennioides . . . . 19.

No groove separates tip of upper lip from snout. No knob on tip of upper lip. Anterior half of maxillaries separated from skin of preorbital by a groove. Mouth more oblique. Two anal spines. Belly partially or entirely scaled. . . . . 20.

19
Complete lateral line with 60–72 scales, average 65 (highest in specimens from southern and southeastern Ohio). Transverse scale rows below central portion of dorsal fin to lateral line 7–10, average 8.5. Symphysis of upper jaw usually with a tip. Scattered rusty-red scales on back and sides of adult males, poorly developed or absent (Ohio River drainage and portions of Lake Erie tributaries).
EASTERN GREENSIDE DARTER "Allegheny" type *Etheostoma blennioides blennioides* (Rafinesque) .... (p. 656)
Complete lateral line with 51–63 scales, average 59. Transverse scale rows below central portion of dorsal fin to lateral line 6–8, average 7. Symphysis of upper jaw usually without a tip. Adult males normally have a few to many rusty red scales scattered over the back (Lake Erie and portions of its tributaries).
CENTRAL GREENSIDE DARTER "Prairie" type *Etheostoma blennioides pholidotum* (Miller) .... (p. 660)

20
Lateral line absent or with fewer than 8 pored scales. Scale rows of lateral series usually 32–37 (extremes, 30–39). Dorsal spines usually 6 (extremes, 5–8). Gill covers rather broadly connected by membrane across isthmus. Opercles partly scaled. Belly scaleless from pelvic insertion to half the distance to anal fin insertion. Basic coloration consists of shades of olives and browns. Breeding male pelvic fins greatly enlarged, reaching beyond anal fin insertion. Maximum length less than 1.8″ (4.6 cm).
LEAST DARTER *Etheostoma microperca* (Jordan and Gilbert) .... (p. 694)
Either lateral line is complete, *or* incomplete and with more than 10 pored scales. More than 39 scales in the lateral series (rarely a rainbow or orangethroat darter has only 39) 21.

21
Gill covers broadly connected across isthmus with membrane, the free edge of membrane forming a broad angle or gentle arc (fig. 163 A). Least distance from membrane angle to snout tip usually contained less than 1.6 times in head length; this distance usually extending from snout to, or beyond, anterior edge of opercle (except rarely in some large variegate darters) .... 22.
Gill covers not at all or only slightly connected across isthmus with membrane thus resulting in an acute angle at junction of gill covers (fig. 151 A); distance from this angle to snout tip usually contained 1.7 or more times in head length and usually extends from snout tip to the anterior half or middle of preopercle. .... 24.

22
Mouth slightly subterminal. Snout rounded. Head triangular; ventral surface of head broad and flattish. Dorsal spines usually 10–13, rarely 9. Length of longest dorsal spine, when measured into head, usually extends from snout to posterior half of eye or beyond. Lateral line complete, or with only an occasional pore missing. Posterior edge of tail straight or slightly emarginate. Lower jaw partly included by snout. Humeral scale inconspicuous, not notably larger or darker than adjacent scales; its greatest depth not equal to eye diameter. Dorsal fins of adults contain red; in breeding males the tips of spines are without whitish knobs. .... 23.
Mouth terminal; lower jaw frequently longer than upper jaw. Head long and narrow; more compressed than triangular. Length of longest dorsal spine, when measured into head, usually extends from tip of snout only to the anterior half of eye. Lateral line incomplete; the pores ending beneath the soft or spinous dorsals. Posterior edge of tail rounded. Humeral scale black, large and differing conspicuously from the smaller, browner adjacent scales; its greatest depth about equal to eye diameter. No scales on cheeks or opercles. No reds or blues on breeding males, but there are prominent, whitish knobs on the tips of the dorsal spines.
BARRED FANTAIL DARTER *Etheostoma flabellare flabellare* Rafinesque .... (p. 691)

23

Cheeks and opercles scaled. Dorsal spines usually 10–11, rarely 9 or 12. Membrane connecting gill covers across isthmus has its free edge almost straight or gently curved (fig. 163 A); least distance from its free edge to tip of snout extending from snout tip to center of opercle. The 5–6 quadrate, dark saddle-bands across back usually inconspicuous. Basic body color greenish; lighter and more brownish and silvery in smallest young. Body of breeding male has 10 or more broad, dark-green vertical bands which encircle body. Maximum total length, 2.5″ (6.4 cm).
EASTERN BANDED DARTER *Etheostoma zonale zonale* (Cope) .... (p. 661)

Cheeks and opercles scaleless. Dorsal spines 12–13, occasionally 11. Membrane connecting gill covers across isthmus has its free edge forming an angle or arc (fig. 165 A); least distance from this angle to snout tip usually extending from snout tip to anterior third of opercle (not that far in largest specimens). Four prominent saddle-bands; the first before dorsals; second between dorsals; third at posterior half of soft dorsal; fourth on caudal peduncle. Posterior half of body with 5–7 bluish-green bars between which are orange-red spots; spots absent in smallest young. Adults have a horizontal orange-red bar, surrounded by yellow, extending from base of pectoral fin to before anal. Breeding adult has much red and blue. Maximum total length, 4.5″ (11 cm).
VARIEGATE DARTER *Etheostoma variatum* Kirtland .... (p. 664)

24

Cheeks usually almost fully covered with deeply embedded scales. Scales in lateral series usually 55–60 (extremes 53–62); lateral line short, ending under spinous dorsal, with 27–42 scales of the lateral series without pores. Dorsal spines usually 8–9 (extremes, 7–11). Body slender, like that of johnny darter; its greatest depth 5.4–6.8 times in standard length. Body of young mottled; these mottlings becoming 10–13 quadrate blotches along sides, which in breeding males are bluish-green with the interspaces rusty-red.
IOWA DARTER *Etheostoma exile* (Girard) .... (p. 680)

Cheeks naked except for occasional scale around eye. Lateral line complete, or ending under soft dorsal with fewer than 25 scales in the lateral series without pores. Usually 10–14 dorsal spines. .... 25.

25

Lateral line usually complete, except for an occasional missing pore (pores may be difficult to see). Scales in lateral line usually more than 52, rarely less than 50. Dark, horizontal, wavy lines usually present between scale rows, especially on posterior half of body. Males with carmine or red spots (scales) scattered over sides of body. Body scaled behind pelvic insertion. .... 26.

Lateral line incomplete, the pored scales ending beneath soft dorsal fin (pores usually conspicuous). Scales in lateral series usually 40–50, rarely more than 52. No dark, horizontal lines between scale rows. .... 27.

26

Snout sharply pointed. Dorsal profile from nape to tip of upper lip straight, or with only a slight angle above eye. Head shape similar to that of blackside darter. Lateral line scales usually about 60 (extremes, 56–63). Caudal slightly rounded. Young have sides mottled and the dark, horizontal lines between the scale rows indistinct, as are the two light areas, one above the other, on the caudal fin base. In large young and adults, the soft dorsal and caudal fins are broadly bordered with light, and there is no black subterminal bar.
SPOTTED DARTER *Etheostoma maculatum* Kirtland .... (p. 668)

Snout blunt. Dorsal profile from nape to tip of upper lip curving sharply downward from above eye to snout tip. Head shape similar to that of the rainbow darter. Lateral line scales usually about 55 (extremes, 50–60). Caudal squarish or slightly emarginate. Small young with 10–13 squarish blotches along the lateral line; the dark, horizontal lines between scale rows indistinct, as are the two circular light areas on caudal fin base. In larger young and adults, the soft dorsal and caudal fins are distinctly and broadly bordered with a dark band which is edged with whitish.
BLUEBREAST DARTER *Etheostoma camurum* (Cope) .... (p. 671)

27

Belly behind pelvics normally scaleless; this naked strip extending about halfway from pelvic insertion to anal fin origin, leaving a bridge of scales before anus similar to that of the river darter. Dorsal spines usually 12–13 (extremes, 11–14). No distinct blotches or bars on sides of young or females. Young male has a suffusion of golden -yellow on body; this color in breeding male is intensely brilliant. Some males have 5–7 narrow, light blue, vertical bars on the sides; these are most intense posteriorly. A whitish or orangish bar at caudal fin base, conspicuous at all sizes, is often separated by a dark, median line, giving the appearance of 2 circular "holes." No greens or reds. Maximum length, 1.8″ (4.6 cm).
TIPPECANOE DARTER *Etheostoma tippecanoe* Jordan and Evermann .... (p. 676)

Belly normally scaled. Dorsal spines usually 10–11 (extremes, 9–12). Sides of body heavily mottled, and usually with 5–13 blotches or bars, even in small young. .... 28.

28

Gill covers slightly connected with membrane. Distance from tip of snout to angle of membrane extends from snout tip to well beyond eye (fig. 170 A). Body deepest beneath middle portion of spinous dorsal fin. Usually 13 pectoral rays. The 5–7 vertical, dark bars on posterior half of body usually separated from each other and not widened appreciably along the lateral band; these bars most distinct in adult male; least so in female and small young. Adult male with much red, and orange, especially between the dark bars on body and on anal fin. Throat of breeding male blue-black. Infraorbital canal complete (fig. 170 B).
RAINBOW DARTER *Etheostoma caeruleum* Storer .... (p. 683)

Gill covers connected without a membrane. Distance from tip of snout to gill-cover angle extends from snout tip to within the eye, or to slightly beyond its posterior edge (fig. 171). Body deepest immediately before spinous dorsal. Usually 12 or fewer pectoral rays. The 5–7 blotches on posterior half of body usually triangulate in shape, tending to merge together along the lateral band; forming bars only in some large males. Dark spots on scales form short, horizontal rows; most conspicuous anteriorly. Breeding male has orange but never deep red, and is without orange or red on the anal fin (only green and light yellow). Throat of breeding male orange; never blue-black. Infraorbital canal disconnected beneath eye, leaving an isolated group of 4 pores before eye (fig. 171 B).
NORTHERN ORANGETHROAT DARTER *Etheostoma spectabile spectabile* (Agassiz)(p. 687)

Family 25 Sciaenidae DRUM .... (pp. 697–700)

Combination of: a rounded tail; a carp- or sunfish-like shape; dorsal fins of 8–9 (extremes, 7–10) spines and more than 25 soft rays; first anal spine shorter than eye diameter, the second much longer than snout length and extremely stout; a complete lateral line which continues across the tail (pores often difficult to see).
FRESHWATER DRUM *Aplodinotus grunniens* Rafinesque .... (p. 697)

Family 26 Cottidae SCULPINS .... (pp. 701–10)

1. Lateral line complete. Preopercle spine long and usually curved into a half-circle. Skin sometimes covered with prickles. Body torpedo-shape. Saddle-bands faint or absent.
SPOONHEAD SCULPIN *Cottus ricei* (Nelson) .... (p. 701)
Lateral line incomplete, usually terminating beneath second dorsal fin. Preopercle spine short, stout and only slightly curved. Skin rarely prickly. Body tadpole-shape. Saddle-bands present and usually brownish-black. .... 2.

2. Combination of: a deeper, more robust body and a blunter, more rounded snout; saddle-bands better developed and more clear-cut, a band or bands under the first dorsal fin often present and usually conspicuous; incomplete lateral line averaging longer, usually containing 20–24 pores, rarely fewer; 11 pores along the operculomandibular canal; the series of mottlings along the lower sides above the anal fin usually inconspicuous; anal papilla of adult male moderate in size, never as long as the first anal ray.
CENTRAL MOTTLED SCULPIN *Cottus bairdi bairdi* Girard .... (p. 704)
Combination of: a more slender body, the caudal peduncle especially slender; more pointed snout; saddle-bands under first dorsal fin never well developed, those under second dorsal poorly developed usually; incomplete lateral line averaging shorter, usually containing 13–18 pores, rarely 19 or 20; 10 pores along the operculomandibular canal; the series of mottlings along the lower sides above the anal fin usually conspicuous and comparatively well-defined; anal papilla of adult male long, sometimes exceeding in length the first anal ray. (Some of the above characters were first indicated by Robins, 1954.)
NORTHERN MOTTLED SCULPIN *Cottus bairdi kumlieni* (Hoy) .... (p. 708)

Family 27 Gasterosteidae STICKLEBACK .... (pp. 536–38)

Combination of: 4–6 dorsal spines which are not connected to each other by membrane, but with each spine posteriorly connected to the back with its own membrane; pelvic fins with a large spine and only 1 or 2 weakly developed rays; no scales; caudal peduncle extremely slender, its least depth less than eye diameter.
BROOK STICKLEBACK *Culaea inconstans* (Kirtland) .... (p. 536)

Natural Hybrids: A List, Their Abundance, Characteristics and Some Apparent Causes of Hybridization

The keys for identification of species and subspecies and the analyses of these with their accompanying drawings are sufficiently detailed to identify all normal individuals, excepting the smaller young with undeveloped characters. The keys and species analyses, however, will not "key out" hybrids or aberrant individuals.

A hybrid may be suspected whenever a specimen to be identified "keys out" to two or more species rather than to only one, or whenever the first impulse of the identifier is to identify the hybrid as belonging to one species (usually one of the parents) but, unsatisfied, next considers it to belong to another species (usually the other parent).

As defined here, hybrids are the offspring of a male of one full species and a female of another full species. In this report the term hybrid does not apply to offspring of a male of one subspecies and female of another subspecies, both forms of which belong to the same species, such as the offspring between scaly johnny and central johnny darters. Such offspring are here called intergrades, despite the fact that they, too, are hybrids because they are the product of two recognizably different genetic backgrounds.

Characteristics of Hybrids

The offspring between one full species and another full species are called F_1 hybrids. The offspring between an F_1 hybrid and an individual of either parent species are called backcrosses or B_1 hybrids.

The F_1 hybrids of many plants, and animals other than fishes, often follow simple Mendelian laws of inheritance as regards dominance and recessiveness. In fishes a blending of the parent characters normally occurs (Hubbs, 1940B: 205–9 and Hubbs, Hubbs and Johnson, 1943: 72). This blending in fish hybrids greatly aids identification whenever one or more characters of one parent differs greatly from the same character in the other parent; such as when one parent had a very large mouth, the other parent a very small one, with the result that the F_1 offspring have mouths intermediate in size and in which there is little or no overlap in size between the hybrids and either parent. Even the identification of a hybrid combination not previously seen is relatively easy when several non-overlapping characters are present and especially if one is well acquainted with the specific characters of both parents.

Identification of F_1 hybrids is difficult or impossible, however, whenever all of the specific characters of one parent are but slightly different from those of the other parent, and especially if there is also present a great degree of individual variability in these characters. We have long suspected that hybridization in nature may occur between the various closely related and superficially similar species of carpsuckers, the black and golden redhorses, the hornyhead and river chubs, rosefin and northern redfin shiners, sand and mimic shiners, blacknose and pugnose shiners, and rainbow and orange-throat darters.

Realizing that simple blending of characters is the rule among fishes, Hubbs and Kuronuma (1942: 291) in 1942 introduced a method of evaluating hybrid characters to which later Hubbs, Hubbs and Johnson (1943: 7) gave a formula. In this formula the rating of a character of one putative species was given the value of 0, the same character in the other putative species was given the value of 100. This character in the F_1 hybrid, since it is more or less intermediate, usually has a value near or at 50 (extremes 35–65). Whenever several of these intermediate characters are totaled (called "summation of characters" by Van Denburgh, 1920: 10), their average is usually close to, or is, 50. This method of evaluating hybrids, or modifications of it, has been used frequently in recent years (Hubbs and Miller, 1943; Hubbs and Hubbs, 1947; Trautman, 1948).

Recent unpublished studies of large numbers of hybrids of a single combination, together with the parent species, have shown that some characters are not exactly intermediate, but consistently favor one parent to a slight degree; such characters seldom exceed a value of 65.

In addition to intermediacy in characters, the F_1 hybrids usually differ from full species, and intergrades between two subspecies of the same species, in their greatly unbalanced sex ratios and degree of sterility. Among sunfishes, for instance, between 75%–100% of the F_1 hybrids are males, and these hybrid males and the few hybrid females are usually sterile (Hubbs and Hubbs, 1933: 629–38). The sterile F_1 hybrid males build nests, guard them, and mate with females of full species, or with hybrid females but normally produce no offspring.

A notable exception to the above is the carp-goldfish F_1 offspring. Considerable circumstantial

Natural and Artificial Hybrid Combinations Recorded from Ohio or Elsewhere

One parent species	Other parent species	Third parent species	Where recorded: Ohio	Elsewhere	Page
Lampreys					
Silver lamprey	Ohio lamprey		x	x	145, 151
Herrings					
Gizzard shad	Threadfin shad			x	201, 205
Salmon					
Chinook salmon	Coho salmon			x	213, 215
Trouts					
Rainbow trout	Brook trout			x	223, 226
Brown trout	Brook trout			x	226, 220
Brook trout	Lake trout			x	226, 228
Whitefishes					
Cisco	Lake whitefish		x	x	231, 236
Pikes					
Grass pickerel	Chain pickerel		x	x	246, 248
Grass pickerel	Northern pike			x	246, 248
Chain pickerel	Northern pike		x	x	246, 248
Northern pike	Muskellunge		x	x	250, 253
Minnows (exotics)					
Carp	Goldfish		x	x	260, 263
Minnows (native)					
Hornyhead chub	Striped shiner		x	x	270, 338
Hornyhead chub	Common shiner		x	x	270, 341
Hornyhead chub	Stoneroller minnow		x	x	270, 399
River chub	Blacknose dace		x	x	274, 292
River chub	Longnose dace		x	x	274, 294
River chub	Creek chub		x	x	274, 299
River chub	Striped shiner		x	x	274, 338
River chub	Common shiner		x	x	274, 341
River chub	Stoneroller minnow		x	x	274, 402
Streamline chub	Gravel chub		x		282, 285
Blacknose dace	Longnose dace		x	x	292, 294
Longnose dace	Creek chub		x	x	294, 299
Longnose dace	Common shiner (*see* Ross and Cavender, 1977: 777–78)		x		294, 341
Longnose dace	Stoneroller minnow		x	x	294, 299
Creek chub	Redbelly dace			x	299, 311
Creek chub	Redside dace		x	x	299, 311
Creek chub	Rosyside dace		x		299, 312
Creek chub	Striped shiner		x	x	299, 338
Creek chub	Common shiner		x	x	299, 341
Southern redbelly dace	Redside dace		x	x	307, 311
Southern redbelly dace	Rosyside dace		x		307, 312
Southern redbelly dace	Striped shiner		x		307, 311
Southern redbelly dace	Common shiner		x		307, 341
Southern redbelly dace	Stoneroller minnow		x		307, 399
Redside dace	Striped shiner		x		311, 338
Redside dace	Common shiner		x		311, 341
Rosyside dace	Common shiner		x		312, 341

One parent species	*Other parent species*	*Third parent species*	*Where recorded*		
			Ohio	*Elsewhere*	*Page*
Silver shiner	Striped shiner		x		321, 338
Rosyface shiner	Striped shiner		x	x	327, 338
Rosyface shiner	Common shiner		x	x	327, 341
Rosyface shiner	Spotfin shiner		x		327, 357
Redfin shiner	Rosefin shiner		x		332, 333
Striped shiner	Stoneroller minnow		x	x	338, 399
Common shiner	Stoneroller minnow		x	x	338, 399
Steelcolor shiner	Spotfin shiner		x		352, 358
Bigmouth shiner	Sand shiner			x	361, 365
Bigmouth shiner	Mimic shiner			x	361, 369
Sand shiner	Mimic shiner		x	x	365, 369
Bullhead minnow	Bluntnose minnow		x	x	390, 396
Fathead minnow	Bluntnose minnow		x		393, 396
Suckers					
Bigmouth buffalofish	Smallmouth buffalofish		x		406, 413
Common white sucker	Longnose sucker			x	459, 461
Catfishes					
Blue catfish	Channel catfish			x	472, 475
Channel catfish	Flathead catfish			x	475, 491
Brown bullhead	Black bullhead		x	x	485, 488
Brindled madtom	Tadpole madtom		x	x	506, 509
Basses					
Striped bass	White bass			x	540, 543
Striped bass	White perch			x	540, 546
Sunfishes					
White crappie	Black crappie		x	x	551, 554
Black crappie	Rockbass			x	554, 557
Black crappie	Largemouth blackbass			x	554, 571
Black crappie	Warmouth			x	554, 574
Black crappie	Bluegill			x	554, 581
Rockbass	Warmouth			x	557, 574
Rockbass	Bluegill			x	557, 581
Smallmouth blackbass	Spotted blackbass		x	x	562, 567
Smallmouth blackbass	Largemouth blackbass		x	x	562, 571
Spotted blackbass	Largemouth blackbass		x		567, 571
Smallmouth blackbass	Spotted blackbass	Largemouth blackbass	x		562, 567, 571
Largemouth blackbass	Warmouth			x	571, 574
Largemouth blackbass	Bluegill			x	571, 581
Warmouth	Green sunfish			x	574, 578
Warmouth	Bluegill		x	x	574, 581
Warmouth	Redear sunfish			x	574, 596
Warmouth	Pumpkinseed			x	574, 600
Green sunfish	Bluegill		x	x	578, 581
Green sunfish	Orangespotted sunfish		x	x	578, 585
Green sunfish	Longear sunfish		x	x	578, 589
Green sunfish	Redear sunfish		x	x	578, 596
Green sunfish	Pumpkinseed		x	x	578, 600
Bluegill	Orangespotted sunfish		x	x	581, 585
Bluegill	Longear sunfish		x		581, 589
Bluegill	Redear sunfish		x	x	581, 596

One parent species	Other parent species	Third parent species	Where recorded		
			Ohio	Elsewhere	Page
Bluegill	Pumpkinseed		x	x	581, 600
Orangespotted sunfish	Longear sunfish		x	x	585, 589
Orangespotted sunfish	Pumpkinseed		x	x	585, 600
Orangespotted sunfish	Redear sunfish			x	585, 596
Longear sunfish	Redear sunfish			x	589, 596
Longear sunfish	Pumpkinseed			x	589, 600
Redear	Pumpkinseed		x	x	596, 600
Perches-Darters					
Sauger	Walleye		x	x	604, 609
Dusky darter	Logperch			x	620, 640
Dusky darter	Greenside darter			x	620, 658
Dusky darter	Rainbow darter			x	620, 685
Dusky darter	Orangethroat darter			x	620, 688
Blackside darter	Logperch darter		x	x	622, 640
Slenderhead darter	Blackside darter			x	622, 628
Slenderhead darter	Logperch darter			x	628, 640
River darter	Logperch darter			x	632, 640
Channel darter	Logperch darter			x	635, 640
Logperch darter	Greenside darter			x	640, 658
Logperch darter	Orangethroat darter			x	640, 688
Banded darter	Rainbow darter			x	663, 685
Bluebreast darter	Tippecanoe darter		x		673, 678
Rainbow darter	Orangethroat darter		x	x	685, 688

and some factual (Matsui, 1931: 135–37) evidence indicates that these offspring and their backcrosses are almost or entirely as fertile as are the parent species. Outside Ohio waters fertile F_1 offspring occur in many families of fishes, notably among the Cyprinodonts. In this family as many as five species have been incorporated into a single individual.

Abundance of Hybrids

The literature contains relatively few references to hybrids prior to 1925. When mentioned, hybrids were frequently considered to be full species and were so described. The green-pumpkinseed sunfish hybrid, described by McKay (1881: 89) as *Lepomis euryorus*, is an example. Fishermen, however, have long recognized some hybrids as such, calling them "mules," "crosses," "jumbos," "mongrels" and "bastards."

Since 1925 a greatly increased number of hybrid combinations have been recognized. In Ohio the abundance varies from one known F_1 individual in a combination such as the one redfin-chain pickerel specimen, to many F_1 individuals such as the green-pumpkinseed sunfish hybrids, which are numerous locally. Because of the abundance, and practicability of rearing some hybrid combinations, they have attained a degree of economic importance (Ricker, 1948: 84–96).

Causes of Hybridization

Formerly it was generally believed that hybrids were largely accidents of nature, occurring without rhyme or reason. Recent studies have demonstrated that certain environmental factors favor an increase in amount of hybridization, especially such factors as overcrowding, removal of isolating barriers, changes in environmental conditions, and presence of much submarginal habitat for one or both parent species (Hubbs and Miller, 1943: 376).

Sunfish hybrids offer an excellent example. These hybrids may be very few in number in one lake, whereas in a neighboring lake hybrids comprise more than 75% of the sunfish population. Investigations of those lakes containing many sunfish hybrids disclose that one or more of the following factors prevail:

(1) Favorable conditions for survival and growth result in overcrowding (Bailey and Lagler, 1938: 605).

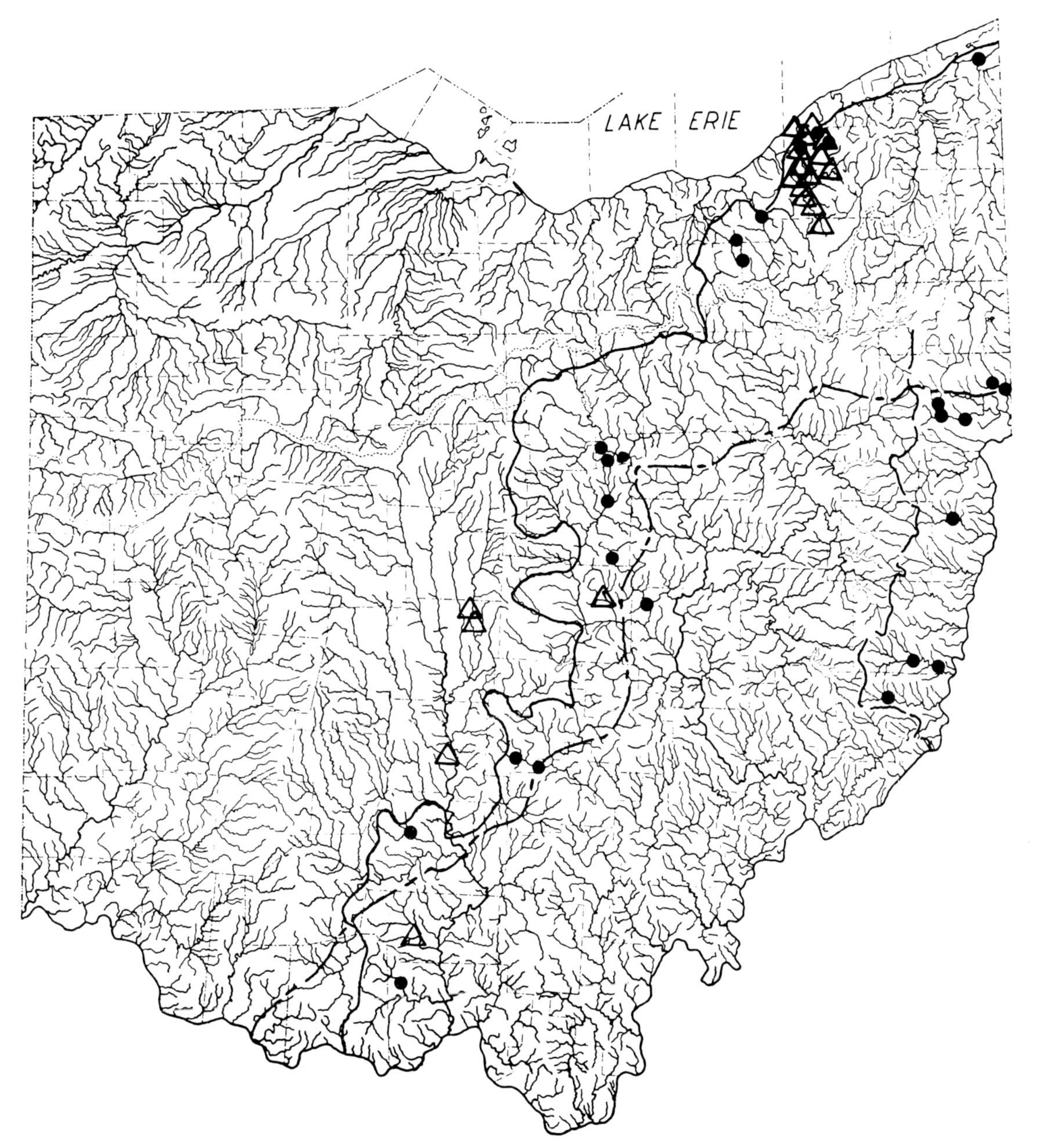

MAP VIII. Common and/or striped shiner × rosyface shiner hybrids *Notropis cornutus* and/or *chrysocephalus* × *Notropis rubellus*
~ Allegheny Front Escarpment
Glacial boundary
Flushing Escarpment
△ Since 1952

(2) Adults of two or more species are forced to spawn in overcrowded areas or where submarginal spawning conditions prevail for one or more species.

(3) One parent species is very numerous, the other is represented by only a few individuals, resulting in difficulty of the latter to find mates (Hubbs, Walker and Johnson, 1943: 19–20 and Hubbs and Hubbs, 1947: 164–65).

(4) Another species of sunfish, formerly absent, is introduced.

(5) A rapid change in ecological conditions oc-

curs, such as the rapid silting-over of the bottom and/or subsequent killing of the large aquatic plants.

Some stream sections contain many hybrids of a given combination whereas other sections do not. Investigations of such hybrid-producing waters often disclose that hybridization is usually influenced by unfavorable stream gradients, and/or recently increased turbidity and siltation have resulted in the increase in amount of submarginal habitats.

In my unpublished studies of some stream-inhabiting hybrid combinations, I found that the common, and/or striped, -rosyface shiner F_1 hybrid apparently is the result of the parent species spawning in waters more swiftly flowing than are those waters in which each species normally spawns. Since such swiftly-flowing waters are restricted to certain physiographical conditions which in Ohio form a distinct pattern, the hybrids likewise show this same distributional pattern. As shown on map VIII, the common, and/or striped, -rosyface shiner hybrids are restricted to those very high gradient streams which lie immediately east of the Allegheny Front and Flushing escarpments.

The blackside-logperch darter F_1 hybrid is apparently the result of the parent species spawning in stream sections where riffles are poorly developed, their bottoms are largely silt-covered, and their currents are less swift than in those riffles in which these species normally spawn (Trautman, 1948: 171–73).

Occasional fish collections contained large series of hybrids (in one instance several hundred), dozens of one parent species, but only a few or no individuals of the other parent species. In each case submarginal conditions prevailed for the parent species which was the least numerous, and it seemed probable that the more numerous parent species was "swamping out," or eliminating through hybridizing, the other parent species; *see* Distribution under smallmouth buffalofish.

Hybridization, Years 1955–80—There was a gratifying increase in our knowledge of natural hybridization among the freshwater fishes of North America. In Ohio, as elsewhere, several heretofore unrecognized interspecific and intergeneric hybrid combinations were recorded. Many small and large hybrid populations were discovered. Principal reasons for recognition of the new hybrid combinations and for the increase in natural hybrid populations were: (1) increased investigations of natural and artificial hybridization and (2) continued increase in the eroding of the original, natural environment resulting in a more artificial one.

During the period investigators made many attempts to artificially hybridize combinations of species, the results ranging from considerable success to none. Egg fertility or viability of young ranged widely between one cross and its reciprocal; the male of species "A" and female of species "B" resulted in viable young whereas in the reciprocal cross the eggs were infertile or the resultant young not viable.

Families of fishes found in Ohio in which one or more hybrid combinations have been recorded, either from Ohio or elsewhere, are as follows:

Petromyzontidae: *see* Hubbs and Trautman, 1937.

Coregonidae: *see* Koelz, 1929; Minckley and Krumholz, 1960.

Salmonidae: natural and artificial hybrids were reported between species of salmons, brook and brown trouts (Brown, 1966) and brook and lake trouts (Slastenenko, 1954).

Esocidae: natural and artificial hybrids were reported between the various species of this family (Bailey, 1938; Black and Williamson, 1947; Eddy, 1941; Eddy and Surber, 1943; Embody, 1918; Greeley, 1938; Greeley and Bishop, 1933; Hubbs, C. L., 1955; Raney, 1957; Underhill, 1939; Weed, 1927; *see* also p. 120).

Cyprinidae: A large amount of information accumulated during this period relative to natural and artificial hybridization between cyprinids. Some previously unrecorded interspecific and intergeneric hybrid combinations were captured in Ohio or elsewhere (Bailey and Gilbert, 1960; Cross and Minckley, 1960; Distler, 1968; Gilbert, 1961; Hubbs, C. L., 1955; Hubbs, Clark, 1959; Hubbs and Bailey, 1952; Hubbs and Laritz, 1961; Hubbs and Strawn, 1957; Pflieger, 1965; Raney, 1940C; *see* also p. 120).

Catostomidae: Several combinations of cyprinid hybrids were reported, as were mass hybridizations between species in the genus *Catostomus* and between *Catostomus* and the genera inhabiting the western United States (Hubbs and Hubbs, 1947; Hubbs, Hubbs and Johnson, 1943; Hubbs and Miller, 1953; Nelson, 1968; *see* also p. 120).

Siluridae: A few hybrid combinations of catfish species have been recorded, such as between species of *Noturus* and species of *Ictalurus*. Mass hybridization was recorded between species of *Ictalurus* (Giudice, 1966; Taylor, 1969; Trautman, 1948; *see* also p. 120).

Centrarchidae: Much information was accumulated in this family relative to natural and artificial hybrids, both interspecific and intergeneric. Many previously unknown hybrid combinations were recorded. With few exceptions all possible hybrid combinations among the Lepomines were recognized.

Probably the most outstanding among the intergeneric hybrids was the "Bluebass" (Saldana, 1971). This hybrid between the bluegill and the largemouth blackbass was captured in Puu Ka Ele and Morita reservoirs on the Hawaiian Island of Kauai. To my knowledge no natural hybrids between these species have been recorded in North America, although young have been artificially produced. It appears to be unknown whether the Hawaiian hybrids can reproduce.

It is intriguing to speculate as to why these two native American species should naturally hybridize in Hawaii, where they were rather recently introduced, and have not done so in their original range in America, where opportunities seem to be so abundant. This brings to mind that in their native Asia the intergeneric carp and goldfish hybridize with extreme rarity, whereas hybridization occurs abundantly between these two species in America, where they have been introduced.

Hybridization of the various species of the genus *Micropterus*, both natural and artificial, has recently received much attention. The Illinois Natural History Survey (1968:74) reported that W. F. Childers artificially produced hybrids between the largemouth and the smallmouth. Their hybrid young successfully spawned at one year of age, whereas in Illinois the young of the largemouth or the smallmouth normally do not spawn until 2 years old.

On April 16, 1970, George Billy, Samuel Leach, Orion Johnson and I captured several largemouth × spotted hybrids in a trap net set in Lake White, Pike County, Ohio. These specimens were intermediate in coloration and scale counts, and had a small patch of teeth on the tongue. However, they were more similar in body form to that of the largemouth. Examination of the scales indicated the rapid growth usual among hybrids. Billy had for several years past captured similar hybrids in Lake White.

Another interesting development in Ohio has been the production of "Hodgen" hybrids. Charles F. Hodgen has two large ponds at his home near Franklin, Warren County, Ohio. In 1960 he drained one of these ponds restocking it with what he believed were all smallmouth blackbass, captured in nearby streams, but included was at least one female spotted bass. On May 8, 1962, he observed a female spotted blackbass spawning with a male smallmouth. He introduced blackbass young into a pond, catching some the next year, which by their rapid growth indicated hybrid vigor. As adults they and their young have produced well.

On December 4, 1968, he gave to the state collections 16 mature fishes of three age groups which were obviously hybrids between the above species. These preserved specimens have been examined by Stephen Taub and me; we found normal appearing gonads and a sex ratio of 1:1.

In the spring of 1969, Hodgen graciously gave breeder smallmouth × spotted blackbass hybrids for a project sponsored by the Ohio State University, Ohio Cooperative Fishery Unit and the Hebron National Fish Hatchery, the latter providing space and assistance. Young obtained from these breeders were stocked in four private ponds, maturing in 1972 and some of them producing offspring in one pond.

Hodgen has kindly permitted me to conduct observations upon the hybrids in his ponds. He had removed some smallmouth × spotted blackbass hybrids from one pond, placing them in a large pond containing smallmouth and largemouth blackbasses. In 1971 and since we have found specimens 12″–20″ (30–50.8 cm) long in this pond which had teeth on the tongue as well as characteristics of all three species, with largemouth blackbass characters sometimes predominating. These fishes gave every indication of being tri-hybrids.

The hybrids appear to have advantages over individual species of blackbasses. They offer a substitute for the largemouth blackbass–bluegill combination in farm ponds, grow more rapidly than do either parent species, and may aid in controlling over-populations of bluegills because of their voracious appetites; and because they frequently jump out of the water and have greater fighting qualities, they offer more sport to the fisherman than does the largemouth. Obviously further experimentation with bass hybrids is needed and a desirable strain of hybrids developed. However, these voracious feeders may be too easily overfished (Birdsong and Yerger, 1967; Childers, 1967; Childers and Bennett, 1961; Hubbs, C. L., 1920; Hubbs, Carl, 1955; Hubbs and Hubbs, 1931; Hubbs and Hubbs, 1932A; Hubbs and Hubbs, 1932B; Hubbs and Hubbs, 1933; Hutchins, 1969; Krumholz, 1950; Lagler and Steinmetz, Jr., 1957; Luce, 1937; Moenkhaus, 1911; Natural History Survey, 1968; Ricker, 1948; Thompson, 1935; West and Hester, 1966; *see* also p. 120).

Percidae: Since 1950 there has been sustained interest in natural hybrids, and in the artificial hybridization between many species of Percidae. Much valuable information has been accumulated. A large number of artificial hybrids have been produced, some of which were thought to be impossible a few years ago. In 1959 Clark Hubbs (50) wrote that Strawn and he had concluded from their experiments that "any hybrid combination can be made between any two species of darter" (Distler, 1968; Hubbs, 1959; Hubbs and Laritz, 1961A and 1961B; Hubbs and Strawn, 1957A and 1957B; Whitehead and Wheeler, 1966; *see* also p. 120).

Part VI

Preserved Fish Collections and Numbers of Fishes Examined

Fortunately there exists a large number of preserved fishes, the majority of which are deposited in the following collections:

Institution housing collection	*Abbreviation of institution collection*	*Years when bulk of material was collected*	*Principal collectors*	*Approximate number of fishes examined by me*
California Academy of Sciences*	CAS	1870–1900	Philip H. Kirsch Seth E. Meek C. Kendall U. S. Fish Commission	500
Chicago Natural History Museum	CNHM	1860–1900	Charles H. Gilbert Seth E. Meek A. J. Woolman J. W. Milner	650
Cincinnati Society of Natural History	CSNH	1860–1900	Charles Dury James A. Henshall Josua Lindall Charles H. Gilbert	500
Marietta College	CM	1925–1935	H. R. Eggleston Edward Warner	100
Oberlin College Museum	OCM	1880–1895	L. M. McCormick Philip H. Kirsch	400
Ohio State Museum	OSM	1920–1938	James S. Hine Robert B. Foster Edward L. Wickliff Milton B. Trautman	9,500
Ohio State University†	OSU	1939–1969	Edward L. Wickliff Edward C. Kinney, Jr. Ted Cavender	67,000
Ohio State University Museum of Zoology	OSUM	1970–1980	Mary A. Trautman Milton B. Trautman	145,000
Stanford Natural History Museum	SNHM	1870–1900	Charles H. Gilbert Philip H. Kirsch Seth E. Meek U. S. Fish Commission	300
University of Akron	UA	1925–1939	Walter C. Kraatz	300
University of Michigan Museum of Zoology	UMMZ	1920–1950	Walter Koelz J. Hilary Deason Carl L. Hubbs Milton B. Trautman	10,500
United States National Museum	USNM	1850–1901	J. W. Milner Spencer F. Baird U. S. Fish Commission	3,500

* The CAS Collections of Ohio fishes have been transferred to the UMMZ collections.

† The Ohio State University collection has been presented to the Ohio Historical Society and was merged with the Ohio State University Museum of Zoology collection at Columbus in 1970.

Survey records indicate that in addition to the 93,250 preserved specimens mentioned above, I also identified between 50,000 and 60,000 preserved specimens belonging to other institutions and to private individuals. After identifying these specimens, they were returned to their owners or discarded at the owners' request. Many thousands of specimens from the Wayne, Ashland, and other county and stream surveys were reluctantly discarded because of lack of permanent storage space, time and money necessary to properly catalog and permanently preserve them. Many thousands of fishes, collected by Edward L. Wickliff during the 1920–25 surveys, had to be discarded because of too long preservation in formalin.

During the 1925–50 period I made more than 2,000 collections, collecting in more than 1250 of the 1420± townships.* While collecting I normally released more than 90% of the catch, often saving none of some species, one or a few individuals of other species, and saving larger series only of those species needed for special study. The survey records show that during the 25-year period as many as 15,000 individuals were released during a 3-hour collecting period in one locality, and a total of between 400,000 and 600,000 fishes were captured, identified and returned to the water.

I also examined many thousands of Lake Erie fishes, while they were in the fish boats or in the fish houses, and Ohio River fishes in the nets and live boxes of the river commercial fisherman. In addition many hundreds were examined which were caught by sport fishermen in all parts of the state.

I have incorporated likewise in this report a relatively few locality records of fishes which I did not identify myself but which were adequately described to me by others. These records included only those species which could not be confused with other species, such as the paddlefishes, large muskellunges, and blue catfishes and alligator gars too large to belong to other species.

During the early years many specimens, then considered as belonging to a single species, were identified upon the spot and returned to the water or if preserved were soon discarded. Later the supposed single species were found to be a complex of two or more species. When this situation was realized, those distribution records of such complexes without specimens were discarded whenever doubt existed as to their specific identity.

During a final check up of all distributional records, those were discarded which did not fit into the distributional pattern for their species if they were not substantiated by specimens. This procedure was considered necessary despite the high probability that most of these records were of correctly identified strays.

* The number of Ohio townships decreased as some townships were incorporated into municipalities.

Methods by Which Fishes and Data Were Collected

The fishes upon which this report is based were collected primarily with the aid of Common Sense seines of 4′ and 6′ (1 and 2 m) in length, and bag seines of 15′ and 20′ (4.6 and 6.1 m) in length, the latter containing a long bag in their centers. Thousands of fishes were captured with the aid of gill, trammel, fyke and trap nets, and trawls, trot lines and angling. Hundreds were taken after having been stunned or killed by excessive amounts of domestic, industrial and soil pollutants; others, dying or dead were picked up along Lake Erie beaches.

The majority were taken during the warmer months and in the daylight hours. Representative samples were taken in the winter and at night.

The following data were usually secured at the time the collection was made: exact locality, such as name of stream, pond or other waters, the stream system and drainage in which it belonged, and the county and townships (section when possible); water conditions, including degree of turbidity, amounts and types of pollutants evident; present water level, and estimate of level during the normal low periods and the height during flood crests, the latter determined by observing the debris in the bank vegetation; an estimation of widths and depths of waters during normal conditions in summer; gradient, whether high, moderate or low; types of bottom and percentage of each type; absence of aquatic vegetation, or amounts of each kind if present; kinds of land vegetation bordering the waters such as an elm-ash-maple, or willow-pioneer weed, or sycamore-elm association; type of valley and flood plain, whether narrow or wide; fishing gear used; names of collectors; time of day; the day, month and year; pertinent remarks such as life history; habitat and behavior data, and color descriptions of living fishes.

In the 1840–1950 period, recorded collections were made at more than 2,500 localities, *see* "Distribution of Collections" map (map IX). Only a single collection was made in about 40% of the localities, and two or more were made upon different days or years in the remaining 60%. Altogether more than 6,000 recorded collections were made at the 2,500 stations, of which fewer than 300 collections were made prior to 1900; fewer than 50 between 1901–19;

MAP IX. Distribution of collections made in Ohio between 1840 and 1955. Each dot represents a locality from which one to several hundred collections were made

more than 2,900 between 1920–38; more than 2,000 between 1939–50. At least one species of fish was recorded at each collecting station, except for about 35 where the waters were so polluted at the time as to be apparently devoid of fish life.

At each of approximately 800 stations a period of two hours or less was spent in investigations; more than two hours were spent at each remaining station.

Years 1955 to 1980. More than 1,000 collecting stations were investigated during this period, some during several dozen days by many individuals and/or organizations and for several hours per day. It is impossible to estimate how many part or entire

MAP X. Distribution of collections made since 1955. Each triangle represents a locality from which one to several dozen collections were made

day collections were made but the total number was obviously more than 2,000.

Areas Thoroughly Investigated and Some Results

County collections: Before 1951, the number of collecting stations per county ranged from 15 (Harrison County) to 64 (Ottawa County), with a county average of slightly more than 29. More than 32 separate collections were made in each county. Counties containing fewer than 20 collecting stations had so few because, like Highland, Medina, Shelby and Warren, they were adjacent to counties which had been thoroughly investigated and it was assumed that their fish faunas were comparatively well known; and/or like Harrison or Cuyahoga their faunas had become pauperized and aberrant because of excessive pollution and because so few suitable collecting sites remained; and/or watershed counties like Crawford, Huron, Morrow, Seneca and Wyandot because their waters lacked a wide diversity of habitats. Greater emphasis was placed upon a stream system as a unit than upon political boundaries.

Ottawa County: Since 1918 many persons and agencies, including faculty and students of Ohio State University, and members of the Ohio Department of Natural Resources and of the U.S. Fish and Wildlife Service, have investigated its waters. A county total of 92 species and four additional subspecies of fishes have been recorded. In an area of about 9 square miles surrounding South Bass Island (Langlois, 1949: 1–57) a total of 80 species and two additional subspecies were recorded during the 1939–53 period, of which 30 species were captured chiefly in the gear of the commercial fishermen, these including such species as silver lamprey, lake sturgeon, cisco and whitefish; 45 species might be captured annually with experimental gear, hand seines and trawls and included minnows, darters, sunfishes, bullheads, trout-perch, and young of the deeper water species; 25 species were so few in numbers and/or so difficult to capture as to be taken upon fewer than eight days, such as the black redhorse and spoonhead sculpin; several species apparently had become extirpated after 1945 or were so reduced in numbers as to have been impossible to capture, such as the pugnose shiner; others like the alewife and orangespotted sunfish invaded the area after 1939.

Since 1951 many collections have been made annually in Ottawa County, and especially in the vicinity of the Bass Islands by faculty and students of the Stone Laboratory. Three exotic species have been added, making a total of 95 species for the county. Of these 18 species have not been recorded since 1955 and several others have decreased greatly in abundance.

Auglaize County: The majority of the collections were made after 1920, first by Edward L. Wickliff, later principally by Clarence Clark (1942: 1–165) and myself. This county was selected partly because it contained remnants of till- and lake-plain, prairie and Black Swamp types of streams. A total of 63 species and two additional subspecies were recorded. Since 1951 only one species, the exotic striped bass, has been added, thereby increasing the total to 64 species.

Scioto County and the lower Scioto River: Before 1951, 97 species and an additional subspecies were recorded for the county. Between October 1, 1939 and September 30, 1940 members of the Ohio Department of Natural Resources conducted daily fyke and trap-net operations in a 2-mile (3-km) stretch of the Scioto River in the northern part of the county. Frequent collecting with hand seines and other gear was also done. A total of 75 species and an additional subspecies were recorded from the 2-mile (3-km) stretch of river. By 1977 there were recorded 104 species of fishes.

Scioto River drainage: No other stream drainage in Ohio has been so thoroughly investigated during the past 100 years as has this one. Although much attention had been given the Scioto River prior to 1950, a considerably increased amount has been given since. Between the years 1960 and 1965 the Ohio Division of Natural Resources, through granting funds, made it possible for several students and others to assist me in the investigations of central Ohio streams (known as COSS). As many as 50 days a year were spent in collecting and in field studies. A comparison between the present and past distribution and abundance of fishes was thus made possible. Various studies were also made of individual species. To date, 125 species have been recorded from this drainage.

Big Darby Creek: This stream rises in the gently rolling hills of Logan County and flows southward into Pickaway County, a distance of more than 75 miles (121 km), to enter the Scioto River a mile above Circleville. It is a free-flowing stream, remaining unimpounded by large dams. There has been comparatively little drastic channelization.

Because of the research and teaching possibilities of this stream, a 1/4-mile (0.4-km) stream section was designated in 1925 as an intensive study area. This section is about 1/2 mile (0.8 km) south of Fox, in Jackson Township, Pickaway County and immediately above the State Route 104 bridge. Throughout the 1925–50 period more than 125 days were spent by others and me investigating this section. Collections were made during every month of the year, the majority in April, May, September and October. By 1951 this short stream area had yielded 81 species of fishes of which 19 were collected on fewer than 4 days and were considered to be strays.

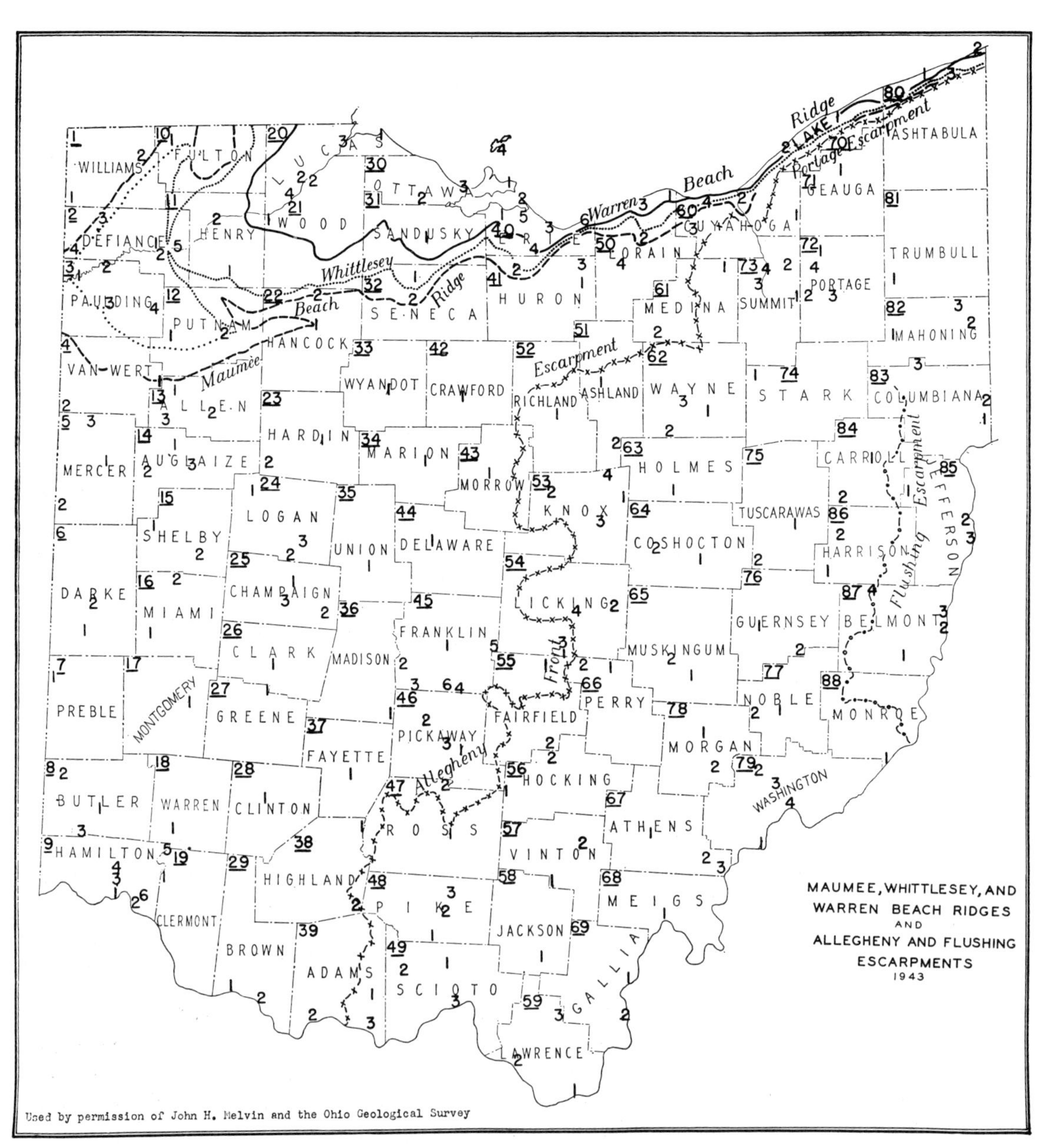

MAP XI. Principal ridges and escarpments, and list of numbered counties, villages and other localities mentioned in the literature or near which important collections were made

Map XI. Principal Ridges and Escarpments, and List of Numbered Counties, Cities, Villages and Other Localities Mentioned in the Literature or Near Which Important Collections Were Made

A Amsterdam, *85*-1*; Antwerp, *3*-1; Apple Creek, *62*-1; Armstrongs Mills, *87*-1; Ashland, *51*-1; Ashtabula, *80*-1; Athens, *67*-1; Auglaize, *13*-1; Aurora, *72*-1; Avon Lake, *50*-1.

B Bellaire, *87*-2; Belpre, *79*-1; Berlin Center, *82*-1; Beverly and Waterford, *79*-2; Blue Creek, *39*-1; Bridgeport, *87*-3; Brinkhaven, *53*-1; Brunersburg, *2*-1; Buckeye Lake, *54*-1; Buckland, *14*-1; Bucyrus, *42*-1.

C Caldwell, *77*-1; Cambridge, *76*-1; Canal Fulton, *74*-1; Carrollton, *84*-1; Castalia and Venice, *40*-1; Cecil, *3*-2; Cedar Point, *40*-2; Celina, *5*-1; Chagrin Falls, *60*-1; Chardon, *71*-1; Chesapeake, *59*-1; Chillicothe, *47*-1; Cincinnati, *9*-1; Circleville, *46*-1; Clarksfield, *41*-1; Cleveland, *60*-2; Clifton, *27*-1; Cloverdale, *12*-1; Columbus, *45*-1; Coney Island Resort, *9*-2; Conneaut, *80*-2; Coolville, *67*-2; Coopersville, *48*-1; Coshocton, *64*-1; Cuyahoga Falls, *73*-1.

D Darbyville, *46*-2; Dayton, *17*-1; Defiance, *2*-2; Delaware, *44*-1; Delphos, *4*-1.

E Edgerton, *1*-1; Elmwood, *9*-3; Elyria, *50*-2.

F Fairport and Painesville, *70*-1; Farmer, *3*-3; Fayette, *10*-1; Findlay, *22*-1; Flushing, *87*-4; Fly, *88*-1; Fort Jefferson, *6*-1; Fort Loramie, *15*-1; Fort Recovery, *5*-2; Fort Seneca, *32*-1; Fox, *46*-3; Fredericktown, *53*-2; Fredericktown, (St. Clair), *83*-1; Freeport, *86*-1; Fremont, *31*-1.

G Gallipolis, *69*-1; Georgesville, *45*-2; Glenford, *66*-1; Gnadenhutten, *75*-1; Grand Rapids, *21*-1; Greenfield, *38*-1; Greenville, *6*-2.

H Handen, *57*-1; Hamilton, *8*-1; Hamler, *11*-1; Hanover, *54*-2; Harrisburg, *45*-3; Hawkins (now Ira), *73*-3; Hebron, *54*-3; Hicksville, *2*-4; Higginsport, *29*-1; Hinckley, *61*-1; Hockingport, *67*-3; Howard, *53*-3; Hudson, *73*-2; Huron, *40*-3.

I Independence, *2*-5; Ira (formerly Hawkins), *73*-3; Ironton, *59*-2.

J Jelloway, *53*-4.

K Kent, *72*-2; Kenton, *23*-1; Killbuck, *63*-1; Kings Creek, *25*-1; Kings Mills, *18*-1; Kingsville, *80*-3.

L Lakeside, *30*-1; Lakeview and Russels Point, *24*-1; Laurelville, *56*-1; Leesville, *84*-2; Lima, *13*-2; Little Cedar Point, *20*-1; Lockbourne, *45*-4; Lockland, *9*-4; Lodi, *61*-2; Lorain, *50*-3; Loudonville, *51*-2; Loveland, *9*-5; Lowell, *79*-3; Lucasville, *49*-1; Ludlow Falls, *16*-1.

M Manchester, *39*-2; Mansfield, *52*-1; Marietta, *79*-4; Marion, *34*-1; Marysville, *35*-1; Maumee, *20*-2; McConnelsville, *78*-1; Mechanicsburg, *25*-2; Milan, *40*-4; Milford, *19*-1; Millersport, *55*-1; Mingo Junction, *85*-2; Mount Gilead, *43*-1; Mount Sterling, *36*-1; Monroeville, *41*-2.

N Napoleon, *11*-2; Neagley, *83*-2; Nellie, *64*-2; Newark, *54*-4; Newcomerstown, *75*-2; New Paris, *7*-1; Newton Falls, *81*-1.

O Oak Harbor, *30*-2; Oak Hill, *58*-1; Oakwood, *3*-4; Oberlin, *50*-4; Olive Green, *77*-2; Olmstead Falls, *60*-3; Ottawa, *12*-2; Otway, *49*-2; Oxford, *8*-2.

P Painesville and Fairport, *70*-1; Paulding, *3*-3; Peninsula, *73*-4; Perrysburg, *21*-2; Philo P. O. (Taylorsville), *65*-1; Piketon, *48*-2; Piqua, *16*-2; Poland, *82*-2; Pomeroy, *68*-1; Port Clinton, *30*-3; Portsmouth, *49*-3; Put-in-Bay, *30*-4.

R Raccoon Island, *69*-2; Ravenna, *72*-3; Redbank, *9*-6; Reynoldsburg, *45*-5; Ripley, *29*-2; Rockbridge, *56*-2; Rockford, *5*-3; Rocky River, *60*-4; Rome (Stout P. O.), *39*-3; Ross P. O. (Venice), *8*-3; Roundhead, *23*-2; Russels Point and Lakeview, *24*-1.

S Salem, *83*-3; Sandusky, *40*-5; Senecaville, *76*-2; Shadeville, *45*-6; Shreve, *62*-2; Sidney, *15*-2; Sinking Springs, *38*-2; Spencerville, *13*-3; Springfield, *26*-1; St. Clair (Fredericktown), *83*-1; Steubenville, *85*-3; St. Marys, *14*-2; Stockport, *78*-2; Stout P. O. (Rome), *39*-3; Streetsboro, *72*-4; Sugar Grove, *55*-2.

T Tappan, *86*-2; Taylorsville (Philo P. O.), *65*-1; Thornport, *66*-2; Tiffin, *32*-2; Toledo, *20*-3.

U Upper Sandusky, *33*-1; Urbana, *25*-3.

V Van Buren, *22*-2; Venice (Ross P. O.), *8*-3; Venice and Castalia, *40*-1; Vermilion, *40*-6.

W Wakeman, *41*-1; Wapakoneta, *14*-3; Washington, C. H. *37*-1; Waterford and Beverly, *79*-2; Waterloo, *59*-3; Waterville, *20*-4; Waverly, *48*-3; West Liberty, *24*-2; West Unity, *1*-2; Wilmington, *28*-1; Willoughby, *70*-2; Willshire, *4*-2; Wooster, *62*-3.

Y Yellow Bud, *47*-2; Youngstown, *82*-3.

Z Zaleski, *57*-2; Zanesfield, *24*-3; Zanesville, *65*-2.

---

* The italicized number is the underscored number in the upper left hand quarter of the county which it represents. Following that is the number representing the city, village or other locality.

Since 1955 Big Darby Creek, and especially the study area, has been ever increasingly investigated by members of many organizations, ichthyologists and students from within and outside Ohio. No stream system in the state has been more thoroughly investigated. Faculty and students of Ohio State University utilize it regularly for investigative purposes as do those from other universities such as John Carroll, Bowling Green State and Ohio Wesleyan. To date, 92 species of fishes have been recorded from the study area of which 15 have not been collected since 1951 and 13 are considered to be strays. An additional 8 species, not recorded in the study section, have been taken elsewhere in Big Darby, giving a total of 100 species for the system.

Olentangy River: This southward flowing stream of central Ohio has several small cities and the northern half of Columbus within its watershed. Its confluence with the Scioto River is in the center of Columbus. Because of its proximity to Ohio Wesleyan and Ohio State universities, considerable research of its fauna and flora has been conducted during the past century. Despite the large number of people living along its banks and within its watershed, it contains a goodly fish population. The stream is much fished by people of all ages.

The Olentangy River compares favorably with Big Darby Creek in the length and size of its watershed. However, there are striking differences in the composition of the fish species. The paddlefish, buffalofishes and other large river species have not been recorded, presumably because of dams situated in the Scioto River a short distance below the confluence of the Olentangy, these dams preventing vernal migrations upstream. Apparently because of dams, human population density, and a more northern location in Ohio, only 71 species have been recorded for the Olentangy, in comparison with 100 species recorded in Big Darby Creek.

Salt Creek: This stream system drains portions of Hocking, Vinton, Jackson and Ross counties in southcentral Ohio, entering the Scioto River in western Ross County. It has been regularly investigated since 1920 and the northern branch was thoroughly studied between 1972–74 by Andrew White, Bruce McLean and associates. To date, 84 species have been reported from the system.

Franklin County: In 1899 and 1900 E. B. Williamson and R. C. Osburn investigated its waters. After 1920 much additional collecting was done by Edward L. Wickliff, several fishery workers and me. A total of 91 species had been recorded, of which, after 1920, ten were considered to have become extirpated and ten to be strays. By 1977 the recorded number of species was 100.

Pike County: Robert B. Foster and I collected a total of 95 species in this county between 1925–34. Many collections have been made since by others and me. At present the recorded number of species is 102.

Wayne County: Its waters were investigated by Ralph V. Bangham and his students, principally in 1935–45. A total of 61 species and an additional subspecies were obtained. By 1977 there were recorded 64 species of fishes.

Ashland County: Mary A. Trautman (*née* Mary Auten) and her students collected fishes during the 1935–40 period, obtaining 63 species and an additional subspecies. By 1977 there were recorded 68 species of fishes.

Lake Erie and tributaries from Lorain County eastward to the Pennsylvania border: This area has been investigated at intervals since J. P. Kirtland began his studies of the zoology of Ohio about 1835. Since then personnel from federal and state organizations, universities and private individuals have studied and observed its fish fauna. Since 1971 Andrew White and associates have conducted intensive studies of the area, which continues (White et al., 1975). To date, 106 species and an additional 5 subspecies have been recorded for the area.

Ohio River: Since 1792 the states of Kentucky and later West Virginia have claimed ownership to all of the Ohio River to its present northern shoreline, despite impoundings which later flooded a considerable portion of land originally belonging to the state of Ohio. Recently the United States Supreme Court ruled that the northern boundary of the Ohio River is to be the 1792 low water mark, at which time Kentucky became a state. This returns to Ohio that portion which consists of 80 to 100 feet (24–30m) southward from the present Ohio shoreline. The ruling gives to the state of Ohio jurisdiction over that portion and also fish protection including our migrating fishes which spawn and summer in Ohio streams.

Columbiana, Defiance and Lucas counties: While residing at Ann Arbor, Michigan, between 1934 and 1938 I, assisted by students and Louis W. and the late Bernard R. Campbell, of Toledo, conducted investigations of many Ohio streams, principally in Columbiana (68 species recorded before 1950; 72 after), Defiance (74 species recorded before 1950; the county has been intensely investigated at times since 1887; *see* Trautman and Gartman [1974]; 76 species recorded after 1951) and Lucas (95 species recorded before 1950; 104 after).

Blacklick Creek: After 1935 Edward L. Wickliff conducted a series of investigations on this central Ohio stream. At least 40 species could be recorded annually.

Other streams: Between 1925–45 I periodically investigated the West Branch of Beaver Creek in Columbiana County, Sunfish Creek in Pike County, Auglaize River in Auglaize County, and several prairie- and till-plain streams, primarily to observe smallmouth blackbass production and changes in abundance of the fish fauna caused by changing ecological conditions.

Comparisons between Various Waters

Before 1951 a total of 160 species and 12 additional subspecies were recorded for all Ohio waters, of which a total of 122 species and seven additional subspecies were listed as occurring, or as having occurred at some time during the 1750–1950 period, in the Lake Erie drainage of Ohio, and 141 species and three additional subspecies in the Ohio River drainage.

By 1980 a grand total of 166 species and 13 additional subspecies had been recorded for all Ohio waters. The increase of six species and one subspecies since 1957 was the result of the introduction of exotic species, inadvertent omission of a species from the first edition, elevation from subspecific to specific rank and recognition of an undescribed species.

Before 1951 the Ohio waters of Lake Erie and its connecting bays and harbors (omitting flowing waters) yielded a total of 93 species and six additional subspecies, of which 30 are classified as strays from streams. These strays included such typical stream species as the creek chub and striped shiner. By 1980, 98 species and six additional subspecies had been recorded for the Ohio waters of Lake Erie and its connecting bays and harbors.

Before 1951 the Maumee River system contained a total of 93 species and two additional subspecies. By 1980, 103 species and two additional subspecies had been recorded. Since 1900 the rapidly increasing rates of turbidity and siltation presumably caused the extirpation of at least six species, drastic reduction in numbers of 27 species and/or the endangering of 19 additional species, all native.

Before 1951 the Ohio River, without its tributaries, yielded a total of 93 species and an additional subspecies. By 1980, 103 species and an additional subspecies, of which about 30 were present during droughts or in winter, and some of which were strays, had been recorded.

Before 1951 the Muskingum River and tributaries yielded 106 species and two subspecies. By 1980, 120 species and a subspecies had been recorded.

The above indicate that:

Each county, carefully investigated, contained more than 60 species of fishes. Counties situated along the divide between the two drainages contained the smallest number of species. Counties bordering Lake Erie and the Ohio River contained the largest number of species.

Short sections, and entire stream systems, in the Ohio drainage contained larger numbers of species than did similar waters in the Lake Erie drainage, provided that they were not heavily polluted. A larger number of species was recorded for the Ohio drainage than for the Lake Erie drainage.

Factors Influencing Fish Distribution and Abundance

To understand fish distribution and abundance it is necessary to realize that each species has its own specialized complex of many ecological niches or habitats, the sum total of which comprises its environment. Wherever its environmental conditions prevail, a fish species under normal conditions can successfully repel all other fish species (including close competitors) from at least part of its specialized environment.

The numbers of a species are usually controlled by the amount of available environment. This environment, in part or in its entirety, may fluctuate greatly from year to year, or from one series of years to another and this, coupled with the high offspring potentials of most fish species, may result in spectacular increases in abundance of a species in a single year, or during two or more consecutive years; *see* Tippecanoe darter.

It is because species (including subspecies) are so restricted to their particular set of environmental conditions that they cannot follow the popular

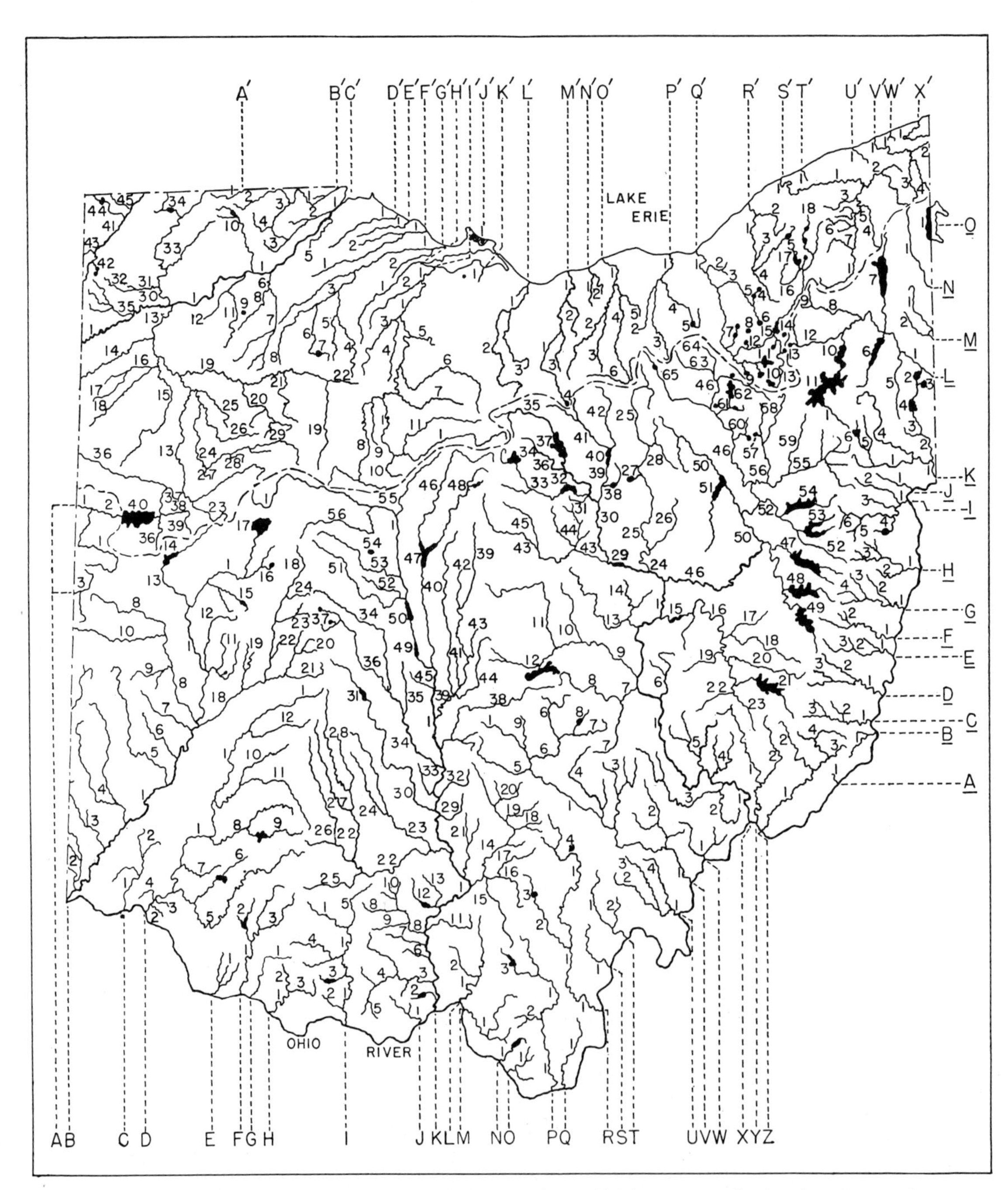

MAP XII. Names of streams mentioned in the literature or from which important collections have been made

Map XII

Ohio River Tributaries

A 1—Wabash R.; 2—Beaver Cr.; 3—Mississinawa R.

B 1—Great Miami R.; 2—Whitewater R.; 3—Indian Cr.; 4—Seven Mile Cr. (right) and Four Mile Cr. (left); 5—Twin Cr.; 6—Bear Cr.; 7—Wolf Cr.; 8—Stillwater R.; 9—Ludlow Cr.; 10—Greenville Cr.; 11—Honey Cr. (below and right), Indian Cr. (left) and Silver Lake (below); 12—Lost Cr.; 13—Loramie Cr.; 14—Loramie Res.; 15—Mosquito Cr. and Kiser Lake; 16—Bokengehalas Cr. and Silver Lake; 17—Indian Lake; 18—Mad R.; 19—Donnels Cr.; 20—Buck Cr. (left); 21—Beaver Cr.; 22—Cedar Run; 23—Kings Cr. (above); 24—Macocheek Cr. (below)

C 1—Mill Cr.; 2—East Br. (below)

D 1—Little Miami R.; 2—Cluff Cr. (above); 3—McCullough Cr. (left); 4—Duck Cr.; 5—East Fork; 6—Stonelick Cr. and Stonelick Lake; 7—O'Bannon Cr.; 8—Todds Fork; 9—Cowan Cr. and Cowan Lake; 10—Caesars Cr.; 11—Andersons Fork; 12—Massies Cr.

E 1—Bull Skin Cr.

F 1—Whiteoak Cr.; 2—Sterling Run and Grant Lake; 3—East Fork

G 1—Straight Cr.

H 1—Eagle Cr.; 2—West Fork; 3—East Fork

I 1—Ohio Brush Cr.; 2—Beasley Fork; 3—Lick Fork and Adams Lake; 4—Little West Fork; 5—Bakers Fork

J 1—Turkey Cr.

K 1—Scioto R.; 2—Pond Cr. and Roosevelt Lake; 3—Scioto Brush Cr.; 4—South Fork; 5—Churn Cr.; 6—Bear Cr. (below); 7—Camp Cr. (below); 8—Sunfish Cr.; 9—Chenoweths Fork; 10—Morgans Fork (below); 11—Beaver Cr.; 12—Peepee Cr. and Lake White; 13—Crooked Cr.; 14—Salt Cr.; 15—Little Salt Cr.; 16—Pigeon Cr. (below); 17—Middle Br.; 18—Queer Cr.; 19—Pine Cr.; 20—Laurel Run; 21—Walnut Cr. (right); 22—Paint Cr.; 23—North Fork; 24—Compton Cr. (right); 25—Rocky Fork; 26—Rattlesnake Cr. (right); 27—Sugar Cr. (between); 28—East Fork (below); 29—Kinnikinnick Cr.; 30—Deer Cr.; 31—Madison Lake; 32—Scipio Cr.; 33—Yellow Bud Cr. (left); 34—Big Darby Cr.; 35—Hellbranch Run (left); 36—Little Darby Cr.; 37—Brush Lake (upper) and Baker Lake (lower); 38—Little Walnut Cr.; 39—Big Walnut Cr.; 40—Alum Cr. (right); 41—Masons Run (below); 42—Little Walnut Cr. (below); 43—Rocky Fork (below); 44—Blacklick Cr.; 45—Scioto Big Run (below); 46—Olentangy R.; 47—Delaware Res.; 48—Whetstone Cr. and Mt. Gilead lakes (right); 49—Griggs Res.; 50—O'Shaughnessy Res.; 51—Mill Cr.; 52—Blues Cr. (below); 53—Bokes Cr.; 54—Richwood Lake (below); 55—Little Scioto R. (right); 56—Rush Cr.

L 1—Little Scioto R.; 2—Rocky Fork

M 1—Pine Cr.

N 1—Storms Cr. and Vesuvius Lake

O 1—Ice Cr.

P 1—Symmes Cr.; 2—Johns Cr.; 3—Black Fork and Lake Jackson

Q 1—Indian Guyon Cr.

R 1—Raccoon Cr.; 2—Little Raccoon Cr.; 3—Lake Alma; 4—Lake Hope

S 1—Campaign Cr.

T 1—Leading Cr.; 2—Little Leading Cr.

U 1—Shade Cr.; 2—West Br.; 3—Middle Br.; 4—East Br.

V 1—Hocking R.; 2—Federal Cr.; 3—Sunday Cr.; 4—Monday Cr.; 5—Clear Cr.; 6—Rush Cr.; 7—Little Rush Cr.; 8—Clouse Lake; 9—Pleasant Run

W 1—Little Hocking Cr.

X 1—Muskingum R.; 2—South Br. Wolf Cr.; 3—West Br. Wolf Cr.; 4—Olive Green (right) and Little Olive (left) crs.; 5—Meigs Cr. (left) and Dyes Cr. (right); 6—Salt Cr.; 7—Moxahala Cr.; 8—Jonathan Cr.; 9—Licking R.; 10—Rocky Fork; 11—North Fork; 12—South Fork and Buckeye Lake (below); 13—Wakatomika Cr.; 14—Little Wakatomika Cr.; 15—Wills Cr. Res.; 16—Wills Cr.; 17—Sugartree Fork; 18—Salt Fork; 19—Crooked Cr.; 20—Leatherwood Cr. (below); 21—Seneca Cr. and Senecaville Res.; 22—Buffalo Fork; 23—Buffalo Cr.; 24—Walhonding R. (below); 25—Killbuck Cr.; 26—Doughty Cr. (right); 27—Shreve Cr. and Beaver Lake (below); 28—Apple Cr. (right); 29—Mohawk Res.; 30—Mohican R.; 31—Pine Run (above); 32—Pleasant Hill Res. (below); 33—Clear Fork; 34—Clear Fork Res. (left); 35—Black Fork (below); 36—Rocky Fork (right); 37—Charles Mill Res.; 38—Odell Lake (above); 39—Lake Fork (right); 40—Mohicanville Res.; 41—Jerome Fork; 42—Muddy Fork; 43—Kokosing R.; 44—Big (right) and Little (left) Jelloway crs.; 45—North Br. (right); 46—Tuscarawas R.; 47—Little Stillwater Cr. and Tappan Res. (below); 48—Brushy Fork and Clendening Res.; 49—Big Stillwater Cr. and Piedmont Res. (left); 50—Sugar Cr.; 51—Beach City Res. and South Br.; 52—Conotton Cr.; 53—Leesville Res.; 54—Atwood Res.; 55—Sandy Cr.; 56—Nimishillen Cr. (right); 57—Myers Lake (above); 58—Middle Br. (right); 59—Little Sandy Cr.; 60—Sippo Cr. (right) and Lake (below); 61—Nimisila Cr. (below) and Comet Lake (right), Luna Lake (left); 62—Portage lakes (left); 63—Wolf Cr. (below); 64—Schocalog Run (right); 65—Chippewa Cr. (left) and Lake

Y 1—Duck Cr.; 2—East Fork

Z 1—Little Muskingum R.; 2—Clear Fork (right); 3—Witten Fork; 4—Cranenest Fork

A 1—Jims Run near Fly, Ohio
B 1—Opossum Cr.
C 1—Sunfish Cr.; —Piney Cr.; 3—Bakers Fork (right)
D 1—Captina Cr.; 2—Bend Fork; 3—North Fork (below)
E 1—McMahon Cr.; 2—Little McMahon Cr. (right); 3—Brush Cr. (right)
F 1—Wheeling Cr.; 2—Campbells Cr.
G 1—Short Cr.; 2—Piney Fork; 3—North Fork (left); 4—Middle Fork (below)
H 1—Cross Cr.; 2—McIntyre Cr.; 3—Salem Cr. (right)
I 1—Big Yellow Cr.; 2—North Fork (below); 3—Brushy Fork; 4—Town Fork and Jefferson Lake; 5—Elklick Run; 6—Elkhorn Cr. (left)
J 1—Little Yellow Cr.
K 1—Little Beaver Cr.; 2—North Fork; 3—Bull Cr. (right); 4—Middle Fork; 5—Cold run (right); 6—West Fork and Guilford Lake (right)
L 1—Mahoning R.; 2—Yellow Cr. and Hamilton Lake (right); 3—Burgess Lake (left); 4—Pine Lake (right); 5—Mill Cr.; 6—Meander Cr. and Res.; 7—Mosquito Cr. and Res.; 8—Eagle Cr.; 9—Silver Cr. (right); 10—Milton Lake (right); 11—Berlin Res.; 12—West Br.; 13—Crystal Lake (above)
M 1—Big Yankee Cr.; 2—Little Yankee Cr. (both of Little Deer System)
N 1—Pymatuning Cr.
O 1—Pymatuning Res.

Lake Erie Tributaries

A′ 1—Bear Cr.
B′ 1—Ottawa R.; 2—Ten Mile Cr.; 3—Prairie Ditch (below and connects with Swan Cr.)
C′ 1—Maumee R.; 2—Swan Cr. (below); 3—South Br. (below); 4—West Fork (below); 5—Grassy Cr. (above); 6—Beaver Cr. (right); 7—Yellow Cr. (below) and Cut-off Ditch (right); 8—West Br.; 9—Hamler Pond (below); 10—Bad Cr. (right) and Delta Lake (above); 11—Turkeyfoot Cr.; 12—School Creek; 13—Auglaize R. (right); 14—Flat Rock Cr.; 15—Little Auglaize R.; 16—Prairie Cr.; 17—Hagerman Cr. (below); 18—Hoaglin Cr. (below); 19—Blanchard R.; 20—Riley Cr. and tribs.; 21—Ottawa Cr. (right); 22—The Outlet (below); 23—Pusheta Cr.; 24—Ottawa R. (below); 25—Plum Cr. (left); 26—Sugar Cr. (below); 27—Little Ottawa R. (between); 28—Lost Cr. (above); 29—Big and Little Hog crs. (below); 30—Tiffin R. (right); 31—Mud Cr.; 32—Lost Cr. (below); 33—Brush Cr. (right); 34—Mill Cr. and Harrison Lake (below); 35—Gordon Cr. and branches (left); 36—St. Mary's R.; 37—Six Mile Cr. (between); 38—East Br. (below); 39—Clear Cr. (right); 40—Lake St. Marys; 41—St. Joseph R.; 42—Lehmans and Little lakes; 43—Fish Cr. (below); 44—Nettle Cr. and Nettle Lake; 45—West Br. (below)
D′ 1—Crane Cr.
E′ 1—Turtle Cr.
F′ 1—Toussaint Cr.; 2—Packard Cr. (right)
G′ 1—Portage R.; 2—Nine Mile Cr. (right); 3—Middle Br. (below); 4—East Br.; 5—South Br. (right); 6—Rocky Fork Cr.; 7—Van Buren Lake; 8—Needles Cr. (right)
H′ 1—Big Muddy Cr.
I′ 1—Sandusky R.; 2—Muskellunge Cr. (below); 3—Wolf Cr. (below); 4—East Br.; 5—Sugar Cr.; 6—Honey Cr.; 7—Sycamore Cr.; 8—Tymochtee Cr.; 9—Little Tymochtee Cr. (above); 10—Little Sandusky R. (right); 11—Broken Sword Cr.
J′ 1—West, Middle and East Harbors
K′ 1—Cold Cr. (right) and Miller Blue Hole (left)
L′ 1—Huron R.; 2—West Br.; 3—East Br. (above)
M′ 1—Vermilion R.; 2—East Fork (right); 3—East Br. (right); 4—Savannah lakes
N′ 1—Beaver Cr.; 2—East Br. (right)
O′ 1—Black R.; 2—West Br.; 3—Wellington Cr.; 4—East Br.; 5—Willow Cr.; 6—West Fork (below)
P′ 1—Rocky R.; 2—Plum Run (right); 3—West Br.; 4—East Br.; 5—Hinkley Lake
Q′ 1—Cuyahoga R.; 2—Mill Cr.; 3—Tinkers Cr. (below); 4—Geauga Lake (above); 5—Aurora Pond (below); 6—Stewart Pond (left); 7—Mud Br. (right) and Wyoga (lower) and Mud (upper) lakes; 8—Crystal Lake (below); 9—Springfield Lake (above); 10—Fritch Lake (left); 11—Mogadore Res. (below); 12—Silver Lake (left); 13—Congress Lake Outlet (left) and Congress Lake (lowest), Muzzy Lake (middle) and Sandy Lake (upper); 14—Brady Lake (below); 15—Rockwell and Pippen lakes (right, and as one); 16—Bridge Cr.; 17—Lake Punderson and Bradley Pond (both to the right, Bradley Pond the farther); 18—Outlet Br. (below) and Alderman Pond (lower) and Aquila Lake (upper)
R′ 1—Chagrin R.; 2—East Br.; 3—Griswold Cr. (below); 4—Aurora Br. (left); 5—Bass Lake (above)
S′ 1—Grand R.; 2—Mill Cr. (below); 3—Trumbull Cr. (below); 4—Rock Cr.; 5—Crooked and Mud crs. (left and across the Grand River); 6—Hoskins and Indian crs. (right); 7—Phelps Cr. (below)
T′ 1—Arcola Cr.
U′ 1—Indian Cr.
V′ 1—Ashtabula R.; 2—Ashtabula Cr. (above); 3—West Br.; 4—East Br. (above)
W′ 1—Creek near Kingsville
X′ 1—Conneaut Cr.

conception that any fish species may be expected to occur (or can be successfully stocked) wherever there is water. An examination of the distribution maps verifies this; *see* the striking distributional patterns of the central mottled sculpin, redbelly and blacknose daces.

An analysis of ecological data was begun in 1940, and has continued as more data were collected. It is obvious from this analysis that escarpments, lake- and till-plains, unglaciated plateau and other physiographic features exerted a profound influence upon fish distribution and abundance, and that there was a correlation between physiography and stream gradients.

After studying the gradients of a few central Ohio streams, the importance of stream gradients to fish distribution was realized, so in 1941, assisted by George Borman, the plotting of all streams except the smallest tributaries was undertaken. U.S. topographical maps were used to plot gradients in feet per stream mile, by dividing the distance in stream miles between two contour lines into the numbers of feet of elevation between these contour lines.

In 1942 I published a paper relative to the correlation of stream gradients with fish distribution and abundance, suggesting that under average conditions, the absence or presence and abundance of a fish species in a given stream section was largely governed by the speed of flow, and that speed of flow, through its ability to erode, scour, dig and deposit various materials, determined to a large degree: (1) the number of riffles per stream mile, their depths, widths, and types of bottoms; (2) number of pools per stream mile, their depths, widths, the presence or absence of currents, and types of bottoms; (3) amount of silt deposited during various water levels; (4) amount and kinds of bank undercutting and resultant tree root exposure; (5) absence, or presence and abundance of macro- and micro-vegetation; (6) absence, or presence and abundance of backwaters, oxbows, sloughs and overflow ponds (Trautman, 1942: 211–23).

In addition to the actual gradient, the speed of flow is sometimes affected by the depth of water since within limits the rate of flow increases as the depth increases. Because of the effects of various rates of flow, water depths and stream widths upon fish distribution and abundance, it becomes necessary to establish ratings for five stream sizes as follows:

Average Stream Widths*		Average Riffle Depths		Low		Moderate		High	
Feet	Meters	Feet	Meters	Feet/Mile	Meters/Kilometer	Feet/Mile	Meters/Kilometer	Feet/Mile	Meters/Kilometer
1.0′- 15.0′	03.m- 4.7m	0.3′- 3.0′	0.1m-0.9m	1.0′-10.0′	0.20m-1.9m	12.0′-20.0′	2.3m-3.8m	25.0′+	4.8m+
16.0′- 30.0′	4.8m- 9.2m	0.5′- 3.0′	0.2m-0.9m	1.0′- 6.0′	0.20m-1.1m	7.0′-18.0′	1.3m-3.4m	20.0′+	3.8m+
31.0′- 45.0′	9.3m-13.8m	1.0′- 3.0′	0.3m-0.9m	1.0′- 5.0′	0.20m-1.0m	6.0′-10.0′	1.1m-1.9m	12.0′+	2.3m+
46′-100.0′	13.9m-30.6m	1.5′- 5.0′	0.5m-1.5m	0.5′- 4.0′	0.10m-0.8m	5.0′- 8.0′	1.0m-1.5m	9.0′+	1.7m+
101.0′+	30.7m+	1.5′-10.0′	0.5m-3.0m	0.1′- 1.0′	0.02m-0.2m	2.0′- 4.0′	0.4m-0.8m	5.0′+	1.0m+

* Stream sizes were determined by striking an average when the stream was at normal summer level.

Other factors affecting fish distribution and abundance:

(1) Physiography and the types of bedrocks and soils: for an example, *see* Trautman (1942: 219–22) concerning lamprey distribution east of the Flushing Escarpment.

(2) Degree and availability of fertility in soils: streams draining fertile soils usually contained faster growing and/or larger fish populations than did those streams draining soils of low fertility.

(3) Soil, domestic and industrial pollutants: below normal fish populations occurred when detrimental pollutants were present, even though stream gradient and other factors were favorable.

(4) Season of year: the absence or presence and abundance of fish species in a given stream section varied according to the seasons, because most of the Ohio fish species are migratory to some extent. Such species, and especially the adults, moved from lower to higher gradients (and usually from larger to smaller streams) with the increase of water temperature in late winter or spring; later, in late spring or summer, the adults and many young began drifting downstream to winter in lower (or at least as low)

gradients. Because of this annual movement into smaller streams and higher gradients some species as the creek chub and smallmouth blackbass were far more numerous in winter in the Ohio River than they were in that stream in summer.

(5) Population pressure from competitors: an example is the smallmouth blackbass which wintered in larger numbers in gradients of less than 5.0′ (1.5 m) per mile (1.6 km) wherever the spotted blackbass was absent than it did in similar gradients where the latter species was present.

The Distribution Maps

Note: Future locality records for each species can be marked on their respective distribution maps and, since these records will be inserted in pencil or in ink, will readily show any changes in distributional patterns, new county and new drainage records, which have been obtained since 1950. Especially interesting will be the records of such invading species as the orangespotted sunfish and white perch, and by using different symbols for subsequent years a graphic record of range extension and speed of invasion will be had.

The symbols on the distribution maps indicate rather accurately the localities where one or more individuals of a species were recorded on one or more days.

Many symbols closely clustered usually denote a center of abundance.

Relative abundance and distribution of many species show a correlation with such physiographic features as escarpments, lake- and till-plains, unglaciated plateaus, large or small lakes, and large or small streams.

Changes in abundance of some species during part or all of the 1850–1950 period are shown by the use of different symbols for the various periods.

Degree of subspecific intergradation throughout sections of the state is demonstrated by the use of various symbols.

The distribution maps fail to portray accurately the relative abundance of a species because each symbol has the same numerical value and may either represent one, hundreds, or thousands of individuals.

The maps do not indicate seasonal changes in abundance of the more migratory species, such as the creek chub which spawns in spring in the smaller creeks and brooks and winters chiefly in larger waters including the Ohio River.

The maps portray rather accurately the distributional patterns for the various species as they existed during the 1920–50 period, but do not indicate the vast changes in distributional patterns of many species prior to 1900, because of the comparatively few numbers of early preserved specimens and literature records. An example is the redbelly dace which unquestionably was more widespread in its Ohio distribution prior to 1900 than it has been since, because habitat conditions for it were more widespread during the early years. As a general rule, those species least tolerant to turbid waters, other pollutants, silt-covered bottoms, ditching and dredging have become more restricted in, or were extirpated from, their Ohio ranges, whereas those species most tolerant to such conditions have become more widespread and have recently invaded areas not previously occupied since 1750.

For further details as to which records have been accepted for use in the distribution maps, *see* "Preserved fish collections and numbers of fishes examined." *See* map X for collections since 1955.

Insert Maps

Each Ohio distribution map has with it a small insert map which shows the approximate present and/or original range in North America of the species or subspecies indicated on the larger map. The range of monotypic species is presented in black. When a species is divided into subspecies the range of the subspecies under discussion is in black whereas the ranges of other subspecies are outlined. In such cases the combined black and outlined areas constitute the entire range of the species in that portion of North America which is shown on the map.

These ranges are generalized, and because there are sections of North America where relatively little is known concerning the distribution of fishes, the limits of range in such sections must be indefinite to a greater or lesser degree. As a general rule the farther the limit of range is from Ohio the greater is the possibility of inaccuracy. This is especially true of northern Canada and Mexico.

The same factors which have caused recent changes in the distributional patterns of many fish species in Ohio have likewise caused drastic recent changes in part or all of the range of some species. Some native species are extending their ranges in one or more directions. In others their ranges are becoming smaller, and in the case of the harelip sucker it may have become extinct.

Some species or subspecies are rather uniformly distributed over much or all of their range; others have large unoccupied areas within their range caused by a lack or a destruction of their habitat.

A few species occur only in isolated colonies which are sometimes separated from each other by considerable distances.

Many native game fishes and other species have been widely introduced into many areas outside of their original or recent range. Only in a few instances have attempts been made to indicate on the insert maps where these introductions were made but mention of these introductions usually was referred to in the legends accompanying the maps. The legends also give other facts essential to an understanding of the ranges.

In order to have these ranges as accurate as possible I have made use of the literature and of the information accumulated by me and many co-workers, especially Reeve M. Bailey, Carl L. Hubbs, Clark Hubbs (Texas), George A. Moore (Oklahoma and general), Royal D. Suttkus (Gulf States), W. Ralph Taylor (catfishes), and Edward S. Thomas.

The Illustrations

With few exceptions there is an outline drawing representing each fish species and subspecies, or a drawing of both sexes of those species in which sexual dimorphism is marked. Exceptions are those species or subspecies which are so similar in general appearance that a drawing of them would be duplication. In addition, there are many illustrations containing anatomical drawings of some part or parts of a fish.

How the Illustrations Were Made

There were between 20 and 40 different counts and measurements taken of ten to 50 individuals of each species or subspecies, except for those species in which fewer than ten individuals were available. Counts and measurements were usually made with the aid of magnification and accuracies within half a millimeter were attempted upon such body parts as length of eye and snout. Particular care was given measurements which were of special taxonomic importance. The specimens usually ranged in size and age from medium- or large-sized young to large, old adults, with the mean frequently about the size of the average adult. After all counts and measurements had been taken the mean value of each was obtained, and that specimen whose counts and measurements came closest to all the means was selected for drawing. Proportional dividers were used to transfer measurements of the specimen to that of the drawing. The drawings from which the reproductions were made were with few exceptions between 12.0″ (30.5 cm) and 14.0″ (35.6 cm) in total length.

Color descriptions and chromatophore patterns were recorded when possible from living fishes, or from dead fishes as soon after their death as practical. The fishes were then preserved in a formalin solution consisting of one part commercial formalin to 7–10 parts of water. A few weeks after preservation in formalin, another description of the then-faded colors and chromatophore patterns was taken, after which the specimens were transferred (after washing 48 hours in water) into a 70% solution of alcohol and distilled water, in which solution they remained permanently. A third set of descriptions was taken, usually within a year, after the specimens had been placed in the alcohol solution. When necessary, separate descriptions were taken of different age groups and/or sexes of the same species. Later, those colors and chromatophore patterns most persistent and indicative of the species were stippled in the drawings. I made all counts, measurements and color descriptions.

After the specimen to be drawn was selected, Mrs. E. R. Weeks or I sketched in its outlines and squamation after which I gave the drawing a thorough examination, modifying it when necessary. Mrs. Weeks assisted with those illustrations on which her initials, E. R. W., appear. All of the stippling was done by me. No conscious attempt was made to include highlights in eyes or on the body, yet the patterns of such species as the Tippecanoe darter give the illusion of having highlights.

Some may consider the lack of colors in the illustrations a handicap, and so it is in those rather few species whose colors in life are taxonomically reliable. However, coloration is of little value in many species, and in some it hides those characters which are taxonomically the most important. In addition a colored plate usually portrays only one sex, or a color phase of one sex during one period of the year, ignoring the three to six quite different color phases of the other sex, or age groups during various periods of the year.

Instructions upon How to Use the Illustrations

The fish drawings are actually diagrammatic illustrations of the species, or of one or both sexes of the species, because all of the counts, measurements and chromatophore patterns are *average* for their species. All measurements of curved surfaces, such as the snout measurement, are actual measurements and are not foreshortened as they frequently are in photographs, paintings and artistic drawings. Therefore, the average length of eye or other body length may be obtained from the drawings by accurately measuring those body parts with dividers, after which the measured body part may be divided into any other body part, thereby obtaining the average proportions between these two parts.

To Identify a Specimen:

(1) Determine the family to which it belongs. If you are not sure, use the "Key to Families of Ohio Fishes."

(2) Leaf through the pages containing the drawings and text for that family, comparing your specimen with the drawings of every species in the family until the one most like your specimen is found.

(3) Use the species analysis given under that species, checking each statement under *Characters*, including both those under *General* and *Specific* (and *Subspecific*).

(4) If the above characters agree with your specimen, check all statements under *Differs*. These statements stress the trenchant characters separating this species from others most closely resembling it, or which superficially resemble it. This section usually exposes errors made previously.

(5) The section on *Most like* and *Superficially like* emphasizes those species which must be eliminated.

(6) Coloration often aids in verifying the identification, and especially in those species in which coloration may be of taxonomic significance.

(7) *Lengths* and *weights* indicate growth rates, and may aid identification. They may disclose a misidentification, especially if your specimen is considerably larger than the maximum length and/or weight given for the species. An example is a fish 4.0″ (10 cm) in length which you identified as a Tippecanoe darter, but whose maximum length is given as only 1.8″ (4.6 cm).

(8) The section on *Hybridizes* is of especial importance when difficulty is had in satisfactorily relegating a specimen to a given species, especially when your specimen appears to fit into no species description satisfactorily but appears to fit partially two or more species. When this occurs your specimen may be a hybrid. This section gives the hybrid combinations recorded from Ohio waters. The chapter on "Natural Hybrids," pp. 119–26, will also be helpful.

Identification, Distribution and Habitat of the 166 Species and 13 Additional Subspecies of Fishes

SILVER LAMPREY

Ichthyomyzon unicuspis Hubbs and Trautman

Fig. 1

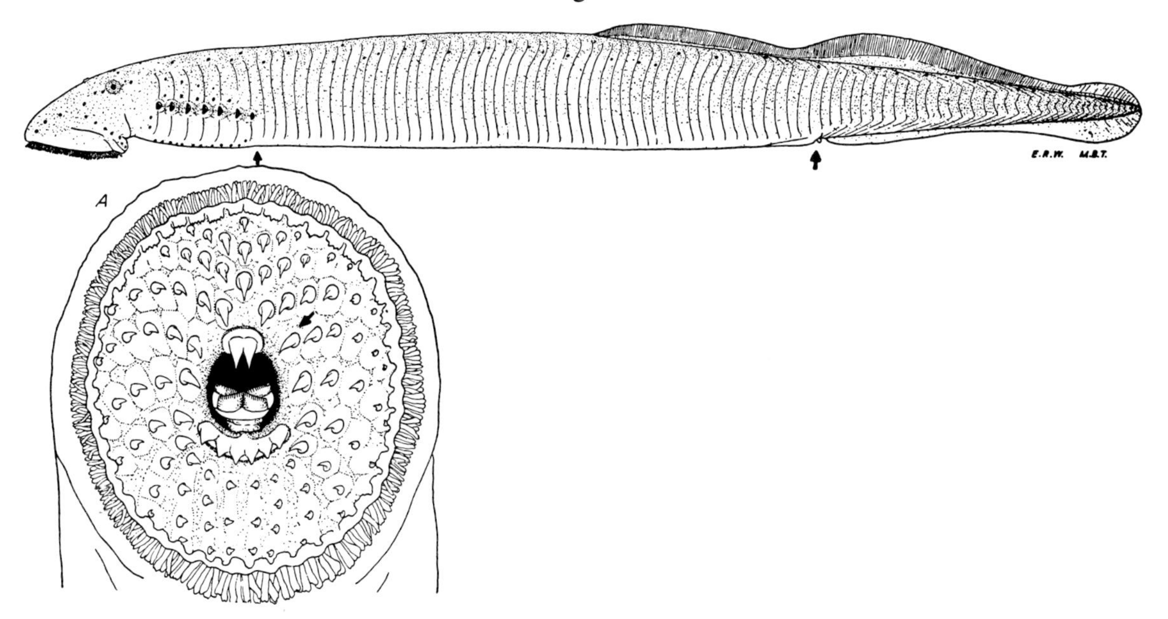

Upper fig.: Lake Erie, Ottawa County, O.
April 20, 1940. 253 mm TL, 10.0" TL.*
Adult female. OSUM 3014.†

Anterior arrow points to the first counted myomere after the last gill opening; posterior arrow points to the last counted myomere before the anus; all myomeres including, and between, these two are counted.

Fig. A: view of mouth; arrow points to one of the 1-pronged (unicuspid) teeth of the circumoral series; also note teeth radiating from the center.

Identification

Adult characters: Dorsal fin not divided. Lateral teeth of circumoral series sharp, long, and with only 1 point or prong, fig. A. Myomeres usually 50–53, rarely to 56, upper figure. Jawless mouth (sucking disc) may be expanded wider than body width. Parasitic on fishes.

Differs: Sea, American brook and least brook lampreys have dorsal divded into 2 fins. Northern brook lamprey has the outer teeth of the radiating series too small to be readily seen, or are hidden by skin. Mountain brook and Ohio lampreys have some or all lateral teeth 2-pointed (bicuspid) and the myomere count is usually more than 55.

Most like: Ohio lamprey.

Ammocoetes differ from adults: Lack teeth. Eyes

* TL. = Total Length; SL. = Standard Length.

† In January, 1956 the trustees of The Ohio Historical Society accepted as a gift from the trustees of The Ohio State University the fish collections of the University, then housed at the Franz Theodore Stone Laboratory at Put-in-Bay, Ohio. They have been removed from the Stone Laboratory and are now in The Ohio State Museum, where they are being integrated with the fish collection there. They will be available to qualified persons for study by 1957 after which all specimens having collection numbers preceded by OSU or OSM will be at The Ohio State Museum at Columbus.

not functional. *See* verbal key and distribution of the various species for tentative identification of ammocoetes.

Coloration: *Adults* and *ammocoetes* light tan or silvery-tan, the spawning *adults* becoming increasingly darker as spawning advances, until near spawning completion, they are blue-black.

Lengths: *Ammocoetes*, maximum 7.0″ (18 cm) TL. Newly transformed individuals 6.0″ (15 cm) TL. Spawning *adults* 10.0″–14.0″ (25.4–35.6 cm) TL.

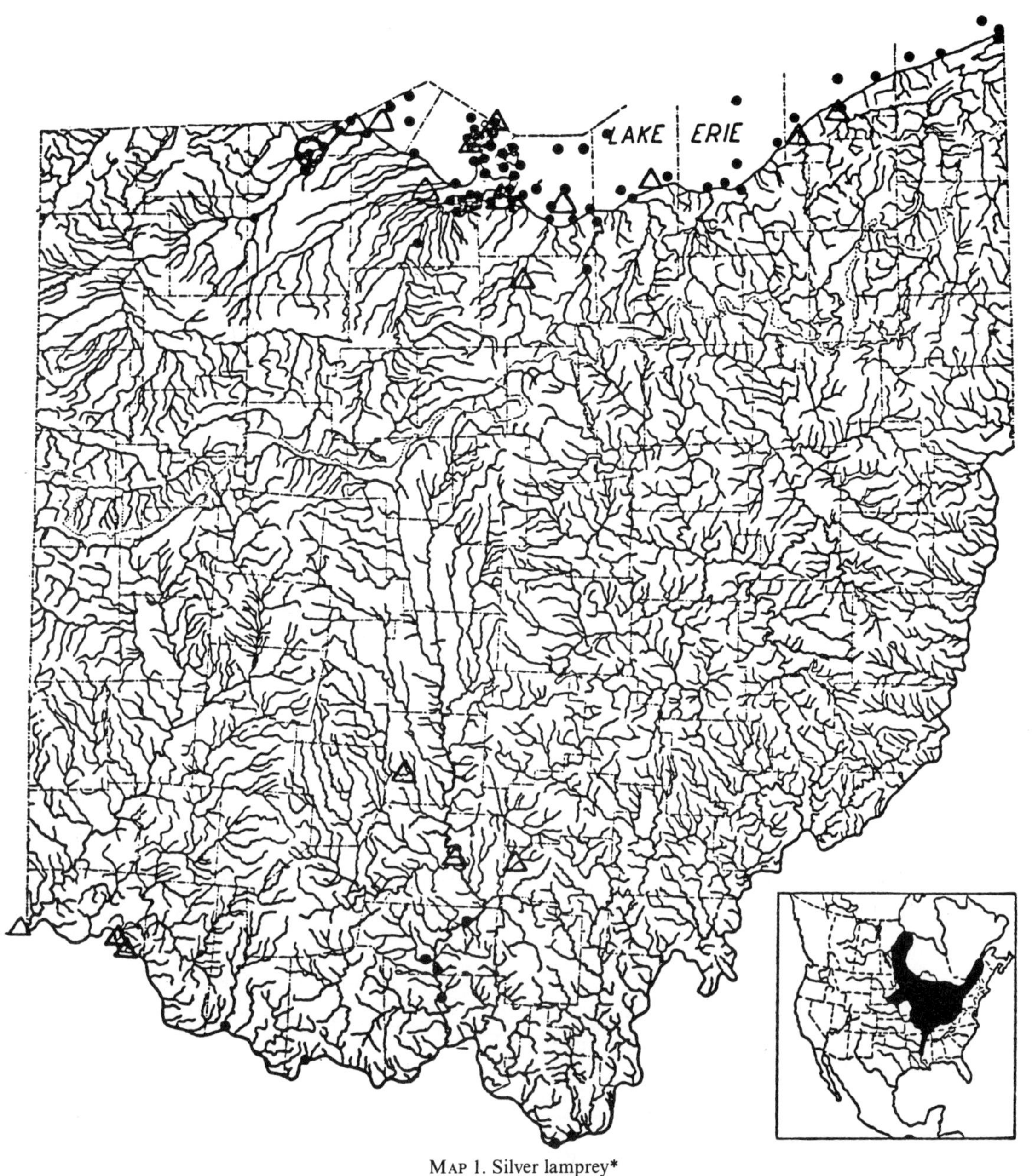

MAP 1. Silver lamprey*

○ Type locality, Swan Creek, Lucas County, O. Locality records. ● Before 1955. △ 1955–80.
Insert: No apparent recent change in range size.

* Ohio distribution shown on the large map in each case; general distribution indicated on the small insert map of North America. A separate map is used where some subspecies are discussed. With each map the name of the species or subspecies is repeated, together with the appropriate legends for the Ohio map and the North American *insert*.

Hybridizes: With Ohio lamprey, *see* Hubbs and Trautman (1937: 14–18).

Distribution and Habitat

Ohio Distribution—Kirtland (1841B:343) referred to this species when he wrote in 1841 that a parasitic lamprey, 12.0″–15.0″ (30–38 cm) long, was present in Lake Erie. A decade later he (1851Q:205) stated that "we have found it very abundant in the waters of Lake Erie" and "During the month of May, immense numbers of this Lamprey ascend the streams in the vicinity of Cleveland." McCormick (1892:9) on May 22, 1887, observed a spawning run; this one was in the Vermilion River. Other less reliable literature references previous to 1900 also refer to its great abundance in Lake Erie and tributaries.

Between 1934–39, Bernard R. and Louis W. Campbell, and I observed the spawning runs of the silver lamprey in Swan Creek, Lucas County, and especially as they attempted to surmount the South Avenue dam in Toledo at night. Below this dam, on June 2, 1935, I observed about 20 individuals spawning in 14 nests.

All evidence, including statements from many commercial fishermen, indicates that the Lake Erie population of silver lampreys has declined drastically since 1900. A commercial fisherman and fish buyer of Venice, Charles Nielsen, told me that before 1900 silver lampreys were so numerous in Sandusky Bay and adjacent Lake Erie that his father was able to sell hundreds of them to biological supply houses; that as late as 1920 a dozen lampreys could usually be found attached to fishes taken from one net, but that after 1945 lampreys were so uncommon that only an occasional one was taken. In 1951 several commercial fishermen told me that for the first time they did not see a single silver lamprey during the spring fishing season.

In the Ohio drainage, the silver lamprey was present chiefly in the Ohio River and lower portions of its larger tributaries. In this drainage it was in competition with its close relative, the Ohio lamprey. Rivermen claimed that during the past 50 years, parasitic lampreys had decreased greatly in abundance in the Ohio River.

On, or a few days previous to, May 25, 1952, some bait-seiners found a spawned-out silver lamprey (OSUM:9312) that was clinging to a rock in Sunfish Creek, about 1 1/4 miles above its confluence with the Scioto River, in Newtown Twp., Pike County. Obviously the lamprey had spawned in the immediate vicinity. Since 1930 fishermen have told me at various times of seeing large lampreys clinging to rocks in this section of Sunfish Creek during late May. This appears to be the only definite spawning locality for the species in the Ohio River drainage.

Habitat—In order to understand the habitat requirements of parasitic lampreys it is necessary to know their life history. In the spring, after the water temperature reaches 50°F (10°C) or more, adult silver lampreys begin to migrate upstream to spawn in the nests they build in the sand- and gravel-bottoms of riffles of moderate-sized streams which have moderate gradients. After hatching, the young dig into bars and beds containing a combination of sand, dark muck, and organic debris which is free from clayey silt. There they remain an unknown period of time, inhabiting their semicircular tunnels, and from the upstream opening of which they thrust their heads in order to feed upon the downstream-drifting microscopic food. In spring, those silver lampreys which have transformed from the ammocoete to adult stage, and which then are 4.0″–6.0″ (10–15 cm) long, begin to drift downstream into larger waters, to parasitize the more abundant fish supply. In this respect all parasitic lampreys differ from nonparasitic or brook lampreys in that the latter do not feed after transforming but migrate immediately upstream to spawn, instead of migrating downstream to feed (*see* northern brook lamprey, under Habitat). The newly transformed silver lampreys enter Lake Erie in late April, May, and June, remaining until the following April or May, when as mature adults, 9.0″–14.0″ (23–36 cm) long, they migrate upstream to spawn.

Such a life history requires rather clear waters in order to capture prey; clean stream bottoms of sand, organic debris, and gravel which are not covered by silt or otherwise polluted, for ammocoete development; and no dams in the streams to prevent migration. Such conditions prevailed in Ohio waters prior to 1875, and then these lampreys were abundant; since then the ever-increasing turbidity, siltation and other pollutants, and dams have resulted in destroying their habitats or preventing migration, thereby drastically reducing their numbers.

Although the silver lamprey was well known for many years, it remained technically undescribed until 1937 (Hubbs and Trautman, 1937:53–65).

Years 1955–80—Statements by such commercial fishermen as the late Charles Nielsen indicate that the numerical status of the silver lamprey has continued to decrease throughout this 25-year period, as shown by the fewer numbers of captured specimens and the decreased numbers of scars on fishes. The latter is significant when it is realized that scars on fishes in Lake Erie are made by both the silver and sea lampreys, the fewer number of scars suggesting that both species were present in small numbers throughout the period and the silver lampreys were possibly decreasing in abundance. During late May and June of 1972 Andrew White and his associates (1975:48) found adult silver lampreys in spawning condition in the lower Chagrin River. Discussing abundance of the silver lamprey with Ohio River fishermen, I learned that they believe the silver and Ohio lampreys have decreased in the river during the past decade. On July 28, 1964 Roger Troutman, Frank Bolin and I found a dying subadult, 7.5″ (19 cm) TL, in the Scioto River immediately above the city of Chillicothe in Ross County. It and many species of fishes were dying from, or were killed by, pollution. Since 1970, Andrew White and others have seen or collected specimens from Salt Creek in Vinton County. Recently members of the Ohio Division of Wildlife removed a silver lamprey from a fish taken in an experimental trap net in Deer Creek, Pickaway County. For additional information relative to life histories, *see* R. H. Morman, 1979.

NORTHERN BROOK LAMPREY

Ichthyomyzon fossor Reighard and Cummins

Fig. 2

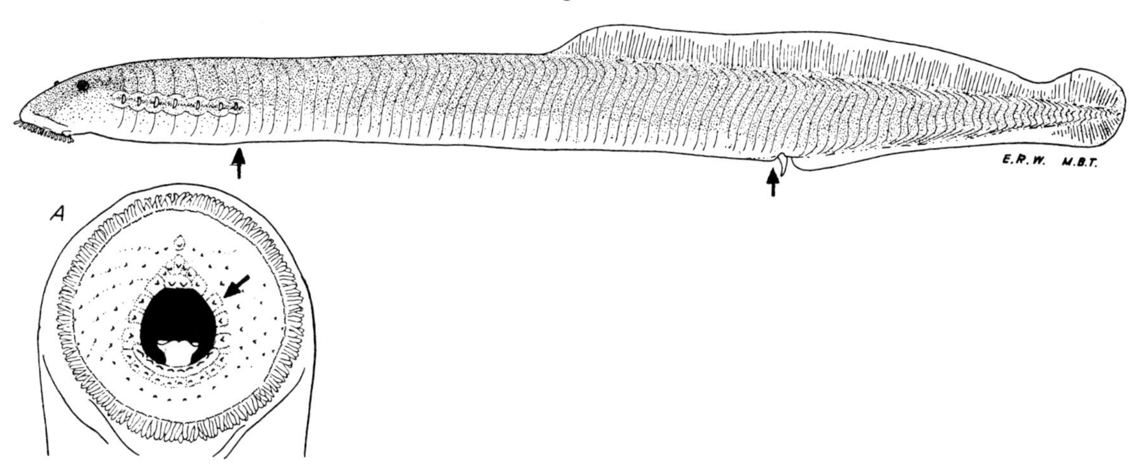

Upper fig.: West Branch, St. Joseph River, Williams County, O.

May 16, 1941. 127 mm TL, 5.0″ TL.
Adult male. OSUM 3189.

Anterior arrow points to the first counted myomere after the last gill opening; posterior arrow points to the last counted myomere before the anus; all myomeres including, and between, these two are counted.

Fig. A: view of mouth; arrow points to one of the 1-pronged teeth of the circumoral series; note the small and partly hidden teeth radiating from the center.

Identification

Adult characters: Dorsal fin not divided. Lateral teeth of circumoral series blunt, small, and with only one prong, fig. A. Radiating teeth blunt, small, many hidden by skin. Myomeres usually 49–55, extremes 47–56, upper fig. Jawless mouth very small, the expanded opening not as wide as body. Not parasitic on fishes. Alimentary tract reduced to a strand in adults.

Differs: Silver lamprey has long, sharp teeth, and very wide sucking mouth. *See* silver lamprey.

Most like: Silver lamprey when it is transforming and does not have the sucking disc and teeth fully developed.

Ammocoetes differ from adults: *See* silver lamprey.

Coloration: *Ammocoetes* and recently transformed *adults* light tan and silvery. Spawning *adults* dark olive-tan, becoming blue-black as spawning comes to an end.

Lengths: Maximum for *ammocoetes* and *adults* 7.0″ (18 cm) TL. *Adults*, usually 5.0″–6.8″ (13–17.3 cm) TL.

Distribution and Habitat

Ohio Distribution—On May 16, 1941, I collected six northern brook lampreys (OSUM:3189) in the West Branch of St. Joseph River, Williams County. This constitutes the first recorded instance of its presence in Ohio, as well as in the Maumee River drainage. On May 21, 1941, I collected three more (OSUM:3209) at the same locality.

On May 14, 1950, G. G. Acker and his class from Bowling Green State University collected six specimens (OSUM:9219) in Phelps Creek, a western tributary of the Grand River, Ashtabula County. On April 28, 1951, my wife and I visited the high-gradient tributaries which enter the Grand River from the west, finding migrating or hiding northern brook lampreys in Indian, Haskins, Crooked and Trumbull creeks (OSUM:9221). All localities mentioned above were visited because studies of Ohio

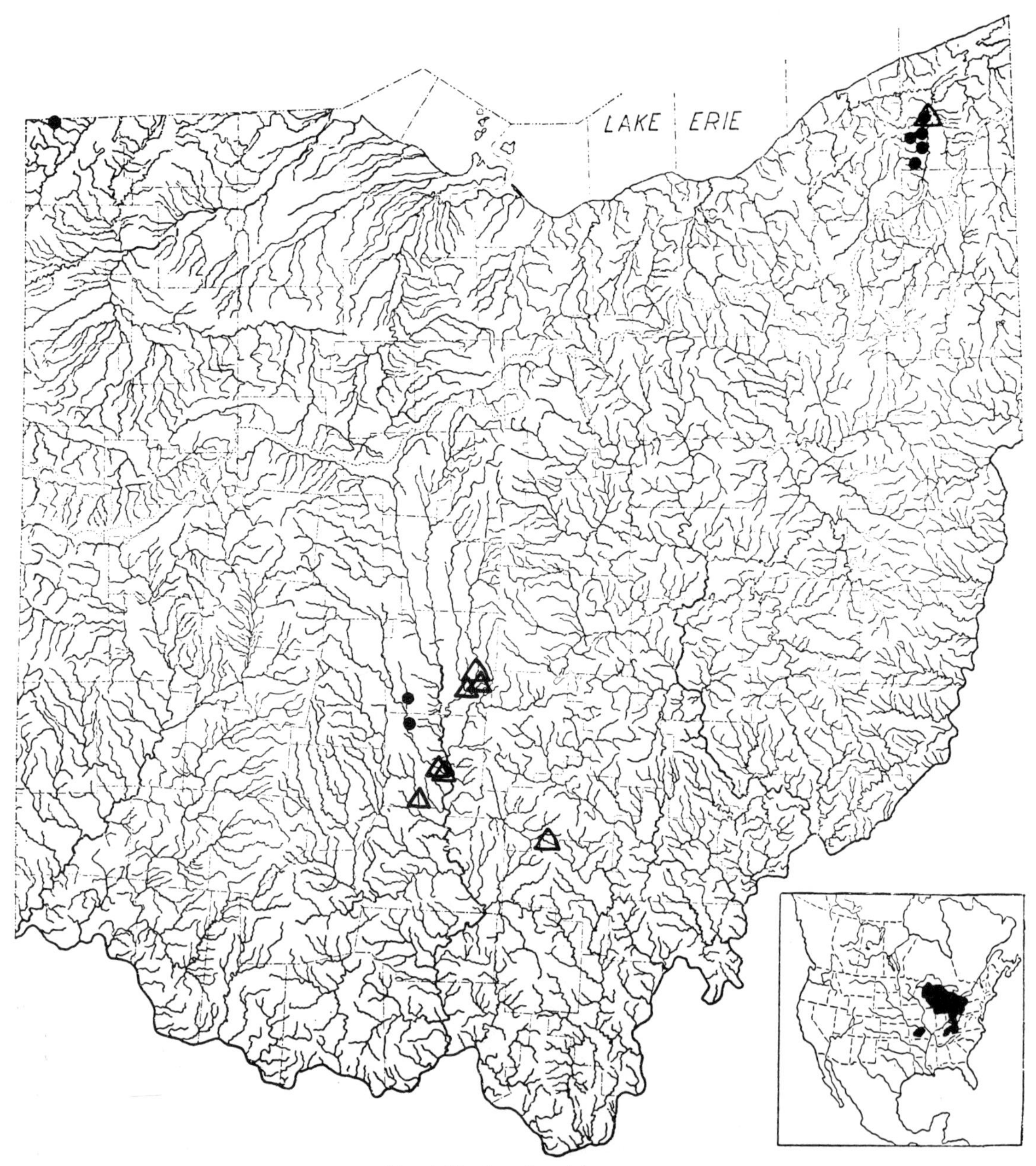

MAP 2. Northern brook lamprey

Locality records. ● Before 1955. △ 1955–80.
Insert: No apparent recent change in range size.

stream gradients indicated the possibility that lampreys might occur in them.

On May 22, 1956, Donald I. Mount and I collected an almost-spent female northern brook lamprey (OSUM:9500) on a sand and gravel riffle of Big Darby Creek in Scioto and Darby townships, Pickaway County. This riffle is situated in a short stretch of river in which the gradient is about 15′ (4.6 m) per mile. This capture constitutes the first record for this species in the Ohio River drainage in Ohio; it has been taken in this drainage in the Tippecanoe River in Indiana (Leach, 1940:21–34).

Hellbranch Run enters Big Darby Creek about 8 miles (13 km) upstream from the above-mentioned riffle. Hellbranch has all of the qualifications for an excellent lamprey spawning stream; it has a gradient of about 17 feet per mile (3.2 m/km) from its mouth to 4 miles (6 km) upstream, a bottom of sand, gravel, boulders, the stream usually flows throughout the year and it has remarkably clear waters for a central Ohio stream. On May 24, in a pool below a riffle which appeared to be ideal for spawning lampreys, I found portions of the white notochords (the fairy necklaces of the hill people of southern Ohio) of five lampreys and part of the skin and flesh of another. Apparently these lampreys had spawned during the previous week. I presume they were *fossor*.

Habitat—As with the parasitic lampreys, some knowledge of the life history of brook lampreys is necessary before their habitat requirements can be understood (*see* silver lamprey under Habitat). The northern brook lamprey, as with other brook lampreys, requires two distinctly different habitats, one for spawning adults, the other for ammocoete development. Spawning adults require small, clear, high-gradient brooks having riffles with bottoms of sand and gravel; after hatching the ammocoetes need permanent waters of lower gradients having bottoms containing bars and beds of mixed sand and organic debris. It is therefore necessary that the high-gradient brooks flow into low-gradient waters having the above characteristics. The most favorable waters are those containing few predacious fishes large enough to feed upon lampreys, for both ammocoetes and adults are very vulnerable to predation by fishes. Unlike parasitic lampreys, brook lampreys after transformation do not go downstream to feed upon fishes, but after leaving their burrows, migrate upstream to spawn. As with all species of lampreys, they die immediately after spawning.

Years 1955–80—As stated above, the first recorded specimen of the northern brook lamprey in the Ohio River drainage of Ohio was taken May 22, 1956, in Big Darby Creek, northwestern Pickaway County. Two days later in an upstream tributary, Hellbranch Run, in southwestern Franklin County, portions of five notochords were taken, presumed to be from individuals of this species.

Realizing that one or more populations of this species must exist in the middle section of the Scioto drainage, extensive efforts were made to find them. On May 5, 1959, William L. Pfleiger captured an adult in Blacklick Creek, eastern Franklin County, and since then we have taken many adults and some ammocoetes from that stream and near its mouth in Big Walnut Creek. A few years ago the late Alvah Peterson gave me two ammocoetes which he captured September 19, 1938, in Blacklick Creek while collecting larval invertebrates. From the locality and myomere counts, these ammocoetes are presumed to be of this species.

On May 4, 1968, the late G. Acker and his class from Bowling Green State University collected nine spawning adults on a riffle of Big Darby Creek, located one mile south of Fox and upstream from the bridge on State Route 104. Although investigated in 1969, none were observed. On May 11, 1970, Carolyn (Cooper) Jenkinson, Mary Trautman and I obtained three which were largely spent. This riffle is the one in which, since 1925, a larger number of seinings and a greater number of hours of observations have been made than in any other locality in Ohio, dramatically illustrating the extreme difficulty involved in locating populations of non-parasitic lampreys. It would appear to be impossible that so vast an amount of investigation over a 45-year period could have been made without taking a single adult or ammocoete.

On May 6, 1969, Joanne E. Stillwell and James K. Bissell collected two adults on a riffle in Deer Creek, Pickaway County, the third major tributary of the Scioto River system in central Ohio in which the species is now known to occur. Undoubtedly other, and in some instances larger, populations exist elsewhere in Ohio waters and especially in Lake Erie tributaries, which have been little investigated.

On the evening of April 20, 1972, the then turbid Olentangy River in Franklin County was slightly overflowing its banks. Big Run, entering the Olentangy from the east and flowing through Whetstone Park, was very clear with the water level slightly above spring normal. Before 1930 Big Run near its mouth contained more than 20 species of fishes, but by 1957 it had been partly modified into a storm sewer. I have recorded since then only a few individuals of six species of fishes. It becomes intermittent or is completely without water shortly after a rain.

On this evening while walking along the stream about 100′ (30.5 m) from its mouth, where the moving water of the creek merged into the backed-up Olentangy River, I saw a lamprey swimming about in the clear water. With field glasses I noted that it was about 6″ (15 cm) long, brown in color and that it had a single dorsal fin. Getting a small seine at my home nearby and accompanied by my wife, I went to where the lamprey was last seen, unsuccessfully looking for it. Since *Ichthyomyzon fossor* is the only *Ichthyomyzon* recorded in the upper half of the Scioto River drainage I am assuming it was a stray of that species.

OHIO LAMPREY

Ichthyomyzon bdellium (Jordan)

Fig. 3

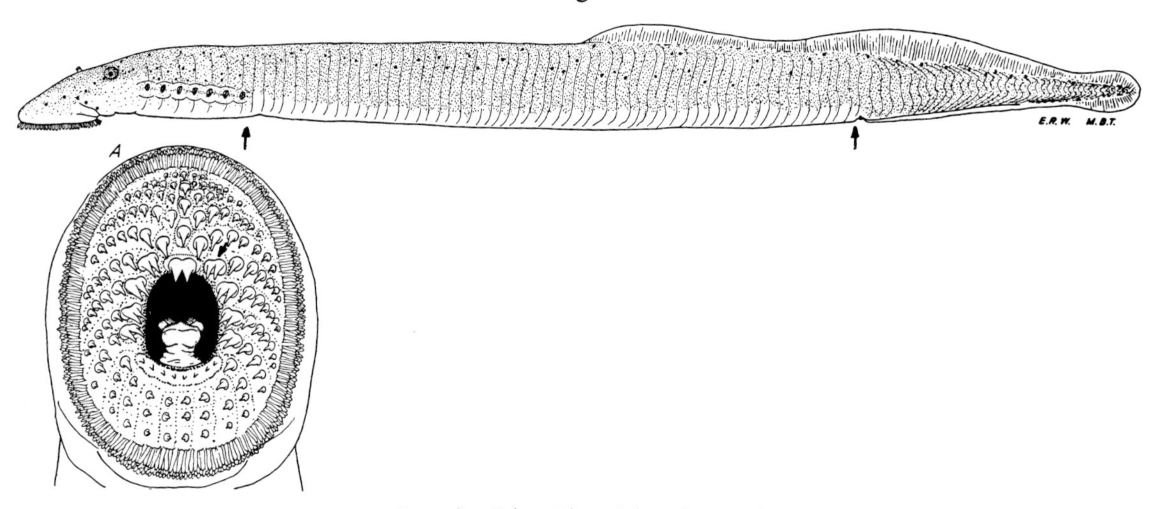

Upper fig.: Scioto River, Scioto County, O.

July 12, 1940. 224 mm TL, 8.8″ TL.
Female. OSUM 2896.

Anterior arrow points to first counted myomere after the last gill opening; posterior arrow points to last counted myomere before anus; all myomeres including, and between, these two are counted.

Fig. A: view of mouth; arrow points to one of the 2-pronged (bicuspid) teeth of the circumoral series; note the sharply-pointed teeth radiating from the center.

Identification

Adult characters: Dorsal fin not divided, although some individuals have fin deeply notched. At least one of the lateral teeth of the circumoral series is 2-pronged (bicuspid), fig. A. Radiating teeth long and sharp. Myomeres usually 55–61, extremes 53–62, upper fig. Jawless mouth may be expanded as wide as body. Parasitic. Alimentary tract of adults functional, except in spawning individuals.

Differs: Silver and northern brook lampreys have only 1-pronged circumoral teeth. Mountain brook lamprey is not parasitic; has smaller mouth. *See* silver lamprey.

Most like: Mountain brook and silver lampreys.

Ammocoetes differ from adults: *See* silver lamprey.

Coloration: *See* silver lamprey.

Lengths: As in silver lamprey.

Hybridizes: Rarely, *see* silver lamprey.

Distribution and Habitat

Ohio Distribution—During the 1920–50 survey, Ohio lampreys were examined which came from the Ohio River as far upstream as Lawrence County, and from the lower Scioto River; in addition, many commercial and sport fishermen described to me parasitic lampreys of this species or the silver lamprey, which were taken in the Ohio River as far upstream as Captina Creek, Belmont County, and in the lower portions of the Muskingum, Leading, Ohio Brush, Little and Great Miami rivers. During this survey the Ohio lamprey appeared to be considerably more numerous in southern Ohio than was the silver lamprey. Commercial fishermen were unanimous in the opinion that parasitic lampreys have decreased greatly in abundance in the Ohio River in recent years.

The Ohio lamprey was first described by Kirtland (1838:170; more fully in 1841B:343, as *Petromyzon argenteus*) from a specimen supposedly taken in the

MAP 3. Ohio lamprey

○ Type locality, Great Miami River. Locality records. ● Before 1955. △ 1955–80.
Insert: A probable decrease in range size since 1900, a result of destruction of habitat.

Great Miami River and was attached to a walleye. (Later, Kirtland, 1851Q:206, stated that the lamprey and walleye came from the Ohio River; for a more complete explanation, *see* Hubbs and Trautman, 1937:86, under *Petromyzon argenteus.*)

Habitat—Similar to the silver lamprey (*see* that species under Habitat) ascending rather small streams to spawn, the newly transformed sub-adults descending into the lower courses of the larger rivers or into the Ohio River, where there is a sufficient population of large fishes upon which to prey.

Years 1955–80—On February 28, 1965, an Ohio lamprey attached to a sucker was taken in the Ohio River at Marietta, Washington County. In October of 1969, Andrew White obtained from Salt Creek, about 2 miles (3 km) south of the Hocking County line in Vinton County, a recently metamorphosed Ohio lamprey which was attached to a white sucker.

During a 3-year-investigation of the entire Ohio River by members of the Department of Biology of the University of Louisville, during which more than 200 collections and many observations were made, only seven lampreys of this species were noted. From this evidence and that of fishermen with whom I conversed, it appears that the population in the Ohio River of the Ohio lamprey, and silver lamprey as well, is continuing to decrease.

MOUNTAIN BROOK LAMPREY*

Ichthyomyzon greeleyi Hubbs and Trautman

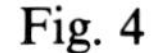

Fig. 4

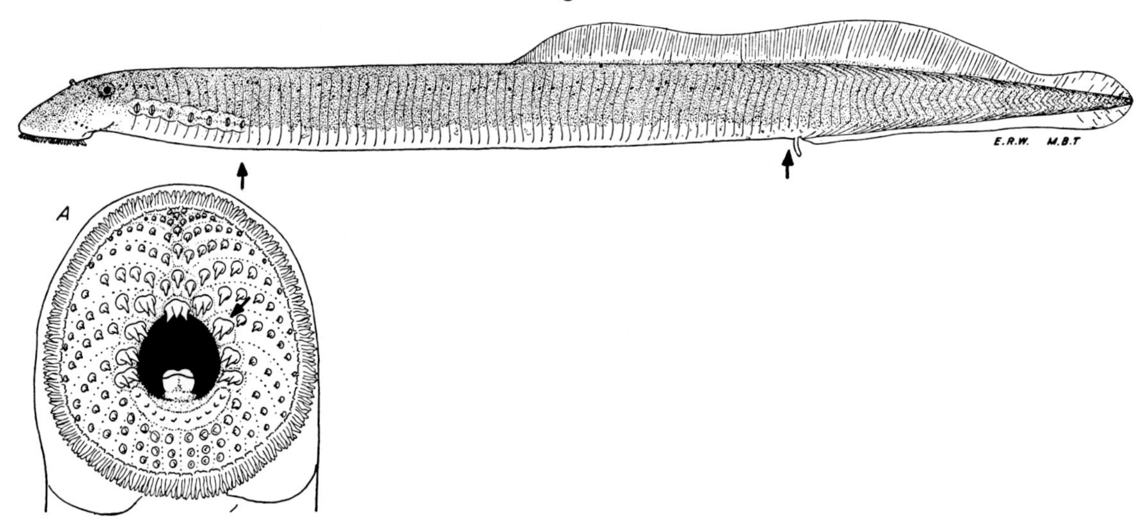

Upper fig.: Mahoning River, Portage County, O.

May 10, 1941. 127 mm TL, 5.0″ TL.
Adult male. OSUM 3132.

Anterior arrow points to first counted myomere after the last gill opening; posterior arrow points to last counted myomere before anus; all myomeres including, and between, these two are counted.

Fig. A: view of mouth; arrow points to one of the 2-pronged circumoral teeth; note sharply-pointed teeth radiating from the center.

Identification

Adult characters: Dorsal fin not divided although some individuals have fin deeply notched. At least one of the lateral teeth of circumoral series is 2-pronged, fig. A. Radiating teeth well developed, especially for a non parasitic species. Myomeres usually 55–60, extremes 53–62, upper fig. Jawless mouth small, cannot be expanded as wide as body. Alimentary tract non-functional and reduced to a strand in adults.

Differs: Ohio lamprey has a larger mouth and is parasitic. American brook and least brook lampreys have no radiating series of teeth. Northern brook lamprey has all circumoral teeth 1-pronged, and all are very blunt.

Most like: Ohio lamprey.

Ammocoetes differ from adults: *See* silver lamprey.

Coloration: *Ammocoetes* light silvery-tan, often bi-colored. Recently transformed *adults* are distinctly bi-colored with olive-tan above, silvery-tan below. Spawning *adults* become a progressively darker blue-black as spawning progresses.

Lengths: Maximum for ammocoetes 7.0″ (18 cm) TL. *Adults*, 5.0″–7.0″ (13–18 cm) TL. Largest *adult*, 7.8″ (20 cm) TL.

Distribution and Habitat

Ohio distribution—In 1841 Kirtland (1841B:-473–75) described *Ammocoetes concolor* from ammocoetes collected in Franklin County (central Ohio, *see Lampetra aepyptera*) and the Mahoning River (presumably near his home, *see* hollow circle in Mahoning County). It appears evident that the ammocoetes from the Mahoning River were mountain brook lampreys, because as indicated below, this species is the most widely distributed and

---

* Allegheny brook lamprey in 1957 edition of *Fishes of Ohio*.

MAP 4. Mountain brook lamprey

○ No preserved specimens. ● Specimens preserved. Locality records. ● Before 1955. △ 1955–80.
Insert: Presumably a decrease in range size since 1900.

abundant brook lamprey of the upper Allegheny system, including the Mahoning River.

In 1901 Williamson (1901:165) recorded that "Herbert McCane has taken a specimen of *Ichthyomyzon concolor* which is preserved in the Salem High School collection." This specimen was collected in a small tributary of the Mahoning River southwest of Salem in Columbiana County (*see* hollow circle in Columbiana County). It presumably was a mountain brook lamprey. In 1931 I visited Salem High School, finding there a few dried-up fish specimens collected by Williamson, but I did not find the lamprey. Also, upon several occasions when I visited the upper Mahoning River tributaries from whence came this specimen, it was apparent that these tributaries had contained suitable lamprey habitats before they were heavily polluted and dammed; in fact, residents described lampreys to me which they had seen in the cleaner portions of these tributaries.

On May 18, 1940, Louis H. Swart, of Warren, Ohio, collected three mountain brook lampreys (two preserved in OSUM:2105) from the West Branch of the Mahoning River in Paris Township, Portage County. These are the first specimens known to be extant. In this stream on May 10–11, 1941, in the townships of Ravenna and Shalersville, Portage County, my wife and I found large colonies of spawning adults; also a migrating adult in Paris Township. Since 1941 we have found adults in this stream during several different years.

In the upper Allegheny River system of northwestern Pennsylvania, Raney (1938:8–9) found this species to be widely distributed and locally abundant; included were localities which were only a short distance from the Ohio border. In 1939 he (1939B:111–12) observed them breeding in Lawrence County, Pennsylvania, the county adjacent to Mahoning County, Ohio.

From the above records, statements of residents, and particularly from the studies of stream gradients, it can be assumed with certainty that the mountain brook lamprey was widely distributed in the Allegheny and upper Ohio river systems in Ohio before these tributaries were heavily polluted or dammed. Conditions were particularly favorable for this species in the following streams, and it may be still present in some: Columbiana County—West Fork of Little Beaver, especially Cold Creek; Little Bull, Big Bull, and Elk creeks of the Middle Fork; North Fork in northeastern Unity Township; Mahoning County—Yellow, Mill (Beaver and Boardman townships), Meander and Mill (Goshen and Berlin townships) creeks, all tributaries of the Mahoning River; Trumbull County—Big and Little Yankee creeks, both tributaries of Shenango River, and Eagle Creek, tributary of the Mahoning River.

Habitat—*See* northern brook lamprey under Habitat.

Years 1955–80—In the spring of 1964 Roy A. Stein (now of the Department of Zoology, Ohio State University) and Dennis Stewart found spawning mountain brook lampreys in Eagle Creek, Trumbull County, a specimen of which is preserved in the state collections (OSUM 13856). This constitutes a slight extension of its known range in Ohio. According to Stein, the presence of the species in Eagle Creek has been known by a few persons for many years. It seems to have been absent every other year, and it is possible that one or more year classes have been eliminated. The population still exists in the West Branch of the Mahoning River, although it appears to be decreasing in numbers because of the partial destruction of its habitat.

SEA LAMPREY

Petromyzon marinus Linnaeus

Fig. 5

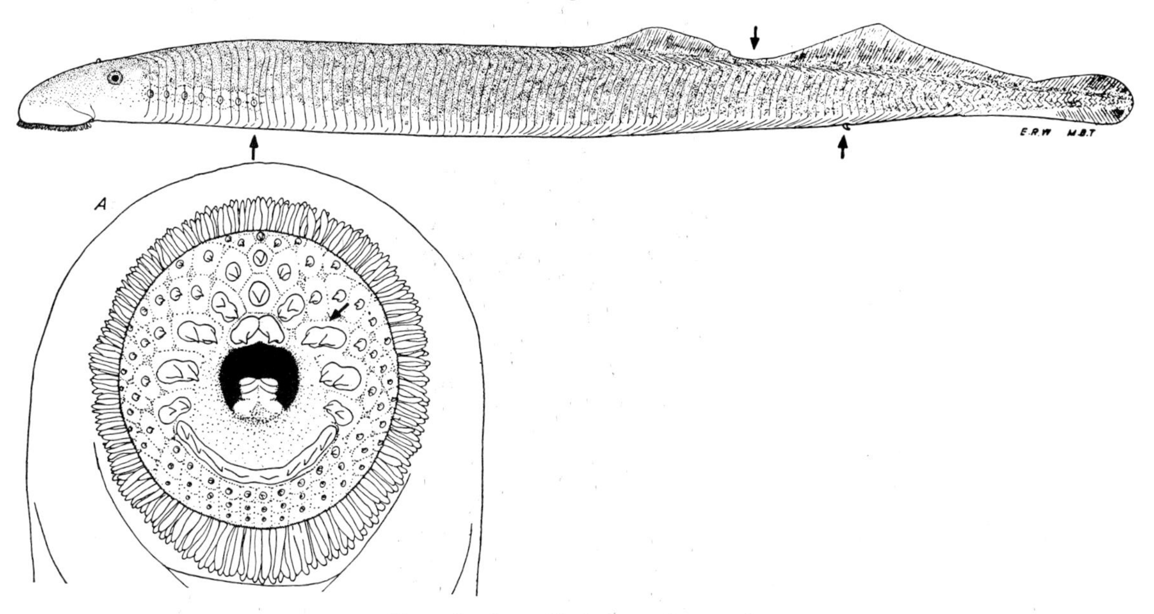

Upper fig.: Swan Creek, Lucas County, O.

April 26, 1935. 418 mm TL, 15.7" TL.
Adult female. OSUM 8579.

Upper arrow points to notch that completely separates the dorsal fin. Anterior arrow points to first counted myomere after the last gill opening; posterior arrow points to last counted myomere before anus; all myomeres including, and between, these two are counted.

Fig. A: view of mouth, showing greatest constriction in diameter possible; when fully expanded mouth is as wide as body (all lamprey species have ability to so constrict mouth diameter). Arrow points to one of the 2-pronged circumeral teeth.

Identification

Adult characters: A wide notch completely separates the dorsal fin. At least one tooth of the circumoral series 2-pronged; usually all are, fig. A. Radiating teeth sharp, large and long. Myomeres usually 65–76, extremes 63–80, upper fig. Jawless mouth large, can be expanded as wide as body. Parasitic. Alimentary tract of adults functional except in spawning individuals. Body more boldly mottled and spotted with dark chocolate than is any other species.

Differs: Least brook and American brook lampreys, the only other species with a divided dorsal fin, are not parasitic, smaller, and do not have radiating teeth.

Most like: Silver and Ohio lampreys when large; American brook lamprey when recently transformed.

Ammocoetes differ from adults: *See* silver lamprey.

Coloration: *Ammocoetes* are tan above, lighter below, sometimes mottled with darker. Non-spawning *adults* are dark tan or brown, usually blotched heavily with chocolate. Spawning *adults* are blue-black.

Lengths: Maximum for *ammocoetes*, 7.0" (18 cm) TL. *Adults*, 6.0"–25.0" (15–63.5 cm) TL. Spawning *adults*, usually 12.0"–25.0" (30.5–63.5 cm) TL.

Distribution and Habitat

Ohio Distribution—Although the sea lamprey presumably was present in the lower St. Lawrence

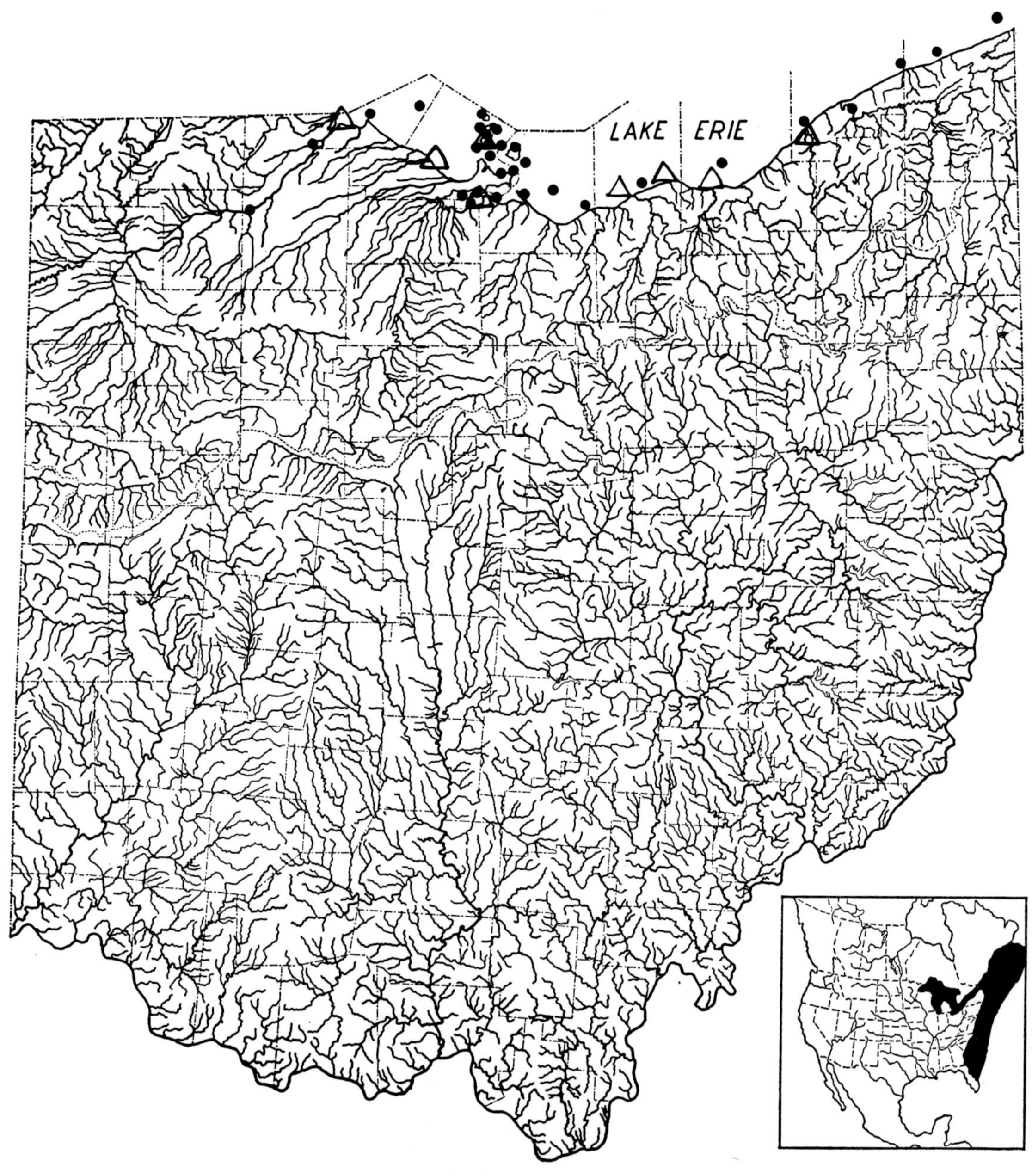

MAP 5. Sea lamprey

Locality records. ● Before 1955. △ 1955–80.
Insert: Recent extension of range into Great Lakes waters above Niagara Falls.

drainage since early post-glacial times, it was prevented from migrating into the upper Great Lakes until 1829 because of its inability to surmount Niagra Falls. In 1829 the completion of the Welland Canal made it possible for the sea lamprey to migrate from Lake Ontario into Lake Erie. For some unexplained reason this lamprey failed to take advantage of this migration route, for it was not until November 8, 1921, almost a century later, that the first specimen was recorded for Lake Erie. This specimen was captured in Canadian waters near Merlin, Ontario (Dymond, 1922:60). The first specimens taken in the Ohio waters of Lake Erie were captured in the autumn of 1927 (Osburn, Wickliff, Trautman, 1930:170).

The first indication of a spawning run in a Lake Erie tributary was that of an adult, collected May 8, 1932, in the Huron River near Flat Rock, Michigan. The first indication of a spawning run from Lake Erie into an Ohio stream was that of an adult, collected May 8, 1934, by Herbert Campbell in Swan Creek at the South Avenue dam in Toledo. The next year, Bernard R. and Louis W. Campbell collected two specimens at the same locality, and on June 2, 1935, I observed two sea lampreys attempting to build a nest on a riffle below the dam.

The sea lamprey has remained uncommon in Lake Erie since its introduction (*see* Habitat). This was not true for the remaining lakes, for after the first specimens were recorded from Lakes Huron and Michigan in 1936, the species increased rapidly in numbers until it had become immensely abundant; by 1946 it had invaded Lake Superior, where it also rapidly increased in abundance (Hubbs and Pope, 1937:172–75; Shetter, 1949: 160–76; Applegate, 1950:2–7).

Habitat—As indicated under silver lamprey (*see* Habitat) all parasitic lampreys have three essential requirements: (1) streams without unsurmountable dams having riffles of clean sand and gravel in which adults spawn; (2) beds of sand mixed with organic debris, or muck that are comparatively free of smothering silts or other injurious pollutants for ammocoete development; (3) large waters containing an abundant fish supply suitable for prey for the transformed lampreys. Lake Erie supplies the third requirement, but its tributaries virtually lack the other two; therefore, the sea lamprey remains uncommon in Lake Erie waters. On the other hand, Lakes Huron, Michigan, and Superior with their clear tributaries abundantly contain all requirements; hence the sea lamprey is abundant (Trautman, 1949: 1–9).

Years 1955–80—During this 25-year period, there was an increased intensity in fishery investigations in Lake Erie. From the evidence obtained by state and federal personnel and other individuals, it is the general opinion that the Lake Erie population of sea lampreys continues to be very small in number and may be decreasing, at least in Ohio waters.

In early June of 1972 Andrew White and associates (White et al., 1975:48) found adult sea lampreys in spawning condition in the lower Chagrin River.

For biology of larval sea lampreys in a Lake Superior, Michigan, tributary, *see* Manion and McLain, 1971:35.

LEAST BROOK LAMPREY*

Lampetra aepyptera (Abbott)

Fig. 6

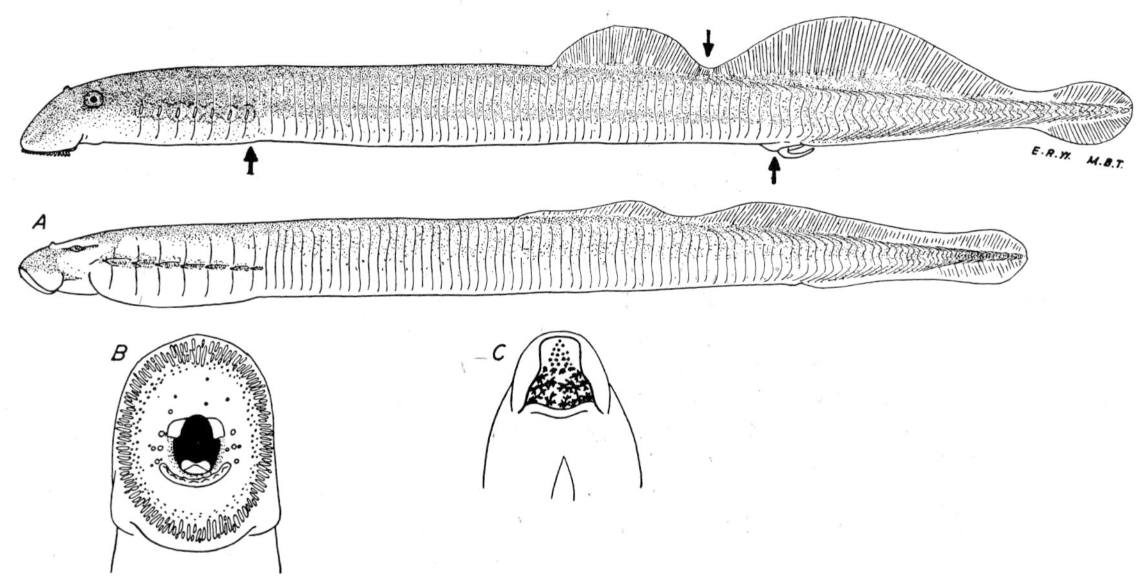

Upper fig.: Little Scioto River, Jackson County, O.

April 4, 1943. 130 mm TL, 5.0″ TL.
Adult male. OSUM 4729.

Upper arrow points to notch that separates the dorsal fin. Anterior arrow points to first counted myomere after the last gill opening; posterior arrow points to last counted myomere before anus; all myomeres including, and between, these two are counted.

Fig. A: Shade Creek, Meigs County, O.

May 25, 1939. 78 mm TL, 3.0″ TL.
Ammocoete or larva. OSUM 263.

Superficially the ammocoetes of all species of lampreys in Ohio are similar to this one; note undeveloped eye; flap or hood partly surrounding the oblique mouth.

Fig. B: view of mouth of adult; note the widely separated supraorals; lack of radiating teeth, and lack of visible teeth on posterior field of sucking disc other than the double row of marginal teeth (in fig. A of American brook lamprey, the arrow points to an additional row of teeth in the posterior field).

Fig. C: view of mouth of ammocoete, note hood surrounding mouth except posteriorly; lack of teeth; presence of fimbrae within the oral opening.

Identification

Adult characters: A wide notch completely separates the dorsal fin. Except for the supra- and infraorals, teeth are too small to be readily seen, many are hidden by skin, and the teeth are in clusters, not in radiating series, fig. B. Myomeres usually 55–60, extremes 54–62, ave. 58, upper fig. Jawless mouth small, usually not as wide as body when fully expanded. Not parasitic. Alimentary tract nonfunctional and reduced to a strand in adults.

Differs: Other species with divided dorsals are the sea and American brook lampreys; these normally have more than 62 myomeres.

Most like: American brook lamprey.

Ammocoetes differ from adults: *See* silver lamprey.

Coloration: *Ammocoetes* are light silvery-tan,

* Ohio brook lamprey in 1957 edition of *Fishes of Ohio.*

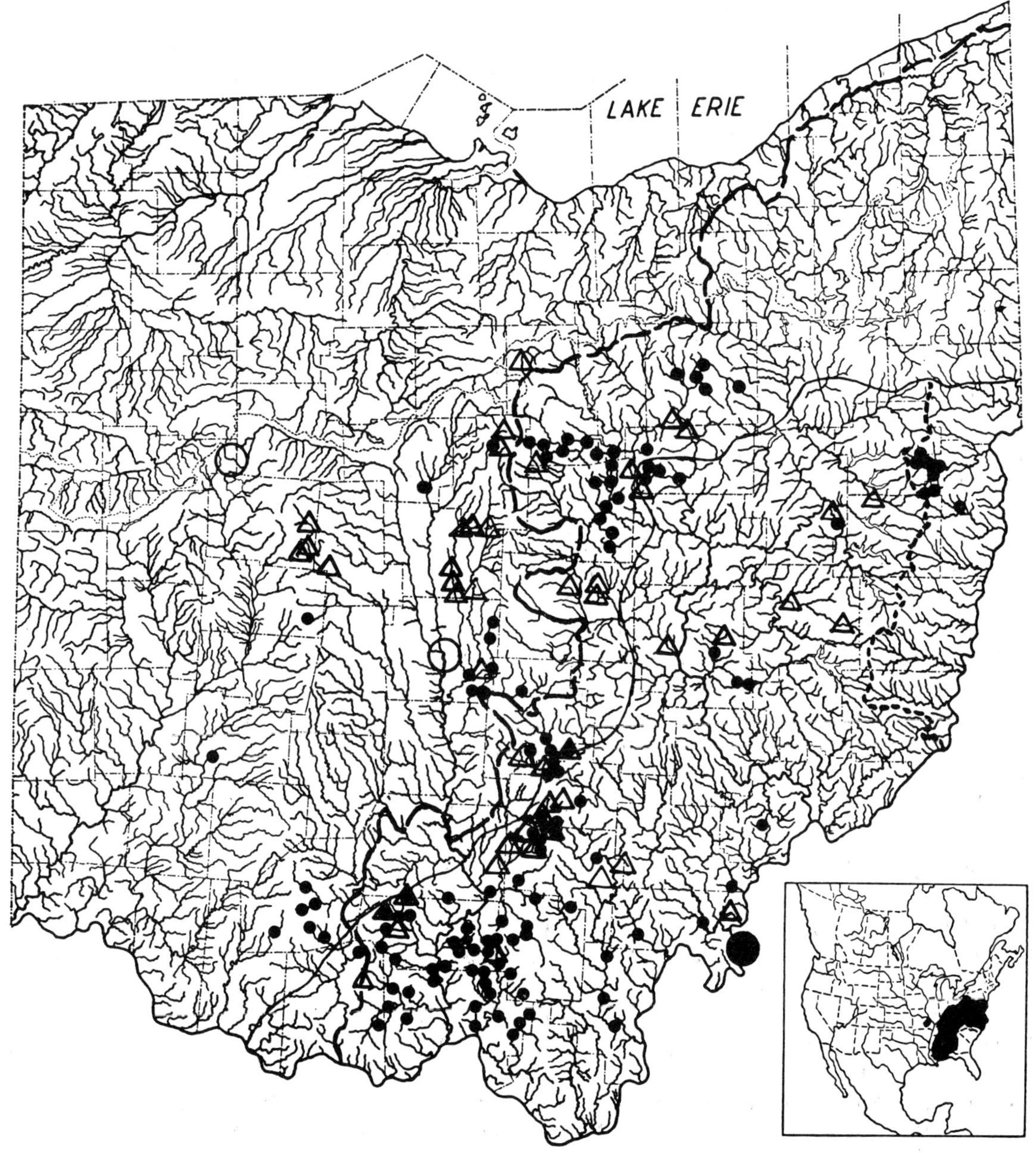

Map 6. Least brook lamprey

○ Specimens not identified as to species.
● Type locality, Ohio River.
• Specimens identified. • Before 1955. △ 1955–80.
Allegheny Front Escarpment.
~ Glacial boundary.
Flushing Escarpment.
Insert: Obvious decrease in range size in Ohio since 1900, a result of destruction of habitat.

occasionally slightly bicolored. Before spawning *adults* are dark tan above, lighter below. Spawning *adults* are blue-black, especially as death approaches.

Lengths: Maximum for *ammocoetes*, 7.0″ (18 cm) TL. *Adults*, 4.8″–7.0″ (12–18 cm) TL. This species averages the smallest of all the species of lampreys in Ohio.

Distribution and Habitat

Ohio Distribution—In the Ohio drainage during the 1925–50 period, the least brook lamprey was numerous only in those areas immediately to the east of the Allegheny Front Escarpment where the maximum relief was between 400–500′ (122–152m) (*see* map VI), and where the streams were the least disturbed by man. There were two principal population centers along this Front; one was in Pike, Scioto and Jackson counties and was in unglaciated territory; the other was in Richland, Knox and Holmes counties, and was mostly in glaciated territory. These centers had much in common; both contained many clear brooks of high gradients and with sand and gravel bottoms, and were suitable for spawning adults; they also had larger, low-gradient waters suitable for ammocoete development.

The species was present only in relict populations in unglaciated eastern Ohio which was least disturbed by glacial action, for here suitable high-gradient spawning streams were virtually absent. It is believed that before mine waste pollution became severe, streams east of the Flushing Escarpment contained large populations of these lampreys, for these small, high-gradient streams should have been favorable for spawning adults, and the then unpolluted Ohio River suitable for ammocoete development.

Although principally an Appalachian species, the least brook lamprey has penetrated into glaciated territory for a considerable distance beyond the line of glaciation. Since 1925, isolated populations have been found in such widely separated localities as the Little Darby headwaters in Champaign County, and Gladys Run in Greene County. On April 5, 1951, I saw an adult brook lamprey, either of this species or the American brook lamprey, in the Scioto River headwaters of Auglaize County (*see* hollow circle in Auglaize County). I was unsuccessful in obtaining the specimen.

Between 1925–50 there was a general decrease in abundance of the least brook lamprey in this state, and during that period populations in many streams were extirpated. This decrease was caused by destruction of lamprey habitats and by the ever increasing demand for both adults and ammocoetes as bait. From observations during the period, statements by many residents, and early literature references to stream conditions prior to 1900, it is abundantly evident that formerly there was far more habitat available for this and other species of brook lampreys than there has been recently, and that formerly the least brook lamprey was more widely distributed and abundant. The literature relative to the 19th century contains many references to the permanency and clearness of Ohio brooks and their suitability for lampreys; an example is Peters Run, which in 1825 (Lee, 1892:273–75) "turned the wheels of Couger's Flouring Mill" but which by 1875 had been eliminated by the city of Columbus. Another central Ohio stream, now called Dry Run for obvious reasons, had a permanent flow before 1840. At that time it flowed through the farm of Joseph Sullivant (now West Columbus; *see* hollow circle in Franklin County), and it was from this stream that Sullivant collected ammocoetes which Kirtland (1841B:473–75 and 1851R:79) used, in part, for his description of *Ammocoetes concolor* (*see* mountain brook lamprey, and Hobbs and Trautman, 1937:24–27). These ammocoetes may have been of this species or of the northern brook lamprey.

Habitat—The least brook lamprey spawned in very small brooks of high-gradients having riffle-bottoms of sand and gravel. Because of the small size of these brooks they reached temperatures above 50°F (10°C) very early in spring, usually in late March or early April; therefore the species was usually the first of the Ohio lampreys to spawn. The ammocoetes required permanent waters of lower gradients having bottoms containing beds of mixed sand and organic debris, suitable for ammocoete development (*see* northern brook lamprey, under Habitat).

Ammocoetes of all species were particularly vulnerable to siltation, for the silting over of the beds of sand and organic debris destroyed their habitat. Mine wastes and other pollutants had similar effects.

The large black circle on the Ohio River, near Portland, Meigs County, is the type locality of the species (Abbott, 1860:466).

Years 1955–80—This species was the lamprey most frequently encountered in Ohio during the 25-year period, and adults and/or ammocoetes were found in more than thirty previously unknown localities. The majority of the localities were situated east of the Allegheny Front Escarpment; however, relict populations have been discovered a considerable distance to the west of it, in Delaware, Union and Logan counties.

Although several former populations have been eliminated during this 25-year period, others continue to survive in small numbers despite such adverse conditions as increased silting, ditching and intensive collecting of adults and ammocoetes for bait. The numbers of lampreys collected for bait rose sharply during the period, and in some localities, especially southern Ohio, collecting was so intensive as to eliminate or greatly reduce the numbers of a relict population.

In the first edition of this report this species was known as the Ohio brook lamprey. More recently it has been renamed the least brook lamprey (Bailey et al., 1960:6; Bailey et al., 1970). Hubbs and Potter (1971:48–49) in their discussion of the phylogony and taxonomy of lampreys have placed the species *aepyptera* in the monotypic genus *Okkelbergia* Creaser and Hubbs, i.e., *Okkelbergia aepyptera.*

AMERICAN BROOK LAMPREY

Lampetra appendix (DeKay)*

Fig. 7

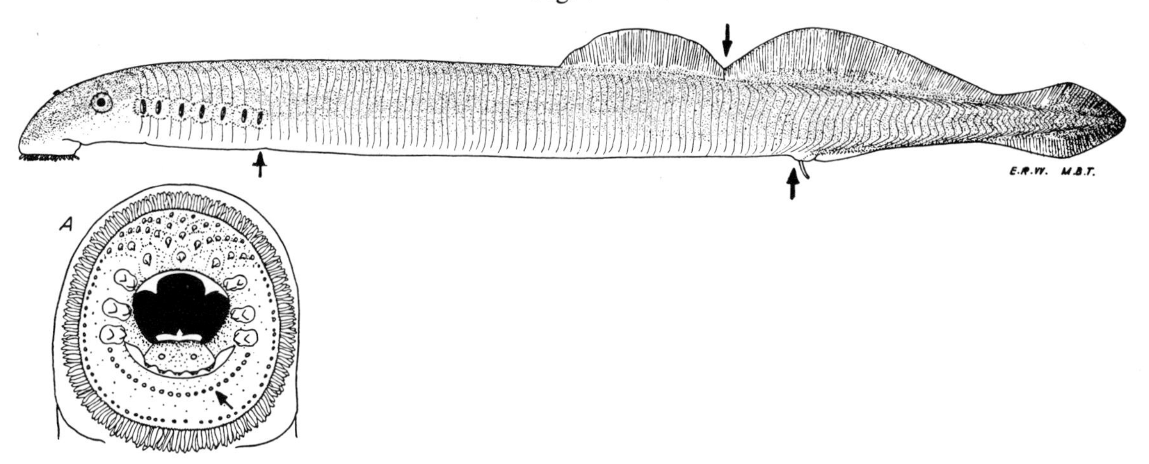

Upper fig.: Chagrin River, Geauga County, O.

May 6, 1940. 151 mm TL, 6.0″ TL.
Adult male. OSUM 2057.

Upper arrow points to notch which separates dorsal fin. Anterior arrow points to the first counted myomere after the last gill opening; posterior arrow points to last counted myomere before anus; all myomeres including, and between, these two are counted.

Fig. A: view of mouth; arrow points to row of teeth (sometimes visible only with aid of magnification) on posterior field of sucking disc, other than the double row of marginals.

Identification

Adult characters: A wide notch completely separates dorsal fin. Supra- and infra-orals, and two pronged laterals, are moderate in size; other teeth small, many hidden by skin. A row of teeth on posterior field usually discernible; *see* arrow, fig. A. Myomeres usually 64–70, extremes 63–73, upper fig. Jawless mouth can be expanded almost as wide as body. Not parasitic. Alimentary tract nonfunctional and reduced to a strand in adults.

Differs: Least brook lamprey has fewer than 62 myomeres and no additional tooth row on posterior field. Sea lamprey has long sharp teeth, in radiating series.

Most like: Least brook lamprey.

Ammocoetes differ from adults: *See* silver lamprey.

Coloration: *Ammocoetes* are dark tan above, silvery-tan below. Before spawning *adults* are dark tan above and lighter below, and sometimes somewhat bicolored. Spawning *adults* are blue-black, darkest when about to die.

Lengths: Maximum for *ammocoetes*, 8.0″ (20 cm) TL. *Adults*, 5.3″–7.7″ (13–19 cm) TL. Largest *adult* 8.0″ (20 cm) TL.

Distribution and Habitat

Ohio Distribution—The American brook lamprey was recorded only in glaciated territory, and its largest populations occurred only in streams of high gradients which drained areas of high maximum relief. One large population spawned in the upper portion of the Mad River (Logan and Champaign counties), some of whose headwaters arise on Campbell Hill, the highest elevation (1550′ [472 m]) in the state, and whose gradient from source to mouth of King's Creek averages 12.6′/mile (2.39 m/km). The other large spawning population centered in the Chagrin River and its East Branch (Geauga and Lake counties) whose gradients aver-

* *Lampetra lamottei* in 1957 edition of *Fishes of Ohio*. Formerly in the genus *Entosphenos*.

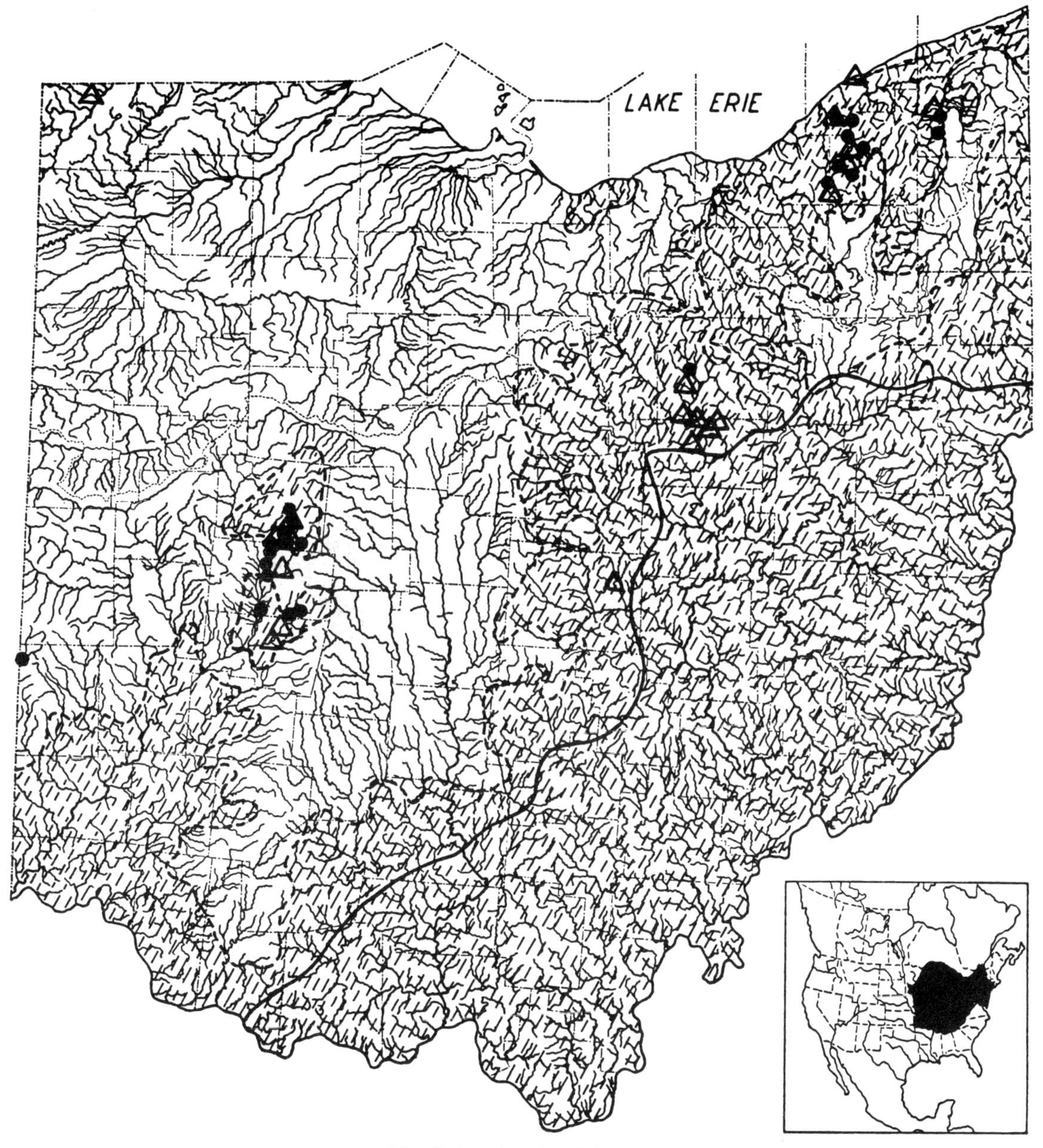

MAP 7. American brook lamprey

Locality records. ● Before 1955. △ 1955–80.
~ Glacial boundary. Shaded area has maximum relief of over 200 feet (70 m).
Insert: A possible decrease in range size in Ohio since 1900.

aged 43.0 and 35.3 feet/mile (8.17 m and 6.71 m/km) respectively. The Whitewater River population (Darke and Preble counties) appeared to be a relict one; it had a gradient of 9.1′/mile (1.7 m/km) and drained an area whose maximum relief was less than 200′ (70 m). The apparently small population in Killbuck Creek (Wayne County) competed with the least brook lamprey, and the Grand River (Ashtabula County) populations with the northern brook lamprey.

It appears probable that before 1900 this lamprey was present in the high-gradient streams of northwestern Ohio, such as the St. Joseph River in Williams County, and Swan Creek in Lucas County. Relict populations may still occur there. No American brook lampreys were found in unglaciated Ohio, even though maximum relief was over 300′ (91.4m) and many high-gradient streams appeared suitable. Its absence might have been due entirely, or in part, to the presence and population pressure of the least brook lamprey.

Habitat—This brook lamprey had the same basic habitat requirement needs as did the other brook lampreys (*see* northern brook lamprey, under Habitat). However, some ecological differences were apparent. Comparisons between the American and least brook lampreys showed that the majority of the American brook lampreys spawned in streams of more than 15′ (4.6m) average width, whereas the least brook lampreys spawned in streams whose average width was less than 15′ (4.6m). Since the larger streams remained colder in spring longer than did the smaller ones, the American brook lamprey usually spawned later in spring than did the least brook lamprey. Both the American and northern brook lampreys spawned in streams of the same average width; but when both were present in the same stream, the height of spawning of the American brook lamprey was often over before the northern brook lamprey reached its spawning areas. As an example: On April 28, 1951, my wife and I found the American brook lamprey at its spawning height in the tributaries of the Grand River (Ashtabula County), whereas the northern brook lamprey had only begun its spawning migration, and all adults of this latter species found were unripe, still in the silver-colored stage, were hiding under objects, or were migrating from one hiding place to another.

Years 1955–80—The statement that the Wayne County population (Killbuck Creek) of the American brook lamprey was small has proved to be erroneous. During the 1955–80 period several populations of this species were discovered in the Killbuck Creek drainage of Wayne and Holmes counties.

On April 17, 1964, Richard A. Zura and Linda B. Scothorn (1966:581) collected five adult American brook lampreys in the West Branch of the St. Joseph River and on October 18 of that year captured a transforming adult a short distance upstream. The latter locality is only a little more than a mile (1.6 km) downstream from the riffle where in 1941 I collected six northern brook lampreys (q.v.). Therefore, the previous suggestion has been confirmed that the American brook lamprey was probably present in the St. Joseph River of Williams County.

LAKE STURGEON

Acipenser fulvescens Rafinesque

Fig. 8

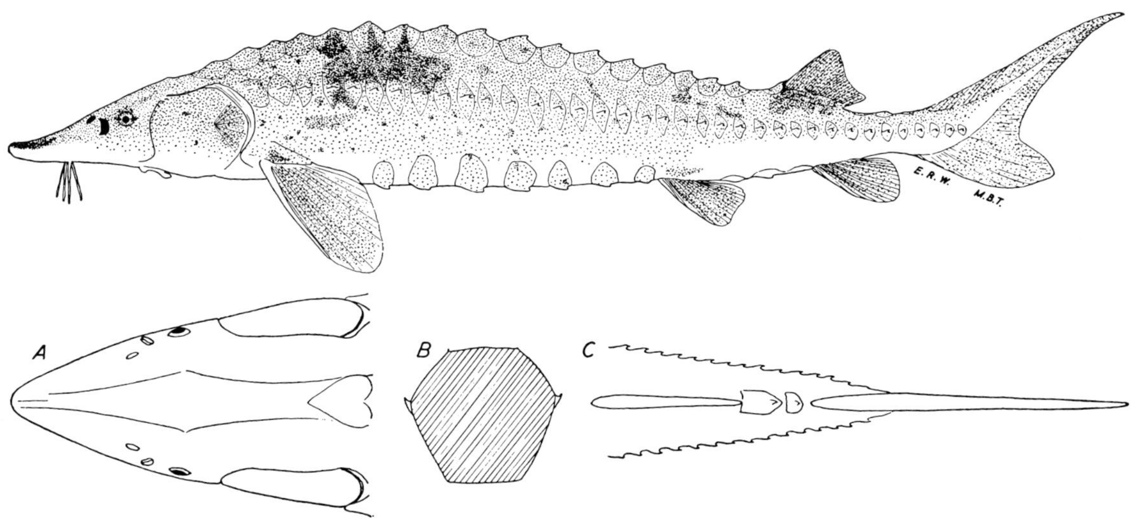

Upper fig.: Lake Erie, Ottawa County, O.

Nov. 13, 1941. 435 mm SL, 20.5″ (52.1 cm) TL.
Immature. OSUM 5033.

Fig. A: dorsal view of head showing even taper of the rather sharp snout.
Fig. B: cross-section of caudal peduncle showing its roundish shape.
Fig. C: dorsal view of caudal peduncle; note that it is *not* completely covered with bony plates.

Identification

Characters: Snout rather conical and sharp, and not notably flattened dorso-ventrally. Caudal peduncle roundish in cross-section, fig. B. Bony plates on caudal peduncle restricted to lateral bands, and to two scales on dorsal ridge, fig. C; these bony plates sharp in young, becoming very blunt or disappearing in large adults. Small young have a filament on the upper lobe of the caudal fin as does the shovelnose, *see* that species.

Differs: Shovelnose sturgeon has a flattened snout, and completely scaled caudal peduncle. Paddlefish has a longer, thinner snout. Catfishes have pectoral and dorsal spines. Gars are completely scaled.

Most like: Shovelnose sturgeon.

Coloration: Olive-yellow, gray or bluish dorsally, milky- or yellow-white ventrally; many individuals sharply bicolored. *Young* under 20.0″ (50.8 cm) in length have four dusky blotches, one on each side of back behind the nape and one on each side of back below the dorsal fin.

Lengths and weights: Usual length 20.0″–55.0″ (50.8–140 cm), weight 2 lbs–60 lbs (1–27 kg). Specimens over 100 lbs (45.4 kg) taken frequently in Lake Erie. In July, 1929, one was taken in Lake Erie near Huron which weighed 216 lbs (98 kg). Maximum length about 8.0′ (244 cm), weight over 300 lbs (136 kg).

Distribution and Habitat

Ohio Distribution—Rafinesque (1820:147), Kirtland (1851–I:229), Jordan (1882:766–68), Henshall (1888:76), Kirsch (1895A:317 and 327), Fowler (1919:52) and many others testified to the former presence of this sturgeon in Lake Erie, in the Ohio River upstream as far as western Pennsylvania, and in the larger inland rivers of Ohio. Many of these

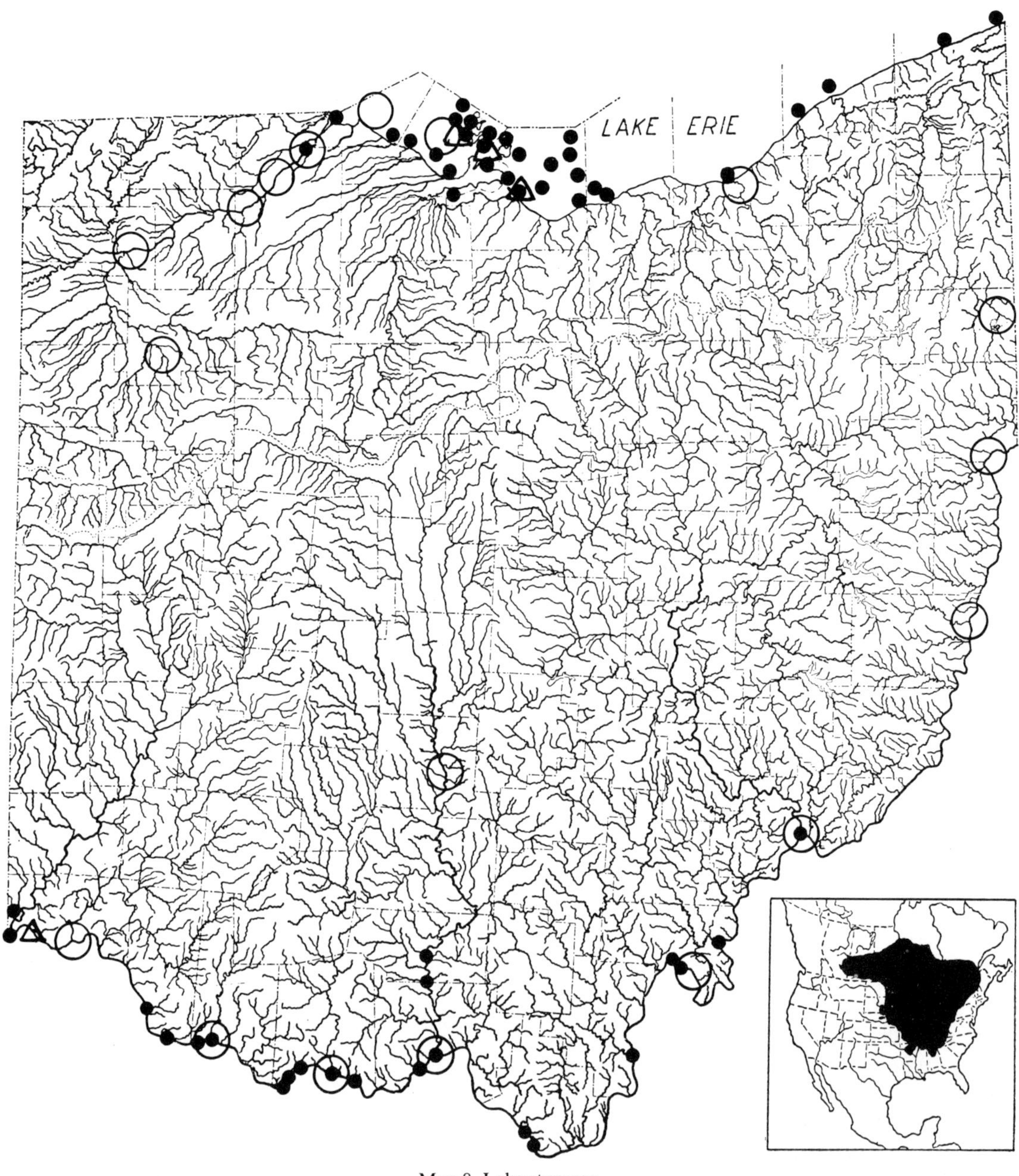

MAP 8. Lake sturgeon

Locality records. ○ Before 1916. ● 1916–50. △ 1955–80.
Insert: Northern and western limits poorly defined.

authors reported upon its former great abundance, especially in Lake Erie and the Ohio River.

In Lake Erie waters before 1850, sturgeons were taken chiefly on set-lines, with seines along shore, or with spears as they ascended streams in late winter and spring on their spawning migrations. In these early days only a portion of the fishes caught were utilized; these were smoked for food, rendered for oil, or their bladders converted into isinglass. Since these huge and then almost worthless fishes greatly damaged fishing gear, they were sometimes thrown upon the beach to rot or were fed to hogs. George H.

Borman told me that his father, John F. Borman, had seen fishermen place sturgeons in large piles and set fire to them.

After 1854, with the advent of pound and fyke nets, and later gill nets, sturgeons were more and more utilized. In 1885, 531,250 lbs (240,971 kg) of sturgeons were taken in the Ohio waters of Lake Erie; the flesh was eaten, the eggs converted into 237,155 lbs (107,572 kg) of caviar; 6,485 gallons (24,548l) of oil were rendered; and the bladders were converted into 277 lbs (126 kg) of isinglass (Smith and Snell, 1891:241–42).

Before 1850 sturgeons were particularly abundant in Lake Erie waters, but even at that early date dams were preventing some of them from reaching their upstream spawning grounds. Kirtland (1850A:l), living in Ohio between 1810–77, recorded that before 1850 "Sturgeon and Muskellonge often run up the Cuyahoga [River] several miles"; but that by 1850 the "Sturgeon had nearly forsaken this shore of the Lake" Erie, about the mouth of the Cuyahoga River. Smith and Snell (1891:248) likewise reported an early decrease of sturgeons in the Maumee River where "Sturgeon once ran up the river by hundreds as far as the rapids above Perrysburg, but at present [1885] . . . are absent"; also that by 1885 they were "very much scarcer" than formerly in western Lake Erie waters near Toledo. Kirsch (1895A:317) mentioned that formerly they were "very abundant" in the Maumee River near Waterville.

After 1890 the yearly poundage of sturgeons began to drop sharply. During the 1939–49 decade only an average of 7,296 lbs (3309 kg) per year were reported as taken from Ohio waters of Lake Erie (Shafer, 1950:1).

Kirtland (1851 I:229) and Henshall (1888:76) noted its abundance in the Ohio River before 1900, and Zeisberger (Hulbert and Schwarze, 1910:73) its occurrence in the Muskingum River before 1780. The late Dr. Howard Jones, of Circleville, told me that before 1870 this species migrated in fair numbers up the Scioto River as far as Circleville; that stragglers were then found as far upstream as Columbus, and in such tributaries as Big Darby and Big Walnut creeks. The older, commercial fishermen agreed that before 1915 the "Bull-nosed Sturgeon" (as this species was almost universally called by Ohio River fishermen) was very common in the Ohio River. At no time, however, was it as abundant as was the shovelnose. (Accepted verbal or published records of sturgeons, observed in these localities before 1916, are represented by hollow circles on the distribution map.)

After 1916, after the river was impounded, the sturgeon began to decline sharply in numbers and since then a fisherman seldom catches more than two in a day, and these rarely exceed 30 lbs (14 kg) in weight.

Habitat—This sturgeon feeds chiefly over a clean bottom of sand, gravel and rocks, and especially where small mollusks, crustaceans and insect larvae are abundant. It avoids soft, muddy bottoms, especially those containing few mollusks or insects. It formerly migrated, apparently in large numbers (Bean, 1892:89), far upstream into inland Ohio waters to spawn. The Auglaize River and its tributaries seemingly were favored spawning areas, for seven farmers told me that their fathers or grandfathers had seen sturgeons spawning in these waters before 1880. Spawning occurred in late May and early June, usually during corn planting time, and I was shown hand-forged spears which were carried by the farmer while he was planting corn. When he heard a sturgeon splashing (called breaching) on a riffle, he used the spear to capture it. While on the riffles the sturgeons, 4′–6′ (122–183 cm) in length, frequently jumped high into the air. The Ottawa River in Putnam County was a particularly favored spawning stream; R. Clevenger told me that his grandfather speared many sturgeons in this stream, 3 miles (5 km) southeast of Kalida, and that he had observed sturgeon 5′ (152 cm) in length, spawning on riffles only a foot (30.5 cm) deep and 10′ (3 m) wide.

The decline in sturgeon abundance appears to have been caused chiefly by inability of the fish to reach its spawning grounds because of dams; by having the former spawning habitat destroyed by silting, pollution or drainage; and by destruction of the great quantities of mussels and gastropods in both the streams (Clark and Wilson, 1912) and Lake Erie. Overfishing might have been a major factor in decreasing the numbers of this species because of the many years in which it may be captured before it spawns once; it apparently does not begin to spawn before it has reached 20 years of age (Harkness, 1924:23) and until it is 48″ (122 cm) or more in length (Schneberger and Woodbury, 1944:35–37).

Years 1955–80—In Lake Erie the lake sturgeon

continued to decline in numbers throughout this period (Van Meter and Trautman, 1970: 67; White et al., 1975:49). Russell L. Scholl of the Ohio Division of Wildlife reports that in the waters of Lake Erie in Erie County, Larry Davis recently "handled in the neighborhood of 50 sturgeon, five of which he termed 'big'." From 1957 to 1959 personnel of the University of Louisville did not take a "rock" sturgeon in their investigations of the Ohio River (ORSANCO, 1962: 71). Woodrow W. Goodpaster, during the summer of 1971, watched a man using hook and line land a fairly large lake sturgeon in the Ohio River between Cincinnati and the Indiana State line.

For life history and fisheries *See* Harkness and Dymond, 1961.

SHOVELNOSE STURGEON

Scaphirhynchus platorynchus (Rafinesque)

Fig. 9

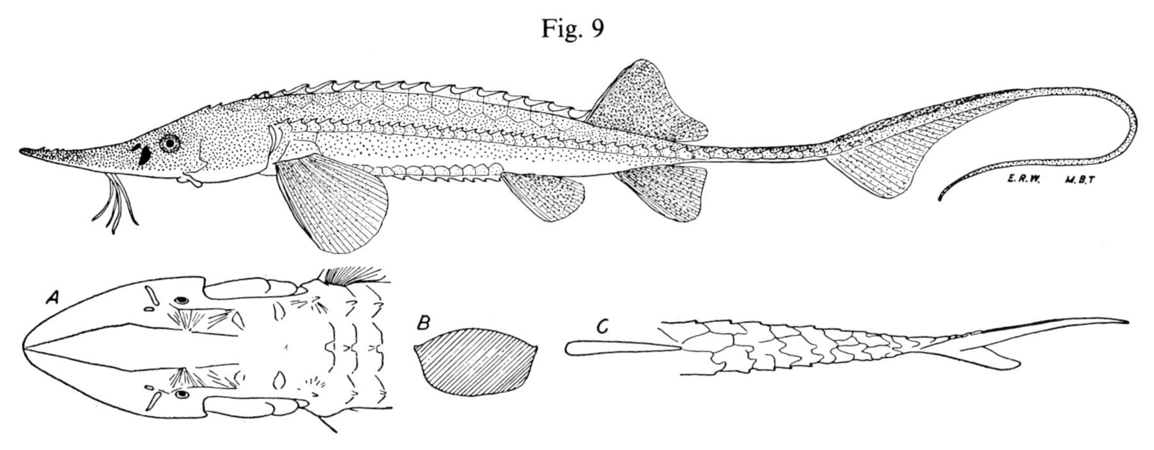

Upper fig.: Missouri River, Monona County, Iowa.

July, 1937. 75 mm SL, 4.8″ to end of long filament.

Immature. Loaned for drawing by Walter Aitken.
Note long caudal filament on this very small specimen; adults lack these. Body proportions similar to those of adults.

Lower three figs.: Ohio River, Lawrence County, O.

Aug. 24, 1939. 485 mm SL, 23.1″ (58.7 cm) TL.
Adult female. OSUM 1076.

Fig. A: dorsal view of head showing a rather broad snout.
Fig. B: cross-section of caudal peduncle, showing that it is notably flattened dorso-ventrally (compare with roundish cross-section of lake sturgeon).
Fig. C: dorsal view of caudal peduncle; note that it is completely covered with bony plates.

Identification

Characters: Snout shovel-shaped, very thin dorso-ventrally, and very broad laterally, fig. A. Caudal peduncle flattened dorso-ventrally, not roundish in cross section as is that of the lake sturgeon, fig. B. Caudal peduncle covered with bony plates, fig. C. Bony plates on dorsal ridge and lateral line keeled, and with sharp points on both young and adult.

Differs: Lake sturgeon has a roundish snout and roundish caudal peduncle. *See* lake sturgeon.

Coloration: Brown, olive, or gray dorsally, whitish ventrally. Many *adults* strongly bicolored. No bold black blotches on body.

Lengths and **weights:** Usual length 18.0″–30.0″ (45.7–76.2 cm), weight 1 lb–5 lbs (0.5–2.3 kg). In 1930, one taken from Ohio River near Pomeroy weighed 9 lbs (4.1 kg). Largest specimen, 32.0″ (81.3 cm), weight 10 lbs (4.5 kg).

Distribution and Habitat

Ohio Distribution— Published statements by Rafinesque (1820:145–46), Kirtland (1851H:233), Jordan (1882:769), Henshall (1888:77), and Bean (1892:72) indicate its former abundance in the Ohio River, and occurrence as far upstream as western Pennsylvania. The older fishermen agree that this sturgeon was abundant upstream as far as Washington County until about 1910; that as many as 75 could be taken in a day on a trotline baited with worms, and especially during spawning runs in late February and March when the river was rising. After ponding of the river began, about 1911, thereby partially stopping the spawning run, the fishermen reported a drastic decrease in abundance.

Between 1925–50 this sturgeon was taken in the Ohio River, most often between the Ohio-Indiana line and Scioto County, and in that section I have seen 20 specimens taken on trotlines in a day. The

Map 9. Shovelnose sturgeon

○ Identified from photograph. ● Identified from specimens.

Insert: Recent withdrawals from the upper Mississippi and Ohio rivers and presumably from the Rio Grande; western limits poorly defined.

species was uncommon or rare between Scioto and Meigs counties. The hollow circle in Belmont County represents the locality where a specimen was taken about 1925; I saw the readily identifiable picture of this fish.

Habitat—Before ponding of the river began, about 1911, the shovelnose apparently fed much over the clean sand and gravel bottoms of chutes and bars, or wherever there was considerable current and a clean bottom. Several fishermen reported that it seemed to congregate wherever there were large quantities of mollusks and snails. The species appeared to avoid the tributary streams; only one fisherman reported taking one in the lower Scioto River, Scioto County, and another fisherman of taking one in the lower Muskingum River, Washington County.

Years 1955–80—The shovelnose sturgeon remained a rather common species in the Ohio River from Portsmouth westward throughout the previous 1925–50 period. Since 1950, however, a marked decrease in abundance has presumably occurred, according to many Ohio River fishermen. This decrease is likewise reflected in the researches by the personnel of the University of Louisville during their intensive 1957–59 investigations, for only one (ORSANCO, 1962:145) was captured. For distribution *see* Bailey and Cross, 1954; food, Held, 1969.

PADDLEFISH

Polyodon spathula (Walbaum)

Fig. 10

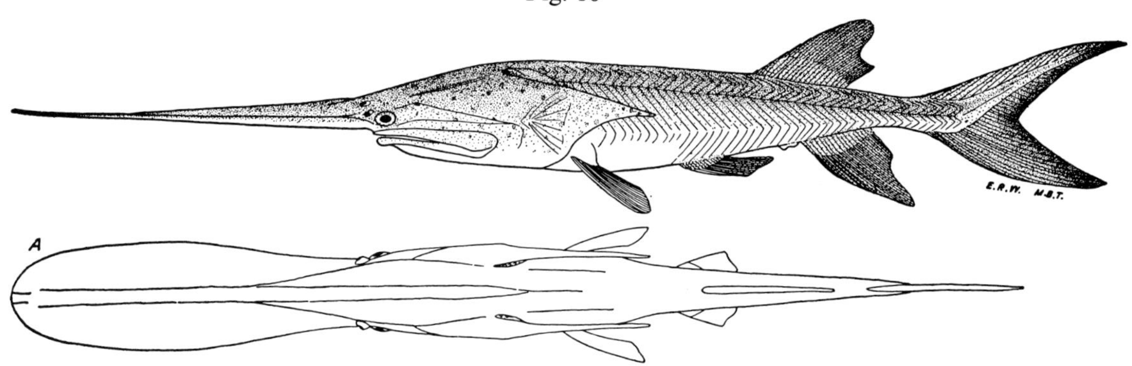

Upper fig.: East St. Louis Waterworks, Illinois

Summer, 1911 or 1912.
Immature.

220 mm SL, 10.0″ (25.4 cm) TL.
OSUM 9315.

Adults similar in body proportions to this small young.
Fig. A: dorsal view; the spatulate bill is slightly wider than in large adults.

Identification

Characters: Long, paddle-shaped snout which is longer than remainder of head. No scales. Hundreds of long gill rakers on each gill arch. Caudal fin heterocercal.

Differs: Gars have scales. Sturgeons have bony plates on body and four long barbels in front of mouth. Catfishes have mouth barbels, pectoral and dorsal spines.

Most like: No other Ohio fish closely resembles it.

Coloration: Bluish-gray, bluish-white or bluish-olive dorsally, lighter and silvery ventrally.

Lengths and **weights:** Usually 20.0″–48.0″ (50.8–122 cm) long, weight 2 lbs–20 lbs (1–9 kg). Largest specimen, 60.0″ (152 cm) long, weight 184 lbs (83.5 kg).

Distribution and Habitat

Ohio Distribution—Zeisberger (Hulbert and Schwarze, 1910:74) noted the presence of the paddlefish in Ohio waters before 1780; others, including Clemens (1827: 201–05 and pl.), Kirtland (1844:22), Henshall (1888:76), and Fowler (1919:53) noted its presence or abundance before 1900 in the upper Ohio River and its larger tributaries in Ohio and Pennsylvania. Klippart (1874:6–7) and Lee (1892:298–99) stressed its abundance before 1880 in the Scioto River at Columbus, before construction of the state dam. Ohio River fishermen agreed that previous to 1915 the paddlefish was abundant in the Ohio River, but that shortly after its impoundment (about 1915) the species decreased markedly in abundance. A decrease in abundance was evident in the river during the 1925–50 period, and in that period the species was most numerous west of Portsmouth in the Ohio River, and in the Scioto River. Several men told me that they saw paddlefish which were taken in the Ohio River above Marietta before 1925. It appeared to be absent in the Muskingum River.

Of the two reliable Lake Erie records, one is of a specimen 5′4″ (163 cm) long, that is mounted and now at Oberlin College, which was taken "in a pound net near Vermilion, in 1874" (McCormick, 1892:9); the other record was of a paddlefish seen by Max Morse (1903:24) "in the Post Company's Fish House, Sandusky, Ohio, in August" of 1903. Morse stated also that "From the fishermen, I learned that the fish is seen at irregular intervals but not commonly." Greene (1935:24), discussing its presence in Lake Erie, concluded that although it may have entered through the early Maumee-Wabash con-

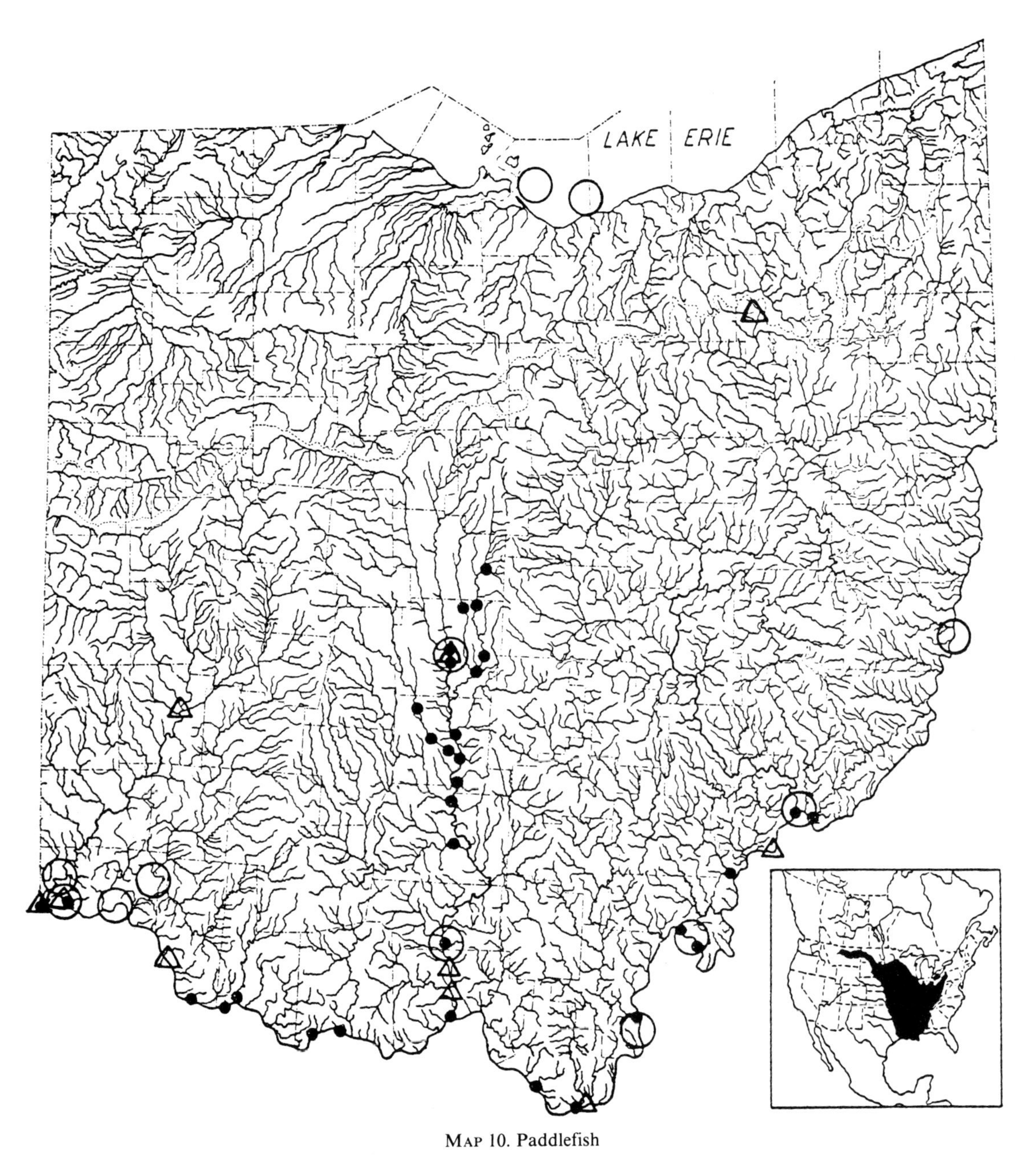

MAP 10. Paddlefish

Locality records. ○ Before 1926. ● 1926–55. △ 1955–80.

Insert: There are large unoccupied areas within the range of this species; stragglers may be expected outside of this range; apparently, a recent withdrawal from the upper Mississippi and Ohio rivers.

nection, it probably invaded the lake through the canals. Accumulated evidence since 1935 favors a pre-Columbian invasion, involving this and several other southern fish species, all of which were present in Lake Erie in relict populations before the building of the canals. Apparently a small paddlefish population existed in Lake Erie waters until at least 1903; possibly its extirpation was caused by dams preventing its upstream migration and thereby preventing it from spawning, and/or by a destruction of its stream spawning habitat.

Habitat—The paddlefish is an inhabitant of the

sluggish pools, back-waters, bayous and oxbows of large rivers, where the gradient is low. Although apparently highly migratory, it seems rarely to move upstream during floods as do many other fish species, but migrates during lower stages of water when currents are weaker. At such times dams are effectual barriers, and it may be for this reason that the paddlefish is absent from the Muskingum, Miami, and Scioto rivers above the first dams. The species is captured in nets, presumably while feeding near the bottom, or with floating gill nets near the surface. Although it apparently migrates mostly near the surface (Evermann, 1902:279) it also feeds near the surface, probably when the light-loving plankton upon which it feeds congregates there.

Years 1955–80—In Ohio this endangered species has shown a decrease in number of individuals taken during this period. Three men reported that they had captured one each from the lower Scioto River, Scioto County. On July 26, 1976, Mr. and Mrs. Ted Stanley, while fishing with nightcrawlers at the first riffle below the Greenlawn Avenue Dam in south Columbus, snagged a paddlefish, the hook entering the cheek. It was 31.75″ (80.65 cm) TL and weighed 5 lbs (2 kg). Personnel of the Ohio Department of Wildlife have taken none during recent years in their annual test-netting operations of Ohio streams. On August 10, 1929, the late Charles F. Walker, the late James S. Hine, and I captured five in one seine hawl at Davenport Puddle, a permanent oxbow of the adjacent Scioto River, in southwestern Pickaway Township, Pickaway County. Between August 17–19, 1972, several of us gill-netted and seined this pond, unsuccessfully attempting to capture a paddlefish.

Investigations of the Ohio River between 1957 and 1959 (ORSANCO 1962:70) disclosed that the species was frequently taken by commercial fishermen in the river below Louisville, but that in the river opposite the state of Ohio, individuals were taken only adjacent to Gallia and Washington counties. Before 1940 I saw many specimens that were captured in the Ohio River from Portsmouth to the Indiana line. I have heard of none recently taken in this stretch of river, except at the power plant in southwestern Hamilton County, where several small specimens have been reported captured near the screen.

In the *Akron Beacon Journal*, June 19, 1977, there appeared an article written by Phil Dietrich, outdoors writer for the *Journal*, stating that Reuben Simmons while crossing a bridge over Pigeon Creek at Collier Road in Summit County, saw a "paddlefish while crossing the creek." It had "beached itself in shallow water and died." The specimen was not preserved, but pictures were taken of it. Pigeon Creek is a tributary of the Muskingum River system.

The paddlefish is a highly migratory species, occasionally surmounting dams during periods of high water. Stranded individuals have been seen or captured in small headwater creeks such as upper Big Walnut Creek in Delaware County before that stream was impounded by Hoover Dam. On the other hand the species could have been obtained by a fisherman in some other state and transported to Summit County, where the fisherman may have learned that it was an endangered, protected species and threw it into Pigeon Creek.

Discarding unwanted fishes into Ohio streams has been recorded several times. On August 8, 1976, Scot Adkins, 13 years old, found an Atlantic sharpnose shark, *Rhizoprionodon terraenovae*, weighing 12 lbs (5 kg) in Big Run, Whetstone Park, north Columbus. Big Run is a tributary of the Olentangy River. The fish was lodged against tree roots in a bend of the tributary. Recent rains had raised the waters of both Big Run and Olentangy River. The shark could have been dropped from a bridge crossing Big Run, 100′ (30.5 m) upstream, or could have floated in from the flooding Olentangy River. This entirely marine species is numerous and frequently captured along the coasts of the southeastern United States.

ALLIGATOR GAR

Lepisosteus spatula Lacepède

Fig. 11

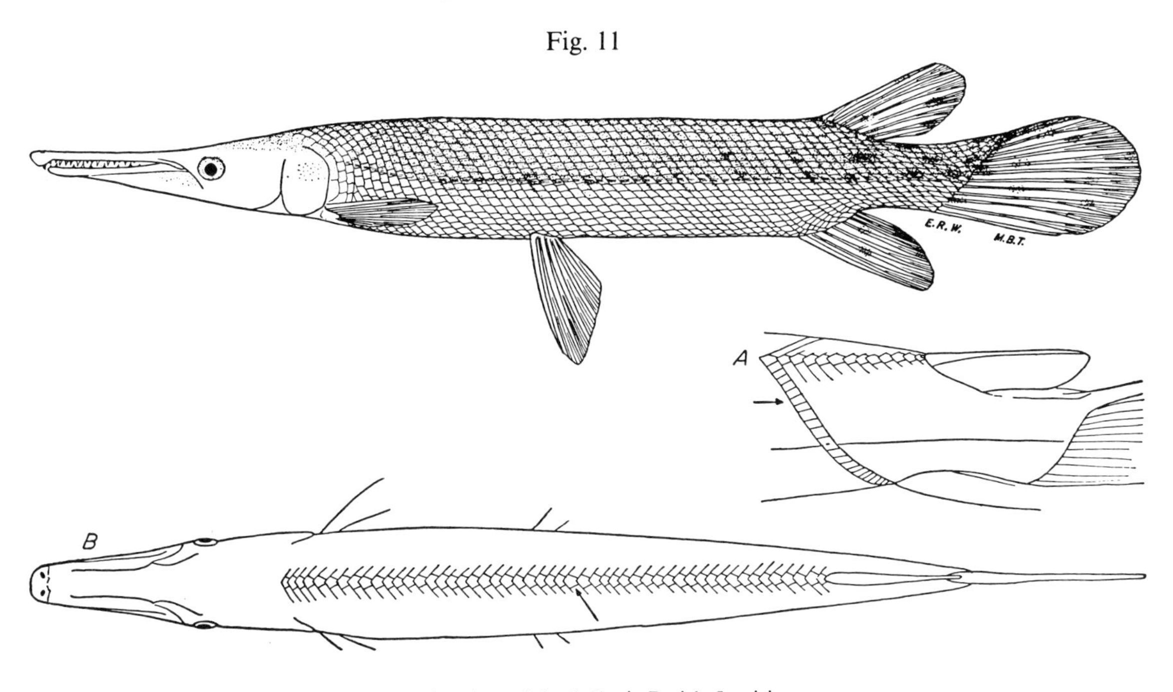

Upper fig.: Avery Island, Iberia Parish, Louisiana.

June 3, 1937. 470 mm SL, 21.5″(54.6 cm) TL.
Immature. OSUM 9268.

Fig. A: arrow points to tranverse row of scales; all these scales are counted including the mid-dorsal scale and anal scale-plate.
Fig. B: dorsal view; the arrow pointing to the mid-dorsal row of scales which begins behind the head and extends to dorsal origin. Note the very wide and short snout.

Identification

Characters: Scales not overlapping: rhombic in shape. Snout duck-shaped, short and very broad, the eye diameter contained 1.5 or more times in width of upper jaw at nostrils. Transverse scale row of 23–25 scales, fig. A. Dorsal ridge contains 50–52 scales, fig. B. Ohio drainage only. Extremely rare in Ohio.

Differs: Other gar species have narrower snouts. Paddlefish has no scales.

Most like: Shortnose and spotted gars.

Coloration: *Adults*—Brown, olive or greenish dorsally, lighter ventrally, sides spotted and mottled, especially posteriorly. *Young* 10.0″—30.0″ (25.4-76.7 cm) long, greenish or brownish dorsally, pale yellow or white ventrally, sides mottled anteriorly and with large, black spots posteriorly, and on anal, dorsal, and caudal fins.

Lengths and **weights:** An Ohio specimen, examined by J. P. Kirtland (before 1845) was 5′8″ (173 cm) long and 25.0″ (63.5 cm) in circumference. Maximum length and weight, outside Ohio waters, are 9′8.5″ (295.9 cm) and 302 lbs (137 kg); from a specimen taken in Belle Island Lake, Vermilion Parish, *Louisiana.*

Distribution and Habitat

Ohio Distribution—Although I have seen no alligator gars taken from Ohio waters, I have accepted the species as a fish occurring in the Ohio River upon the evidence given below.

Kirtland's (1844:18–19, Pl. 1, fig. 2; and 1851J: 149) excellent description and drawing leaves no doubt as to the former presence of the alligator gar in the upper Ohio River. The specimen was 5′ 8″ (173 cm) long; was taken above Cincinnati "up the

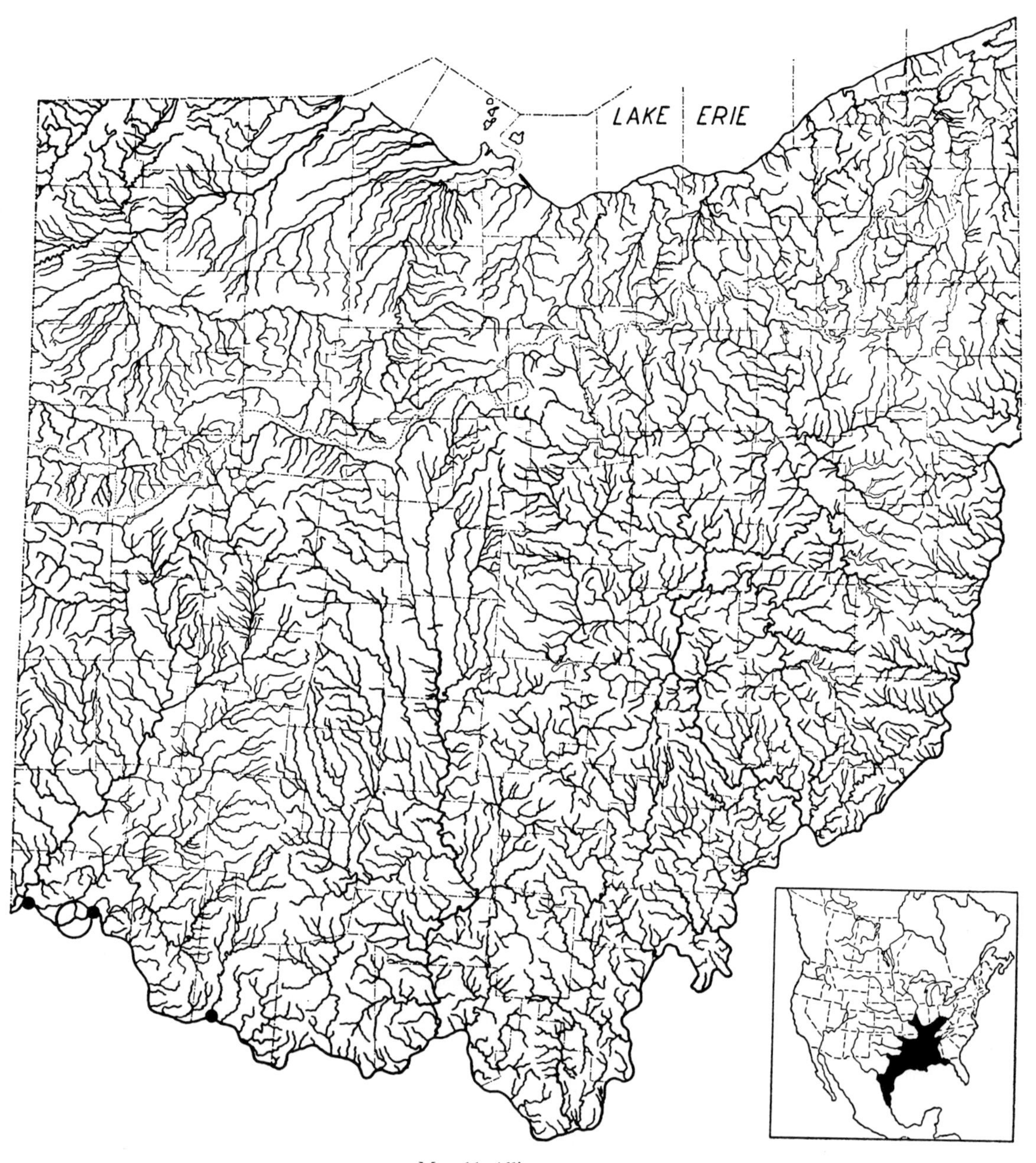

MAP 11. Alligator gar

Locality records. ○ Before 1901. ● 1901–50.
Insert: May be expected farther than indicated up the Missouri and Mississippi rivers, and into Mexico.

river, towards Pittsburg;" and when the drawing of the mounted specimen was made, it was in the Western Museum of Cincinnati. (Later the specimen was destroyed by fire.) Kirtland also stated that a Mr. Silsbee gave him some "bony scales of one taken about four years since" (1838–40). These scales were taken from a specimen presumably larger than the one recorded above. This latter gar was caught at the mouth of Mill Creek, Cincinnati.

In January, 1931, the late Charles Dury, former director of the Cincinnati Museum of Natural History, told me that before the river was dammed (before 1915), he had seen a few specimens of alligator gars taken from the river in Hamilton County, and that he considered it to be "not uncommon."

In May, 1931, an old river fisherman showed me a huge alligator gar scale which he had taken from a fish over 5′ (152.4 cm) in length. He caught the fish, about 1928, in the river at North Bend, Hamilton County. He likewise knew of several more which had been taken by other fishermen between 1910–30.

Harry Brookbank told me that between 1920–40 he took a few alligator gars in the Ohio River near the mouth of White Oak Creek, Brown County, and that during the summer of 1946 a fisherman caught one near the mouth of that creek which was more than 4′ (122 cm) in length. Mr. Brookbank's accurate description of the fish left no doubt as to its identification. The accompanying map gives the localities of those records which are accepted.

Habitat—Chiefly inhabits the bayous, back waters, or oxbows and adjacent lowland lakes of the large, warm, southern rivers.

Years 1955–80—There was no authentic report of the alligator gar having been taken in the inland waters of Ohio or in the Ohio River opposite that state during this period. None was taken in the entire Ohio River by the University of Louisville personnel during their 1957–59 investigations, although the species was known to occur in the lower reaches of that stream (ORSANCO, 1962:70).

For description of postlarval, *see* Moore, Trautman and Curd, 1973:343–44.

SHORTNOSE GAR

Lepisosteus platostomus Rafinesque

Fig. 12

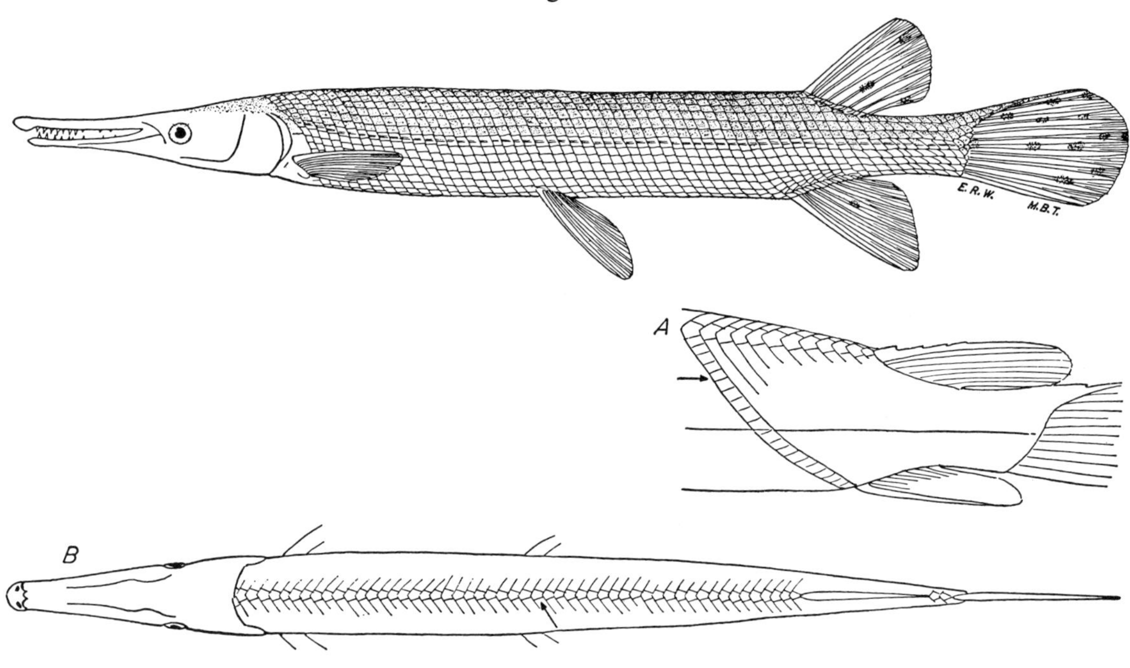

Upper fig.: Scioto River, Scioto County, O.

April 5, 1940. 500 mm SL, 23.1″ (58.7 cm) TL.
Adult. OSUM 1859.

Fig. A: arrow points to transverse row of scales; all these scales are counted including mid-dorsal scale and anal scale-plate.
Fig. B: dorsal view; arrow pointing to mid-dorsal row of scales which begins behind head and extends to dorsal origin. Note rather narrow and short snout.

Identification

Characters: Scales are not overlapping; rhombic in shape. Snout moderately wide, the eye diameter contained 1.0–1.5 times in width of upper jaw at nostrils. Distance from posterior edge of eye to posterior edge of opercle membrane contained 2.9–3.5 times in entire head length in fishes over 10.0″ (25.4 cm) long; all above characters similar to those of the spotted gar. Transverse scale row of 20–23 scales, ave. 20.8, fig. A. Dorsal ridge contains 50–55 scales, ave. 53.3, fig. B. Lateral scales 59–63, ave. 61.3, all scales counted from head to caudal fin, including the small, ill-formed scales at caudal base. Spots few and usually confined to the posterior half of body, or only on the dorsal, anal, and caudal fins. Ventral surface of head without spotting. Ohio drainage only.

Differs: From other gar species in scale counts, or width of snout. Spotted gar recorded only from Lake Erie drainage.

Most like: Spotted gar.

Coloration: *Adults*—Brown, olive or yellowish-slate dorsally with little mottling and few or no spots except on fins. Sides more yellow, lightest on belly. Spots when present on body usually confined to the posterior third. No spots on top of head or mottling on ventral surface. *Young* less than 10.0″ (25.4 cm) long are similar in coloration to longnose gars of corresponding lengths.

Lengths and **weights:** *Young* of year in Oct.

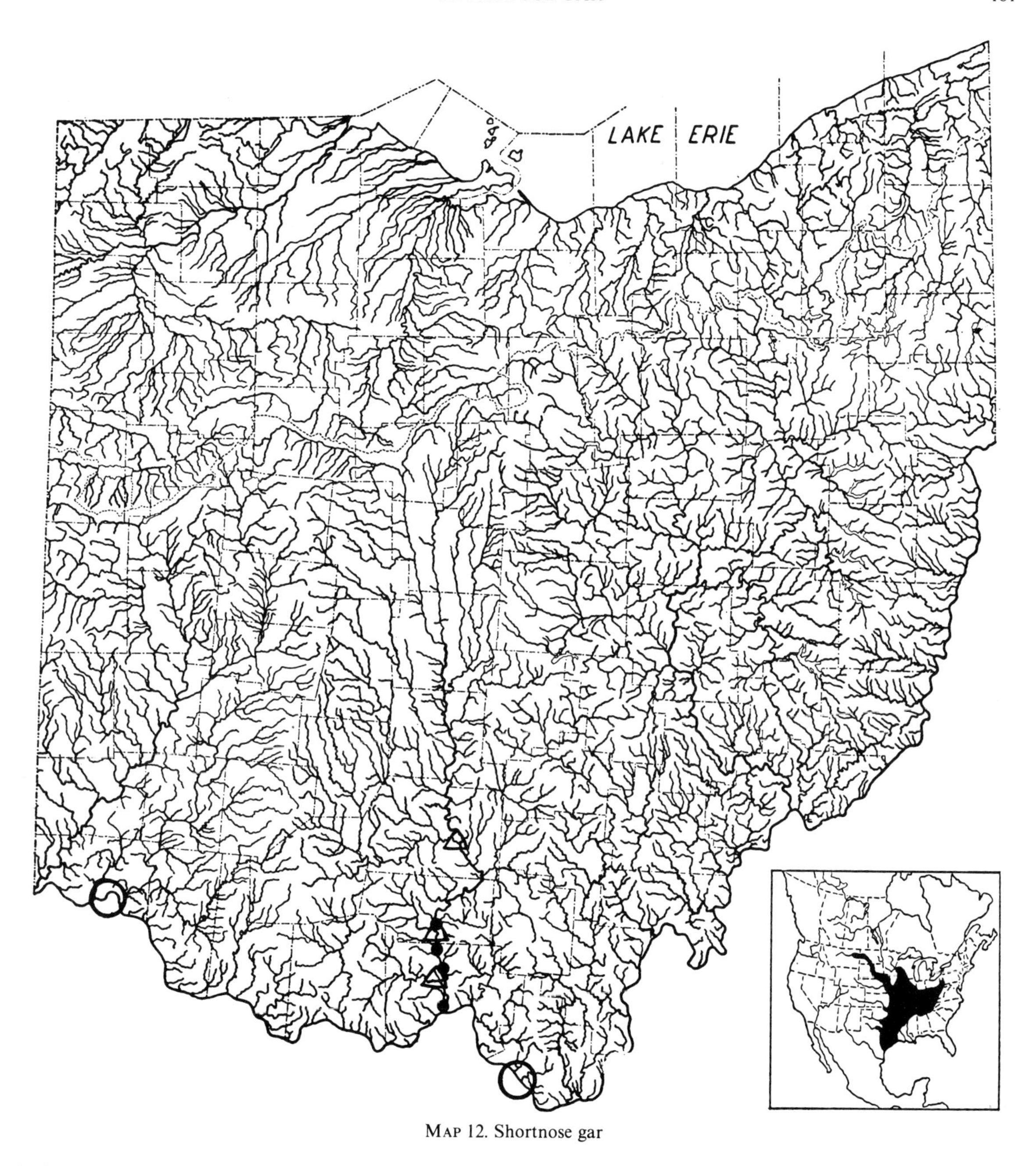

MAP 12. Shortnose gar

○ Literature and verbal records. ● Specimens preserved.
Insert: May range eastward farther into the Gulf States; many literature records unreliable.

7.0″–10″ (18–25 cm) TL. *Adults*, usually 16.0″–30.0″ (40.6–76.2 cm) long, weight 1 lb–5 lbs (0.5–2.3 kg).

Distribution and Habitat

Ohio Distribution—In his description of the shortnose gar in 1820, Rafinesque (1820:72–73) mentioned its occurrence in the Ohio, Miami, and Scioto rivers. In 1899 Osburn (1901:21) recorded the species from the Ohio River at Ironton. The late Charles Dury, in January, 1931, told me that between 1910–30, he saw several specimens that were taken from the Ohio River in Hamilton County; and several fishermen have told me that before

1925 they took an occasional specimen in the lower Muskingum River, and in the Ohio River between Washington and Jefferson counties.

Between 1939–50, a total of 59 specimens (OSUM 241:1859: 3874 and 5698) were taken in four localities in the lower Scioto River or adjacent ponds, principally by John Z. Pelton.

Habitat—The largest populations inhabit lowland lakes, oxbows, and backwaters; smaller populations occur in the still waters of the pools of rivers. In rivers, the species seemingly avoids strong currents and therefore avoids streams of high gradients. Since gars feed by sight, the shortnose seems to prefer clear waters, although at present, they inhabit our silt-laden rivers sparingly. The species does not appear to be adverse to waters which frequently become densely clouded with plankton. It is not found among rooted aquatic vegetation as often as is the spotted gar.

Years 1955–1980—Since 1950 the four overflow ponds along the Scioto River in Pike and Scioto counties, in which 59 shortnose gars had been previously captured, have been largely eliminated, some through construction of highways. Although many attempts to capture the species have been made since 1950 I am aware of the taking of only a few specimens. Harold S. Leach, Jr. of the Ohio Division of Wildlife, test-netting in the Scioto River, captured one east of Chillicothe, Ross County in June 1973 and another below the mouth of Sunfish Creek, Pike County in July 1973.

From 1957 to 1959 personnel of the University of Louisville captured 105 individuals in 31 collections of the entire Ohio River (ORSANCO, 1962:145).

SPOTTED GAR

Lepisosteus oculatus (Winchell)

Fig. 13

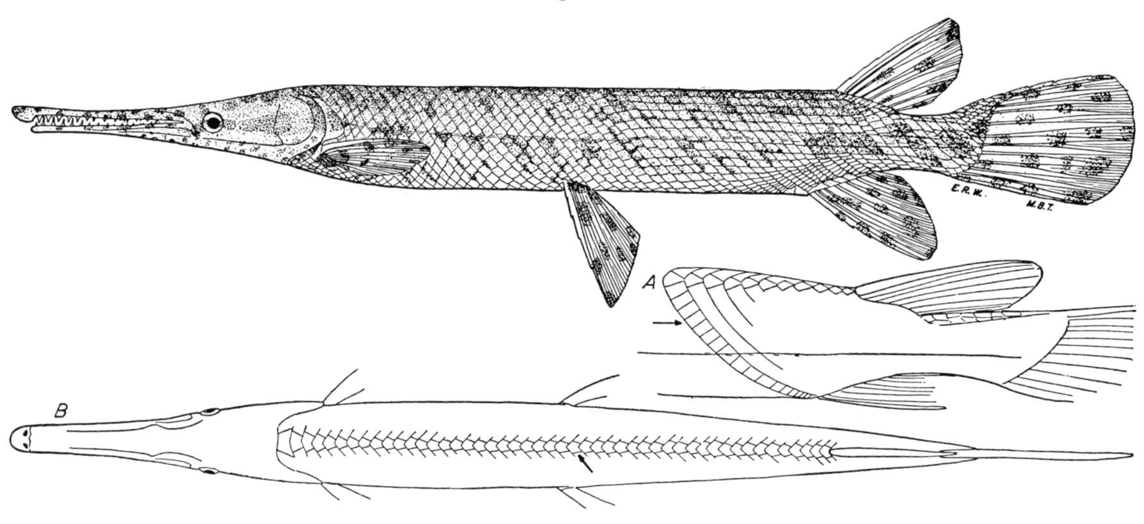

Upper fig.: Sandusky Bay, Erie-Ottawa counties, O.

Oct.—,1931. 434 mm SL, 19.8″ (50.3 cm) TL.
Adult. OSUM 721.

Fig. A: arrow points to transverse row of scales; all these scales counted, including mid-dorsal scale and anal scale-plate.
Fig. B: dorsal view, arrow pointing to mid-dorsal row of scales which begins behind head and extends to dorsal origin. Note rather narrow and short snout.

Identification

Characters: Type of scales, snout width, eye length in upper jaw width, and post-ocular head length in entire head length same as in the shortnose gar. Transverse scale row of 17–20 scales, ave. 19.5, fig. A. Dorsal ridge contains 46–49 scales, ave. 48.2, fig. B. Lateral scales 54–58, ave. 55.8, counting all scales from head to caudal fin. Many spots and blotches over body, fins, and on dorsal and ventral surfaces of head. Lake Erie drainage only.

Differs: *See* shortnose gar.

Most like: Shortnose gar.

Coloration: *Adults*—Dorsally a deep olive-green densely blotched and spotted with darker. Sides lighter with mottlings and spots more distinct. Ventrally yellowish or whitish, and almost invariably with some dark spottings. All fins, and dorsal and ventral portions of head, heavily blotched and spotted. *Young* less than 10.0″ (25.4 cm) long are similar in coloration to longnose gars of corresponding lengths, but mottlings may be more distinct.

Lengths and **weights:** *Young* of the year in Oct., 7.0″–10.0″ (18–25.4 cm) TL. *Adults*, usually 16.0″—-36.0″ (40.6–91.4 cm) long, weight 1 lb–5 lbs (0.5–2.3 kg). Largest specimen 44.0″ (112 cm) long, weight 6 lbs (2.7 kg).

Distribution and Habitat

Ohio Distribution—Kirtland (1851K:157) records the spotted gar as common "about Peach Island at the head of Sandusky Bay" and that "last summer [1850 or '51] a specimen was taken near Cleveland." This Cleveland specimen may have been collected by Spencer F. Baird, for in the U.S. National Museum (3241) there is an undated specimen collected by him at Cleveland, and from which a drawing was made for "The Fishes of North

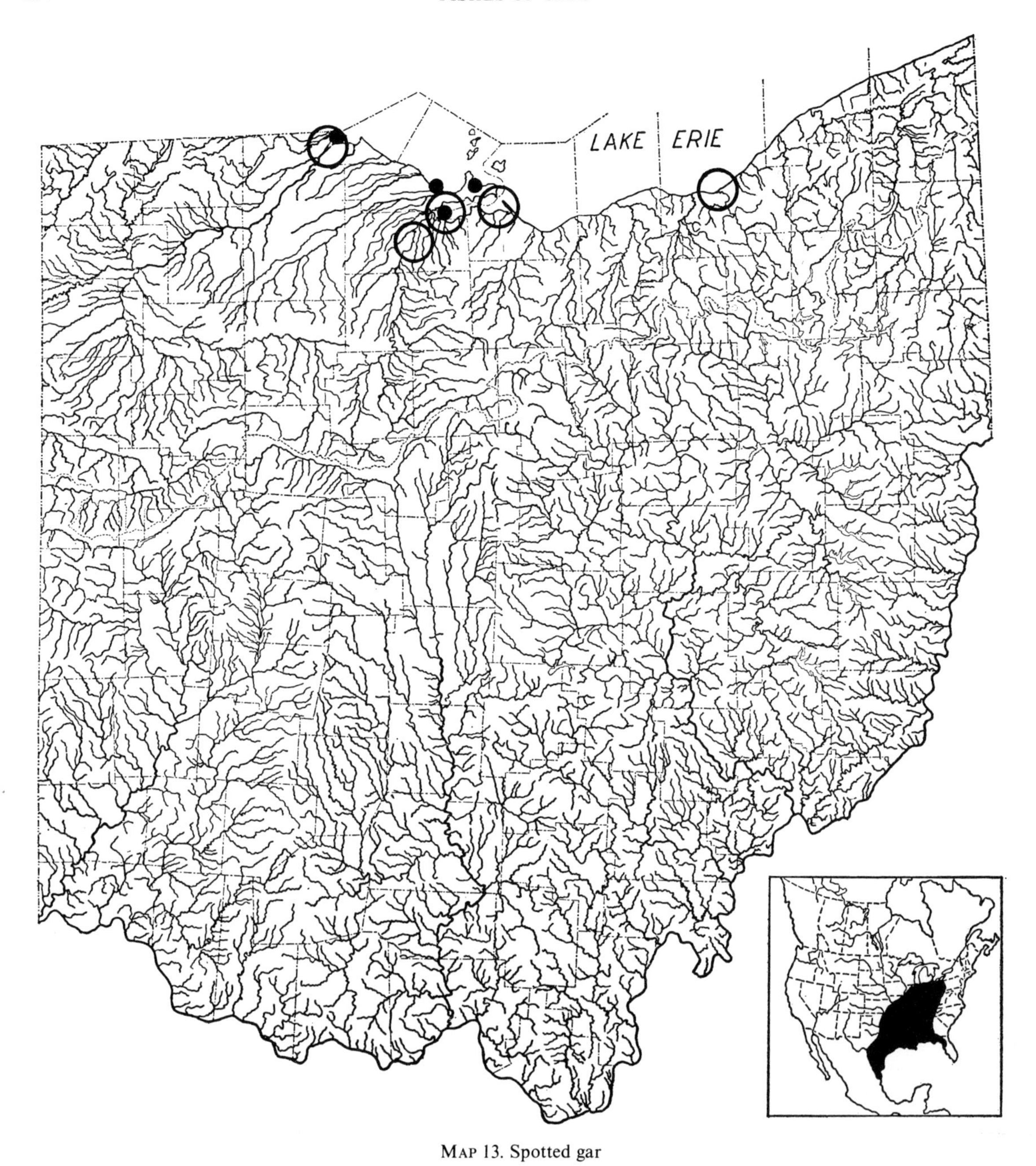

Map 13. Spotted gar

Locality records. ○ Before 1901. ● 1901–50. △ 1955–80.
Insert: May be expected farther west in the larger streams; Mexican range indefinite.

and Middle America" (Jordan and Evermann, 1900:Pt. 4, pl. 42, fig. 49). Henshall (1889:123) records the species from Lake Erie, and in 1931 I saw a specimen in the Cincinnati Society of Natural History, labeled Lake Erie, which the late Charles Dury told me had been taken long ago in the vicinity of Sandusky Bay. McCormick (1892:10) lists it as "Very rare" for Lorain County, but his notes in the Oberlin Museum Library state that "one specimen in Oberlin College Museum labeled 'Lake Erie' is my authority for including this species, as I have neither taken it nor found a fisherman who knows it by sight." Kirsch (1895A:327) in 1893 took a foot-long specimen in the Maumee River at Toledo (USNM76150); C. Rutter on July 31, 1894 collected three at Fremont, presumably in the Sandusky River (USNM126928); and Osburn (1901:21) one in 1899 in Sandusky Bay. In all of the above references this species was misidentified as *L. platostomus*, with which species it was long confused.

The late Charles Lay obtained two specimens (OSUM:721) in September, 1931, from the vicinity of Peach Island in Sandusky Bay, and one specimen each was obtained there during 1946 and 1947 (OSUM:6675 and 7267). On September 13, 1950, Harold Wascko caught one at the mouth of Lacarpe Creek in Ottawa County. Since 1940 K. C. Wahl and Charles Nielsen have obtained a few yearly in Sandusky Bay, and Cameron King told me that a few were taken yearly in Maumee Bay in the few areas where aquatic vegetation is still abundant. These three men claimed that the species had decreased in abundance in recent years.

Habitat—Quiet, clear waters containing a great abundance of aquatic vegetation, such as originally occurred in the bays and harbors of Lake Erie, and in pothole and lowland lakes. In such localities the species decreases in abundance, or disappears with the decrease or disappearance of aquatic vegetation.

Years 1955–80—The spotted gar still exists in small numbers in western Lake Erie and in some of its tributary waters, such as East Harbor and Sandusky Bay (Van Meter and Trautman, 1970:67).

In the 1960 edition of *A List of Common and Scientific Names of Fishes from America and Canada,* the American Fisheries Society (p. 10) placed Cope's specific name of *productus* (1865) in synonomy and substituted Winchell's older name of *oculatus* (1864). Suttkus (1963:71), Moore (1968:39) and other ichthyologists have followed this change.

LONGNOSE GAR

*Lepisosteus osseus** (Linnaeus)

Fig. 14

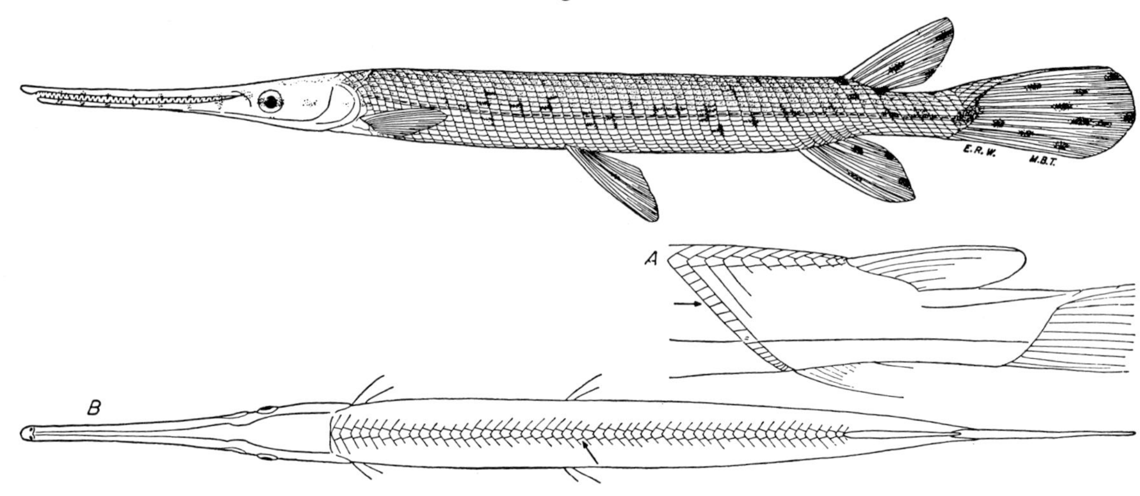

Upper fig.: Hocking River, Athens County, O.

June 28, 1939. 365 mm SL, 17.0″ (43 cm) TL.
Immature. OSUM 1098.

Fig. A: arrow points to transverse scale rows; all these scales counted, including mid-dorsal scale and anal scale-plate.

Fig. B: dorsal view, arrow pointing to mid-dorsal row of scales which begins behind head and extends to dorsal origin. Note long, slender snout.

Identification

Characters: Scales not overlapping; rhombic in shape. Snout beak-like, very long and narrow. Distance from posterior edge of eye to posterior edge of opercle membrane contained 3.5 times in entire head length in fishes more than 4.0″ (10 cm) long; between 3.0–3.5 times in fishes 2.0″–4.0″ (5.1–10cm) long. Width of the narrow upper jaw at nostrils less than eye diameter. Transverse scale row of 17–19 scales, ave. 18.5, fig. A. Dorsal ridge contains 50–52 scales, ave. 51.0, fig. B.

Differs: Other gar species have broader, shorter snouts, and different scale counts.

Most like: Spotted and shortnose gars.

Coloration: *Adults*—Brown, yellow or olive dorsally, yellowish or whitish ventrally. Some specimens from turbid waters lack spots; those from clear, vegetated waters may be as profusely spotted as is the spotted gar. *Young* less than 3.0″ (7.6 cm) long have black or chocolate bellies and dusky lateral bands, with a light yellow band between. *Young* 3.0″–15.0″ (7.6–38.1 cm) long usually retain the dusky lateral band, but have the mid-line of the belly white, bordered on each side with a band of chocolate.

Length and weights: *Young* of year in July, 3.0″–5.0″ (7.6–13cm) long; in Aug., 4.0″–8.0″ (10–20 cm) long; in Oct., 10.0″–15.0″ (25.4–38.1 cm) long. *Adults*, usually 24.0″–40.0″ (60.9–102 cm) long; weight 1 lb–7 lbs (0.5–3.2 kg). Largest specimen, 4′6″ (137 cm) long, weight 14 lbs (6.4 kg).

Distribution and Habitat

Ohio Distribution—The rather large number of literature records and preserved specimens from less than 100 collections indicate that before 1900 the longnose gar was widely distributed and locally abundant throughout the larger waters of Ohio. It is of extreme interest to note that the species began

* For geographic variations within this species, *see* Bailey, Winn and Smith, 1954: 117–18.

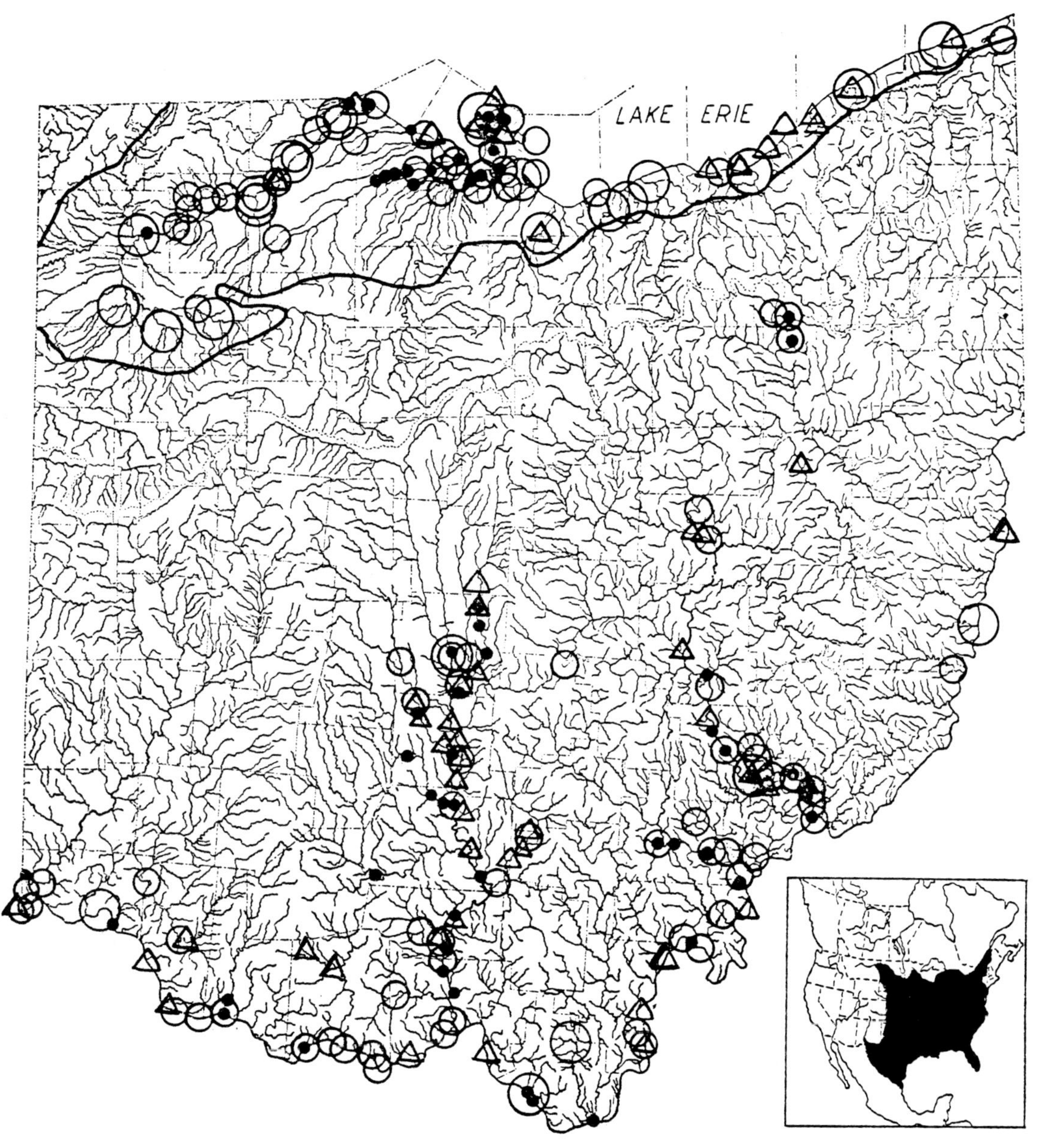

MAP. 14. Longnose gar

Locality records. ○ Before 1901. ○ 1901–35. ● 1936–50. △ 1955–80.
~ Lake Maumee Beach Ridge.
Insert: Large unoccupied areas within the range caused by absence or recent destruction of habitat; Mexican range indefinite.

very early to show a decrease in abundance, for in 1851, Kirtland (1851 L:141) wrote, "This species is now common in the Ohio River and Lake Erie but has greatly diminished in numbers in recent years." A gradual diminution in abundance, as indicated by the distribution map, continued throughout the 1925–50 period. Between 1925–30 the species could be seen basking near the water's surface in almost every long pool in the Maumee River between Defiance and Toledo, and occasionally hundreds could be seen in one pool. Since 1940 only a few have been noted in this entire stretch of river. In 1898 Williamson and Osburn (1898:13) found the species to be "common in the Scioto River south of Columbus where they may be frequently seen in schools." As late as 1925 I noted schools of basking gars in this section of river, but after 1935 the species was uncommon there and since 1945 has occurred only as strays.

The distribution map illustrates well this gar's preference for low-gradient streams, such as the Maumee and Scioto rivers, and its possible absence from the upper half of the higher-gradient Miami system. The absence of longnose gars from Indian Lake, Logan County, where before 1920 habitat conditions were particularly favorable, argues for its absence in the upper Miami system. But its absence from Lake St. Marys and its Wabash and St. Marys outlets is more difficult to comprehend. Until at least 1900 conditions in the greatly vegetated, log-strewn Lake St. Marys should have been extremely favorable, but in more recent years turbidity may have been a detrimental factor.

Habitat—As with the other species of gars present in Ohio waters, the longnose occurs in greatest abundance in the clearest waters of low- or basic-gradient streams, in oxbow lakes, overflow ponds, and in the harbors and bays of Lake Erie. Unlike the other species, however, the longnose is more tolerant to current and is frequently seen feeding or swimming about in moderately swift water. It occurs among aquatic vegetation whenever that is present, but such vegetation is not as important to the longnose as it is to the spotted gar. Its preference for feeding in clear water is shown by the fact that wherever gars are present in abundance and the water is very clear, trotline fishermen have difficulty in keeping these fishes from stripping the bait from their hooks, even on moonlit nights, and at times it becomes necessary to wait with trotline fishing until the water becomes more turbid or for moonless nights.

Years 1955–80—The general Ohio distributional pattern for the longnose gar, existing before 1950, remained essentially unchanged during the present period. The species decreased in numbers in those areas where the amounts of submerged and surface-floating vegetation and/or abundance of its prey decreased, or where some types of pollution and/or turbidity increased. A gradual decrease in numbers of young was observed about South Bass and Gibraltar islands of western Lake Erie and especially in Gibraltar Bay and Squaw Harbor, with the corresponding decrease in amount of aquatic vegetation. Presumably, surface-floating vegetation was quite important to the fry and small young because whenever wave action was present in the open water, no gars were observed there but were found basking at the surface in those quiet pools surrounded by dense masses of floating aquatics. If the open waters of the bay were unruffled, then small gars could be seen at the surface there also. When the young were between 25 and 100 mm TL, their growth was very rapid, some growing as rapidly as a centimeter in a few days.

In the medium- or large-sized streams an increase in numbers of adults was noted whenever the turbidity had been notably decreased. After the reservoir above the Hoover Dam had begun to function as a partial silting basin, the population of gars in Big Walnut Creek below the dam increased markedly, as could be observed when one drifted downstream in a boat. During early June of 1963 a group of approximately forty gars was found on, and adjacent to, a riffle on Big Walnut Creek below the Hoover Dam. Most of these gars were seen spawning at the edges of the riffles amid the stalks of water-willows, which were growing in slowly flowing water 10″–20″ (25–51 cm) deep. The remaining individuals were observed spawning in the more quiet waters of the riffle itself over a substrate of gravel and boulders and also in an adjacent embayment where the waters were 12″–24″ (30–61 cm) deep. A few eggs were found adhering to stones and vegetation. The spawning adults were 22″–30″ (56–76 cm) TL. On June 8, 1962 several hundred young were seen basking near the surface of an embayment of the Scioto River opposite Circleville; Roger Burnard and I captured 113 individuals 40–75 mm TL.

For age and growth *see* Netsch and Witt (1962:251–62).

BOWFIN

Amia calva Linnaeus

Fig. 15

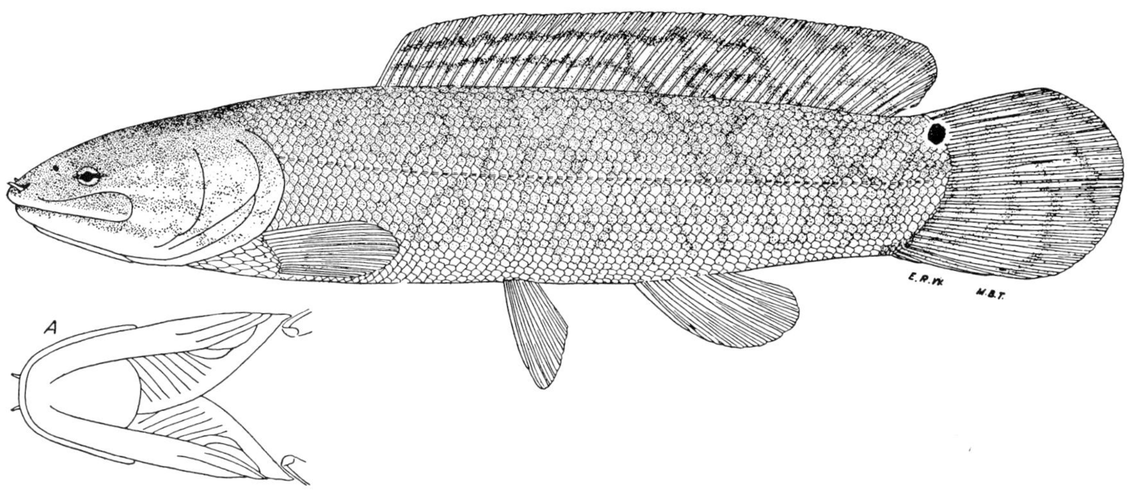

Upper fig.: East Harbor, Ottawa County, O.

Aug. 12, 1940. 440 mm SL, 21.4″ (54.4 cm) TL.
Adult female. OSUM 1891.

Note black spot at upper edge of caudal base; this spot is very distinct in breeding males, less distinct or absent in females and young.

Fig. A: ventral view of head; note the large, somewhat triangular gular plate, situated between the lower jaws anterior to the branchiostegals.

Identification

Characters: Only species of fish in Ohio having a large gular plate occupying the anterior half between the lower jaws, fig. A. The long dorsal fin is not connected to caudal fin and of more than 45 rays. Scales overlap as in most fish species. Color predominantly deep green. Catfish-like in shape. No spines.

Differs: No other species of fish in Ohio has a large gular plate. Catfishes and burbot have one or more barbels.

Most like: No other Ohio fish closely resembles it.

Coloration: Dark olive- or yellow-green dorsally, sides a lighter green, and a cream-green belly. Dorsal fin dark green with two broken, longitudinal, olive bars. Lower fins a vivid green. Spot at upper edge of caudal base most prominent in *adult* males.

Lengths and **weights:** *Young* of year in Oct., 5.0″–9.0″ (13–23 cm) long. *Adults*, usually 15.0″–25.0″ (38.1–63.5 cm) long, weight 1 lb 4 oz to 5 lbs (0.6–2.3 kg). Largest *male*, 27.0″ (68.6 cm) long, weight 5 lbs (2.3 kg); largest female, 31.0″ (78.7 cm) long, weight 8 lbs 8 oz (3.9 kg).

Distribution and Habitat

Ohio Distribution—From the early records it is apparent that before 1900 the bowfin was abundant in Lake Erie (Kirtland, 1851M:109), was present in its tributary streams, and in such glacial pothole lakes as Ladd and Nettle in northwestern Ohio. It seems to have been absent from the Portage Lakes of Summit County, despite an abundance of presumably favorable habitat, and the presence in these lakes of the majority of its associates. It seems logical to assume that if these associates were able to invade these lakes, the bowfin likewise should have been able to invade them.

Between 1925–50 the bowfin occurred sparingly in the Maumee drainage, but was abundant in the

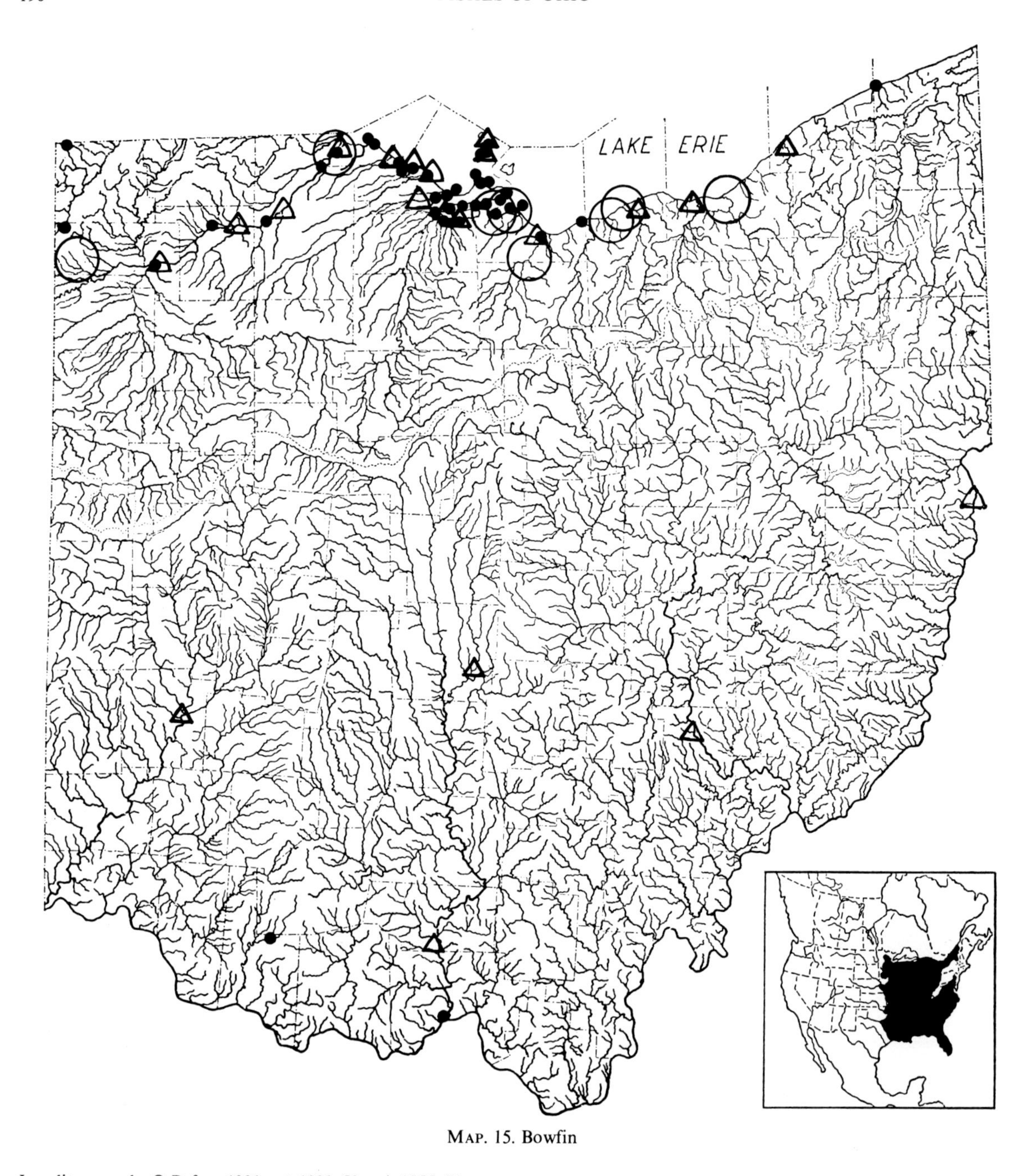

MAP. 15. Bowfin

Locality records. ○ Before 1901. ● 1901–50. △ 1955–80.

Insert: Large unoccupied areas within the range caused by absence or recent destruction of habitat; sparingly introduced west of the original range.

larger Lake Erie marshes and harbors between and including Maumee and Sandusky bays, particularly the latter. Shafer (1950:1) reported the Ohio commercial catch of bowfins for Lake Erie waters for 1949 to be 4,441 lbs (2014 kg).

Jordan (1882:777) indicated that in 1882 the bowfin was present in that portion of the Ohio River which borders Ohio; this statement is strengthened by the occurrence of the species in southern Indiana (Hay, 1894:168–69) and its possible occurrence in Pennsylvania (Fowler, 1919:53). During the 1925–50 period I saw two bowfins taken in the Ohio drainage, one in White Oak Creek, the other in an overflow pond of the Scioto River. In addition several fishermen told me of one that was taken in the Muskingum River near Marietta. It is possible that these strays were inadvertently introduced into southern Ohio with introductions of more desirable species from Lake Erie waters.

Habitat—The largest populations occurred in the bays, marshes, and harbors of Lake Erie wherever the water was the least turbid and there was the greatest abundance of rooted aquatic vegetation. Lesser populations occurred in the small, glacial lakes which likewise contained clear water and an abundance of vegetation, and in the low- or basic-gradient portions of streams and their backwaters, oxbows, and adjacent lowland lakes. The bowfin was not adverse to waters made cloudy by the abundance of plankton but normally occurred sparingly or as strays in waters habitually turbid with clayey silts. It displayed the greatest decreases in abundance in those Ohio waters which formerly were clear and contained much vegetation, but which during the survey had become silty and almost vegetationless.

Years 1955–80—The populations of bowfins in Sandusky Bay and in the remainder of the western Lake Erie region showed no appreciable change during the present 25-year period.

It was previously suggested that the Ohio River populations might be strays that were inadvertently introduced into southern Ohio waters. Since 1957 sufficient additional information has accumulated to indicate that a small population exists in the upper Ohio River drainage. Nelson and Gerking (1968:22) recorded its presence in the White, Wabash and Ohio drainages of Indiana. Twenty-two specimens were taken at five collecting localities during the investigations of the entire Ohio River between the years 1957–59 (ORSANCO, 1962:145). On June 1, 1962 William Shipman took an adult in Big Walnut Creek, Madison Township, Franklin County, the farthest inland record. On April 11, 1963 Howard McClay captured an adult in an overflow pool of the Scioto River in Camp Creek Township, Pike County. On June 17, 1967 Virgil Wilson caught an adult in the Mad River near Dayton, Montgomery County. On May 27, 1973 Dotty Wise caught one in the Muskingum River, a short distance above the towns of Malta and McConnelsville, Morgan County. In addition there were several newspaper records of individuals taken in Ohio drainage streams.

AMERICAN EEL

Anguilla rostrata (Lesueur)

Fig. 16

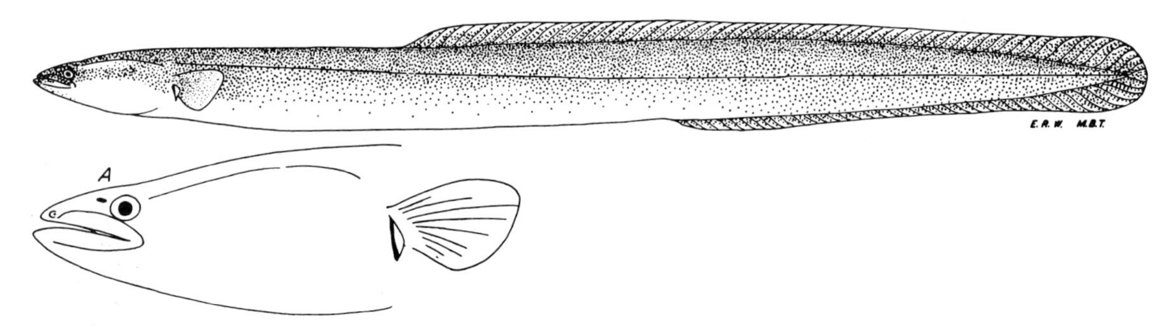

Upper fig.: Scioto River, Scioto County, O.
April 8–15, 1945. 845 mm SL, 33.4″ (84.8 cm) TL.
Female. OSUM 1858.

Fig. A: lateral view of head; note jaws, and small gill cleft before pectoral fin.

Identification

Characters: *No* pelvic fins. The one gill-opening a short, narrow slit immediately in front of pectoral fin. Jaws present, fig. A. Dorsal, caudal and anal fins continuous, about 60 rays from origin of dorsal to posterior tip of caudal. Scales so exceedingly small as to give the impression that body is scaleless.

Differs: Lampreys have circular, jawless mouths; no pectoral fins. Burbot has pelvic fins; a chin barbel. Catfishes have spines.

Superficially like: Lampreys and burbot.

Coloration: Dorsally yellow (young less than 14.0″ [35.6 cm] long and larger fishes from turbid waters), brown, or chocolate-brown (large adults from clear waters). Sides somewhat lighter. Ventrally light yellow, yellow-tan, or whitish. A sharp demarcation between the darker sides and lighter ventral surface, giving a decided bicolored effect. Fins same color as adjacent body parts.

Lengths and **weights:** Smallest Ohio specimen, 11.0″ (27.9 cm) long. Usually 15.0″–40.0″ (38.1–102 cm) long. Fishes 15.0″–22.0″ (38.1–55.9 cm) usually weigh 2 oz–1 lb (57g–0.5kg); 24.0″–30.0″ (60.9–76.2 cm) usually 1 lb–3 lbs (0.5–1.4 kg); 30.0″–40.0″ (76.2–102 cm) usually 3 lbs–5 lbs (1.4–2.3 kg). Largest specimen, 52.0″ (132 cm) long, weight 7 lbs, 8 oz (3.4 kg).

Distribution and Habitat

Ohio Distribution—It is believed that the eel was absent from Lake Erie waters before the completion of the Welland Canal in 1829. By 1844, Kirtland (1844:235) had heard rumors of eels in Lake Erie waters, but it was not until some time later, (date unknown) that he examined an eel which was caught in the Cuyahoga River at Cleveland. In Kirtland's personal copy of Rafinesque's "Ichthyologia Ohiensis," now in the Case Western Reserve University Library, he wrote a marginal note concerning the capture of this Cuyahoga River specimen. This specimen may have been the same one which Garlick (1857:126) mentioned having been taken "last year [1856]" in the Cuyahoga River, Garlick believing that it "undoubtedly found its way from Lake Ontario by the Welland Canal."

In 1820 Rafinesque (1820:143) recorded the eel as present in the Ohio River drainage as far upstream as Pittsburgh, and Kirtland (18510:189) in 1851 as ascending the Ohio drainage to "even its head during spring."

In 1878 the Michigan Fish Commission planted young eels, taken from the upper Hudson River, in southern Michigan waters (Potter, 1879:19), and in 1882 the Ohio Fish Commission began the same practice, liberating 128,100 elvers in waters

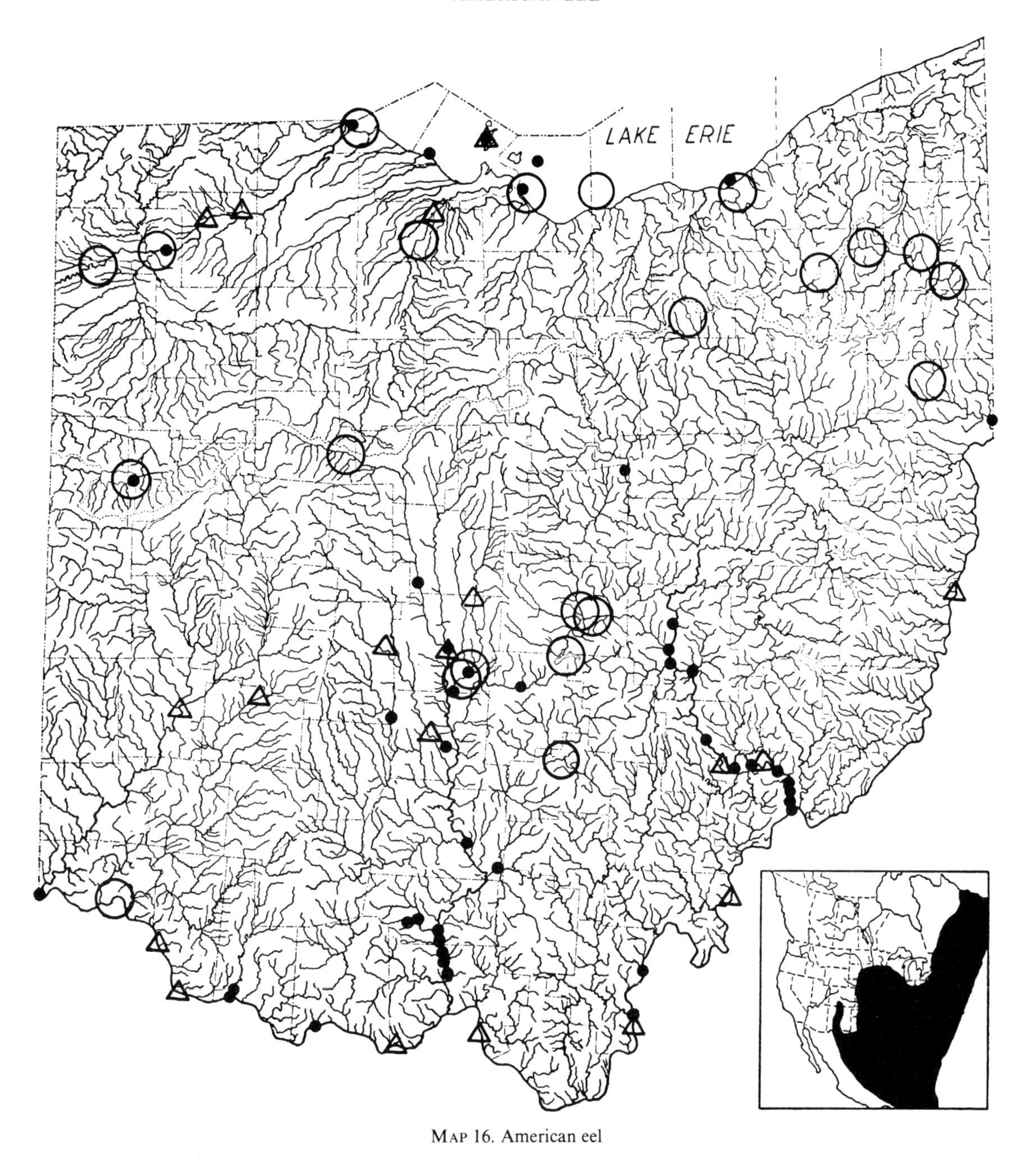

Map 16. American eel

○ Before 1910, literature and verbal records. ● After 1910, specimens. △ 1955–80.
Insert: Stragglers may be found in the upper Great Lakes; large unoccupied areas exist within the range, especially in the highlands.

throughout Ohio. For more than a decade thereafter elvers, mostly from the Hudson River, were liberated in Ohio waters. In 1887 the annual Ohio Fish Commission reports mentioned the capture of eels in many Ohio localities, especially in the Lake Erie drainage, where until these plantings were made, the species had been rare. The majority of the eels appear to have been taken during the 1895–1910 period. Cameron King told me that his father and he caught one or two bushels of eels daily during 1902 in Maumee Bay, and Frank Redding that he took many eels in the Sandusky River below the dams at

Fremont, between 1895 and 1910. In 1897 Williamson and Osburn (1898:36) reported that fishermen found the eel to be "not rare" in Franklin County.

My father and several of his fisherman friends have related to me that between 1895–1910 the eel was rather common in central Ohio. A favorite method of capturing them was to procure a large barrel, bore about 12 holes in the sides, each hole about 2″ (5 cm) in diameter, around which on the inside was tacked the top end of a long stocking, leaving the remainder to hang down inside the barrel. After cutting off the toe of the stocking the barrel was partly filled with old cheese, meat, offal, and garbage, and sunk in a deep pool. The eels, scenting food, squirmed through the holes in the barrel and through the length of the stockings. When the eels were inside the barrel the stockings collapsed, thus preventing escape. As many as 12 eels were taken in a night by this method.

I saw five eels which were captured between 1920–50 in the Lake Erie drainage, and heard of several more. During the same period the eel was rather numerous in the Ohio River as far upstream as Marietta, in the lower Scioto River, and especially in the Muskingum River upstream as far as Zanesville. In June of 1939, I saw five eels which were caught on a trotline in a single night in the Muskingum River near McConnelsville.

Habitat—In Ohio, the eel occurred most frequently in moderate- or large-sized streams where there was an abundance of food, such as living or dead fishes, crayfishes, and garbage. The species was extremely tolerant to turbid waters, presumably finding its food principally by scent rather than by sight. It also was very active on the darkest nights. During the day it sometimes partially or completely buried itself in the mud, sand and gravel, emerging at dusk to begin feeding.

Years 1955–80—Eels have been reported captured annually in the inland waters of Ohio throughout this period. Strays continue to be reported from Lake Erie (Van Meter and Trautman, 1970:74). H. Ronald Preston told me that during 1968 and 1969 the species had been captured in four of nine collection stations in the Ohio River between Belmont and Clermont counties.

Before 1900, thousands of elvers were liberated into Ohio waters, *see* Ohio Distribution above; and although eels are supposedly long-lived, it seems most improbable that individuals from early plantings are still alive. Rather, it appears that this persistent migrant reached Lake Erie waters through the Welland Canal (Trautman, 1960) and that those taken in the Ohio River drainage circumnavigated the dams, possibly during floods.

The early plantings of elvers appeared to be successful. If elvers could be obtained in sufficiently large numbers, they might profitably be reintroduced into inland Ohio waters, and possibly into larger farm ponds having an overcrowding of small bluegills, as an additional fish for food and sport.

SKIPJACK HERRING

Alosa chrysochloris (Rafinesque)

Fig. 17

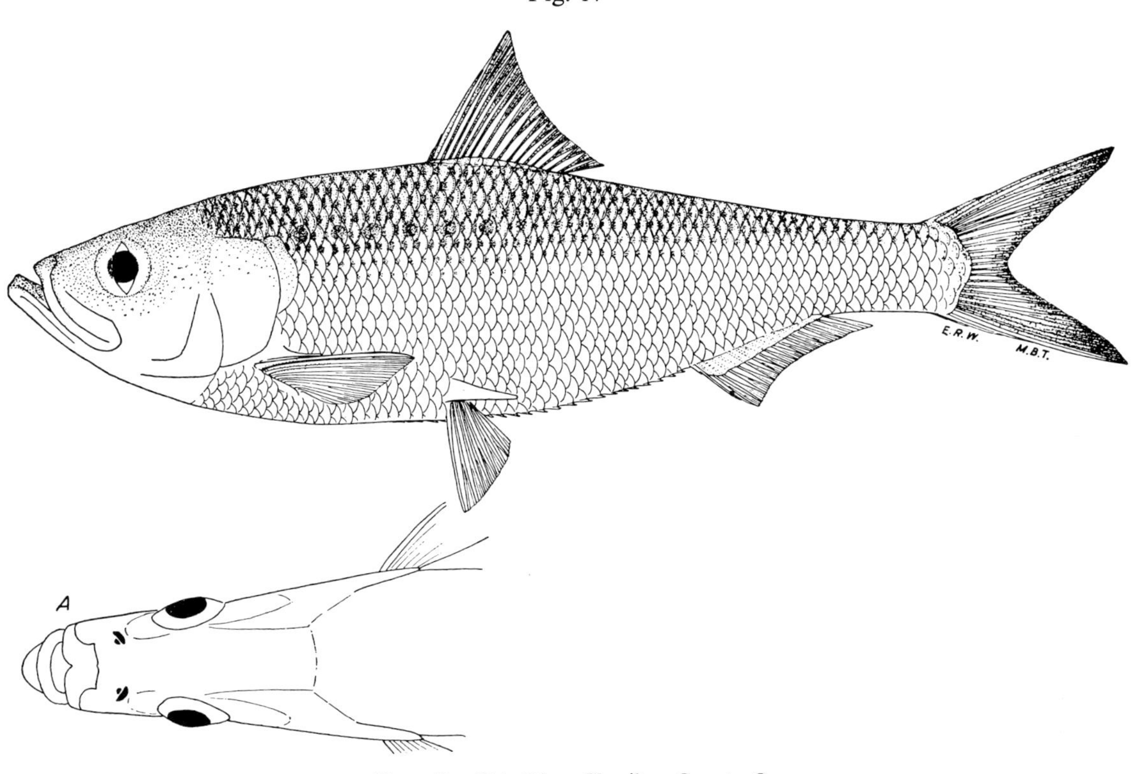

Upper fig.: Ohio River, Hamilton County, O.

Oct. 16, 1942. 130 mm SL, 6.3" (16 cm) TL.
Immature. OSUM 5209.

Note pelvic axillary process and adipose eyelids.

Fig. A: dorsal view of head; note extremely protruding lower jaw.

Identification

Characters: *General*—Pelvic axillary process present. Eyes partly covered, anteriorly and posteriorly, with adipose eyelids. Mouth large, lower jaw protruding. Dorsal fin insertion in front of, or directly over, pelvic fin insertion. Coloration bluish-silvery. *No* adipose fin. *Specific*—Posterior end of upper jaw extending past middle of eye. Gill rakers on lower angle of first gill arch fewer than 30. Ohio Drainage only.

Differs: Whitefish, ciscoes, trouts, and smelt have an adipose fin. Gizzard shad has an included lower jaw, and a greatly elongated posterior dorsal ray. Minnows have no adipose eyelids or pelvic axillary process. Alewife has smaller mouth and more gill rakers.

Most like: Alewife.

Coloration: Deep bluish-silver dorsally, sides less bluish and more silvery, belly silver- or milk-white; entire body with a silvery sheen and blue-gold reflections. A row of 1–9 dusky spots extending from upper angle of gill cleft backward along upper sides; those spots nearest gill cleft usually the largest. Scales on back with dusky blotches at their bases.

MAP 17. Skipjack herring

Locality records. ● Before 1955. △ 1955–80.

Insert: Strays or small numbers of this highly migratory species may be found outside the indicated range, especially in spring.

Lengths and **weights:** *Young* of year in Aug., 1.0″–4.0″ (2.5–10 cm) long; in Oct., 5.0″–8.0″ (13–20 cm). *Adults*, usually 12.0″–16.0″ (30.5–40.6 cm) long, weight 8 oz–1 lb, 4 oz (227g–0.6 kg). Largest specimen, 21.0″ (53.3 cm) long, weight 3 lbs, 8 oz (1.6 kg).

Distribution and Habitat

Ohio Distribution—The statements of Rafinesque (1820:90) that "It seldom goes as far as Pittsburgh [in the Ohio River] and does not run up the creeks"; of Kirtland (1851D:117) that "It is found occasionally in the Ohio River and some of the larger tributaries, but never in the waters of Lake Erie"; of Henshall (1888:79) that it is "Abundant in the Ohio River" indicate that the skipjack herring was present and abundant in the Ohio River and the lower portions of its larger tributaries prior to 1900. Throughout the 1925–50 period the species was present in larger numbers in the Ohio River between Marietta and the Indiana state line; in fewer numbers in the lower Scioto and Muskingum rivers; according to the rivermen, individuals were occasionally taken in the Ohio River between Marietta and the Pennsylvania state line.

Jordan's (1882:873) statement that it "escaped through the canals into Lake Erie" lacks confirmatory evidence and is unacceptable. It is absurd to expect this deep- and swift-water inhabiting species to migrate across Ohio through the sluggish canals when it does not penetrate far inland in the largest unobstructed streams of the Ohio drainage. Furthermore, had the skipjack herring succeeded in invading Lake Erie it presumably should have established itself in these large waters.

Habitat—The skipjack herring appears to avoid the more turbid waters as much as possible. Avoidance of turbid waters was demonstrated many times by the presence of many individuals in the clearer waters about the mouths of tributaries when the Ohio River was turbid. The species fed in large, swiftly swimming schools which forced the huge schools of emerald and mimic shiners to crowd together near the water's surface. Once the minnows were closely crowded together the skipjack dashed in among them, forcing the minnows to rise to the water's surface where they could be captured readily. The species often congregated in large numbers in the swift waters below the dams of the Ohio River where they preyed upon the immense numbers of minnows segregated there.

The species was universally known as the "skipjack" because of its frequent leapings into the air to capture the jumping minnows. It readily took natural and artificial baits, leaping spectacularly in the air and dashing about with great speed when hooked. When taken with the aid of a fly rod and light tackle it ranked among the finest of Ohio game fishes.

Years 1955–80—During the past 20 years the distribution of this species has remained unchanged in Ohio, except for small inland migrations up the Scioto River and some of its tributaries in Ross, Pickaway and Franklin counties. These fishes were reported, with few exceptions, between the years 1958 and 1961 by fishermen who sent me specimens for identification. Some fishermen reported catching as many as eight in a day. Douglas Albaugh has shown me pictures of specimens taken since 1960 at Luke's Chute, Muskingum River, Morgan County. The species remained rather abundant in the Ohio River, where at times it might be readily observed chasing and capturing small fishes near the water's surface.

I follow the American Fisheries Society (Bailey et al., 1970:15) in changing the generic name from *Pomolobus* to *Alosa*; its scientific name, therefore, is *Alosa chrysochloris* (Rafinesque). For some unknown reason the colloquial name for the skipjack in Pike, Scioto and other Ohio River counties is "McKinley shad."

ALEWIFE

Alosa pseudoharengus (Wilson)

Fig. 18

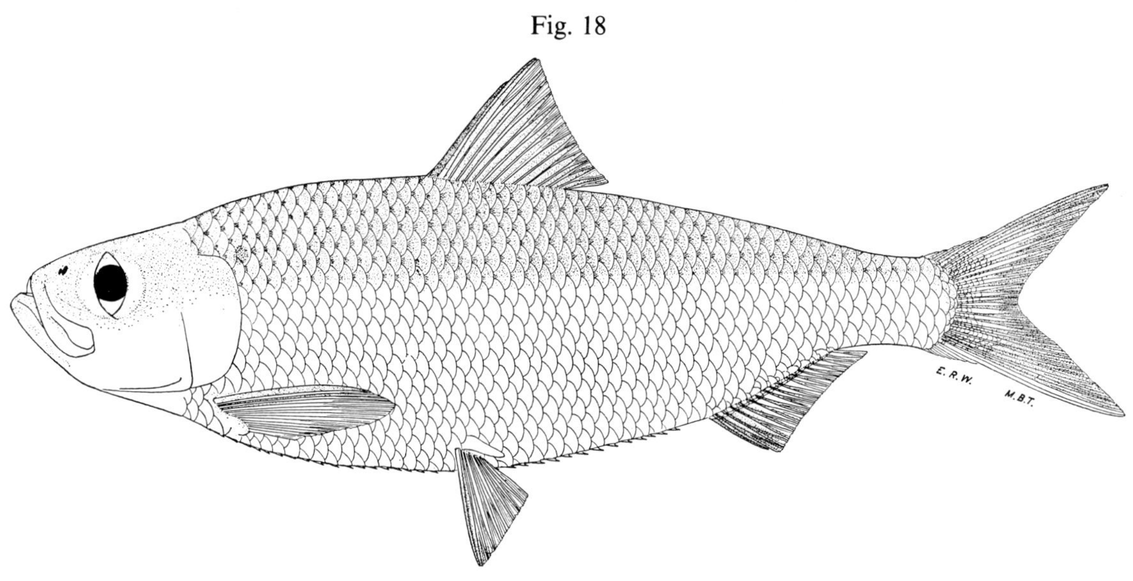

Lake Erie, Ottawa County, O.

Dec. 26–31, 1940.
Adult female.

185 mm SL, 9.0″ (23 cm) TL.
OSUM 3046.

Note that scales on the mid-line of the belly are modified into scutes, giving a serrated or toothed appearance.

Identification

Characters: *General—See* skipjack herring. *Specific*—Posterior end of upper jaw not reaching to middle of eye. Gill rakers on lower angle of first gill arch more than 30. Lake Erie Drainage only.

Differs: *See* skipjack herring.

Most like: Skipjack herring.

Coloration: Similar to skipjack herring, except that golden reflections appear to be lacking.

Lengths and **weights:** Largest specimen, 11.0″ (27.9 cm) long, weight 10 oz (283 g). Commercial fishermen claim to have seen longer and heavier individuals.

Distribution and Habitat

Ohio Distribution—Despite the highly migratory tendency of the marine populations, the alewife of Lake Ontario did not migrate into Lake Erie until rather recently and then, presumably through the Welland Canal. The first definite Lake Erie record is of a specimen 7 3/4″ (19.7 cm) long, taken in Ontario waters near Long Point, in September, 1931 (Dymond, 1932:32). The first record for Lake Huron is of a fish taken in the northern half of the lake in Ontario waters near Duck Island, in March, 1934 (MacKay, 1934:97).

Between 1932–40 several commercial fishermen told me about a "new sawbelly or shad" which was "good to eat," and which they caught infrequently in the Ohio waters of the eastern half of Lake Erie. The majority of these fishes were taken following storms in December, when hundreds were sometimes caught. A few were also taken in shallow water in late May and June. From their description it was apparent that these fishes were alewives. Between December 26–31, 1940, Kenneth H. Doan caught an alewife, 9″ (23 cm) TL, in a gill net set within a few hundred yards of South Bass Island, Ottawa County; this appears to be the first preserved specimen from Ohio waters (OSUM: 3046). On December 27, 1942, following a high wind storm, several fishermen found their gill nets, set near Vermilion

MAP 18. Alewife

Locality records. ● Before 1955. △ 1955–80.

Insert: Recent extension of range above Niagara Falls into Lake Erie and lower Lake Huron; future occupancy of upper Great Lakes expected.

and Lorain, which they had previously given up as lost. In these nets were two kinds of "sawbellies," of which one gizzard shad and two alewives (OSUM:6059) were sent to me. The fishermen estimated that they had taken more than 400 lbs (181 kg) of "the kind that are good to eat" in the nets which they had recovered. Between 1942–50 there were infrequent reports of the taking of alewives in Ohio waters of Lake Erie; by the spring of 1953 it was very numerous about the Bass Islands.

Habitat—Remains in the deeper waters of Lake Erie throughout much of the year, coming into shallower water, or ascending streams to spawn in June or July.

Years 1955–80—Since 1933 the alewife has invaded all of the upper Great Lakes, where it has become at times immensely abundant. The species was first recorded in Lake Erie in 1931, but has been present only in moderate-size populations. During the 1955–80 period the alewife population in Lake Erie fluctuated from year to year, and although numerous in certain localities at times, it never approached the overwhelming abundance that it has in the upper Great Lakes (White et al., 1975:52). I follow others in changing the generic name from *Pomolobus* to *Alosa*; i.e., *Alosa pseudoharengus* (Wilson) (Bailey et al., 1970:15).

On October 13, 1970 H. Ronald Preston, Ted M. Cavender, George Billy, Samuel Leach, I and others collected an alewife in a lock on the Ohio River near Greenup, Kentucky. On October 14, 1970 Preston collected nine in that river at the Captain Anthony Meldahl lock, 2 miles (3.2 km) east of Foster, Bracken County, Kentucky, and opposite Clermont County, Ohio. It is possible that these fishes are the result of introductions of this species into Clayton Lake, a reservoir on the New River near Radford, Virginia (unpublished MS. Cavender and Preston).

EASTERN GIZZARD SHAD

Dorosoma cepedianum (Lesueur)

Fig. 19

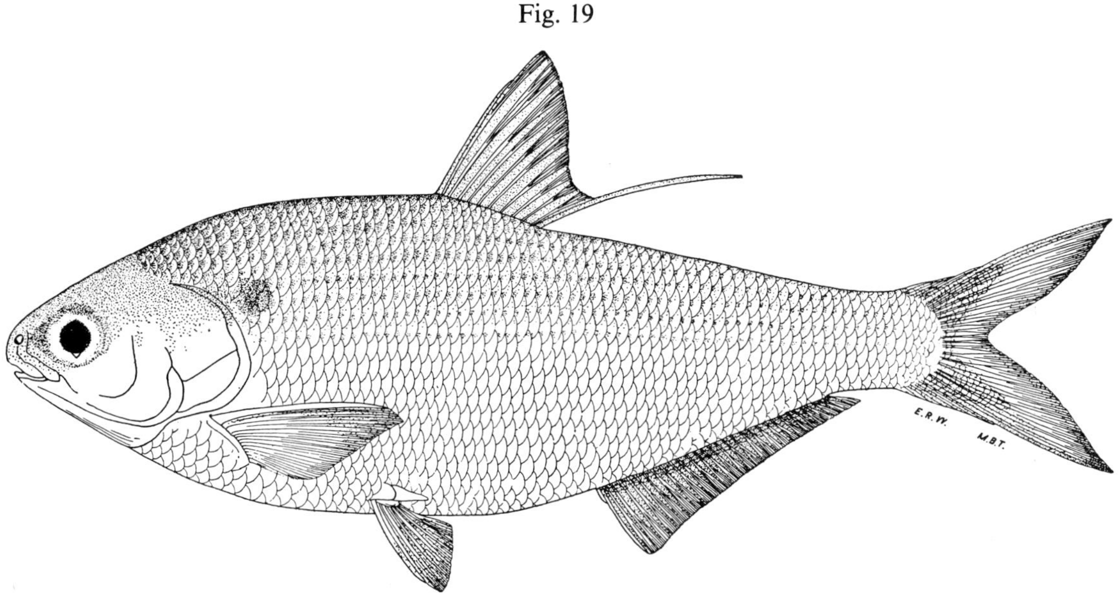

Meander Lake, Trumbull County, O.

May 8, 1942.
Adult male.

223 mm SL, 10.8″ (27.4 cm) TL.
OSUM 4878.

Note elongated last dorsal ray (may be broken off in some specimens) and dusky spot on body behind upper gill cleft.

Identification

Characters: General and Generic—Last ray of dorsal fin greatly elongated (unless broken off as it sometimes is in large young and adults, or not developed as it is in very small young). Pelvic axillary process present. Eyes partly covered, anteriorly and posteriorly, with adipose eyelids. Mouth subterminal, lower jaw included. Upper jaw length contained more than 3.2 times in head length. Snout bluntly rounded. *No* adipose fin. Coloration bluish-silvery. *Specific*—Dorsal fin rays usually 12, range 11–13. Anal fin rays 27–36, usually more than 29. Many melanophores along base of anal fin in young less than 30 mm TL. Lateral line scales usually 58–65, range 52–70. Dark postopercular spot as large as eye or larger. Anterior origin of dorsal insertion distinctly behind pelvic insertion.

Differs: Whitefish, ciscoes, trouts, and smelt have adipose fins. Alewife and skipjack herring have large mouths with the lower jaw projecting. Minnows have anal fins of fewer than 14 rays.

Most like: Threadfin shad. **Superficially like:** Alewife; mooneye; goldeye.

Hybridizes: with threadfin shad (Minckley and Krumholz, 1960).

Coloration: Bluish dorsally, sides more silvery, belly milk-white, with a silvery sheen and bluish and greenish reflections over head and body. A dusky spot, usually smaller than eye, on body behind upper angle of gill cleft; this spot very pronounced in small young, usually distinct in half-grown fishes, may be absent in large adults.

Lengths and **weights:** *Young* of year in July, 1.0″–3.0″ (2.5–7.6 cm); Sept., 2.0″–4.5″ (5.1–11.4 cm); Nov., 2.5″–6.0″ (6.4–15 cm). Around 1 year, 4.0″–9.0″ (10–23 cm). *Adults* of Lake Erie Drainage, 12.0″–16.0″ (30.5–40.6 cm) long, weight 10 oz–1 lb 8 oz (283 g–0.7 kg); Ohio Drainage, 14.0″–18.0″ (35.6–45.7 cm) long, weight 1 lb–3 lbs (0.5–1.4 kg). Largest specimen 20.5″ (52.1 cm) long, weight 3 lbs 7 oz (1.6 kg). Size and weight vary greatly in various bodies of water; marked dwarfing occurs in some localities.

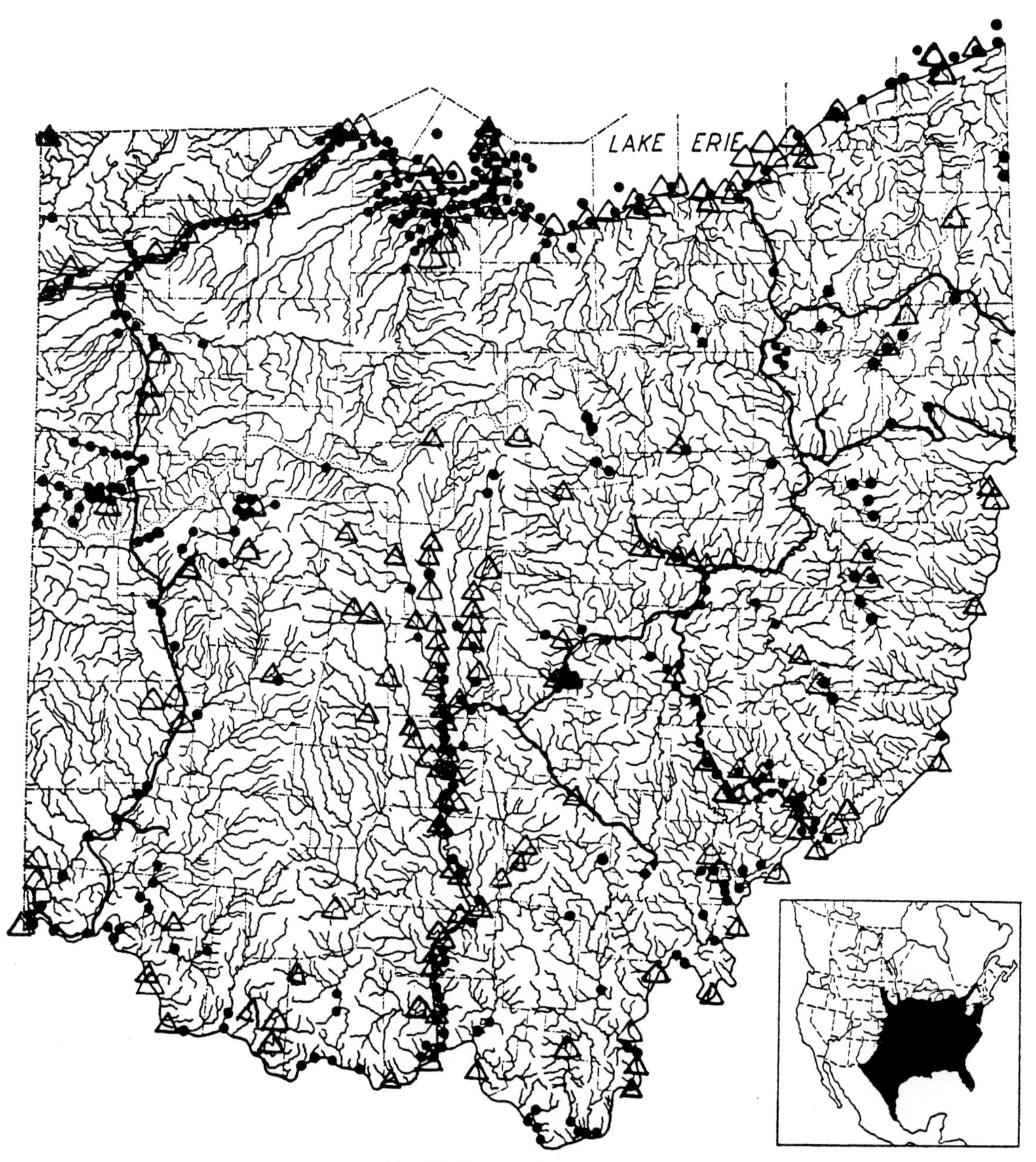

MAP 19. Eastern gizzard shad

Locality records. • Before 1955. △ 1955–80. ~ Canals.
Insert: Formerly absent from large areas in the highlands within its range; many recent and successful introductions within and without its range; Mexican range indefinite.

Distribution and Habitat

Ohio Distribution—There has been much speculation (Miller, 1950:395) as to when the eastern gizzard shad invaded most of Ohio. Rafinesque in 1820 (1820:91; as *Dorosoma notata*) claimed that it came to the Falls of the Ohio River in spring and disappeared in fall. Kirtland in 1838 (1838: 195; as *Chatoessus ellipticus*) stated that it was "frequently exposed for sale" in the Cincinnati markets; that older fishermen informed him (1844:237) that it was unknown about Cincinnati until the "last twenty years" (about 1824) because (Kirtland, 1850C:2) it was a recent "emigrant from more southern waters" and "Since its first appearance it seems according to their [fishermen's] statements, to have increased rapidly in numbers, as it now abounds in the Ohio and Miami Rivers and Dayton Canal;" that it "found its way into Lake [Erie] through either the Dayton and Maumee or the main Ohio canals" and that four specimens were taken in Lake Erie, in November, 1848, near the mouth of the Cuyahoga River. Evermann and Bollman (1886: 339) found the species abundant in the Monongahela River in 1885. Since at least 1892 it has been well distributed in both drainages in Ohio, and abundant in most of the larger streams, reservoirs, lakes, and in Lake Erie.

There appears to be no good reason why this fish should not have been present in the upper Ohio River in pre-Columbian times, or should not have used the Maumee glacial outlet, (Greene, 1935: 16–17) or later crossed from the Ohio into the Lake Erie systems via the watershed marshes which connected both drainages. Although it appears to favor turbid waters where there is an abundance of phytoplankton, it also inhabits clear waters, and although usually inhabiting large waters, spawning adults migrate into small ditches of low gradient on both sides of the watershed where later their young are abundant. Before 1900, watershed marshes emptied into both the Ohio and Lake Erie drainages, thereby making it possible for the species to migrate from one drainage to the other. It seems illogical to assume that formerly the species did not migrate up the larger rivers of the Ohio system as they do today, but did migrate up the newly-constructed canals.

It is my opinion that the species must have been present at least in small numbers in portions of Lake Erie waters before the advent of the canals, but were overlooked, as they are today, until a cyclic peak of abundance occurred and/or there was a winter kill; that the species has increased in abundance in recent years because of a decrease in numbers of other fishes, increased phytoplankton production, or other causes.

Habitat—The eastern gizzard shad was most numerous in lakes, oxbows, sloughs, impoundments, or large streams where the gradient was basic or low. It was tolerant of clear and turbid waters if phytoplankton production were high; fluctuated greatly in abundance, in some years becoming incredibly abundant in the bays and harbors of Lake Erie and in the larger inland impoundments. It winter-killed readily, especially the young (Kirtland, 1844:237; Trautman, 1928:78).

Years 1955–80—The distribution of the eastern gizzard shad in Ohio has remained unchanged since 1950 with the largest concentrations continuing to exist periodically in great numbers in sections of Lake Erie, the Ohio River and the lower gradients of their larger tributaries. Annual fluctuations in population numbers have been periodically noted (Van Meter and Trautman, 1970:68). Large autumnal killings of young of the year still occur in parts of Lake Erie (White et al., 1975:52–53) and in some impoundments, especially those newly created.

Spectacular permanent decreases in abundance also have taken place. At Buckeye Lake between 1925–50, immense numbers of young and some adults could be noted in fall (Trautman, 1940A:110–11). At times several acres of water would be 60% covered with dying and dead yearling shad. Since 1955 no large fall or winter concentrations of dead young have been noted, and normally none or only a few were seen in a day. Several factors presumably have been at least partially responsible for this decrease. The rooted aquatics have been largely eliminated, with only relict patches still present in sheltered areas, thereby eliminating the habitat of eggs and young. The algae-festooned stone wall known as the "North Bank," extending from Buckeye Lake Park westward several miles to the western end of the Lake has been eliminated by the erection of a steel wall in front of it. In spring, at times, this "North Bank" had millions of shad eggs adhering to the algae, but this spawning habitat is now gone. On evenings in late May and until July, before 1950, when the surface of the Lake was unruffled, as great an area as an estimated 10 acres (4 hectares) containing possibly 5–15 young per square foot (50–160 young/m^2) could be seen, the closely packed shad swimming just below the water's surface. Today one seldom sees even a tiny concentration of young shad, possibly because of the almost continuous wave action created by the many speedboats. Can it be that considerable mortality to fry may be caused by the propellers of boats traveling at high speed? Phytoplankton and zooplankton, the food of the shad, appear to be as abundant or more so than before 1950, the greatly enriched waters at times containing so dense a plankton population as to appear green.

In autumn, before 1950, concentrations of gulls,

ducks, and other species of waterfowl remained for days feeding avidly upon this readily available food supply (Pirnie, 1935:155). Since 1950 only infrequently have birds been observed eating young, dead shad. Throughout the 1920's and early 1930's hundreds of boats per day containing 1–5 fishermen could be seen crappie fishing, a fair proportion of them catching their daily limit of 25. During June and July of these years of crappie abundance, Edward L. Wickliff and I examined stomachs of many trap-netted crappies, finding little in their stomachs except numbers of shad fry and young. Since 1950, coinciding with the shad decrease, crappie fishing has likewise decreased until only a few or no boats are found, their occupants fishing for crappies.

For importance as fish food, *see* Wickliff (1933:275–77); for the biology of the shad, *see* Miller (1960:273–88) and Caroots (1976); for distributional references, *see* Nelson and Rothman, 1973.

THREADFIN SHAD

Dorosoma petenense (Günther)

Fig. 20

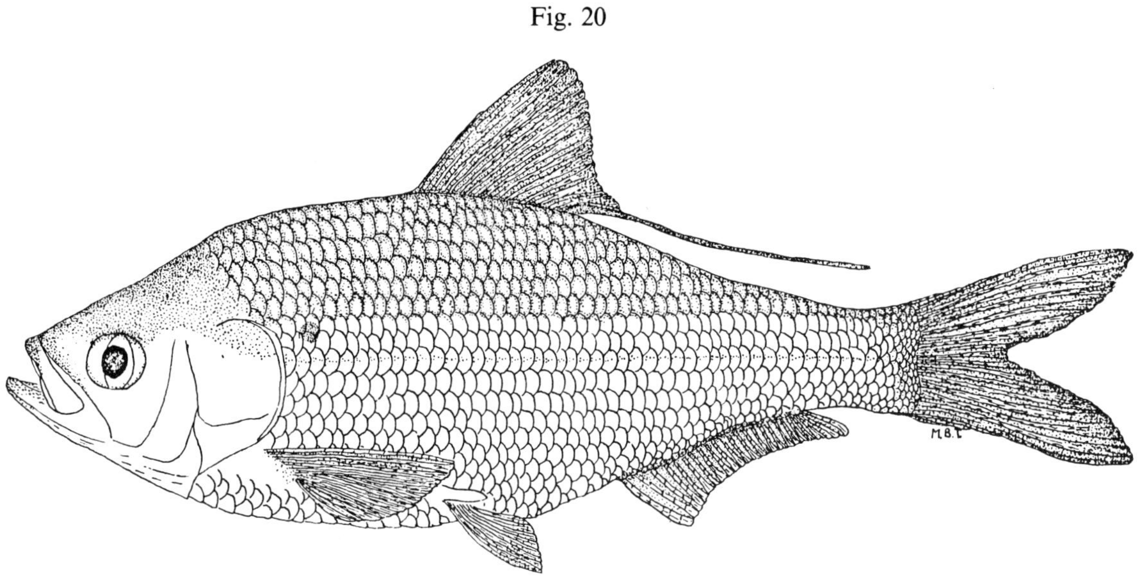

Ohio River, adjacent to Clermont County, O.

October 14, 1970.
Young.

79 mm SL, 4.1″ (10 cm) TL.
OSUM 18414.

Identification

Characters: *General* and *Generic—See* eastern gizzard shad. *Specific*—Dorsal fin rays usually 14 or 15. Anal fin rays 17–27, usually 20 to 25. Few or no melanophores along base of anal fin in young less than 30 mm TL. Lateral line scales fewer than 50, usually 40–43. Dark postopercular spot smaller than eye diameter. Anterior origin of dorsal insertion over pelvic insertion.

Differs: Whitefish, ciscoes, trouts, and smelt have adipose fins. Alewife and skipjack herring have large mouths with the lower jaw projecting. Minnows have anal fins of fewer than 14 rays.

Most like: Eastern gizzard shad.

Coloration: Bluish dorsally, sides more silvery, belly milky-white, with silvery sheen and bluish and greenish reflections over head and body. Small, dark postopercular spot, sometimes absent in adults. Dorsal fin dusky, overcast with yellowish olive; caudal fin with middle portion of each lobe yellow fading distally and basally; anal fin very yellow; paired fins yellow basally.

Lengths: Three to 5″ (7.6–13 cm) in length at 1 year; maximum length 7″ (18 cm) to 8″ (20 cm); few individuals appear to exceed 2 years of age.

Hybridizes: with eastern gizzard shad (Minckley and Krumholz, 1960).

Distribution and Habitat

The threadfin shad apparently was not present in the Ohio River until recently, neither Forbes and Richardson recording it in 1920 for the Mississippi and Ohio rivers adjacent to Illinois, nor O. P. Hay in 1894 for the Ohio River bordering Indiana. It was not until 1957 that the species was first captured in Indiana, in the Little Blue River, a tributary of the Ohio, in Crawford County (Nelson and Gerking, 1968:23). By 1962 (ORSANCO, p. 64) it was present in the Ohio River upstream as far as Maysville, Kentucky (Aberdeen, Brown County, Ohio) and was considered common to abundant from Louisville, Kentucky, downstream. On October 14, 1971, H. Ronald Preston collected a specimen in the Ohio River at the Captain Anthony Mendahl Lock

Map 20. Threadfin shad

Locality records. △ After 1971.

and Dam, one mile downstream from Chilo, southern Washington Township, Clermont County, Ohio, from which the above figure was drawn.

Recently the species has invaded, or was introduced into, such widely distant states as Virginia, Missouri (Pflieger, 1971:322–23), Arizona and California (Parsons and Kimsey, 1954).

In the northern portions of its present range large numbers are sometimes winter-killed, as also occurs with the eastern gizzard shad.

For many years the threadfin was in the genus *Signalosa* but recently has been synonomized with the genus *Dorosoma* (Miller, 1950).

GOLDEYE

Hiodon alosoides (Rafinesque)

Fig. 21

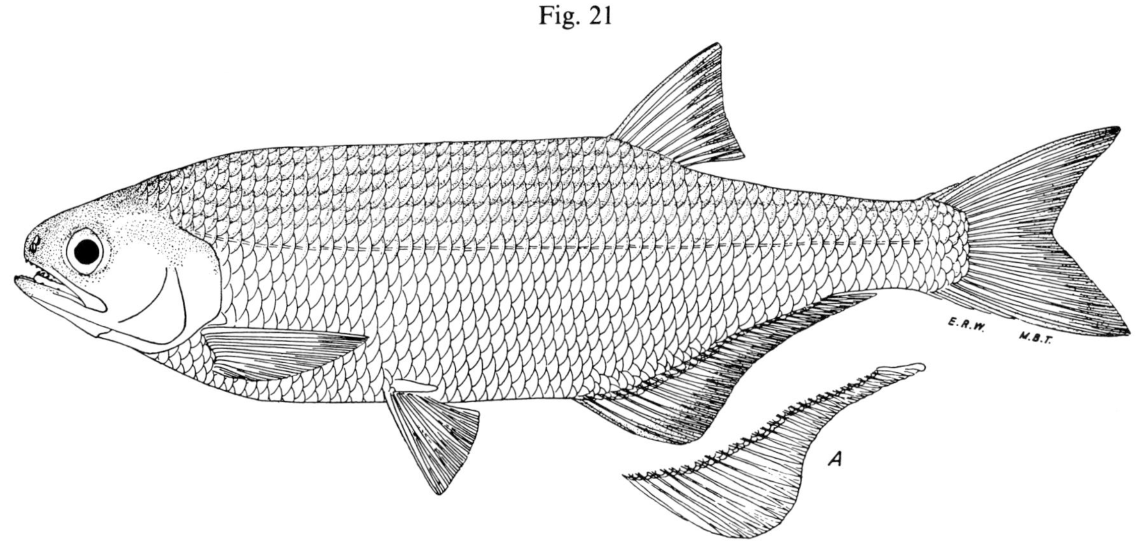

Upper fig.: Hocking River, Athens County, O.

June 28, 1939. 212 mm SL, 10.3″ (26.2 cm) TL.
Immature. OSUM 1105.

The long scale-like process immediately above the pelvic fin base is called the pelvic axillary process; it is present in herrings, whitefishes and trouts.

Fig. A: Muskingum River, Washington County, O.

May 7, 1940. 295 mm SL, 14.0″ (35.6 cm) TL.
Adult male. OSUM 1842.

Anal fin of male with the anterior rays elongated.

Identification

Characters: *General*—Pelvic axillary process present. Eyes partly covered, anteriorly and posteriorly, with adipose eyelids. Canine teeth on tongue and jaws. Gill rakers few, short and knob-like. Coloration intensely silvery. *No* adipose fin. Anal fin of adult males with the elongated anterior rays forming a lobe, and the base of all rays thickened. *Specific*—Dorsal rays 9–10. Dorsal fin inserted slightly behind anal fin origin. Some gold color in iris of eye.

Differs: Whitefish, ciscoes, trouts and smelt have an adipose fin. Skipjack herring, alewife and eastern gizzard shad have the dorsal fin over the pelvic fins. Minnows lack teeth on tongue and jaws. Mooneye has a more anterior insertion of dorsal fin, and more dorsal rays.

Most like: Mooneye.

Coloration: Steel-blue on back, sides silvery, belly milk-white; entire body with a silvery sheen. Some gold color in eye.

Lengths and weights: *Young* of year in Oct., 5.0″–6.5″ (13–17 cm) long. *Adults*, usually 15.0″–17.0″ (38.1–43.2 cm) long, weight 1 lb–2 lbs (0.5–0.9 kg). Largest specimen, 20.0″ (50.8 cm) long, weight 3 lb 2 oz (1.4 kg) (female).

Distribution and Habitat

Ohio Distribution—Although described in 1819 by Rafinesque (1819:421) from specimens taken in

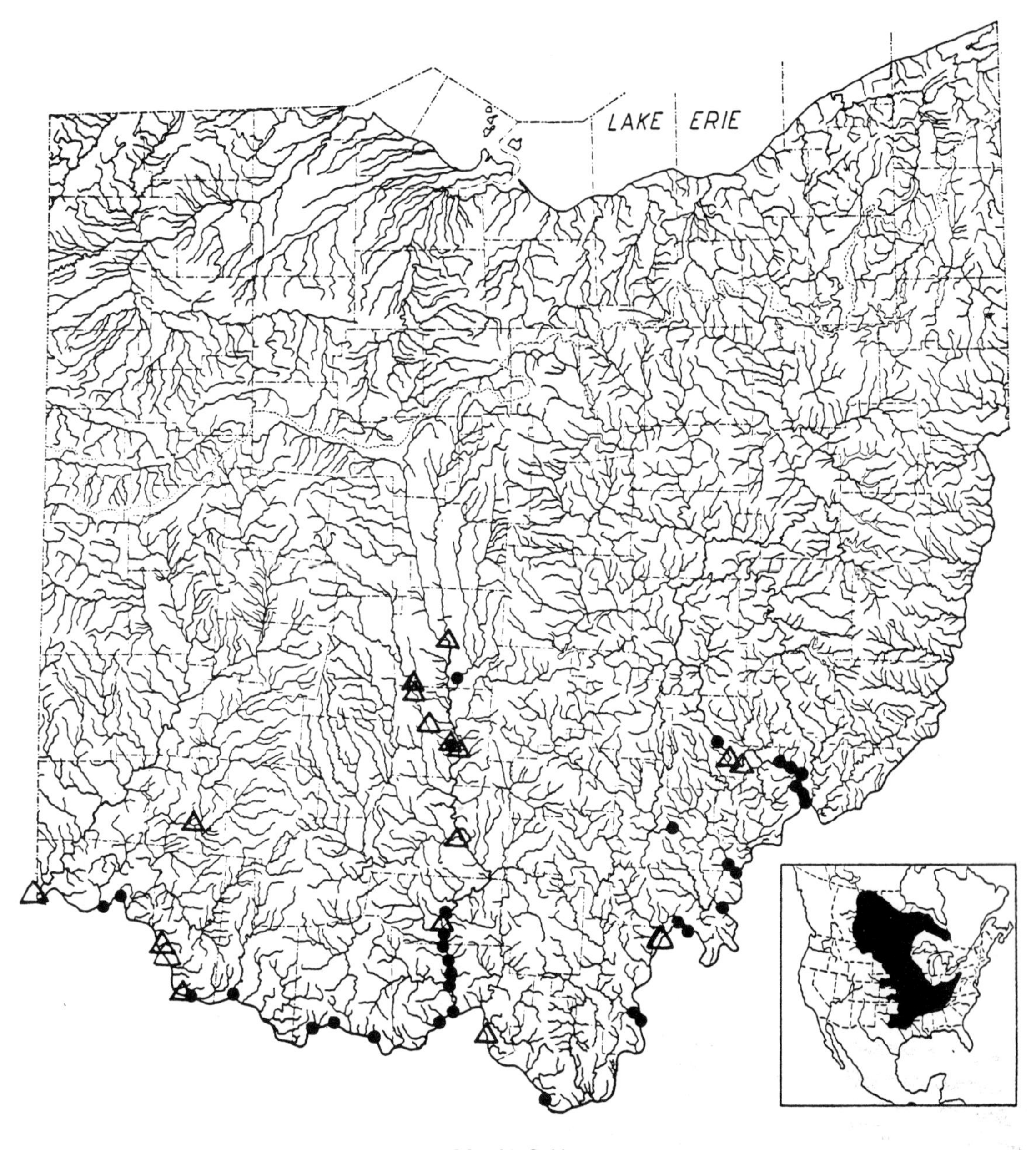

MAP 21. Goldeye

Locality records. ● Before 1955. △ 1955–80.
Insert: Northwestern limits of range indefinite; recent withdrawal from the upper Ohio River.

the lower Ohio River, nothing was known concerning the goldeye in Ohio waters until 1888, when Henshall (1888:79) first recorded it as occurring in the Ohio River (presumably from the vicinity of Cincinnati). The older fishermen of the Ohio River insist that since 1900 the goldeye has been as numerous as, or more numerous than, the closely-related mooneye; since 1920 I have found it to be more numerous than was the mooneye. Since 1920 the species was the most numerous in the lower Scioto River, and in the Ohio River between the mouth of the Scioto River and the Indiana state line; it was less numerous in the lower Muskingum River and in the Ohio River between the mouths of the Muskingum and Scioto rivers. Several Ohio River fishermen have told me that they have seen an occasional goldeye that was taken in the Ohio River between the mouth of the Muskingum River and the Pennsylvania line, especially in spring. In spring, stragglers were taken as far inland as central Ohio.

The goldeye appears to be far more tolerant of turbid water than is the mooneye. Tolerance to turbidity is indicated by its presence, often in large numbers, in the muddy streams of the Missouri system and those of the province of Manitoba. It is therefore possible that the clear-water inhabiting mooneye was formerly more numerous in the upper Ohio Drainage than was the goldeye, but that since the waters of this drainage have become more turbid, the turbid-tolerant goldeye has become the most numerous.

Habitat—The goldeye appears to be far more tolerant to turbidity caused by clays in suspension than by industrial pollutants; it was far more numerous in the turbid Scioto River and Ohio River below Portsmouth, than it was in the more industrially polluted upper Ohio River near the Pennsylvania line. The species congregated wherever small fishes were abundant, such as in the swift waters below the dams on the Ohio River. In the Scioto River it was most numerous in the deeper pools where there was considerable current.

Years 1955–80—With the exception of the Little Miami River locality, the distributional pattern of the goldeye remained the same as it was in 1950.

Evidence accumulates indicating that small numbers of goldeyes migrate upstream during certain years. In September and October of 1958 a fisherman took nine goldeyes in the Little Miami River near Morrow, Warren County; presumably early in 1959 an upstream migration took place in Big Darby Creek, fishermen reporting catches of as many as 12 fishes taken in that stream system between June and September of that year.

The species remained fairly common in the Ohio River. During their 1957–59 investigations of the river, personnel of the University of Louisville recorded the goldeye in 57 of 341 collections. During the 1968–69 investigations of the river by H. Ronald Preston, the species was collected in the river opposite Scioto and Clermont counties.

For description of eggs and early development of young, *see* Battle and Sprules (1960:248–65); for age and growth, *see* Martin (1952:38–49).

MOONEYE

Hiodon tergisus Lesueur

Fig. 22

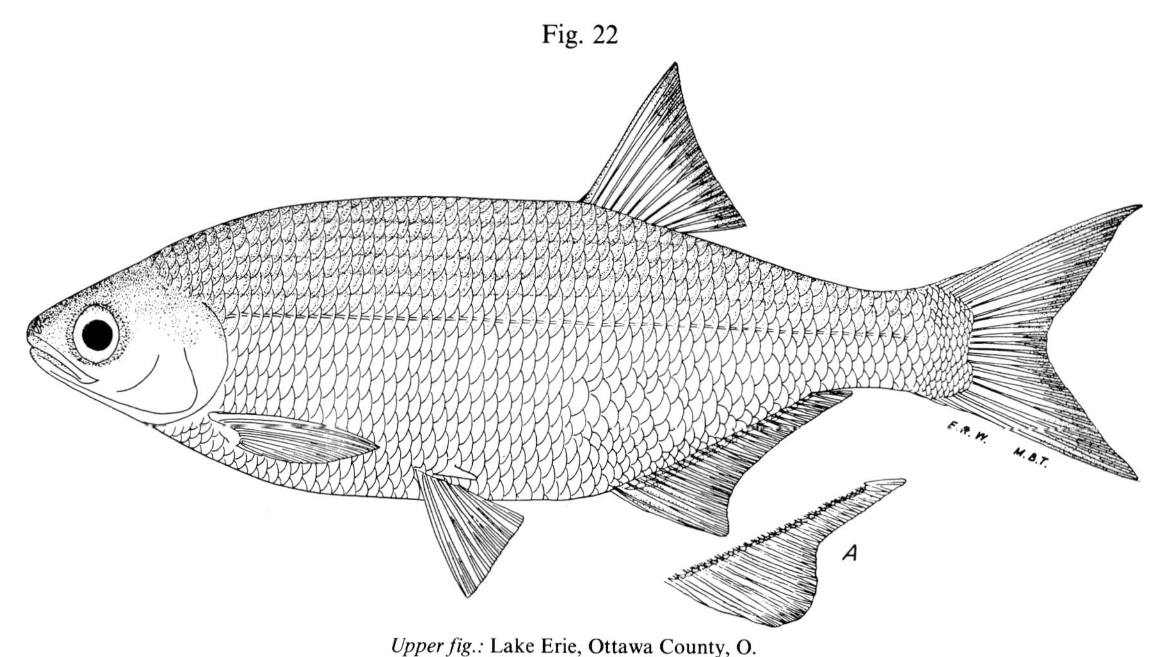

Upper fig.: Lake Erie, Ottawa County, O.

Nov. 4, 1940.
Immature.

197 mm SL, 9.7″ (25 cm) TL.
OSUM 2915.

The irregular squamation of the scales above the anal fin is characteristic of the species.

Fig. A: Lake Erie, Ottawa County, O.

June 5, 1942.
Adult male.

248 mm SL, 12.3″ (31.2 cm) TL.
OSUM 5028.

Anal fin of male with the anterior rays forming a lobe. The specimen has 29 anal rays instead of 25 as illustrated.

Identification

Characters: *General—See* goldeye. *Specific*—Dorsal rays 11–12. Dorsal fin inserted slightly in front of anal fin origin. Iris of eye silvery. More slab-sided and deeper bodied than is the goldeye.

Differs: *See* goldeye.

Most like: Goldeye.

Coloration: Like goldeye, but usually more silvery, with less blue dorsally and without gold color in iris.

Lengths and **weights:** *Young* of year in Oct., 4.5″–6.5″ (11.4–17 cm) long. *Adults*, usually 11.0″–15.0″ (27.9–38.1 cm) long, weight 12 oz–2 lbs (340 g–0.9 kg). Largest specimen, 17.5″ (44.5 cm) long, weight 2 lbs 7 oz (1.1 kg) (female, Ohio drainage).

Distribution and Habitat

Ohio Distribution—Literature references concerning the mooneye before 1900 are in agreement with Kirtland's (1847:338–39) statements that "This fish abounds both in Lake Erie and the Ohio River. It is not very highly valued for food." Because the mooneye was considered a "trash" fish, early commercial fishermen made no particular effort to capture it, and since it was included in the total "trash" fish catch no accurate estimate of the Lake Erie poundage sold per year before 1900 appears available. Kirsch (1895A:330) in 1893 found the species "very abundant" in the Maumee River at Defiance and Grand Rapids. Other early literature references and verbal statements of old fishermen indicate that

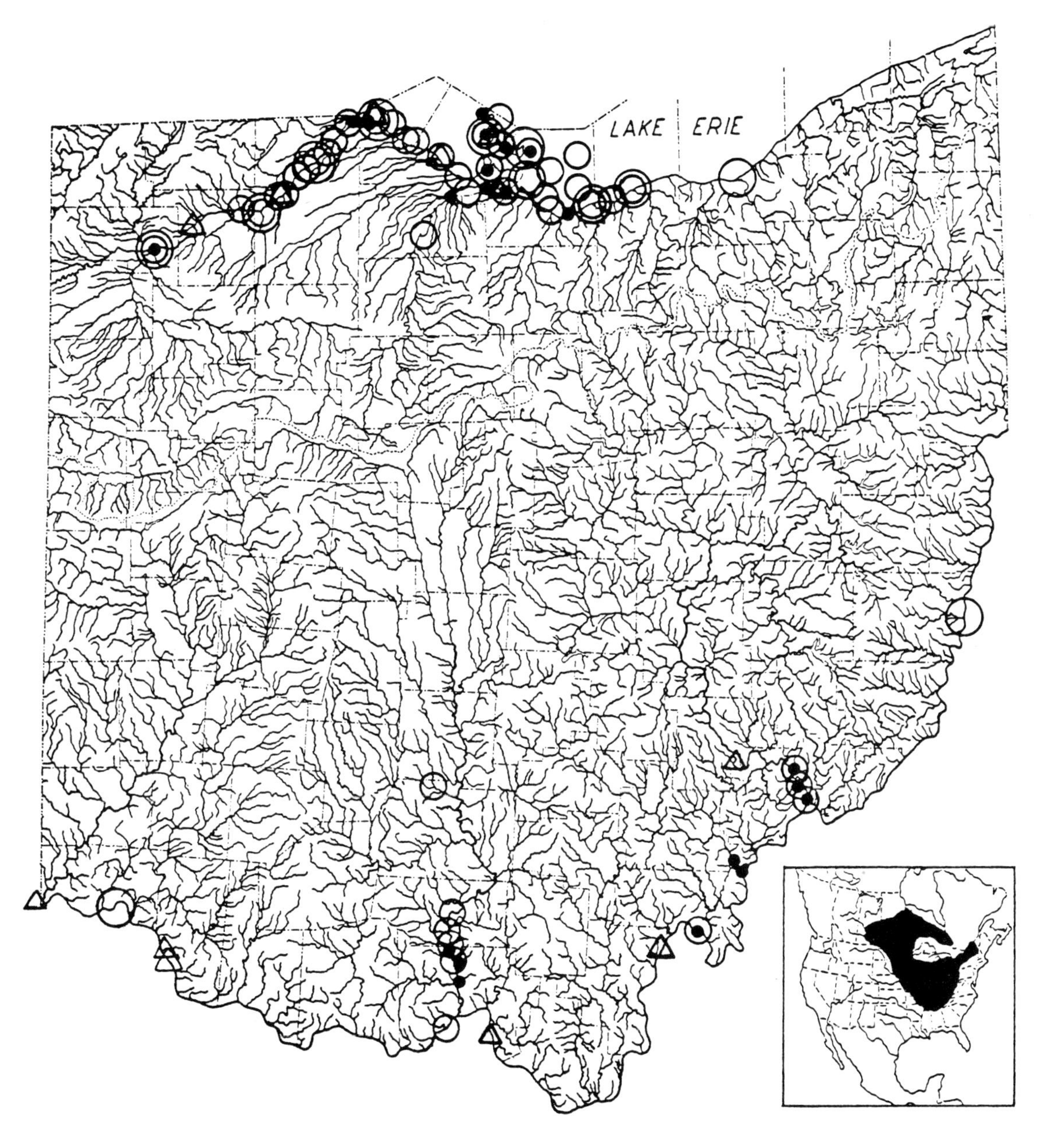

MAP 22. Mooneye

○ Literature records before 1901. Locality records: ○ 1901–35. ● 1935–55. △ 1955–80.

Insert: Large unoccupied areas within the range caused by absence or recent destruction of habitat, may be expected in the northern halves of Lakes Michigan and Huron.

in spring the mooneye migrated in considerable numbers up the larger, clearer streams of Ohio. Between 1920–35 it was still numerous in the Maumee River from Grand Rapids to the mouth, and in the lower Scioto River; but since then it has decreased markedly. According to Henshall (1888:79) and from verbal reports of old fishermen, it was "common" in the Ohio River before 1900, but has decreased greatly since. During rather extensive field work between 1920–35 I did not see a specimen taken from the Ohio River from Adams County westward, and only a few from that county eastward. It can be concluded that the species has decreased in abundance since 1850, particularly in those waters that have become very turbid, such as the Maumee River, and the Ohio River from Scioto County westward. It still remains rather abundant throughout the Ohio waters of Lake Erie.

Habitat—Abundant only in the clearest, largest waters of Ohio where there is an abundant supply of small fishes upon which it feeds. Although often found in non-flowing waters, it feeds mostly in swift waters, such as occur below dams.

Years 1955–80—It is not known how great a decrease in population size of the mooneye occurred in Lake Erie waters during this period because no annual poundage records appear to be available (Van Meter and Trautman, 1970:69). However, a decrease in numbers did occur as indicated by investigations such as that conducted in the vicinity of Cleveland, Cuyahoga County (White et al., 1975: 53). This investigation resulted in the collection of no specimens. Only one record was obtained from the inland waters of the Ohio River drainage. Personnel of the University of Louisville obtained specimens from the entire Ohio River in 29 of their 341 collections. H. Ronald Preston captured individuals in the Ohio River opposite Scioto County during his 1968–69 investigations.

CHINOOK SALMON

Oncorhynchus tshawytscha (Walbaum)

Note: No figure of the chinook was drawn because this species and the coho salmon are superficially very similar. The chinook has spotting on *lower* lobe of caudal fin, and black pigment along bases of the loose, conical teeth.

Identification

Characters: *General—See* Family Salmonidae. *Generic*—Anal rays, 13–20. *Specific*—Anal rays 15 or more; black spottings usually well-defined on lower lobe of caudal fin; black pigment along bases of the not rigid, conical and moderately sharp teeth. Mouth terminal, a pronounced kype or hook on the anterior portion of the upper jaw in mature males; branchiostegals, 13–19; rakers on first gill arch, 18–30, rough and widely spaced. Lateral line scales usually 130–165. Pyloric caeca, more than 100, usually 130–195. Flesh pink, red or white, whitish during spawning. Spawns once and dies, sexual maturity being attained during their 3rd to 7th year, usually during the 4th or 5th.

Differs: *See* coho salmon.

Most like: Coho salmon. **Superficially like:** Brown, rainbow, brook and lake trouts.

Coloration: Larger young and unripe adults usually have the dorsal surface a dark greenish-blue or blackish, becoming more silvery on the sides and immaculately silvery on the ventral surface. The back, dorsal fin and *both* lobes of the caudal fin are spotted. As spawning approaches the coloration becomes more dull and less silvery and the individuals frequently become faintly reddish, rusty and/or grayish in hue. *Young* less than 4″ (10 cm) long have parr marks very strongly developed, extending almost completely across sides of body; are usually wider than the interspaces.

Lengths and **weights:** *Adults*, usually 20.0″–55.0″ (50.8–140 cm) in length, although individuals may be dwarfed if spending their entire lives in fresh water. The jacks (males maturing during 2nd or 3rd year) may be less than 15″ (38 cm) long. Fishes living their entire lives in fresh water may weigh from 1 to 20 lbs (0.5–9.1 kg); those living part of their lives in the ocean may attain a weight of more than 100 lbs (45.4 kg) (heaviest weight recorded, 126 lbs) (57.2 kg).

Hybridizes: with coho salmon.

Distribution and Habitat

Original Distribution—In North America the chinook or king salmon originally inhabited the coastal streams entering the Pacific Ocean from Monterey, California, to Alaska. The fingerlings or smolts migrate downstream into the Pacific Ocean to remain there until approaching maturity, after which they return to their native stream to spawn and die.

Introductions and Distributions in Ohio—The chinook appears to have been first introduced into Ohio waters in 1875, Klippart (1878:9) stating that "The salmon planted in streams flowing into the lake [Erie] have frequently been taken with 'hook and line;' but in one instance many were taken in a net and shipped to market. Of those planted in the Great Miami in 1875, and again in 1876, nothing definite has been learned." Potter (1878:70), then superintendent of the Toledo Fish Hatchery, obtained from the hatchery at Northville, Michigan, "about 30,000 young California [Chinook] salmon. On May 15 [1876 or 1877] he planted half of these in the Maumee River 12 miles (19 km) above Toledo, the remainder in the Portage River at Elmore. On October 9, 1877 Potter received from Livingston Stone, Redding, California, 250,000 California salmon eggs. After hatching, the young were planted in December in the Coshocton, Muskingum, Whetstone (Olentangy), Tuscarawas rivers, Castalia Spring (Cold Creek), Maumee (River) Rapids, Huron River and at Put-in-Bay, Lake Erie.

Since 1890 there have been occasional plantings of chinooks, mostly in the Lake Erie drainage. A revival of planting chinooks occurred in the early 1930's. In 1933, 130,000 between 5″ and 6″ (13 cm and 15 cm) long were released in Lake Erie waters (Anonymous D:1979:18). Since 1970 commercial fishermen in both Ohio and Ontario waters of Lake Erie have reported captures (White et al., 1975:55).

For identification of young, *see* Trautman, 1973; history of salmon in the Great Lakes, 1850–1970,

For identification of young, *see* Trautman, 1973; history of salmon in the Great Lakes, 1850–1970,
see Parsons, 1973.

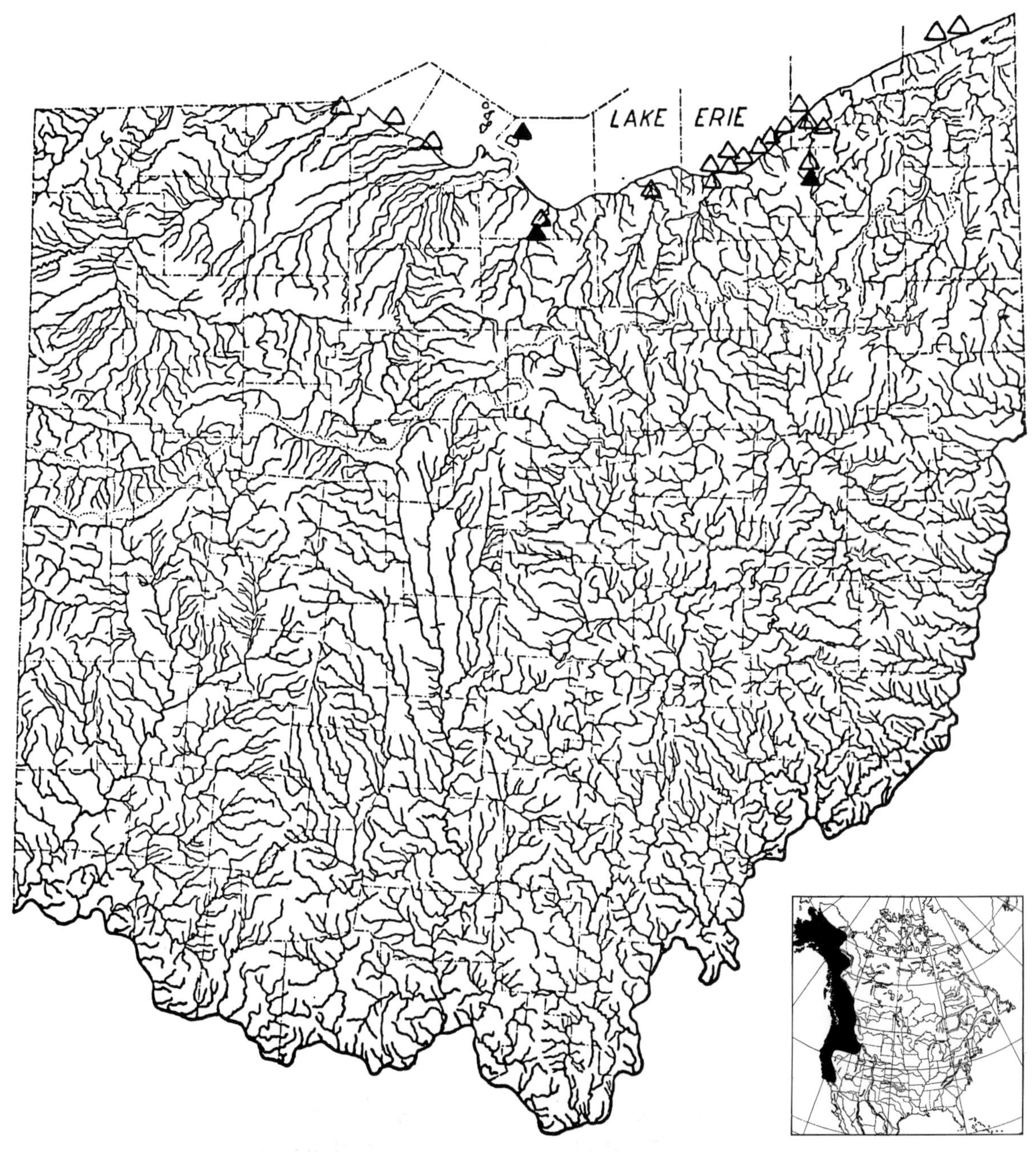

Map 23. Chinook salmon

Locality records. ▲ Planted. △ Recaptured.
Insert: Original freshwater range in North America.

COHO SALMON

Oncorhynchus kisutch (Walbaum)

Fig. 23

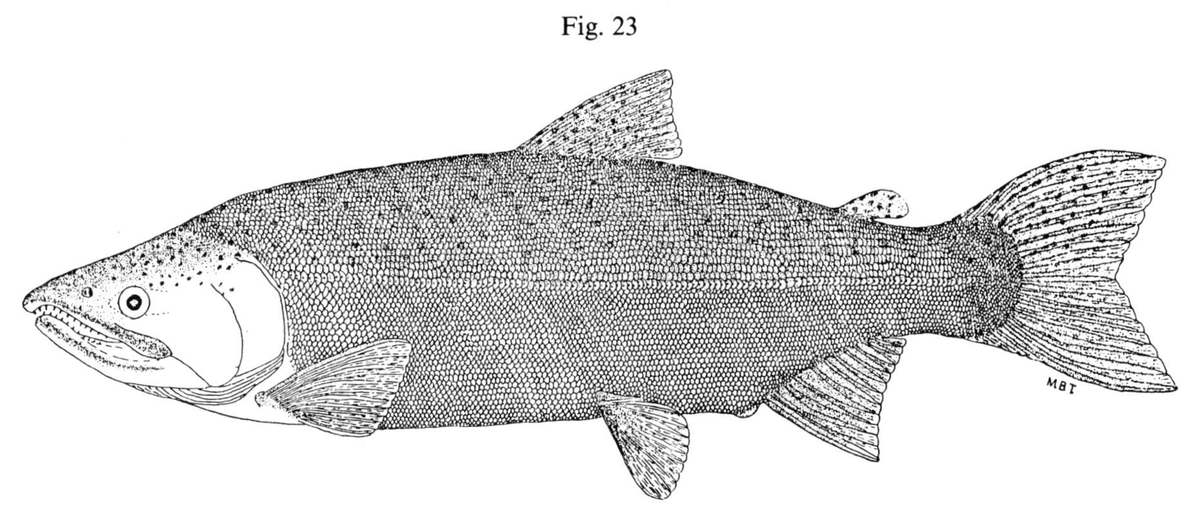

Lake Erie, near Cooley Canal, Jerusalem Twp., Lucas County, O.

Nov. 11, 1969.
Mature male.

520 mm SL, 24″ (61 cm) TL.
OSUM 16772.

Identification

Characters: *General—See* Family Salmonidae. *Generic*—Anal rays, 13–20. *Specific*—Anal rays 15 or fewer; black spottings normally faint or absent on lower lobe of caudal fin; no black pigment along bases of the firmly-set, needle-like teeth. Mouth terminal, a pronounced kype or hook on the anterior portion of the upper jaw in mature males; branchiostegals, 11–15; rakers on first gill arch, 19–25, rough and widely spaced. Lateral line scales usually 121–140. Pyloric caeca fewer than 100, usually 45–85. Flesh pink to red, turning whitish during spawning. Spawn once and die, usually at the end of the third summer.

Differs: *See* chinook salmon.

Most like: Chinook salmon. **Superficially like:** Brown, rainbow, brook and lake trouts.

Coloration: Larger young and unripe adults have the dorsal surface a metallic blue, becoming more silvery on the sides, and immaculately silvery on the ventral surface. The back, usually the dorsal fin and upper lobe of caudal fin are spotted. As spawning approaches the coloration becomes duller and less silvery; when through spawning and/or about to die, may be a mottled dark gray. *Young* less than 4″ (10 cm) long have parr marks very strongly developed, extending almost completely across sides of body; are usually narrower than the interspaces; an orange tinge usually present on pectoral, pelvic and anal fins; white on anterior margin of anal fin.

Lengths and **weights:** *Adults* usually 20.0″–40.0″ (50.8–102 cm) long, except the jacks (males maturing during 2nd year), which may be less than 12″ (30 cm) long. Weights of fishes living their entire lives in fresh water may range from 1–8 lbs (0.5–3.6 kg). Weights for individuals having access to the sea have average weights of 6–12 lbs (2.7–5.4 kg), the maximum record being 31 lbs (14.1 kg).

Hybridizes: with chinook salmon.

Distribution and Habitat

Original distribution—In North America the coho or silver salmon originally inhabited the coastal streams entering the Pacific Ocean from northern California to Alaska. The fingerlings or smolts migrate downstream into the Pacific Ocean to remain there until approaching maturity, after which they return to their native stream to spawn and die.

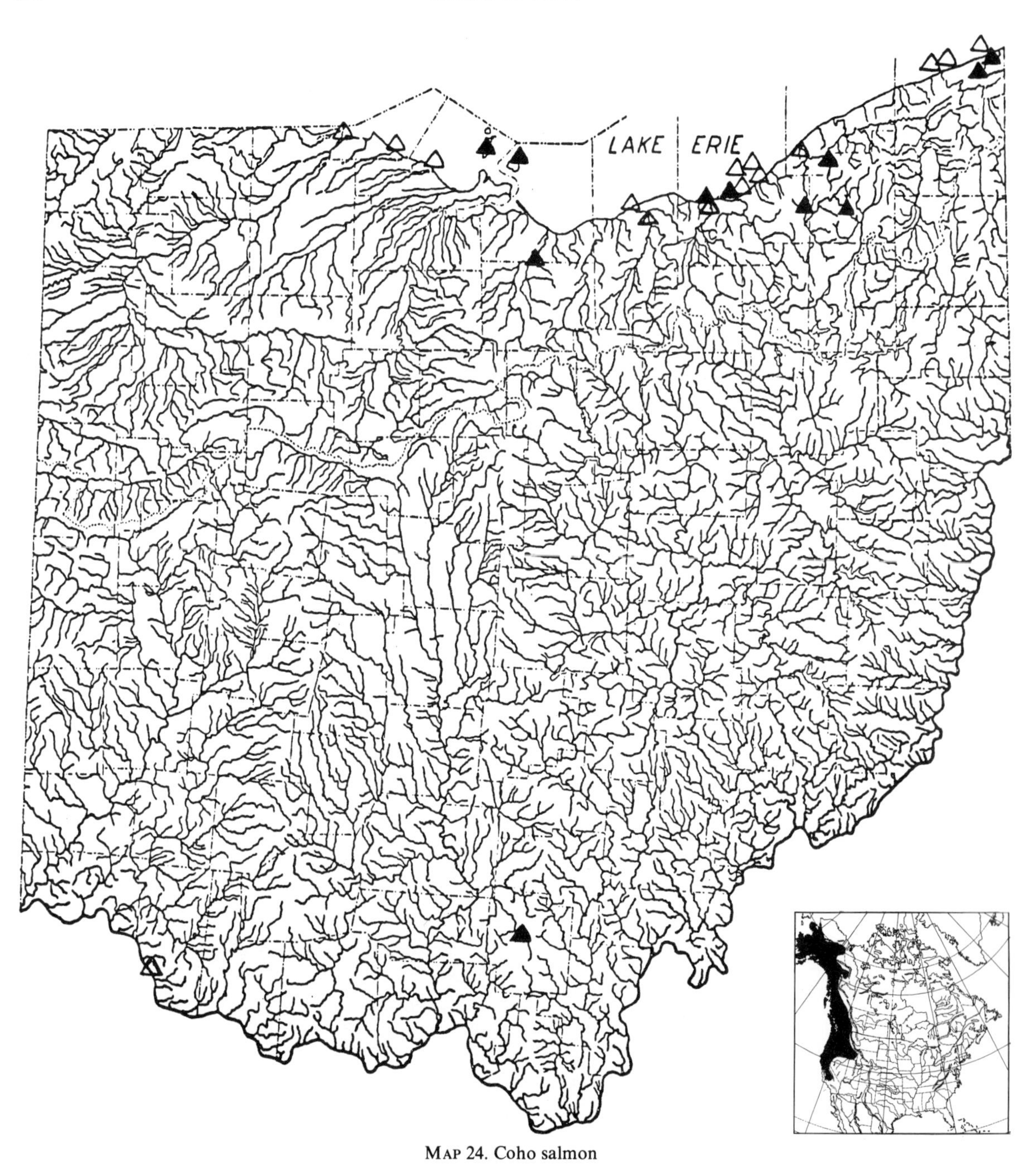

Map 24. Coho salmon

▲ Some introductions since 1970. △ Some recaptures since 1970.
Insert: Original freshwater range in North America.

Introductions and Distributions in Ohio—The coho appears to have been introduced into Ohio waters about 1876, when the first chinook salmons were introduced. A series of re-introductions in Ohio occurred between 1930 and 1934. In Anonymous C (1934) it is stated that 50,000 silver (coho) and 130,000 chinook salmons were stocked in Lake Erie waters in 1933. The eggs, obtained from the Northwest, were hatched and fry reared to between 5″ and 6″ (13 and 15 cm) long at the Castalia Trout Club, Castalia, Ohio. The young were planted in western Lake Erie, Sandusky Bay or adjacent streams. On March 25, 1933, the late T. H. Langlois and I assisted in planting 8,500 cohos in Little

Pickerel Creek, Erie County, near its mouth. They appeared to be in excellent condition, and some were observed moving downstream and entering Sandusky Bay. Although commercial and sport fishermen were alerted, no coho was reported from these plantings.

The introductions of cohos into Lake Erie waters were revived between 1968 and 1970 when more than one-half million fry and fingerlings were planted. Several hundred thousands have been released since in Lake Erie tributaries and in the lake (Anonymous D, 1979:18). Some matured and were captured in Lake Erie and in streams where presumably they had been introduced. Specimens captured in western Lake Erie near the outlet of the St. Clair River possibly had migrated from Lake Huron, where huge numbers had been introduced.

There appear to be several possible reasons why the 1960's introductions were somewhat successful: (1) before 1955, there were huge populations of predacious blue pikes and walleyes plus several other species of large fishes whose numbers recently have been greatly decimated (Van Meter and Trautman, 1970), resulting in possibly greatly decreased population pressure; (2) recently two forage fishes, the alewife and smelt, have invaded Lake Erie and become abundant, offering increased food possibilities; and (3) many of the recent coho introductions were made in the larger and less polluted streams in eastern Ohio instead of Lake Erie. Commercial fishermen in Ohio and Ontario waters report captures of this species, sometimes in fair numbers (White et al., 1975:55). Maintenance of the population in Ohio appears to be entirely by restocking.

On October 6, 1971 Andrew White obtained an adult coho in the Rocky River, one mile (1.6 km) above its confluence with Lake Erie, in Cuyahoga County. This specimen (OSUM 18994) is particularly interesting because, to our knowledge, no young were planted in that stream. In 1976 Margulies and Burch captured a coho in the Ohio River adjacent to Clermont County, Ohio.

For identification of young, see Trautman, 1973B; history of salmon in the Great Lakes, 1850–1970, *see* Parsons, 1973.

Note: Published too late (Lee Emery, Range extension of pink salmon (*Oncorhynchus gorbuscha*) into the lower Great Lakes [Fisheries 6, no. 2 (1981): 7–10]) for inclusion in this edition with complete data is the report that the pink salmon (*Oncorhynchus gorbuscha*) has expanded its range from initial plantings in Lake Superior in 1956 into each of the Great Lakes, where it is now the only self-sustaining species of salmon. "Ohio commercial trapnetters reported catching 8–10 fish at three locations on September 12, 13, and 14, 1979, in the western basin of Lake Erie. . . . Other pink salmon caught by anglers in Lake Erie were reported to state conservation agencies. The largest spawning run in U.S. tributaries to Lake Erie was reported in Walnut Creek, Pennsylvania, where anglers caught 18 fish. All fish caught in U.S. tributaries to Lake Erie were taken between September 12 and October 14. Anglers reported that spawning was completed by the first week in October."

BROWN TROUT

Salmo trutta Linnaeus

Fig. 24

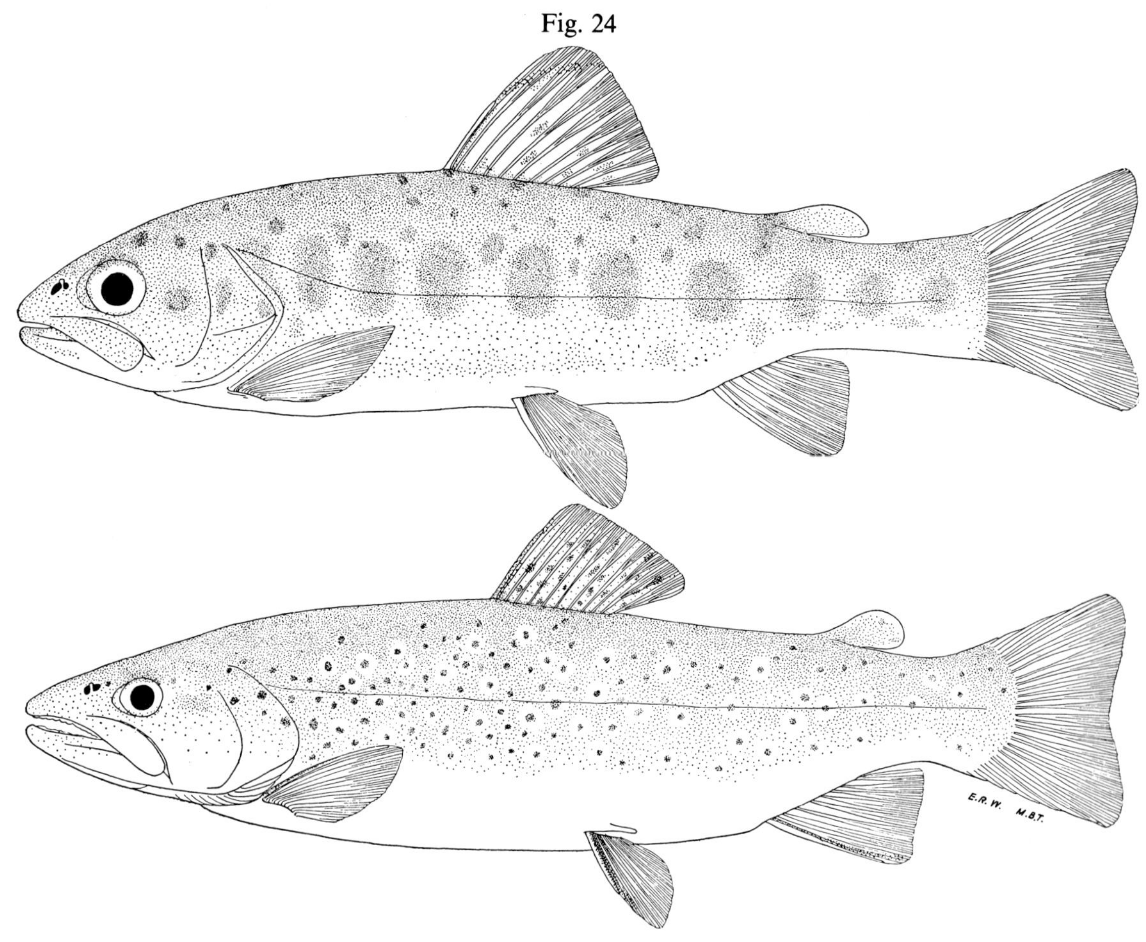

Upper fig.: Castalia Trout Club Hatchery, Cold Creek, Erie County, O.

Sept. 21, 1943. 77 mm SL, 3.6″ (9.1 cm) TL.
Immature. OSUM 6778.

Note the rectangular parr marks along the side which are present on fingerling trout. Compare differences in coloration, in eye, mouth and fin lengths, in body proportions, between this young and the adult.

Lower fig.: Cold Creek, Erie County, O.

April 30, 1940. 320 mm SL, 14.8″ (37.6 cm) TL.
Adult male. OSUM 1805.

Identification

Characters: *General*—Adipose fin and pelvic axillary process present. Scales very small, more than 100 in the lateral series. Mouth very large; the posterior end of upper jaw extending to, or past, the posterior edge of eye. Sharp teeth on tongue and jaws. No spines on fins. *Specific*—Few or no spots on top of head or on caudal fin. Back with or without spots, but lack definite vermiculations. Sides with many orange and red spots, surrounded by a bluish ring. Adipose fin usually without a

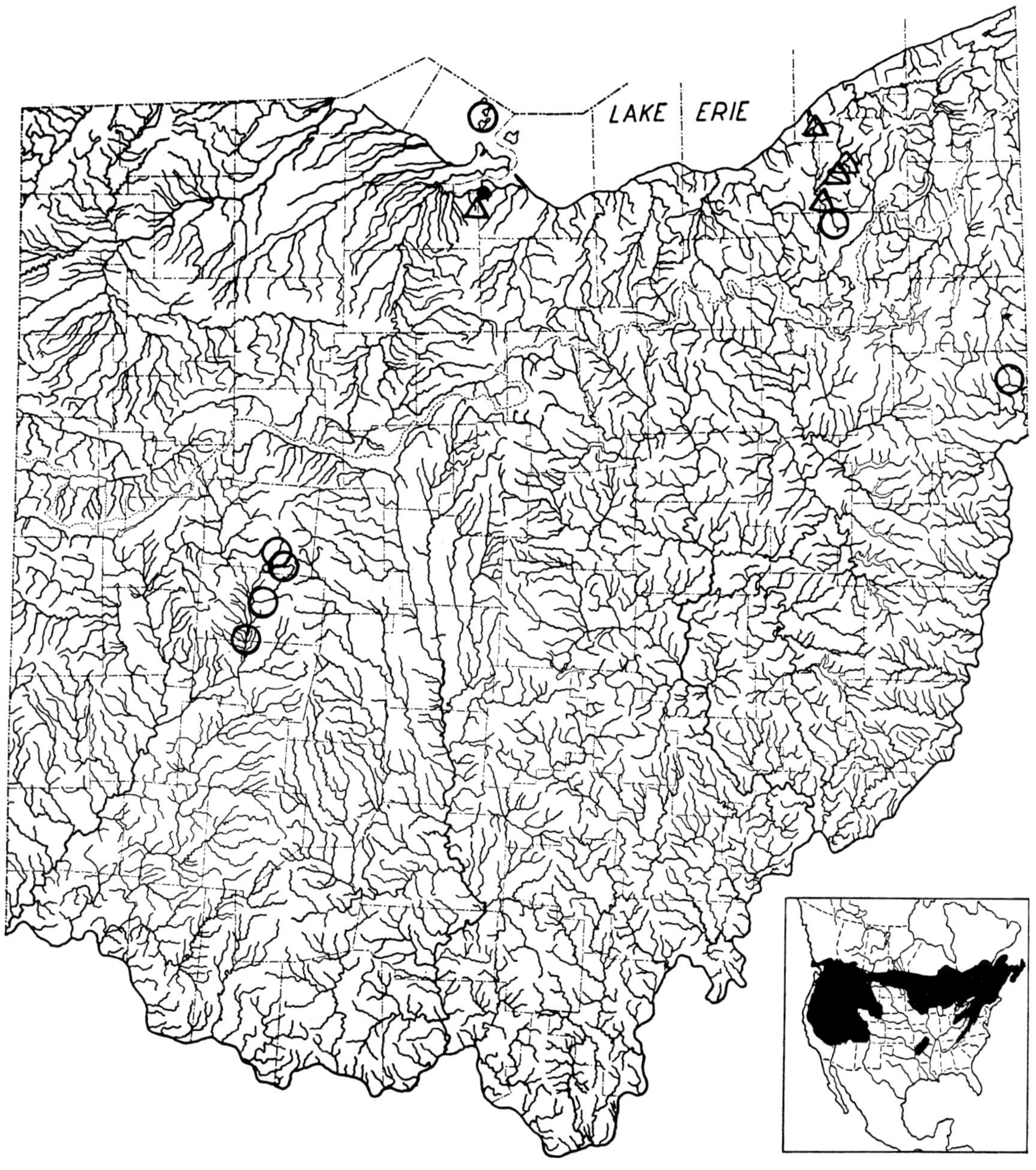

MAP 25. Brown trout

○ One or more captured more than a month after release.
Permanently established. ● Before 1955. △ 1955–80.
Insert: Black area indicates where introduction of this old-world species has been most successful; extensively introduced elsewhere, both to the north and south, with many of these introductions successful in local areas.

blackish margin. Dorsal fin insertion usually closer to snout tip than to caudal base. Anal rays 9, including the first unbranched ray. Introduced.

Differs: Mooneyes, skipjack herring, alewife, and eastern gizzard shad have no adipose fin. Whitefish and ciscoes are unspotted and lack large teeth on tongue and jaws. Smelt is unspotted, has more than 14 anal rays. Rainbow trout has many spots on head and tail. Brook and lake trouts have vermiculated backs.

Most like: Rainbow trout.

Coloration: *Adults*—Greenish- or yellowish-olive dorsally, lighter on sides, yellowish or silvery-white on belly. Dark spots on back and sides, many of which are orange or red, circled with light blue. No red stripe along sides; or milk-white, anterior pelvic and anal rays, lower figure. *Young*—Brownish-olive or yellowish dorsally, lighter and silvery on sides, belly light yellowish or white. Few spots dorsally, sides with 8–12 rectangular parr marks. Spots often absent on fins and lower sides. Body usually with a yellowish cast, upper fig.

Lengths and **weights:** *Adults*, usually 7.0″–22.0″ (18–55.9 cm). Largest specimen, about 32.0″ (81.3 cm) long, weight 13.5 lbs (6.12 kg), caught in Cold Creek, Castalia Trout Club, Erie County, by J. Harris.

Hybridizes: naturally and artificially produced elsewhere with brook trout (Brown, 1966:600).

Distribution and Habitat

Ohio Distribution—The first shipment of brown trouts (as eggs of the Lock Leven trout, *Salmo levenensis*, which was formerly believed to be a distinct species) into the United States by a federal agency appears to have been in 1885 (Smiley, 1889:28–29). Part of this shipment was sent to the Northville Hatchery in Michigan from which fry were distributed later for planting into native waters. It is possible that within a few years some brown trouts from this or later shipments were introduced into Ohio waters, for the Northville Hatchery frequently sent shipments of other species into Ohio for stocking purposes (*see* rainbow and brook trouts). Since 1900 many plantings of brown trouts have been made into various waters, especially in the northeastern quarter of the state. These fishes or eggs were obtained from federal agencies or private firms, principally through requests of sportsmen, or were planted by the state conservation department.

About 1930 brown trouts were introduced into Cold Creek, in Margaretta Township of Erie County, by the Castalia Trout Club, and about the same time they were introduced into the springs and runways which were controlled by the Zanesfield Rod and Gun Club, Inc., in Jefferson Township of Logan County. On November 25, 1931 I assisted in liberating about 125 brown trouts in Cedar Run, Champaign County. The last two mentioned waters are part of the Mad River system.

Of all the many plantings of brown trouts into Ohio waters, only the introduction into Cold Creek has been markedly successful, for in that stream the species maintained its numbers through natural production. In the Zanesfield Rod and Gun Club waters natural propagation may have occurred occasionally. Elsewhere the species disappeared within a short time after the last planting, except that rarely a stray individual would be taken by a fisherman several months after planting, or a few would survive in a private spring pool. The hollow circles on the distribution map represent localities from which I have seen specimens of brown trouts which were taken by fishermen. The only specimen (OSUM 1138) which I have seen from Lake Erie was taken in an experimental gill net, which was set near South Bass Island, June 23, 1939.

Habitat—The brown trout inhabits cold, and clean streams, ponds and lakes in which the water temperature seldom goes as high as 65° F (18° C), and only rarely and for very short periods reaches 75° F (24° C) during the very warmest period of the year; where other fish species and individuals are rather few in numbers; where the deep pools contain "cut banks" and other cover in which it may hide; where there is an abundance of aquatic and land insects and other invertebrates upon which to feed. It feeds chiefly at dusk or at night, and this is especially true of the larger fishes. The lack of continuous cold water throughout the year is the principal reason why Ohio has so few trout waters; in most of the Ohio waters in which brown trout has been planted the water temperature would reach 70° F (21° C) on several days in May, and 80° F (27° C), or more in August.

Years 1955–80—Introductions of this species continue to be made in some of the spring-ponds, lakes, and streams, in some waters repeatedly or annually. The majority of the introductions were planted in the spring-ponds and streams of northern and northeastern Ohio. It is only in the exceptionally cool waters that one or more individuals survive throughout the summer.

We (White et al., 1975:55–56) state that during the survey of the Cleveland Metropolitan area, "approximately 50 adults escaped into the Chagrin River drainage from a trout club in Geauga County. Occasional specimens are captured by sportsmen fishing in the Chagrin and it is assumed that such specimens are the results of similar escapes from stocked ponds. One specimen was collected in the East Branch, Chagrin River."

RAINBOW TROUT

Salmo gairdneri Richardson

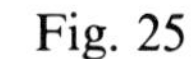
Fig. 25

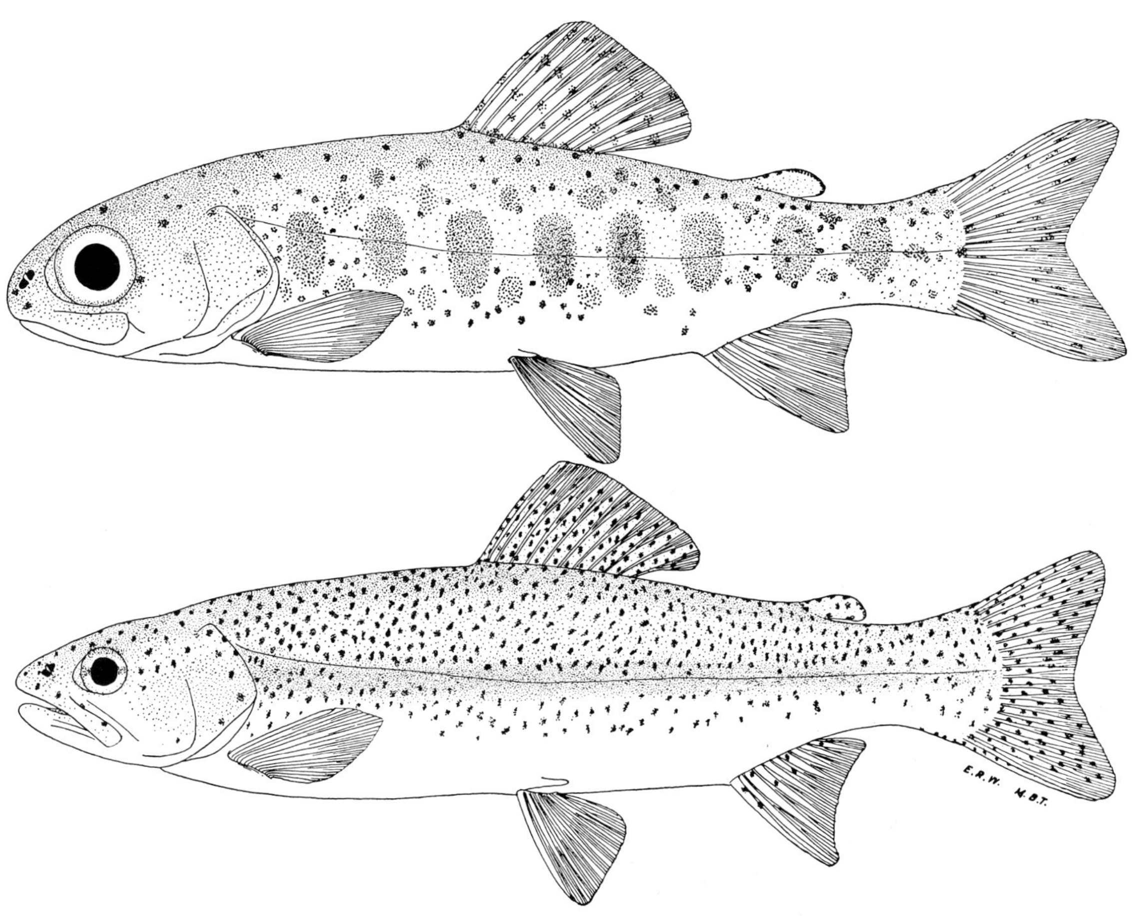

Upper fig.: Mad River, Logan County, O.

April 6, 1942. 65 mm SL, 3.1″ (7.9 cm) TL.
Immature. OSUM 4754.

Note rectangular parr marks along sides. Compare differences in coloration; in length of eye, mouth and fins, and in body proportions, between this young and the adult.

Lower fig.: Macochee Creek, Champaign County, O.

July 10, 1939. 215 mm SL, 10.2″ (25.9 cm) TL.
Adult. OSUM 319.

Identification

Characters: *General—See* brown trout. *Specific*—Back, top of head, dorsal, and caudal fins of large young and adults contain many, small, dark or black spots, and a broad, pink or red band along the sides. No vermiculations on back. Adipose fin usually spotted. Dorsal fin insertion usually midway between snout tip and caudal base. Anal rays 10–12, occasionally 9 in young with undeveloped first ray. Introduced.

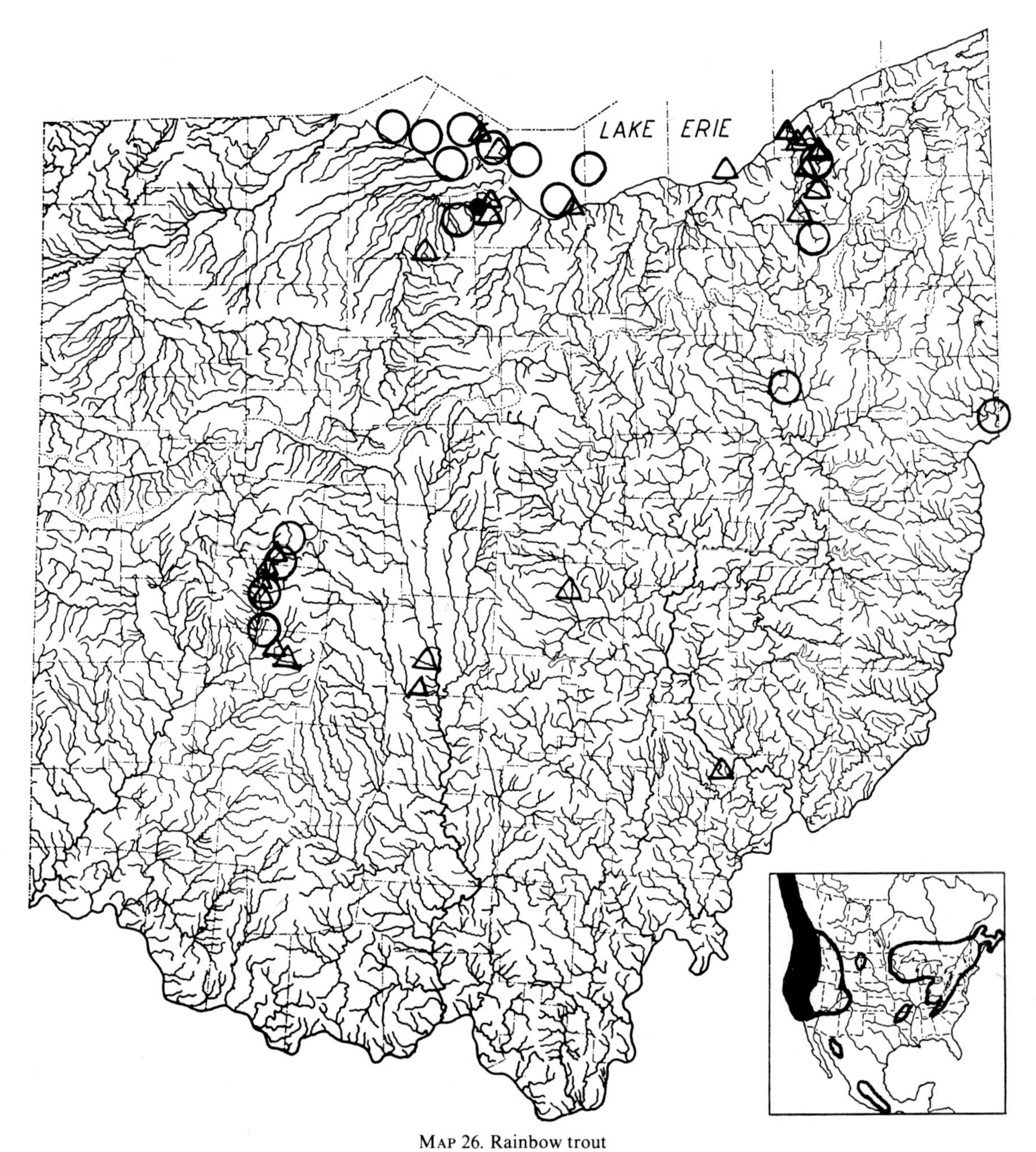

MAP 26. Rainbow trout

○ One or more captured after release or escaping from Cold Creek.
Permanently established. ● Before 1955. △ 1955–80.

Insert: Black area represents approximate original range; outlined areas where introductions have been most successful; extensively introduced elsewhere throughout North America with many introductions locally successful.

Differs: *See* brown trout. Brown trout has fewer or no spots on top of head and caudal fin, and coloration more yellowish.

Most like: Brown and brook trouts.

Coloration: *Adults*—Bluish- or silvery-olive dorsally, sides more silvery with a pink or reddish band, and white or silvery on belly. Many distinct, blackish spots on back, sides, top of head, and on dorsal, caudal, and adipose fins. No red spots on sides or back; no white on anterior rays of pelvic and anal fins, lower fig. *Young*—Bluish- or olive-silvery dorsally, silvery on sides, milk-white on belly. Many spots on back, lower sides, some on top of head, and on dorsal and caudal fins. Adipose fin often edged with dusky. Between 8–12 rectangular parr marks along sides. Body usually decidedly silvery, upper fig.

Lengths and **weights:** *Adults*, usually 7.0″–20.0″ (18–50.8 cm). Largest specimen, about 33.0″ (83.8 cm) long, weight 10 lbs 8 oz (4.8 kg), taken in Cold Creek, Erie County.

Hybridizes: naturally and artifically with brook trout.

Distribution and Habitat

Ohio Distribution—Apparently the first known introduction of the exotic rainbow trout into Ohio waters was in 1884 when in May of that year Smith (1885:436) "obtained 68,000 rainbow trout fry from Michigan; 5,200 of these were planted in the Blue Jacket [of Logan County] a tributary of the [Great] Miami River, and some have been caught there weighing over one-half pound each." In the winter of 1885–86, 3,000 rainbow eggs were sent from the Northville Michigan Hatchery to B. F. Ferris (Clark, 1887:398) of the Cold Creek Trout Club, Castalia (name changed to Cold Creek Sporting Club in 1886 and to Castalia Trout Club in 1890) and 2,000 to S. B. Smith of Zanesfield (not Zanesville, *see* Smith, 1885:436); in addition yearlings and/or 2-year-olds were sent to M. P. Hammond, of Howard (75 fishes) and J. L. Delano, of Mt. Vernon, (100 fishes). Presumably these trouts were liberated into waters in the vicinity of the above-mentioned towns.

Since 1886 many shipments of rainbows have been liberated into Ohio waters by federal and state agencies, fishing clubs and private citizens. In 1932 and for several years thereafter the Mad River drainage was rather heavily stocked by the conservation department, and large plantings were made in Sandusky Bay and Lake Erie by the conservation department and Castalia Farms; in addition, unknown numbers escaped from Cold Creek into Lake Erie.

Although the rainbow presumably was liberated into Cold Creek waters before 1900, it was not until 1926 that large numbers were planted in that stream. Since 1940 it has greatly outnumbered the brown and brook trouts. Cold Creek is the only stream in Ohio where the rainbow propagates naturally and is permanently established. About 1930 the Zanesfield Rod and Gun Club began liberating large numbers of rainbows into its spring-pools and runways (some fishes of which escaped into the Mad River), and occasionally some natural propagation occurred. Elsewhere throughout Ohio the species has failed to establish itself permanently, despite repeated liberations of hundreds of fishes, and in most waters the species disappeared almost immediately after liberation.

The hollow circles on the distribution map represent localities from which I have seen stray rainbows which were taken by fishermen. The earliest record for Lake Erie appears to be that of a specimen captured in 1920 in Canadian waters (Dymond, 1932:33).

Habitat—The rainbow trout inhabited the same types of cold, clear waters as did the brown trout (*see* that species, under Habitat); however, the rainbow preferred a swifter current, fed more readily in the daylight, and was not so secretive. Like the brown, introductions of rainbows into all except a few Ohio waters were unsuccessful because these waters became too warm during the summer months.

Years 1955–80—Introductions of rainbow trouts were made in many Ohio waters during this period, often repeatedly in the same waters. The results were mediocre except for a few spring-ponds or streams sufficiently cool in summer to enable some of them to survive. This appears to be the species of trout most tolerant of our waters. Occasionally, individuals are captured in the Ohio waters of Lake Erie (Van Meter and Trautman, 1970:68). Recently many thousands have been released in the Lake Erie tributaries in northeastern Ohio (Anonymous D, 1979:18).

Recently an increasing number of groups and individuals have been buying catchable-sized rainbows, introducing them into their private waters in the early fall after the water temperature has become suitable. Fishing continues from the time of introduction until late spring, frequently by fishing through holes cut in the ice-covered water. For several years the Southeast Conservation Club of central Ohio has stocked their private waters in Franklin County, thereby supplying an additional incentive for winter fishing.

BROOK TROUT

Salvelinus fontinalis (Mitchill)

Fig. 26

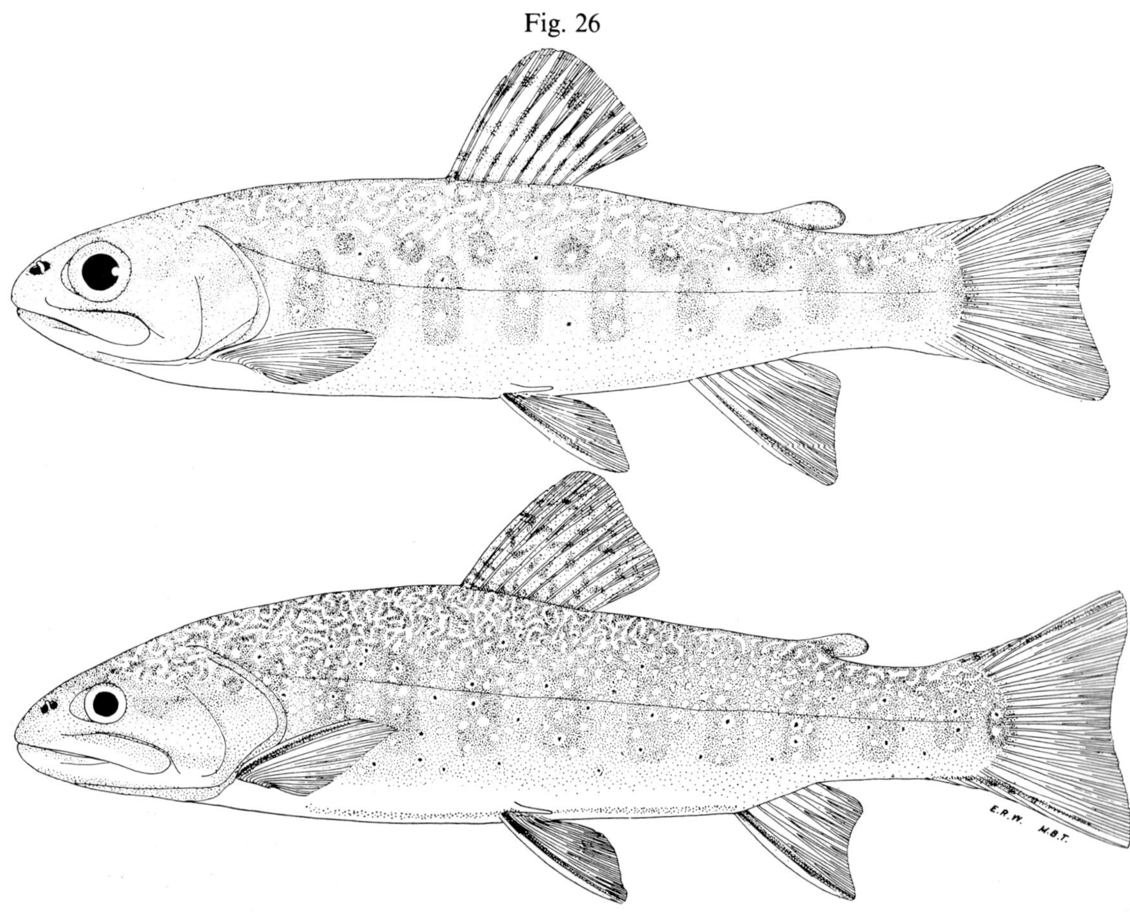

Upper fig.: Miller Blue Hole, Sandusky County, O.

Aug. 18, 1941. 68 mm SL, 3.2″ (8.1 cm) TL.
Immature. OSUM 4245.

Note rectangular parr marks along sides. Compare differences in coloration, and in lengths of eye, mouth, and fins, and in body proportions, between this young and the adult.

Lower fig.: Same locality, date and OSUM number.

Adult male. 180 mm SL, 8.6″ (22 cm) TL.

Identification

Characters: *General—See* brown trout. *Specific*—Top of head and back heavily vermiculated. No black or brown spots on head, back, adipose or caudal fins. Sides of large young and adults with brilliant blue and red spots. Anterior rays of pectoral, pelvic, and anal fins milk-white, bordered posteriorly with dusky, remainder of fins yellowish or reddish. Belly red in breeding males. Tail slightly emarginated, not deeply forked. Usually more than 200 scales in the lateral series. Formerly native in northeastern Ohio.

Differs: *See* brown trout. Brown trout lacks milk-white anterior pelvic and anal rays. Lake trout has deeply forked tail and no blue or red spots.

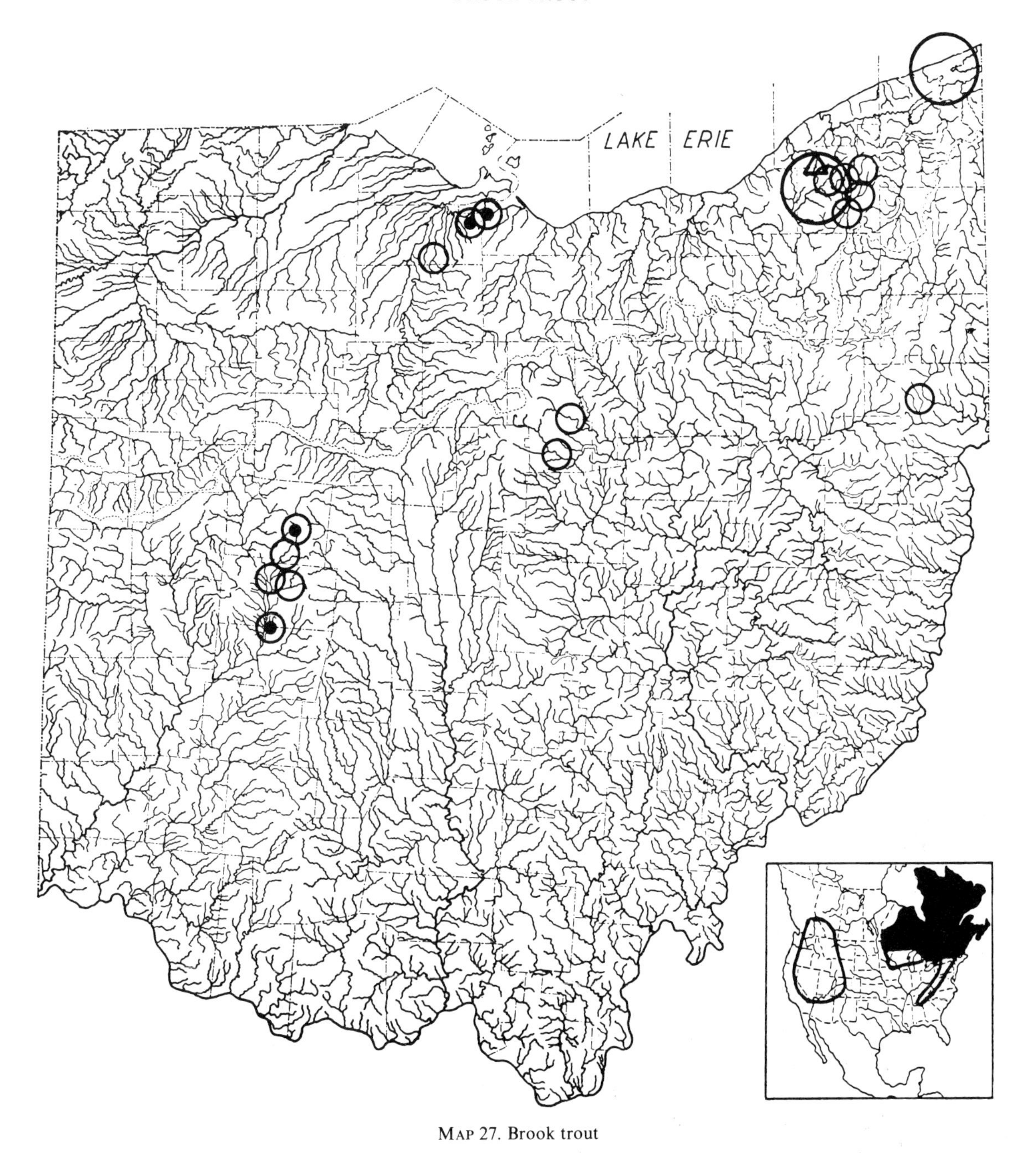

MAP 27. Brook trout

○ Kirtland records, before 1850.
○ Some localities where introduced between 1880–1950.
● More or less permanently established. △ 1955–80.
Insert: Black area represents approximate original range; outlined areas where introductions have been most successful; extensively introduced elsewhere with some local introductions successful.

Most like: Lake trout.

Coloration: *Adults*—Back olive-green heavily vermiculated with lighter, top of head and dorsal fin also vermiculated. Sides lighter with many blue and red spots, the parr marks sometimes faintly evident. Lower sides usually with an orange or reddish band. Mid-line of belly white, except in breeding males; in these the whole ventral surface of body is reddish. Anterior rays of pectoral, pelvics and anal white, lower fig. *Young*—Back olive, lighter and more silvery on sides, belly whitish. Between 8–12 parr marks on sides, also a few yellow and blue spots. No black spots, upper fig.

Lengths and **weights:** *Adults*, usually 5.0″–18.0″ (13–46 cm) long. Largest specimen, about 21.0″ (53.3 cm) long, weight 3 lbs (1.4 kg), taken in Cold Creek, Erie County.

Hybridizes: naturally and artificially produced elsewhere with brown trout (Brown, 1966:600); with lake trout (Slastenenko, 1954: 652–59), and rainbow trout.

Distribution and Habitat

Ohio Distribution—In 1838 Kirtland (1838:195, as *Salmo fontinalis*) wrote that "The speckled trout are to be found in Ohio in only two streams, a small creek in Ashtabula County, and a branch of the Chagrin River, in Geauga County. They also exist in the headwaters of the Allegheny, in Pennsylvania, but never run down into the Ohio [River]." In 1844 and in 1851 Kirtland wrote that the small creek in Ashtabula County was "in Kingsville" (1851T:69) and entered Lake Erie "at Kingsville" (1844:305). In 1930 the banks of this creek still contained remnants of a northern flora of the yellow birch-hemlock-white pine association, and this stream and its watershed gave every indication of formerly having contained a habitat suitable for brook trouts. At the same time several tributaries of the two branches of the Chagrin River in Geauga County likewise appeared to have contained formerly a suitable habitat. By 1945 almost all vestiges of these habitats had been destroyed. In 1857 Garlick (1857:82–83) stated under brook trout that "In a few streams in northeastern Ohio they were found in abundance, thirty or forty years since, and a few are yet to be found on the head waters of the Chagrin River."

Apparently Kirtland was not aware of the exacting habitat requirements of this cold-water species, for he expressed surprise that it was not "diffused into the small streams in the State of Ohio." Since Kirtland's time many well-meaning persons, also not understanding the habitat requirements of the species, have believed likewise that trouts should be "diffused into the small streams" of the state; consequently much futile stocking of brooks has been done. When the cold- and clear-water requirements of the brook trout are understood, it becomes obvious why, despite the many plantings throughout the state, the species has failed to establish itself, except in the few localities mentioned later.

All evidence indicates that the brook trout was absent from Cold Creek and the upper Mad River system, until introduced by the white man. Its absence from Cold Creek in Erie County suggests that: (1) the postglacial westward invasion from the Appalachian uplands had not progressed far enough westward along the south shore of Lake Erie to have reached Cold Creek; or (2) this westward invasion had been stopped by the unsuitable conditions in the northern Ohio lowland; or (3) no invasion of brook trouts occurred through the Maumee or neighboring outlets into Lake Erie; or (4) this invasion occurred in early postglacial time when Cold Creek was still covered by the impounded waters of one of the postglacial lakes; or (5) before the springs which form Cold Creek were in existence. At any rate, the brook trout was absent from Cold Creek until 1868. In that year Mr. Hoyt, of the Castalia Milling Co., was induced by E. Sterling, of Cleveland, to obtain some brook trout eggs. These eggs were hatched successfully, presumbaly using the technique developed by Theodatus Garlick (1857:82–98; and Harris, Bond and Post, 1884:5–7) in 1853. Sometime after the trouts had hatched they escaped from their enclosure into Cold Creek to become the dominant species of trout until 1926. After 1926 the large numbers of rainbows which were liberated in the stream, resulted in that species becoming the most numerous. This change from brooks to rainbows was made in the Cold Creek waters, which were controlled by the Castalia Trout Club, and Castalia Farms, because of the periodic decimation in numbers of the brook trout by diseases, usually furunculosis. The late Webb Saddler told me that disease "once completely killed out the stock" of brook trout.

In 1882 Smith (1885:435–36) obtained fry and eggs from Cassopolis, Mich., and Caledonia, N. Y., placing them in newly-constructed ponds near Zanesfield, Logan County. A few weeks later some fry escaped from the ponds into the Mad River system, and in 1883–84 several trouts were caught in Macocheek Creek and in the ditch connecting Smith's ponds and the Mad River. In 1927 I found brook trouts propagating naturally in Cedar Run of Champaign and Clark counties, and the older residents of the vicinity informed me that this species had been present in Cedar Run for many years.

Possibly these trouts were descendants of the 1882 plantings.

About 1900, brook trouts were introduced into the ponds and runways now controlled by the Zanesfield Rod and Gun Club. This property is adjacent to the Mad River and the village of Zanesfield. Brooks were the important species in these waters until about 1930 when rainbows were first heavily stocked. Brooks have propagated naturally in these waters, and some have escaped into the Mad River system.

Brook trouts have been present for many years in the Miller Blue Hole, of Townsend Township, Sandusky County, where they have propagated naturally.

Since the first introductions in 1868, many plantings of brooks have been made throughout Ohio waters and especially during the 1880–95 and 1932–34 periods. The hollow circles on the distribution map show a few of these many localities where brooks have been planted.

Habitat—The brook trout inhabits the same general types of cold, clean waters as does the brown (*see* brown trout under Habitat), except that the brook normally requires even colder waters, and will inhabit in larger numbers the small creeks and small ponds. The brook hides quickly under "cut banks" and other cover when frightened as does the brown, but the brook feeds more freely in the daylight, is less cautious, and consequently can be caught more readily. The brook usually inhabits waters which flow less swiftly than those inhabited by the rainbow.

Years 1955–80—Plantings of brook trouts into a few of the waters of Ohio continued during this period. On the whole, these plantings were less successful than introductions of the brown and rainbow trouts (White et al., 1975:56).

LAKE TROUT

Salvelinus namaycush (Walbaum)

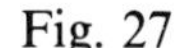

Fig. 27

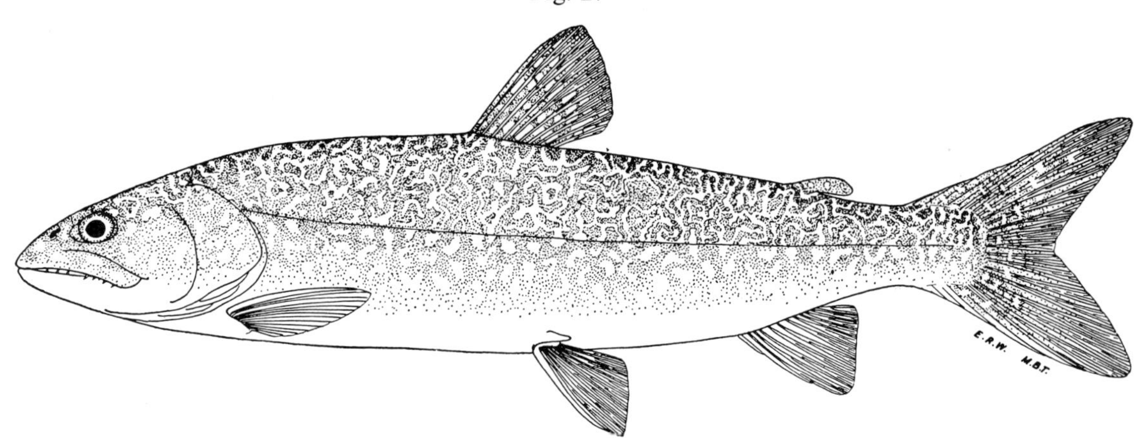

Pine Lake, Charlevoix County, Michigan.

Aug. 10, 1926.
Adult.

400 mm SL, 18.5″ (46.9 cm) TL.
OSUM 9269.

Identification

Characters: *General*—*See* brown trout. *Specific*—Only species of trout in Ohio with a deeply forked tail. Top of head and back heavily vermiculated with shades of gray. Sides with an occasional light spot which is tinged with reddish, but no bluish spots. Lower fins unicolored, the first rays not sharply different in color from other rays. Usually fewer than 210 scales in the lateral series.

Differs: *See* brown trout. Other trouts have squarish or slightly emarginated tails.

Most like: Brook trout.

Coloration: Dorsally olive- or brownish-gray, heavily vermiculated with lighter grays. Sides lighter than back, more silvery, vermiculations fainter, and with some spotting. Belly light or silvery, without red. Dorsal and caudal fins vermiculated. Lower fins unicolored.

Lengths and **weights:** *Adults*, usually 15.0″–36.0″ (38.1–91.4 cm) long, weight 1 lb–20 lbs (0.5–9.1 kg). Commercial fishermen give maximum length at about 4′ (122 cm), weights between 40–54 lbs (18.1–24.5 kg).

Hybridizes: with brook trout.

Distribution and Habitat

Ohio Distribution—Literature references before 1900 denote the presence of a moderately large population of lake trouts in the deeper waters of eastern Lake Erie, and of strays in the shallower western end. Kirtland (1851P:10) noted the capture before 1851 of several from the lake near the mouths of the Cuyahoga and Rocky rivers, and of the considerable numbers which were captured each winter by fishing in water 70′–80′ (21 m) deep in the eastern section of the lake; in 1874 Milner (1874:38) stated that "In the shallow waters of Lake Erie, in the western part of the lake, they are scarcely found at all, though numerous in the deeper portion, east of the city of Cleveland"; in 1885 Smith and Snell (1891:295) recorded that only 5,200 lbs (2,359 kg) of lake trouts were taken that year in the extreme eastern end of Lake Erie as compared to 3,660,000 lbs (1,660,148 kg) of sturgeons and 2,011,425 lbs (912,367 kg) of walleyes, and that (1891: 264) "Very few lake trout are ever seen west of Huron [Ohio]." Since 1900 Greeley (1929:169) reported that lake trouts were of minor importance to the commercial fisheries of eastern Lake Erie although "Good catches are sometimes made in late fall." Between 1925–30 fair catches were made occasionally in the Lake Erie waters of Ashtabula County with gill-nets which were primarily set for blue pike; during that period I saw as many as 15 trouts, ranging between 3–15 lbs (1.4–6.8 kg) in weight, taken in one net. After 1930 a decrease in abundance became ap-

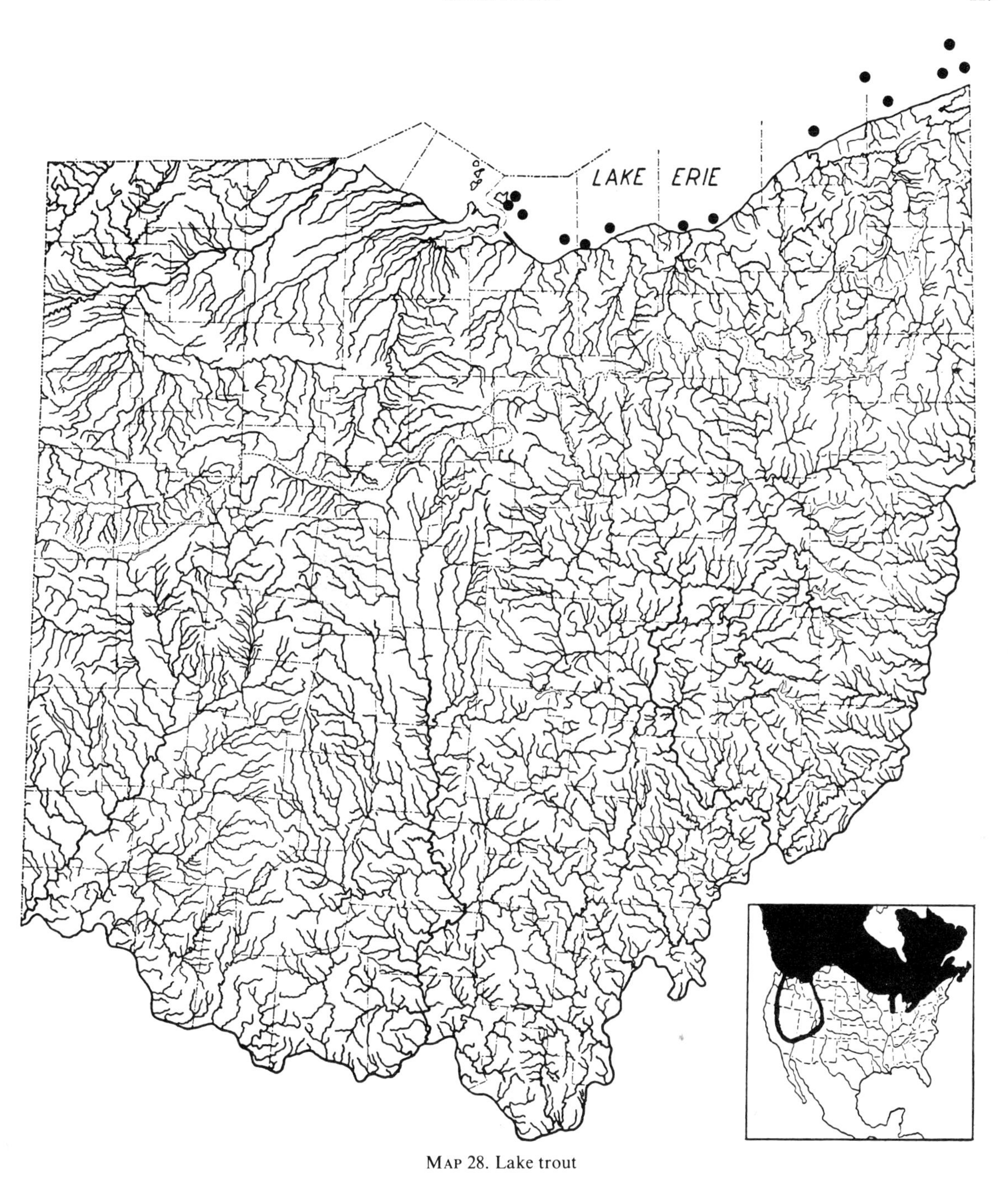

MAP 28. Lake trout

● Locality records before 1955.

Insert: Black area represents approximate original range; outlined area where introductions have been most successful; northern limits of range indefinite.

parent and since 1940 relatively few have been taken. Since 1925 I have seen three individuals which were taken in western Lake Erie near Kelleys Island, of which two were greatly emaciated, having thin bodies, sharply-ridged backs, and bulky heads, all factors usually associated with submarginal conditions. Lake Erie constitutes the extreme southern range of the species in this section of the country (Rostlund, 1952:260–61).

Habitat—Primarily confined to the deeper waters of the eastern half of Lake Erie, although venturing into shallower waters in late fall to spawn and to feed about the rocky and gravelly reefs, and returning to deeper waters in early spring.

Years 1955–1980—The decline in the population size of the lake trout in Lake Erie, apparent since 1850, continued during this period. By 1975 it appeared to be very rare, or possibly extirpated, from the Ohio waters of the Lake (Van Meter and Trautman, 1970:68; White et al., 1975:58).

For a general life history of the lake trout, *see* Eschmeyer (1957).

GREAT LAKES CISCO

Coregonus artedii artedii* Lesueur

Fig. 28

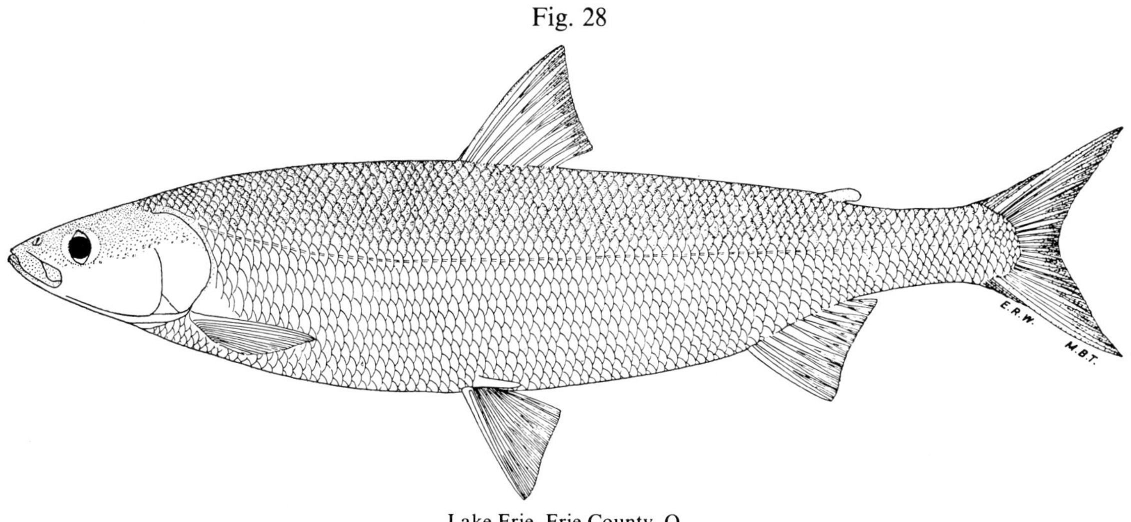

Lake Erie, Erie County, O.

Nov. 27, 1941.
Adult female.
Note terminal mouth.

340 mm SL, 13.8″ (35.1 cm) TL.
OSUM 5008.

Identification

Characters: *General*—Adipose fin, adipose eyelids, and pelvic axillary process present. Scales in lateral series 70–90. Tongue and jaws without large, sharp, canine teeth. Coloration bluish-silvery. No spines on fins. *Specific*—Mouth terminal, the snout sharp. Gill rakers on entire first gill arch 41–53. *Subspecific*—Body torpedolike in shape, its depth usually more than 3.7 times in standard length. Longest pelvic ray usually contained more than 1.7 times in distance between origins of pelvic and anal fins. Lateral line scales usually more than 79. Back a deep bluish. Entire fish bluer and less silvery (hence the commercial fisherman's name of "blueback") than is the Lake Erie cisco.

Differs: Mooneye, goldeye, skipjack herring, alewife, and eastern gizzard shad have no adipose fin. Trouts and smelt have large canine teeth. Whitefish has a rounded snout and subterminal mouth.

Most like: Lake Erie cisco and whitefish.

* Many ichthyologists, especially formerly, restrict the genus *Coregonus* to include only some of the whitefishes, placing the ciscoes in the genus *Leucichthys*.

Coloration: Dorsal half of head and body steel-blue with a silvery and emerald-green overcast. Ventral half silvery and milk-white with greenish reflections.

Lengths and weights: *Adults*, usually 11.0″–15.0″ (27.9–38.1 cm) long, weight 9 oz–1 lb 9 oz (255 g–0.7 kg). Largest specimen 19.0″ (48.3 cm) long, weight 2 lbs 9 oz (1.2 kg). Commercial fishermen give maximum length at about 20.0″ (50.8 cm), weight about 3 lbs (1.4 kg).

Intergrades: Abundantly with Lake Erie cisco; often the bulk of the catch consists of intergrades.

Hybridizes: not uncommonly with the whitefish. These hybrids are called "mule whitefish" by commercial fishermen, and hybrids more than 5 lbs (2.3 kg) are not unusual. Koelz (1929:539) records a male from Lake Erie, Canadian waters, which weighed 11 lbs 15 oz (5.4 kg). I have seen a mounted specimen taken about the Bass Islands, which weighed 13 lbs (5.9 kg), and many others 15.0″–24.0″ (38.1–60.9 cm) in length and 2–5 lbs (0.9–2.3 kg) in weight. In shape they resemble the Lake Erie cisco, but their mouth-shape and gill raker count are intermediate between the cisco and whitefish.

LAKE ERIE CISCO

Coregonus artedii albus* Lesueur

Fig. 29

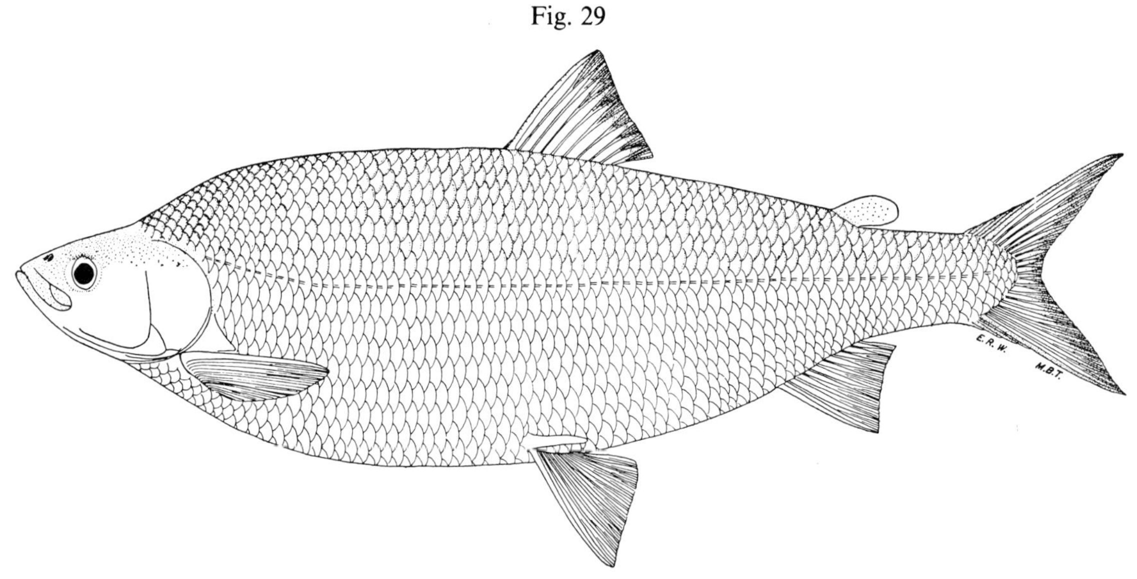

Lake Erie, Erie County, O.

Nov. 27, 1941.
Adult female.
Note terminal mouth.

410 mm SL, 19.5″ (49.5 cm) TL.
OSUM 5009.

Identification

Characters: *General* and *specific*—*See* Great Lakes cisco.

Subspecific—Body depth deep and slab-sided, like body of whitefish. Body depth is usually less than 3.7 times in standard length. Longest pelvic ray usually contained less than 1.8 times in distance between origins of pelvic and anal fins. Lateral line scales usually fewer than 79. Color less bluish and more silvery.

Differs: *See* Great Lakes cisco.

Most like: Great Lakes cisco and whitefish.

Coloration: Dorsal half of head and body light steel-blue with much silvery. Ventral half milk-white and silvery.

Lengths and **weights:** *Adults*, usually 11.0″–18.0″ (27.9–45.7 cm) long, weight 11 oz–3 lbs (312 g–1.4 kg). Largest preserved specimen (OSUM 5009) is 19.5″ (49.5 cm) long, 3 lbs 12 oz (1.7 kg) in weight. The two largest specimens measured and weighed, were 21.5″ (54.6 cm) long, and weighed 5 lbs 8 oz (2.5 kg). Commercial fishermen report maximum lengths of 24.0″–25.0″ (60.9–63.5 cm), and weights of 6–8 lbs (2.7–3.6 kg).

Intergrades: *See* Great Lakes cisco.

Hybridizes: with whitefish, *see* Great Lakes cisco.

Distribution and Habitat

Ohio Distribution—The following general statements apply to the cisco as a species:

As in the case of the whitefish, the first cisco fishery of importance about Lake Erie was established about 1815 as a seine fishery in the shallow waters of Maumee Bay and in the Detroit River (Scott, 1951:13). However, at that time ciscoes were classed as rather worthless "soft fish" and were often thrown away (Smith and Snell, 1891:253). It was not until after 1870, after the pound nets had become numerous, that the cisco began to be of major commercial importance. By 1885 it was the most important species of the commercial fishery, and that year the total catch for Ohio ports of 15,037,000 lbs (6,820,668 kg) was the largest catch to date. In that year the heaviest runs occurred

* *See* footnote on p. 231.

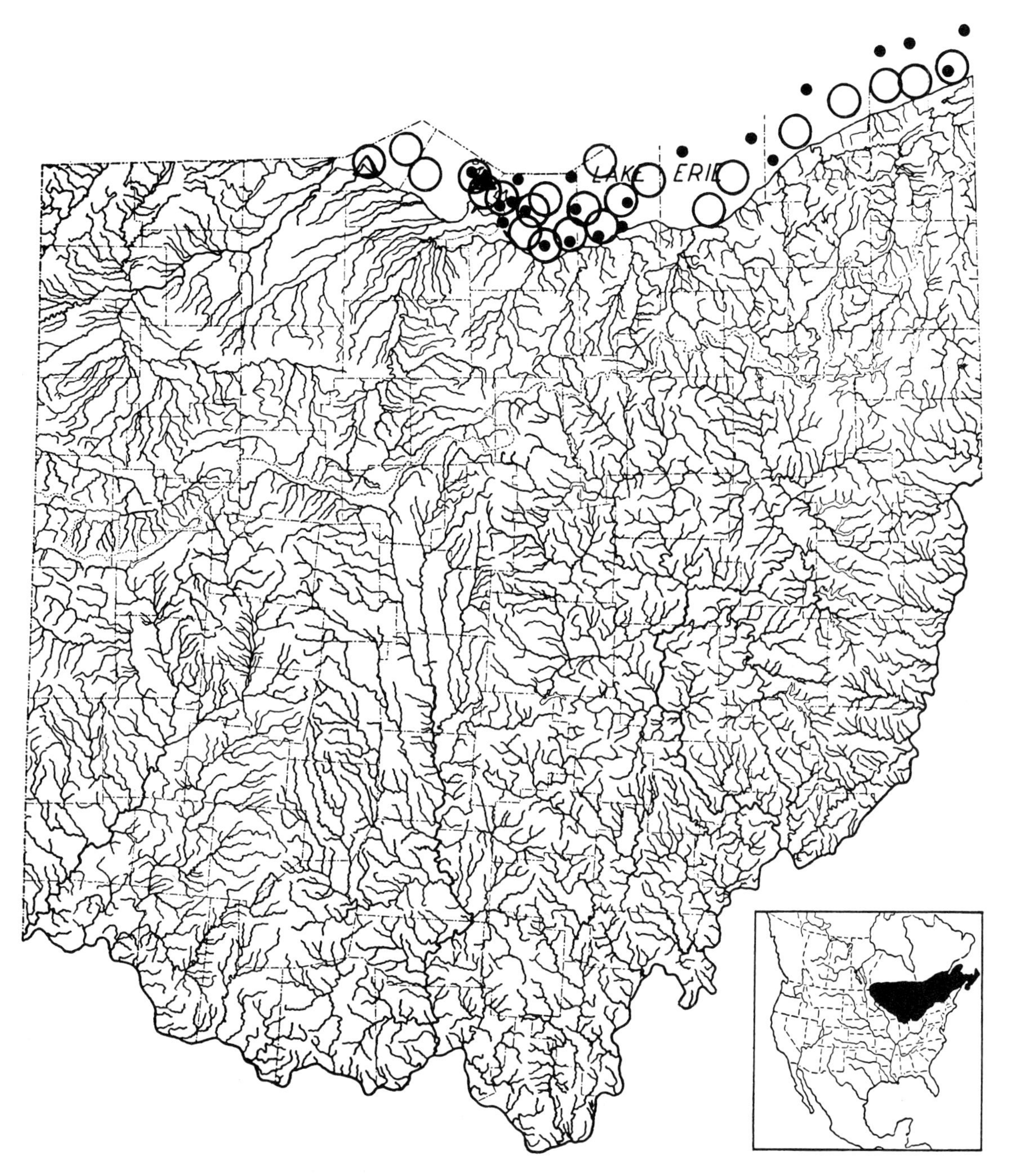

MAP 29a and 29b. Great Lakes and Lake Erie ciscoes

○ Lake Erie cisco. ● Great Lakes cisco. Locality records. Before 1955. △ 1955–80.
Insert: Northern and northwestern limits of range of the species indefinite, these limits depending in part upon interpretation of the relationships with other ciscoes.

between November 12–22, Smith and Snell (1891:241 and 262) reporting that "One firm at Put-in-Bay took out 11 tons [10 metric tons] of herring from their seven nets on the 16th of November and left several tons of them. The next day they were full again. . . . Five tons [4.5 metric tons] per day was not an uncommon catch, and Mr. Bower says that some claimed to have taken 10 tons [9.1 metric tons] from one net in a single day's fishing. At 'Kellys' Island alone between 25 [23 metric tons] and 30 tons [27 metric tons] of herring were dumped overboard or allowed to escape for lack of a market."

Koelz (1926:592) records great fluctuations in the amount of the annual catch for all of Lake Erie during the 1885–1924 period. Van Oosten (1930:211) states that during the 1913–24 period the total catch for Lake Erie ranged between 13,547,485 lbs (6,145,035.8 kg) (1916) and 48,822,520 lbs (22,-145,523 kg) (1918); in 1924 the total catch was 32,200,661 lbs (14,605,974 kg), dropping to 5,657,-000 lbs (2,565,972 kg) in 1925. The decline in abundance continued until 1937, when in that year Langlois and Edmister (1939:2b) gave the Ohio catch as only 1,523 lbs (690.8 kg), and Scott (1951:13) gave the Canadian catch as only 99,400 lbs (45,087 kg). Between 1938–44 the annual Ohio catch fluctuated between 8,594 lbs (3,898 kg) in 1941 (Gray, 1942) and 76,919 lbs (34,890 kg) in 1939 (Edmister, 1940). In 1945 a notable increase occurred, Gray (1946) recording the Ohio catch as 677,212 lbs (307,178 kg); for 1946 Gray (1947) recorded the Ohio catch as 1,578,570 lbs (716,027.3 kg) and Scott (1951:13) for the Canadian catch as 9,525,000 lbs (4,320,467 kg). After 1946 another decline occurred; in 1949 the Ohio catch was 57,289 lbs (25,986 kg) (Shafer, 1950:1). Van Oosten (1930:204–14) believed that the 1925 population collapse was caused by overfishing; Langlois (1946:101–04) attributed the collapse to other unfavorable environmental factors.

No group of eastern North American freshwater fishes displays greater variations in body proportions than does the Coregonidae, not only between two isolated populations of supposedly the same subspecies in two different bodies of water, but there is also considerable variation in body proportions from year to year in the population of a small, isolated lake. Hile (1937:123) has shown that in some small lakes in Wisconsin the "populations assigned to the same subspecies showed differences as great or greater than those which now separate some of the subspecies of *L. artedii.*" Realizing this I hesitate to attribute subspecific status to the two quite different types of ciscoes inhabiting Lake Erie and prefer to call them "types." Following is a discussion of these two types.

The *Coregonus artedii artedii* type is comparable in body proportions to *C. artedii artedii* of Lakes Huron and Ontario; it has a terete body like that of the smelt; a distinctly bluish back; is of smaller average size and weight (seldom more than 3 lbs [1.4 kg]); occupies the deeper waters of the eastern two-thirds of Lake Erie; since 1907 and probably throughout historic time, it has been more numerous than has the other type; is called the "spike or blueback herring" by the commercial fishermen.

The *Coregonus a. albus* type is slab-sided like the whitefish; silvery, and of larger average size and greater weight (before 1907, examples of 4–7 lbs [1.8–3.2 kg]; occupies the shallower waters; is called the "jumbo herring or cisco" by commercial fishermen who greatly prefer it to the "blueback" as food.

Commercial fishermen separate the more typical examples without difficulty, but experience great difficulty with the intermediates (as they do with "gray pikes" which are the intermediates between blue pikes and yellow walleyes).

Three species of Ohio fishes have two recognized subspecies in Lake Erie: blue pike and walleye; central and northern mottled sculpins; eastern and central quillback carpsuckers. The first named of each of these three pairs occupy the deeper waters as does the *C. a. artedii* type; the last three named occupy the shallower habitat, as does the *C. a. albus* type.

Habitat—*Coregonus artedii artedii* type.

As in Lakes Ontario and Huron, this type inhabits moderately deep water throughout most of the year; perhaps concentrating at times in the "deep hole" area (Van Oosten, 1930:209); invading the shallow Ohio waters east of Kelleys Island only in very small numbers since 1930.

Habitat—*Coregonus artedii albus* type.

According to the older commercial fishermen of western Lake Erie, the "jumbo herring" and intermediates were the types of ciscoes taken in the ice fishery about the Bass Islands between 1890–1907 (Doan, 1944:69). About these islands much ice fishing was done in bays where the water was less than 10′ (3 m) in depth, by fishing through a hole in the ice and using a pearl button as a lure. The greatest amount of ciscoes caught by one man in a day in this manner was 300 lbs (136 kg). After 1907, ice fishing for ciscoes declined rapidly; but fair gill-net and trap-net catches were made near the islands until about 1917, after which these catches declined drastically.

Since 1939 the majority of the ciscoes caught

about the islands of western Lake Erie, which I have examined, have been intermediates or of the *albus* type; the majority of those taken during spawning seasons on the shoals in Erie and Lorain counties have been intermediates, with the remainder consisting mostly of the *artedii* type and a few of the *albus* type. Farther eastward in Ashtabula County, the majority were *artedii* and the remainder intermediate, with few of the *albus* type.

Years 1955–80—As may be noted under Ohio Distribution, great fluctuations in the sizes of the annual catches of ciscoes in Lake Erie have taken place since at least 1885. Since 1885, also, there has been a continuous, although fluctuating, decline. Between 1938 and 1944 the annual catch per year varied from 8,594 lbs (3,898 kg) in 1941 to 76,919 lbs (34,890 kg) in 1939. In 1945 the Ohio catch rose to 677,212 lbs (307, 178 kg), a notable increase, and in 1946 to a surprising 1,578,570 lbs (716,027 kg), thereafter declining again. This huge 1945–46 increase apparently was the result of one, or at the most two, very successful spawning seasons. By 1954 the annual Ohio catch had declined to 48,815 lbs (22,142 kg), dropping to *5* lbs (2.3 kg) in 1967 and to *62* lbs (28 kg) in 1968 (Anonymous B, 1968:6; White et al., 1975:58). *See* lake whitefish, p. 236.

LAKE WHITEFISH

Coregonus clupeaformis (Mitchill)

Fig. 30

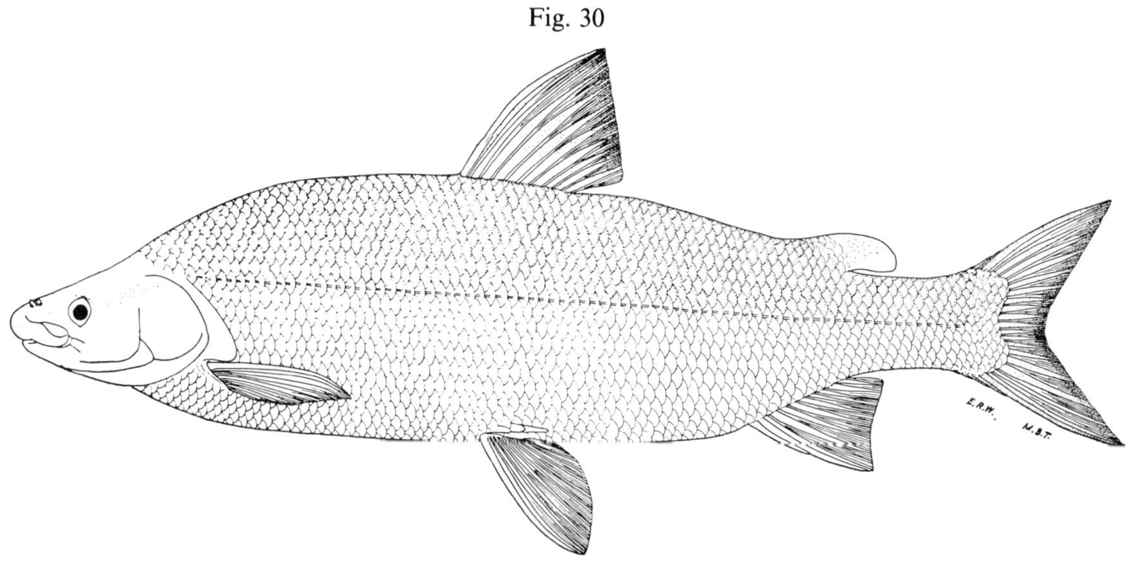

Lake Erie, Ottawa County, O.

Nov. 29, 1941.
Adult male.

375 mm SL, 17.5″ (44.5 cm) TL.
OSUM 5010.

Note subterminal mouth.

Identification

Characters: *General—See* Great Lakes cisco. *Specific*—Mouth definitely subterminal, the snout rounded. Gill rakers on entire first gill arch 25–30. Scales in lateral line usually 73–86, extremes 73–93.

Differs: *See* Great Lakes cisco. Ciscoes have terminal mouths and more than 41 gill rakers on the first gill arch.

Most like: Ciscoes.

Coloration: Dorsal half of head and body olive-green or olive-blue with a silvery overcast. Sides silvery; ventral half of body milk-white. Whitefishes have a more greenish cast whereas ciscoes have a more bluish cast. Male whitefish has scales tuberculated during the spawning season.

Lengths and **weights:** *Adults*, usually 17.0″–22.0″ (43.2–55.9 cm) long, weight 1 lb 8 oz–4.0 lbs (0.7–1.8 kg). In 1920 or 1921, Roy Webster, of South Bass Island, caught four whitefishes near Starve Island, Ottawa County, which totaled 52 lbs (23.6 kg). Van Oosten and Hile (1949:232) record a specimen 28.8″ (73.2 cm) long, weight 10.75 lbs (4.88 kg). Commercial fishermen report maximum lengths of 30.0″–36.0″ (76.2–91.4 cm), and weights of 13–15 lbs (5.9–6.8 kg).

Hybridizes: with ciscoes. *See* Great Lakes cisco.

Distribution and Habitat

Ohio Distribution—Presumably the whitefish has been present in Lake Erie since early post-glacial times (Rostlund, 1952:262). Klippart (1877A:34) quotes Charles Carpenter as stating that "Mr. U. S. Webb tells me a tradition of Delaware County was that, one hundred years ago [about 1777], the Delaware and Wyandot Indians were in the habit of visiting this vicinity [particularly Sandusky Bay, etc.] for picking up the whitefish thrown up on the beaches by north-east storms."

Until 1848 the only important whitefish fishery about Lake Erie was in the Detroit River (Smith and Snell, 1891:261) and it was the general opinion that whitefishes were not numerous in Lake Erie

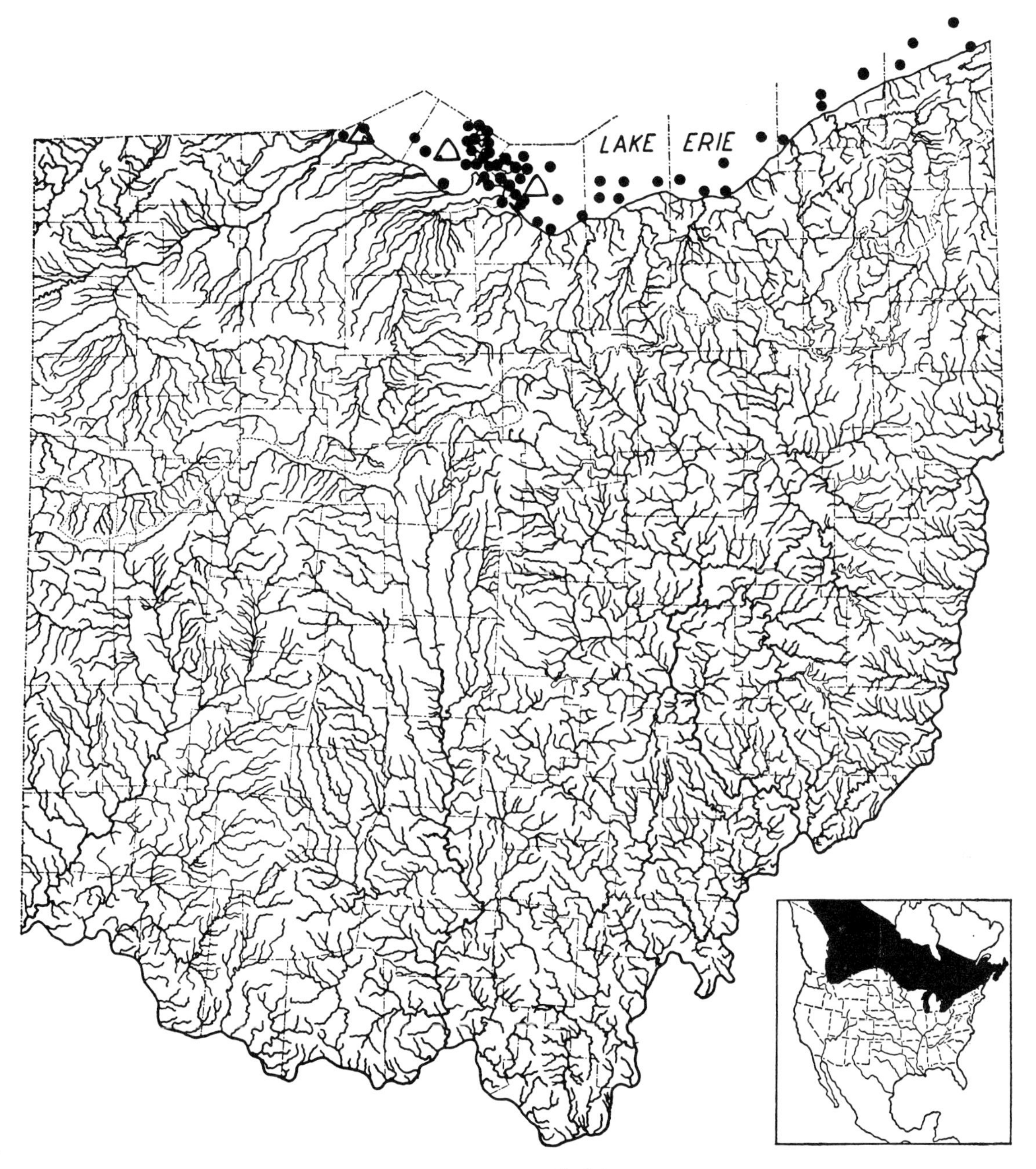

MAP 30. Lake whitefish

Locality records. ● Before 1955. △ 1955–80.
Insert: Northern and western limits indefinite; has been introduced successfully into many waters, especially mountain lakes.

itself. Kirtland (1838:195) shared that opinion for he pointed out, that although an occasional whitefish was taken in Lake Erie within the limits of Ohio, the species was too uncommon to be of commercial importance. In 1846 the capture in a seine in Sandusky Bay (Klippart, 1877A:31) of about 100 whitefishes caused much speculation. Later, in the autumn of 1848, Mr. and Mrs. Charles Webb saw great numbers of fishes swirling at the surface near Kelleys Island, and by trolling succeeded in snagging a whitefish. Realizing that the species might be present in considerable numbers about the island, Benjamin Deeley set gill nets in the autumn of 1851, which "caught well." At the same time a pound net (pound net introduced into Lake Erie waters in 1850), set in the connecting channel between East Harbor and Lake Erie, caught "emmense quantities" of whitefishes. In 1852 a pound net set in North Bay of Kelleys Island produced 1–4 tons (0.9–3.6 metric tons) of whitefishes per haul (Klippart, 1877A:34). With the realization of the whitefish abundance about the Bass Islands, and rapid increase in the number of pound nets, the whitefish fishery rapidly attained considerable commercial importance. In 1885, 1,249,000 lbs (566,536.9 kg) were brought into Ohio ports (Smith and Snell, 1891:241). Between 1940–49 the 10-year average was 704,315 lbs (319,472 kg) annually (Shafer, 1950:1), the extremes in poundage ranging from 164,504 lbs (74,617.8 kg) (1944) to 1,616,459 lbs (733,213.5 kg) (1948).

Until 1890 whitefishes spawned in the Detroit River and Maumee Bay. After 1890 a sharp decrease in the size of the Detroit River spawning run occurred; by 1900 the run apparently had stopped entirely. About 1900 the ever increasing silt load of the Maumee River began smothering the Maumee Bay spawning areas, causing the annual take in Maumee Bay to decline until by 1918 only 10–20 whitefishes were taken daily by such commercial fishermen as Cameron King and his father. This decline has continued, and since 1940 only an occasional stray has been taken. In 1950 the species still spawned in Lake Erie as far west as Niagara Reef, eastern Ottawa County; a few may still spawn about West Sister Island, in Lucas County.

Habitat—Throughout the warmer months the whitefishes chiefly inhabited waters of 40′ (12 m) or more in depth, and the bulk of the population was in the eastern half of the lake. In fall an inshore and/or western migration began, its speed of advance depending upon the rapidity with which the water temperature declined. The first whitefishes, usually males, arrived on the reefs about the Bass Islands when the water temperature reached about 50° F (10° C). The incoming fishes sometimes swam near the water's surface, and on those windless days without waves they were seen swirling near the surface. Spawning began in earnest after the water temperature had dropped below 45° F (72° C). Paul Webster has captured spawning whitefishes as early as October 18 and as late as December 15 (last day of commercial fishing season). By the time the spring fishing season opened, on March 15, all except a few of the whitefishes had left the shallow flats of western Lake Erie for the deeper waters. In May an inshore movement from the deep waters of eastern Lake Erie into the shoals took place, individuals remaining on the shoals until June, after which the species remained in the deep waters until fall (Koelz, 1929:538).

Koelz (1931:373) in 1931 gave subspecific rank to the whitefish of Lake Erie, naming it *Coregonus clupeaformis latus*. I prefer not to accept the subspecific status for the Lake Erie population until more extensive research upon the assumed morphological characters of the Coregonidae has been made.

Years 1955–80—As may be noted under Ohio Distribution, the poundage of whitefishes brought into Ohio ports prior to 1900 was more than a million lbs (453,592.4 kg) annually, whereas the 10-year average between 1940 and 1949 was only 704,315 lbs (319,472 kg), except for a peak of 1,616,459 lbs (733,213.5 kg) in 1948. In 1954 the annual catch was 145,449 lbs (65,974.6 kg), dropping to an all-time low of 699 lbs (317 kg) in 1963, to 809 lbs (367 kg) in 1964, rising to 5,555 lbs (2,520 kg) in 1965, then dropping to 1,157 lbs (524.8 kg) in 1968 (Anonymous B, 1968:8; Van Meter and Trautman, 1970:68; White et al., 1975:59).

It is becoming increasingly obvious that violent fluctuations in annual abundance of a fish species, except in those relatively few cases of overfishing, is an indication of unstable and/or environmental changes as Langlois (1946:101–64) formerly suggested. *See* also silver chub, a non-commercial species.

RAINBOW SMELT*

Osmerus mordax (Mitchill)

Fig. 31

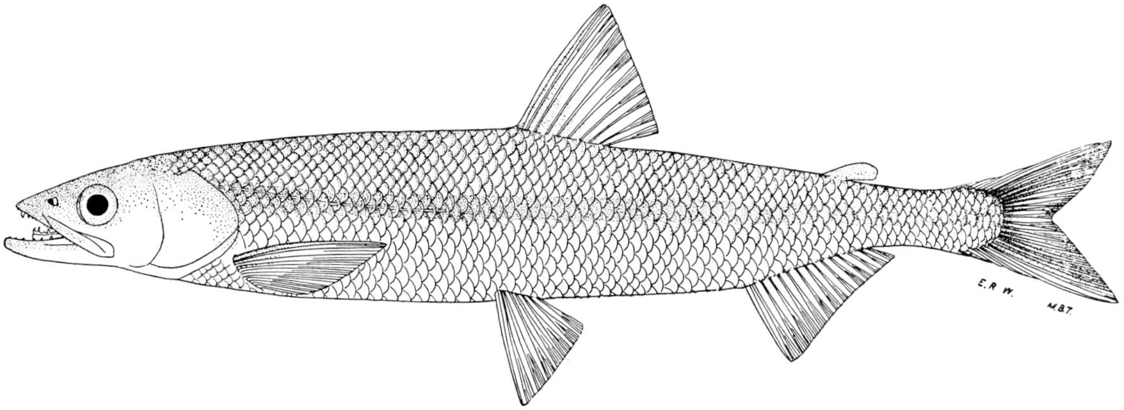

Lake Erie, Erie County, O.

July 24, 1939.
Female.

198 mm SL, 9.0″ (23 cm) TL.
OSUM 1428.

Identification

Characters: Adipose fin present. Teeth on tongue and jaws. Mouth large; the posterior end of upper jaw extending beyond posterior edge of eye. Anal rays 13–17. Lateral line scales 60–74. Coloration silvery. No pelvic axillary process.

Differs: Mooneye, goldeye, skipjack herring, alewife, and eastern gizzard shad have no adipose fin. Whitefish, ciscoes and trouts have a pelvic axillary process. Trout-perch lack large teeth and have ctenoid scales.

Most like: Mooneye and ciscoes.

Coloration: Predominantly silvery; dorsal half of head and body steel-blue; sides lighter; ventral half silvery-white. A faint, dark lateral band present in some large individuals.

Lengths and **weights:** *Young* of year in July, 1.0″–2.0″ (2.5–5.1 cm). Around 1 year old, 3.0″–6.0″ (7.6–15.2 cm). Around 2 years old and mature, 7.0″–10.0″ (18–25.4 cm) long, weight 2 oz–4 oz (57–113 g). Largest specimens 12.0″–14.0″ (30.5–35.6 cm) long, weight 5 oz–8 oz (142–227 g).

Distribution and Habitat

Ohio Distribution—Van Oosten (1937A:161) stated that the single introduction of smelt in 1912 into Crystal Lake, Benzie County, Michigan, Lake Michigan drainage, "is considered the source of all the smelt now found in the Great Lakes except Ontario" and that from this introduction the species spread throughout the upper lakes so that by 1932, spawning runs of smelt occurred as far south as Lake St. Clair (Van Oosten, 1937A:166). The first smelts were taken in Lake Erie in 1932 from the Ontario waters near Port Dover. The first specimens (UMMZ 103359) from the Ohio waters of Lake Erie were captured on June 30, 1936, in gill nets 16 miles (26 km) NNE of Vermilion, Ohio. Later in 1936, additional specimens were taken in various localities, ranging from western Lake Erie east to the Ohio-Pennsylvania line, and at that time it was apparent that the species had become well established. Since 1936, smelt have been taken annually in Lake Erie waters, and since 1949 spawning runs have occurred each spring in several small Lake Erie tributaries of northeastern Ohio, particularly in Lake (Arcola Creek) and Ashtabula (Indian Creek) counties.

Habitat—Smelt inhabited the deeper waters of Lake Erie at all seasons, as indicated by trap and gill-net captures. Adults were also found in shallow waters in spring; some going up small tributaries in northeastern Ohio to spawn. Post-larval and small

* American smelt in 1957 edition of *Fishes of Ohio*.

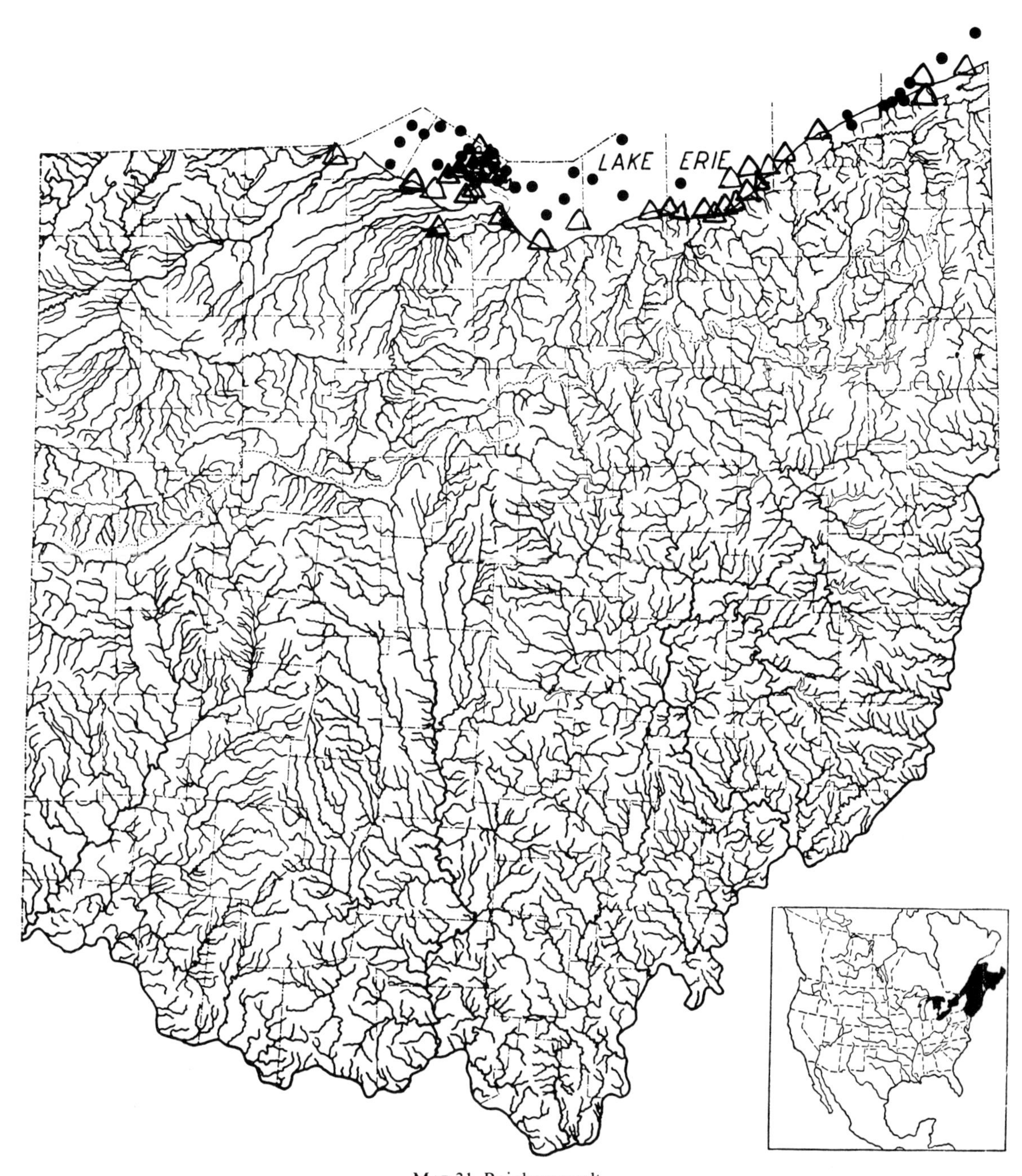

MAP 31. Rainbow smelt

Locality records. ● Before 1955. △ 1955–80.

Insert: Recent extension of range in Great Lakes above Niagara Falls presumed to be the result largely of introduction; populations in Lakes Michigan and Superior may be increasing.

fry were taken about the gravel bars around Gibraltar and South Bass islands in June and July during several years between 1940–50; more than 50 were taken in one seine haul on June 25, 1942, these ranging from 0.8″–1.3″ (20–34 mm) in total length. Although there appear to be no recorded observations of smelt spawning in lakes, it is difficult to believe that these small fry, captured about the islands, were hatched in small tributaries on the mainland, and that they had traversed the open waters of the lake to inhabit the shallow bars about the islands. Rather it seems more probable that the adults spawned over the bars where normally there is a current, or in the narrow connecting passage between the lake and Terwilliger's Pond where seiches (Krecker, 1928:1–22) produced almost continuous currents. In August of several years, young 2.0″–3.0″ (5.1–7.6 cm) in total length were taken in the open lake, in depths of more than 20′ (6.1 m).

Years 1955–80—The annual numbers of rainbow smelts inhabiting Lake Erie appeared to fluctuate considerably (White et al., 1975:59). During some years between April and June heavy mortality, presumably post spawning, occurred. Then sandy beaches, such as at Bay Point in Ottawa County, contained windrows of dead smelt, deposited by wave action. Windrows were as long as 1/8 of a mile (200 m), widths 5′ to 15′ (1.5–4.6 m), and maximum depths of a foot (0.3 m). Bathing beaches were rendered unusable until long trenches were dug, the fishes bulldozed into them and covered with sand.

Goodly-sized spawning runs were noted in such tributaries as Turkey, Rock, Conneaut, Whitman, Redbrook and Wheeler creeks and the Ashtabula River, all in Ashtabula County and Arcola Creek in Lake County.

In 1970 Van Meter and Trautman (1970:68) followed D. E. McAllister (1963:15) in considering *mordax* to be conspecific with *eperlanus*, we employing *Osmerus eperlanus mordax* (Mitchill). McAllister has informed me since that it is his present opinion that *O. eperlanus* is specifically distinct from *O. mordax*; that the rainbow smelt of the North Pacific and Arctic is *O. m. dentex*, and that the eastern North American population is *Osmerus m. mordax* (Mitchill). In the first edition of this report the rainbow smelt was called the American smelt.

CENTRAL MUDMINNOW

Umbra limi (Kirtland)

Fig. 32

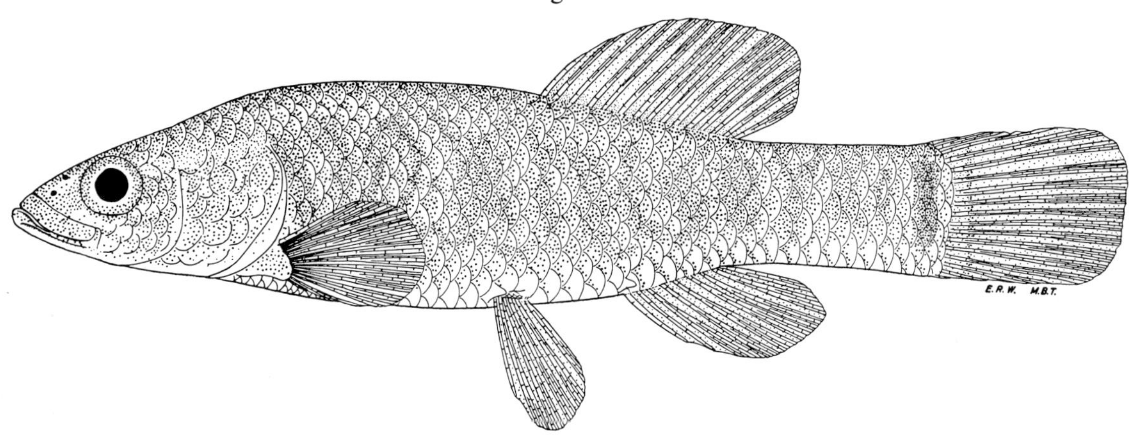

Deer Run, Stark County, O.

May 28–29, 1941.
Young female.

48 mm SL, 2.3″ (5.8 cm) TL.
OSUM 3239.

Identification

Characters: No groove separating tip of upper lip and snout. Top of head, cheeks and opercles scaled. Tail rounded. Pelvic fin insertion closer to caudal base than to tip of upper lip. Anterior insertion of dorsal fin over, or just behind, pelvic fin insertion. A vertical dusky bar at caudal base, its width over half the diameter of eye. No lateral line, and fewer than 50 scales in the lateral series. Fins without spines.

Differs: Pikes have forked tails and more than 90 scales in lateral series. Killifishes, topminnow and mosquitofish have a deep, wide groove separating upper lip and tip of snout. Trout-perch has an adipose fin. Pirate perch has the anus far before anal fin origin. Suckers and minnows have scaleless heads. Darters have dorsal spines.

Most like: Killifishes, topminnow, pirate perch and pikes.

Coloration: Dorsal half of head and body dark olive-green or brown, mottled with dark browns. Sides lighter, the mottling more distinct, occasionally vague barring present. Ventral half of body a dirty yellow-white. All fins brownish. A prominent caudal bar.

Lengths: *Adults*, usually 2.0″–4.0″ (5.1–10 cm) long. Largest specimen 5.2″ (13 cm) long. For growth rate, *see* Applegate (1943: 92–96).

Distribution and Habitat

Ohio Distribution—Records of early occurrence and published accounts of environmental conditions prior to 1900 suggest that the mudminnow was formerly widespread and abundant throughout glaciated Ohio, except for the southern third of the state, and that it was absent from unglaciated Ohio except in a few localities where it had penetrated a few miles beyond the edge of glaciation. Although no records of its occurrence exist, it was probably widespread throughout the Auglaize, Portage and Sandusky river systems before 1850, before wholesale ditching, dredging, draining and increased turbidity had lowered the water table, stopped the flow of many springs, destroyed much of the aquatic vegetation, greatly increased the turbidity of the water, and covered the muck, peat and organic debris of the bottoms with silt.

During the 1925–50 surveys the mudminnow decreased greatly in abundance in many localities such as about Buckeye Lake, or disappeared entire-

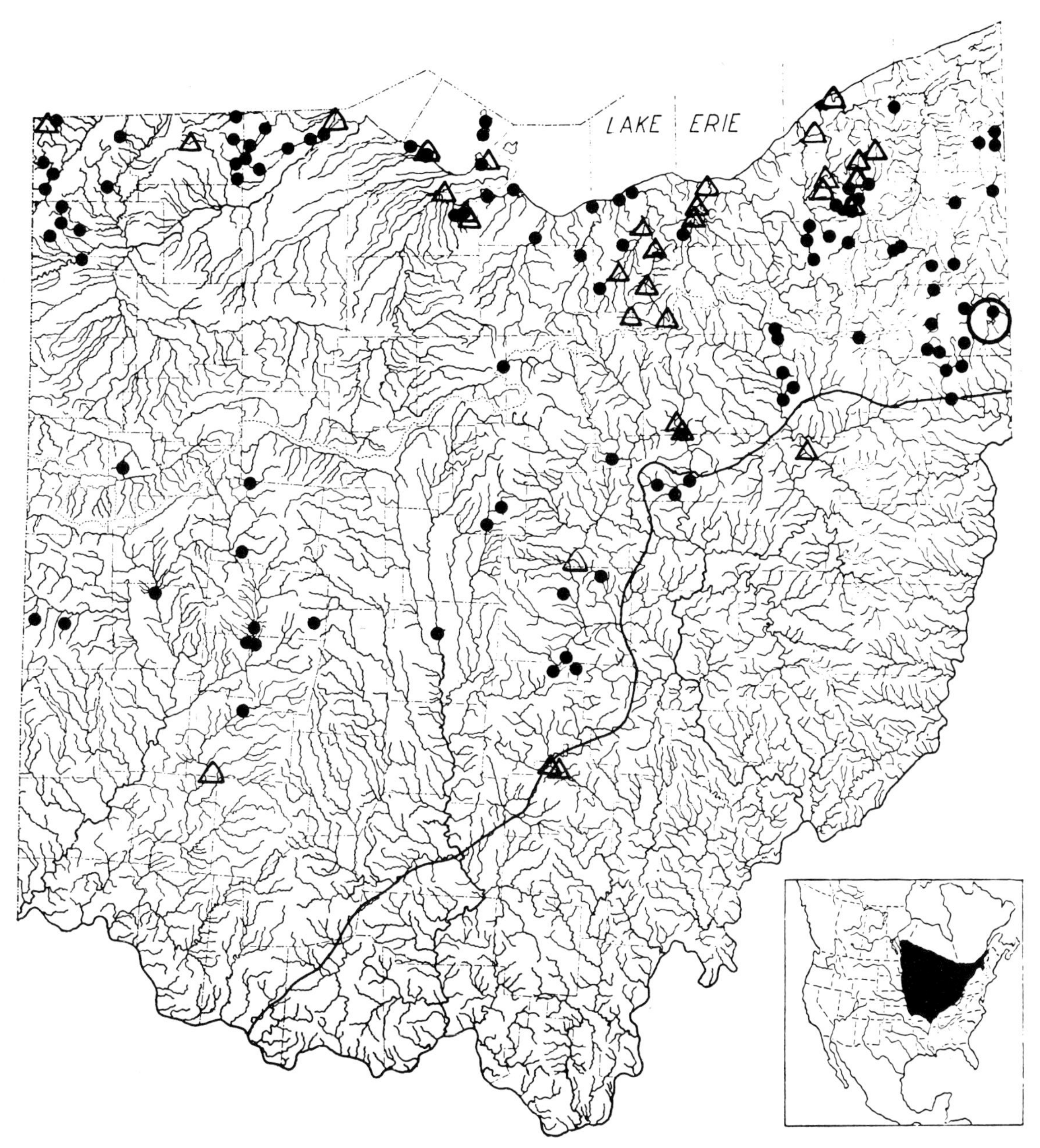

MAP 32. Central mudminnow

○ Type locality, Yellow Creek, Mahoning County.
Locality records. ● Before 1955. △ 1955–80. ~ Glacial boundary.

Insert: Possible recent withdrawals from many sections along the southern edge of range; northern and northwestern limits of range indefinite; large areas within the southern half of range recently have become unfavorable for the species.

ly as in Franklin County. The changes mentioned in the first paragraph caused these decreases in abundance or extirpation. The mudminnow was described in 1841 by Kirtland (1841A:277; as *Hydrargira limi*) from specimens collected in Yellow Creek, in the village of Poland, now Mahoning (in 1841 it was Trumbull) County.

Habitat—The mudminnow was present in abundance only in the undisturbed clear-water areas of basic or low gradients where the bottoms were of organic debris, muck, peat, and there was aquatic vegetation. A soft bottom seemed necessary because the species burrowed, tail first, into the bottom to hide, rest, or aestivate during droughts. It seemed to be partial to the cooler waters. It migrated in March, April and May into small streams of moderate gradients to spawn (Adams and Hankinson, 1928:386). Relict populations, obviously upon the verge of extirpation, were found in disturbed areas of turbid water which lacked aquatic vegetation and had bottoms of yellow clays.

Years 1955–80—As predicted, several northern Ohio populations were extirpated, chiefly because of channeling, ditching and dredging operations (White et al., 1975:59–61).

Two new county records are of particular interest because they are located east of the glacial boundary. On August 22, 1962, F. A. Bolin collected specimens from a drainage ditch in Lawrence Township, Tuscarawas County, and on March 21, 1969 Andrew White and Franklin A. Bolin located a population in Goodhope Township, Hocking County, in a permanent pool that was occasionally covered by Hocking River flood waters.

For relationship between the Umbridae and Esocidae, *see* Nelson (1972).

GRASS PICKEREL*

Esox americanus vermiculatus Lesueur

Fig. 33

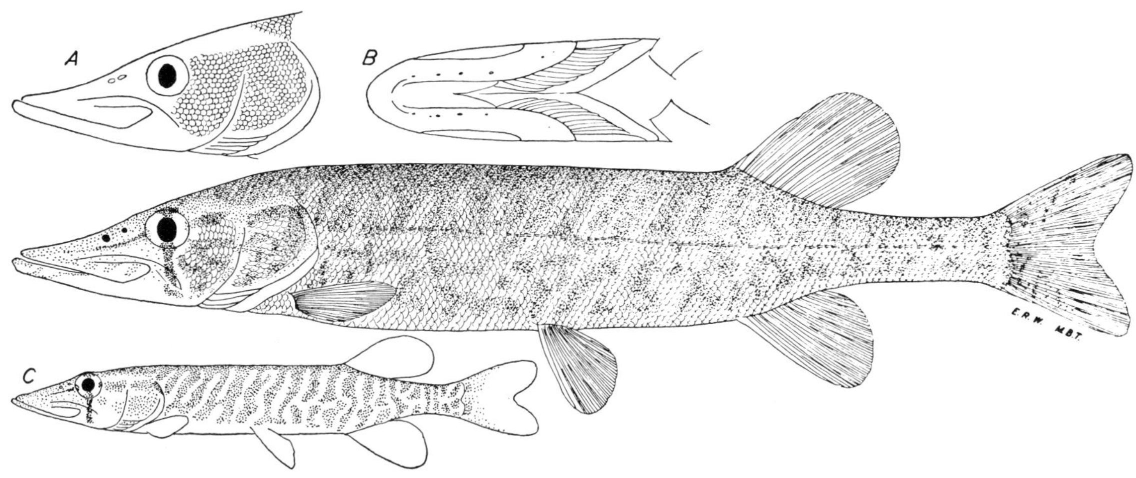

Central fig.: Chagrin River, Geauga Co., O.

April 17, 1940. 145 mm SL, 6.7" (17 cm) TL.
Adult female. OSUM 2011.

Fig. A: lateral head view showing completely scaled cheek and opercle.
Fig. B: ventral head view showing 12 branchiostegals on each side, and 4 sensory pores on each lower jaw.
Fig. C: Mosquito Creek, Trumbull Co., O.

Oct. 10, 1939. 86 mm SL, 4.1" (10 cm) TL.
Immature. OSUM 3025.

Note oblique bars on side and prominent tear-drop.

Identification

Characters: *General*—All or upper half of cheeks and opercles scaled; no scales on top of head. Tail deeply forked. No groove separating tip of upper lip from snout. Pelvic fin insertion closer to caudal base than to tip of upper lip. Dorsal fin without spines and inserted far back, mostly over anal fin base. Large canine teeth in jaws. More than 90 scales in lateral series. *Specific*—Opercles and cheeks fully scaled, fig. A. Prominent vertical tear-drop extending downward from eye. Branchiostegals usually 11–13; sensory pores on each lower jaw fewer than 7, usually 4, fig. B. Snout rather short; distance from tip of upper jaw to center of eye usually less than distance from center of eye to posterior edge of opercle. Lateral line scales usually fewer than 110. Adults usually less than 12.0" (30.5 cm) long.

Differs: Mudminnow, killifishes, topminnow, and mosquitofish have scales on top of head. Gars, bowfin, suckers and minnows have no scales on head. Pirate perch has dorsal base over pelvics. Trouts, smelt and trout-perch have adipose fins. White bass, all sunfishes, walleyes, and darters have dorsal spines. Chain pickerel has dark reticulations on sides; pike and muskellunges have half-scaled opercles.

Most like: Chain pickerel.

Coloration: Olive- or yellow-brown dorsally; sides lighter and boldly barred or mottled with darker browns; ventral surface yellowish or whitish. A very distinct tear-drop. Fins plain or vaguely blotched. No white or black spots, or longitudinal chain-like reticulations.

* Colloquial names: mud and redfin pickerel. Central redfin pickerel in 1957 edition of *Fishes of Ohio*.

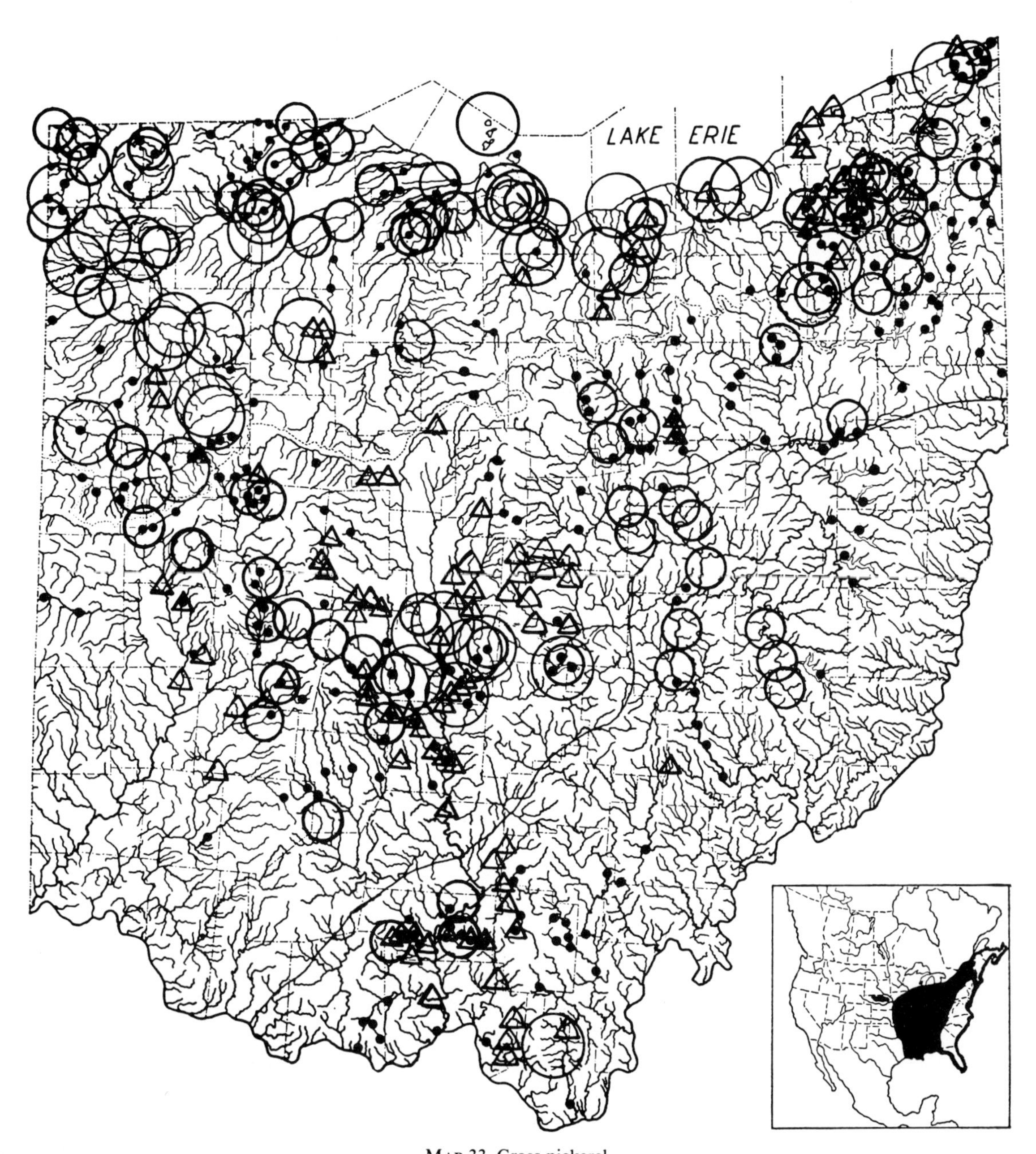

MAP 33. Grass pickerel

Locality records. ○ Before 1901. ○ 1901–21. ● 1922–55. △ 1955–80. ~ Glacial boundary.

Insert: Black area represents range of this subspecies, outlined area of other subspecies; a rather broad band of intergradation exists wherever two subspecies meet; large unoccupied areas occur in mountainous regions.

Lengths and **weights:** *Young* of year in July, 1.5″–3.5″ (3.8–8.9 cm); in Oct., 3.5″–5.5″ (8.9–14 cm) long. *Adults*, usually 5.5″–10.0″ (14–25.4 cm) long, weight 1 oz–6 oz (28–170 g). Largest specimen, 15.0″ (38.1 cm) long, weight 14 oz (397 g).

Hybridizes: with chain pickerel and northern pike.

Distribution and Habitat

Ohio Distribution—Before 1900 Ohio contained an abundance of marshes, springs, bog ponds, oxbows, and low-gradient streams which contained clear waters and much aquatic vegetation; these waters were prime habitat for the grass pickerel. It is assumed therefore that the species was formerly more numerous and more generally distributed than it has been since 1900, after the majority of such waters had been drained, or had become turbid and almost without vegetation. The former abundance and a more general distribution are substantiated by such statements as those of Kirtland (1854A:78; as *Esox umbrosus*) "Lake Erie, and some of its tributaries"; McCormick (1892:24) for Lorain County "Common: found in head waters of most streams and among the pads of spatter-docks in the bayous"; Kirsch (1895A: 530) "Common throughout the Maumee basin"; Williamson and Osburn (1898:37) for Franklin County "A very common species, especially in weedy and clear streams"; also by preserved specimens which were collected before 1900 in Cuyahoga, Lorain, Defiance, Putnam, Allen, Franklin, and Hamilton counties, and in the Ohio River (CAS 7160; Bates, one specimen).

During the 1925–50 period the largest populations in Ohio occurred in the low-gradient streams, marshes and ponds which were on the Lake Erie Ohio River watershed, from Ashland County northeastward to Ashtabula County. The largest populations in unglaciated Ohio occurred in those partly-filled valleys of the old Teays stage drainage of southern Ohio (map II) which were later flooded by proglacial Lake Tight (Wolfe, 1942: fig. 1); in this section the species was abundant locally in the vegetated backwaters, oxbows and lowland streams. Small populations were present in the low-gradient, unglaciated streams from Noble County northward to Tuscarawas County.

By 1925 the grass pickerel was no longer "common throughout the Maumee basin" as it had been in 1893, for by 1925 it had become uncommon or absent in many sections. A continued decrease in abundance occurred during the 1925–40 period, and since 1940 I have seen only one specimen taken in the Maumee River itself; however, moderate populations still were present in the sandy, vegetated "Oak Openings" tributaries of Fulton and Lucas counties. Likewise its status in Franklin County has changed from "very common" in 1897 to a few relict populations and strays in 1950.

Habitat—The largest populations inhabited clear, densely vegetated waters of base- or low-gradient streams, springs, marshes, oxbows, overflow and pothole ponds, and inland lakes whose bottoms were composed chiefly of organic debris. Its numbers decreased, or it was extirpated, from such waters whenever ditching or increased turbidity destroyed the aquatic vegetation and the formerly clear waters. Because of its tendency to migrate, occasional specimens were found in such unfavorable conditions as muddy waters flowing over silted bottoms and it was these occasional individuals which have given rise to the general misconception that this pickerel is tolerant to turbid waters and clay-silt bottoms, and to its colloquial name of "mud pickerel." There seems to be a marked interspecific competition between this pickerel and the other species of pikes, for the grass pickerel was rare or absent in any locality where another species of pike was abundant.

Years 1955–80—The grass pickerel, which inhabits in largest numbers the low-gradient or static, vegetated waters, has maintained its numbers in Ohio wherever such habitats have been largely unmolested. The species decreased in numbers, or became extirpated, wherever ditching, dredging or other forms of channelization destroyed its habitat (White et al., 1975:61). Such practices were rather general in the Maumee River drainage of northwestern Ohio, and in that area there appeared to be the greatest population decrease.

For many years prior to 1950 the western subspecies, *Esox americanus vermiculatus*, was almost universally known as the grass pickerel; but at the time the first edition of this report was published, the English name had been changed to the redfin pickerel. In the 1960 edition of *A List of Common and Scientific Names of Fishes from the United States and Canada* (Bailey et al., 1960:12) the English name was changed back to grass pickerel. Unfortunately, the eastern form, *Esox americanus americanus*, retained the name of redfin pickerel, thereby giving one species two English names.

For relationships between *Esox americanus americanus* and *Esox americanus vermiculatus* and between the Umbridae and Esocidae, *see* Nelson, 1972.

CHAIN PICKEREL

Esox niger Lesueur

Fig. 34

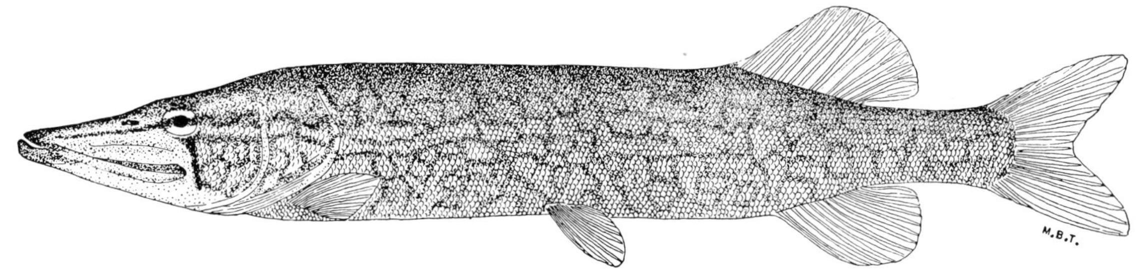

Long Lake, Summit County, O.

April 28, 1950.
Breeding female.

445 mm SL, 20.0″ (50.8 cm) TL.
OSUM 9124.

Identification

Characters: *General—See* grass pickerel. *Specific*—Similar to grass pickerel in having fully scaled cheeks and opercles, and a prominent teardrop. Young under 9.0″ (23 cm) long barred and mottled very similarly to grass pickerel; but more than 10.0″ (25.4 cm) long the bars are replaced by the distinctive longitudinal chain-like reticulations or vermiculations. Branchiostegals usually 14–16; sensory pores fewer than 7. Snout moderately long; distance from tip of upper jaw to center of eye usually greater than distance from center of eye to posterior edge of opercle. Lateral line scales usually 112–135. Adults usually 12.0″–25.0″ (30.5–63.5 cm) long.

Differs: *See* grass pickerel; it lacks dark reticulations along the sides.

Most like: Grass pickerel.

Coloration: *Adults* and *young* over 10.0″ (25.4 cm) long are olive- or yellow-green dorsally; the sides lighter, more greenish-yellow, and with dark chain-like reticulations; ventral surface milk-white often tinged with yellow. All fins unspotted. *Young* under 8.0″ (20 cm) long are barred and mottled quite like adult grass pickerel.

Lengths and **weights:** Hatchery raised *young* of year in Sept., 3.0″–8.0″ (7.6–20 cm) long. *Adults*, usually 12.0″–26.4″ (30.5–67.1 cm), weight 7 oz–3 lbs (198 g–1.4 kg).

Hybridizes: with grass pickerel and northern pike.

Distribution and Habitat

Ohio Distribution—Daniel C. Armbruster, in a letter of Feb. 19, 1952, informed me that 250 adult chain pickerels were introduced into Long Lake, Coventry Township, Summit County in the spring of 1935; also that in the fall of that year an additional 450 adults were liberated in the same lake. Since 1935 the species has been present in this lake in moderate numbers.

Habitat—The chain pickerel inhabited all depths of water in Long Lake, and was found most often about submerged brush and aquatic vegetation.

The various species of pikes of the genus *Esox* are known to hybridize with one another (Weed, 1927:10–12; Eddy, 1944:38–43). A specimen from Long Lake (OSUM:8481) appears to be a hybrid between the chain pickerel and the grass pickerel, for it has an intermediate lateral line scale-count of 110, an intermediate color pattern, and other intermediate characters. The grass pickerel was present in Long Lake in 1935 when the chain pickerel was introduced. A small population of grass pickerel was still present in 1950.

Years 1955–80—This species has continued to maintain itself in Long Lake, Summit County, where it was first introduced into Ohio. During the 1955–80 period the species was stocked in 52 water areas in 25 counties (as indicated by black triangles on the distribution map), according to Clarence F. Clark, formerly in Wildlife Research of the Ohio Division of Wildlife.

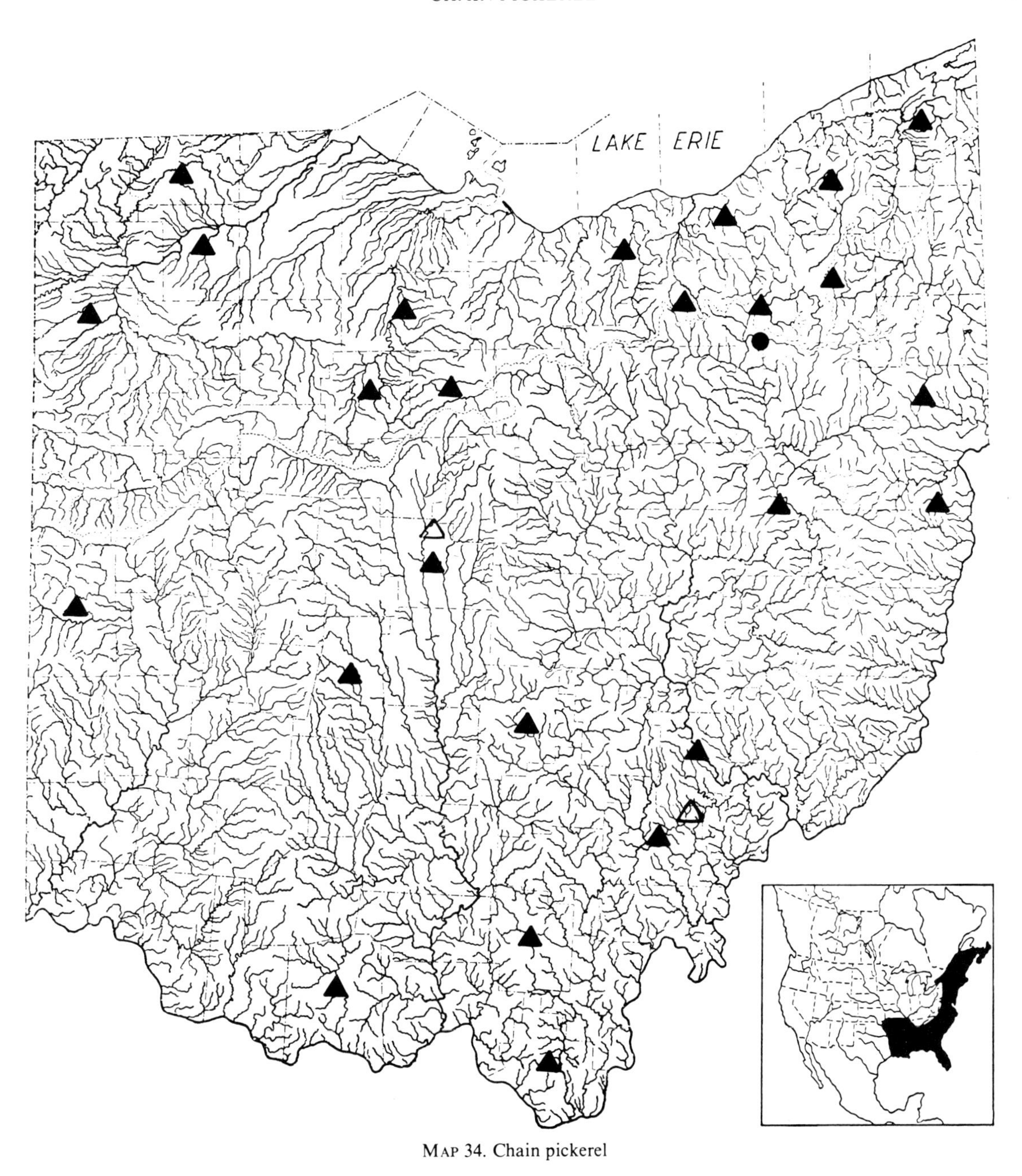

Map 34. Chain pickerel

Locality records. ● Before 1955. △ 1955–80. ▲ Counties stocked since 1954.
Insert: Many successful introductions outside its original range as far west as the state of Washington.

NORTHERN PIKE

Esox lucius Linnaeus

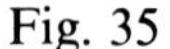

Fig. 35

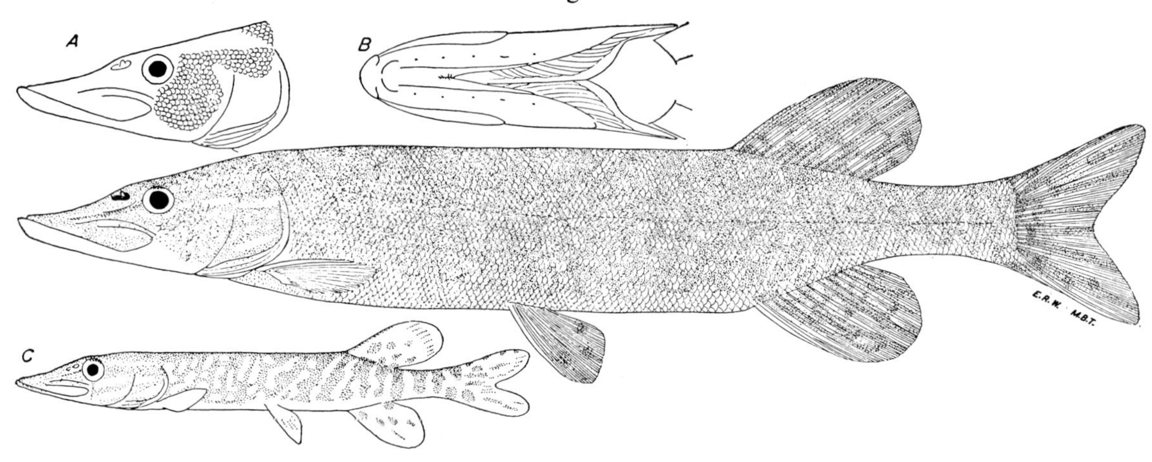

Central fig.: Lake Erie, Ottawa County, O.

Last half of March, 1942. 400 mm SL, 18.5″ (46.9 cm) TL.
Adult male. OSUM 5019.

Fig. A: lateral head view showing completely scaled cheek and half-scaled opercle.
Fig. B: ventral head view showing 15 branchiostegals on the gill cover, and 5 sensory pores on each lower jaw.
Fig. C: Portage River, Ottawa County, O.

Aug. 17, 1942. 156 mm SL, 7.2″ (18 cm) TL.
Immature. OSUM 5317.
Note oblique bars on side and lack of a prominent tear-drop.

Identification

Characters: *General—See* grass pickerel. *Specific*—Opercles scaleless on their lower halves, cheeks fully scaled, fig. A. Conspicuous white or yellow-green spots scattered over body and the vertical fins in fishes over 11.0″ (27.9 cm) long; the spots on sides tending to form oblique rows. No pronounced tear-drop. Branchiostegals usually 14–16, sensory pores on each lower jaw usually fewer than 7, fig. B.

Differs: *See* grass pickerel. Grass and chain pickerels have fully scaled opercles. Muskellunges have half scaled cheeks.

Most like: Muskellunges and pickerels.

Coloration: *Adults* and *young* more than 12.0″ (30.5 cm) are olive- or brownish-green dorsally; the sides with many white or yellow-green spots as large as pupil of eye, which tend to form oblique rows; belly yellowish-white. Vertical fins orange-yellow with bold streakings and spots of dusky. *Young* less than 11.0″ (27.9 cm) long have oblique bars, fig. C.

Lengths and **weights:** *Young* of year in Sept., 6.0″–10.0″ (15–25.4 cm). *Adults*, usually 19.0″–37.0″ (48.3–93.9 cm) long, weight 2 lbs–12 lbs (0.9–5.4 kg). Commercial fishermen give maximum lengths and weights ranging from 37.0″–48″ (93.9–122 cm) long and 12 lbs–30 lbs (5.4–13.6 kg); they claim that the larger fishes were captured before 1920.

Hybridizes: with muskellunge, chain and grass pickerel. *See* Eddy (1944A).

Distribution and Habitat

Ohio Distribution—Kirtland (1844:233–34) recorded the pike as inhabiting "Lake Erie, the Ohio and most of their Tributaries" prior to 1844; Klippart (1877:82) recorded it in 1877 as occurring in "all our streams, reservoirs and Lake Erie." The presence before 1900 of the pike in Lake Erie and its tributaries is unquestioned because of the presence of preserved specimens and reliable published ac-

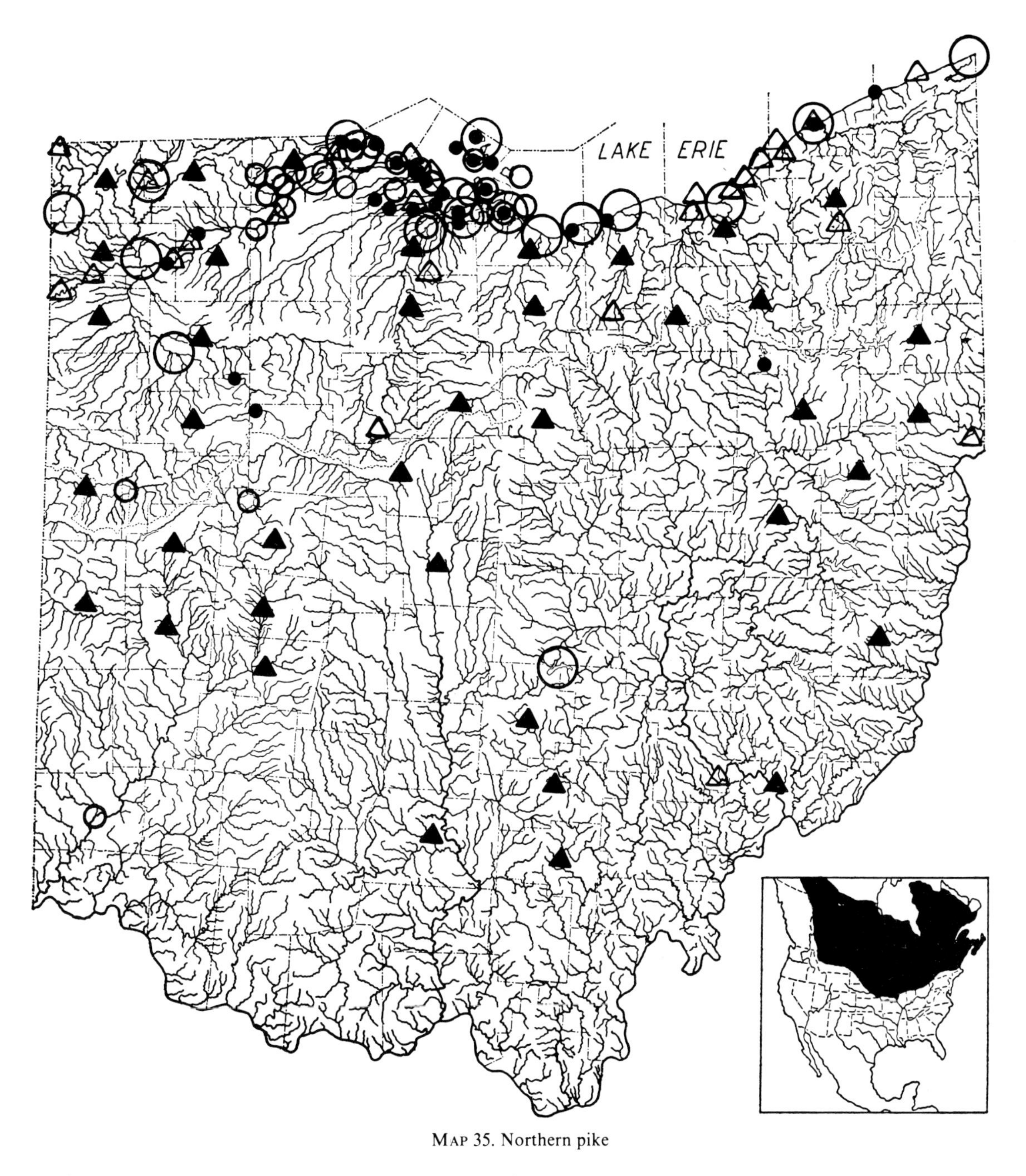

MAP 35. Northern pike

Locality records. ○ Before 1910. ○ 1910–33. ● 1943–50. △ Since 1952. ▲ Counties stocked.
Insert: Western and northern limits indefinite; widely introduced south and west of its original range. Also a native of Eurasia.

counts. The occurrence of the northern pike in the Ohio drainage streams of the southern third of the state is questioned because of lack of specimens and because the Ohio muskellunge has been, until recently, almost universally known as the pike and all except a few of the published accounts recorded the muskellunge as pike. The only unquestioned Ohio drainage records are those of Osburn (1901:72) for a pike at Buckeye Lake in 1900; records of specimens for Lake St. Marys and the Great Miami River in 1930; a 1938 record for Indian Lake. All of these records are presumably of introduced individuals, and in none of these localities has the species become established. The Summit County record is of an introduction made in 1950.

Despite lack of early, positive records for the Ohio drainage, it is assumed that the species was present before 1900 in those Ohio River tributaries which were situated on the Ohio River-Lake Erie watershed. The extensive marshes along this watershed drained into both systems during periods of high water, and especially in spring when the pikes were in these marshes for the purposes of spawning. It was therefore possible for these marsh-spawning pikes to enter the Ohio drainage. This assumption is substantiated further by the fact that the pike was formerly present in the upper portions of the Ohio River tributaries in Indiana (Gerking, 1945:76–77) and northwestern Pennsylvania (Fowler, 1919:65).

From the statements of early writers (Kirtland, 1838:194; and 1854A:78–79) and from commercial fishery reports it is apparent that before 1900 the pike was an abundant species in Lake Erie waters. Its exact abundance is unknown, however, because almost invariably the poundage of pikes and muskellunges was totaled together in the early catch reports. These reports show that the pike and muskellunge were among the very first species to attain considerable commercial importance in the Lake Erie fisheries, primarily because they could be readily seined in the shallow waters, speared on the riffles or through the ice, and caught on hook and line (Kirtland, 1851B:87). These methods of capture were the important ones prior to 1850; after 1855 the newly-introduced gill and pound nets became increasingly important as methods of capture. In 1885 the combined poundage of pike and muskellunge for the Ohio waters of Lake Erie was 263,840 lbs (119,676 kg) (Smith and Snell, 1891: 241). A marked decrease in poundage occurred after 1885. By 1900, according to Cameron King, the maximum poundage of pike from two-pound nets in Maumee Bay was 1,800 lbs (816.5 kg) and 2,200 lbs (997.9 kg); by 1910 the maximum poundage had decreased to 500 lbs (227 kg); after 1910 the decrease in poundage caught in Maumee Bay was rapid. By 1945 the total poundage for Lake Erie waters was only 3,764 lbs (1707 kg) (Shafer, 1950:1). Before 1885 the pikes and muskellunges, speared through holes in the ice or caught under the ice with seines, were stacked in wagons and transported to Cleveland, Toledo and Detroit for sale.

Many factors appear to be responsible for the decrease in abundance of the pike and muskellunge in Lake Erie waters during the 1820–1950 period. As Kirtland (1850A:1) indicated, the decrease before 1850 was chiefly caused by dams preventing migration of the fishes to their spawning grounds; by 1900 overfishing may have become a factor. Since 1900 the prime factors for the decrease in abundance unquestionably have been the preventing of the fishes from reaching their spawning grounds because of dams or polluted waters; ditching, draining and diking of streams and marshes; greatly increased amount of turbidity and rate of siltation on the bottoms with the resultant decrease in amount, or elimination of, aquatic and flooded land vegetation.

Habitat—Chiefly an inhabitant of bays, marshes, and pools of low-gradient streams where the waters were clear and there was considerable aquatic and flooded land vegetation. The species began migrating during winter and early spring, going from the deeper waters of Lake Erie into the adjacent, shallower marshes, and upstream into the headwater marshes. In such waters the species deposited its spawn about aquatic and flooded land vegetation, immediately before, during, or shortly after the ice break-up. After spawning, a migration toward deeper waters was begun.

Years 1955–80—The breeding populations of the pike in Lake Erie and its tributaries continued to decrease during this period, and those populations were eliminated wherever dredging and draining were intensive (White et al., 1975:61). The most successful spawning stream during the period appeared to be the Tiffin River in northwestern Ohio. This population is in danger of extinction in the future because the U.S. Soil Conservation Service has been planning to channelize that portion of the stream containing the best pike waters. If this plan is effected, channelization will completely destroy the natural environment, thereby eliminating this prize game fish and many other species from these waters.

During the 1952–80 period the Ohio Division of Wildlife introduced or reintroduced young into 39 counties in Ohio, stocking 74 lakes and reservoirs and two streams as indicated by black triangles on the distribution map. These stocked fishes offered sport for fishermen by their capture; however, because of the deteriorating environment throughout much of Ohio, reproduction was successful in only four counties where the species was stocked.

GREAT LAKES MUSKELLUNGE

Esox masquinongy masquinongy Mitchill

Fig. 36

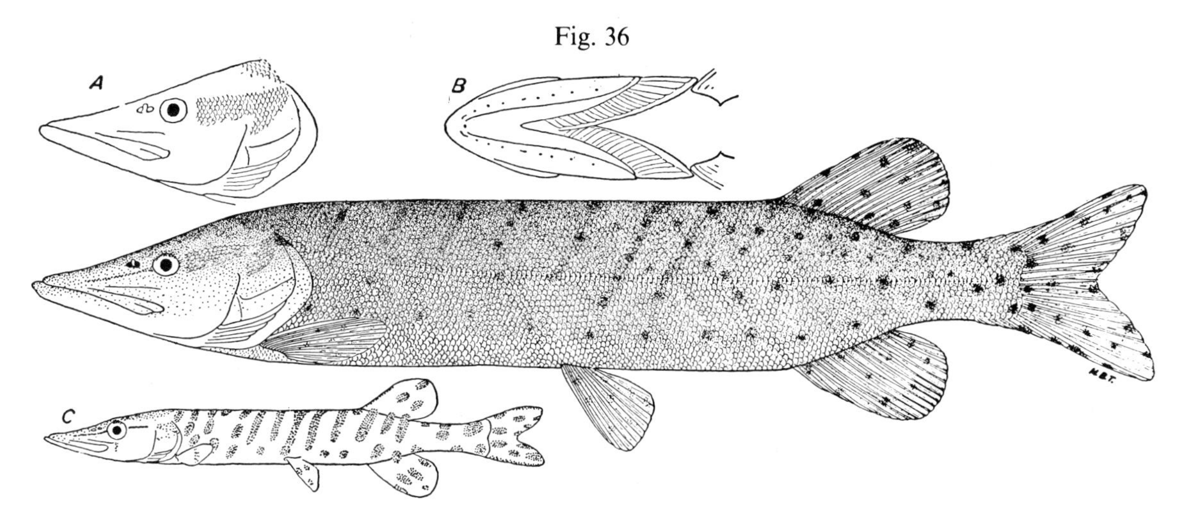

Central fig.: Sandusky Bay, Erie County, O.

Oct. 16, 1941. 640 mm SL, 29.3″ (74.4 cm) TL.
Adult male. OSUM 5031.

Fig. A: lateral head view showing half-scaled cheek and opercle.
Fig. B: ventral head view showing 18 branchiostegals on the gill cover, and 9 sensory pores on each lower jaw.
Fig. C: Pigeon Lake, Ontario, Canada.

Fall, 1940. 148 mm SL, 6.6″ (17 cm) TL.
Immature. OSUM 83.
Note the tendency for bars on side to be broken.

Identification

Characters: *General—See* grass pickerel. *Specific*—Both cheeks and opercles partly or entirely scaleless on their lower halves, fig. A. Dark or black (not white) spots scattered over part or all of the sides in fishes more than 12.0″ (30.5 cm) long; the spots tending to form oblique rows. Dark spots on vertical fins. Branchiostegals usually 17–19, sensory pores on each lower jaw usually 7–11, fig. B. *Subspecific*—Scales usually confined to upper halves of cheeks and opercles, and with little tendency to encroach downward on the anterior halves of the opercles. The dusky spots on sides tend to form rows; these rows always present on the posterior half of the fish, and often on the anterior half. Dorsal, caudal, and anal fins with dusky spots.

Differs: *See* grass pickerel; pickerels and pike have fully scaled cheeks.

Most like: Ohio muskellunge.

Coloration: *Adults* and *young* more than 12.0″ (30.5 cm) long are greenish-, yellowish-, or brownish-olive dorsally; the sides a lighter yellow-green with golden reflections, and with dusky spots which tend to form oblique rows; belly white or yellowish. Vertical fins olive or olive-green, usually with dusky spots. *Young* less than 7.0″ (18 cm) long are yellow-green dorsally, whitish ventrally, with oblique bars of olive-green on the sides, fig. C.

Lengths and weights: Fishes 15.0″–24.0″ (13.1–60.9 cm) long usually weigh 2 lbs–5 lbs (0.9–2.3 kg); 24.0″–30.0″ (60.9–76.2 cm), weight, 3 lbs, 8 oz–8 lbs (1.6–3.6 kg); 30.0″–45.0″ (76.2–114 cm), weight 5 lbs–25 lbs (2.3–11.3 kg). Commercial fishermen report maximum weights of Lake Erie specimens of 35–40 lbs (15.9–18.1 kg). For length-growth relationships in muskellunges, *see* Krumholz (1949:42–48), Williams (1948:10–11, 15) and Schloemer (1936:185–93).

Hybridizes: with northern pike.

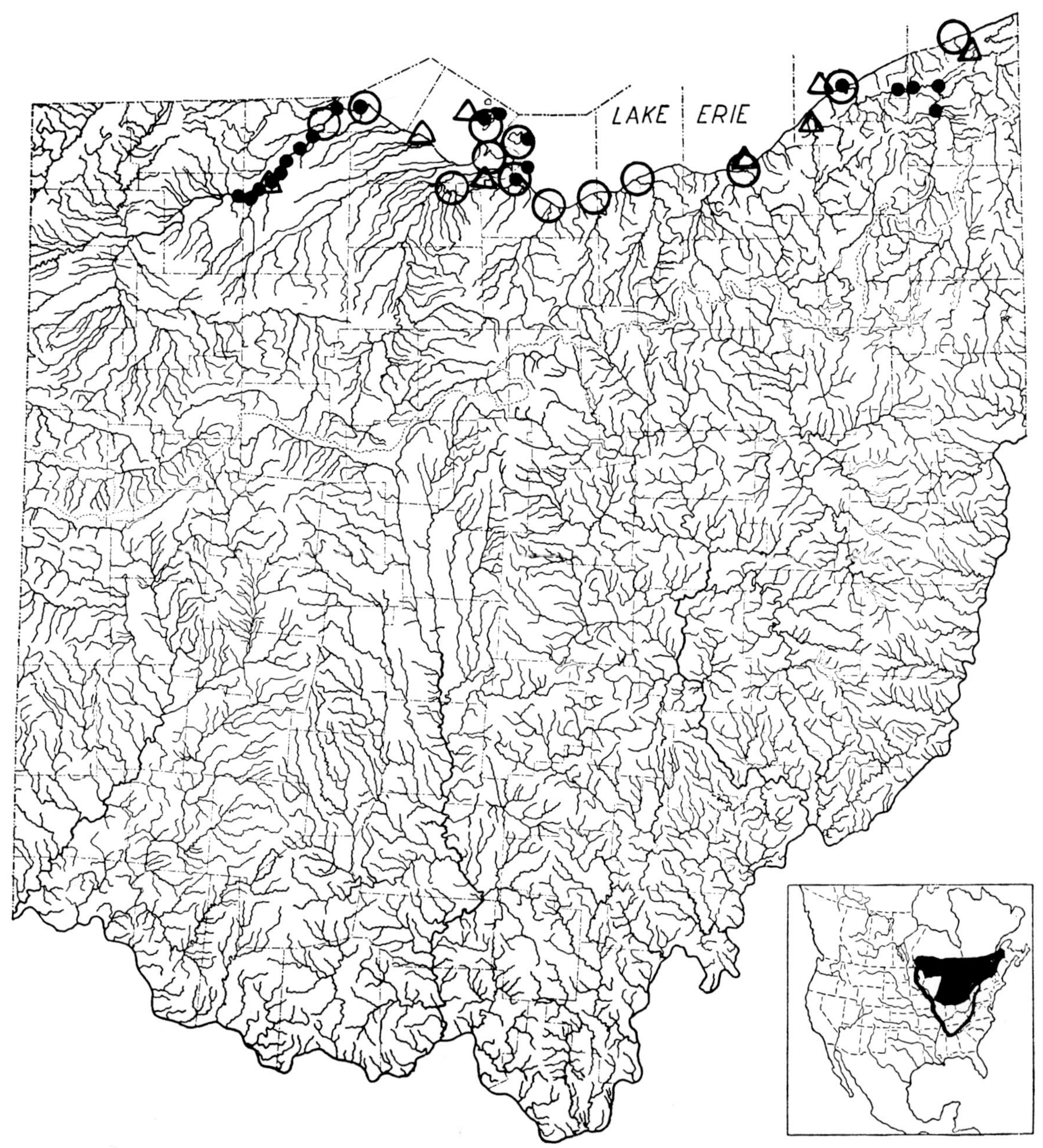

MAP 36. Great Lakes muskellunge

Locality records. ○ Before 1900. ● After 1920. △ 1955–80.

Insert: Black area represents range of this subspecies, outlined area of others; has been successfully introduced into waters inside and outside of its range.

Distribution and Habitat

Ohio Distribution—This species was abundant before 1830 in Lake Erie, its bays and some of its tributaries, and especially from Sandusky Bay westward. Klippart (1877:31) recorded it as abundant in Lake Erie in 1830; Smith and Snell (1891:266) stated that "muscallunge were also very abundant" about 1830 in Sandusky Bay.

The muskellunge was one of the first species to

become commercially important. Until 1850 the bulk of the commercial catch was taken by spearing the fishes as they passed beneath holes in the ice or swam over riffles, and by hauling seines in the marshes, streams, and shallow bays, either beneath the ice or after it had disappeared. Many were taken in late winter and spring when they ascended or descended streams. After 1855 fyke, pound and trap nets, set in the bays and open lake, became increasingly important as a method of capture. Fishing with hook and line has long been practiced, but it is one of the least effective methods of capture.

This species appears to have been one of the first of the fishes in Ohio to show a decrease in abundance. Kirtland (1851A:61) reported that "forty years since (1811) this fish was far more abundant than at present (1850)"; also (Kirtland, 1850A:1) that by 1850 the "Muskellunge has become scarce, and no longer seeks the mouth of the rivers to deposit its spawn." Confusion exists in the early fishery reports as to the relative abundance of the muskellunge and the pike because the poundage of the two species was always totaled together. After 1850 the pike was the more numerous of the two. Smith and Snell (1891:241) gave 263,840 lbs (119, 676 kg) as the total muskellunge-pike poundage that was captured in the Lake Erie waters of Ohio in 1885. In 1892 McCormick (1892:24–25) reported that in Lorain County, fishermen said that they "used to be more common than now"; Kirsch (1895A:330) in 1895 stated that "Fishermen on the lower course of the Maumee River say that formerly the maskalonge was very abundant in that stream but that now one is seldom taken there. They are also decreasing in Lake Erie." In 1941 Cameron King, for many years a commercial fisherman, told me that between 1902–05 as many as 100 muskellunges were caught in Maumee Bay in a day, but since then the species has decreased in numbers until at present it appears to be absent. Between 1930–50 I noted the continued decrease in abundance of the species. By 1950 the Great Lakes muskellunge was so reduced in numbers in Ohio waters as to be in danger of extirpation.

Habitat—Chiefly an inhabitant of Lake Erie, its bays and the deeper portions of its larger tributaries throughout the warmer months. Ascended streams and migrated into the marshes in late winter and early spring before, during, and shortly after the ice break-up, to spawn in clear, shallow waters containing much herbaceous land and aquatic vegetation, after which the adults returned to deeper waters. As late spring and summer droughts dried up the shallow waters, the fry and larger young drifted into the deeper waters of the marshes, bays and open lake. Formerly the clear streams of the Maumee system were particularly suited for spawning purposes for these low-gradient streams widely overflowed their banks in spring, submerging prairie and swamp forest vegetation and producing favorable conditions for deposition of eggs and growth of fry. Several residents of Putnam and Hancock counties have told me that their fathers or grandfathers speared muskellunges and pikes in the small, marshy creeks during early spring.

Many agricultural and urban practices have contributed to the destruction of the habitat of the muskellunge, such as the ditching of streams; draining of marshes; isolating the marshes from Lake Erie by means of dikes; increasing the turbidity of the water and the siltation rate in the stream and bay bottoms; increasing other types of pollution; damming of streams; elimination of vegetated, overflow areas. Before 1910 the intensive commercial fishery may have been a factor contributing to its decrease in abundance; however, it appears most probable that today the muskellunge population would be no larger than it is, had not a single fish been captured commercially during the past century, because of the present, almost total destruction of its habitat.

Years 1955–80—During the period this subspecies of the muskellunge has continued to decrease in abundance in the Lake Erie waters of Ohio, where only infrequently an individual was taken (Van Meter and Trautman, 1970:69; White et al., 1975: 63). Russel L. Scholl of the Ohio Division of Wildlife reports that since 1966 one was taken at Fairport, Lake County in an experimental gill net, one by Charles Muller at South Bass Island, Ottawa County, and three in Sandusky Bay in Ottawa and Erie counties. Andrew M. White reports that a fisherman caught one in 1970 west of Cleveland Harbor and in 1972 one was taken by an angler near the mouth of the Chagrin River, Lake County. Another was reported as captured in Ashtabula River, Ashtabula County. It is possible that one or more of these individuals were *Esox masquinongy ohioensis*, which subspecies has been introduced into many counties in waters tributary to Lake Erie. *See* map 37.

OHIO MUSKELLUNGE

Esox masquinongy ohioensis Kirtland

Fig. 37

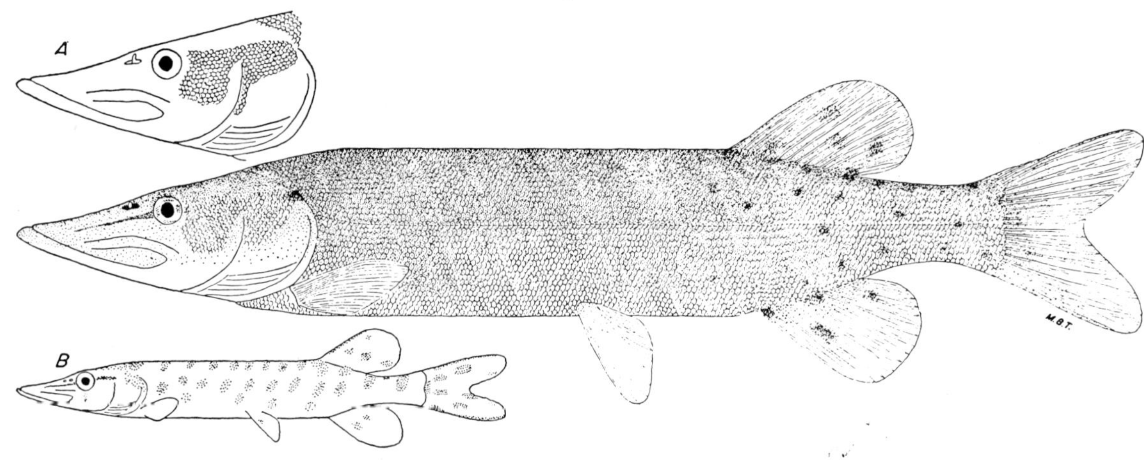

Central fig.: Shade Creek, Meigs County, O.

June 26, 1939. 510 mm SL, 24.3″ (61.7 cm) TL.
Immature male. OSUM 1101.

Fig. A: lateral head view showing half-scaled cheek and opercle; note downward extension of scales along the anterior margin of opercle, a frequent occurrence in this subspecies.

Fig. B: Crooked Creek, Pike County, O.

Aug. 20, 1940. 141 mm SL, 6.8″ (17 cm) TL.
Immature. OSUM 3011.

Note that in this small specimen the oblique bars have already broken up into spots.

Identification

Characters: *General—See* grass pickerel. *Specific—See* Great Lakes muskellunge. *Subspecific*—Scales of upper halves of cheeks and opercles often encroaching downward, the encroachment on the opercles often occurring as a vertical bar on the anterior edge, fig. A. Dusky spots on sides often confined to the posterior third of the body and the vertical fins, and usually do not tend to form oblique rows.

Differs: *See* grass pickerel. Pickerels and pike have fully scaled cheeks.

Most like: Great Lakes muskellunge.

Coloration: *Adults* and *young* more than 12.0″ (30 cm) long are greenish-, yellowish-, or brownish-olive dorsally; the sides a lighter olive-green, yellow or golden, and with some dusky spots; belly white or yellowish. Vertical fins olive or olive-green, usually with a few dusky spots. Fishes taken from turbid waters may be a light yellow dorsally, silver ventrally, and with few or no dusky spots. *Young* less than 9.0″ (23 cm) long are yellow-green dorsally, whitish ventrally, with oblique rows of spots along the sides.

Lengths and **weights:** One *young* of year in May (UMMZ 105,722; Sunfish Creek, Pike County), 31 mm SL, 1.6″ (4.1 cm) TL; one in July (OSUM 431; Scioto Brush Creek, Scioto County), 98 mm SL, 4.6″ (12 cm) TL; one in Aug. (OSUM 3011; Crooked Creek, Pike County), 141 mm SL, 6.8″ (17 cm) TL; one June fish slightly over one year old, 14.5″ (36.8 cm) TL. *Adults*—Comparable to those lengths and weights given for the Great Lakes muskellunge. Most specimens, taken by angling in recent years, weighed 5 lbs–15 lbs (2.3–6.8 kg), with 42 lbs (19.1 kg) as maximum weight. Old rivermen give maximum weights of 40–80 lbs (18.1–36.3 kg). There are many old records of "pike" of more than 40 lbs (18.1 kg) in weight. Kirtland (1854A:79) recorded that it "attains a weight of 31 1/4 lbs"; Hildreth

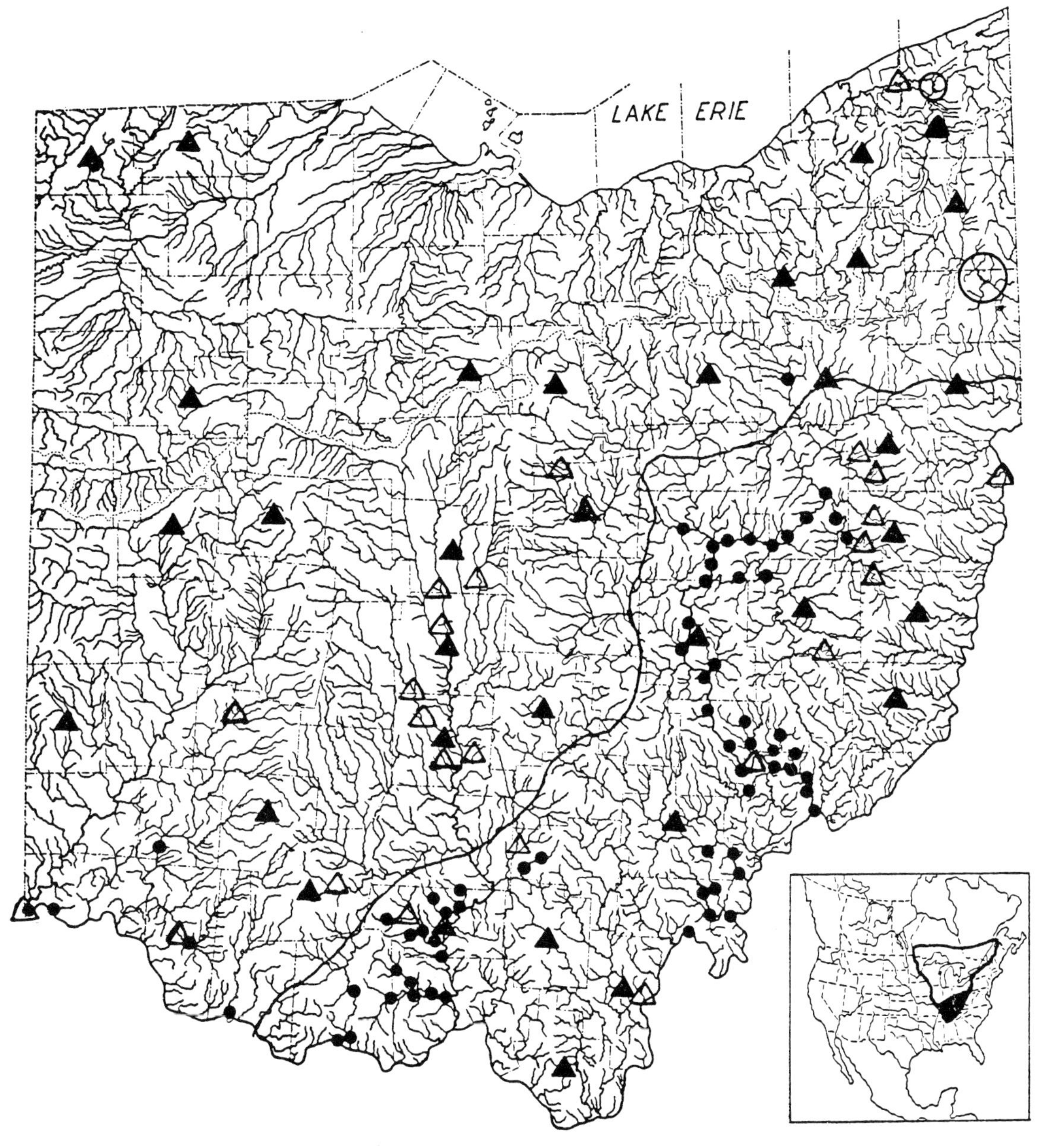

MAP 37. Ohio muskellunge

Locality records. ○ Type locality. ○ Introduced since 1930. ● After 1900. △ After 1955. ▲ Counties stocked. ~ Glacial boundary.

Insert: Black area represents range of this subspecies, outlined area of others; widely introduced, often unsuccessfully, as far west as California.

(1848:498–99) that "Judge Gilbert Devoll took a pike in the Muskingum which weighed nearly one hundred pounds, on the 2nd day of July, 1788," which was "about six feet in length."

Hybridizes: with northern pike.

Distribution and Habitat

Ohio Distribution—Hildreth (1848:498–500), while writing concerning the "Natural production of the rivers" when "Ohio was first occupied by the whites" (before 1803), stated that the pike, as this species was then called, was "the king of fish in the western waters." From his descriptions of the Indian and white man's several methods of capturing this species, it is obvious that the Ohio muskellunge was then rather widespread and abundant in Ohio drainage waters.

Kirtland (1850A:1) in 1850 mentioned that "While the tributaries of Lake Erie and the Ohio River were unobstructed by dams and were not swept by seines, they abounded with large and valuable species [including this one], which, in their vernal migrations crowded in immense shoals on the ripples." By 1850 great changes had occurred with "the finny tribes" because "All the migratory species have been excluded from the Mahoning River by the construction of dams." From such references it is obvious that in early pioneer times the Ohio muskellunge was a numerous and much prized food fish which was present in most of the larger streams of southern Ohio. However, doubt exists as to its exact abundance before 1909, because early Ohioans did not distinguish between this species and the pike (*Esox lucius*). Until recently this muskellunge was almost universally known as the "pike" despite the fact that in 1854, Kirtland (1854A:78–79) pointed out the differences between the Ohio and Great Lakes muskellunges and the pike. In 1854 he described the Ohio muskellunge, naming it *Esox ohiensis*, using as a type, a specimen from the Mahoning River.

In recent years the species has shown a steady decrease in numbers. It was fairly numerous as late as 1930, for between 1927–30 I saw at least 15 barns in Pike and Scioto counties upon whose sides were nailed the heads of 5–25 "pikes," all caught in a single season in Sunfish and Scioto Brush creeks. Today few Ohio fishermen catch more than five muskellunges a year and then only through considerable fishing effort.

The pioneers captured this species chiefly with the use of spears, guns, hooks and lines and hand-drawn seines. The majority were captured during the vernal, upstream migration and especially after the mill dams were built, which concentrated the fishes below those dams. Later, funnel nets and fish traps were used also and until 1930 an effective method was to "run set lines." These lines were 5′–10′ (1.5–3 m) long, tied at one end to a tree limb overhanging deep water, the other end having a large hook baited with a sucker or catfish, 1/4 to 1 1/4 lbs (0.1–0.6 kg) in weight, or a large sunfish. After 1930, the usual method of capture has been with rod and reel and using natural or artificial baits. In 1935 this subspecies was introduced into Lake Erie waters in Ohio when 2000 fry, obtained from the Chautauqua Hatchery in New York, were placed in the Grand River in Austinburg Township, Ashtabula County (*see* small hollow circle).

Habitat—The largest populations occurred in those sections of streams in unglaciated Ohio, such as Wills, Shade, Sunfish (Pike County) and Scioto Brush creeks, which have gradients of less than 3′/mile (0.57 m/km), have considerable aquatic vegetation, such as cow and white water lilies and cattails; have long pools often deep and narrow, and with much submerged brush and timber. This muskellunge, like all pikes, has a decided preference for clear water, sandy bottoms and considerable aquatic vegetation.

Years 1955–80—During this period, the abundance of the Ohio muskellunge appeared to have decreased or remained unchanged in the few areas in those unglaciated portions of the Scioto and Muskingum river systems which were the centers of its abundance in Ohio prior to 1950.

According to information obtained from Clarence F. Clark, the Ohio Division of Wildlife released upon one or more occasions young in 44 reservoirs and lakes and 12 streams, in 33 counties in Ohio, as indicated by black triangles on the distribution map. Fishermen caught as much as 10% of the released fishes in some reservoirs, which represents good stocking success of this much-prized trophy fish.

COMMON CARP

Cyprinus carpio Linnaeus

Fig. 38

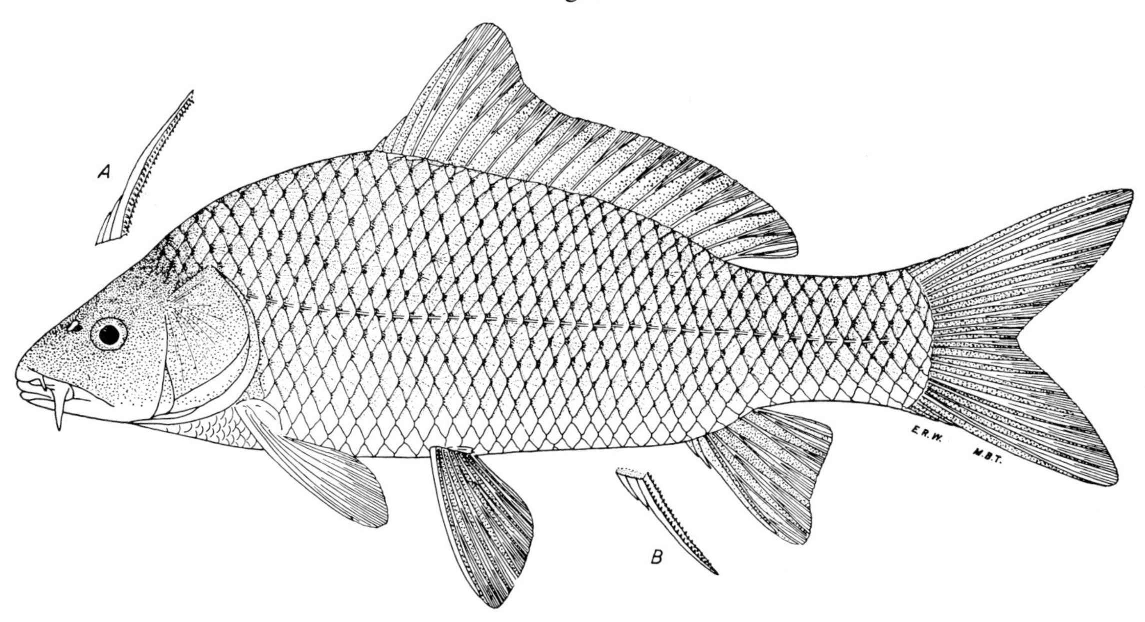

Central fig.: Lake Erie, Ottawa County, O.

April 23, 1943. 295 mm SL, 14.8" (37.6 cm) TL.
Male. OSUM 3879.

Fig. A: dorsal spinous-ray showing double row of serrations on its posterior edges.
Fig. B: anal spinous-ray showing double row of serrations on its posterior edges.

Identification

Characters: *General*—Dorsal fin with one serrated spinous-ray, fig. A and more than 16 soft rays. Anal fin with one serrated spinous ray and 5–6 soft rays. Teeth only in the throat, on the pharyngeal arches; none in the mouth. *Generic* and *Specific*—Upper jaw bearing two fleshy barbels on each side, the posterior one the largest. Complete lateral line with 32–38 scales, except in those partly-scaled or scaleless individuals known as "half-scaled," "mirror" or "leather" carp. A dark spot at the scale bases. Gill rakers on the first arch 21–27. Pharyngeal teeth in three rows of 1, 1, 3–3, 1, 1, with the three teeth in each main row molar-like. Mouth almost horizontal.

Differs: Goldfishes lack barbels on the jaw, have fewer lateral line scales. All other species of minnows and suckers have no spinous ray on dorsal and anal. White bass, blackbasses, and drum have more than one spine in the dorsal fin.

Most like: Goldfish. **Superficially like:** Buffalofishes, carpsuckers.

Coloration: Dorsally a slaty- or golden-olive. Upper sides lighter and more bronzy-golden; lower sides golden-yellow with some silvery. Ventrally a yellowish-white. Scales on back and sides each with a dark basal spot and each outlined with darker at their edges, giving a cross-hatched appearance. Generally a dark bar on body immediately behind gill cleft, and a vertical dusky bar on caudal base; both are most prominent on the small young, less noticeable or absent on adults. Dorsal and caudal fins gold-olive; remaining fins more yellowish, becoming a bright yellow-orange in breeding adults.

Lengths and **weights:** Much length variation in *young* of year in Oct., dwarfed fishes or those hatched late 0.7"–1.3" (1.8–3.3 cm); those growing rapidly

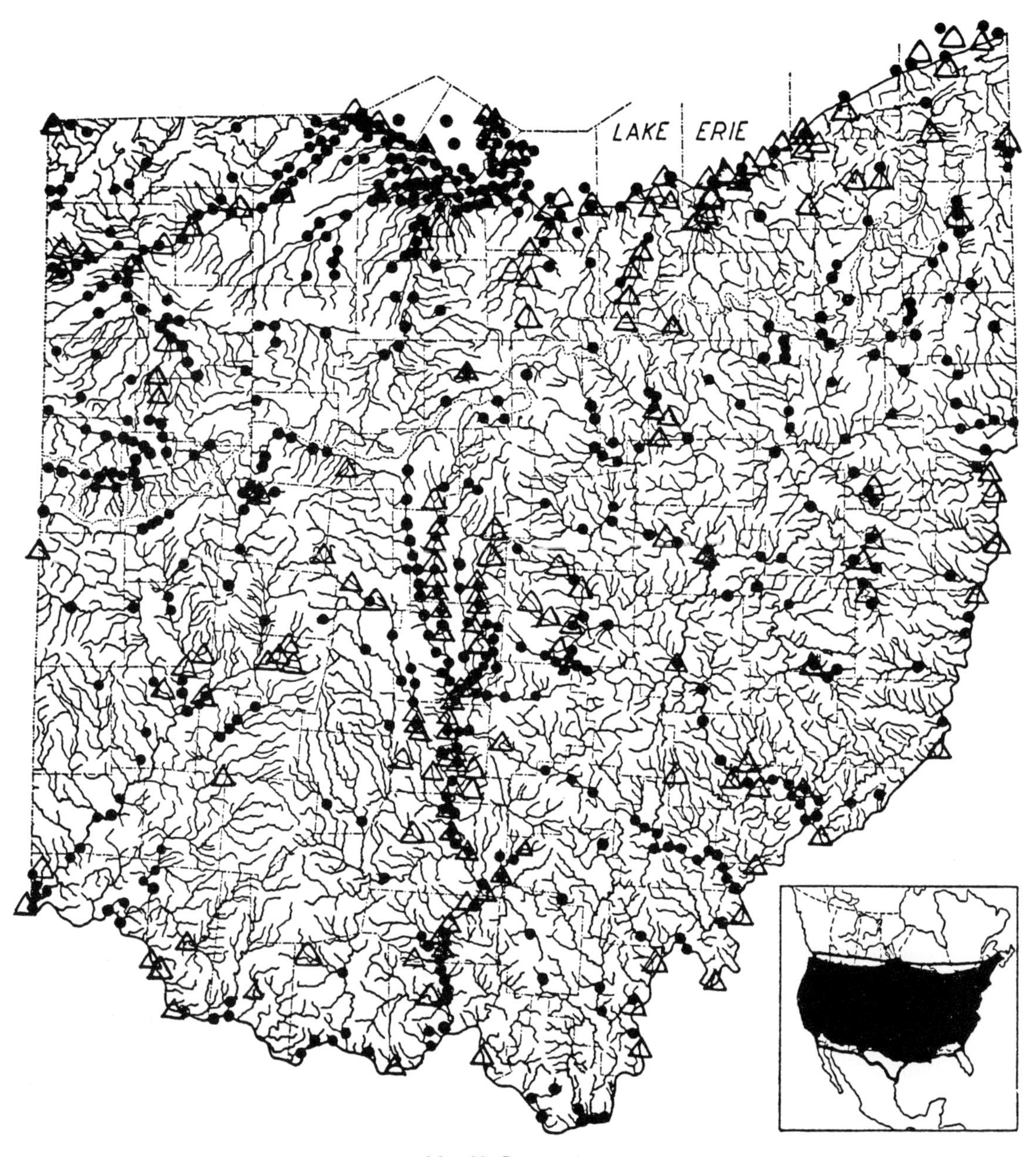

Map 38. Common carp

Locality records. ● 1880–1955. △ 1955–80.

Insert: Black area represents section where introductions were most successful, outlined area where introductions were few and/or were only locally successful.

or hatched early 4.0″–8.0″ (10–20 cm) long. *Adults,* usually 12.0″–30.0″ (30.5–76.2 cm) long, weight 14 oz–20 lbs (397g–9.1 kg). Roy Thompson weighed one, taken in western Lake Erie, of 58 lbs (26.3 kg). Commercial fishermen give maximum lengths of 40.0″–48.0″ (102–122 cm), weights 40–60 lbs (18.1–27.2 kg).

Hybridizes: abundantly with goldfish; the hybrids are fertile and back crosses abundant. In some seine hauls in Sandusky and Maumee bays,

30–90% of the carp-goldfish catch consists of hybrids, with the remainder carp, or goldfish, or both.

Distribution and Habitat

Ohio Distribution—The first carps apparently were introduced into Ohio waters in the fall of 1879, when the U.S. Fish Commission sent shipments of 16 to 39 yearlings to each of six applicants in the Cincinnati and Fremont areas (Smiley, 1884: 974–76; Cole, 1905:547). In 1880 more than 100 applicants, scattered throughout Ohio, received carps from the Federal Government, and yearly introductions on a large scale continued until at least 1886. In 1896, Ohio applicants received the last shipments from the Federal Government for stocking purposes.

On December 8, 1880, the Fish Hatchery Superintendent for the State of Ohio, Emery D. Potter (1881: 16–17), received from the Federal Government 750 mirror carps, 4.0″–6.0″ (10–15 cm) long. Distribution of carps by the state first began in the spring of 1881, when young carps were planted in the Maumee River and Ten Mile Creek, and others distributed for private pond stocking to persons in widely separated localities in the state (Howell, 1882:11). In 1882 more than 125 applicants, widely scattered throughout Ohio, received carps, and the annual distribution of thousands throughout the state continued for more than a decade.

The majority of the fishes distributed before 1885 were placed in private ponds where many escaped into the streams, some within a few days after planting them (Smiley, 1884:782–97). In 1881, carps were reported captured in Ten Mile Creek, and by 1885 were taken in streams in widely separated localities throughout Ohio. By 1890 they were well distributed throughout the state but not abundant. By 1893 Kirsch (1895A:328) found them "Very abundant in the Maumee River at Toledo, Ohio, and in west end of Lake Erie," but at that time they were not sufficiently numerous to be mentioned in the commercial catch. In 1899, 20 years after the initial stocking, 3,633,679 lbs (1,648,209 kg) were taken from Lake Erie waters. Since 1900 they have been most abundant in western Lake Erie, the Maumee and Scioto river systems, and in reservoirs throughout the state. They have been least numerous in the cooler waters of northeastern Ohio and in waters draining the lands of low fertility in eastern Ohio. Many fishermen insist that in some streams carps are not as abundant as formerly; these statements coming almost entirely from areas where land fertility has most markedly decreased or where recently the amount of organic pollution has been noticeably lessened.

Habitat—The carp was most abundant in basic- or low-gradient, warm streams, lakes, reservoirs or overflow sloughs containing an abundance of organic matter, either contributed by sewage or by biologic conversion of inorganic fertilizers from fields. The species foraged to some extent upon rooted aquatic vegetation or its by-products, but was not dependent upon vegetation or rootlets except for deposition of eggs. It was tolerant to all types of bottom, and clear or turbid waters, except in cases of excessive turbidity where proper development of the food chain was chiefly prevented. It was rare or absent in clear, cold waters, and streams of high gradients.

Years 1955–80—As observed prior to 1950, the common carp during the 1955–80 period attained its greatest numerical abundance in the static and low-gradient waters throughout inland Ohio, and especially in the enriched waters of the Maumee and Scioto river systems and in reservoirs. The species also maintained its numbers in the shallow waters of western Lake Erie and in Sandusky Bay, where it was of considerable commercial importance (Van Meter and Trautman, 1970:69), as it was to a lesser extent in the Ohio River (ORSANCO, 1962:146).

Although well distributed throughout Ohio, none or only strays were recorded in those portions of streams having the highest gradients, in waters containing much acid mine wastes or of little fertility (White et al., 1975:65).

GOLDFISH

Carassius auratus (Linnaeus)

Fig. 39

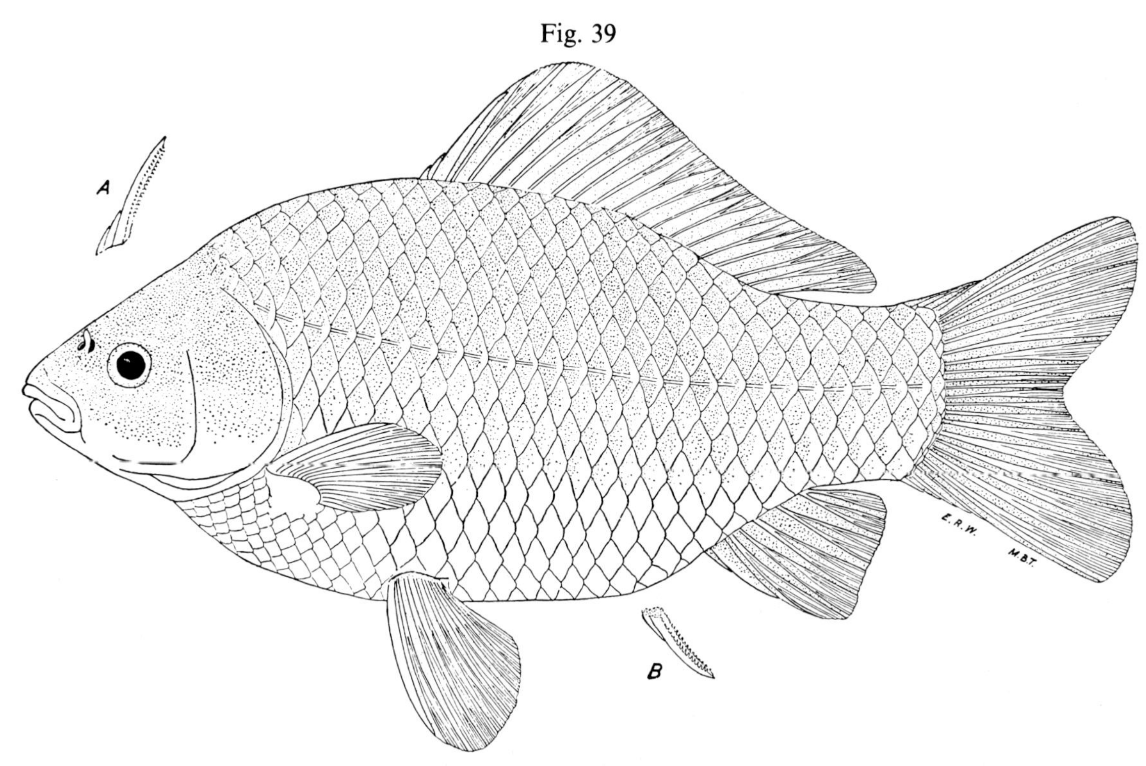

Central fig.: Lake Erie, Ottawa County, O.

April 26, 1942. 220 mm SL, 11.4″ (28.9 cm) TL.
Male. OSUM 5024.

Fig. A: dorsal spinous-ray showing double row of serrations on its posterior edges.
Fig. B: anal spinous-ray showing double row of serrations on its posterior edges.

Identification

Characters: *General—See* carp. *Generic* and *Specific*—Upper jaw without barbels; if small barbels are present the fish is a carp-goldfish hybrid or a carp. Complete lateral line with 25–31 scales. No dark spot at each scale base and no prominent cross-hatching effect. Gill rakers on the first arch 37–43. Pharyngeal teeth in one row of 4–4, and are not molar-like. Mouth rather oblique.

Differs: *See* carp.

Most like: Carp. **Superficially like:** Buffalofishes, carpsuckers.

Coloration: Some individuals are all scarlet, red, pink, silvery, brown, white, gray or black; others have a combination of two or more of these colors. Occasional "fancy" varieties are found with comet- or double-tails or pop-eyes. Most wild goldfishes have reverted to their natural coloration which is: Dorsally a slaty- or brownish-olive with a bronzy sheen. Sides lighter and more yellowish. Ventrally a yellow-white or white. Dorsal and caudal fins slaty-olive or brown, remaining fins light brownish or transparent-white.

Lengths and **weights:** Much length variation in *young* of year in Oct.; dwarfed fishes or those hatched late 0.6″–1.3″ (1.5–3.3 cm) long; those growing most rapidly or hatched early 1.4″–5.0″ (3.6–13 cm). Adults, usually 10.0″–16.0″ (25.4–40.6 cm) long, weight 12 oz–3 lbs 8 oz (340 g–1.6 kg). Individuals from small ponds spawn at a length of 5.0″ (13 cm). Specimens more than 18.0″ (45.7 cm)

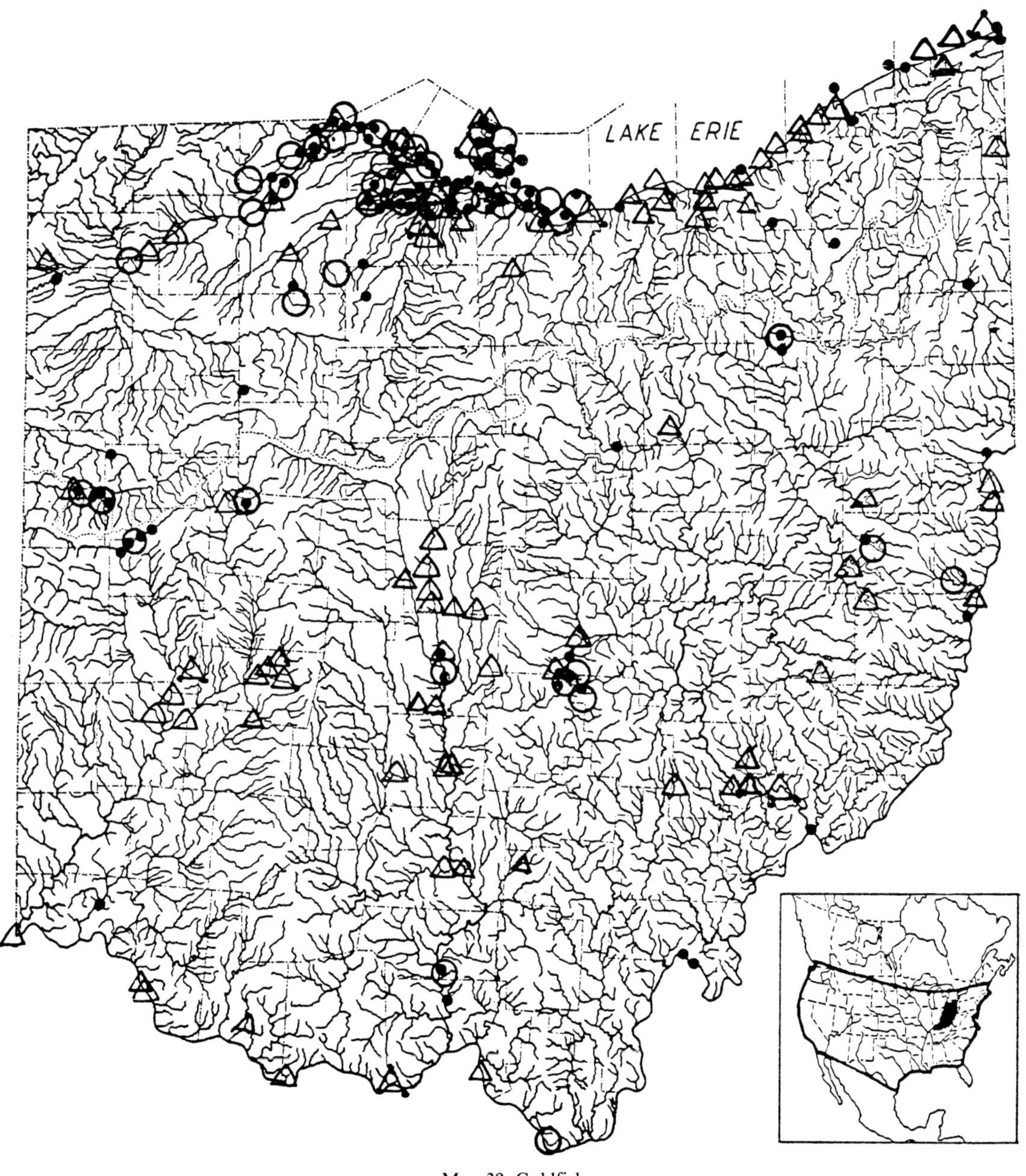

MAP 39. Goldfish

● Goldfish locality records. ○ Goldfish-carp hybrids, locality records. △ 1955–80.
Insert: Black area represents section of great population density, outlined area where introductions were fewer and/or were locally successful; widely introduced, often inadvertently, by bait fishermen throughout most of the United States.

long almost invariably show some hybrid characters.

Hybridizes: *see* carp.

Distribution and Habitat

Ohio Distribution—The first definite record of the exotic goldfish for the public waters of Ohio was in 1888, the species having escaped from private ponds into the canal basin near Elmwood, Hamilton County (Henshall, 1888:79). It is possible, however, that the species had escaped or had been introduced into Ohio waters before this, for as early as 1850 Kirtland (1850D:29) suggested that the goldfish be introduced into ponds and streams; likewise it was an ornamental fish for many years previous to 1888. In 1885–86, 11 Ohio applicants received goldfishes from the federal government for stocking (McDonald, 1887:393).

Despite its wide distribution and early introduction, the goldfish at present is not as widely abundant over the state as is the carp; for the goldfish is abundant only in the shallower waters of western Lake Erie, in the low-gradient streams tributary to western Lake Erie, and in a few larger, inland reservoirs. Hybrids may be expected wherever the carp and goldfish occur together, and in some areas in and near Lake Erie the number of hybrids is greater than the number of both parent species.

Habitat—The goldfish has about the same ecological requirements as has the carp, but the former species appears to be more specialized. The goldfish seems to be less tolerant to moderate or high gradients, cool water, great turbidity, and rapid siltation, domestic and industrial pollutants. It often reaches greater numbers than does the carp in shallower water and denser aquatic vegetation and seems to be more dependent upon dense vegetation.

Years 1955–80—Throughout this period goldfishes in Ohio were most numerous in the western basin of Lake Erie, including Sandusky Bay, where they were of some commercial value (Van Meter and Trautman, 1970:69; White et al., 1975:65). In inland Ohio the species has been recorded in an additional 20 counties (*see* map 39) and has increased considerably in numerical abundance in some inland localities. Introductions and reintroductions into Ohio waters continued at an increasing rate, especially in those localities where goldfishes were raised for forage, bait or aquarium purposes. Some aquarium fishes were released into Ohio waters by individuals.

The goldfish-carp hybrid population was large in the western Lake Erie region. In inland Ohio we captured upon several occasions one or more hybrids in a given locality before a "pure" goldfish was taken. Later in the same locality "pure" goldfishes usually outnumbered the hybrids. Apparently, the first goldfishes to invade an area did not or could not find other goldfishes with which to mate, and therefore mated with carps. The hybrid population thus produced was sufficiently numerous, apparently, so that one or more could be captured. Later when the goldfishes in an area became sufficiently numerous, they spawned among themselves, after which we could capture "pure" goldfishes. *See* orangespotted sunfish.

During the 1957–59 investigations of the entire Ohio River (ORSANCO, 1962:146) by William M. Clay and his associates, the goldfish was taken in 27 of the 341 collections.

GOLDEN SHINER

Notemigonus crysoleucas (Mitchill)

Fig. 40

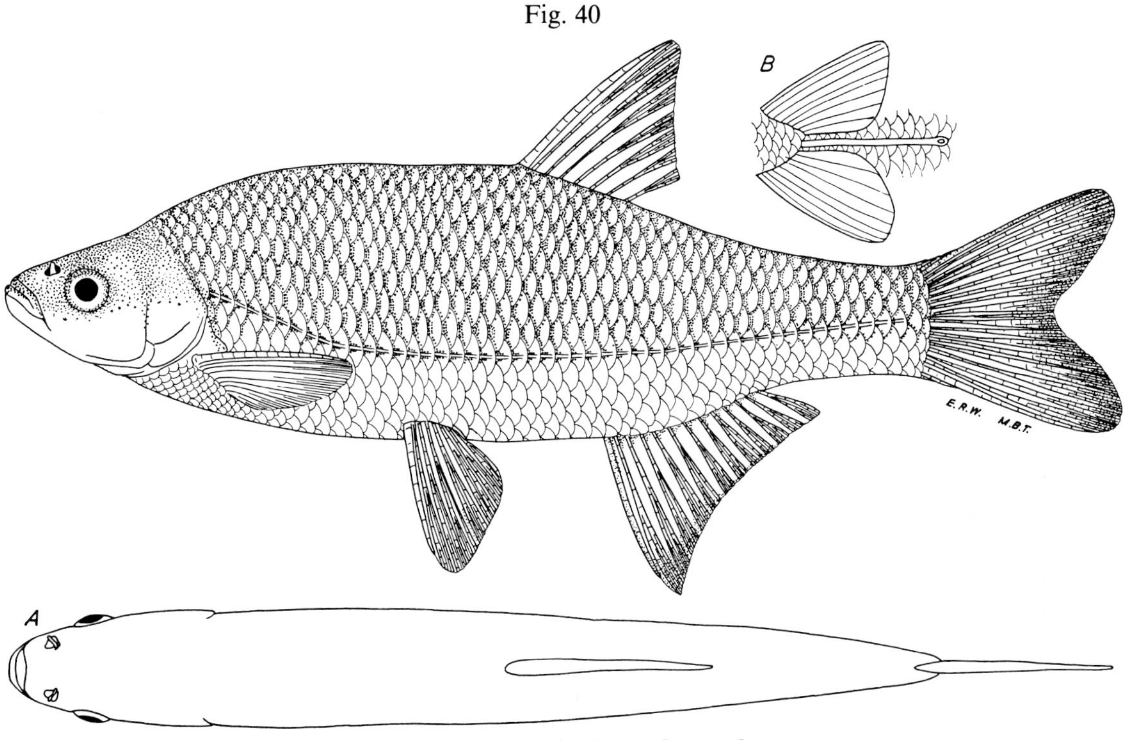

Central fig.: Lake Erie, Ottawa County, O.

Feb. 4, 1942. 135 mm SL, 6.4″ (16 cm) TL.
Male. OSUM 6730.

Fig. A: dorsal view showing the narrow width of the body.

Fig. B: ventral view from pelvics to anus; note the naked strip along midline of belly over which the scales do not cross. No other species of minnow in Ohio has this character.

Identification

Characters: *General*—A single, spineless dorsal fin of 8 developed rays, *except* in the pugnose minnow which has 9, and in the bullhead, fathead and bluntnose minnows which have a rather stout half-ray immediately preceding the first developed ray. Teeth on the pharyngeal arches in one or two rows, the main row of each arch with 4–6 teeth only. No teeth in mouth. Cycloid scales. Pelvics abdominal. Body scaled. Head without scales. Belly never serrated. No adipose fin. No dorsal, anal, or pelvic spines. No pelvic axillary process. No adipose eyelids. *Generic* and *Specific*—Midline of belly from pelvic fins to anus naked, the scales not crossing this naked strip, fig. B; in small young this character may be difficult to see. Anal fin rather deeply falcate, especially in adults, and usually contains 11–13 rays, extremes 10–5. Complete lateral line greatly decurved anteriorly, of 44–54 scales. Body decidedly slab-sided, fig. A. Coloration of large young and adults a distinctive golden-yellow. Teeth, 5–5, and hooked.

Differs: No other species of minnow in Ohio has a naked strip from pelvics to anus. Suckers have more than 8 dorsal rays. All species of herrings, mooneyes, gizzard shad, and whitefishes have adipose eyelids.

Most like: Redfin, rosefin, and emerald shiners.
Superficially like: Common and steelcolor shiners.

Coloration: Dorsally golden underlaid with olive-green. Sides more golden and with some

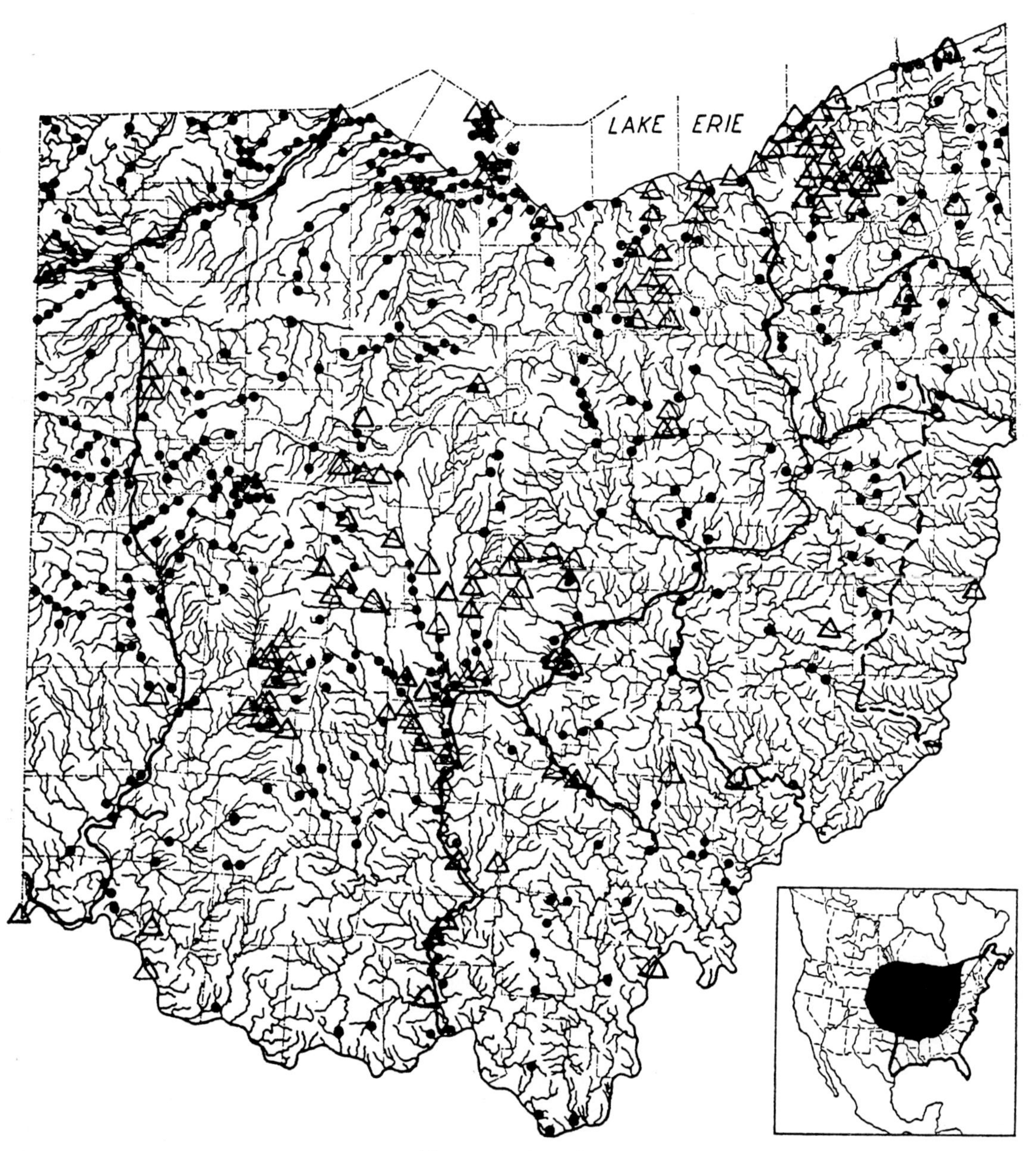

Map 40. Golden shiner

• Before 1955. △ 1955–80. ~ Canals. ⁄ \ Flushing Escarpment.
Insert: Black area represents range of this subspecies, outlined area of another subspecies (Bailey, Winn and Smith, 1954:123–24); widely introduced westward, often inadvertently by bait dealers and fishermen (Miller, 1952:32–33).

silvery reflections. Ventrally yellowish or yellow-silvery. Fins light olive, yellow, or in the small young, silvery. *Young* less than 3.0″ (7.6 cm) long generally have a dusky lateral band; this band becoming faint in larger young and obsolete in adults. Smallest young have a silvery appearance similar to most species of shiners. *Large adults* are often a dark golden.

Lengths and **weights:** *Young* of year in Oct., 0.7″–4.0″ (1.8–10 cm) long. *Adults*, usually 3.0″–7.0″ (7.6–18 cm). Largest specimen, 10.5″ (26.7 cm) long, weight 12 oz (340 g).

Distribution and Habitat

Ohio Distribution—Rafinesque (1820:92) and later writers indicated that until 1900 the golden shiner was widely distributed and locally abundant in Ohio. Until 1935 it was most abundant in the low- or base-gradient, less turbid and usually weedy waters of northwestern Ohio, and in the waters draining the prairies of west-central Ohio. It was least numerous in the two tiers of counties bordering the Ohio River, particularly avoiding the high-gradient streams east of the Flushing Escarpment.

Since 1900, marked decreases in abundance, and extirpation in small sections have occurred, particularly in the harbors, marshes and tributary streams of western Lake Erie, including the Maumee River system, and in the prairie areas of west-central Ohio. As an illustration: in 1898, Osburn and Williamson (1898:13) recorded it as "generally common" in the waters of the western half of Franklin County; but between 1926–32 only several small populations remained and since 1940 only strays, or relict populations, have been found. These decreases in abundance obviously resulted from destruction of their habitat, chiefly through drainage, increased turbidity, siltation, industrial and domestic pollutants. On the other hand, invasion or introduction of golden shiners into the many newly constructed farm ponds and larger impoundments, especially in eastern Ohio, has resulted in many recent, local increases in abundance. During the past century there has been a gradual shifting of abundance from low-gradient, sluggish, clear streams and marshes to headwater impoundments and farm ponds.

Habitat—The largest populations occurred in the quiet waters of low- or base-gradient streams, and in sloughs, oxbows, canals, ponds, and lakes having relatively clear waters, bottoms chiefly of organic debris and/or sand, and much aquatic vegetation. When large populations were found in turbid, weedless waters, the turbidity was almost invariably caused by a large planktonic pulse. Small populations occurred in weedless waters which infrequently were silt laden and had a clayey silt bottom; but only strays or relict populations occurred in habitually silt-laden waters, and especially where silt deposition on the bottom was rapid. Formerly the canals may have been an important factor in Ohio distribution, particularly southward, for the abandoned remnants of canals which were weedy and had clear waters still contained large populations.

Years 1955–80—The golden shiner remained rather well distributed in Ohio despite marked decreases in abundance or extirpation of former populations throughout the state. Despite efforts to obtain more individuals, fewer than seven specimens were taken in each of 36 localities, and in only four localities during the period were more than seven individuals obtained. Observations indicate that the species was on the verge of extirpation in many areas, caused primarily by modifications of the environment through channelization, dredging, ditching and destruction or elimination of the aquatic vegetation (White et al., 1975:65–66).

During the period increasing numbers of this species were propagated throughout Ohio for use as bait, for introductions into farm ponds or other waters and as food for fishes in hatcheries. Some propagated golden shiners escaped into nearby waters, but only in a few instances did this result in the establishment of a population.

HORNYHEAD CHUB

Nocomis biguttatus (Kirtland)

Fig. 41

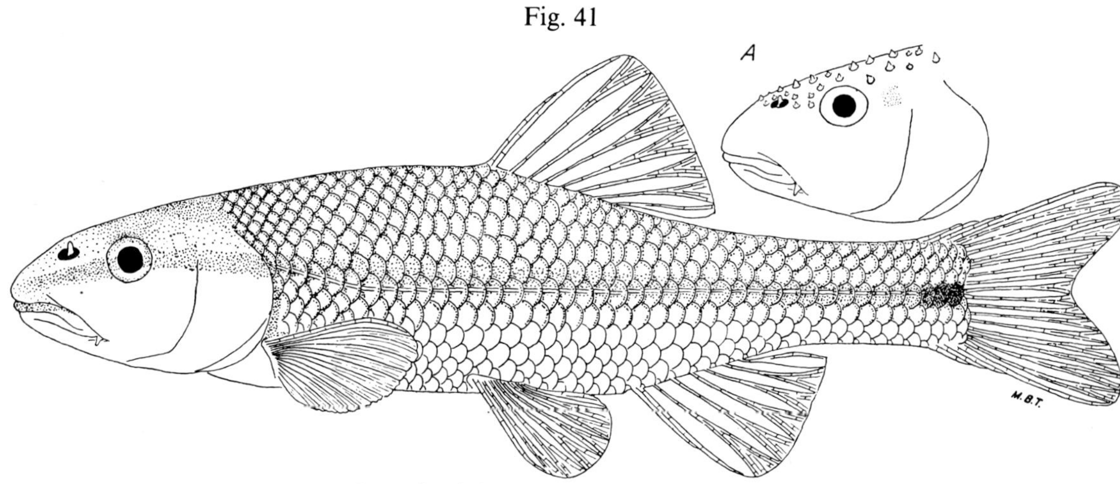

Lower fig.: Cold Run, Columbiana County, O.

May 28, 1941. 130 mm SL, 6.2" (16 cm) TL.
Breeding male. OSUM 3212.

Fig. A: lateral view of head; note barbel at posterior angle of mouth and in this breeding male, the tubercles and squarish post-ocular spot. Same fish as above.

Identification

Characters: *General—See* golden shiner. *Generic*—A small barbel at posterior end of upper jaw, fig. A. Tip of upper jaw separated from snout by a deep groove. Scales in the complete lateral line 57 or fewer. Teeth 1,4–4,1 or 4–4, or some combination of these. Constructs mound nests; males have large cephalic breeding tubercles; breeding coloration pink-rosy, orange and bluish. *Specific*—General coloration brownish. Eye small, contained 1.0 (small young)–2.2 (large adults) times in length of upper jaw. Mouth almost terminal, large and slightly oblique. Snout rather short, the shortest distance (least suborbital width) between the lower, bony edge of eye and the bony edge of snout (does not include groove or upper jaw) usually contained 2.0–2.5 times in the postorbital length (distance from posterior edge of eye to posterior edge of operculum membrane); likewise, the distance from tip of snout to anterior edge of eye is usually contained 1.2 (large adults)–1.7 (small young) times (avg. 1.3) in the postorbital length. Dusky or brownish spot at caudal base very distinct in young, faint or absent in large adults. Caudal fin dark orange or reddish in young. Breeding male has tubercles on head extending as far back as nape (fig. A). Breeding adults have a squarish red or carmine spot behind eye (fig. A). Teeth, usually 1,4–4,1; occasionally a tooth on an inner row is missing.

Differs: River chub has a longer snout. All other minnows with barbels have a larger eye or more lateral line scales. Remaining species of minnows and all suckers lack barbels.

Most like: River and silver chubs. **Superficially like:** Creek chub and tonguetied minnow.

Coloration: Dorsally a dark olive- or brownish-yellow with light coppery reflections. Sides light olive and more yellowish. Ventrally pale yellow or white. Scales of back and sides with darkened free edges, giving a cross-hatched appearance. A dusky band, encircling snout, crossing opercles and sides of body to tail, usually prominent in young, less so or obsolete in adults. Caudal spot most distinct in young. Dorsal and caudal fins light olive or brownish, the caudal often flushed with reddish, especially in the young; remaining fins faintly olive or transparent. Heads of *breeding males* and upper

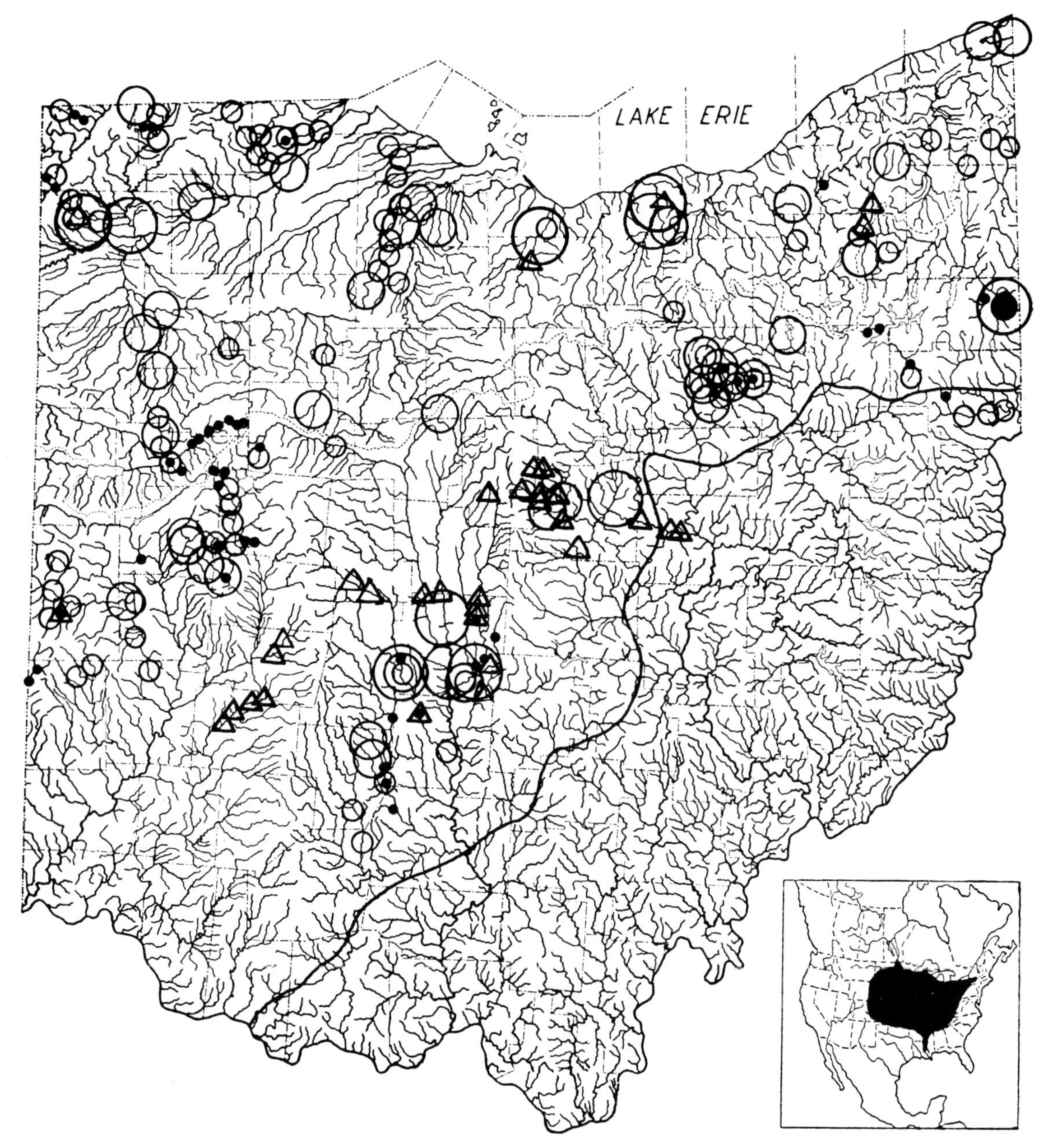

MAP 41. Hornyhead chub

Locality records. ○ Before 1901. ○ 1901–24. ○ 1925–38.
● 1939–52. △ 1955–80. ● Type locality, Yellow Creek, Mahoning County.
~ Glacial boundary.
Insert: Unoccupied areas within range appear to be increasing in size and number because of destruction of habitat.

part of the body suffused with pinks, rose and blues, the tubercles a pallid white.

Lengths and **weights:** *Young* of year in Oct., 1.0″–3.0″ (2.5–7.6 cm) long. *Adults*, usually 4.0″–7.0″ (10–18 cm). Largest specimen, 9.3″ (24 cm) long, weight 6 oz (170 g). The largest fishes almost invariably are males as is customary with all species of minnows in which the male builds a nest and guards the eggs. A smaller species than the river chub. *See* Lachner (1946:209–10).

Hybridizes: with common and striped shiners, stoneroller minnow.

Distribution and Habitat

Ohio Distribution—References before 1925 to the hornyhead chub in Ohio are mostly a composite of this species and the river chub, for it was not until 1926 (Hubbs, 1926:28–30) that both were recognized as distinct species. Literature records comprising both species (Jordan, 1882:861–62; McCormick, 1892:20; Kirsch, 1895A:329 "in every stream examined;" Osburn, 1901:64) indicate that before 1900 the two species, lumped together, were abundant and widespread in distribution, and were extensively used for bait and food. Preserved specimens of the hornyhead from the Maumee, Huron, Black, Scioto, and Mahoning river systems, and reliable literature records (Kirtland, 1841B:344–45, description of species, type locality, Yellow Creek, Mahoning County, Ohio; Parker, Williamson and Osburn, 1899:27) demonstrate its widespread distribution before 1900 throughout glaciated central and northern Ohio.

Between 1925–50 the hornyhead was found throughout glaciated Ohio, except in the extreme southwestern part. It penetrated unglaciated territory only for a few miles in Columbiana County, in the glacial outwash streams which were in reality glaciated. Throughout the 1925–39 period there was a marked decrease in abundance in the Maumee system and streams tributary to Sandusky Bay; after 1939 none was found in the Maumee River itself or in the Sandusky Bay tributaries. Between 1925–30 the hornyhead was numerous in Big Darby Creek, near Fox, Pickaway County, after which it decreased in abundance until 1934 when the last specimen was taken, despite yearly seinings since. By 1950 the species occurred in more or less relict populations in rather widely separated localities.

Habitat—The largest populations occurred in small or medium-sized streams having a moderate or sluggish current, a sandy, gravelly bottom, and usually where there was some aquatic vegetation such as attached algae, pondweeds, lizard's-tail, turtlehead, and water-willow. Clayey silts were tolerated if they did not entirely prevent plant growth, and cover the stones and boulders with silt so that there was a decrease in abundance of the aquatic insect larvae. The hornyhead preferred smaller-sized streams than did the river chub, and as indicated by its more westerly distribution, appeared to be more tolerant of sluggish currents, turbidity and silted stream bottoms. For food habits and biology, *see* Lachner (1950:229–36; 1952:433–66).

Years 1955–80—Throughout most of the state the populations of this inhabitant of clean sand and gravel substrates continued to decline in abundance (White et al., 1975:68), except for a few localities which appear unique. One of these was Big Walnut Creek in northern Franklin County, immediately below the Hoover Reservoir Dam. This reservoir acted as a settling basin for silt and organic matter except during brief periods of exceptionally high water. Also, the dam contained three outlets, one near its top, another in the middle and the third, which was more frequently used, near the bottom. Water drawn from the lower outlet was usually less turbid than that taken from near the top or when water overflowed the dam lip. After the dam began functioning, it became evident that for a few miles downstream the sand and gravel substrate contained less silt than formerly. Shortly thereafter, a marked increase in the numbers of hornyhead chubs occurred in this section below the dam. During the central Ohio stream surveys in the 1960's my assistants and I found a goodly population of hornyhead chubs below the reservoir dam as did Cavender and Crunkilton (1976:113) in the 1970's. A somewhat similar condition occurred below O'Shaughnessy Dam on the Scioto River. This section contained less silt, but since the stream system drained some of the richest, most heavily fertilized cornlands of Ohio and contained considerable sewage enrichment, the boulders, gravel and sand became heavily coated with diatoms during colder weather and much algae during warmer. During this period only one specimen was collected below this dam, in contrast to the abundance below Hoover Dam.

Another locality where this species maintained its numbers or increased in abundance was the Kokosing River system in Knox County. This system drains a rolling terrain containing much meadow and wooded land and comparatively few heavily cultivated fields. The larger populations of chubs were in those headwaters where there was clean sand and gravel.

During the 20-year period no stream in Ohio was seined so frequently or so intensively as was the lower Big Darby system, yet only one specimen was collected, that on July 4, 1962, in Muhlenberg Township, Pickaway County.

For many years prior to 1951 the hornyhead and river chubs were in the genus *Nocomis* (Osburn, Wickliffe and Trautman, 1930:172). Bailey in 1951 (192) merged *Nocomis* and other genera in the genus *Hybopsis*, where they remained for several years. Recent studies have demonstrated that the hornyhead and river chubs are sufficiently distinct from other chubs to warrant reinstating them in the genus *Nocomis*; *see* Davis and Miller (1967:1–37) on brain and gustatory structures; Reno (1969:771) on the cephalic lateral-line system; Jenkins and Lachner (1971:3–11) on interpretation of *Nocomis* and *Hybopsis*. The hornyhead therefore becomes *Nocomis biguttatus* (Kirtland).

RIVER CHUB

Nocomis micropogon (Cope)

Fig. 42

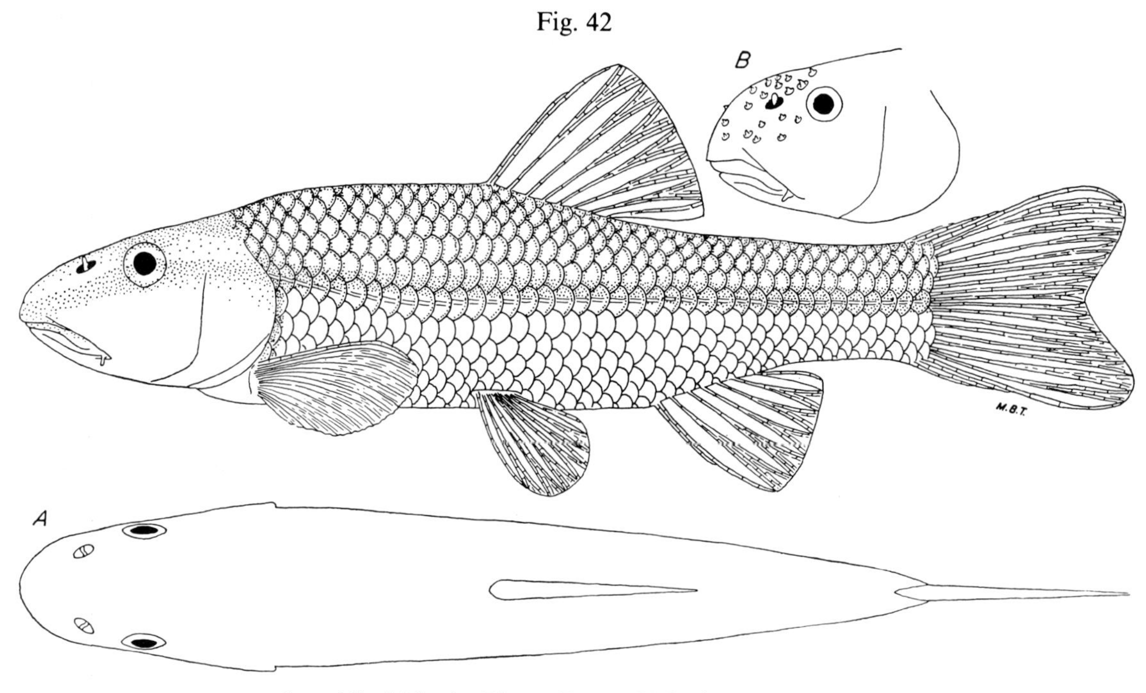

Central fig.: Mahoning River at Portage-Mahoning County line, O.

Nov. 17, 1939. 130 mm SL, 6.4″ (16 cm) TL.
Female. OSUM 1304.

Fig. A: dorsal view; note body shape.

Fig. B: Same collection as above but another specimen.

Male. 165 mm SL, 7.7″ (20 cm) TL.

Lateral view of head showing the swollen snout and tubercles of the male.

Identification

Characters: *General—See* golden shiner. *Generic—See* hornyhead chub. *Specific*—General coloration brownish. Eye small, contained 1.0 (small young)–2.2 (large adults) times in length of upper jaw. Snout overhangs the large, slightly oblique mouth. Snout rather long, the shortest distance (least suborbital width) between the lower, bony edge of eye and the bony edge of snout (does not include groove or upper jaw) usually contained 1.3–2.0 times in postorbital length (distance from posterior edge of eye to posterior edge of operculum membrane); likewise, the distance from tip of snout to anterior edge of eye is usually contained 0.8–1.2 (ave. 1.0) times in the postorbital length. Caudal spot usually faint or absent; prominent only in small young. Caudal fin slatish or olive, and is reddish or orange only in fishes living in polluted waters. Breeding male has the tubercles on the head restricted to the snout and interorbital region, fig. B. No carmine spot behind eye of breeding adults. Teeth 4–4.

Differs: *See* hornyhead chub. Hornyhead has a shorter snout.

Most like: Hornyhead and silver chubs. **Superficially like:** Creek chub and tonguetied minnow.

Coloration: Similar to hornyhead chub, except

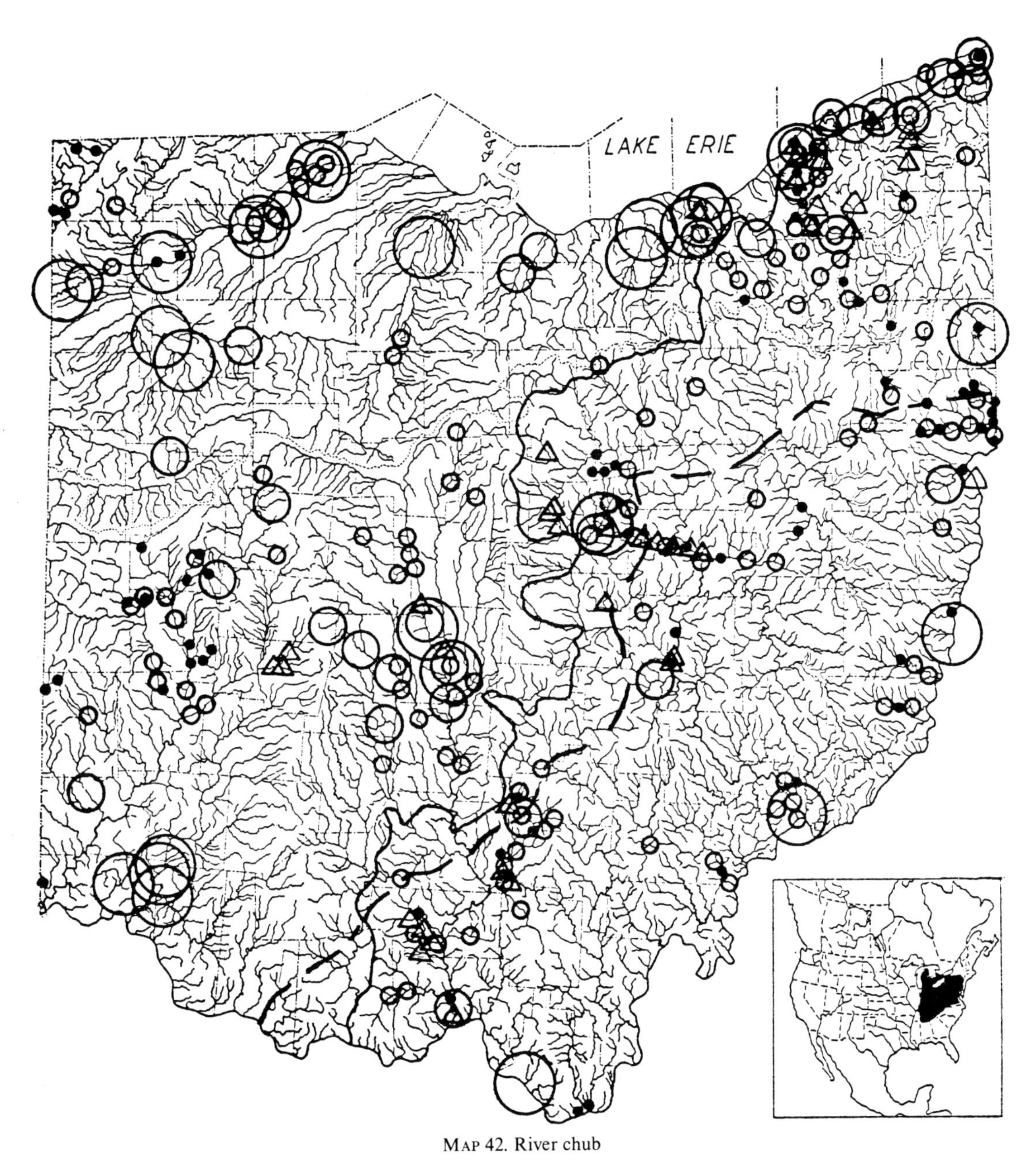

MAP 42. River chub

Locality records. ○ Before 1901. ○ 1901–24. ○ 1925–37. ● 1938–52. △ 1955–80.
~ Allegheny Front Escarpment. /\ Glacial boundary.
Insert: Unoccupied areas within range appear to be increasing in size and number because of destruction of habitat.

that the lateral band averages less conspicuous, the caudal fin is slate and never flushed with red although it may have an orange tinge on fishes from polluted waters, the crimson postocular spot is missing, and the caudal spot usually absent even in small young.

Lenghts and **weights:** *Young* of year in Oct., 1.0″–3.5″ (2.5–8.9 cm) long. *Adults*, usually 4.0″–8.5″ (11–22 cm) long. Largest specimen, 11.3″ (28.7 cm) long, weight 10 oz (283 g). Largest fishes invariably males. A larger species than the hornyhead.

Hybridizes: with blacknose and longnose daces, creek chub, striped and common shiners and stoneroller minnow.

Distribution and Habitat

Ohio Distribution—As stated under hornyhead chub, the literature records before 1925 were a composite of the river and hornyhead chubs, these records indicating that the two species, lumped together, were abundant, widespread in distribution and extensively used for bait and food. Preserved specimens of the river chub from the Ohio River, and from the Maumee, Black, Sandusky, Rocky, Chagrin, Scioto, Muskingum, Mahoning and Little Miami systems, and reliable literature records (Williamson and Osburn, 1898: lower fig., pl. 15; Parker, Williamson and Osburn, 1899:27) demonstrate its widespread distribution in Ohio before 1900.

During the 1925–38 surveys the river chub was widely distributed in Ohio, the largest populations occurring along the line of glaciation, and in that area of high relief immediately east of the Allegheny Front Escarpment from Cuyahoga to Ashtabula counties. It was presumably absent from that portion of unglaciated Ohio which extended from Jackson to Carroll counties, where the streams of base- or low-gradients had been little affected by glaciation and consequently had few well-defined gravel riffles. Since 1938 the river chub has largely or entirely disappeared from the Maumee (two wintering records totaling three specimens) and Sandusky river systems, and from the central Ohio portions of the Scioto River system. Until 1930 it was still present in small numbers in Big Darby Creek near Fox, Pickaway County, but yearly seinings since have failed to capture it.

The hornyhead and river chubs (Reighard, 1943:397–423; on breeding) and the tonguetied minnow, inhabited clear streams having bottoms of clean gravel and sand, and all three built huge nests of pebbles or small gravel. Probably because of similarities in habitats and nest building these three species were presumably competitors and were therefore probably intolerant of each other as suggested by the distribution maps. As examples: in Wayne and Auglaize counties the hornyhead was the dominant; in Columbiana and southern Ashland counties the river chub was dominant; in the Mad River system the tonguetied minnow was the most numerous except near its confluence with the Great Miami where stragglers of the river chub occurred. In Franklin County, the river chub was dominant between 1920–30 but after 1930 the hornyhead became the dominant.

Habitat—The largest populations of river chubs occurred in streams which were of moderate or large size, had moderate or high gradients, whose bottoms were of gravel, boulders and bedrock, and where there was little or no aquatic vegetation except water willow and attached algae. Populations decreased greatly in abundance or disappeared entirely wherever there was an increase in turbidity and siltation, and especially when it resulted in smothering over of the stones and boulders of the riffles, thereby resulting in a decrease in abundance of the aquatic insect larvae beneath and around the stones and boulders. For food habits and biology, *see* Lachner (1950:229–36; 1952:433–66).

Years 1955–80—As with the other species of *Hybopsis* and *Nocomis* present in Ohio, this one generally continued to decline in abundance throughout the period (White et al., 1975:68). Only three centers of abundance containing moderate populations were found. One was in the Walhonding River with its population center in Coshocton County and downstream from the center of abundance of the hornyhead chub in Knox County. This distributional pattern is expected because the river chub is primarily an inhabitant of larger streams and more swiftly flowing riffles, where there are more gravel and boulders and less sand. A moderate population appeared to be maintaining itself in the high-gradient Chagrin River in northeastern Ohio. A third population was present in the Scioto River and especially its glacial-outwash tributaries south and east of the Allegheny Front Escarpment, in the unglaciated, hilly country where the terrain is largely wooded or is in brush or meadowland. It should be noted that the Walhonding and Chagrin rivers populations were also east of the Escarpment.

On October 18, 1960, 11 adults were captured in the Scioto River below the O'Shaughnessy Dam in Franklin County at a time when the substrate was heavily coated with diatoms and general conditions appeared unfavorable. Since 1960 we have failed to obtain more specimens from this section.

The river chub has been recently reinstated in the genus *Nocomis* and becomes *Nocomis micropogon* (Cope); *see* hornyhead chub.

SILVER CHUB

Hybopsis storeriana (Kirtland)

Fig. 43

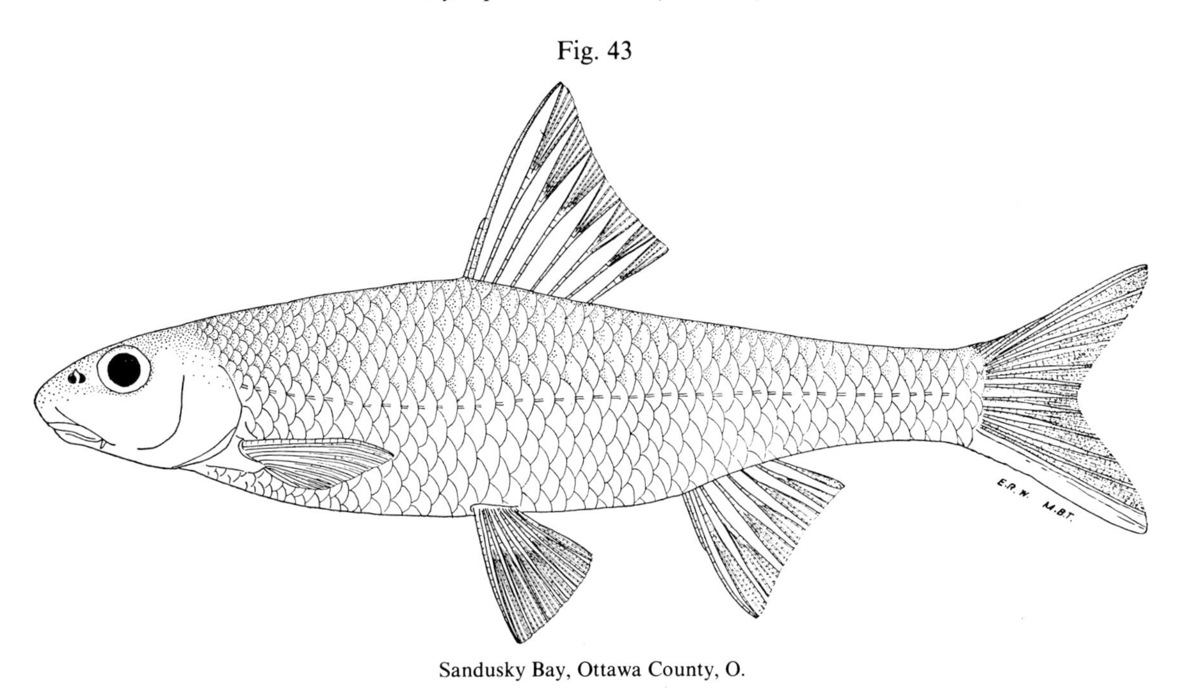

Sandusky Bay, Ottawa County, O.

Aug. 17, 1939.
Female.

100 mm SL, 4.8" (12 cm) TL.
OSUM 457.

Identification

Characters: *General—See* golden shiner. *Generic*—A small barbel at posterior end of upper jaw. Tip of upper jaw separated from snout by a deep groove. Scales in complete lateral line 57 or fewer, usually fewer than 50. Teeth 1,4–4,1 or 4–4, or some combination of these. Does not construct a mound nest; males have no large cephalic breeding tubercles or cephalic swellings or crests; coloration of both sexes silvery. *Specific*—General coloration silvery. Eye moderately large, contained 0.4–0.9 (1.0 in smallest young) times in upper jaw length. Mouth rather small and considerably overhung by the snout. Bony interorbital space contained 0.7–1.4 times in length of short snout. Dorsal fin forward, the distance from dorsal origin to caudal base much greater than distance from dorsal origin to snout. No spots on body, except for occasional blackish cysts of parasites. Lateral band faint or absent, never encircling snout. Ventral edge of caudal fin milk-white in large young and adults, and as it is in the spottail shiner. Teeth 1,4–4,1.

Differs: Hornyhead and river chubs have smaller eyes and a light brownish coloration. Bigeye chub has a bold lateral band encircling snout and a more posteriorly inserted dorsal fin. Streamline, gravel and speckled chubs have spotting over the body. Other minnows having barbels have more lateral line scales. Remaining species of minnows lack barbels. Suckers have more dorsal rays. Mooneye and herrings have pelvic axillary processes. Trouts, whitefishes, smelt, and trout-perch have an adipose fin.

Most like: Hornyhead, river and bigeye chubs. **Superficially like:** Streamline chub, river and common shiners.

Coloration: Dorsally a pale greenish-olive overcast heavily with silver. Sides very silvery. Ventrally a milk-white. All fins transparent, the ventral edge of caudal fin and the two or three adjacent rays

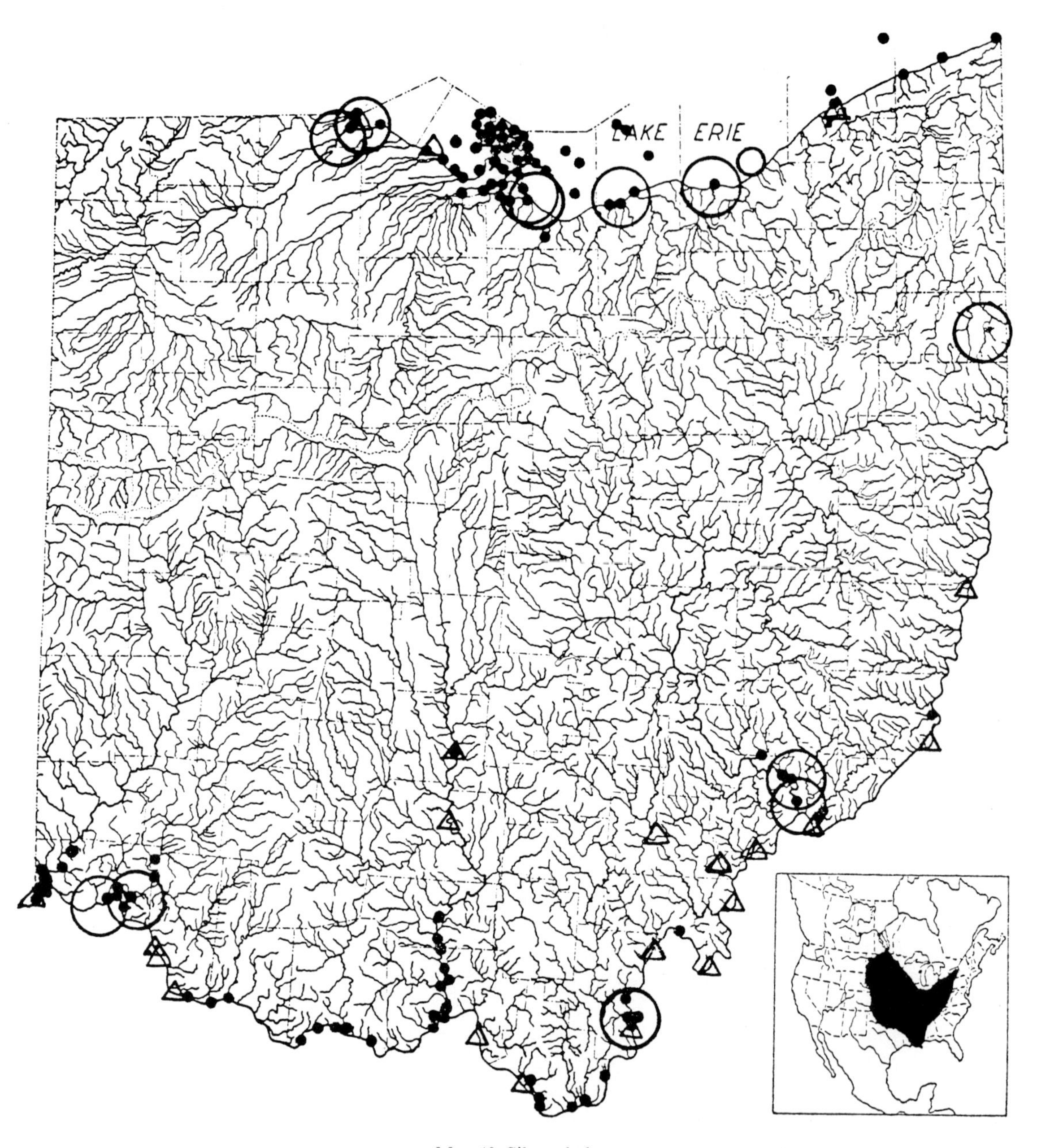

MAP 43. Silver chub

Locality records. ○ Before 1900. ○ Type locality, Lake Erie, most probably near Cleveland. ● After 1901. △ 1955–80.
Insert: Western limits of range indefinite.

milk-white, a distinctive feature in large young and adults.

Lengths and **weights:** *Young* of year in Oct., 1.0″–3.0″ (2.5–7.6 cm) long. Around **1** year, 2.0″–4.5″ (5.1–11 cm). *Adults*, usually 4.0″–7.0″ (10–18 cm). Largest specimen, 9.1″ (23 cm) long, weight 6 oz (170 g).

Distribution and Habitat

Ohio Distribution—From the writings of Kirtland (1847:30–31, includes description of species; type locality, Lake Erie), Jordan (1882:843), and McCormick (1892:20), it is evident that the silver chub was at least as abundant in Lake Erie in the

1847–1925 period as it has been since. Kirtland and Jordan stated that the species seldom occurred in streams tributary to Lake Erie, and this has been true during the 1925–50 surveys. McCormick stated that "On July 8, 1891, after a storm, a great many large ones were thrown on the beach by the waves"; this was undoubtedly a spawning mortality, for upon many occasions during their spawning seasons in June and July between the years 1925–50 I have seen large numbers of spawned adults dead along Lake Erie beaches. It is from these instances of spawning mortality, results of deep-water trawlings, and studies of stomach contents of predatory fishes, that realization of the great abundance of the silver chub in Lake Erie was obtained.

Less is known concerning its early abundance in the Ohio River drainage. Henshall (1889:78) stated that in Hamilton County in 1888 it was "Common in Little Miami River and Clough Creek," and there are preserved specimens, collected in 1888 by C. H. Gilbert and J. A. Henshall from the Ohio River in Hamilton County, and Raccoon Island, in Gallia County, and from the Muskingum River, Lock 2, Washington County. Evermann and Bollman (1886:338) noted that Baird and Kirtland collected a specimen, in August, 1853 in Yellow Creek, Ohio. In the California Academy of Sciences (2967) collections there are two specimens labeled "*Erinemus storerianus*, Yellow Creek, O." Evermann and Bollman (1886:339) stated likewise that in 1885 it was abundant in the Monongahela River, at Monongahela City, Pa. These early records indicate a widespread distribution in the Ohio drainage previous to 1900.

Between 1925–50, the silver chub was quite numerous in the Ohio River between the Indiana state line and Marietta, and in the lower Muskingum and Scioto rivers. On October 18, 1928, Robert B. Foster and I took a stray at the mouth of Big Darby Creek, Pickaway County, at a time when pollution in the Scioto River had forced a great multitude of fishes into the clean waters of Big Darby. This is the farthest inland record.

Habitat—The silver chub occurred in greatest abundance in Lake Erie in waters from 3′–60′ (1–18 m) in depth, in large, deep waters of low- or base-gradient streams which had rather clean and usually gravelly bottoms. It was essentially a pool species when the bottom was not covered by flocculent silts; where pool bottoms were so covered it resorted to riffles. When the large streams were very turbid or were depositing unusually large amounts of silt, it would temporarily migrate into clearer streams of higher gradients, such as from the lower-gradient Great Miami into the higher gradients of the Whitewater River of Hamilton County. When the waters were very clear it retired into deep water. If the assumption is correct that this species requires a clean bottom of gravel and sand, it should not be as abundant in the silty-bottomed Ohio River recently as it was before 1900 when large areas of clean sand and gravel existed. The Ohio River and Lake Erie populations appear to differ morphologically. Most of the Ohio River specimens are more streamlined in appearance, have less body depth at the dorsal origin, and their heads are less triangular. The snouts of many Ohio River adults are more bulbous and appear to overhang the upper lip more than do the snouts of the Lake Erie adults. Despite the present, turbid waters of the Ohio River, preserved specimens from that stream are more heavily pigmented and less silvery than are Lake Erie specimens, and the usually faint band along the sides and encircling the snout is more distinct than it is in Lake Erie specimens. If found to be subspecifically distinct, the Ohio River population becomes *Hybopsis storeriana lucens* Jordan (1880:238–39).

Years 1955–80—Until 1953 the silver chub apparently remained as abundant in Lake Erie waters as it had been for many years previously. After 1953 a drastic decrease in numbers occurred, almost simultaneously with the catastrophic decline in numbers of mayfly (*Hexagenia*) larvae and the subsequent rise in numbers of midge (Chironomidae) and sludge worm (Oligochaete) larvae (p. 32) (White et al., 1975:69). Trawls, made since 1965 in the same localities in western Lake Erie and at the same seasons of the year, resulted in usually obtaining none (one specimen, from South Bay, Middle Bass Island, July 11, 1968), whereas before 1950 similar trawls captured dozens and often hundreds of silver chubs. Obviously, this decrease cannot be attributed to over-fishing because comparatively few were ever taken for food or bait. Its decrease must be attributed to other factors, the most logical of which seems to be a marked deterioration in the amount of suitable habitat caused by recent modifications in Lake Erie. Scott and Crossman (1973 : 421) state that "The present status of the species in Lake Erie is in doubt but obviously it is rare."

Changes in abundance in the rivers of the Ohio drainage are not as clear-cut because this deep-water-inhabiting species is not readily caught with hand seines. Despite this, fewer specimens have been captured since 1955 in the Scioto drainage under comparable conditions than were taken prior to 1940. However, the finding during 1964 of 11 dead specimens killed by pollution in the Scioto River immediately above Chillicothe and of one partially decomposed specimen at the mouth of Big Darby Creek indicates that this deep-water-inhabiting species may be present well upstream toward Columbus. Specimens were collected as far upstream in the Hocking River as Athens by members of Ohio University (Zahuranec, 1962:-842–43).

Similar to the other species in Ohio in the genera *Nocomis* and *Hybopsis*, the silver chub inhabits the bottoms of streams and, as with the others, reaches its greatest numerical abundance over substrates of clean gravel and sand and appears to be rather susceptible to pollutants of diverse types.

The silver chub continued to be numerous in the Ohio River, the 1957–59 investigations reporting 11,789 specimens (ORSANCO, 1962:62).

NORTHERN BIGEYE CHUB

Hybopsis amblops amblops (Rafinesque)

Fig. 44

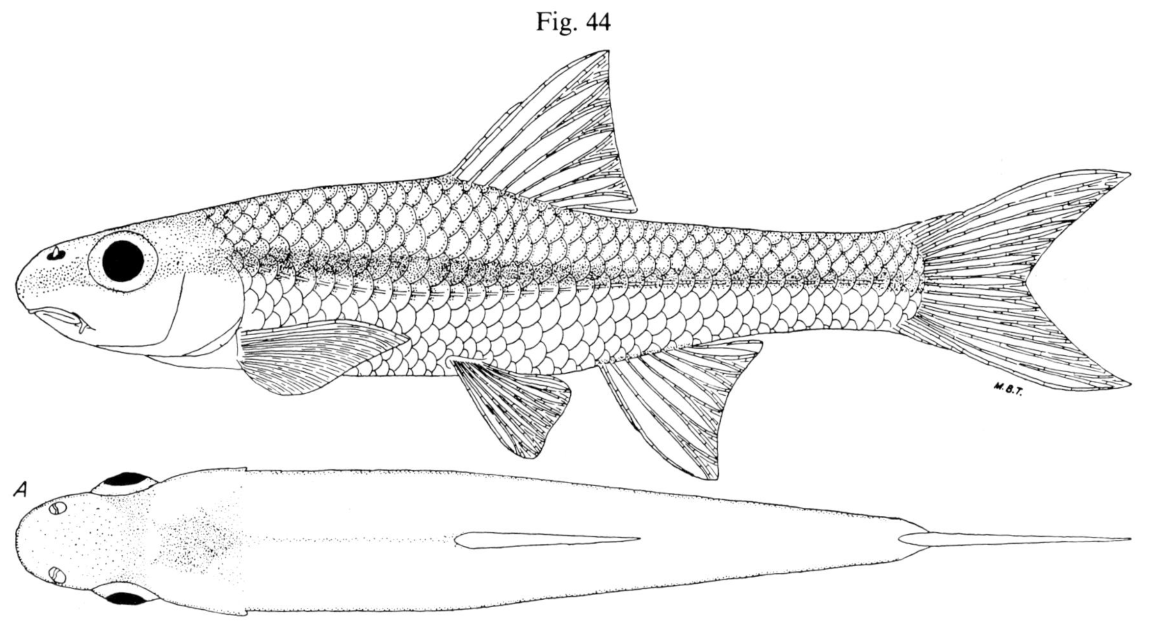

Upper fig.: Cold Run, Columbiana County, O.

May 28, 1941. 69 mm SL, 3.4″ (8.6 cm) TL.
Male. OSUM 3213.

Fig. A: dorsal view showing color pattern and distinctive shape.

Identification

Characters: *General—See* golden shiner. *Generic—See* silver chub. *Specific*—Coloration silvery with a bold, black lateral band which projects forward to cross opercles and encircles snout. Eye large, contained 0.4–0.9 times in upper jaw length. Mouth small, considerably overhung by the snout. Bony interorbital space contained 0.7–1.4 times in snout length. Dorsal fin origin usually equidistant between the distance from dorsal origin to caudal base and distance from dorsal origin to snout. No black spots on body. Teeth, 1,4–4,1.

Differs: *See* silver chub. Silver chub has dorsal origin farther forward and lacks black lateral band.

Most like: Streamline, gravel, speckled and silver chubs. **Superficially like:** Suckermouth minnow, mimic, sand and blacknose shiners.

Coloration: Dorsally an olive-straw, the scales edged with darker giving an appearance of wavy, longitudinal lines rather than cross-hatching. Ventrally silvery or milk-white. A black band, as wide as eye, encircles snout, crosses opercles and continues along sides to tail. Fins transparent. Specimens from clear waters often very straw-colored, heavily overlaid with silvery and with a striking, black lateral band; those from turbid waters pale, the lateral band barely evident. *Male*—Has minute tubercles on dorsal half of head.

Lengths: *Young* of year in Oct., 1.0″–2.3″ (2.5–5.8 cm) long. Around **1** year, 1.5″–2.5″ (3.8–6.4 cm). Breeding *adults*, usually 2.3″–3.5″ (5.8–8.9 cm) long. Largest specimen, 3.9″ (9.9 cm) long.

Distribution and Habitat

Ohio Distribution—Many preserved specimens and reliable published records prior to 1900 (Henshall, 1888:78; Kirsch, 1895A:329; Williamson and Osburn, 1898:34, Pl. 19; Parker, Williamson and Osburn, 1899:27; Osburn, 1901:63) indicate the former widespread distribution and abundance of

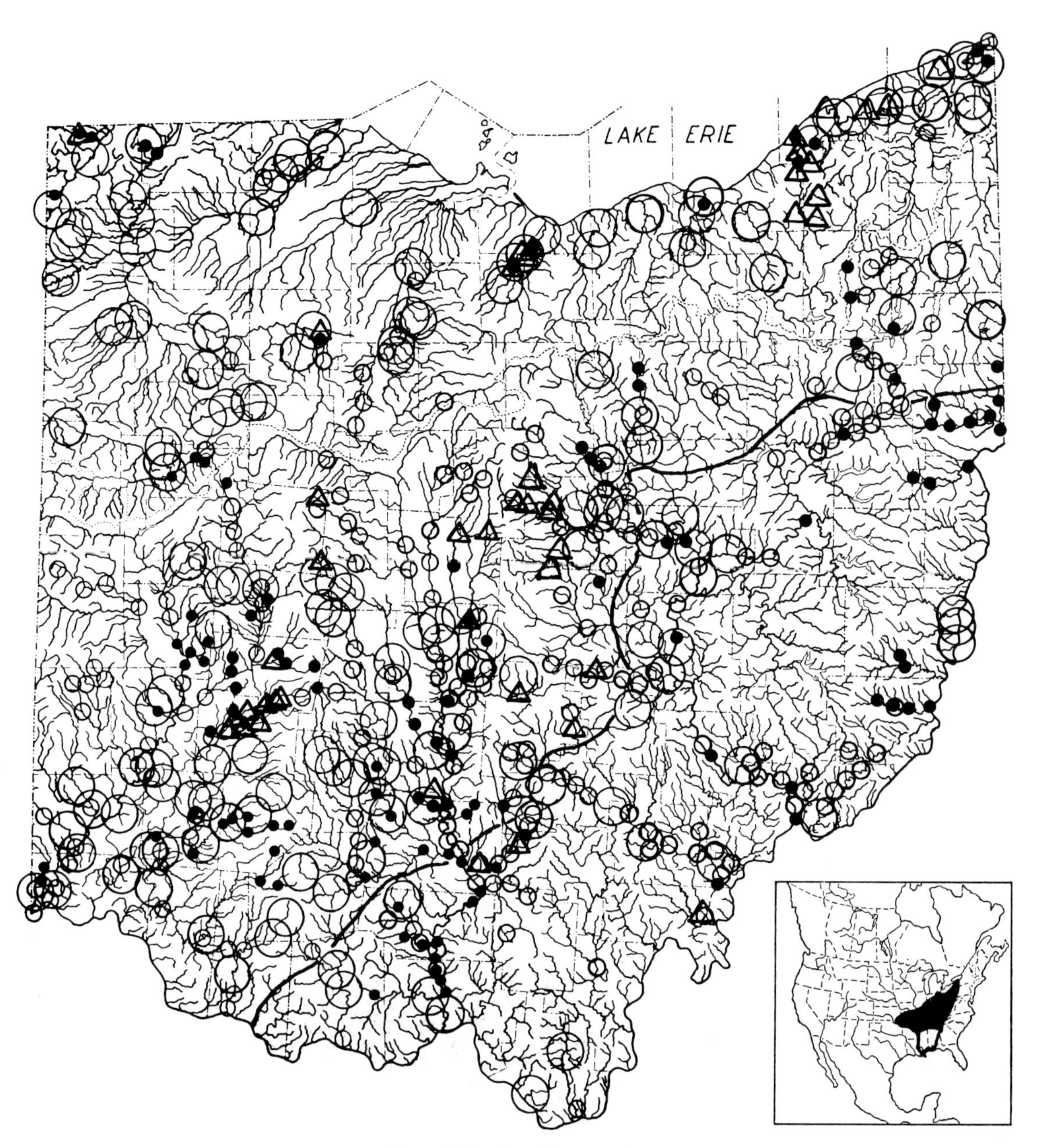

MAP 44. Northern bigeye chub

Locality records. ○ Before 1924. ○ 1925–38. ● 1939–54. △ 1955–80. ~ Glacial boundary.
Insert: Black area represents range of this subspecies, outlined area of another subspecies whose validity has been questioned recently by Bailey, Winn and Smith (1954:124); unoccupied areas within range of species appear to be increasing in size and number.

this species in Ohio. Since 1900, a marked decrease in abundance and extirpation of populations have occurred in several localities, particularly in northwestern Ohio where in 1893 Kirsch stated that the bigeye chub "no doubt inhabits all the streams" of the Maumee system of Ohio and Indiana (Gerking, 1945; map 27). During the 1925–39 surveys the species was widespread throughout the Maumee River system, but disappeared after 1939 from the main stream and was present only in relict populations or as strays in some of the headwaters. Until 1939 it was present in the Sandusky River drainage;

extensive collecting since has failed to produce a specimen.

Between 1925–50 the largest populations occurred along the line of glaciation and in the headwater streams of the Great and Little Miami rivers. It was conspicuously absent in streams of highest gradients such as those of the upper Chagrin River systems; also in that portion of the unglaciated area of southeastern Ohio where the mature, base-gradient streams were little affected by glacial action. However, small populations were present in those rejuvenated, glacial outlet streams of the southeastern section such as the Muskingum and Hocking rivers, which had been rejuvenated by glacial action and therefore had moderate gradients and well-defined riffles.

Habitat—The largest populations occurred in moderate- or small-sized streams in which the gradients were moderate and where there was much clean sand and fine gravel in the pools and/or riffles. Streams of high gradients had low populations, especially if the swift current prevented accumulation of sand and fine gravel on the bottoms. Likewise, streams of low gradients normally had low populations except those few which had so little siltation that the sand and fine gravel of their bottoms were not covered with silt.

In some prairie streams of west-central Ohio between 1925–50 I observed changes which were caused by increased siltation upon the stream bottoms, with subsequent decreases in the sizes of the bigeye chub populations. Until 1925 the bottoms of these streams were mostly of sand and gravel and had little clayey silt, and in them goodly populations of bigeye chubs were present in both the pools and riffles; between 1930–40 the pool bottoms of these streams became silted, and these chubs were then confined chiefly to the riffles; after 1940 the riffle bottoms became silted also, there was further decrease in the size of the chub population, or the species disappeared. Unquestionably the total Ohio population of bigeye chubs has decreased greatly since 1900.

Years 1955–80—Before 1940 moderate to large populations of the bigeye chub existed in the majority of the streams of central Ohio; but recently less than two dozen localities have been found to contain populations, and these appear to be very small. Since 1960 this condition appears to exist throughout Ohio, not a single large concentration of individuals being noted, despite determined efforts to locate one (White, et al., 1975:69).

The sandy and gravelly substrate of small- or moderate-size streams containing little or no silts is the preferred habitat of this species. This habitat is disappearing because of ditching, increased drainage, unfavorable land use and silt covering the substrate. Many streams have become intermittent, and this lack of continuous streamflow disrupts or otherwise prevents normal development of aquatic invertebrates, the principle food of the bigeye chub. Stoppage or drastic reduction in streamflow prevents fishes from remaining in suitable situations, trapping some and forcing others downstream into unfavorable conditions. If the reduction in habitat continues, it will only be a relatively short period of time before most or all of the remaining remnant populations in Ohio are eliminated.

OHIO STREAMLINE CHUB*

Hybopsis dissimilis dissimilis (Kirtland)

Fig. 45

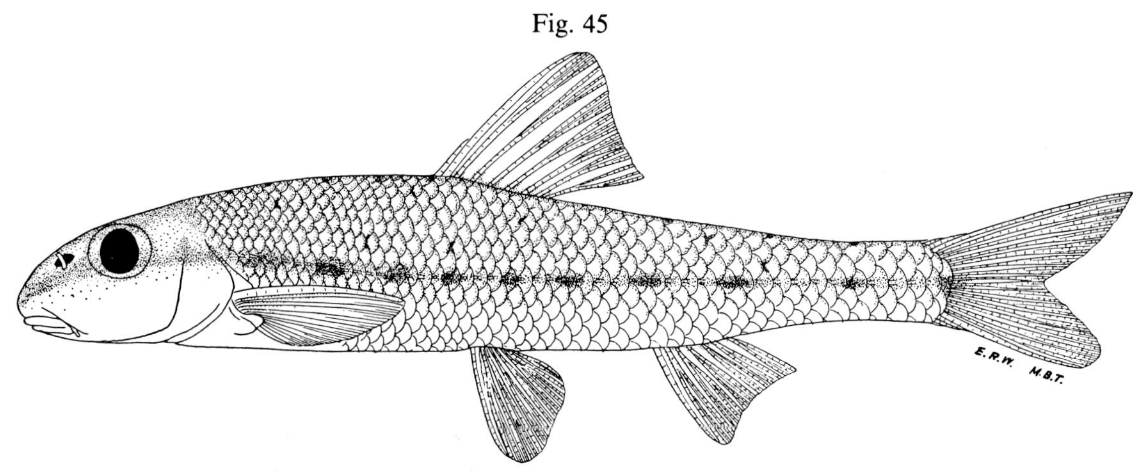

Muskingum River, Muskingum County, O.

Oct. 19, 1939.
Male.

64 mm SL, 3.2″ (8.1 cm) TL.
OSUM 924.

Identification

Characters: *General—See* golden shiner. *Generic—See* silver chub. *Specific*—Upper jaw length seldom as long as diameter of eye. Posterior end of lower jaw not reaching a point directly below anterior edge of eye. Bony interorbital space usually contained 1.5–2.4 (1.3 in smallest young) times in the moderately long snout. Preorbital length shorter than, or equal to, postorbital length of head. Lateral line scales usually 44–47, rarely 43. The broken lateral band consisting of 7–11 oblong, blackish spots, including the prominent caudal spot; all these spots faintest in the largest adults. Several light spots alternate with darker spots along the dorsal ridge; these best seen when looking downward through the water as the fish rests on the bottom. Spottings on back mostly X- or W-shaped; seldom roundish. Teeth, 4–4.

Differs: Gravel chub has a longer snout, lacks the lateral spots. Speckled chub has roundish spots on back. *See* silver chub.

Most like: Gravel and speckled chubs. **Superficially like:** Suckermouth minnow, river and bigeye shiners.

Coloration: General coloration silvery, the back with an undercast of greenish-olive and emerald, with the dark scale edges sometimes giving a cross-hatched effect. Predorsal ridge with 3–7 small, light dots, postdorsal ridge with 6–11, these most prominent when fish is alive. Sides intensely silvery, except for broken lateral band that encircles snout and crosses opercles, and continues along sides of body from head to tail in a series of 5–11 oblong, blackish spots. These spots often connected by a dark gray band. Caudal spot usually of same size and color intensity as other lateral band spots. Ventrally a milk-white. Fins unspotted, transparent or silvery. *Breeding male*—Head covered with many minute tubercles, including the lips and branchiostegals; breast and scales on anterior half of body sometimes containing minute tubercles.

Length: *Young* of year in Oct., 1.3″–2.4″ (3.3–6.1 cm) long. Around **1** year, 2.0″–2.8″ (5.1–7.1 cm). *Breeding adults*, usually 2.5″–4.0″ (6.4–10 cm). Largest specimen, 4.2″ (11 cm) long.

Hybridizes: with gravel chub.

Distribution and Habitat

Ohio Distribution—In 1841 Kirtland (1841B: 341–42 and Pl. 4; as *Luxilus dissimilis*) described and adequately figured the "Spotted Shiner," after presumably having obtained two specimens from

* Ohio spotted chub in 1957 edition of *Fishes of Ohio*.

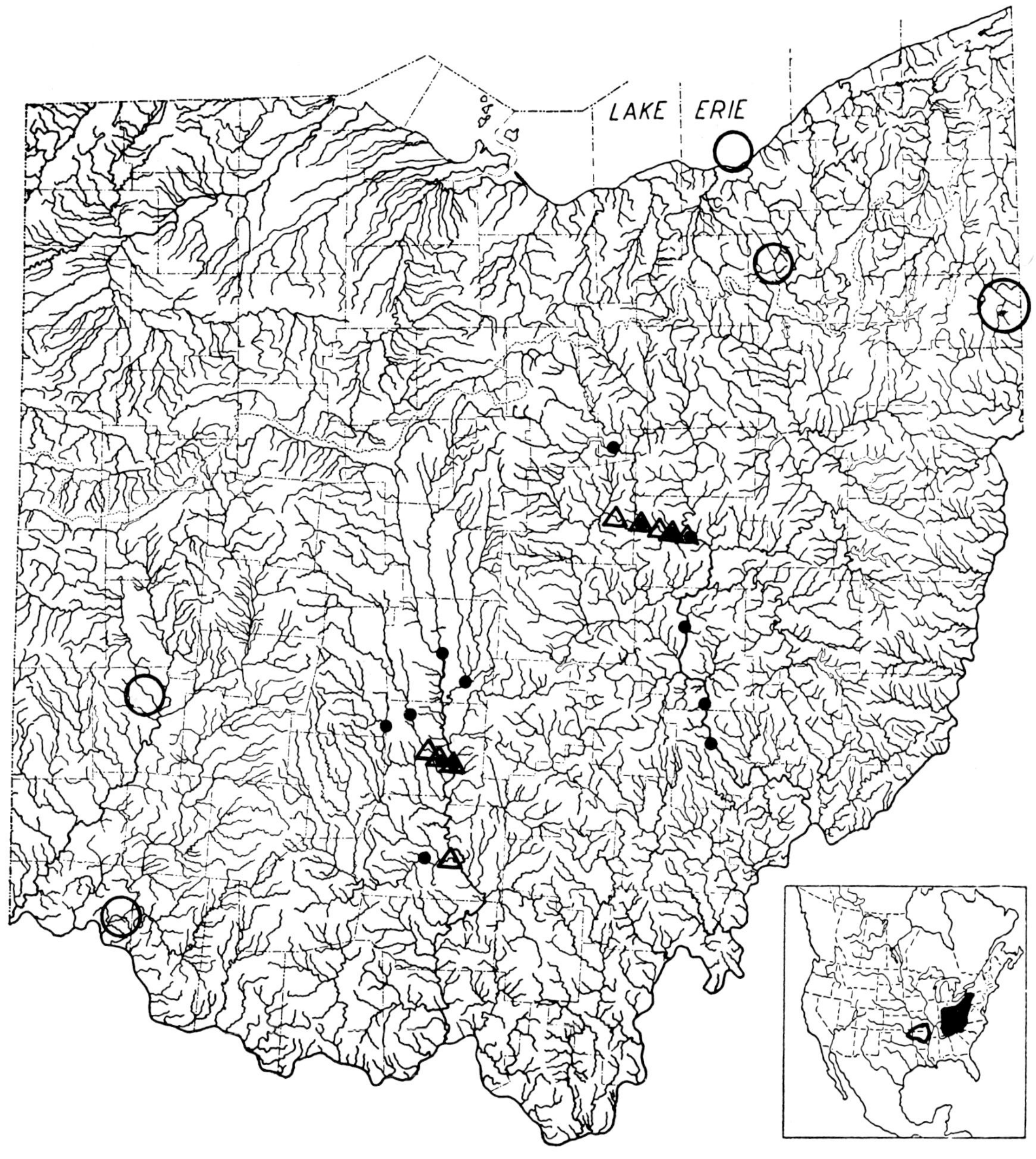

MAP 45. Ohio streamline chub

Locality records. ○ Type locality, Mahoning River, Mahoning County.
○ Before 1900, without preserved specimens. ● 1897-1955, with specimens. △ 1955–80.
Insert: Black area represents range of nominate species, outlined area of a disjunct subspecies (Hubbs and Crowe, 1956:7).

the Mahoning River near Youngstown. In his description of the species he stated that after obtaining these two fishes he "found several dead specimens upon the shore of Lake Erie, near Cleveland." Apparently Kirtland later realized that he was mistaken about its occurrence in Lake Erie, for in 1851 he (1851Q:189) omitted all reference to Lake Erie, stating only that "It inhabits the Mahoning River, near Youngstown"; and adding the significant ecological statement that it "is usually found in deep water at the foot of riffles."

About 1927 Carl L. Hubbs and I realized that in

Ohio the streamline chub and an undescribed species, the gravel chub, had until then been considered as a single species. Because of this, only those records can be accepted which have either preserved specimens, adequate descriptions or identifiable photographs to substantiate them. The following records therefore must be excluded: Henshall's (1888:79) early records for the Little Miami River and O'Bannon Creek in Hamilton County; Osburn's (1901:62) 1899 record for *H. dissimilis* in the Stillwater Creek near Dayton, and 1900 record for the Cuyahoga River at Hawkins (now Ira, Summit County). Williamson and Osburn's record of 1897 for Big Walnut Creek in Franklin County is acceptable because there are specimens (USNM 64790 and CAS 8869).

Between 1925–50 the streamline chub disappeared from several stream sections, coincidental with the covering over of the gravel bars and riffles with clayey silts. Successful spawnings during several years were noted in small sections in both the Muskingum and Scioto river drainages, resulting in a temporary and marked increase in abundance in these sections. As an example: During many seinings between 1925–47 fewer than a dozen individuals were collected in Big Darby Creek near Fox. In 1948 a successful spawning occurred, and between July–October a dozen young or more could be taken in an hour of seining. There are no acceptable records for the Lake Erie drainage.

Habitat—The streamline chub inhabited the riffles and bars in streams of moderate size, where a favorable current had deposited coarse sands and gravels and kept them comparatively free of silt, and where the waters were 1′–4′ (0.3–1.2 m) deep in summer, and slightly deeper in winter. The largest numbers of streamline chubs were found usually about the bars and deeper waters at the foot of riffles, and in the deeper, gravel-bottomed pools which were present in the main portions of riffles. The species seemed to avoid aquatic vegetation. It disappeared promptly from a riffle when the gravel became heavily coated with silt or other pollutants. The habitats of the streamline and gravel chubs seemed to be very similar, and this similarity must have resulted in competition between the two. Like the gravel chub, the streamline must have been far more abundant before 1850 than it was after 1925 (*see* gravel chub, under Habitat).

Years 1955–80—Throughout the 1925–50 period there were moderate populations of streamline chubs present in several localities in the Muskingum and Scioto river systems. These localities were investigated during the 1955–80 period, and the species was taken only in the Walhonding drainage of the Muskingum system and in Paint and Big Darby creeks of the Scioto system.

The history of varying abundance of the streamline chub in lower Big Darby Creek is well documented because that section has been seined, and there have been observations made by others and me annually since 1925. Despite more than 200 investigations of this section since 1925, no streamline chub was collected until 1939, when one was taken, and not until 1942 was another captured. Apparently successful spawnings by a small population of adults took place in 1947 and 1948, for in late summer and fall of 1948 dozens of young and small adults could be captured during a few hours of seining. The species remained fairly numerous until 1950; then no more were taken until 1960 when a few were captured. By 1962 many were again present, and for the first time a specimen was taken in the Scioto River about a mile downstream from the mouth of Big Darby. These marked fluctuations in annual abundance also occur in several other species in lower Big Darby Creek; *see* Scioto madtom and Tippecanoe darter. Apparently, during most years submarginal conditions prevail for these species in this stream section, and it is only during exceptional years that there is a fair population of young present.

In 1841 Kirtland described *Luxilus dissimilis* and gave it the name of spotted shiner. In the literature the names spotted shiner or spotted chub have been in common usage for this species, as it was in the 1957 edition of this report, until recently changed to streamline chub (Bailey et al., 1960:14).

EASTERN GRAVEL CHUB

*Hybopsis x-punctata trautmani** (Hubbs and Crowe)

Fig. 46

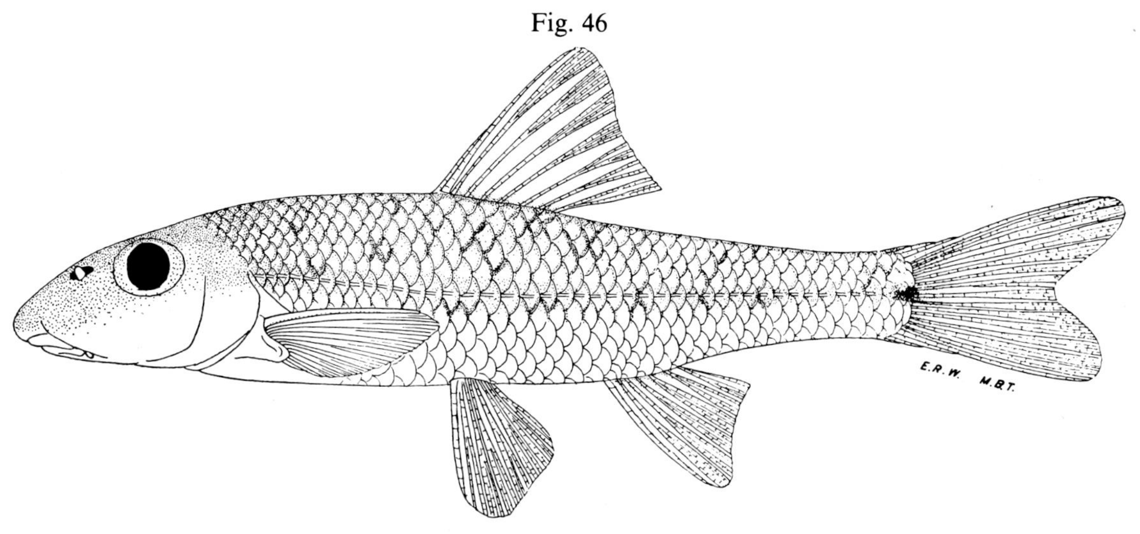

Walhonding River, Coshocton County, O.

Oct. 20, 1939.
Adult.

70 mm SL, 3.5″ (8.9 cm) TL.
UMMZ 177278.

Identification

Characters: *General—See* golden shiner. *Generic—See* silver chub. *Specific*—Upper jaw length usually equal to eye diameter. Posterior end of lower jaw usually falling considerably short of the point directly below the anterior edge of eye. Bony interorbital space usually contained 1.5–2.4 times in the long snout. Preorbital length longer than postorbital head length, except in the smallest young. Lateral line scales 38–43, usually 42 or fewer. No blackish, oblong spots along lateral line. Back and sides usually contain a few to many W- or X-shaped markings; these markings sometimes absent in large adults. No spotting on dorsal ridge. Teeth, 4–4.

Differs: *See* silver chub. Streamline chub has blackish spots along lateral line and more scales.

Most like: Streamline and speckled chubs. **Superficially like:** River, bigeye and mimic shiners.

Coloration: General coloration silvery, the back with an undercast of greenish- or brownish-olive, with the dark scale edges usually giving a cross-hatched appearance. Sides a clouded silvery. Ventrally milk-white. Back and sides usually with few or many W- or Y-shaped markings, and without rounded spots, except for the often prominent caudal spot. Lateral band when present, pale gray and ill-defined. Fins unspotted, transparent or somewhat silvery. *Breeding male*—Head covered with many minute tubercles, including lips and branchiostegals; breast and scales on anterior half of body sometimes containing tiny tubercles.

Lengths: *Young* of year in Oct., 1.1″–2.4″ (2.8–6.1 cm) long. Around **1** year, 1.7″–2.8″ (4.3–7.1 cm). *Breeding adults*, usually 2.5″–3.8″ (6.4–9.7 cm). Largest specimen, 3.9″ (9.9 cm) long.

Hybridizes: with streamline chub.

Distribution and Habitat

Ohio Distribution—It was not realized until rather recently that the species formerly known as the spotted shiner (*Luxilus dissimilis*, Kirtland) was a complex of two species, the gravel and streamline chubs. Because of this complex and lack of adequate descriptions, photographs or specimens, all of the early records are unidentifiable as to species,

* Hubbs and Crowe, 1956:4 and 7.

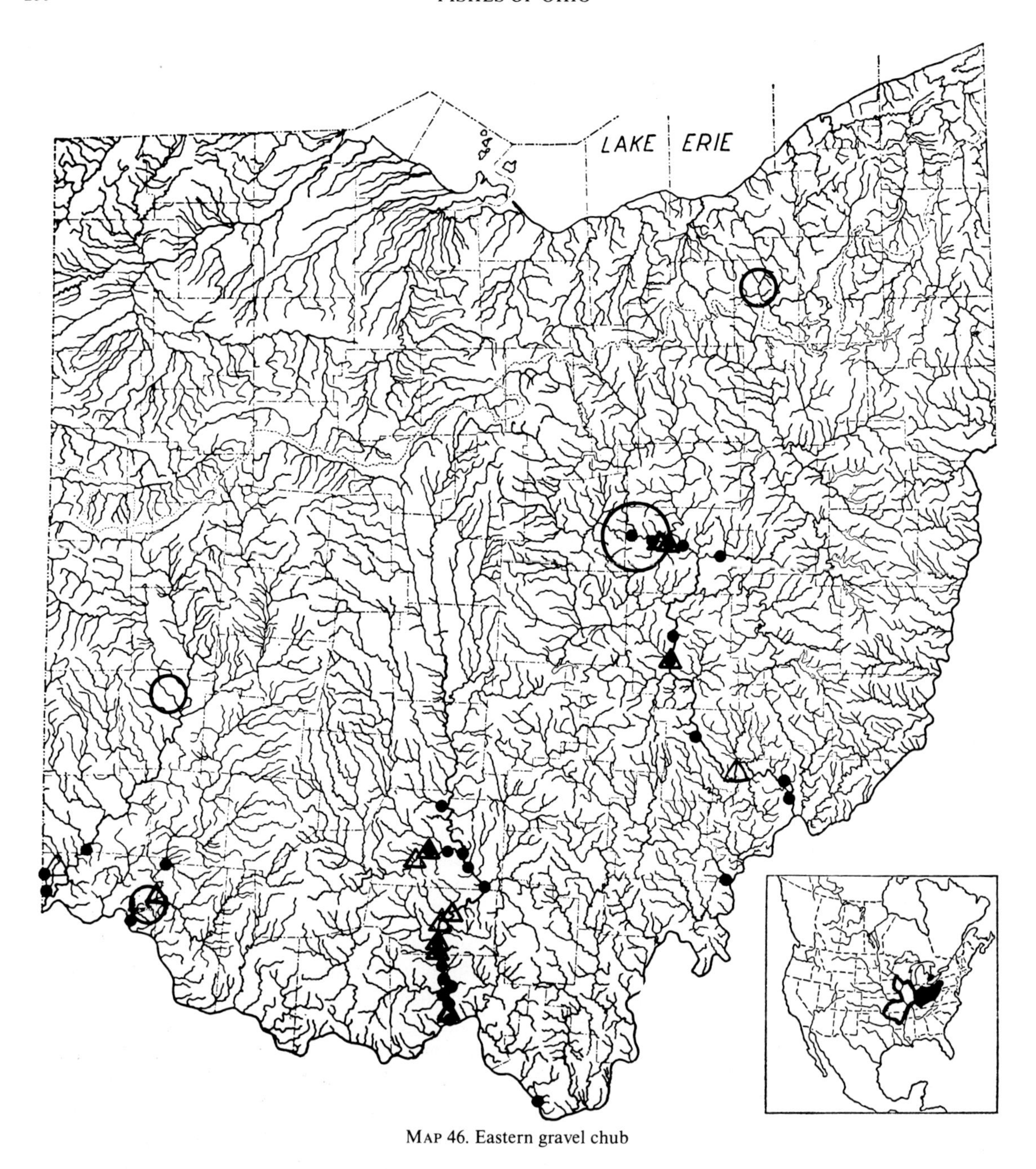

MAP 46. Eastern gravel chub

Locality records. ○ Type locality, Walhonding River, Coshocton County.
○ Before 1900, without preserved specimens.
● 1920–55, with specimens. △ 1955–80.
Insert: The two disjunct black areas represent the range of this subspecies, outlined area of another; unoccupied areas within range appear to be increasing in size and number; recently the species may have become extirpated from the small black area.

except the Williamson and Osburn record of 1897 for Big Walnut Creek, of Franklin County; specimens from this collection were preserved and are streamline chubs. The remaining records, all unidentifiable, are given by Osburn (1901:62) as follows: Little Miami River and O'Bannon Creek in Hamilton County, Stillwater Creek near Dayton and the Cuyahoga River at Hawkins; *see* Ohio streamline chub, under Distribution.

During the 1925–50 period the gravel chub was

found in moderately large populations only in sections of the Walhonding, Muskingum, Scioto, Little Miami and Whitewater rivers. Small populations or strays were found in the Tuscarawas, Hocking, Paint and Deer creeks. None was found in the Scioto system north of Ross County, despite much collecting in Pickaway and Franklin counties. A decrease in abundance or extirpation of the species was recorded for several sections during the 1925–54 surveys. It disappeared from many riffles and bars immediately after the sand and gravel of the stream bottoms became heavily silt-covered.

No acceptable Ohio records exist for the gravel chub in the Lake Erie drainage. The species has been taken in the tributary waters of the Great Lakes, however, in the Thames River of Ontario which empties into Lake St. Clair.

Habitat—The gravel chub chiefly inhabited the large sand and gravel riffles and bars of moderate- or large-sized streams, where the current had deposited coarse sands and gravels and had kept these comparatively free of clayey silts and other injurious pollutants, and where the water depth was between 1′–4′ (0.3–1.2 m) in summer and 2′–6′ (0.6–1.8 m) in winter. Wherever silt was lacking the gravel chub was present in the largest numbers in the slower-flowing, deeper waters, but where silting was rapid the species was forced to abandon this habitat and could then be found in shallower and swifter waters which still contained a suitable sand and/or gravel habitat. When this latter habitat became silt-covered, the relict population was forced into the swifter waters where the bottoms consisted of large gravels and boulders, or the species disappeared entirely from that section. It avoided rooted aquatics and the larger species of algae and aquatic mosses; in fact little vegetation grew among the sand and gravel habitats frequented by the gravel chub.

Competition between the gravel and streamline chubs, especially while feeding, seemed to be rather keen. I frequently observed both species feeding within a few inches of each other. However, whenever habitat conditions seemed particularly favorable for both species, the gravel chub inhabited the deeper, slower-flowing waters, whereas the more active streamline chub frequented the shallower, more swiftly-flowing portions.

I believe that both the gravel and streamline chubs were far more widespread and abundant in Ohio prior to 1850 than they have been since, because their habitats undoubtedly were more extensive before the agricultural and industrial activities of the white man had eliminated so many of them. (*See* pp. 23–24).

Years 1955–80—During the 1925–50 year period, as indicated above, there were moderately large populations of this species in more or less continuous sections of the Walhonding-Muskingum, Scioto, Whitewater-Great Miami, and Little Miami river systems. During this period investigations of these rivers resulted in locating populations only in the Walhonding-Muskingum river system and in the lower Scioto River and its tributary Paint Creek. Moderately large populations were found during low-water periods in the lower Scioto River in Pike and Scioto counties. In the Walhonding-Muskingum river system, goodly populations were present in the Walhonding River, and only in two localities downstream in the Muskingum River.

The gravel and streamline chubs were sparingly present on the same riffles in the Walhonding River in Coshocton County. On these riffles the two species occupied the same types of substrates apparently, except that the larger-stream-inhabiting gravel chub occupied the deeper, faster portions of the riffles, whereas the streamline chub, usually an inhabitant of smaller streams, occurred in the shallower, more slowly moving sections of the same riffle.

OHIO SPECKLED CHUB

Hybopsis aestivalis hyostoma (Gilbert)

Fig. 47

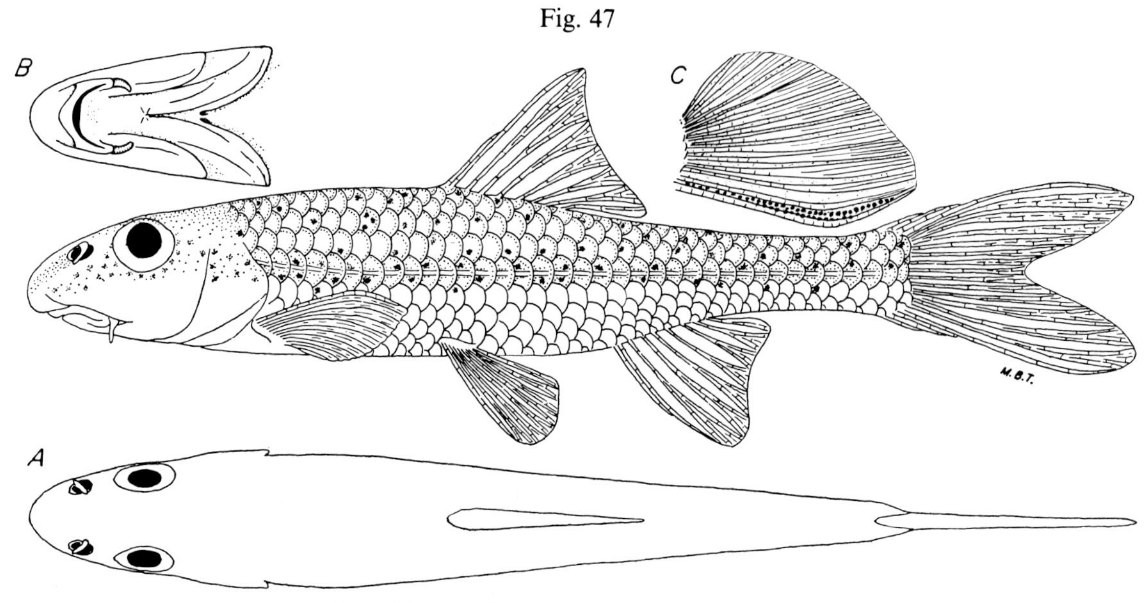

Central fig.: Muskingum River, Washington County, O.

Oct. 18, 1939. 55 mm SL, 2.8″ (7.1 cm) TL.
Male. OSUM 857.

Fig. A: dorsal view, note position of eyes.
Fig. B: ventral view of head showing long barbels at posterior ends of mouth, and the overhanging snout.
Fig. C: pectoral fin of male, showing tubercles.

Identification

Characters: *General—See* golden shiner. *Generic—See* silver chub. *Specific*—Upper jaw length often greater than eye diameter. Posterior end of lower jaw usually extending beyond a point directly below anterior edge of eye. Bony interorbital space contained 1.5–2.4 times in snout. Snout long, bulbous, and greatly overhangs the inferior mouth. Preorbital length usually as long, or longer, than postorbital length. Roundish, blackish spots usually conspicuously and abundantly scattered over back and sides. Teeth, 4–4.

Differs: *See* silver chub. Streamline and gravel chubs have W- or X-shaped markings.

Most like: streamline and gravel chubs.

Coloration: Dorsally a light olive- or brownish-yellow overlaid with much silvery. Sides quite silvery. Ventrally milk-white. Back and sides generally profusely spotted with round, blackish spots. Minute spots tend to form a band which encircles snout, and extends across the opercles and sides of body to the tail; this band often not apparent while fish is alive, often very prominent in preserved specimens. Fins transparent, the lower fins with some silvery. *Breeding male*—Has many minute tubercles on head and breast. Large breeding females sometimes contain these tubercles, but in the females the tubercles are less well developed.

Lengths: *Young* of year in Oct., 1.1″–1.8″ (2.8–4.6 cm). *Breeding adults*, usually 1.8″–3.0″ (4.6–7.6 cm); rarely more than 3.0″ (7.6 cm) long.

Distribution and Habitat

Ohio Distribution—There are four locality records before 1929: Ohio River at Raccoon Island,

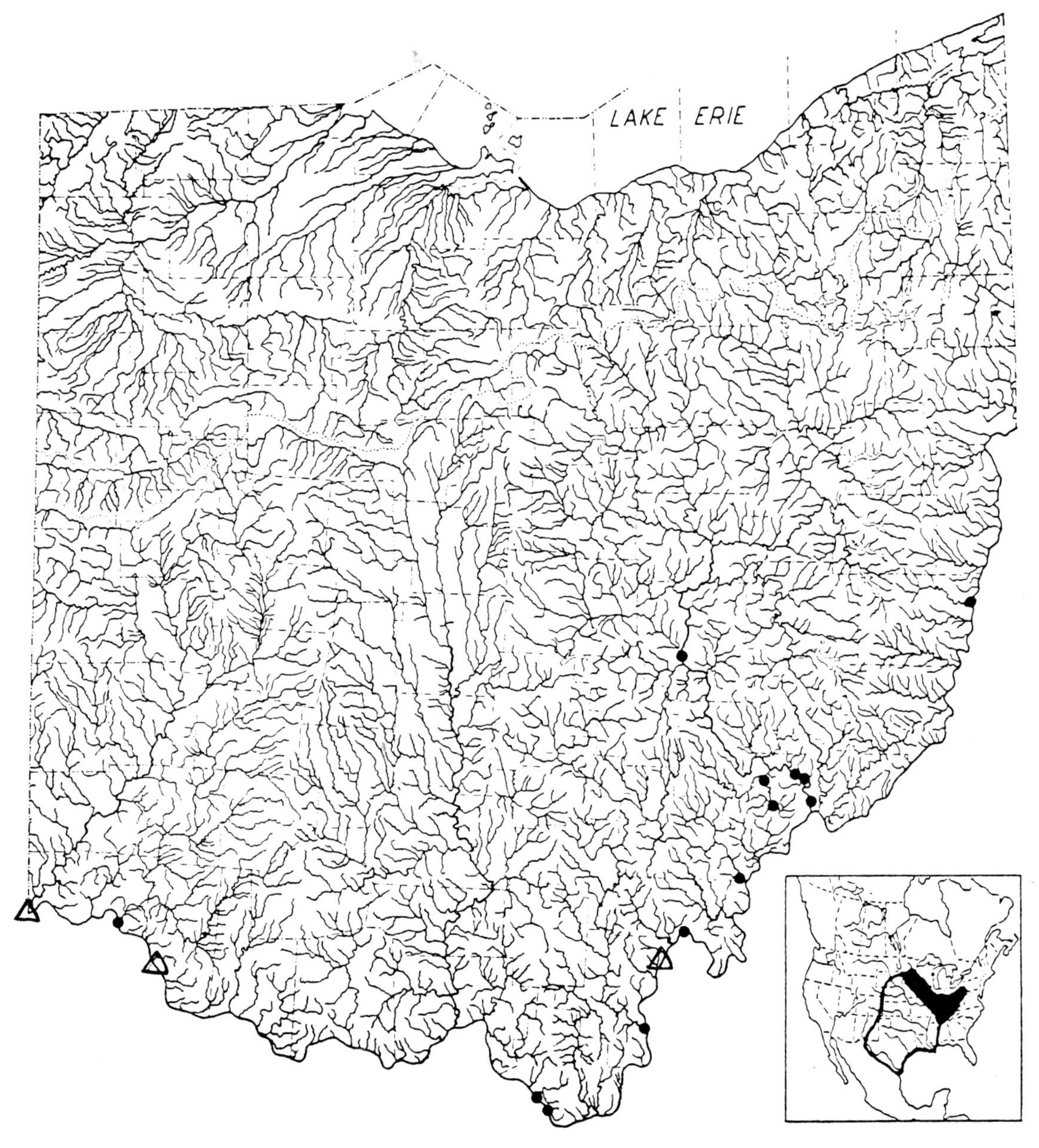

MAP 47. Ohio speckled chub

● Locality records until 1954. △ 1955–80.
Insert: Black area represents range of this subspecies, outlined area of other subspecies; area of intergradation, and its width between the subspecies indefinite, as are the western and Mexican limits of range.

Gallia County, in 1888 (Henshall, 1889:124); Ohio River at Ironton, Lawrence County, in 1899; Ohio River at Bellaire, Belmont County, in 1900 (Osburn, 1901:62); South Branch of Wolf Creek, Washington County, on July 13, 1921 (collected by Edward L. Wickliff). All other collections were made by me since 1929. These later collections indicate that a fairly large population existed in the lower third of the Muskingum River and that smaller populations were present in the Ohio River and lower portions of its tributaries between Lawrence and Washington counties. The only collection west of the Scioto River was of an individual taken in the Ohio River, Hamilton County, in 1942. It appears highly significant, that despite much effort, no specimens of this large-river, sand-inhabiting species were taken from the more turbid Scioto River, presumably because during the 1925–54 surveys this stream, unlike the Muskingum, lacked large areas of silt-free, clean sand and gravel. It is quite possible that previous to 1925 the speckled chub was more abundant and widespread than it has been since, before the sand bottoms of the Scioto and other Ohio drainage rivers were covered with silt.

Habitat—The speckled chub inhabited the clean sand and fine gravel bottoms of the swifter portions of large rivers, usually remaining in waters more than 4′ (1.2 m) in depth during the day, especially if these waters were clear, and venturing into the shallows only on dark nights. Upon three nights, specimens were collected in shallow water until the bright moon began to shine, after which none could be taken.

Years 1955–80—Although concerted daytime attempts were made during the period to capture the speckled chub, none was taken. Habitat conditions for the species in the Ohio River opposite the State of Ohio appear to have deteriorated. Reasons for failing to capture it in the lower Muskingum River are not known, except that no intensive, concerted efforts were made during dark, moonless nights, similar to those made prior to 1950.

During the 1957–59 investigations of the entire Ohio River by William M. Clay and associates, a total of 82 specimens in seven collections were taken (ORSANCO, 1962:147). Between 1976 and 1978 Margulies and Burch captured the species in three of their five collecting stations in the Ohio River adjacent to the state of Ohio.

WESTERN BLACKNOSE DACE

Rhinichthys atratulus meleagris Agassiz

Fig. 48

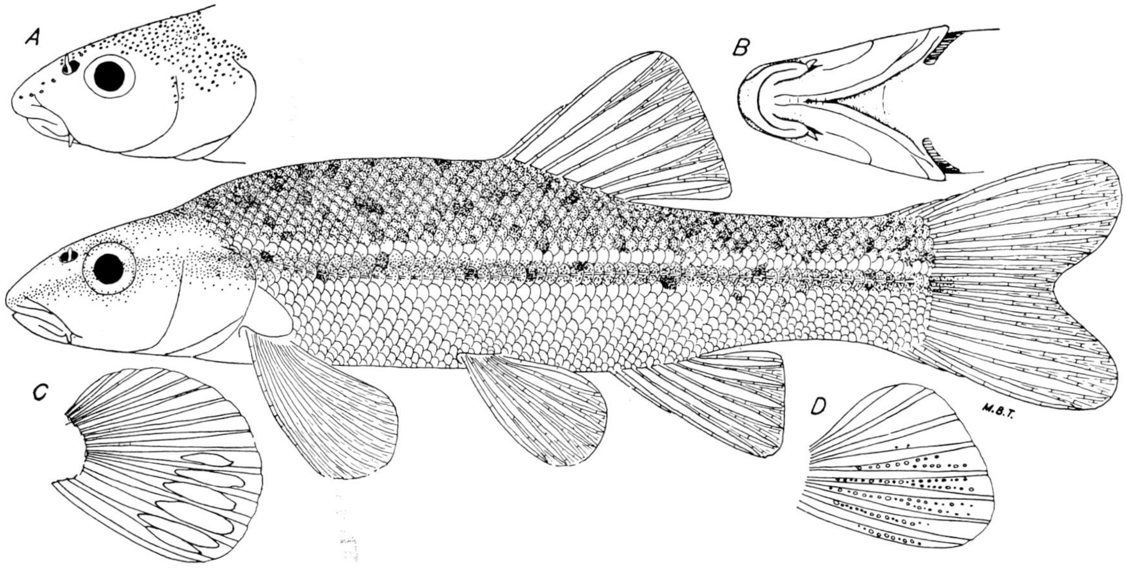

A B C D M.B.T.

Central fig.: Buck Run, Hocking County, O.

April 30, 1938. 59 mm SL, 2.8" (7.1 cm) TL.
Breeding male. OSUM 2438.

Fig. A: lateral view of head, showing distribution of minute tubercles.
Fig. B: ventral view of head showing the snout slightly overhanging the upper lip.
Fig. C: pectoral fin; note breeding pads.
Fig. D: pelvic fin; note tubercles.

Identification

Characters: *General—See* golden shiner. *Generic*—A small barbel at posterior end of upper jaw, figs. A and B. Tip of upper jaw *without* a groove separating it from tip of snout; in this character this genus differs from *Hybopsis* and *Nocomis*. Scales in complete lateral line 56–70. Teeth, normally 2,4–4,2. *Specific*—Tip of upper lip about on a level with the lower edge of eye. Snout scarcely projecting beyond the oblique and subterminal mouth, figs. A and B. Preorbital length of head contained 1.1 (large adults)–1.7 (small young) times in the postorbital length. Eye contained 3.8 (small young)–5.3 (large adults) times in head length. Dark lateral band usually contrasts sharply, both above and below, with the silvery sides.

Differs: Longnose dace has a longer snout and smaller eye. Other minnows with barbels have fewer lateral line scales and except for the tonguetied minnow have a groove separating tip of upper lip from remainder of snout. Suckers have more than 8 dorsal rays.

Most like: Longnose dace. **Superficially like:** Redbelly dace and stoneroller minnow.

Coloration: Dorsally a dark greenish- or bluish-olive. Sides lighter with a dusky band, about as wide as eye, encircling snout and extending across opercles and sides of body to tail. Usually a narrow, light streak separates the dusky lateral band from the dark back. Lower sides usually with some yellowish, belly milk-white. Many dusky scales scattered over back and sides, resulting in a mottled appearance. Fins light olive or transparent. In the breeding male the lateral band is suffused with pink or red, and the lower fins contain much yellow, with the pectoral fins having the pads in their centers a deep orange or orange-red, fig. C. Head of breeding

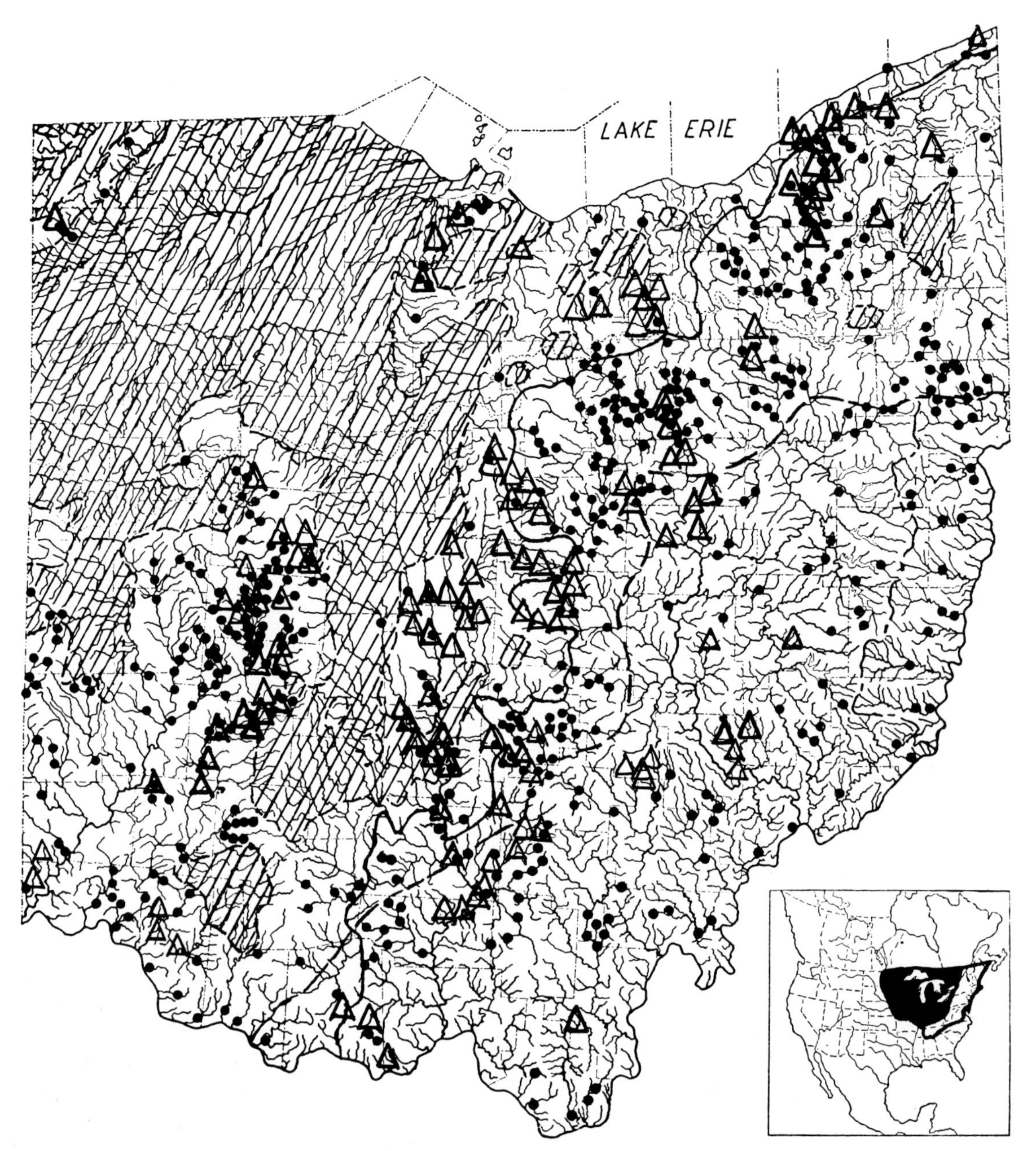

MAP 48. Western blacknose dace

Locality records. • Before 1955. △ 1955–80. ~ Allegheny Front Escarpment.
/ \ Glacial boundary. – — Flushing Escarpment.
Shaded area has maximum relief of 100′ (30.5 m) or less.

Insert: Black area represents range of this subspecies, outlined area of other subspecies; northern range limit indefinite; introduced outside original range, usually inadvertently by bait fishermen.

male with tubercles, fig. A; pelvic fins with tubercles, fig. D. Breeding female has the lateral band suffused with yellow.

Lengths: *Young* of year in Oct., 0.8″–2.0″ (2.0–5.1 cm) long. Around **1** year, 1.0″–2.4″ (2.5–6.1 cm). *Adults*, usually 2.0″–3.8″ (5.1–9.7 cm); seldom reaching 4.0″ (10 cm) long.

Hybridizes: with river chub, longnose dace.

Distribution and Habitat

Ohio Distribution—Before 1900 the blacknose dace presumably was an abundant inhabitant of Ohio brooks, excepting perhaps the streams of lowest gradients and in areas where maximum relief was less than 100′ (30.5 m). Its avoidance, since 1925 at least, of streams in areas of low relief is remarkable; the comparatively few records in these areas (shaded portion on distribution map) consist of strays as in Big Darby Creek in central Ohio, or of small colonies adjacent to areas of higher relief as in western Ohio.

Since 1900 drastic decreases in abundance and elimination of colonies, have occurred in many sections of higher relief (unshaded portion); by 1950 large populations flourished only in the upper Mad River drainage, along the glacial boundary, and immediately to the east of the Allegheny Front Escarpment. During the 1925–54 period decreases in abundance or elimination of populations were obviously caused by drastic habitat modifications, which converted the former sand and gravel bottom, forest and brush bordered, permanently flowing, unpolluted, spring-fed, clear-water brooks into surface-fed, intermittent, flash-flood, turbid, silt-bottomed, deforested ones. Many other brooks became polluted with domestic, industrial and mine wastes; others were ditched or completely eliminated. In many of the still-suitable brooks this extremely vulnerable species faced extirpation because of the excessive seining by bait dealers.

Habitat—An inhabitant of moderate- and high-gradient brooks whose waters were clear except for brief periods, which had a permanent flow, sand and gravel bottoms, well-defined riffles for spawning purposes, pools containing deep holes, undercut banks, brush and roots for safety from enemies, and shade during much of the day. The blacknose inhabited these waters throughout the year, but in submarginal brooks, especially intermittently-flowing ones, the species was forced to retreat downstream into larger waters and especially during drought and winter. Because of this, small numbers or strays were found in the larger streams.

Ohio specimens appeared to be typical *R. a. meleagris*, except those individuals collected east of the Flushing Escarpment (especially from tributaries of Sunfish Creek in Belmont County). These Belmont County specimens had a tendency toward the larger mouth and head shape of the southern blacknose dace, *Rhinichthys atratulus obtusus* Agassiz.

Years 1955–80—Between 1955 and 1979 the blacknose dace was collected throughout much of Ohio and in more than 125 localities. Many additional localities were likewise investigated, in which none was found but which before 1950 contained thriving populations. In these latter localities it was obvious that the former favorable habitat of this species had been largely or entirely destroyed (White et al., 1975:70–71).

Invariably, the most flourishing populations were found in those permanently flowing brooks of high gradient which were turbid only for brief periods of time. Rarely were a few individuals or strays taken in areas having a maximum relief of 100′ (30.5 m) or less; *see* map 48.

For some aspects of reproduction in this subspecies *see* Tarter (1969:454–59).

LONGNOSE DACE

Rhinichthys cataractae (Valenciennes)

Fig. 49

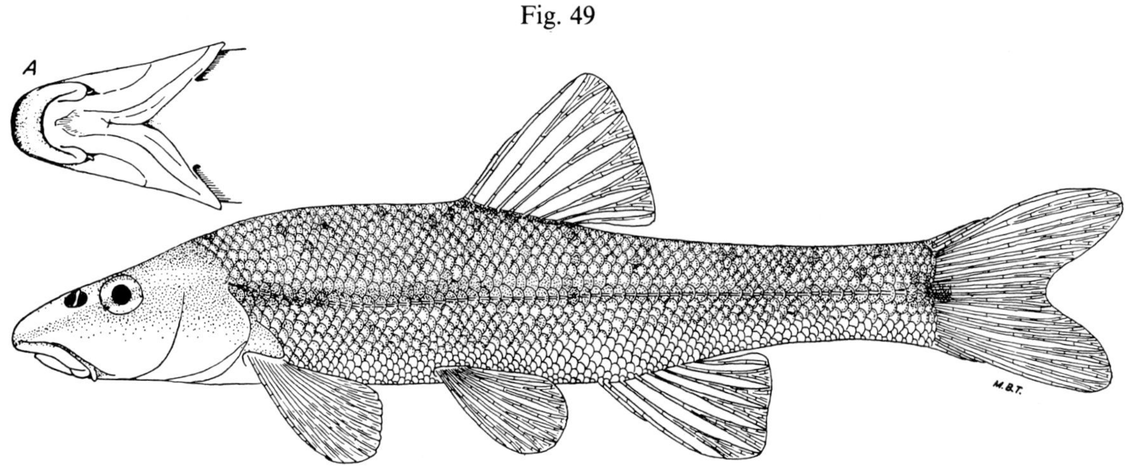

Lower fig.: Lake Erie, Erie County, O.

April 19, 1940. 68 mm SL, 3.3″ (8.4 cm) TL.
Breeding female. OSUM 2015.

Fig. A: ventral view of head; note the overhanging snout.

Identification

Characters: *General*—See golden shiner. *Generic*—*See* blacknose dace. *Specific*—Tip of upper lip far below the level of the lower edge of eye. Snout projecting far beyond the almost horizontal mouth, fig. A. Preorbital length of head usually contained 0.7 (large adults)–1.2 (small young) times in the postorbital length. Eye contained 4.8 (small young)–6.3 (large adults) times in head length. Dusky lateral band poorly defined, often absent.

Differs: *See* blacknose dace. Blacknose dace has a larger eye, less overhanging snout.

Most like: Blacknose dace. **Superficially like:** Tonguetied and stoneroller minnows.

Coloration: Dorsally a dark greenish- or brownish-olive. Sides somewhat lighter, with or without a vague lateral band; when present, band is about as wide as eye, and then encircles snout, crosses opercles and extends backward across body to tail. Ventrally a milk-white or silvery. Many dusky scales scattered over back and sides resulting in a mottled appearance. Fins light olive or transparent. A caudal spot usually present. Breeding males have an orange or rosy blush to upper edge of upper lip; a red spot above pectoral fin origin; the fins, especially their basal portions, flushed with orange or rose.

Lengths: *Young* of year in Nov., 1.0″–2.3″ (2.5–5.8 cm) long. Around **1** year, 1.5″–2.5″ (3.8–6.4 cm). *Adults*, usually 2.5″–4.5″ (6.4–11 cm) long.

Hybridizes: with river chub, blacknose dace, creek chub, stoneroller minnow and common shiner.

Distribution and Habitat

Ohio Distribution—The collection of this species (UMMZ 63030, 3 sp.; OSUM 1, 1 sp.), in August, 1853, by S. F. Baird in Yellow Creek, Mahoning County, constitutes the first record for the state, and the only positive Ohio River drainage record. The longnose dace probably was more widely distributed in eastern Ohio before 1860 than this record indicates; it still occurs in the upper Ohio drainage in northwestern Pennsylvania (Raney, 1938:31, map 42). In the National Museum (USNM 36592) are ten large specimens; concerning these the museum catalog contains the following "Marietta, Ohio, William Holden, Entered October 13, 1884."

The first Lake Erie drainage record for Ohio is

MAP 49. Longnose dace

Locality records. ● Before 1955. △ 1955–80.
Insert: Some isolated populations along, or south of, southern border of range; northern limit of range indefinite.

that of Osburn (1901:61), who collected a specimen on August 2, 1900 in a tributary of the Grand River near Painesville. Since 1929 the species was found to be fairly numerous in winter in the Chagrin River drainage, and along the shores of Lake Erie from Erie County eastward. Much shallow water seining and trawling in the deeper waters about the Bass Island region of western Lake Erie failed to produce a specimen.

All recent collections were from seinings made during the cooler months. Apparently few remained in the Chagrin River system during the heat of summer. In fall they migrated into the Chagrin River from the lake, and by January were common, to remain so until April. They appeared to be likewise absent from the shores of Lake Erie during the heat of summer, for we were able to collect them only between the months of September and late May. Apparently they moved into offshore waters for the summer.

Habitat—In inland Ohio the longnose dace was most numerous only in the swiftly-flowing riffles and pools with considerable current of the high-gradient portions of the Chagrin River system, where the stream bottom was composed chiefly of gravel, boulders and bed rock. In Lake Erie it inhabited the gravelly, boulder-strewn beaches, often hiding under the larger stones and boulders, and remaining in the shallow waters even during a heavy surf.

Years 1955–80—Since 1970 Andrew White and his associates have conducted intensive investigations along the shores of Lake Erie in the vicinity of Cleveland, and in the Rocky, Cuyahoga and Chagrin rivers. Seinings along the shores and lower portions of the rivers have produced occasional specimens. A local population has been present since at least 1970 in a limited portion of the East Branch of the Chagrin River, which is normally isolated from Lake Erie by a dam at Willoughby, except possibly during high floods (White et al., 1975:71).

NORTHERN CREEK CHUB

Semotilus atromaculatus atromaculatus (Mitchill)

Fig. 50

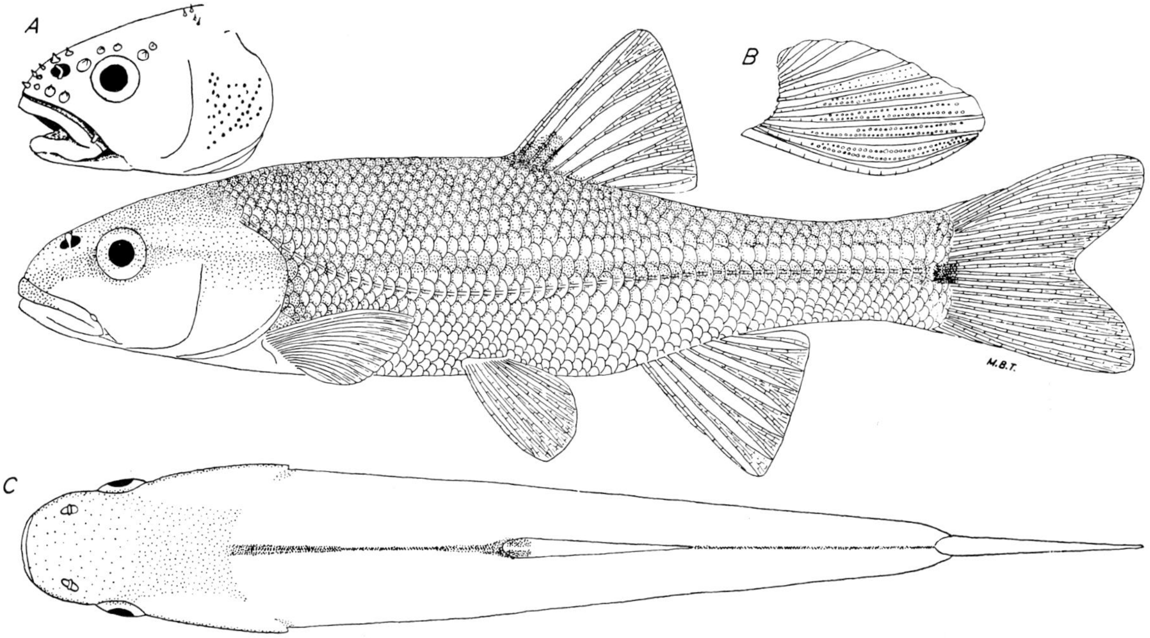

Central fig.: Mad River, Logan County, O.

April 6, 1942. 78 mm SL, 3.8″ (9.6 cm) TL.
Breeding female. OSUM 4743

Fig. C: dorsal view of the above; the head shape and dorsal stripe which enlarges at the dorsal fin origin are very characteristic.

Figs. A and B: Sunfish Creek, Monroe County, O.

May 2, 1940. 163 mm SL, 7.7″ (20 cm) TL.
Breeding male. OSUM 2039.

Fig. A: lateral view of head; note the small, flap-like barbel well in advance of the posterior end of the upper jaw in the groove between the jaw and preorbital portion of snout; also breeding tubercles of male.

Fig. B: pectoral fin of breeding male showing tubercles.

Identification

Characters: *General—See* golden shiner. *Generic* and *Specific*—A small, flap-like barbel which is well in advance of the posterior end of the upper jaw, and is hidden or nestles in the deep groove separating upper jaw and preorbital portion of snout, fig. A. Barbels must be looked for carefully, because they usually are hidden in the groove; occasionally the barbel is absent on one side; rarely on both sides. Tonguetied minnow is the only other species in Ohio having barbels in a similar position. A deep groove separates tip of upper lip from remainder of snout. Jaws equal and mouth terminal. Mouth large, the posterior end of upper jaw extending well beyond the anterior edge of eye. Tip of upper lip extending to, or above, anterior edge of eye. A prominent dusky blotch at base of anterior dorsal fin, a characteristic feature of this species. Complete lateral line with 50–60 scales. Predorsal stripe and bluntly-pointed head distinctive. Teeth, variable, usually 2,5–4,2 or 2,4–5,2; occasionally 2,4–4,2 or 2,5–5,2; rarely with only 1 tooth in the inner row.

Differs: Western tonguetied minnow lacks the

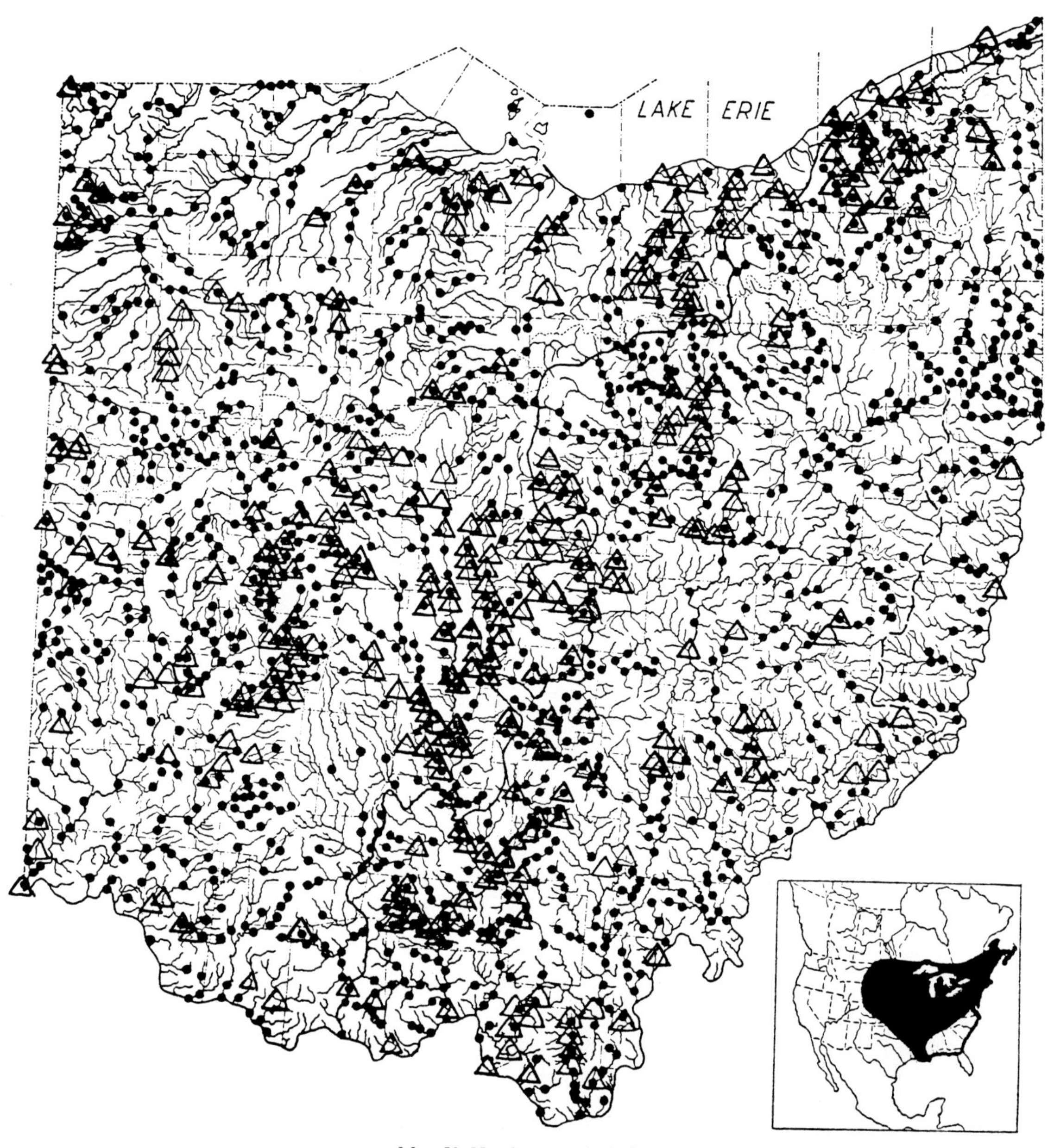

MAP 50. Northern creek chub

Locality records. ● Before 1955. △ 1955–80. ~ Allegheny Front Escarpment. /\ Flushing Escarpment.
Insert: Black area represents range of nominate subspecies, outlined area of a nominal subspecies whose validity is questioned by Bailey, Winn and Smith (1954:124); widely introduced, often inadvertently by bait fishermen, into many western states.

groove between tip of upper lip and snout. Other species of minnows lack barbels or have them at the posterior end of the upper jaw. Suckers have more than 8 dorsal rays.

Most like: Redside and rosyside daces, hornyhead and river chubs. **Superficially like:** Western tonguetied minnow; redfin and rosefin shiners.

Coloration: Dorsally olivaceous with a steel-blue and silvery overcast, a dusky spot at base of anterior dorsal rays. Sides lighter and more silvery. Ventrally milk-white. A distinct but usually small, oblong caudal spot. Fins transparent, with a silvery-bluish cast. A dusky band encircles snout, crosses opercles and extends backward along body to tail; this band most distinct in young, or in fishes from clear or vegetated waters; least distinct or absent in old fishes or those from turbid waters. *Breeding male*—Has large tubercles on dorsal half of head, and tiny tubercles on cheeks and opercles, fig. A; minute tubercles on free edges of some body scales, especially those anteriorly and on pectoral, fig. B, pelvic, anal and caudal rays; head and body contains blues, greens and rose. *Breeding female*—Less brilliant coloration.

Lengths and **weights:** *Young* of year in Oct., 1.3″–2.8″ (3.3–7.1 cm) long. Around **1** year, 1.8″–3.5″ (4.6–8.9 cm). *Adults*, usually 3.0″–8.0″ (7.6–20 cm). Fishes 8.0″–10.0″ (20–25.4 cm) long uncommon. Largest specimen 11.9″ (30.2 cm) long, weight 12 oz (340 g). All fishes more than 9.5″ (24 cm) long have been males.

Hybridizes: with river chub, longnose, redbelly, redside and rosyside daces, striped and common shiners.

Distribution and Habitat

Ohio Distribution—The generally abundant and ubiquitous creek chub occurred at least as strays, in every Ohio stream capable, even for only short periods, of sustaining fish life; small populations or strays were found in almost every pond, slough, reservoir and lake. The species avoided Lake Erie, although strays have been found about the Bass Islands, and a small specimen was taken in the open waters of the lake. It was essentially a small creek or brook species throughout spring, at which time the greatest population densities occured in areas of high relief such as are in the upper Great Miami River drainage, along the Allegheny Front Escarpment, and east of the Flushing Escarpment. After the spawning season most of the adults, subadults and fry moved downstream into larger waters. Fair-sized populations were present from October to March in the Ohio River, and especially about the mouths of its small tributaries. The chub occurred in the fewest numbers in streams of the lowest gradients, or those which were heavily polluted, extremely turbid, or had heavily silted bottoms.

Habitat—An abundant inhabitant, particularly in spring, of the brooks, creeks and small streams having gradients between 10′–75′/mile (1.9–14 m/km) which had well scoured bottoms of sand, gravel, boulders and bedrock, well defined riffles, and pools with holes sufficiently deep and/or with brush, roots or other cover sufficient for retreat in times of danger. The majority of the population moved into the deeper pools of larger streams during droughts and in winter, seemingly semi-hibernating during the coldest weather under cutbanks, among piles of leaves lying on the stream bottom, and under stones. Most individuals, especially adults, moved into smaller streams at the approach of spring, after the water temperature had reached 40° F (4° C).

The creek chub, in addition to being one of the most prized bait minnows, was sought after as a pan fish chiefly by boys, except in streams east of the Flushing Escarpment where pollutions and high gradients had limited the fish fauna. In these streams the creek chub was much sought after as a pan fish by adult fishermen. A century ago Kirtland (1850E:213) wrote that "during winter, they are frequently taken in great numbers by cutting a hole through the ice and fishing with a hook baited with pork." Kirtland considered them to be an excellent pan fish.

Years 1955–80—The creek chub was present, at least as strays, in all except the most turbid of streams or those which were heavily polluted otherwise. The species continued to be one of the most numerous present in most of the smaller streams, especially those whose substrates of sands, gravels, boulders or bedrocks normally were usually free of silts. Large-sized populations survived in many diverse situations, from the smallest of isolated pools to the largest of streams, including the Ohio River. It often increased greatly in numbers after a stream was channelized or dredged (Trautman and Gartman, 1974:168). It sometimes became a dominant species in some of the larger channelized streams, whose former fish fauna consisted of goodly numbers of diverse species but after channelizing had become pauperized.

For nest-entry behavior of female creek chubs *see* Ross, M. R., 1976:378–80; nest building, Ross, 1977.

WESTERN TONGUETIED MINNOW

Exoglossum laurae hubbsi Trautman

Fig. 51

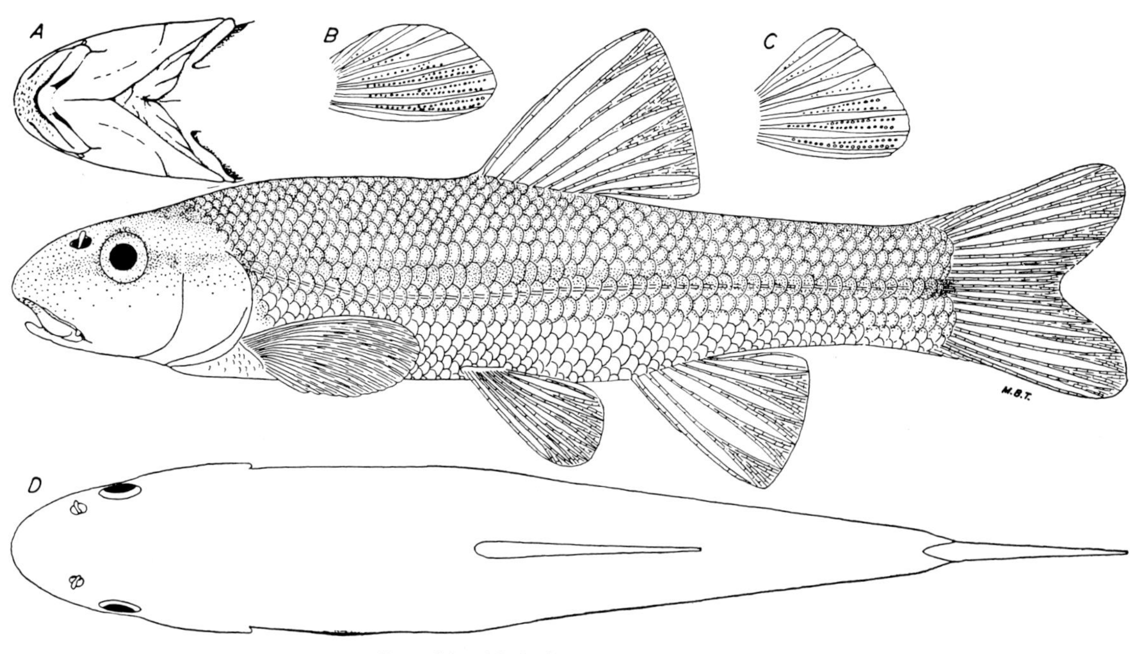

Central fig.: Mad River, Logan County, O.

April 6, 1942.
Male.

100 mm SL, 4.9″ (12 cm) TL.
OSUM 4747.

Fig. A: ventral view of head; note peculiarly shaped lower jaw.
Fig. B: tuberculated pectoral fin of male.
Fig. C: tuberculated pelvic fin of male.
Fig. D: dorsal view.

Identification

Characters: *General—See* golden shiner. *Generic* and *Specific*—A small, flap-like barbel which is well in advance of the posterior end of the upper jaw, and is hidden or nestles in the deep groove separating upper jaw and preorbital portion of snout; this barbel similar to that of the creek chub, *see* that species. Tip of upper jaw *not* separated from snout by a groove, and the tip is considerably below lower edge of eye. Lower jaw definitely included; also highly modified in that the fleshy lobes, which in most minnows entirely cover the lower jaw, in this species cover only the posterior two-thirds of each side, leaving the anterior third covered only with a horny sheath, fig. A. Complete lateral line with 48–52 scales. Snout more pointed, fig. D, than is snout of creek chub.

Differs: *See* creek chub. Creek chub has a terminal, larger mouth. **Superficially like:** Stoneroller minnow; creek, hornyhead and river chubs.

Coloration: Dorsally a slaty-olive; sides lighter; under-parts a pellucid white. The entire fish has a deep violet- or bluish-purple and green overcast which gives this species a peculiar and most distinctive appearance. Fins transparent, tinged with olive or silvery in the larger fishes. *Breeding male*—Has minute tubercles on the pectoral, fig. B, and pelvic, fig. C, fins.

Lengths and **weights:** *Young* of year in Oct., 1.3″–2.3″ (3.3–5.8 cm) long. Around **1** year, 1.8″–2.5″ (4.6–6.4 cm). Some breeding *adults* only 2.7″

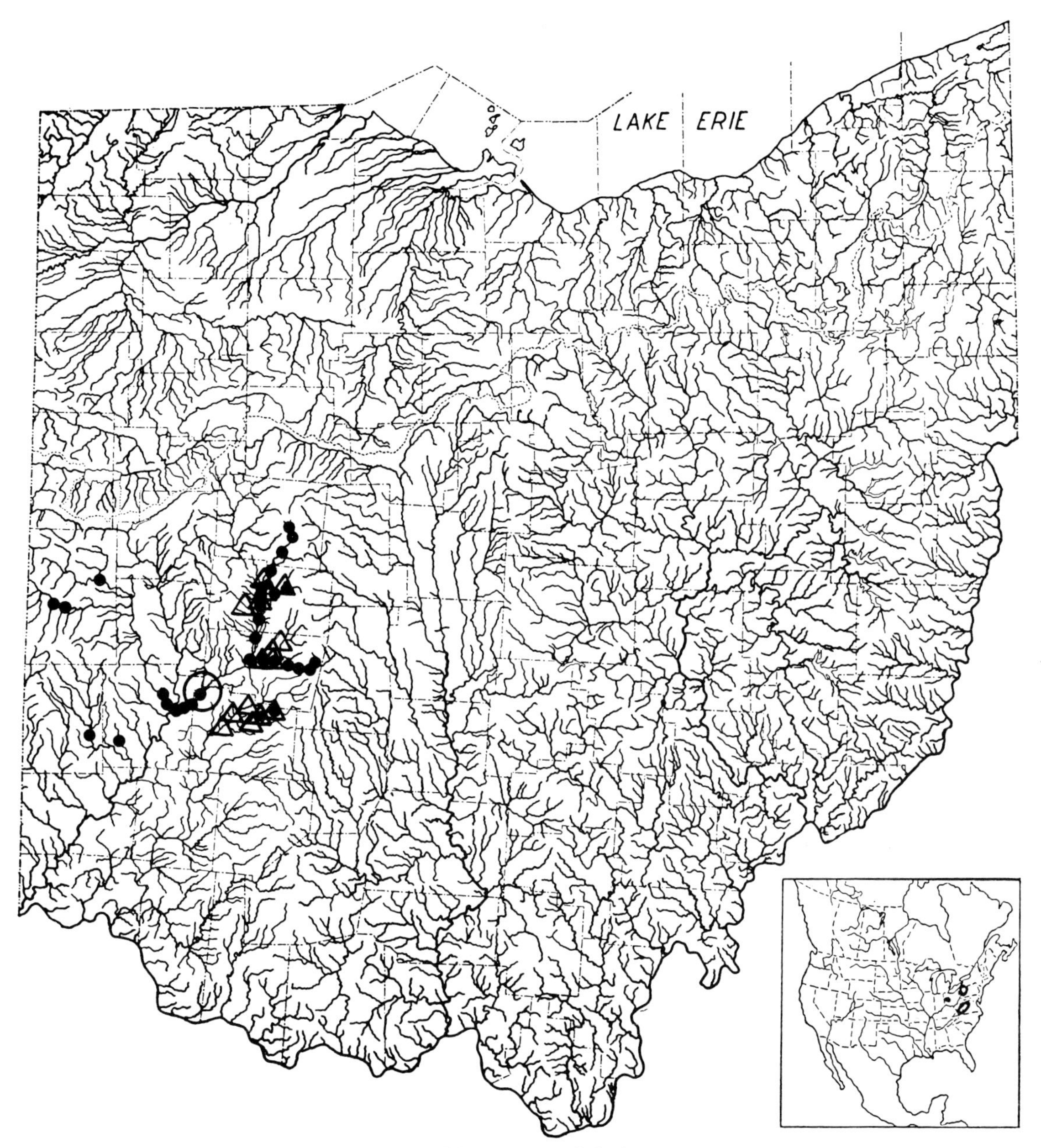

MAP 51. Western tonguetied minnow

Locality records. ○ Type locality, Mad River, Greene County. ● After 1927. △ 1955–80.
Insert: Black area represents range of this subspecies, the two outlined areas the disjunct range of the nominate form.

(6.9 cm) long; usually 3.0″–5.8″ (7.6–15 cm) long. Largest specimen, 6.1″ (15 cm) long; weight 1.5 oz (42 g).

Distribution and Habitat

Ohio Distribution—In the Stillwater River near Dayton on August 15, 1899, R. C. Osburn collected six minnows, presumably of *Parexoglossum l. hubbsi*, which he (1902:64–65) recorded later as *Exoglossum maxillingua* (Trautman, 1931A:1). Since 1927 many specimens have been taken from the upper tributaries of the Great Miami system. In June, 1952, Frank Steinbach found the species to be not uncommon in Massie's Creek above the Cedarville Falls, in eastern Cedarville Twp., Greene County, which is in the Little Miami River system.

Habitat—The western tonguetied minnow was most numerous in streams normally of considerable clarity and seldom turbid, where the gravel and pebble bottoms were almost, or entirely, free from clayey silts, the gradients were about 8′ per mile (1.5 m/km) [range 3′–20′/mile (0.6–3.8 m/km)], the numbers of other fish species and especially of individuals were low, and where the river and hornyhead chubs were rare or absent (*see* river chub). The western tonguetied minnow spawned and inhabited moderately-flowing riffles and pools, building large nests each containing as much as a bushel of small roundish pebbles. The nests were rectangular in shape, 16″–18″ (41–46 cm) in width and 1′–4′ (0.3–1.2 m) in length, and sometimes had one end butting against a large boulder or log or the nest was between two or more large boulders. This minnow usually built its nest with the long axis at right angles to the current, unlike those of the river and hornyhead chubs, whose nests were circular or with their long axis parallel with the current.

Years 1955–80—Between 1952 and 1955 Brown (1960:50–51) conducted fishery investigations on the headwaters of the Little Miami River, including Massie Creek, in Clark and Greene counties, collecting 1 to 10 tonguetied minnows at 6 of his 17 collecting stations.

In 1931 I described *Parexoglossum hubbsi* (Trautman 1931A:1–11; type loc., Mad River, Greene County). At that time the then recently described eastern tonguetied minnow, *Parexoglossum laurae* (Hubbs, 1931:1–12) was known only from the Kanawha River above the falls, and the largest specimen of *hubbsi* was only 63 mm SL. Since then the eastern tonguetied minnow has been found in the upper Allegheny River system of New York and Pennsylvania, and in the Lake Ontario basin in the Genesee River in New York above the falls (Raney, 1941B:272); also the maximum size of the western tonguetied minnow has been found to equal that of the eastern tonguetied minnow. The findings of the eastern subspecies inhabiting streams other than those of the Kanawha River system and examination of many additional western tonguetied minnows indicate that the western subspecies is less distinct from the eastern form than was formerly supposed; in fact, it appears that the eastern and western forms may be inseparable. Until a thorough study is made, it seems more conservative to retain *hubbsi* as a subspecies of *Exoglossum laurae*.

In the 1957 edition of this report this species was *Parexoglossum laurae hubbsi*, western tonguetied chub. In the third edition of *Common and Scientific Names of Fishes* (Bailey et al., 1970:19) the genus *Parexoglossum* was placed in synonomy and included with the barbless cutlips minnow in the genus *Exoglossum*, as *Exoglossum laurae hubbsi*, western tonguetied minnow.

In 1968 and 1969 Stephen H. Taub, then with the Ohio Cooperative Fishery Unit, began a life-history study of this minnow in Ohio. He investigated and attempted to collect specimens at all of the stations where the species had been found since 1927, including those of Brown, succeeding in recording them only in Kings Creek and adjacent Mad River downstream to the Route 36 bridge west of Urbana.

Taub's investigations have centered on the 9-mile-long (14.5–km) Kings Creek in Champaign County. Taking water temperatures at least weekly throughout the summers of 1968–69, he recorded a maximum of only 70° F (21° C), which is below the average maximum temperature of an Ohio stream. He observed that the waters were seldom turbid except below a gravel operation conducted throughout 1968 and until mid-1969. The substrate appeared to be especially suited for this species, being composed chiefly of small gravels such as used by it in nest-building. The species was most numerous in those stream sections in which the banks were wooded and undercut and where well-defined riffles and alternating pools were present. It seemed to be intolerant of recently disturbed banks and channelization. Nest-building was most active when water temperatures approached 60° F (16° C), usually during mid-May. Taub's preliminary conclusion was that possibly fewer than 10,000 individuals were to be found in Ohio in 1970.

SUCKERMOUTH MINNOW

Phenacobius mirabilis (Girard)

Fig. 52

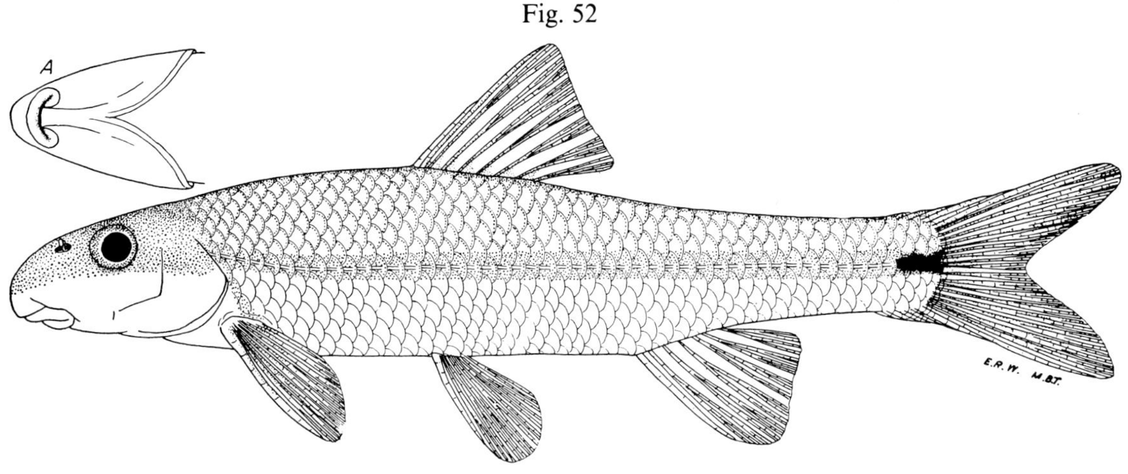

Lower fig.: Fish Creek, Williams County, O.

July 13, 1939. 72 mm SL, 3.5″ (8.9 cm) TL.
Male. OSUM 355.

Fig. A: ventral view of head; note distinctive lower jaw.

Identification

Characters: *General—See* golden shiner. *Generic* and *Specific*—No barbel about mouth. Fleshy lips of lower jaw restricted to lobes on their posterior halves, fig. A, thereby superficially resembling the lower jaw of the western tonguetied minnow. Tip of upper jaw separated from remainder of snout by a deep groove. Mouth inferior; the snout overhanging it. Scales in the complete lateral line 43–51. Anal rays 7. A black band usually extending along the sides. Alimentary tract equal to head and body length. Peritoneum silvery. Teeth 4–4.

Differs: All species of chubs have barbels. Stoneroller minnow lacks fleshy lips entirely. Remaining species of minnows have normal-appearing lower jaws. Suckers have more than 8 dorsal rays, and anal fin situated more posteriorly.

Superficially like: Bigeye, streamline and speckled chubs, small suckers; western tonguetied and stoneroller minnows. A most distinctive species.

Coloration: Dorsally yellowish- or slaty-olive with a pronounced silvery sheen and a sharp, slate line along dorsal ridge. Sides silvery, except for the dusky band, as wide as eye, which encircles snout, extends across opercles and continues backward along sides, to end in an oblong, deep-black caudal spot. This band sometimes faint or absent in large specimens from turbid waters. Ventrally milk-white with much silvery. Fins transparent, the lower ones sometimes whitish. *Breeding male*—Has tiny tubercles on the dorsal half of head and predorsal region of body; also on pectoral and pelvic fins.

Lengths: *Young* of year in Oct., 1.5″–2.8″ (3.8–7.1 cm) long. Around **1** year, 2.0″–2.9″ (5.1–7.4 cm). *Adults*, usually 2.5″–4.0″ (6.4–10 cm). Largest specimen, 4.8″ (12 cm) long.

Distribution and Habitat

Ohio Distribution—Before 1800 this plains-inhabiting species presumably ranged only as far eastward as the Mississippi River, except perhaps for outlying and/or existing relict, eastern populations. It was not until 1876 that the species was first recorded for Illinois (Nelson, 1876:46); also in that year David S. Jordan collected a specimen in the White River of Indiana (USNM 117360). It was not recorded for Ohio until 1920, when specimens were

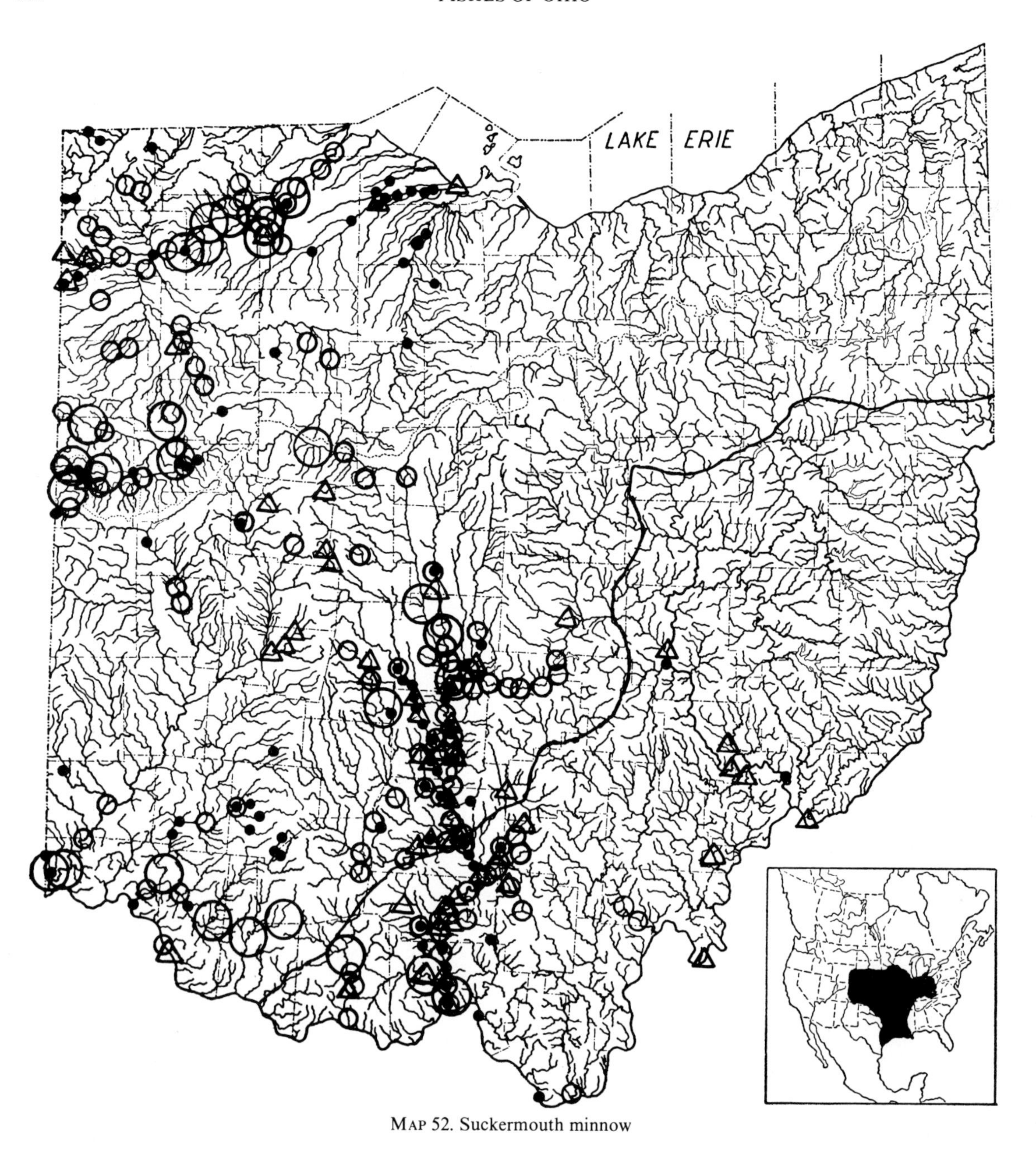

MAP 52. Suckermouth minnow

Locality records. ○ 1920–25. ○ 1926–38. ● 1939–54. △ 1955–80. ~ Glacial boundary.
Insert: The northeastern extension of the range presumably represents a recent eastward invasion and/or an expanding and merging of existing relict, northeastern populations.

taken in the Maumee and Scioto river systems. Presumably, between 1909 and 1920 it had migrated across Indiana, following the eastward extension of its habitat, made possible through conversion of virgin prairie and forest into vast cornfields which changed the clear, gravel- and sand-bottomed prairie-type streams into turbid plain-type ones with silty bottoms. By 1922 it was abundant and well distributed throughout the Maumee River drainage in Ohio; by 1938 its range extended in isolated populations in the Ohio River drainage as far eastward as Meigs County; also to Buckeye Lake in the Muskingum River drainage where it possibly had been introduced inadvertently from the bait buckets of fishermen. By 1950 it had become widespread in the streams tributary to Sandusky Bay, and was present in isolated populations in the Muskingum River. If present agricultural practices continue it may become rather well distributed throughout the remainder of glaciated Ohio; but should be restricted in the unglaciated portion to isolated populations, except for such glacial outlets as the Muskingum River. Like other invading species, it became unusually abundant in a given locality a few years after first invading that locality, after which its numbers decreased noticeably as the species became a permanent part of the fauna.

Habitat—The suckermouth minnow chiefly inhabitated in largest numbers the riffles of those portions of streams whose waters were usually turbid and rich in organic material, and whose gradients were sufficiently high so that the gravel on the riffles remained comparatively free from silt. It usually reached large population densities in large streams and rivers which were rich in organic material, and especially when competition pressure from other riffle species was low. Populations were low or absent in the clearest streams, in streams of high gradients, and in normally turbid streams whose gradients were too low to prevent silt accumulations from covering the stones on the riffles. The suckermouth occurred in large numbers throughout the Maumee and Scioto river systems, the two most turbid stream drainages in the state.

Years 1955–80—Between 1920 and 1950 the eastward invasion of the suckermouth minnow across Ohio was rapid. Since 1950 little eastward progress has been observed, and the species has failed to materially extend its range in both glaciated and unglaciated territory; *see* map 52.

As has been stated above, this species often became abundant in a given locality a few years after invasion, after which its numbers usually decreased. Declines in abundance of established populations continued after 1950, in some localities to a considerable degree. The species never has been present in as large numbers in glacial outlet streams, such as the lower Muskingum River, as it has been at times in glaciated portions of the Scioto and Maumee drainages.

SOUTHERN REDBELLY DACE

Phoxinus erythrogaster (Rafinesque)

Fig. 53

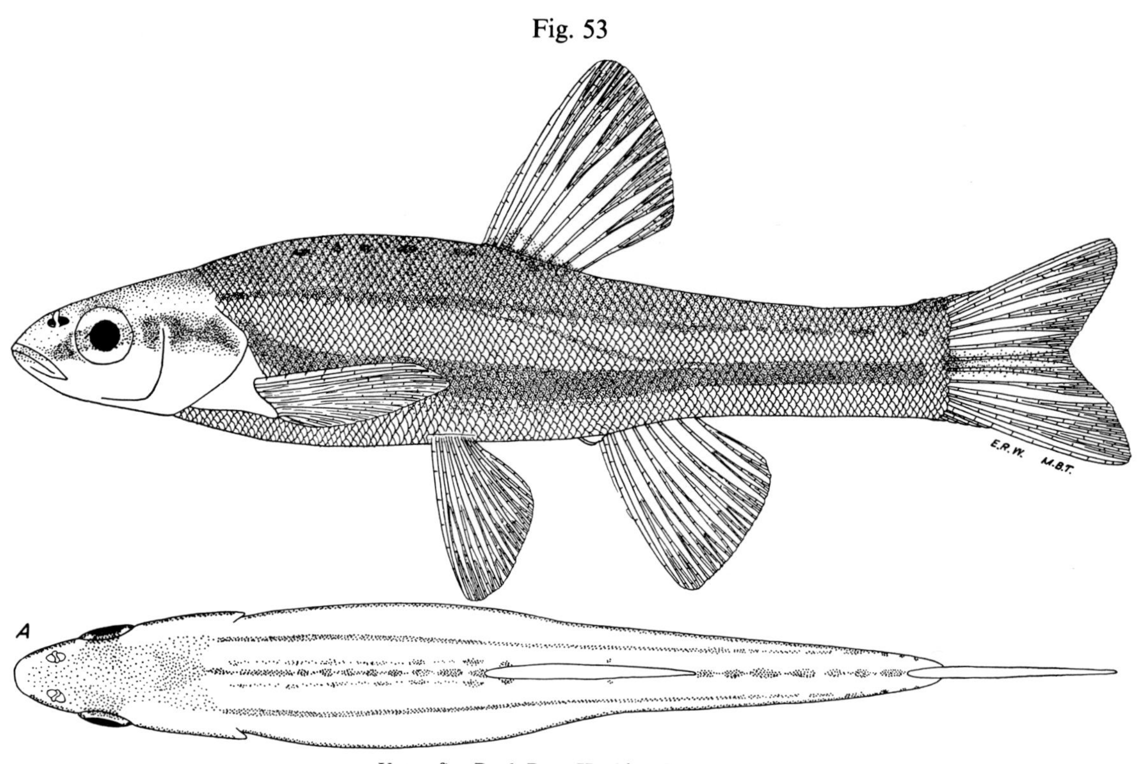

Upper fig.: Buck Run, Hocking County, O.

April 30, 1938.
Male.

55 mm SL, 2.6″ (6.6 cm) TL.
OSUM 2440.

Fig. A: dorsal view; note distinctive color pattern.

Identification

Characters: *General—See* golden shiner. *Generic* and *Specific*—Body scales so small they normally cannot be seen without the aid of magnification; between 70–95 scales in lateral series, usually more than in any other species of minnow in Ohio. Mouth small. Jaws equal. Length of upper jaw contained 3.2–4.0 times in standard head length. Intestine long, with two crosswise coils and a loop. Peritoneum black. No barbel. Two dusky, longitudinal bands along each side, separated by a light or yellowish band. Teeth, 5–5.

Differs: Other species of Ohio minnows have fewer lateral line scales, or only one lateral band.

Most like: Small blacknose dace. **Superficially like:** Small creek chub.

Coloration: Dorsally a deep greenish-olive with a series of dark blotches along the dorsal ridge, and sometimes an additional row of blotches paralleling the ridge predorsally, fig. A. A dusky longitudinal band, often broken into blotches posteriorly separates back and upper sides; below this is a silvery-yellowish band which begins on the snout and continues backward to caudal fin. Immediately below this light band is a dusky one, which encircles snout and extends backward across opercles and sides to the tail. Lower sides and belly silvery, white, or flushed with yellow, red, or crimson. Fins transparent olive or flushed with yellow. *Breeding male*—Exceedingly brilliant, with the back a bright greenish-olive; the two black bands on the sides separated by white or yellow; the cheeks silvery-white; the ventral surface of head and body a deep

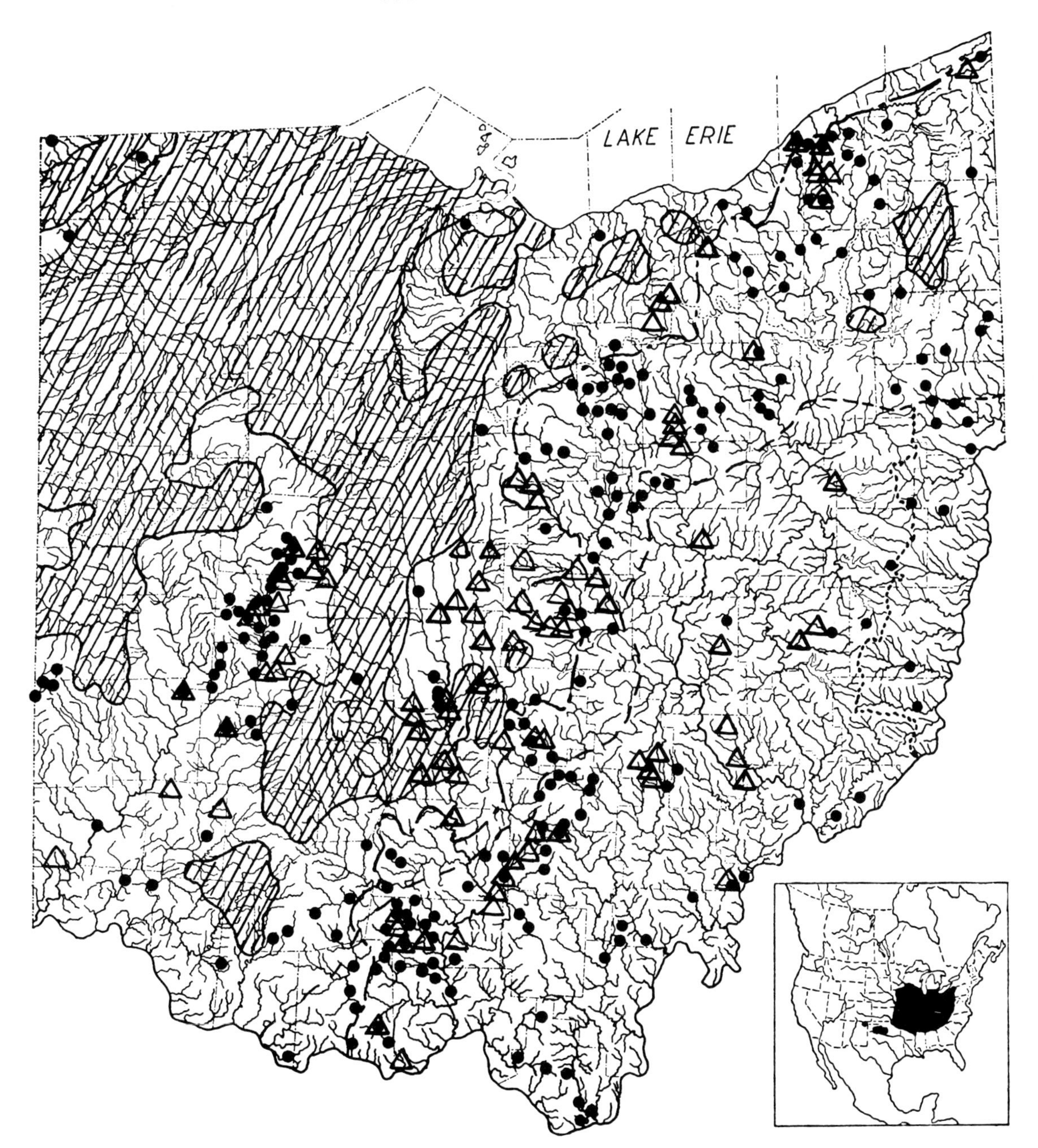

MAP 53. Southern redbelly dace

Locality records. • Before 1955. △ 1955–80. – — Allegheny Front Escarpment.
/\ Glacial boundary. . . . . . . Flushing Escarpment.
Shaded areas have maximum relief of less than 100′ (30.5 m).
Insert: Unoccupied areas within range increasing in size and number; widely introduced outside original range, usually as a bait or forage fish; the New Mexico population may be the result of introduction.

carmine; the fins flushed with yellow; a carmine spot on base of dorsal fin. *Female*—Less brilliantly colored and with less red.

Lengths: *Young* of year in Oct., 0.7″–1.5″ (1.8–3.8 cm) long. Around **1** year, 1.0″–1.8″ (2.5–4.6 cm). *Adults*, usually 1.5″–2.8″ (3.8–7.1 cm). Largest specimen, 3.0″ (7.6 cm) long.

Hybridizes: with redside and rosyside daces, striped and common shiners and stoneroller minnow.

Distribution and Habitat

Ohio Distribution—From the statements of Kirtland (1844:24; 1850:213) and other early writers, our knowledge of former conditions in Ohio, and of the environmental requirements of the redbelly dace, it is obvious that before 1900 this dace was well distributed and abundant in most of the brooks of moderate and high gradients in Ohio. The greatest populations presumably occurred along the glacial boundary, along the Allegheny Front Escarpment and immediately to the east of it, and in the Mad River system. The species was least numerous in brooks of that portion of unglaciated Ohio which were least disturbed by glacial action, and in the till-plains of the northwestern section. It may have avoided those sections of Ohio in which the maximum relief was less than 100′ (30.5 m). It was a close associate of the blacknose dace; *see* that species.

The literature concerning early conditions in Ohio abounds with references to the many springs, some of remarkable size (Brown, 1817:297), and of the "small but durable brooks of the most excellent water" (*Western Intelligencer*, 1811:2) which occurred throughout the then largely forested state. Clear and shaded waters were the prime habitat of the redbelly dace. Because of the recent drastic lowering of the water table and subsequent drying up of innumerable springs, these formerly clear, wooded brooks and streams of "durable" waters have become intermittent streams flowing between treeless and brushless banks and fields and subjected to flash floods and turbid waters. These conditions destroyed the dace habitat; consequently, it has been extirpated from innumerable brooks and from large sections of Ohio, of which Franklin County is an example. In this county in 1897, Osburn and Williamson (1898:12–13) recorded this dace as "occurring in abundance in brooks flowing into the Scioto [River] from the west"; in 1925 a survey of these brooks disclosed that Big and Breckenridge runs in Jackson Township still contained relict populations; in 1953 a single redbelly was collected in Big Run by Edward Wulkowicz.

Habitat—The largest populations of redbelly dace occurred in permanent brooks of clear waters which were not subjected to frequent flooding, which flowed between wooded banks and contained long pools of moving water, and which had "cut banks" overhung by vegetation. These "cut banks" appeared to be very important as places of refuge. It was of paramount importance that these brooks contained sufficient water throughout the year, for unlike the creek chub and some other headwater species, few redbelly dace migrated downstream in summer; instead they remained in the headwaters throughout the year.

When frightened and especially if "cut banks" were absent, the redbelly dace formed closely packed schools in mid-water, instead of scattering to hide under objects as many of the other brook species do. Schooling dace were extremely vulnerable to capture by bait seiners, and in a few hours two commercial bait seiners could capture 75% of the dace population in a half-mile of stream. In frequently seined streams the dace populations were very low, even in those brooks containing a large amount of suitable habitat. It was only when visiting a brook protected from seiners and having much suitable habitat, that one realized how numerous this species could be, and how abundant it must have been formerly.

Years 1955–80—Although the redbelly dace has been taken in more than 60 localities in Ohio since 1955, relatively few were in, or adjacent to, the shaded areas in the above map. Such areas have a maximum relief of less than 100′ (30.5 cm). The collecting locality farthest inside a shaded area was on the riffle of Big Darby Creek 100′ (30.5 m) above State Route 104, southeast Jackson Township, Pickaway County. An individual was collected there April 8, 1958, and was the only one captured during hundreds of days of collecting on this riffle during the past 50 years by me or others. Presumably it was a stray.

I have investigated tributaries which were well inside the shaded areas and which environmentally appeared to be suitable for the redbelly dace, but found none. Although such tributaries were probably suited for the redbelly dace, they were too isolated by streams of lower gradients and possibly never were invaded by this species.

During this period populations present before 1950 were observed to have become extirpated as the streams were disturbed by channelization, ditching, tree removal along banks and/or increased turbidity (White et al., 1975:73).

In the first edition of *The Fishes of Ohio* the southern redbelly dace was called *Chrosomus erythrogaster* Rafinesque. Recently, Banarescu (1964:336) synonomized *Chrosomus* with the Eurasian genus *Phoxinus*. This was accepted by the committee of the American Fisheries Society (Bailey et al., 1970:70). The scientific name therefore becomes *Phoxinus erythrogaster* (Rafinesque).

REDSIDE DACE

Clinostomus elongatus (Kirtland)

Fig. 54

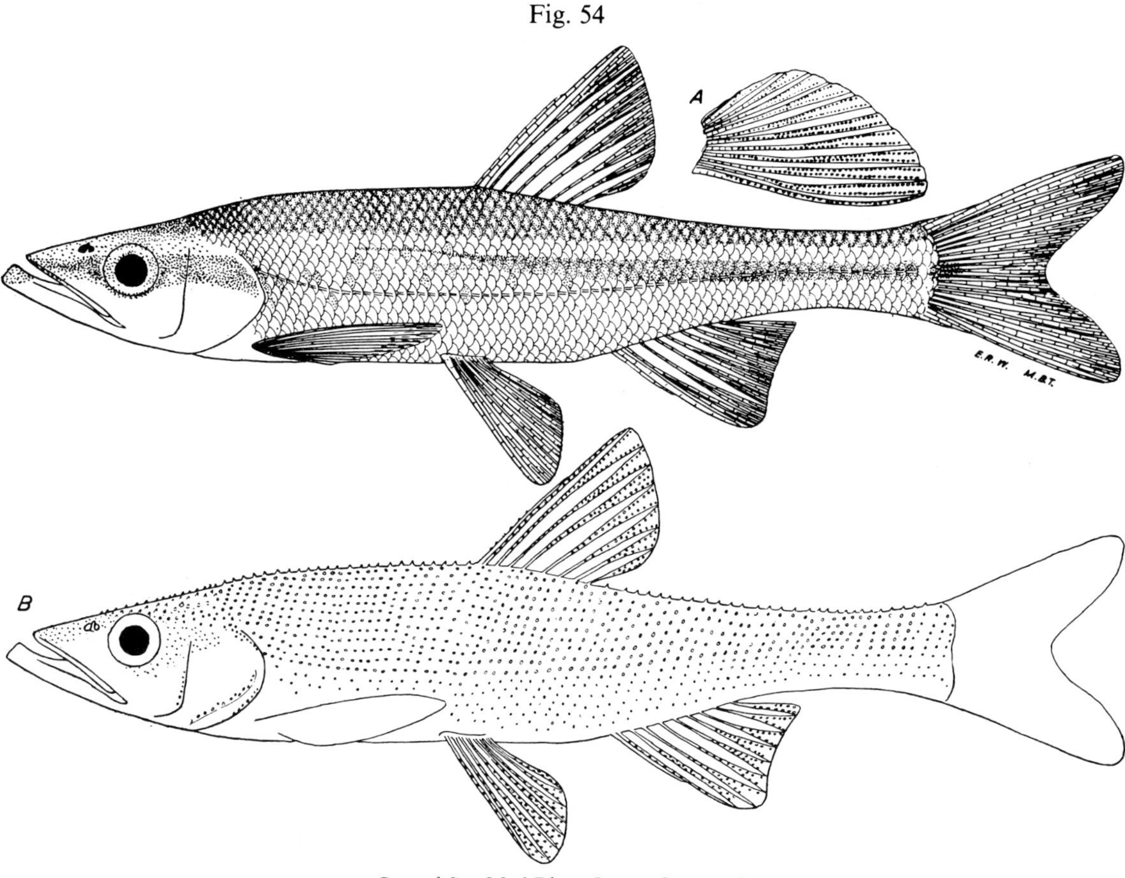

Central fig.: Mad River, Logan County, O.

May 30, 1940.
Breeding male.

67 mm SL, 3.3″ (8.4 cm) TL.
OSUM 2120.

Fig. A: tuberculated pectoral fin of male.
Fig. B: Male; showing tubercle distribution on head, body and fins.

Identification

Characters: *General—See* golden shiner. *Generic*—Mouth larger than in any other genus of minnows in Ohio; the upper jaw contained less than 2.8 times in head length and usually less than 2.5. The prominent lower jaw extends beyond the upper jaw in all except the smallest young of the rosyside dace. No barbels. Lateral line complete. Intestine short with only a single loop. Peritoneum silvery. Teeth normally 2,5–5,2. *Specific*—Lateral line scales 59–70. Body depth usually more than 4.5 times in standard length, extremes 4.4 (adults)–5.4 (small young). Caudal peduncle depth usually contained 2.7–3.4 times in head length. Adults with carmine-scarlet on sides.

Differs: Rosyside dace has fewer lateral line scales and a deeper body. Creek chub has a barbel. Redbelly dace has a small mouth. Other Ohio minnows have a barbel and/or larger scales. Trouts, smelt, and whitefish have adipose fins. Skipjack herring, mooneye and goldeye have adipose eyelids.

Most like: Rosyside dace and creek chub.

Coloration: Dorsally olive-green, heavily overcast with emerald and steel-blue. A white, (small young) yellow, or golden-yellow (adults) band ex-

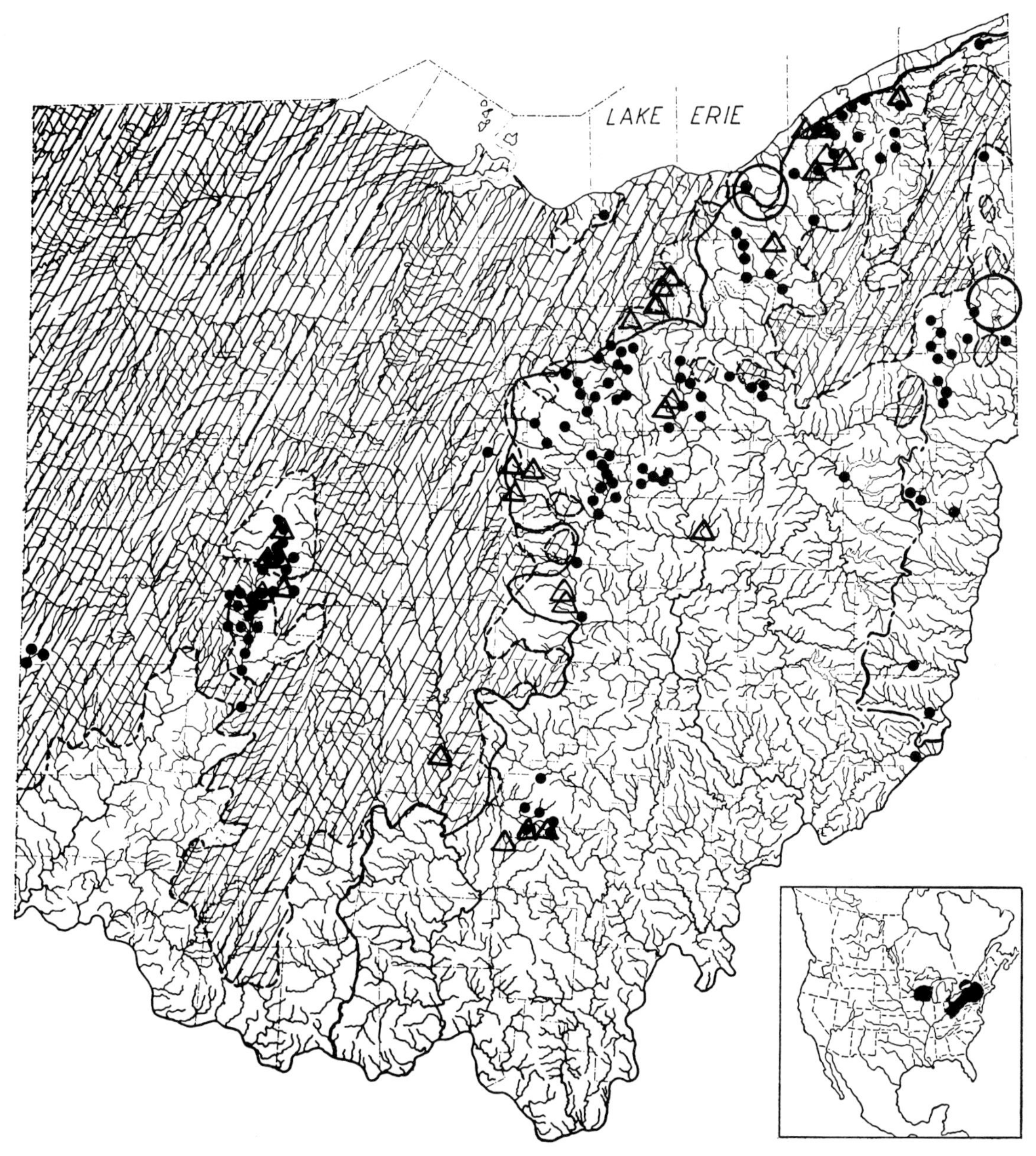

MAP 54. Redside dace

○ Type localities, tributaries of Lake Erie near Cleveland and Mahoning River, Mahoning County.
Locality records. ● Before 1955. △ 1955–80.
~ Allegheny Front Escarpment. – — Flushing Escarpment.
Shaded areas have maximum relief of less than 200′ (60.9 m).
Insert: Unoccupied areas in the disjunct black areas increasing in size and number.

tends from tip of snout backward between the darker back and lateral band to the tail. Lateral band is dusky on tip of lower jaw and extends backward across snout, cheek and opercle; almost immediately behind the gill-cleft the band becomes white (young), orange (young males or females), carmine or scarlet (adult males), continuing backward thus to beneath the dorsal fin, after which the band becomes dusky again, remaining so backward to the tail. Ventral half of head and body silvery-

and milk-white. Fins transparent. *Breeding male*—Very brilliant coloration; back bright steel blue; a golden-yellow band along upper sides; anterior half of lateral line bright scarlet; tubercles abundantly distributed over head, body and fins, figs. A and B. *Small young*—Lack scarlet coloration except for a small spot in upper angle of gill-cleft and in some individuals, a blush of scarlet over the 10 anteriormost scales of the lateral line.

Lengths: *Young* of year in Oct., 1.3″–2.4″ (3.3–6.1 cm) long. Around **1** year, 1.4″–2.8″ (3.6–7.1 cm). *Adults*, usually 2.5″–3.8″ (6.4–9.7 cm). Largest specimen, 4.5″ (11 cm).

Hybridizes: with creek chub, redbelly dace, striped and common shiners.

Distribution and Habitat

Ohio Distribution—During the 1925–50 surveys there were two centers of redside dace abundance in Ohio. One was the clear waters of the upland streams of the Mad River system, the other the upland streams along the upper half of the Allegheny Front Escarpment and immediately to the east of it. A small population existed in similar streams in Hocking County. Relict populations which decreased drastically in abundance during the surveys, existed in the upper headwaters of the Whitewater River in Darke and Preble counties, and in the hill streams east of the Flushing Escarpment. From studies of its ecological requirements it is obvious that its distributional range in Ohio must have been far more widespread before 1900 than it was after. Previous to 1900 it should have been present and abundant in streams along the upper two-thirds of the Allegheny Front Escarpment from Hocking County northeastward, extending for more than 30 miles (48 km) east of that escarpment, as well as for a considerable distance to the west of it; it should have been present in many streams in northeastern Ohio south to the glacial boundary, and should have been particularly abundant in the hill streams east of the Flushing Escarpment. Populations may have been present in some St. Joseph and Maumee river tributaries in Williams and Fulton counties of northwestern Ohio. It should have been absent from the till-plain streams of the Maumee River drainage and the lowland streams of west-central Ohio.

During the 1925–50 surveys many redside populations decreased drastically in abundance, or disappeared entirely, and in every case it was obvious why the species had so decreased or disappeared. Coal mine pollution was the principal cause east of the Flushing Escarpment; elsewhere it was chiefly agricultural practices which changed the streams from clear water, clean-bottomed creeks with wooded banks to turbid, silt-bottomed ones with barren, eroding banks, thereby eliminating the redside habitat. In a few localities the introduction of domestic or industrial pollutants was the primary cause.

Habitat—The redside dace frequented brooks and small streams whose waters were normally very clear, whose gradients were moderate or high, and whose bottoms were composed of clean gravel, sand, or bedrock. Organic debris may or may not have been present; there was little or no clayey silt. The pools usually had a moderate current flowing through them, and many pools contained brush or roots which the redsides used as emergency cover. The banks were shaded invariably by brush and trees. This species seemingly preferred the cooler waters, and it may have been the warmer temperatures which were primarily responsible for its absence in extreme southern Ohio.

Years 1955–80—Attempts to locate unrecorded populations of the redside dace in Ohio counties other than those heretofore recorded were unsuccessful (White et al., 1975:73), except in the following two instances.

On July 26, 1964 K. Roger Troutman and James M. Norrocky discovered a population in Little Mill Creek, Coshocton County.

On April 24, 1964 Donald I. Mount captured an adult in Big Darby Creek, Pickaway County. The presence of this individual in this locality is most surprising because no population of this dace has been found in the Big Darby drainage, which lies wholly in an area whose maximum relief is less than 200′ (60.9 m) (*see* distribution map). The redside dace is sold by bait dealers, and it is highly probable that this Big Darby specimen was an escape from a fisherman's bait container or was released by the fisherman. Because of the isolation of this record, this explanation appears to be somewhat more plausible than that the fish was a stray.

ROSYSIDE DACE

Clinostomus funduloides Girard

Fig. 55

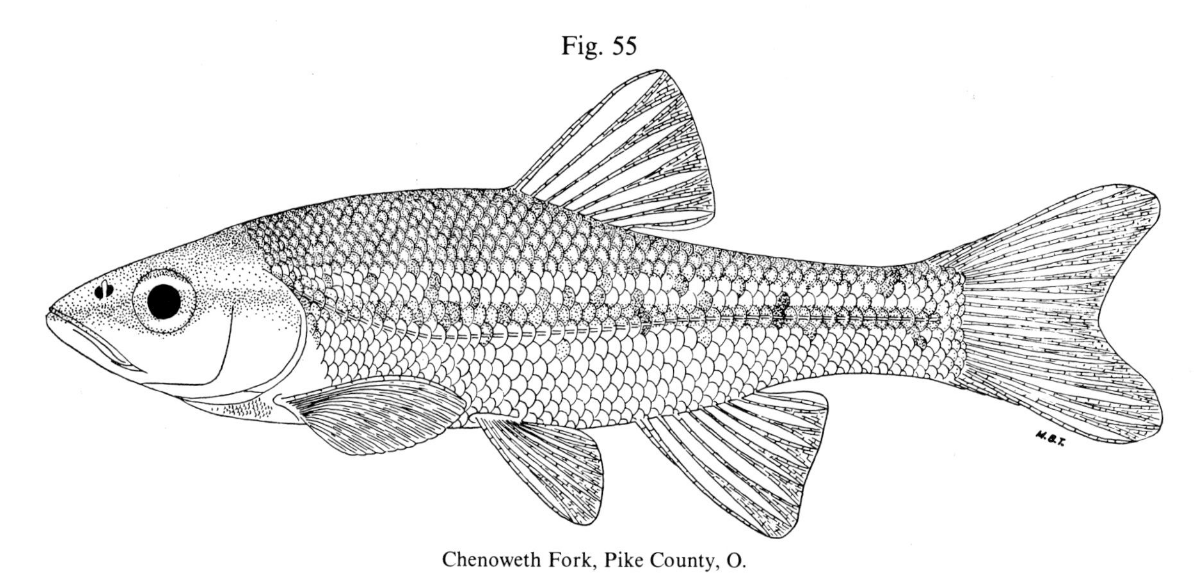

Chenoweth Fork, Pike County, O.

May 3, 1930. 65 mm SL, 3.3″ (8.4 cm) TL.
Male. OSUM 509.

Distribution of tubercles on the breeding male similar to that of the redside dace, figs. A and B.

Identification

Characters: *General—See* golden shiner. *Generic—See* redside dace. *Specific*—Lateral line scales 48–57, usually more than 51. Body depth usually contained 3.9 (adults)–4.7 (small young) times in standard length. Caudal peduncle depth usually contained 2.0–2.7 times in head length. Adults with rose on sides.

Differs: *See* redside dace. Redside dace has more lateral line scales, and a more slender body.

Most like: Redside dace and creek chub.

Coloration: Similar to redside dace, except that the colors are not so brilliant and the pattern not so sharply defined; the color on the sides is less carmine and more rosy; the posterior half of the lateral band is less dusky and less well defined; usually there are more dark scales along the sides. *Breeding male*—Heavily tuberculated over head, body and fins, as is the redside dace; *see* that species, figs. A and B.

Lengths: *Young* of the year in Oct., 1.0″–2.0″ (2.5–5.1 cm) long. Around **1** year, 1.4″–2.4″ (3.6–6.1 cm). *Adults*, usually 2.2″–3.5″ (5.6–8.9 cm). Largest specimen, 4.3″ (11 cm) long.

Hybridizes: with creek chub, redbelly dace and common shiner.

Distribution and Habitat

Ohio Distribution—The presence of the rosyside dace in Ohio waters was first noted in 1929. Since then this dace has been found in the upland, clear-water, limestone-bottomed streams, which are 600′ (183 m) or more above sea level, of the Blue Grass Region of unglaciated, southern Ohio.

During the 1929–50 period several isolated populations of rosyside dace decreased greatly in numbers, disappeared or there occurred notable fluctations in abundance. These changes were directly related to changes or fluctuations in the amount of suitable habitat.

It is interesting to contemplate upon the possible length of time that the rosyside daces have occupied their present stations in Ohio. The present distribution in these unglaciated, upland streams suggests that the stations are too widely separated and isolated from each other by unfavorable lowland streams for this highly non-migratory species to have be-

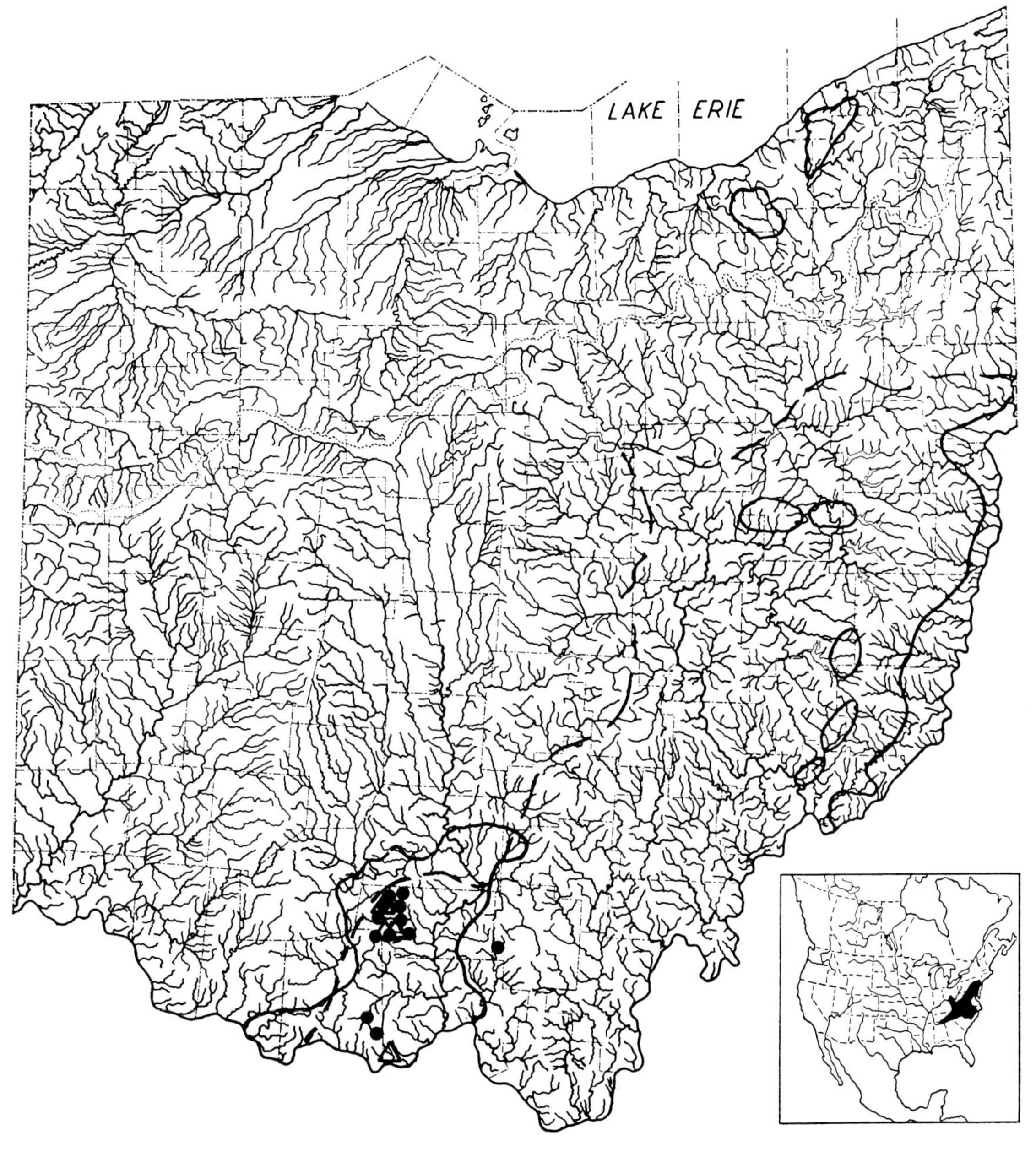

MAP 55. Rosyside dace

Locality records. ● Before 1955. △ After 1955. /\ Glacial boundary.
Enclosed areas have maximum relief of over 500′ (152 m).
Insert: Southwestern limits of range indefinite.

come established recently through the migration of strays from a central source. Rather it suggests that these relict populations have been present in their long-undisturbed tributaries for a great period of time. Likewise, the Ohio populations are at present separated by the Ohio River from populations inhabiting the upland streams of eastern Kentucky and West Virginia; this river at present seems to be an effective barrier separating the two. It is possible that the Ohio populations may have been isolated from the Kentucky and West Virginia ones since as long ago as the Teays Drainage Stage (Stout and

Lamb, 1938: map p. 59) which was previous to the formation of the present Ohio River, when the Teays River and its tributaries flowed northward from West Virginia and Kentucky into Ohio, including a tributary, the Portsmouth River. This river rose in Kentucky and flowed northward through the present Blue Grass Region of Ohio to join the Teays in Pike County. Later an early glacier blocked the Teays, bringing into existence the "Cincinnati River," a very large stream which could have isolated the Ohio from the Kentucky populations. During the Lake Tight stage (Wolfe, 1942: Fig. 1) the rosyside dace may have been restricted to those portions of the upland tributaries which were more than 900′ (274 m) above sea level. It seems significant that at present all of the known Ohio stations for the rosyside dace are in headwaters of the ancient Teays system, headwaters which presumably have not been greatly modified since Teays time.

Habitat—The rosyside dace was numerous only in the permanent sections of those small, upland streams which were more than 600′ (183 m) above sea level, which had bottoms of gravel, boulders and bedrock consisting of limestones and shales (Bownocker, 1920), and whose waters were normally very clear and of a deep bluish cast. The rosyside dace and bigeye shiner were close associates and both species disappeared from waters when they became habitually turbid; *see* bigeye shiner.

Years 1955–80—In this period several clear, upland limestone streams were investigated in attempts to discover heretofore unrecorded populations of the rosyside dace. Only one, in Scioto County, was located outside the range as given on map 55. In a few localities no individuals were found where before 1950 the species had been present, and it is possible that these former populations have been extirpated. Within the recorded Ohio range of this dace only Chenoweth Fork, a tributary of Sunfish Creek in western Pike County, maintained a thriving population.

For many years the scientific name of the rosyside dace was *Clinostomus vandoisulus* (Valenciennes), and this name was used in the first edition of this report. Recently, Lachner and Deubler (1960:358–60) have shown that the name *vandoisulus* is inapplicable and suggested the first available name, which is *funduloides* Girard. The scientific name therefore is *Clinostomus funduloides* Girard (Bailey et al., 1960:13), and this name is used in the present edition. Some recent authors, such as Moore (1968:63), have merged *Clinostomus* with *Richardsonius*, thus becoming *Richardsonius funduloides* (Girard).

PUGNOSE MINNOW

Notropis emiliae emiliae (Hay)

Fig. 56

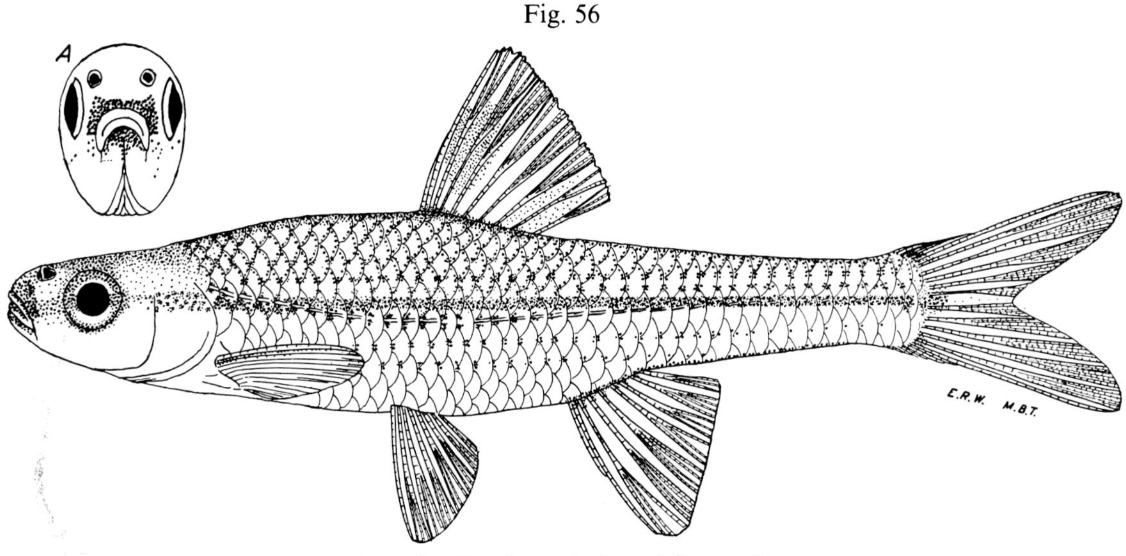

Lower fig.: East Reservoir, Summit County, O.

Aug. 9, 1939. 48 mm SL, 2.3″ (5.8 cm) TL.
Male. OSUM 419.

Fig. A: frontal view of head showing distribution of the breeding tubercles of the male.

Identification

Characters: *General—See* golden shiner. *Generic* and *Specific*—Only species of minnow in Ohio which normally has *9* fully-developed dorsal rays; aberrant specimens with only 8 are rare. Mouth extremely small and nearly vertical. Peritoneum silvery with darker specklings. Lateral line incomplete with 37–39, extremes 36–40, scales in the lateral series. Large young and adults with one or two faintly dark or dusky areas on dorsal fin. Teeth 5–5, rarely 5–4.

Differs: Very much like the pugnose shiner in size and in having a small, very oblique mouth, but the shiner differs in having a black peritoneum; a transparent dorsal fin; 8 dorsal rays; fewer lateral line scales. Other species of minnows in Ohio have only 8 dorsal rays. Small chubsuckers have less oblique mouths; other suckers have 10 or more dorsal rays.

Most like: Pugnose shiner; small chubsuckers; small golden shiner and blackchin shiner.

Coloration: Dorsally olive-straw overlaid with silvery; the dark edgings of the scales giving a pronounced crosshatched appearance. Sides yellow-silvery or silvery, with usually a faint band before and after eye, but which normally is more distinct as a lateral band along the sides. Ventrally silvery with a faint emerald sheen. Fins of young and adult female silvery. *Adult male*—Has a dusky blotch between the first four dorsal rays and another between the last three rays, leaving the middle two rays transparent or whitish (while a fish is in the water the two dusky dorsal blotches give the delusion of two separate dorsal fins); posterior blotches on dorsal fin vary greatly; in some males they are very intense; in others absent. Many tiny, very sharp tubercles surround the mouth of breeding males, fig. A.

Lengths: *Young* of year in Oct., 1.0″–1.7″ (2.5–4.3 cm) long. Around **1** year, 1.3″–2.0″ (3.3–5.1 cm). *Adults*, usually 1.5″–2.3″ (3.8–5.8 cm). Largest specimen, 2.5″ (6.4 cm) long.

Distribution and Habitat

Ohio Distribution—This inconspicuous species was not described until 1880 (Hay 1880:507–08) and was not recorded as occurring in Ohio until 1893

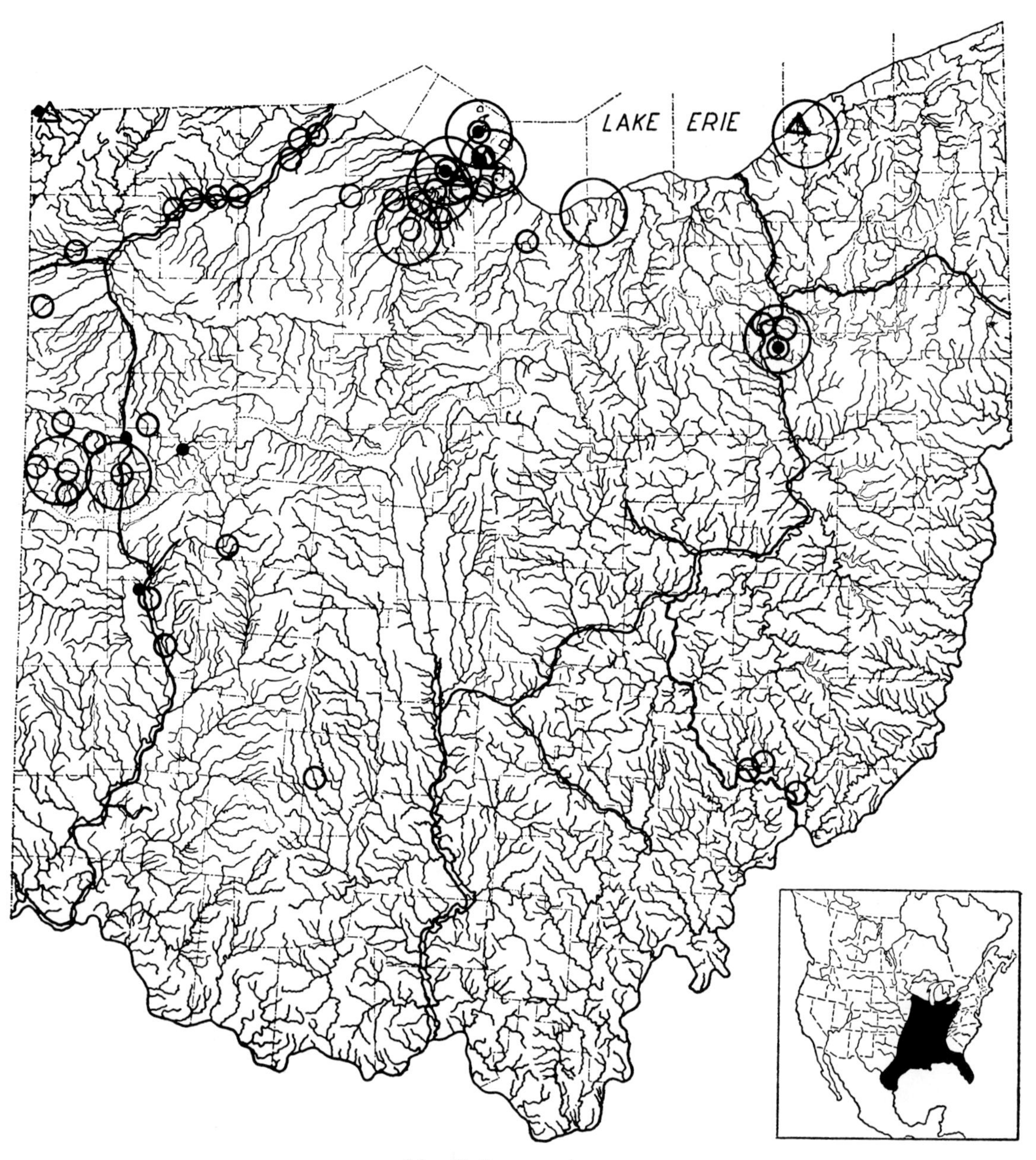

MAP 56. Pugnose minnow

Locality records. ○ Before 1901. ○ 1901–38. ● 1939–54. △ 1955–80. ~ Canals.
Insert: Unoccupied areas in at least the northeastern section of the range increasing in size and number.

(Kirsch 1895A:329). Preserved specimens, published statements, distribution of the species in surrounding states, and our knowledge of its environmental requirements indicate that the pugnose minnow should have been rather well distributed in Ohio before 1900, and probably occurred in the largest numbers about aquatic vegetation in the clear, prairie-type, low-gradient streams, the bayous, oxbows, canals and glacial lakes, and in the harbors and bays bordering Lake Erie. Between 1825–1900 the canals, with their aquatic vegetation and quiet waters unquestionably produced a suitable habitat; between 1920–30 it was found in canals in Mercer, Henry and Summit counties.

Between 1920–30 it was abundant in Put-in-Bay Harbor, in East, Middle and West harbors, in Sandusky Bay and streams entering it, in the Maumee River in Henry and Wood counties, in the streams of Mercer County and Portage Lakes; small populations existed in the lower Muskingum River system. Since 1930 it has drastically decreased numerically or disappeared from most of these localities and at present appears to be threatened with possible extirpation from Ohio waters.

Habitat—The pugnose minnow occurred abundantly only in the clearer waters of low or base gradients where there was aquatic vegetation and the bottom was of sand or organic debris. Gerking's (1945:53) perplexing statement that during the 1940–43 surveys in Indiana, the pugnose was found in turbid waters without vegetation is contrary to the general conception of its environmental requirements. This contradiction can be explained when it is realized that before their final disappearance, relict populations persisted in Ohio for several years after almost all of the aquatic vegetation had disappeared, and after turbidity and siltation had become great. This persistence of small populations to exist for a time after only submarginal conditions remain may have been operating in southern Indiana during the 1940–43 fish surveys.

Years 1955–80—During the past 25 years, despite a concerted effort to capture it, the pugnose minnow was taken only at Nettle Lake in extreme northwestern Ohio and in the Chagrin River near its mouth, Lake County (White et al., 1975:73–74). There remained some submerged aquatic vegetation in both localities. Before 1952 several isolated populations existed in widely separated areas in Ohio, such as in the Muskingum, Maumee and Portage river systems and in East Harbor. Since many of these localities were investigated and no pugnose minnows found, it may be assumed that these populations have been extirpated.

The pugnose minnow has been in the genus *Opsopoeodus* since its description. Recently, Carter R. Gilbert and Reeve M. Bailey (1972:1–34) suggested that *Opsopoeodus* be synonomized with *Notropis* despite its having nine dorsal rays instead of eight as in *Notropis*. They described a peninsular Florida race, the Ohio form becoming *Notropis e. emiliae* (Hay). All members of the genus *Notropis* are commonly known as shiners. *Notropis emiliae* for the time being must remain the pugnose minnow because *Notropis anogenus* is currently known as the pugnose shiner. Roundnose shiner might be a substitute common name for *N. emiliae*, "round" being a synonym of "pug."

COMMON EMERALD SHINER

Notropis atherinoides atherinoides Rafinesque

Fig. 57

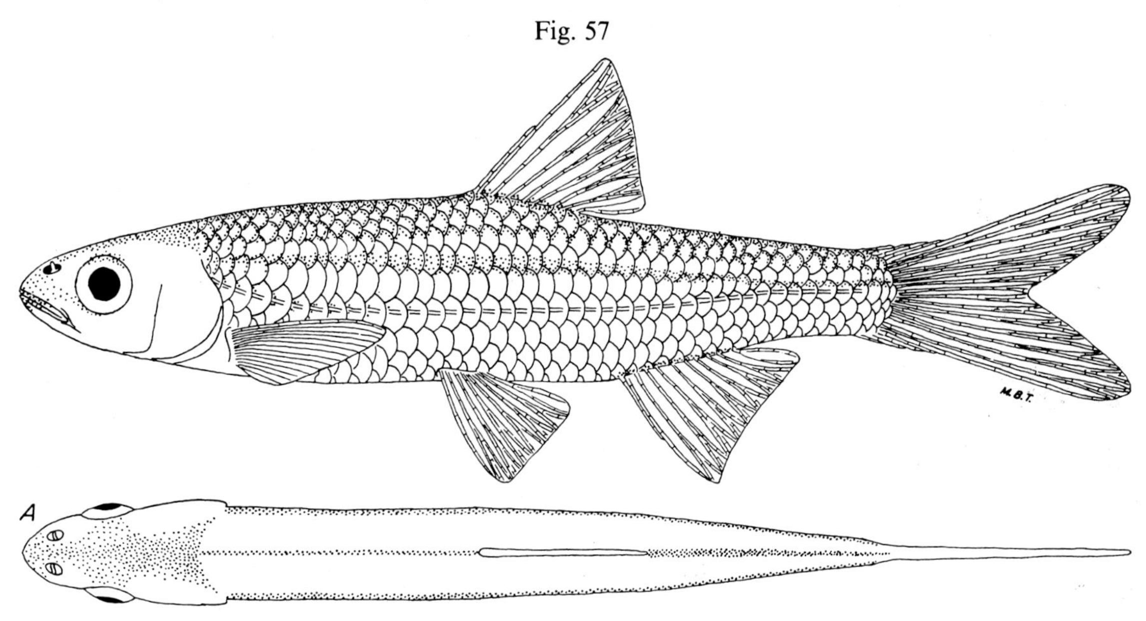

Upper fig.: Lake Erie, Lake County, O.

April 16, 1940. 77 mm SL, 3.7″ (9.4 cm) TL.
Adult female. OSUM 1981.

Fig. A: dorsal view; note color pattern.

Identification

Characters: *General*—Dorsal fin of 8 developed rays, with the anterior half-ray an inconspicuous splint which is bound tightly to the first developed ray. Teeth in the throat in one or two rows on each of the two pharyngeal arches, and never more than six teeth in a row. No teeth in mouth. Lateral line scales 32–48, normally less than 42 except in redfin and rosefin shiners. Lower jaw with fleshy lobes; lobes normal, not restricted or absent. Alimentary tract less than twice the length of head and body. Body with cycloid scales. Head scaleless. Pelvic fins abdominal. No barbels, adipose fin, adipose eyelids, pelvic axillary process, dorsal, anal or pelvic spines; belly not serrated as in herrings. *Specific*—Anal rays normally 10–13, rarely 9. Dorsal fin transparent, without spots. Body silvery with emerald reflections; mid-dorsal stripe faint or absent. Snout rounded and short; preorbital length usually contained 1.5 or more times in postorbital length of head. Dorsal origin over, or slightly behind, pelvic origin. Body slab-sided, fig. A.; width of body immediately behind head usually contained 1.8–2.4 times in height of dorsal fin. Body deepest in region of dorsal insertion. Complete lateral line with 36–40 scales; fewer than 22 predorsal scales. Teeth, usually 2,4–4,2.

Differs: Silver shiner has two dusky crescents between nostrils; a larger eye. Rosyface shiner has a sharper-pointed, longer snout; dorsal origin is more posterior. Redfin and rosefin shiners have more lateral line scales; more predorsal scales; a black blotch on fin at dorsal origin. Other shiners normally have 9 or fewer anal rays.

Most like: Silver and rosyface shiners.

Coloration: Dorsally olive-silvery with an emerald, blue, and green overcast; streak along dorsal ridge faint and slender, or absent. A greenish-emerald band usually extends along upper sides from upper angle of gill cleft to the tail. Sides very silvery, with an emerald green band evident when viewed from some angles. Ventrally silvery and milk-white. Fins transparent, lower fins with some

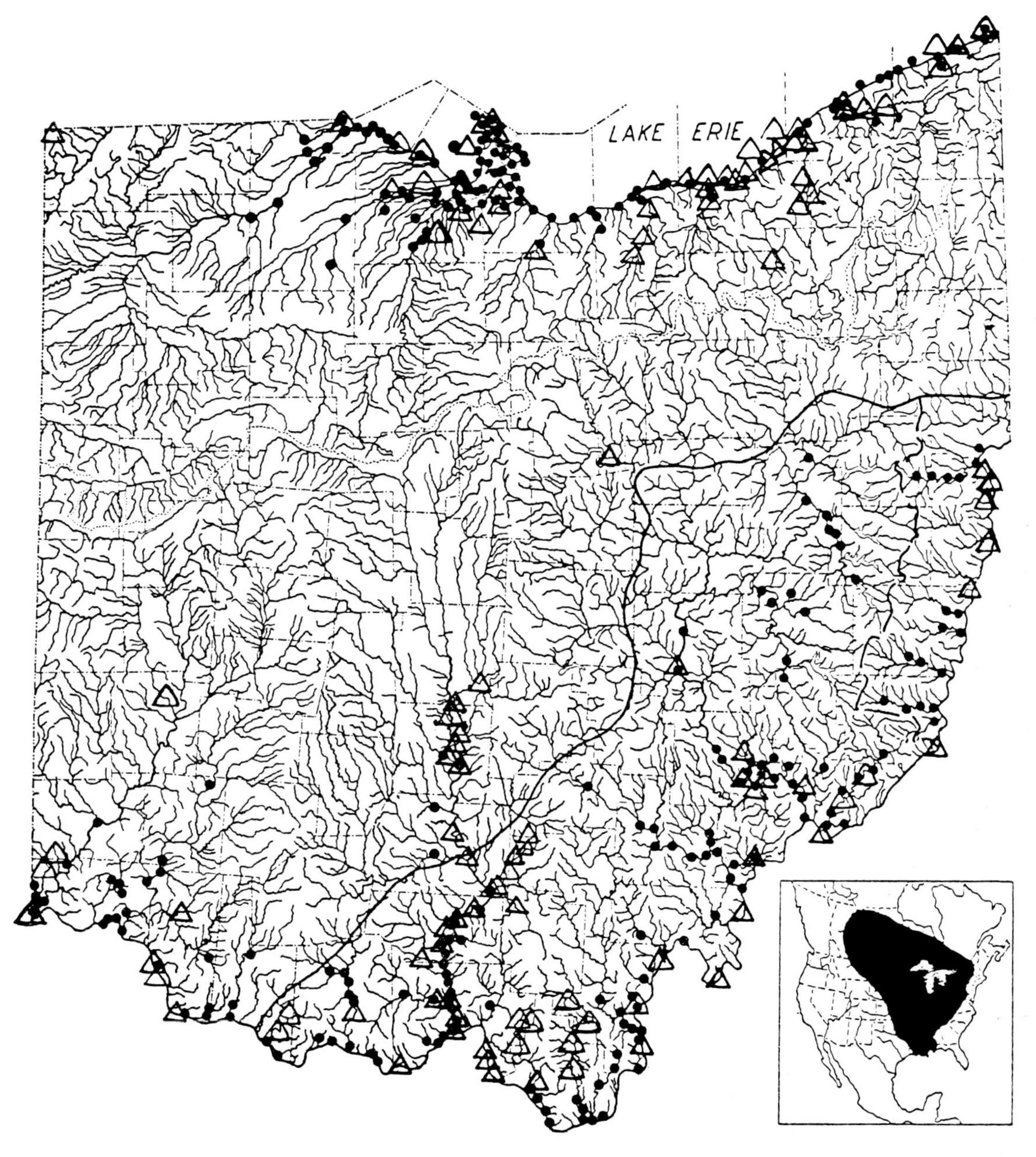

MAP 57. Common emerald shiner

Locality records. ● Before 1955. △ 1955–80. ~ Glacial boundary. /\ Flushing Escarpment.
Insert: Black area represents range of the nominate subspecies; outlined portion of the upper Great Lakes of a nominal subspecies.

white. *Breeding male*—Has tubercles extremely small and chiefly confined to upper surface of pectoral fins.

Lengths: *Young* of year in Oct., 1.0″–2.3″ (2.5–5.8 cm) long. Around **1** year, 1.3″–2.8″ (3.3–7.1 cm). *Adults*, usually 2.5″–3.3″ (6.4–8.4 cm). Largest specimen, 3.9″ (9.9 cm) long.

Distribution and Habitat

Ohio Distribution—The few preserved specimens that were taken before 1900 establish the early presence of the emerald shiner in Lake Erie, the Ohio River, and a few of their tributaries. The majority of the early literature references must be ignored because of the confusion which existed until recently between the emerald, silver, and rosyface shiners. Osburn's (1901:56–57) inland locality records before 1901 presumably are based upon mistaken identification of the silver for the emerald shiner; for specimens which he and Williamson collected in 1897, now in OSUM, and identified by them as emerald shiners, are silver shiners, as is the photograph of a silver shiner in their joint thesis (Williamson and Osborn, 1898: pl. 7) which is labelled "*Notropis atherinoides.*" As further proof, Williamson and Osburn did not record the silver shiner for Franklin County, although their specimens proved that the species was there before 1900.

Between 1920–50 the emerald shiner was immensely abundant in Lake Erie where occasionally, and especially in fall, the large, dense schools darkened the water in the boat slips, harbors, and mouths of rivers. The species was abundant also in the Ohio River, sometimes crowding in huge schools below dams, in association with the channel mimic shiner. Fair numbers were normally present in the mouths of streams tributary to Lake Erie and the Ohio River. Normally the species ascended for a much greater distance up the Ohio River tributaries than it did those of Lake Erie; this was likewise true of the trout-perch. The presence and abundance of the emerald shiner in any inland locality usually fluctuated drastically from one season to another, except in the base- or low-gradient, mature streams tributary to the Muskingum River in the unglaciated area from northern Noble to Tuscarawas counties, where a small population was always present.

Normally the emerald shiner avoided fast currents, but at times, often in fall and chiefly as large adults, it migrated into streams of high gradients. This was particularly true of the high-gradient streams east of the Flushing Escarpment; possibly it invaded these streams so as to escape from the highly polluted Ohio River. At times, often in fall, huge schools appeared at the mouths of streams. Kirtland (1954B:44–45, as *Alburnus nitidus*) records a concentration which occurred on October 9–12, 1853, in the shallow and rapid water of the mouth of the Rocky River; fortunately Kirtland's statements are accompanied by a recognizable figure of the emerald shiner.

Habitat—It is most abundant in large, deep, sluggish rivers and in Lake Erie, exhibiting a preference for clear water. The type of bottom seems to be of little importance to this mid-water and surface-swimming species, for it seldom comes in contact with the bottom, even during the spawning season. It tends to remain in the mid-section of deep water during the daytime whenever the water is clear, or during high wave action. The species comes to the surface at dark, in summer often in great schools, to feed upon the small midges and other flying insects that hover just above the water's surface. It avoids rooted aquatic vegetation.

Emerald shiners vary considerably in body proportions and fin length. Specimens from different year classes or for different localities in Lake Erie vary almost as much from one another as do Lake Erie specimens from those from the Ohio River. However, Ohio River specimens are usually less deep and more terete, have more prominent-appearing eyes, usually have shorter, more rounded fins, and appear to be more robust; whereas Lake Erie specimens are more slab-sided, have more sharply-pointed, fragile-appearing fins, and as a whole seem to be more frail; in these respects they resemble *Notropis atherinoides acutus*, the lake emerald shiner of Lake Michigan.

Years 1955–80—During this period the Ohio distribution of the emerald shiner remained relatively unchanged from that prior to 1950. It sporadically invaded the Scioto River drainage, occasionally in large numbers, as far upstream as Franklin County.

Unlike some other species inhabiting Lake Erie, this shiner has shown little or no decrease in numbers during recent years, and it remains one of the most abundant species present in that lake (White et al., 1975:74). During the 1957–59 investigations of the entire Ohio River (ORSANCO, 1962:147), 428,463 emerald shiners were tabulated, the largest number of individuals recorded for any species during this investigation.

SILVER SHINER

Notropis photogenis (Cope)

Fig. 58

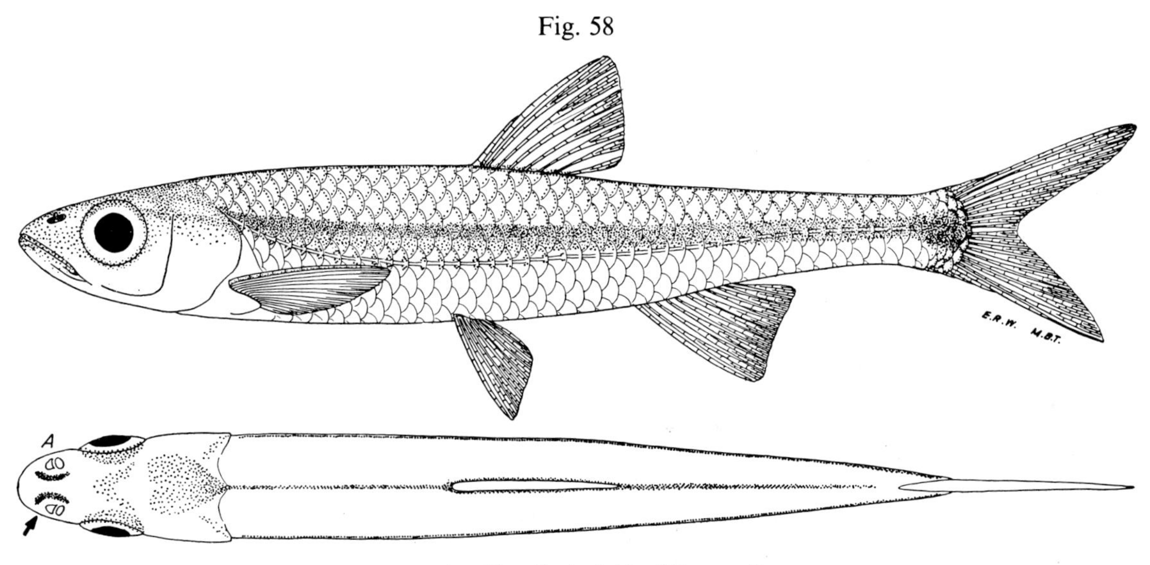

Upper fig.: Clear Fork, Ashland County, O.

Sept. 9, 1939. 76 mm SL, 3.5″ (8.9 cm) TL.
Adult male. OSUM 519.

Fig. A: dorsal view showing color pattern; note arrow pointing to the crescent-shaped group of melanophores between the nostrils.

Identification

Characters: *General—See* emerald shiner. *Specific*—Anal rays normally 10–13, rarely 9. Dorsal fin transparent without spots. Body silvery, with some bluish reflections, the mid-dorsal stripe generally distinct, black, and sometimes fairly broad. Two dark crescents of melanophores occur between the nostrils, fig. A.; these are faint or absent in smallest young; faint in some of the largest breeding adults. Snout more pointed and longer than in the emerald shiner; its length usually contained less than 1.5 times in the postorbital length of head. Eye large; contained 2.7 (small young)–3.5 (large adults) times in head length. Dorsal origin over, or slightly behind, pelvic origin. Body wider and roundish; its width behind head usually contained 1.2–1.9 times in dorsal fin height. Body deepest midway between head and dorsal insertion. Complete lateral line usually with 36–40 scales; fewer than 22 predorsal scales. Teeth, usually 2,4–4,2.

Differs: *See* emerald shiner. Emerald shiner has a smaller eye, more slab-sided body; rosyface shiner has a smaller eye, the dorsal origin decidedly behind pelvic origin.

Most like: Emerald and rosyface shiners. **Superficially like:** smelt.

Coloration: Dorsally greenish-, yellowish-, or slaty-olive heavily overcast with silver and steel-blue; a rather wide, very black stripe along dorsal ridge, fig. A. Sides silvery and milk-white; ventrally the same. Fins transparent, the lower ones sometimes whitish. *Breeding male*—Tubercles too small to be seen without magnification; the largest tubercles are on the upper surface of the pectorals and smaller ones on the head and on the scales of the forepart of body.

Length: *Young* of year in Oct., 1.5″–2.4″ (3.8–6.1 cm) long. Around **1** year, 2.0″–3.0″ (5.1–7.6 cm). Adults, usually 2.7″–4.3″ (6.9–11 cm). Largest specimen, 5.2″ (13 cm) long.

Hybridizes: rarely with striped shiner.

Distribution and Habitat

Ohio Distribution—There are a sufficient number of preserved specimens taken between 1854–1925, plus reliable published records (despite inability of most of the early workers to recognize

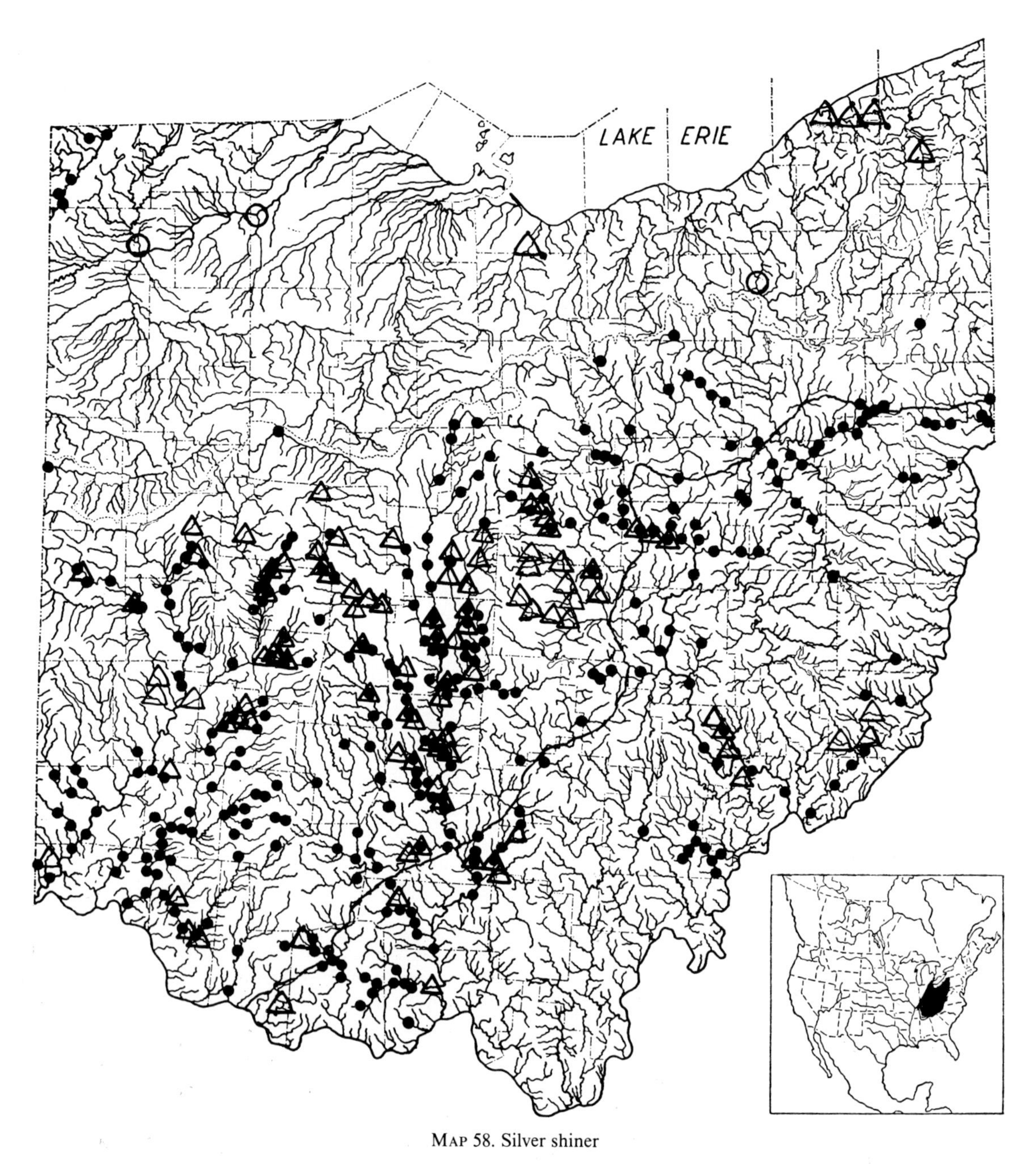

MAP 58. Silver shiner

○ Literature records before 1901, without specimens.
Locality records. ● After 1920. △ 1955–80. ~ Glacial boundary.
Insert: No obvious recent change in limits of range.

the species in the *photogenis-rubellus-atherinoides* complex) to justify the statement that between 1854–1925 the distributional pattern of the silver shiner in the Ohio River drainage was essentially the same as it was during the 1925–1950 surveys. Kirsch's records (1895A:329 as *N. arge*) for the Tiffin River at Brunersburg and the Maumee River at Grand Rapids probably are correct, for in 1894 the Maumee River still contained a rather clear-water fish fauna. Osburn's (1901:57) record for the Cuyahoga River, Summit County, is questioned because in 1901 he misidentified this species, as illustrated by the misidentification of the Franklin County material (Williamson and Osburn, 1893:31–32, Pl. 7; *see* rosyface shiner). McCormick's statements (1892:19) and my identification of his material, demonstrate that his Lorain County records of *N. photogenis* were based upon specimens of *N. atherinoides* and *N. rubellus*.

Since 1920 the largest populations have occurred in the Great and Little Miami river systems and in a broad band paralleling the line of glaciation. Small populations were present in high-gradient streams entering the Ohio River between Athens and Columbiana counties, and in the St. Joseph River system in northwestern Ohio. In 1929 I took a silver shiner (OSUM 292) in the Wabash River, Mercer County, presumably a stray from Indiana. (Gerking, 1945: map 34). The silver shiner must have been present and probably was abundant in the Ohio River before that stream was impounded and when it was less turbid and less polluted. During the 1920–50 period the species decreased markedly in numbers in many localities, especially in those portions where turbidity and siltation had increased greatly.

Habitat—The silver shiner was abundant in moderate- and large-sized streams which had relatively clear waters throughout most of the year, moderate or high gradients, and clean gravel and boulder-strewn bottoms. In such streams the species was most numerous on the deep, swift riffles and in the swifter eddies and currents of the pools immediately below such riffles. It usually avoided heavily silted bottoms and rooted aquatic vegetation. It normally occurred in schools and when feeding frequently jumped into the air to capture flying insects.

Years 1955–80—As may be noted on map 58, there has been little change in the distribution of this species in the Ohio River drainage since 1955. It remains a prominent part of the fish fauna in many moderate and large-sized streams.

On June 25, 1900 R. C. Osburn collected this species, recording it as *Notropis arge* (Cope), in the Cuyahoga River at Hawkins. He considered it to be rare at this locality. In the 1957 edition of this report (p. 342) I questioned the record because the specimens were lost and elsewhere Osburn had misidentified the species (*see* above). Although there were records for the Maumee drainage in northwestern Ohio, there were none for the northeastern portion of the state.

Since June 19, 1975 Andrew White and associates have captured many specimens in the Grand River and its tributaries, in Lake and Ashtabula counties. In 1976 Ted Cavender and class captured this species in the Vermilion River, Erie County, at a collecting station where for many years instructors have taken classes from the Franz Theodore Stone Laboratory for the purpose of collecting fishes. It is unknown when and how this species invaded these Lake Erie tributaries. Between July 29 and August 8, 1971 the silver shiner was captured in the Grand River drainage of Ontario, Canada (Gruchy, Bowen and Gruchy, 1973B:1379–82). These recent captures suggest that R. C. Osburn's record from the Cuyahoga River probably is valid.

POPEYE SHINER

Notropis ariommus (Cope)

Fig. 59

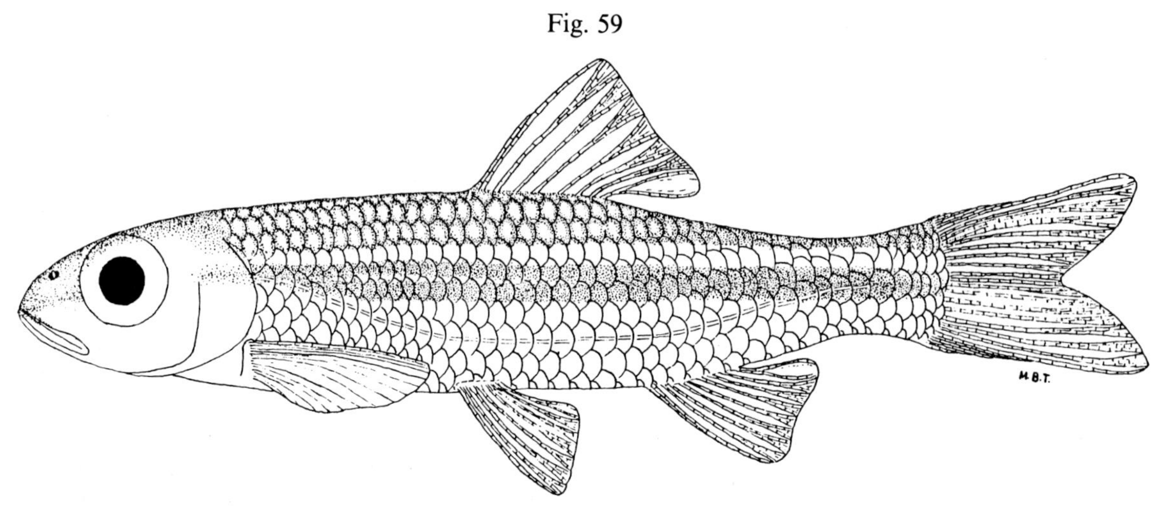

Maumee River, 2 1/2 mi. above Antwerp, Paulding County, O.

Aug. 16, 1893.
Adult.

57 mm SL, 2.7″ (6.9 cm) TL.
OSUM 16166.

Identification

Characters: *General—See* Family Cyprinidae. *Generic*—Dorsal fin of 8 developed rays, the anterior half-ray an inconspicuous splint bound tightly to the first developed ray. Lower jaw with fleshy lobes. Body with cycloid scales. Head scaleless. No barbels, adipose fin, adipose eyelids, pelvic axillary process, dorsal, anal or pelvic spines. *Specific*—Tip of lower jaw dusky; a dusky or blackish lateral band extending from tip of snout posteriorly, very faint across opercles and body to approximately beneath dorsal origin, thence more distinct to caudal base. Lateral line complete with 33–37 scales; in the two Ohio specimens the lateral line bends rather sharply downward along the central half of the body. Predorsal scales fewer than 18; body circumferential scales usually 24 or 25. Mouth large and terminal; upper jaw length about as long as eye length. Eye very large, proportionately larger than in any other Ohio species of *Notropis*, usually contained 2.5–3.1 times in head length. Peritoneum silvery, uniformly overlaid with brown pigment. Anal rays 9, rarely 8 or 10. Teeth 2,4–4,2 (*see* Gilbert, 1969:474–92).

Differs: *See* bigeye and blackchin shiners. Bigeye shiner has 8 anal rays, rarely 7; 1,4–4,1, rarely 0,4–4,0 teeth. Blackchin shiner has incomplete lateral line, 8 anal rays, rarely 7; 1,4–4,1, rarely 0,4–4,0 teeth; silvery peritoneum.

Most like: Bigeye shiner. **Superficially like:** Blackchin, blacknose, pugnose, mimic and sand shiners.

Coloration: Dorsally olive-yellow, overlaid with silvery, the scales of the back with broad, dark edges. A row of silvery scales immediately dorsal to the lateral band, between the dark scales of the back and lateral band. Ventrally silvery and milk-white. Fins transparent.

Lengths: *Adults*, usually 2.0″–2.5″ (5.1–6.4 cm) TL.

Distribution and Habitat

Ohio Distribution—Prior to publishing the first edition of *The Fishes of Ohio* in 1957, I had examined all collections of Ohio fishes that were known to me without finding Kirsch's two specimens of *N. ariommus*, taken August 16, 1893, in the Maumee River, Paulding County, Ohio. I had carefully looked for them in the University of Michigan Museum of Zoology collections prior to 1940, at which time they were deposited, unbeknown to me, in the California Academy of Science collections. However, when in 1950 I examined the California

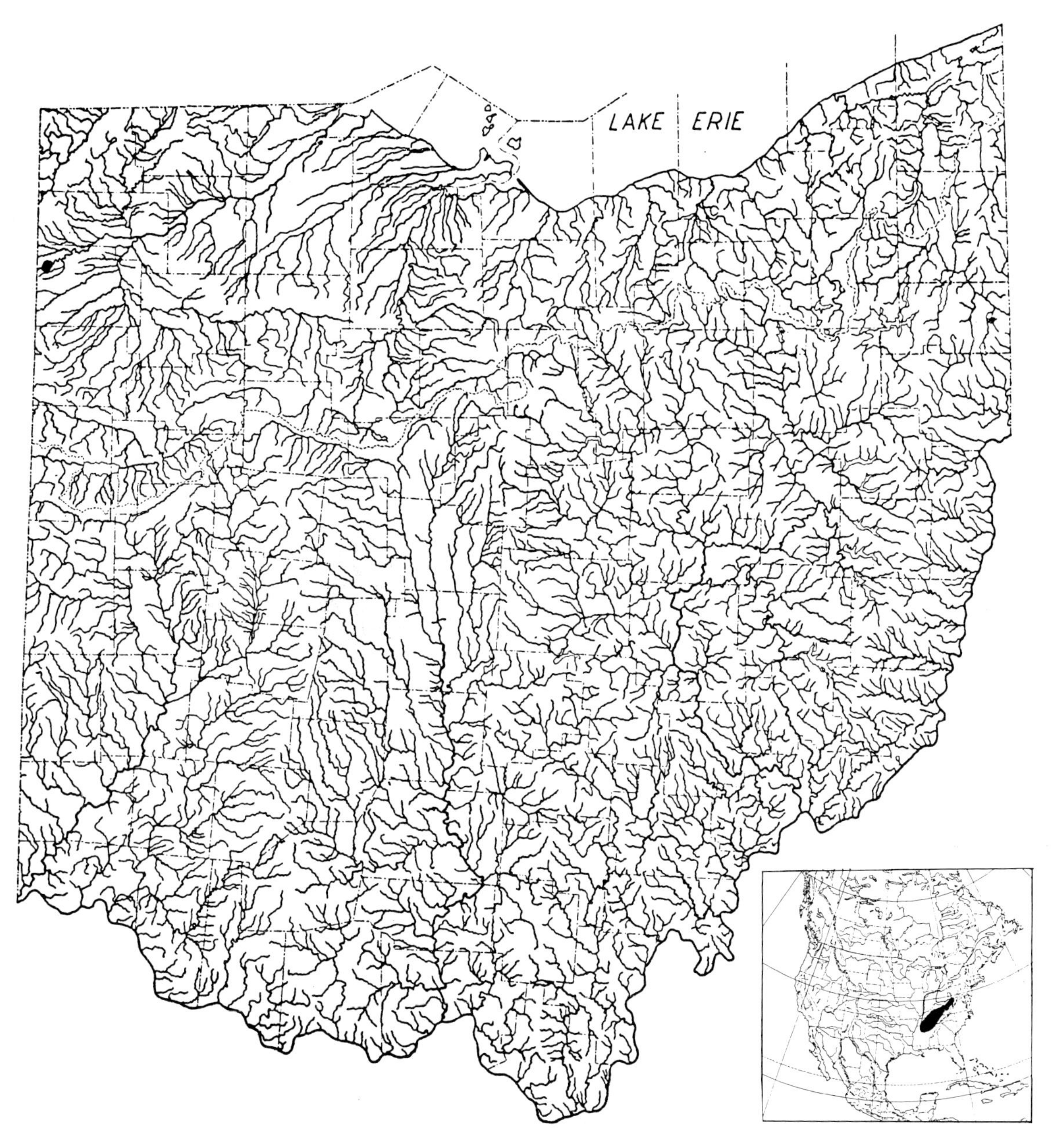

MAP 59. Popeye shiner

● Before 1900.
Insert: former range outlined; present range in black.

Academy's collections, the two specimens of *N. ariommus* had been sent a few years previously to the Museum of Zoology at Ann Arbor. I found no record of this transfer at the Academy.

Since Kirsch did not give the tooth count and erroneously gave the anal fin ray count as *8*, and especially since I had seen specimens of *Notropis boops* which Kirsch had collected during the same year in the Maumee system in Ohio, I assumed that the two specimens likewise were *Notropis boops*, the bigeye shiner. Fortunately, Carter R. Gilbert (1969:485) discovered these two specimens at Ann Arbor. One of these is now deposited in the Museum of Zoology of the University of Michigan

(UMMZ 187285), and the other has been presented to the Ohio State University collection (OSUM 16166). No specimens have been collected in the Great Lakes drainage or in the Ohio River tributaries entering from the north during the past 75 years.

Habitat—As the huge eye suggests, the popeye shiner is an inhabitant of waters of great clarity. Originally the Maumee River was such a system, draining a level or slightly rolling terrain in which the prairie and woodland flora held erosion to a minimum. Its original fish fauna was composed of such species as the sturgeon, bowfin, mooneye, muskellunge, northern, greater and river redhorses, harelip sucker, hornyhead, river and bigeye chubs, pugnose minnow, silver, rosyface, and bigeye shiners, northern madtom, pirate perch, pumpkinseed sunfish, sauger, walleye, and channel, gilt, and sand darters, all of which required "clear water" and/or much aquatic vegetation. The disappearance of the popeye shiner from the Maumee system was coincidental with the increase in soil erosion (*see* Habitat under harelip sucker).

ROSYFACE SHINER

Notropis rubellus (Agassiz)

Fig. 60

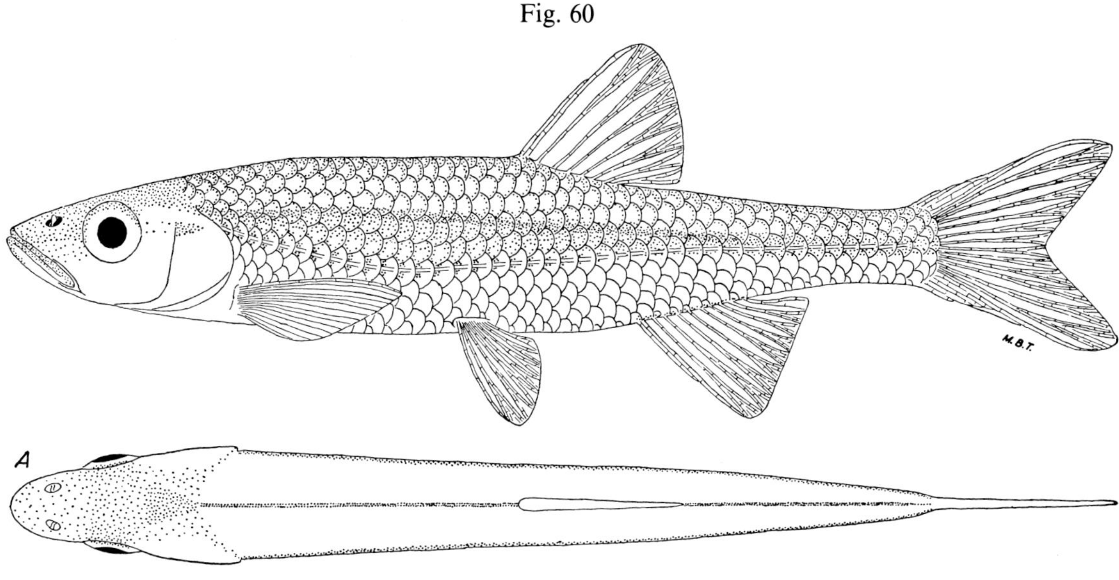

Upper fig: East Branch, Rocky River, Medina County, O.

April 15, 1940. 64 mm SL, 3.1″ (7.9 cm) TL.
Adult male. OSUM 1944.

Fig. A: dorsal view showing color pattern.

Identification

Characters: *General—See* emerald shiner. *Specific*—Anal rays normally 10–13, rarely 9. Dorsal fin transparent; without spots. Body olive-blue and silvery; breeding males rosy anteriorly; breeding females less so. Mid-dorsal stripe often faint and narrow, or absent; distinct only in some adults. No dark crescents between nostrils as in the silver shiner. Snout sharply pointed; its length contained less than 1.5 times in postorbital head length. Dorsal origin decidedly behind pelvic origin. Body roundish; its width behind head usually contained 1.1–1.8 times in height of dorsal fin. Eye small; contained 3.5–4.3 times in head length. Body deepest midway between head and dorsal fin. Complete lateral line usually with 36–40 scales; fewer than 22 predorsal scales. Teeth, usually 2,4–4,2.

Differs: *See* emerald shiner. Emerald shiner has a shorter snout; more slab-sided body; dorsal origin over pelvic origin.

Most like: Emerald and silver shiners. **Superficially like:** Rosefin and redfin shiners.

Coloration: Dorsally olive-blue or greenish-olive with an overcast of blue and silver; back usually darker than that of the emerald and silver shiners. Mid-dorsal stripe usually faint or absent, or narrow if distinct. Sides silvery with a lavender sheen; a lateral band of lavender apparent at some angles. Ventrally silvery-white. Fins transparent. *Breeding male*—Similar to above description, except that back is more intensely blue; dorsal head surface bluish-green with a blush of orange-red; a narrow scarlet bar extending on body from upper gill cleft downward to base of pectoral fin and with bright rosy lips, snout, sides of head, and a rosy blush often extending across breast and forepart of body. Tiny tubercles cover head and some scales of the predorsal region, and upper surfaces of pectoral rays. *Female*—Contains some rosy.

Lengths: *Young* of year in Oct., 1.0″–2.0″ (2.5–5.1 cm). Around **1** year, 1.3″–2.5″ (3.3–6.4 cm). *Adults*, usually 1.8″–3.0″ (4.6–7.6 cm). Largest specimen, 3.5″ (8.9 cm) long.

Hybridizes: with striped, common and spotfin shiners.

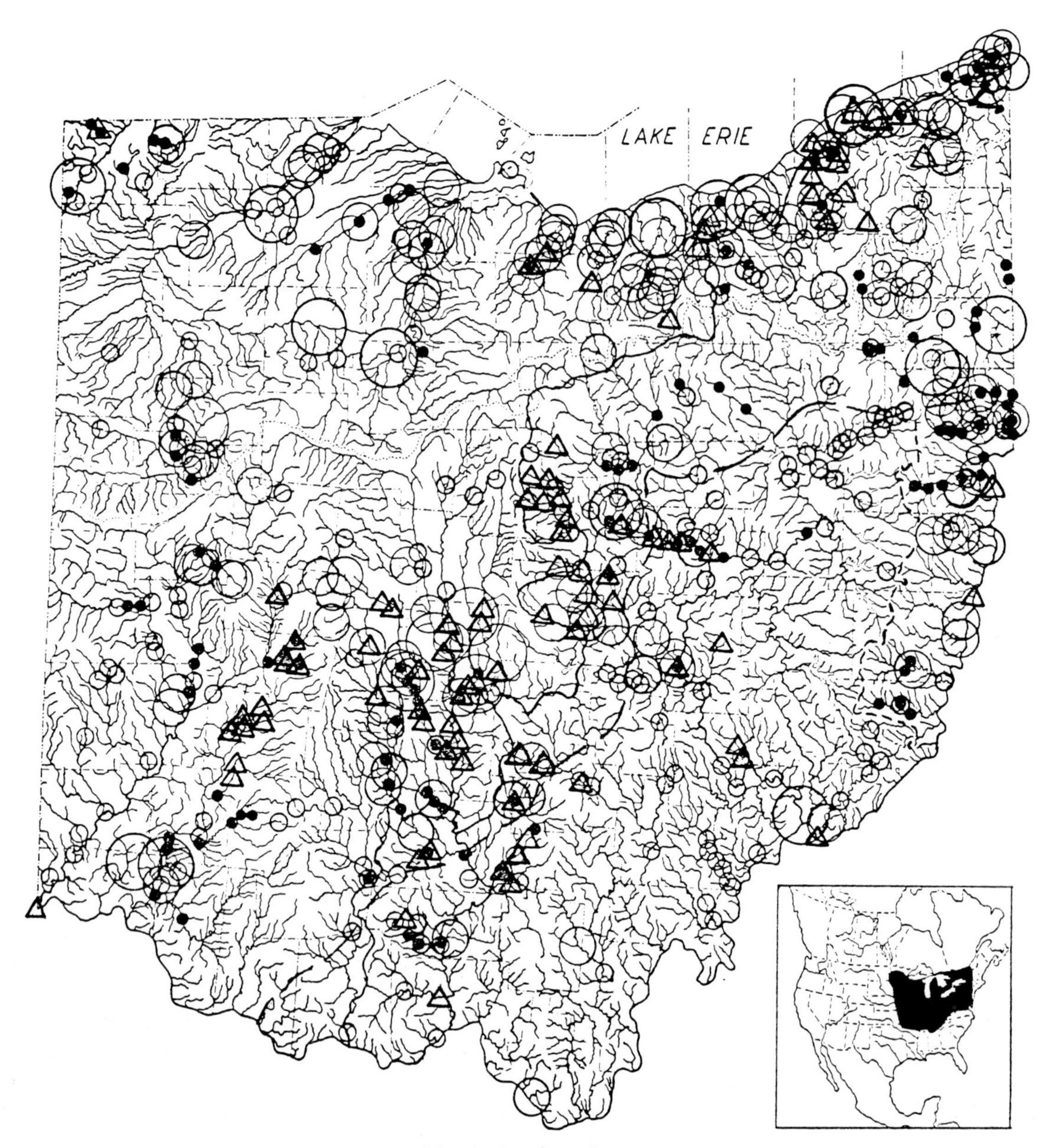

MAP 60. Rosyface shiner

Locality records. ○ Before 1901. ○ 1901–24. ○ 1925–38. ● 1939–54. △ 1955–80.
~ Allegheny Front Escarpment. /\ Glacial boundary. ------- Flushing Escarpment.
Insert: Western and southern limits of range indefinite.

Distribution and Habitat

Ohio Distribution—It is evident from preserved material and reliable literature references that during the 1890–1938 period the rosyface shiner was widely distributed and abundant throughout most of the Ohio stream systems. Since 1938 it has decreased markedly in numbers, or has become extirpated in such streams as the Sandusky, Hocking, Maumee, and lower portions of the Muskingum and Great Miami rivers. Throughout the entire period its greatest population densities were in those areas of high relief such as are present in the upper Great Miami drainage, along the Allegheny Front

Escarpment and immediately to the east of it, and east of the Flushing Escarpment. It was absent or occurred in small numbers in the low- and base-gradient, glaciated streams of the till-plains and prairie areas, and in the base-gradient, unglaciated streams of the southeastern section. It was accidental in the Ohio River and in Lake Erie.

Habitat—The rosyface was abundant in moderate-sized streams which had relatively clear waters throughout most of the year, rather high gradients, and clean bottoms of sandy gravel, gravel, boulders, or bedrock. It spawned over sandy gravel, gravel, and bedrock in the fast current of rather shallow portions of riffles, and over such nests as those of the hornyhead and river chubs. It wintered in the deeper riffles and pools, and at that season was found in the largest inland streams. The rosyface was an associate of the silver shiner throughout the year and the two frequently spawned on the same riffles; however, the rosyface spawned chiefly in May, whereas the silver spawned principally in June and early July; also the rosyface spawned in smaller streams than did the silver shiner. The rosyface was intolerant to turbid waters and bottoms covered with silt, and marked decreases in abundance were observed in several stream sections which had become increasingly turbid and silted during the 1928–50 surveys.

Years 1955–80—During this period the rosyface shiner could be found in numbers in Ohio only in the less turbid and higher-gradient streams adjacent to the Allegheny Front Escarpment and the glacial boundary (White et al., 1975:76). Although relict populations surely were present in the Maumee River system, only one was found. Observations indicate that many populations throughout the state had become extirpated or that only a few were taken in a given locality where, before 1950, the species was very numerous.

NORTHERN REDFIN SHINER

Notropis umbratilis cyanocephalus (Copeland)

Fig. 61

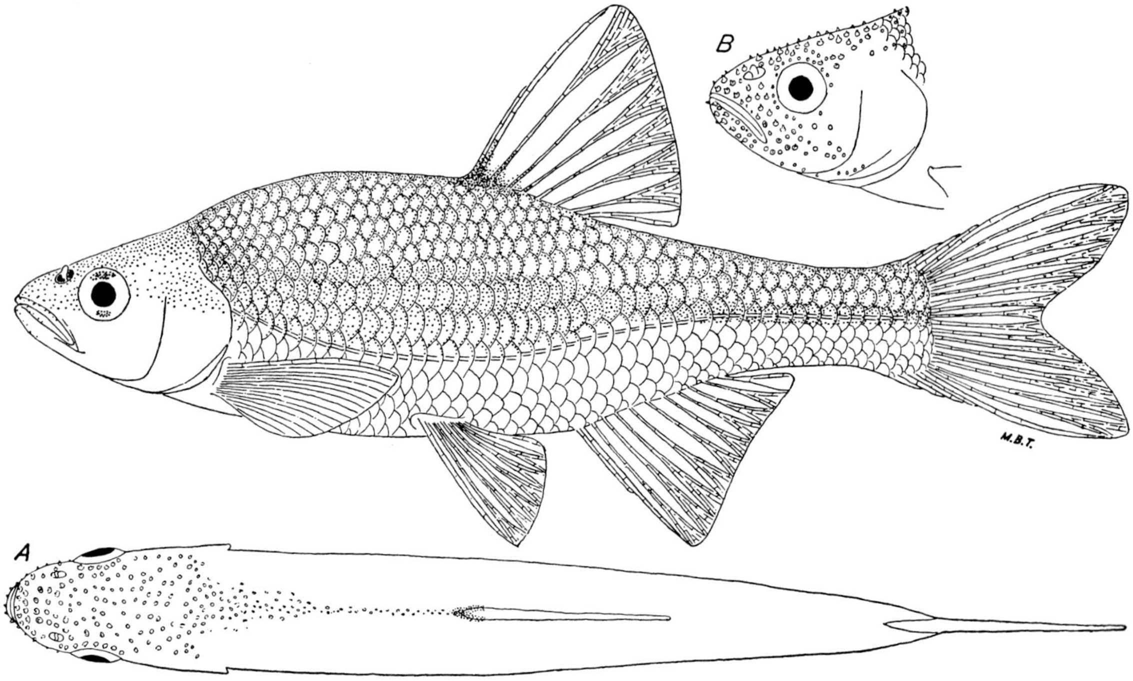

Central fig.: Mill Creek, Fulton County, O.

May 16, 1941. 58 mm SL, 2.8" (7.1 cm) TL.
Male. OSUM 3176.

Fig. A: dorsal view showing the dark spot at dorsal origin; also tubercle distribution of the male.

Fig. B: Auglaize River, Auglaize County, O.

July 9, 1941. 49 mm SL, 2.5" (6.4 cm) TL.
Breeding male. OSUM 3344.

Lateral view of head showing tubercle distribution of male; note tubercles on cheeks.

Identification

Characters: *General—See* emerald shiner. *Specific*—Anal rays normally 11–13, occasionally 10. Dorsal fin with a prominent blotch at base of anterior rays; least noticeable on small young. Body bluish-silvery; breeding adults with red on fins. Body compressed and slab-sided; its width usually contained more than 1.8 times in depth. Head length usually less than body depth. Saddle-bands across back usually faint or absent in living fishes; often absent in preserved specimens. Complete lateral line with 41–46 scales, extremes 40–48. Predorsal scales usually more than 24. Breeding male has tubercles on *cheeks*, fig. B. Teeth, normally 2,4–4, 2.

Differs: Rosefin shiner averages fewer anal rays; a more slender body; bolder saddle-bands; has no tubercles on cheeks. Emerald, silver and rosyface have fewer lateral line scales; no dusky blotch on dorsal fin. Common shiner has dorsal fin farther forward; steelcolor and spotfin have a posterior blotch on dorsal fin; all three normally have 9 or fewer anal rays as do all other species of shiners.

Most like: Rosefin shiner. **Superficially like:** Emerald, rosyface, steelcolor and spotfin shiners.

Coloration: Dorsally bluish or brownish-blue,

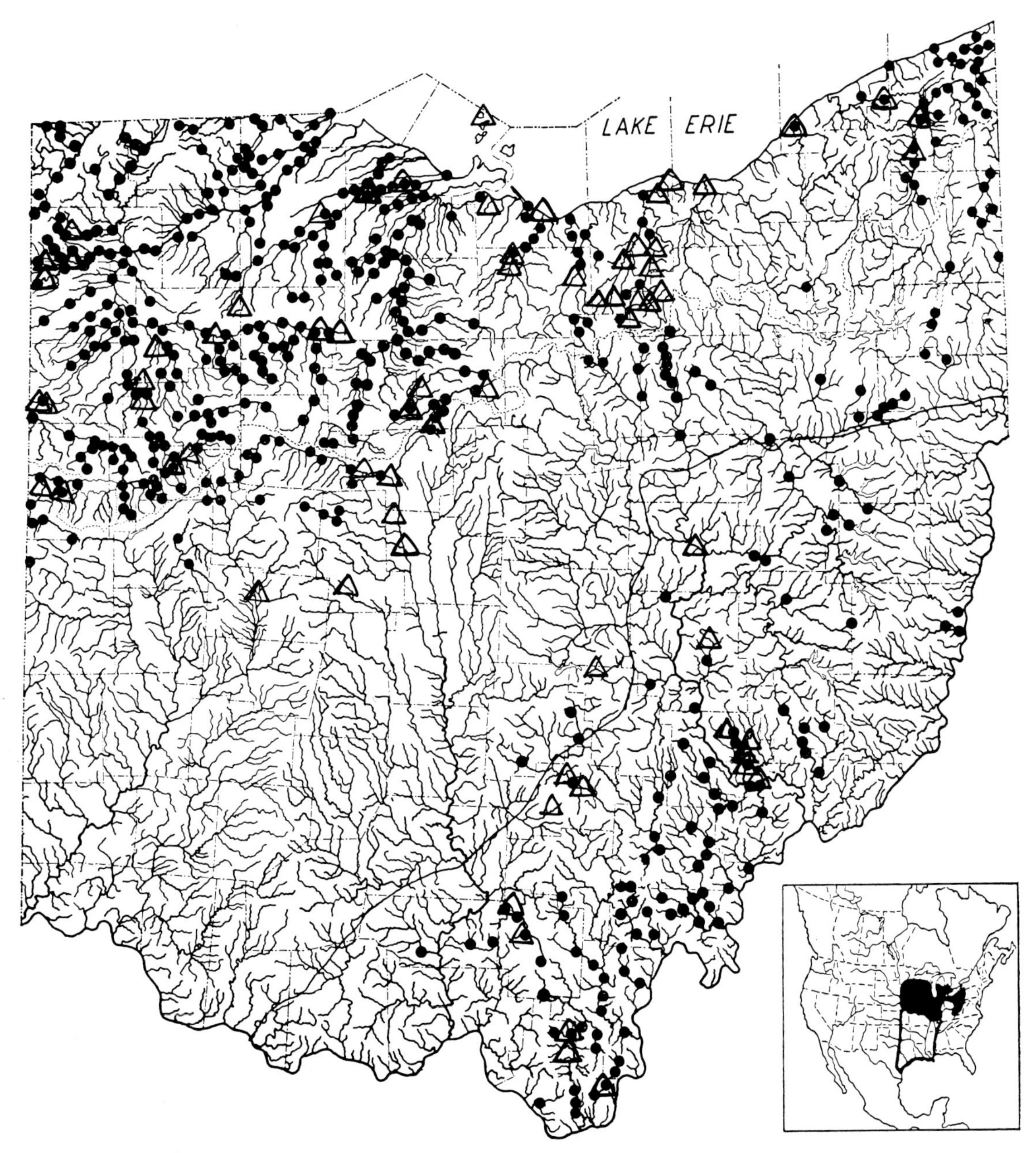

MAP 61. Northern redfin shiner

Locality records. ● Before 1955. △1955–80. ~ Glacial boundary.
Insert: Black area represents range of this subspecies, outlined area of the nominate subspecies.

overlaid with steel-blue and silvery. A dusky spot at base of anterior dorsal rays. Sides lighter and more silvery. Ventrally silvery and milk-white. Fins of non-breeding fishes transparent with some bluish and silvery. When present, the 4–11 saddle-bands are usually faint, except in breeding adults. *Breeding male*—Extremely greenish-blue on back; head brilliantly bluish dorsally; a distinct gray-blue streak along the dorsal ridge. All fins flushed with pink or red. Tubercles on dorsal half of head, lower jaw, and *cheeks*, fig. B; on predorsal scales of back, fig. A; sometimes on scales of sides; on basal third of the pelvic, anal and caudal fins. *Female*—Less brilliantly colored; some females contain a few tubercles.

Lengths: *Young* of year in Oct., 0.7″–2.0″ (1.8–5.1

cm) long. Around **1** year, 1.3″–2.2″ (3.3–5.6 cm). *Adults*, usually 1.8″–2.9″ (4.6–7.4 cm). Largest specimen, 3.2″ (8.1 cm) long.

Hybridizes: with rosefin shiner.

Distribution and Habitat

Ohio Distribution—It was not until after 1925 that the redfin and rosefin shiners were generally recognized as distinct species; fortunately sufficient preserved material exists of both species to warrant the statement that during the past century there has been no drastic change in the general distributional pattern of either species in Ohio. During the 1925–50 period the largest populations of redfins occurred in the low-gradient streams of the till-plains of northwestern Ohio, in the low-gradient streams of Trumbull and Ashtabula counties, and in unglaciated streams of low or base gradients between the line of glaciation and the Ohio River. It particularly avoided the high-gradient streams from Cuyahoga County to Ashtabula County. Since 1925 it decreased in abundance in those localities where increased turbidity, and especially silting of stream bottoms, was greatest.

Habitat—The redfin was abundant in streams of all sizes which normally had clear waters, low or moderate gradients, sandy, gravelly bottoms, and some aquatic vegetation. It spawned over sand and gravel in rather sluggish riffles and in pools having currents, apparently utilizing the swifter riffles only when the slower ones and the pools had their bottoms silt-covered. It was essentially a pool species after spawning and displayed a preference for submerged aquatic vegetation. When not spawning it was rather tolerant of turbidity and silted bottoms, and displayed marked decreases in abundance in a locality only after the faster riffles became silt-covered.

The redfin and rosefin shiners occurred together occasionally in the same stream section. In Bokes Creek, Union County, both species were present in about equal numbers, and in that stream on July 8, 1928, Robert B. Foster and I found the redfin spawning over a sandy bottom in the sluggish currents of riffles and pools, and on the same day the rosefin was spawning in the fastest riffles over a bottom of glacial gravel which was mostly limestone. In every instance the two species while spawning were widely separated ecologically.

Years 1955–80—The only evidence of extension, if such, of the range in Ohio of the redfin during this period was a collection of 19 specimens, taken by Donald E. Veth and me, April 3, 1963, in Buck Run, Big Darby drainage in southern Union county. The low-gradient and other habitat conditions at the point of capture appeared to be favorable for this species. Schools of rosefins were found both above and below the stream section where the school of redfins was found.

A mature individual was captured April 25, 1967, by G. F. Ahrens, my wife and me in the Mad River, Champaign County. At the time many trout fishermen were in this section of the Mad River, some using minnows as bait. It is quite possible that the redfin collected was brought there as bait and escaped. The conditions where this specimen was captured appeared unfavorable to this species.

Another adult was taken July 3, 1969, by the ichthyology class and me in a pond on North Bass Island, Ottawa County. This pond was connected with Lake Erie during periods of high water. Annual collections made in this pond before and since have resulted in the capture of no other specimens. This specimen was most probably a stray from some inland stream or an escape from a minnow bucket (White et al., 1975:76).

In the spring of 1969 Franklin F. Snelson, Jr., and I examined the material from Bokes Creek, Union County, he agreeing that some of the mature males were quite possibly hybrids between the redfin and the rosefin. Both species have been observed spawning in this stream, a few spawning schools in close proximity to each other. Some of the schools of both species appeared to be spawning under submarginal conditions in this disturbed stream.

Hunter and Wisby (1961:113–15) found the redfin shiner in Wisconsin utilizing occupied and unoccupied green sunfish nests for egg depositions. I and others have noted the utilization of sunfish nests in Ohio, not only of the green sunfish but also of the longear and the orangespotted sunfishes. The substrate in the sunfishes' nests was markedly similar to that over which the species spawned otherwise. For a redescription of the redfin shiner *see* Snelson and Pflieger, 1975:231–49. For a discussion of the red shiner (*Notropis lutrensis*), see page 366.

ROSEFIN SHINER

Notropis ardens (Cope)

Fig. 62

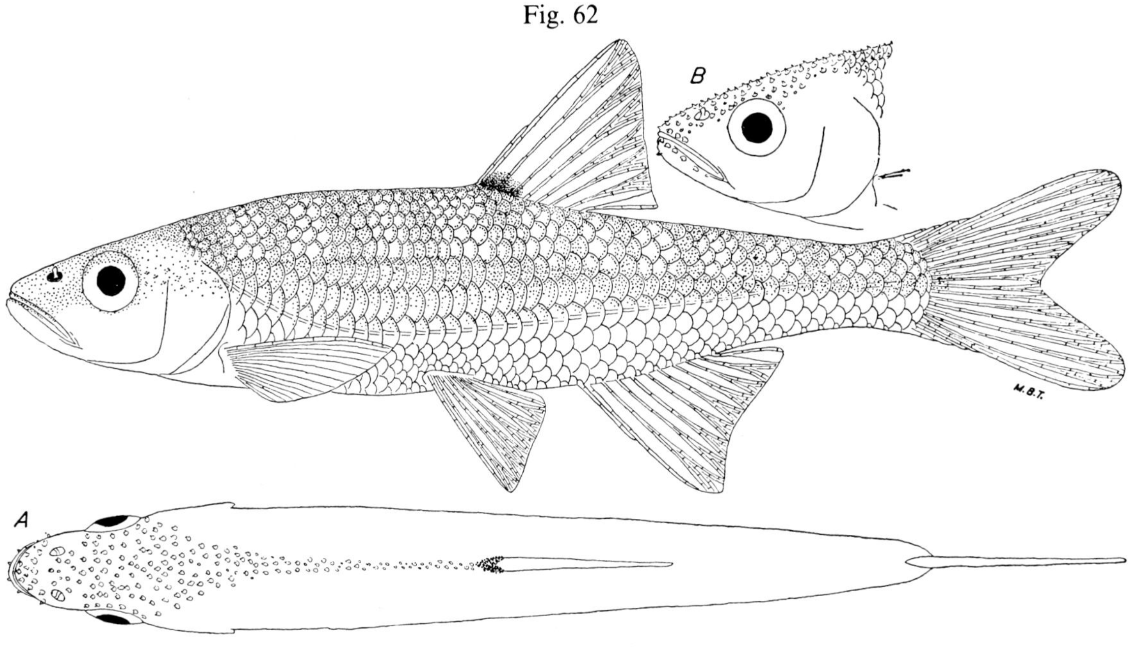

Central fig.: Beaver Creek, Clark County, O.

May 28, 1940.
Breeding male.

63 mm SL, 3.0″ (7.6 cm) TL.
OSUM 2072.

Fig. A: dorsal view, showing the dark spot at dorsal origin; also tubercle distribution of the male.
Fig. B: lateral view of head showing tubercle distribution of the male; note absence of tubercles on cheeks.

Identification

Characters: *General—See* emerald shiner. *Specific*—Anal rays 9–11; most often 10, rarely 12. Dorsal fin with a prominent blotch at base of anterior rays, similar to that of the redfin shiner. Body bluish-silver; the 4–11 dark saddle-bands usually prominent in large young and adults in life; these bands often present in preserved specimens. Breeding adults with rosy fins. Body less slab-sided than is body of the redfin; its width usually contained less than 1.8 times in depth. Head length usually equal to, or greater than, body depth. Complete lateral line of 41–46 scales, extremes 40–48. Predorsal scales usually more than 24. Breeding male *without* tubercles on cheeks, fig. B. Teeth, normally 2,4–4,2.

Differs:—*See* redfin shiner. Redfin averages more anal rays; a deeper body; has saddle-bands faint or absent; tubercles on cheeks.

Most like: Redfin shiner. **Superficially like:** Emerald, rosyface, steelcolor and spotfin shiners.

Coloration: Similar to redfin shiner, except that the 4–11 dark gray-blue saddle-bands are usually more distinct; the fins are more scarlet or rosy, and not so dark red. *Breeding male*—Lacks tubercles on cheeks; tubercles on the lower jaw and chin are more or less confined to a row along each jaw.

Lengths: *Young* of year in Oct., 0.7″–2.0″ (1.8–5.1 cm) long. Around **1** year, 1.3″–2.3″ (3.3–5.8 cm). *Adults*, usually 1.8″–3.0″ (4.6–7.6 cm). Largest specimen, 3.5″ (8.9 cm) long.

Hybridizes: with redfin shiner.

Distribution and Habitat

Ohio Distribution—As with the redfin (*see* that species) the distributional pattern of the rosefin in

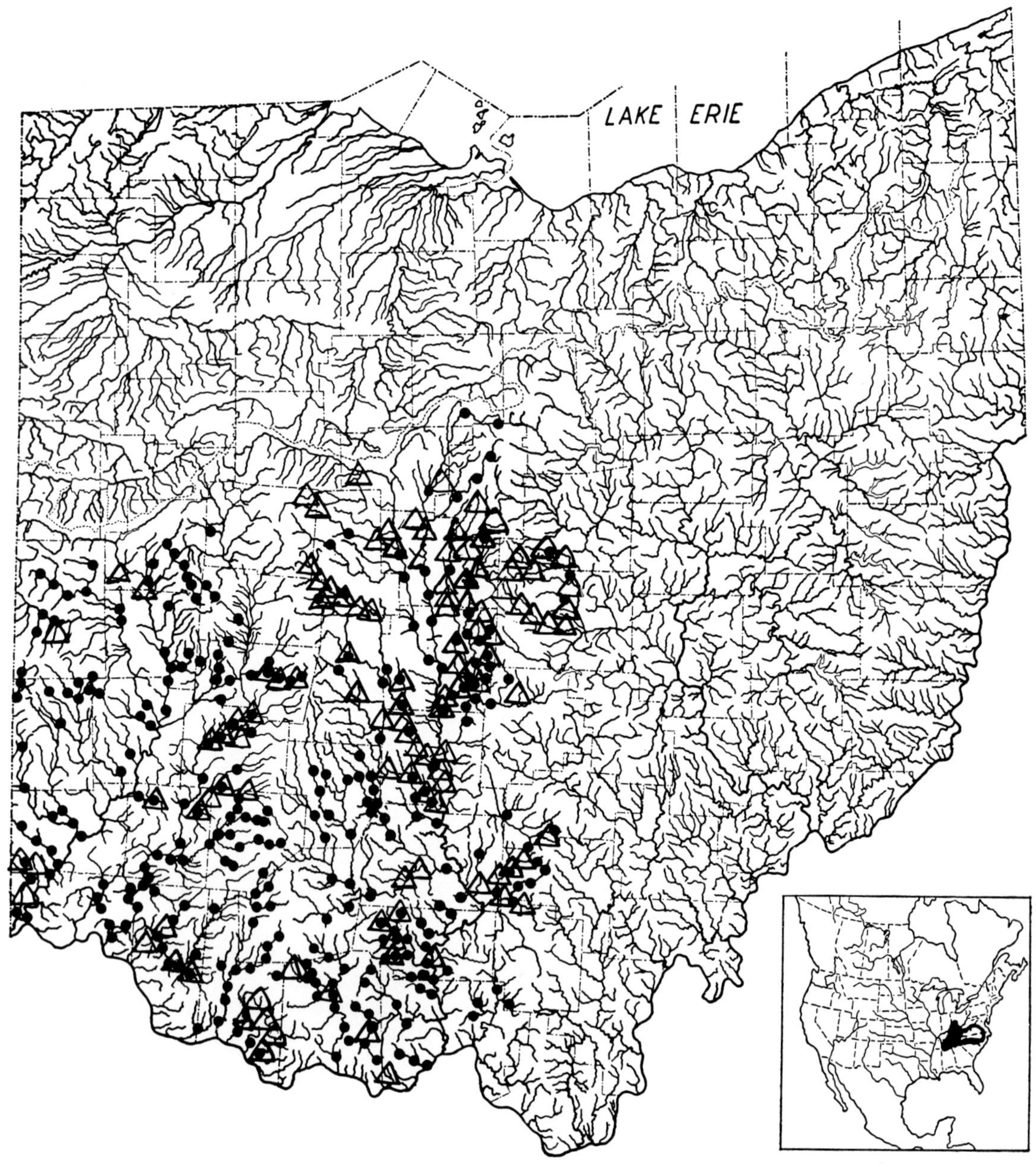

MAP 62. Rosefin shiner

Locality records. ● Before 1955. △ 1955–80.
Insert: Black area represents range of this subspecies, outlined areas of the nominate subspecies.

Ohio displayed no apparent change during the past century. Throughout the 1925–50 period the largest populations occurred in the clear-water, limestone-bottomed streams of the unglaciated, bluegrass region of Adams, Scioto, and the western half of Pike counties.

Habitat—The rosefin was abundant only in very clear-water streams and brooks which had moderate and high gradients, and limestone bedrock or limestone-gravel bottoms in which clayey silt deposits were greatly restricted or absent even in the deeper pools. It spawned in the faster currents of riffles and pools, remaining about the spawning localities throughout the warmer periods of the year, and retiring into deeper, more quite pools during midwinter. It likewise inhabited, in smaller numbers, larger streams of various types if they contained gravel riffles and did not have too much clayey silts on their stream bottoms. The species appeared intolerant of very turbid waters and where siltation was rapid. *See* redfin for occurrence of both species in the same stream section.

Years 1955–80—As noted on the distribution map above, no change is evident in the distribution of this species in Ohio during this period. As is expected, populations decreased in those areas where turbidity or other pollutant conditions increased considerably.

See redfin shiner for discussion of hybridization between the rosefin and redfin shiners in Bokes Creek, Union County.

The trinomial *lythrurus* has been omitted because this name for the rosefin may prove to be untenable (Snelson and Pflieger, 1975: 231–49).

CENTRAL STRIPED SHINER

Notropis chrysocephalus chrysocephalus (Rafinesque)

Fig. 63

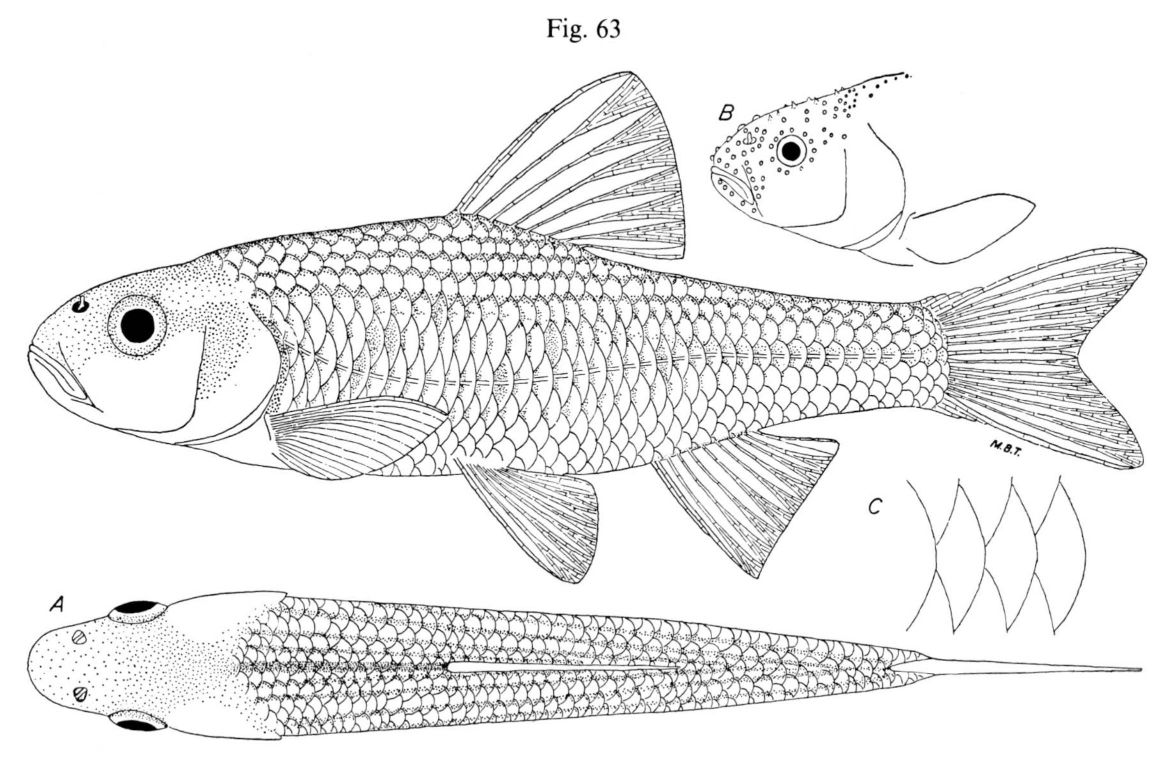

Central fig.: North Fork, Paint Creek, Fayette County, O.

Nov. 21, 1939. 107 mm SL, 5.2″ (13 cm) TL.
Male. OSUM 1322.

Fig. A: dorsal view; note pairs of dusky lines converging posteriorly, and the broad dorsal stripe from head to dorsal origin.

Fig. B: Scioto River, Scioto County, O.

Feb. 6, 1940. 135 mm SL, 6.5″ (17 cm) TL.
Male. OSUM 2988.

Fig. B: lateral view of head showing tubercle distribution of male.
Fig. C: anterior lateral line scales showing their great depth.

Identification

Characters: *General—See* emerald shiner. *Specific*—Anal rays normally 9, extremes 8–10. Unspotted dorsal fin far forward; distance from dorsal origin to caudal base extending from dorsal origin considerably beyond the snout; usually a greater distance than the length of the eye. Scales in the anterior half of the lateral line greatly elevated; their height 2.0–3.0 times greater than their exposed width, fig. C. Snout rounded. Eye large; 2.5 (smallest young)–4.7 (large adults) times in head length. Body slab-sided. Complete lateral line with 37–40 scales. Peritoneum dark brown. Teeth, 2,4–4,2. *Subspecific*—Predorsal scales usually 11–21, rarely as many as 25 (carefully count all scales from occiput to dorsal origin which *cross* the dorsal ridge, including those in which most of a scale lies to one side of the ridge). In large young and adults dark wavy lines between the scale rows of the back converge posteriorly, each wavy line on one side of the back meeting its corresponding line on the other side, thereby forming several narrow, long V's which are conspicuous when viewed from above,

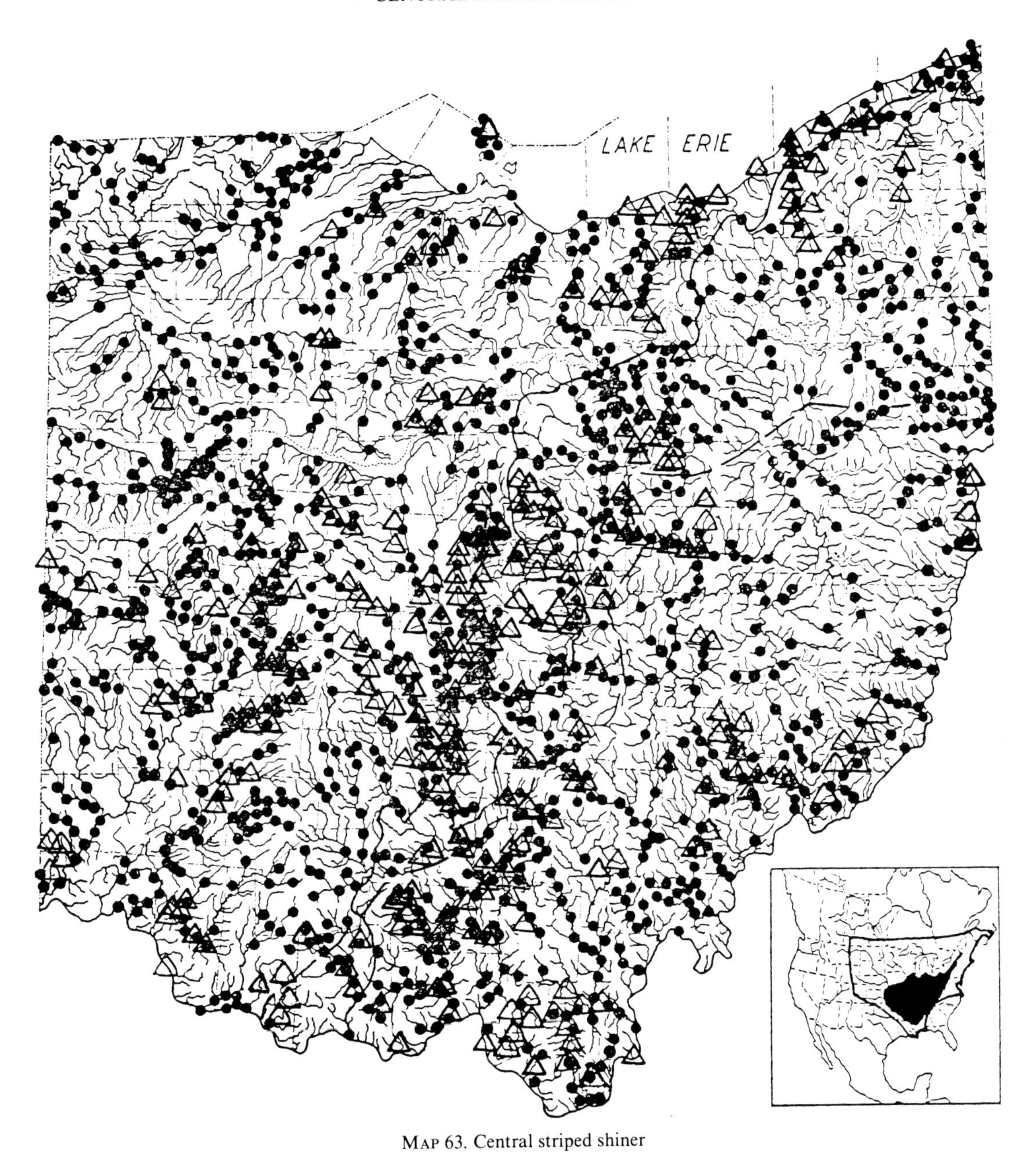

MAP 63. Central striped shiner

Locality records. ● Before 1955. △ 1955–80. ~ Allegheny Front Escarpment. /\ Glacial boundary.

Insert: Black area represents range of this subspecies, outlined area of other subspecies and of range of common shiner.

fig. A. Intergrades or hybridizes abundantly with the common shiner.

Differs: Common shiner has more than 24 predorsal scales; lacks the V's on the back. No other species of shiner in Ohio has a combination of: 9 anal rays; greatly elevated anterior lateral line scales; a dark brown peritoneum. The silvery minnow has a more subterminal mouth; a very long intestine; a blackish peritoneum.

Most like: Common shiner. **Superficially like:** Silvery minnow, silver chub, steelcolor and spotfin shiners.

Coloration: Dorsally greenish-, bluish-, or slaty-olive with a bluish-silver overcast. Mid-dorsal stripe prominent, broad and usually brownish or slaty-blue. Dark wavy lines usually present between the scale rows of the back. Sides bluish-silvery in adults; mostly silvery in young. Sides with scattered dark scales. Ventrally silvery- or milk-white. Fins of nonbreeding fishes transparent or tinged with white. *Breeding male*—Bright blue on dorsal half of head and back; sides of head and body with blues, pinks and rose; distal third of dorsal, caudal and anal a deep pink-rose; remaining fins paler. First ray of pectoral dusky. Large tubercles on top of head, snout, and lower jaw. *Female*—Less brilliantly colored; without tubercles (*Breeding behavior: see* Raney, 1940:1–14).

Lengths: *Young* of year in Oct., 1.0″–2.5″ (2.5–6.4 cm) long. Around **1** year, 1.5″–3.5″ (3.8–8.9 cm). *Adults*, usually 3.0″–7.0″ (7.6–18 cm). Specimens more than 8.0″ (20 cm) long uncommon. Largest specimen, 9.3″ (24 cm) long (*see* Marshall, 1939: 148–54).

Hybridizes: with hornyhead, river and creek chubs, redbelly and redside daces, rosyface and silver shiners and stoneroller minnow.

Distribution and Habitat

Ohio Distribution—Preserved specimens and statements like those of Kirtland (1847:26–27 as *Rutilus playgrus*) and Osburn (1901:55–56) leave no doubt as to the widespread distribution and great abundance of the striped shiner, as a species, in Ohio previous to 1900. Throughout the 1925–50 surveys the striped shiner was one of the six most universally distributed species of fishes in Ohio, and was presumably present, at least as strays in every Ohio stream.

The largest populations of central striped shiners occurred in regions of considerable maximum relief, such as in the area drained by the upper half of the Great Miami River system, along the southern half of the Allegheny Front Escarpment, and in the area adjacent to the Ohio River from Lawrence County to the Pennsylvania state line. The smallest populations occurred in lowland, sluggish, turbid streams such as were present in the Maumee River drainage. This subspecies was rare or present as strays in Lake Erie and in the Ohio River during late spring and early summer.

Habitat—During the spawning season the largest populations of central striped shiners were present in the brooks and smaller streams where the gradients were moderate or high, the waters normally clear, the bottoms primarily of gravel, boulders, bedrock and sand, and where brush or other escape cover was present in the pools. After the spawning season many adults and young moved downstream to winter in the larger, deeper waters of lower gradients, and during winter or in drought periods many were present in the Ohio River.

The central striped shiner differed ecologically from the common shiner in that the former seemingly preferred warmer waters, it moved downstream in larger numbers after the spawning season and occupied deeper and larger waters in winter, and seemed to be more tolerant to turbid waters, silt upon the stream bottoms and lower stream gradients. Populations comprised entirely of central striped shiners without visible evidence of intergradation with the common shiner were present only in the southern third of Ohio.

Years 1955–80—Before 1955, when the above was written, *Notropis cornutus chrysocephalus* (Rafinesque) and *Notropis cornutus frontalis* (Agassiz) were considered to be conspecific. In 1960 Carter R. Gilbert completed, at the University of Michigan, his doctoral dissertation, "A systematic and geographical study of the subgenus *Luxilus* (Family Cyprinidae)," in which he suggested that the heretofore conspecific *N. cornutus chrysocephalus* and *N. cornutus frontalis* be considered specific. In 1961 Gilbert (181–92) published a condensed version of his 1960 dissertation. Despite criticism, such as R. J. Miller (1968:640–47), the Committee on Names of Fishes of the American Fisheries Society (Bailey et al., 1970:21) accepted Gilbert's conclusions, designating the common names of striped shiner to *N. chrysocephalus* and common shiner to *N. cornutus*. Hoping to retain some taxonomic stability in fishes, I follow the conclusions of the committee.

I based my former conspecific conclusions on the Ohio material collected prior to 1955, which is preserved in the OSUM. I have reexamined this material and much of that collected in Ohio since. Besides the characters previously employed, I attempted to utilize those suggested by Gilbert (1961:185–87) and especially the greater pigmented area of the chin and gular areas in *chrysocephalus*.

The pigmentation character was found to be unreliable in those areas where both forms were sympatric and was found to be more reliable the farther the *chrysocephalus* populations were removed from those of *cornutus*. However, in southern Ohio, where only *chrysocephalus* was found, there were specimens without considerable pigmentation, whereas in northeastern Ohio, where both forms were found, some individuals of *cornutus* occasionally contained considerable pigment.

The greatest instability, and the largest numbers

of intermediates (hybrids?) occurred in those stream sections in which the greatest and/or most rapid ecological changes were taking place. In such sections an increase in the number of *chrysocephalus* almost invariably occurred. In Auglaize County, on the southern edge of *cornutus* influence, the populations consisted almost entirely of hybrids and *chrysocephalus*.

Between 1971–73 Andrew White and others investigated the streams of the Cleveland area (White et al., 1975). Large numbers of *chrysocephalus*, *cornutus*, their hybrids or intergrades were collected from the Chagrin and Rocky rivers. From these specimens I took counts and measurements, and recorded the amount of pigment on chin and gular areas. The data indicated that the Rocky River headwaters contained small populations of what appear to be typical *cornutus*, the Chagrin River containing larger populations, especially in the upper Chagrin and its Aurora Branch. The lowest portion of both streams contained what appear to be more or less typical *chrysocephalus*. However, the lower Chagrin River specimens contained 2 to 5 more predorsal scales than did those from the lower Rocky River. Many of the specimens from the lower Rocky River contained as few predorsal scales as did *chrysocephalus* from southern Ohio, outside the present range of *cornutus*.

In 1975 Miles M. Coburn submitted to John Carroll University a M.S. thesis entitled "Evidence for a subspecific relationship between the Common and Striped Shiners in the Chagrin River, northeastern Ohio." He concluded that: (1) mean scale counts exhibited a cline throughout the drainage with only one exception; (2) the degree of intergradation is closely related to the physiography of the river, presence of stream obstructions and probably the amount of canopy; (3) the upper Chagrin and Aurora Branch contain nearly pure populations of *cornutus*; (4) the East Branch demonstrates complete intergradation; (5) melanophore patterns do not fall into two groupings as Gilbert (1960) stated, but are highly individualistic and inconsistent when compared to scale counts; (6) the two taxa may become morphologically indistinct in the future.

To me, *cornutus* and *chrysocephalus* appear to be in that period of evolution in which they are in between subspecies and species. They appear not to have obtained the status of "biological" species but rather appear to be "potential" or "incipient" species. When in this stage, they may attain specific immunity at some time in the future or may retrogress, depending upon conditions.

Some species of animals have several recognizable subspecies, such as the wide-ranging song sparrow, *Melospiza melodia* (Checklist of North American Birds, 1957:630–37), which contains 31 recognized subspecies. Some of these subspecies are so similar in size and color as to be difficult to recognize, whereas others are strikingly different. It appears illogical to assume that throughout their wide range as a species the many subspecies of song sparrows will become definitive species within a short period of time. It appears more probably that, if specific status occurs at all, it will take place within only one or a few subspecies at a time and throughout a long period of time.

In the case of *chrysocephalus* and *cornutus* specific status may not be attained simultaneously but may occur only at one or more isolated points along the broad front of intergradation, and will spread from such a point or points throughout the sympatric range, until genetic immunity is attained. If genetic immunity has been attained at one or more points and not throughout the broad front of integradation, it therefore depends upon where the researcher has obtained the largest amount of data as to whether he decides as to its specific or subspecific identity.

For a discussion of hybridization versus intergradation between striped and common shiners, *see* Gilbert (1961:181–92); Coburn (1975).

COMMON SHINER

Notropis cornutus (Mitchill)

Fig. 64

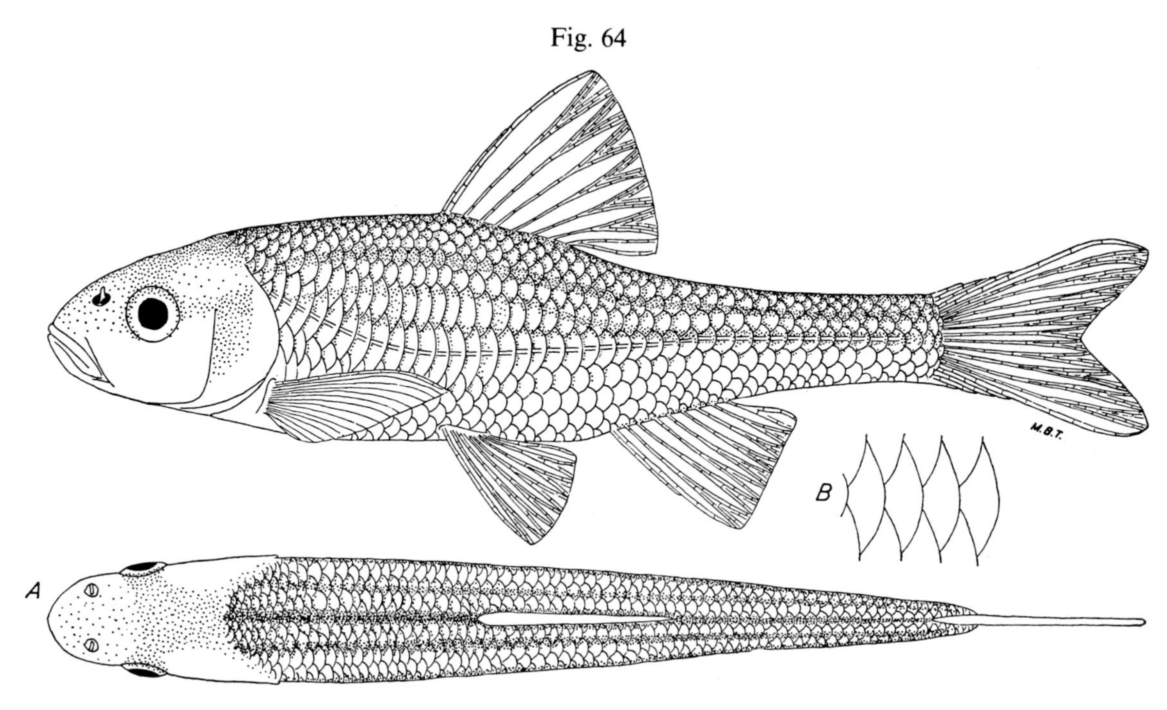

Upper fig.: East Branch, Rocky River, Medina County, O.

April 15, 1940. 79 mm SL, 3.7″ (9.4 cm) TL.
Male. OSUM 1942.

Fig. A: dorsal view, note the dusky lines remaining parallel with one another and not converging posteriorly. Note also the broad dorsal stripe which extends from the head to the tail.

Fig. B: anterior lateral line scales, showing their great depth.

Identification

Characters: *General—See* emerald shiner. *Specific—See* central striped shiner. *Subspecific*—Predorsal scales usually 25–32, extremes 22–36; carefully count all scales from occiput to dorsal origin which *cross* the dorsal ridge, including those in which most of a scale lies to one side of the ridge. The one or two wavy lines between the scale rows on each side of the back lie parallel to the dorsal ridge; therefore the corresponding scale rows on each side of back do not merge posteriorly as they do in the central striped shiner, fig. A. Intergrades or hybridizes abundantly with the central striped shiner.

Differs: *See* central striped shiner.

Most like: *See* central striped shiner.

Coloration: Dorsally greenish-, bluish-, or slaty-olive with a bluish-silver overcast. A very pronounced gray-blue mid-dorsal stripe, about half as wide as eye diameter. One or two gray-blue wavy lines run between the scale rows on each side of the mid-dorsal ridge, and parallel to it; when viewed from above they show as 3–5 parallel lines, fig. A. Sides bluish-silvery, mostly silvery in young; sides with some dark scales. Ventrally silvery and milk-white. Fins whitish or transparent. *Breeding adults*—Similar to those of the central striped shiner, except that the colors are usually more subdued; the reds are pinkish rather than rosy.

Lengths: *Young* of year in Oct., 1.0″–2.5″ (2.5–6.4 cm) long. Around **1** year, 1.3″–3.2″ (3.3–8.1 cm). *Adults*, usually 2.8″–6.0″ (7.1–15 cm). Specimens

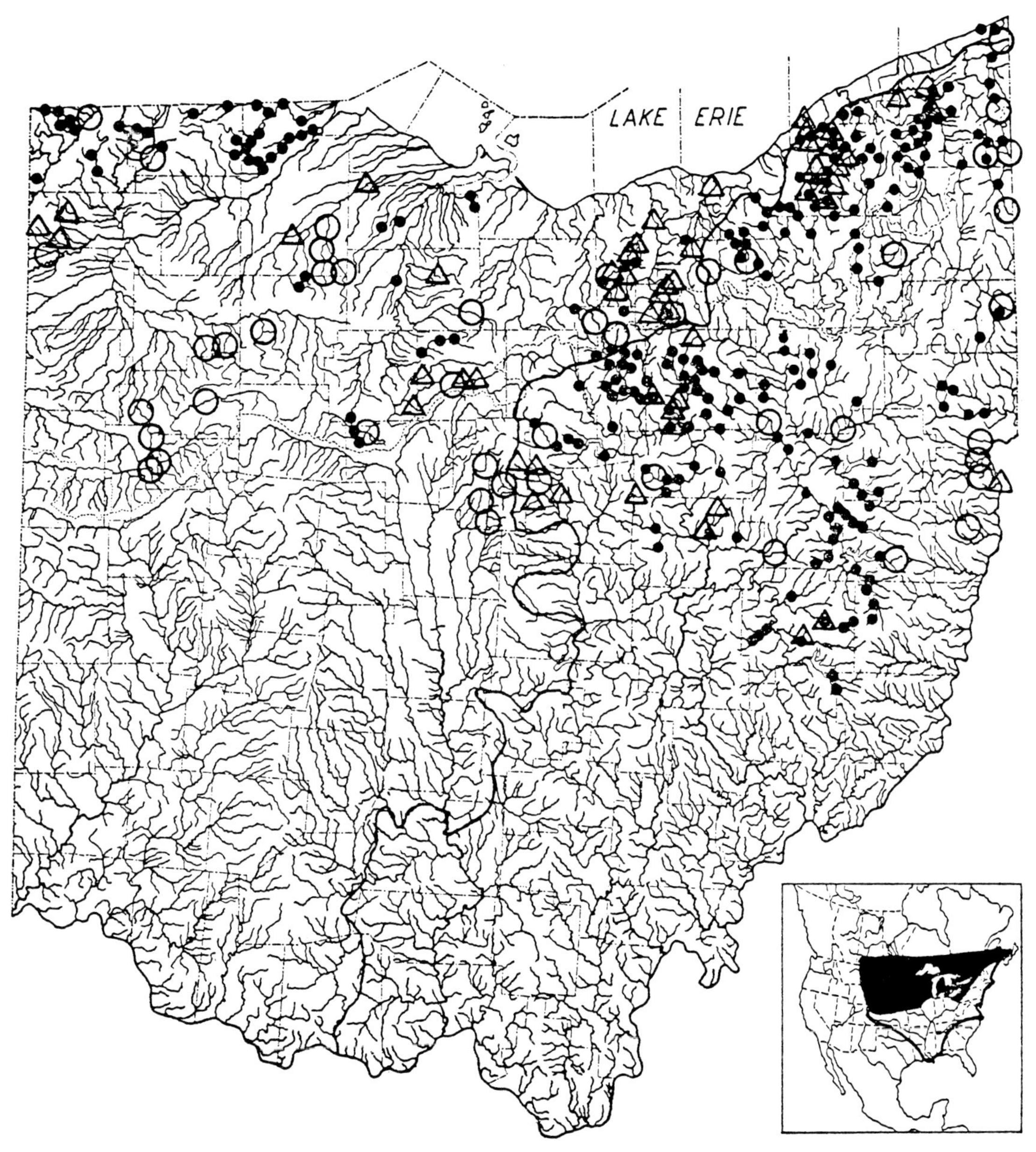

MAP 64. Common shiner

Locality records. Largely consisting of intergrades between striped and common shiners. This species, with few or no recognizable intergrades: ● Before 1955. △ 1955–80.
~ Allegheny Front Escarpment.

Insert: Black area represents range of this species, outlined area of striped shiner; northern limit of range indefinite.

more than 6.5″ (17 cm) long are uncommon. Largest specimen, 8.2″ (21 cm) long.

Hybridizes: with hornyhead, river and creek chubs, longnose, redbelly, redside and rosyside daces, rosyfaced shiner and stoneroller minnow.

Distribution and Habitat

Ohio Distribution—During the 1925–50 surveys the largest populations of common shiners occurred in small streams in those areas of considerable maximum relief located in extreme northwestern Ohio north of the Maumee River, along the northern half of the Allegheny Front Escarpment and immediately to the east of it, and in the highlands from Ashland County southeast to Harrison County.

Habitat—The largest populations consisting of predominantly typical common shiners were found only in those isolated portions of spring-runs and brooks of moderate or high gradients where the waters were usually clear, cool and shaded by brush or trees, where the stream bottoms were of clean sand, gravel, bedrock or organic debris, and where clayey silts were almost absent. This species seemed less tolerant than was the central striped shiner to warm and turbid waters and to silted bottoms, and seemed to be slightly more tolerant to submerged aquatic vegetation.

The post-glacial history of Ohio suggests that both the northern and central striped shiners were present in some unglaciated portions of the Ohio and/or Mississippi river drainages during glacial times; that in post-glacial times an invasion of both subspecies into glaciated Ohio occurred; that prior to 1800 conditions were more favorable to the common shiner than they have been since, because after 1800 agricultural practices caused the elimination of many springs, the draining of surface waters, the removal of brush and trees along brook banks thereby allowing the unshaded waters to become warmer and turbidity of the waters to increase; that after 1900 conditions for the common shiner became increasingly unfavorable, and became more favorable for the central striped shiner; that if this trend continues the central striped shiner eventually will replace the common shiner in those streams now dominated by the common shiner.

Years 1955–80—On the whole during this period the range in Ohio of the common shiner retreated slightly northeastward. The size of the populations within its Ohio range continued to decrease under unfavorable environmental conditions. Hybridization, or intergradation, was evident in most of the populations, including those in the smaller tributaries.

The Chagrin River populations remained large, and the common shiner was present in small numbers even in the larger and lower section of that stream. Andrew White and his associates collected many specimens of common and striped shiners and their hybrids during their investigations of that stream. Examining these specimens, we found that those which appeared to be striped shiners were found to have on the average two to five more predorsal scales than did the striped shiners from the lower section of the Rocky River. In fact, near the mouth of the Rocky River many specimens of striped shiners contained as few predorsal scales as did populations in central Ohio outside the range of the common shiner (White et al., 1975:77).

In those areas where only *cornutus* or *chrysocephalus* is present, identification of a hybrid between a parent of one of these species and another species, such as *Notropis rubellus*, is relatively simple. This is not true of the Chagrin River system, where populations of *cornutus, chrysocephalus* and their hybrids occur. We have found that in many instances it was impossible to ascertain with any degree of certainty which parent, either *cornutus* or *chrysocephalus* or both, was principally involved.

RIVER SHINER

Notropis blennius (Girard)

Fig. 65

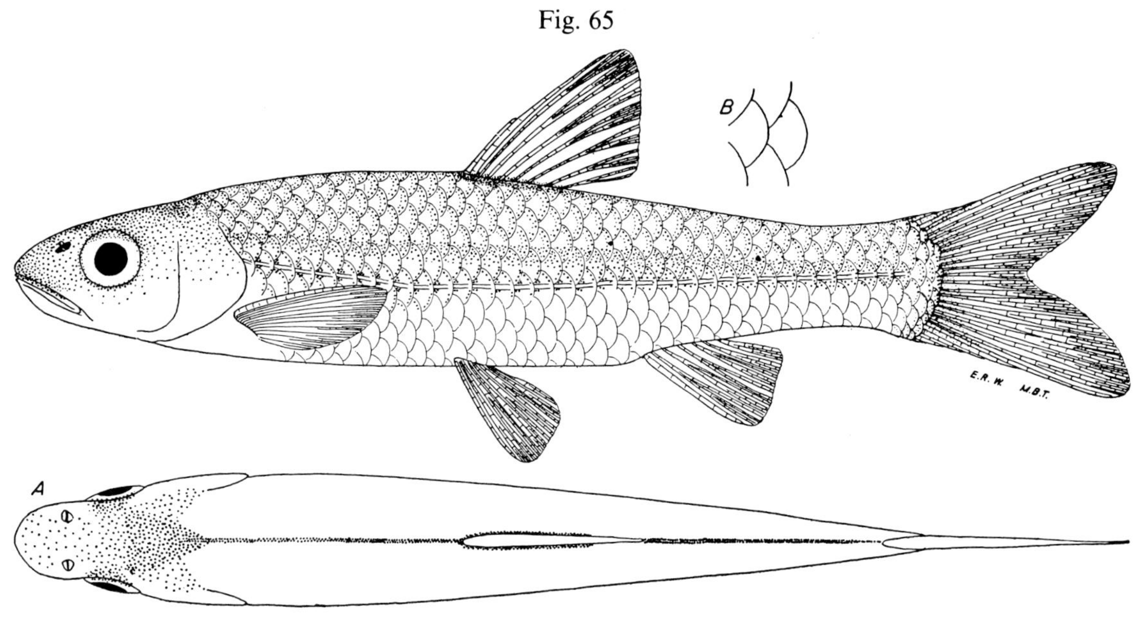

Central fig.: Ohio River, Lawrence County, O.

Aug. 22–23, 1939. 57 mm SL, 2.8″ (7.1 cm) TL.
Male. OSUM 586.

Fig. A: dorsal view; note the dark dorsal stripe completely encircling the dorsal fin base.
Fig. B: anterior lateral line scales showing that the exposed height of each scale is less than twice its width.

Identification

Characters: *General—See* emerald shiner. *Specific*—Anal rays normally 7, rarely 8. No prominent lateral band in life; a dark band sometimes appearing in preserved specimens. Anterior lateral line scales not elevated; less than 2.0 times as high as wide, fig. B. Mid-dorsal stripe rather broad, uniform in width, and completely *surrounds* base of dorsal fin, fig. A. Mouth large; upper jaw normally longer than the eye. Eye small; usually contained 3.5–4.8 times in head. Complete lateral line with 34–37 scales. Fewer than 17 predorsal scales. Peritoneum silvery. Teeth, 2,4–4,2; 1,4–4,1; combination of these.

Differs: The superficially similar sand shiner has the mid-dorsal stripe *not* surrounding base of dorsal fin; has a smaller mouth. Among those species of shiners normally possessing 8 anal rays there may be found occasional individuals having only 7 anal rays; such individuals may be eliminated through use of the following characters: the ghost and mimic shiners have greatly elevated scales along the lateral line; bigmouth shiner has the mid-dorsal stripe not surrounding base of dorsal fin, and a more horizontal mouth; blackchin, bigeye, blacknose, and pugnose shiners have dusky lateral bands; spottail shiner has a black caudal spot; spotfin shiner has dusky blotches on webbings between the last three dorsal rays.

Most like: Sand, mimic, ghost, and bigmouth shiners. **Superficially like:** Silvery and silverjaw minnows.

Coloration: Dorsally olive-straw, heavily overlaid with silvery. Sides silvery; a faint plumbeous lateral band evident at some angles, often becoming prominent in preserved specimens. Ventrally silvery- or milk-white. Fins transparent. *Breeding adults*—Only slightly more brilliant than nonbreeding adults and large young. *Breeding male*—

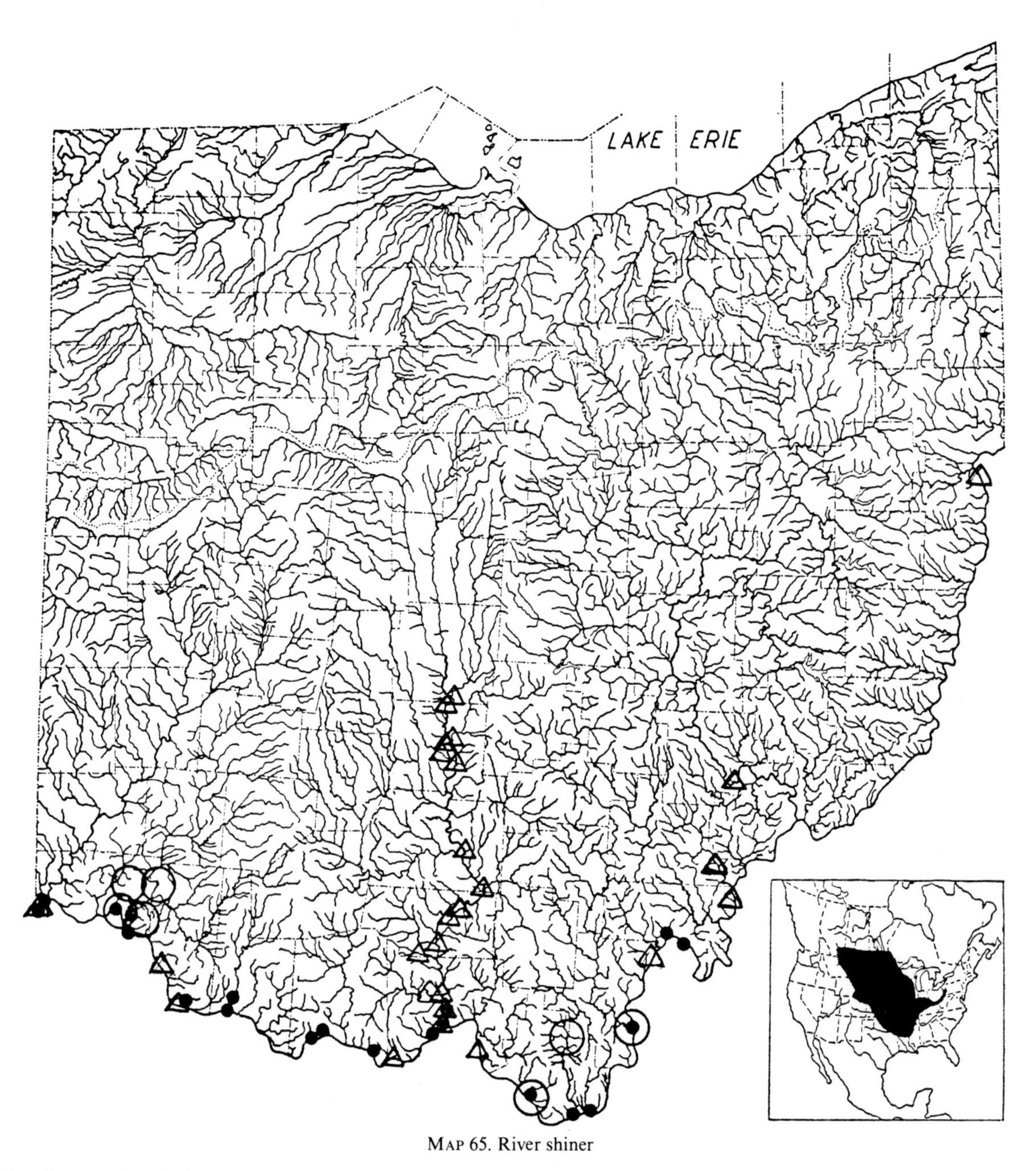

MAP 65. River shiner

Locality records. ○ Literature records without specimens. ● Preserved specimens. • Before 1955. △ 1955–80.

Insert: Northern and northwestern limits of range indefinite; many large areas within range where species is absent, largely because of a lack of habitat.

Has very tiny tubercles on the anterior rays of the pectoral fins.

Lengths: *Young* of year in Oct., 0.8″–2.2″ (2.0–5.6 cm) long. Around **1** year, 1.0″–2.5″ (2.5–6.4 cm). *Adults*, usually 2.0″–4.0″ (5.1–10 cm). Largest specimen, 5.2″ (13 cm) long.

Distribution and Habitat

Ohio Distribution—The river shiner was first recorded for Ohio by Henshall (1888:78 as *N. jejunus*) with the statement "Common in Little Miami River and Bloody Run." In the Cincinnati Museum of Natural History in 1931 I found three collections made by Henshall which were labeled "*N. jejunus*". These collections contained only two river shiners, the remaining specimens were sand and mimic shiners. The two river shiners were collected in Little Miami River, Hamilton County. The catalog books of the Stanford University Natural History Museum contain records of three collections of *N. jejunus;* unfortunately these collections were discarded before I could see them. One collection was from Raccoon Bar in the Ohio River, Gallia County; another from Clough Creek, Hamilton County; a third from the Little Miami River at Redbank, Hamilton County. The collections were made about 1888 by Henshall and Charles Gilbert, and the fishes were identified by Gilbert. There are two collections in the USNM (41018 as *N. jejunus*, 3 sp.; 41035 as *N. dinemus*, 1 sp.) which were taken by Henshall and Gilbert at Raccoon Island and are of this species. In 1899 Osburn (1901:56) found the river shiner to be "Common in the Ohio River and Ice Creek at Ironton and in John's Creek at Waterloo," and six specimens from Ice Creek, taken May 31, 1899 have been preserved (OSUM 7). In 1885 two specimens were taken in the Monongahela River, Lock 9, at Monongahela City, Pa., by Evermann and Bollman (1886:338); these the authors compared with Forbes' type of *N. jejunus* and found them to be identical with the type. From the Ohio and Pennsylvania records it is assumed that previous to 1900 the river shiner was present in the Ohio River between the Indiana and Pennsylvania state lines.

Throughout the 1925–50 surveys the river shiner was sufficiently numerous in the Ohio River from Scioto County to the Indiana state line to be important as a bait minnow, and I have taken as many as 200 in one haul of a collecting seine. This shiner occurred commonly in the Ohio River tributaries only at their mouths, and despite intensive seining none was taken in the Scioto River farther upstream than 4 miles (6.4 km) from its mouth.

Habitat—The river shiner usually remained in waters deeper than 3′ (0.9 m) throughout the day whenever these waters were clear, venturing into shallower waters only at night when they were clear or in the daytime only when the waters were turbid. Although found over silty bottoms it occurred in the largest numbers over gravel and sand bars, spawning throughout the summer until late August over the sand and gravel.

Years 1955–80—Between 1925–50 the river shiner was collected in the Scioto River upstream only a distance of approximately 4 miles (6.4 km) above its confluence with the Ohio River. During 1962 and 1963 a marked upstream migration in the Scioto River occurred, I and several assistants taking specimens as far upstream as Big Walnut Creek, Franklin County. On October 22, 1960 14 specimens, and on May 31, 1961 3 more, were taken by members of Ohio University in the Hocking River, 5.1 miles (8.2 km) upstream from its confluence with the Ohio River, all of which were published by Zahuranec (1962:842–43). On June 15, 1973 Douglas Albaugh and I collected a river shiner from the Muskingum River at Luke's Chute, Morgan County.

Irregular invasions or great increases in numerical abundance have occurred, probably the result of several causes, including a decrease in amount of habitat or highly successful spawning. Note the great increase in numbers of the bullhead minnow in the Scioto drainage after 1935 and its absence, except for a few specimens, after 1950; also the 1958 and 1961 northward invasions of the skipjack herring in that drainage.

During the 1957–59 investigations of the entire Ohio River, the river shiner was found to be an abundant species; 1,416 specimens being recorded (ORSANCO 1962:62).

SPOTTAIL SHINER

Notropis hudsonius (Clinton)

Fig. 66

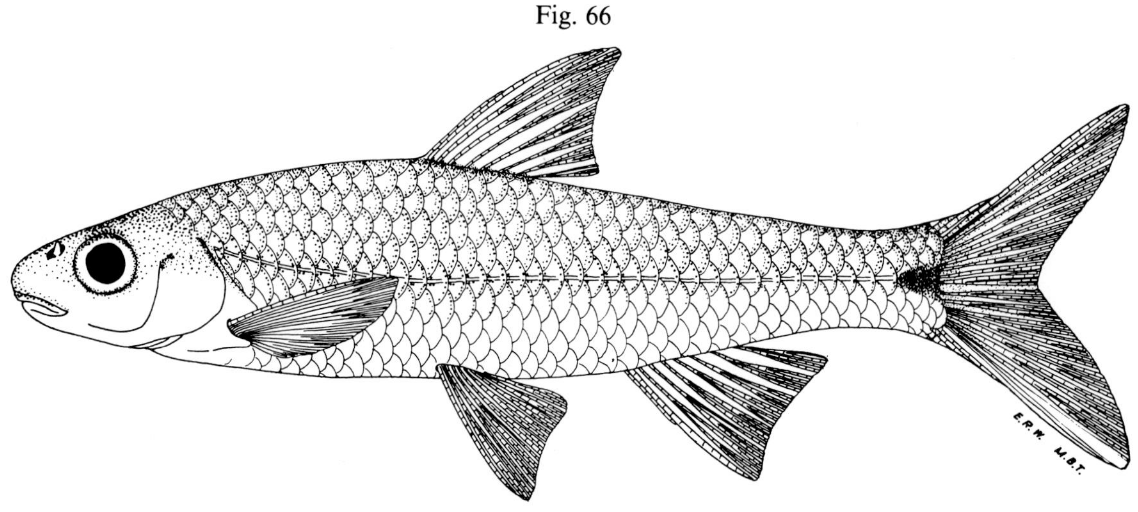

Lake Erie, Ottawa County, O.

Feb. 19, 1944.
Female.

74 mm SL, 3.8″ (9.7 cm) TL.
OSUM 6075.

The ventral edge of the caudal fin is whitish.

Identification

Characters: *General—See* emerald shiner. *Specific*—Usually a very distinct caudal spot, except in some large adults from very turbid waters in which the spot is faint or absent. Body very silvery. Ventral edge of caudal fin of large young and adults generally milk-white, as is the ventral edge of the silver chub. Dorsal and anal fins of adult more deeply falcate than in any other species of shiner. Dorsal fin far forward; usually the distance from dorsal fin origin to caudal base extending from dorsal origin to *considerably* beyond tip of snout. Eye large, 2.9 (young)–3.7 (adults) times in head length. Eye diameter longer than upper jaw length. Scales in complete lateral line usually 36–40. Predorsal scales fewer than 20. Peritoneum silvery, speckled with darker. Anal rays normally 8, rarely 7. No dark lateral band evident in life; such a band may be present in preserved specimens. Teeth, variable, usually 2,4–4,2; all combinations occur between this and 0,4–4,0.

Differs: No other species of Ohio shiner has combination of: a black caudal spot; dusky lateral band absent; dorsal fin so far forward. Silvery minnow lacks the caudal spot; has a long intestine and black peritoneum. Silver, river, and hornyhead chubs possess a barbel on posterior end of upper jaw.

Most like: Young resemble river, mimic and ghost shiners. **Superficially like:** Silver and hornyhead chubs; silvery minnow.

Coloration: Dorsally greenish, or bluish-olive overlaid with whitish-silvery. Sides silvery. Ventrally silvery- and milk-white. Usually a bold black caudal spot. Ventral edge of caudal fin usually milk-white; remaining fins transparent. *Breeding adults*—Similarly colored; the male has tiny tubercles on the dorsal half of head and basal portion of the anterior pectoral rays.

Length: *Young* of year in Oct., 2.0″–3.0″ (5.1–7.6 cm) long. Around **1** year, 2.5″–3.3″ (6.4–8.4 cm). *Adults*, usually 2.3″–5.0″ (5.8–13 cm). Largest specimen, 5.8″ (15 cm) long.

Distribution and Habitat

Ohio Distribution—During the 1925–50 surveys the spottail shiner was abundant along the shores and about the islands of Lake Erie; it occurred in smaller numbers about the mouths of the tributary

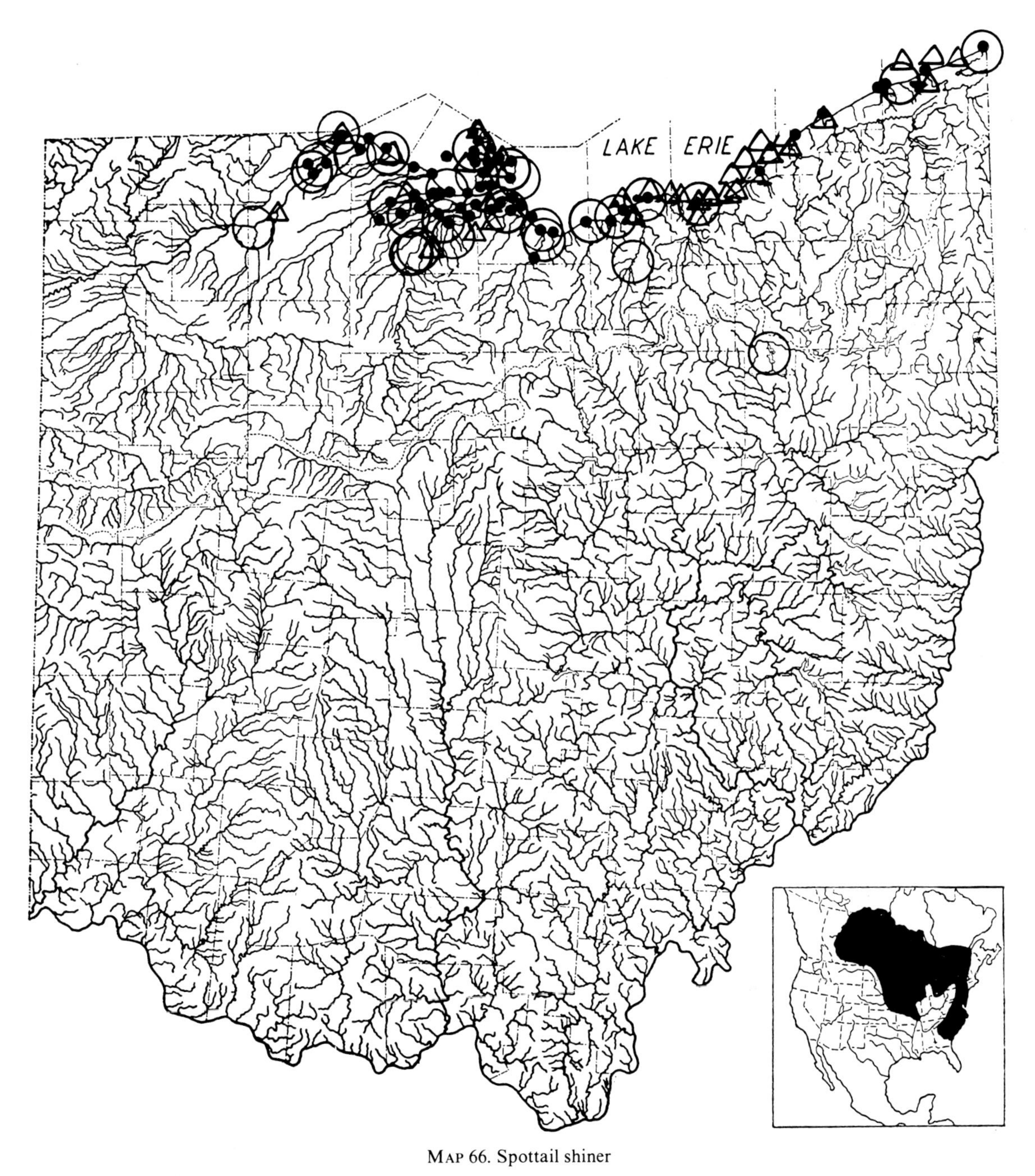

MAP 66. Spottail shiner

Locality records. ○ Before 1926. ● 1926–52. ● Before 1955. △ 1955–80.
Insert: Northern limit of range indefinite.

streams, usually penetrating upstream only to the first well-defined riffles. Apparently the species penetrated farther upstream before 1925 than it did later, for the two collections farthest inland were made before 1900, one from the Black River in central Lorain County (CNHM 1080; two specimens) and the other from the Maumee River at Grand Rapids in Wood County (CNHM 1978; five specimens). Possibly the species penetrated farther inland when the tributary waters were less silt-laden and otherwise less polluted. There was a definite decrease in numbers in Sandusky Bay during the 1925–50 period, for before 1930 it was one of the most numerous species in the survey seine hauls whereas after 1945 it occurred uncommonly and was absent from several areas. The spottail shiner probably entered the Lake Erie drainage through the Maumee-Wabash and other glacial outlets during early post-glacial times.

On July 31, 1926, Walter C. Kraatz collected two spottails (OSUM 636) in East Reservoir, Summit County, and on August 17, 1926 he collected another (OSUM 9218) in the same reservoir "from a sandy beach." East Reservoir is one of the Portage Lakes, which are in the Ohio drainage; elsewhere in this drainage the spottail has been taken in three glacial lakes in the Wabash system of Indiana (Gerking, 1945: Map 40) and in western Pennsylvania (Evermann and Bollman, 1886:336–37 and Fowler, 1919:60).

Habitat—In Lake Erie the spottail occurred in greatest abundance in clear waters, where the depth was between 3′–60′ (0.9–18 m), and the bottom was of sand or gravel. It seemingly avoided bottoms of flocculent clayey silts and turbid waters, and its recent decrease in numbers in the tributary streams, Sandusky Bay and Maumee Bay is attributed to increased silting and turbidity in these waters.

Years 1955–80—Since 1955 the spottail shiner appears to have maintained its numerical abundance in Lake Erie (Van Meter and Trautman, 1970:71; White et al., 1975:43) and has continued its sporadic short-range migrations up those Lake Erie tributaries which habitually were the least turbid.

It is interesting to contemplate why the emerald and spottail shiners apparently have maintained their numbers in Lake Erie since 1955, whereas the silver chub has not. The emerald shiner chiefly inhabits the surface- and mid-waters of the lake, and the species has been rarely observed in direct contact with the substrate during any period of its life cycle. The spottail shinner is primarily a mid-water inhabitant of Lake Erie and repeatedly has been observed over clean sand substrates and avoiding silty ones. The silver chub definitely tends to remain immediately above or upon the substrate as do the other species of the genus *Hybopsis* and *Nocomis* inhabiting Ohio waters. It is possible that the changing conditions of the substrate since 1955 which presumably resulted in the profound changes in the invertebrate fauna (p. 32) have likewise adversely affected the silver chub.

BLACKCHIN SHINER

Notropis heterodon (Cope)

Fig. 67

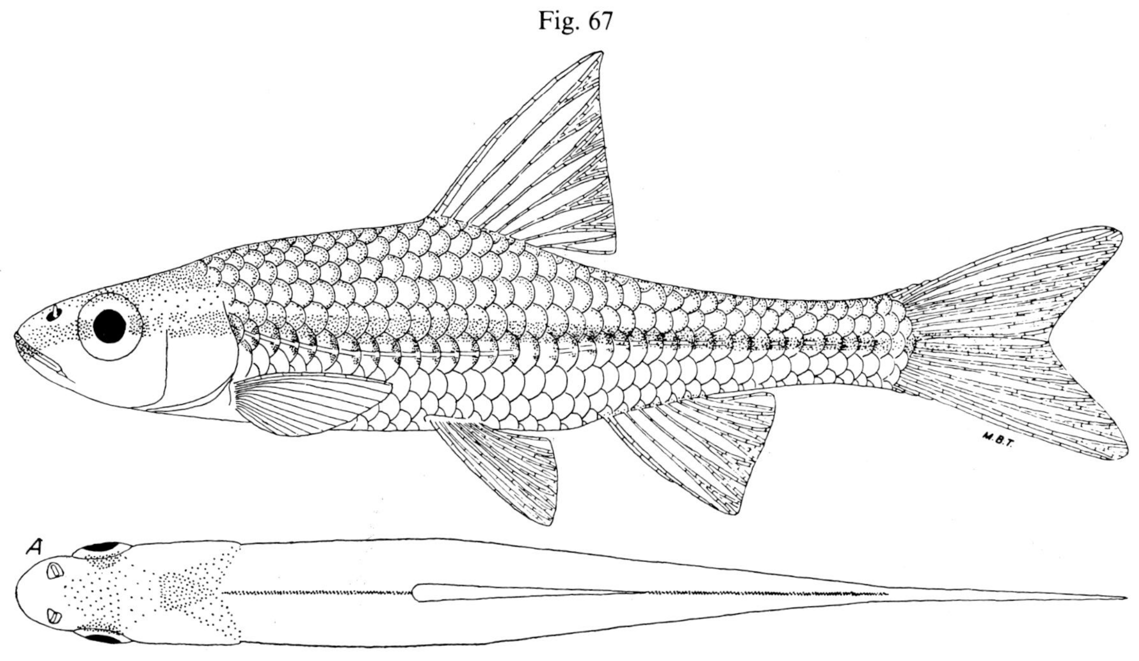

Upper fig.: East Harbor, Ottawa County, O.
July 17, 1933.
Breeding female.
45 mm SL, 2.2″ (5.6 cm) TL.
OSUM 3089.

Fig. A: dorsal view showing shape and color pattern.

Identification

Characters: *General—See* emerald shiner. *Specific*—Tip of lower jaw dusky, a dusky or black band extending from tip of snout, backward across opercles and body to caudal fin. Lateral line incomplete, with 34–38 scales in the lateral series. Predorsal scales fewer than 15. Mouth moderately large; upper jaw usually not quite as long as eye diameter. Eye diameter usually contained 3.0 or more times in head length in adults; 2.5 times in very small young. Peritoneum silvery. Anal rays 8, rarely 7. Teeth, usually 1, 4–4,1; rarely 0,4–4,0.

Differs: Only other species of shiners in Ohio having black-tipped, lower jaws are the bigeye and pugnose shiners; these have black peritoneums and complete lateral lines.

Most like: Bigeye shiner. **Superficially like:** Pugnose, blacknose, mimic, and sand shiners.

Coloration: Dorsally olive-yellow, overlaid with silvery, the scales of the back with broad, dark edges. Upper sides silvery; the scales immediately above the dark, lateral band lack dark edges, thereby contrasting sharply with the dark-edged scales of the back. A broad, usually black, lateral band extends from tip of lower jaw backward across snout, opercle and body to tail. Scales of lateral line have a dusky spot immediately above and below each pore, giving a dark, cresecent-shaped mark to the posterior edge of each scale. Ventrally silvery and milk-white. Fins transparent. *Breeding adults*—Similarly colored; the male has tiny, almost microscopic, tubercles on dorsal surface of head and on the upper surface of the pectoral rays.

Lengths: *Young* of year in Oct., 0.7″–1.4″ (1.8–3.6 cm) long. Around **1** year, 1.0″–2.0″ (2.5–5.1 cm). *Adults*, usually 1.6″–2.5″ (4.1–6.4 cm). Largest specimen, 2.8″ (7.1 cm) long.

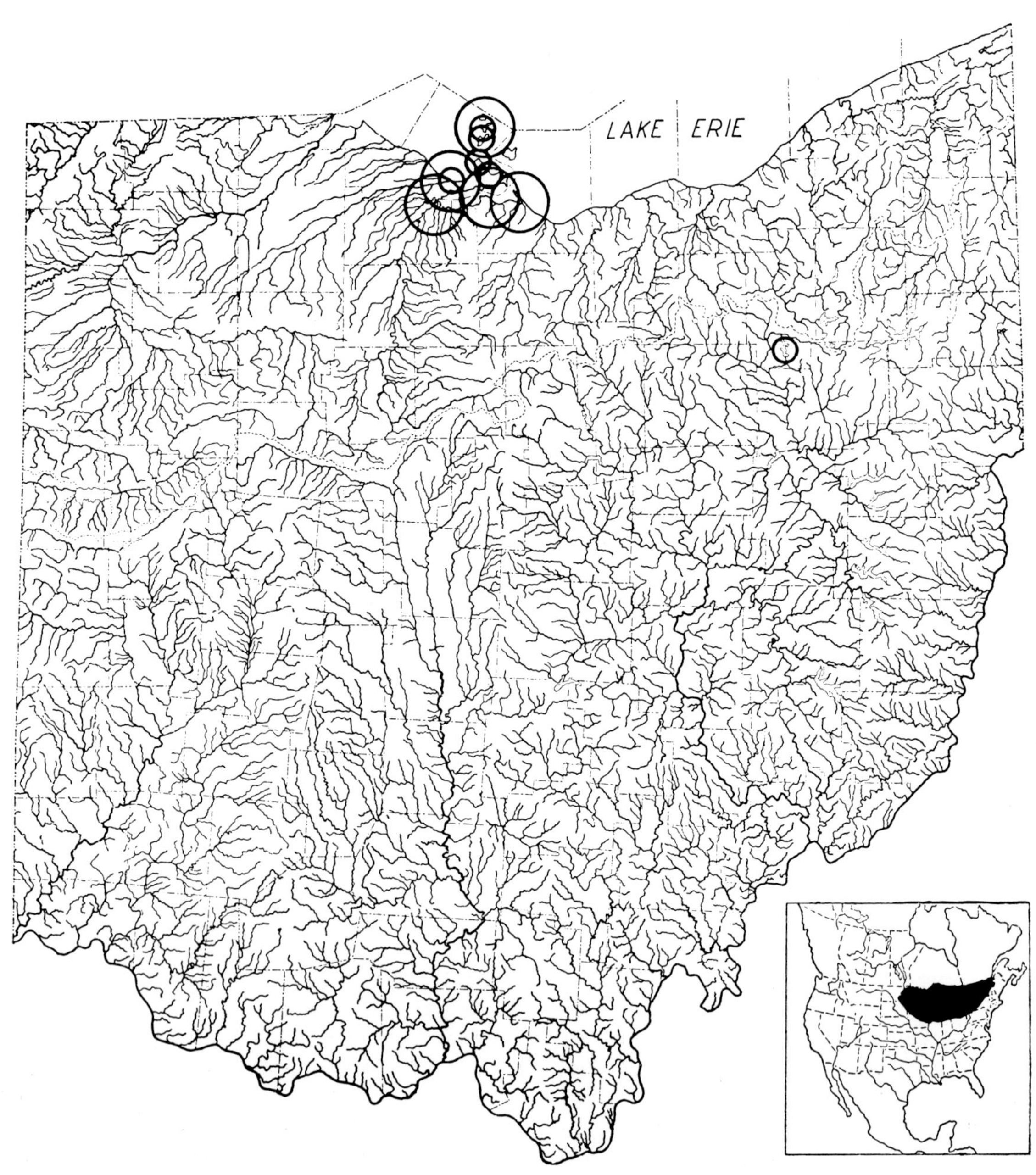

MAP 67. Blackchin shiner

Locality records. ○ Before 1901. ○ 1901–40.
Insert: Unoccupied areas within southern half of range (at least) increasing in size and number.

Distribution and Habitat

Ohio Distribution—In Sandusky Bay in 1899 Osburn (1901:52) collected the first recorded Ohio specimens of the blackchin shiner. Between 1900–30 the species was numerous in East, Middle, and West harbors, at the mouth of the Portage River, about South and Middle Bass islands and in some of the Portage Lakes. After 1930 the blackchin decreased rapidly in abundance; since 1940 none has been taken in Ohio waters although many attempts to capture it were made. It is possible that after 1950 the species was absent from Ohio waters. Between 1933–38 Raney (1938:33) collected specimens in Conneaut Lake, in Crawford County, Pa., which county is adjacent to Ashtabula County, O. It is possible that before 1900 the blackchin occurred in many other glacial pothole lakes of northern and western Ohio, and in some of the low-gradient, clear-water, vegetated streams of the Black Swamp and the prairie areas.

Habitat—The blackchin populations were largest in glacial pothole lakes and the bays and harbors of Lake Erie where the waters were very clear, the bottoms were of clean sand, gravel or organic debris, and where there were dense beds of submerged aquatic vegetation. It disappeared almost immediately when waters became turbid, the bottoms silt-covered, and the aquatic vegetation vanished.

Years 1955–80—Although persistent and frequent attempts since 1955 were made to capture the blackchin shiner in the island, harbor and bay regions of western Lake Erie, none was taken. It is probable that the species has become extirpated from Ohio waters (Van Meter and Trautman, 1970:71).

BIGEYE SHINER

Notropis boops Gilbert

Fig. 68

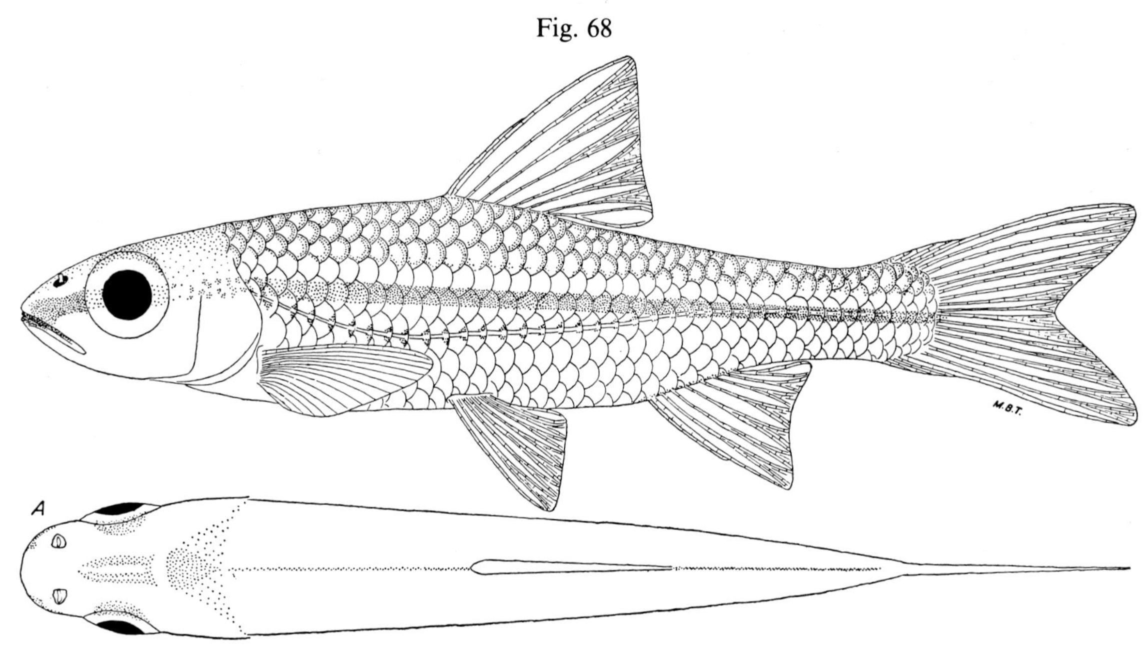

Upper fig.: Chenoweth Fork, Pike County, O.

May 29, 1940. 45 mm SL, 2.3″ (5.8 cm) TL.
Male. OSUM 2090.

Fig. A: dorsal view showing shape and color pattern.

Identification

Characters: *General—See* emerald shiner. *Specific*—Tip of lower jaw dusky; a dusky or black band extending from tip of snout backward across opercles and body to caudal fin. Lateral line *complete*, with 34–38 scales. Predorsal scales fewer than 17. Mouth large and terminal; upper jaw usually as long as the very large eye. Eye diameter usually contained 2.0–3.0 times in head length. Peritoneum *black*. Anal rays 8, rarely 7. Teeth, usually 1,4–4,1; rarely 4–4.

Differs: *See* blackchin shiner. Blackchin shiner has an incomplete lateral line and a silvery peritoneum.

Most like: Blackchin shiner. **Superficially like:** Pugnose, blacknose, mimic and sand shiners.

Coloration: Essentially the same color pattern as that of the blackchin shiner, except that in the bigeye the lateral line scales lack the crescent-shaped marks along the posterior edge of each pored scale. *Breeding adults*—Have same coloration as non-breeding adults; the male has tubercles present along the free edges of some of the anterior body scales; upper surfaces of the rays of the pectoral fins roughened into serrated ridges.

Lengths: *Young* of year in Oct., 1.0″–1.5″ (2.5–3.8 cm) long. Around **1** year, 1.3″–2.4″ (3.3–6.1 cm). *Adults*, usually 1.8″–2.8″ (4.6–7.1 cm). Largest specimen, 3.6″ (9.1 cm) long.

Distribution and Habitat

Ohio Distribution—The fewer than 100 recorded fish collections made prior to 1900 indicate that before 1900 the bigeye shiner probably was well-distributed throughout the western half of Ohio, especially in the Maumee River drainage (UMMZ 63040; CAS 9186), and that as early as 1897 it had become uncommon in central Ohio (Osburn and Williamson, 1898:13–14). The more than 2000 col-

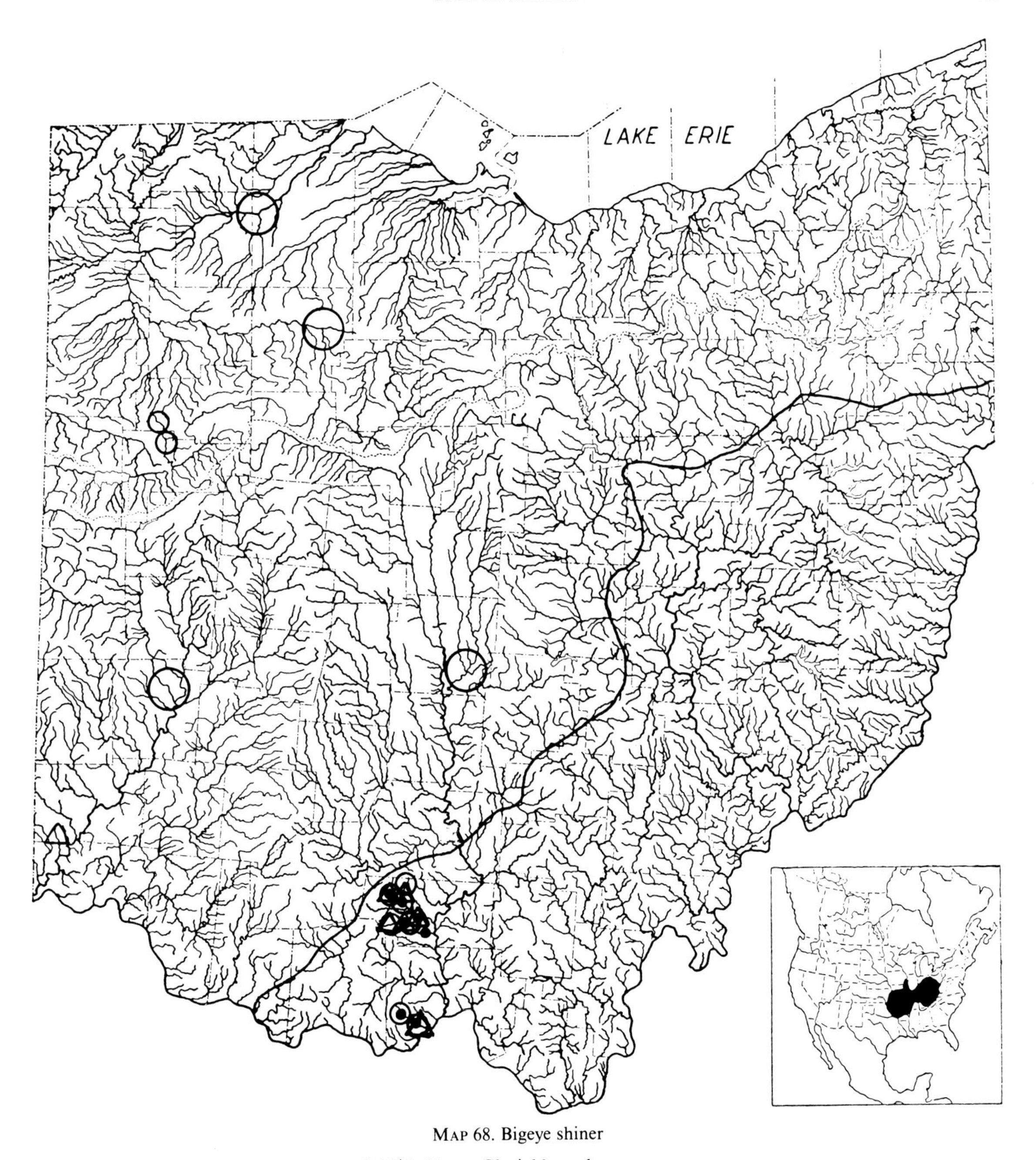

MAP 68. Bigeye shiner

○ Before 1900. ○ 1921–41. ● 1942–54. △ 1955–80. ~ Glacial boundary.
Insert: Unoccupied areas within northeastern section of range increasing in size and number.

lections made since 1925 show that between 1930–41 a relict population was still present in the Auglaize River at the Auglaize-Allen county line, but that since 1941 none has been found there, or elsewhere in western Ohio.

In 1927 I found the bigeye inhabiting the clear-water, limestone-bottomed streams above the 600′ (183 m) contour line in the unglaciated portions of Pike and Scioto counties, where it may have been present since interglacial times.

Habitat—The bigeye was an inhabitant of the clearest of streams in which the waters were turbid for only brief intervals during floods, the stream bottoms were of sand, gravel, bedrock and small amounts of organic debris, and where silty clays were virtually absent. In these streams this shiner fed by sight, capturing its animal food chiefly in mid-water and from the surface. On summer evenings it fed extensively upon the small insects which hovered a few inches above the water, jumping gracefully into the air to capture them.

The bigeye was present in streams as long as these were essentially of the clear-water type, disappearing when the waters became silt-laden and the stream bottoms became covered with silt. As an example: In 1928, Morgan's Fork, in Pike County, contained a large population of bigeye shiners. One of its tributaries became turbid because of silt which entered from a newly made hill cornfield. Before the tributary became turbid it contained a large population of shiners; these disappeared as soon as the stream became turbid. A few years later erosion was stopped, after the cornfield had been allowed to revert to a brushy field, whereupon the waters became clear, the gravel of the stream bottom became free of silt, and the bigeye again became abundant.

Years 1955–80—During this period the bigeye shiner was definitely recorded in only one locality in unglaciated territory, despite attempts to locate other populations, especially in the Maumee and upper Scioto river systems. On July 27, 1973 two specimens were collected in Paddy's Run Creek, a tributary of the Great Miami River in Crosby Township, Hamilton County (Bauer, Branson, Colwell, 1978:147). The population in Chenoweth Fork of western Pike County continued to flourish, as did the smaller population in Turkey Creek, southwestern Scioto County.

STEELCOLOR SHINER

Notropis whipplei (Girard)

Fig. 69

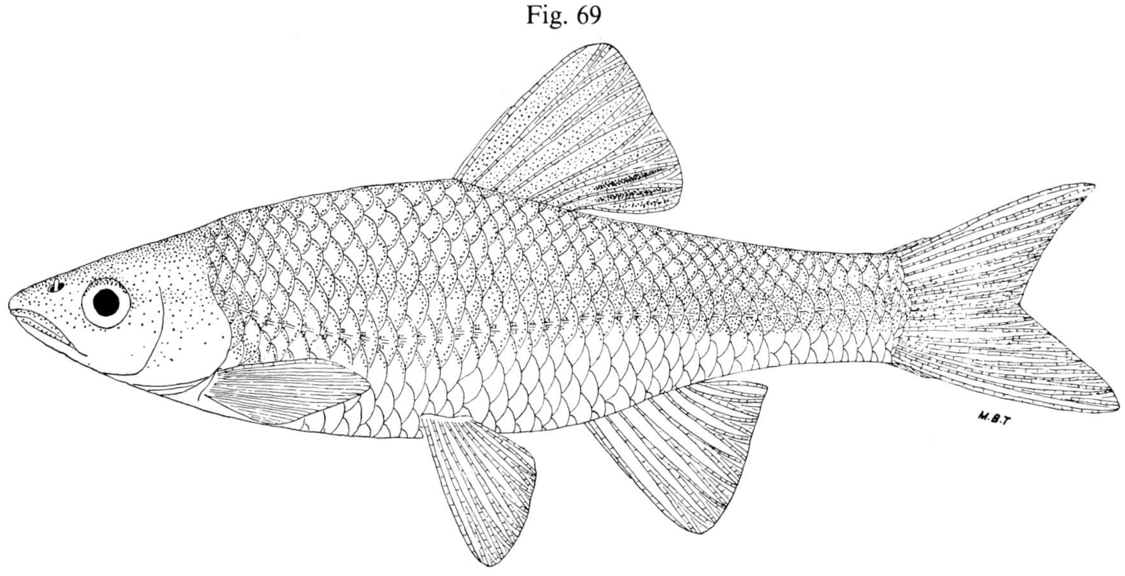

Scioto River, Scioto County, O.

June 22, 1940. 77 mm SL, 3.7" (9.4 cm) TL.
Breeding female. OSUM 2527.

Note position of the lateral band on the caudal peduncle, and the melanophores on the webbing of the anterior half of the dorsal fin.

Identification

Characters: *General—See* emerald shiner. *Specific*—Anal rays, 9, rarely 8 or 10. A dusky blotch on the webbing between last three rays of the dorsal fin (among Ohio shiners only the spotfin has a similar blotch; in both species this blotch is faint or absent in small young). In large young and adults the webbing between the anterior dorsal rays is rather thickly speckled with tiny black spots (melanophores); in breeding adults this webbing is densely covered with these spots. General coloration bluish-silvery, overlaid with olive-yellow (in large specimens this coloration is distinctive of this species and the spotfin). Snout pointed and head triangular in lateral outline. Eye small; usually contained 3.8–4.5 times in head length; as low as 3.0 times in smallest young. Dorsal fin origin about equidistant between caudal base and tip of snout. Distal edges of dorsal and anal fins straight or slightly convex. In life, the dark lateral band is usually indistinct or absent, but in preserved specimens this band becomes visible, especially along the caudal peduncle where it is centered over the lateral line scales and consequently, is equidistant between the dorsal and ventral surfaces of the peduncle, *see* fig. (this band is usually evident to the unaided eye, but is best seen with aid of magnification). Body of adult averages deeper than does that of the very similar spotfin. Peritoneum silvery, speckled with darker. Teeth, 1,4–4,1 or 1,4–4,0.

Differs: The very similar spotfin has 8 anal rays; dark specklings usually few or absent on webbing of anterior dorsal fin, except in breeding adults; in preserved specimens, the center of the dark lateral band on the caudal peduncle lies below the pores of the lateral line. No other species of shiner in Ohio has a dusky blotch on the posterior half of the dorsal fin.

Most like: Spotfin shiner. **Superficially like:** Redfin, rosefin, striped and common shiners.

Coloration: Dorsally bluish-silvery with considerable olive-yellow, the pronounced cross hatching on scales of back and sides diamond-shaped; this color combination and diamond-shaped cross-hatching are characteristic of this species and the

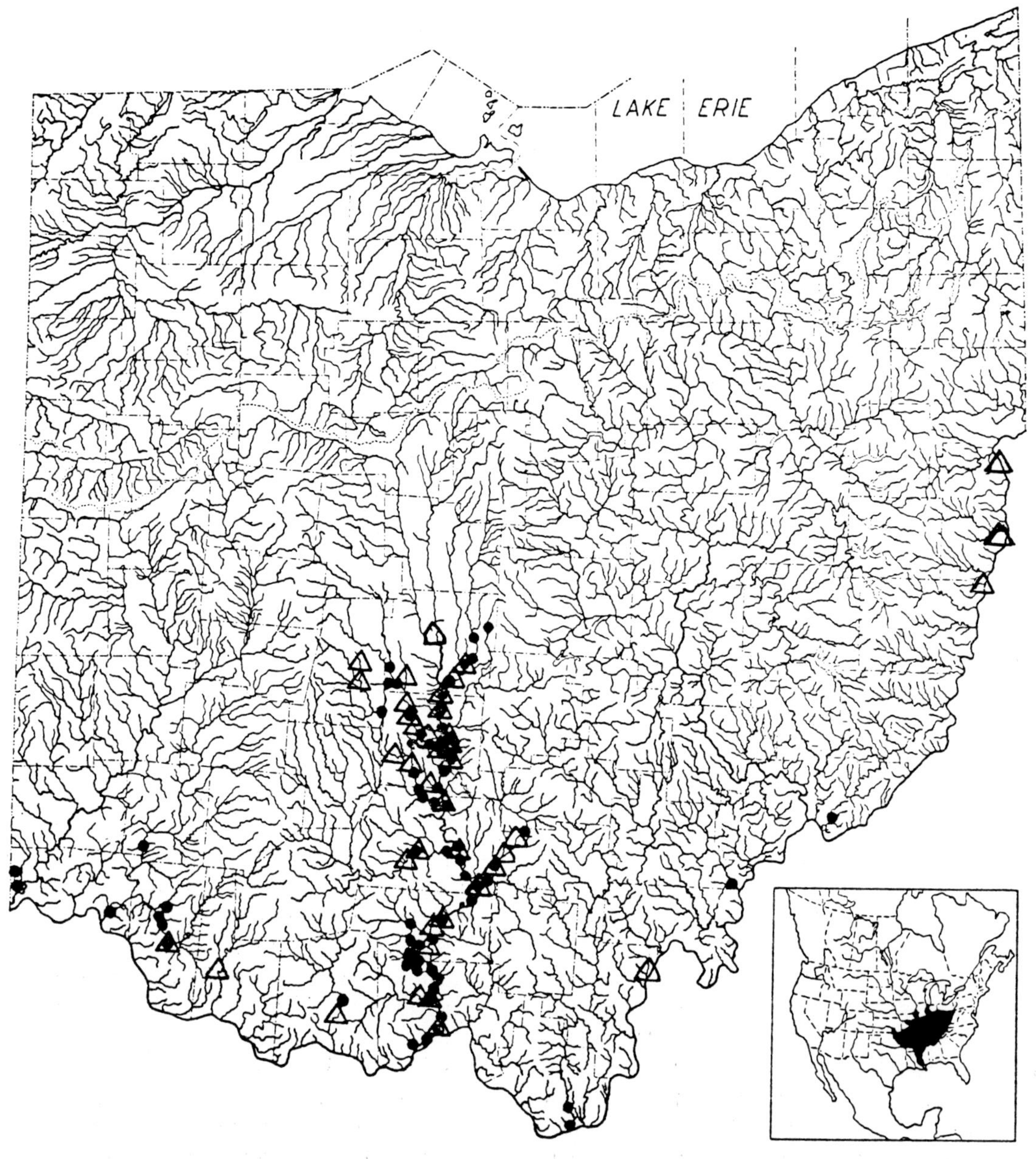

MAP 69. Steelcolor shiner

Locality records. ● Before 1955. △ 1955–80.
Insert: Limits of range, except northeastern section, rather indefinite.

spotfin. Sides bluish-silvery with less yellow. Ventrally pale bluish and milk-white. Lateral band steel-blue if present in life; becomes dusky and pronounced in preserved specimens. A dusky bar on body immediately behind gill cleft which extends from upper angle of gill cleft to pectoral fin base. A dusky blotch on webbing between the last three rays of dorsal fin; anterior webbing usually with scattered specklings. All fins of large young and adults usually have some olive, overlaid with milk-white and yellow; fins of small young usually transparent. *Breeding male*—A brilliant steel-blue overlaid with yellow; much milk-white and a little yellow on the fins; small but sharply-pointed tubercles on snout and dorsal half of head; a row of smaller ones along lower jaw; tiny tubercles on scales of predorsal region and along lateral line; minute tubercles on some rays of all fins except the caudal.

Lengths: *Young* of year in Oct., 1.0″–2.5″ (2.5–6.4 cm) long. Around 1 year, 1.6″–3.0″ (4.1–7.6 cm). *Adults*, usually 2.3″–4.5″ (5.8–11.4 cm). Largest specimen, 5.3″ (13 cm) long.

Hybridizes: with spotfin shiner.

Distribution and Habitat

Ohio Distribution—Until 1938, the steelcolor shiner was confused with the closely related spotfin shiner, and since no specimens collected before 1920 have been found, no statement concerning the presence, abundance, and distribution in Ohio of the steelcolor shiner prior to 1920 can be made. Since 1920, the steelcolor shiner has been present in the Scioto River and lower portions of its tributaries from Franklin County southward; and although sometimes as abundant in a locality as was the spotfin, it rarely was more abundant and frequently was less so. The steelcolor was also present, usually in small numbers, in the lower courses of streams tributary to the Ohio River from the Indiana state line to Lawrence County, but east of Lawrence County occurred only in two collections as strays. Curiously, not even a stray was found in the Ohio River although numerous attempts to obtain specimens were made, especially about the mouths of streams. Repeated collections in the lower Muskingum River likewise failed to produce a specimen.

Habitat—The largest populations occurred in those sections of the larger streams of the Scioto River system where there was considerable timber and brush in the water, a moderate gradient, and where turbidity and siltation were lowest, such as in the lower portion of Big Darby Creek. They were taken over the clean gravel and boulders of moderately fast riffles and likewise in pools having a silted bottom and no current.

No two species of Ohio fishes appeared to have fewer recognizable habitat differences, especially while spawning, than did the steelcolor and the spotfin. Both spawned under seemingly the same conditions, about brush and usually underneath logs and timber as described by Hankinson (1930:73–74). On June 9, 1942, William Doherty, Mary A. Trautman and I found a log in Big Darby Creek near Fox under which both species were spawning. The partly submerged log was about 15′ (4.6 m) long, 18″ (46 cm) in diameter, and its base was buried in the stream bank. The bottom of the log was raised about a foot (30 cm) from the gravel bottom of the stream. Near the outer end of the log, a group of perhaps 30 steelcolor shiners were spawning, depositing their eggs on the underside of the log, while 10′ (3.0 m) nearer the bank and in a slower current a slightly larger group of spotfins were likewise spawning. Although both groups were disturbed repeatedly by us, and adults of both species were collected (OSUM 5041 and 5042), the two groups would form quickly after each disturbance, and never was an individual of one species taken in the spawning group of the other.

Ecological reasoning and examinations of their present ranges suggest that at one time the two species had been separated from each other sufficiently long to produce genetic isolation, and that more recent developments, presumably postglacial, have brought the two in contact. At present the steelcolor's range lies mostly southwest of Ohio whereas the spotfin's extends much farther northeastward into the St. Lawrence River drainage, the two occupying the same area from Ohio to Alabama and Missouri. It is possible that the steelcolor is the more recent invader.

I have found no evidence of mass hybridization between the two species, the only suggestion that some hybridization may have taken place is that spotfins outside the Ohio range of the steelcolor such as in the upper Muskingum and Lake Erie river drainages, very rarely have 9 anal rays, more often 7; whereas in areas where both occur together the spotfins show a higher percentage of 9 anal rays, and some tend to have slightly deeper bodies or have the lateral band character poorly defined. Specimens with 9 anal rays also are frequent in the Wabash River system in western Ohio, which may be significant because Gerking (1945: Map 46) has recorded the steelcolor shiner in the Wabash system in Indiana within a few miles of the Ohio line.

Years 1955–80—During 1959 and 1960 W. L. Pflieger conducted a field study of the steelcolor and spotfin shiners. He found that since 1959 there had occurred no extension of the central Ohio distributional range of this species (*see* map 69). Since 1954, however, its numerical abundance throughout the limited range in Ohio has been maintained, despite what appears to be severe population pressure from the closely related spotfin shiner.

H. Ronald Preston and his associates collected the species in the Ohio River opposite Belmont County, between the years 1968 and 1970.

SPOTFIN SHINER

Notropis spilopterus (Cope)

Fig. 70

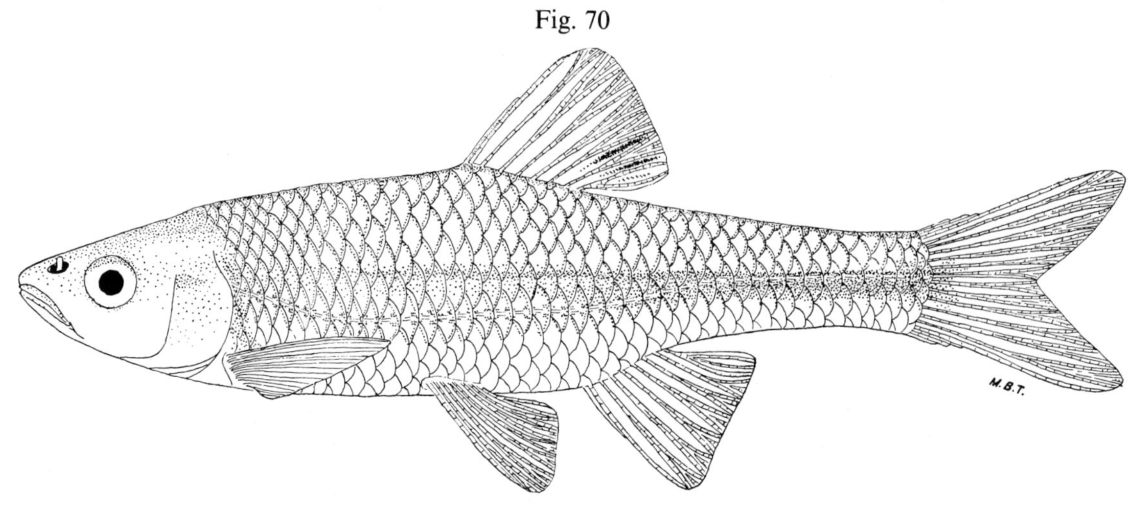

Walhonding River, Coshocton County, O.

Oct. 20, 1939. 74 mm SL, 3.6″ (9.1 cm) TL.
Adult male. OSUM 990.

Note position of the lateral band on the caudal peduncle, and the lack of melanophores on the webbing of the anterior half of the dorsal fin.

Identification

Characters: *General—See* emerald shiner. *Specific*—Similar to those of the steelcolor shiner, except that the spotfin normally has only *8* anal rays, rarely 9; the dark specklings on the webbing of the anterior half of the dorsal fin are few or absent except in large or breeding adults; the body of the adult averages more slender; in preserved specimens the lateral band on the caudal peduncle lies chiefly below the pores on the lateral scales and below the center of the peduncle, *see* fig. Apparently a slightly smaller species than the steelcolor. *See* steelcolor shiner for similar characters.

Differs: *See* steelcolor shiner.

Most like: Steelcolor shiner. **Superficially like:** Emerald, redfin, rosefin, striped and common shiners.

Coloration: Very similar to that of the steelcolor in life, as is distribution of the breeding tubercles. Differs primarily in the absence or fewness of dark specklings on the webbing of the anterior half of the dorsal fin of all except breeding adults.

Lengths: *Young* of year in Oct., 0.9″–2.4″ (2.3–6.1 cm) long. Around **1** year, 1.6″–2.8″ (4.1–7.1 cm). *Adults*, usually 2.0″–4.0″ (5.1–10 cm). Largest specimen, 4.8″ (12 cm) long.

Hybridizes: with steelcolor and rosyface shiners.

Distribution and Habitat

Ohio Distribution—From specimens collected, and from literature references by Henshall (1888:78), McCormick (1892:18), Kirsch (1895A: 329) and Osburn (1903:54–55), it is obvious that the steelcolor shiner (recorded by them as *whipplii*) was wide ranging and abundant in Ohio before 1920. Since 1920 the species has been present, often abundant, throughout most of the streams of the state, including the Ohio River. It occurred, usually in smaller numbers in sloughs, ponds, reservoirs and lakes; it was often numerous in the shallow, more protected or weedy waters of Lake Erie, such as in the harbors and about the Bass Islands.

Habitat—The apparently aggressive spotfin was tolerant of a surprisingly large variety of habitats. The species occurred in the greatest abundance in streams of base- or low-gradients, such as those in northwestern Ohio from Sandusky Bay westward,

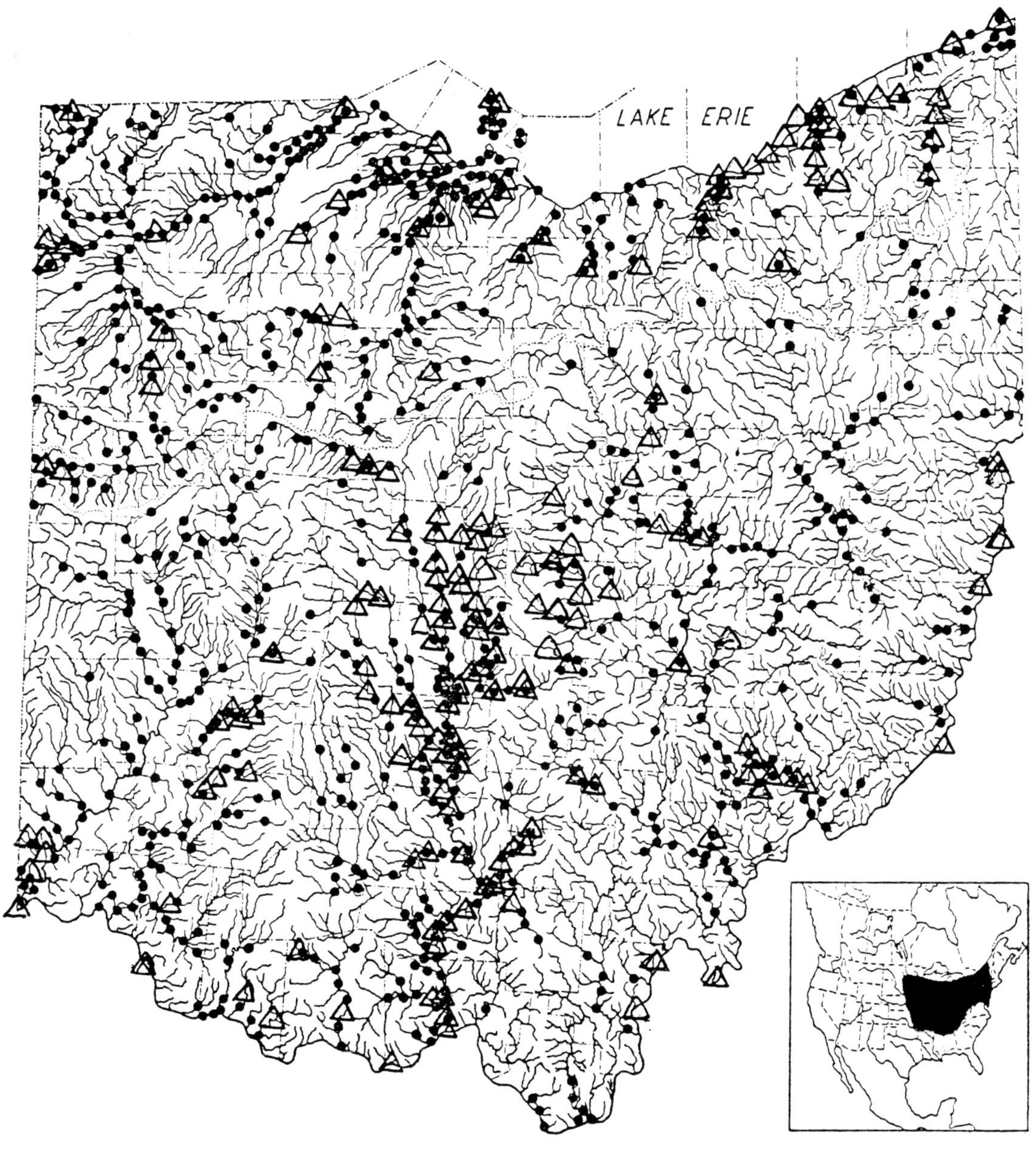

MAP 70. Spotfin shiner

Locality records. ● Before 1955. △ 1955–80.
Insert: Range limit may extend farther westward than indicated.

and in the Scioto River system, and usually was the most numerous shiner where the waters were turbid, or there was considerable siltation, domestic or industrial pollutants. Because of its method of egg deposition on the underside of objects, it appeared to be less affected by silted bottoms than were most species of minnows, and under such conditions usually would be the most numerous spawning species of shiner. It occurred likewise in large numbers about submerged aquatic vegetation. It was less numerous in moderate- and high-gradient streams of clear water, or where there were a large number of other fish species and individuals (*see* steelcolor shiner).

Years 1955–80—Hundreds of collections and observations of the spotfin shiner have been made since 1950, these indicating that the species has remained one of the widely distributed and abundant cyprinids throughout much of Ohio. As stated above, this minnow has a high degree of tolerance to turbidity, siltation and many other types of pollutants.

It remained numerous in the shallow waters of Lake Erie and especially in the western basin (Van Meter and Trautman, 1970; White et al., 1975:79). The reasons are not known why it was so uncommon in the Ohio River, as has been indicated by the 1957–59 investigations of that stream (ORSANCO, 1962:147).

Pflieger (1965:1–8) obtained many F_1 spotfin × steelcolor shiners by isolating breeding males of one species with breeding females of the other. He succeeded in rearing many hybrid young.

A specimen (OSUM 1579) taken in Scioto County in 1939, which I had previously identified as a natural hybrid, was compared with the hybrids raised by Pflieger, this comparison proving correct my earlier identification of the presumed natural hybrid.

For spawning, *see* Gale and Gale, 1977.

CENTRAL BIGMOUTH SHINER

Notropis dorsalis dorsalis (Agassiz)

Fig. 71

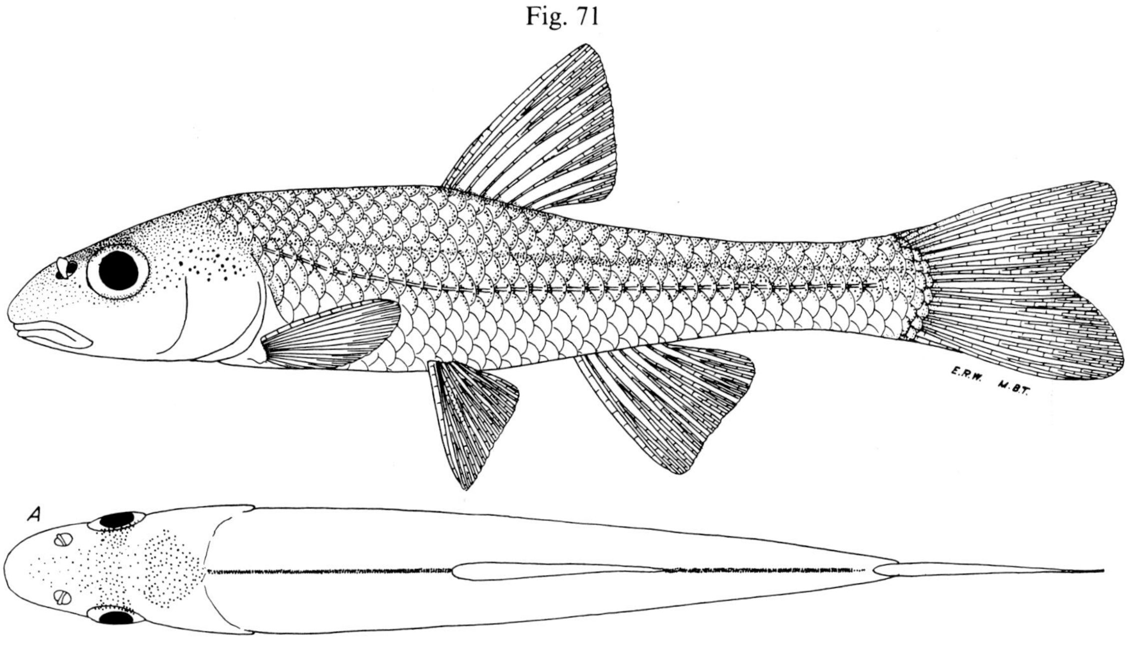

Upper fig.: Black River, Lorain County, O.

Sept. 14, 1929. 51 mm SL, 2.5″ (6.4 cm) TL.
Adult female. OSUM 9316.

Fig. A: dorsal view showing body shape, and dark mid-dorsal stripe that does not surround base of dorsal fin.

Identification

Characters: *General—See* emerald shiner. *Specific* —Anal fin rays normally 8, rarely 7 or 9. Head flattened on its ventral surface, so that head is triangular in cross-section. Shape of head, body proportions, and silvery coloration strikingly similar to those of the silverjaw minnow. Mouth horizontal and large; length of upper jaw longer than eye diameter except in smallest young. Complete lateral line with 36–39 scales of the usual shape, their exposed surfaces less than 2.0 times as high as wide. Peritoneum silvery, with darker specklings. Intestine not as long as combined length of head and body. Teeth, normally 1,4–4,1.

Differs: The similar-appearing silverjaw minnow has large cavernous spaces on the ventral surface of the head, *see* silverjaw minnow, fig. B. Other species of shiners with 8 anal rays have *either* a distinct lateral band; *or* greatly elevated lateral line scales; *or* spot at caudal base; *or* spot on dorsal fin. Chubs have barbels. Silvery minnow has a longer intestine; a black peritoneum.

Most like: Silverjaw minnow; sand, river, and mimic shiners; young of silver chub.

Coloration: Dorsally pale olive-straw heavily overlaid with silvery, the scales narrowly dark-edged. Sides silvery. Ventrally silvery- and milk-white. Fins transparent. *Breeding male*—Has minute tubercles scattered over the head, and scattered tubercles on some of the scales on the anterior half of body. The basal half of the pectoral rays are enlarged, their upper surface serrated.

Lengths: *Young* of year in Oct., 1.1″–2.0″ (2.8–5.1 cm) long. Around **1** year, 1.3″–2.5″ (3.3–6.4 cm). *Adults*, usually 2.0″–2.8″ (5.1–7.1 cm). Largest specimen, 3.0″ (7.6 cm) long.

Hybridizes: with sand shiner. No Ohio specimens.

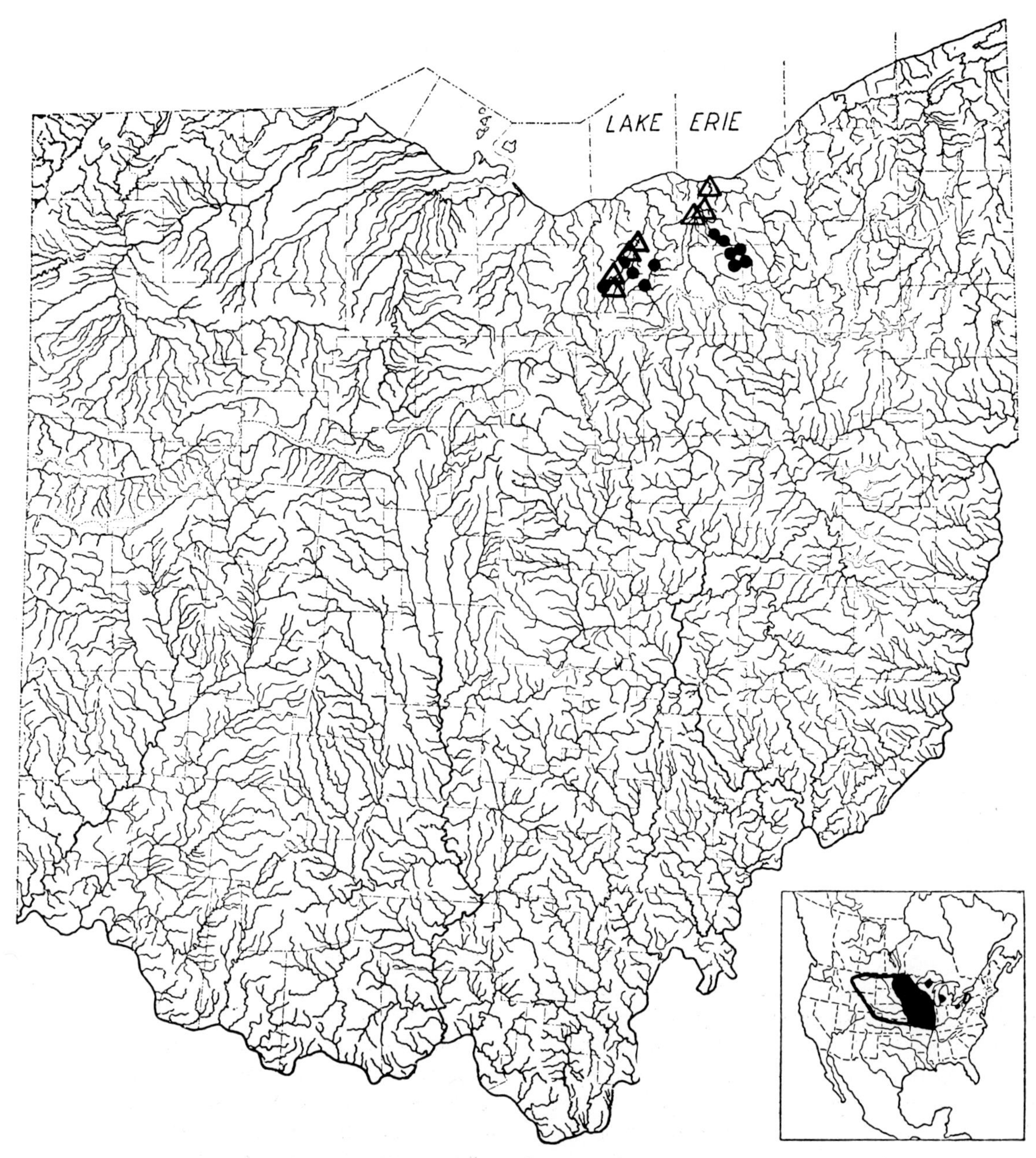

MAP 71. Central bigmouth shiner

Locality records. ● After 1922. △ 1955–80.

Insert: The four disjunct black areas represent the present range of the nominate subspecies, the large outlined area to the west represents a nominal subspecies, the small one to the east another.

Distribution and Habitat

Ohio Distribution—On August 7, 1922, E. L. Wickliff collected the first known specimen (OSUM 214) of the bigmouth shiner in Ohio; it was taken in the West Branch of Black River, 4 miles west of La Grange, Lorain County. Between 1922–34 the species was taken several times in the upper halves of the Black and Rocky rivers. On September 14, 1928, James S. Hine and I found it to be abundant in the West Branch of Black River, Brighton Twp., Lorain County, collecting more than 150 specimens (OSUM 163) in an hour. It was never abundant in the East Branch of Rocky River, and the largest number taken was a total of 23 in 2 collections

(OSUM 286 and 287), on June 2, 1929. Since 1940 none has been taken in the Black River and only one on April 15, 1940, in the Rocky River (OSUM 1941). The populations in both stream systems may be on the verge of extirpation.

It seems logical to assume that the bigmouth shiner was at least as numerous between 1888–91 as it was in 1928, despite the fact that McCormick (1892:18) did not record the species for Lorain County, probably confusing it with his *Notropis deliciosus* or *Notropis stramineus*. At Oberlin College in 1931 I examined some unpublished notes of McCormick's, which contained an unannotated list of Lorain County fishes, including *Notropis gilberti* = *Notropis dorsalis*. It is therefore possible that McCormick did capture and recognize the bigmouth shiner but did not publish his findings.

Habitat—The bigmouth shiner and silverjaw minnow were superficially very similar, both having a triangulate head, a similar mouth, the same color pattern and the same adult size. Both species occupied the same type of habitat, and the methods of schooling, feeding, resting, and retreat when frightened were the same (*see* silverjaw minnow). The bigmouth remained the dominant species in a stream section as long as it maintained a large or fair sized population; once it became rare the silverjaw became abundant.

The history of the bigmouth and silverjaw in Ohio may be as follows: a xerothermic period occurred during late post-glacial times (Transeau, 1935:435 and 1941:208; Conant, Thomas and Rausch, 1945:66) which favored the eastward invasion into Michigan, Ohio, Pennsylvania and New York (Hubbs and Lagler, 1947:67–68) of the plains-inhabiting bigmouth. Later a retreat westward occurred, leaving isolated populations in the East, one of which was in northern Ohio.

The present range of the silverjaw lies principally south and southwest of Ohio extending northeastward only into northeastern Ohio and northwestern Pennsylvania and New York. The species appears to have continued to advance northeastward during the past 50 years, and sometime before 1922 the silverjaw reached the Ohio range of the bigmouth and encircled it. Since 1928 I have noted the shrinking in size of the Ohio range of the bigmouth, and invasion and great increase in numbers of the silverjaw into territory formerly occupied by the bigmouth. If this invasion of the silverjaw continues it will be only a matter of time before the bigmouth is extirpated from Ohio waters.

Years 1955–80—Prior to 1952 the bigmouth shiner was numerous in the West Branch of the Black River in Lorain County, whereas in the East Branch only single specimens were captured in each of two localities, suggesting that in the East Branch the species was on the verge of extirpation (White et al., 1975:81).

On the other hand, the silverjaw minnow, a species remarkedly similar in form and habits to the bigmouth shiner, never has been recorded in the West Branch, but has been recorded several times in the East Branch, first by McCormick (1892:19) in 1892, twice by E. L. Wickliff (unpublished) in 1922, and twice in the same locality by Mary A. Trautman (née Auten) and students in 1939, from the headwaters in Ashland County. The absence of the silverjaw from the West Branch may be explained by the establishment before 1850 of dams above the junctions of both branches (Drake, 1850:371) before the silverjaw had invaded the East Branch.

Since 1952 no bigmouth shiners have been taken in the East Branch despite the thorough and recent investigations of the fishes of Lorain County by Bernard J. Zahuranec. Zahuranec found that the silverjaw minnow had increased in numbers in that branch, his four largest collections of that species containing between 87 and 171 individuals.

The bigmouth shiner has continued to be present in the West Branch and was found to be abundant in a locality in Pittsfield Township, Lorain County, by Zahuranec and me. Recently, Andrew White and associates have captured the species in the lower half of the Rocky River in Cuyahoga County.

The observation above that the bigmouth remains numerous in a locality so long as no or few silverjaws are present has been strengthened by the investigations since 1952.

SAND SHINER

Notropis stramineus stramineus (Cope)

Fig. 72

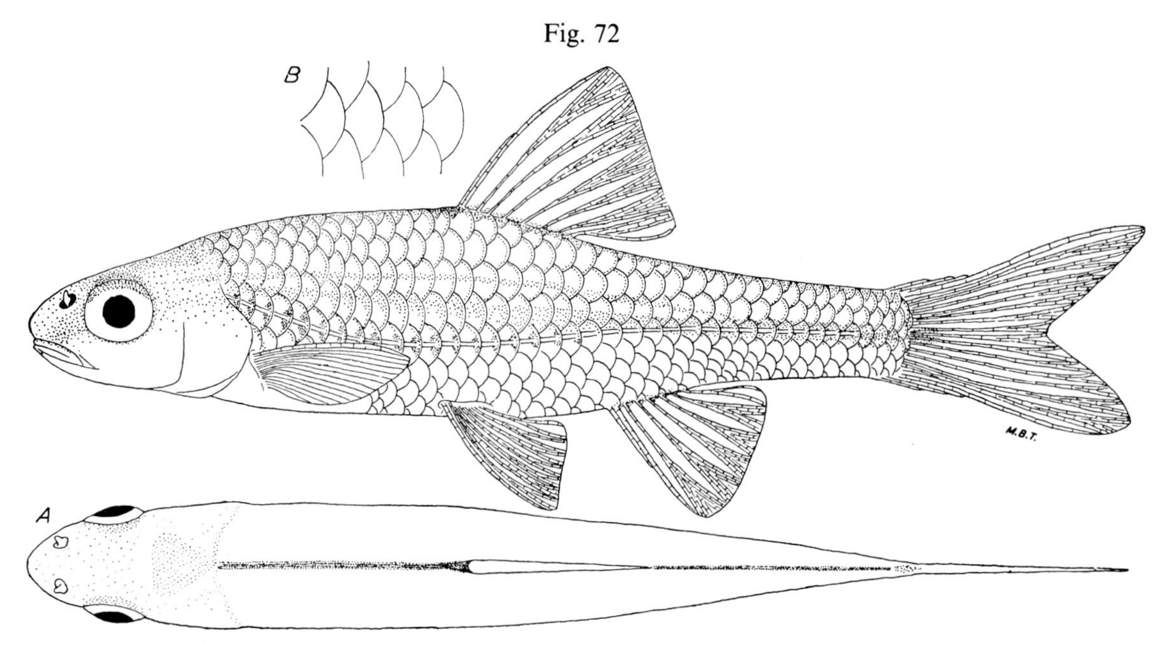

Central fig.: Scioto River, Scioto County, O.

April 24, 1940.
Adult female.

49 mm SL, 2.4″ (6.1 cm) TL.
OSUM 2494.

Fig. A: dorsal view, showing dark, bold mid-dorsal stripe which widens at dorsal fin origin.
Fig. B: anterior lateral line scales; their height is less than twice their exposed width.

Identification

Characters: *General—See* emerald shiner. *Specific*—Normally 7 anal rays, rarely 8; the rays more robust than in the mimic shiner and the webbing often heavily tinged with milky-white. The 34–38 scales in the complete lateral line are of the usual shape; their exposed heights are less than 2.0 times their widths, fig. B. Mid-dorsal stripe usually prominent although sometimes variable in width, but almost invariably expanded into a wedge-shaped spot at dorsal origin, fig. A. This stripe does not extend around the base of the dorsal fin, as in the river shiner. Mouth small; upper jaw length usually shorter than eye diameter, except in small young. Eye large; usually contained less than 3.5 times in head length. Dorsal fin origin often slightly behind pelvic origins. Peritoneum silvery. Teeth, 4–4.

Differs: River shiner has the mid-dorsal stripe surrounding base of dorsal fin; a smaller eye; larger mouth. Other species of shiners have more than 7 rays normally; or elevated lateral line scales; or bold, dusky lateral bands.

Most like: River shiner. **Superficially like:** Mimic and bigmouth shiners.

Coloration: Dorsally a light olive- or straw-yellow with a silvery overcast, the scales faintly dark-edged. Sides silvery, a faint bluish lateral band sometimes evident. Two dark spots, often quite pronounced, one immediately above, the other below, the pore of each lateral line scale on the anterior half of body. Ventrally silvery- and milk-white. Fins transparent or tinged with white. *Breeding adults*—More strongly tinged with straw color; fins, especially the anal, whitish basally. *Breeding male*—With microscopic tubercles on head; basal halves of pectoral rays thickened and serrated, as sometimes are the pelvic rays.

Lengths: *Young* of year in Oct., 1.3″–1.6″ (3.3–4.1 cm) long. Around **1** year, 1.4″–2.3″ (3.6–5.8 cm). *Adults,* usually 1.5″ (3.8 cm) (with ripe eggs)–2.8″ (7.1 cm). Largest specimen, 3.2″ (8.1 cm) long.

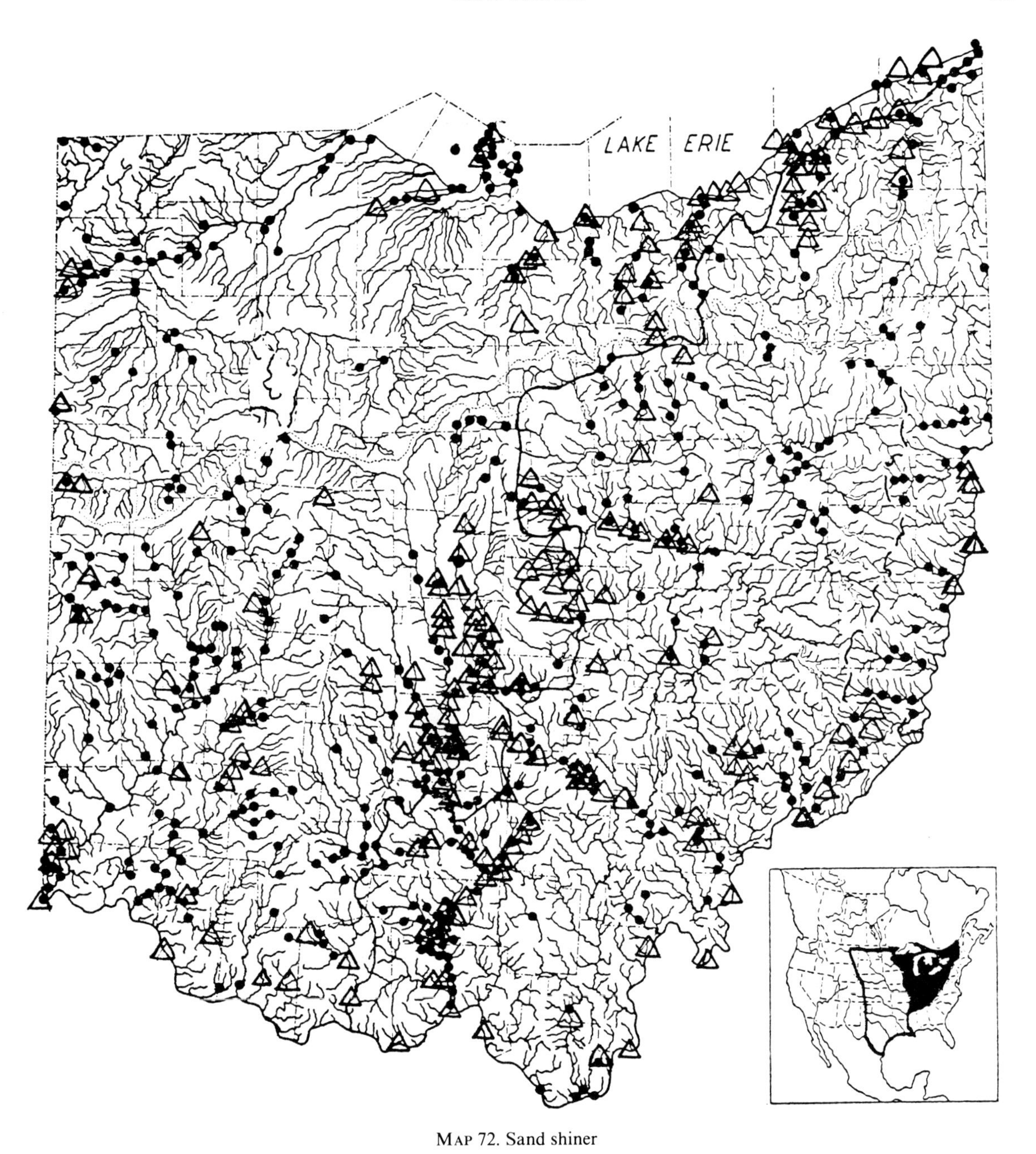

MAP 72. Sand shiner

Locality records. ● Before 1955. △ 1955–80. ~ Allegheny Front Escarpment. /\ Flushing Escarpment.
Insert: Black area represents range of this subspecies, outlined area of another subspecies; zone of intergradation between subspecies appears to be quite broad.

Hybridizes: with bigmouth and mimic shiners; no Ohio specimens.

Distribution and Habitat

Ohio Distribution—Until 1928 the sand shiner was hopelessly confused with several other species of shiners (Hubbs and Greene, 1928: 375–79), and therefore literature references previous to 1928 are of little value. Preserved specimens taken before 1900 are from Ottawa, Erie, Lorain, Cuyahoga, Defiance and Hamilton counties. Presumably, this inhabitant of clear-water, sand-bottomed streams was at least as widespread and abundant in Ohio before 1900 as it was between 1920–35.

Since 1920 the largest populations have occurred in areas of high maximum relief, such as in the upper Great Miami River drainage, along the Allegheny Front Escarpment, and east of the Flushing Escarpment; the smallest populations in the low-gradient streams of northwestern Ohio and the mature, unglaciated streams of the southeastern section. During the 1920–50 period a drastic decrease in abundance occurred in the Maumee and Auglaize rivers, for before 1935 the species was taken in almost every collection made in these rivers, whereas since 1935 the species was found in abundance only at Grand Rapids, Wood County, and a few other localities where the sand and gravel were still relatively free from silt. It was abundant in the shallow waters of Lake Erie, and rare or absent in the Ohio River.

Habitat—The largest populations were present on the sand- and gravel-bottomed riffles and in those pools having considerable current, of the clear-water, moderate- and large-sized streams of moderate and high gradients; smaller populations occurred in smaller streams of the same type. Large populations also were present about the exposed, sand- and gravel-bottomed beaches of Lake Erie. The species was absent or rare in low-gradient streams, except in sections where the sand was not silt-covered. It was seldom found among rooted-aquatic vegetation, and seemed to be surprisingly tolerant to some inorganic pollutants, such as mine wastes, provided these pollutants did not cover the sand and gravel.

In 1930 Osburn, Wickliff and Trautman (1930:173) recorded both *Notropis stramineus* (Cope) and *N. missuriensis* (Cope) as occurring in Ohio. Recent investigations indicate that only one form is present.

Years 1955–80—The distributional pattern throughout Ohio of the sand shiner remained relatively unchanged since 1952 (*see* map 72). The species retained its numerical abundance or increased in numbers in those sections of high maximum relief such as those adjacent to the Allegheny Front Escarpment (White et al., 1975:81)

Summerfelt and Minckley (1969:452) point out that in Kansas the abundance of the sand shiner varied inversely with the abundance of the red shiner, *Notropis lutrensis* (Baird and Girard)*. Something of the same inverse relationship exists between the sand and redfin shiners in Ohio. Whenever both were present in the same stream section, the sand shiner usually dominated the sandy and least silted substrates of riffles, and the redfin was numerous only in the more sluggish portions of the stream where some silting of the sandy-gravelly substrate occurred. If, however, the sand shiner was absent, the numbers of redfins were larger usually, and the species then occupied those cleaner substrates normally inhabited by the sand shiner.

This inverse relationship was not noted whenever the sand shiner and the rosefin shiner occupied the same stream section. The latter is a close relative of the redfin shiner, but unlike that species the rosefin inhabits the more rapidly flowing portions of riffles and appears to compete strongly with the sand shiner for living space if the substrate is sandy or of a sandy gravel. If the substrate is of bedrock, the numbers of rosefins then greatly outnumber the sand shiner. Apparently, interspecific competition was a factor in limiting the numbers of both species when both were present. When one species was absent, the other might be quite abundant.

Recently Bailey and Allum (1962:64) have shown that formerly the sand shiner was "incorrectly known as *Notropis blennius* (Hubbs, 1926:43), [and] has recently been called *Notropis deliciosus*, a name that now must be placed in the synonymy of *Notropis texanus* (Suttkus, 1958:307–18). The sand shiner next becomes *Notropis stramineus* (Cope)."

For some aspects of the life history of the sand shiner *see* Summerfelt and Minckley (1969:444–53).

---

* In 1928 I began a systematic study of the redfin and rosefin shiners in Ohio, noting a marked similarity between some individuals of the redfin and red shiner (*Notropis lutrensis*). The latter's range is largely west of the Mississippi River. At the time I thought that possibly the more silt-tolerant red shiner had migrated eastward into Ohio, as had the orangespotted sunfish. Cavender (1981:18) has collected redfin shiners recently in northern Ohio, stating that the red shiner "has been widely sold as an aquarium fish in Ohio."

NORTHERN MIMIC SHINER

Notropis volucellus volucellus (Cope)

Fig. 73

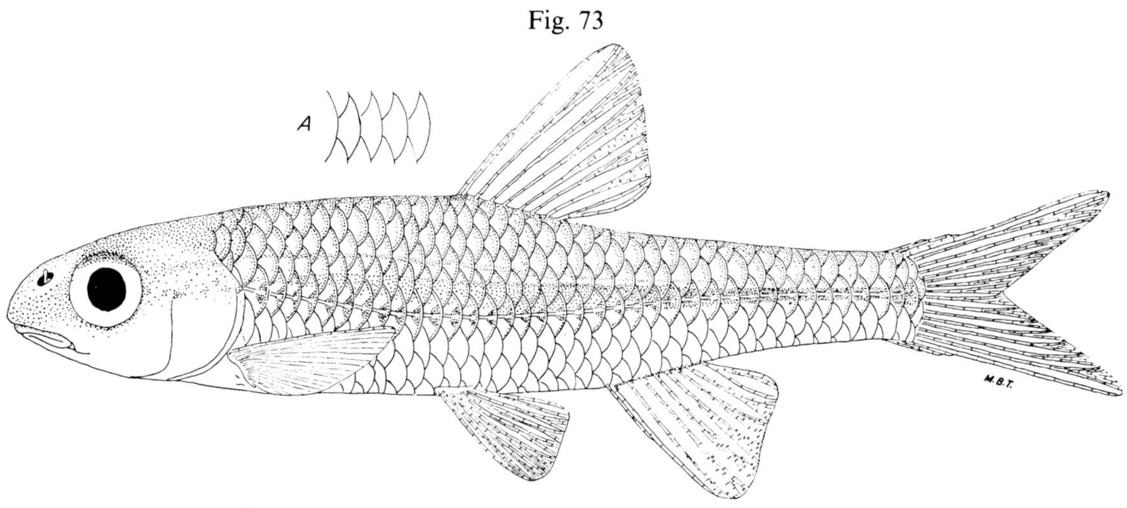

Lower fig.: Fish Creek, Williams County, O.

July 13, 1939. 48 mm SL, 2.4″ (6.1 cm) TL.
Adult female. OSUM 346.

Fig. A: anterior lateral line scales; their height is more than twice their exposed width.

Identification

Characters: *General—See* emerald shiner. *Specific* —Anal rays normally 8, extremes 7–9; these rays thinner and frailer than those of the sand shiner, and the webbing transparent with little or no whitish. The complete lateral line contains 33–38 scales; those in the anterior half of the line greatly elevated, their exposed heights more than 2.0 times their widths, fig. A. Mid-dorsal stripe often faint and irregular in outline; it does not surround base of dorsal fin. Length of upper jaw shorter than diameter of the moderately large eye. An ill defined, often inconspicuous, lateral band usually present on sides of body, with only a faint suggestion of a band across opercles and snout; lateral band may be quite dark in specimens taken from clear, weedy waters, or may be absent in specimens taken from turbid waters. Peritoneum silvery. Intestine about equal in length to standard length. Infraorbital canal complete, extending around lower half of eye and ending anteriorly between nostril and eye. Teeth, 4–4. *Subspecific*—Lateral band normally rather prominent as are the black edges of the scales of the back. Body averages more slender; its depth usually contained more than 4.5 times in standard length, its width usually more than 1.7 times in its depth. Depth of caudal peduncle averages more than 2.5 times in head length. Dorsal fin height averages more than 2.5 times in predorsal length. Anterior lateral line scales usually about 3.0, extremes 2.5–3.8, times as high as wide, fig. A. Scales in complete lateral line usually 35–37, extremes 34–38. Intergrades between northern and channel mimic shiners present in extreme southern Ohio.

Differs: Channel mimic shiner has a less developed and less intense color pattern; a deeper and wider body; deeper caudal peduncle; less elevated scales. Ghost shiner has little or no color pattern; a deeper body and an interrupted infraorbital canal. Other species of shiners in Ohio with 8 anal rays have the lateral line scales less than 2.0 times as high as wide. Chubs have barbels.

Most like: Channel mimic and ghost shiners. **Superficially like:** River, bigmouth, and sand shiners.

Coloration: Dorsally greenish-, bluish-, or olive-straw overlaid with silver, the scales narrowly dark-edged. Mid-dorsal stripe absent, faint, or ill-defined. Sides silvery with a faint, dark lateral band

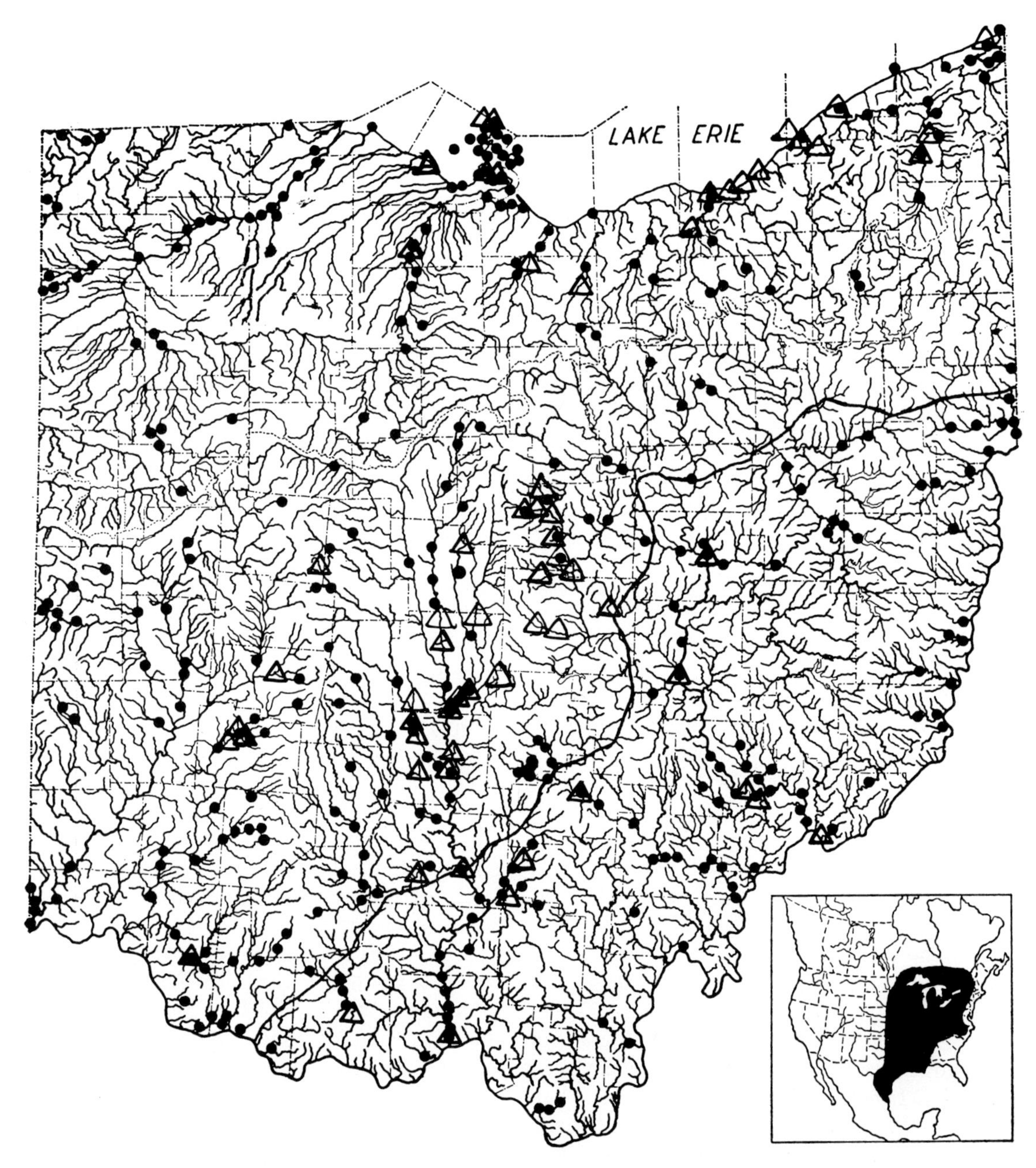

MAP 73. Northern mimic shiner

Locality records. ● Before 1955. △ 1955–80. ~ Glacial boundary.

Insert: The southwestern segment of the population may be subspecifically or specifically distinct; large unoccupied areas occur within the southern portion of the range; western and Mexican boundaries indefinite.

which is usually absent or very poorly defined on snout and opercles. The dark spots above and below each pore on the lateral line scales are inconspicuous. Ventrally silvery- or milk-white. Fins transparent. *Breeding adults*—Little brighter than non-breeding fishes, the males with microscopic tubercles on the dorsal surface of head, and with the pectoral rays slightly thickened and serrated on their dorsal surfaces.

Lengths: *Young* of year in Oct., 0.8″–1.6″ (2.0–4.1 cm) long. Around **1** year, 1.2″–2.5″ (3.0–6.4 cm). *Adults*, usually 1.5″–2.8″ (3.8–7.1 cm). Largest specimen, 3.0″ (7.6 cm) long. For life history, *see* Black (1945:297–324).

Hybridizes: with bigmouth and sand shiners; no Ohio specimens.

Distribution and Habitat

Ohio Distribution—Until 1928 the mimic shiner, as a species, was hopelessly confused with other species of shiners (*see* sand shiner); therefore the literature references previous to 1928 are of little value. Preserved specimens from Hamilton, Mercer, Defiance, Lucas, Lorain and Cuyahoga counties indicate its presence in Ohio before 1900. There is no evidence to indicate that the species was not as abundant and widespread in distribution in Ohio before 1900 as it has been since; in fact, conditions should have been more favorable before 1900 and the species should have been more abundant then.

Between 1920–50 large populations of northern mimic shiners were found along the shores and islands of Lake Erie and throughout most of inland Ohio, except in the most mature streams of unglaciated Ohio and in the lake plains and Black Swamp streams of the northwestern quarter. Since 1935 there have been marked decreases in abundance in those stream sections where silting became most rapid, notably in portions of the Maumee River system.

Habitat—The largest populations inhabited rather clear, moderate- and large-sized streams of low and moderate gradients where the bottom was a sandy silt, gravel, or clayey silt, provided that silt deposition was very slow. Fair-sized populations occurred in small, clear low-gradient, headwater streams, especially those streams along the Lake Erie-Ohio River divide which formerly were marshy. This species is more tolerant to rooted aquatics than is the sand shiner. When the northern mimic and sand shiners occurred in the same stream sections the two species usually showed decided habitat preferences, the mimic shiner inhabiting the more slowly flowing riffles or pools with little or no visible current and having bottoms containing some silt among the sand and gravel, whereas the sand shiner occupied the swifter flowing riffles and pools where the sand and gravel were more free of silt. Only small populations of mimic shiners occurred in low gradient streams which were frequently turbid and where silt deposition was rapid. Its Lake Erie habitat was not as well differentiated from that of the sand shiner as was its stream habitat, but the mimic shiner showed a decided tendency to occupy more silty bottoms and more sheltered locations than did the sand shiner, and unlike the sand shiner it occurred in fair or large numbers in waters as deep as 35′ (11 m).

Years 1955–80—As stated above, the decrease in abundance throughout Ohio of the mimic shiner was noted throughout the 1920–50 period and especially wherever silting of the substrate of streams was pronounced, as it was in the Maumee system. Since 1950 such decreases in abundance have continued, and, except for rare occasions, no large numbers of individuals have been captured in any locality. None was taken in the Maumee River system despite determined efforts to capture it; however, relict populations must still occur in that system.

Before 1950 the mimic shiner was numerous and often abundant about the Bass Islands of western Lake Erie, where frequently more than a hundred individuals could be taken in less than an hour. Since 1950 usually fewer than eight mimic shiners were taken during a collection, and the largest number captured in a 2-hour period was only 32. From these data and many observations, it is assumed that the species was not as numerous after 1950 as it had been previously (White et al., 1975:83).

CHANNEL MIMIC SHINER

Notropis volucellus wickliffi Trautman

Fig. 74

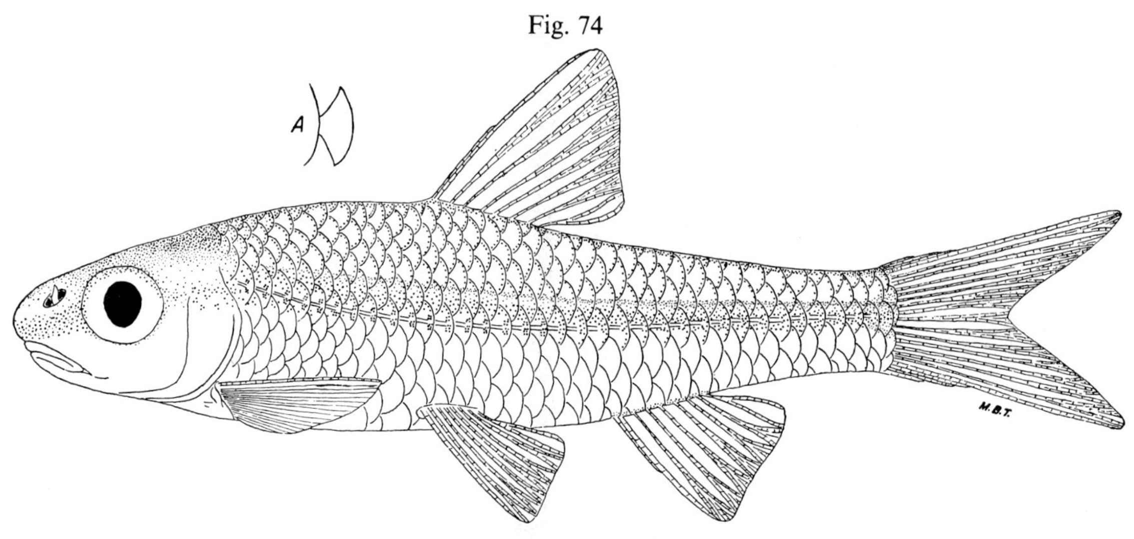

Lower fig.: Ohio River, Lawrence County, O. Aug. 22–23, 1939. Adult female. 47 mm SL, 2.3″ (5.8 cm) TL. OSUM 583.

Fig. A: anterior lateral line scale; its height is more than twice its exposed width.

Identification

Characters: *General—See* emerald shiner. *Specific—See* northern mimic shiner. *Subspecific—* Compared with the northern mimic shiner the color pattern of the channel mimic shiner is less prominent and the lateral band is more vague, or is absent. Body averages deeper and wider, giving a turgid, robust appearance; body depth usually contained less than 4.5 times in standard length; body width averaging less than 1.7 times in depth. Depth of caudal peduncle averages less than 2.5 times in head length. Dorsal fin height averages less than 2.4 times in predorsal length. Anterior lateral line scales less elevated, normally about 2.5, extremes 2.1–3.2, times as high as wide, fig. A. Scales in complete lateral line usually 33–36, extremes 32–37. Intergrades present.

Differs: *See* northern mimic shiner.

Most like: Northern mimic and ghost shiners. **Superficially like:** River, bigmouth and sand shiners.

Coloration: Similar to northern mimic shiner in color pattern of adults, and tubercle distribution of breeding males, but differs in being generally paler, with fewer and/or smaller melanophores along the lateral band.

Lengths: *Young* of year in Oct., 0.8″–1.6″ (2.0–4.1 cm) long. Around **1** year, 1.2″–2.5″ (3.0–6.4 cm). *Adults,* usually 1.5″–2.8″ (3.8–7.1 cm). Largest specimen, 3.0″ (7.6 cm) long.

Distribution and Habitat

Ohio Distribution—The range in Ohio of this subspecies (Trautman, 1931B:468–74) is confined to the lower courses of the Ohio River tributaries, and to the Ohio River itself where it often abounds in tremendous numbers as far upstream as Belmont County. During the spawning season, from June to August, all except a few of the Ohio River specimens examined were typical channel mimic shiners; but during the colder months there were also specimens of intergrades and northern mimic shiners. Likewise, in the lower courses of the Ohio River tributaries, the spawning populations were mostly channel mimic shiners; but progressing up-

MAP 74. Channel mimic shiner

○ Type locality, GreatMiami River mouth, Hamilton County. Locality records. ●After 1920. △ 1955–80.

Insert: Range of this subspecies may extend farther up the Missouri and down the Mississippi rivers; for range of the species *see* **Insert**, map 73.

stream the numbers of relatively few intergrades and northern mimic shiners increased rapidly until only typical northern mimic shiners occurred. During the colder months, or during late summer drought periods, the lower portions of the Ohio River tributaries contained mixed populations, which in some instances were mostly intergrades or intergrades and northern mimic shiners. Obviously after the spawning season, there was a downstream drifting of northern mimic shiners and intergrades, and this was greatest during the drought years.

Habitat—As the name "Channel" implies, this subspecies is an inhabitant of the deeper waters of the Ohio and Mississippi rivers and lower portions of their tributaries. It inhabits the midwaters of the large pools, often where there is no visible current; it likewise congregates at times, together with the emerald shiner, in immense schools in the swift currents below dams. The largest numbers were taken with hand seines over the large gravel bars, particularly during the spawning season; but smaller schools likewise were taken in turbid waters over a soft mud bottom. Because of its tolerance to various bottom types and its abundance, it is assumed that this subspecies is fairly tolerant to present turbid conditions. Nothing is known of its abundance before 1900 when the river was not ponded, the waters were less turbid, and the sand and gravel riffles and bars were not covered with silt.

Years 1955–80—Unlike the inland and Lake Erie populations of the mimic shiner, the present taxon displayed no diminution in numbers in the Ohio River during the 1955–80 period. Large schools often consisting entirely of this shiner could be observed whenever such predators as the skipjack herring herded them into the shallows in an attempt to capture some. William Clay and associates, recording the 1957–59 investigations of the entire Ohio River, stated that they captured 50,241 individuals in 99 collections of this form (ORSANCO, 1962:148). The erection and operation of six high-lift dams since 1950 apparently has had no adverse effect upon the population.

It is unknown why the population of the Ohio River form of this shiner continues to thrive, whereas the inland and Lake Erie populations of *Notropis volucellus* are decreasing in numbers. I know of no observations relative to the requirements of breeding adults, development of eggs, or growth of young for the Ohio River subspecies.

In a communication from Bruce H. Bauer to me of September 20, 1978 he wrote: "Most people here [University of Tennessee] feel that *Notropis wickliffi* warrants specific status. Recent collections from the Duck River, a large Tennessee River tributary, have yielded large collections of *N. volucellus* that are readily divisable into two taxa, one of which fits *wickliffi*. We have also taken the *wickliffi* type in the Mississippi and other large rivers in western Tennessee." Several other ichthyologists have informed me of their conviction that *wickliffi* is specifically distinct, to which I agree.

GHOST SHINER

Notropis buchanani Meek

Fig. 75

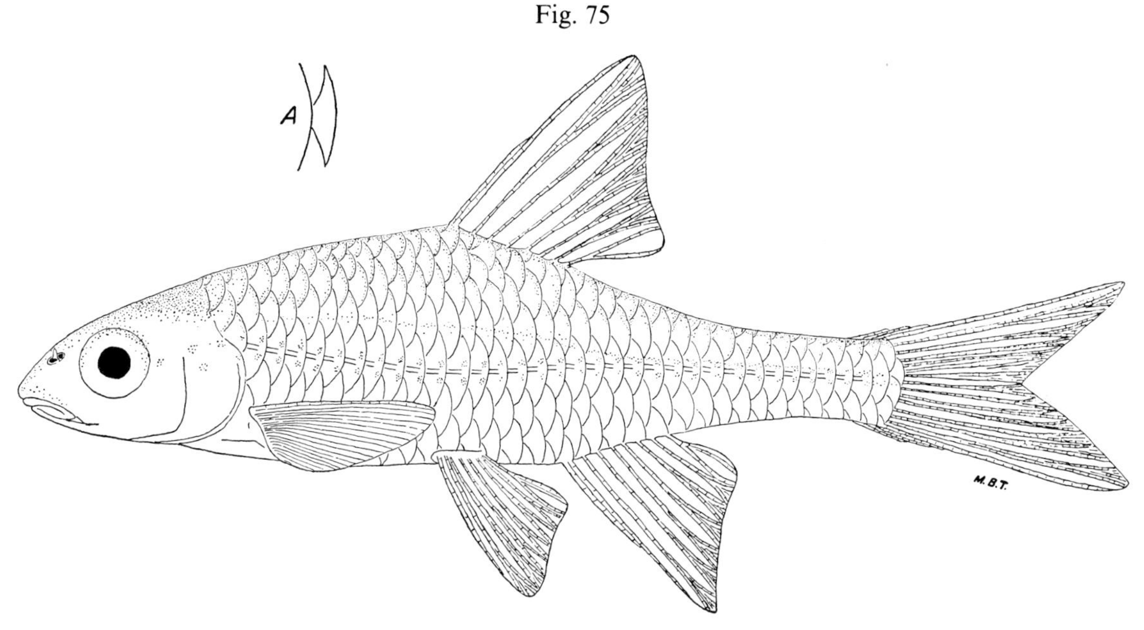

Lower fig.: Muskingum River, Washington County, O.

May —, 1940. 36 mm SL, 1.8" (4.6 cm) TL.
Adult female. OSUM 2869.

Fig. A: anterior lateral line scale; its height is more than three times as high as wide.

Identification

Characters: *General—See* emerald shiner. *Specific*—Anal rays normally 8, extremes 7–9, the rays very frail. Complete lateral line with 30–35 scales, those anteriorly greatly elevated; exposed portions always more than 3.0 times as high as wide, and usually about 4.0 times, fig. A. Dark specklings (melanophores) few and often apparently absent; the milky-white general coloration suggesting the name ghost shiner. Mouth small; usually 3.9–4.1 times in head length. Eye large; less than 3.8 times in head length; its diameter longer than length of upper jaw. Body depth equal to, or greater than, head length; body depth and head length contained fewer than 3.9 times in standard length. Depth of caudal peduncle contained more than 2.8 times in head length; length of caudal peduncle 3.8–4.3 times in standard length. Dorsal fin high; more often falcate than straight distally; its height contained 1.6–2.1 times in the predorsal length. Peritoneum silvery. Length of intestine about equal to standard length. No canals connecting infraorbital pores. Teeth, 4–4.

Differs: Mimic shiner has larger and darker specklings; less body depth; shorter fins; infraorbital canal connected by pores. Other species of shiners having 8 anal rays have the anterior lateral line scales less than 2.0 times as high as wide. Chubs have barbels.

Most like: Mimic shiner. **Superficially like:** River and sand shiners.

Coloration: Dorsally very pale olive, heavily overlaid with silvery and milky-white; virtually lacking the dark edgings on the scales. Sides and ventral parts silvery and milk-white. *Breeding male*—Has microscopic tubercles on dorsal surface of head; dorsal surface of the anterior pectoral rays slightly enlarged and roughened.

Lengths: *Young* of year in Oct., 0.8"–1.5" (2.0–3.8 cm) long. Around **1** year, 1.1"–2.3" (2.8–5.8 cm). *Adults*, usually 1.3"–2.3" (3.3–5.8 cm). Largest specimen, 2.5" (6.4 cm) long.

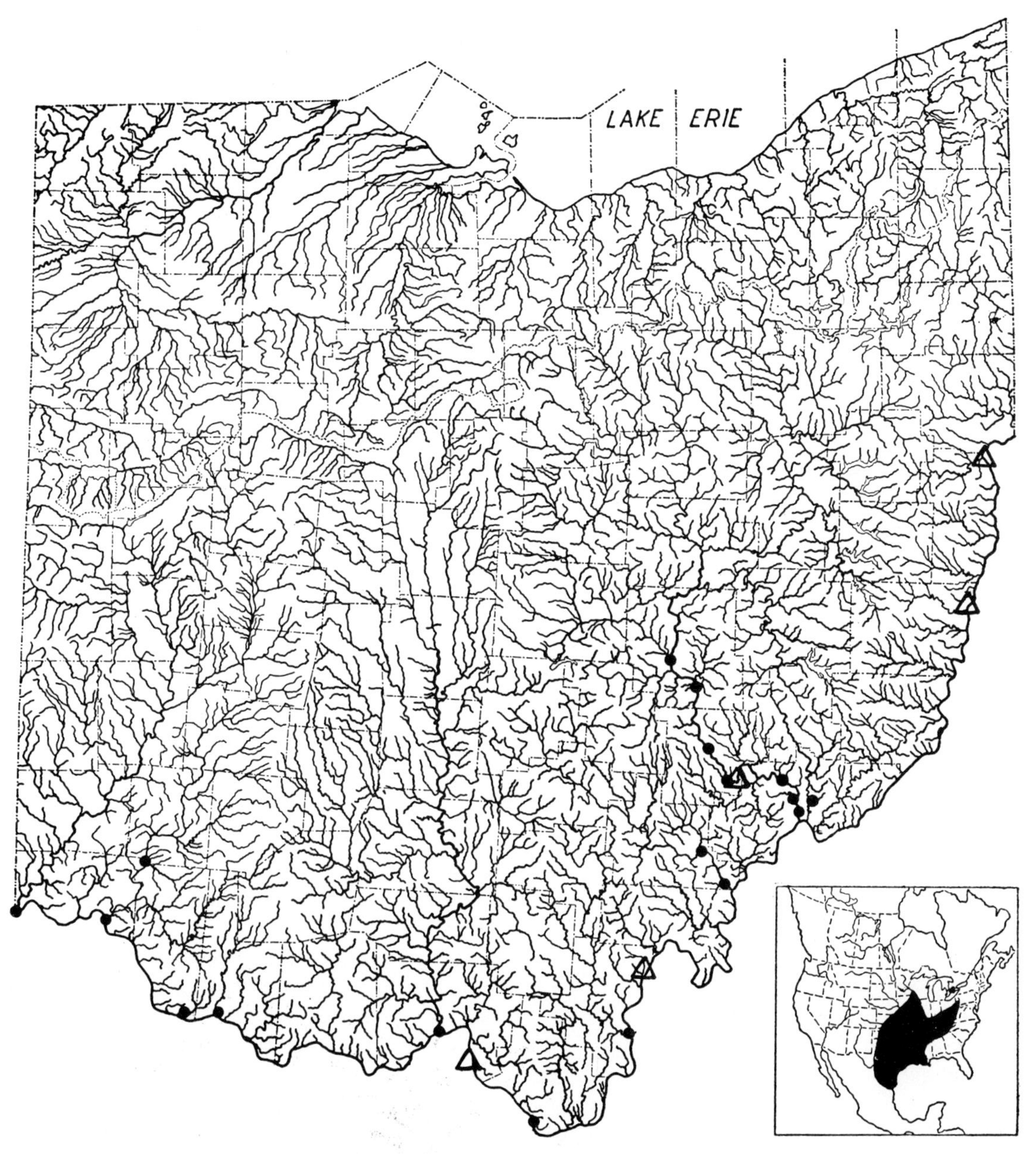

MAP 75. Ghost shiner

Locality records. ● Before 1955. △ 1955–80.

Insert: Unoccupied areas in the northeastern section of the range increasing in size and number; Mexican limits of range indefinite.

Distribution and Habitat

Ohio Distribution—Between 1920–50 the largest numbers of ghost shiners were taken in the quiet waters of the ponded Muskingum River, from its mouth upstream to Zanesville. Elsewhere, in the lower portions of the Ohio River tributaries and in the Ohio River itself, the species was present only in small numbers or as strays. It seemed to be virtually absent from the Scioto River.

The ghost shiner was first recorded as occurring in Ohio waters in 1930 (Osburn, Wickliff, Trautman, 1930: 173) when at that time and until recently, it was considered a subspecies of *Notropis volucellus*. Recently W. Ralph Taylor (Bailey, 1951:223) discovered that in the ghost shiner the infraorbital canal is undeveloped, whereas it is completely developed in the mimic shiners; consequently the ghost shiner is now considered to be specifically distinct. Because of failure to recognize the ghost shiner until rather recently, nothing is known concerning its distribution or abundance before 1920; however, its habitat requirements suggest that before 1900 its habitat was more prevalent in southern Ohio than it has been recently.

Habitat—The ghost shiner definitely sought clear, quiet waters, and a clean sand and gravel bottom, and usually where there were some submerged aquatics, such as pondweeds. The largest concentrations were found invariably in the quieter, clearer waters of an embayment or mouth of a small brook when the Muskingum River was very turbid. This apparent intolerance of turbidity and current may be the reason for its absence in the unimpounded and more turbid Scioto River, and rarity in the Ohio River.

Years 1955–80—A total of 34 specimens was captured in the Muskingum River, Morgan County, in June of 1966, the only collection made in inland Ohio during the 20-year period. Presumably it is present elsewhere in the Muskingum River. H. Ronald Preston and his associates have taken specimens from four localities in the Ohio River between the years 1968 and 1970.

BLACKNOSE SHINER

Notropis heterolepis Eigenmann and Eigenmann

Fig. 76

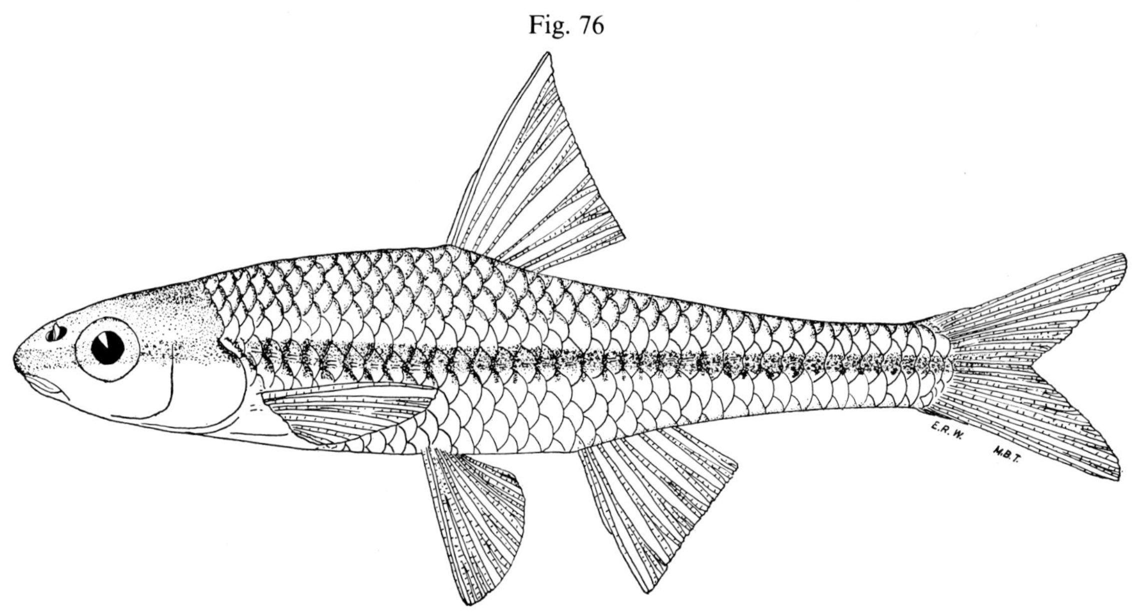

Ludlow Creek, above the Falls, Miami County, O.

About 1930.
Adult male.

52 mm SL, 2.4″ (6.1 cm) TL.
OSUM 3606.

Identification

Characters: *General—See* emerald shiner. *Specific*—Lower jaw not tipped with dusky. A dusky or black band extending from tip of snout backward across opercles and body to caudal fin. The dark posterior borders of the scales in, and immediately below, the lateral band expanded to form black, crescent-shaped bars, a distinctive characteristic of the species. Lateral line incomplete, lacking pores posteriorly. Scales in lateral series, 34–38. Snout slightly overhangs the small, almost horizontal mouth. Upper jaw length usually much shorter than eye diameter. Peritoneum silvery. Anal rays 8, seldom 7. Teeth, 4–4.

Differs: Blackchin and bigeye shiners have larger and terminal mouths; chins tipped with black. Pugnose shiner has an extremely small, oblique mouth; chin tipped with black; a black peritoneum. Mimic shiner has greatly elevated lateral line scales. River and sand shiners rarely have 8 rays; those specimens which do, lack the bold lateral band. Chubs with dusky bands have barbels.

Most like: Blackchin and bigeye shiners. **Superficially like:** Bigeye chub; mimic, sand and pugnose shiners.

Coloration: Dorsally gray- or yellow-olive, overlaid with silvery, the scales boldly edged with black. Upper sides silvery, the scales immediately above the dark lateral band lack dark edges, thereby contrasting sharply with the dark-edged scales of the back. Chin whitish. A broad, black lateral band extends from tip of snout backward across opercles and sides of body to tail; the dark, posterior borders of scales in the lateral band and immediately below it are expanded to form crescent-shaped bars. Ventrally silvery and milk-white. Fins transparent; sometimes a trace of olive in dorsal and caudal fins. *Breeding male*—Has microscopic tubercles on dorsal surface of head; dorsal surfaces of pectoral rays thickened and roughened.

Lengths: *Young* of year in Oct., 1.1″–1.8″ (2.8–4.6 cm) long. Around **1** year, 1.5″–2.5″ (3.8–6.4 cm). *Adults*, usually 1.7″–2.5″ (4.3–6.4 cm). Largest specimen, 2.7″ (6.9 cm) long.

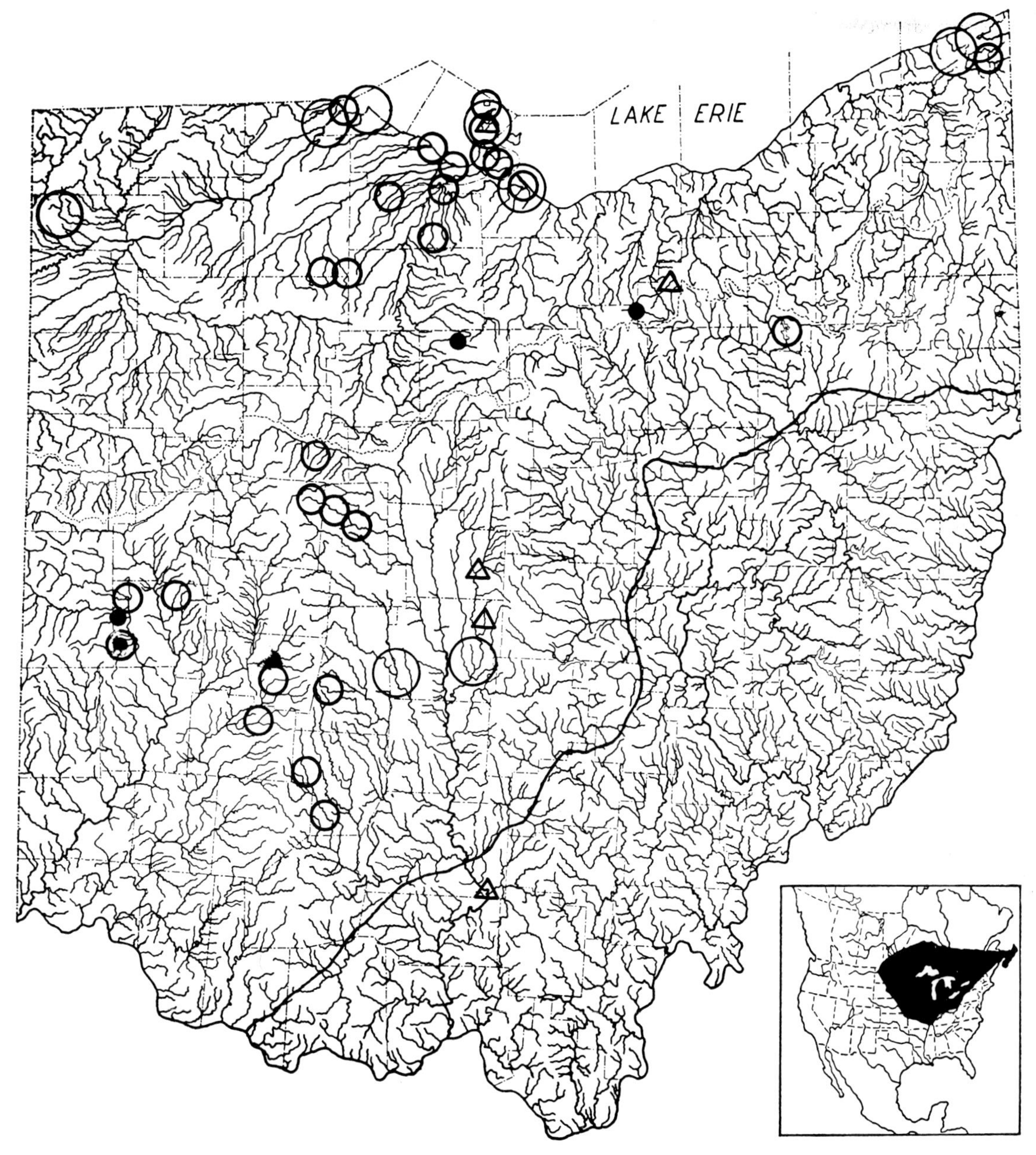

MAP 76. Blacknose shiner

Locality records. ○ Before 1901. ○1901–38. ●1939–54. △ 1955–80. ~ Glacial boundary.
Insert: Unoccupied areas throughout the southern portion of range increasing in size and number; northern limits of range indefinite.

Distribution and Habitat

Ohio Distribution—When the nine reliable records from the fewer than 100 recorded collections, made prior to 1900, are compared with the 31 records from more than 2000 collections made since 1925, it may be assumed that the blacknose shiner was far more abundant and widespread throughout glaciated Ohio before 1900 than it has been since. During the 1925–50 surveys it was obvious that the species was rapidly becoming greatly reduced in numbers in, or extirpated from, many areas where it had been abundant before 1935. During the 1925–35 decade the blacknose was abundant in East and Middle harbors and about the Bass Islands such as in Squaw Harbor at South Bass Island; since 1945 none has been taken in these localities although annual attempts to capture the species have been made. It now appears possible that the blacknose may become absent from Ohio waters within a few years, if it has not already disappeared.

Williamson's (1901:168) record of the blacknose (as *Notropis cayuga*) for Cold Run, Columbiana County, must be questioned because of lack of specimens and the distinct possibility that these fishes were sand or mimic shiners rather than blacknose shiners.

Habitat—The blacknose shiner occurred in largest numbers in the glacial lakes, harbors and bays of Lake Erie, and in the pools of small, clear prairie streams of low gradients where the waters were usually very clear, there was some or much aquatic vegetation, and the bottoms were of clean sand, gravel, marl, muck, peat, or organic debris. The species disappeared rapidly when these waters became more turbid, the bottoms began to silt over with clay, and the aquatic vegetation began to be reduced in amount or disappeared.

Years 1955–80—As stated above, by 1950 the blacknose shiner in Ohio had been reduced to relict populations and had been presumably extirpated in many streams. Since then the sizes of the populations have decreased in most areas, or the populations have been extirpated.

In 1957 George J. Phinney discovered a population of these shiners in Rocky Fork Creek, a tributary of Big Walnut Creek in Franklin County where the species still exists. In 1960 Thomas Thatcher collected one in an unamed tributary of Big Walnut Creek, 1 mile (1.6 km) north of Galena in Delaware County. In 1962 Richard Zura collected five specimens at Put-in-Bay, Ottawa County, where none had been captured since 1945. In 1963 four of us collected a stray in the Scioto River in southeastern Ross County. In 1964 Bernard J. Zahuranec captured three specimens in an unnamed tributary of the East Branch of the Black River, Medina county. Obviously the species is in danger of extirpation in Ohio.

PUGNOSE SHINER

Notropis anogenus Forbes

Fig. 77

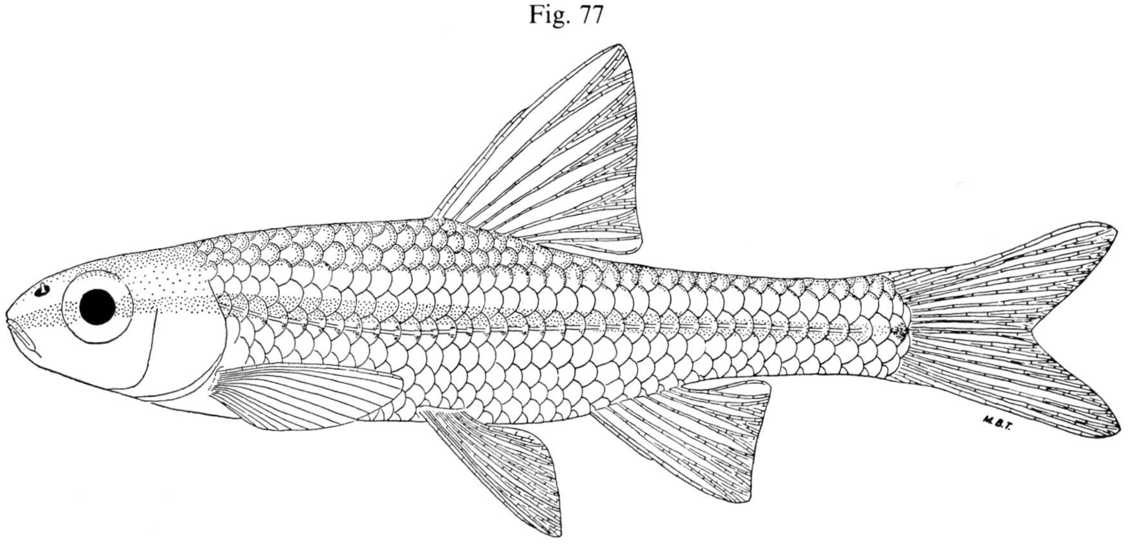

East Harbor, Ottawa County, O.

Aug. 27, 1929.
Adult male.

34 mm SL, 1.7″ (4.3 cm) TL.
OSUM 1784.

Identification

Characters: *General—See* emerald shiner. *Specific*—Chin normally tipped with dusky; a dark band extending backward across snout, opercles and body to tail. Mouth so oblique as to be almost vertical, and very small. Upper jaw length contained 4.5–5.1 times in head length. Scales in the complete lateral line 34–37. Peritoneum *black*. Anal rays often 8, rather frequently 7. Teeth, 4–4.

Differs: No other species of shiner has a small, almost vertical mouth; dusky chin; complete lateral line; a definite lateral band; black peritoneum. The very similar-appearing pugnose minnow has 9 dorsal rays, incomplete lateral line, and silvery peritoneum.

Most like: Pugnose minnow. **Superficially like:** Blackchin and bigeye shiners.

Coloration: Dorsally olivaceous or olive-straw, overlaid with silvery, the scales dark-edged. Tip of chin dusky. A blackish band extends backward across snout, opercles and body to tail. Scales immediately above lateral band yellowish-silvery and without dark edges, thereby contrasting sharply with the dark-edged scales of the back. Ventrally silvery-white. Fins transparent. *Breeding male*—Has microscopic tubercles on top of head; dorsal surface of pectoral rays thickened and roughened, as is sometimes the first pelvic ray.

Lengths: *Adults* usually 1.3″–1.9″ (3.3–4.8 cm) long. Largest specimen, 2.2″ (5.6 cm) long.

Distribution and Habitat

Ohio Distribution—The pugnose shiner was first recorded for Ohio in 1894 (not in 1898 as recorded by Osburn, Wickliff and Trautman, 1930:172), when on July 23, 1894, C. Rutter collected a specimen at Port Clinton, and on July 28, two specimens at Lakeside. Rutter's specimens are in the U. S. National Museum (69524, 1 sp.; 58625, 2 sp.). In August, 1898, Henry B. Ward reportedly collected 30 pugnose shiners at Put-in-Bay, (South Bass Island) from which he obtained the types of two species of Myxosporida. These fishes were identified by W. C. Kendall, who noted that they "did not agree in all details with the descriptions of that species" (Ward, 1919:50). Apparently the fishes were not preserved. Although environmental conditions at Put-in-Bay presumably were favorable for

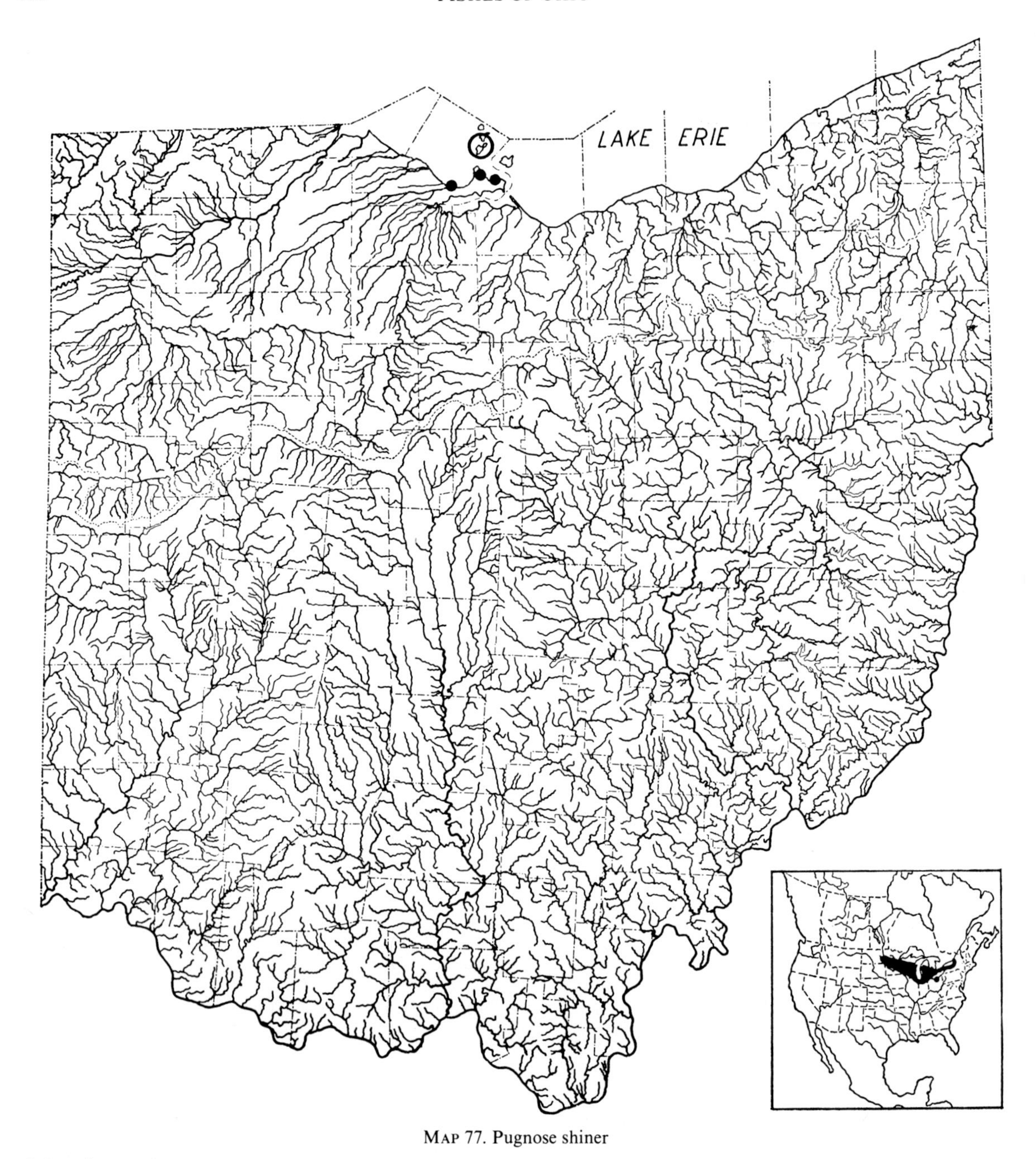

MAP 77. Pugnose shiner

○ Questionable literature record. ● 1894–1931, specimens.

Insert: Now only in isolated areas within its former known range, several of these areas appear to be decreasing in size and number.

the pugnose shiner in 1898, the record is questioned because of Kendall's statement and because the superficially similar pugnose minnow occurred there in abundance. Hubbs (1926:40) questioned the Put-in-Bay record because the figure (Ward, 1919:Pl. 5) "scarcely suggests this species"; I found identification of the figure beyond that of a cyprinid to be questionable. Until 1929 the pugnose shiner could be taken rather readily in East Harbor; the last specimens were collected there in 1931. Since then none has been taken in Ohio waters, although extensive collecting for it has been done.

This post-glacial relict probably occurred elsewhere in glaciated Ohio before 1900, such as in

Sandusky and Maumee bays and in inland pothole lakes. In Indiana, Gerking (1945:65) recorded it from Hamilton Lake, Whitley County, which is in the Maumee River drainage a few miles west of the Ohio-Indiana line.

Habitat—The pugnose shiner occurred only in the very clear waters of glacial lakes, and in streams of low gradient, where there was a profuse abundance of aquatic vegetation and the bottom was of clean sand, marl, or organic debris. The species appeared to be extremely intolerant to turbidity of the water which presumably was a primary factor in its extirpation from Ohio waters.

Years 1955–80—No specimens of the pugnose shiner were taken during this period, despite numerous attempts to capture it about the island region, bays and marshes of western Lake Erie. It is assumed that the species has become extirpated from Ohio waters.

For a map of the present disconnected range in the northern United States and Canada of this rare cyprinid, *see* Bailey (1959:fig. 1).

SILVERJAW MINNOW

Ericymba buccata Cope

Fig. 78

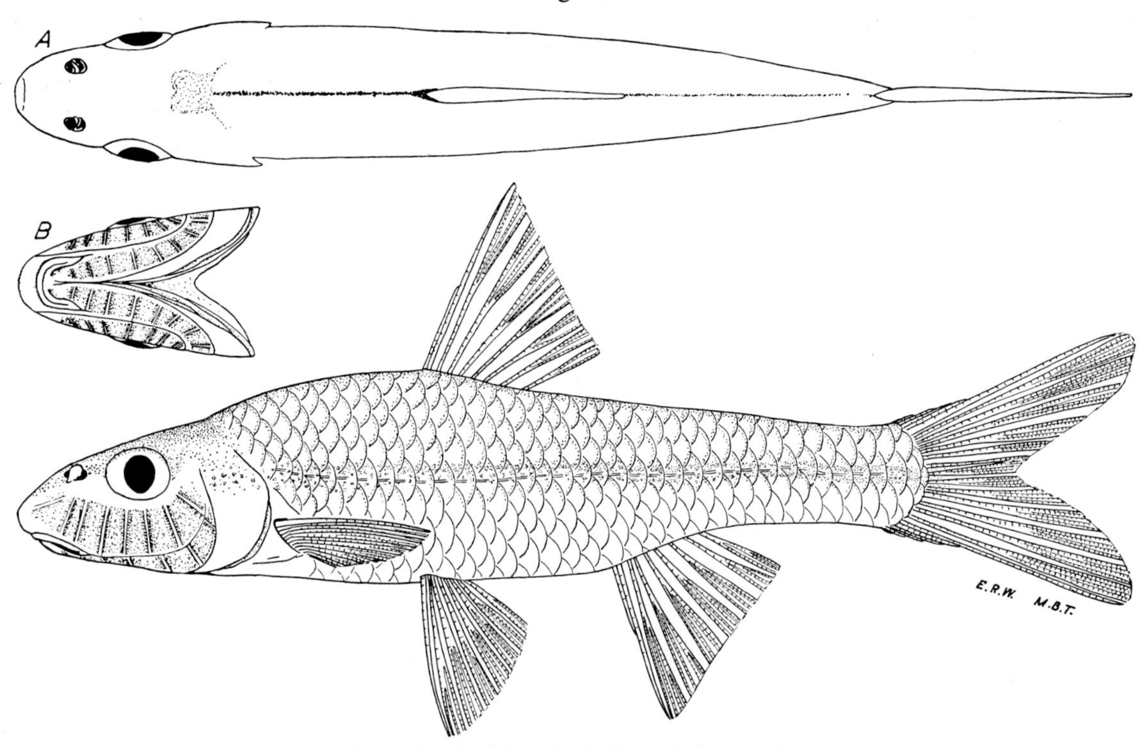

Lower fig.: Rockford Creek, Hancock County, O.

June 27, 1939.
Male.

54 mm SL, 2.8″ (7.1 cm) TL.
OSUM 1750.

Fig. A: dorsal view; note body shape and mid-dorsal stripe.
Fig. B: ventral view of head; note the large cavernous spaces.

Identification

Characters: *General*—Similar to the genus *Notropis* (*see* emerald shiner) except that the present genus has large cavernous spaces on the ventral and lower sides of the head, fig. B. *Specific*—Head flattened ventrally so that in cross-section it is triangular; this head shape and silvery coloration result in a striking resemblance to the bigmouth shiner. Anal rays normally 8, very rarely 7–9. Snout slightly overhanging the almost horizontal mouth. Upper jaw usually about as long as eye diameter. Mid-dorsal stripe narrow but distinct predorsally; less distinct or almost absent behind dorsal. Scales in complete lateral line 31–36; of the usual shape. Teeth, 1,4–4,1 or 4–4.

Differs: The superficially similar bigmouth shiner lacks the cavernous spaces on the head; has a larger mouth; bolder mid-dorsal stripe; a greater average number of lateral line scales. All other species of Ohio minnows lack the large cavernous spaces.

Most like: Bigmouth shiner. **Superficially like:** River and sand shiners; young of the silver chub.

Coloration: Dorsally pale olive-yellow, overlaid with much silvery. Sides less olivaceous and more silvery; no distinct lateral band in life. Ventrally silvery- and milk-white. Fins transparent. *Breeding male*—Has dorsal surface of pectoral rays thickened and roughened, as is sometimes the first pelvic ray.

Lengths: *Young* of year in Oct., 1.0″–1.8″ (2.5–4.6 cm) long. Around **1** year, 1.2″–2.5″ (3.0–6.4 cm).

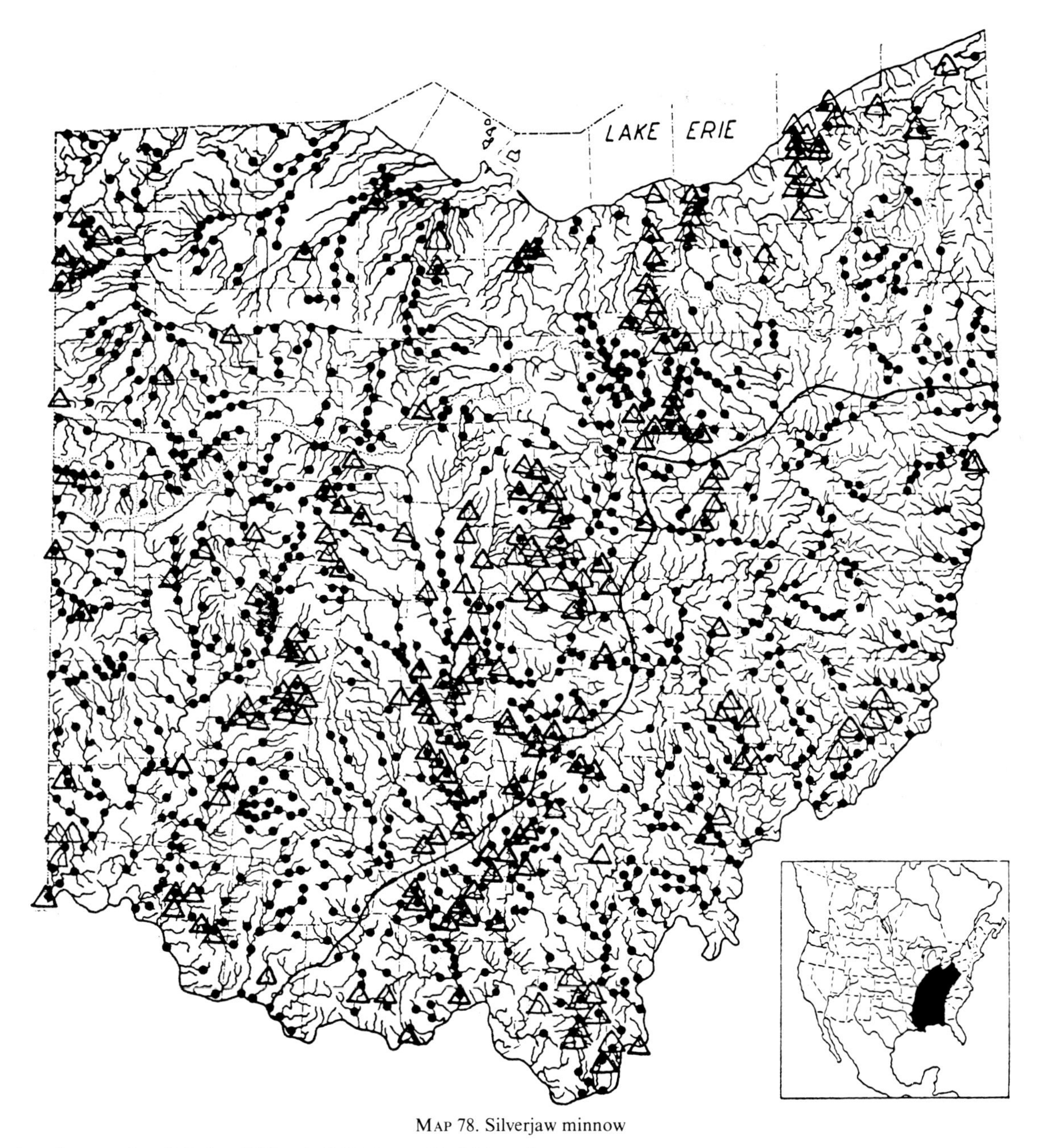

MAP 78. Silverjaw minnow

Locality records. ● Before 1955. △ 1955–80. ~ Glacial boundary.
Insert: May be extending its range northeastward.

Adults, usually 1.7″–3.0″ (4.3–7.6 cm). Largest specimen, 3.8″ (9.7 cm) long.

Distribution and Habitat

Ohio Distribution—Apparently the first capture of a silverjaw minnow in Ohio, until now unpublished, was of a specimen collected in 1853 by S. F. Baird in Yellow Creek of the Ohio River drainage (near Poland); it is preserved in the USNM (7001). The species was described 12 years later (Cope, 1865:88). In 1882 Jordan (1882:855) reported the presence of the species in the Ohio drainage but failed to give a definite locality so that it was not until 1888 that the first published Ohio drainage record appeared (Henshall, 1888:78). In 1884 Seth E. Meek collected specimens (USNM 36803) in the Lake Erie drainage, publishing locality records in

1889 (Meek, 1889:437). As indicated by the many preserved specimens and literature records, the species was abundant and widespread throughout Ohio during the 1892–1925 period, except perhaps in the northeastern corner.

Since 1920, the silverjaw has been one of the most abundant minnows in Ohio. It was particularly numerous along a broad band on both sides of the line of glaciation; it occurred only in moderate or small numbers in the low-gradient streams of the prairie- and till-plain areas of westcentral and northwestern Ohio; before 1925 it was almost absent from the high-gradient and/or cooler streams from Cuyahoga to Ashtabula counties but since then has become more numerous and more generally distributed.

Habitat—The silverjaw is a sand-inhabiting species, reaching greatest abundance only in brooks and small streams of moderate gradients where the sands on the bottoms of the pools, bars, and riffles are relatively free from a covering of clayey silts. It has a surprisingly high tolerance to turbidity, domestic and industrial pollutants, especially mine wastes, provided these do not cover the sand. Since 1925 the species has shown the greatest decrease in abundance in the low-gradient, till-plain, and prairie streams. In several of these streams, I have noted the change in status from large populations before 1930, when the sands were still relatively free of silts, to small populations after 1940, after the sands were mostly covered with silts. It was uncommon or absent, even where sand was present, in the cooler and/or high-gradient streams, such as are in Geauga County; it occurred only accidentally in Lake Erie, inland lakes, ponds, and the Ohio River. *See* the bigmouth shiner for its status in Lorain County.

Years 1955–80—As stated above, the silverjaw minnow was abundant throughout Ohio before 1950 with the possible exception of the northeastern section. Since 1950 the species has remained widespread throughout the state; however, on the whole, its numbers have markedly decreased. In some localities, especially those where the sandy substrate has been covered with silt or other pollutants, it was extirpated or greatly reduced in numbers. Exceptions were the East Branch of the Black, Rocky, upper Cuyahoga and Chagrin rivers, where the species increased in numbers. (*See* bigmouth shiner; White et al., 1975:83.)

It apparently occurred only as a stray in the entire Ohio River. Investigations between 1957–59 indicated that the species was captured in only nine localities with a total of 17 individuals (ORSANCO, 1962:147). Clay (1975:191–92) found it to be one of the most abundant minnows in the sandy-bottom streams of eastern Kentucky.

For some aspects of the life history and ecology of the silverjaw minnow, *see* Ross, D. F., (1972:1–78).

MISSISSIPPI SILVERY MINNOW

Hybognathus nuchalis nuchalis Agassiz

Fig. 79

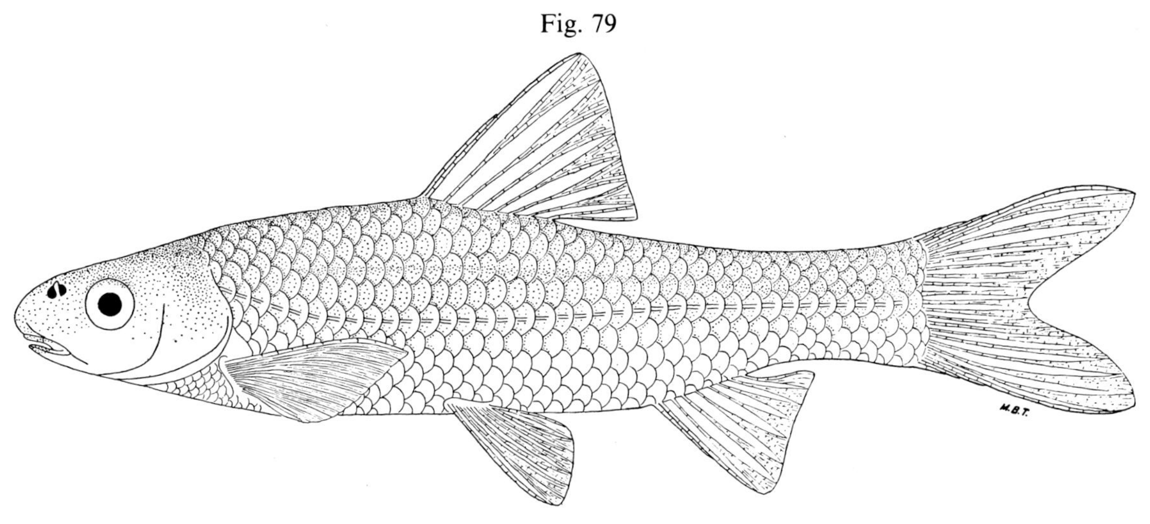

Little Miami River, Hamilton County, O.

Collected before 1900.
Adult male.

80 mm SL, 3.9″ (9.9 cm) TL.
OSUM 625.

Identification

Characters: *General*—Similar to the genus *Notropis*, (*see* emerald shiner) except that the present genus has a small knob on the tip of the lower jaw inside the mouth, and a much longer intestine, which is 3–10 times the body length. *Specific*—Dorsal fin well forward; distance from dorsal origin to caudal base usually much greater than distance from dorsal origin to tip of snout. Snout slightly overhangs the almost horizontal mouth. Eye and mouth about equal in length, each contained 3.6 (small young)–4.8 (adults) times in head length. Head rather subconical. Mid-dorsal stripe broad and distinct. Anal rays normally 8. Scales 35–38 in the complete lateral line. Peritoneum dusky. Teeth, 4–4. (Characters taken principally from Ohio specimens which were collected before 1900.)

Differs: Common shiner has a less pointed head; elevated anterior lateral line scales; like all other shiners lacks the long intestine and knob on tip of lower jaw. Silver and all other chubs have barbels.

Most like: Common, river and spottail shiners; silver chub; bullhead minnow.

Coloration: Dorsally pale olivaceous, overlaid with much silvery. No dark edgings to scales. A broad, slaty mid-dorsal stripe. Sides whitish-silvery. Ventrally milk-white. Fins transparent. (Color description taken from fresh Indiana and Illinois specimens.)

Lengths: Of 9 preserved Ohio specimens, 3.3″—4.0″ (8.4–10 cm) long. Forbes and Richardson (1920:114) give 6.0″ (15 cm) as maximum length.

Distribution and Habitat

Ohio Distribution—All existing records of the occurrence of the silvery minnow in Ohio were made before 1900; these were published by Henshall (1889:124) or are in the catalog book of the Cincinnati Society of Natural History. The catalog records are: "Little Miami (Loveland); Clough Creek, Hamilton County; White Oak Creek." One Ohio collection of five (OSUM 2) and another of four (OSUM 625) silvery minnows have been preserved. Both collections were obtained from the Cincinnati Society of Natural History. All of the many attempts to collect the silvery minnow since 1900 have failed.

The records of former occurrence and absence of

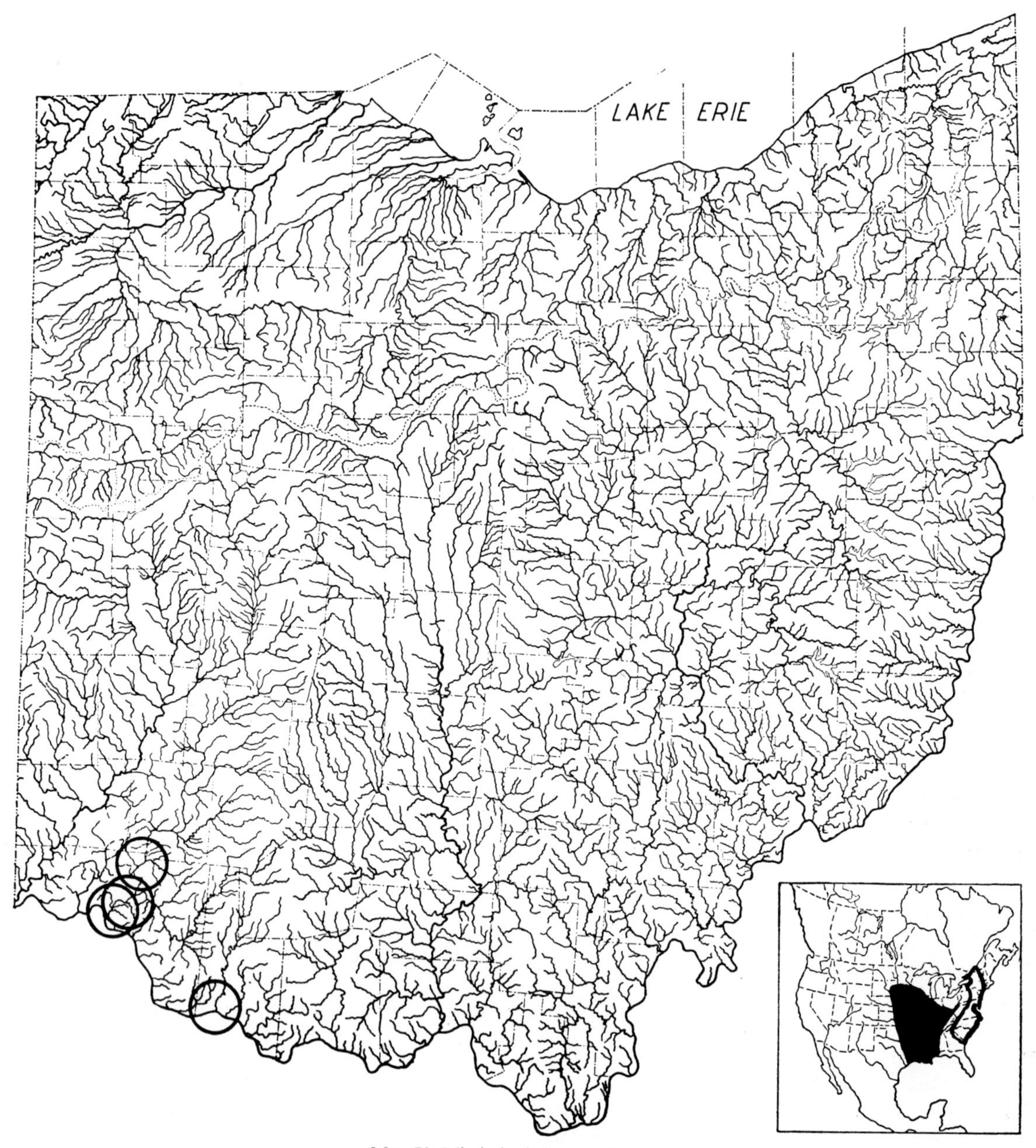

MAP 79. Mississippi silvery minnow

○ Before 1900.

Insert: Black area represents range of nominate subspecies; has been retreating from northeastern portion of its range; outlined area represents range of another subspecies.

recent records from Pennsylvania (Cope, 1881:91; Fowler, 1919:58; Raney, 1939A:46), West Virginia (Raney, 1947:12), and northeastern Kentucky (Clark: 1937), suggest that the silvery minnow is retreating from the northeastern portion of its range. It likewise appears to be retreating from southeastern Indiana, remaining locally abundant only in the lower Wabash River system (Gerking, 1945:67).

Habitat—My observations of the silvery minnow outside of Ohio, and published accounts by others, indicate that the largest populations occur only when there is little or no current, such as in lowland lakes, oxbows, bayous and the quiet portions of pools in streams, where the waters are rich in phytoplankton, and the gravelly-sand, muck, or debris-covered bottoms are not smothered by silt. Both Mississippi and eastern (*H. n. regius*) silvery minnows have been found spawning only where the rate of siltation of the bottom is low, and usually where there is much organic debris or vegetation (Raney, 1939C: 676). Such spawning habitat is uncommon in southern Ohio today. Probably the early disappearance of the silvery minnow from southern Ohio and much of the remainder of the upper Ohio River drainage, was caused by increased turbidity of the waters and disappearance, through smothering or covering with silt, of aquatic vegetation and the muck and organic debris in the lowland waters.

Years 1955–80—Efforts were made during this period to locate a relict population of silvery minnows, especially in southwestern Ohio, but without success. It therefore can be assumed that the species has been extirpated from Ohio, as it appears to have been from much of the upper Ohio River drainage. Clay (1975:190) states that "the silvery minnow is abundant in the lower Ohio River, from Henderson County [Kentucky] downstream, but above this point it is sparse," continuing that the farthest upstream locality is the locks at Louisville.

Examinations of the localities where the species was captured in Ohio before 1900, the Little Miami River, White Oak Creek, and Clough Creek (sometimes spelled "Cluff") suggest that formerly these streams normally carried little silt in suspension. For much of its length Clough Creek has substrates of bedrock, gravel, sand, some muck, and very little silt, so that its waters were or are seldom turbid. Examination of the literature (Cook, 1959:96–97; Cross, 1967:149; Fingerman and Suttkus, 1961:466; Forbes and Richardson, 1920:115; Metcalf, 1966:134–35; Pflieger, 1975:169 and others) indicate that this minnow inhabits a wide range of diverse habitats including ox bows and other static waters, swift, clear streams, moderate-sized to large rivers, some silt-laden as is the Mississippi; that the substrates generally consist of gravel, sand, organic debris, and mud; that the species feeds primarily near or on the substrate, and sometimes in large compact schools. Because of such tolerance to diverse habitats, it is difficult to understand why it has disappeared from the upper Ohio River and its tributaries.

NORTHERN BULLHEAD MINNOW

Pimephales vigilax perspicuus (Girard)

Fig. 80

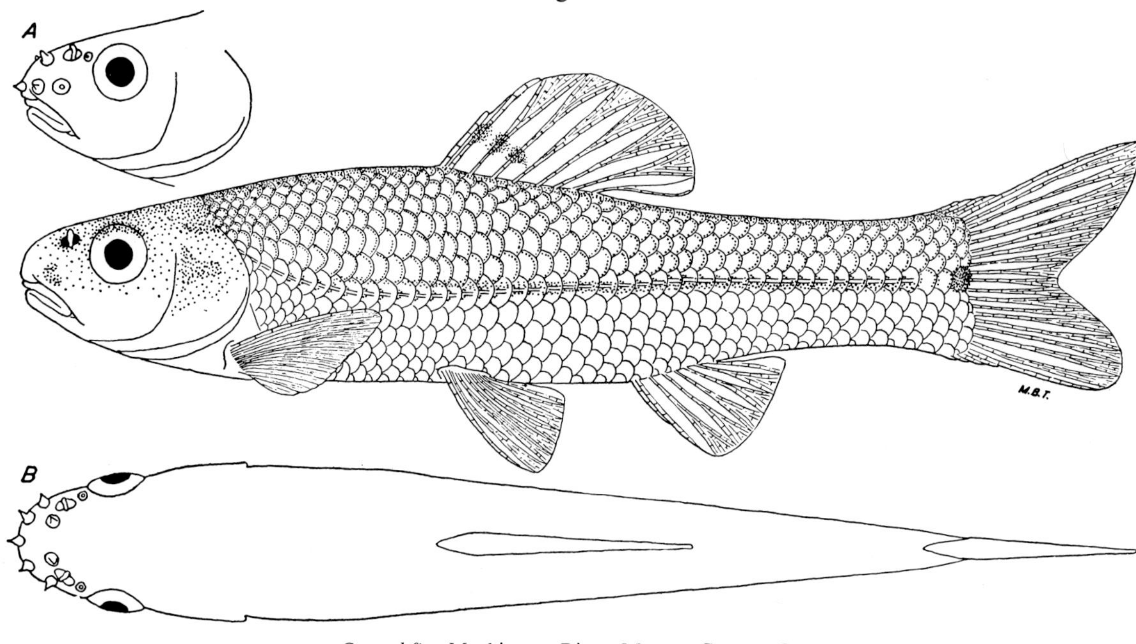

Central fig.: Muskingum River, Morgan County, O.

June 29, 1939. 58 mm SL, 2.8″ (7.1 cm) TL.
Male. OSUM 311.

Figs. A and B: same collection as above.

Breeding male. 66 mm SL, 3.2″ (8.1 cm) TL.

Fig. A: lateral view of head; note tubercles.
Fig. B: dorsal view; note body shape and tubercles.

Identification

Characters: *General* and *generic*—In addition to the usual 8 developed dorsal rays, there is a rather stout, blunt-tipped, half-ray situated before the first developed ray; this half-ray and the first developed ray are connected together with membrane (membrane and half-ray least developed in smallest specimens). Dorsal surface of head and back before dorsal origin notably flattened; scales of predorsal are small, irregular and crowded in appearance. Usually more than 20 scales along the predorsal ridge from head to dorsal origin. A dark, often vague, blotch on each of the first two or three developed dorsal rays; each blotch situated about a third of the length of the ray above its base; these blotches may be absent in small young. Scales in lateral series 41–49. Anal rays 7, rarely 8. Teeth, usually 4–4. *Specific*—A crescent-shaped spot on side of snout between anterior third of upper jaw and eye; an irregularly shaped, triangular spot on upper opercle behind eye; these dark markings least apparent on breeding or very large adults, and on smallest young. No lateral band, although there may be a narrow, dusky line along the caudal peduncle. A prominent, blackish caudal spot. Mouth almost terminal, and moderately oblique. Peritoneum *silvery*. Intestine S-shaped, but not coiled. Lateral line complete.

Differs: Bluntnose and fathead minnows have a black or brown peritoneum; longer intestine; lack the dark crescent on side of snout. Rosefin and

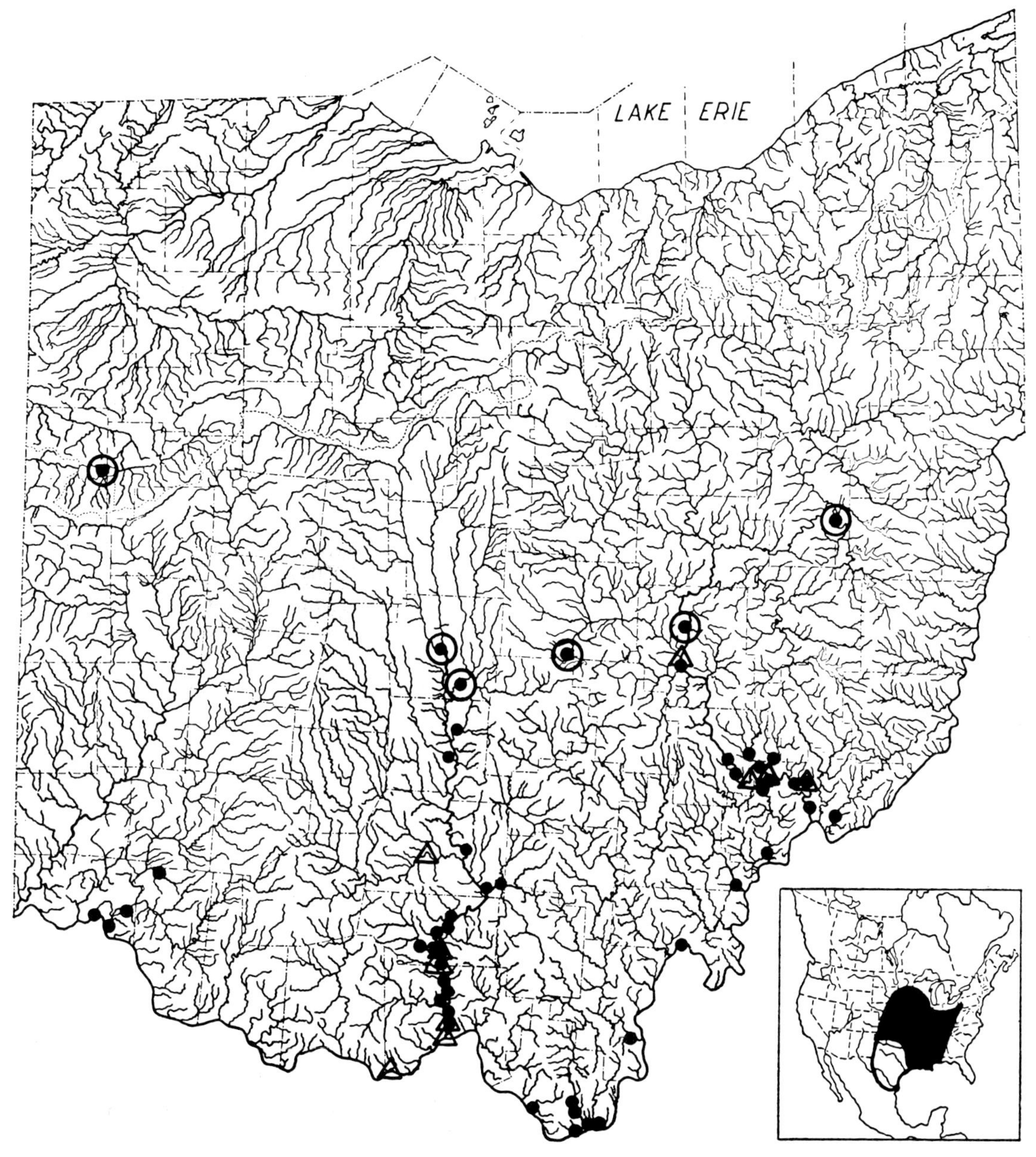

MAP 80. Northern bullhead minnow

Locality records. Before 1955: ⊙ Only one specimen. ●Several or many specimens. △ 1955–80.
Insert: Black area represents range of this subspecies, outlined area of another.

redfin shiners have more anal rays. Stoneroller minnow lacks fleshy lobes on lower jaw. Chubs have barbels.

Most like: Fathead and bluntnose minnows. **Superficially like:** Rosefin and redfin shiners, stoneroller minnow, young of hornyhead and river chubs.

Coloration: Dorsally greenish- or yellowish-olive, overlaid with much silvery, the scales faintly dark-edged. Sides lighter and more silvery. When present, lateral band is restricted to a thin line which is above the lateral line pores anteriorly, crossing beneath the lateral line below the dorsal fin, and continuing as a darker and slightly wider band to the tail. A bold, blackish caudal spot. A dark crescent on side of snout, and an irregularly-shaped triangular spot on opercles behind eye. Ventrally silvery- and milk-white. Fins pale olive or transparent, except for the dusky spots on each of the two or three anterior dorsal rays. *Breeding male*—Has two rows of large, sharply-pointed tubercles on snout, figs. A and B; a moderately well-developed pad on the flattened area of back before the dorsal fin; pectoral rays thickened and slightly roughened.

Lengths: *Young* of year in Oct., 0.8″–2.0″ (3.0–5.1 cm) long. Around **1** year, 1.2″–2.5″ (3.0–6.4 cm). *Adults*, usually 1.5″–3.0″ (3.8–7.6 cm). Largest specimen, 3.7″ (9.4 cm) long.

Hybridizes: with bluntnosed minnow.

Distribution and Habitat

Ohio Distribution—There are few reliable records of the occurrence of the bullhead minnow in Ohio waters before 1900. In the USNM (41028) are two specimens, collected by Gilbert and Henshall in the Muskingum River at Beverly. Henshall (1888:78) in 1888 found it "Common in O'Bannon Creek," a tributary of the Little Miami River, Clermont County. In the Cincinnati Society of Natural History are specimens labeled, "*Cliola vigilax*, Little Miami River and Muskingum River, Henshall"; possibly some of these specimens may have come from O'Bannon Creek. There is an identifiable photograph by Williamson and Osburn (1898: Pl. 14), labeled "*Cliola vigilax.*" This photograph is of the only specimen collected during their Franklin County survey, and came from Big Walnut Creek.

Prior to 1935 comparatively few bullhead minnows were taken in collections; by 1940 it had increased so greatly in numbers, especially in the lower Scioto River, that occasionally hundreds were captured in one seine haul; by 1945, it was of common occurrence in the tanks of commercial bait dealers in Columbus. Despite the recent increase in abundance and unintentional introductions into many sections by bait fishermen there appears to have been no notable recent extension of its range in Ohio.

Upon several occasions I found the bullhead minnow in the bait pails of fishermen at Buckeye Lake, so it was not surprising to capture a specimen there during seining operations. On September 3, 1927 a specimen was collected in Lake St. Marys, Mercer County (Hubbs and Black, 1947:31) which likewise probably was an escape from the minnow bucket of a fisherman. The St. Marys specimen constitutes the only Lake Erie drainage record for Ohio.

Habitat—An inhabitant of sluggish pools of the larger inland streams of the Ohio River drainage and their connecting backwaters and bayous, where there was little or no current, little or no rooted aquatic vegetation, and the bottom was of silt-covered gravel, silt or deep mud. In such situations in the lower Scioto River, it sometimes outnumbered the total number of all other species of minnows in the seine hauls. It appeared to be highly tolerant to turbidity and the rapid deposition of silt. It avoided strong currents and I found only one specimen, obviously a stray, in the main current of the Ohio River (Hubbs and Black, 1947:31).

Years 1955–80—Since 1950 the populations of the bullhead minnow throughout its range in the Ohio River tributaries have on the whole decreased in numbers, especially in the Scioto River system. A brief history of the abundance of this minnow in the Scioto River system follows.

The earliest known record in the Scioto drainage is that of an individual taken in 1897 by E. B. Williamson and R. C. Osburn in Big Walnut Creek, Franklin County (Osburn, 1901:50). Until 1935 only an occasional specimen was captured. After that a phenomenal increase took place, and until 1945 hundreds could be collected daily in many localities in Ross, Pike and Scioto counties. During this period of abundance large numbers were sold as bait. After 1950 it again became rare in the Scioto drainage, and the largest number taken in one collection since was only 14, in Pike County on September 12, 1963, and only after several hours of seining.

During the 1935–50 period the number of individuals taken in the Muskingum River system was low in comparison with the large numbers that could be captured in the Scioto River system. Since 1950, however, the species has apparently retained its numerical abundance in the former better than it has in the latter.

Before 1950 I saw only one specimen taken in the Ohio River. During the intensive surveys conducted on that stream between 1957 and 1959, only 14 individuals from five collections were taken (ORSANCO, 1962:149). Margulies and Burch reported the capture of a specimen in the Ohio River adjacent to Jefferson County, Ohio in 1976.

NORTHERN FATHEAD MINNOW

*Pimephales promelas promelas** Rafinesque

Fig. 81

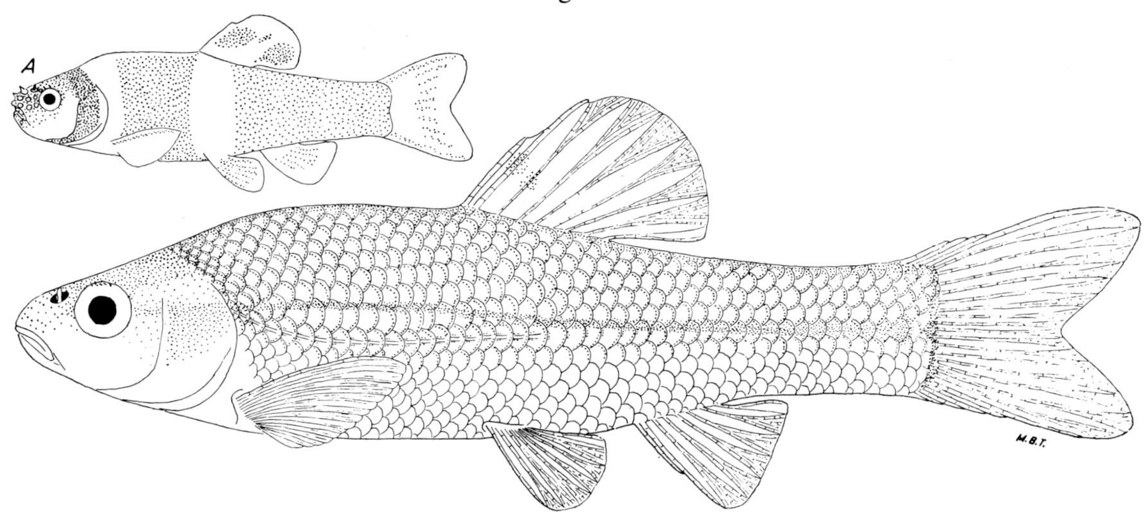

Lower fig.: Six Mile Creek, Auglaize County, O.

July 8, 1941. 49 mm SL, 2.3″ (5.8 cm) TL.
Adult male. OSUM 3266.

Fig. A: Hussy Creek, Auglaize Co., O.

July 8, 1941. 50 mm SL, 2.4″ (6.1 cm) TL.
Breeding male. OSUM 3250.
Note breeding tubercles and breeding color pattern.

Identification

Characters: *General—See* bullhead minnow. *Specific*—Lateral line *incomplete* (may be complete in specimens in some areas outside Ohio). Mouth terminal and decidedly oblique. Peritoneum brownish-black. Intestine long and coiled one or more times. Dusky band usually absent on snout and opercles, except in some small young; lateral band across body distinct only in some small young, and in adults from clear and/or weedy waters. Body depth usually contained less than 4.5 times in standard length. A narrow, dark, vertical bar often present at base of caudal fin.

Differs: Bluntnose minnow has a subterminal and almost horizontal mouth; a bold lateral band; complete lateral line; a more silvery coloration. Bullhead minnow has a complete lateral line; dark crescent on side of snout; short intestine; silver peritoneum. *See* bullhead minnow.

Most like: Bluntnose and bullhead minnows. **Superficially like:** Creek chub, redbelly dace and pugnose minnow.

Coloration: Dorsally olive or olive-yellow, tinged with copperish or purplish in the larger fishes, and yellow-silvery in small young. Sides lighter with more yellow or tan. *See* under Characters for description of lateral band and caudal bar. Ventrally whitish-yellow or silvery. Fins tinged with yellow-olive, sometimes with silvery; dorsal fin with a dusky spot on each of the two or three anteriormost rays, these spots absent in smallest young. *Breeding male*—Has a blackish head except for the lighter cheeks and light tubercles; a golden-copper band encircling body immediately behind head, another such band beneath the dorsal fin; remainder of body darker olive with copperish and purplish reflections, fig. A. Snout of large breeding male contains

* For geographical variations, *see* Taylor, 1954: 42.

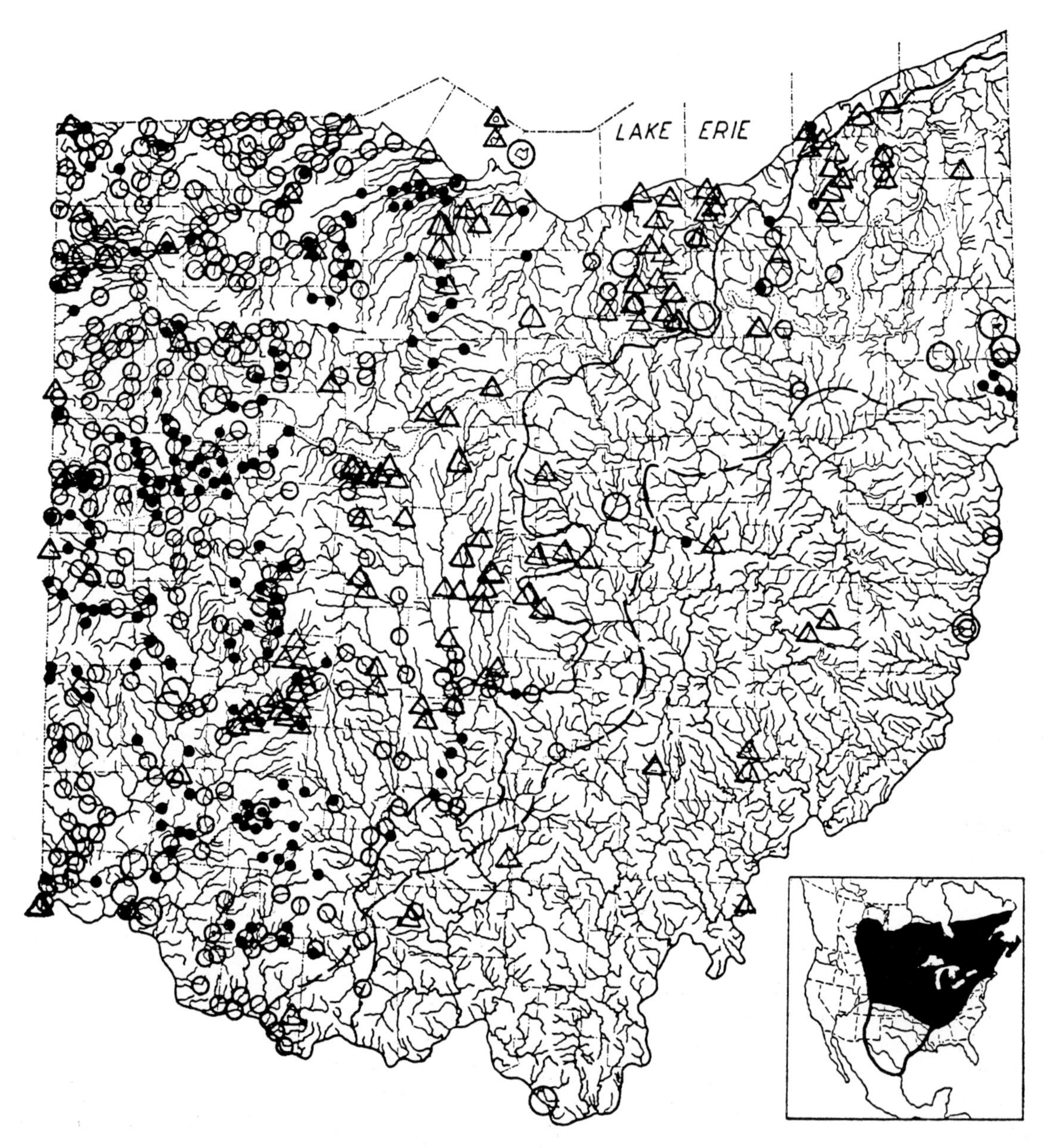

MAP 81. Northern fathead minnow

Locality records. ○ Before 1901. ○ 1901–38. ● 1939–54. △ 1955–80. ~ Allegheny Front Escarpment. /\ Glacial boundary.

Insert: Black area represents range of this subspecies, outlined area of another; zone of intergradation between the two subspecies appears to be rather indefinite as do the limits of the range in Mexico; widely introduced by bait dealers, fishermen and organizations throughout the United States outside its original range.

two rows of sharply-pointed tubercles, with a few tubercles between the two rows suggesting a third row; also two tubercles along each side of lower jaw, fig. A. The thick spongy predorsal pad on back is well developed, as it is in the bluntnose (pad used by male for cleaning the eggs which the female had deposited on the under side of stones and other objects); this spongy mass so thick as to completely hide many scales. Dorsal surface of anterior pectoral rays contain many microscopic, sharply-pointed serrations or tubercles.

Lengths: *Young* of year in Oct., 0.5″ (Aug.–Sept.

hatchings)–2.5″ (May-June hatchings) (1.3–6.4 cm) long. Around **1** year, 1.0″–3.0″ (2.5–7.6 cm). *Adults*, usually 1.6″–3.0″ (4.1–7.6 cm). Largest specimen, 3.5″ (8.9 cm) long. For life history and propagation, *see* Markus, 1934:116–222; Hubbs and Cooper, 1936: 83–87; Wascko and Clark, 1948: 1–16.

Hybridizes: with bluntnose minnow.

Distribution and Habitat

Ohio Distribution—Previous to 1901 the fathead presumably was generally distributed in Ohio only in the two westernmost tiers of counties; elsewhere it appeared to have been present only in relict colonies along the Ohio River and in northeastern Ohio. It was first recorded for the state about 1840 by Kirtland (1841B: 467) who found three specimens in a spring-run on his farm in Boardman Township, Mahoning County (Trumbull County in 1840).

Agricultural practices since 1901, resulting in increased turbidity, increased silting of stream bottoms, and/or reduction in numbers of other fish species and individuals in some waters, have aided the establishment, increase and abundance of the fathead elsewhere in glaciated Ohio. Recent plantings by the conservation department and escapes from the minnow buckets of fishermen, have aided likewise in establishing the species in some localities.

By 1926, the fathead had invaded, or had greatly increased numerically in the third westernmost tier of counties; by 1935 it had become well established in central Ohio, especially in Franklin and Union counties; by 1950 it had become well established and abundant in many of the Sandusky Bay tributaries of northcentral Ohio. In the 1925–50 period it failed to become widely distributed east of the Allegheny Front Escarpment, despite plantings of the species by the conservation department and inadvertent plantings from the minnow buckets of fishermen.

In the future it may become numerous in some impoundments and farm ponds in unglaciated territory. Since 1925 this important forage and bait minnow has been propagated by the conservation department and bait dealers, and hundreds of thousands were seined annually from Ohio streams to be used and sold as bait.

Habitat—The fathead minnow was most abundant in muddy brooks, small muddy creeks, ponds and small lakes, where population pressure from other fish species was not great. The fathead and bluntnose minnows were competitors, and the fathead occurred in greatest population densities *only* where the bluntnose was absent or comparatively few in numbers. The fathead was extremely tolerant to both very clear and very turbid waters. It was also very tolerant to extremes in pH; some of the largest populations occurred in highly turbid, alkaline streams where few other fish species, especially the bluntnose, were present, and also in very clear, highly acid bog-ponds which contained few other fish species and individuals. The fathead was uncommon or absent in streams of moderate and high gradients, and in most of the larger and deeper impoundments.

Years 1955–80—The range and centers of abundance of the fathead minnow in Ohio west of the glacial boundary and the isolated populations to the east of that boundary appear to have remained largely unchanged since 1952. During the present period the species was collected only a few times in unglaciated territory east of the boundary (White et al., 1975: 83–84).

Many thousands of fatheads were propagated and sold annually, and the species was found in the bait containers of fishermen in every part of the state. Many of these minnows escaped into public waters, were released by fishermen or were introduced into newly created impoundments. It is therefore surprising that, despite such introductions, the species did not become more widely dispersed in unglaciated Ohio.

BLUNTNOSE MINNOW

Pimephales notatus (Rafinesque)

Fig. 82

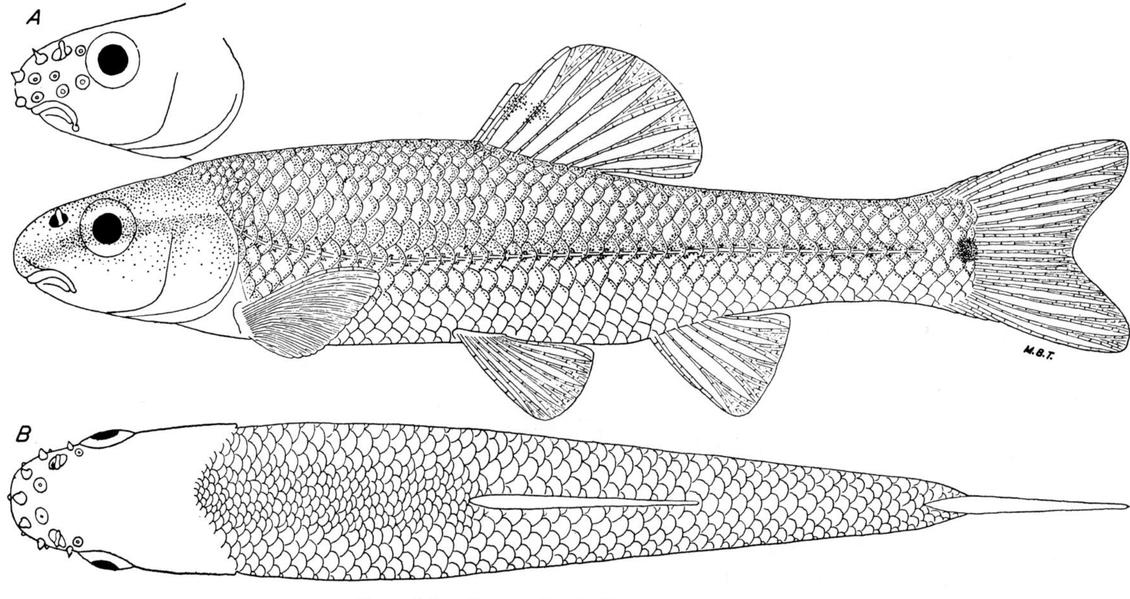

Central fig.: Ottawa Creek, Hancock County, O.

May 14, 1941. Male. 60 mm SL, 2.8″ (7.1 cm) TL. OSUM 3157.

Figs. A and B: Same collection as above.

Breeding male. 67 mm SL, 3.2″ (8.1 cm) TL.

Fig. A: lateral view of head; note breeding tubercles; also the tubercle-like papilla at posterior angle of mouth.
Fig. B: dorsal view; note body shape, breeding tubercles, crowding and irregularities of the predorsal scales.

Identification

Characters: *General—See* bullhead minnow. *Specific*—Lateral line *complete.* Snout slightly overhangs the almost horizontal mouth. Peritoneum brownish-black. Intestine long and coiled one or more times. Dusky band usually conspicuous as it encircles snout and crosses opercles; lateral band rather wide and conspicuous, as are the black caudal spot and cross-hatching of scales of the back and sides. Body depth usually contained more than 4.5 times in standard length.

Differs: Fathead minnow has a terminal and oblique mouth; incomplete lateral line; a more yellowish-golden coloration. Bullhead minnow has a dark crescent on side of snout; short intenstine; silvery peritoneum. *See* bullhead minnow.

Most like: Fathead and bullhead minnows. **Superficially like:** Bigeye and gravel chubs, and blacknose shiner.

Coloration: Dorsally yellowish-, greenish-, or bluish-olive, heavily overlaid with silvery; scales dark-edged giving back and sides a pronounced cross-hatched appearance. A dusky band usually present around snout and crossing opercles, extending backward as a conspicuous lateral band to the very black caudal spot. Specimens from clear and/ or vegetated waters are boldly marked; those from turbid waters often pale, lacking the band and cross-hatching. Ventrally silvery- and milk-white. Fins transparent in young, more olive in adults; the dorsal normally with a dusky blotch on each of the two or three anterior-most rays in adults. The combination of cross-hatching; back flattened predor-

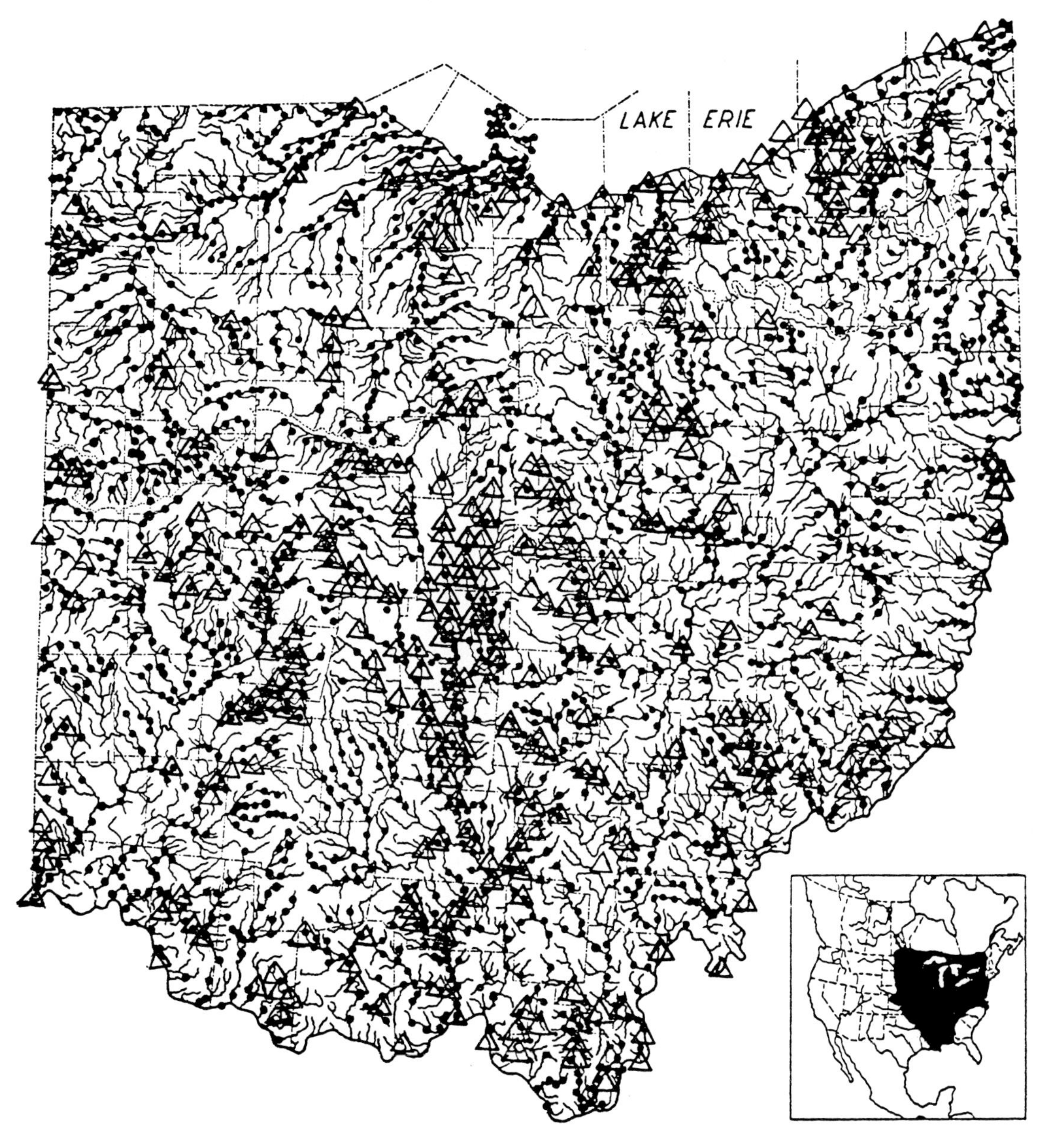

MAP 82. Bluntnose minnow

Locality records. ● Before 1955. △ 1955–80.
Insert: Widely introduced outside its original range, both inadvertently by fishermen and by introductions through the efforts of individuals and organizations.

sally and with crowded scales; lateral band; subterminal mouth; general silvery coloration makes this abundant minnow readily recognizable. *Breeding female*—May have a suffusion of yellow or light-tan. *Breeding male*—Has a blackish head and dusky-bluish body anteriorly; three rows of breeding tubercles across snout which in large males are sharp-pointed. Large males also have a fleshy barbel-like papilla at the posterior end of the upper jaw, figs. A and B. Spongy pad on flattened, predorsal portion of back well developed; so thick as sometimes to hide the scales completely, *see* flathead minnow, under Coloration. Dorsal surface of anterior pectoral rays contain many microscopic, sharply-pointed tubercles.

Lengths: *Young* of year in Oct., 0.5″ (August-

September hatchings) (1.3–7.1 cm)–2.8″ (May hatchings) (1.3–7.1 cm) long. Around **1** year, 1.0″–3.3″ (2.5–8.4 cm). *Adults*, usually 1.6″–3.5″ (4.1–8.9 cm). Largest specimen, 4.3″ (11 cm) long. For life history and propagation, *see* Van Cleave and Markus, 1929:530–39; Hubbs and Cooper, 1936: 83–87; Wascko and Clark, 1948:1–16.

Hybridizes: with fathead and bullhead minnows.

Distribution and Habitat

Ohio Distribution—From the many preserved specimens collected before 1900 and the testimony of early writers, it is apparent that this most important forage and bait minnow was as widely distributed throughout Ohio waters before 1900 as it has been recently. Since 1930 it has been taken in more than 2,000 collections, more often, in larger numbers and in more localities, than has any other species of fish; after 1925 it was propagated by the conservation department and distributed throughout Ohio; it was also propagated by bait dealers. Millions of bluntnose minnows were seined annually from the streams and sold and used as bait.

Habitat—The bluntnose was a most plastic species, occurring in all types of waters, except the deeper waters of Lake Erie and the Ohio River. Its outstanding distributional and numerical success appeared to be the result of such favorable factors as its ability to survive, at least in small numbers, in areas where there was severe competition from a large number of other species and/or of individuals; ability to inhabit waters whose gradients ranged from 0′–100′/mile (0m/km–19m/km); its high degree of tolerance to turbidity and other organic and inorganic pollutants; its ability to inhabit and to spawn in all types of waters except the deepest and largest; its habit of migrating up the smallest, temporary brooks to spawn from which a large proportion of adults and young succeeded in returning to deeper waters before these brooks stopped flowing; its method of spawning, in which the female deposited her eggs on the underside of stones, logs, tin cans, boards, leaves, and other objects where they could not be readily silted over; the male's method of guarding the eggs, and his frequent cleaning of the eggs with the spongy tissue on his nape which kept the silt from suffocating them; the long spawning season which extended from early April to early September (mostly May to July).

The largest populations occurred in streams and lakes of moderate size which drained lands rich in organic matter and had high phytoplankton populations, and where the fathead minnow was absent and the number of other species and individuals was rather low. It occurred in the smallest numbers wherever its close relative, the fathead minnow, was abundant; in waters where turbidity and rate of siltation were excessive; where the pH was extremely alkaline or acid; where the stream gradients were more than 50′/mile (9.5 m/km); where there was an unusually large concentration of fishes of other species which preyed upon it or were its competitors.

Years 1955–80—This ubiquitous species was captured at every collecting station in inland Ohio which supported fish life, except upon very rare occasions. Normally it was present in moderate numbers and was seldom abundant.

The species was widespread and often common in the shallower waters of Lake Erie (White et al., 1975:84) and especially in the island region of the western third of the lake (Van Meter and Trautman, 1970:72). It was moderately numerous in the Ohio River (ORSANCO, 1962:148).

OHIO STONEROLLER MINNOW

Campostoma anomalum anomalum (Rafinesque)

Figs. 83 and 84

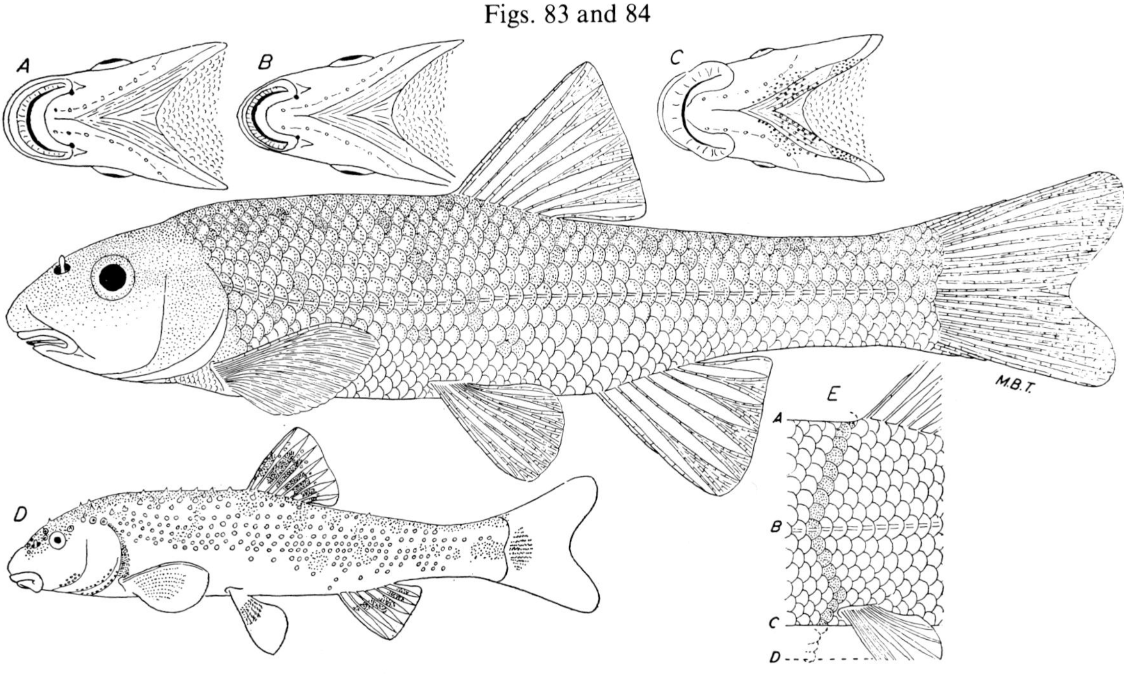

Central fig.: Sunfish Creek, Monroe County, O.

Sept. 29, 1939. 77 mm SL, 3.7″ (9.4 cm) TL.
Male. OSUM 737.

Fig. A: ventral view of head, showing the large mouth of the non-breeding Ohio stoneroller.

Figs. C, D and E: East Fork, Buck Creek, Champaign County, O.

April 18, 1942. 115 mm SL, 5.7″ (14 cm) TL.
Breeding male. OSUM 4786.

Fig. C: Ventral view of head, showing the enlarged lips of the breeding male Ohio stoneroller.

Fig. D: lateral view, showing tubercle distribution, and color pattern of the fins of the Ohio stoneroller. Central stoneroller has similar tubercle distribution and color pattern.

Fig. E: dorsal-pelvic fin region, showing method of counting the scales encircling body immediately before dorsal and pelvic origins. The line of darkened scales is counted, beginning at the mid-dorsal scale (A), counting downward along one side of body to lateral line (B), continuing to pelvic origin (C) and on to mid-line of belly (D), thence around the other side of body in the same manner to, but not including, the mid-dorsal scale which had been counted previously.

Fig. B: West Branch, St. Joseph River, Williams County, O.

Oct. 24, 1939. 60 mm SL, 3.0″ (7.6 cm) TL.
Male. OSUM 1218.

Ventral view of head of non-breeding central stoneroller; note the narrow mouth and more pointed snout.

Identification

Characters: *General—See* golden shiner. *Generic and specific*—A cartilaginous sheath replaces the fleshy lips which cover the jaws in most species of minnows, *Central Figure*; this sheath widest and most conspicuous on the lower jaw, fig. B. Mouth subterminal and almost horizontal. Alimentary tract contained 1.5 (smallest young)–9.0 (adults) times in length of head and body; intestine usually coiled about the air-bladder (*see* Kraatz, 1924:265–98). Peritoneum black. Basic coloration

MAP 83. Ohio stoneroller minnow

Locality records.
○ No Ohio stoneroller influence apparent during spawning season.
○ Mostly intergrades between Ohio and central stonerollers.
● Localities from which large series of intergrades were examined.
● Ohio stonerollers during spawning season.
△1955–80. ~ Glacial boundary.
/\ Flushing Escarpment.
Insert: Black area represents range of this subspecies, outlined area of other subspecies.

tan or light brownish, with little or no silvery. Darkened scales present over most of back and sides; often no distinct lateral band; no distinct caudal spot. Anal rays 7. Teeth 4–4. *Subspecific*—Scales in complete lateral line usually 45–50, extremes 43–51. Scale rows around body immediately before dorsal and pelvic origin, usually 38–42, extremes 35–43, fig. E. Width of gape usually 3.8–5.0, extremes

3.6–5.1, times in length of head, fig. A; gape width greatest in breeding males, fig. C. Intergrades abundantly with central stoneroller.

Differs: Central stoneroller has a greater average number of scales and a smaller gape. All other species of minnows in Ohio have the *upper* jaw covered with a fleshy lip; lack coiling of the intestine about the air bladder. Suckers have well-developed lips and more than 8 dorsal rays.

Most like: Central stoneroller. **Superficially like:** Tonguetied minnow; young of hornyhead and river chubs; small suckers, especially the white sucker.

Coloration: Dorsally olive- or yellowish-tan with the silvery sheen faint or absent. Sides lighter and more yellowish. Many dark scales scattered, usually singly or in pairs, over back and sides. Ventrally whitish or yellowish. Fins transparent in small young; olive or yellow in the larger fishes. General coloration of large young and non-breeding adults tan or muddy-yellow; small young more silvery; those coming from clear and/or weedy waters may have a dusky lateral band. *Breeding male*—Bright yellow-tan with much tan, orange and black in the fins, fig. D; lips a livid white, hence the colloquial name of "Tallowmouth Minnow." Sharply pointed and whitish tubercles cover most of the dorsal surface of head and body, those on dorsal fin, branchiostegals, and elsewhere are smaller, fig. D.

Lengths: *Young* of year in Oct., 1.1″–2.5″ (2.8–6.4 cm) long. Around **1** year, 1.3″–3.0″ (3.3–7.6 cm). *Adults*, usually 2.0″ (breeding Females)–6.0″ (usually breeding males) (5.1–15 cm). Largest specimens, 7.0″(18 cm) long. Males normally longer than are the females.

Hybridizes: with hornyhead and river chubs, longnose and redbelly daces, striped and common shiners.

Distribution and Habitat

Ohio Distribution—After reviewing the literature, Osburn (1901:90) stated that the stoneroller as a species was a very abundant and widely distributed minnow in Ohio prior to 1900. Since 1920, the species has been present in every Ohio stream capable of sustaining fish life. It was most abundant in those stream sections, anywhere in the state, where the bottom harbored an abundance of micro-plants and small animals upon which the stoneroller fed; it was rare or absent only in streams where the oxygen had been depleted, or where pollutants or clay siltation had greatly decreased the amount of its food. Only small populations were present in the Ohio River, and it was not taken in Lake Erie.

Although specimens having the characters of the Ohio stoneroller have been taken throughout Ohio, except in the extreme northwestern section, evidence of intergradation was virtually absent only in unglaciated Ohio. It was completely absent in breeding adults only east of the Flushing Escarpment, and in a narrow band along the Ohio River in unglaciated territory. Intergradation between breeding adults was marked northwest of a line that extended from Butler County northeastward through Pickaway, Ashland to Ashtabula counties; intergradation became pronounced, although still leaning towards this subspecies, in the headwaters of the Wabash, Miami and Scioto river systems of the Ohio River drainage, and the headwaters of the Cuyahoga, Chagrin and Grand in the Lake drainage. In northeast Ashtabula County, the Ohio stoneroller occurred in almost typical form north almost to Lake Erie. Intergradation elsewhere in the Lake drainage is discussed under central stoneroller.

As with most stream fishes in Ohio, considerable downstream movement occurred after the spawning season, resulting in the spawning intergrades of the headwaters moving downstream as far south as Pickaway County, to winter in the spawning range of the Ohio stoneroller.

Habitat—The largest populations were present where there was available the combination of many small streams of moderate and high gradients having sandy-gravel bottoms in which to spawn, moderate sized streams of moderate gradients in which to summer, and lower gradients in the larger streams in which to winter. It was essentially an inhabitant of riffles, and in the streams of low gradient was virtually restricted to riffles; however, it became a pool species in those streams of very high gradients in which there was considerable current in the pools.

Intergradation.— The Ohio stoneroller, subspecies *anomalum*, differs from the central stoneroller, subspecies *pullum*, in having lower average scale counts and a wider gape (*see* key for Characters). Since the characters of the two forms widely overlapped, an accurate indication of subspecific status of a population in any section could be obtained only from examinations of two series, each of 15 or more specimens, one taken in the spawning season, the other in winter. The distribution maps of the two subspecies show the spawning ranges of each subspecies (solid circles) and of intergrades (hollow circles).

Years 1955–80—As a species, the stoneroller minnow was captured in almost as many seining localities throughout inland Ohio as was the bluntnose minnow (White et al., 1975:84). As with the latter, it was usually present in moderate numbers and was seldom rare or abundant at any collecting station.

The typical subspecies of the Ohio stoneroller, *see* above, continued to dominate the southern and northeastern portions of Ohio, including the Ohio River. In that stream it was fairly numerous (ORSANCO, 1962:146).

The central stoneroller or its intergrades, *see* below, continued to occupy the Maumee River system and the Lake Erie tributaries to the east.

CENTRAL STONEROLLER MINNOW*

Campostoma anomalum pullum (Agassiz)

* For figure of this species *see* figs. 83 and 84

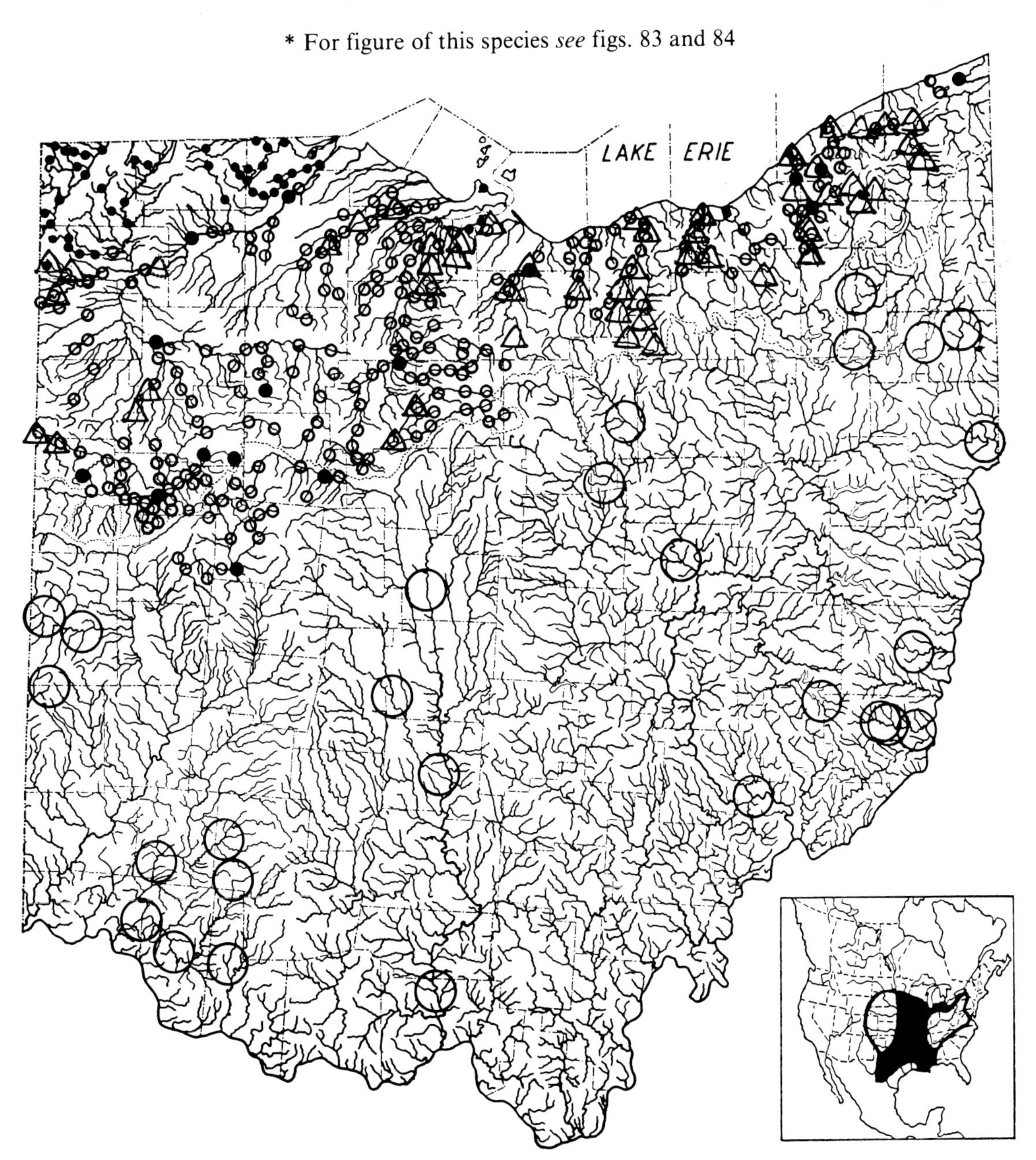

MAP 84. Central stoneroller minnow

Locality records.
○ No central stoneroller influence apparent during spawning season.
○ Mostly intergrades between central and Ohio stonerollers.
● Localities from which large series of intergrades were examined.
● Central stonerollers during spawning season.
△ 1955–80.
Insert: Black area represents range of this subspecies, outlined areas of other subspecies.

Identification

Characters: *General, generic* and *specific—See* Ohio stoneroller. *Subspecific*—Scales in complete lateral line usually 50–56, extremes 49–64. Scale rows around body immediately before dorsal and pelvic origins usually 40–50, extremes 39–54, fig. E. Width of gape usually 4.8–5.4, extremes 4.6–5.6, times in length of head, fig. B; gape width greatest in breeding males.

Differs: Ohio stoneroller has fewer average number of scales and a wider gape; *see* Ohio stoneroller.

Most like: Ohio stoneroller. **Superficially like:** Tonguetied minnow; young of hornyhead and river chubs; small suckers.

Coloration: Similar to that of Ohio stoneroller; *see* that subspecies.

Lengths: *Young* of year in Oct., 1.2″–2.6″ (3.0–6.6 cm) long. Around **1** year, 1.3″–3.2″ (3.3–8.1 cm). *Adults*, usually 2.0″ (breeding females)–6.3″ (usually breeding males) (5.1–16 cm). Largest specimen, 7.5″ (19 cm) long.

Hybridizes: *see* under Ohio stoneroller minnow.

Distribution and Habitat

Ohio Distribution—Populations with only central stoneroller characters occurred in a narrow band north of the Maumee River, in the harbors of the Ottawa County peninsula, and in a few brooks in Erie and Sandusky counties which emptied into Sandusky Bay. Elsewhere throughout the Lake drainage, and in a small portion of the upper Miami River system of the Ohio River drainage, spawning populations were composed mostly of intergrades which leaned toward this subspecies. Occasional specimens of apparently typical central stonerollers were found throughout the intergrading territory, and especially in the Maumee River after the spawning season.

The central stoneroller, whose range is westerly, presumably invaded the Lake drainage through the Maumee and/or other glacial outlets, penetrating across the present watershed into the upper portions of the Ohio River tributaries. The Ohio stoneroller presumably was present in part of the upper Ohio River drainage during glacial times, penetrating into both drainages of glaciated Ohio after the last glacial retreat. The meeting of the two subspecies has produced a wide, triangular-shaped area of intergradation, with its base in southwestern and apex in northeastern Ohio.

Habitat—Similar to that of the Ohio stoneroller.

BLUE SUCKER

Cycleptus elongatus (Lesueur)

Fig. 85

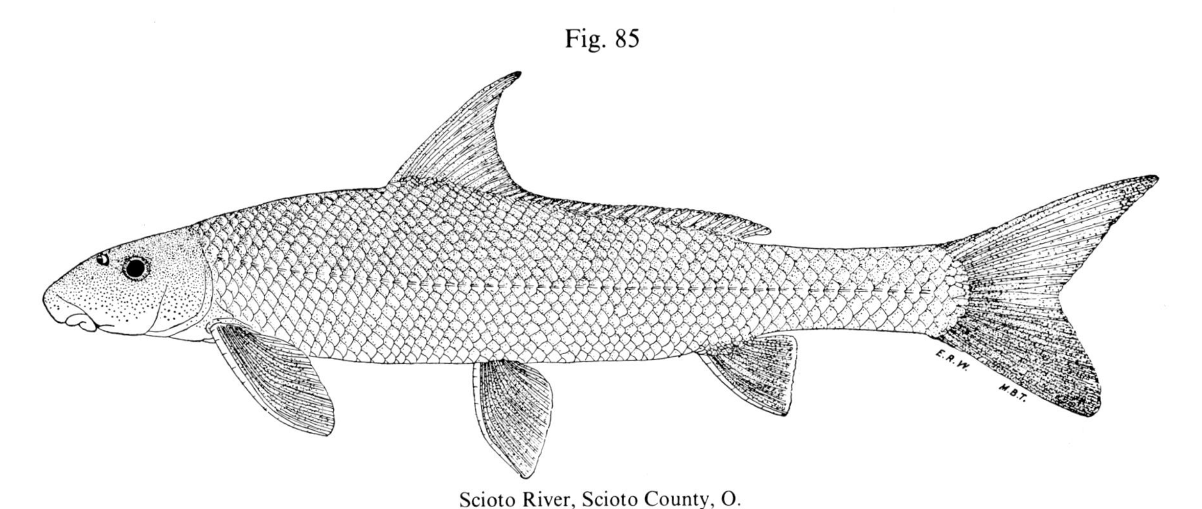

Scioto River, Scioto County, O.

Jan. 12, 1940.
Immature male.

185 mm SL, 9.4″ (24 cm) TL.
OSUM 3009.

Identification

Characters: *General*—The fleshy lips are papillose, plicated or striated. Anal fin far back; distance from its anterior insertion to caudal base usually more than 1.8 times the distance from anal fin insertion to posterior edge of opercle. Teeth only in throat, borne on two pharyngeal arches, each arch with 15 or more teeth. Cycloid scales on body. No scales on head. No spines in fins. No adipose fin. No pelvic axillary process. *Specific*—Scales in the complete lateral line, 53–59. Dorsal fin of 30–37 rays. Eye closer to posterior edge of opercle than to snout tip. Body depth 4.0 times or more in standard length. Lips strongly papillose. Color dark bluish.

Differs: Other suckers have fewer dorsal rays, or if they have as many rays, are deeper bodied. Carp and goldfish have a dorsal and anal spine. Walleyes and blackbasses have many dorsal spines.

Most like: Black buffalofish. **Superficially like:** Black redhorse, white and longnose suckers.

Coloration: Dark olive-blue or bluish-slate dorsally; bluish-white ventrally, the lips white. All fins very dark slaty-blue. Breeding *adults* blue-black.

Lengths and weights: *Young* of year, Oct. to Jan., 5.0″–9.5″ (13–24 cm) long, weight 1.0–3.5 oz (28–99 g). October fishes with one annulus, 12.0″–14.0″ (30.5–35.6 cm) long, 9.0 oz–12.0 oz (255–340 g). *Adults*, 16.0″–19.0″ (40.6–48.3 cm) long, usually weigh 1 lb, 8 oz–2 lbs, 8 oz (0.7–1.1 kg); 20.0″–30.0″ (50.8–72.2 cm) long, weight 3 lbs–10 lbs (1.4–4.5 kg), with ripe females the heaviest. Rivermen give maximum lengths at 36.0″–40.0″ (91.4–102 cm), weights 12 lbs–15 lbs (5.4–6.8 kg).

Distribution and Habitat

Ohio Distribution—The blue sucker has been present throughout the 19th century in the Ohio River from the Indiana state line to the Pennsylvania state line, and during that period has been considered as a rather common, well-known, and highly-valued food fish. Rafinesque (1820:119) found it before 1820 "in the Ohio as far as Pittsburgh;" Kirtland (1851E:349) before 1851 as "rather common in the Ohio River at Cincinnati, during certain seasons, especially the Spring and Autumn;" Kirtland also stated that it was preferred to any other sucker species as an article of food, and that it "is never found in Lake Erie or its tributaries." Henshall (1888:77) considered it to be "Not uncommon in the Ohio River" before 1888.

Between 1925–50 the blue sucker was most frequently taken in the Ohio River between the Indiana state line and city of Portsmouth, and in the lower portions of the Great Miami, Scioto, and

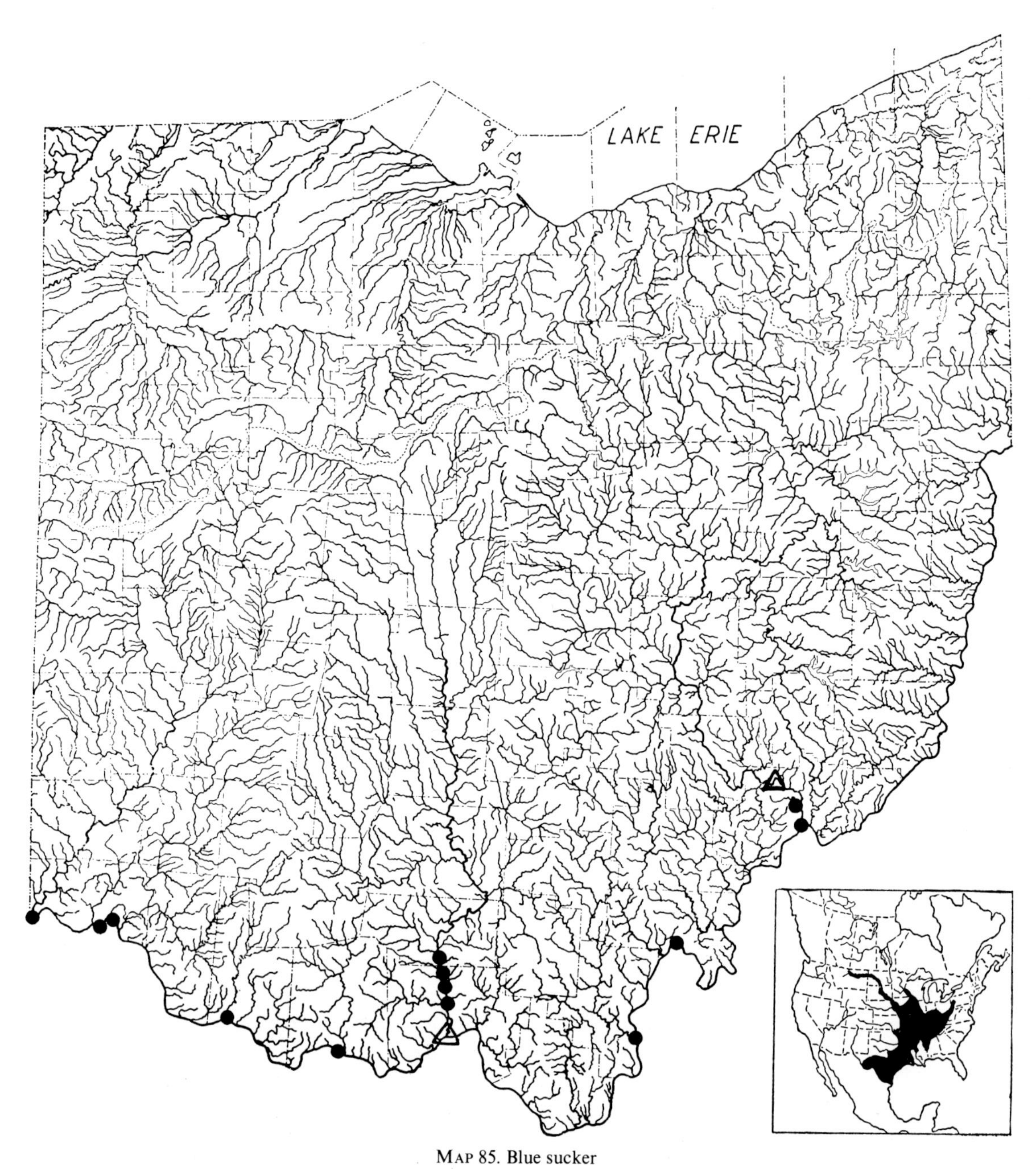

MAP 85. Blue sucker

Locality records. ● Before 1955. △ 1955–80.
Insert: May ascend western rivers farther upstream than indicated; Mexican range limit indefinite.

Muskingum rivers. It was taken less frequently in the Ohio River between Portsmouth and Marietta. Although I saw no specimens, I interviewed several fishermen who had taken an occasional specimen of this distinctive sucker during the 1925–50 period in the Ohio River between the city of Marietta and the Pennsylvania state line. These fishes were taken usually in spring during a high stage of water when industrial pollution was at its minimum.

Habitat—Principally an inhabitant of the deeper waters of the Ohio River, especially those channels and pools having a moderate current, and of similar situations in the lower portions of its tributary streams. It was captured chiefly in trap nets and with trot lines during its spring and fall migrations.

The rarity of this species in the Missouri River and other highly turbid streams, and its abundance in the clearer, cleaner sections of the Ohio and Mississippi rivers, argues for its preference for clear waters. Its recent apparent decrease in abundance in the Ohio River, similar to its decrease in the Illinois River of Illinois (Forbes and Richardson, 1920:66) suggests its intolerance to turbidity and other pollutants.

Years 1955–80—During this period I was informed by personnel of the Ohio Division of Wildlife and others that an occasional specimen was taken in inland Ohio, with an experimental trapnet or otherwise, in the lower portions of the Scioto and Muskingum rivers. This species is seldom taken by fishermen using hook and line.

Personnel of the University of Louisville captured the species only in five of 341 collections in the entire Ohio River (ORSANCO, 1962:67 and 148).

BIGMOUTH BUFFALOFISH

Ictiobus cyprinellus (Valenciennes)

Fig. 86

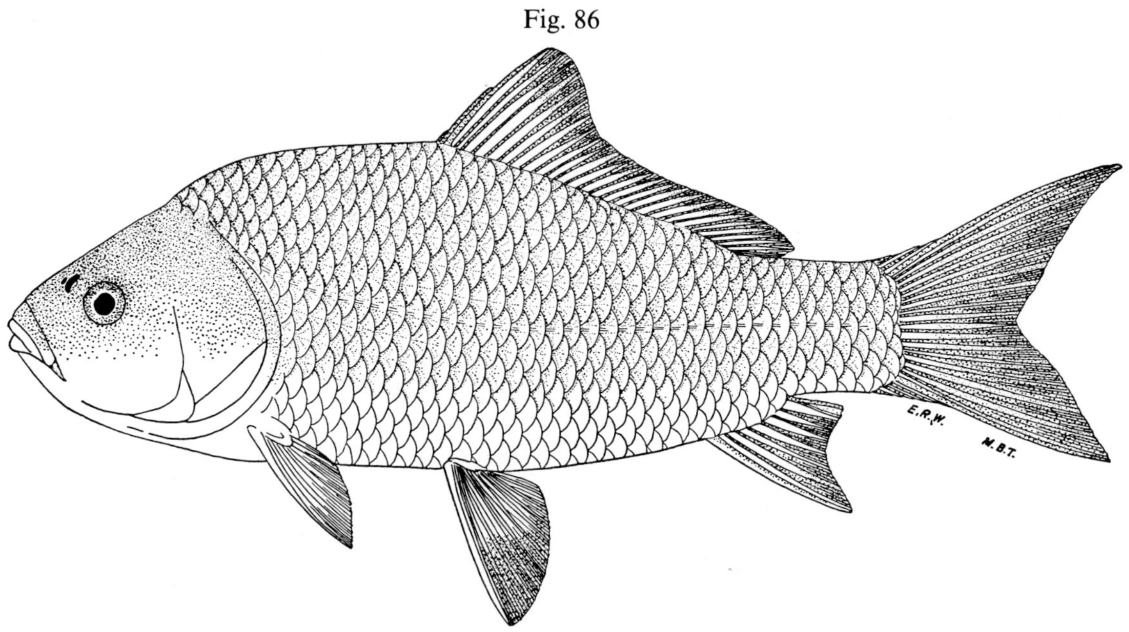

Overflow pond beside the Scioto River, Franklin County, O.

July 9, 1928.
Immature.

210 mm SL, 10.6″ (26.9 cm) TL.
OSUM 117.

Identification

Characters: *General—See* blue sucker. *Generic—* Dorsal fin of 24–32 rays. Subopercle broadest at its middle, with its free (posterior) edge forming an even curve. In specimens over 7.0″ (18 cm) long the distance from eye to free angle of preopercle usually contained 1.0 or more times in the distance from eye to upper angle of gill cleft. Anterior fontanelle closed or much reduced; posterior fontanelle open. Lateral line scales 35–43. Coloration brownish. *Specific*—Mouth more terminal than in any other species of sucker; mouth very oblique and large. Tip of upper lip about on the level with lower edge of eye. Lips thin and only faintly striated.

Differs: Other suckers have subterminal mouths. Carp, goldfish, walleyes, white bass, blackbasses and drum have one or more dorsal spines.

Most like: Other buffalofishes and carpsuckers, carp and brown-colored goldfish.

Coloration: Slaty- or olive-bronze dorsally; sides lighter and more olive-yellow; yellow and whitish ventrally. Fins uniformly light brownish-slate. Fishes from turbid waters often very pale and yellowish; from very clear waters often quite olive-blue.

Lengths and **weights:** *Young* of year in Sept., 1.7″–4.0″ (4–10 cm) long. Around **1** year, 5.0″–7.0″ (13–18 cm). *Adults*, usually 15.0″–30.0″ (38.1–76.2 cm) long, weight 2 lbs–30 lbs (0.9–13.6 kg). Rivermen give maximum lengths of 35.0″–40.0″ (88.9–102 cm), weights 40 lbs–60 lbs (18.1–27.2 kg). I weighed one of 43 lbs (19.5 kg), taken in the lower Muskingum River. Outside Ohio waters Harlan and Speaker (1951:60) report weights up to 80 lbs (36.3).

Hybridizes: with smallmouth buffalofish.

Distribution and Habitat

Ohio Distribution—The bigmouth buffalofish was not definitely recorded for Ohio until 1878 (Klippart, 1878:109–10), because of inability of early naturalists to recognize this species. In 1882, Jordan (1882:805–06) mentioned that it "abounds

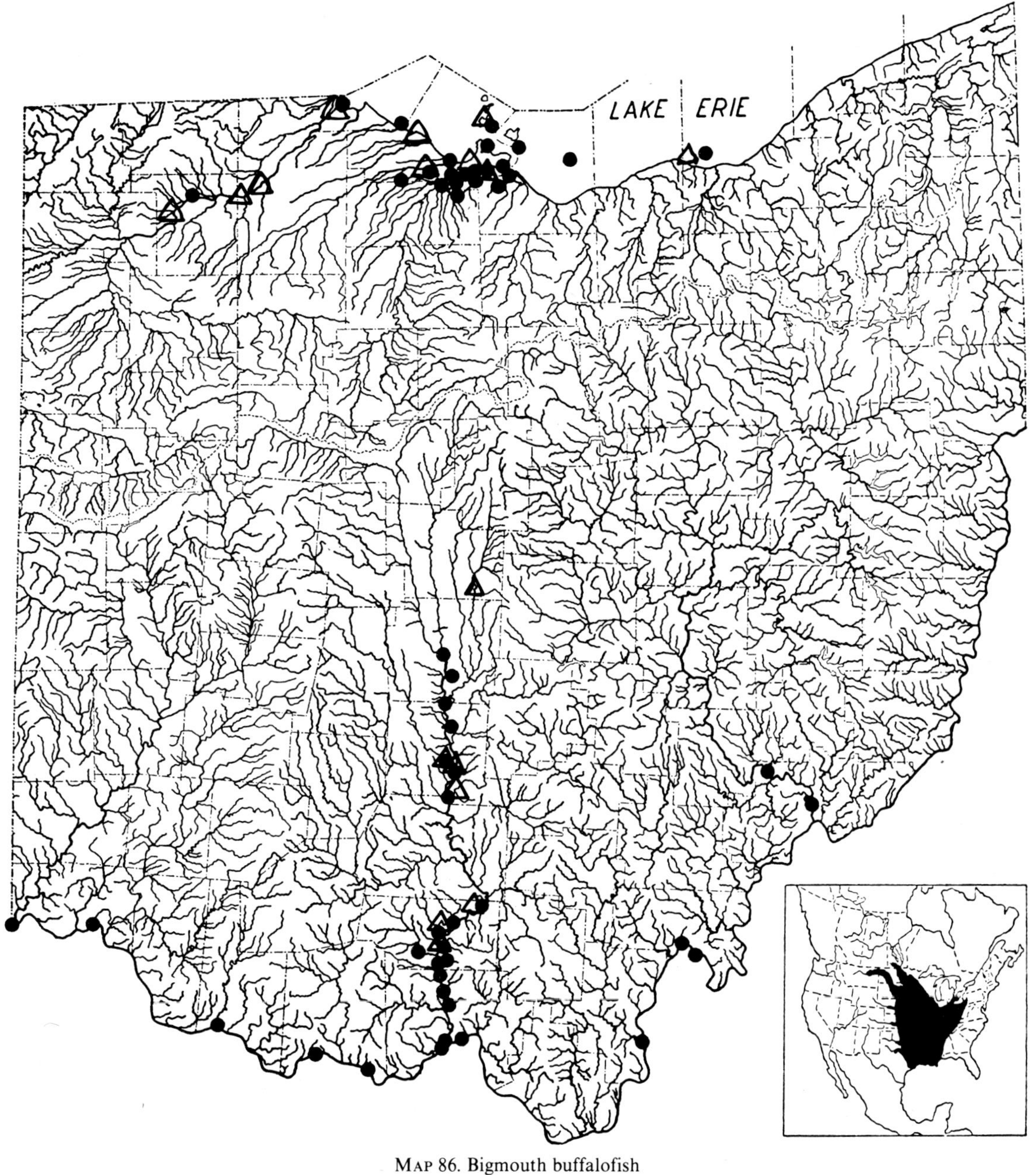

MAP 86. Bigmouth buffalofish

Locality records. ● Before 1955. △ 1955–80.

Insert: Original range limits somewhat uncertain because of confusion of this species with other buffalofish species, and because of early introductions; western limits of range indefinite.

in the Ohio River and its larger tributaries." Later Henshall (1888:77) stated that the species was very common in the Ohio River.

Between 1925–50 the species was taken in larger numbers than were either of the other two species of buffalofishes. In the Ohio drainage it was most numerous in the Ohio River from Cincinnati to Portsmouth, and in the Scioto River and its overflow ponds from Portsmouth to Columbus. It occurred in smaller numbers in the Ohio River between Portsmouth and Marietta and in the lower Muskingum River. Although I saw no specimens, reliable river fishermen accurately described this species to me, stating that since 1900 they had occasionally taken it in the Muskingum River as far upstream as Zanesville, and in the Ohio River between Marietta and the Pennsylvania state line.

According to the older fishermen the species has maintained, or increased, its numbers in the Ohio River since 1890; I have noted no drastic change in abundance since 1925.

The bigmouth apparently was at least sporadically present in Lake Erie before 1900, for there is a specimen in the Harvard Museum of Comparative Zoology, collected by Kirtland at Rockport, Cuyahoga County, Ohio (Hubbs, 1930B:427) in November of 1854. About 1920, plants of buffalofishes were made by the federal government in western Lake Erie and Sandusky Bay; by 1924 bigmouth buffalofishes were being taken about South Bass Island and Sandusky Bay (Hubbs, 1926:19); since 1924 their numbers have increased so that in 1949 more than 5,000 lbs (2,268 kg) (Shafer, 1950:2) were taken in western Lake Erie, and fishermen reported annually capturing a few as far east as Cleveland. By 1942 it was present in the Maumee River as far upstream as Henry County. It is surprising that after 1854 there were no more records until 1920, after plantings were made by the federal government.

Habitat—The bigmouth buffalofish was often abundant in the shallow, turbid, overflow ponds and lowland lakes along the large southern Ohio streams, and like the two other species of buffalofishes, it was also numerous in the deep pools of the Ohio River and its larger tributaries. The bigmouth appeared to be very tolerant of turbid water, and seemingly was a direct competitor with the carp, with which species it associated. In Lake Erie waters the bigmouth occurred in greatest abundance in the upper half of Sandusky Bay, where the waters were often extremely turbid. In spring, from late March and until May, it spawned in the tributary streams of Sandusky Bay, even ascending the smallest ditches, and individuals, 25 lbs (11.3 kg) or more in weight, were captured in shallow ditches which were only a few feet in width.

The sustained abundance of the bigmouth in the Ohio River since 1925 suggests that this species, with its large, terminal mouth (and therefore probably different feeding habits) may be better suited to present, turbid water conditions and rapid silting of the stream bottoms than is the smallmouth buffalofish with its smaller and inferior mouth.

Years 1955–80—The species appears to have maintained its numerical abundance throughout the 1955–80 period, showing no marked changes in its Ohio distributional pattern. Small populations appear to have become established in western Lake Erie and in Hoover Reservoir, eastern Delaware County.

It was captured in 23 of 341 collections in the entire Ohio River (ORSANCO, 1962:149) during the 1957–59 surveys.

BLACK BUFFALOFISH

Ictiobus niger (Rafinesque)

Fig. 87

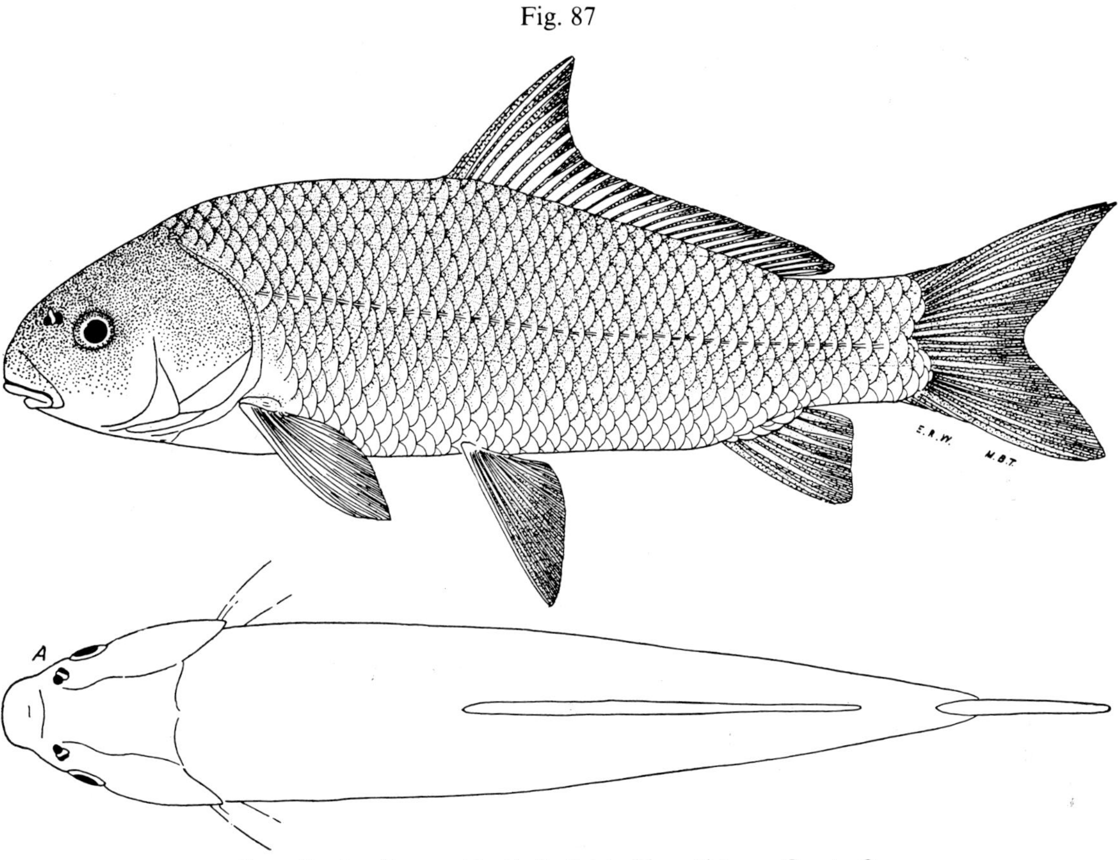

Upper fig.: Overflow pond beside the Scioto River, Pickaway County, O.

Aug. 10, 1929. 276 mm SL, 13.7″ (34.8 cm) TL.
Immature male. OSUM 364.

Fig. A: dorsal view; note width of head between gill covers.

Identification

Characters: *General—See* blue sucker. *Generic—See* bigmouth buffalofish. *Specific*—Mouth subterminal and almost horizontal; the tip of the lower lip far below the lower margin of eye. Lips moderately full and striated. Eye rather small; in fishes over 7.0″ (18 cm) long the eye is contained 5.1–7.4 times in head length; 2.0–2.5 times in snout length, and is usually equal to, or shorter than, length of upper jaw. Head thick at opercles; the greatest width at opercles usually contained 4.7–5.5 times in standard length, fig. A. Body usually more slender than in any other species of buffalofish; its depth in standard length usually contained 2.9–3.4 times, extremes 2.6–3.6. Fishes less than 12.0″ (30.5 cm) long may be very difficult to separate from the smallmouth buffalofish.

Differs: *See* bigmouth buffalofish. Smallmouth buffalofish usually has a shorter upper jaw in comparison to eye length, and a greater body depth. Bigmouth buffalofish has a terminal mouth.

Most like: Smallmouth buffalofish. **Superficially like:** Carpsuckers, carp and brown-colored goldfish.

Coloration: Slaty-blue or slaty-bronze dorsally with a greenish overcast. Sides more olive-brown. Yellow and white ventrally. Fins a uniform slate or

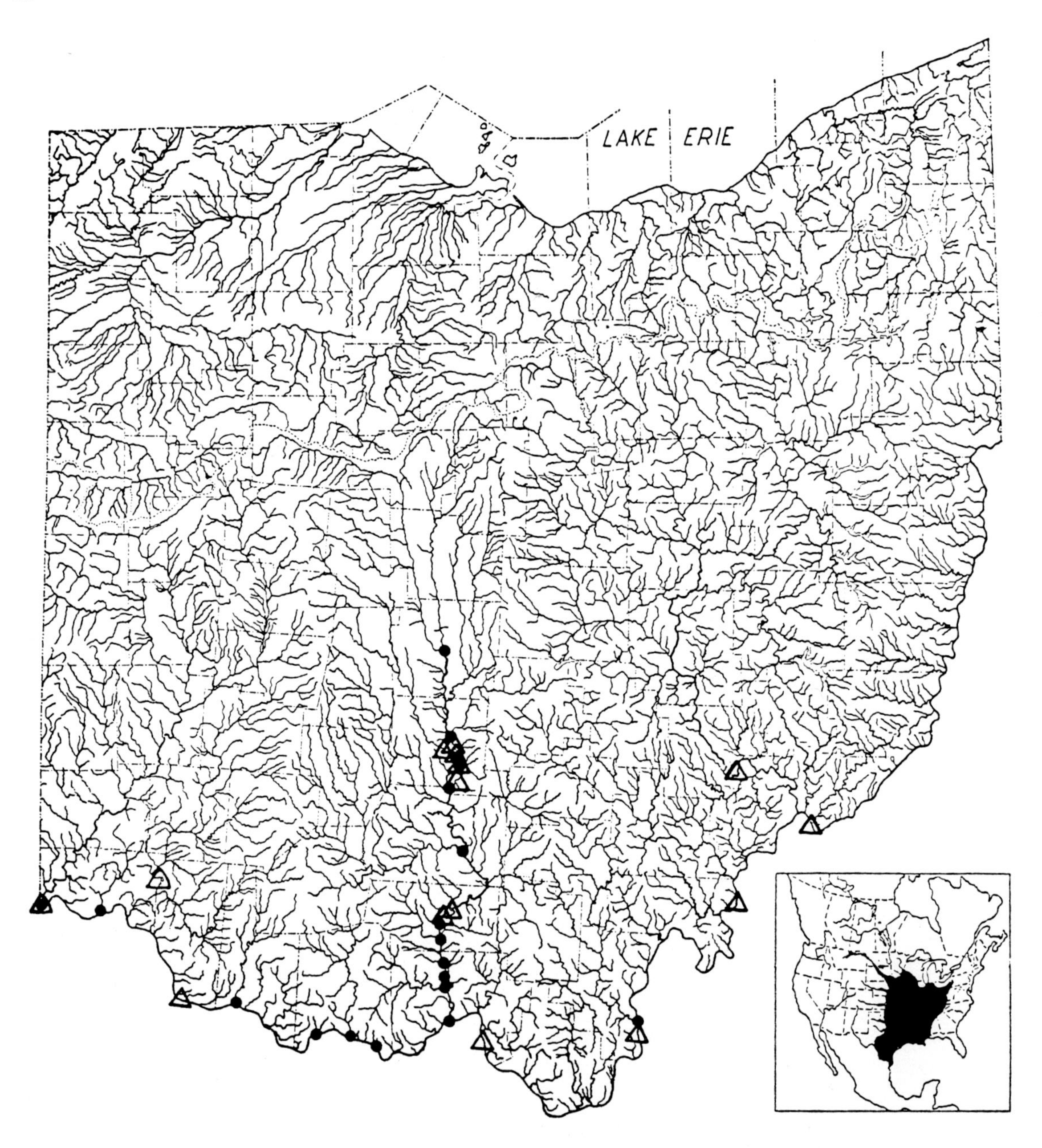

MAP 87. Black buffalofish

Locality records. ● Before 1955. △ 1955–80.
Insert: Introduced into lower Great Lakes and elsewhere outside of its original range.

slaty-olive. Normally the darkest and most slaty-colored of the buffalofishes; light-colored only when taken from very turbid waters. Spring *males* often blackish.

Lengths and **weights:** *Young* of year in Sept., 1.5″–4.0″ (3.8–10 cm) long. Around **1** year, 5.0″–8.0″ (13–20 cm). *Adults*, usually 15.0″–26.0″ (38.1–66.0 cm) long, weight 1 lb–10 lbs (0.5–4.5 kg). Largest specimen, 28.0″ (71.1 cm) long, weight 15 lbs, 8 oz (7.0 kg).

Distribution and Habitat

Ohio Distribution—Formerly the black buffalofish was thoroughly confused with the smallmouth buffalofish; therefore little is known concerning the presence of the former in Ohio waters before 1890. Those few fishermen who I believe were capable of recognizing this species during the 1900–25 period, insist that the black buffalofish was as numerous in the Ohio River before 1925 as it has been since. Since 1925, I have found the black to be almost as numerous as was the bigmouth buffalofish in the Ohio River between Gallipolis and the Indiana state line, and in the Scioto River as far upstream as Columbus. A few fishermen claimed that they have taken an occasional specimen in the Ohio River between Gallipolis and the Pennsylvania state line, usually during spring floods.

Habitat—Black buffalofishes were frequently found in association with the other two buffalofishes; however, its preferred habitat seemingly is intermediate between the preferred habitats of the bigmouth and that of the smallmouth. Occasionally it was abundant in turbid, mud-bottomed, shallow, overflow ponds and sloughs. Harry Brookbank, of Brown County, and other reliable river fishermen stated that during the summer, following the great Ohio River flood in the spring of 1937, the black buffalofish was more abundant than it had been previously.

Some commercial fishermen of the Ohio River considered the black buffalofish to be the hybrid between the bigmouth and smallmouth buffalofishes, because the black buffalofish usually occupied an intermediate habitat and its external morphological characters approach intermediacy, especially when young. I have seen a few specimens, less than 15.0″ (38.1 cm) in total length, which appeared to be intermediate in characters and these may have been hybrids.

Years 1955–80—The Scioto River population showed no changes in numerical abundance during the 25-year period. In 1962 two test nets set for 4 days in the Scioto River, short distances above and below the mouth of Big Darby Creek, Pickaway County, by Paul Scowden, Timothy Hood and Ray Riethmiller, members of the Ohio Department of Natural Resources, and I produced one to 12 specimens, 12″ to 18″ (30–46 cm) in length; per day, per net. On June 14, 1973, Jerry Boyd and Dave Christman, members of the Ohio Department of Natural Resources, lifted experimental trap nets in the Muskingum River, Luke's Chute, Morgan County, Douglas Albaugh and I accompanying them. One of the nets contained a black buffalofish. Recently a specimen was reported captured in the Little Miami River, Hamilton and Clermont counties.

During the 1957 to 1959 Ohio River investigations by the personnel of the University of Louisville, only seven specimens were recorded for the entire river (ORSANCO, 1962:149). During his 1968 and 1969 investigations of the river, H. Ronald Preston captured the species in four of his nine collecting stations bordering the State of Ohio. Apparently the species has decreased in numbers in the river since 1950.

SMALLMOUTH BUFFALOFISH

Ictiobus bubalus (Rafinesque)

Fig. 88

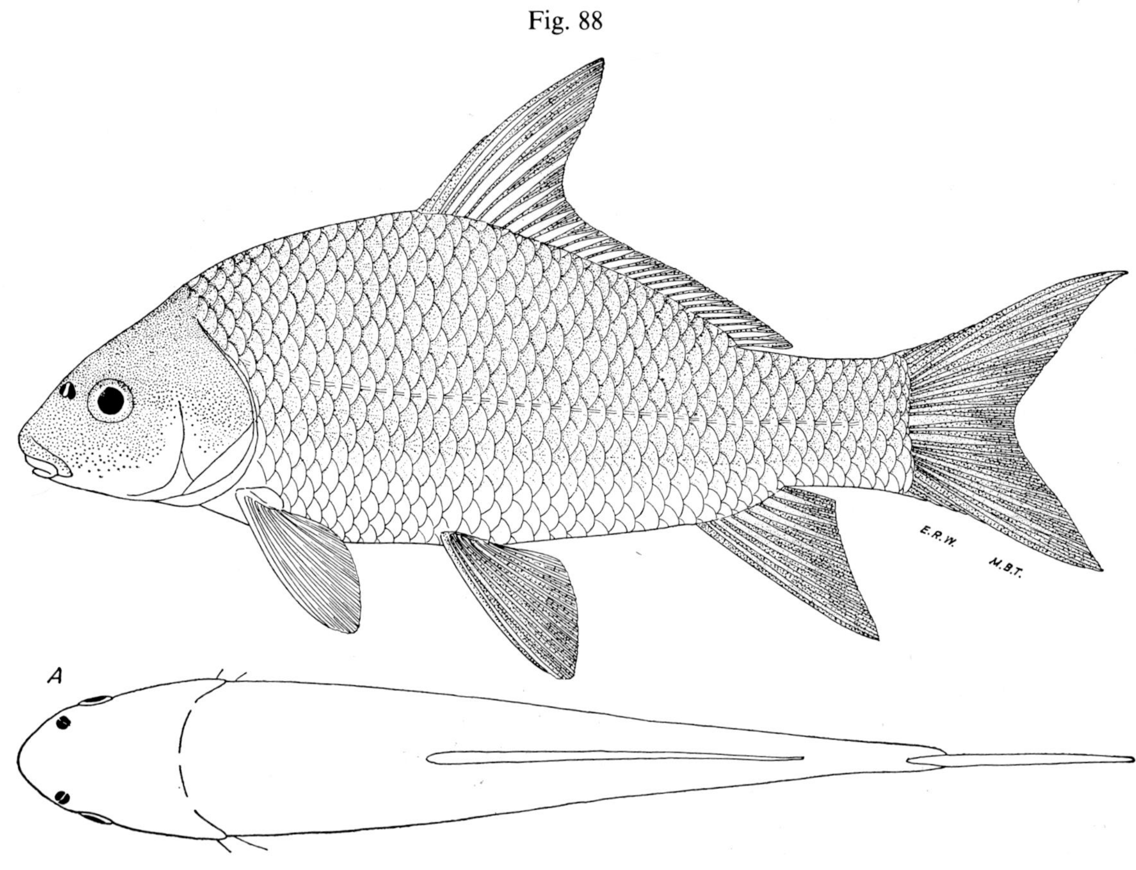

Upper fig.: Scioto River, Scioto County, O.

April 7–15, 1940. 231 mm SL, 12.3″ (31.2 cm) TL.
Immature. OSUM 1857.

Fig. A: dorsal view; note head shape.

Identification

Characters: *General—See* blue sucker. *Generic—See* bigmouth buffalofish. *Specific*—Mouth subterminal and almost horizontal, the tip of the lower lip far below the lower margin of the eye. Lips moderately full and striated. Eye moderately large; in fishes more than 7.0″ (18 cm) long the eye is contained 4.4–5.9 times in head length; 1.5–2.0 times in snout length, and usually is equal to, or longer than, the length of the upper jaw. Head moderately thick at opercles; the greatest width at opercles usually contained 5.2–6.1 times in standard length, fig. A. Body deep, especially in large fishes; its depth in standard length usually contained 2.4–2.8 times, extremes 2.2–3.0 times. Fishes less than 12.0″ (30.5 cm) may be very difficult to separate from the black buffalofish.

Differs: *See* bigmouth buffalofish. Black buffalofish usually has a longer upper jaw in comparison to eye length, and a more slender body.

Most like: Black buffalofish. **Superficially like:** Carpsuckers, carp and brown-colored goldfish.

Coloration: *Young* and small *adults*—Slaty-bronze or slaty-olive dorsally, sometimes with a bluish overcast. Sides lighter, more bronze or golden. Yellowish or white ventrally. Fins uniformally brownish-slate. As the species grows larger it

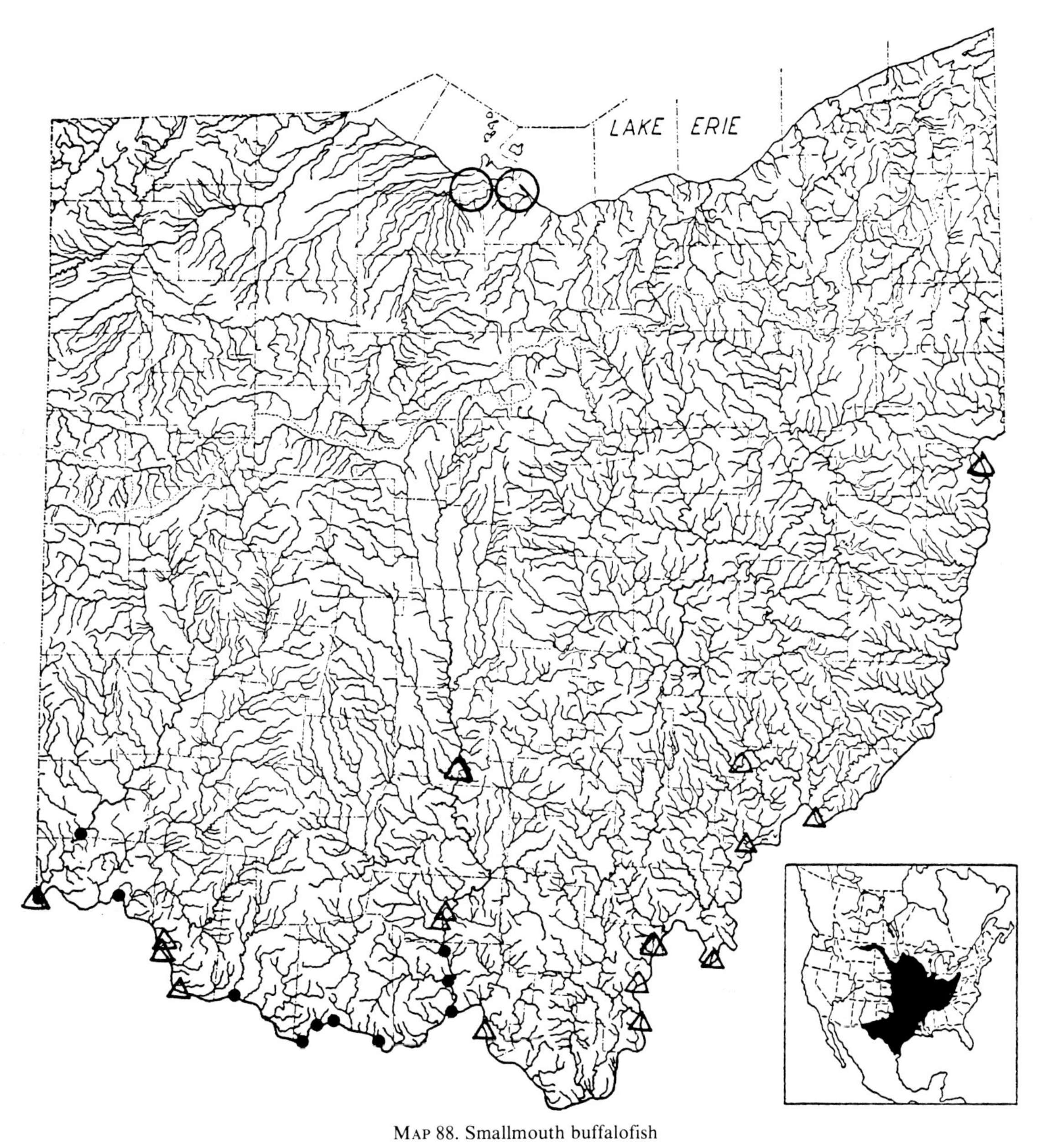

MAP 88. Smallmouth buffalofish

Locality records. ○ Hybrids only. ● Before 1955. △ 1955–80.
Insert: Introduced into lower Great Lakes and elsewhere outside of its original range.

tends to become the lightest colored of all the buffalofishes and especially if taken from turbid waters; these individuals may be a light yellow-olive dorsally with considerable silver-white ventrally.

Lengths and **weights:** *Young* of year in Sept., 1.5″–4.0″ (3.8–10 cm) long. Around **1** year, 5.0″–7.5″ (13–19 cm). *Adults*, usually 15.0″–28.0″ (38.1–71.1 cm) long, weight 1 lb, 8 oz–20 lbs (0.7–9.1 kg). River fishermen give maximum lengths of 30.0″–38.0″ (76.2–96.5 cm), weights 30 lbs–40 lbs (13.6–18 kg). Schoffman (1944:3–9) records a Redfoot Lake, Tenn., specimen as 33.0″ (83.8 cm) long, weight 25 lbs., 8 oz. (11.6 kg).

Hybridizes: with bigmouth buffalofish.

Distribution and Habitat

Ohio Distribution—In 1820 Rafinesque (1820: 112) described what presumably was the smallmouth buffalofish, giving it the name of *Catostomus bubalus*; but his description is so vague as to cast doubt upon whether he had the smallmouth or one of the other species of buffalofishes in mind (Hubbs, 1930A:12). In the description Rafinesque mentioned that the smallmouth was a common species in the Ohio River and was taken as far upstream as Pittsburgh. Kirtland's (1851W:341) statement that in 1845 the smallmouth inhabited the Ohio River is unquestioned because his drawing and verbal description clearly indicate that he had this high-backed species in mind. In 1878 Kpippart (1878:111) mentioned that it "formerly was exceedingly abundant in the Scioto and some other streams emptying into the Ohio," but that by 1878 it had become of "less frequent occurrence." In 1888 Henshall (1888:77) stated that it was abundant in the Ohio River.

During the 1925–50 period I saw many specimens which were taken in the Ohio River from the Indiana state line to Portsmouth and in the lower portion of the Scioto and Great Miami rivers. Those fishermen who correctly recognized this species told me of one or more individuals which they had taken in the Ohio River between Portsmouth and the Pennsylvania state line prior to 1930. It was the almost universal opinion of these fishermen that the smallmouth had decreased in abundance during the past 50 years and especially after the installation of dams in the Ohio River. They believed that the ponding of the Ohio River, subsequent reduction of current and elimination of "chutes" had been adverse factors.

Between 1939–50 many fishermen reported capturing in Sandusky Bay an occasional buffalofish which was deep bodied and had the undershot mouth of a quillback, and which obviously was not a bigmouth buffalofish. On May 4, 1950 fishery workers and I took one of these high-backed individuals (OSUM 9129) of about 14.0″ (35.6 cm) in total length from an experimental net which was set near the mouth of Sandusky Bay in Erie County. On October 13–14, 1952 Robert Cummins and Warren Handwork took six specimens (OSUM 3919) of between 10.5″–18.0″ (26.7 cm–45.7 cm) in length, in the upper part of Sandusky Bay in Ottawa County. Those seven specimens have been studied critically by Carl L. Hubbs and myself. We consider them to be, beyond any reasonable doubt, hybrids between the bigmouth (subgenus *Megastomatobus*) and the smallmouth or black buffalofishes (subgenus *Ictiobus*) for they clearly show the intermediate characters between these subgenera. In the hybrids, the width of the base of the pharyngeal arch is intermediate between the thin, frail base of the bigmouth and the widely-flanged, heavy base of the other two species; the length of the teeth on the arch is intermediate between the long, thin teeth of the bigmouth and the shorter, heavier teeth of the other two; the length and number of gill rakers on the first and second gill arches are intermediate between the longer and more numerous gill rakers of the bigmouth and the shorter and fewer gill rakers of the other two. Our specimens are hybrids between the bigmouth and the smallmouth (and not the black buffalofish) because of the sharp predorsal ridge and deep body of the hybrids; such a deep-bodied hybrid could only result from a cross in which one parent was the deep-bodied smallmouth. Hybrids between the bigmouth and black buffalofishes would have the roundish bodies of their parents.

In the past 25 years Carl Hubbs and I identified many buffalofishes without finding a hybrid, although we were constantly on the lookout for them. The hybridization of buffalofishes in Sandusky Bay can be explained logically as follows: A small population of bigmouths was present, at least sporadically, in Sandusky Bay before 1920. Between 1920–30 plants of buffalofishes were made in western Lake Erie waters which presumably consisted of bigmouths and smallmouths. The shallow, often turbid waters of Sandusky Bay offered a suitable habitat for the bigmouth, but conditions were submarginal for the smallmouth which normally inhabits deeper, clearer waters having a current. Consequently the bigmouths increased in abundance, the smallmouths did not, and some of the few smallmouths spawned with the abundant bigmouths thereby producing hybrids. We have noted the presence of hybrids, or of mass hybridization, following the introduction of a species not native into a body of water containing a closely related species which had been long established, and especially when the conditions were favorable for one species and submarginal for the other.

Habitat—The smallmouth usually inhabited in large numbers only the deeper, swifter and, whenever possible, the clearer waters of the larger rivers. During floods it tended to remain in the permanent river channels and did not wander over flooded lands in numbers as did the other two species, and it was seldom captured in the shallow, turbid and smaller overflow ponds left by retreating waters. Fishermen stated that the largest numbers of smallmouths were taken in the Ohio River during those years in which the waters remained clearer than normal for long periods of time.

Years 1955–80—The abundance and distributional pattern of the smallmouth buffalofish in inland Ohio and in the Ohio River appears to have remained unchanged except that a few specimens were taken on August 18 and 19, 1972 at Davenport Pond, southern Pickaway County, by a group from John Carroll and Ohio State universities. As it was previous to 1950, this sucker was the most numerous of the three species of buffalofish inhabiting the Ohio River. In their 1957–59 investigations of the entire Ohio River personnel of the University of Louisville captured this species in 83 of the 341 localities investigated, with a total of 504 specimens, whereas only 155 bigmouth buffalofish and seven black buffalofish were taken (ORSANCO, 1962:148–49). In his 1968–69 investigations of the Ohio River, H. Ronald Preston captured the smallmouth buffalofish in six of his nine localities bordering the state of Ohio.

EASTERN QUILLBACK CARPSUCKER

Carpiodes cyprinus cyprinus (Lesueur)

Fig. 89

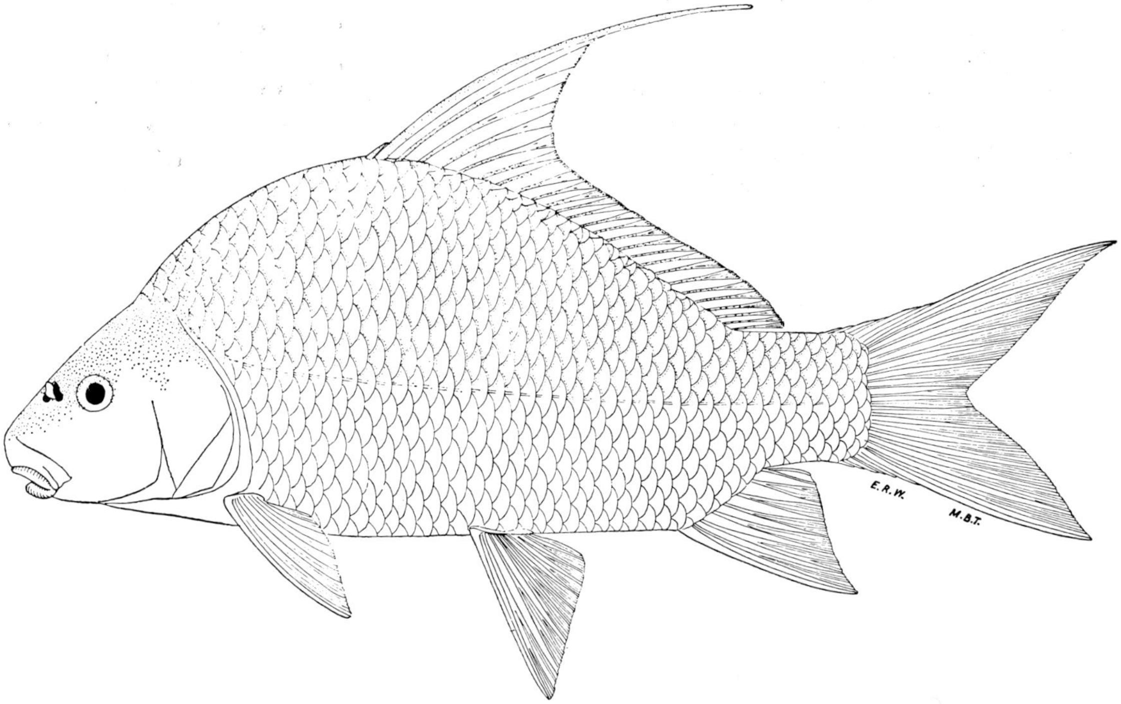

E.R.W. M.B.T.

Lake Erie, Ottawa County, O.

Summer, 1939.
Immature.

200 mm SL, 10.3″ (26.2 cm) TL.
OSUM 1184.

Identification

Characters: *General—See* blue sucker. *Generic—* Dorsal fin of 22–30 rays. Subopercle broadest *below* its middle, at the free angle of its posterior edge; this angle giving the subopercle a triangular appearance. In specimens more than 7.0″ (18 cm) long the distance from eye to free angle of preopercle usually is contained 1.0 or less times in the distance from eye to upper angle of gill cleft. Anterior fontanelle open as is the posterior fontanelle. Lateral line scales 33–42. Coloration silvery. *Specific—No* small knob present at tip of lower jaw (*see* highfin carpsucker, arrow in fig. A pointing to knob). Tip of lower lip considerably in advance of anterior nostril, with the nostrils above the posterior third of the lower jaw, or behind it. Snout long, usually contained 3.0–3.5 times in head length. When not broken off, the longest anterior rays of the depressed dorsal fin are longer than half the length of the dorsal fin base. *Subspecific*—Body deep and more sunfish-like in shape; its depth in standard length usually contained 2.3–2.7 times, extremes 2.2–3.4. Eye in large young and adults usually smaller, and contained 5.4–7.0 times extremes 4.0–8.8, in head length; eye larger in young less than 4.0″ (10 cm) long and is usually contained more than 4.5 times in head length; eye length in snout length usually 1.6–2.7 times in large young and adults, 1.2–2.7 in small young. Caudal peduncle depth in standard length usually 5.9–7.9 times with an average of 6.9. Small young are difficult or impossible to separate from small central quillbacks (Trautman, 1956:38).

Differs: Central quillback has a larger eye and more slender body. Other carpsuckers have a knob

MAP 89. Eastern quillback carpsucker

Locality records. ● Before 1955. △ 1955–80. ~ Warren Beach Ridge.
Insert: Black area represents range of this subspecies, outlined area of another; northwestern range limits indefinite.

at tip of lower jaw. Buffalofishes have an evenly curved posterior edge to the subopercle *without* a distinct angle, and are less silvery. Carp, goldfish, walleyes, white bass, blackbasses, and drum have one or more dorsal spines.

Most like: Central quillback and other carpsuckers. **Superficially like:** Carp, goldfish and buffalofishes.

Coloration: Bronze-olive dorsally, each scale with silvery, greenish and bluish reflections. Sides whiter and more silvery. Ventrally milk- or yellowish-white. Anterior and distal margins of dorsal fin edged with black; remainder of dorsal, also caudal and anal, light olive-slate. Remaining fins white, sometimes tinged with yellow or orange. Tip of snout and lips often milk-white.

Lengths and **weights:** *Young* of year in Nov., 2.0″–5.0″ (5.1–13 cm) long. *Adults*, usually 12.0″–24.0″ (30.5–60.9 cm) long, weight 14 oz–8 lbs (397g–3.6kg). Lake Erie commercial fishermen give maximum lengths for the "Shad" as 24.0″–26.0″ (60.9–66.0 cm), weights 9 lbs–12 lbs (4.1–5.4 kg).

Distribution and Habitat

Ohio Distribution—Since 1900 the eastern quillback carpsucker has been moderately numerous in the Ohio waters of Lake Erie, occurring in typical form throughout the deeper waters of the eastern two-thirds. Evidence of intergradation between this subspecies and the central quillback was apparent in specimens from the western third of the lake, and especially those specimens which came from the shallower waters and Sandusky Bay. With few exceptions those specimens taken in the lower courses of the Lake Erie tributaries between the Warren Beach Ridge and the Lake, were obviously intergrades.

Authorities differ as to the abundance before 1900 of the eastern quillback in Lake Erie. Kirtland (1847:274–76 and 1851S:373, as *Sclerognathus cyprinus*) considered it to be rare or uncommon, as did McCormick (1892:14, as *C. thompsoni*); but Jordan (1882:811, as *C. thompsoni*) and Jordan and Evermann (1896, Pt. 1:167) believed it was abundant in Lake Erie, particularly in Sandusky Bay and at Toledo.

Considerable confusion existed during the past century relative to the specific and subspecific status of the Lake Erie population. Kirtland (1847:274–76) considered this population to be conspecific with the Ohio River population in 1847. In 1882 Jordan (1882:811) considered the population to be specifically distinct, referring to the Lake Erie population as *C. thompsoni* and that of the Ohio River as *C. velifer*, as also did Henshall (1889:124), McCormick (1892:14) and Jordan and Evermann (1896, Pt. 1:167).

In 1930 Osburn, Wickliff and Trautman (1930:171) considered the Lake Erie drainage and Ohio River drainage populations to be identical, and followed Hubbs (1926:21) in using Lesueur's specific name of *cyprinus*, instead of Rafinesque's *velifer* which had long been used. But in 1934, in a mimeographed list of Ohio fishes, Wickliff and Trautman (1934:1) gave subspecific status to the Lake Erie population, calling it *C. cyprinus thompsoni*, and leaving the Ohio River population as *C. cyprinus cyprinus*; Trautman (1940B:5; 1946:7; 1950:7) followed this procedure in subsequent lists.

In 1941 Hubbs and Lagler (1941:41) believed the two populations to be identical; but in 1947 they (1947:50) gave subspecific status to both populations. In addition they correctly pointed out that specimens from the Atlantic coast were identical with those from the Great Lakes, and therefore *thompsoni* (type locality, Lake Champlain) must be synonymized with the older name *cyprinus* (type locality, Elk River and streams entering Chesapeake Bay). As a result, the Lake Erie population became *Carpiodes cyprinus cyprinus*, the name long used for the Ohio River population.

Habitat—The eastern quillback chiefly inhabited the moderately deep waters of Lake Erie where wave action was not severe and the bottoms were of sand, sandy gravel, sandy silt, mucky silt, or clayey silt. The largest, fastest growing and fattest specimens that I have seen came from waters 15′–25′ (4.6–7.6 m) in depth, in Lorain and Cuyahoga counties, where the bottom seemed to be largely of a sandy silt.

Years 1955–80—This subspecies of the quillback demonstrated little or no change in abundance in Lake Erie during this period (Van Meter and Trautman, 1970; White et al., 1975:86).

CENTRAL QUILLBACK CARPSUCKER

Carpiodes cyprinus hinei Trautman

Fig. 90

Lower fig.: Scioto River, Scioto County, O.

Aug. 14, 1940. 272 mm SL, 14.0″ (35.6 cm) TL.
Female. OSUM 1889.

Fig. A: ventral surface of head; note lack of knob on tip of lower jaw (*see* highfin carpsucker for knob on lower jaw).

Identification

Characters: *General—See* blue sucker. *Generic* and *specific—See* eastern quillback. *Subspecific—* Body moderately slender; its depth in standard length usually contained 2.7–3.0 times, extremes 2.5–3.9. Eye in large young and adults averaging larger, usually contained 4.4–5.6 times, extremes 3.0–5.8, in head length and usually less than 4.5 times in fishes less than 4.0″ (10 cm) long and more than 4.8 times in large adults; eye length in snout length usually 1.2–1.7 times in large young and adults. 1.0–1.3 in small young. Caudal peduncle depth in standard length usually 6.4–8.4 times with an average of 7.4. Small young are difficult or impossible to separate from small eastern quillbacks (Trautman, 1956:38).

Differs: *See* eastern quillback.

Most like: Eastern quillback and other carpsuckers. **Superficially like:** Carp, goldfish, and buffalofishes.

Coloration: Similar to eastern quillback.

Lengths and **weights:** *Young* of year in Nov., 2.0″–5.0″ (5.1–13 cm) long. *Adults*, usually 12.0″–20.0″ (30.5–50.8 cm) long, weight 12 oz–4 lbs (340 g–1.8 kg). Maximum lengths of 24.0″–26.0″ (60.9–66.0 cm) and weights of 8–9 lbs (3.6–4.1 kg) occur in those impoundments, such as Indian and Buckeye lakes, where growth is exceedingly rapid, the fishes very fat and deep bodied. Dwarfing frequently occurs in this very plastic genus, resulting in maximum lengths of 12.0″ (30.5 cm) and weights of 10 oz (283.5 g) in some waters; these individuals approach the body proportions of the plains carpsucker, *Carpiodes forbesi* (*see* Trautman, 1956:39).

Hybridizes: *See* river carpsucker.

Distribution and Habitat

Ohio Distribution—Since 1900 the quillback as a species, has been the most widely-distributed and generally the most abundant species of carpsucker

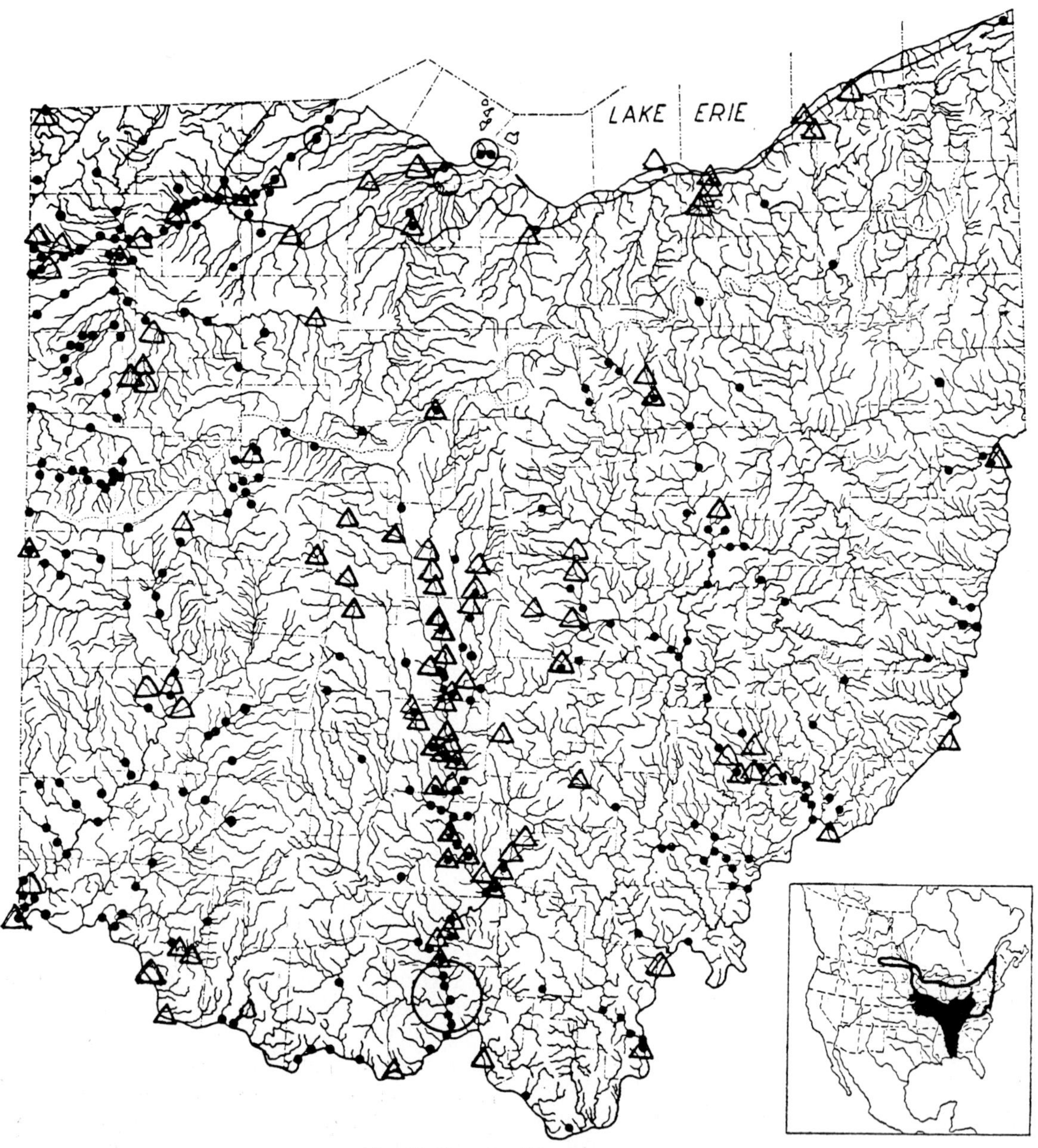

MAP 90. Central quillback carpsucker

○ Type locality, Scioto River, Scioto County.
○ Large series of intergrades examined.
● Records after 1880. △ 1955–80. ~ Warren Beach Ridge.
Insert: Black area represents range of this subspecies, outlined area of another; southwestern range limits indefinite; reports of specimens taken as far southwestward as Oklahoma, and from a few Gulf Coast tributaries.

in the inland waters of Ohio and in Lake Erie. In the Ohio River, however, the present subspecies was taken by commercial fishermen in much fewer numbers than was the river carpsucker, and in slightly fewer numbers than was the highfin. In every county of inland Ohio the quillback was presumably present, at least as strays, but was very numerous only in the low-gradient streams of northwestern and southern Ohio. It occurred in the fewest numbers in the northeastern section of Ohio.

Preserved specimens, collected before 1900, indicate its early presence in the Little Miami and Ohio rivers of southwestern Ohio, in the Scioto drainage of central Ohio, and in various localities in the Maumee drainage of northwestern Ohio; the few literature references which appear reliable, indicate its presence elsewhere in the state. However, no accurate estimate of early abundance for inland Ohio appears possible because of the almost hopeless and extensive confusion existing in the literature before 1926 concerning the various species of carpsuckers (Hubbs, 1926:21). Most literature references previous to 1926 refer to the quillback as *Carpiodes velifer*, the name now applied to the highfin (*see* eastern quillback).

Few North American genera of fresh-water fishes show such diverse morphological types as do the carpsuckers. Presumably these types are the result of environmental conditions, and because of these differences subspecific, and less often specific, identification of some specimens is difficult or impossible.

Central quillbacks, living in large waters of relatively low turbidity having an abundance of food, such as Buckeye Lake, grow very rapidly, are excessively fat, deep-bodied and small-eyed, and because of these characteristics they resemble the eastern carpsucker, often to a remarkable degree. On the other hand individuals from turbid waters containing little food, and others heavily parasitized, grow slowly, are terete and large-eyed and resemble *Carpiodes forbesi*, a form inhabiting turbid streams west of the Mississippi River. In 1934 Wickliff and I (1934:1) included *forbesi* in the Ohio list, as I did in my 1940 Ohio fish key (1940:5). Upon realizing that the few specimens which had been called *forbesi* were in reality emaciated *C. cyprinus*, I omitted this species from the later keys (Trautman, 1946:7; 1950:7).

Habitat—The central quillback chiefly inhabited the lower gradient portions of large- and medium-sized streams, the largest populations occurring in those streams having bottoms of a sandy silt or sandy muck and where the amount of infertile, clayey silt was small. Spawning adults ascended small creeks in spring and have been found in base- and low-gradient, dredged ditches of less than 10′ (3.0 m) in average width. Their young could be found in such low-gradient, small streams, as well as in larger waters, throughout the summer; in some ditches in Paulding and Van Wert counties the young were so numerous that more than a peck (8.8 l) could be taken in one haul of a seine 15′ (4.6 m) in length. Both young and adults began drifting downstream in late summer and early fall, wintering in large waters of low gradient. In winter many of necessity inhabited turbid waters and clayey silt bottoms, presumably because of a lack of clearer waters and cleaner bottoms; but where clear waters and clean bottoms did occur, large concentrations of quillbacks likewise occurred. This subspecies was seldom found among aquatic vegetation (Gerking, 1945:38).

Years 1955–80—Intergradation between the Lake Erie population of *Carpiodes c. cyprinus* and the inland population of *C. c. hinei* appear to be confined to the vicinity of the mouths of streams entering Lake Erie (White et al., 1975:86–87). A few miles above these mouths fairly typical *hinei* may be found in moderate numbers.

Usually there are surprisingly slight morphological differences between specimens of *C. c. hinei* occurring throughout inland Ohio. However, the slowest-growing individuals are quite terete, especially if heavily parasitized, whereas the most rapidly growing individuals are round-bodied, such as those specimens taken from Buckeye Lake. Although specimens taken from Buckeye Lake approach the body depth of *C. c. cyprinus* from Lake Erie, the former are more round-backed whereas the latter are definitely keeled. If *C. c. hinei* were an environmental response, greater variation should be expected within the inland population. I have seen many adult specimens collected by the Illinois Natural History Survey and taken from streams in Illinois which are markedly similar to inland Ohio specimens.

This inland subspecies showed no marked changes in abundance or distributional pattern during this period. The largest populations continued to inhabit the lower gradients of the larger inland streams and in some portions of the Scioto and Maumee systems were abundant.

In 1968 and 1969 H. Ronald Preston collected the species in five of his nine Ohio River collecting stations that bordered the state of Ohio.

NORTHERN RIVER CARPSUCKER

Carpiodes carpio carpio (Rafinesque)

Fig. 91

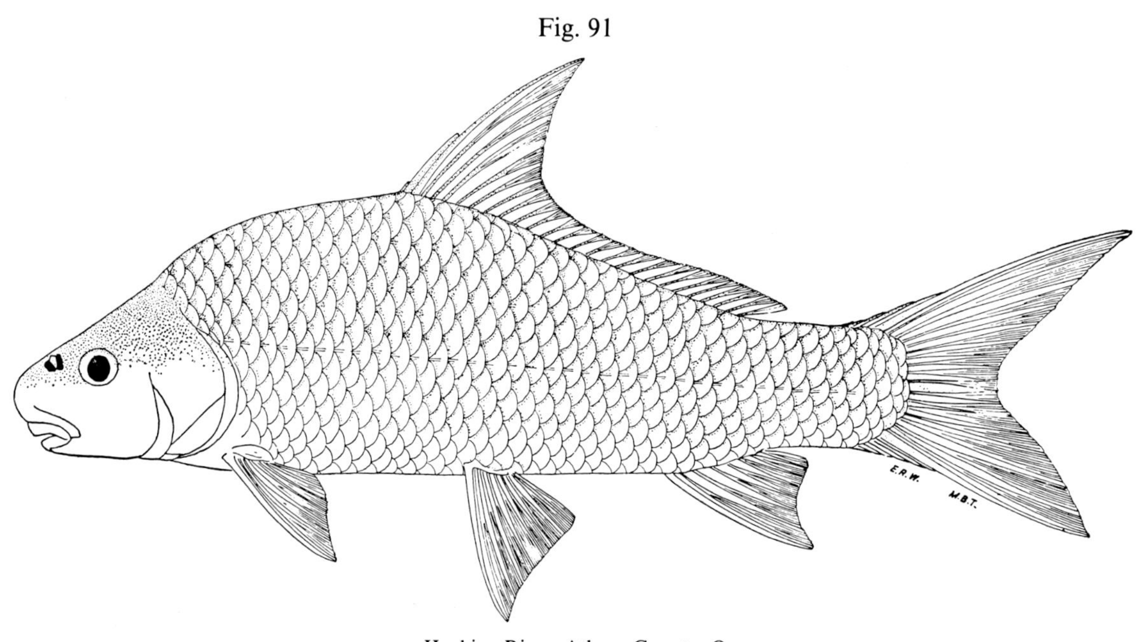

Hocking River, Athens County, O.

June 27, 1939.
Breeding female.

290 mm SL, 15.3″ (38.9) TL.
OSUM 1099.

Identification

Characters: *General—See* blue sucker. *Generic—See* eastern quillback. *Specific*—A small knob present on tip of lower jaw, *see* highfin carpsucker, fig. A; except in small young in which the knob may be undeveloped. Tip of lower lip scarcely or not at all in advance of the anterior nostril. Snout in adults pointed, the small eye giving the false impression that this species has a longer preorbital length of head, in comparison to its postorbital length, than does the highfin carpsucker. Anterior dorsal rays average shorter in length than those of any other carpsuckers; they usually are less than two-thirds the length of the dorsal base. Body moderately slender; the depth usually contained 2.7 (young)–3.3 (large adults) times in standard length. Eye of large young and adults usually contained 3.0 (young less than 4.0″ [10 cm] long)– 6.2 (adults) times in head length. Large adults readily identifiable; young less than 3.0″ (7.6 cm) long difficult or impossible to separate from young highfin carpsuckers.

Differs: Quillbacks lack knob on tip of lower jaw. Highfin carpsucker has a larger eye and deeper body. Buffalofishes lack the angle on posterior edge of subopercle. Carp, goldfish, walleyes, white bass, blackbasses, and drum have one or more dorsal spines.

Most like: Highfin and quillback carpsuckers, buffalofishes, carp and goldfish.

Coloration: Bronze- and greenish-olive dorsally, each scale with much silvery reflection. Sides whiter and more silvery. Ventrally milk-white or with a yellowish cast. Anterior and distal margins of dorsal fin edged with black; remainder of dorsal, caudal, and anal light olive-slate. Lower fins white or tinged with orange. Tip of snout and lips milk-white.

Lengths and **weights:** *Young* of year in Nov., 2.0″–5.0″ (5.1–13 cm) long. *Adults*, usually 13.0″–24.0″ (33.0–60.9 cm) long, weight 14 oz– 7 lbs (397 g–3.2 kg). Rivermen state that fishes more than 6 lbs (2.7 kg) in weight are uncommon. Largest specimen, 25.0″ (63.5 cm) long, weight 10 lbs, 4 oz (4.6 kg).

Hybridizes: possibly with highfin and quillback.

MAP 91. Northern river carpsucker

Locality records. ○ Presumably this species. ● After 1880. △ 1955–80.
Insert: Black area represents range of this subspecies, outlined area of another; area of intergradation between subspecies probably very broad; limits of range in Mexico indefinite.

Although I have seen adults with most of their specific characters intermediate between two species of carpsuckers, I hesitate to call them hybrids because the species of no other genus of Ohio fishes are more variable in their morphological characters, or approach each other more closely (*see* Hubbs, Hubbs, and Johnson, 1943:8–11).

Distribution and Habitat

Ohio Distribution—During the 1925–50 period the river carpsucker was confined chiefly to the Ohio River from the Indiana state line upstream to Washington County, and to the lower portions of its larger tributaries. It appeared to be more common in the lower Scioto River than in any other tributary, and in this river well-marked upstream movements occurred during May and downstream ones in late August and early September. Upon a few occasions during these movements I have, in one night, taken dozens of adults in gill nets placed across the Scioto River in Pike County. In April, commercial fishermen netted considerable numbers in the Ohio River, especially opposite Brown and Adams counties; in this river the species outnumbered the other two species of carpsuckers. Commercial fishermen were among the few in Ohio who could identify the adults; some of these fishermen considered this species to be a hybrid between the other two species of carpsuckers.

Although sub-adults and adults were identifiable almost at a glance, many young under 4.0″ (10 cm) in total length could not be satisfactorily separated from the young of the highfin carpsucker; because of this, the locality dots on the distribution map represent only localities where larger specimens were collected or observed. Young, believed to be of this species, were taken as far inland in the Scioto system as northern Pickaway County, and in the Muskingum system as far upstream as southern Muskingum County.

The few preserved specimens collected before 1900 from the Ohio River (CAS 3978; CSNH, no number) substantiate statements by Jordan (1882:809–10) and Henshall (1888:77) as to its former presence in that river.

On September 4, 1927 Robert B. Foster and I collected four young (OSUM 72; UMMZ 86020), 2.5″–4.0″ (6.4–10 cm) in total length in Bad Creek, Henry County, about 5 miles (8 km) above its confluence with the Maumee River (*see* hollow circle). These young were originally identified by Carl L. Hubbs, myself, and others as river carpsuckers. As river carpsuckers they comprised the first record for the species in the Great Lakes drainage (Hubbs, 1930B:428). Since at present some doubt exists relative to our ability to identify such small individuals, it is possible that these specimens may be highfin carpsuckers. The presence of the river carpsucker in the Great Lakes drainage is surprising, and our failure to find more specimens in subsequent years suggests that: (1) individuals of this species were included in shipments of buffalofishes which were introduced into western Lake Erie waters between 1920–30, and which have failed to become established; (2) an Ohio drainage collection became mixed with the Bad Creek collection; (3) a relict population has long existed in these waters. I consider the first suggestion as the most logical.

Habitat—The largest populations of river carpsuckers occurred only in the deeper waters of the Ohio River from Washington County downstream, and in the deeper, longer pools of the lower portions of its tributaries. Small numbers of adults were found in the larger lowland ponds; presumably these had become trapped while spawning among the submerged portions of trees and brush, and/or while the river was in flood; in fall their young often were very numerous in these ponds. This species was more confined to larger rivers than were the other two species of carpsuckers.

Hubbs and Lagler (1947:50) and Gerking (1945: 39) stress the presence of this species in silty and turbid waters. Admittedly this is true, since at present most of our larger rivers are silt laden, and if the species is to be present at all, it must occasionally inhabit silty waters. It seems significant, however, that in Ohio, the largest, fastest-growing river carpsuckers came from large, deep overflow ponds whose waters were notably less silt laden than were the rivers; also, there was agreement among rivermen that river carpsuckers were much more abundant before 1910, when the bottom of the Ohio River was less silt covered, than they have been since.

Years 1955–80—During this period the abundance and distributional pattern of the river carpsucker appeared to remain the same in inland Ohio as it had been prior to 1950, except for upstream extensions of its range in the Scioto, Hocking and Muskingum rivers.

Ohio River fishermen informed me that this was the most numerous of the three species of carpsuckers, as it had been before 1951. This view is strengthened by the 1957–59 investigations of the river by personnel of the Biology Department of the University of Louisville (ORSANCO, 1962:148).

HIGHFIN CARPSUCKER

Carpiodes velifer (Rafinesque)

Fig. 92

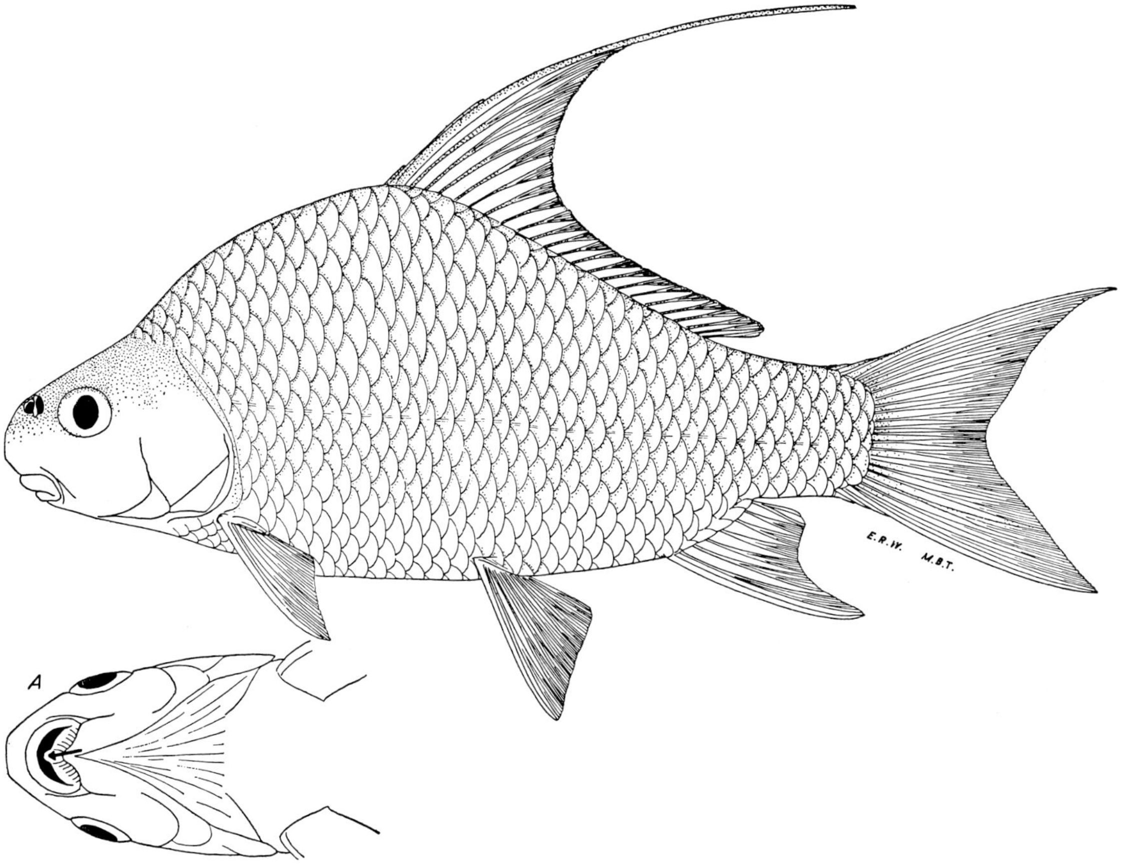

Upper fig.: Scioto River, Scioto County, O.

Nov. 6, 1939. 230 mm SL, 12.6″ (32 cm) TL.
Female. OSUM 1851.

Fig. A: ventral view of head; note arrow pointing to knob on tip of lower jaw.

Identification

Characters: *General—See* blue sucker. *Generic—See* eastern quillback. *Specific*—A small knob present on tip of lower jaw, *see* fig. A, except in small young in which knob may be undeveloped. Tip of lower lip usually below, or posterior to, the anterior nostril; only occasionally in adults is the tip of lower lip in advance of the anterior nostril. Adult snout rounded, and because of the large eyes appears to be very short, giving the false impression that this species has a shorter preorbital length of head, in comparison to its postorbital length, than does the river carpsucker. Anterior dorsal rays of large young and adults, when unbroken, longest of any species of carpsuckers; they are usually longer than the dorsal base. Body of large young and adults so deep as to be sunfish-like in shape; its depth usually contained 2.0 (large adults)–3.0 (small young) times in standard length. Eye contained 3.0 (young less than 3.0″ [7.6 cm])–4.8 (adults) in head length. Adults readily identifiable; young less than 3.0″ (7.6 cm) long difficult or impossible to separate from young river carpsuckers.

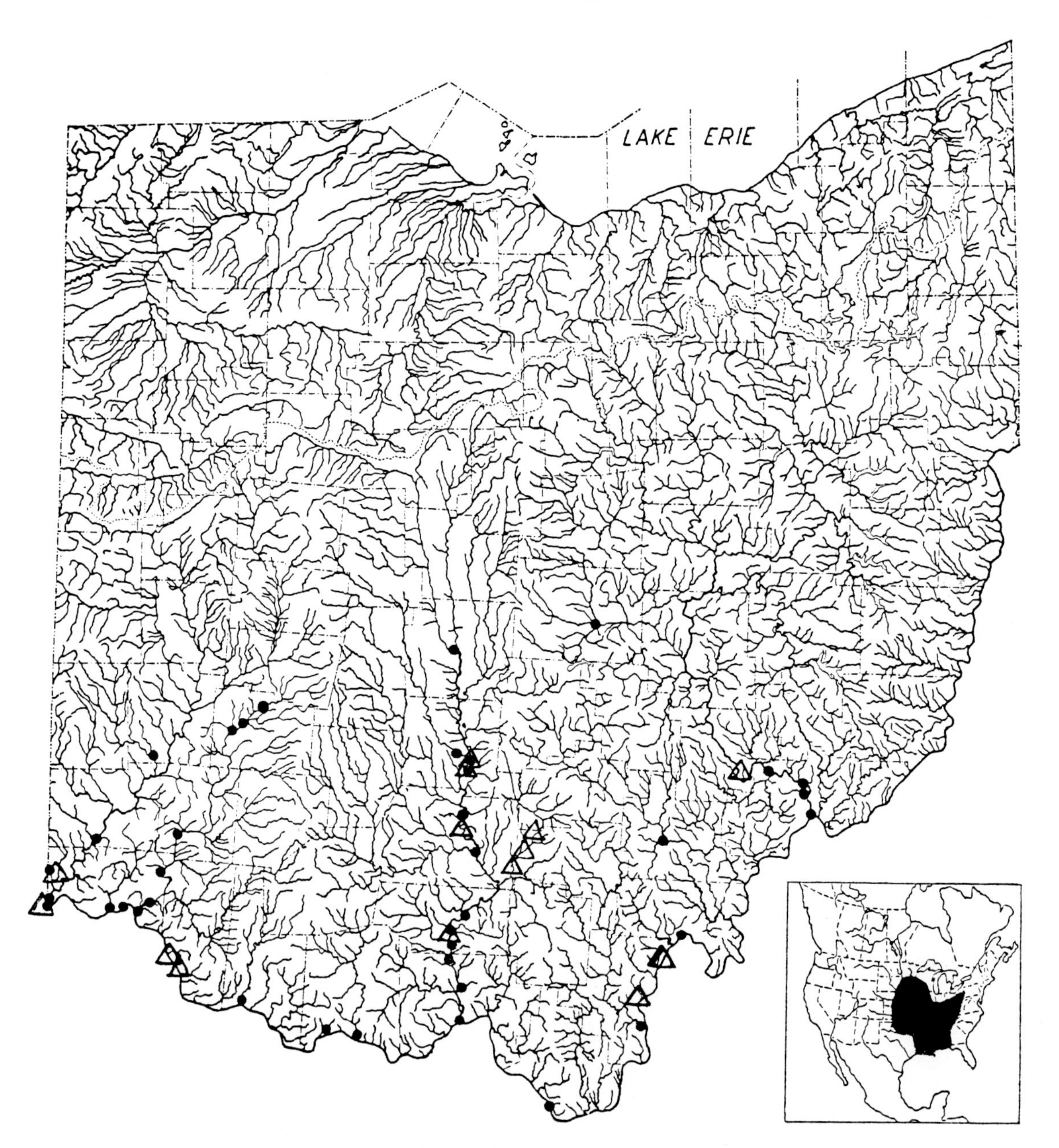

MAP 92. Highfin carpsucker

Locality records. ● After 1925. △ 1955–80.

Insert: Western limits of range indefinite, partly because of confusion of this species with other species of carpsuckers.

Differs: *See* river carpsucker. River carpsucker is more slender-bodied, has a smaller eye.

Most like: River carpsucker, quillback, buffalofishes, carp and goldfish.

Coloration: Same as river carpsucker.

Lengths and **weights:** *Young* of year in Nov., 2.0″–4.0″ (5.1–10 cm) long. *Adults*, usually 11.0″—14.0″ (27.9–35.6 cm) long, weight 8 oz–1 lb, 12 oz (227 g–0.8 kg). Rivermen state that fishes more than 14.0″ (35.6 cm) long are uncommon, and give maximum weights of between 2–3 lbs (0.9–1.4 kg). Many spawning females are only 9.0″ (23 cm) long.

Largest specimen, 15.0″ (38.1 cm) long, weight 2 lbs (0.9 kg).

Hybridizes: *See* river carpsucker.

Distribution and Habitat

Ohio Distribution—Between 1925–50 the highfin carpsucker was present in the Ohio River from the Indiana state line to Washington County, and in the lower halves of its larger tributaries. It appeared to be as numerous in these lower halves as it was in the Ohio River; in that river it was less numerous than was the river carpsucker. The highfin, together with the quillback and river carpsuckers, migrated up the lower Miami, Scioto and Muskingum rivers during May, and downstream in late August and September. Commercial fishermen recognized the large adults as a distinct species, calling them the "Hump-backed Carp." Although the adults were readily recognizable, young under 3.0″ (7.6 cm) total length were difficult or impossible to separate satisfactorily from the young of the river carpsucker; because of this the dots on the distribution maps represent only those localities where the larger and readily identifiable highfins were observed or taken.

Considerable confusion existed before 1900 concerning the various species of carpsuckers; as a result the older literature records are mostly unreliable. The most reliable records appear to be those of Jordan (1882:812–13, as *Carpiodes cutisanserinus* and *C. difformis*) and Henshall (1888:70, as *C. difformis*), both of whom stated that this species was common or abundant in the Ohio River. Their statements, and especially that of Henshall, are strengthened by specimens collected in the Ohio and Muskingum rivers previous to 1900 (CSNH, no number; SNHM 4025 and 4084), and by the statements of old commercial fishermen who knew the highfin and who stated that they and their fathers took the species abundantly before 1900, before the river was ponded. Likewise these men said that until 1920 the "Hump-back" was numerous in the Ohio River from Washington County upstream to the Pennsylvania state line. It likewise was present, at least formerly, in the Ohio River tributaries in Pennsylvania, for Cope's (1871:481) type of *cutisanserinus* from the Kiskiminitas River is a highfin (type No. 6649, in the Philadelphia Academy of Natural Science; in 1950 Henry W. Fowler graciously re-examined the type for me and sent me a large photograph of it). Fowler (1919:63 as *difformis*) also examined specimens obtained by Cope from the Youghiogheny River.

Until 1930 (Osburn, Wickliff, Trautman, 1930:171) the Ohio literature referred to the highfin as *C. difformis* or *C. cutisanserinus*, and *C. velifer* was used as the name of the quillback. Since 1930 the specific name of the quillback has been *cyprinus*.

Habitat—The largest populations of highfins occurred in the moderately deep waters of the Ohio River and the lower halves of its larger tributaries. Smaller populations were present in the lowland ponds, oxbows and sloughs. The highfin differed from the river and quillback carpsuckers in that it was not as restricted to the deepest, largest waters as was the river carpsucker, nor did it normally inhabit the smaller streams and creeks as did the quillback.

Rafinesque (1820:114) called this species "Sailor fish, Flying fish and Skimback" because of its curious habit of "skimming" along the surface with part of its back and its dorsal fin exposed, and because it frequently jumped above the water's surface. I have noted that this skimming and jumping occur most often on quiet evenings in late spring and early summer. The habit was indulged in by all species of carpsuckers, but the highfin appeared to skim and jump more frequently than did the others, and especially more than did the river carpsucker.

Years 1955–80—In those sections of the lower Scioto River system where the most intensive collecting was done during this period, as well as prior to it, the abundance and distributional pattern of the highfin carpsucker appeared to remain unchanged except possibly for a somewhat larger population in the Scioto above Circleville, where since 1955 the waters have been more inhabitable for fishes than they were previously. It is assumed that the abundance and distributional pattern elsewhere in inland Ohio remained largely unaltered.

In the Ohio River the species continued to be the least numerous of the three species of carpsuckers, according to statements made by fishermen and from the results of the 3-year investigations of that stream by personnel of the Biology Department of the University of Louisville (ORSANCO, 1962:148).

SILVER REDHORSE

Moxostoma anisurum (Rafinesque)

Fig. 93

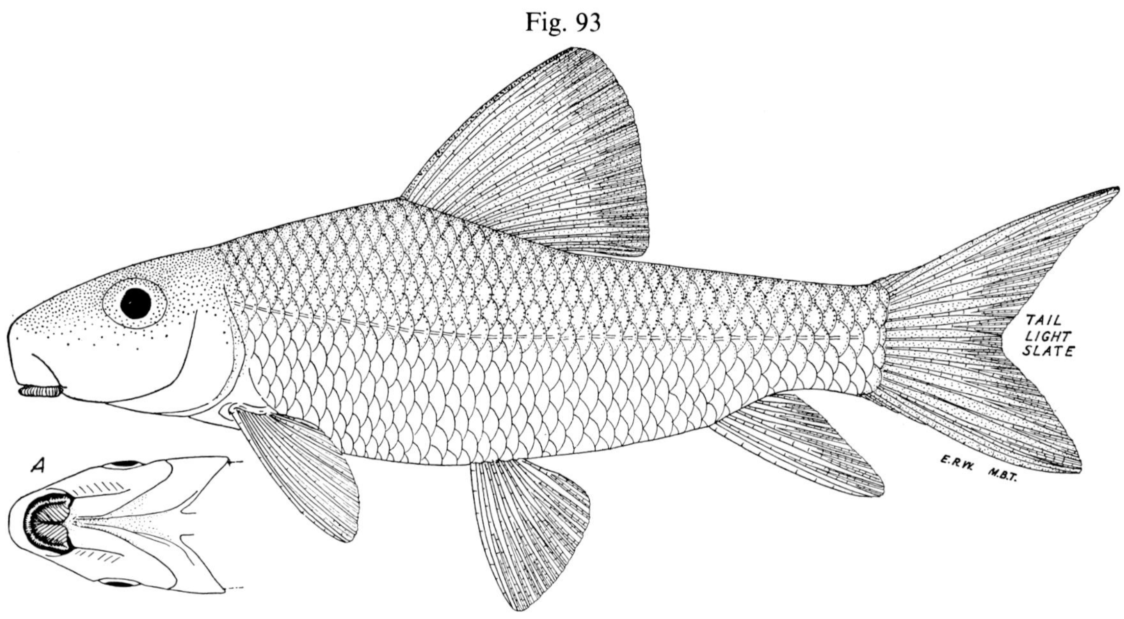

Upper fig.: Killbuck Creek, Wayne County, O.

Oct. 3, 1940. 146 mm SL, 7.5″ (19 cm) TL.
Immature. OSUM 2159.

Fig. A.: ventral view of head; note narrow upper lip and fullness and shape of lower lips.

Identification

Characters: *General—See* blue sucker. *Generic—* Dorsal fin of 11–18 rays. Scales 50 or fewer in the *complete* lateral line. Head rounded between eyes. Upper lips separated from tip of snout by a wide groove. Lower lips large, the halves widely joined together. Air bladder in three parts. *Specific—* Dorsal rays usually 15–16, extremes 13–18. Distal margin of dorsal rounded outward (convex) in adults; straight in young. Length of longest dorsal ray usually equal to distance from dorsal origin to space between the eyes. Tail slate-colored in life (broken blood vessels may give the false impression of a reddish-colored tail). Body silvery, the scales of back without dark spots at their bases. Lower lips very full, and usually broken up into papillae. Body deep for this genus, usually contained 2.8 (adults)–3.7 (young) times in standard length. Pharyngeal teeth very thin and compressed.

Differs: Other species of redhorses and spotted sucker normally have fewer dorsal rays and more slender bodies. Hog sucker is depressed between the eyes. White and longnose suckers have more lateral line scales. Chubsuckers have no lateral lines. Buffalofishes, carpsuckers, carp and goldfish have more dorsal rays. Minnows have the anal fin farther forward.

Most like: Black and golden redhorse.

Coloration: Mostly silvery and white. Bronzy- or greenish-olive dorsally, the scales often dark-edged but lacking dark spotting at their bases. Sides silvery sometimes with faint, golden reflections. Silver or milk-white ventrally, fins light-slate and white.

Lengths and **weights:** *Young* of year in Oct., 2.0″–5.0″ (5.1–13 cm) long. Around **1** year, 4.0″–8.0″ (10–20 cm). Adults, usually 11.0″–22.0″ (27.9–55.9 cm) long, weight 8 oz–5 lbs (227 g–2.3 kg). Largest specimen, 25.0″ (63.5 cm) long, weight 8 lbs, 4 oz

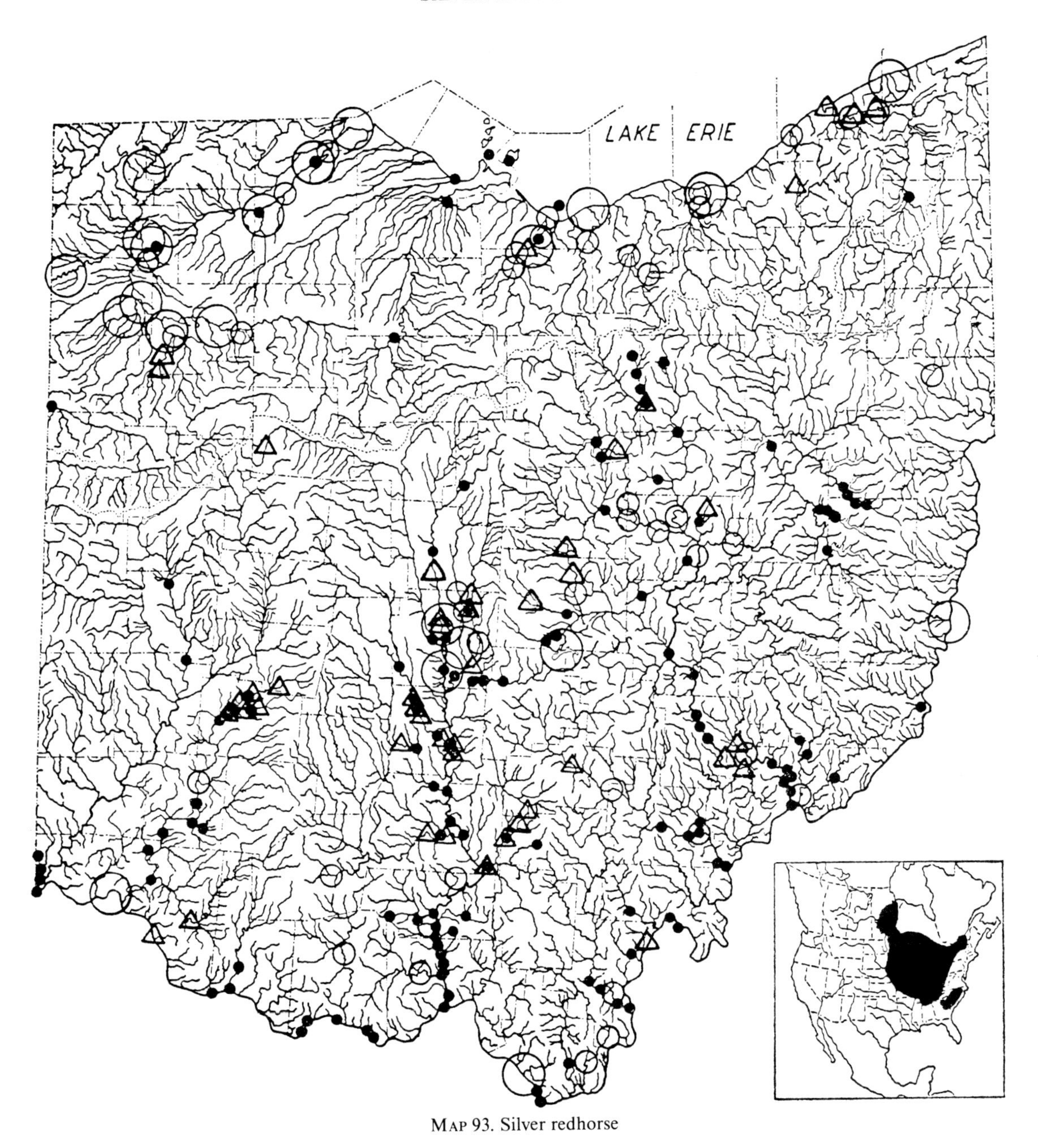

MAP 93. Silver redhorse

Locality records. ○ 1885–1900. ○ 1901–20. ● 1921–54. △ 1955–80.
Insert: Strays may be expected in the lower Mississippi River, especially after floods.

(3.7 kg). Ohio rivermen give maximum weights of 7–10 lbs (3.2–4.5 kg); Lake Erie fishermen, 5–8 lbs (2.3–3.6 kg).

Distribution and Habitat

Ohio Distribution—Rafinesque, Kirtland, Jordan and other 19th century writers failed to separate satisfactorily the various species of redhorse suckers; consequently we have few reliable published records of the silver redhorse previous to 1900. Because of this unreliability only records based upon preserved specimens and the most reliable published accounts are used; these records are represented on the distribution map by the use of large, hollow circles.

When the fewer than 100 collections of fishes made between 1885–1900 are compared with the more than 2,400 collections made between 1925–55, it becomes apparent that the silver redhorse was more numerous and possibly more widespread before 1900 than the species has been since; also that it has decreased greatly in abundance in the Lake Erie tributaries, especially in the Maumee system. Observations, made by experienced fishermen since 1900 and by me since 1925, indicate such a decrease in abundance. This decrease has been caused presumably by increased turbidity and siltation, for the silver redhorse is especially intolerant to increased turbidity and siltation.

Although the distribution map shows no records for some counties, this highly migratory species unquestionably has been present in every county at some time during the present century.

Habitat—Chiefly an inhabitant of large streams, in which it showed a preference for the long, deep pools of those sections where gradients were low or basic, and where rates of silt sedimentation and/or detrimental industrial pollutants were at a minimum. It was conspicuously rare or absent in those stream sections, otherwise favorable, where the gradient was not sufficiently high to scour the stream bottoms and to prevent rapid siltation.

Years 1955–80—Throughout this period considerable numbers of silver redhorses were sometimes captured wherever intensive collecting was undertaken in medium- and large-sized streams, and especially by employing electric shockers and gill and fyke nets (White et al., 1975:87). As stated above, this highly migratory species unquestionably was present in every county of the state, at least in small numbers. The species was found to be as numerous during this present period as it had been between 1925–55.

Before 1935 I accompanied commercial fishermen as they lifted their hoop and wing nets in the Ohio River and also assisted them in "running" their trotlines. At that time the silver redhorse was rather numerous and occasionally was captured in considerable numbers. It was, therefore, a surprise to learn that only 11 individuals were recorded for the entire Ohio River during the thorough 1957–59 investigations of that stream (ORSANCO, 1962:149). Margulies and Burch collected the species in the Ohio River adjacent to Gallia County, Ohio, in 1976. For distributional range for the species of *Moxostoma, see* Jenkins, 1970.

BLACK REDHORSE

Moxostoma duquesnei (Lesueur)

Fig. 94

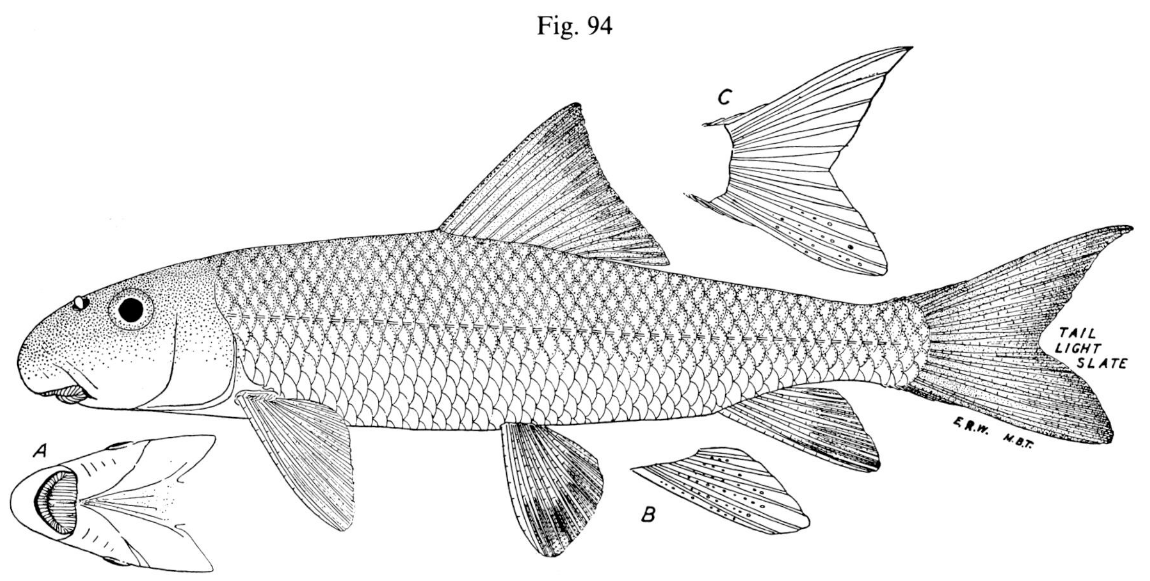

Central Fig.: North Fork Paint Creek, Fayette County, O.

Nov. 21, 1939. 157 mm SL, 7.7″ (20 cm) TL.
Immature. OSUM 1318.

Fig. A: ventral view of head; note overhanging snout and the posterior edges of lower lips forming a straight line.

Figs. B and C: Muskingum River, Washington County, O.

April 16, 1940. 265 mm SL, 13.2″ (33.5 cm) TL.
Breeding, tuberculated male. OSUM 1828.
Tuberculated anal (B) and caudal (C) fins.

Identification

Characters: *General—See* blue sucker. *Generic—See* silver redhorse. *Specific*—Dorsal rays usually 12–13, extremes 11–15. Length of longest dorsal ray shorter than distance from dorsal origin to head above eye; in larger specimens extending only to nape. Similar to silver and golden redhorses in having no distinct dark scale bases and a slate-colored tail. Body in life darker, not as silvery as in silver redhorse or as bronze as in the golden redhorse. Distal edge of dorsal slightly falcate in large young and adults; often straight in small young. Scales in lateral line 44–47, extremes 42–49. Almost invariably one or both pelvic fins have ten rays. Least depth of caudal peduncle usually contained more than 1.7 times in its length. Eye small, usually more than 2.2 times in snout length of large young and adults; about 1.8 times in snout length of small young. Lips usually striated. Body slender, usually the depth is contained more than 4.0 times in standard length. Posterior edges of lower lips form a straight line, fig. A. Anal (B) and caudal (C) fins tuberculated in males; the snout has no tuberculations.

Differs: Silver redhorse has more dorsal rays and a deeper body. Golden redhorse has fewer lateral line scales and normally only 9 pelvic rays. Other redhorses have red tails in life. *See* silver redhorse.

Most like: Golden redhorse. **Superficially like:** Spotted sucker.

Coloration: Dorsally gray- or brownish-olive with silvery-blue reflections, the scales usually dark edged but lacking dark bases. Sides lighter and more bluish-silvery, occasionally with faint, bronze reflections. Ventrally silvery- or milk-white. Fins

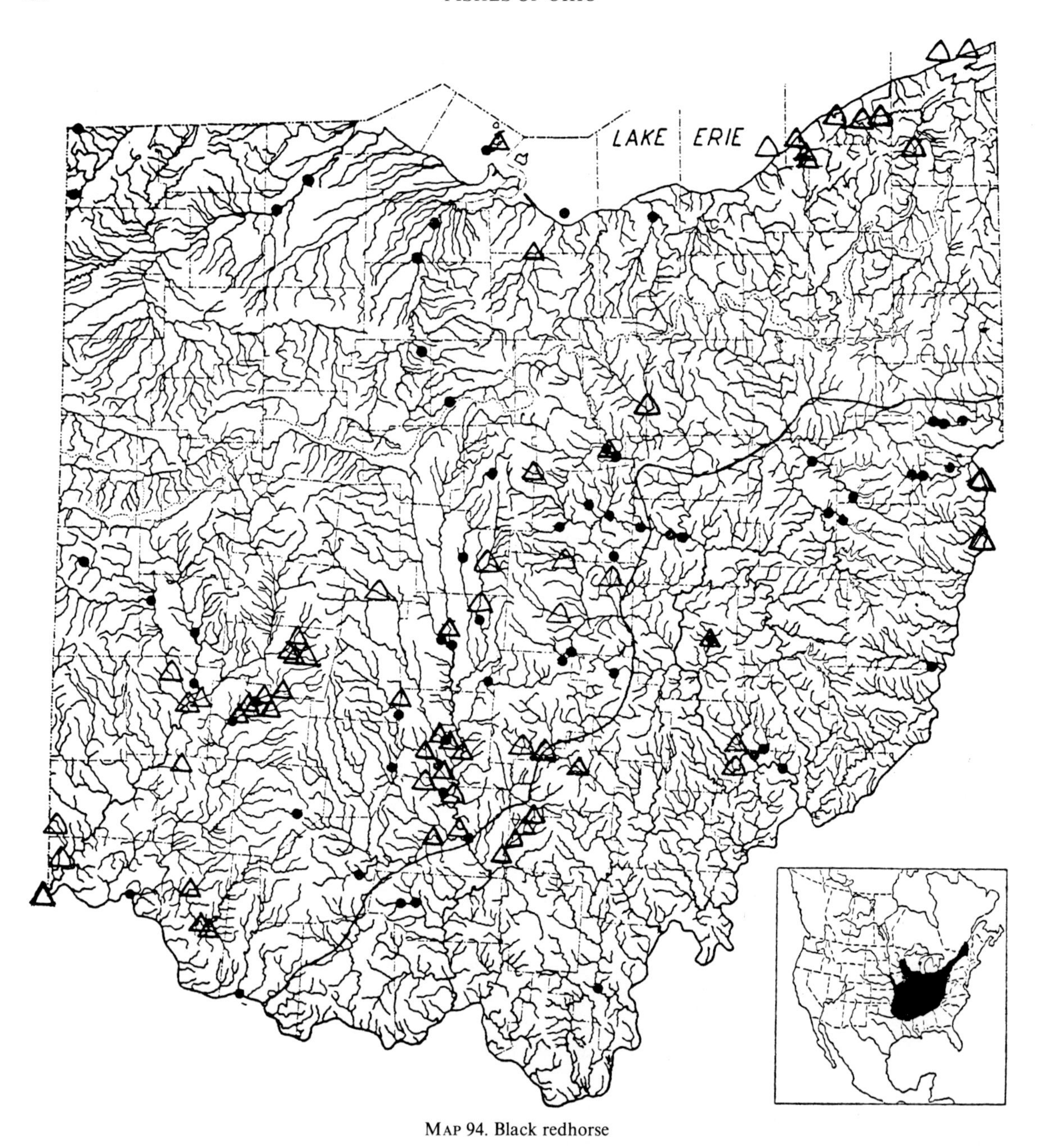

MAP 94. Black redhorse

Locality records. ● Before 1955. △ 1955–80. ~ Glacial boundary.
Insert: Rare or absent in streams of low gradients throughout its range.

light slate and whitish, except in specimens taken from streams containing such inorganic wastes as coal-mine pollution; these often have orange-tinged lower fins.

Lengths and **weights:** *Young* of year in Oct., 2.0″–3.5″ (5.1–8.9 cm) long. *Adults* usually 10.0″—–15.0″ (25.4–38.1 cm) long, weight 6 oz–1 lb, 8 oz (170g–0.7 kg). Largest specimen, 17.3″ (43.9 cm) long, weight 2 lbs, 4 oz (1.0 kg).

Distribution and Habitat

Ohio Distribution—Since 1930 the black redhorse has been present in moderately-sized populations along the line of glaciation in Ohio, with the largest population centered in the Walhonding River system. The species appeared to be uncommon in the Miami and Maumee river systems.

It appears remarkable that this sucker was not definitely recorded in northeastern Ohio, particularly in that area from Cuyahoga County northeastward to Ashtabula County, and in such high-gradient streams as the Chagrin River; however, between 1940–51 I saw redhorses spawning in the fast waters of the Chagrin system which certainly were this species, for with binoculars I saw that they were darker in coloration than are the other species of slate-tailed redhorses and that they lacked tubercles on the snout, such as are present on the male golden redhorse. Unfortunately, I was unable to capture these spawning suckers.

It was not until 1930 that the black redhorse was fully recognized as a species (Hubbs, 1930:19–24); therefore little is known concerning its distribution and abundance in Ohio before that year. Ecological reasoning indicates that this fast-and clean-water inhabiting species should have been rather numerous before 1900, especially in the swift Ohio River tributaries east of the Flushing Escarpment. Old residents relate that previous to 1900 these tributaries had large migrations of spawning redhorses.

Habitat—Primarily an inhabitant of the swifter-flowing portions of moderate- or large-sized streams having normally clear waters and bottoms of gravel, bedrock and sand, and where siltation was at a minimum. It appeared to be less tolerant of low gradients, turbidity and siltation than was the golden redhorse.

Years 1955–80—It was only when diversified collecting gear was utilized for several days or weeks in clear and swift streams that the numerical status of the black redhorse could be reasonably ascertained. An example was the investigation of the upper Little Miami River and its tributaries including Massie Creek, in Greene and Clark counties. Between the years 1952 and 1957 Brown (1960: 53–54) and others investigated these streams, employing electric shocker and other devices. They found that during 1957 the Little Miami River produced a total of 564 redhorses of the genus *Moxostoma*, of which an estimated 20% (112) were black redhorses. In Massie Creek this species "appeared to dominate [in greater numbers than any other *Moxostoma* species] in certain sections." From such types of collecting it may be assumed that this elusive species was more numerous and widespread in its Ohio distribution in both drainages than is indicated by map 94.

In the first edition of this report I stated that I had seen what I believed were black redhorses spawning in the fast waters of the Chagrin River. This statement has been recently proven to be correct by the capturing of many specimens in the Chagrin River by Andrew White and his associates (White et al., 1975:87).

For a life-history study *see* Bowman, 1970: 546–59.

GOLDEN REDHORSE

Moxostoma erythrurum (Rafinesque)

Fig. 95

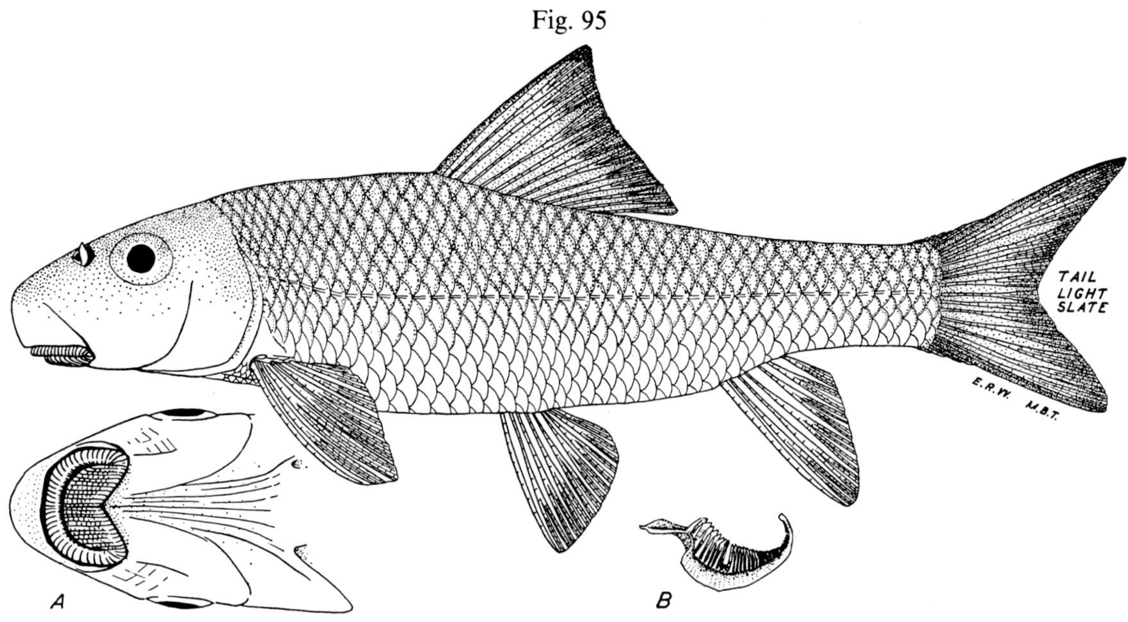

Upper fig.: Killbuck Creek, Wayne County, O.

Oct. 3, 1940. 151 mm SL, 7.6″ (19 cm) TL.
Immature. OSUM 2158.

Fig. A: ventral view of head; note coarse lips and angle formed by the posterior edges of the 2 halves of the lower lip.
Fig. B: pharyngeal arch; note the thin, compressed teeth with a central tooth missing; (one or more teeth are usually missing).

Identification

Characters: *General—See* blue sucker. *Generic—See* silver redhorse. *Specific*—Dorsal rays usually 12–14, extremes 11–15. Length of longest dorsal ray shorter than distance from dorsal origin to head above eye, or in the larger specimens extending only to junction of occiput and nape. Similar to silver and black redhorses in having a slate tail and no distinct dark scale bases. Body in life usually with a bronze-golden cast. Distal edge of dorsal slightly falcate in large young and adults; straight in small young. Scales in lateral line usually 39–42, extremes 37–44. Rays in both pelvic fins normally 9. Least depth of caudal peduncle usually contained less than 1.6 times in its length. Eye moderate or large, usually less than 2.2 times in snout length of large young and adults; about 1.4 times in snout length of small young. Lips striated. Body moderately deep, the depth usually contained less than 4.0 times in standard length. Posterior edges of lower lips form an angle, fig. A. Snout of breeding male contains tubercles, as do the anal and caudal fins.

Differs: *See* silver and black redhorses.

Most like: Black and silver redhorses.

Coloration: Dorsally grayish- or bronzy-olive, the scales golden-green at their bases but without dark spotting; the entire dorsal half usually with a bronzy or golden cast in large young and adults; silvery cast in small young. Sides lighter with a bronze-silvery cast. Ventrally silvery- and milk-white, the lips white, the snout grayish-white. Dorsal and caudal fins a light slate, remaining fins whiter or with an orange tinge. Fishes taken from polluted waters may have all fins orange or reddish-orange.

Lengths and **weights:** *Young* of year in Oct., 2.5″–4.5″ (6.4–11 cm) long. *Adults*, usually 11.0″–18.0″ (27.9–45.7 cm) long, weight 8 oz–2 lbs, 8 oz (227 g–1.1 kg). Dwarfed individuals breed when

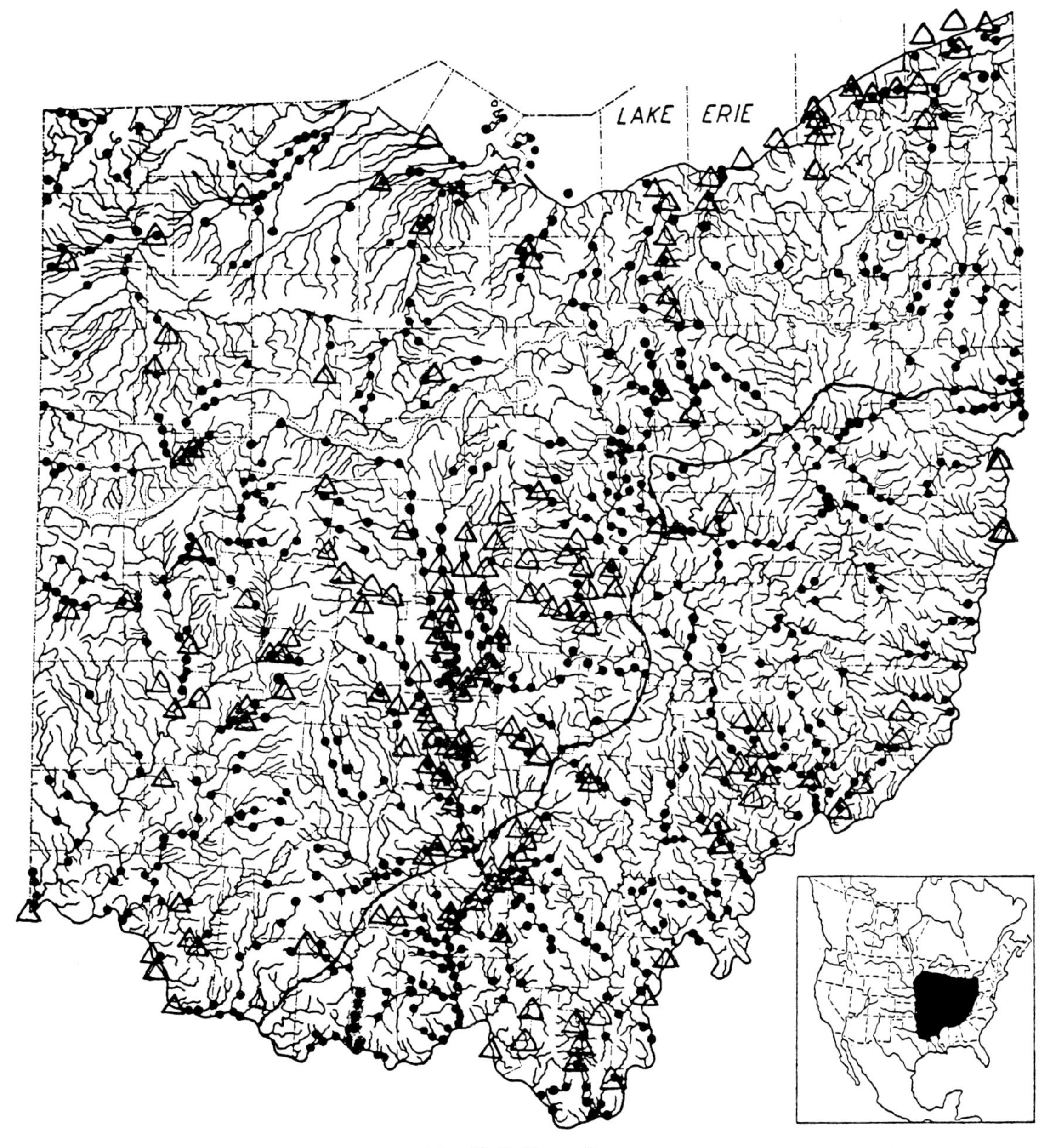

MAP 95. Golden redhorse

Locality records. ● Before 1955. △ 1955–80. ~ Glacial boundary.
Insert: Western limits of range indefinite.

only 9.5″ (4 cm) long. Maximum length 26.0″ (66.0 cm), weight 4 lbs, 8 oz (2.0 kg) (Lake Erie). Heaviest fishes were females.

Distribution and Habitat

Ohio Distribution—Since 1920 the golden redhorse has been the most widely distributed and most abundant species of redhorse (*Moxostoma*) inhabiting Ohio streams. Undoubtedly the species occurred, at least sporadically, in every permanent tributary in the state. Because of its highly migratory nature it penetrated, during freshets, the most industrially-polluted streams. The seasonal concentrations of the golden redhorse rather closely paralleled those of the hog sucker, but unlike the latter species, the former did not spawn habitually in the very smallest of tributaries or those of the highest gradients. Avoidance of high gradients is aptly demonstrated by the comparative fewness of records in the Chagrin River system and other streams of high gradients in the region from Cuyahoga County northeastward into Ashtabula County. Like the hog sucker, the golden redhorse was intolerant of domestic and industrial pollutants, continuous turbidity and rapid silting of stream bottoms; consequently its numbers were low in the mine-polluted areas of southeastern Ohio, the rapidly silting prairie and lake-plain streams of northwestern Ohio, and the industrially-polluted Great Miami River from Dayton downstream to its mouth.

Although literature references are almost wholly unreliable, specimens from more than 25 localities which are preserved in museums indicate that this redhorse was widely distributed throughout Ohio during the 1853–95 period. There is no obvious reason to assume that it was not as abundant and widespread throughout the 19th century as it has been since.

Habitat—The largest numbers were present in the relatively-clear streams of median gradients where the riffles were composed of sand, gravel, boulders and bedrock, and where the pools were relatively free from rapidly accumulating silt. The species appeared to be rather tolerant to a moderate amount of turbidity and siltation, for it was present in such waters in moderate numbers. As a species it ascended the smaller streams in spring, moved downstream during the summer and fall, and wintered in the larger streams. It was seldom found in waters having an abundance of aquatic vegetation, and only small numbers were present in the larger inland lakes and the open waters of Lake Erie.

Years 1955–80—During this period the golden redhorse remained the most abundant representative of the genus *Moxostoma* in the inland waters (White et al., 1975:89). It is assumed that this highly migratory species was present at times in all except a few of the streams of Ohio. Decreases in numerical abundance or extirpation from a stream or stream section were observed whenever silting or organic or inorganic wastes had markedly increased in amounts. Channelization and dredging greatly modifies the waters, changing the fish fauna from that of larger food and game fishes to smaller, forage species or causes the formerly flowing waters to become intermittent, or in some cases eliminating the stream entirely.

The highly migratory golden redhorse began moving upstream toward its spawning grounds during early and mid-spring. Spawning was observed from late April until late May and occurred after water temperatures reached 50° F (10° C) or above. Although migrating and spawning have been observed many times, two instances remain outstanding. Between May 15–18, 1931 Robert Foster and I netted the Scioto River near the Pike-Scioto County line upstream a short distance from the mouth of Camp Creek. During the daylight hours the gill and trammel nets caught many dozens of these redhorses, considerably fewer after dark. During daylight hours few were found in Camp Creek; these usually were weakened, dying or dead. As darkness approached, hundreds entered Camp Creek, we counting approximately a thousand during a half-hour period. Spawning conditions were ideal. There were goodly currents over riffles, and through the shallow pools the substrate was of sand, gravels and stones producing an almost continuous spawning habitat, the water so clear as to be able with the assistance of our head jack lights to observe spawning. Although more or less distributed, it was noted that spawning usually occurred when one male was closely parallel to a female on one side and another male on the other, all vibrating simultaneously. We captured several dozen, fin clipping them, and the next day recapturing a few in our nets in the Scioto River. This spawning was unusual in that it apparently involved only one species of sucker.

On September 13, 1930 Robert Foster and I set two trammel nets entirely and two gill nets partly across the Scioto River at the same locality. The night was extremely dark until after 3 a.m. We moved as rapidly as possible in a boat from one bank to the other, releasing the fishes downstream. A massive downstream movement of Moxostomines was in progress, consisting of a fair number of silver and river redhorses and a large number of golden redhorses. They were migrating within 1.5′ (45 cm) of the water's surface, and rarely was one taken deeper. When a sufficient number became entrapped, the top of the net sank a foot (30 cm) or more beneath the surface, thereby allowing others to swim over it. We captured and released about 500 that night, probably only a moderate number of those moving downstream.

Desiring to learn more about the migratory habits, the late T. H. Langlois and I persuaded the then Ohio Division of Conservation to have a crew of men trap-netting the river daily at the above locality for an entire year. Their results indicated a considerable semi-annual movement of suckers and many other species.

Investigation of the entire Ohio River between the years 1957–59 indicated that this was the most numerous species of *Moxostoma* in that stream (ORSANCO, 1962:149; Clay, 1975:112–13).

NORTHERN SHORTHEAD REDHORSE

Moxostoma macrolepidotum macrolepidotum (Lesueur)

Fig. 96

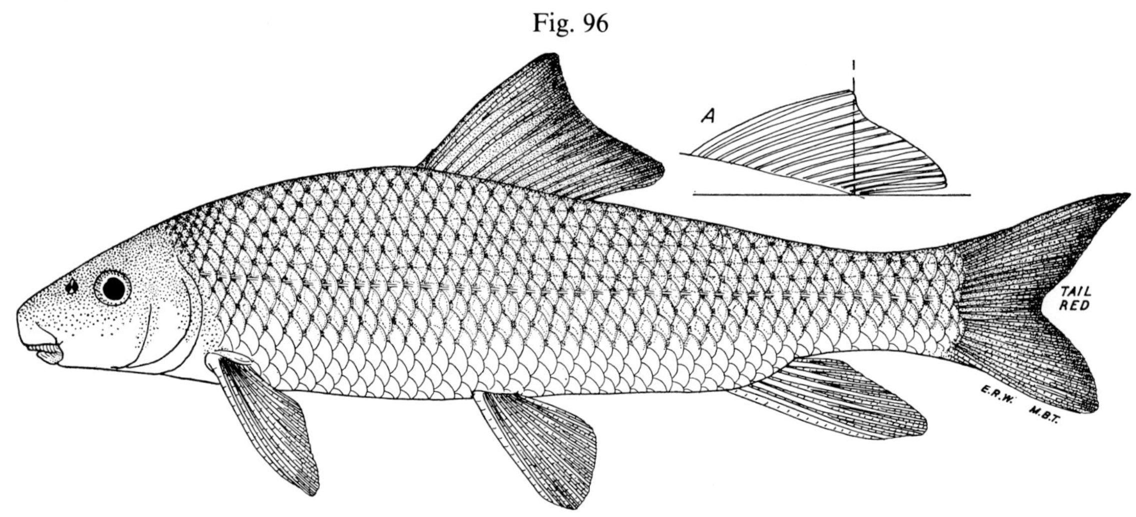

Lower fig.: Lake Erie, Ottawa County, O.

Dec. 26, 1940. 205 mm SL, 10.2″ (25.9 cm) TL.
Immature. OSUM 3047.

Fig. A: dorsal fin; note that the tip of the anterior rays falls far short of extending backward to the tip of the posterior rays.

Identification

Characters: *General—See* blue sucker. *Generic—See* silver redhorse. *Specific and subspecific*—As with the Ohio shorthead, greater and river redhorses, this sucker has a red or pink tail in life and distinct, dark spots at the scale bases; except in very small young in which both characters may be undeveloped. Head subconical and very short, usually contained 4.3–5.4 times in standard length in large young and adults; 3.5–4.0 times in young less than 3.0″ (7.6 cm) long. Dorsal rays usually 12–14, extremes 11–15. Dorsal fin falcate, except in small young in which the distal edge may be almost straight. Longest anterior dorsal rays not extending to end of last rays when depressed, fig. A. Mouth small, the posterior edges of the two halves of the lower lip forming a straight line; *see* Ohio shorthead redhorse, fig. A. Pelvic fins normally with 9 rays each. Body depth averaging 3.9 times in standard length, extremes 3.5–4.3. Lateral line scales 42–44, extremes 40–47. Tubercles confined on breeding male to anal and caudal fins.

Differs: From other redhorses, except Ohio shorthead redhorse, in having a combination of red tail, short head and falcate dorsal. *See* silver redhorse.

Most like: Ohio shorthead redhorse.

Coloration: Dorsally an intense olive-yellow, with golden, greenish, and coppery reflections, the scales with dark spots at their bases. Sides lighter and more bronzy. bventrally milk- or yellowish-white. Dorsal, caudal, and anal fins of adults brilliant red; paler and pinker in young. Pelvics and pectorals yellowish-pink, sometimes flushed with crimson.

Lengths and **weights:** *Young* of year in Oct., 2.0″–4.0″ (5.1–10 cm) long. *Adults*, usually 13.0″—18.0″ (33.0–45.7 cm) long, weight 12 oz–3 lbs (340 g–1.4 kg). Fishes between 2–3 lbs (0.9–1.4 kg) in weight mostly females. Largest specimen, 24.4″ (61.9 cm) long; weight 4lbs, 2 oz (1.9 kg). Lake Erie fishermen give maximum weights of 4–6 lbs (1.8–2.7 kg).

Distribution and Habitat

Ohio Distribution—Despite confusion existing before 1925 relative to the comparative abundance of the various species of suckers, it is apparent that this redhorse has been abundant for the past century

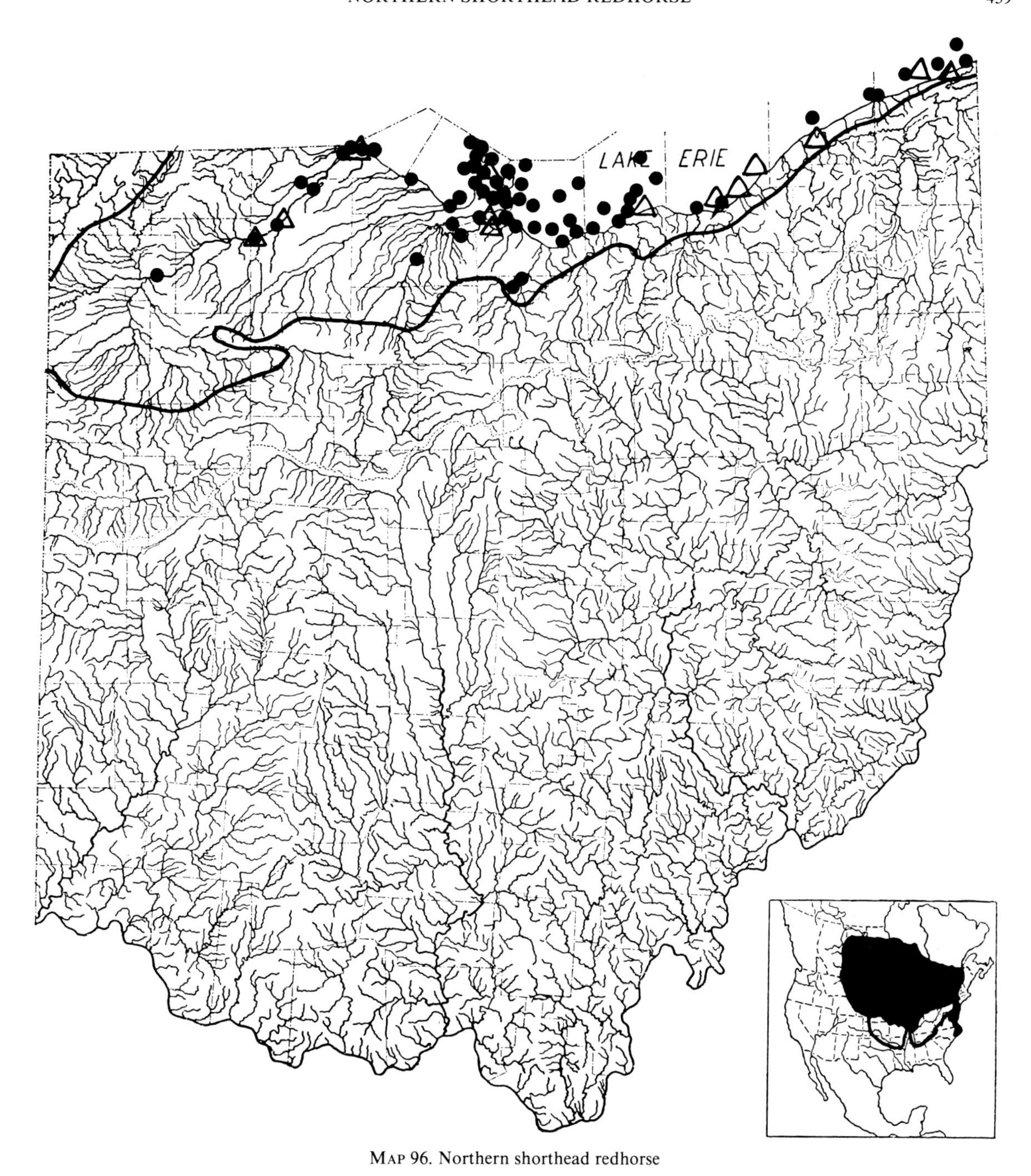

MAP 96. Northern shorthead redhorse

Locality records. ● Before 1955. △ 1955–80. ~ Lake Maumee Beach Ridge.
Insert: Black area represents range of this subspecies, outlined area of other nominal subspecies, northern limits of range indefinite.

in Lake Erie, Sandusky and Maumee bays, and in the lower courses of streams tributary to Lake Erie and especially the Maumee River (Kirsch, 1895A:328). The older commercial fishermen believe that the species was more abundant before 1925, in such now turbid waters as Sandusky and Maumee bays and the Maumee River, than it has been since. Kirtland (1851G:309) accurately described and figured the species in 1851, although he did confuse it with other sucker species.

Since 1925 the shorthead redhorse and white sucker have comprised almost the total commercial catch of "mullets" for Lake Erie; of this total fewer than 30% were shortheads. During the 1925–50 period, small spring migrations into Lake Erie tributaries occurred, and it was the general opinion of fishermen that these runs were larger before 1930 than after 1940.

Habitat—The northern shorthead primarily inhabited the rather shallow and clearer waters of Lake Erie where there was a comparatively clean and silt-free bottom of sand, gravel or bedrock. Smaller populations or strays were present in the turbid waters of bays and streams which had heavily silted bottoms, and especially in spring. As with other species of redhorses, this one was readily killed by domestic and industrial pollutants.

Years 1955–80—In Ohio during this period the northern shorthead redhorse was confined almost entirely to Lake Erie, where it showed no notable changes in numerical abundance until recently. It continued to contribute significantly to the commercial fishery (Van Meter and Trautman, 1970:72). Evidence obtained since 1970, however, indicates that the species has begun to decline in numbers, as have several other species of fishes that were formerly abundant in Lake Erie (White et al., 1975:89).

The specific name of *aureolum*, long associated with this species, was recently replaced by *macrolepidotum*. The Lake Erie population is now scientifically known as *Moxostoma m. macrolepidotum* (Lesueur) (Bailey et al., 1960:18). For spawning behavior, *see* Burr and Morris, 1977.

OHIO SHORTHEAD REDHORSE

Moxostoma macrolepidotum breviceps (Cope)

Fig. 97

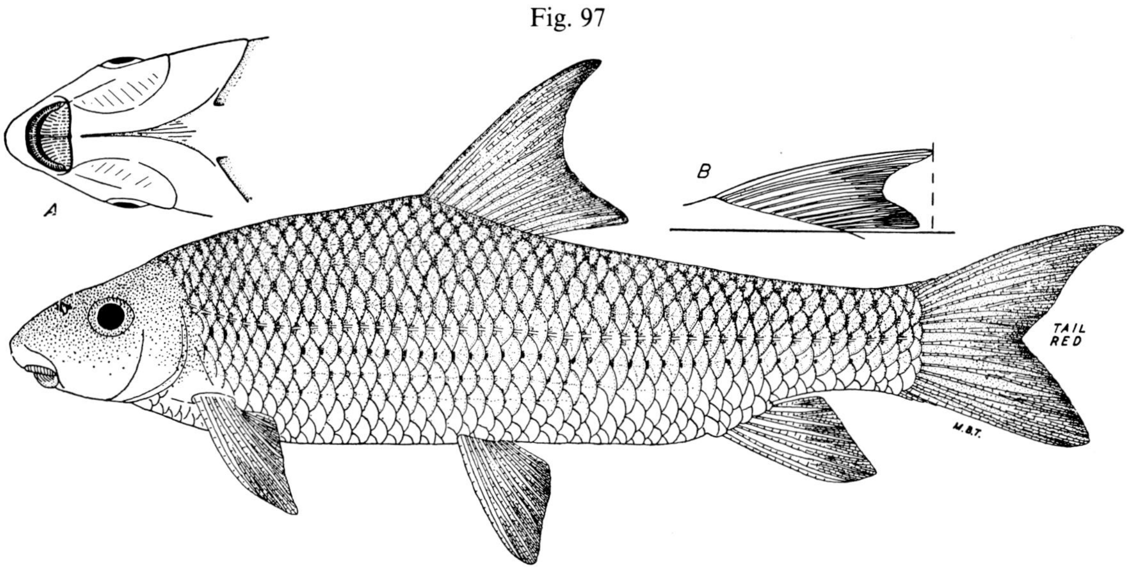

Lower fig.: Scioto River, Scioto County, O.

Aug. 5, 1940. 140 mm SL, 6.9″ (18 cm) TL.
Immature. OSUM 2899.

Fig. A: ventral view of head; note pointed snout and the posterior edges of the lower lips forming a straight line.
Fig. B: dorsal fin; note that the tip of the anterior rays extends to or beyond the tip of the posterior rays.

Identification

Characters: *General—See* blue sucker. *Generic—See* silver redhorse. *Specific and subspecific*—Similar to northern shorthead redhorse, *except* that one or both pelvic fins contain 10 rays; body averages slightly deeper with an average depth of 3.7, extremes 3.3–4.2; dorsal fin averages more deeply falcate, with the tips of the longest anterior rays in adults usually extending to or beyond the tips of the posterior rays, fig. B.

Differs: *See* northern shorthead redhorse.

Most like: Northern shorthead redhorse.

Coloration: Similar to that of the northern shorthead, except that the dorsal and caudal are more scarlet and less dark red.

Lengths and weights: *Young* of year in Oct., 2.0″–4.0″ (5.1–10 cm) long. *Adults*, usually 11.0″–17.0″ (27.9–43.2 cm), weight 11 oz–2 lbs, 8 oz (312 g–1.1 kg). Largest specimen, 19.0″ (48.3 cm) long; weight about 3 lbs (1.4 kg). Rivermen give maximum weights of 3 lbs, 8 oz–4 lbs (1.6–1.8 kg). Apparently does not attain so large a size as does the lake-inhabiting northern shorthead.

Distribution and Habitat

Ohio Distribution—The Ohio shorthead redhorse was rather numerous during the 1925–50 period in the Ohio River between Marietta and the Indiana state line and in the Walhonding, Muskingum and lower half of the Scioto rivers. Throughout the period the largest populations were present in the clearer waters of the Muskingum and Walhonding drainages; the more turbid waters of the lower Scioto River contained a smaller population; the species was present in small numbers or as strays in the industrially-polluted Great Miami River. The older rivermen believed that the Ohio shorthead redhorse was far more numerous in the Ohio River prior to 1910 before the river was ponded when there was little industrial pollution, than it has been since, and that before 1915 it was still fairly numer-

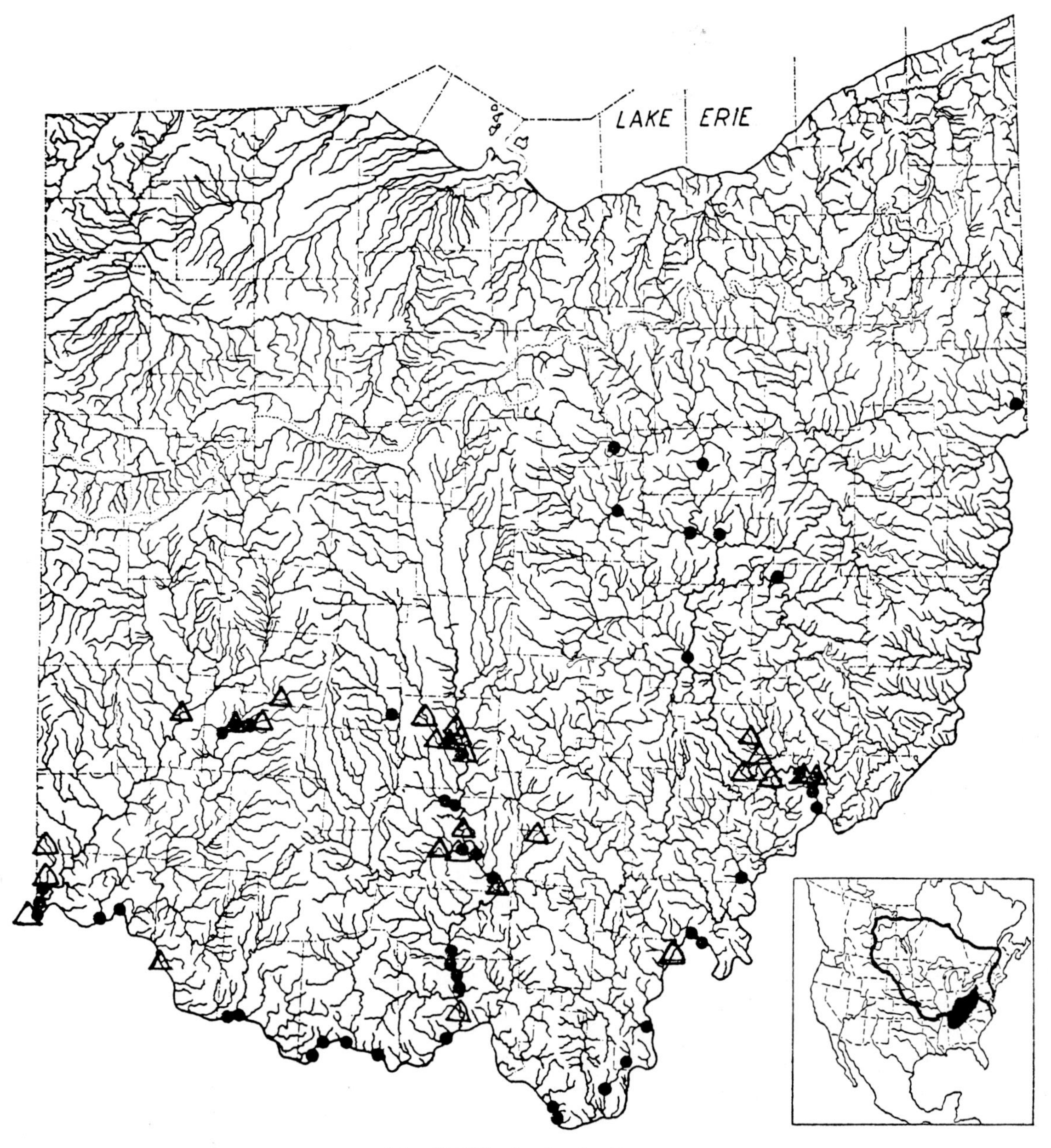

Map 97. Ohio shorthead redhorse

Locality records. ● Before 1955. △ 1955–80.
Insert: No apparent recent change in range size.

ous in the Ohio River between Marietta and the Pennsylvania state line.

Habitat—The Ohio shorthead redhorse was most numerous in those larger streams which had a moderate- or swiftly-flowing current, relatively clear water throughout most of the year, and pools containing bottoms of sand and gravel which were not smothered beneath a thick layer of clayey silt; the Walhonding River in Coshocton County was such a stream. In spring, spawning adults were also found in quite small, swiftly-flowing streams. The species seemed to be particularly susceptible to most industrial pollutants and was killed by pollutants which apparently had little effect upon such resistant species as the spotted and smallmouth blackbasses.

The range of the Ohio shorthead redhorse includes all of the Ohio River basin except the Wabash River and the Ohio River from the mouth of the Wabash downstream to the Mississippi River. In the Wabash and lower section of the Ohio River the northern shorthead replaces the Ohio shorthead redhorse. These two redhorses are currently considered to be species, and this concept appears valid when Ohio shorthead redhorses from the Muskingum River and the Ohio River above Portsmouth are compared with northern shortheads from Lake Erie. However, individuals collected in the Ohio River near the Indiana state line are more intermediate in their characters, and it seems quite probable that when a sufficient number of specimens have been collected from the Ohio River between the Indiana state line and the mouth of the Wabash, they will prove that the northern shorthead and Ohio shorthead redhorse are only subspecifically distinct (Trautman and Martin, 1951:8–9).

Years 1955–80—The Ohio shorthead redhorse ("Southern Shorthead"-Clay, 1975:113) was another Moxostomine whose numerical abundance could be ascertained only when diversified collecting gear was used in an area for several days or weeks (Brown, 1960:54). After utilization of such gear, it became apparent that this inhabitant of rather deep waters was more numerous and widespread in its Ohio distribution than was generally supposed. Its numerical abundance decreased wherever water quality deteriorated but increased where water quality improved, as it did in the Scioto River in the vicinity of the confluence with Big Darby Creek after water quality in the Scioto had improved.

Between 9 a.m. and noon of the heavily overcast day of May 7, 1963 Donald Veth and I observed an estimated 1,000 suckers, genera *Hypentelium* and *Moxostoma*, spawning in Big Darby Creek, central Darby Township, Pickaway County, water temperature 58° F (14° c). In this group were eight adult Ohio shorthead redhorses spawning in 10″–15″ (25–38 cm) of water on a rather rapidly flowing riffle overshadowed by trees. Spawning occurred over the same types of redds, consisting of gravels 1/4″–2″ (0.6–5 cm) in diameter, and apparently in the same types of habitat as were those of the spawning golden redhorses which surrounded them. It may have been the overcast sky and shade from the trees which aided in producing this spawning activity at this time of day; usually such activity is confined to the twilight hours or night.

As suggested previously, further study (Jenkins, 1971) has shown that *breviceps* is conspecific with *macrolepidotum*, thereby becoming *Moxostoma macrolepidotum breviceps* (Cope).

GREATER REDHORSE

*Moxostoma valenciennesi** Jordan

Fig. 98

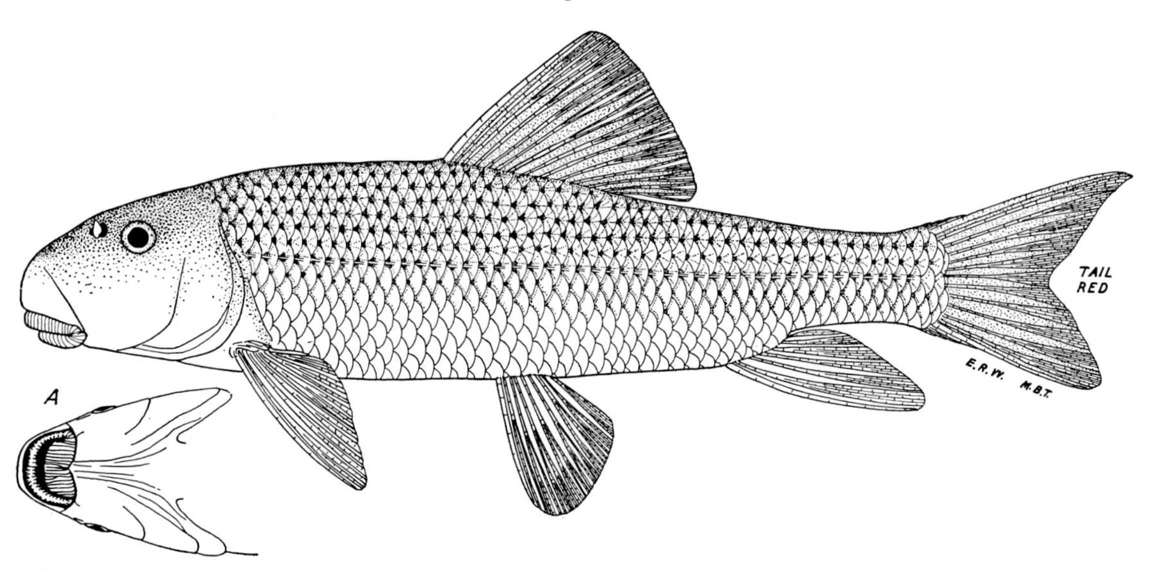

Upper fig.: Gordon Creek, Paulding County, O.

Aug. 31, 1930. 260 mm SL, 12.7″ (32.3 cm) TL.
Young adult. OSUM 668.

Fig. A: ventral view of head; note the slight angle or indentation at junction of the posterior edges of the two lower lips.

Identification

Characters: *General—See* blue sucker. *Generic—See* silver redhorse. *Specific*—Similar to the shorthead and river redhorses in having a red or pink tail in life and distinct, dark spots at the scale bases (both characters faint in small young). Head bulky and long, the forehead and occipital regions much rounded, the head usually contained less than 4.3 times in standard length of yearlings and adults; 3.0–3.8 times in small young. Mouth large, lips coarse and deeply striated, the posterior edges of the two halves of the lower lip forming a slight angle. Distal edge of dorsal fin usually very convex in adults, almost straight in young. Eyes small, usually contained 4.0 or more times in fishes less than 7.0″ (18 cm) long; 5.0 or more times in fishes 8.0″–14.0″ (20–35.6 cm); 7.0 or more times in fishes over 17.0″ (43.2 cm) long. Breadth of pharyngeal arches generally less than their depths, the teeth moderately compressed and comb-like (*see* golden redhorse, fig. B). Tubercles on breeding males confined to anal and caudal fins.

Differs: Silver, black and golden redhorses and spotted sucker have slate tails. Shorthead and Ohio redhorses have shorter heads. River redhorse has a larger eye and squarish pharyngeal teeth.

Most like: River redhorse.

Coloration: Dorsally a brownish-or greenish-olive with a rich, bronze overcast, the scales with

* In 1844 Valenciennes (Cuvier et Valenciennes, 1844:457, Pl. 517) described *Catostomus carpio*. Jordan (1886:73), realizing that Rafinesque (1820:56) had previously used the name *Catostomus carpio* for another species, offered the substitute name of *valenciennesi* for *carpio*. In 1868 Günther (1868:20) wrongly used *carpio* for the species now called *Moxostoma anisurum*, and his action was followed by others such as Jordan (1878:115 and 118). However, in 1942 Legendre (1942:227–33), while attempting to identify his recently found copper redhorse, realized that *valenciennesi* was not *anisurum*, and believing that his newly discovered sucker was described previously by Valenciennes, he used the substitute name *valenciennesi* for it. Later, in 1952, Legendre (1952:VI) realized that *valenciennesi* was not the copper redhorse but was the species which Hubbs (1930 A:24–28) described in 1930 as *Moxostoma rubreques*. Legendre therefore placed *rubreques* in the synonomy of *Moxostoma valenciennesi*, the greater redhorse (*see* that species). That left the copper redhorse without a specific name, so Legendre (1952:VI) gave it the current name of *Moxostoma hubbsi*.

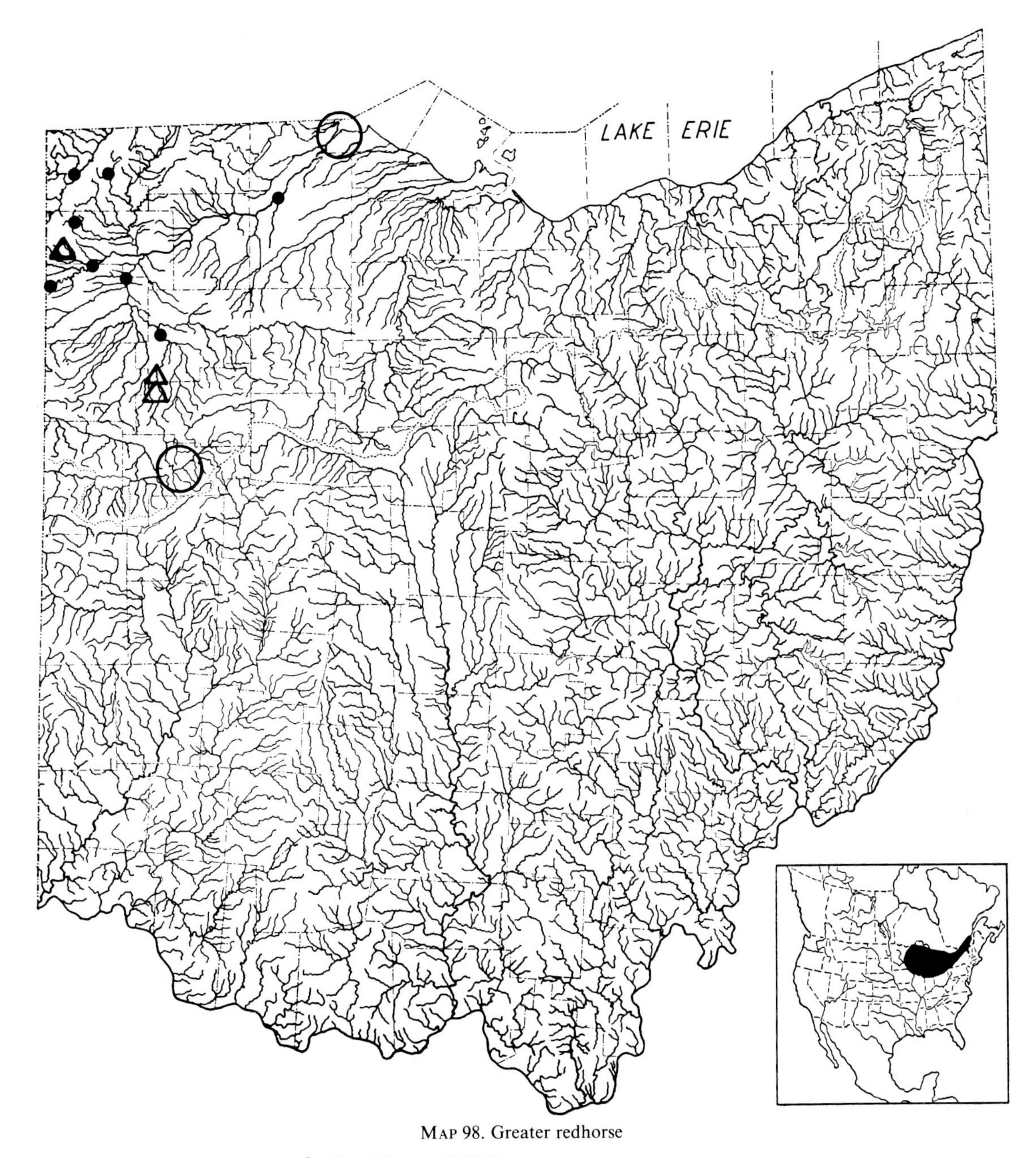

MAP 98. Greater redhorse

Locality records. ○ Before 1900. ● After 1920. △1955–80.
Insert: Range may have extended farther southwestward formerly; size of unoccupied areas within range appears to be increasing.

dark spots at their bases. Sides lighter and more golden. Ventrally a golden- or milk-white. Dorsal, caudal, and anal fins a deep lake-red, lightest in small young. Anterior rays of pelvic and pectoral fins whitish, remainder of pelvics light reddish; pectorals olive.

Lengths and **weights:** *Young* of year in Aug., 1.5″–3.0″ (3.8–7.6 cm) long. Largest specimen seen, a decomposing fish, was 24.5″ (62.2 cm) long; estimated weight 3–5 lbs (1.4–2.3 kg). Hubbs (1930A: 17) believes that the species attains a weight of "12 or 16 pounds." (5.4–7.3 kg).

Distribution and Habitat

Ohio Distribution—In the CAS (18141) there is a gill-netted specimen, 110 mm in standard length, labeled "Toledo, O." This specimen was probably collected by P. H. Kirsch or A. J. Woolman between 1890–1900, and probably was captured either in the lower Maumee River or in Maumee Bay. In the USNM (76155) there are two specimens, 133 mm and 150 mm in standard lengths, labelled *Moxostoma aureolum*, which were taken by Philip Kirsch on Aug. 4, 1893 in the Auglaize River at Wapakoneta, Auglaize County.

Since 1925 the greater redhorse appears to have been confined in Ohio to the Maumee system. On the distribution map are four dots, representing localities on the Maumee River, and on Gordon Creek in Paulding County, Lost Creek in Definace County, and St. Joseph River in Williams County; in these localities preserved specimens have been obtained. The four remaining dots represent localities where I found large, dead fishes too decomposed for preservation but still identifiable; all of these fishes presumably had been killed by soil or industrial pollutants. On Sept. 2, 1933 in the Maumee River near Waterville, Lucas County, I found about two dozen of these decomposing suckers, some of which were slightly more than 2′ (70 cm) in length. The gill chambers of these suckers were packed with clay mixed with mucous; presumably the clay had caused asphyxiation.

I have shown specimens of the greater redhorse to commercial and sport fishermen who insisted that they knew this species well; some stated that between 1890–1920 the species was rather numerous in the Maumee River; others that it was formerly present in the Sandusky, Vermilion and Grand rivers, or in Lake Erie near Toledo, Sandusky, Huron, Lorain, Conneaut, or about the Bass Islands. Other fishermen who did not see the preserved specimens accurately described the species to me, stressing particularly its huge head and weight of as much as 10 lbs (4.5 kg). This circumstantial evidence suggests that the species was present in Lake Erie and its tributaries before 1920 and that in recent years it has been extirpated from much of its former range in Ohio.

In 1929 I sent to Carl L. Hubbs a small specimen of a greater redhorse which I had found in an old collection at the Ohio Division of Conservation; with the collection was a label upon which was written "Ohio River Drainage." Hubbs (1930A:25) mentioned this specimen (UMMZ 85862) in his description of the greater redhorse. This Ohio River drainage record I now consider unacceptable because: of the lack of a more definite locality and other essential data; of the possibility that the specimen was taken in the Lake Erie drainage and placed in the wrong bottle; the species has not been found since in the Ohio River drainage in Ohio.

Habitat—From my observations of the greater redhorse in such clear-water streams as the Au Sable River in Michigan, and the observations of Hubbs and Lagler (1947:52) and others, it is apparent that the species primarily inhabits large streams having clear waters throughout most of the year and bottoms of clean sand, gravel or boulders. The dead specimens I have found, with their gills impacted with clay or otherwise injured, indicate a high degree of intolerance to excessive turbidity and chemical pollutants; this evidence suggests that it is these adverse factors which have been responsible for the decrease in abundance of the species in Ohio waters during the past century.

Years 1955–80—The decrease in the numerical abundance of the greater redhorse, apparent prior to 1950, has continued since. After fruitlessly searching for a relict population, especially in the Maumee River system, one was found in 1967 by Darrell Allison and Eric Angle in the Auglaize River, Putnam and Allen counties south of Fort Jennings. In June of 1968 there were 21 fishes taken; 3 more were captured in July of 1969, all between 10.25″ and 20.75″ (26.04 and 52.71 cm) TL.

This population exists in the Auglaize River from its confluence with the grossly polluted Ottawa River upstream an undetermined distance. It occupies a seemingly precarious position in this stream section because of possible increased pollution from the cities of Delphos and Wapakoneta, oil wells and increased enrichment from the huge amounts of commercial fertilizers, liquid and solid, which are annually applied on the soils of this intensively farmed section.

Darrell Allison advised me that between June 12–16, 1978 fyke nets employed by the Ohio Division of Wildlife captured and released 11 greater redhorse suckers from the Auglaize River, Putnam County which were 12″–19″ (30–48 cm) TL; 10 individuals in Allen County, 12.5″–17.5″ (31.8–44.5 cm) TL. This small population continues to survive and will so long as there is no marked increase in amount of pollutants.

On July 18, 1973, during our investigations of Gordon Creek, Donald K. Gartman and I met two brothers, Paige and Richard Craig, along the North Branch near Cicero. A few moments before our arrival, one of the brothers had caught a greater redhorse, 20″ (51 cm) TL. All of us immediately seined several hundred yards of the stream, Gartman and I reseining it a few weeks later, without seeing or capturing another. The North Branch at Cicero is at the base of the Fort Wayne end moraine and is quite small for Moxostomine spawning purposes. However, the gradient and substrate of sand and gravel was suitable, a condition found only in the undredged headwaters of the otherwise low gradient Gordon Creek (Trautman and Gartman, 1974:171).

RIVER REDHORSE

Moxostoma carinatum (Cope)

Fig. 99

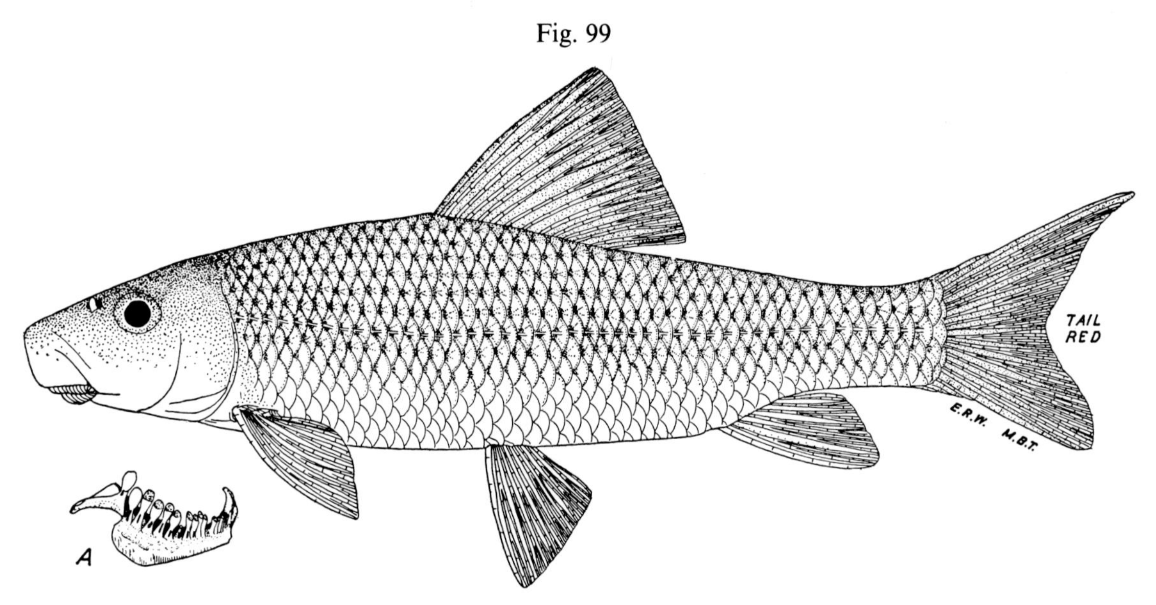

Upper fig.: Walhonding River, Coshocton County, O.

Oct. 20, 1939. 125 mm SL, 6.0″ (15 cm) TL.
Immature. OSUM 995.

Fig. A: left pharyngeal arch showing the heavy molar-like teeth; note that the anteriormost two are just developing and have not attained their grinding surfaces, others have grinding surfaces, and that two teeth are missing.

Identification

Characters: *General—See* blue sucker. *Generic—See* silver redhorse. *Specific*—Similar to greater redhorse in having a red tail in life; dark spots at scale bases; a long, bulky head which is usually contained fewer than 4.3 times in standard length of yearlings and adults, and 3.0–3.8 times in small young; a large mouth, with coarsely striated lips, the posterior edges of the two halves of the lower lips forming a slight angle. The distal edge of the dorsal is usually straight or slightly convex. The eye is larger than in the greater redhorse, and is contained 4.0 or fewer times in fishes less than 7.0″ (18 cm) long; 5.0 or fewer times in fishes 8.0″–14.0″ (20–35.6 cm); 7.0 or fewer times in fishes more than 17.0″ (43.2 cm) long. The head is rather flat between the eyes, the occiput not highly arched. Breadth of pharyngeal arches as great as, or greater than, their depths in cross section; the teeth large, squarish and with large grinding surfaces, fig. A. Tubercles on snout of breeding male; also on anal and caudal fins.

Differs: *See* greater redhorse.

Most like: Greater redhorse. **Superficially like:** Golden redhorse.

Coloration: Similar to greater redhorse except that the tail of adult is a lighter red, and the pelvics are less reddish and more olive.

Lengths and **weights:** *Adults*, usually 13.0″–24.0″ (33.0–60.9 cm) long; weight 1 lb–7 lbs (0.5–3.2 kg). Individuals more than 4 lbs (1.8 kg) common. Largest specimen, 29.0″ (73.7 cm) long; weight 10 lbs, 8 oz (4.8 kg). Rivermen give maximum weights of 12–14 lbs (5.4–6.4 kg).

Distribution and Habitat

Ohio Distribution—The river redhorse was not recognized as a distinct species until 1870 because of its superficial resemblance to other suckers in the confusing group of Moxostomine fishes. It was not until 1882 that the species was known to be present in Ohio waters, when Jordan (1882:831–32; as *Placopharynx carinatus*) wrote that a skeleton was

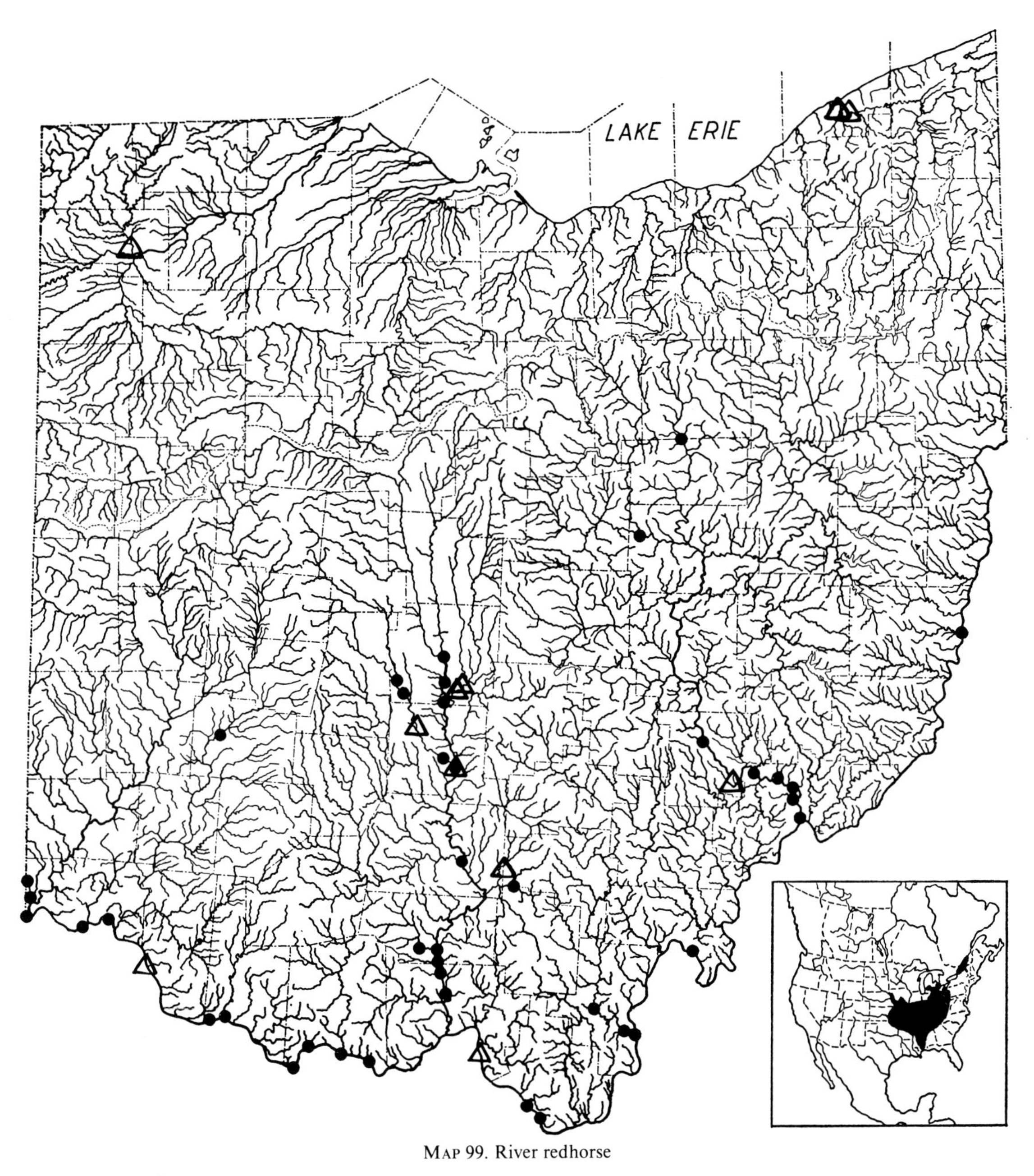

MAP 99. River redhorse

Locality records. ● Before 1955. △ 1955–80.

Insert: Bones in debris from former Indian village sites suggest that during late pre-Columbian times the now disjunct Mississippi River drainage and St. Lawrence River drainage populations were connected by populations inhabiting Lakes Erie and Ontario and their tributary waters.

found by "Dr. J. M. Wheaton in the Scioto River near Columbus." Despite the lack of factual evidence, it may be assumed that the species has been long present in that portion of the Ohio River bordering the state of Ohio and in its larger tributaries within the state.

During the 1925–34 period the river redhorse was most numerous in the Ohio River from Portsmouth to the Indiana state line and in the lower portions of the Scioto and Muskingum rivers; in these sections I captured with the aid of gill-nets and traps, as many as 100 adults in a 24-hour period. The species was less numerous in the upper halves of the larger tributaries. It was quite uncommon in the Ohio River between Portsmouth and the Pennsylvania state line, presumably because of the presence of the large amounts of industrial pollution. The species appeared to be less numerous throughout the state after 1940 than it was before 1930.

This redhorse was probably more abundant and widespread in its Ohio drainage distribution than the records on the distribution map indicate; it was seldom captured in the smaller collecting seines. Its presence in a locality was often unsuspected until a mass killing of fishes by pollution occurred, then sometimes many individuals were found. The records for Greene, Vinton, Belmont and Wayne counties are of individuals killed by pollution.

Published references to the presence of the river redhorse in the Lake Erie drainage of Ohio previous to 1900 are questioned because of lack of specimens and because of the confusion then existing relative to the redhorse suckers. McCormick (1892:16) in 1892 stated that the river redhorse was "common among other mullets" in the Lake Erie waters of Lorain County, and that "The body seems more compressed and the outlines have different curves." This latter statement is not suggestive of the river redhorse, for it has a terete body and the usual redhorse body outline. McCormick's statements are weakened further by his unpublished notes which I examined at Oberlin College, and in which was the following "I have looked in vain for it [river redhorse] among the other mullets at Lorain."

Habitat—Since 1925, the largest populations have occurred in the deeper waters of the Ohio River and lower portions of its larger tributaries; smaller numbers were present in the upper portions of these tributaries, especially during the spawning season. Since 1940, the largest inland populations appear to have been centered in those sections of the Muskingum River which were the least polluted and had the least turbid waters. The Scioto River, with its more turbid waters, contained smaller numbers, and the heavily polluted Great Miami River contained few. From this avoidance of polluted areas, it is apparent that the river redhorse like most Moxostomine species, is intolerant to much turbidity and siltation.

Years 1955–80—During this period only five river redhorses were recorded for the Scioto River drainage of central Ohio despite much collecting. It is surprising that only one individual of this deep-water-inhabiting redhorse was taken in the trap nets and other collecting gear in that portion of the Scioto River near the mouth of Big Darby Creek, and especially since the improved water quality there apparently resulted in increased numbers of other sucker species of the genus *Moxostoma*. Note above the numbers captured before 1957.

What evidence was obtained elsewhere in inland Ohio suggested that this redhorse was decreasing in abundance. Those reliable commercial and sport fishermen who recognized this species stated that it had decreased in abundance in the Ohio River since 1950. This assumption is strengthened by investigations made by personnel of the University of Louisville, they recording only 40 specimens, and those from only four localities in the entire Ohio River (ORSANCO, 1962:149).

After the 1957 edition of *The Fishes of Ohio* was published, I became aware of the 1893 capture by Philip Kirsch of this species in the Maumee drainage in Ohio. On August 18, 1893 Kirsch collected fishes in the Tiffin River, about 2 miles (3.2 km) above its confluence with the Maumee River, Defiance County. Three specimens of *Moxostoma* were preserved, and originally were in the collections of Indiana University, IU 6479. They were catalogued as *M. anisurum*. However, only one of the three specimens is *anisurum*, the others are, one *M. erythrurum* and one juvenile *M. carinatum*. In Kirsch's (1895:328) listing of the genus *Moxostoma* only the species *anisurum* is listed from the above collection, he stating that this species was "Not scarce." These misidentifications are readily understood because at that time great confusion existed as to speciation within the genus *Moxostoma*. There is no reason to question this record, because about that time specimens were collected in Michigan, in other tributaries of western Lake Erie. At present the specimen is in the University of Michigan, Museum of Zoology collection.

Goslin (1943:50; as *Placopharynx carinatus*) records having found maxillary bones and part of the pharyngeal arch of the river redhorse at a prehistoric Indian village site at Fairport, Lake County, near the mouth of the Grand River. In 1938 Carl L. Hubbs, E. C. Case and I identified these bones as belonging to the river redhorse. In 1974, while routinely identifying the John Carroll University collections, Andrew White found an individual of this species which may have been collected in 1969 from some stream in northeastern Ohio by

Edwin J. Skoch and class. At the time, White and I were inclined to question the locality because collections were being made in the Ohio River drainage and the possibility existed that during student class use this specimen had been misplaced.

In June of 1977 White and several students began a thorough investigation of the Grand River from the Ashtabula-Lake County line downstream. They made 24-hour-trap and gill-net collections at nine collecting stations within 3 miles (4.8 km) of the mouth of the Grand River, obtaining such pollution-tolerant species as carp, goldfish, bullheads and white suckers, but no individual of any species of the genus *Moxostoma*. All of these stations were below the outlets of two large chemical plants.

Trap and gill nets were set on June 29 between the outlets of the first and second chemical plants. On June 30, White and his daughter Pamela, Frank Banks, Neil Bernstein, Miles Coburn and I removed from this gear, taken from the trap, besides other species, 30 silver suckers, 2 black redhorses, 6 golden redhorses, 1 northern shorthead redhorse and 2 river redhorses. The largest river redhorse was an adult male, 450 mm SL (OSUM 35500) with the breeding tubercle scars still evident on snout and anal fin; the other, a juvenile male, 261 mm SL (JCU 1807). Presumably, the river redhorse has been present in the Grand River since pre-Columbian times.

Apparently the section of the Grand River where *M. carinatum* was captured was well suited formerly for this species. The substrate consisted of bedrock boulders and gravels which should have been a favorable habitat for snails and other aquatic invertebrates, presumably the principal foods of this redhorse. Today this section appears to be unsuited for pollution-intolerant Moxostomines. There were discarded automobile tires and other rubbish, and the boulders and refuse along the banks were stained a reddish-brown. There were windrows of a granular substance. It is possible that these Moxostomines had drifted downstream since spawning farther upstream, and had concentrated above the outlet of the lower chemical plant.

HARELIP SUCKER

Lagochila lacera Jordan and Brayton

Fig. 100

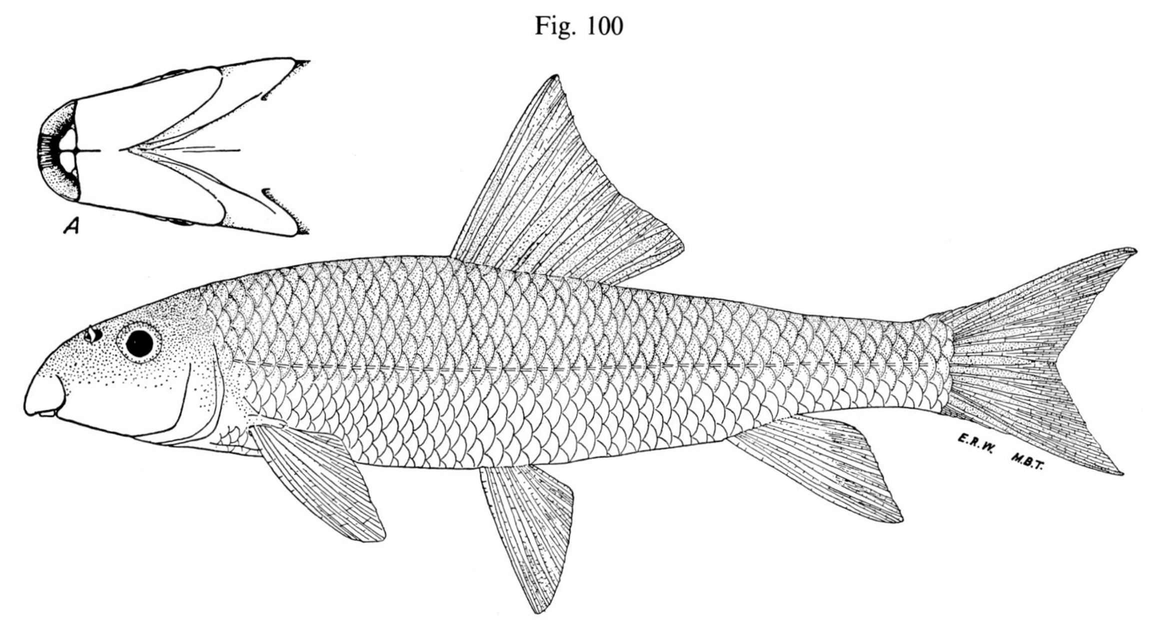

Lower fig.: Blanchard River, Putnam County, O.

In 1893, presumably Aug. 8. 105 mm SL, 5.5″ (14 cm) TL.
Immature. OSUM 1.

Fig. A: ventral view of head; note peculiar lips, and their unstriated lower halves.

Identification

Characters: *General—See* blue sucker. *Generic* and *specific*—Similar to the redhorses except for the mouth. There is no groove separating the tip of the upper lip from the snout. The halves of the lower lip are completely separated into distinct lobes, these halves are smooth or only slightly roughened, fig. A. No distinct dusky spots on the scale bases. No specimens have been taken in Ohio or elsewhere during the past 75 years.

Differs: Other species of suckers have a deep groove separating the tip of upper lip and snout.

Superficially like: Black and golden redhorses.

Coloration: No accurate color description apparently exists. Because of lack of dusky scale bases, the tail should be slate, and the general coloration similar to that of the silver, black and golden redhorses.

Lengths and **weights:** The three specimens collected in Ohio, are 115 mm SL, 140 mm TL (OSUM:1); 102 mm SL, 127 mm TL (SNHM: 5158); 116 mm SL, 143 mm TL (CNHM:1847). Klippart (1878:105) states that the fish collected in the Scioto River at Columbus "weighed several pounds." Jordan (1882:833) referring to the "large specimen" sent to him by Klippart, gives maximum lengths of "1 to $1^1/_2$ feet." (30 to 45 cm).

Distribution and Habitat

Ohio Distribution—In April of 1878, Klippart (1878:105) sent "a fine large specimen" of harelip sucker to Jordan (1882:833) which was captured "at Columbus, just below the State dam" in the Scioto River. Klippart informed Jordan that this species was well known to Columbus fishermen who called it "May Sucker" because it spawned in May (as do the species of the closely-related genus *Moxostoma*). Presumably Jordan gave this large specimen to the Cincinnati Society of Natural History, for in the Museum catalog book is the

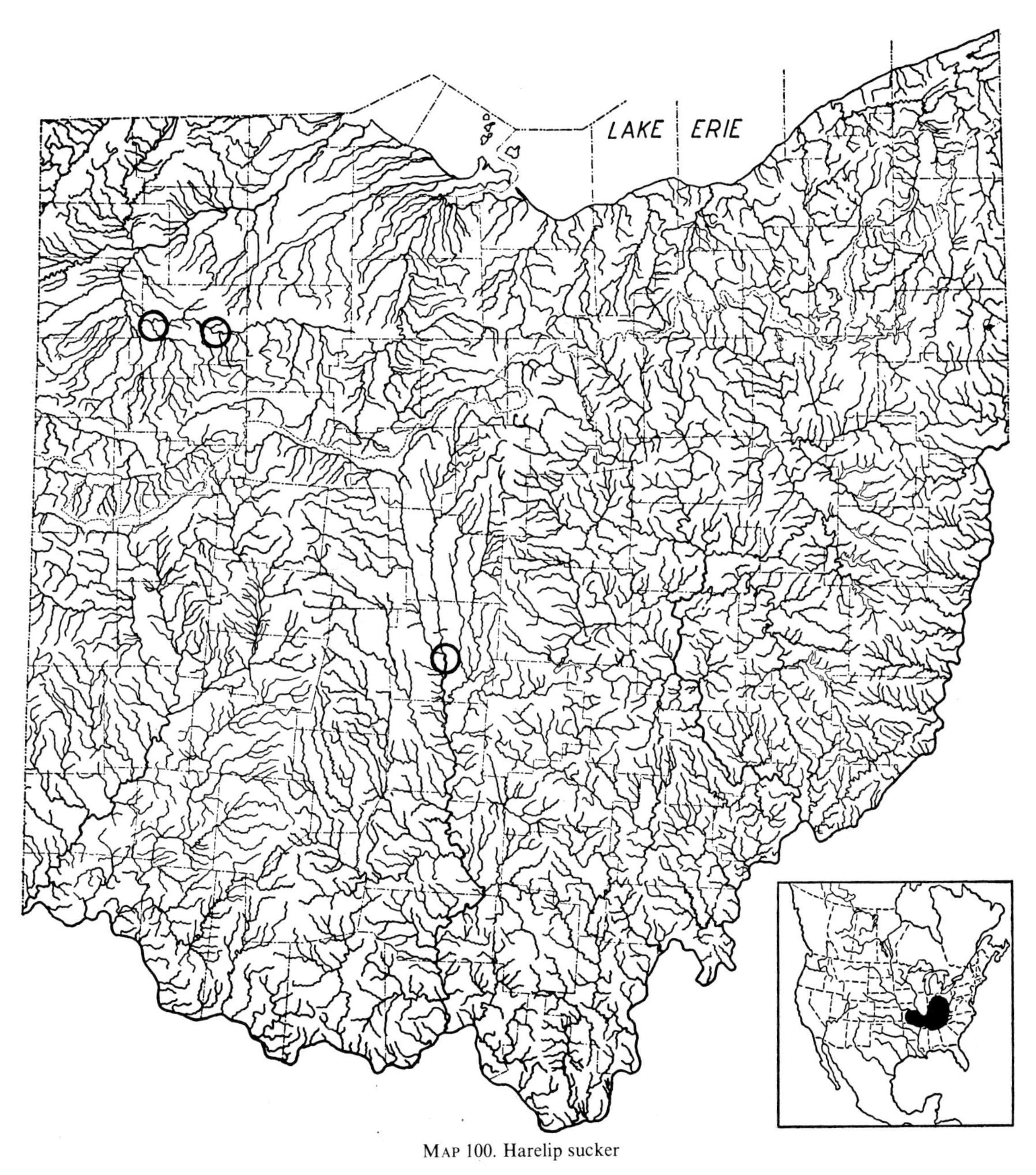

MAP 100. Harelip sucker

Locality records. ○ Before 1900.
Insert: Black area surrounds the scattered localities where the species was formerly collected.

following: "Identified by Jordan and sent to him by J. H. Klippart. Scioto River near Columbus. April, 1878." The specimen appears to be no longer extant.

On August 8, 1893, Kirsch (1895A:328) collected "many small specimens" in the Blanchard River at the village of Ottawa, Putnam County, of which two (CNHM:1847 [236] and OSUM:1) are preserved. On August 9–10, 1893 he collected "one specimen, 5 inches long" in the Auglaize River near Cloverdale, Putnam County. It likewise is preserved (SNHM:5158). Since about 1900, no specimens have been taken in Ohio or elsewhere, and the species may be extinct throughout its range. Its former range extended from Georgia, Alabama and

Arkansas northward to Indiana and Ohio. In Indiana this sucker was called pea-lip sucker (Evermann and Jenkins, 1888:45).

Habitat—Kirsch (1895A:324–26) described the stream sections where he collected his specimens as having bottoms of limestone bedrock, coarse gravel, sand, and a hard *whitish* clay. Since 1925 the stream bottoms in these localities have been covered with fine silts, and it is only when one removes the layer of silt that the bedrock, gravel, sand, or hardpan of white clay is exposed. Between 1850–1900 those portions of the Scioto and Maumee systems which originally were of the clear-water, prairie-stream type, have been converted into turbid streams of the western plains-type. It is this destruction of clear water and clean stream bottoms which apparently has been the principal factor in the extirpation of the harelip in Ohio waters. The harelip, with its small, specialized mouth and closely bound gill covers must have been particularly susceptible to asphyxiation through impacting about the gills by colloidal clays.

Years 1955–80—Although increased efforts were made to capture this sucker in Ohio and elsewhere, no specimens were obtained, and it is now generally assumed that the species is extinct.

Jenkins (1971) theorizes that this species was trophically specialized and fed by sight, unlike other catostomids which were more adapted to taste feeding, and that even prehistorically this species probably was in delicate balance with its environment.

Upon several occasions I have found massive killings of moxostomine suckers in a stream below a field in which a farmer had been cultivating previous to a hard rain storm. From this field rivulets were depositing water heavily laden with silt into the stream, causing heavy turbidity. Examination of the dead suckers disclosed that the gills and entire gill-chambers were compacted with silt, heavily saturated with mucous. Obviously these suckers had died of asphyxiation. The small and unusually shaped mouth of the harelip sucker restricted the passage of water across the gills more than do the larger mouths of species of *Moxostoma*, and together with the broad isthmus and restricted gill openings, resulted in decreased ability to flush out mucous and sediment from the gill chambers. Because the harelip was presumably a sight feeder, the plowing of the prairie soils and resultant converting of the waters of streams from clear to heavily sediment-laden should have been sufficient to cause extinction of the species.

NORTHERN HOG SUCKER*

Hypentelium nigricans (Lesueur)

Fig. 101

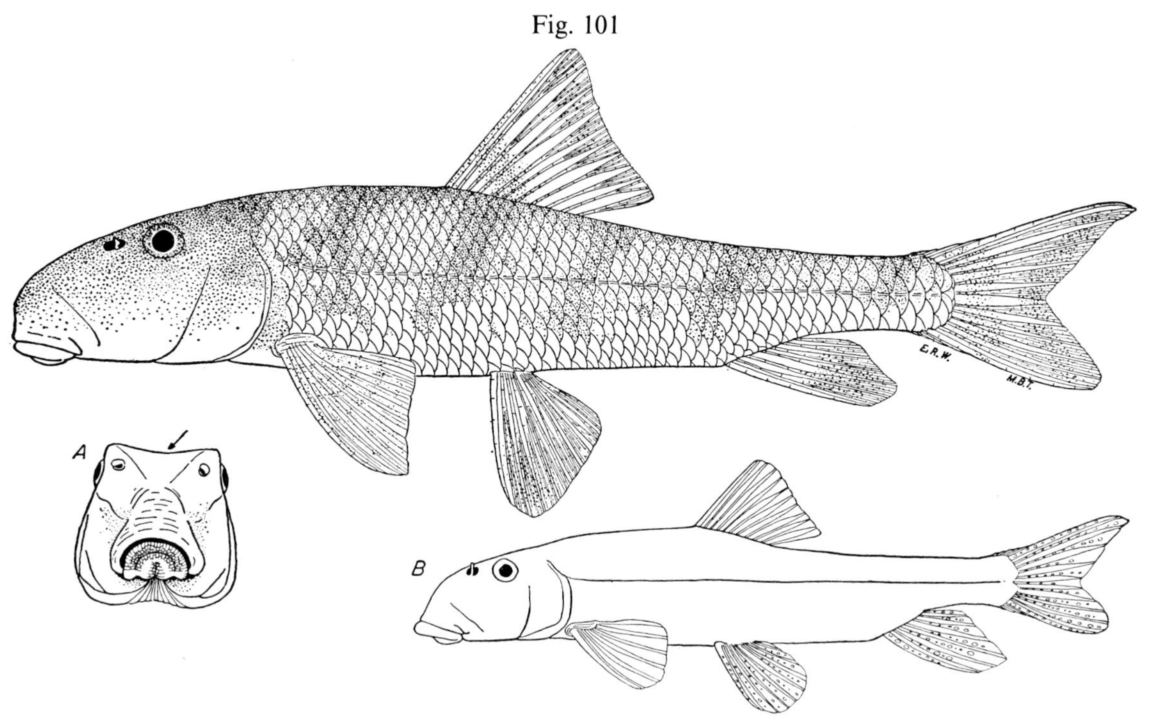

Upper fig.: Walhonding River, Coshocton County, O.

Sept. 30, 1939. 140 mm SL, 6.9″ (18 cm) TL.
Female. OSUM 770.

Fig. A: frontal view of head, the arrow pointing to the depression between the eyes. No other species of sucker in Ohio is depressed between the eyes.

Fig. B: Little Bull Creek, Columbiana Co., O.

May 12, 1942. 173 mm SL, 8.5″ (2.2 cm) TL.
Breeding male. OSUM 5323.
Note breeding tubercules on pelvic, anal, and caudal fins.

Identification

Characters: *General—See* blue sucker. *Generic* and *specific*—Head wide and strongly depressed between the eyes, the orbital rims elevated, fig. A. Lips heavily papillose. Body with 4–6 dark, oblique saddle-bands. Complete lateral line of 46–54 scales. Air bladder in two parts. Breeding male has tubercles on pelvic, anal, and caudal fins, fig. B; in old males there are minute tubercles on some body scales. Breeding females often have small tubercles on the fins.

Differs: No other species of sucker has the head depressed between the eyes.

Superficially like: Small specimens resemble small redhorses and chubsuckers.

Coloration: Dorsally a dark bronzy- or brownish-olive, lightest in the smallest young, with 4–6 dark chocolate, brown, or dusky oblique saddle-bands. Sides lighter and more bronzy-golden, the snout lighter, lips white, and the groove between upper lip and snout black. Ventrally pale yellowish- or milk-white. Dorsal and caudal fins light olive, remaining fins whitish, often tinged with yellow or olive. *Young*—Sometimes sharply bicolored.

* Scales on caudal peduncle too few: 16 usually encircle peduncle.

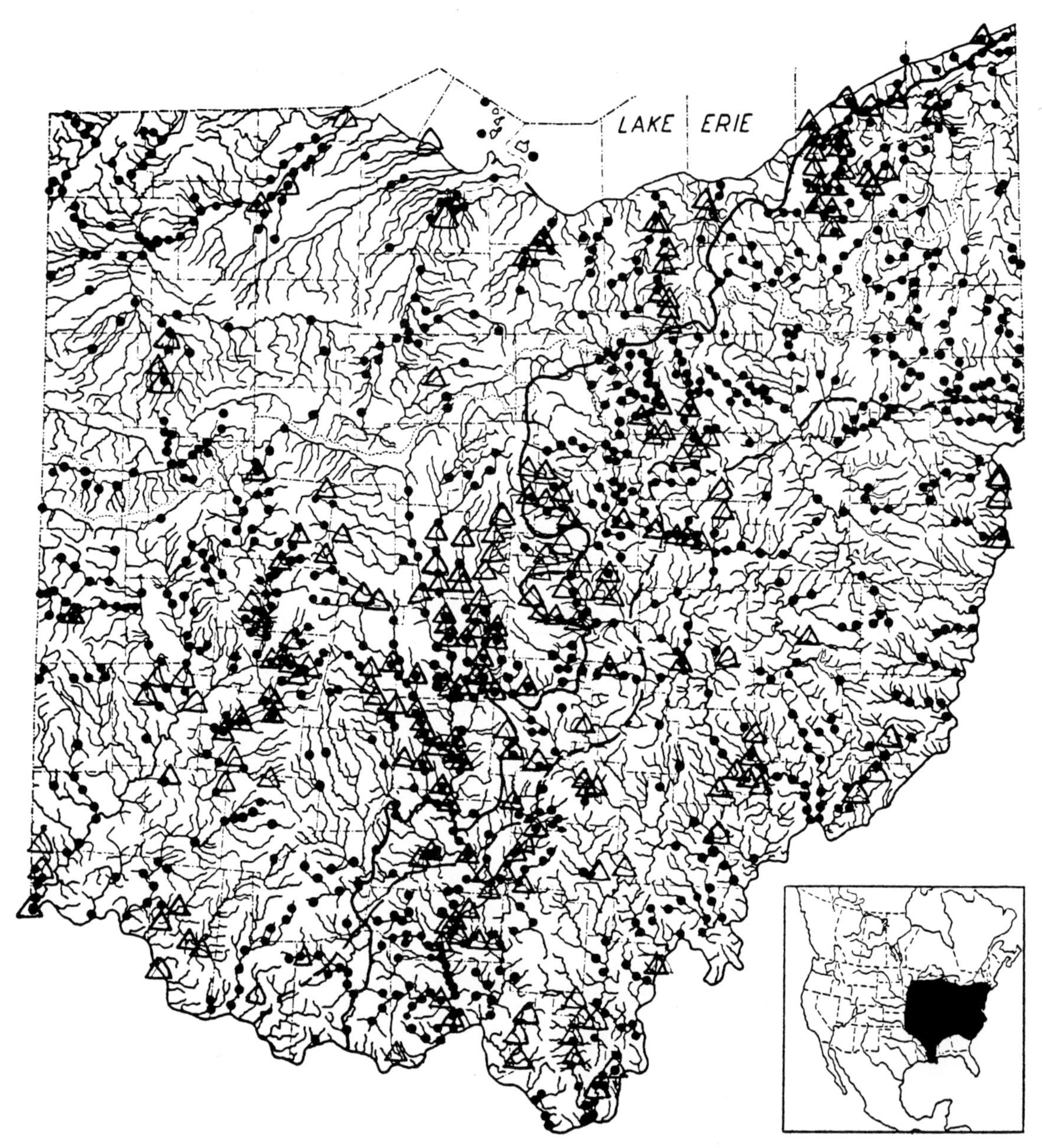

MAP 101. Northern hog sucker

Locality records. ● Before 1955. △ 1955–80. ~ Allegheny Front Escarpment. /\ Glacial boundary.
Insert: Western limits of range indefinite.

Lengths and **weights:** *Young* of year in Oct., 2.0″–3.5″ (5.1–8.9 cm) long. *Adults*, usually 7.0″—14.0″ (18–35.6 cm) long; weight 2 oz–1 lb 8 oz (57 g–0.7 kg). Dwarfing is common, especially in small streams of high gradients; in such streams males breed when only 4.0″ (10 cm) long. Largest specimens were taken in the Ohio River and as strays in Lake Erie. Largest specimen, from Lake Erie near Kelleys Island, was 22.5″ (57.2 cm) long; weight 4 lbs, 3 oz (1.9 kg). Lake and rivermen give maximum length of 24.0″ (60.9 cm); weight, 5 lbs (2.3 kg).

Distribution and Habitat

Ohio Distribution—Although the highly-migratory northern hog sucker superficially appeared to be distributed rather uniformly throughout the streams of Ohio, it occurred in abundance only in certain localities at certain seasons of the year. During the spawning season from late March to early June, the species congregated wherever streams had moderate- or high-gradients and well-defined, stony riffles. It was particularly abundant along the line of glaciation, along the Allegheny Front Escarpment, in the Ohio River tributaries from Lawrence County to the Pennsylvania state line, in the upper Great Miami system, and in the morainal sections of Williams and Defiance counties. The smallest populations occurred in the low-gradient streams of the Black Swamp area of northwestern Ohio, and streams of low gradients in southeastern Ohio, especially those containing considerable mine pollution.

After the spawning season many adults and some of the young began to drift downstream into larger waters which often had lower gradients. They moved downstream throughout late summer and fall until they reached the larger, deeper waters where they wintered, and where they remained until the water temperature rose to above 40° F (4° C), after which they began their vernal upstream migrations.

This inhabitant of clear, clean, swiftly-flowing waters was an abundant species before 1900 as indicated by museum specimens and the literature. Since 1900 it has been one of the most abundant of Ohio suckers.

Habitat—Throughout the warmer half of the year the largest numbers were found in those stream sections where the gradients were moderate or high, the well-defined riffles had bottoms of clean sand, gravel, boulders and bedrock, and the pool bottoms were comparatively free of clayey silts. During the winter the largest numbers were found in deeper and more quiet waters, and only the young and smaller adults remained in the faster waters. Small populations occurred in the larger impoundments, and only strays were present in small impoundments and in Lake Erie.

The northern hog sucker was intolerant to habitually turbid waters, to most of the domestic and industrial pollutants, and especially to those stream sections of low gradients where silt and other pollutants tended to settle upon the stream bottom. Individuals were found in the swiftest-flowing portions of the riffles, where their stream-lined bodies, widely-spread pectoral fins, and the downward push of the current upon their depressed heads held them in position (even recently killed specimens will remain on the bottom without being swept downstream if properly placed). The species avoided profuse amounts of aquatic vegetation.

Years 1955–80—The distribution map of the highly migratory northern hog sucker gives the impression that it is quite uniformly distributed throughout inland Ohio, which is true as far as locality records are concerned. It is not true, however, as regards numbers of individuals in any locality, because the numbers of individuals range from abundant in moderate- and high-gradient streams with comparatively little turbidity to few individuals or migrating strays in waters of low gradient and rather high turbidity. From general observations it is believed that this once abundant species in Ohio decreased in numbers during this period. Relatively few large spring concentrations were observed, such as were frequently noted prior to 1950.

COMMON WHITE SUCKER

Catostomus commersoni commersoni (Lacepède)

Fig. 102

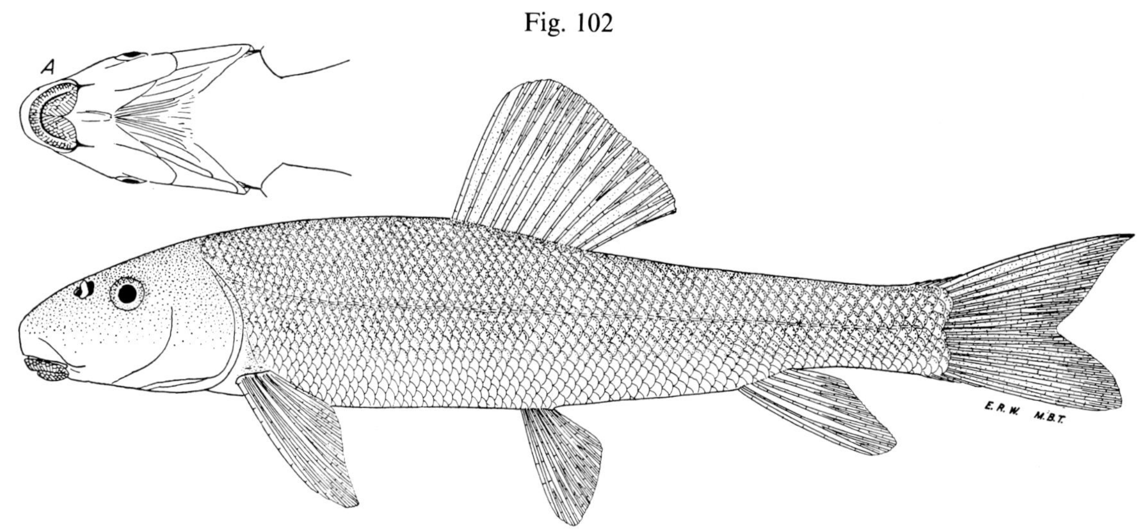

Lower fig.: Vermilion River, Ashland County, O.

Aug. 7, 1939. 163 mm SL, 8.0″ (20 cm) TL.
Female. OSUM 228.

Fig. A: ventral view of head; note papillate lips and the deep angle produced by the posterior edges of the halves of the lower lip.

Identification

Characters: *General—See* blue sucker. *Generic—* More than 50 scales in the *complete* lateral line. Head rounded between the eyes. Tip of upper lip separated from snout by a deep groove. Lips papillose, the lower lip large, the posterior edge of the two halves forming a deep angle, fig. A. Air bladder in two parts. Dorsal fin usually contains 10–13 rays, extremes 9–14. *Specific*—Lateral line with 55–85 scales. Rounded snout projects slightly or not at all beyond tip of upper lip. Lower lip not flaring greatly. Posterior angle of mouth extending backward only to anterior edge of nostrils. A distinct rosy lateral band on the male only when the individual is in the act of spawning.

Differs: Longnose sucker has a greater number of lateral line scales. Redhorses, northern hog sucker, spotted sucker and chubsuckers have fewer lateral line scales. Carp and goldfish have a dorsal spine; other minnows have the anal fin farther forward.

Most like: Longnose sucker. **Superficially like:** Black redhorse and spotted sucker.

Coloration: Dorsally olive- or brownish-slate, the scales generally margined with darker. Sides more silvery. Ventrally white or milk-white. Snout and lips a pellucid-white. Dorsal and caudal fins light slate, remaining fins whiter, except in polluted waters where they may be tinged with orange. Although normally an obscurely colored species, the male during the spawning act is fantastically brilliant, for the back is olive with a bright lavender sheen, the lateral band pink or reddish, and between the back and lateral band is another band of whitish-yellow overcast with pink.

Lengths and **weights:** *Young* of year in Oct., 2.0″–4.5″ (5.1–11 cm) long. *Adults*, usually 10.0″–20.0″ (25.4–50.8 cm) long; weight 6 oz–4 lbs (170 g–1.8 kg). Dwarf males in small streams of high gradients, or from over-populated ponds, breed when only 4.5″ (11 cm) long; dwarfs from high-gradient streams often have a higher than average scale count. The largest specimens occur in the Ohio River and Lake Erie, and in the larger impoundments. Largest specimen, 25.0″ (63.5 cm) long; weight 5 lbs, 3 oz (2.4 kg). Commercial

MAP 102. Common white sucker

Locality records. ● Before 1955. △ 1955–80.

Insert: Black area represents range of this subspecies, outlined area of another; band of intergradation between the two subspecies appears to be very broad; northern limits of range indefinite; introduced widely, often inadvertently by bait fishermen, outside of this range and especially in the southwest.

fishermen give maximum length of 28.0″–30.0″ (71.1–76.2 cm); weight 6–8 lbs (2.7–3.6 kg). For life history, *see* Reighard (1920) and Stewart (1926).

Hybridizes: with longnose sucker; no Ohio specimens.

Distribution and Habitat

Ohio Distribution—Kirtland (1851C:317) stated that "This species abounds in every permanent stream, lake and pond in Ohio;" other early writers

(Osburn, 1901:34), and preserved specimens, testify to the universal distribution and great abundance of the white sucker in Ohio before 1900.

Since 1925 the white sucker has been one of the six fish species occurring most often in collections. The largest populations occurred in Lake Erie, the larger impoundments, the Ohio River, and the larger inland streams of both drainages. It was present in smaller numbers in the smaller impoundments and smaller streams, even those brooks which flowed intermittently.

Habitat—Throughout winter the bulk of the population remained in the deeper, larger streams of low gradients, and in the deeper waters of the impoundments. Some individuals remained in these waters to spawn; others ascended the streams when the waters reached a temperature of 40° F (4° C) or more, spawning from March to June in streams of all sizes and all gradients, when the water temperatures were usually between 50°–68° F (10°–20° C) (extremes 43°–74° F [6°–23° C]). Migration and spawning occurred mostly at night and at dusk. The species was most generalized in its habits, tolerating all gradients from basic to the highest; appeared to be more tolerant to turbidity, siltation, and other organic and inorganic pollutants than was any other species of sucker; could survive in waters low in oxygen; it tolerated dense aquatic vegetation. It congregated in large numbers in those waters rich in nutrients.

Years 1955–80—The rather evenly distributed white sucker continued to be one of the most numerous, if not the most numerous, Catostomid throughout Ohio waters. It remained numerous in Lake Erie, contributing significantly to the commercial fishery (Van Meter and Trautman, 1970:73; White et al., 1975:90). In the inland waters it showed no substantial change and may have increased slightly in abundance, seemingly because of its high tolerance to gradients ranging from basic to high, to considerable turbidity, to many types of organic and inorganic pollutants, and to low oxygen requirements. Brown (1960:54–55) after his thorough investigation of the headwaters of the Little Miami system, stated that it was generally the most abundant sucker and was the most abundant non-cyprinid species collected with shocker devices in Massie Creek during 1956–57. The 1957–59 studies of the Ohio River (ORSANCO, 1962:148–49) of 19 species of Catostomids indicated that the number of specimens taken (627) was surpassed only by the golden redhorse (744) and river carpsucker (1,157).

For embryonic stages *see* Long and Ballard, 1976.

EASTERN LONGNOSE SUCKER

Catostomus catostomus catostomus (Forster)

Fig. 103

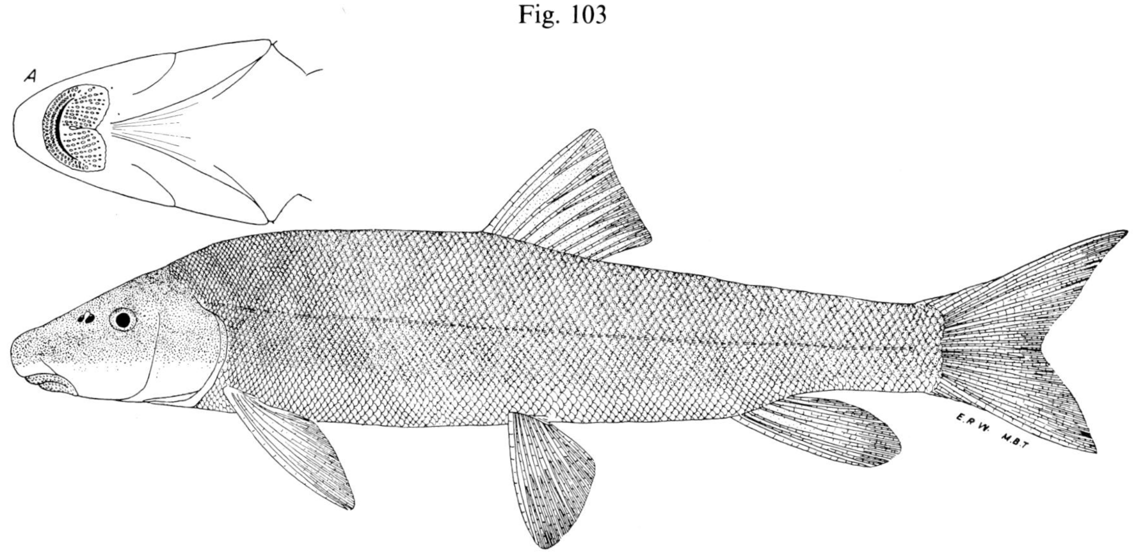

Lower fig.: Lake Erie, Erie County, O.

July 3, 1941. 343 mm SL, 16.5″ (41.9 cm) TL.
Female. OSUM 3876.

Fig. A: ventral view of head; note papillate lips, the wide flaring of the lower lip and the deep angle produced by the posterior edges of the halves of the lower lip.

Identification

Characters: *General—See* blue sucker. *Generic—See* white sucker. *Specific*—Lateral line with more than 85 scales. Bulbous snout projects far beyond the tip of upper lip. Lower lips flaring backward, fig. A. Posterior angle of mouth extends backward beyond anterior nostril. Adult male in spring with a rosy lateral band.

Differs: *See* white sucker.

Most like: White sucker.

Coloration: Back a dark olive-slate, the scales sharply outlined with darker. Sides somewhat lighter. Ventrally milky-white. The abrupt change on the lower sides between the dark sides and the white ventral surface give a decided bicolored effect. Dorsal and caudal fins light slate; remaining fins milk-white or transparent. *Adult spring male*—Has a rosy lateral band.

Lengths and weights: The four preserved specimens range from 14.0″–18.0″ (35.6–45.7 cm) long; weight 1–3 lbs (0.5–1.4 kg). Lakemen give maximum length of 24.0″ (60.9 cm), weight of slightly more than 4 lbs (1.8 kg) (OSUM: 3876; PSUM:728; UMMZ:9719).

Hybridizes: with white sucker; no Ohio specimens.

Distribution and Habitat

Ohio Distribution—In 1878 Klippart (1878:109) recorded the longnose sucker as occurring in the Ohio waters of Lake Erie; in 1882 Jordan (1882:817) considered it to be "quite abundant in Lake Erie;" in 1889 Henshall (1889:124) briefly mentioned its presence in the lake. Jordan's statement that it is abundant in Lake Erie and its possible presence in the Ohio River drainage in Ohio is questioned because of lack of sufficient evidence.

Commercial fishermen have been long aware of the presence of a small population in the Ohio waters of Lake Erie. Those fishermen fishing east of Lorain record capturing as many as a dozen individuals in a year; those fishermen fishing western Lake Erie record the capture of 1–4 in 15 years.

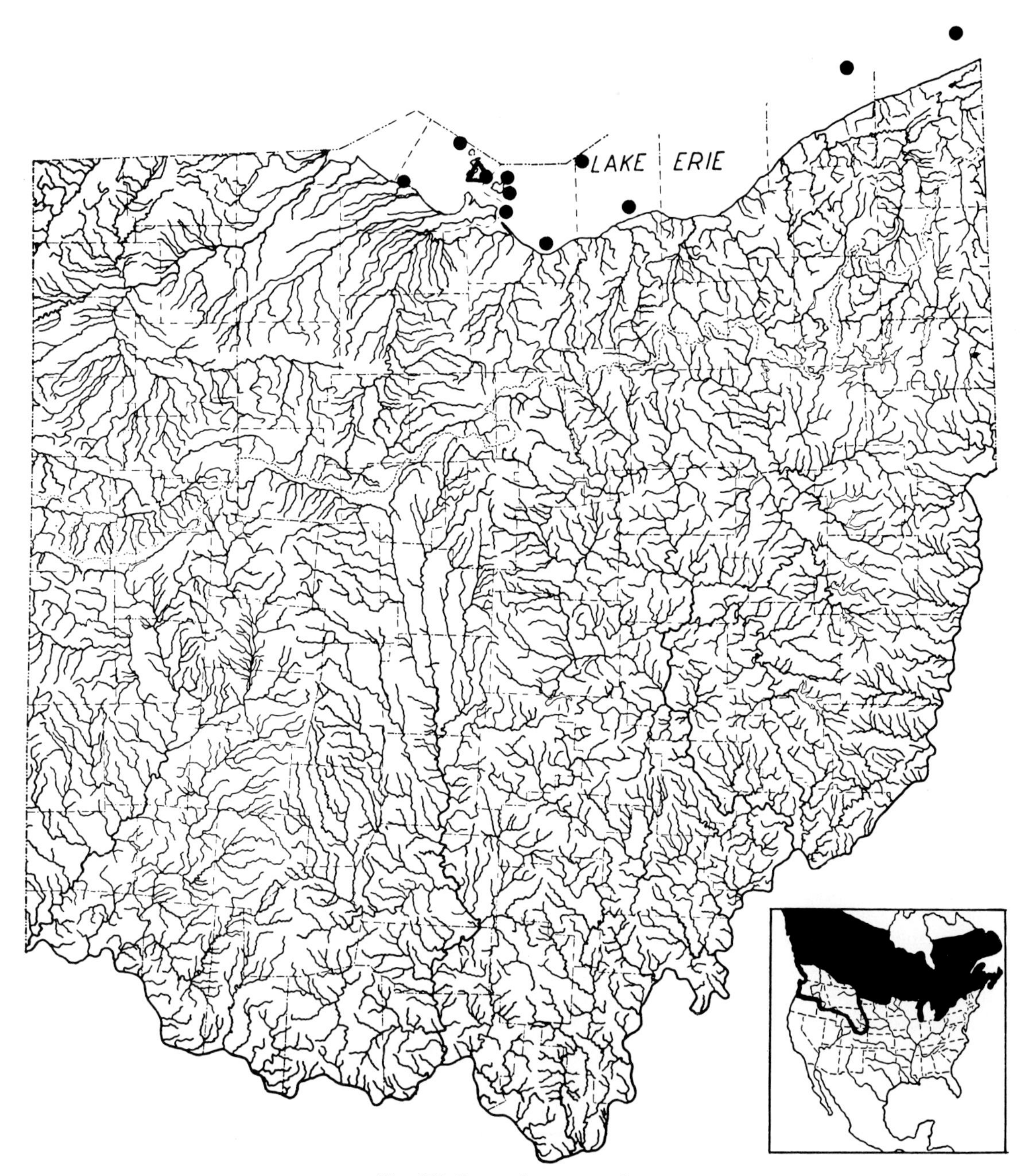

MAP 103. Eastern longnose sucker

Locality records. ● After 1920. △ 1955–80.

Insert: Black area range of this subspecies, outlined area of another; northern limits of range indefinite.

Among these fishermen the species is called redside sucker because of the red lateral streak, or sturgeon sucker because of its long snout.

Habitat—The longnose sucker occurred in the deeper and cooler waters of Lake Erie throughout most of the year, coming into waters of less than 25′ (7.6 m) in depth only in the spring, and presumably for the purpose of spawning about the reefs.

Years 1955–80—Commercial fishermen occasionally captured longnose suckers throughout the waters of Lake Erie as they had formerly but in decreasing numbers. This possible decrease in abundance may be attributed entirely or in part to decreased commercial fishing intensity in recent years (Van Meter and Trautman, 1970:73; White et al., 1975:90).

SPOTTED SUCKER

Minytrema melanops (Rafinesque)

Fig. 104

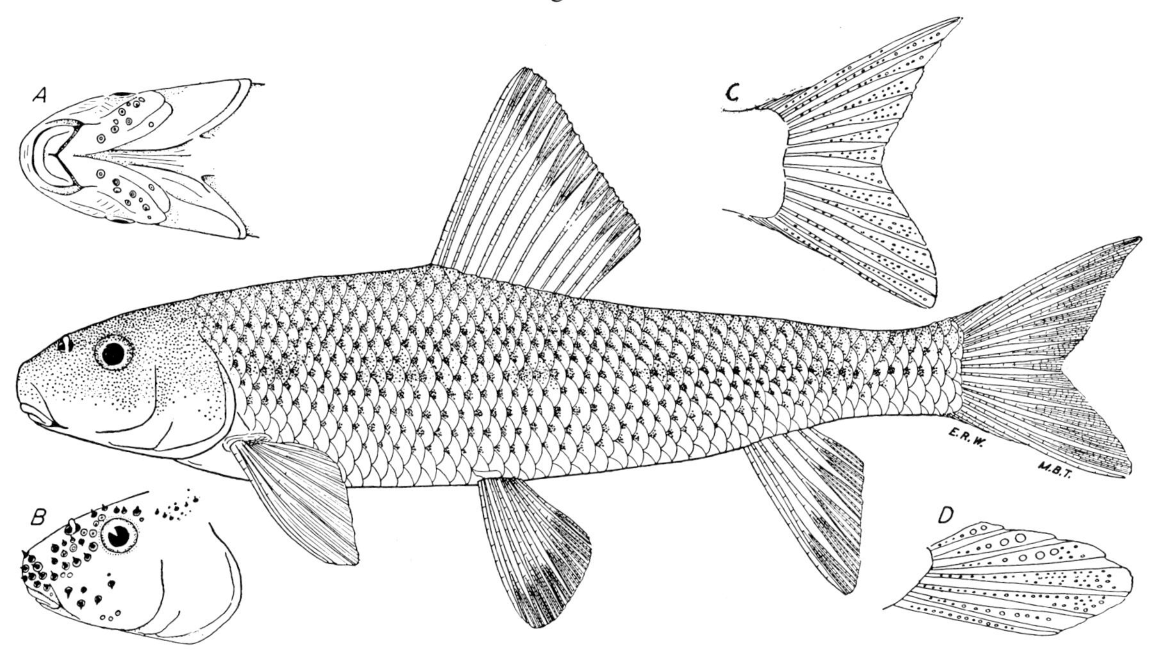

Central fig.: St. Marys River, Mercer County, O.

Oct. 14, 1941. 146 mm SL, 7.1″ (18 cm) TL.
Immature. OSUM 3525.

Figs. A, B, C, D: Lake Erie, Ottawa County, O.

June 11, 1942. 260 mm SL, 12.4″ (31.5 cm) TL.
Breeding male. OSUM 5030.
Showing tubercle distribution on the breeding male; ventral view of head (A); lateral view of head (B); caudal fin (C); anal fin (D).

Identification

Characters: *General—See* blue sucker. *Generic* and *specific*—Lateral line incomplete or absent. Scales in lateral series usually 43–45, extremes 42–47. A black spot at each scale base, these forming longitudinal streaks that are most prominent on the sides of large young and adults. Spots in small young faint, undeveloped, or largely restricted to region above anal base. Dorsal fin usually of 11–12 rays, extremes 10–13. Distal edge of fin often slightly concave in adults, straight in young. Lips thin and striated.

Differs: All other suckers except chubsuckers have developed lateral lines. Chubsuckers lack dusky spots at scale bases, and have fewer scales in the lateral series. Minnows have the anal fin farther forward.

Most like: White sucker. Small young lacking distinct spots sometimes resemble redhorses, especially the black redhorse.

Coloration: Dorsally dark greenish- or brownish-olive lightening to a more bronzy-green on upper sides. Sides lighter and more silvery, the squarish, black spots more distinct than those on the back. Ventrally milk-white and silvery. Dorsal, caudal, and anal fins slaty-olive; remaining fins whitish. *Breeding male*—Has a narrow, chocolate-gray lateral band extending from snout tip to tail; immediately above this band another of light grayish-

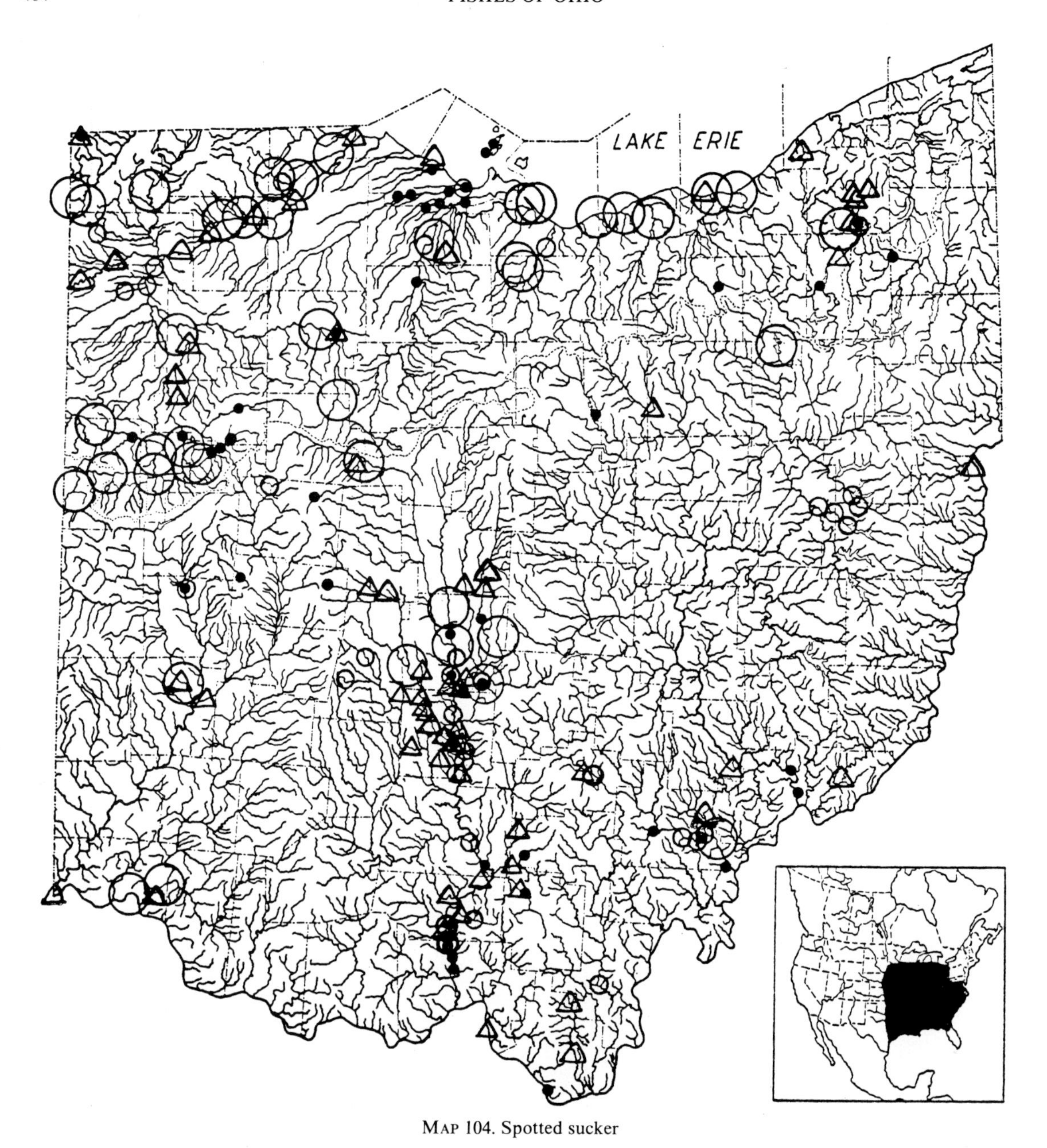

MAP 104. Spotted sucker

Locality records. ○ Before 1925. ○ 1925–38. ● 1939–54. △ 1955–80.
Insert: Introduced outside its normal range, usually inadvertently by bait fishermen.

pink; above this another of dark lavender with the back itself a lighter lavender.

Lengths and **weights:** *Young* of year in Oct., 2.0″–4.0″ (5.1–10 cm) long. *Adults*, usually 9.0″—–15.0″ (23–38.1 cm), weight 6 oz–1 lb, 12 oz (170 g–0.8 kg). Dwarf males breed when 6.0″ (15 cm) long. Largest specimen, 17.7″ (44.9 cm) long, weight about 3 lbs (1.4 kg).

Distribution and Habitat

Ohio Distribution—Klippart (1878:107), Jordan (1882:825) and others indicate that before 1900 the spotted sucker was widely distributed throughout the state. Kirtland (1851F:413) and Jordan (1882:825) stated that it was very numerous in Lake Erie; this may have been particularly true of the bays, harbors and marshes with their clean, sandy bottoms and aquatic vegetation. Before 1850 the prairie streams of Ohio, such as the low-gradient, gravelly-bottomed Blanchard River, and the base-gradient streams of unglaciated Ohio such as the Stillwater River of Tuscarawas and Harrison counties should have been particularly well suited for this species.

Since 1920 the spotted sucker has been rather locally distributed in Ohio. As indicated by the distribution map there has been a pronounced decrease in abundance, and possible extirpation in some areas, such as in the Maumee drainage. During the 1920–50 period it occurred only as strays in Lake Erie. Presumably the decrease in abundance of this species began prior to 1920 and was caused primarily by destruction of its habitat.

Habitat—The largest populations were present in those lakes, overflow ponds, sloughs, oxbows, and those stream sections, which had low or basic gradients, clean sandy, gravelly or hard clay bottoms and where the siltation rate was extremely low. This small-mouthed sucker with the closely-bound gill-covers seemed to be particularly intolerant to turbid waters, various industrial pollutants, and to lake and stream bottoms covered with flocculent clay silts.

Years 1955–80—During this period the distributional pattern and numerical abundance of the spotted sucker appeared to remain essentially unchanged in Lake Erie and inland Ohio (White et al., 1975:91). Most of the individuals noted were, as had been observed prior to 1950, in static or low-gradient portions of streams and in oxbows where silt deposition was not rapid. As previously also, one or only a few specimens were usually taken at any time. The highest number captured was 20; these were taken in trap nets in the Auglaize River, Putnam County, near Fort Jennings.

Apparently, habitat conditions during this period were rather favorable for this species in the Ohio River, for it was the fourth most numerous Catostomid taken during the investigations of that stream between 1957 and 1959 (ORSANCO, 1962: 68, 149). Prior to 1950 I took only one specimen in the Ohio River (Lawrence County) during rather intensive collecting with various types of gear.

H. Ronald Preston did not capture the species in the Ohio River opposite the state of Ohio during his 1968–69 investigation; Margulies and Burch obtained a spotted sucker in the Ohio River adjacent to Jefferson County, Ohio, in 1976.

WESTERN LAKE CHUBSUCKER

Erimyzon sucetta kennerlyi (Girard)

Fig. 105

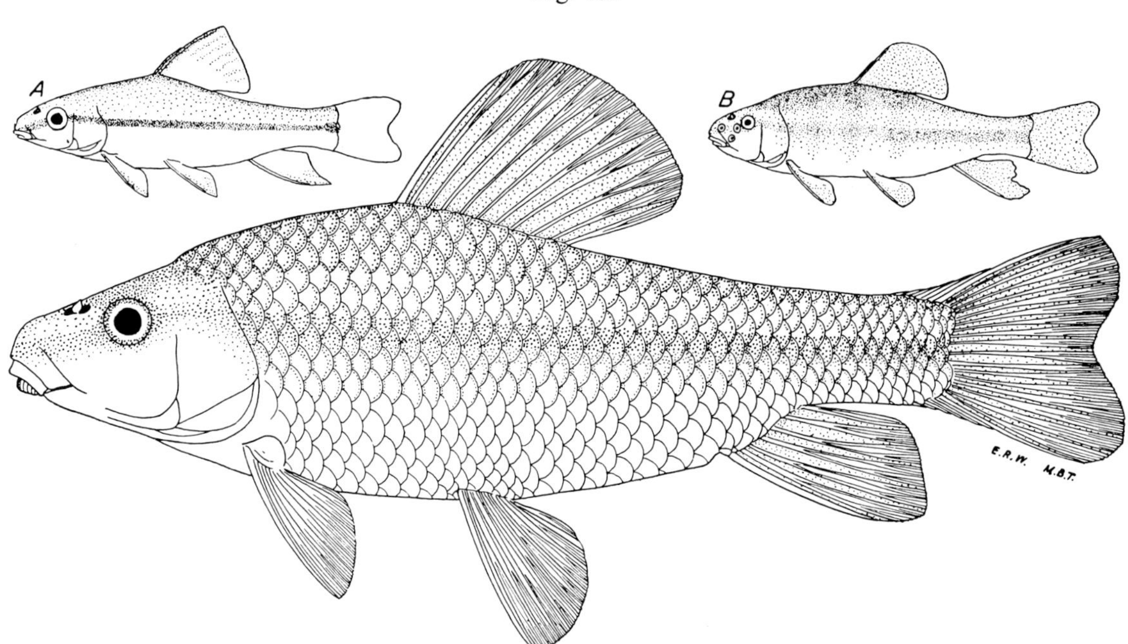

Lower fig.: Mosquito Creek, Trumbull County, O.

Oct. 10, 1939. 155 mm SL, 5.7″ (14 cm) TL.
Female. OSUM 3019.

Fig. A: Fish Creek, Williams County, O.

July 13, 1939. 39 mm SL, 1.9″ (4.8 cm) TL.
Immature. OSUM 343.
Note distinctive color pattern and marked resemblance to some minnows and shiners.

Fig. B: Buckeye Lake, Perry County, O.

April, 1939. 155 mm SL, 7.5″ (19 cm) TL.
Breeding male. Specimen accidentally destroyed.
Note color pattern and three tubercles on side of snout; also falcated anal fin.

Identification

Characters: *General—See* blue sucker. *Generic*—No lateral line. No black spots at scale bases. Distal edge of dorsal fin decidedly convex in adults, straighter in young. Air bladder in two parts. Tip of upper lip separated from snout by a groove. Breeding males with three tubercles on each side of snout and a falcate anal fin, fig. B. *Specific*—Dorsal rays usually 11–12, extremes 10–13. Lateral scale rows usually 35–37, extremes 33–40. Body depth usually contained 3.3 times or less in standard length in large young and adults. Small young with an intense black band from snout tip to tail and anterior edge of dorsal fin black, these black markings becoming faint or absent in large young.

Differs: Other species of suckers have lateral lines, except the spotted sucker and creek chubsucker. The spotted has black spots at the scale bases, the creek chubsucker fewer dorsal rays and more scales in the lateral series.

Most like: Creek chubsucker. Small young frequently mistaken for minnows or shiners.

Coloration: Dorsally a deep olive- or greenish-bronze with a pronounced bronze overcast and with the dark scale edges giving a cross-hatched appear-

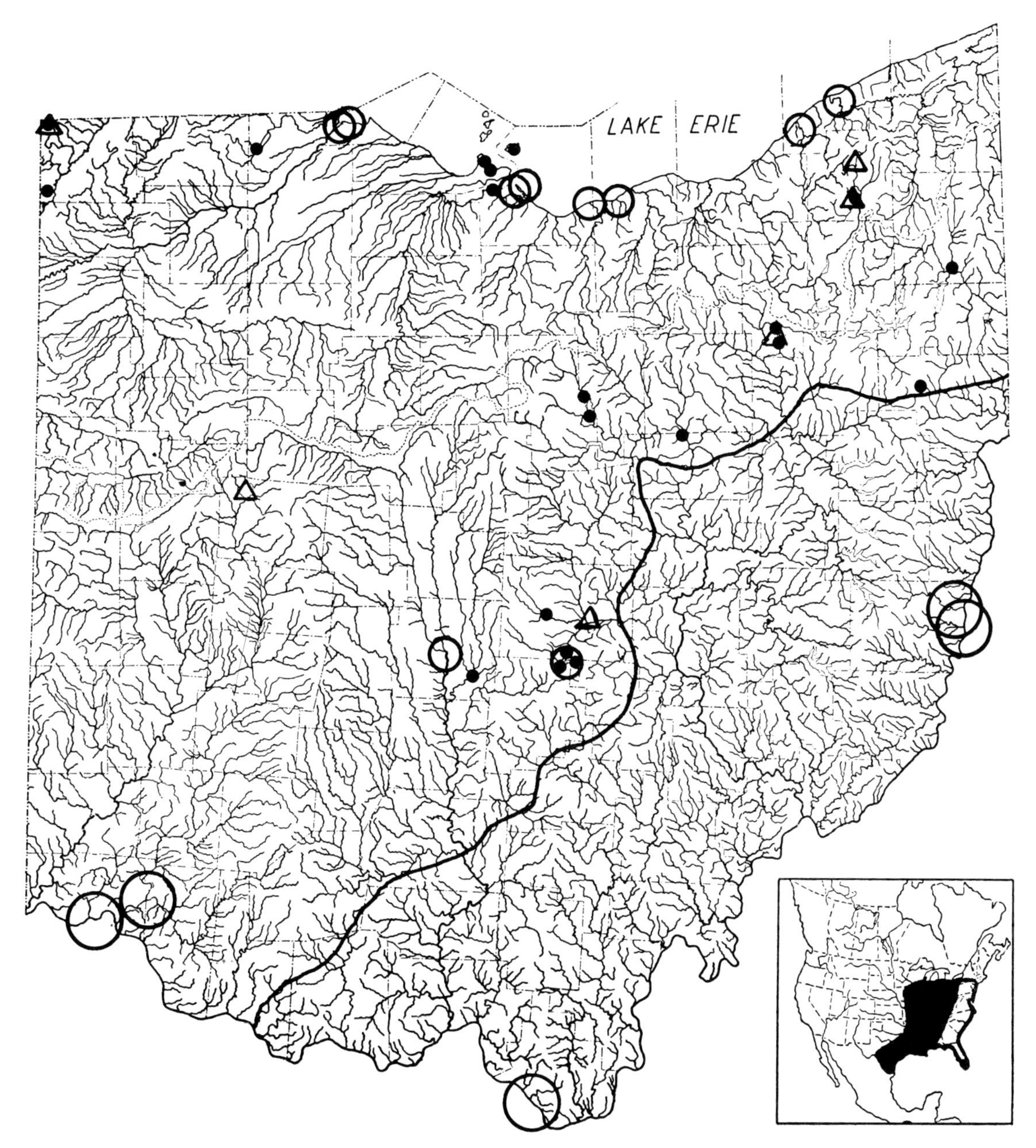

MAP 105. Western lake chubsucker

○ Questionable records. ○ Before 1901. ● 1901–54. △ 1955–80. ~ Glacial boundary.
Insert: Black area represents range of this subspecies, outlined area of the nominate subspecies (Bailey, Winn and Smith, 1954: 123).

ance. Sides lighter and more golden, a faint lateral band sometimes present in large young and adults, which may become broken into bars in large adults. Ventrally olive-yellow, most silvery in smallest young. Dorsal, caudal, and anal fins olive-yellow, remaining fins paler or whitish. *Breeding adults*—Sometimes show 6–8 vague, broken, ventral bars besides the broken lateral band, fig. B. *Small young*—Yellowish-silver with a black band extending from snout to tail and anterior rays of dorsal fin black.

Lengths and **weights:** *Young* of year in Oct., 1.7″–3.3″ (4.3–8.4 cm) long. *Adults*, usually 5.0″–10.0″ (13–25.4 cm) long, weight 2 oz–10 oz (57 g–283 g). Largest specimen (Buckeye Lake), 11.5″ (29.2 cm) long, weight about 14 oz (397 g).

Distribution and Habitat

Ohio Distribution—The Hamilton County records of Henshall (1889:77) for 1888, and Lawrence and Belmont county records of Osburn (1901:36) for 1899 and 1900 are questioned because of lack of specimens, confusion in identification by authors of this species and the creek chubsucker, and present seemingly unsuitable habitat. It is of course possible that before the Ohio River was ponded this species may have been present. Records made before 1900 along Lake Erie are in part substantiated by specimens or reliable published descriptions, and since the habitat for the species was still present there after 1925, these records are accepted. The species was abundant in Sandusky Bay before 1900 (Osburn, 1901:361); before 1877 in Buckeye Lake and in the Scioto River (Klippart, 1877:85–86 and specimen in UMMZ collection in 1854). Since 1920 populations have been found where there formerly were large marshes or glacial pothole lakes, such as at Buckeye and Portage lakes. The species should have been present in the original marshes and lakes now flooded by Lake St. Marys and Indian Lake, but I have seen no specimens. Small populations also occurred in those clear streams of fair size near old pothole lakes, especially when they contained much aquatic vegetation. Because of unfavorable conditions several populations greatly decreased in abundance or disappeared during the 1925–50 period; such as at East Harbor where before 1930 there was a large population, whereas since 1945 few individuals have been observed.

Habitat—The lake chubsucker inhabited pothole lakes and larger streams of glaciated Ohio, often in widely separated, relict populations. It appeared to be highly intolerant of turbidity and siltation and usually occurred most abundantly where there was much submerged aquatic vegetation and where the bottoms were of sand or fine gravel. It appeared intolerant of high gradients in streams. Because this species and its close relative, the creek chubsucker, occupy rather markedly different habitat niches, there appears to be little competition between the two. Both occurred in different habitat niches in the same stream systems of northwestern Ohio.

Years 1955–80—During early historic time the lake chubsucker apparently was a rather widely distributed species in glaciated Ohio, occurring principally in the pothole ponds and low-gradient, vegetated streams. Several formerly known populations appear to have been extirpated from the state since 1950. None was found during the 1955–80 period in such localities as Buckeye Lake, Sandusky Bay and East Harbor, where previously it had been numerous. The species remained present in some suitable areas in northeastern Ohio (White et al., 1975:91). The only new county record of which I am aware was obtained by Rex Worley and others of the Ohio Division of Wildlife, who about 1964 collected a lake chubsucker in a trap net set in Indian Lake near Lake Ridge, Stokes Township, Logan County (*see* above relative to Indian Lake).

WESTERN CREEK CHUBSUCKER

Erimyzon oblongus claviformis (Girard)

Fig. 106

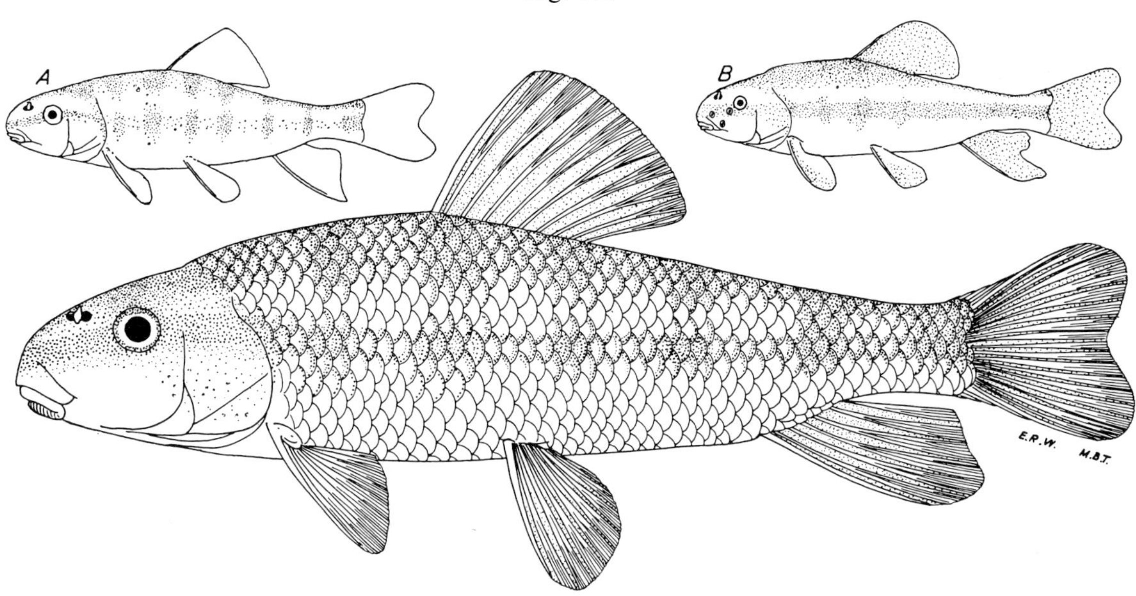

Lower fig.: Auglaize River, Auglaize County, O.

Oct. 15, 1941. 98 mm SL, 4.7″ (12 cm) TL.
Female. OSUM 4086.

Fig. A: Greenville Creek, Darke County, O.

Nov. 7, 1940. 45 mm SL, 2.3″ (5.8 cm) TL.
Immature. OSUM 2246.
Note distinctive color pattern and resemblance to some minnows and shiners.

Fig. B: St. Marys River, Mercer County, O.

Oct. 14, 1941. 130 mm SL, 6.3″ (16 cm) TL.
Male. OSUM 3524.
Note color pattern and three tubercles on side of snout; also falcated anal fin.

Identification

Characters: *General—See* blue sucker. *Generic—See* lake chubsucker. *Specific*—Dorsal rays usually 9–10, extremes 8–11. Lateral scale rows usually 39–41, extremes 37–43. Body depth usually contained 3.3 times or more in standard length. A series of 5–8 blotches along the lateral line, with a saddle-band directly above each lateral blotch, these bands and blotches most distinct in young; adults with the lateral blotches more or less confluent.

Differs: *See* lake chubsucker.

Most like: Lake chubsucker. Small young frequently mistaken for minnows or shiners.

Coloration: Dorsally an olive-bronze with a bronze or golden-yellow overcast, and with the dark scale edges giving a cross-hatched appearance. Sides lighter and more golden. Ventrally yellowish or whitish. Between 5–8 saddle-bands crossing back and a blotch on sides directly beneath each band; these most distinct in small young. In *adults*—blotches on the sides form a more or less confluent band.

Lengths: *Young* of year in Oct., 1.5″–2.0″ (3.8–5.1 cm) long. *Adults*, usually 3.0″–6.0″ (7.6–15 cm). Smallest breeding males with tubercles and bifurcate anal fins only 2.4″ (6.1 cm) long. Largest specimen, 7.1″ (18 cm) long.

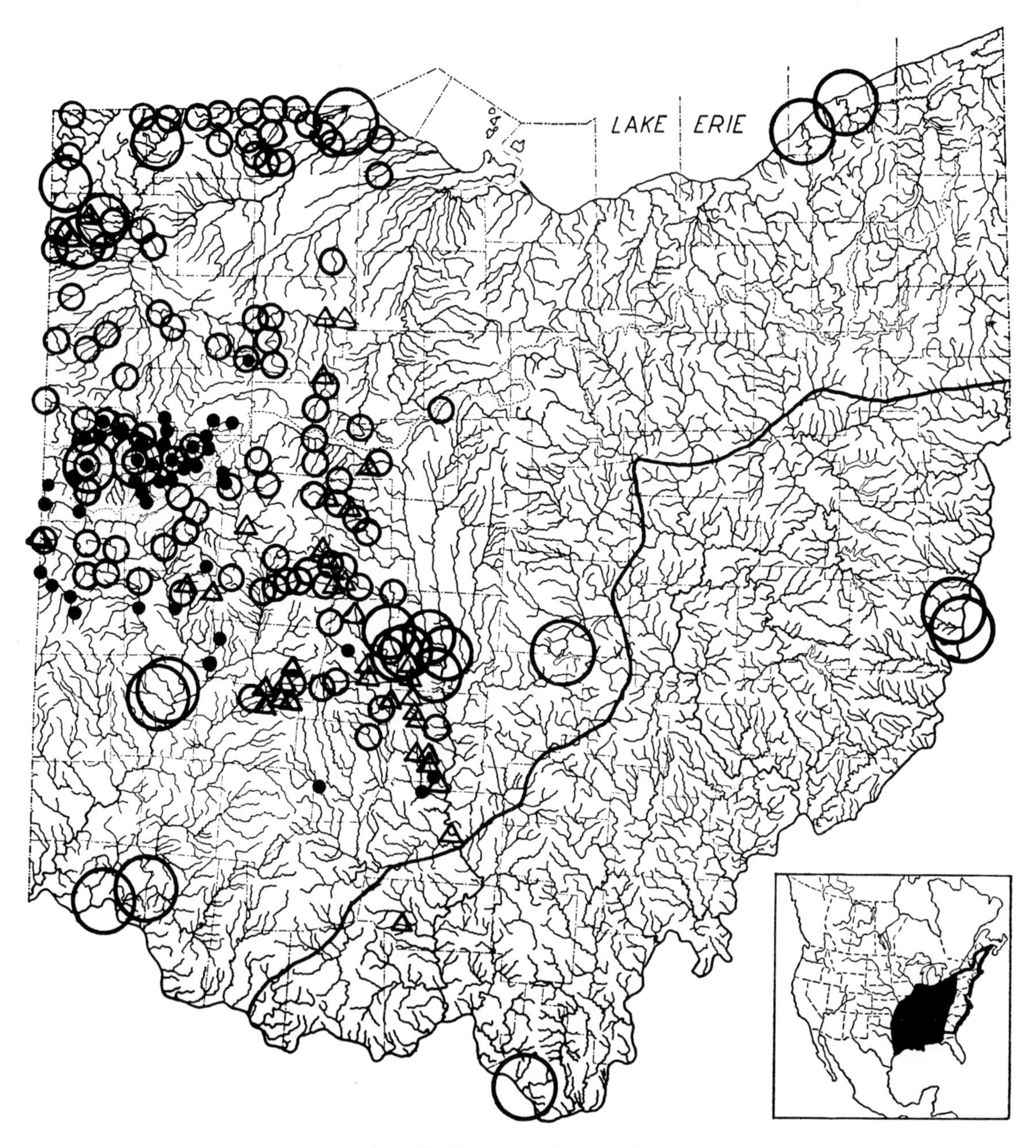

MAP 106. Western creek chubsucker

○ Questionable records. ○ Before 1901. ○ 1901–38. ● 1939–54. △ 1955–80. ~ Glacial boundary.
Insert: Black area represents range of this subspecies, outlined area of another.

Distribution and Habitat

Ohio Distribution—Before 1900, the creek chubsucker was probably confined almost entirely to the clear prairie and till-plain streams of moderate and high gradients in central and northwestern Ohio. Henshall's (1889:77) chubsucker records of 1888 for Hamilton County are without adequate descriptions or specimens; they may have been this species, or the lake chubsucker, or misidentified. Because of lack of adequate evidence, Henshall's records are questioned as are Osburn's (1907:36) for Montgomery, Lawrence, Belmont and Lake counties, Kirsch's (1895A:328) for Toledo, Lucas County, and Williamson's for Licking County (Osburn, 1907:36). Williamson and Osburn (1898:12) in 1897 found this chubsucker to be abundant in the prairie areas west of the Scioto River in Franklin County where today it is rare or absent. Between 1920–39 large populations occurred in the small clear streams of the Scioto tributaries of Madison and Union counties, and the Auglaize tributaries of Hancock and Putnam counties. In these streams many specimens could be taken with little effort; since 1940 the species has decreased markedly in these streams and recent efforts to capture them have largely failed. Apparently this sucker is retreating from the more turbid and silted waters of the Scioto and Maumee river drainages; remnant populations still occur only in the clearest of the old prairie streams such as in Yellowbud Creek, in southern Pickaway County.

Habitat—A stream species, which in spring inhabited the small clear prairie streams of moderate and high gradients. The largest populations occurred in streams whose bottoms were mostly of sand and gravel and which contained little clayey silt. After spawning or in early summer the adults migrated downstream into the larger creeks and remained there throughout fall and winter. Like the redhorses, the creek chubsucker was killed by extremely turbid waters. Upon several occasions I found many dead chubsuckers in a stream section whose water contained much clayey silt. This silt had been washed into the stream from recently cultivated cornfields during a brief summer shower. The sticky silt had packed about the gills of the chubsuckers, suffocating them.

Years 1955–80—As illustrated on distribution map 106 there was little extension in the distributional range of the creek chubsucker in glaciated Ohio during the 1955–80 period. The species was found only in a few localities in the generally turbid Maumee River system, such as in the Blanchard River above the city of Findlay, where turbidity and harmful pollutants were less than in most sections of the Maumee system, and in the headwaters of Gordon Creek, Defiance County (Trautman and Gartman, 1974:171). There must remain other relict populations in this system, especially in those small tributaries having little turbidity and the least amounts of other pollutants.

Populations in some of the smaller, prairie-type tributaries of the Scioto River and in the headwaters of both Miami River systems indicate that the species was numerically holding its own. In fact, it appeared to be increasing in abundance in those localities where there has been an observable decrease in amounts of silt deposition upon the stream's substrate.

BLUE CATFISH

Ictalurus furcatus (Lesueur)

Fig. 107

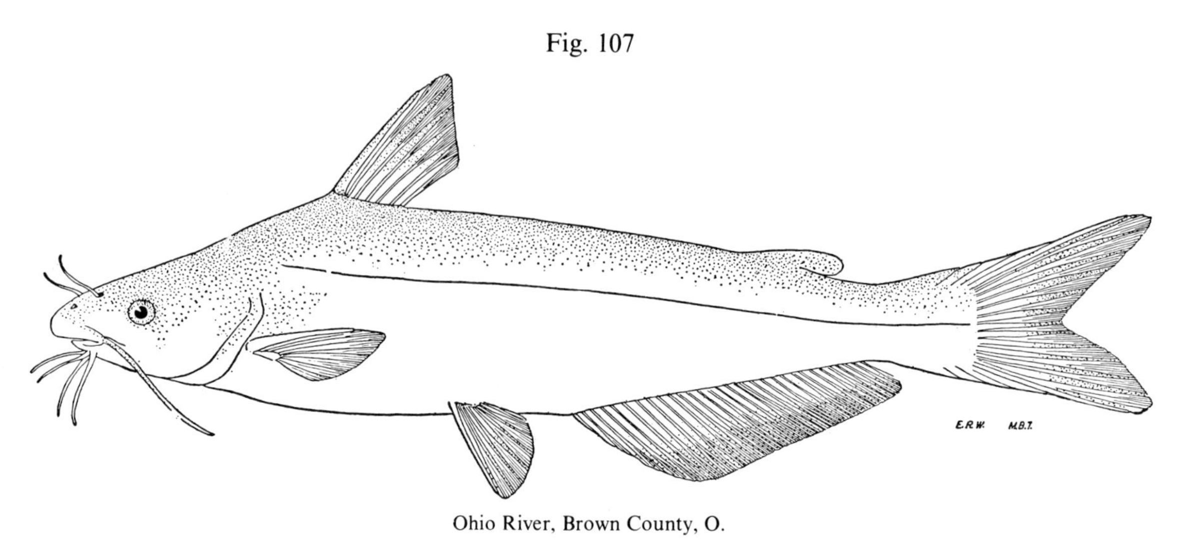

Ohio River, Brown County, O.

April 27, 1931.
Immature.

218 mm SL, 10.6″ (26.9 cm) TL.
OSUM 678.

Identification

Characters: *General*—A sharp, robust spine in the dorsal and pectoral fins; none in pelvic or anal fins. Usually 8 barbels (occasionally one or more missing) encircle mouth. No scales. Adipose fin present. Breeding male has a very wide head and bulging cheeks and opercles. *Generic*—Posterior third of adipose fin a free lobe. Lower jaw not longer than upper jaw, except rarely in old, aberrant bullheads. *Specific*—Tail deeply forked. Upper jaw longer than lower jaw. Anal fin of 30–36 rays, counting rudimentary rays, its distal margin almost straight. Anal fin base long, usually contained 2.9–3.3 times in standard length. Body without black spots. Eyes appearing to be situated in the ventral half of head.

Differs: Channel and white catfishes have fewer anal rays; distal edge of anal rounded; a shorter anal fin base. Other species of catfishes have squarish, rounded, or slightly emarginated caudal fins.

Most like: Channel catfish. **Superficially like:** White catfish.

Coloration: Dorsally usually pale-bluish, with much silvery. Sides lighter. Ventrally silvery-and milk-white. Fins transparent-white with some bluish basally. *Breeding adults* darker; male has blue-black head and the dorsal half of body largely bluish-black. Normally no dark spots on body.

Lengths and weights: Largest Ohio specimen I have seen in life was about 41.0″ (104 cm) long, weight 40 lbs (18.1 kg). Picture of a specimen "about 4 1/2 feet [137 cm] long and weighed 92 lbs [41.7 kg]." *See* text.

Hybridizes: with channel catfish.

Distribution and Habitat

Ohio Distribution—Jordan (1882:785–86) saw "one or two specimens taken at Cincinnati [Ohio River]," and Henshall (1888:77) recorded the species as "Common in the Ohio River." From these statements and those of old Ohio River fishermen, it is obvious that the readily-identifiable "Mississippi" or "White" catfish was present before 1900 in the Ohio River between the Indiana state line and Belmont County. The fishermen are in universal agreement that the species was far more abundant before the Ohio River was ponded (prior to 1911) than it has been since; at least many more fishes were caught before than after ponding. I was shown many pictures of large specimens taken in the Ohio

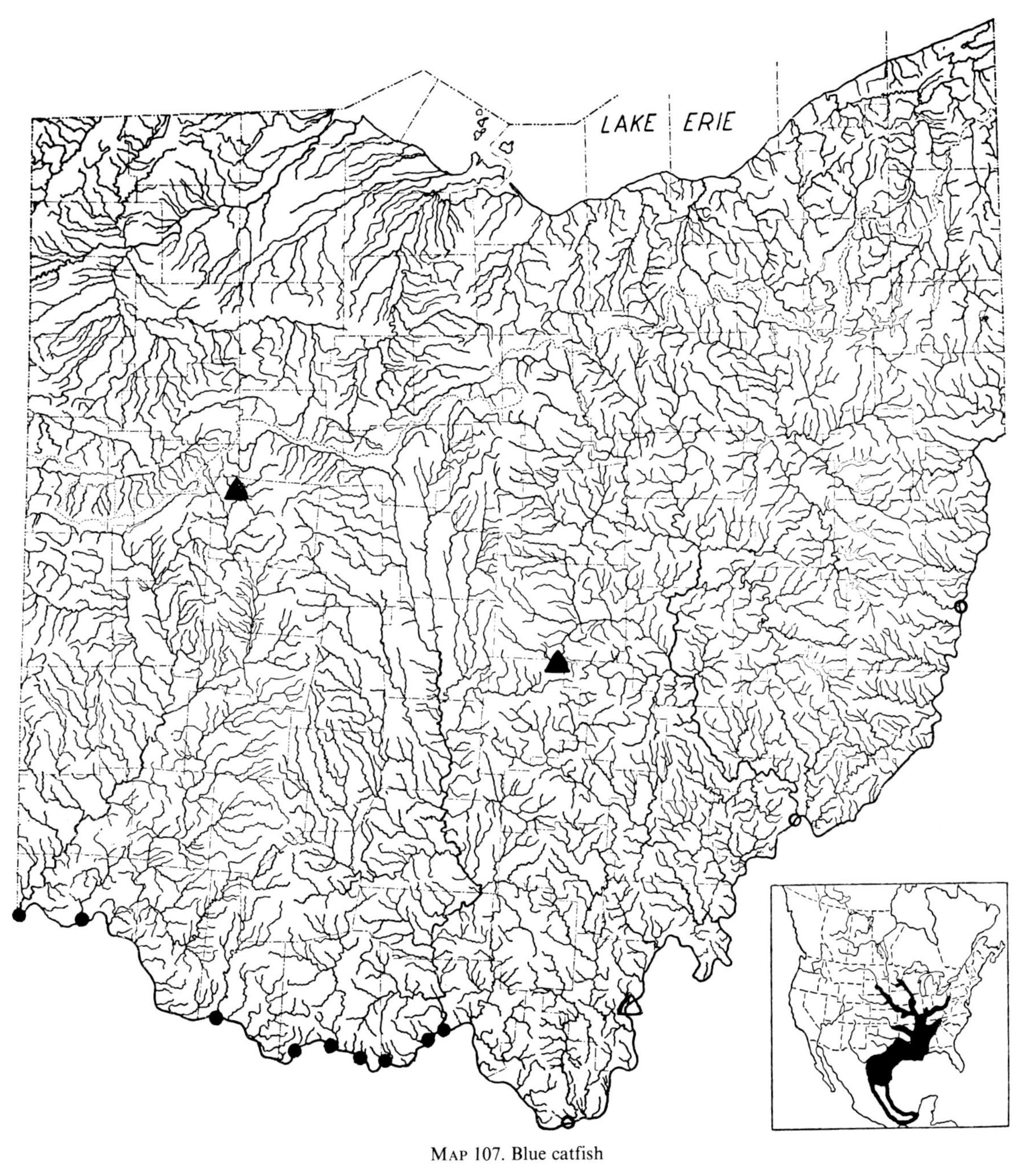

MAP 107. Blue catfish

○ Reliable verbal reports or photographs. ● Specimens examined. △ 1955–80. ▲ Introduced 1975.

Insert: Most of the Mexican segment of the population has been considered to be subspecifically distinct; limits of range in Mexico indefinite.

River between the Indiana state line and Portsmouth. The largest specimen was one which weighed 92 lbs (41.7 kg); it was caught near Higginsport, Brown County, in July, by Harry Brookbank and Charles Glaze and was taken before the river was ponded. Between 1925–50 I examined more than 20 specimens ranging from 1–40 lbs (0.5–18.1 kg) in weight, all taken between the Indiana state line and Portsmouth. The three hollow circles on the Ohio River between Portsmouth and the Pennsylvania state line represent localities where fishermen caught huge catfishes, which from their descriptions were of this species. These fishermen also had taken the species farther downstream and recognized it. All three specimens were caught before the river was ponded.

Habitat—Those fishermen who have fished most persistently for this species before the river was ponded agree that most of its feeding was done in the swiftly-flowing chutes or rapids, and over bars or elsewhere in pools wherever there was a good current and where the bottom was of bedrock, boulders, gravel or sand. When not feeding and in winter it apparently retired into deeper waters where the fishermen took it in hoop nets, sometimes in depths of more than 30′ (9 m). It seemingly avoided the silted bottoms of the most sluggish pools. The ponding of the river and subsequent rapid and deep silting of the bottoms have largely destroyed these habitats.

Before the Ohio River was impounded, jugging was the usual method of fishing for this catfish, and also for the large channel and flathead catfishes. An earthen jug, or gallon tin can, was tied to one end of 9′–15′ (3–5 m) of No. 30 twine. A No. 10 hook was tied to the other end and was baited with a piece of raw beef larger than a baseball, a sucker or bullhead of a pound (0.5 kg) in weight, a chicken less than one-third grown, or a small kitten. Between five and ten jugs with lines and hooks attached were floated down the chutes, the fishermen following behind in a boat.

In Ohio the species is called "Mississippi" or "White Cat," the latter a most appropriate name, for all specimens I have seen were whitish or milk-white, not predominantly bluish as is the channel catfish. I follow the American Fisheries Society recommendation in calling it blue catfish, which to Ohio fishermen is the channel catfish.

Years 1955–80—During this period I heard rumors, or saw newspaper accounts, of a few individuals presumed to have been blue catfishes, caught in the Ohio River between its confluence with the Scioto River and the Ohio-Indiana state line.

According to members of the University of Louisville (ORSANCO, 1962:65) the blue catfish did "not occur in great numbers above Louisville," Kentucky, but became increasingly abundant below that city.

H. Ronald Preston caught none during his 1968 investigation of the Ohio River opposite the state of Ohio but did capture individuals from the river opposite central-eastern Indiana.

In 1975 the Ohio Division of Wildlife introduced the blue catfish into Buckeye and Indian lakes. It is hoped that these introductions are as successful as were the introductions of the channel catfish more than half a century previously.

CHANNEL CATFISH

Ictalurus punctatus (Rafinesque)

Fig. 108

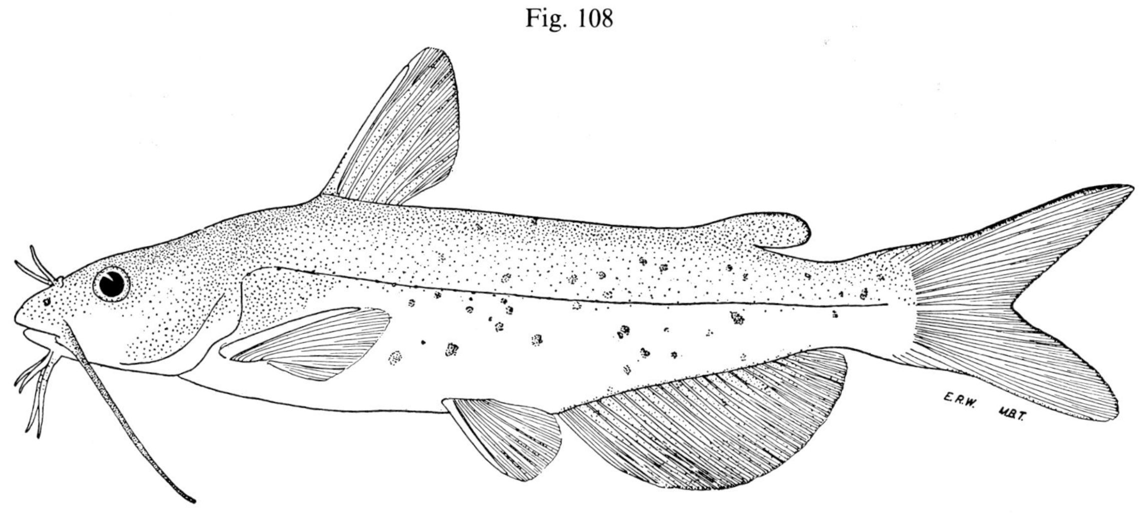

Muskingum River, Washington County, O.

April 4, 1940. 198 mm SL, 10.3″ (26 cm) TL.
Immature. OSUM 1820.

Identification

Characters: *General* and *generic*—*See* blue catfish. *Specific*—Tail deeply forked. Upper jaw longer than lower jaw. Anal fin of 24–30 rays, counting rudimentary rays, its distal margin rounded. Anal fin base rather short, usually contained 3.4–3.7, extremes 3.2–4.0, times in standard length. Body of young, and of some adults weighing as much as 5–9 lbs (2.3–4.1 kg), with few or many black spots. Eyes definitely situated in dorsal half of head.

Differs: Blue catfish has more anal rays; a longer anal fin base; anal fin margin almost straight. White catfish has fewer anal rays; a shorter anal fin base; anal fin margin very rounded. Other species of catfishes have squarrish, rounded, or slightly emarginated caudal fins.

Most like: Blue and white catfishes.

Coloration: *Young* less than 14.0″ (35.6 cm) long (colloquially called squealers, ladycats, spotted, silver cats, and fiddlers) are bluish- or olivaceous-silvery dorsally, and silvery-white ventrally; these have few to many spots on body. Adults 13.0″–24.0″ (33.0–60.9 cm) long (colloquially called silver, channel, and blue cats) are silvery-, olive-, or slaty-blue dorsally and yellow- or milk-white ventrally; all except adult males of this group usually have a few or many spots. Largest adults (colloquially called channel and blue cats [usually males], canal boaters, black warriors and loggerheads) are dark steel-blue with whitish bellies. *Male*—During breeding season has a blue-black head, blackish-blue body dorsally, whitish-blue ventrally. Fins in fishes of all ages correspond in color with that of adjacent body parts.

Lengths and **weights:** *Young* of year in Oct., 2.0″–4.0″ (5.1″10.2 cm) long. Around **1** year, 3.5″–7.5″ (8.9–19 cm) long. *Adults*, usually 11.0″–30.0″ (27.9–76.2 cm) long, weight 12 oz–15 lbs (340 g–6.8 kg). Largest specimens, 33.0″–46.0″ (83.8–117 cm) long, weight 25–30 lbs (11.3–13.6 kg). Most commercial fishermen give maximum weight of between 32–38 lbs (14.5–17.2 kg). Largest on record, 47.5″ (120.6 cm) long, weight 58 lbs (26 kg).

Hybridizes: with blue and flathead catfishes.

Distribution and Habitat

Ohio Distribution—The astute Kirtland (1850H:173, as *Pimelodus coerulescens*) in 1850 commented that because of morphological and color differences in this species, resulting from age, sex,

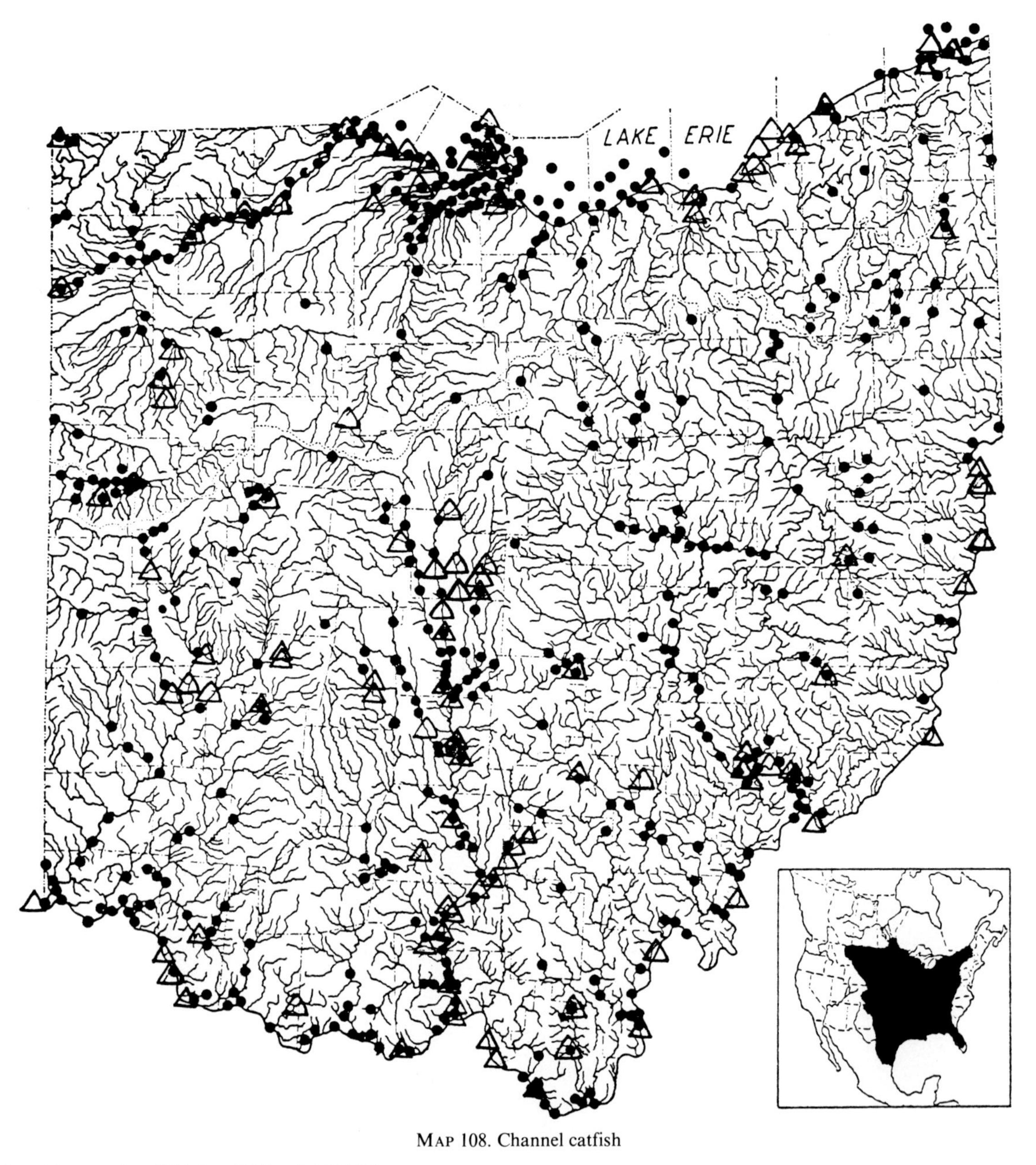

MAP 108. Channel catfish

Locality records. ● Before 1955. △ 1955–80.

Insert: Introduced successfully into many areas outside of this range, both in eastern and western states; range limits in Mexico indefinite.

and habitat, "it would be an easy matter to manufacture half a dozen new species from the varieties captured at one haul of a seine" in the Cuyahoga River. These differences did cause great confusion, with the result that several scientific and common names were given to this species, under the assumption that two to six species were involved. Since we now realize that all of these diverse names obviously referred to the channel catfish, the old published records are considered reliable.

Concerning early distribution and abundance, Kirtland continued with the statement that the species is "extensively diffused through the waters of Lake Erie and the Ohio River," a statement corroborated by many other authors who published before 1900. Kirtland likewise commented concerning the possible effect of dams upon this highly migratory species, writing that this catfish was "decreasing in numbers in many tributary streams as they are becoming obstructed with mill dams."

Between 1920–50 this prized food and game fish was well distributed throughout the larger streams and lakes of Ohio; was particularly abundant in the Ohio River, Lake Erie, and lower courses of their larger tributaries. The 1941–49 yearly average of commercially caught channel catfishes in Lake Erie waters was 725,909 lbs (329,267 kg). The low-gradient portions of the Maumee and Scioto rivers contained especially large populations, despite turbidity, whereas such high-gradient streams as the Mad and Chagrin rivers contained very low populations. Large populations were also present in many of the larger lakes and impoundments. During the 1920–50 period, the Conservation Department introduced large numbers in waters immediately after impoundment, and within a few years many young were being taken. About 1908 the Conservation Department introduced channel catfishes in the long established Indian Lake, which apparently had contained none previously. A few years after this introduction other fishermen and I began catching large numbers of young; since then the species has been one of the important sport fishes of that lake.

Habitat—The greatest populations occurred in the deeper and/or larger waters of low or base gradients which had fairly clean bottoms of sand, gravel, or boulders; also over bottoms of silt, provided the rate of silt deposition was slow. It seldom occurred in dense beds of aquatic vegetation. The adults were highly migratory, ascending surprisingly small streams for the purpose of spawning. The yearlings and subadults seem to be more tolerant to fast currents than are the adults, for considerable numbers winter under boulders in rather swiftly flowing water, or feed, chiefly at night, in such riffles.

Years 1955–80—Throughout this period the channel catfish was present, sometimes in fair- to large-size populations, in the larger inland streams, ponds and reservoirs. It was introduced into many of the larger farm ponds.

The species was common to abundant in the Ohio River (ORSANCO, 1962:149), and also in the shallower waters of Lake Erie (Van Meter and Trautman, 1970:73), where in some localities the species was of considerable commercial importance (Anonymous B, 1968:6; White et al., 1975:91–92).

The annual poundage of catfishes raised for human consumption in the United States increased greatly after 1960, and particularly in southwestern states such as Arkansas and California. Propagation in Ohio of this species for human consumption, if undertaken, will take longer to produce marketable fishes because of the shorter growing seasons, cooler waters and lack of large supplies of water.

WHITE CATFISH

Ictalurus catus (Linnaeus)

Fig. 109

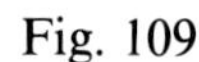

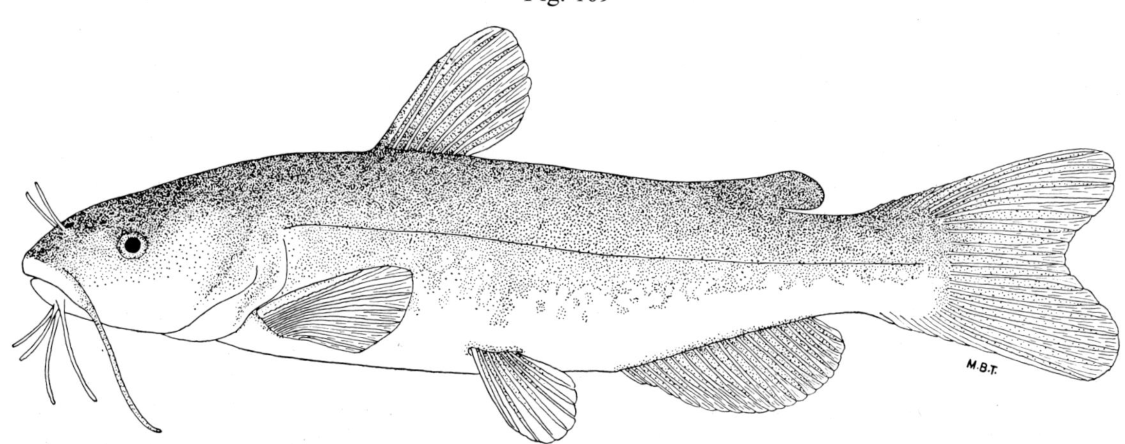

Either the Potomac River in Maryland or the Rappahannock River in Virginia.

Late March or early April, 1954. 253 mm SL, 12.2″ (30.9 cm) TL.
Adult male. OSUM 9349.

Identification

Characters: *General* and *generic—See* blue catfish. *Specific*—Tail moderately forked. Upper jaw longer than lower jaw. Anal fin of 18–24 rays counting rudimentary rays, its distal margin very rounded. Anal fin base short, usually contained 4.3–5.2, extremes 3.2–4.0, times in standard length. No distinct dark spots on body. Eyes situated in dorsal half of head. Width of head averages wider in non-breeding adults and young than in comparable individuals of channel catfishes of the same sex.

Differs: Blue and channel catfishes have more anal rays and a longer anal fin base. Other species of catfishes have squarish, rounded, or slightly emarginated caudal fins.

Most like: Channel catfish. **Superficially like:** Bullheads because of the very wide head.

Coloration: Non-breeding adults and young darkest dorsally, ranging from a milky-gray to a dark blue; sides lighter, sometimes mottled, or with a sharp demarcation between the darker color of the lower sides and whitish ventral surface which give such individuals a distinctly bicolored appearance. Fins lighter in coloration than adjacent body parts and lack the sharply defined dusky borders present in many young and subadult channel catfishes. *Adults*—Males and some females during breeding season have the head a dark bluish-black, dorsal half of body dusky-blue, and ventral half a whitish-blue.

Lengths: *Adults* usually 10.0″–18.0″ (25.4–45.7 cm) long, weights 8 oz–4 lbs, 8 oz (227g–2.0 kg). Maximum size reported to be 24.0″ (60.9 cm).

Distribution and Habitat

Ohio Distribution—Robert Cummins, Jr. has supplied much of the following: According to C. J. Riley, a commercial fisherman and transporter of fishes, 2500 lbs (1134 kg) of white catfishes were released in Sandusky Bay about 1939. Since then many have escaped annually from the Riley Fish Company's holding cars and pens, located near the mouth of the Portage River. Despite these annual introductions the white catfish seems to have failed as yet to establish itself permanently in Lake Erie waters.

During the past 15 years many tons of white catfishes have been liberated in private lakes in inland Ohio in which lakes the public pays for the privilege to fish, such as Sippo Lake east of Canton and Springdale Lake near Cincinnati. During the

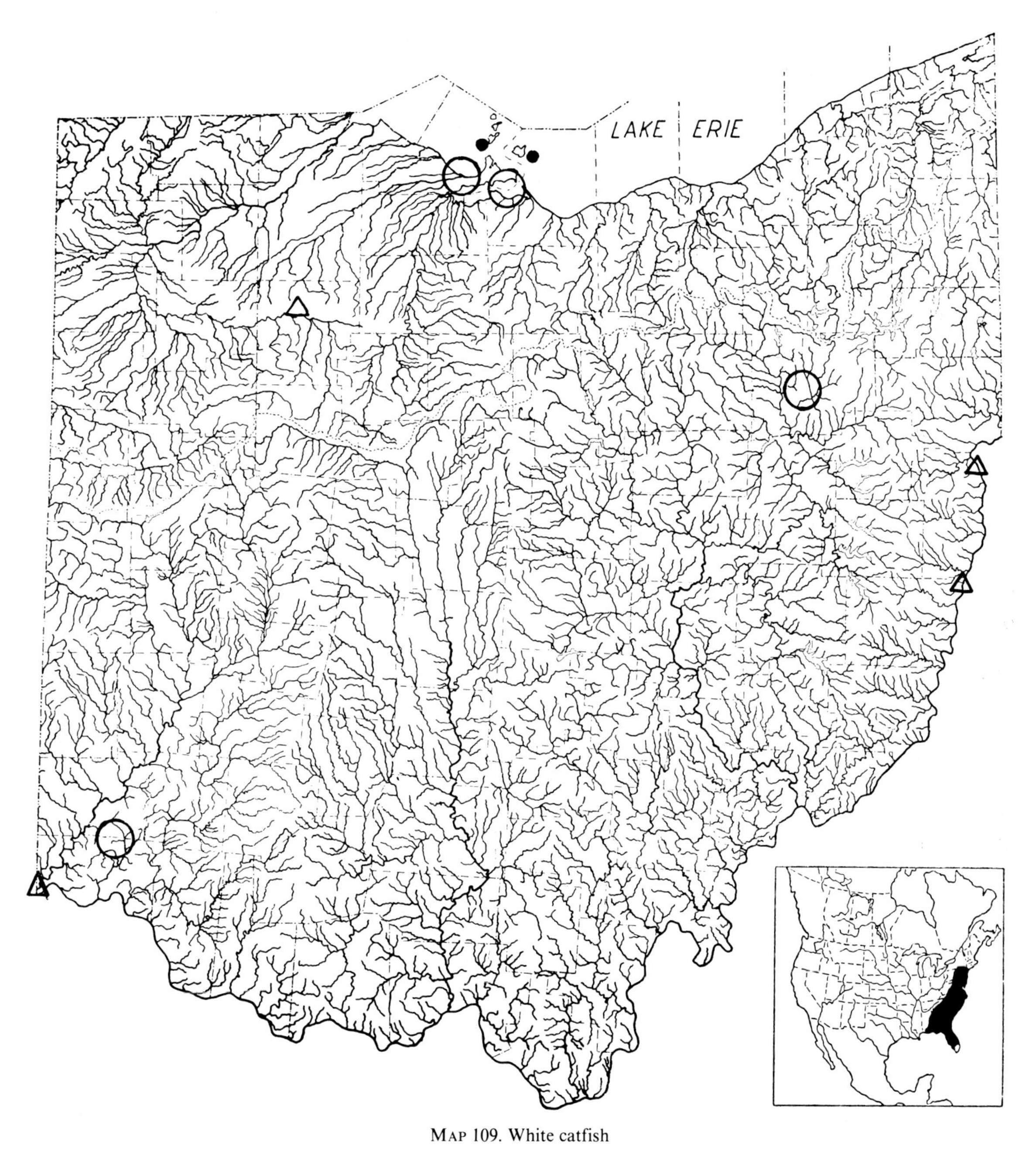

MAP 109. White catfish

○ Some localities where introduced. Reported captures. ● Before 1955. △ 1955–80.
Insert: Introduced into many states outside of its original range.

spring of 1954 the Riley Fish Company transported approximately 10 tons (9.1 metric tons) of these fishes, releasing most of them in the waters of southern Ohio and Indiana. The majority of the catfishes were obtained from Maryland and Virginia, and from such rivers as the Potomac and Rappahannock.

In 1953 Warren D. Handwork examined two specimens which were taken in the experimental nets of the Ohio Division of Wildlife. These nets were set in the vicinity of Kelleys Island, Erie County. C. J. Riley states that he found occasional individuals in his trap nets which were fished south of Green Island, Ottawa County. In the spring of

1954 fishermen brought several fork-tailed catfishes, presumably white catfishes, to the Port Clinton Fish Company for identification.

Owners of the private "pay lakes" consider the white catfish to be an excellent species for their purpose because it bites freely and is of a desirable size. It would not be desirable at present as a Lake Erie commercial species because the white catfish is difficult for the fishermen to separate from the channel catfish, and there is a 14-inch (36–cm) legal limit on fork-tailed catfishes, a length attained by comparatively few white catfishes. If the white catfish did become abundant in Lake Erie waters it presumably would compete, at least for food, with the commercially more desirable channel catfish.

Habitat—In its native range, which includes the coastal streams from Pennsylvania and New Jersey south to Florida, the white catfish inhabits the fresh and slightly brackish waters of streams, ponds and bayous. The species apparently is somewhat migratory. It is rather tolerant to swiftly flowing waters but seemingly prefers a more sluggish current than that preferred by the channel catfish, and appears to be more tolerant to a heavily silted bottom. In fact, the white catfish seems to occupy a habitat somewhat intermediate between that of the channel catfish, which usually inhabits swiftly flowing waters and/or a clean sand, gravel and bedrock bottom, and the bullheads which inhabit more quiet waters that often have soft bottoms of clayey silt and/or organic debris. The white catfish does not occur in as large numbers in the dense beds of aquatic vegetation as does the yellow bullhead, nor does it occur in as large numbers in the small, muddy, shallow ponds as does the black bullhead.

Years 1955–80—Apparently, the white catfish has failed to become permanently established in Lake Erie (Van Meter and Trautman, 1970:73). The species presumably has become established, temporarily at least, in that portion of the Ohio River adjacent to eastern Ohio. Three specimens from the Ohio River, taken by H. Ronald Preston, are catalogued in the state collections as follows:

OSUM 18432: 1 specimen 6 3/4″ (17 cm) TL, taken September 20, 1967 at the Pike Island Locks and Dam, Ohio River, near Yorkville, northeastern Pease Township, Belmont County, Ohio. OSUM 15253: 2 specimens 2 3/4″ (7.0 cm) and 3 1/4″ (8.3 cm) TL, taken September 24, 1968 at the New Cumberland Locks and Dam, Ohio River, near Stratton, Knox Township, Jefferson County, Ohio.

In a communication, dated December 19, 1968 Preston stated that "Several dozen were collected in our rotenone sampling at the New Cumberland Locks and Dam." Since the specimens were small, it appears plausible that they had hatched in the river. Normally, only subadults or adults are released in pay lakes, farm ponds, and so on. This species should be looked for in the tributaries entering the Ohio River in Jefferson and Belmont counties.

On April 30, 1974 a white catfish, weighing 4 lbs, 7.5 oz (2.0 kg) and 20.4″ (51.8 cm) in length was caught by Thomas Michel, in a pond a few miles northwest of Ottawa, Hancock County.

YELLOW BULLHEAD

*Ictalurus natalis** (Lesueur)

Fig. 110

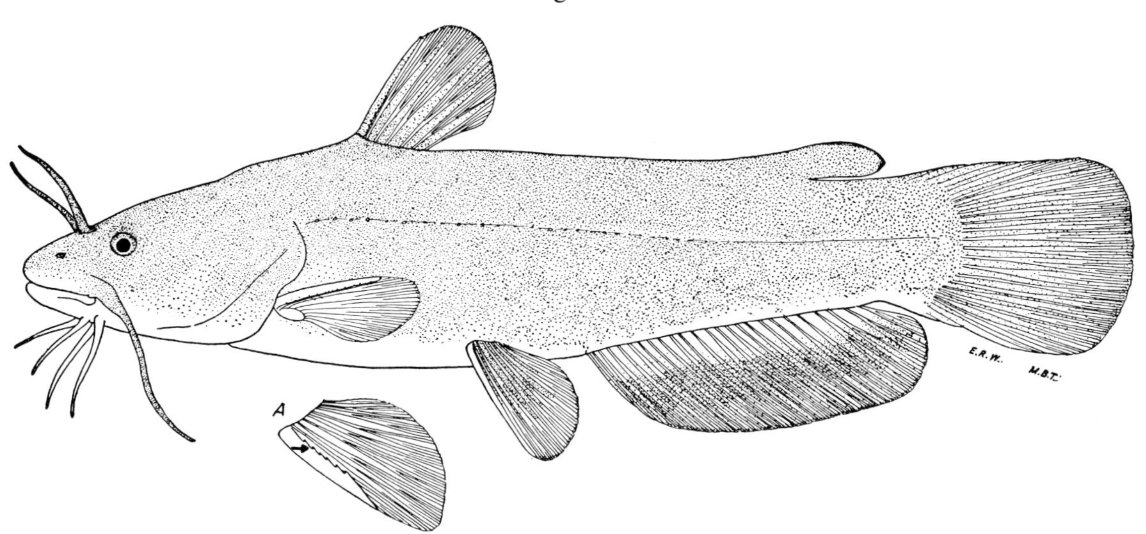

Upper fig.: Sunfish Creek, Monroe County, O.

Sept. 29, 1939. 160 mm SL, 7.5″ (19 cm) TL.
Female. OSUM 738.

Fig. A: left pectoral fin; arrow pointing to small, distinct serrations on posterior edge of pectoral spine.

Identification

Characters: *General* and *generic—See* blue catfish. *Specific*—Tail slightly rounded normally, sometimes straight in small young. Jaws equal or thc upper jaw is slightly the longer; lower jaw rarely the longer. Chin barrels whitish or yellowish, without black spots. Anal fin of 25–26 (extremes 23–28) rays, counting rudimentaries; its distal margin almost straight and usually edged with dusky or black. Anal fin usually with a broad, vague, horizontal, dark band, situated on the central half of the rays and running parallel with base of fin. Posterior edges of pectoral spines with sharp teeth or serrations, fig. A, these becoming blunt in large fishes.

Differs: Black and brown bullheads have black spotting or pigment on basal half of chin barbels; fewer anal rays. Flathead has lower jaw longer than upper jaw. Blue, white and channel catfishes have forked tails. Stonecat and other madtoms have adipose fin without a free lobe.

Most like: Brown bullhead. **Superficially like:** Black bullhead, flathead, stonecat and tadpole madtom.

Coloration: Dorsally yellow-olive (young, or fishes from turbid waters) brown, brownish-black (clear waters with black-muck bottom), or slaty-black (usually when about to hibernate or during hibernation). Sides lighter and more yellowish, sometimes vaguely mottled with darker. Ventral surface of head and body bright yellow, yellow-white or milk-white. Fins the same color as, but usually lighter than, adjacent parts of body; *see* Characters for description of anal fin. *Young* less than 2.0″ (5.1 cm) long often entirely black, except for the whitish belly and under side of head.

Lengths and **weights:** *Young* of year in Oct., 2.0″–3.5″ (5.1–8.9 cm) long. Around **1** year, 2.5″—5.0″ (6.4–13 cm). *Adults*, usually 5.5″–15.0″ (14–38.1 cm), weight 2 oz–2 lbs (57 g–0.9 kg). Largest specimen, 18.3″ (46.5 cm) long, weight 3 lbs, 10 oz (1.6 kg).

---

* Formerly *Ameiurus natalis*; *see* Taylor, 1954:43.

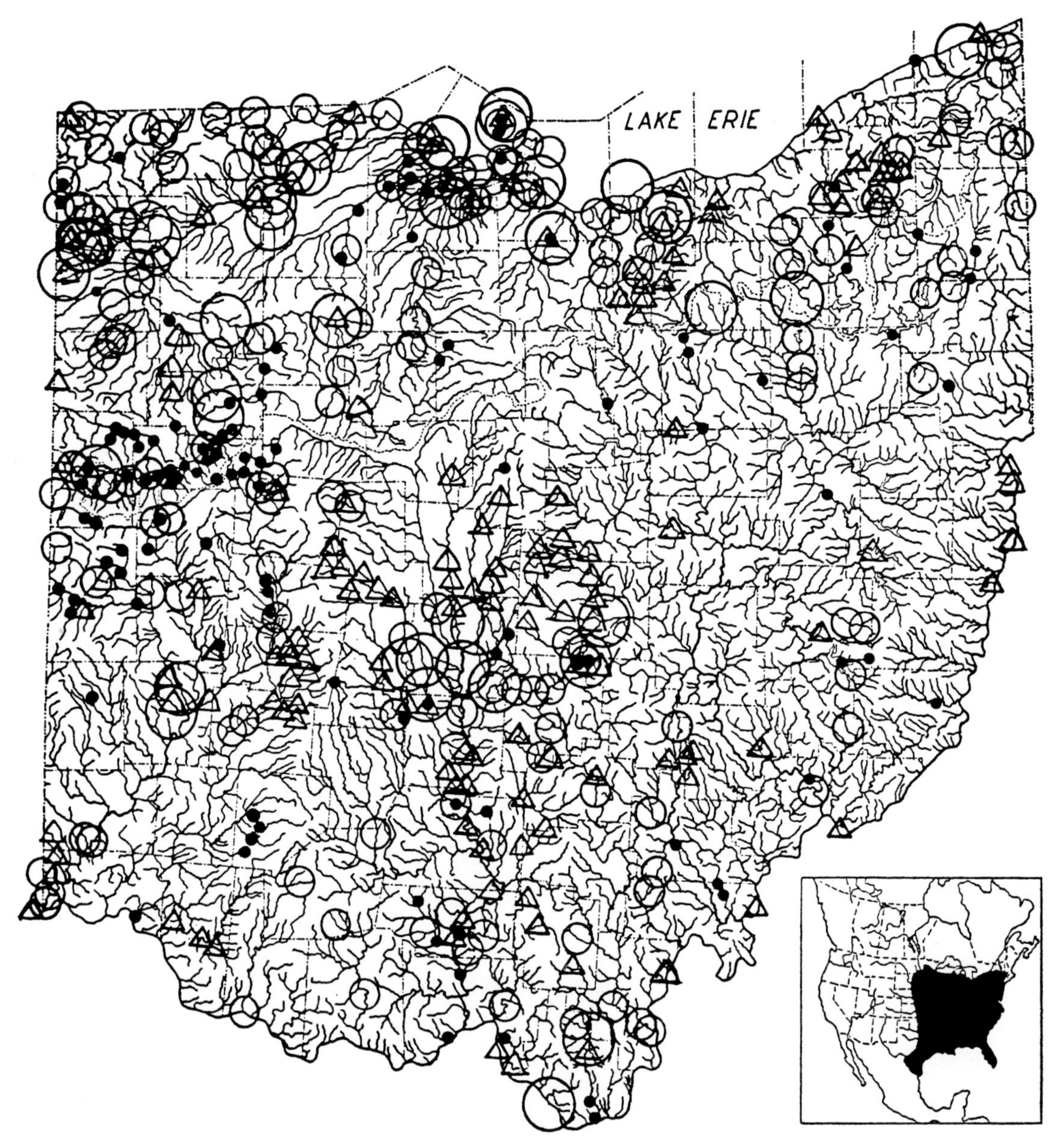

MAP 110. Yellow Bullhead

Locality records. ○ Before 1901. ○ 1901–37. ● 1938–54. △ 1955–80.
Insert: Successfully introduced into many areas outside of its range; limits of range in Mexico indefinite.

Distribution and Habitat

Ohio Distribution—Published records before 1880 are unreliable because of inability of writers to separate this species from the brown and black bullheads. Records between 1880–1900, often substantiated by preserved specimens, indicate that the species was rather well distributed over Ohio, was most abundant in the Lake Erie drainage, and was less numerous in the Ohio drainage, particularly in the unglaciated portion along the Ohio River. Comparisons of collections made before and after 1939, indicate a decrease or lack of records after 1939 in such localities as Sandusky Bay, Maumee River itself, and streams of Franklin County.

Although no definite retreat from considerable sections of its Ohio range is apparent, the species has decreased drastically in abundance in many localities in recent years, and this despite repeated stocking by the State. Between 1907–18 I caught hundreds of bullheads annually in the northeast corner of Indian Lake. Before 1911 more than 90% of these were yellow bullheads. After 1911 intensive stump and log removal from the lake, subsequent decrease in amount of aquatic vegetation which caused increased wave wash and turbidity, had so decreased its habitat that by 1918 less than 50% of the bullheads taken were yellow bullheads. By 1950 apparently less than 10% of the bullheads caught in that section of Indian Lake were yellow bullheads. I (1940A:Pl.1) noted a similar decrease in numbers of this species at Buckeye Lake since 1906, as the amount of aquatic vegetation decreased; and published upon its apparent decrease, between 1887–1938, in Lost and Gordon Creeks (1939A:280 and 287).

Habitat—The largest populations occurred in the shallow portions of large bays, lakes, ponds, and streams where the gradient was basic or low, water clear and aquatic vegetation profuse. It was often present in small low-gradient brooks which contained normally clear waters and some aquatic vegetation. Several diverse bottom types were tolerated, such as gravel, sand, peat or muck. Clayey silts were tolerated, if the clay would not become re-suspended in the water during wave action or increased current. Populations of yellow bullheads appeared to remain higher under adverse conditions if the two interspecific competitors, the brown and black bullheads, were absent.

Years 1955–80—Small or isolated populations of the yellow bullhead existed over much of inland Ohio between 1955–75 (White et al., 1975:93). The species was most numerous in the original prairie areas, such as those in west central Ohio and especially in the smaller streams where there remained aquatic vegetation and silting of the substrate was not severe. The greatest decreases in numbers and extirpation of relict populations were noted whenever silting became excessive, such as it did in much of the Maumee River system. It is not known why the species so drastically decreased, almost to the verge of extirpation, in the older, larger canal reservoirs, such as Indian and Buckeye lakes.

On the whole its numbers decreased in the shallows and marshes of the western third of Lake Erie (Van Meter and Trautman, 1970:73). In the entire Ohio River (ORSANCO, 1962:149) only 121 yellow bullheads were taken during the 1957–59 investigations.

BROWN BULLHEAD

*Ictalurus nebulosus** (Lesueur)

Fig. 111

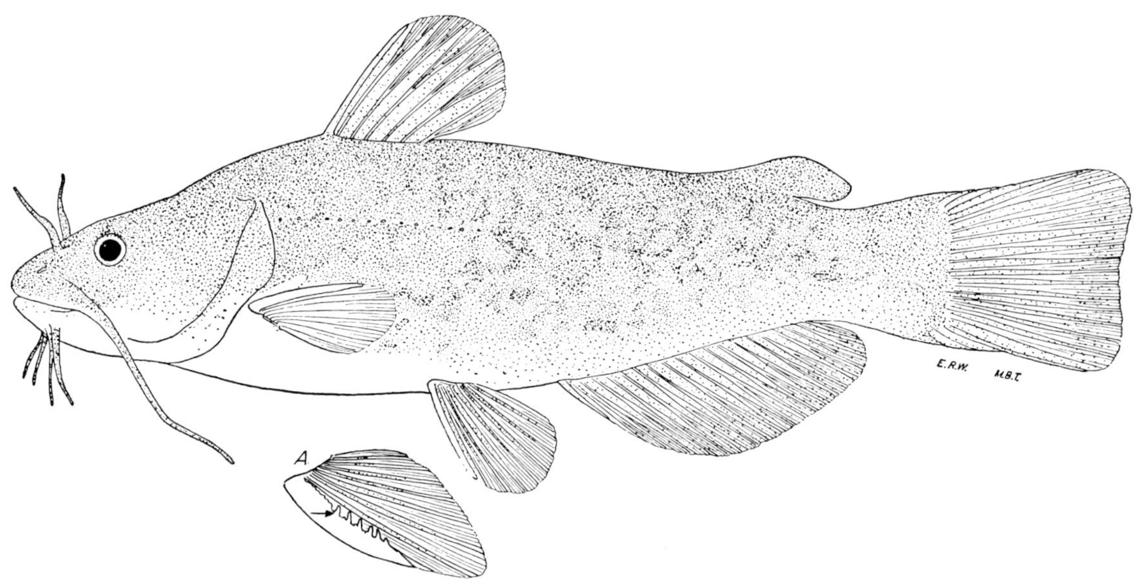

Upper fig.: Chagrin River, Geauga County, O.

Aug. 8, 1939. Female. 142 mm SL, 6.8″ (17 cm) TL. OSUM 400

Fig. A: left pectoral fin; arrow pointing to serrations on posterior edge of pectoral spine.

Identification

Characters: *General* and *generic*—*See* blue catfish. *Specific*—Tail usually slightly emarginate, sometimes straight in small young. Jaws equal or upper jaw slightly longer; lower jaw rarely the longer. Chin barbels gray, black or black-spotted, at least at their bases. Anal fin usually of 22–23 (extremes 21–24) rays, counting rudimentaries; its distal margin usually slightly rounded. Anal fin usually mottled or darkest on basal third or half of fin. Posterior edges of pectoral spines with many sharp teeth (or serrae) which may become blunted in large fishes, fig. A (to note presence, size and sharpness of teeth, grasp anterior and posterior edges of pectoral fin between thumb and forefinger and pull outward; the sharp spines will prick and prevent the finger from moving outward). Body averaging rather slender. Body of adults often conspicuously mottled, especially on sides, and there is no sharp demarcation line between ventral surface of body and lower sides.

Differs: Black bullhead lacks the sharp teeth on posterior edge of pectoral spines, except in small young; has fewer anal rays; is more chubby. *See* yellow bullhead.

Most like: Black and yellow bullheads. **Superficially like:** Flathead, stonecat, and tadpole madtom.

Coloration: Dorsally yellow-, olive-, slaty- or light chocolate-brown with vague, darker mottlings. Sides lighter and often heavily blotched and mottled with dark browns and chocolates. Ventrally bright yellow, yellow- or milk-white. Fins same color as, but usually lighter than, adjacent parts of body; for description of anal fin, *see* Characters. *Young* less than 2.0″ (5.1 cm) long often black except for whitish under-surface of head and white belly.

Lengths and **weights:** *Young* of the year in Oct., 2.0″–4.8″ (5.1–12 cm) long. Around **1** year, 2.7″–6.0″ (6.9–15 cm). *Adults*, usually 6.0″–16.0″ (15–40.6 cm)

* Formerly *Ameiurus nebulosus nebulosus*; Taylor, 1954:43.

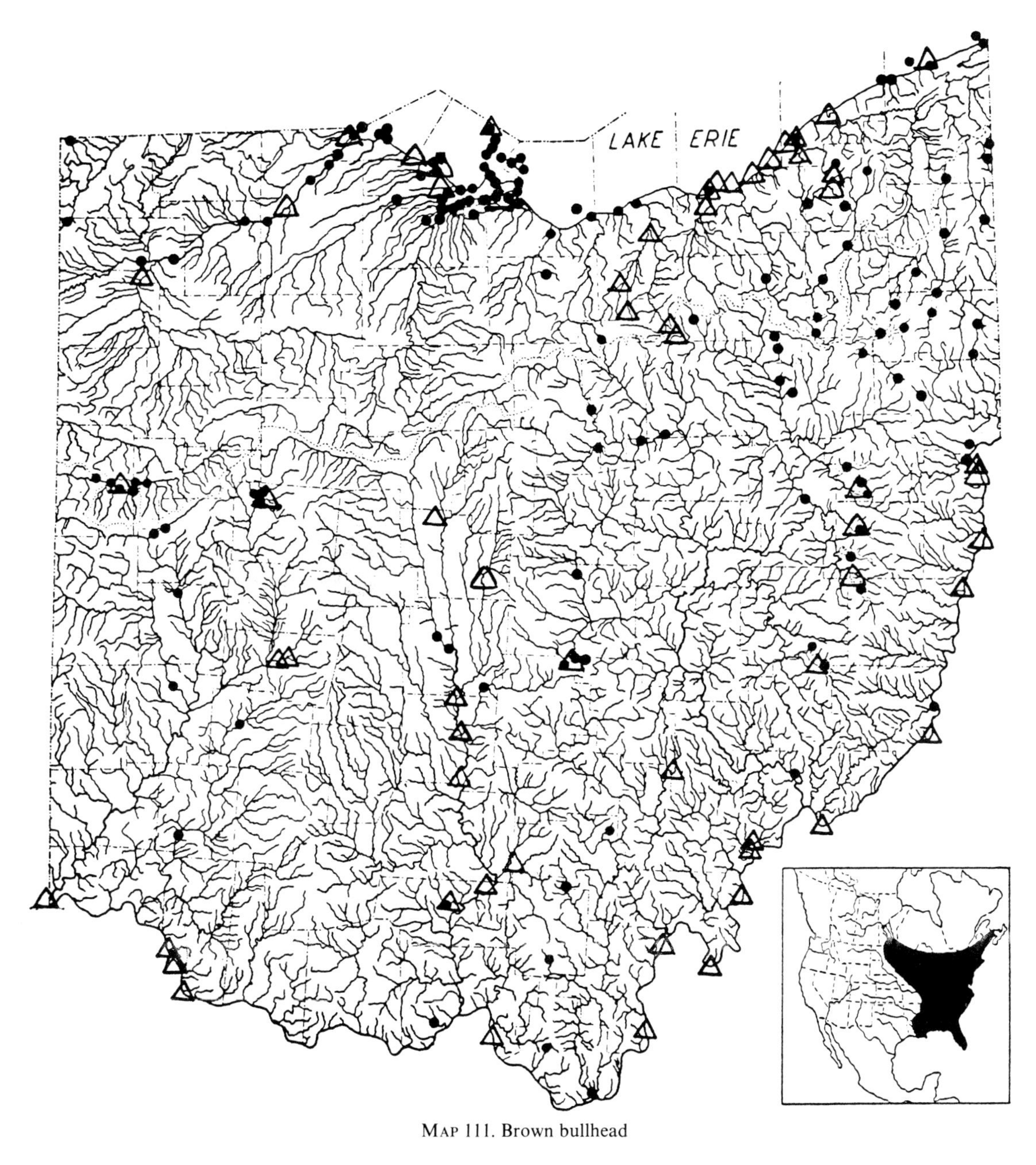

MAP 111. Brown bullhead

Locality records. ● Before 1955. △ 1955–80.
Insert: Successfully introduced into many areas outside of its range.

long, weight 2 oz–2.5 lbs (57 g–1.1 kg). Largest specimen, 18.8″ (47.8 cm) long; weight 31 lbs, 14 oz (1.8 kg).

Hybridizes: with black bullhead, especially in the shallow bays and waters of western Lake Erie, and in small impoundments where only submarginal habitat for one species, or both, exists; also in the Ohio River.

Distribution and Habitat

Ohio Distribution—Between 1920–50 the brown bullhead was abundant only in the clearer waters of Lake Erie and its adjacent bays, harbors and marshes, and in widely separated, post-glacial lakes and impoundments of the northern fifth of the state. Only strays were present in flowing waters, except in the northeastern section where small populations were found in some clear, weedy low-gradient streams. During this period the state conservation department liberated hundreds of thousands of brown bullheads into Ohio waters, but only those introduced into the larger lakes and impoundments became permanently established. It is highly significant that despite the mass introductions of brown bullheads into Pike and Scioto counties, the only recoveries we have taken in more than 425 days of stream and lake surveying in these counties have been from lakes Roosevelt and White. I have never seen a specimen taken from the Ohio River.

During the 1907–50 period I noted decided fluctuations in abundance of brown bullheads in several lakes, particularly Buckeye and Indian lakes. In the latter, the brown bullhead became more abundant after 1915, after a marked decrease in yellow bullheads (*see* that species) had occurred. The brown bullhead remained abundant until about 1930, after which it began to decrease in abundance coincidental with a decided increase in numbers of black bullheads.

Little is known concerning the brown bullhead prior to 1900, for few species of Ohio fishes were more persistently misidentified than were the three species of bullheads. I find that all specimens identified originally as brown bullheads, which were taken in the Ohio drainage, prove to be black or yellow bullheads, as are several specimens collected in the Lake Erie drainage. As examples: Williamson and Osburn (1898:16 and Pl. 14) state that the species was "Found in most of our larger streams" in Franklin County; yet their photograph of a supposed brown bullhead is of a yellow bullhead. Of particular interest is a black bullhead (SNHM 3293), long assumed to be a brown bullhead, that was taken in Mill Creek, Hamilton County, presumably by Henshall between 1885–90. Because of these almost universal misidentifications no early literature records have been used.

Judging from our present knowledge of its exacting habitat requirements in Ohio, I believe that prior to 1900 the brown bullhead may have been rather widespread and abundant in the Lake Erie drainage, particularly in the then clear waters of Sandusky Bay, West Harbor, and clear, low-gradient, vegetated streams such as the Maumee and St. Marys rivers; that it was present in pothole, post-glacial lakes in the Ohio drainage as far south as Buckeye Lake; that it was rare or absent in the southern third of Ohio, at least before the building of the canals.

Habitat—The largest Ohio populations occurred in western Lake Erie waters, and in the deeper, larger inland ponds and impoundments, where the waters were rather clear (at least with few clayey silt particles) and usually cool; there was a moderate amount of aquatic vegetation; the bottom was of sand, gravel or dark muck, with or without larger pieces of organic debris. The brown bullhead tended to remain in deeper water than did the other two bullhead species, and in areas where there was less vegetation than that seemingly required by the yellow bullhead. It appeared to be far less tolerant to turbid waters than was the black bullhead.

Years 1955–80—There was little change in the numerical status of the brown bullhead in inland Ohio during the 1955–80 period. Typical brown bullheads normally were present only in small numbers in most localities, despite introductions from Lake Erie and elsewhere of this species, but there were many hybrids between brown and black bullheads. The numbers of typical brown bullheads continued to decline in the old canal reservoirs, such as Indian and Buckeye lakes, where formerly the brown bullhead had been abundant.

Together the brown and black bullheads and their hybrids were commercially important in Lake Erie (White et al., 1975:93–94). The majority were taken from the western third of the lake and Sandusky Bay. The 1957–59 investigations of the Ohio River indicate that the brown bullhead was the least numerous of the three bullhead species inhabiting that stream (ORSANCO, 1962:149).

Before 1950, as indicated under Hybridizes above, there had been considerable hybridizing and backcrossing between brown and black bullheads in western Lake Erie and in several inland impoundments. After 1951 hybridization in some impoundments became so prevalent that whole populations appeared to be largely F_1 hybrids or backcrosses, and "typical" examples of one or both species were rare or absent.

The 12-acre Sportsmen's Lake in Hardin County is an example of what may happen in inland impoundments. In 1966 the fish population of this lake was rotenoned. Darrell Allison, of the Ohio Division of Wildlife, sent me two bullheads which were hybrids or backcrosses between brown and black bullheads. In a letter (March 1, 1967) he stated that approximately 35,000 other bullheads had been taken which likewise were considered to be hybrids, and, "At the time of the total rotenone removal, no pure strains of either *I. melas* or *I. nebulosus* were found to be present in the lake."

H. Ronald Preston, of the Federal Water Pollution Control Administration, gave the state collections several dozens of bullheads taken in the vicinities of dams along the Ohio River from Pittsburgh to Cincinnati. He is of the opinion that the majority of these bullheads are hybrids or backcrosses, to which I and Ted Cavender agree.

Mass hybridization of bullheads in Ohio may be the result of one or several factors. From original records the pre-Columbian range of the brown bullhead presumably extended along both coasts of Florida, the Atlantic coast northward to New Brunswick, and westward through southern Ontario to Saskatchewan and the northern United States to Minnesota. Because of the absence of old specimens from southern Ohio, it is assumed that originally the brown bullhead was largely or entirely absent from that area. The range of the black bullhead was confined largely to the Mississippi and lowlands including the Ohio River drainage.

There should have been comparatively little hybridization between the black and brown bullheads in late pre-Columbian times because of relative stability of their respective habitats. With the invasion of Europeans there began an increasing disruption of the aquatic environment which tended to decrease or eliminate the clear-water, aquatic vegetation habitat of the brown and to favor the silt-tolerant black bullhead. The destruction of the aquatic vegetation in the shallows and marshes of western Lake Erie and increasing turbidity favored the black and produced unfavorable conditions for the brown, resulting in hybrid swarms in the shallower waters of western Lake Erie. It is from such populations that many hundreds of thousands of bullheads were captured, to be later released in inland Ohio waters, some descending into the Ohio River.

It is intriguing to note that when mass hybridization occurs in the small, silty, largely vegetationless impoundments, the majority of the population resembles black bullheads, and that a large number of "typical" blacks may be present, but there may be few or no "typical" browns. On the other hand, in the deeper waters of the Ohio River the situation appears to be reversed, and backcrosses usually favor the brown rather than the black bullhead.

BLACK BULLHEAD

*Ictalurus melas** (Rafinesque)

Fig. 112

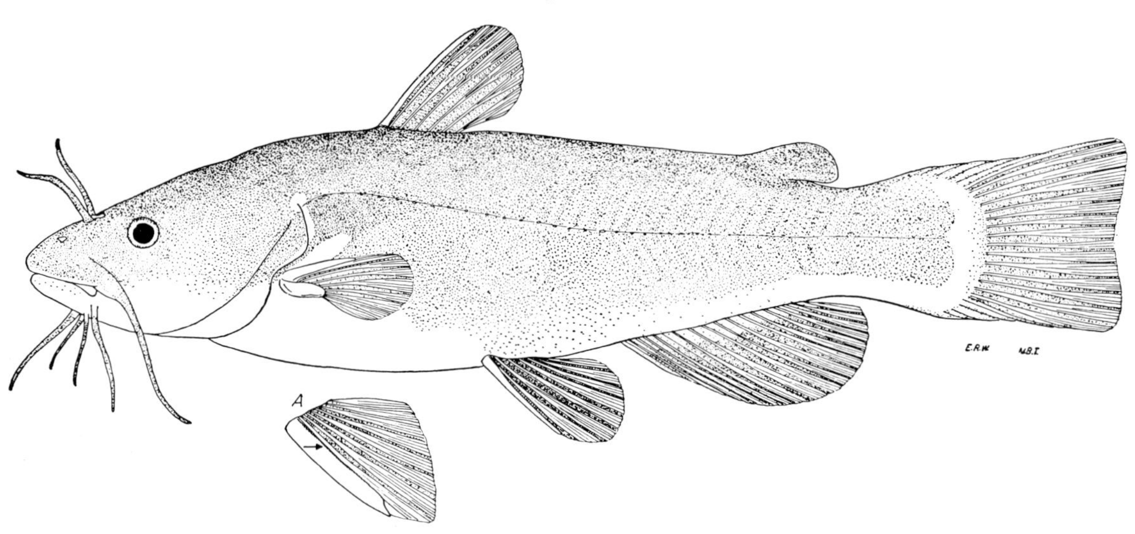

Upper fig.: Lake Erie, Ottawa County, O.

April 9, 1943. 185 mm SL, 8.6″ (22 cm) TL.
Female. OSUM 6668.

Fig. A: Little East Fork, Todds Fork, Clinton County, O.

Oct. 17, 1942. 90 mm SL, 4.3″ (11 cm) TL.
Female. OSUM 5251.

Left pectoral fin; arrow pointing to slightly roughened but not serrated, posterior edge of pectoral spine.

Identification

Characters: *General* and *generic—See* blue catfish. *Specific*—Tail usually slightly emarginate, sometimes straight in young. Jaws equal or upper jaw slightly longer; lower jaw rarely longer. Chin barbels gray, black, or black spotted, at least at their bases; barbel coloration similar to that of brown bullhead. Anal fin usually of 17–21 (extremes 16–22) rays, counting rudimentaries, its distal margin normally rounded. Basal third of anal fin light, webbing and rays of the distal two-thirds darker, usually giving fin a distinctly bicolored appearance; except in young less than 2.0″ (5.1 cm) long in which the entire fin may be black. Posterior edges of pectoral spines without sharp teeth which catch the finger (*see* Characters under brown bullhead) although many young and some small adults (also the hybrids) have the posterior edges somewhat serrated (fig. A). Body chubby and deep at dorsal origin; angle steep from dorsal to snout. Body of adults usually bicolored, with a sharp demarcation between the darker, lower sides and the lighter ventral surface. A light, vertical, caudal bar, usually conspicuous in large young and adults; this bar connects with the light color of the ventral surface.

Differs: *See* yellow and brown bullheads. Yellow bullhead has whitish chin barbels; brown bullhead has the posterior edge of the pectoral spine serrated.

Most like: Brown and yellow bullheads.

Coloration: Dorsally greenish-, yellowish-, brownish-, or slaty-olive. Sides lighter and usually yellower or whiter. Ventrally bright yellow, yellow- or milk-white. Fins normally conspicuously blacker than adjacent body parts; for description of anal fin, *see* Characters. *Breeding male*—Often jet black

* Formerly *Ameiurus melas melas*; *see* Taylor, 1954:43.

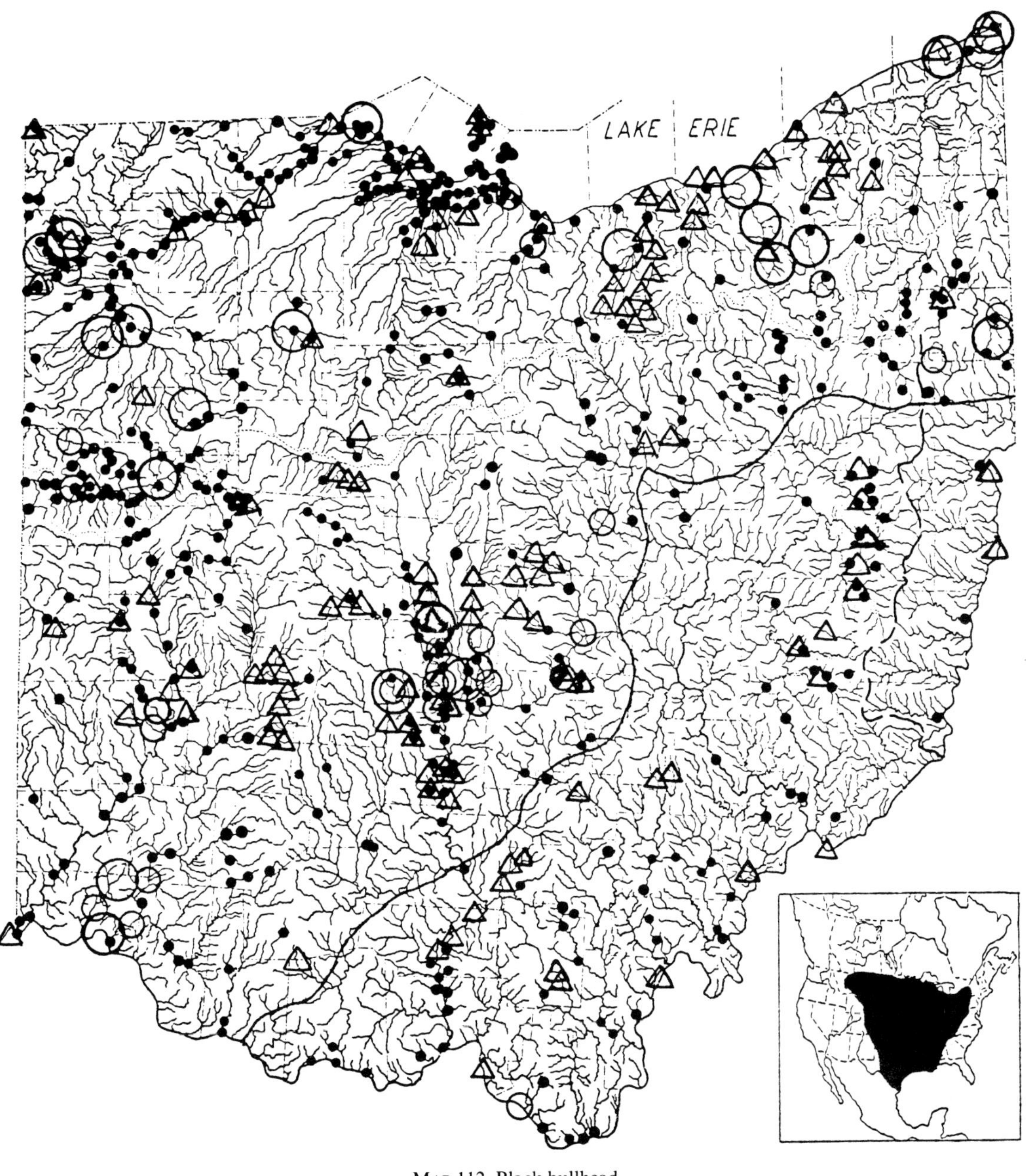

MAP 112. Black bullhead

○ Specimens collected before 1901. ○ Literature references without specimens before 1901. ● 1901–54. △ 1955–80. ~ Glacial boundary. /\ Flushing Escarpment.

Insert: Successfully and sometimes inadvertently introduced into many areas outside of this range; limits of range in Mexico indefinite.

with bright yellow or white belly. *Young* less than 2.0″ (5.1 cm) long, black except for whitish belly.

Lengths and **weights:** *Young* of year in Oct., 1.2″–4.0″ (3.0–10 cm) long. Around **1** year, 1.5″–5.0″ (3.8–13 cm). *Adults*, usually 4.5″ (11 cm) (smaller in small ponds when breeding adults are dwarfed)–12.0″ (30.5 cm), weight 1 oz–15 oz (28–425 g). Largest Specimen, 16.8″ (42.7 cm) long; weight 2 lbs, 12 oz (1.2 kg).

Hybridizes: *see* under brown bullhead.

Distribution and Habitat

Ohio Distribution—Because of the confusion in identification of the three species of bullheads that existed before 1900, it is necessary to separate localities from which there are preserved specimens from those where the published references are unsupported by specimens. From both types of records it appears that before 1900, the black bullhead was fairly well distributed throughout glaciated Ohio, but seemingly was neither so widely distributed nor as abundant as was the yellow bullhead. This is readily understandable; for obviously conditions before 1850 were far more favorable for the yellow than for the black bullhead.

Since 1850 conditions have become increasingly favorable for the black bullhead. This increase in amount of favorable habitat began early, for Kirtland (1850G:141, as *Pimelodus catus*) commented that although by 1850 the species was "common in the rivers, ponds, and lakes of Ohio" with large numbers being taken, it was "rapidly multiplying at that time," especially in "The muddy waters of the canals and reservoirs." As indicated under "Yellow Bullhead," I noted at Indian Lake a reversal from a dominance of yellow bullheads between 1907–11 to a dominance of the black bullheads after 1940. I likewise observed similar reversals, or an increase in abundance of black bullheads, at Buckeye Lake and several other localities.

Between 1920–55 the black bullhead was the most widely distributed and abundant of the three bullhead species. Because of its preference for low gradients, rather warm waters and high tolerances to turbidity and siltation, it was particularly numerous about Sandusky Bay and northwestern Ohio, especially after 1940, and conversely, it was less numerous in the colder, high-gradient waters of the northeastern portion. This preference for low-gradient streams is well illustrated by noting the many locality records in Ottawa and Auglaize counties, and the fewer records in Ashland and Wayne counties, four counties which have been thoroughly investigated since 1935. In unglaciated Ohio it was most numerous in impoundments and in base-gradient streams; least numerous or almost entirely absent in the higher-gradient streams along the line of glaciation and east of the Flushing Escarpment.

Habitat—The largest populations occur in base- and low-gradient portions of small and moderate sized streams; in the impoundments, backwaters, oxbows, and overflow ponds, particularly along the larger rivers; in quarries and farm ponds; and in the shallower and more turbid waters of Lake Erie. The species seemingly prefers silty water and soft mud bottoms, is highly tolerant of many types of industrial and domestic pollutants, and warm waters. It appears incapable of invading in abundance the deeper, cooler, clearer waters, with or without some vegetation, which is the habitat of the brown bullhead, or the very clear water, heavily vegetated habitat of the yellow bullhead. It frequently becomes over-populated in small waters with a consequent dwarfing in size.

Years 1955–80—During this period the black bullhead, with rare exceptions, maintained or increased its numbers throughout its Ohio range, largely because changing conditions in ponds, reservoirs and streams favored this silt-tolerant, vegetationless, shallow-water-inhabiting bullhead (White et al., 1975:94). However, populations in the old canal reservoirs continued to decline for reasons not fully understood.

The black bullhead, or hybrids between it and the brown bullhead, were quite numerous in the shallow waters of western Lake Erie (Van Meter and Trautman, 1970:73). It was the most numerous of the three bullhead species in the Ohio River (ORSANCO, 1962:149). For discussion of hybridization in inland Ohio and the Ohio River *see* the last two paragraphs under brown bullhead.

FLATHEAD CATFISH

Pylodictis olivaris (Rafinesque)

Fig. 113

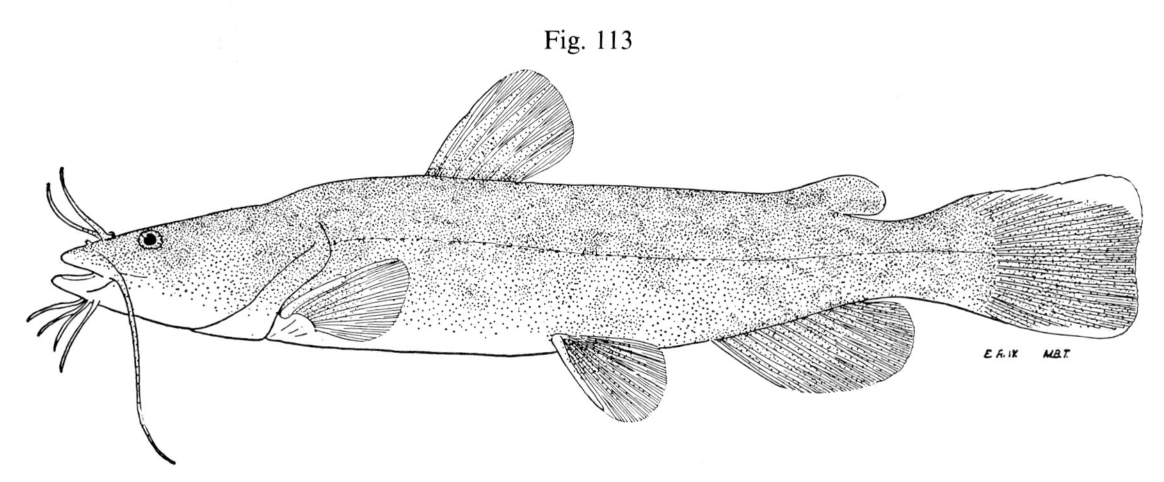

Scioto River, Scioto County, O.

July 25, 1940. 273 mm SL, 12.8″ (33.5 cm) TL.
Male. OSUM 1885.

Note light dorsal and ventral edges of caudal fin.

Identification

Characters: *General—See* blue catfish. *Generic* and *specific*—Lower jaw *always* longer than upper jaw. Posterior margin of tail straight or slightly emarginate; dorsal and ventral edges of tail lighter colored or whitish. Anal fin base very short; fin usually of 14–17, extremes 13–18 rays, counting all rudiments. Head wide and notably flattened between eyes. Adipose fin very large. Pre-maxillary bands of teeth with backward extensions (*see* stonecat, fig. A). General coloration brown, mottled with darker brown and dusky.

Differs: No other species of catfish in Ohio has a flattened head between the eyes; a longer lower jaw (except rarely an old bullhead with a misformed lower jaw).

Most like: Brown bullhead. **Superficially like:** Black and yellow bullheads and stonecat; tadpole madtom may be mistaken for a small flathead.

Coloration: Dorsally yellow- (turbid water), olive-, or dark brown (clear water), with blotches of darker chocolate. Sides lighter, brownish, and mottled with darker. Ventrally yellow or yellowish-white. Fins same color as adjacent body parts. Dorsal and ventral tips of caudal fin white in small young; these areas lighter colored than adjacent parts of fin in larger fishes.

Lengths and weights: *Young* of year in Oct., 2.0″–4.0″ (5.1–10 cm) long. Around **1** year, 3.5″–8.0″ (8.9–20 cm). *Adults*, usually 14.0″–45.0″ (35.6–114 cm) long, weight 1–45 lbs (0.5–20.4 kg). The 52-lb (23.6 kg) flathead from Lake Erie drainage, mentioned in text, was "4 foot" (122 cm) long. Largest specimen, was about 53.0″ (135 cm) long, weight 82 lbs (37.2 kg), taken in Ohio River, Adams County, in 1930.

Hybridizes: with channel catfish.

Distribution and Habitat

Ohio Distribution—Published references and preserved specimens indicate the presence before 1900 of the flathead catfish in the Ohio River and its larger tributaries eastward as far as western Pennsylvania (Fowler, 1919:57). Since 1900 this species, usually called shovelhead cat in southern Ohio, was common and much sought after in the Ohio River upstream as far as Marietta, and in the lower Scioto and Muskingum rivers. Smaller populations occurred in the Little Miami, White Oak, Ohio Brush, Symmes, Raccoon, Shade and Hocking rivers and

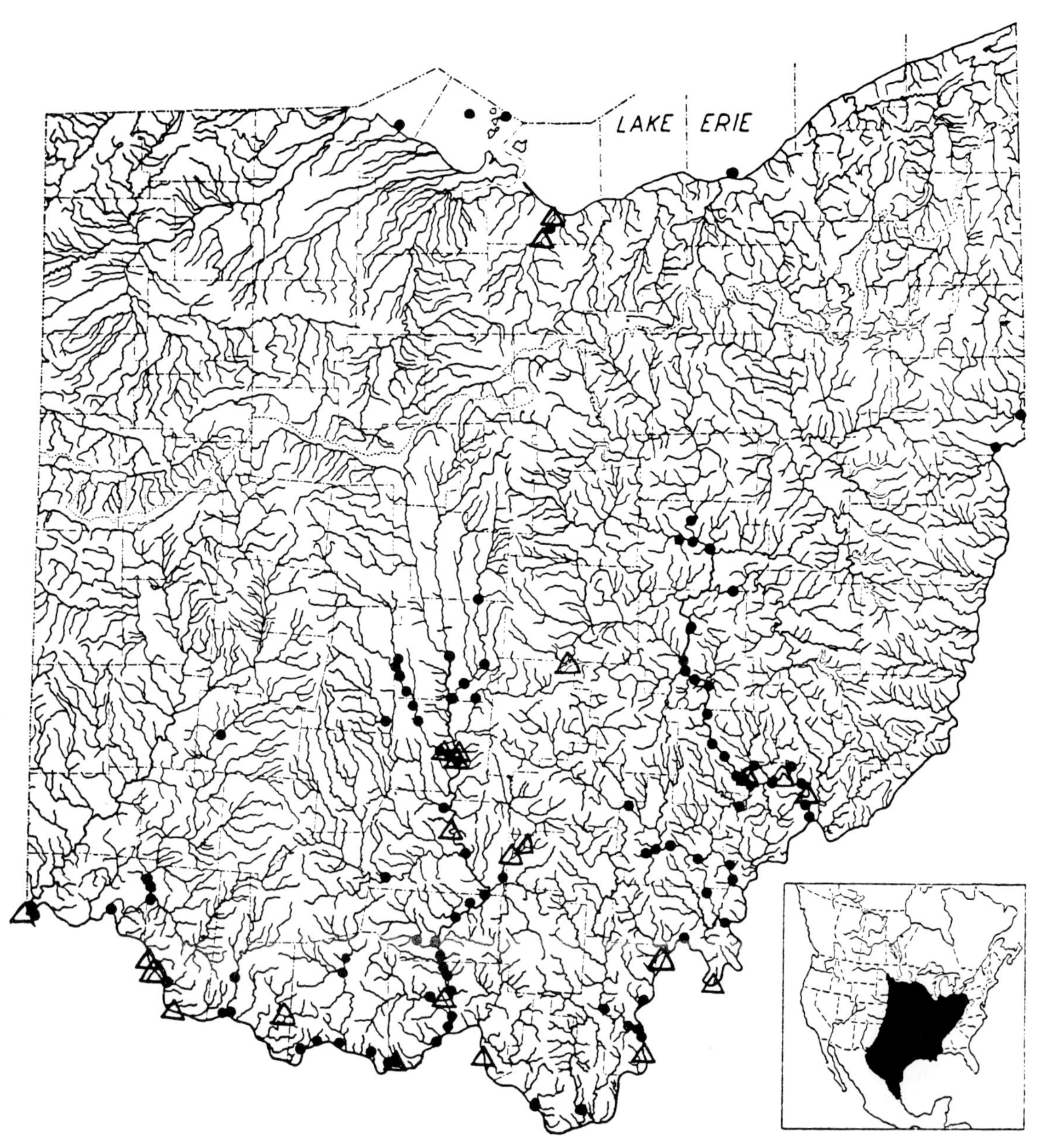

MAP 113. Flathead catfish

Locality records. ● Before 1955. △ 1955–80.
Insert: Sparingly introduced outside of this range; limits of range in Mexico indefinite.

creeks. Its rarity in the Great Miami River system may have been caused by a combination of higher gradients and too much inorganic pollution.

In 1892 McCormick (1892:12, as *Leptops olivaris*) recorded a relict colony of "Mud-cats" from the Ohio waters of Lake Erie, stating that the species was "quite rare." In his unpublished notes, preserved at Oberlin college, McCormick stated that he saw a specimen taken from the pounds (pound nets) in September, 1890. Since 1938, five specimens from Lake Erie waters have been preserved, one of 3 lbs (1.4 kg) in the Cleveland Museum, and one each of 1 lb (0.5 kg), 2.5 lbs (1.1 kg), 2.75 lbs (1.2 kg) and 14 lbs (6.4 kg) in the OSUM (1866;

6664; 8467; and 8471) collections. All of these were from Lake Erie except the largest which came from the Huron River. There appears to be a small population of flathead catfish in the Huron River, where they are taken yearly. Recently, fishes of 52 lbs (23.6 kg) (Meyers, 1947:30) and 47 lbs (21.3 kg) (Kiddney, 1951:27) have been caught in that stream.

Habitat—This fish occurred most abundantly in sluggish, long, deep pools of the low-gradient portions of large streams, and although of necessity, the species was forced to inhabit extremely turbid streams, it is significant that in such streams flatheads were found usually over hard bottoms, or when over silt bottoms, where silt deposition was very slow. The species was taken in the Ohio River on trotlines in water deeper than 50′ (15 m). The fish likewise fed at night on riffles so shallow that the dorsal fins stuck out of the water. I have seen them feeding thus on shallow riffles in the Scioto River. When using jack lights, I have seen large flatheads, with their mouths widely open, lying on the bottom and usually beside a log or other object, in water less than 5′ (1.5 m) in depth. Several Ohio River fishermen have told me that they have seen frightened fish dart into the open mouths of flatheads, to be swallowed immediately. The large numbers of such hiding species as rockbass, spotted blackbass, and small catfishes, found in the stomachs of large flatheads, lend credence to these statements.

Years 1955–80—Prior to 1950 the Scioto River from Chillicothe downstream was noted for its excellent "Shovelhead" (the name almost universally applied to the species in Ohio) fishing. Since 1950 fishing for this species in the Scioto downstream from Chillicothe has largely ceased because of the drastic decrease in its numbers and rank flavor of those taken. This decrease in abundance and strong flavor were caused by pollutants, apparently those coming chiefly from paper mills and atomic wastes.

Unlike those decreasing populations of the lower Scioto, the Muskingum River populations appear to have remained relatively constant during the 1955–80 period. The species continued to be numerous in the Ohio River, as indicated by the 1957–59 investigations (ORSANCO, 1962:150).

In the Lake Erie drainage the relict population (White et al., 1975:94–95) tenaciously remained extant in the Huron River despite one severe case of pollution in 1968 when a broken gasoline line polluted the lower half of that stream. A few individuals were caught by fishermen annually in this river during the last 25-year period.

For age and growth of this species in Iowa *see* Mayhew, 1969:118–21.

STONECAT MADTOM

Noturus flavus Rafinesque

Fig. 114

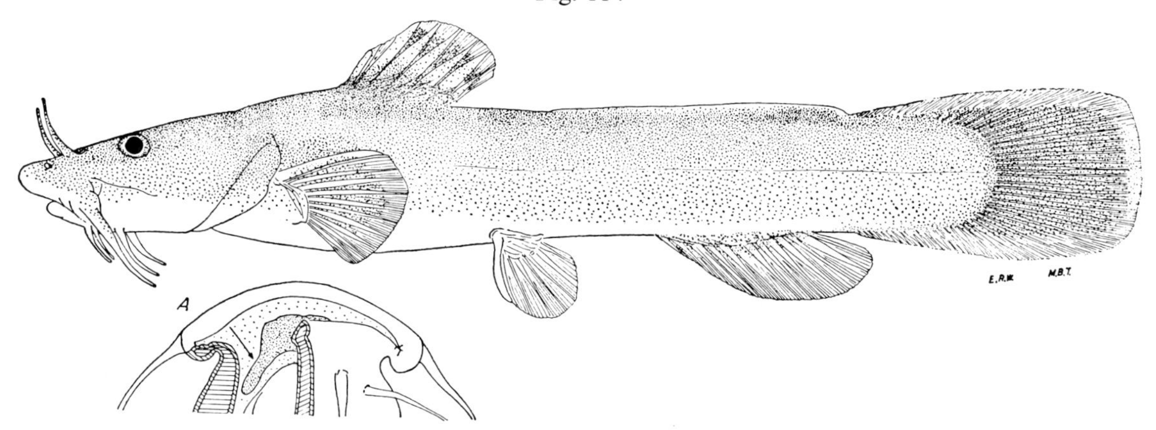

Upper fig.: Clear Fork, Ashland County, O.

Sept. 9, 1939. 160 mm SL, 7.4″ (19 cm) TL.
Male. OSUM 524.

Fig. A: ventral view of anterior portion of head with right half of lower jaw removed to show the right half of the premaxillary band of teeth which is situated on the roof of the mouth. Arrow points to the backward extension of this premaxillary band of teeth. Such an extension of the premaxillary teeth occurs also in the flathead catfish, but is lacking in bullheads and other species of madtoms.

Identification

Characters: *General—See* blue catfish. *Generic—* The long adipose fin is bound to the back *throughout* its entire length and separated from the caudal fin by a notch. *Specific—No* spinelets, serrae or teeth on posterior edge of pectoral spine. Adipose fin low and relatively inconspicuous. Upper jaw much longer than lower jaw. Humeral process (in the figure the outline of the process may be seen lying directly above and parallel to the anterior portion of the pectoral spine) very short and very little, if any, longer than the width of the base of pectoral spine. The four saddle-bands, described under northern madtom (*see* Coloration), are faint when present. Caudal squarish with a light border. Body slender, becoming deeper in largest adults. Body usually bicolored, the dark back and sides contrasting sharply with the light ventral surface, except in individuals from Lake Erie or from turbid, inland waters; these are more unicolored. Premaxillary band of teeth with backward extensions, *see* arrow, fig. A.

Differs: Tadpole madtom has lower jaw almost as long, or as long, as upper jaw, has a high adipose fin, is unicolored and deep-bodied, and the premaxillary teeth lack backward extensions. Other madtoms have spinelets on posterior edge of pectoral spine. Flathead catfish has lower jaw longer than upper jaw. Bullheads, blue, white and channel catfishes and flathead have the adipose fin with a free lobe posteriorly.

Most like: Tadpole madtom. **Superficially like:** Other madtoms, and young of bullheads and flathead.

Coloration: Dorsally yellow-olive (Lake Erie), bluish-olive (streams usually turbid), or blue-black (streams usually clear). Sides lighter. Ventrally yellowish- or milk-white. The darkest-colored fishes are almost invariably bicolored; these have a sharp line of demarcation separating the dark sides from the light ventral surface. Fins somewhat lighter than adjacent body parts. Caudal with a whitish border. Saddle-bands across back faint or absent.

Lengths and **weights:** *Young* of year in Oct., 1.2″–3.2″ (3.0–8.1 cm) long. Around **1** year, 2.2″—–4.0″ (5.6–10 cm). *Adults*, usually 3.8″–9.5″ (9.7–24 cm) long. Largest specimen, 12.3″ (31.2 cm) long; weight 1 lb, 1 oz (0.5 kg).

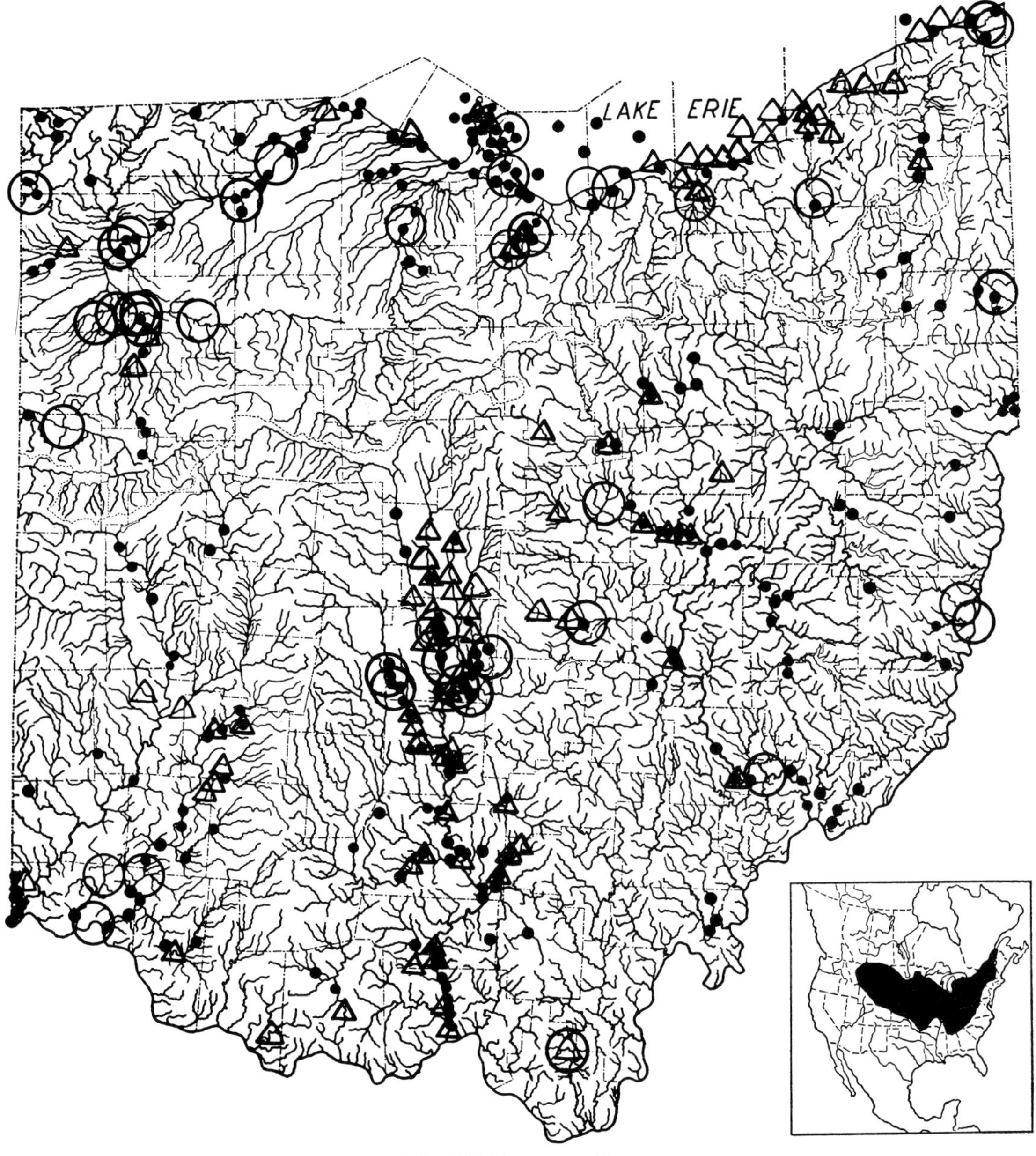

MAP 114. Stonecat madtom

Locality records. ○ Before 1901. ● 1901–54. △ 1955–80.

Insert: Unoccupied areas exist within this range, especially in the western half.

Distribution and Habitat

Ohio Distribution—The general distribution and abundance of the stonecat in the larger waters of Ohio before 1900 are well indicated by the many preserved specimens and literature references. Since 1900 this locally numerous species has shown marked decreases in abundance in the more turbid sections of the inland streams, particularly in the Scioto and Maumee river systems. As an example, in 1893 Kirsch (1895A:327) found it to be numerous in the Auglaize River where since 1925 it has occurred only sparingly. Before 1900, the Great Miami River below Dayton unquestionably contained much suitable habitat, and although preserved specimens and literature references are lacking, it is assumed that the stonecat was then abundant, and that since then the great quantity of industrial pollutants has destroyed that habitat. According to the commercial fishermen, the stonecat was abundant in the Ohio River previous to 1925, and was then caught in numbers to be used as trotline bait; since 1925, after impoundment had eliminated the chutes and silted over the gravel bars, the species occurred only as strays. A large population was present about the limestone reefs and gravel bars of Lake Erie.

Habitat—The largest populations occurred in two types of habitats which superficially appeared to be vastly different, but which in reality were similar ecologically. One was the gravel, boulder and bedrock riffles of the larger streams of moderate gradient which were comparatively free of silt and other pollutants and where the insect, crayfish, and forage fish populations were high. The other habitat was the limestone bedrock and gravel reefs and beaches of Lake Erie where the currents produced stream-like conditions, and consequently there was a high population of aquatic insects, crayfishes, and forage fishes. Although the stonecat is predominantly a riffle species, it is intolerant of the strong currents of high-gradient streams. It also occurred sparingly in low-gradient streams where the sluggish current cannot prevent the silting of riffle bottoms. Young were sometimes numerous among the sparse aquatic vegetation of sand and gravel riffles. The stonecat was an excellent index of smallmouth blackbass abundance, for almost invariably one was abundant only where the other was abundant.

The stonecat is an exception to the general rule that the same species of fish grows to larger size and heavier weight in Ohio River waters than it does in Lake Erie. Stonecats grow to twice the length and three times the weight in Lake Erie than they do in southern inland waters. Likewise, coloration was different; Lake Erie stonecats were light-brown or olive-brown dorsally, dirty-white or yellow-white ventrally, and were quite unicolored; whereas stream stonecats were olive-green or olive-blue dorsally, milk-white ventrally and were distinctly bicolored. Lake Erie fishermen call this species the "Beetle-eye."

Years 1955–80—Throughout this period the stonecat madtom was present in the larger streams of moderate gradients of the Ohio River drainage and especially in the larger glacial outwash streams of the Scioto and Muskingum river systems. The species was often quite numerous in some collecting stations. Marked increases in numbers were noted wherever the waters below dams were relatively clear because the resultant reservoirs acted as settling basins.

During this period only one specimen was captured in the Maumee River system, and although many populations were found to have become extirpated, there undoubtedly remained isolated colonies in sections of those few remaining streams of moderate to large size which were the least silt-laden.

This species continued to be fairly numerous about the islands, reefs and shoals of western Lake Erie (Van Meter and Trautman, 1970:73) and in the Cleveland area (White et al., 1975:95); however, recent collections indicate that the population may be decreasing in numbers. It continued to be largely accidental in the Ohio River, and the University of Louisville biologists investigating that stream between 1957 and 1959 recorded taking only five collections having only 34 individuals in all (ORSANCO, 1962:149).

On July 13, 1968 in the Huron River in Erie County, 18 students and I collected 10 stonecats which were young of the year between 16 and 25 mm SL. They were hiding among the algae growing on the upper surface of large, flat slabs of shale situated in the faster-flowing waters of riffles. The bright green algae was approximately an inch in length and very dense. Many other young were frightened out of the algae but not collected.

For diagnosis and distribution of this species *see* Taylor, 1969:111–28.

MOUNTAIN MADTOM

*Noturus eleutherus** Jordan

Fig. 115

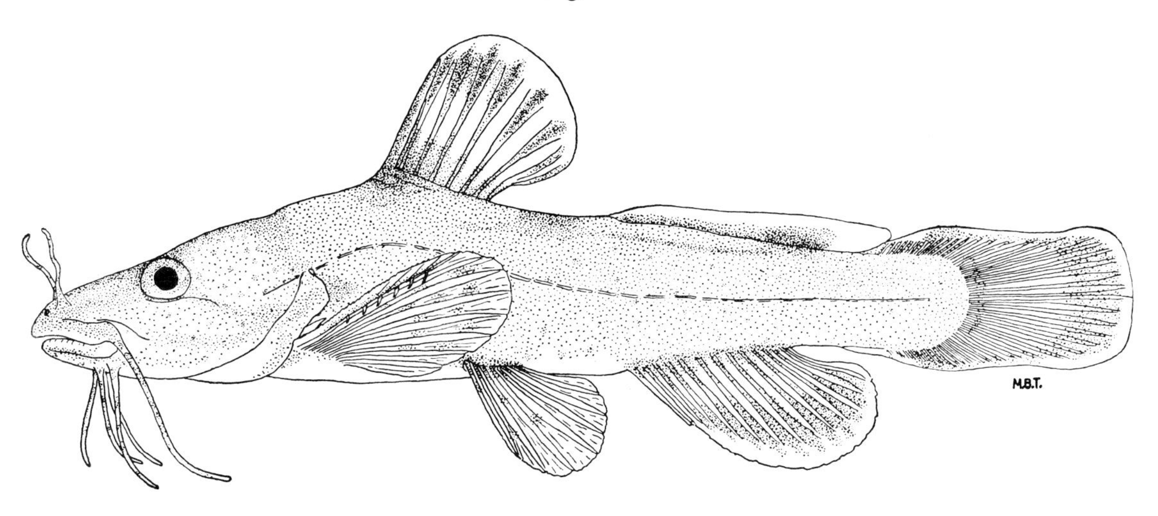

Scioto River, Scioto County, O.

Nov. 16, 1939.
Male.

56 mm SL, 2.7″ (6.9 cm) TL.
OSUM 38.

Identification

Characters: *General—See* blue catfish. *Generic—See* stonecat madtom. *Specific*—Blackish saddle-band on caudal peduncle does *not* extend to distal edge of adipose fin, thereby leaving that edge whitish, in which character it agrees with the northern madtom. Caudal fin *without* a dark mid-caudal bar or crescent. Humeral process (in the figure the outline of the process may be seen lying directly above and parallel to the anterior portion of the pectoral spine) longer than is the width of pectoral spine shaft, but never longer than shaft plus serrae (spinelets, teeth or barbs). Body and caudal peduncle usually more slender, depth of caudal peduncle of large young and adults usually contained more than 1.0 times in snout length. Coloration more uniform, saddle-bars and other body markings usually faint. Like the northern madtom, the present species has 4–8 serrae on each pectoral spine; the low adipose fin is almost separated from the caudal fin and with a tendency to form a free posterior flap; upper jaw is longer than the lower; the distal edge of dorsal fin whitish, below which is a sub-distal, dusky band, most conspicuous in adults; caudal squarish; pre-maxillary band of teeth *without* backward extensions (*see* stonecat, the arrow pointing to an extension).

Differs: Stonecat and tadpole madtoms lack teeth on pectoral spine. Northern madtom has a dark mid-caudal bar and a longer humeral process. Scioto madtom has no or very few melanophores on adipose fin; humeral process absent or very short. Brindled madtom has the blackish saddle-bar extending upward onto the distal edge of the adipose fin.

Most like: Northern and brindled madtoms. **Superficially like:** Stonecat, Scioto and tadpole madtoms.

Coloration: Dorsally olive- or grayish-tan overlaid and mottled with gray. The 4 saddle-bands appear to be absent in some individuals, indistinct in others, and well-defined in a few which usually are medium-sized fishes. When present, the anteriormost band crosses the hind head and nape, another is below the dorsal origin, the third is anterior to the adipose fin, and the posteriormost encroaches upward on the adipose fin. Sides

* Formerly *Schilbeodes eleutherus*; *see* Taylor, 1954:44.

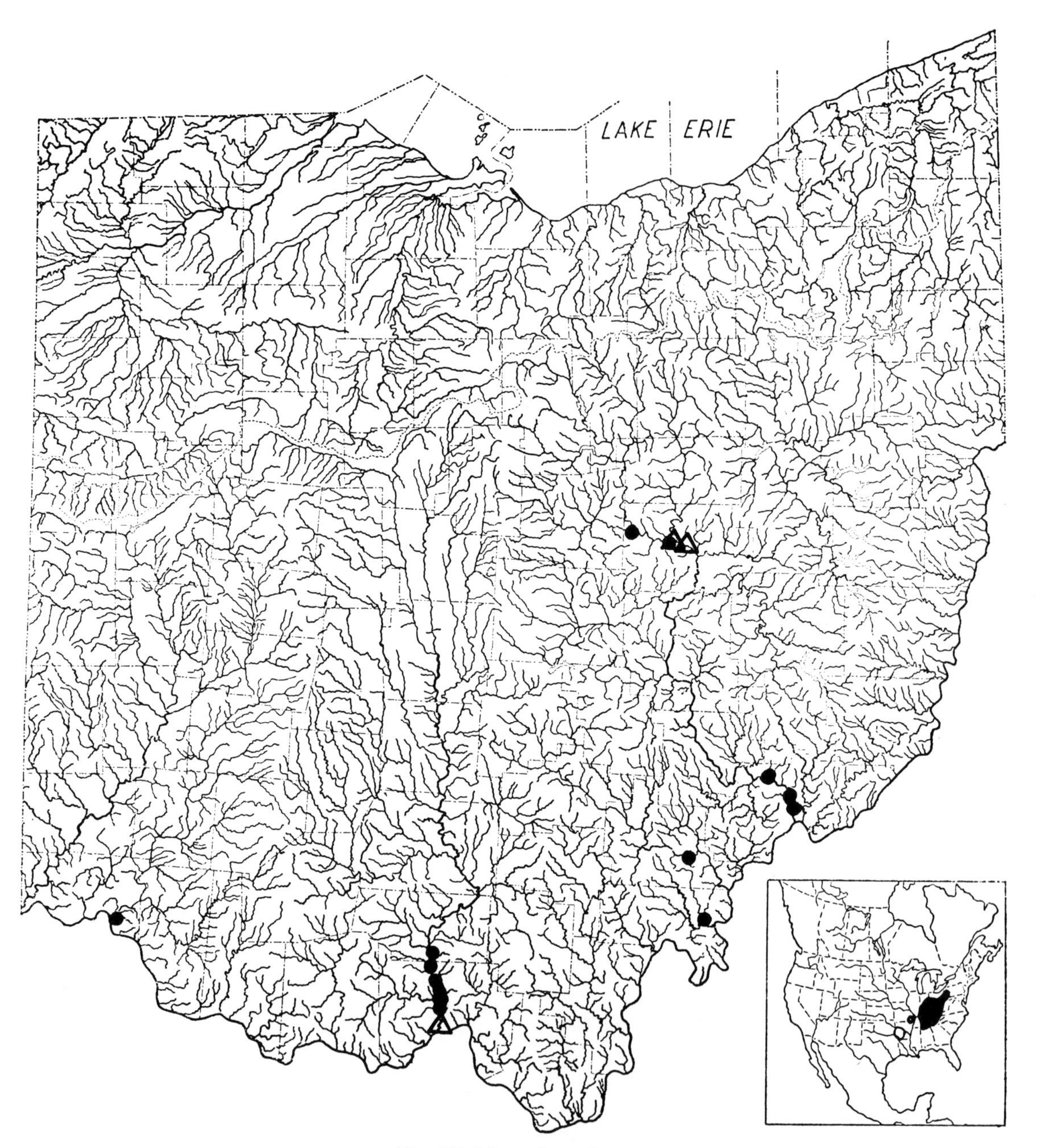

MAP 115. Mountain madtom

Locality records. ● After 1926. △ 1955–80.
Insert: Black area represents range of this form, outlined area of another.

grayish- or whitish-tan, sometimes sprinkled or mottled with dusky. Lips and ventral surface dull white. Dorsal fin usually has a whitish margin and a sub-distal, dusky bar. Caudal fin usually with basal and sub-terminal darkish crescents, but without a dark mid-dorsal bar. This species normally is less prominently marked and more tan than is the northern madtom. *Breeding male*—The broad head and bulging cheeks are blue-black; anterior half of body bluish and dusky.

Lengths: *Young* of year in Oct., 1.0″–2.3″ (2.5–5.8 cm) long. Around **1** year, 1.4″–2.5″ (3.6–6.4 cm). *Adults*, usually 2.2″–3.8″ (5.6–9.7 cm). Largest specimen, 5.0″ (13 cm) long.

Distribution and Habitat

Ohio Distribution—The earliest preserved Ohio specimens of mountain madtom which I have examined were collected by me on July 3, 1927 in Shade Creek, Meigs County, Ohio. (*See* next species for explanation of Osburn's [1901:28] Franklin County record of *eleutherus*.) Since 1927 the mountain madtom has been found most frequently in the lower portions of the Scioto and Muskingum rivers, occupying virtually the same range in southern Ohio as the northern madtom. In fact, the two species were usually found associating together in the same riffle sections and to all appearances were occupying the same type of habitat. This suggests that these two species formerly occupied different ranges but in post-glacial times were brought into close association with each other during and since the redistributional scramble and reinvasion of territory, and that there now exists considerable intraspecific competition.

From observations on the ecological requirements of the mountain and northern madtoms I conclude that these essentially clear-water inhabiting species were far more abundant and more generally distributed throughout Ohio waters before 1900 than since. This must have been especially true of the Ohio River where during the 1920–50 surveys none of these fishes was found. The many literature references (such as U.S. Army, 1935:2; Rafinesque, 1820:53) and statements of old rivermen indicate that in this river before it was ponded there existed clear waters, stony riffles, chutes and bars which should have been prime habitat for these species.

Habitat—Since 1927 the mountain and northern madtoms were numerous only in those portions of the deep, wide and swift riffles of southern Ohio streams which had boulder and gravel bottoms. Specimens were usually taken during daylight in water less than 18″ (46 cm) in depth only when the water was turbid, when clear these species seemed to be absent from these shallows during the day, and then could be captured in shallow water only at night. Presumably they moved into deep water when the day was bright and the water clear. Most of the specimens captured in the daytime were found under stones; but at night, with the aid of a strong jacklight, these madtoms could be seen swimming and feeding among the stones. There appeared to be little competition between these two species and the brindled madtom, for when the three were present in the same stream section, the former two were found in the stony, fast waters, whereas the latter was found among the stones and debris in the pools or sluggish riffles.

Years 1955–80—Since 1950 the mountain madtom was taken only in two of the several localities where it had been taken previously. One locality was the Walhonding River in Coshocton County, where the population appeared to have maintained its numerical abundance. The other was the lower Scioto River in Pike and Scioto counties, where a decrease in numbers was noted, caused, it is believed, by increased amounts of pollutants. In October of 1963, when 12 madtoms were taken in the Scioto immediately above Portsmouth, there were globs of detergent foam floating upon the water's surface; the water had a blackish cast, as it had that same day in the Scioto immediately below Columbus and unlike the grayish-green coloration in the stream above the mouth of Big Darby Creek. Unsuccessful attempts were made to collect this species elsewhere in Ohio and especially in the Muskingum River. Isolated populations undoubtedly occur in this state elsewhere in the larger streams of the Ohio River drainage. For distribution and diagnosis of the species, *see* Taylor (1969:160–67).

NORTHERN MADTOM

*Noturus stigmosus** Taylor

Fig. 116

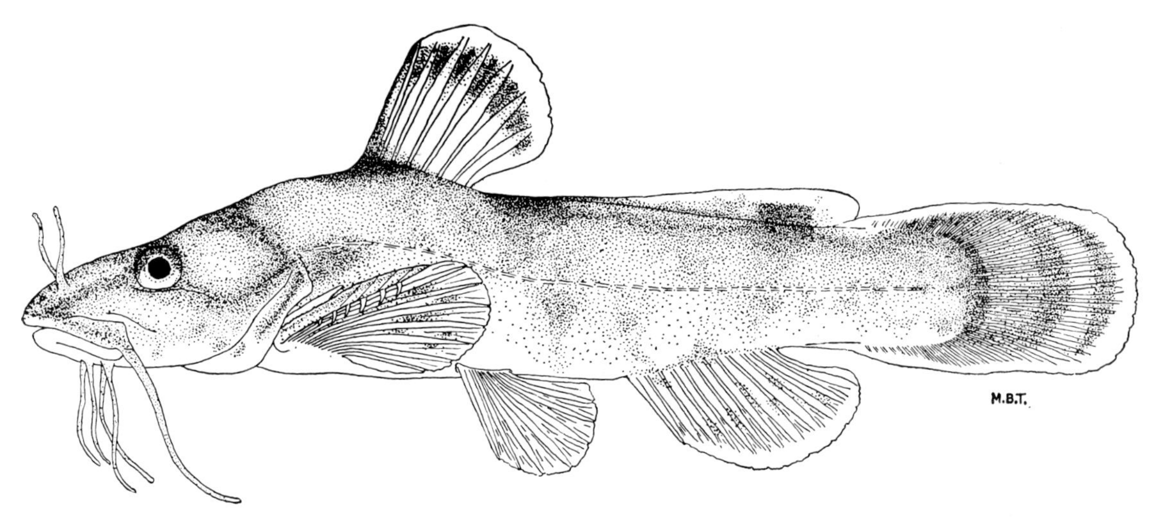

Scioto River, Scioto County, O.

Nov. 15, 1939.
Adult male.

61 mm SL, 2.9″ (7.4 cm) TL.
OSUM 74.

Identification

Characters: *General—See* blue catfish. *Generic—See* stonecat madtom. *Specific*—Like the mountain madtom, the northern has a blackish saddle-band on the caudal peduncle which does *not* extend upward to the distal edge of the adipose fin; for other similarities, *see* mountain madtom, under *Specific*. Caudal fin normally has a distinct, dark mid-caudal bar or crescent. Humeral process (in the figure the outline of the process may be seen lying directly above and parallel to the anterior portion of the pectoral spine) usually longer than width of pectoral spine shaft plus serrae (spinelets, teeth or barbs). Body chubbier, depth of caudal peduncle of large young and adults usually contained 1.1 times or less in snout length. Coloration usually more pronounced, the markings contrasting sharply with the ground color.

Differs: Stonecat and tadpole madtoms lack teeth on pectoral spine. Mountain madtom lacks the dark mid-caudal bar and has a shorter humeral process. Scioto madtom has no or very few melanophores on adipose fin; humeral process absent or very short. Brindled madtom has the blackish saddle-bar extending upward onto the distal edge of the adipose fin.

Most like: Mountain and brindled madtoms. **Superficially like:** Stonecat, Scioto and tadpole madtoms.

Coloration: Dorsally olive-grayish, mottled with dusky and sometimes overlaid with tan. The four saddle-bands usually are rather prominent, the anteriormost band crosses the hind head and nape, another is below the dorsal origin, the third is anterior to the adipose fin origin and the posteriormost encroaches upward onto the adipose fin. Sides grayish, usually heavily mottled with dusky. Lips and ventral surface dull white, the front of the abdomen usually containing round chromatophores (spots); these are absent in the largest individuals. Dorsal fin with a whitish margin and a subdistal dusky bar which is absent only in smallest young. Caudal fin with basal, mid-caudal and subterminal darkish crescents. Pectoral fins usually spotted prominently. *Breeding male*—Has the

* Formerly in the genus *Schilbeodes*; *see* Taylor, 1954:44.

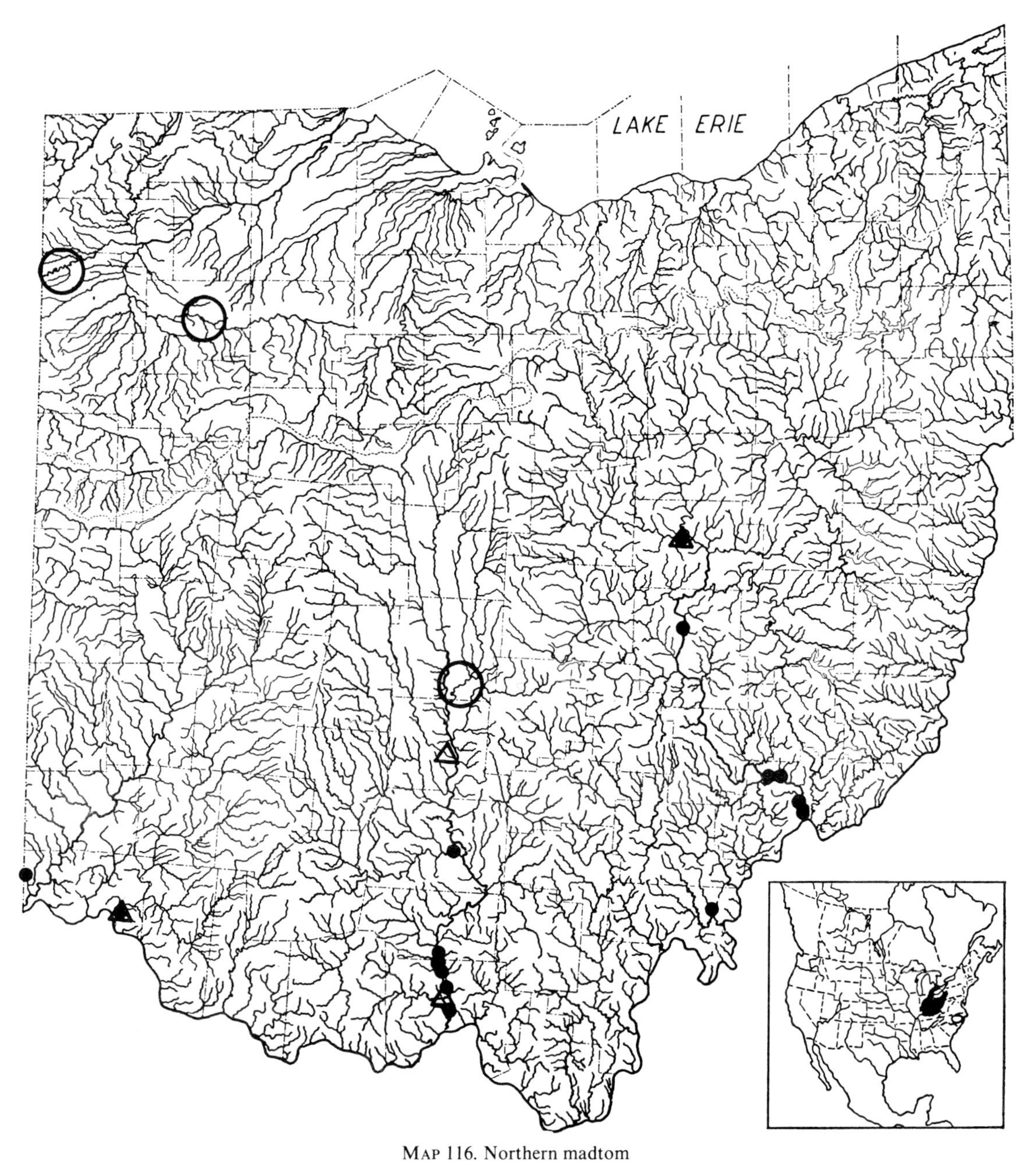

MAP 116. Northern madtom

Locality records. ○ Before 1900. ● After 1926. △ 1955–80.
Insert: Black area represents the range of the form found in Ohio, outlined area represents range of the Atlantic slope population of *N. furiosus.*

broad head and bulging cheeks blue-black or black and the anterior half of body dusky.

Lengths: *Young* of year in Oct., 1.0″–2.3″ (2.5–5.8 cm) long. Around **1** year, 1.4″–2.5″ (3.6–6.4 cm). *Adults*, usually 2.2″–3.8″ (5.6–9.7 cm). Largest specimen, 5.2″ (13 cm) long.

Distribution and Habitat

Ohio Distribution—In 1930 Osburn, Wickliff and Trautman (1930:174) recorded for Ohio, some specimens of madtoms which they identified as *Schilbeodes furiosus*. These fishes had been captured by Trautman in the lower portions of Shade Creek and the Scioto and Muskingum rivers during 1927 and 1929. Later, the discovery of marked sexual differences in the catfishes resulted in the temporary and mistaken belief that specimens previously identified as *eleutherus* were males, and that those identified as *furiosus* were females. As a result *furiosus* was omitted from subsequent state lists.

In his unpublished doctoral thesis William Ralph Taylor (1955:349) demonstrates that a madtom related to *Noturus furiosus* inhabits Ohio waters; because this population was previously known as *furiosus* the name is used here. Taylor shows that the specimen (OSUM 9348)† collected in the Blanchard River on August 8, 1893 by Kirsch (1895A:327) at his Ottawa station in Putnam County, is this species.

On August 16, 1893, at his fishing station "2 1/2 miles (4 km) above Antwerp, Ohio" (Paulding County), Kirsch collected five specimens (IU 6551), presumably in the "short rocky riffles." These madtoms were identified originally as *Noturus miurus*, Kirsch (1895A:327) mentioning that these specimens approached *N. eleutherus* in coloration. Recently Taylor identified them as *N. furiosus*.

Since 1920 many attempts have failed to take this species in the then-turbid Maumee River system in Ohio. The species still inhabits the rather clear Huron River of southeastern Michigan, a Lake Erie tributary (Hubbs, 1930B:432, as *Schilbeodes furiosus*).

On June 25, 1897 Osburn and Williamson (1898:12) collected a madtom in Big Walnut Creek at Lockbourne, Franklin County, which Osburn (1901:28) considered to be the first specimen of *eleutherus* (OSUM 627) taken in Ohio. This specimen is *furiosus*, as Taylor has pointed out.

Since 1927 the northern madtom has been found to be numerous only in the lower portions of the Scioto and Muskingum rivers. *See* under mountain madtom.

† In an exchange of fish specimens with the Chicago Natural History Museum, on February 7, 1953, the Ohio State University fish collection received a madtom, catalogued originally as "1956, *Noturus miurus*, Blanchard River, Ottawa. USFC." Although the original label lacks date, collector's name and the state, the handwriting and position of this entry in the CNHM catalogue point to this specimen having been collected as stated above. I assume that it was so collected.

Habitat—*See* mountain madtom.

Years 1955–80—During this period the northern madtom was captured in only a few localities in Ohio: the Walhonding River in Coshocton County, the lower Scioto River in Pike and Scioto counties, Big Darby Creek at the riffle upstream from Route 104 in Pickaway County, and the Little Miami River, Hamilton County north of Newtown. The Walhonding River population appeared to be stabilized. There was a decrease in numbers in the lower Scioto River as there was in the numbers of the mountain madtom.

On October 22, 1957, Donald I. Mount and I collected two northern madtoms together with a Scioto madtom in "riffle 3" of Big Darby Creek, *see* under Scioto madtom, Ohio Distribution. These are the only northern madtoms taken in Big Darby Creek between the years 1924 and 1973. In May, 1974, Ted Cavender and his class captured one on the same riffle. It may be significant that they were captured in the same small area in which 17 of the 18 Scioto madtoms were taken.

Since 1950 none have been captured in the Great Lakes drainage of Ohio. A large specimen (OSUM 14324) was taken in a trawl in Lake St. Clair, Ontario, Canada, near the origin of the Detroit River by Harry Van Meter on July 25, 1963. The species is still present in the Huron River of southeast Michigan.

Osburn (1901:28) considered this madtom as being *Schilbeodes eleutherus* (Jordan). In 1940 Osburn, Wickliff and Trautman (1930:174) believed it to be conspecific with *Schilbeodes furiosus* (Jordan and Meek) of the southeastern United States. In 1957 synonymized *Schilbeodes* with *Noturus* (1957:151–52), the species then becoming *Noturus furiosus* Jordan and Meek, the name used in the first edition of this report. Taylor in 1969 (184–88) demonstrated that *Noturus furiosus* was confined to the Neuse and Tar river systems of North Carolina and that the population of the Ohio River system, formerly referred to as *furiosus* was undescribed. He therefore described it as *stigmosus* (Taylor, 1969:187). The Ohio population of the northern madtom, therefore, becomes *Noturus stigmosus* Taylor with the type locality as Huron River, Michigan; *see* Taylor (1969:173–80).

SCIOTO MADTOM

Noturus trautmani Taylor

Fig. 117

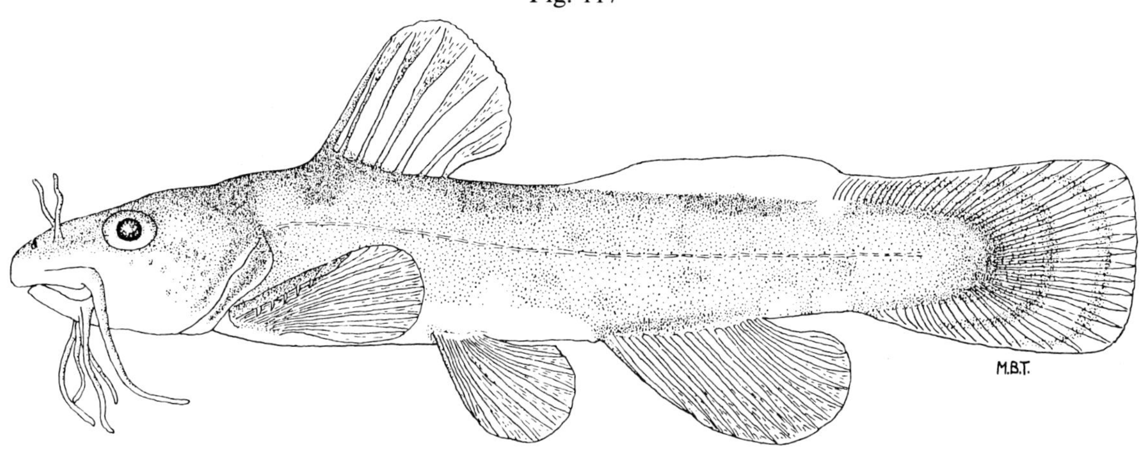

Big Darby Creek, Pickaway County, O.

November 4, 1943.
Male.

44.1 mm SL, 2.2″ (5.6 cm) TL.
UMMZ 187098 (holotype).

Identification

Characters: *General—See* blue catfish. *Generic —See* stonecat madtom. *Specific*—Blackish saddle-band on caudal peduncle does *not* extend onto adipose fin, thereby leaving that fin immaculate. Caudal fin *with* a dark mid-caudal bar or crescent, except in smallest young where it is undeveloped. Humeral process usually absent or very short, although the posterior edge of process may be somewhat roughened. Has typically 9 pelvic rays, 45–51 caudal rays, 6 soft dorsal rays, 5–7 posterior serrae on pectoral spines and 11 preoperculomandibular pores. (For fuller account, *see* original description, Taylor, 1969:156–60.)

Differs: Stonecat and tadpole madtoms lack teeth on pectoral spine. Mountain and northern madtoms have well-defined humeral process, they and brindled madtom have the blackish saddle-band extending upward onto the adipose fin.

Most like: Mountain and northern madtoms. **Superficially like:** Brindled, stonecat and tadpole madtoms.

Coloration: Dorsally dusky olive or dark brown overlaid or mottled with gray. The 4 saddle-bands usually well developed, the anteriormost crossing hind head and nape and extending downward across opercle, the posteriormost three breaking off sharply ventrally on the upper side along a parallel line, the first lying below the dorsal origin, second anterior to the adipose fin and the third below the adipose fin but which *does not* encroach upon it. In some specimens the sides are mottled. A dark brown crescent is present on the posterior portion of the caudal peduncle, often extending forward and upward, to become a dusky blotch on body and rudimentary rays; this may be preceeded by a light area between it and the posteriormost saddle-bar. The caudal fin contains a dark, rather vague, medial crescent and a subdistal one. The ventral third of the head and upper lips are unspotted and milky-white. Ventral surface of body from breast to anus whitish; in some specimens a bridge of chromatophores extends ventrally between the anus and origin of anal fin. Two adjacent, semi-circular whitish areas, often quite conspicuous, are situated ventrally between the posterior end of the anal fin and first procurrent rays of the caudal. The above description taken from specimens preserved and alive, and from three kept in an aquarium for a year.

Lengths: 1.4″–2.4″ (3.6–6.1 cm) long.

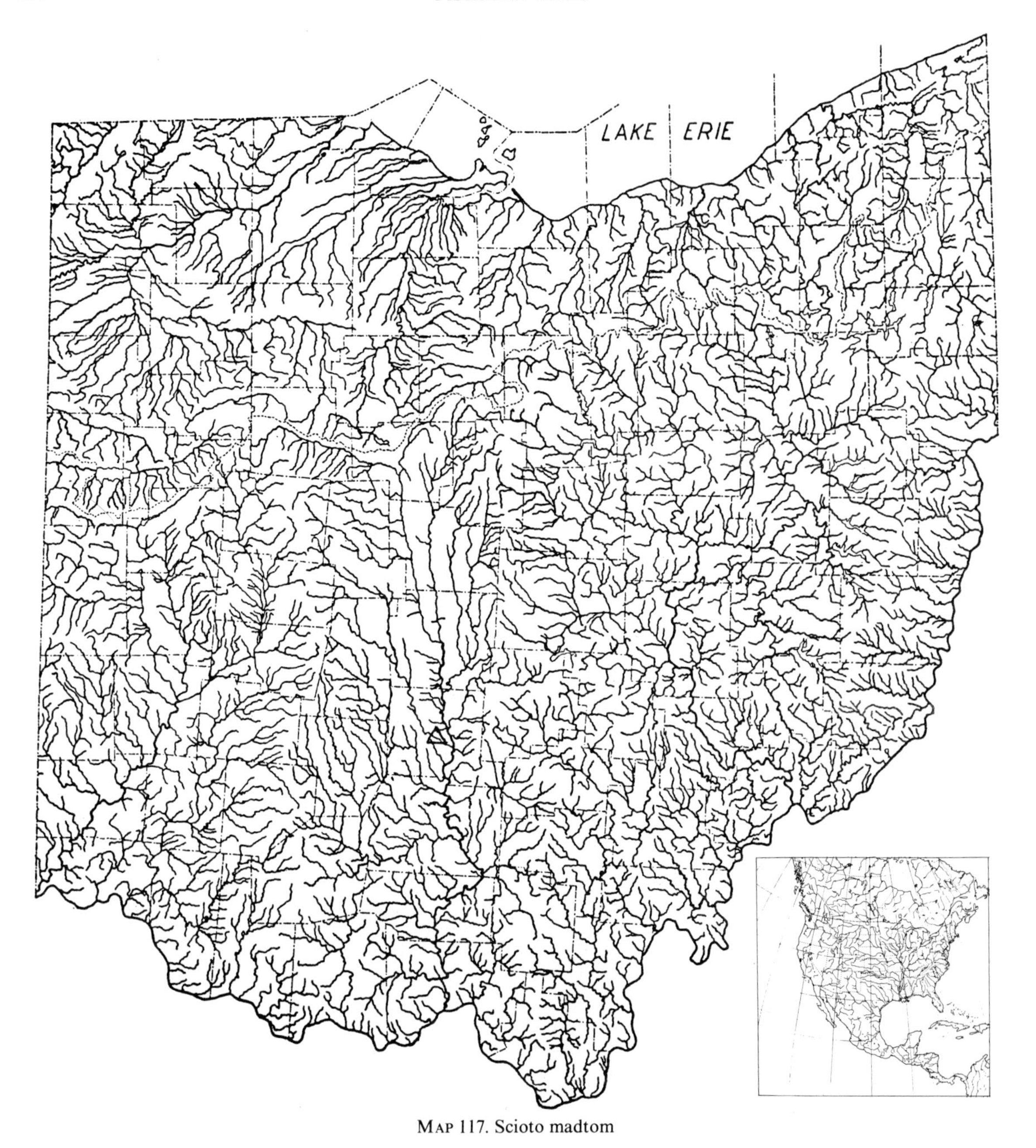

MAP 117. Scioto madtom

△ Collected between years 1943–57; only known locality.

Distribution and Habitat

Ohio Distribution—While seining Big Darby Creek on November 4, 1943, the late Walter Cunningham and I collected two madtoms from "riffle 3" of a series of four riffles which extended 100′ to 200′ (30.5–60.9 m) upstream from the bridge of State Route 104, in Pickaway County, southeastern Jackson Township, one mile south of the village of Fox. One of these madtoms William Ralph Taylor (1969:158–59) later designated as the holotype of this species. During a study of winter conditions in Big Darby Creek, December 30, 1943, my wife and I captured another from the open riffle ("riffle 1") of an otherwise ice- and snow-covered stream. Because we had difficulty identifying these specimens as to species, many attempts were made to capture more, but it was not until November 16, 1945, that another was taken. After that not one was caught until the autumn of 1957, when Donald I. Mount and I captured one each on September 26 and October 3, two each on October 9 and 15, one on October 22, and seven on November 17. None has been taken since, despite the increased and intensified efforts by Raymond F. Jezerinac, E. Bruce McLean, Roger Burnard, James Norrocky, Bruce May, Thomas Thatcher, K. Roger Troutman, Donald Veth, other students and me seining at various hours of the day and night during every month and under many diverse stream conditions.

Since all specimens were taken during fall and winter between September 26 and December 30, it was thought that possibly the population center of this madtom might be upstream and that the individuals obtained had migrated downstream to these riffles to winter. Consequently, many seinings, some involving entire days, were conducted from immediately above "riffle 3" upstream a distance of more than 13 1/2 miles (21.7 km). Using a boat, we drifted downstream stopping to seine every recognizable type of habitat enroute.

Habitat—Seventeen of the 18 specimens taken were captured in the downstream 20′ (6.1 m) of a riffle 60′ (18.3 m) in length, "riffle 3," at a point where the water velocity was decreasing and the substrate was composed of sandy gravel with occasional small stones, the largest about 4″ (10 cm) in diameter. Frequently disturbing this substrate while seining apparently did not permanently discourage its use by the Scioto madtom because 14 individuals were taken during 6 days between late September and mid-November of 1957.

Since 1924 no stream section in Ohio has been seined more assiduously and intensely than have these riffles, and few species of Ohio fishes have been more consistently sought after than this one. Plausible theories as to why more were not taken include: (1) this species normally inhabits unseinable situations such as crayfish burrows or other holes in the banks; (2) it spawns upstream, where this madtom becomes so widely dispersed that it is impossible to collect one; (3) it has become extirpated since 1957.

That all except four of the 18 specimens were taken during the autumn of 1957 suggests that a previous spawning was so successful that 14 specimens could be taken and that since 1957 no such successful spawning has occurred; *see* Tippecanoe darter for drastic fluctuations in annual numbers.

BRINDLED MADTOM

*Noturus miurus** Jordan

Fig. 118

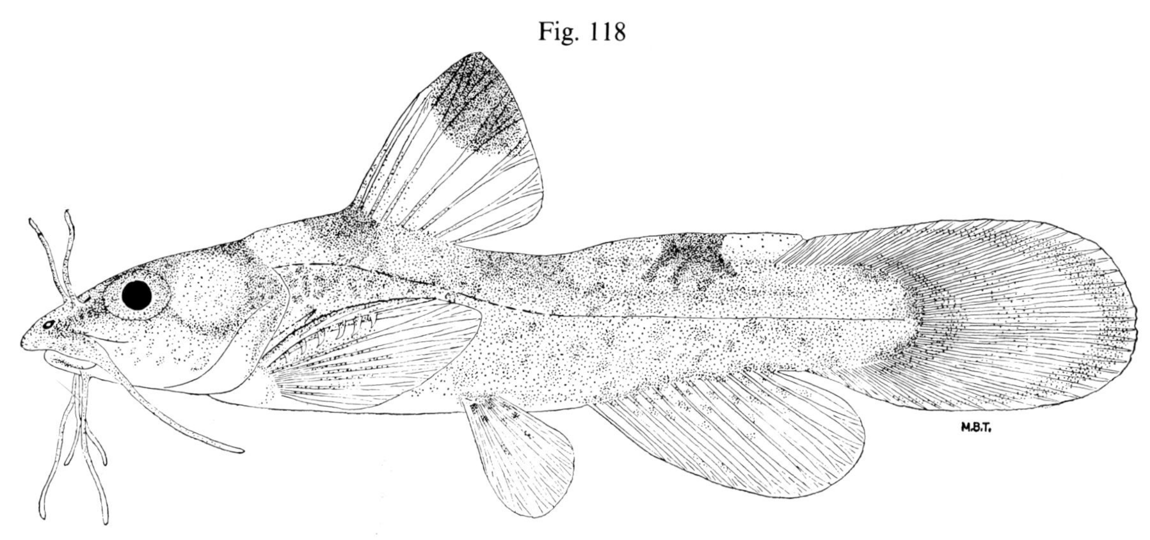

Fish Creek, Williams County, O.

July 13, 1939.
Breeding female.

63 mm SL, 2.9″ (7.4 cm) TL.
OSUM 361.

Identification

Characters: *General—See* blue catfish. *Generic—See* stonecat madtom. *Specific*—Blackish saddle-band on caudal peduncle extends upward across the adipose fin to include the distal edge of that fin. Humeral process (in the figure the outline of the process may be seen lying directly above and parallel to the anterior portion of the pectoral spine) is longer than is the width of the pectoral spine shaft plus serrae (spinelets, teeth or barbs). A roundish, dusky or black spot covers the distal third of the four anterior dorsal rays (least conspicuous in smallest young and largest adults). Caudal fin roundish in shape; body moderately deep; eye rather large and usually contained 2.0 or fewer times in snout length at all ages. The brindled is similar to the mountain and northern madtoms in having 4–8 sharp teeth on the posterior edge of the pectoral spine, four blackish saddle-bands across the back and a longer upper than lower jaw.

Differs: Stonecat and tadpole madtoms lack teeth on pectoral spine. Mountain and northern madtoms have light border to adipose fin, a smaller eye, longer snout, and a more squarish tail. Scioto madtom has no or very few malanophores on adipose fin.

Most like: Mountain and northern madtoms. **Superficially like:** Stonecat, Scioto and tadpole madtoms.

Coloration: Similar to that of the northern madtom, except that the blacks contrast more sharply with the grays and whites, there is less olive and tan, and in the blackish spotting on adipose and dorsal fins.

Lengths: *Young* of year in Oct., 1.0″–2.2″ (2.5–5.6 cm) long. Around **1** year, 1.4″–2.5″ (3.6–6.4 cm). *Adults*, usually 2.2″–3.8″ (5.6–9.7 cm). Largest specimen, 5.2″ (13 cm) long.

Hybridizes: with tadpole madtom, *see* text.

Distribution and Habitat

Ohio Distribution—The large number of brindled madtom records made before 1900, when compared with the relatively few collections of that period, indicate that the species was more widely

* Formerly *Schilbeodes miurus*; *see* Taylor, 1954:44.

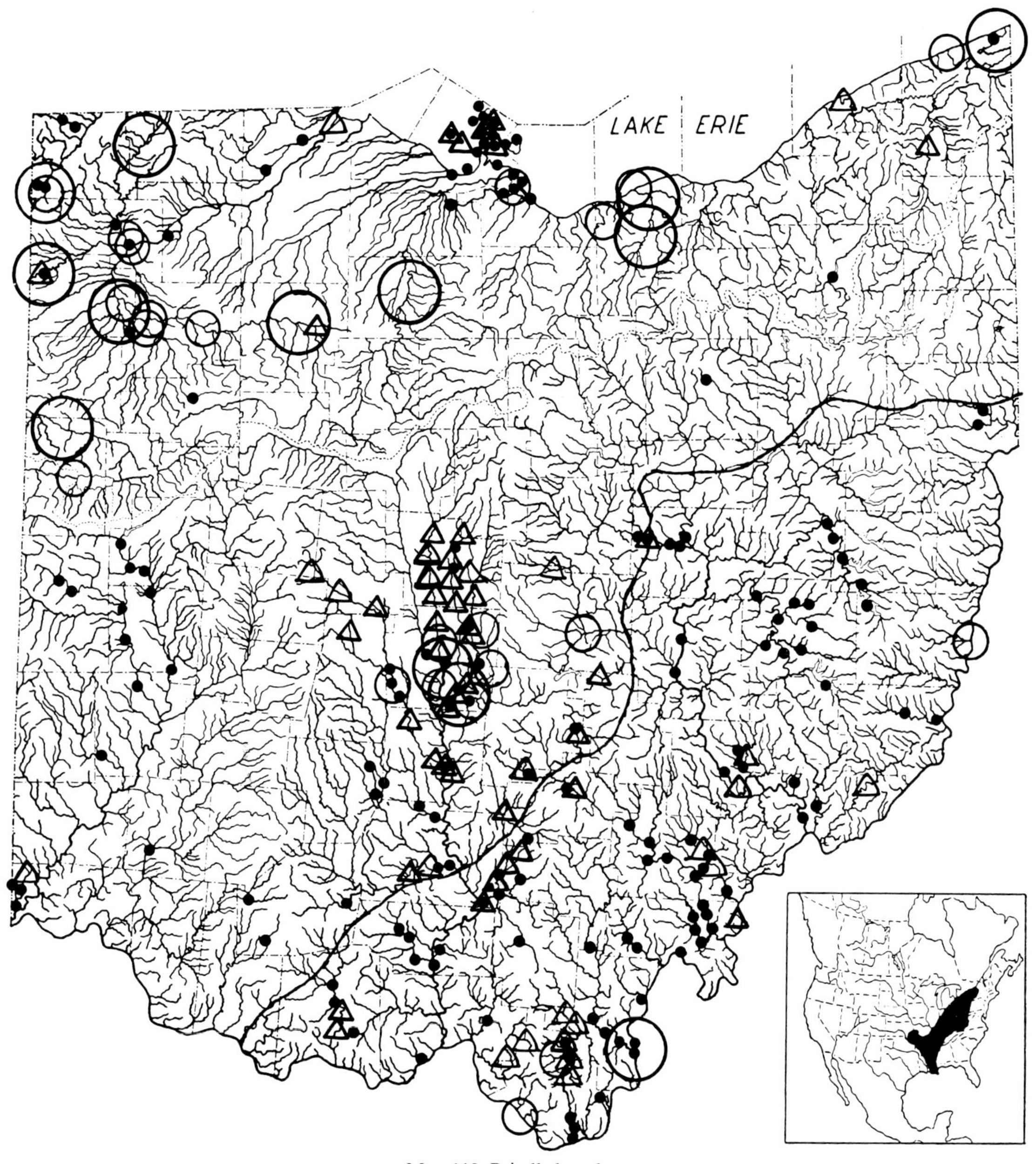

MAP 118. Brindled madtom

○ Before 1901, preserved specimens. ○ Before 1901, from the literature and without specimens.
● After 1901, specimens. △ 1955–80. ~ Glacial boundary.
Insert: No apparent recent changes in limits of range.

distributed and abundant before 1900 than it has been since. During the 1925–50 period the largest populations inhabited unglaciated Ohio; a moderately-sized population was present in Lake Erie; and the species was absent in northeastern Ohio, except for a few relict populations. Between 1893–1950, a decrease in abundance took place in the Maumee River system, for in 1893, Kirsch (1895A:327) found this madtom to be well distributed and locally abundant, whereas since 1940 only a few specimens, usually strays, have been taken. This decrease in abundance is obviously the result of destruction of habitat through the covering-over with clayey silts of the organic debris, and the sand and gravel bottoms.

Habitat—The largest populations occurred in base- or low-gradient streams where the bottom was of sand and organic debris, and where viscous clayey silts were almost or entirely absent. In such waters the species primarily inhabited pools, with little or no current, hiding in the daylight under stones and such organic debris as roots, leaves, twigs and logs; it occurred in smaller numbers in riffles of sluggish or moderate flow, but usually only if sand were present; it was found occasionally in sluggish riffles or pools among such aquatic vegetation as pond-weeds.

In Lake Erie this madtom was most numerous where the bottom contained stones, sand and organic debris, and was somewhat protected from wave action; it also was taken frequently while trawling at depths of between 10′ (3.0 m) and 40′ (12.2 m). It rarely occurred over a soft mud or muck bottom, the habitat of the tadpole madtom; or in the deep, swift riffles, the habitat of the mountain and northern madtoms.

In 1930, *Schilbeodes nocturnus* (Jordan and Gilbert) was recorded for Ohio waters (Osburn, Wickliff, Trautman, 1930:174) on the basis of a specimen collected in western Lake Erie; in 1948 I (1948: 166–74) demonstrated that this specimen is a hybrid between the brindled and tadpole madtoms.

Years 1955–80—During this period the brindled madtom demonstrated no appreciable changes in numerical abundance or in its distribution throughout inland Ohio (White et al., 1975:95). Trawling about the Bass Islands and western Lake Erie region indicated that it may have been as numerous in waters of 10′–30′ (3.0–9.1 m) in depth as it had been before 1950. The species was of rare occurrence in the Ohio River.

The six species of the genus *Noturus* offer splendid illustrations of specialized habitats and degrees of difficulty in collecting the various species.

The tadpole madtom has a highly specialized and easily recognizable habitat niche in static or low-gradient waters having a stablized, soft muck bottom containing brush and branches. By examining a stream section, one could locate such habitat areas readily and, by methodically seining each, obtain a fairly accurate estimation of population abundance.

The stonecat madtom normally occurred among the stones or gravels, or beneath the larger boulders of riffles. It normally inhabitated relatively shallow waters, was captured readily, and the size of the population apparently could be estimated with some accuracy.

The only habitat observed for the Scioto madtom was in the tails of two of a series of four riffles in which the substrate was composed of sandy gravel and a few stones and in water of less than 20″ (51 cm) in depth. It appears that the species should have been captured readily in such a habitat had it occurred in some numbers.

The mountain madtom, an inhabitant of fast-flowing riffles having a substrate of stones and boulders of varying sizes and water depth of more than 18″ (46 cm), was a difficult species to capture, especially in the daytime. With few exceptions, this species could be captured successfully only in late summer, fall or winter during periods of low water. An estimate of its numerical abundance could not be obtained with any degree of accuracy.

The northern madtom gave the impression of competing for living space with the mountain madtom whenever the two occurred in the same riffle; however, the northern sometimes occupied smaller, less deep, and slower-flowing riffles than did the mountain madtom. As in the case of this latter species no accurate estimate of numerical abundance could be ascertained.

The brindled madtom had a readily recognizable habitat niche which largely consisted of a rather firm substrate covered with leaves, twigs, roots, and so forth, usually in sluggish-flowing riffles and in pools, especially along the banks. A fairly accurate estimate of abundance could be ascertained under favorable conditions at any period of the year, as was the case with the tadpole and stonecat madtoms.

For distribution and diagnosis *see* Taylor (1969:190–201).

TADPOLE MADTOM

*Noturus gyrinus** (Mitchill)

Fig. 119

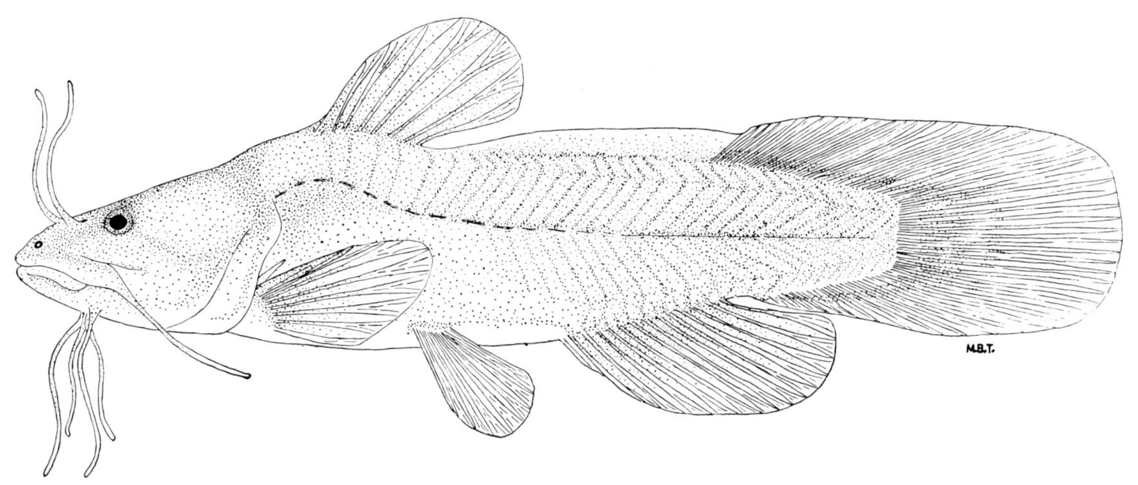

Muddy Creek, Ottawa County, O.

July 23, 1941.
Female.

45 mm SL, 2.3″ (5.8 cm) TL.
OSUM 3388.

Identification

Characters: *General—See* blue catfish. *Generic—See* stonecat madtom. *Specific*—No spinelets, serrae or barbs on the posterior edge of pectoral spine, instead there is a shallow groove between the two posterior edges of the spine. Adipose fin high, the shallow notch between adipose and caudal fins often inconspicuous. Lower jaw as long, or almost as long, as upper jaw. Humeral process (in the figure the outline of the process may be seen lying directly above and parallel to the anterior portion of the pectoral spine) is much longer than is the width of the pectoral spine at base. The four saddle-bands, described under northern madtom (*see* Coloration), are faint when present. Caudal very rounded. Body deep and tadpole shaped. Eye very small. Body and tail never sharply bicolored, the general coloration brownish, and normally there is a dark streak along the lateral line.

Differs: Stonecat has upper jaw longer than lower, body is often sharply bicolored, has a low adipose fin, is slender, and the premaxillary band of teeth has backward extensions.

Most like: Stonecat. **Superficially like:** Other madtoms and young bullheads.

Coloration: Dorsally light-yellow (turbid waters), yellow- or chocolate-brown (bogs having stained waters and black bottoms). Sides lighter, the color gradually fading downward to merge with the yellow-white or white of belly and ventral surface of head. Fins same color as adjacent body parts. The four saddle-bands faint or absent. *Breeding male*—Head and forepart of body dark chocolate or brownish-dusky.

Lengths: *Young* of year in Oct., 0.8″–2.0″ (2.0–5.1 cm) long. Around **1** year, 1.2″–2.5″ (3.0–6.4 cm). *Adults*, usually 1.7″–3.5″ (4.3–8.9 cm). Largest specimen, 4.4″ (11 cm) long.

Hybridizes: with brindled madtom, *see* text of that species.

Distribution and Habitat

Ohio Distribution—Between 1920–50 the tadpole madtom was distributed irregularly over glaciated Ohio, where it was restricted chiefly to low-gradient streams which formerly carried little silt, to clear inland lakes, to marshes, harbors, and bays of Lake Erie, and to abandoned canals or adjacent

* *Schilbeodes mollis* of recent publications; *see* Taylor, 1954: 44.

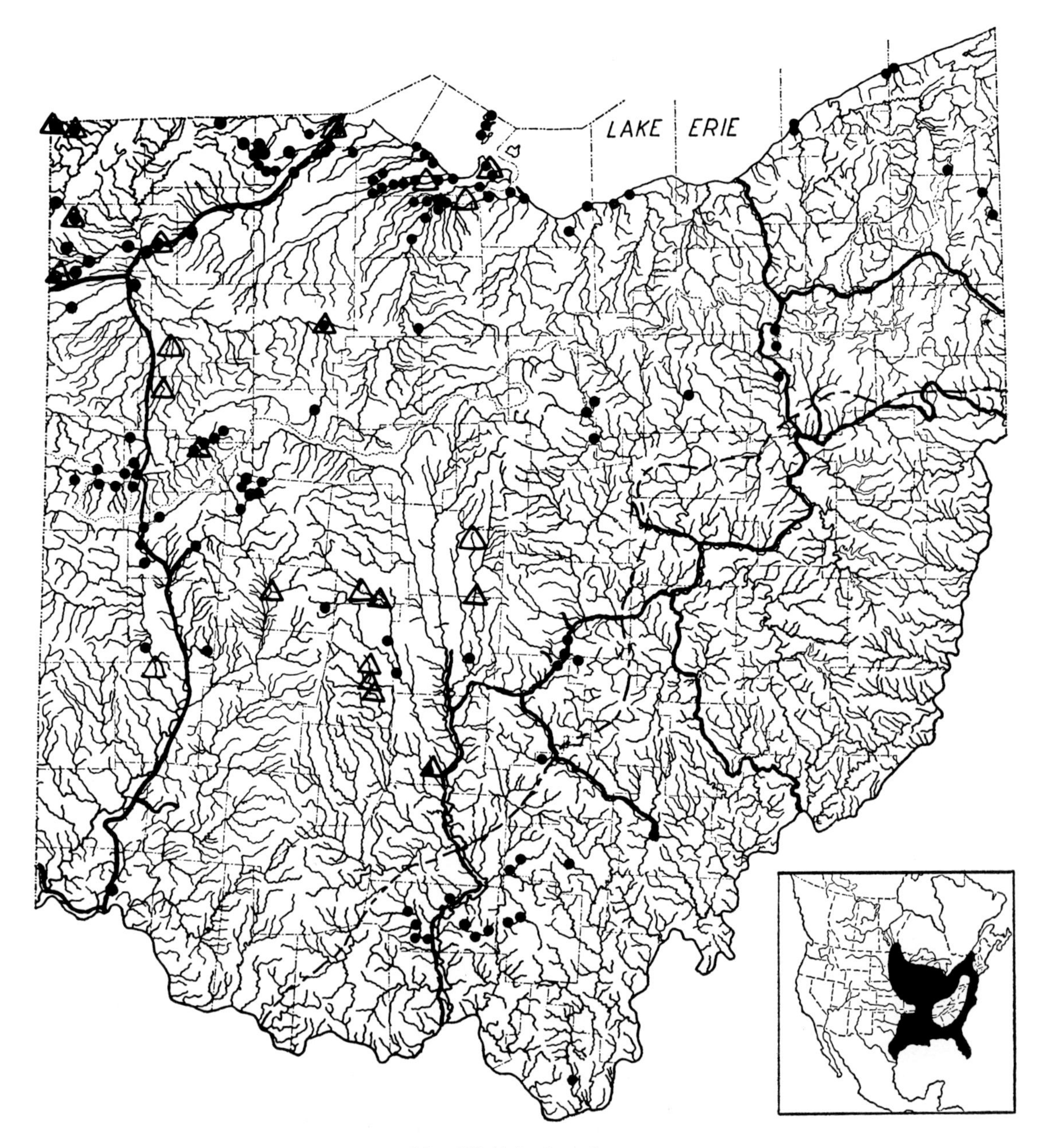

MAP 119. Tadpole madtom

Locality records. ● Before 1955. △ 1955–80. ~ Canals. /\ Glacial boundary.
Insert: May occur in isolated populations in the Appalachians.

streams or reservoirs which formerly supplied water to these canals. When the canals were functioning, they were potential migration routes for this madtom. The isolated population which Henshall (1889:124) recorded for Ross Lake, a basin of the Miami and Erie Canal in Hamilton County, may have been a result of the species moving southward through the canal system. In unglaciated Ohio, the tadpole madtom was restricted to the valleys of the old Teays Stage drainage (map II, above) which were flooded later by proglacial Lake Tight (Wolfe, 1942: Fig. 1). Conditions in these broad, partly filled valleys with their lowland streams containing numerous back waters, oxbows, and considerable aquatic vegetation, produced a habitat for this madtom.

Although literature references and preserved specimens before 1900 are few, the tadpole madtom undoubtedly was more generally distributed and abundant throughout glaciated Ohio before 1900 than it has been since. The draining of marshes, ditching of streams, and increased turbidity and siltation, especially in prairie areas, have destroyed many former habitats. During the past 30 years I have observed the extirpation or great decrease in abundance of many populations through the destruction of the habitat of this species.

Habitat—The most flourishing populations occurred in base- or low-gradient lowland streams, springs, marshes, oxbows, pothole lakes, and protected harbors and bays of Lake Erie, where conditions were relatively stable, the water was usually clear, the bottom was of soft muck which generally contained varying amounts of twigs, logs, and leaves, and where there usually was an abundance of such rooted aquatics as pondweeds and hornwort. The species seemed to be highly intolerant to much turbidity and rapid silting; but despite these, it often clung tenaciously, as a decreasing, isolated population, to areas of increasing turbidity and silting, disappearing only after the last remnants of its habitat had been eliminated.

In 1944 Hubbs and Raney (1944:25–26) believed it necessary to change the specific name of the tadpole madtom, long known as *gyrinus* (of Jordan) to the name *mollis* (of Hermann). Taylor (1954:44) in his recent monographic and largely unpublished studies of North American catfishes has concluded that the name *mollis* is "untenable because of the indefinite and confused original description."

Years 1955–80—As expected, this inhabitant of slowly or sluggishly flowing waters having substrates of soft muck and much organic debris with or without rooted aquatics has decreased in numbers throughout this period. Several relict populations have become extirpated because of ditching, dredging, increased siltation or other destruction of their habitat (White et al., 1975:97). Several of the relict populations continued to exist in the former, natural prairie areas where favorable conditions remained largely unchanged. The number of relict populations continued to decrease in western Lake Erie and adjacent waters. The former large populations about the Bass Islands appear to have become extirpated.

For distribution and remarks concerning the tadpole madtom *see* Taylor (1969:35–54).

PIRATE PERCH

Aphredoderus sayanus (Gilliams)

Fig. 120

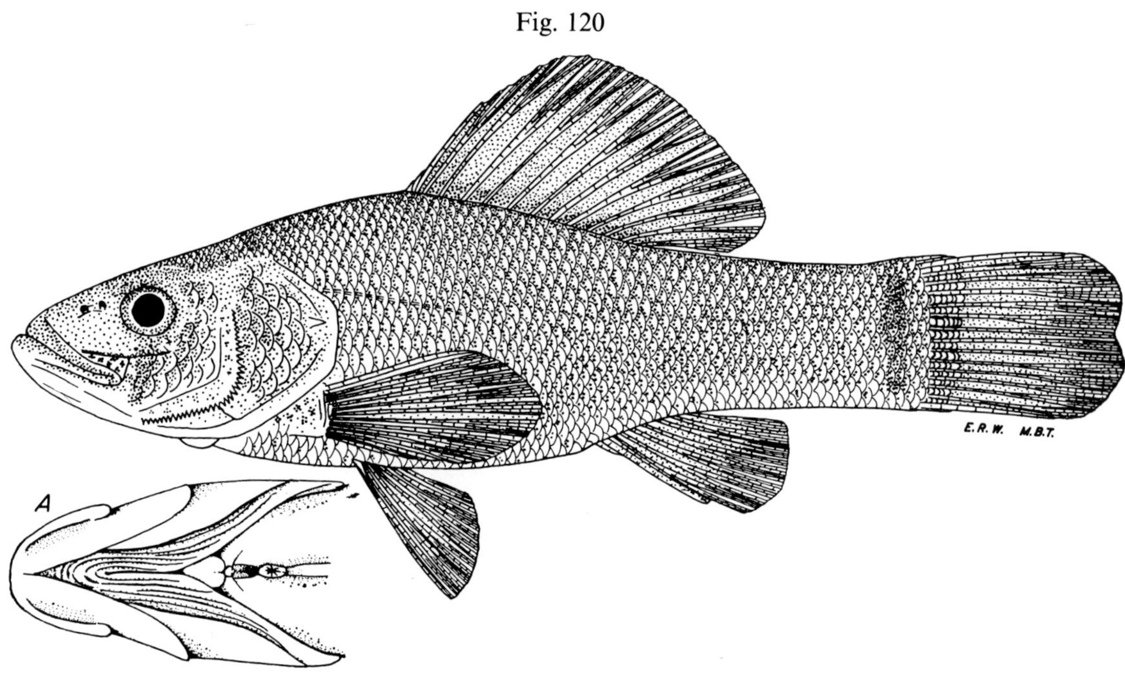

Upper fig.: Auglaize River, Auglaize County, O.

Oct. 15, 1941. 73 mm SL, 3.7″ (9.4 cm) TL.
Female. OSUM 4103.

Fig. A: ventral view of head and breast; note position of anal opening behind the junction of the gill covers.

Identification

Characters: Fishes longer than 1.2″ (3.0 cm) have anal opening between, or anterior to, the pelvic fins, fig. A; smaller fishes have the opening between pelvic fins and anal fin origin, but always well in advance of anal fin origin (in all other species of Ohio fishes the anal opening is immediately in front of the anal fin). Scales strongly ctenoid (rough), including those on cheeks and opercles. Usually three dorsal, two anal, and one pelvic spines; spines short and difficult to find. No adipose fin.

Differs: No other species of fish in Ohio has an anal opening so far in advance of anal fin origin. Trout-perch has an adipose fin. Sunfishes have more than three dorsal spines. Mudminnow and minnows lack spines.

Most like: Green and warmouth sunfishes, mudminnow and young of largemouth blackbass. **Superficially like:** Other species of sunfishes and trout-perch.

Coloration: Dorsally a dark slaty-olive. Sides lighter and more brownish. Ventrally whitish-yellow or white. A distinct, dark, vertical caudal bar, similar to that of the mudminnow. Dorsal and caudal fins slaty-olive, remaining fins lighter and yellower. Head and body with a distinctive blue-purple cast. *Breeding male*—Almost dusky on head and dorsal half of body.

Lengths: of Ohio specimens, 1.5″–4.3″ (3.8–11 cm) long.

Distribution and Habitat

Ohio Distribution—In 1875 Jordan (1877A: 60–61; as the type of *Aphrodedirus cookianus*) recorded the first pirate perch for the Maumee drainage, from Sawyer's Creek in northeastern Indiana. In 1887 Meek (1889:439) collected the first specimen (USNM40104) recorded for Ohio, in Gordon Creek, Maumee River drainage, Defiance County, and on August 1 or 2, 1893, Kirsch (1895A:330) collected one (USNM76077) in the St.

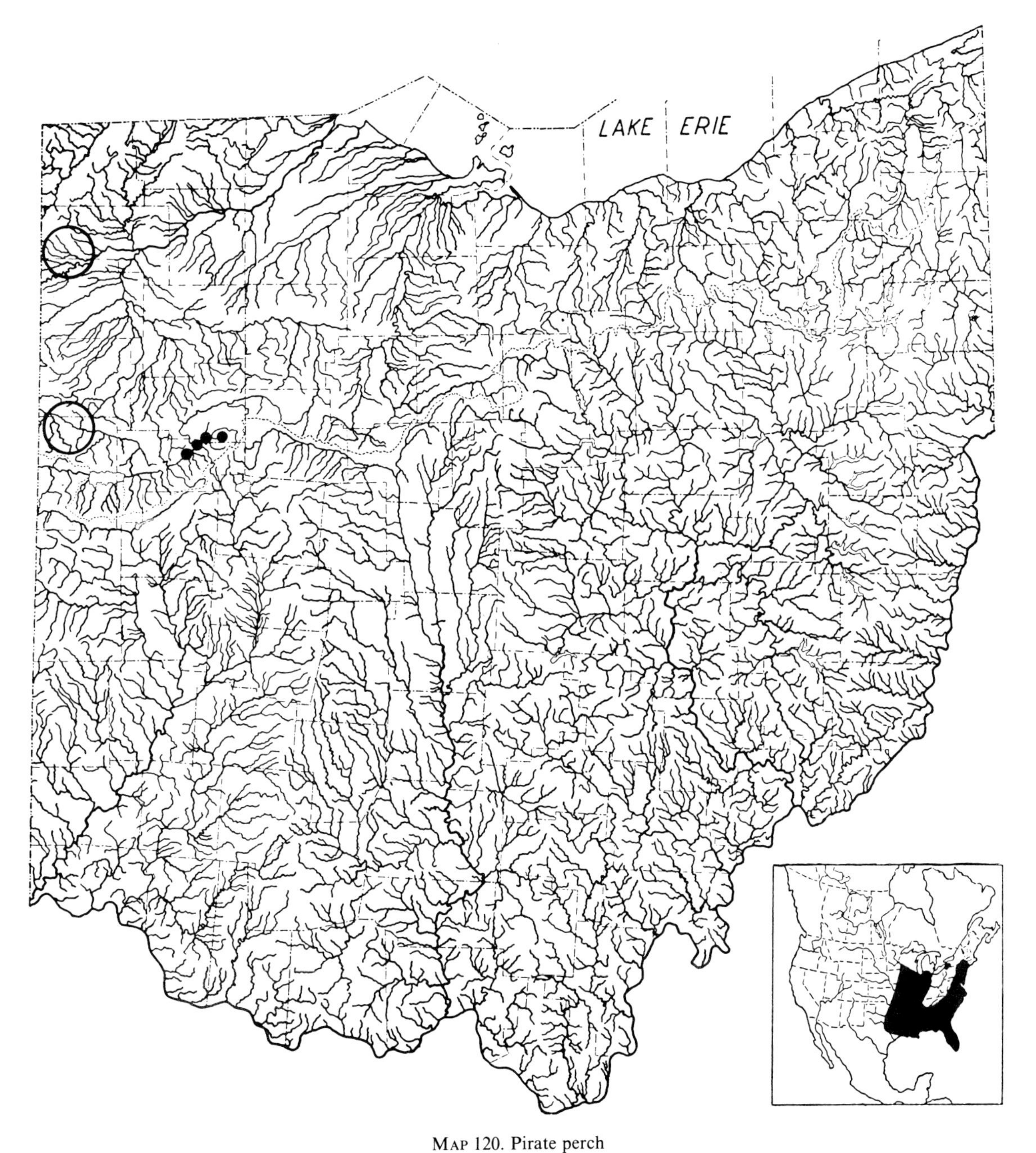

MAP 120. Pirate perch

Locality records. ○ Before 1900. ● After 1939.
Insert: Unoccupied areas within range appear to be increasing in size and number.

Marys River near Rockford, Mercer County. In June of 1954, while examining Ohio fishes which are preserved in the USNM, I was delighted to find almost all of the fish specimens collected by Seth E. Meek in Lost and Gordon creeks including the pirate perch and also to examine P. H. Kirsch's specimen.

In 1889 Henshall (1889:125) recorded the pirate perch for Lake Erie; in the catalogue book of the CSNH is the following: "545, one specimen, Lake Erie, Dr. J. A. Henshall." I could not find this specimen in 1931.

On July 6, 1940, Clarence F. Clark and Ferd Baily collected a pirate perch while removing stranded fishes from an oxbow that was being cut off from the Auglaize River in Auglaize County through dredging operations. On September 10, 1942, specimens were collected from areas in the river that were disturbed by the dredge (Clark, 1949:219–20). Since 1942 Clark and/or I have taken specimens at infrequent intervals. From the above data, literature accounts and my own observations (Trautman, 1939A:275–82) it is obvious that prior to 1900 the Maumee drainage in Ohio, and the lowland streams entering western Lake Erie must have contained much suitable habitat for the pirate perch which since 1900 has been destroyed. Few fish habitats could be so easily destroyed by dredging, ditching, silting and draining. During the 1925–50 period there still existed small, widely-scattered remnants of its habitat in sections of such low-gradient streams as the St. Marys River in Mercer County and Hog Creek in Hardin and Allen counties.

Habitat—The pirate perch inhabited oxbows, overflow ponds, marshes, estuaries, large springs and pools of streams where the gradients were less than 3′/mile (0.9 m/km); where the bottom consisted of soft, dark muck which contained so much decomposing organic material that in walking over it one's feet sank deeply, releasing quantities of gas which was generated by bacterial action. Such a bottom usually had many twigs, leaves, roots and down-timber lying upon it; the pools were bordered and the riffles choked with dense stands of such emergent aquatics as turtlehead, lizard's-tail and waterwillow. The pirate perches lurked about the vegetation and underneath the debris (Becker, 1923:1–4).

Years 1955–80—No records of the capture of the pirate perch in Ohio have been reported for this period. Recently attempts have been made to collect it in the Auglaize River in Auglaize and Allen counties, where it was present until 1950. Despite the fairly adequate habitat conditions still prevailing in restricted areas, these attempts were unsuccessful.

TROUT-PERCH

Percopsis omiscomaycus (Walbaum)

Fig. 121

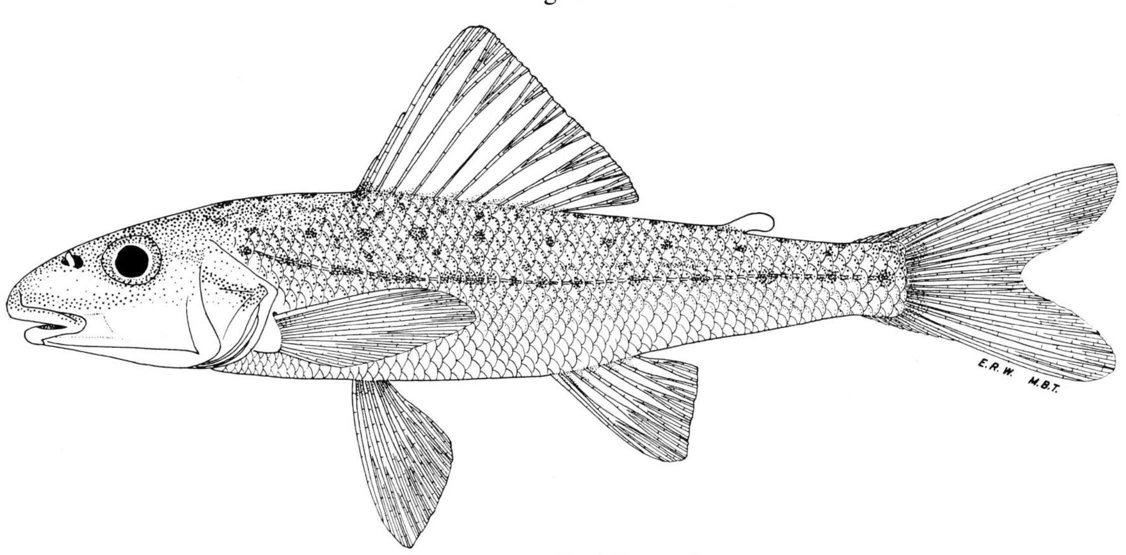

Clear Fork, Ashland County, O.

Sept. 9, 1939.
Female.

84 mm SL, 4.2″ (11 cm) TL.
OSUM 525.

Identification

Characters: An adipose fin. Ctenoid scales which feel rough when fish is stroked from tail to head. Usually two, extremes 1–3, splintlike, weak dorsal spines: one or two anal spines; tiny pelvic spines; all spines so frail and small as to be seen only by careful observation, and usually only after some dissection. The small teeth are in brush-like bands in mouth. Color pale yellowish and silvery, with a series of dark spots along mid-dorsal stripe; a series along lateral line; one or two additional series between these on each side of back.

Differs: Only species of fish in Ohio having combination of: adipose fin and rough (ctenoid) scales. Trouts, whitefish, cisco, and smelt have smooth (cycloid) scales; lack the frail spines. Minnows and pirate perch lack adipose fin. Catfishes lack scales.

Superficially like: Pirate perch; smelt; young cisco.

Coloration: Dorsally pale straw-yellow and silvery with spots; for description of spots, *see* under Characters. Ventral portion of head and body silvery-white, with a yellowish cast. Fins transparent. Body has a pellucid cast.

Lengths: *Young* of year in Oct., 1.5″–3.0″ (3.8–7.6 cm) long. Around **1** year, 2.5″–4.5″ (6.4–11 cm). Around **2** years, 3.0″–5.0″ (7.6–13 cm). Largest specimen, 5.0″ (13 cm) long.

Distribution and Habitat

Ohio Distribution—The records before 1900 are rather contradictory. Kirtland, for some unknown reason, failed to record the trout-perch for either the Lake Erie or Ohio River drainages; Jordan and Evermann (1896, Pt. 1:784) considered the species to be rare in Ohio "streams south of Lake Erie," but stated that it was abundant in Lake Erie and other Great Lakes. McCormick (1892:23) found the trout-perch to be "Common in Lake Erie, Black River and Beaver Creek" of Lorain County; Henshall (1889:79) found it abundant in the Little Miami River, Ohio drainage, as did Osburn, (1901:75) in the Franklin County streams west of the Scioto River. Kirsch (1895A) failed to find it in the Maumee River system of Lake Erie. From these and other published records, and from preserved specimens, it is apparent that before 1900 the Ohio range of the trout-perch must have been similar to

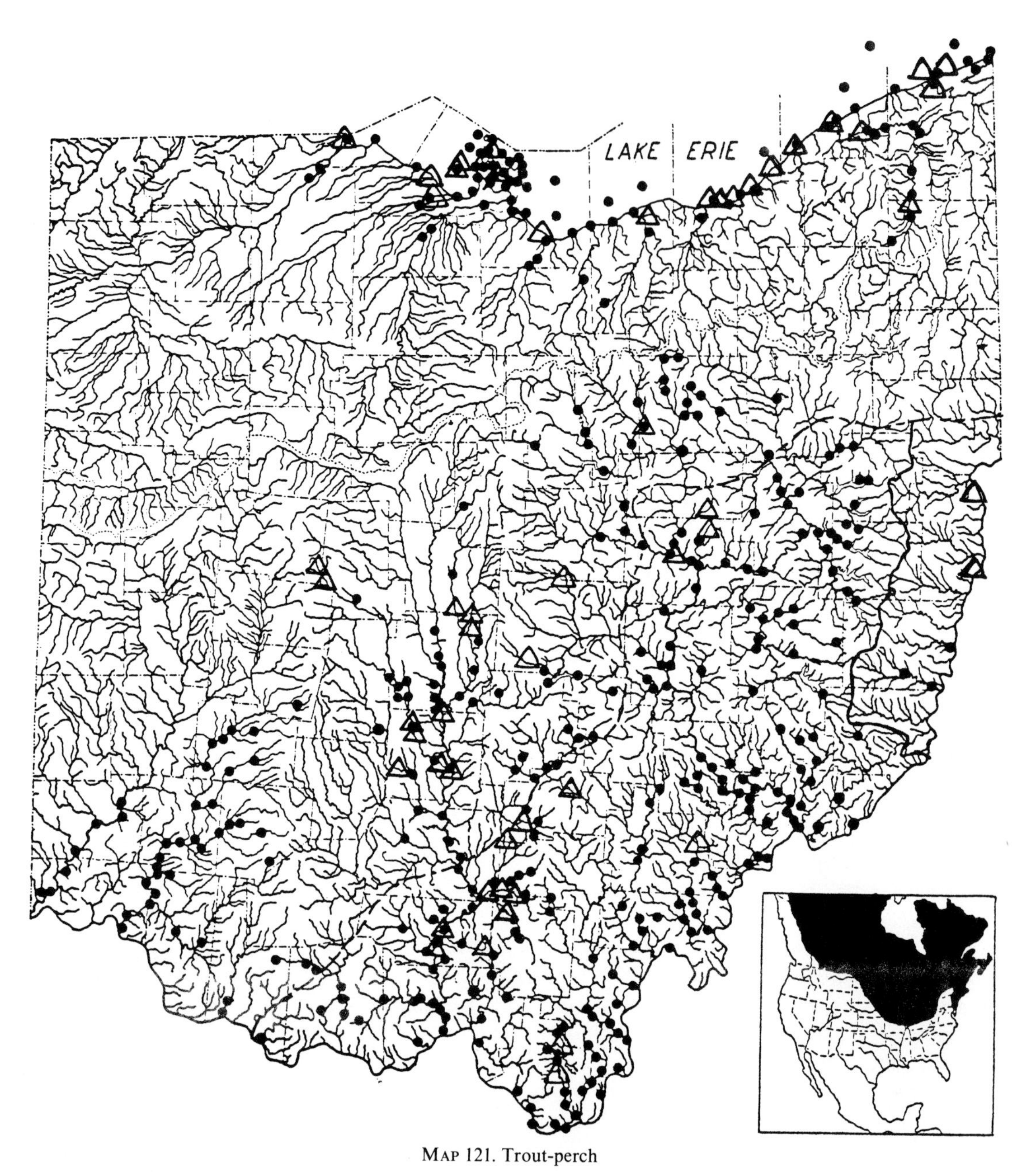

MAP 121. Trout-perch

Locality records. ● Before 1955. △ 1955–80. /\ Glacial boundary. ~ Flushing Escarpment.
Insert: Northern limits of range indefinite.

its range since 1920, and that before 1900 the species was abundant in Lake Erie, and at least locally abundant in the Ohio drainage.

Between 1920–50 the trout-perch was abundant in Lake Erie; but only more or less temporary or sporadic populations of varying sizes occurred in its tributaries. An exception was the lower half of the Grand River which contained a small, more or less stable population. This is interesting, because ecologically the Grand River is more similar to those unglaciated Ohio drainage streams which contain large populations than it is to any other Lake Erie tributary. As a rule, the species penetrated less far upstream in the Lake Erie tributaries than it did in those of the Ohio River, and it was entirely absent from the upper two-thirds of those Lake Erie tributaries from the Sandusky River westward.

In the Ohio River between 1920–50 the trout perch was most numerous between Scioto and Washington counties. The largest inland populations occurred in the low-gradient, unglaciated streams from Noble northward to Tuscarawas County with the greatest density in Guernsey County; moderate populations were present in the unglaciated streams from Hocking southward to Lawrence County; rather small populations inhabited the streams along the line of glaciation; and, as a rule, populations in glaciated territory became progressively smaller the farther they were removed from unglaciated territory. Only strays occurred in high-gradient streams east of the Flushing Escarpment, and there are few records for the Ohio drainage of western Pennsylvania (Raney, 1938:60). Likewise, the few records for southwestern and no records for western Ohio indicate that the species occurs rarely in Indiana; this assumption is well demonstrated by Gerking's (1945:81, Map 73) few Indiana records.

Habitat—In Lake Erie the trout-perch occurred most abundantly over a clean sand or gravel bottom, and was taken at depths of almost 200′ (60.9 m) (Fish, 1929:170). It usually was numerous in waters less than 5′ (1.5 m) in depth only during the late spring-early summer spawning season, or on dark nights when it presumably came inshore for the purpose of feeding.

In inland streams the largest and more permanent populations inhabited the longer and deeper pools of the base- or low-gradient streams where the bottom consisted mostly of sand and fine gravel, and where viscous clays were relatively uncommon. During daylight in such pools, the trout-perch hid under cut-banks, tree roots, organic debris and rubbish, or retired into deep water, coming at dark to feed in the open shallows and even upon the sluggish riffles. Occasional specimens were accidentally caught on trotlines or in nets in the deeper portions of the Ohio River. The species definitely avoided rooted aquatic vegetation, and seemed to be highly intolerant to the clayey silts of the till plains and the more eroded prairies.

Years 1955–80—The overall population of the trout-perch in Lake Erie definitely became smaller after 1950 than it had been previously, the decrease in numbers becoming very apparent after 1960 (Van Meter and Trautman, 1970:74; White et al., 1975: 97). Before 1950 large numbers of spawning adults could be taken occasionally from early summer to midsummer, especially along sandy beaches. Frequently, many post-spawning trout-perches, dead or dying, accumulated in small windrows on the beaches. Since 1960 no such evidences of such large spawning concentrations have been noted.

No change in numerical abundance was observed in inland Ohio populations since 1950. Occasional spawning concentrations were observed; the earliest was May 4, 1961 (OSUM 13858), when a fair-sized school was found spawning in Big Walnut Creek immediately below the Hoover Dam, Franklin County. A fair-sized population was noted in the 1970's by Cavender and Crunkilton (1976:119).

The Ohio River population appeared to be decreasing in size. The 1957–59 investigations (ORSANCO, 1962:150) of the entire Ohio River resulted in capturing the species in only 16 of the 341 collecting stations. H. Ronald Preston did not capture it during his 1968–69 investigations of the river opposite the state of Ohio. During the years 1976 to 1978 Margulies and Burch collected the species in the Ohio River adjacent to Jefferson County, Ohio.

For life history in western Lake Erie *see* Kinney (1950); for life history elsewhere *see* Magnuson and Smith (1963).

EASTERN BURBOT

Lota lota lacustris (Walbaum)

Fig. 122

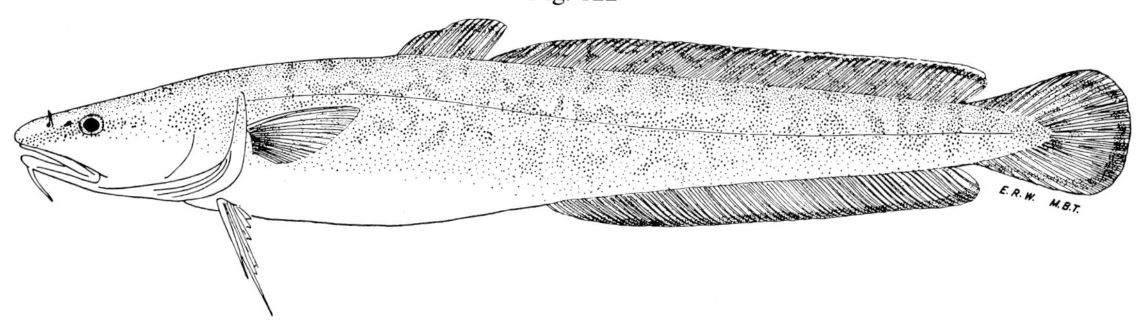

Lake Erie, Ottawa County, O.

April —, 1942.
Male.

554 mm SL, 23.6″ (59.9 cm) TL.
OSUM 5015.

Identification

Characters: *One* barbel near tip of chin. Two dorsal fins, both without spines; second dorsal of more than 65 rays. Anal fin of more than 60 rays. Dorsal, caudal, and anal fins separated from one another. Scales so small as to be almost invisible without magnification. Catfish-like in shape.

Differs: Catfishes have four chin barbels; only one dorsal; dorsal and pectorals have a spine each. Eels and lampreys lack pelvic fins. No other species of fish in Ohio has a median chin barbel.

Superficially like: Eel and catfishes.

Coloration: Dorsally olive-brown or brown, heavily mottled and blotched with chocolate-brown; this mottling most distinct on smallest young, often obsolete in large adults. The lighter-colored sides often contrast sharply with the dark mottlings. Ventrally yellow-white or white, the demarcation between the darker sides and ventral surface giving a bicolored effect. Fins of same general color as adjacent body parts; dorsal and caudal fins mottled; they and anal fins usually have a dusky margin.

Lengths and **weights:** *Young* of the year in Aug., 1.0″–4.0″ (2.5–10 cm) long. Around **1** year 6.0″–8.0″ (15–20 cm). *Adults*, usually 16.0″–32.0″ (40.6–81.3 cm). Fishes 10.0″–15.0″ (25.4–38.1 cm) long usually weigh 5 oz–1 lb 8 oz (142 g–0.7 kg); 16.0″–20.0″ (40.6–50.8 cm) weigh up to 2 lbs, 10 oz (1.2 kg); 20.0″–25.0″ (50.8–63.5 cm) weight up to 5 lbs 8 oz (2.5 kg), or heavier if huge stomach is packed with food; 25.0″–30.0″ (63.5–76.2 cm) weight up to 10 lbs (4.5 kg). Largest specimen, 32.5″ (82.6 cm) long, weight 12 lbs, 5 oz (5.6 kg). Commercial fishermen give maximum weights of between 12–14 lbs (5.4–6.4 kg). According to Cahn (1936:163–65) and Adams and Hankinson (1928:517) spawning occurs in winter; my observations on spawning agree with theirs and with the statement by Fish (1930:9) that young hatch in late May or June.

Distribution and Habitat

Ohio Distribution—The presence of the burbot in Lake Erie has been known since early historic times; however, before 1850 little attention was paid to this species because it was considered to be almost worthless as food. In 1838, Kirtland (1838: 196) noted its presence in Lake Erie, later (1844:25) commenting that this fish, little esteemed as food, was taken in considerable numbers with hooks and seines about Cleveland Harbor. Early commercial reports failed to mention the burbot, for when it was included in the total commercial catch, it was lumped together with such species as suckers and drum under the names "soft fish" or "trash fish." Its value as food fit for human consumption has recently increased; during the 1939– 49 period, a yearly average of 364,260 lbs (165,226 kg) (Shafer, 1950:1) was taken commercially from the Ohio waters of Lake Erie.

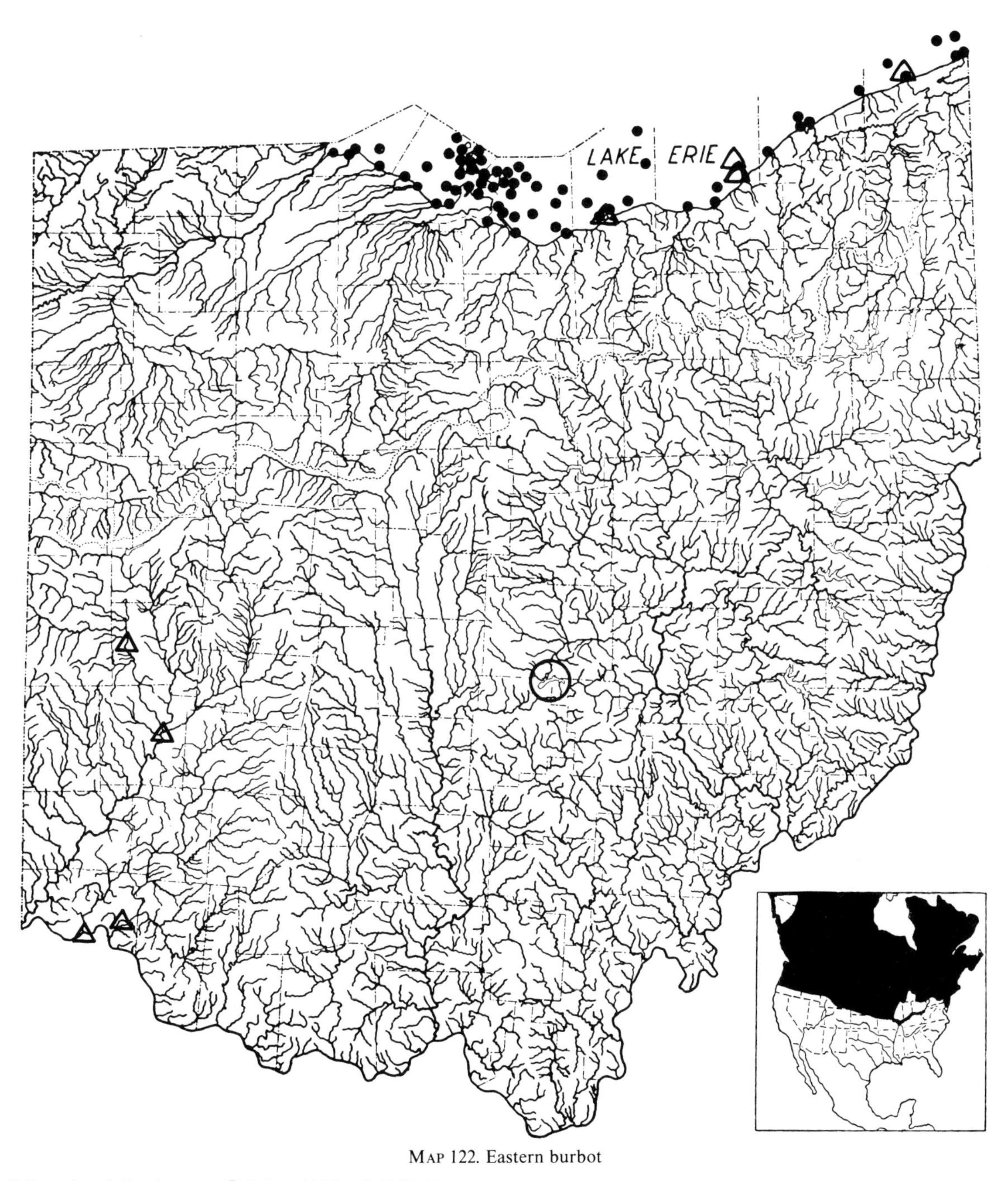

MAP 122. Eastern burbot

○ Introduced. Specimens: ● Before 1955. △ 1955–80.

Insert: Black area represents range of this subspecies; *Lota lota leptura* Hubbs and Schultz occupies the outlined area in northwestern British Columbia and Alaska; large unoccupied areas exist within the southern portion of the range.

It seems improbable that, at any time since 1800, there has been an established population of burbots in the upper Ohio River drainage; ecological conditions in this system have been too unfavorable for this deep- and cold-water species. The early records for this drainage, all outside of Ohio, are based upon solitary individuals (Jordan, 1882:995; Hay, 1894:293; Forbes and Richardson, 1920:331). It has been suggested that these strays entered the Ohio River drainage from the Great Lakes through the canals (Rostlund, 1952:279); but it seems more logical to assume that they strayed from the upper Mississippi into the Ohio River. In 1920, Edward L. Wickliff examined a burbot, caught in Buckeye Lake (hollow circle), which presumably was introduced into that lake from Lake Erie, one of the many hundred thousands of fishes of several species, caught in commerical nets in Lake Erie, which were transported inland and liberated in public waters during the 1915–30 period.

Habitat—A deep- and cold-water species, tending to congregate in the colder, deeper waters of Lake Erie during the warmer months, coming into shallow waters along shore in winter, but apparently not ascending tributary streams (Kirtland, 1851N:161).

The scientific name of the burbot has long been *Lota maculosa* (Lesueur). Recently, J. Murray Speirs (1952:99–103) pointed out that the older specific name *lacustris* of Walbaum, long associated with *Ictalurus punctatus*, belonged to the burbot, hence the name change to *Lota lota lacustris* (Walbaum).

Years 1955–80—Before 1950 the burbot was very numerous or abundant in Lake Erie and was usually an inhabitant of deeper and/or cooler waters. Since 1950 it has decreased markedly in abundance (Van Meter and Trautman, 1970:74; White et al., 1975:97–98), as have other deep- and cold-water species, such as the lake trout, cisco, whitefish and blue pike.

There were four recent inland records. Thomas Thatcher reported the capture, on August 7, 1963, of one in the Little Miami River directly north of Newtown, Hamilton County. Peter Munroe identified one taken May 11, 1968 in the Stillwater River, 1 mile (1.6 km) southwest of Pleasant Hill, Newton Township, Miami County, he writing that "at least one pay lake near the Stillwater River stocks Burbot." On July 13, 1963, V. Larrmore caught a burbot at Dayton in the Great Miami River while fishing below the confluence of the Stillwater River at the Steel Dam. On December 27, 1963, M. G. Kover captured another at the same place. Both were approximately 20″ (50.8 cm) long. They are preserved in the Dayton Museum of Natural History.

As indicated above, it is possible that these individuals strayed from the upper Mississippi River drainage. Or they may have been inadvertently or otherwise introduced. The species is fished for commercially in Lake Erie, and some of the catch is transported inland to be liberated in public waters or in "pay lakes."

William M. Clay and other investigators from the University of Louisville reported the occurrence of the burbot in the Ohio River, stating that "In the spring of 1960, several specimens of burbot were taken by a commercial fisherman in nets near Cincinnati, Ohio, and two of these specimens are preserved in the collections at the Department of Biology, University of Louisville" (ORSANCO, 1962:71).

EASTERN BANDED KILLIFISH

Fundulus diaphanus diaphanus (Lesueur)

Fig. 123

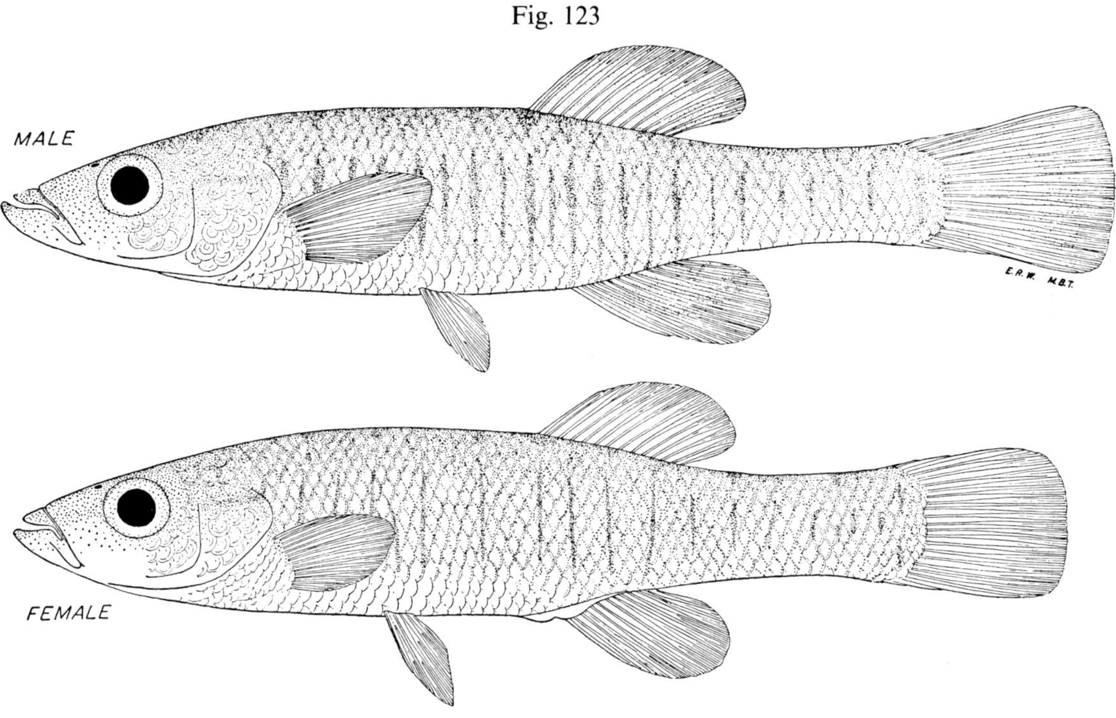

Upper fig.: Little Yellow Creek at mouth, Columbiana County, O.

Aug. 17, 1944. 42 mm SL, 2.0″ (5.1 cm) TL.
Adult male. OSUM 6389.

Lower fig.: same locality, date and OSUM number.

Breeding female. 51 mm SL, 2.5″ (6.4 cm) TL.

Identification

Characters: *General* and *generic*—Large, round scales on top of head, cheeks and opercles. Deep groove separates tip of upper jaw from tip of snout; jaw capable of considerable forward extension because of the deep groove. Tail rounded. Pelvic fin origins closer to tip of snout than to caudal base. Anterior dorsal origin far behind pelvic origins. Jaws with bands of small, sharp teeth. No lateral line. No spines. Anal fin similar in both sexes. Female lays eggs. *Specific*—Dorsal fin origin decidedly in advance of anal fin. Both sexes have dark vertical bands on their sides. *Subspecific*—Dorsal rays usually 13–14. Scales in lateral series usually more than 42. Vertical bands on sides narrower and more regular in shape; those on caudal peduncle often short and not fused into a median, lengthwise stripe; adult males generally have 9 or more vertical bands before dorsal origin; the females have fewer. Introduced.

Differs: Western banded killifish averages fewer dorsal rays and scales in the lateral series, also has a lower average number of vertical bands before dorsal fin origin. Blackstripe topminnow has fewer dorsal rays; dorsal origin is behind anal origin. Central mudminnow has no groove separating tip of upper jaw from snout. *See* central mudminnow.

Most like: Western banded killifish and male blackstripe topminnow. **Superficially like:** Female mosquitofish and mudminnow.

Coloration: Dorsally olivaceous, olive-yellow, or yellow-brown, with a dark brown mid-dorsal stripe and vague mottlings. Sides lighter, yellower, the

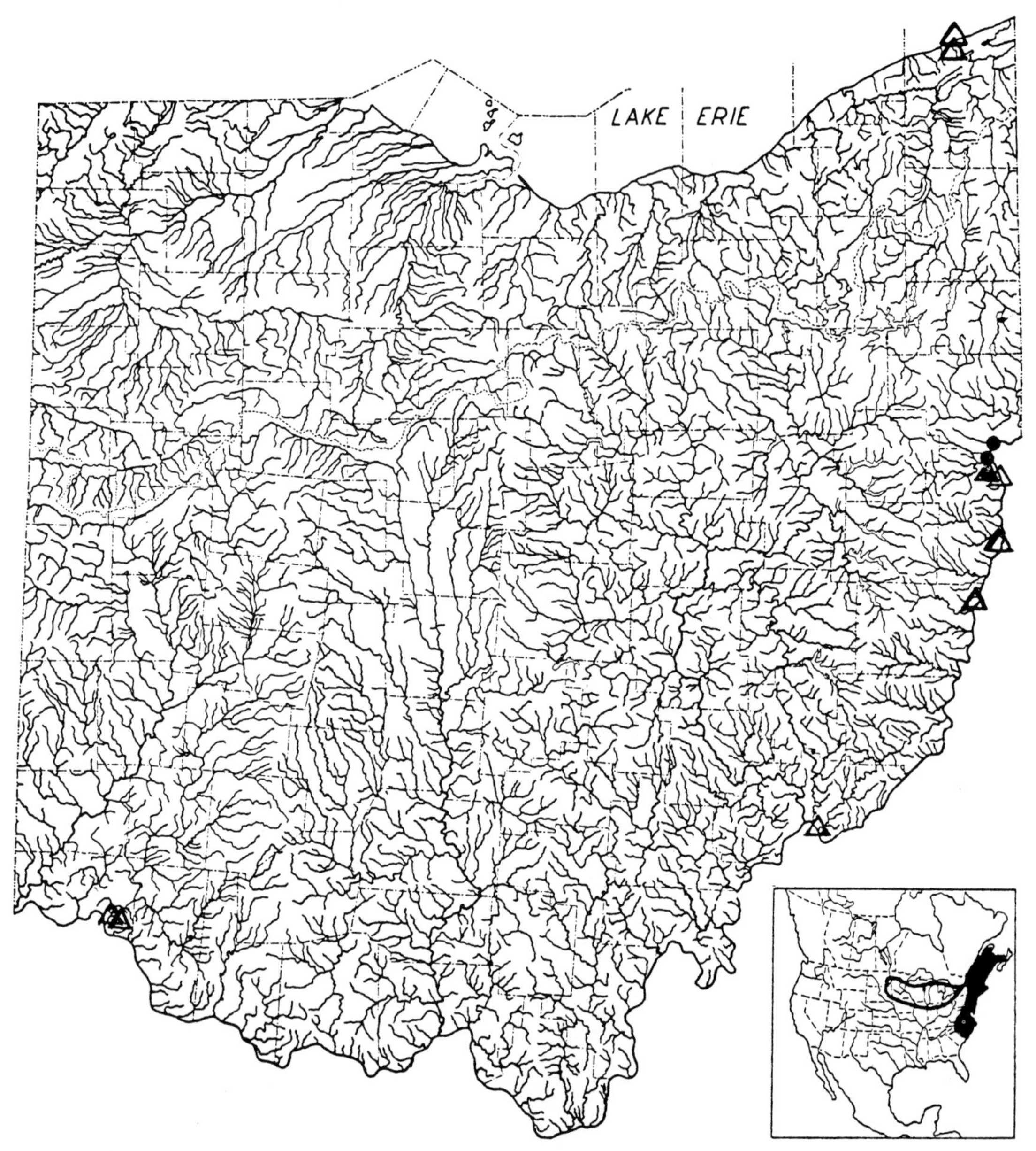

MAP 123. Eastern banded killifish

Locality records. ● After 1943. △ 1955–80.
Insert: Black area represents original range of this introduced form, outlined area is of another subspecies.

bands brown. Ventrally yellow-, or silvery-white. Fins light olive or olive-yellow.

Lengths: *Young* of year in Oct., 0.8″–2.3″ (2.0–5.8 cm) long. Around **1** year, 1.5″–2.8″ (3.8–7.1 cm). *Adults*, usually 2.1″–3.2″ (5.3–8.1 cm). Largest Ohio specimen, 3.3″ (8.4 cm) long; maximum length elsewhere, about 4.0″ (10 cm).

Distribution and Habitat

Ohio Distribution—In 1938, Raney (1938:59) wrote that eastern banded killifishes and mummichogs (*Fundulus heteroclitus*), captured in the Delaware River drainage of eastern Pennsylvania, had been liberated in the upper Ohio drainage of extreme western Pennsylvania, and that both species were then flourishing in the Beaver River system. In a letter to me, dated October 22, 1942, Raney wrote that in some collections made recently in the Ohio River in Beaver County, Pennsylvania, he had identified specimens of both the eastern banded killifishes and the mummichog. Some of the eastern banded killifishes were collected 4 miles west of Midland, which locality is within a few miles of the Ohio-Pennsylvania line.

On August 17, 1944, I collected 93 specimens of the eastern banded killifish at the mouth of Little Yellow Creek, Columbiana County, and on the next day took 20 specimens at the mouth of Big Yellow Creek in Jefferson County. On May 12, 1950, G. G. Acker collected this killifish in Carter Run, a tributary of Big Yellow Creek in Knox and Saline townships, Jefferson County.

Habitat—In Ohio this subspecies was taken in the quiet backwaters at the mouth of streams where the bottom was of sand, gravel, and a few boulders, and there were small beds of a fine-leaved pondweed.

Years 1955–80—This subspecies of the banded killifish, not a native of Ohio, was captured in five additional localities in the vicinity of the Ohio River in the past 25 years.

John G. Eaton and Peter T. Frame presented eight specimens to the state collections (OSUM 13865), taken July 10, 1964, from Clough Creek, Hamilton County. Three year-classes were represented, suggesting successful spawnings. Many young less than 1″ (2.5 cm) TL were captured and released. The Clough Creek population could have resulted from specimens originally introduced into eastern Ohio (*see* above); however, it seems more logical to assume that this population came from some aquarium in the Cincinnati area. This fish is considered a desirable aquarium species.

During 1967–68 H. Ronald Preston obtained specimens from the Ohio River in Belmont County. In 1978 Margulies and Burch obtained specimens in the Ohio River adjacent to central Jefferson County, Ohio.

On July 9, 1975, Andrew White collected specimens (OSUM 33514) in Ashtabula Harbor, Lake Erie, in Ashtabula County. *Fundulus d. diaphanus* appears to be more tolerant of some types of pollutants, requires little or no aquatic vegetation, and may be more tolerant to larger streams and lakes than is the now relict population in Ohio of *F. d. menona*.

WESTERN BANDED KILLIFISH

Fundulus diaphanus menona Jordan and Copeland

Fig. 124

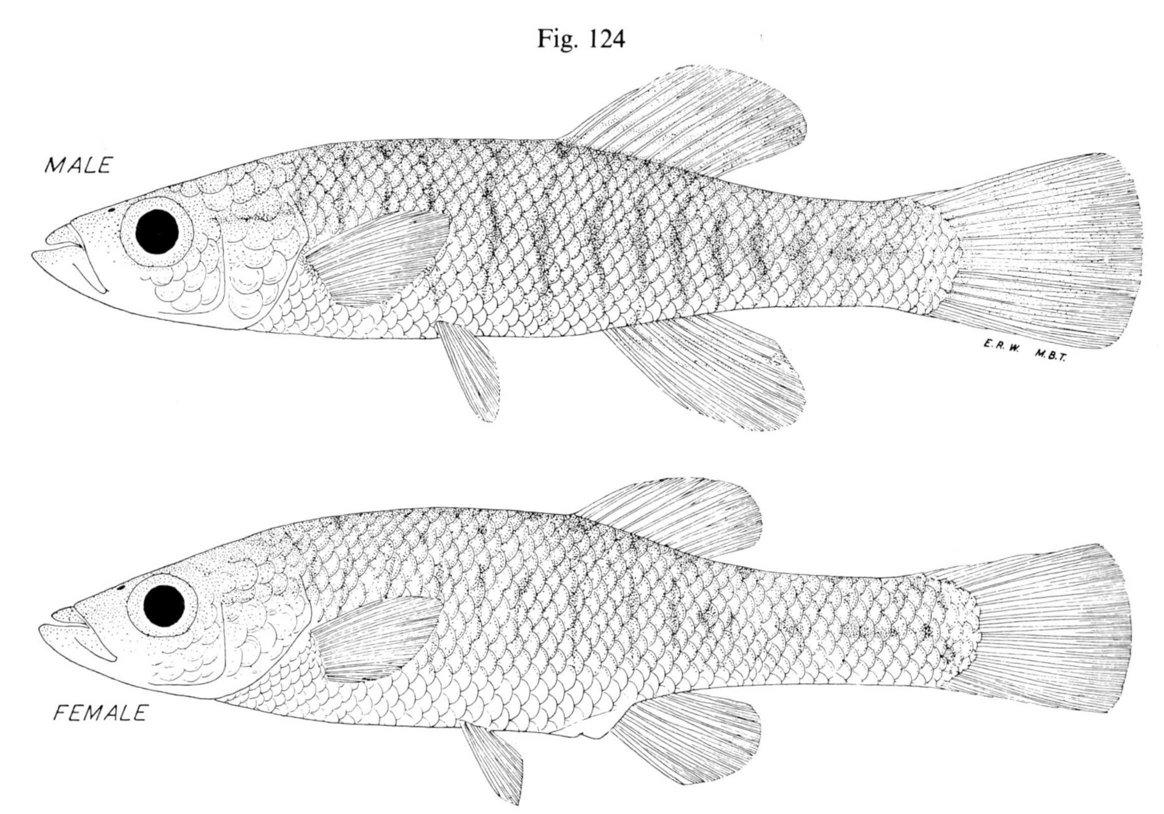

Upper fig.: East Harbor, Ottawa County, O.

Aug. 24, 1933. 42 mm SL, 2.1″ (5.3cm) TL.
Adult male. OSUM 3093.

Lower fig.: same locality, date and OSUM number.

Breeding female. 50 mm SL, 2.4″ (6.1 cm) TL.

Identification

Characters: *General, generic,* and *specific—See* eastern banded killifish. *Subspecific*—Dorsal rays usually 12–13. Scales in lateral series usually fewer than 42. Vertical bands on sides broader and less regular in shape; those on caudal peduncle generally fused together by a median, lengthwise stripe; adult males generally have 8 or fewer vertical bands before dorsal origin; females usually have fewer than 7. Native.

Differs: Eastern banded killifish has a greater average number of dorsal rays and lateral scales, also has a greater average number of vertical bands before the dorsal fin origin. *See* eastern banded killifish and central mudminnow.

Coloration: Similar to eastern banded killifish.

Lengths: *Young* of year in Oct., 0.8″–2.2″ (2.0–5.6 cm) long. Around **1** year, 1.5″–2.5″ (3.8–6.4 cm). *Adults*, usually 1.5″–2.8″ (3.8–7.1 cm). Largest specimen, 3.2″ (8.1 cm) long.

Distribution and Habitat

Ohio Distribution—Few other species of Ohio fishes displayed as great a decrease in abundance during the 1920–50 surveys as did the western banded killifish. During these surveys it became extirpated from such localities as the Portage River, and in others, such as East, Middle and West harbors and about the Bass Islands (McCormick, 1892:23), it decreased so greatly in abundance as to

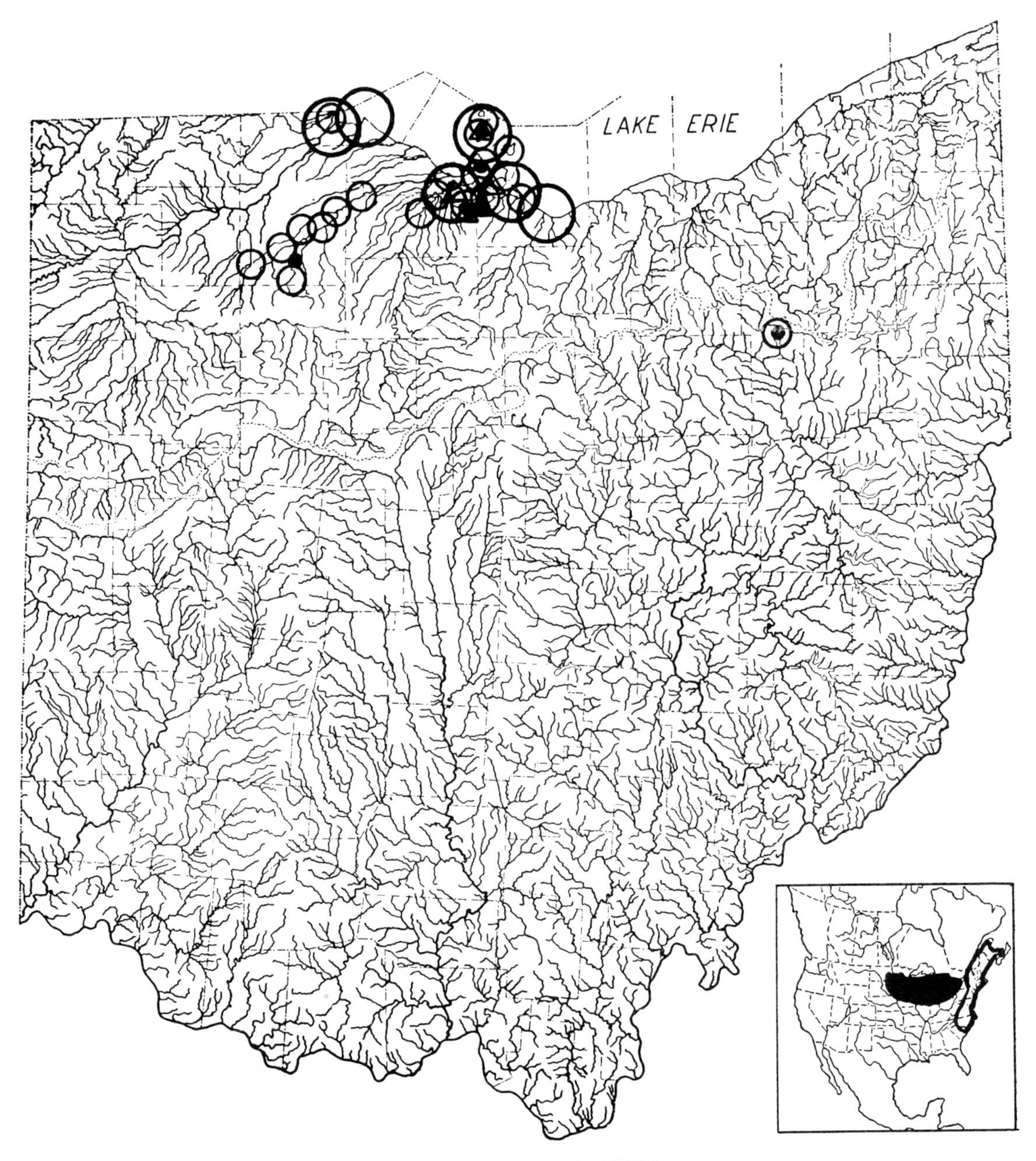

MAP 124. Western banded killifish

Locality records. ○ Before 1900. ○ 1920–30. ● 1931–54. △ 1955–80.
Insert: Black area represents range of this subspecies, outlined area of the nominate form.

verge on extirpation. Only in the relatively undisturbed Miller Blue Hole of Townsend Township, Sandusky County, did the population appear to hold its own during the 1920–50 period. Changes in abundance of this and the other species of topminnows were readily observed, because of their surface-swimming habit which made them most conspicuous.

This killifish should have been far more widespread and abundant throughout the northern half of Ohio before 1900 than it was after 1920, for previous to 1900 there undoubtedly was far more available habitat. This species should have extended far inland in such river systems as the then clearer and more vegetated Maumee and Sandusky, and should have been present in many other glacial lakes other than the Portage Lakes, as it was in similar lakes in Indiana (Gerking, 1945:79) and Pennsylvania (Raney, 1938:59).

Habitat—The largest populations occurred only in very clear streams of low- and base- gradients and in harbors, marshes and glacial lakes, where there was a profuse amount of aquatic vegetation and the bottoms consisted of sand, marl, or organic debris. The smallest populations were found in localities where the bottoms were of clayey silts, and in such localities the species eventually disappeared. *See* blackstripe topminnow for the complementary distribution of these two topminnows and of the recent encroachment of the blackstripe into the former range of the western banded killifish.

Years 1955–80—This subspecies of the banded topminnow, a native of Ohio, continued to decrease in numbers and in the size of its range in the state. It was observed or collected in only three localities during the present period.

On April 19, 1952 David and Richard Zura captured one in Terwillegar's Pond on South Bass Island, Ottawa County. None has been taken since despite thorough collecting annually.

A thriving, small population continues to exist in the Miller Blue Hole, Sandusky County, where classes from the Franz Theodore Stone Laboratory had the opportunity of observing it annually. If in the future the ecology of this particular spring is modified, this native subspecies could be extirpated from Ohio.

On April 15, 1972 Andrew White, Michael Kelty and Cyrus Pourzanjani collected two in Sandusky Bay near Gypsum.

BLACKSTRIPE TOPMINNOW

Fundulus notatus (Rafinesque)

Fig. 125

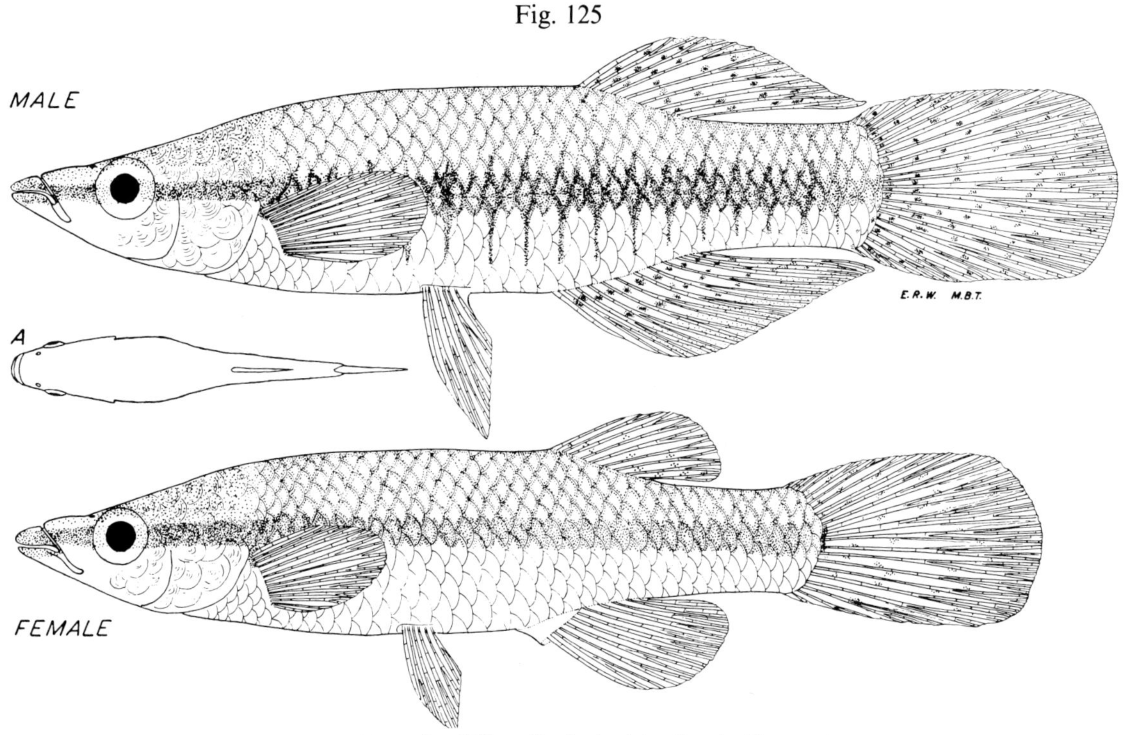

Upper fig.: Willow Creek, Auglaize County, O.

July 9, 1941. 51 mm SL, 2.4″ (6.1 cm) TL.
Adult male. OSUM 3319.

Lower fig.: same locality, date and OSUM number.

Breeding female. 47 mm SL, 2.3″ (5.8 cm) TL.

Fig. A: dorsal view of female; shape characteristic of all Ohio species of killifishes.

Identification

Characters: *General* and *generic—See* eastern banded killifish. *Specific*—Dorsal fin origin slightly behind anal fin origin; rarely directly over. Dorsal rays normally 7–9. Scales in lateral series 29–36. Adults of both sexes have a lateral band which is solid in the female, but in the male tends to break up into a confluent series of short, vertical bars. Adult males with large, posteriorly-pointed dorsal and anal fins.

Differs: Banded killifishes have a greater number of dorsal rays; dorsal origin is before anal origin. *See* central mudminnow.

Most like: Banded killifishes. **Superficially like:** Mudminnow and mosquitofish.

Coloration: Dorsally above lateral band, olive- or brownish-yellow, the scales with dark edges giving a somewhat cross-hatched appearance; Ohio River drainage populations usually with 15–30 black spots scattered over back, *see* text. Both sexes have a black band, beginning on both lips, extending backward across snout, cheeks, and opercles, continuing across body to caudal base; in adult males the lateral band tends to break up into a confluent series of short vertical bars, *see* upper fig. Ventral half of head and body whitish-yellow or silvery-white. Ventral fins light olive-yellow; dorsal, caudal and anal fins black-spotted in adult males, the spots tending to form rows; these fins sometimes faintly spotted in adult females.

Lengths: *Young* of year in Oct., 1.0″–2.0″ (2.5–5.1

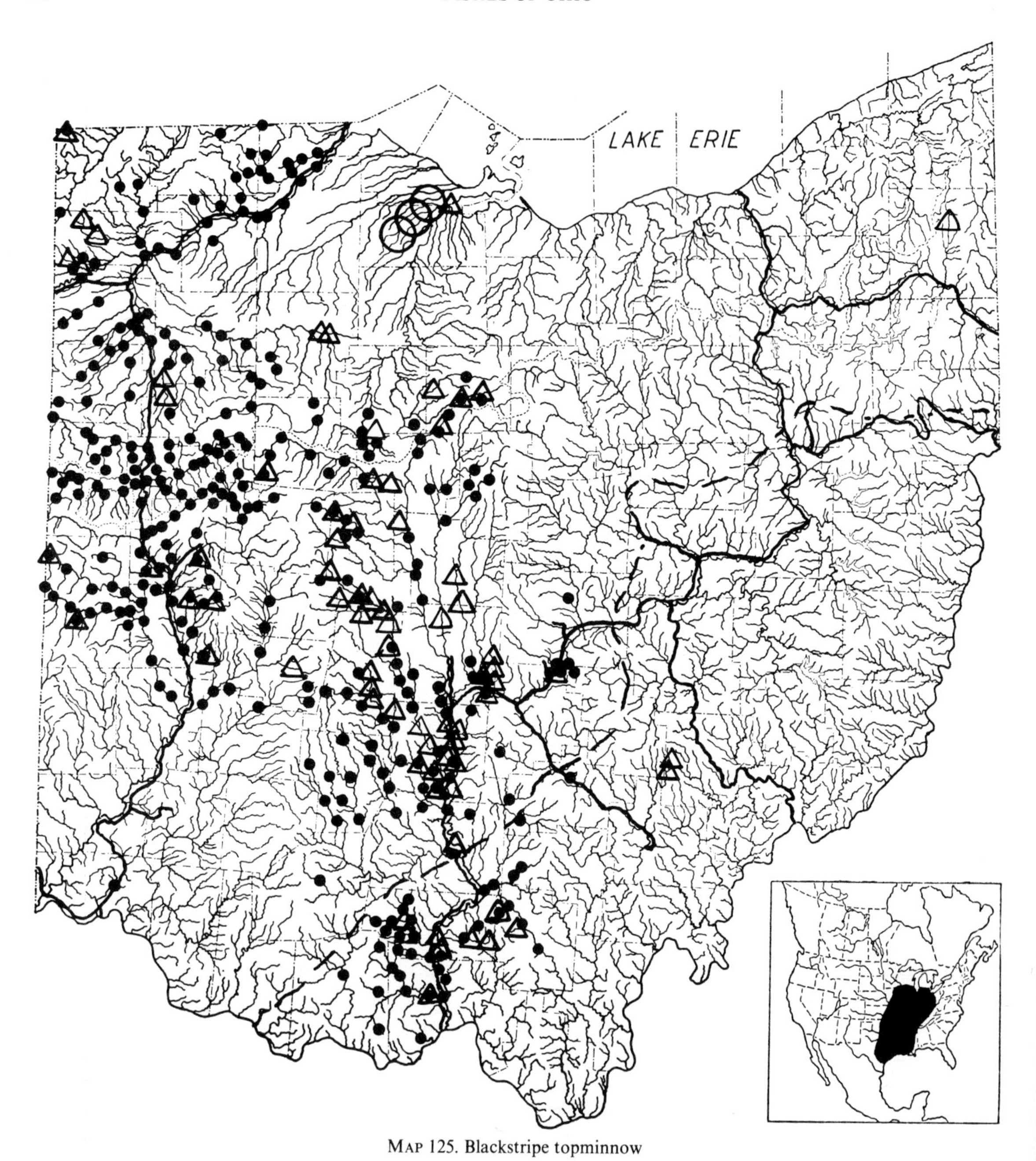

MAP 125. Blackstripe topminnow

○ Range extension after 1940. Locality records. ● 1880–1954. △ 1955–80. ~ Canals. /\ Glacial boundary.
Insert: Eastern limits of range indefinite because of confusion in former identification of the closely related *Fundulus olivaceus* (Storer).

cm) long. Around **1** year, 1.1″–2.5″ (2.8–6.4 cm). *Adults*, usually 1.8″–3.8″ (4.6–9.7 cm). Largest specimen, 3.0″ (7.6 cm) long.

Distribution and Habitat

Ohio Distribution—Until 1940 the Ohio distributional pattern of the blackstripe topminnow was similar to the distributional pattern of the prairies, except that the blackstripe was absent in that area from the mouth of the Maumee River eastward to Erie County, which area was the range of a presumed competitor, the western banded killifish (*see* distributional map of that species). During the 1925–50 period the western banded killifish decreased greatly in abundance or disappeared entirely from much of the above described area thereby lessening any possible population pressure of a closely related species. It may have been this lessening of population pressure and/or a change towards greater turbidity in the streams and/or environmental factors which made possible the invasion of the blackstripe into the former range of the banded killifish. At any rate, on August 12, 1941 W. J. K. Harkness and I took the first known specimen of a blackstripe within the former center of the range of the banded killifish, in Muddy Creek of Sandusky County. By May 16, 1949, the blackstripes had increased sufficiently in that stream so that Gerald G. Acker collected 21 specimens (*see* hollow circles).

Apparently the blackstripe was abundant in the Ohio canals where these canals traversed prairie areas, for the canals contained aquatic vegetation and a favorable habitat. In fact, those sections of the canals extant during the 1925–50 period still contained flourishing populations. These canals possibly served as former migration routes which would explain the isolated Hamilton County record of Henshall (1888:79). His specimens came from Ross Lake, a part of the Miami and Erie canal system. The populations at Buckeye Lake and in the nearby canals seem to have been a result of an eastward migration along the Ohio and Erie canal.

Two rather readily recognizable types of blackstripe topminnows occurred in Ohio. One type appears to have entered Ohio via the Wabash-Maumee glacial outlet (Greene, 1935:16), invading glaciated prairie areas as far south as south-central Ohio. This type is characterized by having a deep, turgid body and pale coloration with few or no spottings on the dorsal half of the body. The other type apparently entered Ohio from a center of distribution in the limestone, bluegrass uplands to the south, perhaps before the formation of the Ohio River, and invaded the clear, limestone, upland streams of the Ohio Bluegrass Region in Pike and adjacent counties. This latter type is characterized by a more slender body and a more vivid coloration; it has spottings on the dorsal half of the body of the adults. It is possible that such morphological and color differences may be caused entirely or in part by the degree of nutrition and turbidity of the water, the more robust, paler form being the result of a greater food supply and more turbid waters. It appears more probable, however, that the two types came from slightly different genetic stocks. Specimens taken near the line of glaciation in Ross and Fayette counties, which likewise is the dividing line between the two types of habitat, show intermediate characteristics.

Habitat—The largest populations occurred in relatively clear waters of lakes, ponds, canals, and in streams of low and base gradients where there was some aquatic (or submerged land) vegetation. The blackstripe apparently was not as dependent upon aquatic vegetation as was the western banded killifish, and the blackstripe was more tolerant to turbid waters. However, marked decreases in abundance occurred in those sections of the Maumee River drainage which showed the greatest increases in turbidity during the 1925–50 surveys.

Years 1955–80—Apparently, the blackstripe topminnow has maintained its numerical abundance throughout much of the Ohio range during the present period. The only exceptional record was the discovery of a population in Burr Oak Reservoir in Morgan County in 1962. It appears logical to assume that this species was stocked in the newly created reservoir, either inadvertently or otherwise, as were other species of fishes.

In August of 1972 the first Canadian specimens were captured in southwestern Ontario, suggesting an increase of a relict population, an extension of range or escapes of importations from the United States (Gruchy, Bowen and Gruchy, 1973A: 683–84).

MOSQUITOFISH

Gambusia affinis (Baird and Girard)

Fig. 126

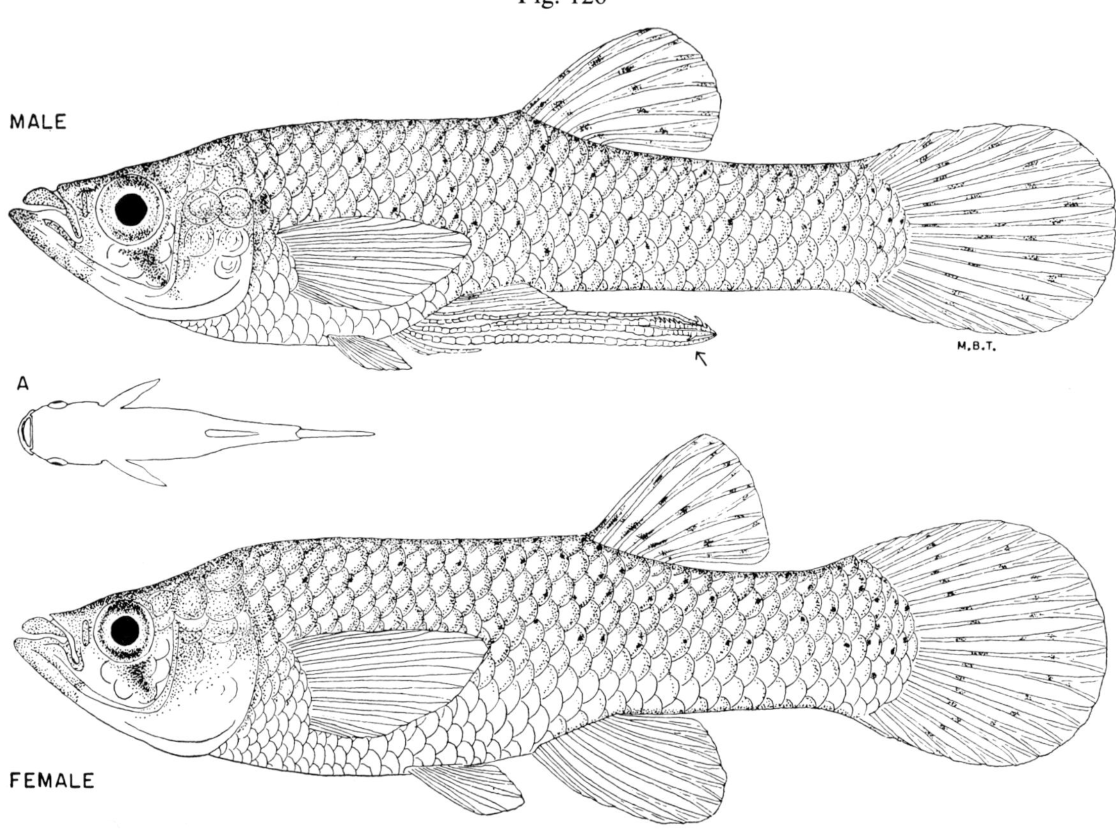

Upper fig.: Horseshoe Pond, Swan Creek Drainage, Lucas County, O.

Sept. 8, 1950. 23 mm SL, 1.2″ (3.0 cm) TL.
Adult male. OSUM 9197.

Lower fig.: same locality, date and OSUM number.

Adult female. 39 mm SL, 2.0″ (5.1 cm) TL.
Fig. A: dorsal view of female.

Identification

Characters: *General*—Similar to those given under eastern banded killifish, except that in this family of killifishes (Poeciliidae) the adult male has the anal fin modified into an intromittent organ, upper fig., arrow; the female bears young alive instead of laying eggs. *Specific*—Dorsal rays 6–8, usually 7. Scales in lateral line series usually 27–30, seldom more. Dorsal fin origin far behind anal fin origin. No bars or bands on sides. Introduced.

Differs: Killifishes and topminnow have dorsal fin origin in front of or only slightly behind anal fin origin; males have unmodified anal fins. Mudminnow has pelvic fin origin nearer caudal base. *See* mudminnow.

Superficially like: Killifishes, topminnow and mudminnow.

Coloration: Dorsally whitish-olive overlaid with silvery, the scales boldly edged with dusky giving back a cross-hatched appearance. A prominent, dusky mid-dorsal stripe. Sides lighter, more silvery, and the scales less prominently dark-edged. Ventrally silvery- and milk-white. Many black spots

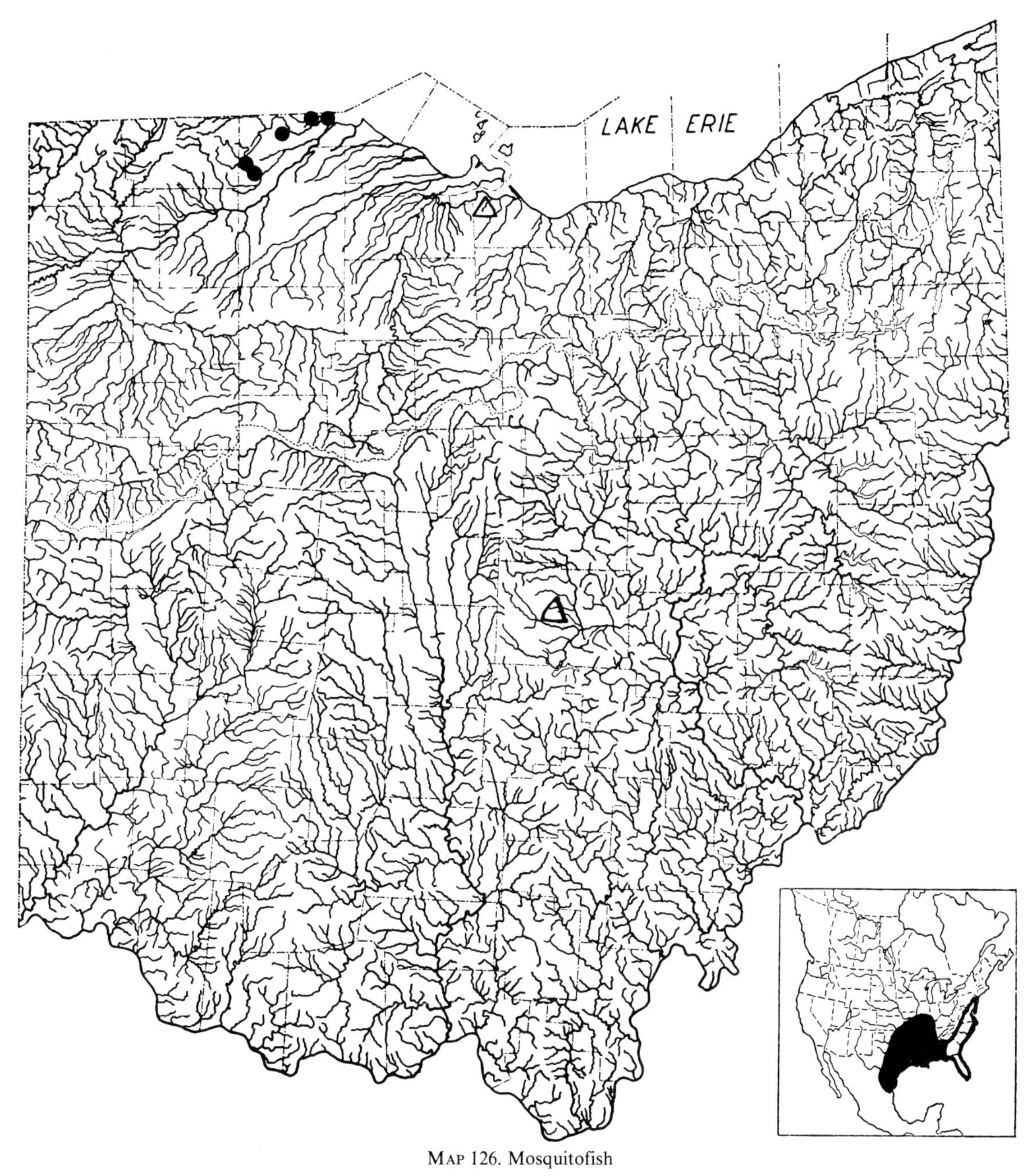

MAP 126. Mosquitofish

Locality records. ● After 1948. △ 1955–80.

Insert: Black area represents the presumed original range of the nominate form, outlined area of another form; the species has been widely introduced in North America outside of its original range, in other continents and islands.

scattered over back and sides. All fins transparent; the dorsal and caudal fins spotted, the spotting tending to form two or three wavy rows. In adult female a bluish bar extends from behind pectoral fin almost to beneath dorsal fin origin.

Lengths: *Adult males*, 0.8″–1.3″ (2.0–3.3 cm) long. *Adult females*, 1.8″–2.3″ (4.6–5.8 cm) long.

Distribution and Habitat

Ohio Distribution—In a mimeographed report, Harold Wascko (1950:1–4) wrote that in the early part of 1947, H. A. Crandell, superintendent of the Toledo Sanitary District, obtained a brood stock of *Gambusia affinis holbrooki* from Florida and one of *G. affinis affinis* from Michigan. Specimens of both subspecies were planted in several localities in the Oak Openings of western Lucas County. The eastern form, *holbrooki*, disappeared after the first winter, but the western form, *affinis affinis*, has been flourishing for four years and has begun to invade areas adjacent to where it was stocked.

Habitat—The species inhabited base- or low-gradient waters such as ponds, small pools, ditches, drains, marshes, and sluggish creeks, occurring in the greatest abundance where there was clear water and aquatic vegetation. As early as 1912, Forbes and Richardson (1920:216–17) called attention to the "injurious competition" resulting when this topminnow invades the range of the blackstripe and starhead (*Fundulus dispar*) topminnows.

Years 1955–80—For several years the Toledo Zoological Society has captured mosquitofishes from private and public waters in the Oak Openings of Lucas County, to hold them over winter in their aquarium building. The following spring they and their young were released into the same waters. It is not known if any of those left over winter outdoors survived, and if they did, how many (Van Meter and Trautman, 1970:74).

Robert W. Alrutz recently informed me that a few years ago the mosquitofish had been stocked in an old spring-fed minnow pond on the Denison University Biological Reserve situated north of Denison in Granville Township, Licking County, and that some had successfully over-wintered.

In the 1957 edition of this report this livebearer was called the western common gambusia. In *A List of Common and Scientific Names of Fishes* (Bailey et al., 1970:31) the common name was changed to mosquitofish. This name is here used.

BROOK SILVERSIDE

Labidesthes sicculus (Cope)

Fig. 127

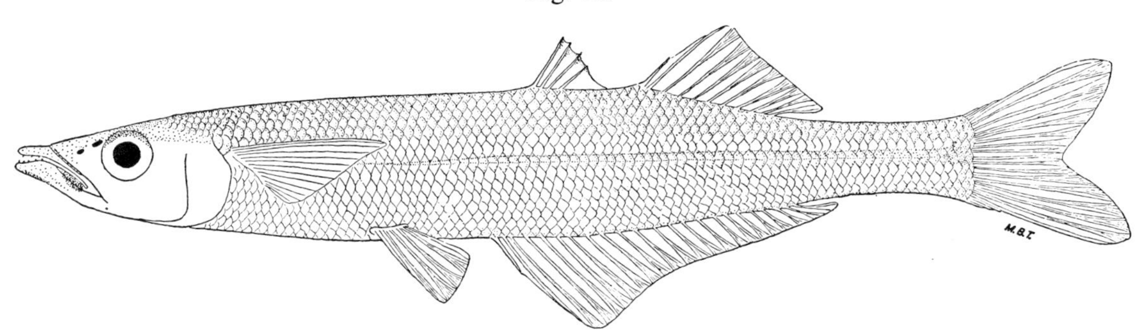

Lake Erie, Ottawa County, O.

June 30, 1942.
Breeding male.

73 mm SL, 3.3″ (8.4 cm) TL.
OSUM 5097.

Identification

Characters: First dorsal fin contains 4–6 weak, flexible spines which bend as readily as do rays. Spinous and soft dorsals well separated, both fins situated *over* the long anal fin; anal fin has one spine and 20–25 rays. Mouth produced into a short beak. Upper jaw may be projected considerably forward because of the deep groove separating tip of upper jaw and snout. Incomplete lateral line of 75–79 scales. Pectoral fin insertion opposite upper angle of gill cleft.

Differs: No other species of fish in Ohio has *both* dorsal fins situated entirely over the long anal fin. Minnows, smelt, cisco, mooneye, goldeye and skipjack herring have only one dorsal fin.

Superficially like: Smelt, silver shiner; possibly other slender shiners.

Coloration: Dorsally a pellucid-straw or greenish-yellow, overlaid with much silvery. At some angles the lateral band may be intensely silvery-white with some bluish. Ventrally silvery-white with some greenish, golden and bluish reflections. Fins transparent, except for the black-tipped dorsal spines.

Lengths: *Young* of year in Oct., 1.8″–3.5″ (4.6–8.9 cm) long. *Adults*, usually 2.8″–3.8″ (7.1–9.7 cm) long. Largest specimen, 4.2″ (11 cm). Observations by Hubbs (1921:274–75) on silverside in Michigan indicate that the species spawns after its first winter of life (around 1 year old), and that only a small remnant lives through the second winter; therefore, the silverside virtually has only a one-year life cycle. These observations agree with those made in Ohio.

Distribution and Habitat

Ohio Distribution—The many early literature records refer to its abundance and capture in more than half of the localities investigated, and plainly indicate that before 1900 the silverside was usually abundant in most of the areas investigated, and presumably was widely distributed over the state. During the 1930–50 period the silverside was far less abundant than formerly, for it was taken only in about 200 of the more than 2000 collections made during that period, and in many localities where it was found, only strays were taken. This decrease in abundance was aptly illustrated in Franklin County where in 1897 Williamson and Osburn (1898:38) found the silverside to be "abundant, and of general distribution," but where in 1930 it was no longer generally distributed or abundant, and where after 1945 only strays or relict populations remained. A similar decrease in abundance occurred in the Auglaize River in Putnam and Defiance counties, where in 1893 Kirsch (1895A:336–37) found the species to be present in every locality seined, but where since 1925 only a few strays have been taken; likewise in Sandusky Bay, where in 1899 Osburn (1901:76) found the species "very abundant, on sandy bottom in shallow water" but where since

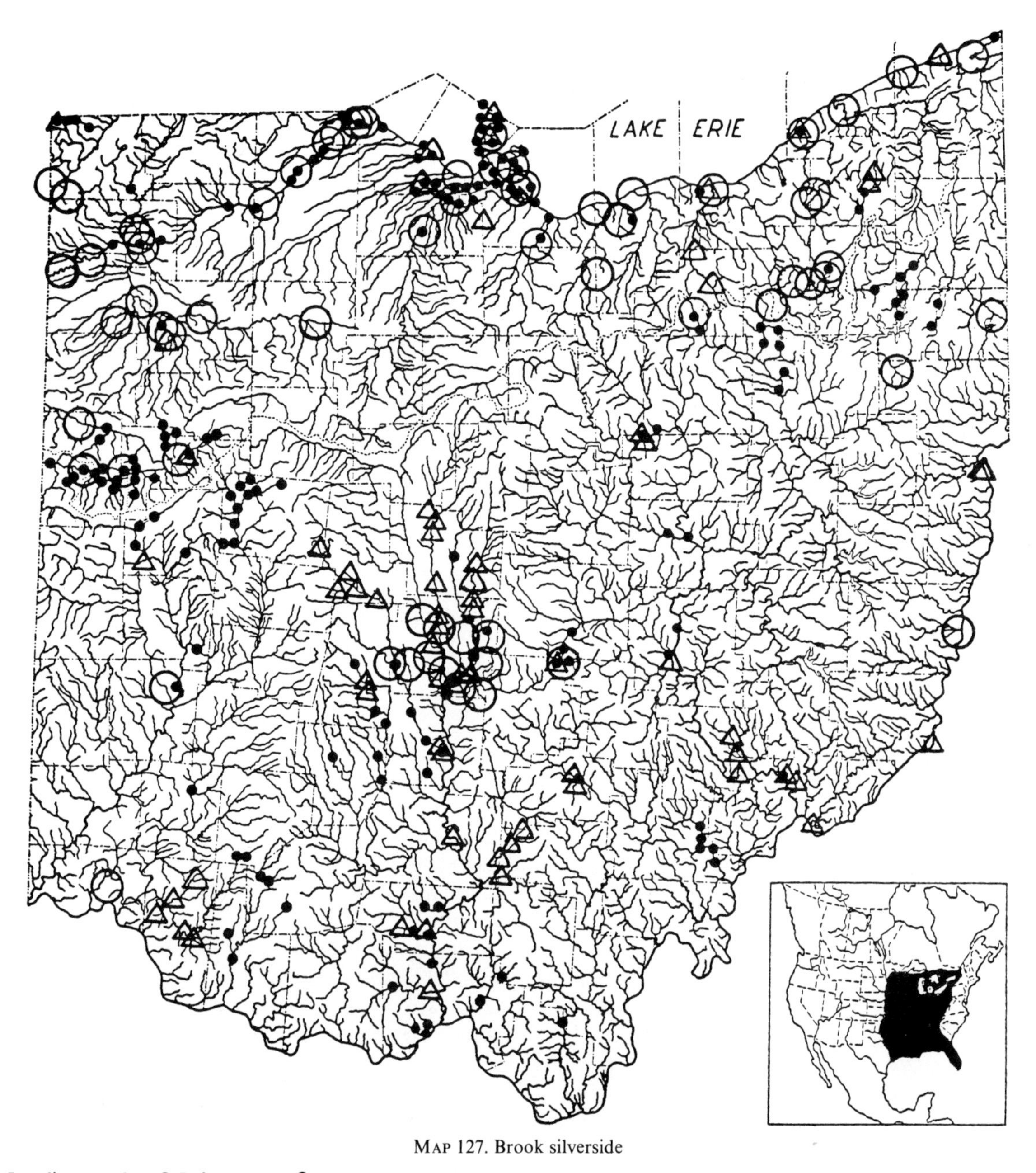

MAP 127. Brook silverside

Locality records. ○ Before 1901. ● 1901–54. △ 1955–80.

Insert: Isolated colonies are present in areas outside this range, especially westward; large unoccupied areas exist within range, some of which are increasing in size.

1945 it has been uncommon. Chief among the factors causing this waning in abundance is the increased turbidity of the waters.

Habitat—The largest populations occurred in glacial ponds and lakes, and about the bays and harbors of Lake Erie where the waters were clearest, and the bottoms were composed of clean sand, gravel or organic muck. Before 1900 the silverside was common in large streams, such as the Maumee, Scioto, and possibly the Ohio River, but since then has become almost absent in these now turbid waters. Since 1940, those individuals inhabiting streams were found principally in the small, clear, upland brooks, such as the limestone streams in western Pike County, or in the clearer backwaters of the recently constructed reservoirs. The usually active silverside often stopped feeding in the daytime when the water became turbid, and then lay listlessly at the surface as it does at night (Hubbs, 1921:267–68), and as do other surface-feeding species such as the bigeye shiner.

Years 1955–80—The number of localities in Ohio where the brook silverside occurred and its numerical abundance have been decreasing since 1930; *see* Ohio Distribution above. These decreases continued after 1951, and several additional relict populations were known to have been extirpated (White et al., 1975:100). Although possibly a few relict colonies may have existed, none was found in the Maumee River system, except at Nettle Lake, Northwest Township, Williams County. Isolated populations occurred in the waters of northeastern Ohio (Van Meter and Trautman, 1970:78).

In the Ohio River drainage the usually widely separated populations appear to have maintained their range size and numerical abundance. Although some previously known colonies were extirpated, with few exceptions the more than 30 collections containing silversides taken since 1955 contained fewer than 10 individuals per collection, whereas before 1934 large numbers were observed or collected at the majority of collecting stations. An exception was Big Walnut Creek, central Ohio, where in Hoover Reservoir and immediately below my assistants and I found moderately large populations, as later did Cavender and Crunkilton (1976:120). H. Ronald Preston informed me that in 1968 or 1969 a few were taken in the Ohio River in the vicinity of dam 15, Belmont County. In 1978 Margulies and Burch captured the species in the Ohio River adjacent to Jefferson County, Ohio.

BROOK STICKLEBACK

Culaea inconstans (Kirtland)

Fig. 128

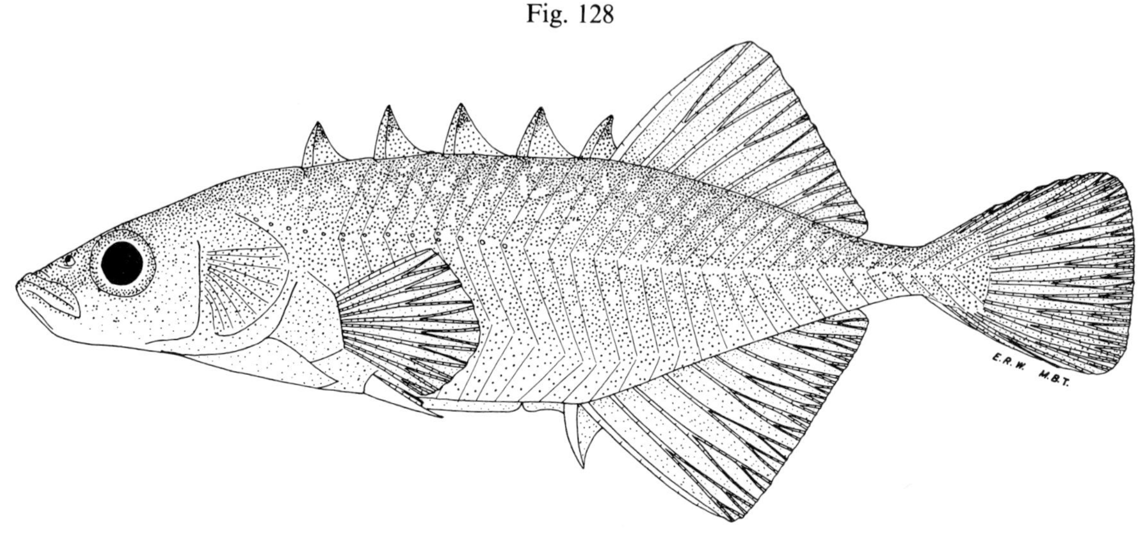

Tippecanoe Creek, Mahoning River drainage, Mahoning County, O.

May 11, 1942.
Breeding male.

44 mm SL, 2.1″ (5.3 cm) TL.
OSUM 4939.

Identification

Characters: The 4–6 stout, dorsal spines are *not* connected to each other with membrane; dorsal spine membranes restricted to triangular sections, each section of which is attached only to the posterior edge of one spine and to the back. No scales. Pelvic fins each with a stout spine and only one or two small rays. Posterior edge of tail straight.

Differs: All other species of fishes in Ohio having dorsal spines have them connected to one another with membrane (webbing).

Superficially like: Young of darters; there is little resemblance to any other species in Ohio.

Coloration: *Adult*—Dorsally yellow-olive or dark olive-green, spotted with lighter. Sides lighter. Ventrally pale whitish-yellow. Dorsal, caudal and anal fins pale olive-green, pelvics light yellow-green; pectorals transparent tinged with olive. *Breeding male*—Much darker; entire head dusky-black; dorsal third of body greenish-black; sides lighter; belly dusky-green. Dorsals and pelvics dusky; caudal and anal fins greenish-dusky; pectorals slate. Spottings on back and sides obscured by dusky flush. *Breeding female*—Similar to breeding male except that duskiness is not so intense. *Young*—Similar to nonbreeding adult, except lighter in color; more silvery; fins more transparent.

Lengths: *Young* of year in Oct., 1.2″–2.0″ (3.0–5.1 cm) long. *Adults*, usually 1.5″–2.3″ (3.8–5.8 cm). Largest specimen, 2.7″ (6.9 cm) long.

Distribution and Habitat

Ohio Distribution—In his catalogue of animals found in Ohio, Kirtland (1838:191–92) in 1838 mentioned that he found this unique species "in a small stream in the village of Poland" (presumably Yellow Creek) and that he had furnished the Boston Academy with a figure and description of it for publication. This description appeared in 1841 (1841A:273–74) and the type locality was given as "ditches and muddy pools in *Trumbull* Co. Ohio." These ditches and pools were probably in the vicinity of Poland, for it was not until 1846 that Mahoning County was formed from portions of Trumbull and Columbiana counties (Howe, 1900:175).

In 1877 Jordan (1877:66) had specimens which were collected in Cuyahoga County, and in 1882 he (1882:998) stated that the species was "very abun-

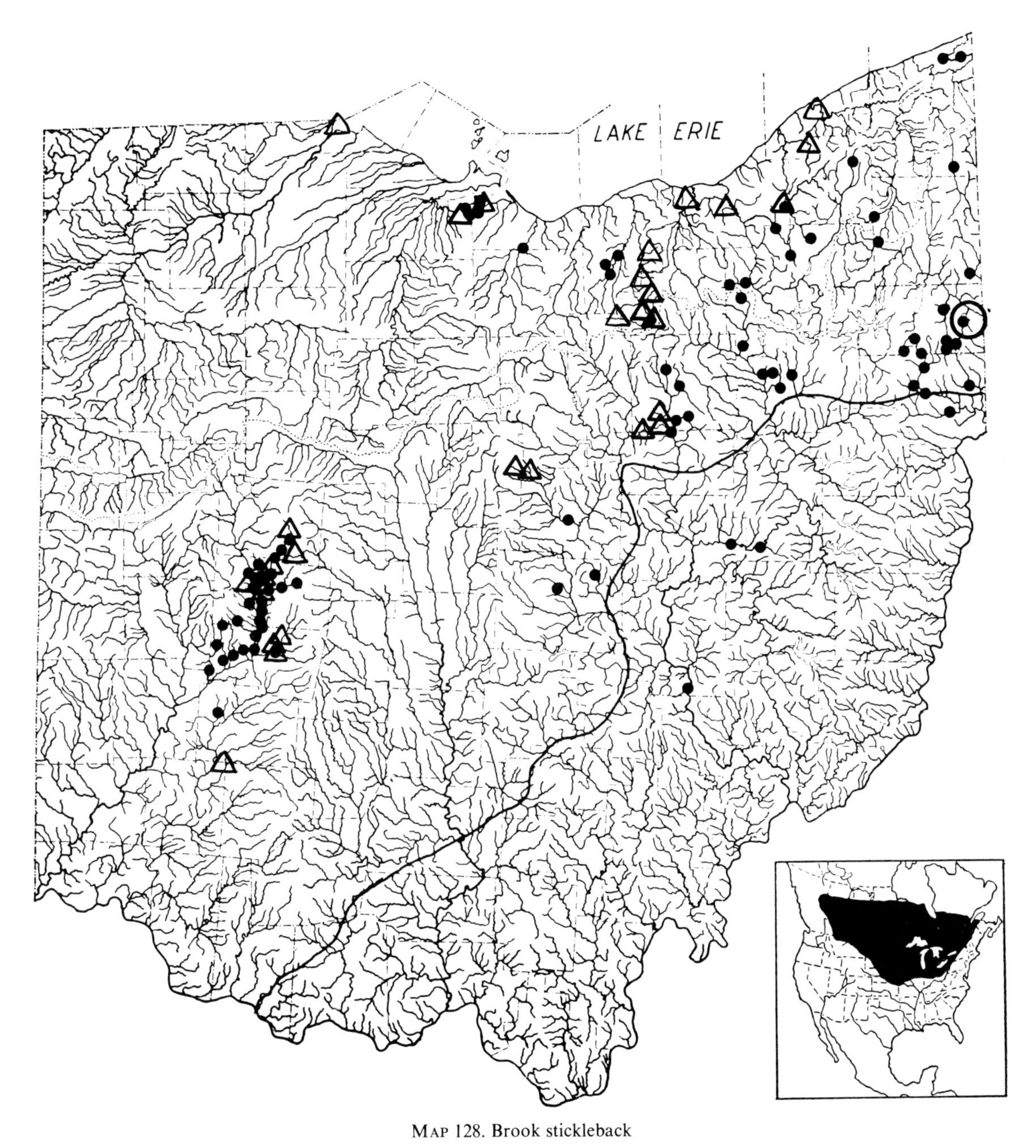

MAP 128. Brook stickleback

○ Type locality, Yellow Creek, Poland, Mahoning County. Locality records. ● Before 1954. △ 1955–80. ~ Glacial boundary.
Insert: Large unoccupied areas occur along southern limits of this range; northern range limits indefinite.

dant in many streams in the northern part of Ohio," which was undoubtedly true for northeastern Ohio. There seemingly are no records from the Black Swamp areas of northwestern Ohio and the adjacent part of Indiana (Gerking, 1945:107).

During the 1925–50 surveys the species was abundant in the upper Mad River drainage, the Miller and Castalia blue holes of Sandusky and Erie counties, and in scattered localities in northeastern Ohio. It was rare in unglaciated Ohio. The Muskingum River record of southern Muskingum County is of a stray that was captured by Robert B. Foster and me on July 2, 1927, when that stream was very turbid and in flood. We captured it in very swift water. Presumably this individual had strayed far from its brook or spring habitat and was being swept downstream when we accidentally captured it. During the 1925–50 surveys many isolated populations of sticklebacks throughout glaciated Ohio either decreased markedly in abundance or were extirpated.

Habitat—This cold-water inhabiting species is on the southern edge of its range in Ohio. The most flourishing populations were present in habitually clear, cold springs and brooks where there was much submerged aquatic vegetation and the bottom was of muck, peat or marl. The species was extremely tolerant to either highly alkaline or highly acid conditions, but was intolerant to turbidity. It was very pugnacious, both with its own kind and with other fish species. Dredging, ditching and draining readily destroyed its habitat, after which it immediately disappeared.

Years 1955–80—The collections of the brook stickleback during this period indicate little change in its distributional pattern prior to 1950. However, noticeable decreases in numbers, or extirpation, of several populations were observed, caused by the partial erosion or destruction of its vegetated, clear-water habitat (White et al., 1975:98). The total Mad River population apparently decreased in numbers, as possibly did the overall size of its range in Ohio.

In 1950 Whitley (44) showed that *Eucalia* Jordan, 1878, was preoccupied by Felder, 1861, in the Lepidoptera. Whitley proposed the generic name *Culaea*; this has been generally accepted (Bailey and Allum, 1962:83; Nelson, 1969:2442–43; Bailey et al., 1970:33). For geographic variation and distribution *see* Nelson (1969).

STRIPED BASS

*Morone saxatilis** (Walbaum)

Fig. 129

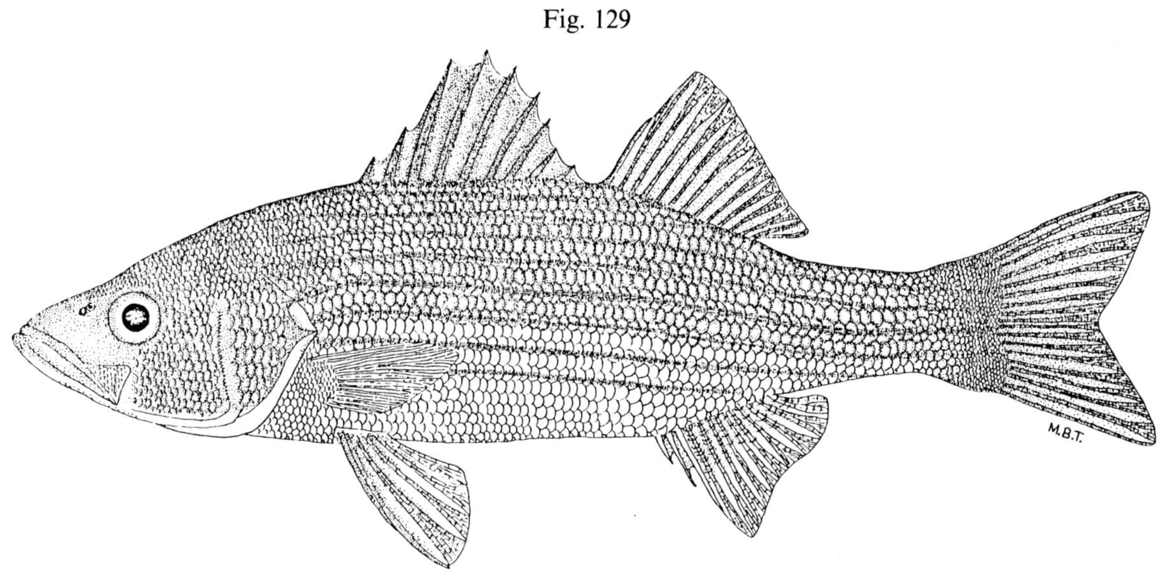

Reared at U.S. National Fish Hatchery, Licking County, O.

Hatched, May, 1968.
Collected, Aug. 27, 1969.
Subadult.

212 mm SL, 10.5″ (26.7 cm) TL.
OSUM 16165.

Identification

Characters: *General—See* Family Serranidae. *Generic*—Dorsal fins well separated, the spinous dorsal normally having 9 spines. A sharp spine present on posterior angle of opercle (earflap). Head mostly covered with scales. *Specific*—Anal rays normally 10–12; the three anal spines graduated, the first not notably shorter or stouter than the second; length of second anal spine usually contained 5 times in head length; lower jaw projecting beyond upper jaw; back evenly arched, body depth 3.45 (young)–4.2 (large specimens) times in standard length; scales in lateral series normally 57–67; a series of 7–9 longitudinal lines along the sides and back; teeth on base of tongue in 2 parallel patches. Introduced.

Differs: White bass has anal spine contained less than 4 times in head length, body depth usually deeper, normally less than 3.5 times in standard length, except in very small young; teeth on base of tongue in a single patch. White perch usually has less than 10 anal rays; second anal spine stout, about equal in length to the third, and much longer than the first; body depth normally greatest immediately before spinous dorsal fin origin. Blackbasses and other sunfishes have dorsal fins broadly joined together, and lack the opercle spine. Drum, walleye, and sauger have only two anal spines.

Most like: White bass. **Superficially like:** White perch and drum.

Coloration: Dorsally deep bluish-silvery with many greenish, silvery and steel-blue reflections. Sides more silvery white, often with a bronzy cast. Ventrally milk-white. Vertical fins dusky-green to black, ventrals dusky-silver, pectorals greenish, except in small young in which all fins are more silvery. Large young and adults have 7–9 longitudinal lines along scale rows of sides and back, these absent in small young, which may have 6–10 dusky vertical crossbars.

Lengths and **weights:** 7.0″–9.0″ (18–23 cm) long, weight 3–6 oz (85–170 g). Subadults and adults, 12.0″–16.0″ (30.5–40.6 cm) long, weight 1–2 lbs

* *Roccus saxatilis* in the past.

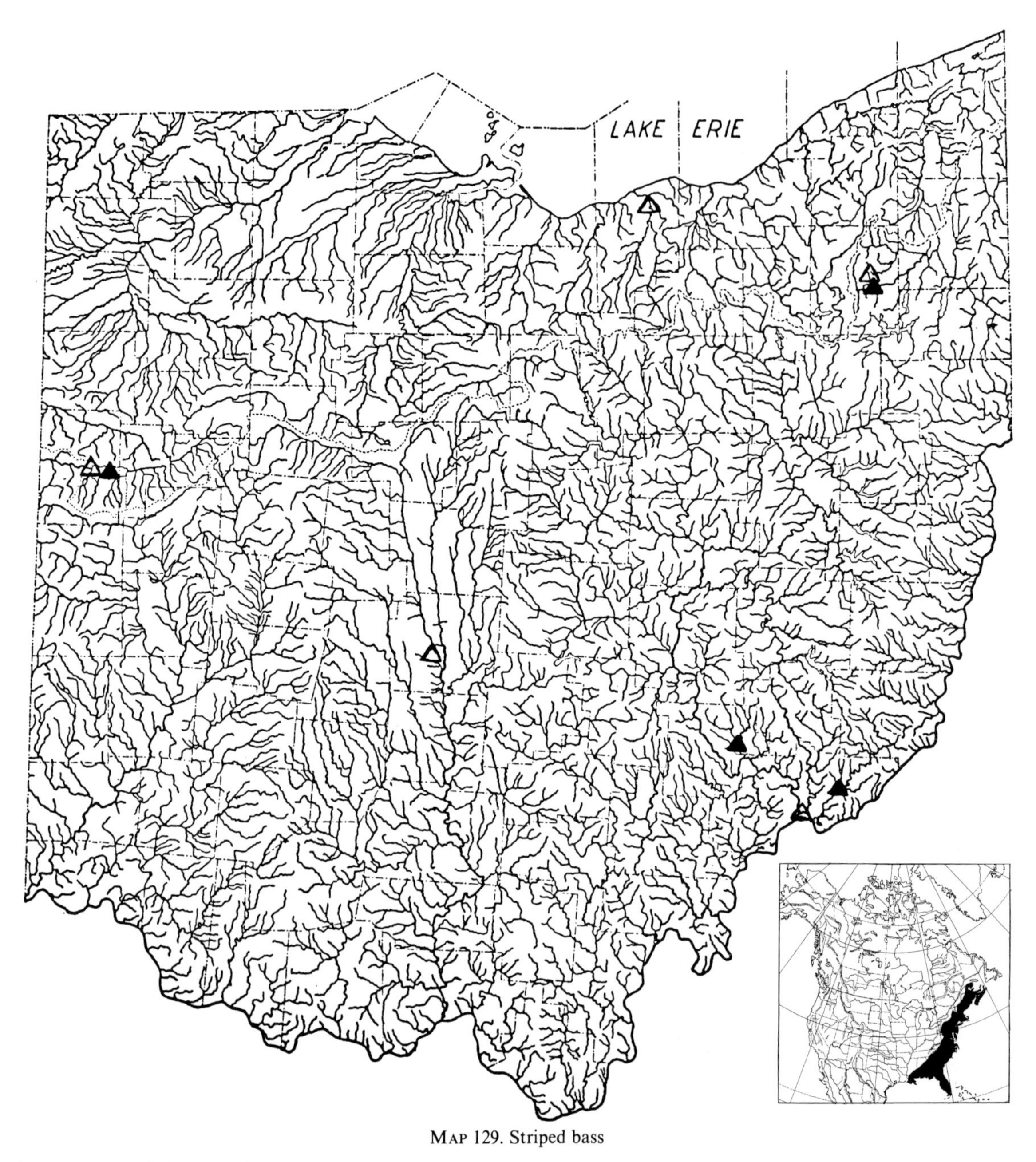

MAP 129. Striped bass

▲ Introduced. △ Recaptured.
Insert: Original freshwater range in North America; widely introduced since.

(0.5–0.9 kg); 20″–25″ (50.8–63.5 cm) long, weight 3.5–7.0 lbs (1.6–3.2 kg); along Atlantic Coast maximum weights range up to 60 lbs (27.2 kg), rarely heavier; reaching heaviest weights only by individuals spending much of their life in salt water.

Two of the few striped bass captured in Lake St. Marys, Ohio, were large, one of 32″ (81 cm) TL and 15 lbs (6.8 kg) in weight, in 1973, and another of 35″ (89 cm) TL and 17.5 lbs (7.9 kg) in weight, in 1974.

Hybridizes: with white bass and white perch. *See* Kerby, 1979.

Distribution and Habitat

Distribution—Original distribution of this coastwise, anadromous species was the Atlantic Ocean from the St. Lawrence River, Canada, south into northern Florida; less numerously and sporadically in the Gulf of Mexico from northern

Florida westward to the Tchefuncta River in Louisiana. Successfully introduced on the Pacific Coast, in San Francisco Bay, California, in 1879, the species now occurs coastwise from Los Angeles County, California, northward to the Columbia River, Washington. "Land-locked" striped bass flourish in some waters of the southeastern United States.

Ohio Distribution—In 1968 eggs were obtained by the Ohio Division of Wildlife from hatcheries in Arkansas and South Carolina and the resultant fry (approximately 81,000) introduced into West Branch Reservoir, Portage County, Lake St. Marys, Mercer and Auglaize counties, Meigs Creek, Morgan and Washington counties, Little Muskingum River, Washington County and elsewhere with discouraging results. Only a few have been captured by fishermen in West Branch Reservoir and fewer than one hundred in Lake St. Marys.

On October 5, 1971 Stephen H. Taub, Bernard Dowler and Martin Bozeman caught in an experimental gill net a striped bass, 5″ (13 cm) TL, in the Ohio River near Marietta, Ohio. During the late 1960's the West Virginia Conservation Department released several thousands in the Ohio River and vicinity of which this specimen is probably one of the first known recoveries.

As in the past, exotic species such as the striped bass continue to be liberated into Ohio waters in the hope that one of these, unlike such native species as the walleye, blue pike and muskellunge, can become established and numerous (as did the carp) in the presently modified and enriched waters. Parsons (1974;1–60) discussed the suitability of the striped bass for introduction into the Great Lakes, stating wistfully that "The western basin of Lake Erie has considerable sport fishing potential for striped bass." I prefer to have the monies spent attempting to increase population sizes of our native species through the re-establishment of as much of their former, favorable environment as is possible.

Habitat—With relatively few exceptions, the large young and adults inhabit the Atlantic, Pacific, and Gulf of Mexico waters of the United States and Canada, spawning in coastwise streams as much as 250 miles (402 km) inland. A few land-locked populations exist, one of the largest being above the Pinopolis Dam in the Santee-Cooper River system of South Carolina.

WHITE BASS

*Morone chrysops** (Rafinesque)

Fig. 130

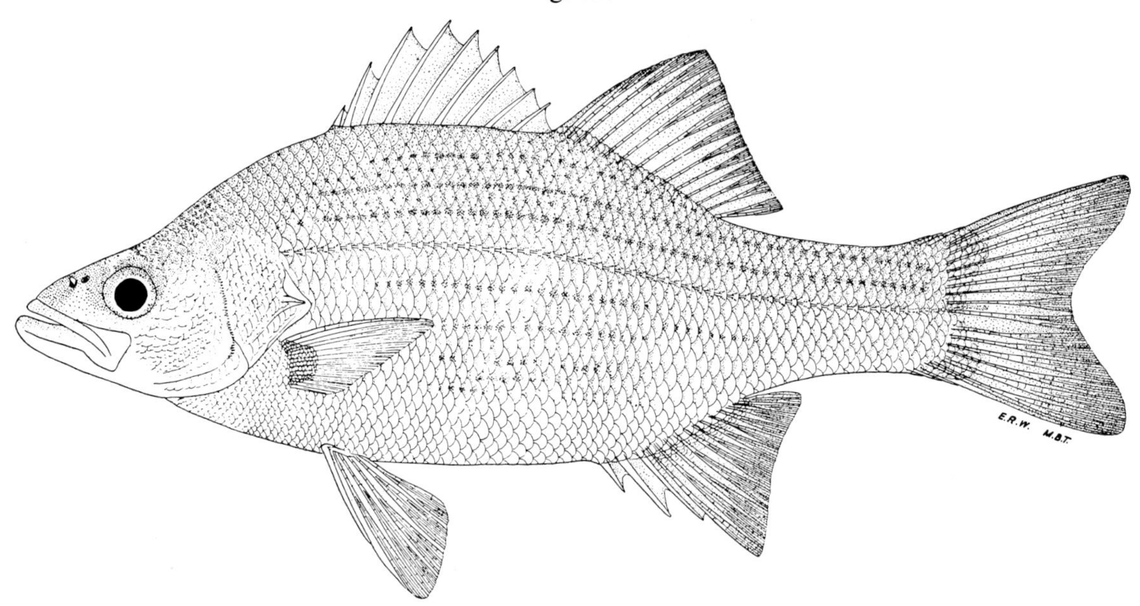

Meander Lake, Mahoning River Drainage, Trumbull County, O.

May 8, 1942.
Adult male.

163 mm SL, 8.1″ (21 cm) TL.
OSUM 4885.

Identification

Characters: *General* and *generic*—Dorsal fins well separated, the spinous dorsal normally having 9 spines. A sharp spine present on posterior angle of opercle (earflap). Head mostly covered with scales. *Specific*—Anal rays usually 11 or 12, rarely 10 or 13; the three anal spines graduated, the first not notably shorter or stouter than the second; lower jaw definitely projecting beyond upper jaw; back evenly arched, body depth greatest under spinous dorsal; mouth rather large, the length of upper jaw normally going 2.2–2.7 times in head; scales in lateral series normally 52–56 (extremes, 51–58); a series of 6–12 longitudinal lines along the sides and back, resulting from the dark spotting on the scales (absent or faint in smallest young); coloration predominantly bluish-silvery.

Differs: White perch has fewer than 10 anal rays, a smaller mouth and lacks the distinct dark longitudinal lines on sides. Blackbasses and other sunfishes have dorsal fins broadly joined together, and lack the spine on the opercle. Drum, walleye, and sauger have only two anal spines. Mooneye, goldeye, whitefish, cisco, smelt have no spines. Striped bass has long second anal spine.

Most like: White perch and striped bass. **Superficially like:** Drum, largemouth blackbass, walleye, sauger.

Coloration: Dorsally deep bluish-silvery with many greenish, golden, silvery and steel-blue reflections. Sides more silvery-white with less gold, green, and blue. Ventrally milk-white. Dorsal, caudal, and anal fins silvery near their bases, more bluish distally. Pectoral and pelvic fins white, milk-white or transparent. For description of dark, horizontal lines, *see* under Characters. *Breeding adults*—Brilliant blue-white when taken from clear waters; more olivaceous when taken from turbid waters.

Lengths and **weights:** *Young* of year in Oct., 2.0″–8.0″ (5.1–20 cm) long. Around **1** year, 5.0″–10.5″ (13–26.7 cm) long, weight 0.5 oz–11 oz

* *Lepibema chrysops* or *Roccus chrysops* until recently.

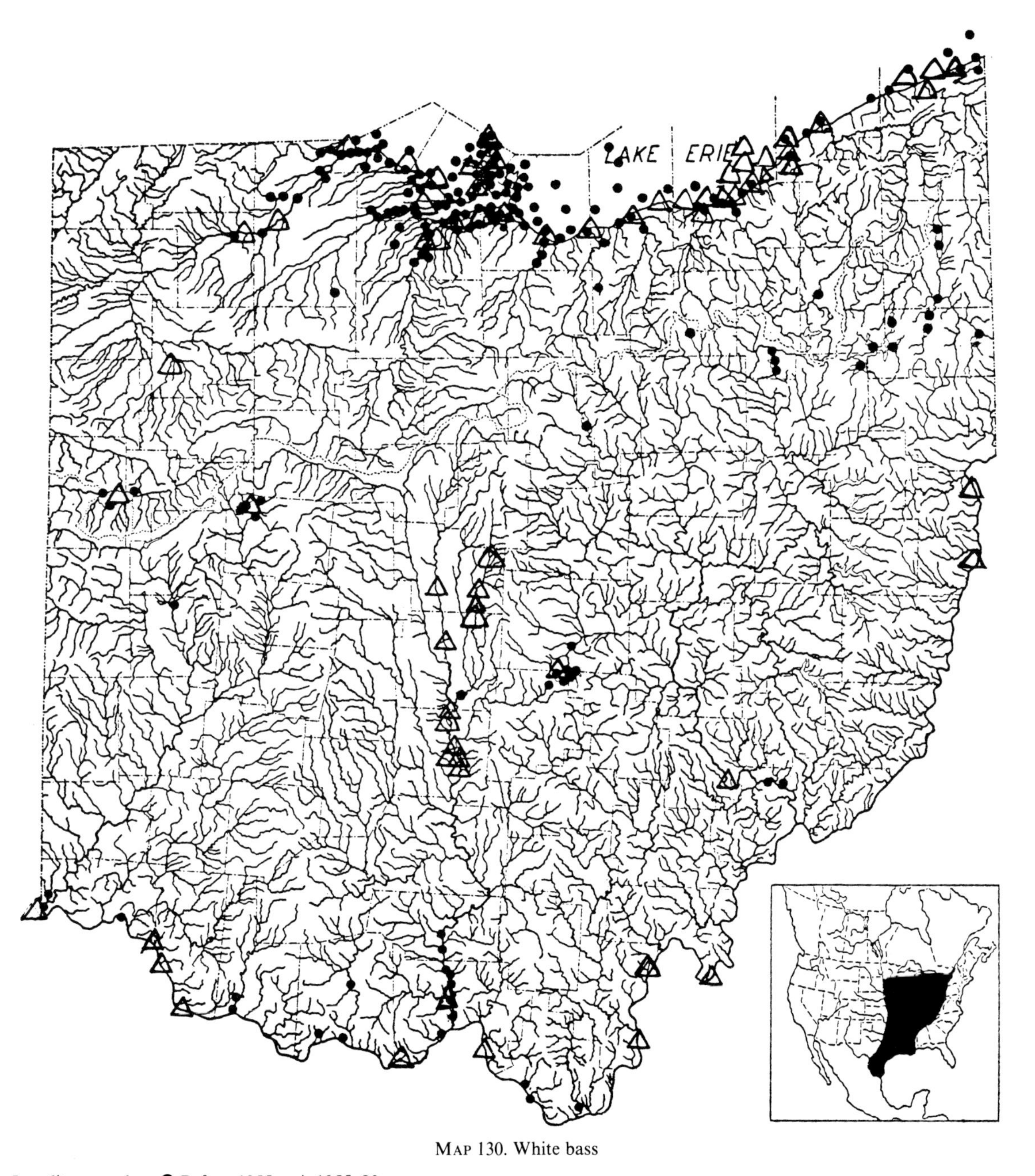

MAP 130. White bass

Locality records. ● Before 1955. △ 1955–80.
Insert: Introduced successfully in several localities within and outside of its range; limits of range in Mexico indefinite.

(14–312 g). *Adults*, usually 7.5″–16.0″ (19–40.6 cm) long, weight 3 oz–2 lbs, 3 oz (85 g–1.0 kg). Largest specimen, 17.8″ (45.2 cm) long, weight 3 lbs 1 oz (1.4 kg). Largest fishes were taken from the large, inland impoundments. For Lake Erie growth rates, *see* Van Oosten (1942:307–34); for Buckeye Lake, *see* Roach (1943:263–66).

Hybridizes: with striped bass, *see* Kerby, 1979.

Distribution and Habitat

Ohio Distribution—In 1830, when commercial fishing in western Lake Erie was conducted chiefly with the use of shore seines and hook and line, "the greater part of the catch consisted of white bass [*Morone chrysops*] which were then most numerous" (Klippart, 1877A:31). Between 1853–60 white bass appeared in "enormous schools" about the Bass Islands and were taken in the newly introduced trap nets, selling "for 6 1/4 cents a hundred pounds" and "As many as 10 tons of them have been thrown overboard at one time for lack of a purchaser." (Smith and Snell, 1891:263); also that "Since 1860 they have been growing scarcer every year, and only a few are now [1885] taken." This decrease in the catch from 1860 to 1885 could have been an actual decrease in abundance, or could have been caused entirely or in part by this exceedingly migratory and schooling species moving out of the range of Ohio fishermen, or into Ohio waters where these fishermen did not fish. Fluctuations in abundance have been noted in recent years, either in only a part of the lake or throughout Lake Erie. Cameron King and his father, fishing Maumee Bay between 1900–41, found white bass to be almost entirely absent during some years, whereas during others, as in 1920–22, they were very abundant. Between 1939–51 I found a somewhat similar condition in the immediate vicinity of South Bass Island. Van Oosten (1942:308) found no progressive decline in abundance of white bass during the 1913–38 period, but did find a yearly fluctuation ranging from 121,124 lbs (54940.9 kg) (1927) to 840,671 lbs (381,322 kg) (1921) a year, and a tendency toward definite cycles. During the 1939–49 period the average commercial catch in Ohio waters of Lake Erie has been 549,510 lbs (249,254 kg) (Shafer, 1950:1).

For the Ohio River drainage, Kirtland (1850J:53) records that "Forty years since [1810] it was very common in the Mahoning River where we saw a few as late as the year 1840." Later writers considered the white bass as having been uncommon in the Ohio drainage (Jordan, 1882:955) or have made no statement concerning it. Between 1920–50 the species was quite uncommon in the Ohio River, and was found only as far upstream as Lawrence County. It was uncommon also in the lower Scioto River, and was almost absent in the other streams such as the ponded Muskingum River. Old rivermen insisted that this migratory fish decreased greatly in abundance immediately following installation of dams in the Ohio River or tributary waters.

Since 1900 many thousands of adults have been introduced or planted in the larger inland impoundments and streams by the conservation department. In some of the clearer and larger impoundments, the species has become established and has offered good fishing, such as in Indian, Buckeye, and several lakes in northeastern Ohio. But in the more turbid lakes, such as St. Marys (Clark, 1942:142) and Loramie, repeated introductions have failed to establish the species. It is still too soon to predict whether it will become permanently established in the large impoundments of the Muskingum Conservancy District. Occasional strays are taken in inland streams such as Big Walnut Creek in Franklin County.

Habitat—The white bass was most abundant in Lake Erie and the larger inland impoundments where the water was clear, bottom firm and not of a soft, flocculent silt, water depth of less than 30′ (9.1 m), and where there was an abundance of small fishes, including its own young, upon which it fed. During warm evenings many individuals moved into water only a few feet in depth in pursuit of small fishes, or to feed upon insects, such as the larger species of mayflies. In spring the white bass migrated up tributary streams, sometimes in large numbers.

Years 1955–80—The annual poundage of white bass brought into the Ohio ports of Lake Erie during the 1955–80 period fluctuated as greatly as it had previously, and it is too soon to predict a decline in abundance such as has occurred with other commercially important food fishes since 1950 (White et al., 1975:100). The years 1955–56 were the last ones in which more than 2 million lbs (907,184.7 kg) were brought into Ohio ports, although as late as 1961 there were reported 1,905,362 lbs (864,257.7 kg). In 1968 only 673,579 lbs (305,530 kg) were reported (Anonymous B, 1968:8). Recently I have not seen the frequent, large catches taken by the sport fishermen about the Bass Islands as I did before 1955 when a fisherman often captured 50–125 white basses in an evening. Spring runs continue in the lower portions of such tributary streams as the Sandusky River (Van Meter and Trautman, 1970:75).

Fair- to moderate-sized populations existed in some of the older and larger reservoirs and impoundments of the Ohio River drainage in which they had been initially or repeatedly introduced. Plantings were made in the newly created impoundments. In the larger inland streams the number of individuals was low, as it had been previously. The recent catches by fishermen in the Scioto River and its tributaries in Franklin and Pickaway counties are assumed to be the result of introductions into such upstream reservoirs as Delaware and Hoover. During the 1968–69 investigations H. Ronald Preston reported the species present in the vicinities of dams in the Ohio River between Meigs and Clermont counties.

For age and growth in Lake Erie *see* Van Oosten (1942:307–34); for reproduction and a good bibliography *see* Riggs (1955:87–110). Much research has been conducted and published recently relative to this important sport and commercial species.

P. J. P. Whitehead and A. C. Wheeler (1966:23) demonstrated that under the rules of nomenclature *Morone* Mitchill has priority over *Roccus* Mitchill. Therefore, *Roccus chrysops* becomes *Morone chrysops* (Rafinesque).

WHITE PERCH

*Morone americana** (Gmelin)

Fig. 131

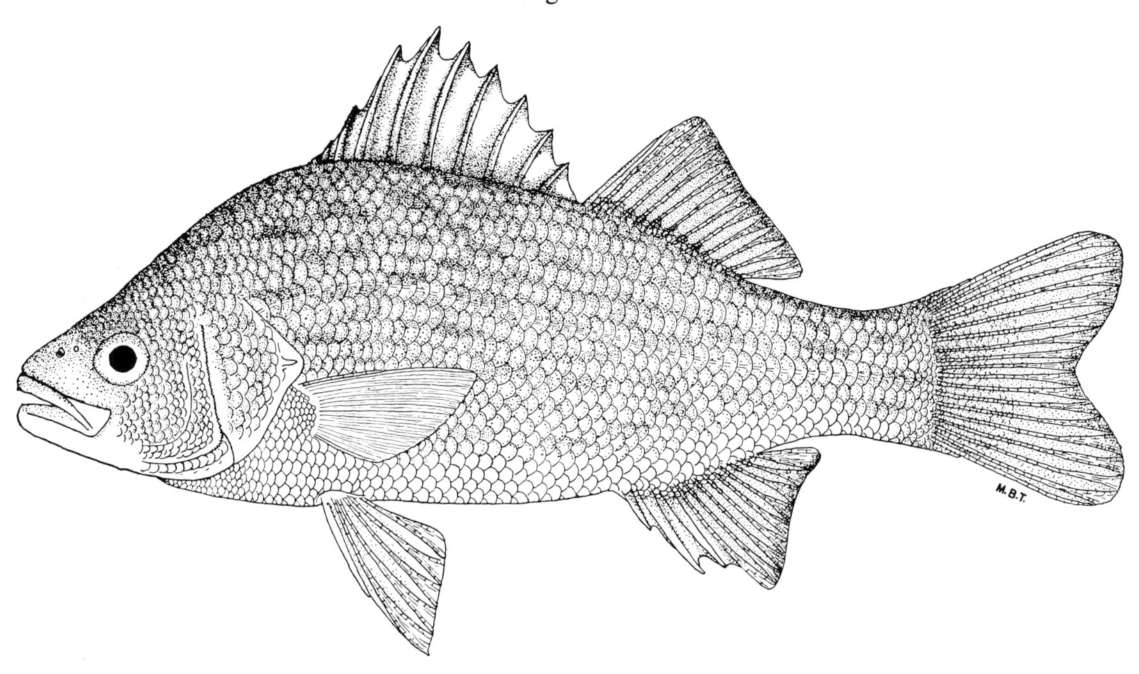

Lake Erie, 8.5 miles west of Erie, Pa.

June 20, 1953.
Adult female.

178 mm SL, 8.5" (22 cm) TL.
OSUM 9322.

Identification

Characters: *General* and *generic*—*See* white bass. *Specific*—Anal rays usually 8 or 9, rarely 10; the three anal spines not graduated, the first much shorter than the long, stout second spine; jaws equal or the upper jaw slightly shorter; body depth normally greatest immediately before spinous dorsal fin, giving fish the body shape of the freshwater drum; mouth rather small, the length of upper jaw usually going 2.7–3.5 times in head length; scales in lateral series 47–51 (extremes, 45–53); no distinct longitudinal rows along sides and back; coloration more olivaceous-silvery. Introduced.

Differs: White bass has more than 10 anal rays, a larger mouth and has dark, longitudinal lines on sides. Striped bass has back not arched; usually more than 10 anal spines.

Most like: White bass. **Superficially like:** Drum, largemouth blackbass, walleye, sauger.

Coloration: Dorsally slaty-olivaceous with many greenish, bluish and silvery reflections. Sides lighter, more silvery. Ventrally milk-white and silvery. The scales of the dorsal half of body bordered with darker, leaving the centers of many scales light, and giving some individuals a mottled appearance or of having light wavy rows on the back. No distinct dark, longitudinal lines on sides as in the white bass, although faint lines are present in some specimens. Dorsal, caudal, and anal fins silvery near their bases, darker distally; pectorals and pelvics mostly silvery-white or transparent.

Lengths and **weights:** 8.3"–9.8" (21–25 cm) long, weights 4–7 oz (113–198 g) (9 specimens). Adults normally 7.0"–12.0" (18–30.5 cm) long, weight 4 oz–1 lb, 4 oz (113g–0.6 kg).

Hybridizes: with striped bass, *see* Kerby, 1979.

---

* *Roccus americanus* until recently.

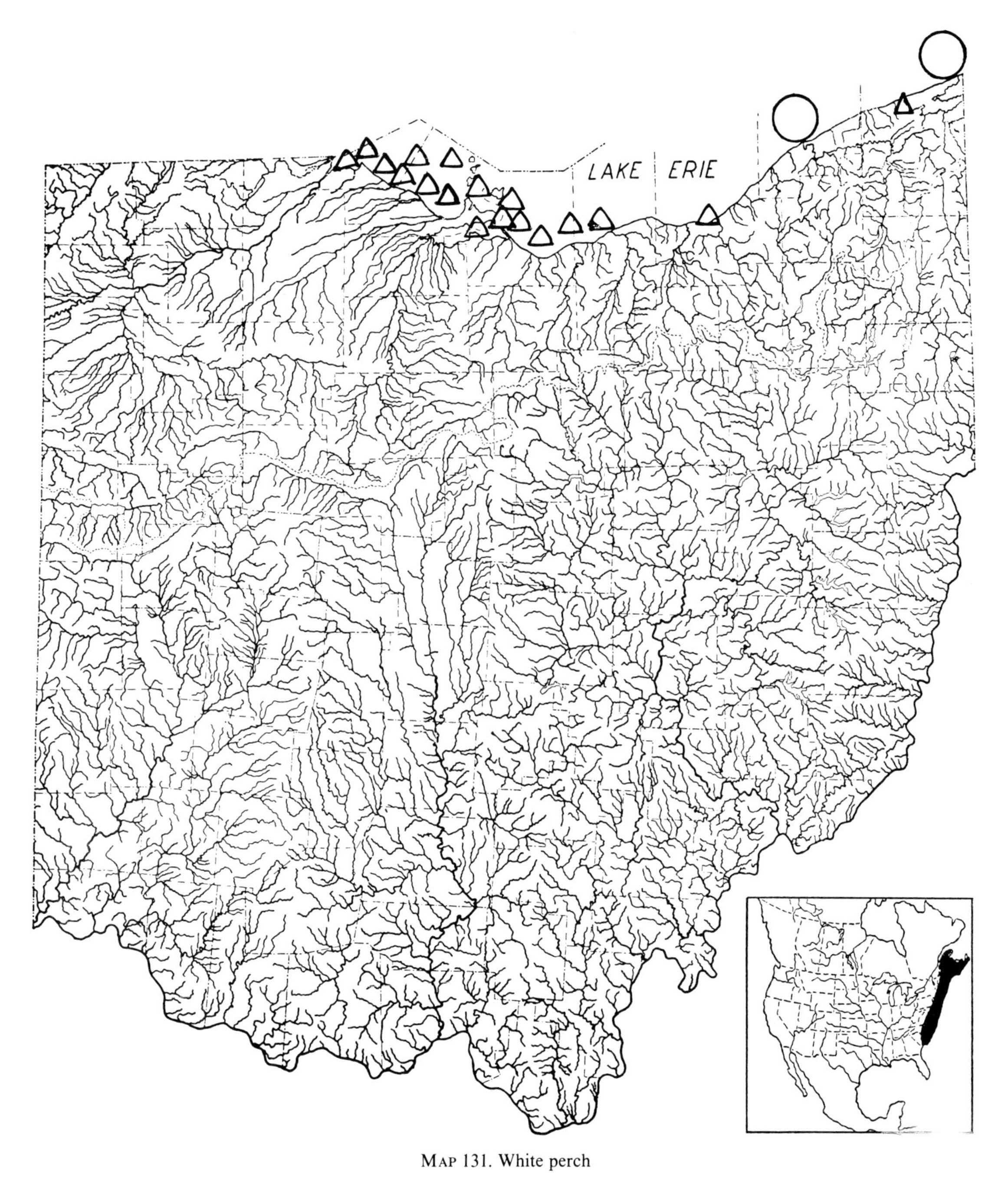

MAP 131. White perch

○ Reported captures before 1955. △ 1955–80.
Insert: Introduced widely within and outside of its original range.

Distribution and Habitat

Ohio Distribution—Originally the white perch ranged along the Atlantic coast from Nova Scotia to Georgia, occurring in salt and brackish waters and in the fresh waters in the lower section of the streams. It was landlocked in many lakes. The species was not present originally in the Great Lakes above Niagara Falls.

On both June 20 and 27, 1953, a single white perch was captured in a pound-net, set in Lake Erie 8.5 miles (14 km) west of Erie, Pa., and about 25 miles (40 km) east of the Ohio-Pennsylvania line. Fred Ralph captured these fishes and gave them to Alfred Larsen, fishery biologist for Pennsylvania. Ralph stated that at the time these specimens were caught, large numbers of white bass were also taken. Larsen sent the June 20 specimen to me for verification, and it is now No. 9322 in the OSUM collection. *See* Larsen (Copeia, 1954:154).

Ohio Fish Management Supervisor Robert Cummins, Jr., interviewed many commercial fishermen during the summer of 1953. One of these fishermen reported that on August 25, he captured a white perch in the Ohio waters of Lake Erie near Conneaut. Another fisherman, Joe Dominish, of Fairport, told Cummins that in early August he saw a white perch in his fish house which had been captured in Lake Erie near Fairport. Cummins saw neither of these fishes.

It is not known when or where the white perch entered, or was introduced into, the Lake Erie drainage. The species often becomes abundant after introduction into waters where it is not native. It is probable that it may become established throughout Lake Erie, hence its inclusion here, despite the lack of specimens from Ohio waters.

Habitat—Throughout its original range the white perch inhabits in greatest numbers the clearer waters of low and base gradients. It is often abundant in small lakes.

Years 1955–80—There appears to be no authentic record of the capture of a white perch in Lake Erie waters between 1953 and 1973, although there were a few unconfirmed reports (Van Meter and Trautman, 1970:75). In March of 1973 Russell L. Scholl, Supervisor of Lake Erie Fisheries Research, ODNR, Division of Wildlife, informed me that on March 13, 1973, a white perch was captured by J. Schwartz, 5 miles (8 km) northeast of Port Clinton, Ottawa County. This specimen was presented to the Ohio State University Museum of Zoology and catalogued OSUM 21840. Another was taken March 21, 1974, in a trap net set in Lake Erie near the Toledo water intake, Lucas County, by Fritz Komorny, and two more in the fall, one near Lorain, the other near South Bass Island.

In May, 1974 Stephen J. Nepszy of the Ontario Ministry of Natural Resources informed Scholl that none had been reported to date from the Canadian waters of Lake Erie. Considerable publicity was given to white perch in Lake Erie during the winter of 1974–75, with the result that a few dozen records were confirmed, both in Canadian and United States waters of Lake Erie. It is unknown at present how abundant the species eventually will become in Lake Erie, and what effect, adverse or otherwise, it will have on its close relative, the white bass.

For establishment of the white perch in Lake Erie, *see* Busch, Davies and Nepszy, 1977.

WHITE CRAPPIE

Pomoxis annularis Rafinesque

Fig. 132

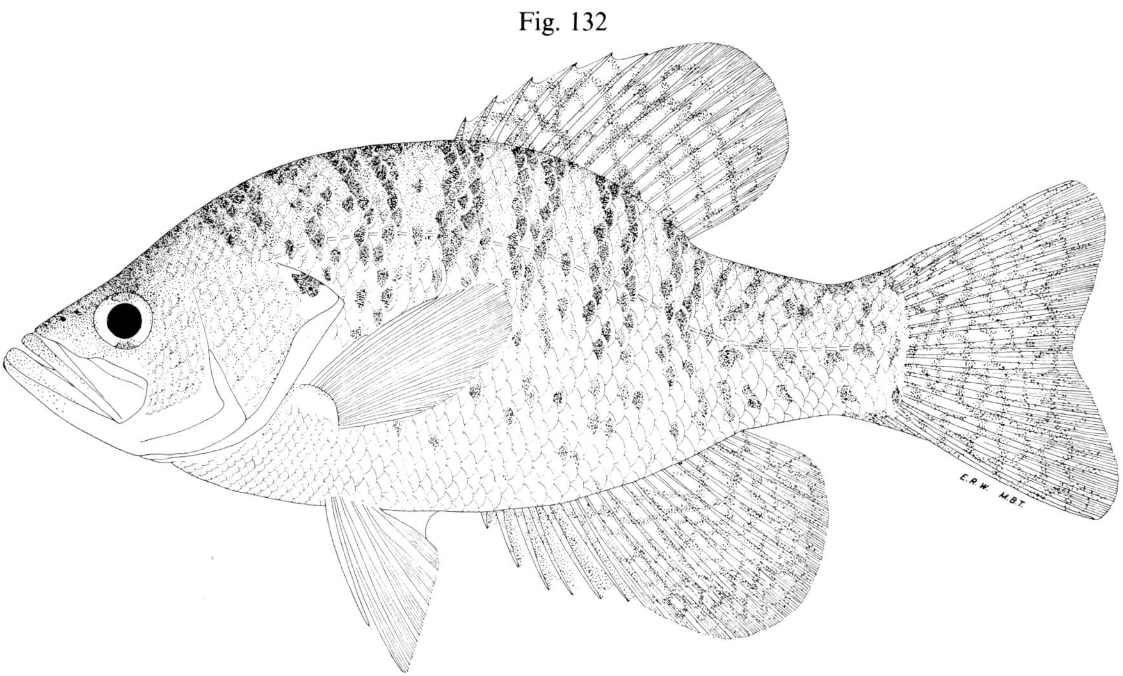

Meander Lake, Mahoning River Drainage, Trumbull County, O.

May 8, 1942. 203 mm SL, 10.5″ (26.7 cm) TL.
Adult male. OSUM 4886.

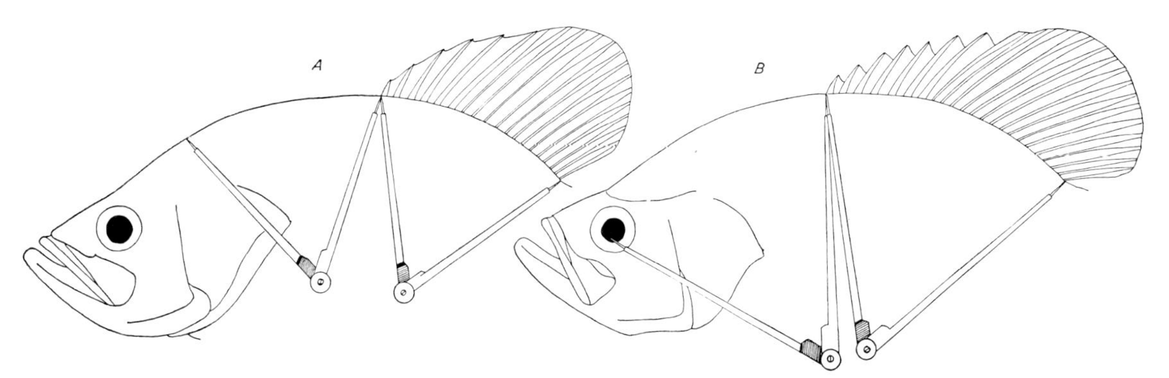

Fig. A: partial outline of white crappie; when the length of the dorsal base is projected forward with the aid of dividers, the dorsal base length extends from the dorsal origin to above the opercle, seldom farther forward.

Fig. B: partial outline of black crappie; when length of dorsal base is projected forward with the aid of dividers, the dorsal base length extends from the dorsal origin to above the eye or cheek.

Identification

Characters: *General*—Spinous and soft dorsals broadly joined together, a distinct notch between the two fins in a few species, notably the largemouth blackbass. *No* spine on posterior angle of opercle as in the white bass. *Generic*—Anal spines normally 6, extremes 5–7. Dorsal spines normally fewer than 10. Branchiostegal rays 7. Preopercle finely serrated. Usually more than 28 long, slender gill rakers on each first arch. *Specific*—Dorsal spines usually 5–6, extremes 4–8. Length of dorsal fin base, when pro-

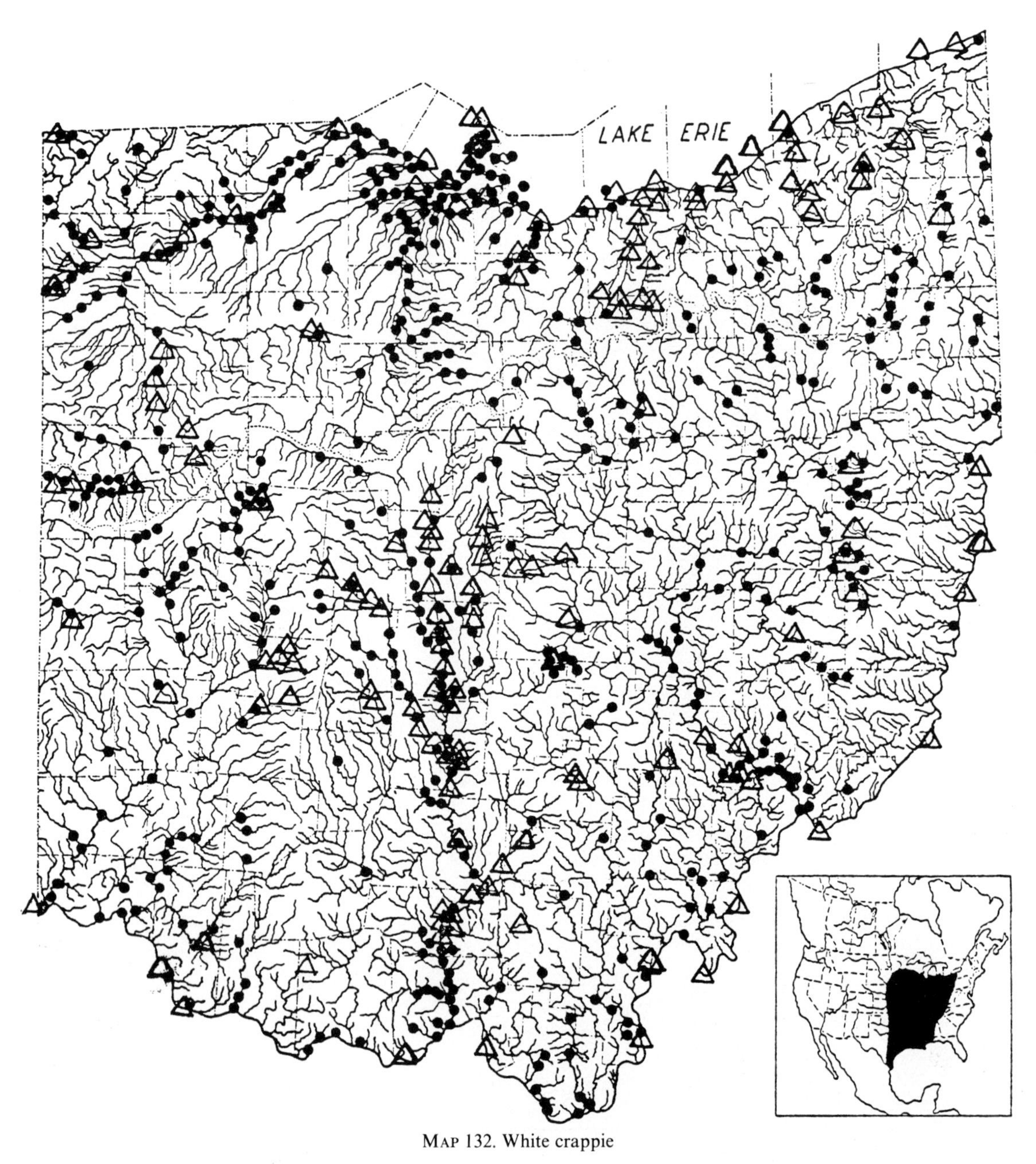

MAP 132. White crappie

Locality records. ● Before 1955. △ 1955–80.
Insert: Widely and successfully introduced into many states outside of this range.

jected forward with dividers, extends from dorsal origin to above opercle, seldomly farther forward, fig. A. Chain-like double bands on back and sides are usually rather regular in shape and spacing. Mouth moderately oblique. Body depth contained usually more than 2.4 times in standard length, except in unusually deep-bodied, large adults.

Differs: Black crappie normally has 7–8 dorsal spines; a longer dorsal fin base. Rockbass has more dorsal spines. Blackbasses, other sunfishes, and white bass have only three anal spines. Drum and walleye have only two anal spines.

Most like: Black crappie and rockbass. **Superficially like:** White bass, blackbasses, other sunfishes.

Coloration: Dorsally olive- or brownish-green

with many bluish, greenish and silvery reflections. Sides lighter and more silvery-white. Ventrally silvery- and milk-white. Back and sides have 5–10 chainlike, double bands of dusky or black; these bands most distinct on the upper sides, absent on ventral surface of body. Dorsal, caudal and anal fins usually heavily vermiculated with darker and/or with oscelli; these markings most prominent on adults, least conspicuous or absent in small young. Considerable variation in amount and intensity of spotting; smallest young often faintly spotted; large adults from turbid waters may almost lack spotting, especially the females; large adults, especially breeding males, from clear and/or vegetated waters may be as heavily spotted as are male black crappies. *Male*—Usually darker and more boldly marked than is female.

Lengths and **weights:** *Young* of year in Oct., 1.0″–3.8″ (2.5–9.7 cm) long. About **1** year, 1.2″–5.0″ (3.0–13 cm). *Adults*, usually 5.0″–14.0″ (13–35.6 cm), long, weight 1 oz–2 lbs. Largest specimen 20.0″ long, weight slightly over 3 lbs (1.4 kg). Dwarfing occurs in over-populated impoundments, or in submarginal habitats; adults in such waters may spawn when only 4.3″ (11 cm) in length, and they seldom attain a length greater than 6.5″ (17 cm). Between 1906–18 white crappies over 2 lbs (0.9 kg) in weight were rather common in Indian Lake.

Hybridizes: with black crappie.

Distribution and Habitat

Ohio Distribution—The white crappie was well distributed over Ohio during the 1920–50 period. Large populations were present in most of the larger ponds, lakes, and impoundments, where it was one of the most sought after of pan fishes by the fishermen. Moderate populations occurred in the bays, marshes, harbors, and shallow waters of western Lake Erie; smaller populations were present in the larger and more sluggish streams and their overflow ponds and oxbows, particularly the Scioto and Maumee rivers. The white crappie was the least numerous in northeastern Ohio, and was rare or absent in high-gradient streams. During the 1920–50 period the state conservation department trapped hundreds of thousands of adults from such lakes as Rockwell and Meander, liberating them throughout Ohio waters. This mass liberation has resulted in stocking most of the suitable waters, and was particularly successful in populating the recently built impoundments.

Less is known concerning its distribution and abundance between 1810–75, because then the white and black crappies were not separated as to species. Kirtland (1850N:69, as *Centrarchus hexacanthus*) obviously referring to both, mentioned their presence in the Ohio and Great Miami rivers, in Sandusky Bay, and Lake Erie at Cleveland. Kirtland also leaves the impression that crappies may have increased in abundance between 1810–50, for he wrote that "Since Pike, Pickerel, and Muskellunge, have become less abundant than formerly, this species has rapidly multiplied." Former commercial fishermen agree that at Buckeye Lake before 1880, crappies were not as abundant as were the bluegills and pumpkinseeds, but that crappies later became far more abundant, especially after 1920. This appears likewise true for Lake St. Marys, where since 1920 the crappies have far outnumbered all other sunfishes.

After 1875 the white and black crappies were usually separated in the literature. Klippart (1877:78) states that before 1877 the white crappie was taken in Lake Erie and abounded in the Scioto River and many inland lakes, and suggests that its Ohio range had been considerably extended through its utilization of the Ohio canals as migrating routes. It was the prevailing opinion in the literature that before 1900 the white crappie was less abundant in northern Ohio than was the black crappie (Jordan, 1882:924–25).

Habitat—The white crappie was one of the most abundant of pan fishes during the survey, apparently because of its tolerance to a wide variety of habitats, and especially toward turbidity and siltation. The most favorable habitats were ponds, lakes, or impoundments of more than 5 acres (2 hectares) in extent, and especially in those where competition and predation from bluegills, pumpkinseeds, and largemouth blackbass was slight. Smaller populations were present in the sluggish pools of moderate or large sized streams of base or low gradients and their larger overflow ponds, bayous, and oxbows. This crappie was found over both soft and hard bottoms, even where turbidity was severe and siltation rapid. It frequently inhabited areas containing aquatic vegetation and congregated about submerged brush, logs, stumps, and tree roots. The white crappie was outnumbered by the black crappie in cool, clear waters having an abundance of submerged aquatics and a clean, hard bottom.

Years 1955–80—As it had been previous to 1950, the white crappie continued to be common or abundant in the shallows, bays and marshes of the western third of Lake Erie and in some of its tributaries, such as the Maumee River (Van Meter and Trautman, 1970:75). It was less numerous in the eastern two-thirds of the state and in northeastern Ohio (White et al., 1975:102).

It was an almost universally distributed species in the flowing and static waters of inland Ohio and was undoubtedly present, usually in considerable numbers, in every county. As in previous years, new impoundments were stocked and older ones restocked. There was a considerable increase in the number of farm ponds in which crappies were liberated. An occasional newly constructed reservoir or large farm pond produced a few very large specimens, 20″ (51 cm) in length and 3 lbs (1.4 kg) in weight, shortly following the initial stockings. Some of the larger fishes were hybrids between the white and black crappies.

Some indication of the large size of the population existing before 1900 in Buckeye Lake (Licking Reservoir) may be obtained from Smith (1898B: 510), he stating "that a total of 1500 lbs. of Crappy or Strawberry Bass were taken commercially in 1894, and that a total of 32,300 lbs. were taken during the same year in Indian Lake." It is questionable whether at present there exists in Buckeye Lake a total of 1,500 lbs (680.4 kg) of crappies of both species or a total of 32,300 lbs (14,651 kg) in Indian Lake.

During the 1955–80 period the population of the white crappie at Buckeye Lake notably decreased, as did the numbers of fishermen fishing for them. Apparently, the gradual changing of the lake bottom from soft muck to clay hardpan, the increase in amount of speedboat activity and the changing environment of the lake in general adversely affected this species. Of all adverse factors, however, the drastic decrease in the numbers of gizzard shad (*see* p. 203), the principle food of the crappies, was the major contributing factor. For life history of the species at Buckeye Lake *see* Morgan (1954:113–44).

The investigations of the entire Ohio River between 1957–59 by personnel of the University of Louisville (ORSANCO, 1962:151), and those of H. Ronald Preston of 1968–70 indicate that the white crappie was a very numerous species in the Ohio River, as did those during 1976–78 of Margulies and Burch.

BLACK CRAPPIE

Pomoxis nigromaculatus (Lesueur)

Fig. 133

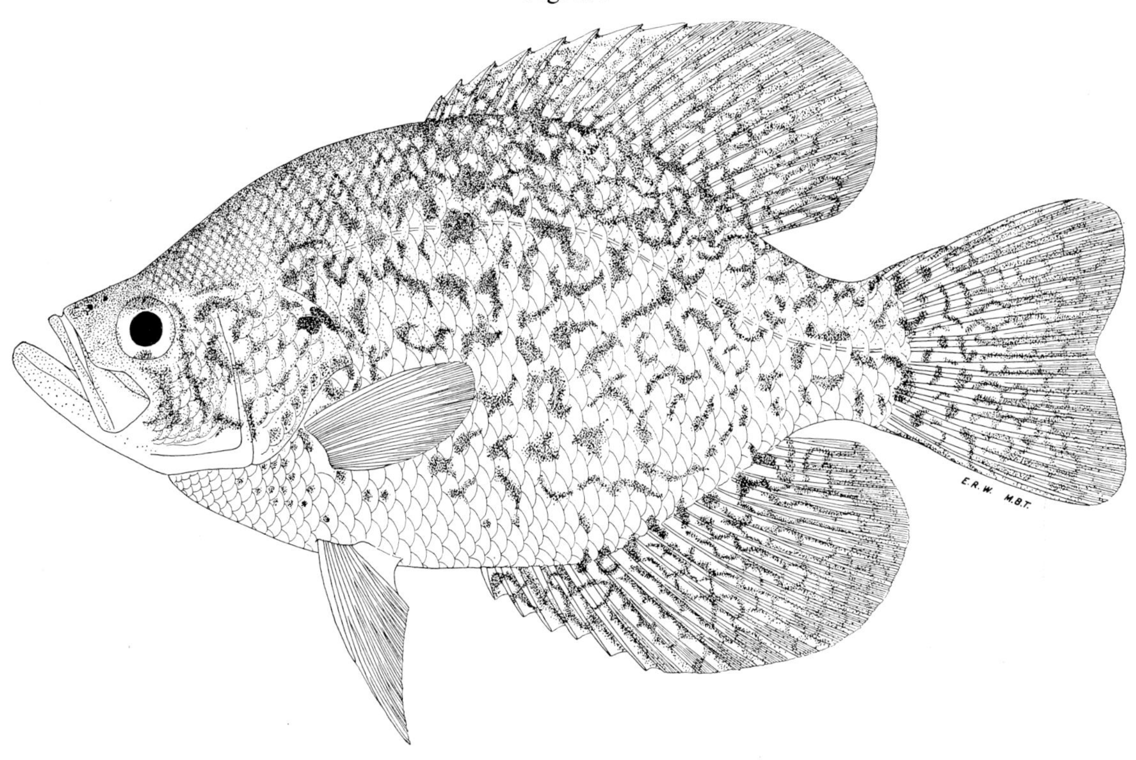

Lake Erie, Ottawa County, O.

Dec. 6–7, 1939.
Male.

190 mm SL, 9.8″ (25 cm) TL.
OSUM 1127.

Identification

Characters: *General* and *generic—See* white crappie. *Specific*—Dorsal spines 7–8, extremes 6–9. Length of dorsal fin base, when projected forward with dividers, extends from dorsal origin to above eye or cheek; *see* fig. 132 B. Body profusely speckled and mottled; occasional specimens have broken and highly irregular chainlike, double bands. Mouth very oblique. Body depth contained less than 2.4 times in standard length, except in smallest young.

Differs: White crappie has fewer dorsal spines; shorter dorsal base; chainlike bands on sides. *See* white crappie.

Most like: White crappie and rockbass. **Superficially like:** Blackbasses, other sunfishes, white bass.

Coloration: Dorsally golden- or olive-green with a heavy overcast of bluish and silvery. Sides more silvery. Ventrally silvery- and milk-white. Head, back, and sides boldly mottled with blue- or slate-black blotches, the markings rather evenly distributed and seldom tending to form chainlike bands on the sides. Vertical fins strikingly vermiculated, some of the vermiculations form light colored oscelli. Paired fins transparent, silvery or whitish. Intensity of mottling varies greatly; fishes from turbid waters are usually rather faintly marked; those from clear, vegetated waters are brilliantly and strikingly colored. In some localities, some crappies have a broad brown or brown-black band, a little less wide than pupil diameter, extending from dorsal fin origin forward along the dorsal ridge to tip of upper lip, continuing across the center of the chin backward on ventral surface of head to throat. Fishes with such bands have been found

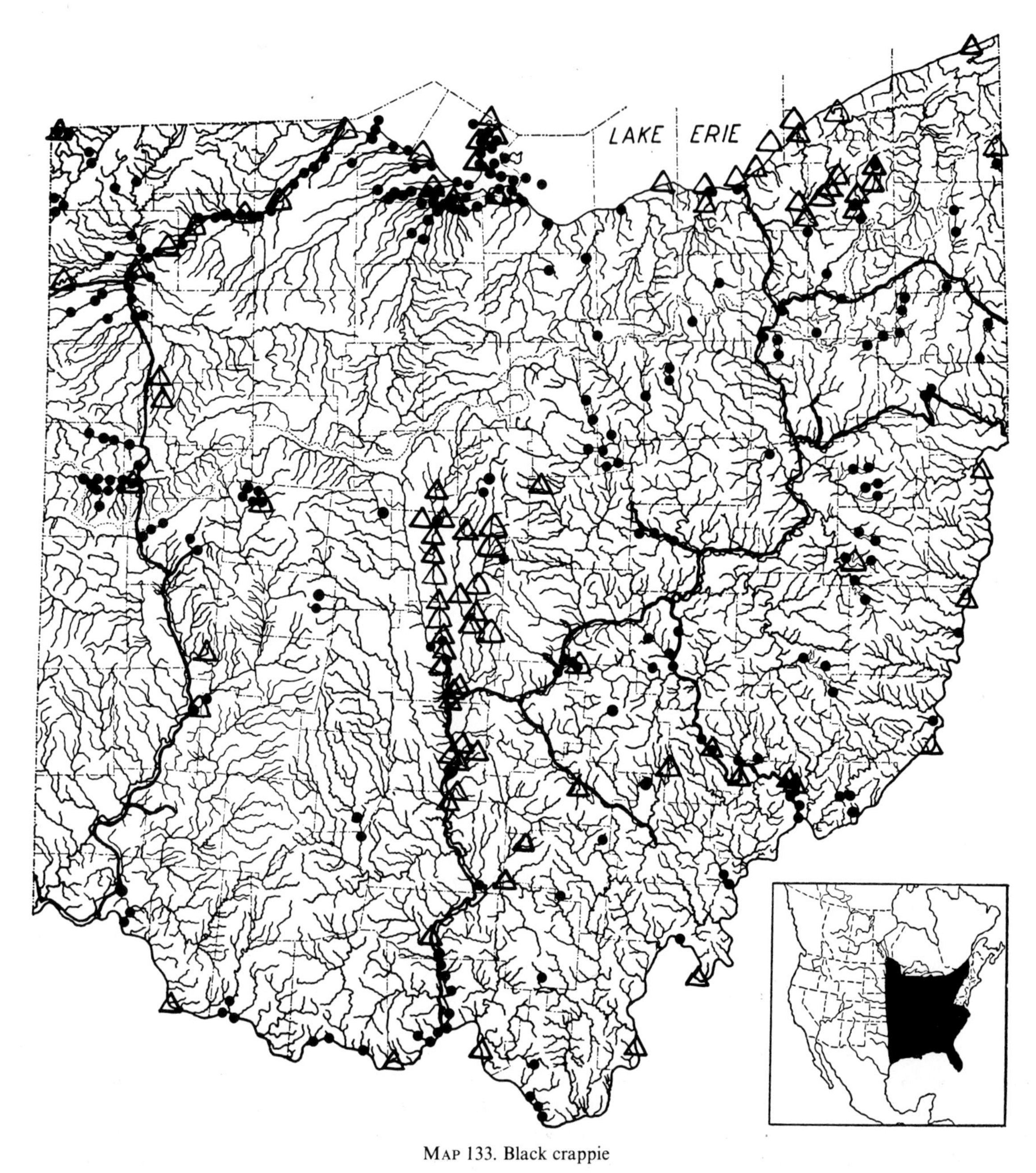

Map 133. Black crappie

Locality records. ● Before 1955. △ 1955–80. ~ Canals.
Insert: Widely and successfully introduced into many states outside of this range.

throughout Ohio, most of them from the clearer, more vegetated waters.

Lengths and **weights:** *Young* of year in Oct., 1.0″–3.0″ (2.5–7.6 cm) long. Around **1** year, 1.1″—–4.5″ (2.8–11 cm) long. *Adults*, usually 5.0″–12.0″ (13–30.5 cm) long, weight 1 oz–1 lb, 12 oz (28 g–0.8 kg). Largest specimen, 18.0″ (46 cm) long, weight about 3 lbs, 8 oz (1.6 kg).

Hybridizes: naturally in Ohio with white crappie; naturally or artificially produced elsewhere with rockbass, largemouth blackbass, warmouth and bluegill sunfishes (Hester, 1970:102).

Distribution and Habitat

Ohio Distribution—The black crappie in the 1920–55 period was not as generally distributed over Ohio as was the white crappie, nor were its populations usually as large; this was especially true of the Ohio drainage. Moderate populations were present in some of the ponds, lakes, and impoundments of inland Ohio, in the bays, marshes, harbors, and shallows of western Lake Erie, and particularly in those waters which were clear and/or contained much aquatic vegetation. Small, more or, less permanent populations inhabited some portions of the larger streams and their overflow ponds, principally the Muskingum River. The species was absent from the most turbid waters and high-gradient streams.

During the 1920–55 period the state conservation department liberated many thousands of adult black crappies throughout Ohio waters, which were taken from such lakes as Meander and Rockwell; despite this mass liberation, permanent populations were absent from many counties and stream systems. There were also notable decreases in abundance in such turbid streams as the Maumee. As an example: most of the collections made in this river between 1920–30 contained black crappies and it then was considered not uncommon; between 1930–38 fewer than half of the collections contained this species; after 1939 it was taken only in a few localities where there were extensive weed beds. In 1893, Kirsch (1895A:336) recorded the "Calico Bass" as present in the Maumee River at every locality investigated; however, since he failed to mention the white crappie, it is assumed that he probably included both species in this statement.

As with many other fish species, the black crappie displayed great fluctuations in abundance throughout the 1920–55 period. These fluctuations may have been of a cyclic nature. Fluctuations were especially great in such impoundments as Buckeye Lake, where during certain years few black crappies would be taken, whereas during others the species appeared to become half as abundant as was the very numerous white crappie. Similar fluctuations in abundance occurred in the Muskingum and Ohio rivers. Such fluctuations may have been an expression of greater survival during some years, which were the result of temporarily favorable conditions.

Less is known concerning the black crappie previous to 1920. From the few reliable literature references and absence of preserved specimens from southern Ohio, there is a possibility that before 1925 the black crappie may have been almost or entirely absent from the Ohio drainage. In the Lake Erie drainage before 1900, it probably occurred only in western Lake Erie, the inland glacial lakes, and the more weedy, base- or low-gradient portions of streams. Because of ecological conditions prevailing prior to 1850, it is possible that the black crappie may have been widespread and abundant in the former Black Swamp; also that after establishment of the Ohio canals, it may have become more widely distributed, especially in the Ohio drainage.

Habitat—Since 1920 fairly stable populations of black crappies occurred only in those ponds, lakes, and impoundments where competition from the white crappie was not too severe; where the waters were usually clear; and where there was an abundance of submerged aquatics, and sandy or muck bottoms. Small populations were present in the base- or low-gradient portions of large streams where similar conditions prevailed. The black crappie seemed to be much less tolerant to silty waters and the rapid depositing of yellow clays upon the bottoms than was the white crappie, and was more dependent upon a greater profusion of aquatic vegetation. (Wickliff and Trautman, 1947:28).

Years 1955–80—At some time in the past the black crappie was stocked in every county of Ohio, and during the present period undoubtedly occurred in every county at least in small numbers. However, it was only in the comparatively few localities where the waters were habitually clear, there was much aquatic vegetation and general conditions were stabilized that it outnumbered the white crappie. There was an overall decrease in the numbers inhabiting the marshes, bays and shallows of western Lake Erie and in the vicinity of Cleveland (White et al., 1975:102) as the general turbidity in the shallows increased and/or the aquatic vegetation decreased. In inland Ohio there likewise appeared to be an overall decrease in numbers coincidental with increasingly unfavorable environmental conditions. Buckeye Lake was an example, *see* under white crappie. The larger populations scattered about the state seemed to vary annually in size more than did the white crappie populations.

The 1957–59 investigations of the Ohio River (ORSANCO, 1962:151) indicated that the numbers of white crappies were three times greater than the numbers of blacks. H. Ronald Preston noted that in the 1968–69 investigations of that river black crappies were collected in seven of the nine stations bordering the state of Ohio. Margulies and Burch captured black crappies at all of their Ohio River stations during the years 1976 to 1978.

NORTHERN ROCKBASS

Ambloplites rupestris rupestris (Rafinesque)

Fig. 134

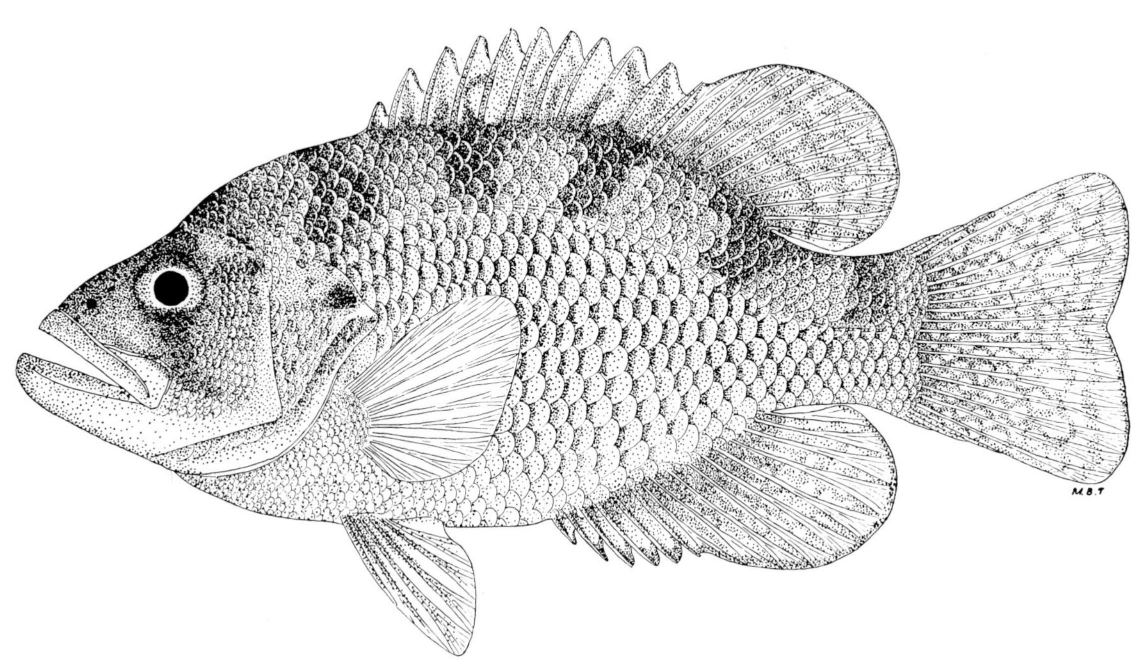

Huron River, Erie County, O.

July 28, 1940.
Female.

158 mm SL, 7.9″ (20 cm) TL.
OSUM 2849.

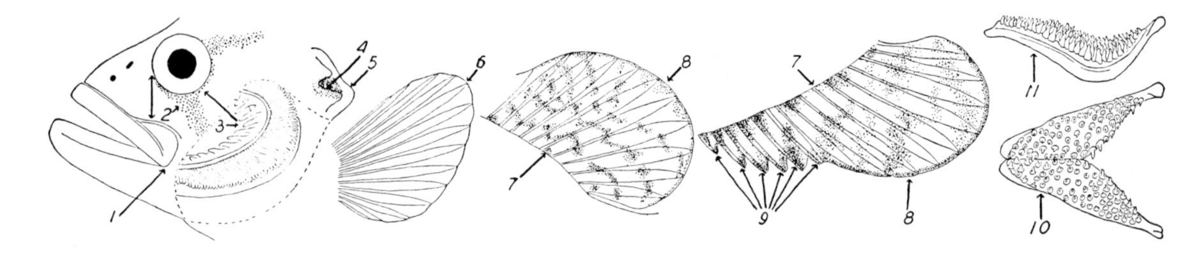

Nos. 1–11

(1) mouth large, posterior end of upper jaw extending well beyond anterior edge of eye.
(2) a faint tear-drop usually present.
(3) dotted portion of gill cover removed, exposing first gill arch; note moderately long gill rakers, the longest longer than diameter of pupil.
(4) opercle flap stiff, the unflexible opercle bone extending posteriorly to the opercle membrane.
(5) opercle flap membrane whitish, without red.
(6) tip of pectoral fin rounded.
(7) soft dorsal and anal fins mottled and vermiculated; usually oscelli are present.
(8) soft dorsal and anal fins margined with dusky or black.
(9) 6 anal spines (black and white crappies normally have the same number).
(10) dorsal view of lower pharyngeal arches showing their moderate width and many teeth.
(11) lateral view of a pharyngeal arch; note the pointed teeth.

Identification

Characters: *General*—*See* white crappie. *Generic*—Anal spines usually 6, extremes 5–7. Dorsal spines 11–12, extremes 10–13. Branchiostegal rays 6. Preopercle smooth or slightly roughened. Usually 15 or fewer gill rakers on each first arch. *Specific*—Dorsal fin base long, when projected forward with dividers, extends from dorsal origin to tip of upper jaw or beyond. *See* Nos. 1–11 for remaining characters.

Differs: Crappies have fewer dorsal spines. Blackbasses and other sunfishes, and white bass have fewer anal spines.

Superficially like: Crappies, warmouth and green sunfishes.

Coloration: Dorsally a dark slaty- or olive-green with bronze and coppery reflections; smallest young more dusky and slaty, sometimes lacking bronze and green. Usually 6, extremes 4–7, dark, broad saddle-bands extend across head and back; these end abruptly at the lateral line. Sides below lateral line lighter, most of the scales having a bold, dusky, basal spot; these spots tending to form longitudinal rows. Ventrally white or bronzy-white. Vertical fins heavily vermiculated with darker; vermiculations on the soft-rayed portions often forming oscelli. Margins of vertical fins usually edged with black. Paired fins olive-transparent, and unspotted. *Adults*—from clear waters boldly marked with black and bronze; those from very turbid waters often yellow-bronze and lacking the black markings. *Breeding males* are blackish.

Lengths and **weights:** *Young* of year in Oct., 0.8″–2.0″ (2.0–5.1 cm) long. Around **1** year 1.1″–3.5″ (2.8–8.9 cm). *Adults*, usually 4.3″–10.5″ (11–26.7 cm) long, weight 1 oz–14 oz (28–397g). Largest specimen, 14.7″ (37.3 cm) long, weight 1 lb 15 oz (0.9 kg). Rockbass confined to small streams are often dwarfed; fishes more than 9.0″ (23 cm) long uncommon in streams; fishes 9.0″–10.5″ (23–26.7 cm) not uncommon about the Bass Islands of Lake Erie.

Hybridizes: with warmouth, bluegill, naturally and artificially (Hester, 1970:102).

Distribution and Habitat

Ohio Distribution—Many preserved specimens, collected between 1853–1900, and many literature references, especially those of Kirtland (1844:239, as *Centrarchus aeneus*) and Osburn (1901:79), leave no doubt as to the general distribution and great abundance of the rockbass in Ohio waters prior to 1900. During that early period, the species was of commercial importance in the island region of western Lake Erie, where an enormous population was then present.

During some period since 1900, the rockbass presumably was present in all Ohio stream systems, including the smallest Ohio River tributaries. Its greatest abundance in inland Ohio was west of the Allegheny Front Escarpment, particularly in the limestone gravel and bedrock streams of the Scioto River drainage of central Ohio, the middle portions of the two Miami River systems, and the upper Sandusky River. The species was moderately abundant east of the escarpment only in the limestone streams of the glaciated portion, and in the unglaciated section only in glacial outlet streams which contained much outwash.

From statements by commercial fishermen and a few literature references, the rockbass seems to have decreased greatly in abundance in the island region of western Lake Erie since 1900, presumably for the same reasons as those which caused the decrease in abundance of the smallmouth blackbass (*see* that species). Between 1939–51 I noted a decrease in numbers of rockbass rising for mayflies on late June evenings in the bay in front of our home on South Bass Island. Before 1944 hundreds were seen rising during a 5-minute period, whereas since 1948 only dozens have been seen during the same interval.

In 1893, Kirsch (1895A:330) found the rockbass in the Maumee drainage to be widespread in distribution and "common." In 1930, it was still fairly well distributed and locally abundant, but since then has disappeared or decreased greatly in numbers in many sections, especially those where silt has succeeded in covering the sand and gravel, such as in many sections of the Maumee and St. Marys rivers. Throughout the 1920–55 period the rockbass was very uncommon in the Ohio River, except during the colder months; but according to many rivermen, they captured it in considerable numbers in their nets in the Ohio River before its impoundment. Only strays or very small populations occurred in inland lakes, and then only in those having the cleanest waters and sand or gravel bottoms.

Habitat—Although rockbass could be found in almost every Ohio stream during the 1920–55 survey, it attained its greatest abundance in the clearer

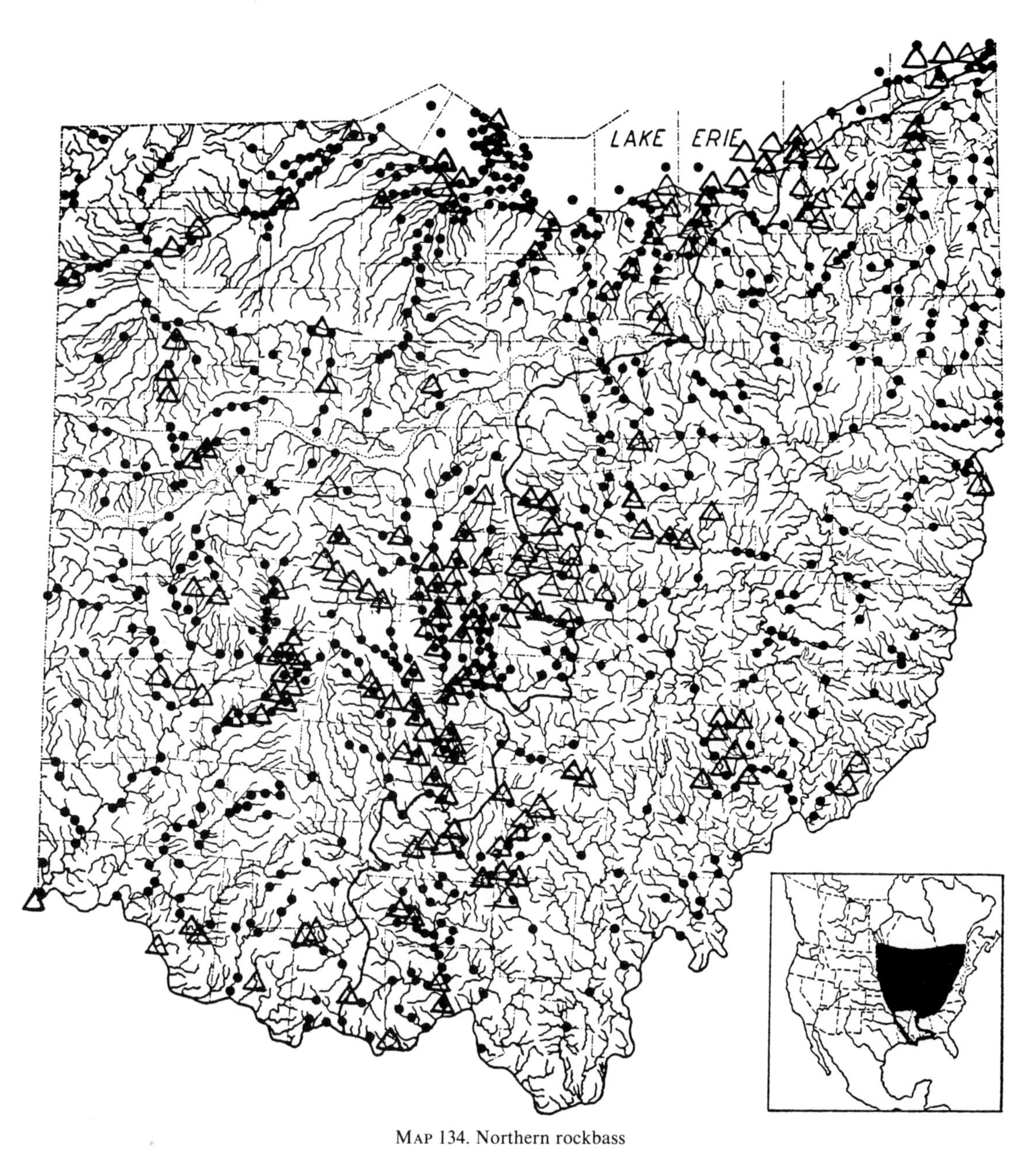

MAP 134. Northern rockbass

Locality records. ● Before 1955. △ 1955–80. ~ Allegheny Front Escarpment.

Insert: Black area represents range of this subspecies, outlined area of other subspecies; widely introduced into several states outside this range.

streams of moderate size and moderate gradients, where the bottom was of boulders, gravel and/or bedrock and usually of limestone, and where there were many "step-offs" and beds of water-willow.

In 1844, Kirtland (1844:239) noted astutely that "During winter it does not migrate." Kirtland lived at a time when migratory fishes, like the smallmouth blackbass, were observed migrating in great "shoals" upstream in spring, and when the absence of migrating rockbass would be most obvious. This non-migratory habit of the rockbass was abundantly observed during the 1920–55 period. It was found that the hibernating numbers in given stream sections were not materially different from the spring numbers, and that in these sections throughout the winter, the hibernating and hiding rockbass could be found under leaves and debris, among tree roots and water-willows, and about the rocks beside the "step-offs." However, immediately preceding and during the spawning season, adult females tended to congregate in certain pools.

Years 1955–80—In the Ohio waters of Lake Erie a fluctuating downward trend in the numbers of rockbasses was apparent throughout the 1955–80 period, continuing the decline which began before the present century (White et al., 1975:102–3). This decline was particularly noticeable in the shallow waters of the western third of the lake, where originally the population size had been very large. Flyfishing about South Bass Island after 1950 rarely produced more than 10 of these sunfishes per evening, whereas previous catches of 10 to 30 per evening were to be expected. Inability to catch the larger numbers may have been the result in part of the almost total disappearance of two large species of mayflies of the genus *Hexagenia* which before 1950 had been so immensely abundant and upon which the rockbasses fed heavily.

In inland waters the abundance and distribution of the species appear to remain relatively unchanged except for decreasing numbers in low-gradient, turbid, or highly enriched streams such as the Maumee and Portage rivers. In the latter the rockbass was taken infrequently after 1960, whereas in the 1930's it was one of the most numerous species taken by fishermen.

The Ohio River population was low, H. Ronald Preston stating that he had captured the species in the Ohio River opposite Ohio only adjacent to northern Jefferson County, as did Margulies and Burch during the years 1976 to 1978.

NORTHERN SMALLMOUTH BLACKBASS

Micropterus dolomieui dolomieui Lacepède

Fig. 135

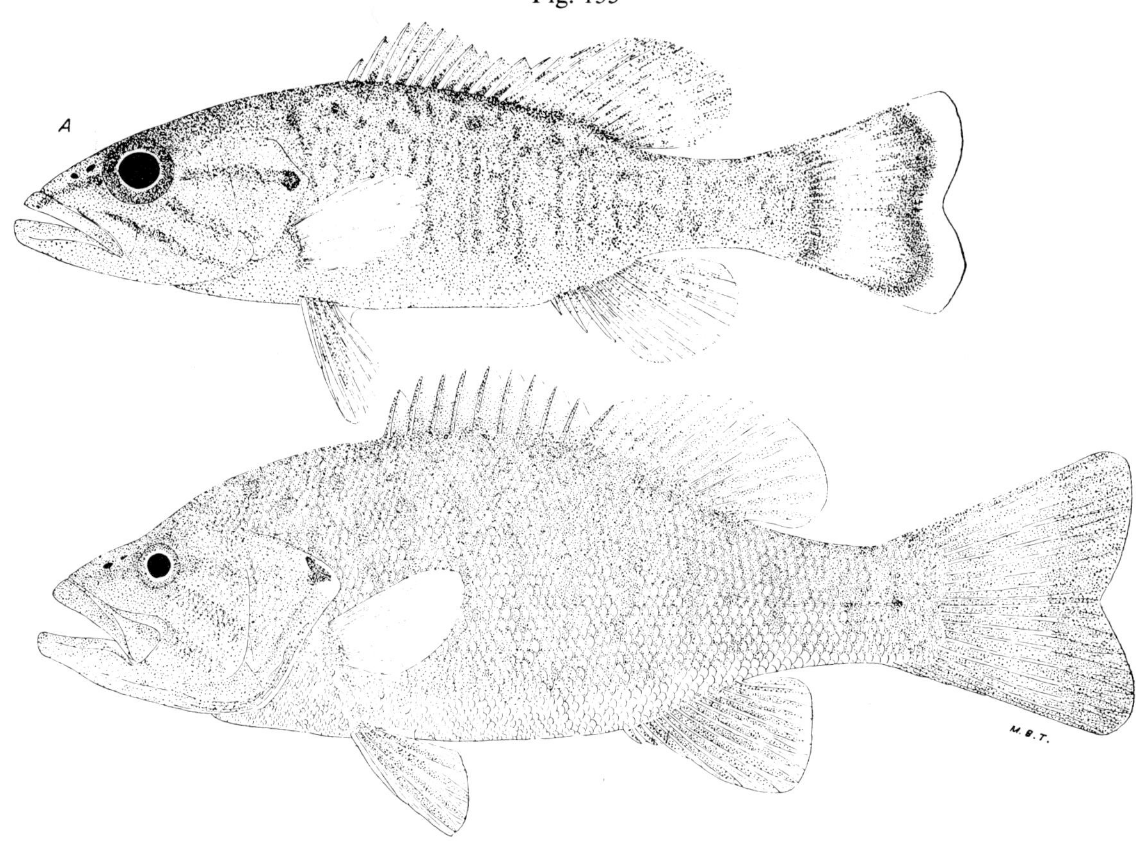

Upper fig.: Deer Creek, Madison County, O.

Nov. 8, 1940. 53 mm SL, 2.6″ (6.6 cm) TL.
Young of year. OSUM 2367.

In comparing the young (upper fig.) with the adult (lower fig.), note in young the larger eye; higher fins; tricolored tail; many faint bars on side.

Lower fig.: Lake Erie, Ottawa County, O.

Aug. 3, 1943. 213 mm SL, 10.0″ (25.4 cm) TL.
Adult. OSUM 6669.

Note in adult the faintly bicolored tail; shape of dorsal fin; size of mouth, the posterior end of which does not extend beyond, or only slightly beyond, the posterior eye margin; the faintly barred sides.

Identification

Characters: *General—See* white crappie. *Generic*—Anal spines 3. Dorsal spines normally 10, rarely 9–11. Body elongate and slender for a sunfish; body depth usually 3.0–4.5, extremes 2.5 (largest females)–5.0 (smallest young) times in standard length. More than 55 scales in the lateral line. *Specific*—Mouth moderately large; posterior end of upper jaw usually extends to beneath center of eye in fishes less than 6.0″ (15 cm) long (upper fig.); to beneath the posterior half of eye in the smaller adults; to slightly beyond the posterior edge of eye in some large fishes (lower fig.). Margin of spinous dorsal a gentle curve. Notch between the two dorsal fins shallow, the shortest spine in the center of the notch (usually the second last) contained 1.2 (largest adults)–1.9 (smallest young) times in length of the

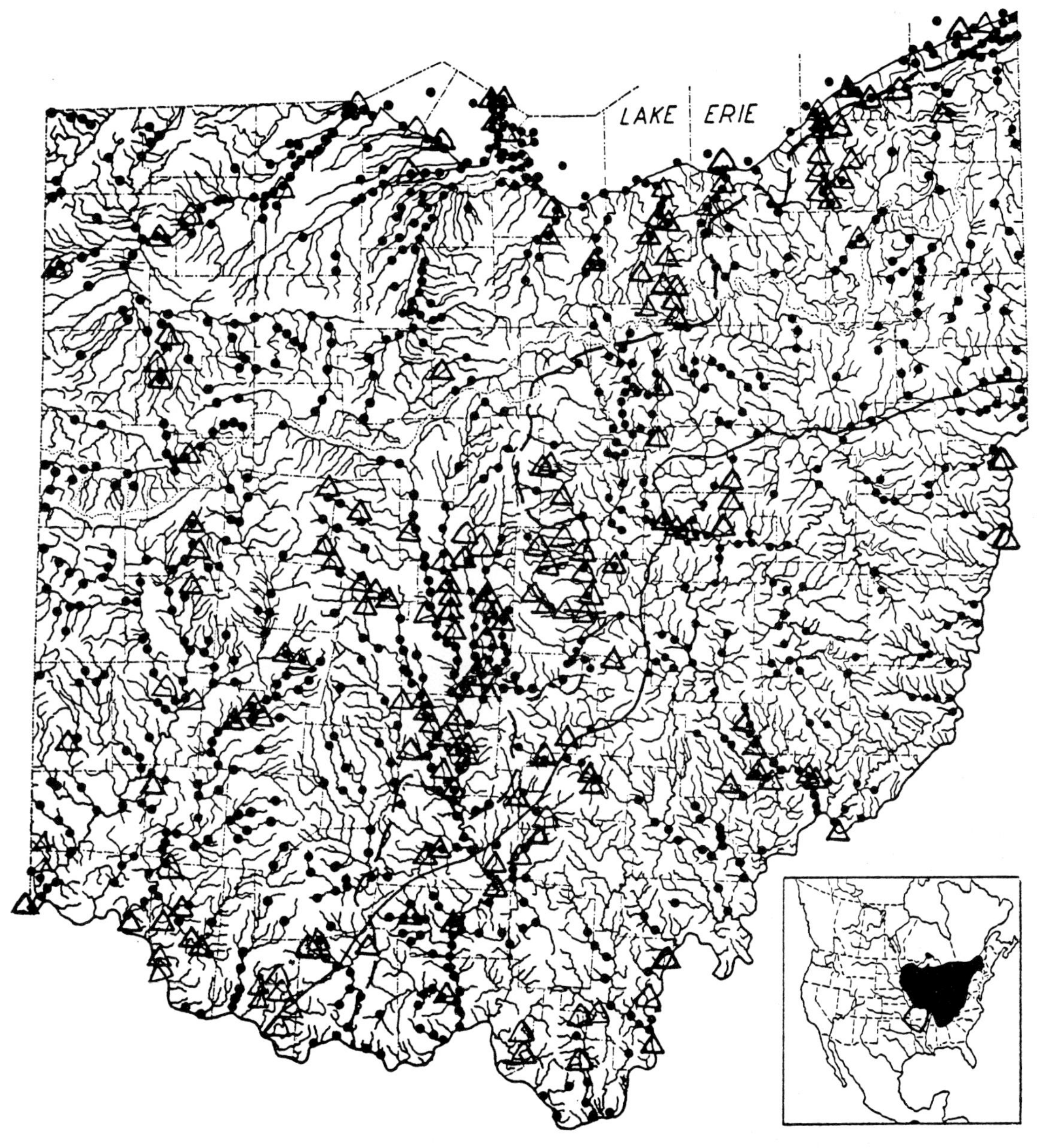

MAP 135. Northern smallmouth blackbass

Locality records. ● Before 1955. △ 1955–80. ~ Glacial boundary. /\ Allegheny Front Escarpment.

Insert: Black area represents range of nominate form; outlined area of other subspecies; widely and often successfully introduced outside of this range in North America and on the continents and islands.

longest spine (third or fourth). Dorsal rays usually 13–15. Anal rays usually 11, extremes 10–12. Complete lateral line with 67–81 scales, rarely less. Pyloric caeca typically unbranched. No longitudinal rows of dusky spots on lower sides.

Differs: Spotted blackbass has longitudinal rows of dusky spots on lower sides; a lateral band; fewer dorsal rays on the average, and fewer lateral line scales. Largemouth blackbass has a lateral band; larger mouth; greatly curved spinous dorsal; fewer lateral line scales. Warmouth, green, other sunfishes, and white bass have fewer lateral line scales. Drum has two anal spines.

Most like: Spotted and largemouth blackbasses. **Superficially like:** Warmouth and green sunfishes, especially the small young.

Coloration: *Adults*—Dorsally varies from light yellow-green to dark olive-green, with bronzy reflections. Sides lighter, more yellowish and with some light slate. Ventrally yellow-white, or white tinged with light slate. Usually three dark streaks on each side of head, radiating backward from snout and eye. Back usually mottled with dark olive-green. Sides with 9 (adults mostly)–16 (young) dark, vertical bars; these never form a definite lateral band. A moderately-sized dusky spot on opercle flap. Vertical fins light olive-slate with some bronze, the tail often slightly bicolored with the posterior third the darker. Paired fins pellucid-white, tinged with olive-yellow. *Breeding male*—Dark greenish-bronze; the back markings and bars on sides blackish. *Young* less than 6.0″ (15 cm) long—has a tricolored tail, with the basal third orange-yellow, the middle third dusky and in the form of a vertical bar, and the distal third whitish. *Fry* are entirely black.

Lengths and **weights:** *Young* of year in Oct., 1.5″–4.5″ (3.8–11 cm) long (hatchery fishes that are force-fed may reach a length of 7.0″ [18 cm] by Oct.). Around **1** year 3.0″–6.5″ (7.6–17 cm). *Adults*, usually 10.0″–18.0″ (25.4–45.7 cm). Fishes 10.0″—–12.0″ (25.4–30.5 cm) long, usually weigh 11 oz–1 lb, 3 oz (312 g–0.5 kg); 12.0″–14.0″ (30.5–35.6 cm), as much as 2 lbs, 2 oz (1.0 kg); 14.0″–16.0″ (35.6–40.6 cm), as much as 3 lbs, 2 oz (1.4 kg). Above 16.0″ (40.6 cm) weight is extremely variable; as examples, a gravid female only 19.0″ (48.3 cm) long weighed 4 lbs, 14 oz (2.2 kg), whereas a male 21.8″ (55.4 cm) long weighed only 3 lbs, 10 oz (1.6 kg). Largest Ohio specimen, 24.5″ (62.2 cm) long, weighed 7 lbs, 8 oz (3.4 kg). Maximum weights of 10–14 lbs (4.5–6.4 kg) recorded outside Ohio.

Hybridizes: with spotted and largemouth blackbasses.

Distribution and Habitat

Ohio Distribution—From many preserved specimens (32 Ohio collections in USNM alone) and innumerable literature references it is abundantly evident that from the time of the arrival of the white man and until about 1900, the smallmouth blackbass was immensely abundant in Lake Erie and most of the inland Ohio streams. It was a valued source of food for both Indian and white man, particularly during those early years whenever game was scarce and crops poor. These men captured the blackbass in many ways; with spears, gigs, bows and arrows, and with jacklights in a canoe or boat (Hildreth, 1848:466–67, as spotted perch); with hooks and lines, seines, nets, weirs, and guns (Kirtland, 1847:30). Blackbass were particularly vulnerable to capture in these early days because of the great clarity of the water, especially to spearing as the fishes migrated upstream to spawn, as they spawned in the shallows, or lay in the pools during the low-water of late summer. After 1810, hundreds of dams blocked the upstream migration, and the blackbasses were then taken in huge numbers with seines in the pools below these dams; then later in the season, when the blackbasses moved downstream, they were caught in cribs and traps placed on the aprons of dams.

Until commercial fishing for blackbasses in Lake Erie was prohibited in 1902, both the largemouth and smallmouth blackbasses were included in the blackbass poundage of fish commercially taken. Both species had become commerically important by 1830, and were taken mostly in shallow waters with hand seines or with hook and line. Kirtland (1851B:87) mentioned that between 1849–51 the waters of Lake Erie near the mouth of the Cuyahoga River were "literally black with fishing boats," the boats containing hook and line fishermen, many of which were fishing only for sport, and that frequent catches of 100 walleyes and blackbasses were taken in a morning by one person. Fisher, Klippart and Cummings (1878:69) reported that "During the season [1877] several tons of blackbass are taken *daily* in the vicinity of the islands in Lake Erie, with hook and line, showing that this mode of fishing is just as destructive as any other." Few, if any, of these island blackbasses were largemouths. For the year 1885, Smith and Snell (1891:241) gave the commerical catch for both blackbasses as 599,000 lbs (271,702 kg) for the Ohio ports of Lake Erie, of which 450,000 lbs (204,117 kg) were taken in the vicinity of Sandusky and the islands. Upon many days between 1865–88, Cooke (Pollard, 1935:100–312), alone or with friends, caught 30–100 smallmouths about the islands with hook and line. After 1885 the annual poundage of commercially caught blackbasses began dwindling so rapidly in amount that in 1902 the Ohio Legislature passed a law prohibiting the commercial capture of blackbasses in the Ohio waters of Lake Erie. Advocates of this law predicted that with commercial fishing pressure removed, the blackbasses would return quickly to their former abundance, which they did not. Instead a continued decline was apparent, even though there were ever increasing restrictions on hook and line fishing until 1951, after which most of the restrictions were removed.

Since 1920 the largest numbers of smallmouths have occurred in western Lake Erie and in streams west of the Allegheny Front Escarpment. The largest numbers in the unglaciated section were in the glacial outlets, such as the Walhonding, Muskingum, and Beaver rivers. Smallmouth numbers were relatively low in the southernmost two tiers of counties where spotted blackbasses were abundant, and relatively few were taken in hoop nets in the Ohio River.

Between 1925–55 I periodically made observations on several streams. Such test streams as the West Branch of Beaver Creek, Columbiana County, showed no changes in smallmouth numbers, amount of turbidity or habitat during the 25 years; streams such as Big Darby, Pickaway County, displayed a moderate deterioration in habitat; and others such as the turbid Maumee River showed definite decreases in abundance and amount of smallmouth habitat.

Recent smallmouth blackbass populations in western Lake Erie were patently far below those of 1885, despite no commercial take, and hook and line fishing pressure far below the average for inland waters. I noted a possible decrease between 1939–51 about South Bass Island, where each year I fished in the same manner, at the same places, and the same period of evening. Before 1944, my catch per hour, and schools of feeding blackbasses observed, were approximately one-third larger than after 1947. It is possible that the following were factors in the decrease in abundance from 1885 to 1951:

It is ecologically sound reasoning to assume that more blackbasses migrated upstream out of the lake, and more young were produced to drift downstream into the lake, before these tributaries were dammed or had become greatly polluted, and before the spawning and nursery grounds were covered with silts, than were produced after these detrimental factors came into existance.

In the extremely clear waters of Whitmore Lake, Michigan, I saw smallmouth eggs hatch at a depth of 22′ (6.7 m), but never to more than 3′ (0.9 cm) in depth in turbid waters whose visibility range was less than 6″ (15 cm) (Hubbs and Bailey, 1938:39). In western Lake Erie smallmouth nests which are in waters of less than 5′ (1.5 m) in depth are in danger of destruction when violent winds expose these nests to drastic wave and undertow action, and especially when this is accompanied by a sudden lowering of the water level. It follows, therefore, that nests in waters deeper than 5′ (1.5 m) are safe from storms, provided these waters remained clear, as presumably they did formerly. But in recent years western Lake Erie tributaries have carried sufficient eroded clays from cornfields into western Lake Erie, so that the accumulation on the bottom, when disturbed by violent wave action, can cause enough turbidity to affect adversely blackbass eggs in the deeper nests.

Habitat—The largest inland populations occurred in those stream sections where about 40% of the stream consisted of riffles flowing over clean gravel, boulder, or bedrock bottoms; where the pools normally had a visible current and a maximum depth of more than 4′ (1.2 m), there was considerable water-willow, and the gradients were between 4′–25′/mile (0.8–4.8 m/km) (Burton and Odum, 1945:188–90; Trautman, 1942:211–18). As the waters warmed to above 40° F (4° C) in spring, adult blackbasses moved upstream to spawn in streams 20′–100′ (6.1–30.5 m) in average width, having gradients of 7′–25′/mile (1.3–4.8 m/km). After spawning, adults and most of the young began drifting downstream, their speed of drift depending upon stream levels; they wintered in larger, deeper waters of gradients usually less than 7′/mile (1.3 m/km). Spawning populations were low or absent in streams having gradients less than 3′/milc (0.6 m/km), such as the St. Marys River, whose gradient, from the Ohio-Indiana line upstream 39.7 miles (63.9 km) averages 1.1′/mile (0.2 m/km); also in streams of very high gradients, such as the East Branch of the Chagrin, which has gradients ranging from 17′–75′/mile (3.2–14.3m/km) (Trautman, *op. cit*).

The largest Lake Erie populations were about the limestone reefs and bars, where there was some current and the water was 3′–20′ (0.9–6.1 m) in depth.

Years 1955–80—The smallmouth blackbass population of Lake Erie, especially in the western third about the Bass Islands and in the vicinity of Cleveland (White et al., 1975:104) declined in total numbers during this period. The decline appeared to be similar to that which had been occurring since before 1900 except that possibly the recent one may have become more accentuated.

The total number of smallmouths caught during a day in summer about the island region definitely appeared to be less than it was before 1945 despite an obvious increase in the total number of sport fishermen. Throughout the last few years, also, the number of blackbasses taken on a fly during a summer's evening about South Bass Island seldom exceeded two or three, whereas before 1950 three to ten could be taken frequently. These latter numbers now seem as remote as do the 30–100 smallmouths taken per day between 1865 and 1888 by Cooke and/or his friends (Pollard, 1935:100–312).

There appeared to be a definite increase in the average lengths and weights of the smallmouths captured since 1960. Such increases sometimes occur when a population of fishes or other animals is decreasing numerically. Possibly the modifications that have taken place in Lake Erie recently and which contributed to the catastrophic decrease in the population of other predatory species such as the blue pike, may have resulted in the increased lengths and weights of the smallmouths, probably through increases in amounts of food and/or available living space.

The distribution of the smallmouth in the inland streams remained largely unchanged during the period. However, the numerical abundance in many localities declined, especially in the lower-gradient, more turbid or channelized waters. The decrease in the numbers of bass fishermen, especially flyfishermen, was more apparent than was the decrease in smallmouth numbers indicated by collections. On summer's evenings during the 1920's and 30's when conditions were favorable, one to ten flyfishermen could be observed along a stretch of stream; and occasionally one had to wait, as custom dictated, for a fishermen to leave a favored pool section before one began fishing it. Since 1960 none or few flyfishermen were encountered where formerly a half-dozen could be seen. Part of this decrease was caused by the recent surge in popularity of the effecitve method of spin-casting. Yet the total numbers of spin-casters were far below those of the former flyfishermen. If we judge by the number of bass fishermen and their catches, we must conclude that the inland smallmouth population was not as large as it had been before 1940.

The 1957–59 investigations (ORSANCO, 1962 :151) of the Ohio River indicate that the smallmouth is the least numerous of the three species of blackbasses inhabiting that stream. During the 1968–69 investigation of the Ohio River, H. Ronald Preston captured smallmouths only at two of the nine collecting stations which border the state of Ohio, Margulies and Burch at all of their collecting stations between 1976 and 1978.

NORTHERN SPOTTED BLACKBASS

Micropterus punctulatus punctulatus (Rafinesque)

Fig. 136

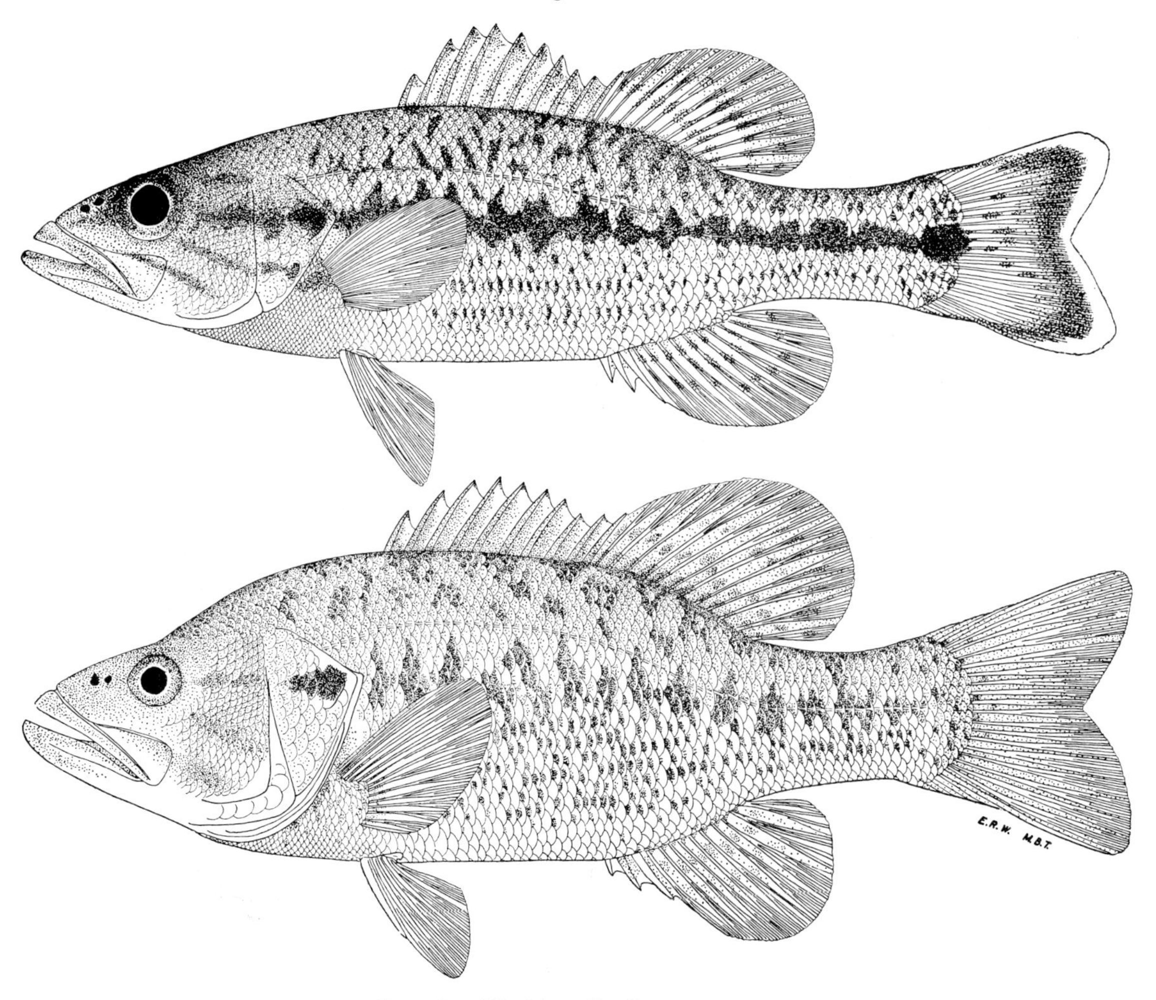

Upper fig.: Ohio River, Hamilton County, O.

Oct. 16, 1942. 70 mm SL, 3.3″ (8.4 cm) TL.
Young of year. OSUM 5222.

In comparing young (upper fig.) with adult (lower fig.), note in young the pike-like, slender body; the larger eye; the higher and more curved spinous dorsal fin; tricolored tail; bold caudal spot; continuous lateral band; both young and adult have longitudinal rows of spots on lower sides, and the posterior end of the upper jaw does not extend beyond the eye.

Lower fig.: Scioto River, Scioto County, O.

Nov. 13, 1939. 255 mm SL, 12.5″ (31.8 cm) TL.
Adult female. OSUM 1148.

Note in adult the deeper body; smaller eye; lower but still moderately-curved spinous dorsal fin; faintly bicolored tail; bold opercle spot; broken lateral band; diamond-shaped spots on back.

Identification

Characters: *General—See* white crappie. *Generic—See* smallmouth blackbass. *Specific—* Mouth moderately large; posterior end of upper jaw usually extends to beneath posterior half of eye in young and most adults; occasionally to slightly

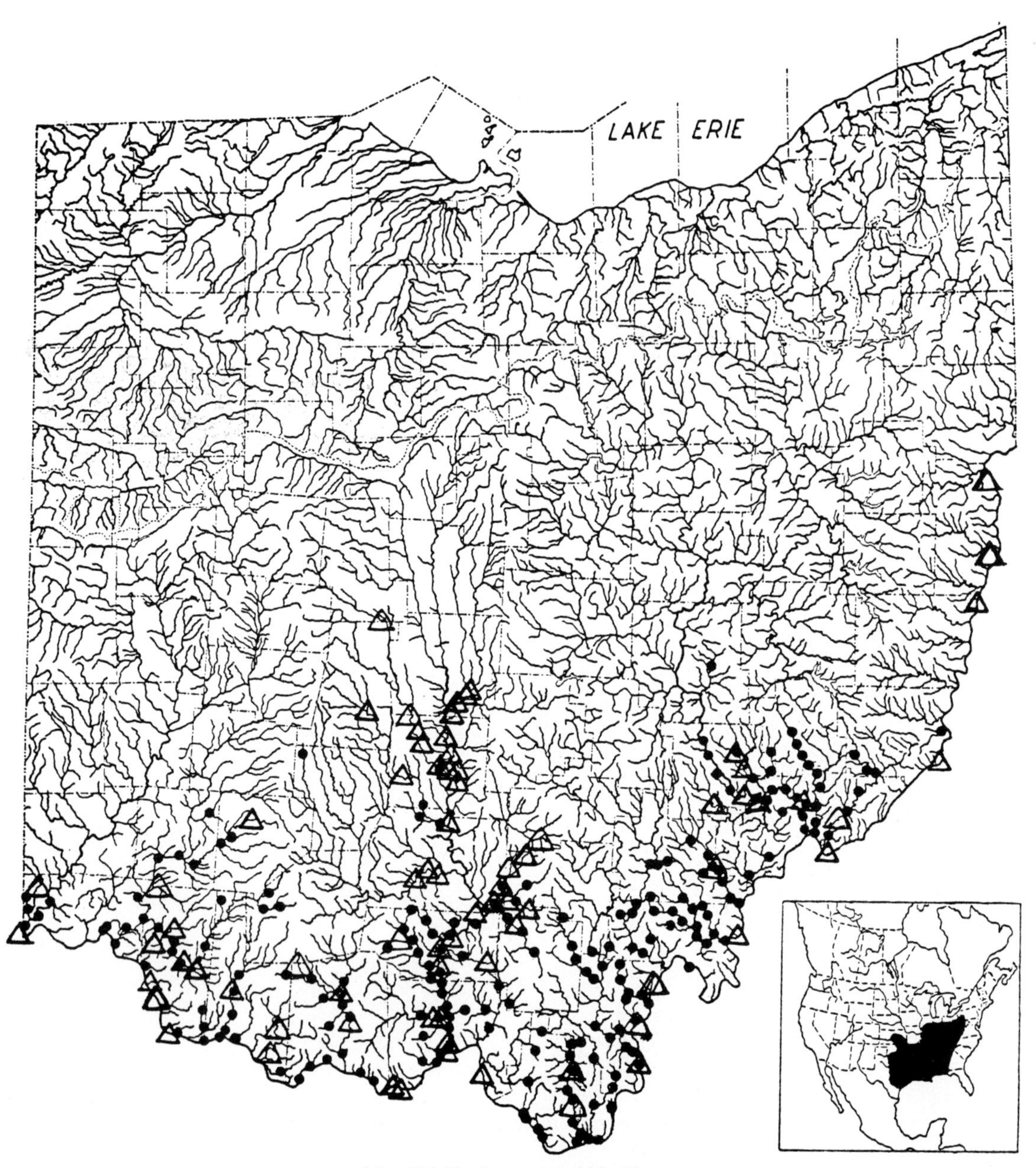

MAP 136. Northern spotted blackbass

Locality records. ● Before 1955. △ 1955–80.
Insert: Black area represents range of the nominate and other nominal forms.

beyond posterior edge of eye in large adults. Margin of spinous dorsal a gentle curve. Notch between the two dorsal fins shallow; the shortest spine in center of notch (usually the second last spine) contained 1.2 (largest adults)–1.9 (smallest young) times in length of the longest spine (third or fourth). Dorsal rays usually 12, extremes 11–13. Anal rays usually 10, extremes 9–11. Complete lateral line usually of 60–68, extremes 55–77, scales. Usually conspicuous longitudinal streaks on sides below lateral line (undeveloped in smallest young), formed by each scale having a dusky base. Lateral band usually broken up into somewhat confluent, short bars; in young this band is more confluent. Opercle spot usually larger and blacker than in the other blackbasses. Young with a conspicuous caudal spot, and a slender, pike-like shape. Pyloric caeca typically unbranched. Patch of teeth on tongue.

Differs: Smallmouth blackbass has no lateral band; has more dorsal and anal rays; more lateral line scales. *See* smallmouth blackbass.

Most like: Smallmouth and largemouth blackbasses. **Superficially like:** Warmouth and green sunfishes.

Coloration: *Adults*—Dorsally yellow- or olive-green, the back with many dark olive markings which are often diamond-shaped. Sides lighter olive-green and more yellowish. A broken lateral band consisting of short vertical bars extends forward as a large, black opercle spot, and continues across cheek to eye as one of the three bars which radiate from snout and eye on each side of the head. Lower sides yellowish-white, most of the scales with dusky bases, resulting in the formation of horizontal lines; it is from these distinctive spots that the species derives its name of spotted blackbass. Vertical fins yellow-olive; the dorsal mottled and spotted with darker; the caudal less mottled, its distal third slightly darker than the remainder, often giving a bicolored appearance to the tail; anal fin lighter than dorsal and mottled. The paired fins are light and without spotting. *Young*—Has a tricolored tail when less than 6.0″ (15 cm) long as does the smallmouth; the tail orange-yellow basally, the middle section a vertical, dusky bar, the distal third whitish. Lateral band usually solid, tending to break up into vertical bars. Caudal spot large and prominent. Spots on lower sides less well developed in large young but may be absent in smallest young.

Lengths and **weights:** *Young* of year in Oct., 1.5″–4.0″ (3.8–10 cm) (abundantly fed hatchery fishes reach a maximum length of 6.5″ [17 cm]). Around 1 year, 3.0″–7.0″ (7.6–18 cm). *Adults*, usually 7.0″–17.0″ (18–43.2 cm) long, weight 5 oz–2 lbs, 8 oz (142 g–1.1 kg). Largest specimen, 20.1″ (51.1 cm) long, weight 3 lbs, 2 oz (14 kg). Hybrids larger.

Hybridizes: with smallmouth and largemouth blackbasses.

Distribution and Habitat

Ohio Distribution—Although the spotted blackbass was described as early as 1819 (Rafinesque, 1819:420, as *Calliurus punctulatuse)* it was not generally recognized by fishery workers as a distinct species before 1927 (Hubbs, 1927:1–15).

Since 1925 the spotted blackbass, during its spring migrations up the Ohio River, outnumbered more than 50 to 1 the smallmouth blackbass in the commercial nets, and the largemouth to spotted ratio was even greater. I have seen as many as 20–80 spotted blackbasses taken in one wing-net in the spring, usually when the Ohio River was sufficiently high so that the nets could be set among the willows along the bank. In fall when the river was exceptionally clear, I have seen schools of 40–70 spotted blackbasses, 8.0″–15.0″ (20–38 cm) in total length, lying under old boats and barges in such localities as the mouth of White Oak Creek, Brown County, and at that season have seen large strings of "Spotties" that were caught with minnows by fishermen.

This blackbass occurred in abundance in the southernmost tier of Ohio counties from Hamilton to Washington, in streams having gradients between 0.5′ and 3.0′/mile (0.1 and 0.6 m/km). Experimental nets, placed in the Scioto River, northern Scioto County, between October 1,-1939–September 30, 1940 by the Conservation Department, produced 334 spotted bass and only 8 smallmouths (Trautman, 1942:217). During the present century many thousands of smallmouth and largemouth blackbasses were propagated, or trapped in Lake Erie, and were liberated in southern Ohio waters, but the first spotted blackbasses were not liberated until 1932, and the numbers stocked have been relatively few. Yet despite the mass liberation of the largemouth and smallmouth (possible competitors), the spotted still far outnumbers them, a splendid illustration of the ability of a species to withstand interspecific competition provided ecological conditions favor it.

The failure of ichthyologists and anglers to recognize the distinctive spotted blackbass before 1927 is surprising. Part of this failure was due to the weight of Henshall's opinion and his (1917:45–47) insistence that there were only two species of blackbasses. His identifications of spotted blackbasses from the Ohio River near Cincinnati, in the CNHM collections, plainly show that some he originally labelled largemouths, others smallmouths, only to later reverse his opinions. Still later, Josua Lindall re-examined these blackbasses, identifying as largemouths most of the spotteds which Henshall had identified as smallmouths.

Although ichthyologists and sport fishermen may not have recognized the spotted blackbass, the commercial fishermen of the Ohio River long believed it to be a distinct species, calling it "Speckled, Yellow or Spotted bass, perch or trout." They agree that it was the most abundant bass in the Ohio River since and before 1900. Since 1927 much has been learned concerning the species (Howland, 1931:89–94; Hubbs and Bailey, 1940:14–22).

Habitat—The spotted blackbass chiefly inhabited moderate- or large-sized streams having gradients of less than 3′/mile (0.6 m/km) and long, sluggish, rather deep pools. It appeared to be more tolerant to turbid waters and a silt bottom than were the largemouths and smallmouths. Upstream migration in spring was pronounced after the water temperature reached 50° F (10° C) or more, and then adults penetrated farther upstream than at any other period of the year, inhabiting the smaller streams having gradients under 8.0′/mile (1.5 m/km), and deep holes if the water was rather clear or shallower waters if they were turbid. In early summer, adults and most of the young drifted downstream, to winter in large, low-gradient streams. On August 17, 1930, I collected the three species of blackbasses within a hundred feet of each other, in Bullskin Creek, Clermont County, each species showing its habitat affinities. The smallmouths were on the riffle, the spotteds were in the sluggish pool, and the few largemouths were in a connecting and weedy oxbow.

Years 1955–80—The distributional range in the Ohio River drainage of the spotted blackbass remained similar to that existing before 1950, except for a northward extension in the Scioto River system into Franklin and Union counties. As far as could be ascertained, the population size within its range remained unchanged.

The species continued to be numerous in the Ohio River, as demonstrated by the 1957–59 investigations (ORSANCO, 1962:151), by specimens obtained by H. Ronald Preston during 1968–69, in seven of the nine river collections that bordered the state of Ohio, and by Margulies and Burch at all of their stations on the Ohio River.

NORTHERN LARGEMOUTH BLACKBASS

Micropterus salmoides salmoides (Lacepède)

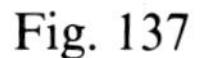

Fig. 137

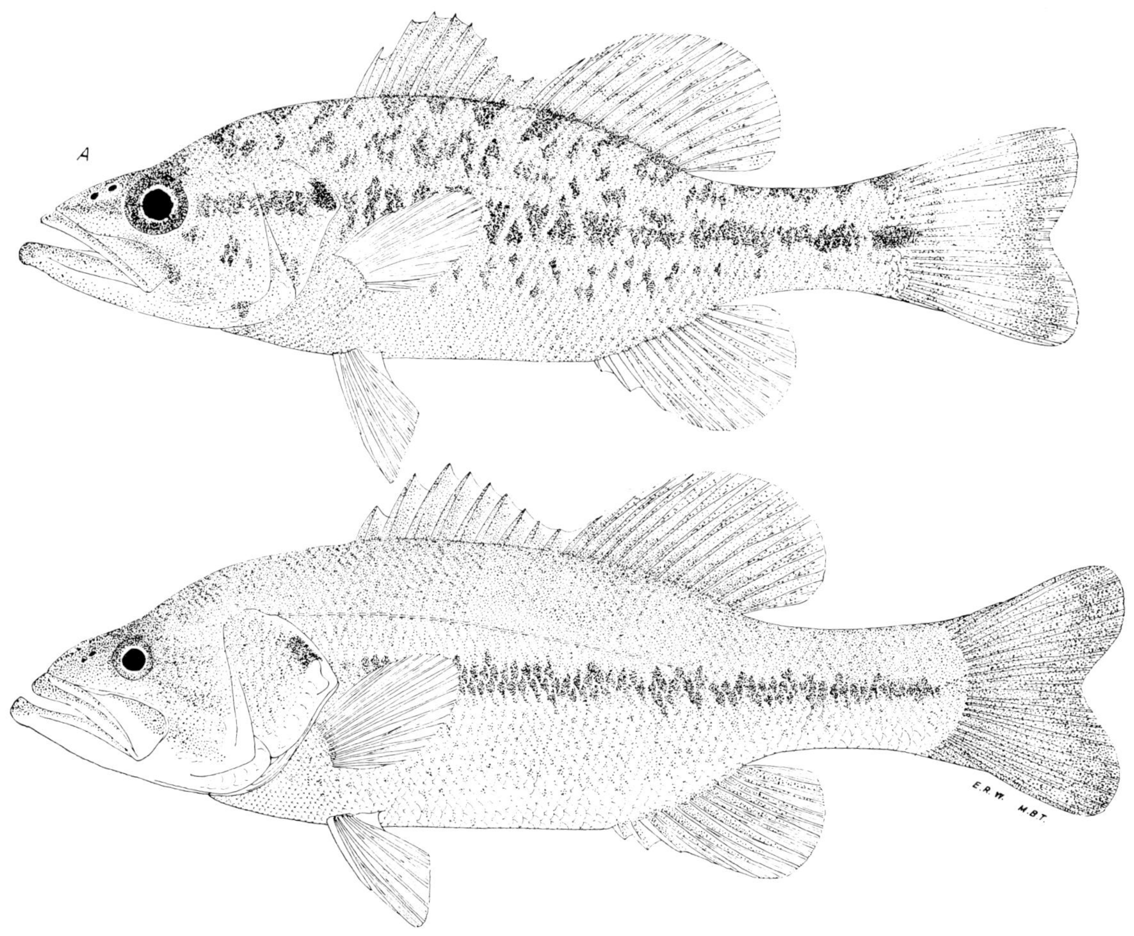

Upper fig.: Walhonding River, Coshocton County, O.

Sept. 30, 1939. 79 mm SL, 3.8″ (9.7 cm) TL.
Young of year. OSUM 21.

In comparing young (upper fig.) with adult (lower fig.), note in young the larger eye; the very high and greatly curved spinous dorsal fin; bicolored tail; dark mottlings on back; broken lateral band; length of upper jaw which does not extend beyond the posterior edge of the large eye.

Lower fig.: Lake Rockwell, Cuyahoga River Drainage, Portage County, O.

Sept. 22, 1943. 216 mm SL, 10.5″ (26.7 cm) TL.
Female. OSUM 5845.

Note in adult the smaller eye; rather high, curved spinous dorsal fin; faintly bicolored tail; few or no markings on back; broken lateral band; length of upper jaw which extends beyond the posterior edge of the eye; lack of longitudinal rows of dark spots on the lower sides.

Identification

Characters: *General—See* white crappie. *Generic—See* smallmouth blackbass. *Specific—* Mouth very large; in fishes more than 6.0″ (15 cm) long the posterior end of the upper jaw extends well beyond the posterior margin of eye; in fishes less than 5.0″ (13 cm) long it extends to the posterior

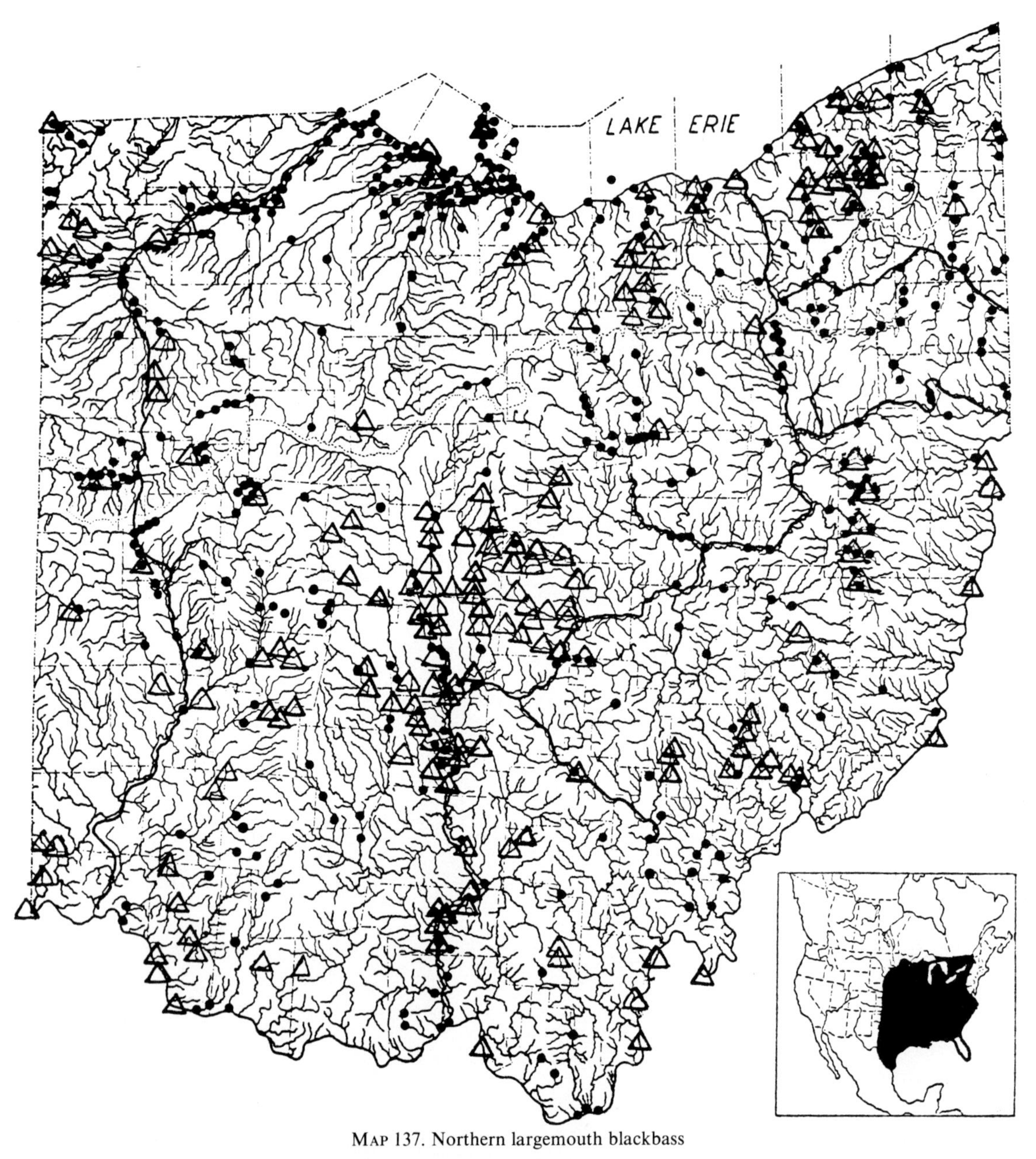

MAP 137. Northern largemouth blackbass

Locality records. ● Before 1955. △ 1955–80. ~ Canals.

Insert: Black area represents range of nominate form, outlined area of other subspecies; widely and often successfully introduced outside of this range in North America and on other continents and islands.

margin of eye. Margin of spinous dorsal fin considerably curved and sickle-shaped. Notch deep between the dorsals; the shortest spine in center of notch (usually the second last spine) usually contained 2.0 (largest adults and smallest young)—3.3 times in length of longest spine (third or fourth). Dorsal rays usually 12–13, extremes 11–14. Anal rays 11, extremes 10–12. Complete lateral line usually of 58–69 scales, rarely more. Lateral band rather uniform in width in adults, lower fig.; more broken in small young, upper fig. Pyloric caeca usually branched at base. No longitudinal rows of dark spots on sides below lateral band as in spotted blackbass.

Differs: *See* spotted and smallmouth blackbasses.

Most like: Spotted and smallmouth blackbasses.

Superficially like: Warmouth and green sunfishes.

Coloration: *Adults*—Dorsally varies from a pale

olive-yellow (warm or turbid waters) to a dark olive-green (clear and/or vegetated waters), normally overlaid with bronze or gold. Back vaguely mottled with darker in some adults. Sides lighter, usually olive-yellow with a silvery sheen; the lateral band usually solid or mostly confluent; considerably broken only in smaller adults. Lower sides with dusky spots on comparatively few scales; these spots fail to form longitudinal rows. Ventrally white or yellow, often milk-white on belly. A moderately-sized black spot on opercle. Three bands radiating backward from snout and eye usually present, although often faint. Fins light yellow-olive, the dorsal and anal sometimes vaguely mottled with darker, the caudal frequently bicolored with darker olive-slate, forming a wide band on the posterior third of the fin. Pelvics with some silvery. *Young*—Coloration more contrasting and with a more silvery sheen than in adults; blackish spotting on back better defined and more intense; basal spots on scales below lateral line more conspicuous; a blackish band extending across snout, opercles, and continuing as a broken lateral band backward to an ill-defined caudal spot. Tail distinctively bi-colored, the posterior third the darker.

Lengths and **weights:** *Young* of year in Oct., 2.0″–5.5″ (5.1–14 cm) long (abundantly fed hatchery fishes reach a maximum of 8.0″ [20 cm]). Around **1** year 3.0″–7.5″ (7.6–19 cm). *Adults*, usually 10.0″—20.0″ (25.4–50.8 cm) long. Fishes 10.0″–12.0″ (25.4–30.5 cm) long usually weigh 7 oz–1 lb (198 g–0.5 kg); 12.0″–15.0″ (30.5–38.1 cm) long usually 9 oz–2 lbs, 7 oz (255 g–2.4 kg). Length-weight relationship varies greatly after fishes reach a length of more than 15.0″ (38.1 cm); weights of fishes 18.0″ (47.7 cm) long range from 2 lbs, 8 oz–3 lbs, 12 oz (1.1–1.7 kg); fishes 20.0″ (50.8 cm) from 3 lbs, 6 oz–4 lbs, 12 oz (1.5–2.2 kg). Largest specimen, 23.5″ (59.7 cm) long, weight 8 lbs, 14 oz (8.9 kg), a decrepit, frayed-finned fish, caught by a small boy fishing with a cane pole baited with a worm, at Buckeye Lake in 1927. Older Ohio fishermen report maximum weights of 9.0–11.0 lbs (4.1–4.99 kg), these fishes taken in the old canal reservoirs before 1915.

Hybridizes: with black crappie, spotted and smallmouth blackbasses, warmouth and bluegill sunfishes.

Distribution and Habitat

Ohio Distribution—Preserved specimens and reliable literature references testify to the presence before 1900 of the largemouth blackbass in the weedy bays, harbors, marshes, and shallows of Lake Erie, the base- and low-gradient portions of its tributaries, and the glacial lakes of its watershed. Klippart (1877A:31, as *Calliurus floridanus*) records that about Sandusky, the largemouth (and undoubtedly the smallmouth also) had attained considerable commercial importance by 1830, and that then the largemouth was very abundant and one of the most sought after species in Sandusky Bay (*see* smallmouth blackbass).

Although specimens and reliable records are lacking, it is believed that the largemouth was present in the Ohio drainage before 1825, or prior to the building of the canals, especially in the glacial lakes of the middle third of the state, and the weedy, prairie streams into which they drained. Published records for the southern third of Ohio are entirely unreliable, because in that area the largemouth and the spotted blackbasses were hopelessly confused (*see* spotted blackbass). However, the species surely was present since its habitat was present in the bayous, oxbows and overflow ponds along the larger rivers. The establishment of the canals, between 1830–50, produced a continuous narrow habitat for the species twice across Ohio, and this unquestionably influenced the later distribution of the largemouth, as the distribution map indicates.

Before 1900, the largemouth was of considerable commercial importance in the larger impoundments, such as Indian, Buckeye and St. Marys lakes, from which thousands of barrels of blackbasses were shipped (Clark, 1951:15). Since the prohibiting of the commercial capture of blackbasses in 1902, the largemouth has become the most sought after and prized game species of Ohio's lakes, ponds and impoundments.

Stocking of this species in Ohio waters began about 1885 or slightly before; since 1920 largemouths have been liberated by the hundreds of thousands throughout Ohio, with every county receiving some. Such mass stocking has established the species in the newly impounded waters, and in isolated ponds, canals, and along the larger rivers. But the fact remains that despite mass liberation the largemouth occurs only sparingly or in isolated colonies in the streams of Ohio, particularly in the southern half. Between 1920–50 few were taken in the hoop nets in the Ohio River, in marked contrast to the abundance of the spotted blackbass.

Kirsch (1895A:331) explained that in the 1893 survey no largemouths were taken in the Auglaize River, but that they were "common" in the

Maumee. This is an example of the avoidance by this species of flowing waters, for the Auglaize has a higher average gradient than does the Maumee. Comparison of Kirsch's collections with those after 1940 indicates a recent decrease in abundance of the largemouth, as would be expected with increased turbidity and the discontinuance of most of the adjacent canal.

Habitat—The largemouth was essentially a species of non-flowing waters, such as occur in ponds, lakes, impoundments, oxbows and overflow ponds, which contained much aquatic vegetation and reasonably clear water. The preferred bottom types were soft muck and organic debris, gravel, sand and hard, non-flocculent clays. Only small numbers occurred in the larger, non-flowing pools of streams, or in non-flowing waters where suspended clays usually kept the water roily, thereby smothering the vegetation and covering the bottom with a flocculent mass of clay. Only strays occurred in high-gradient sections of streams.

Years 1955–80—Throughout this period the overall numerical abundance of the largemouth blackbass continued to decrease in the shallows of Lake Erie (White et al., 1975:105) and especially in its adjacent harbors, marshes and Sandusky Bay (Van Meter and Trautman, 1970:75). The obvious decrease in the amount of submerged aquatics undoubtedly was an adverse factor. The rather drastic decrease in submerged aquatics in East Harbor coincided with the equally drastic decrease in the size of its largemouth population. The disappearance of much of the submerged vegetation in Sandusky Bay was presumably important in the decrease of largemouths, but it is believed that continued turbidity also was a major factor. Coincidental with the decrease in numerical abundance of this species was a corresponding decrease in the numbers of individuals fishing for largemouths. East Harbor and Sandusky Bay were formerly noted for their splendid bass fishing.

Many thousands of farm ponds were constructed throughout inland Ohio during this period, all except a few of which were stocked with largemouths. This resulted in a major increase in the production of largemouths in Ohio and in a corresponding increase in the number of persons fishing for them. Increased production in farm ponds has more than counterbalanced the decreasing production in such formerly outstanding largemouth waters as Buckeye Lake. As an example of the former production at Buckeye Lake, Smith (1898B:510) states that 92,500 lbs (41,957 kg) of "bass" were taken commercially during 1894. For an excellent discussion of causes of recent decrease in the numbers of largemouth blackbasses in Buckeye Lake, *see* Judd, 1971:1–86.

Routine collections from the flowing waters of inland Ohio contained a higher percentage of largemouths during this period than were recorded from the collections taken between 1930 and 1950. This increase in flowing waters may have been due in part to the numbers of largemouths which "escaped" from or left farm ponds during floods.

Between 1928 and 1933 I conducted investigations of the fish life in that section of the Ohio River bordering the state of Ohio. I recorded few largemouths taken commercially or in any other manner, the spotted blackbass greatly outnumbering them. During the 1957–59 investigations of the entire Ohio River (ORSANCO, 1962:151) 258 largemouths and 251 spotted blackbasses were recorded. H. Ronald Preston informed me that during the 1968–69 investigation seven of the nine collecting stations bordering the state of Ohio produced largemouths (*see* spotted blackbass). Margulies and Burch recorded the collecting of specimens from all of their stations in the Ohio River between 1976 and 1978. Environmental modifications favorable to the largemouth must have occurred in the Ohio River since 1950. Possibly the newly constructed high-lift dams may have produced more pond-like or favorable conditions, or a continued influx of largemouths from farm ponds and inland streams may have resulted in their increased numbers.

WARMOUTH SUNFISH

*Lepomis gulosus** (Cuvier)

Fig. 138

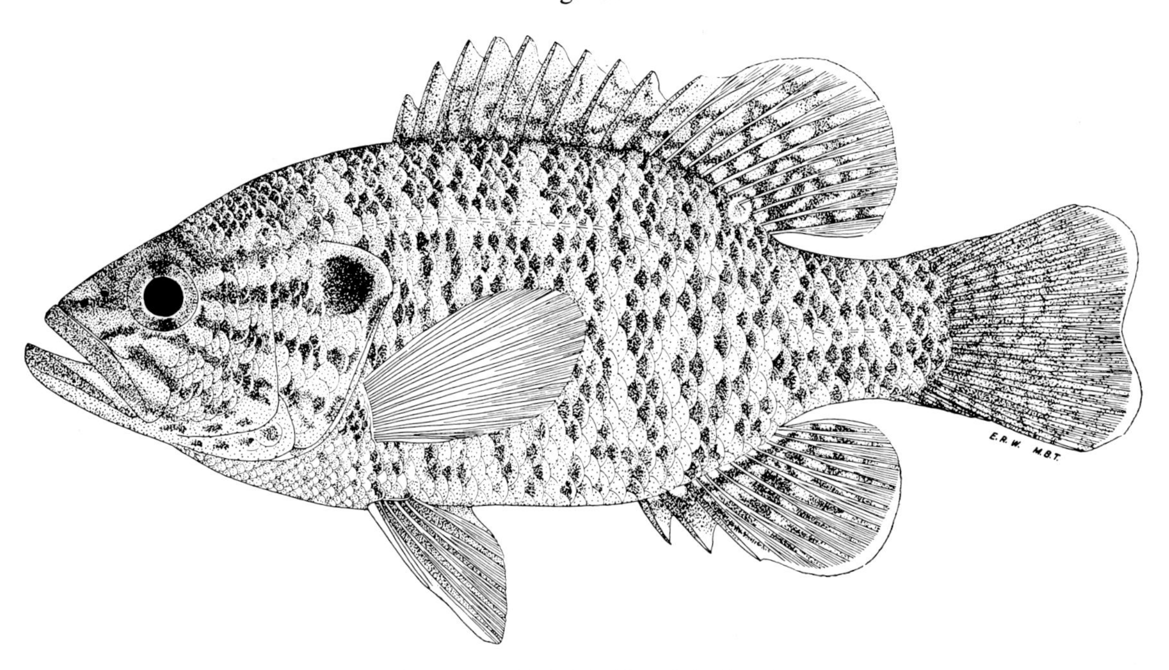

Overflow pond beside Scioto River, Pike County, O.

Nov. 8, 1939.
Female.

120 mm SL, 5.8" (15 cm) TL.
OSUM 1776.

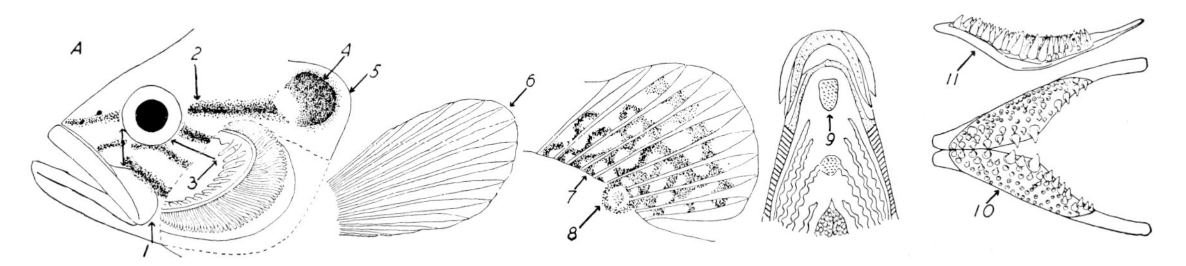

Nos. 1–11

(1) mouth large; posterior end of upper jaw extending well beyond anterior edge of eye, usually to center of eye or beyond in adults.
(2) 3–5 dark bands radiating backward from snout and eye.
(3) dotted portion of gill cover removed, exposing first gill arch; note long gill rakers, usually as long as diameter of pupil.
(4) black opercle flap stiff, the inflexible opercle bone extending posteriorly to the opercle membrane.
(5) tip of opercle membrane whitish, without red.
(6) tip of pectoral fin rounded; fin rather short.
(7) soft dorsal vermiculated; vermiculations usually form light colored oscelli.
(8) breeding male with a brilliant orange spot at base of last three rays; spot absent or ill-defined in breeding female.
(9) teeth on tongue, which no species of the genus *Lepomis* normally possesses.
(10) dorsal view of lower pharyngeal arches showing their narrow width and bluntly conical teeth.
(11) lateral view of a pharyngeal arch; note heavy basal portion of the conical teeth.

---

* Formerly in the genus *Chaenobryttus.*

Identification

Characters: *General—See* white crappie. *Generic* and *specific*—Band of tiny teeth on tongue, *see* 9 above; supramaxilla well-developed, its length greater than the greatest width of the maxilla (upon which the supramaxilla rests); similar otherwise to the genus *Lepomis*. Anal spines 3. Dorsal spines normally 10, rarely 9–11. Complete lateral line of fewer than 45 scales. Body depth usually contained 1.9–2.6, extremes 1.7–2.9 times in standard length. *See* Nos. 1–11 for remaining characters.

Differs: All sunfishes in genus *Lepomis* lack teeth on tongue, except rarely a green sunfish which may have a few; spotted blackbass has patch of teeth on tongue but it has more lateral line scales. Green sunfish normally lacks teeth on tongue; lacks the *dark* radiating bands on side of head; usually contains a dusky spot at base of posterior rays of dorsal and anal fins. Crappies and rockbass have more than three anal spines. Smallmouth and largemouth blackbasses have more lateral line scales.

Most like: Green sunfish. **Superficially like:** Blackbasses, northern longear and redear sunfishes, and rockbass.

Coloration: *Adults*—Dorsally varies from light yellow-olive to dark olive-green, vermiculated with lighter and with dull bluish, purplish and golden reflections. Sides lighter and more orange-yellow. Ventrally greenish-yellow or white, usually with a slaty tinge and/or mottlings. Back and sides with 6–11 chainlike double bands of dark olive or dusky; these bands very prominent in breeding adults and fishes from clear, vegetated waters; absent in fishes from turbid waters. Sides of head with 3–5 dark gray or lavender bands radiating backward from snout and eye. A purplish-lavender sheen over most of body. Soft dorsal, caudal and anal fins boldly vermiculated, the vermiculations usually forming light oscelli; fin margins edged with white or yellow-white. Paired fins unspotted and transparent or olive. *Breeding male*—Has a brilliant orange spot at base of posterior three dorsal rays; slate-colored pelvics. *Young*—Similar to adults but with the vertical fins faintly spotted and vermiculated; smallest young with transparent fins.

Lengths and **weights:** *Young* of year in Oct., 0.8″–2.0″ (2.0–5.1 cm) long. Around **1** year, 1.5″—3.2″ (3.8–8.1 cm). *Adults*, usually 4.1″ (10 cm) (from dwarfed populations)–10.0″ (25.4 cm), weight 1 oz–14 oz (28–397 g). Largest specimen, 11.2″ (28.4 cm), weight 1 lb (0.5 kg).

Hybridizes: naturally in Ohio with bluegill sunfish; naturally or artificially produced elsewhere with black crappie, rockbass, largemouth blackbass and bluegill (Birdsong and Yerger, 1967:62–71; Hester, 1970:102), redear and pumpkinseed sunfishes.

Distribution and Habitat

Ohio Distribution—The only Ohio record for the warmouth sunfish previous to 1920 is of July 28, 1893, when Kirsch (1895A:330) collected specimens in Fish Creek, near Edgerton, Williams County. (Osburn [1901:79] erroneously recorded that Kirsch also took specimens in the Tiffin River at Brunersburg.) In 1920 the warmouth sunfish was found abundantly in Nettle Lake, and since then has been found to be locally abundant in weedy, sluggish streams and lakes in glaciated, northeastern Ohio. Undoubtedly, this sunfish has been present in northern Ohio since early times, possibly invading the Lake Erie drainage through the Maumee-Wabash route, and utilizing the flooded watershed marshes to cross over into the upper Ohio River drainage. Conditions before 1900 should have been favorable for this sunfish in Sandusky Bay, and in the harbors and marshes bordering western Lake Erie, when these areas contained clear waters and an abundance of aquatic vegetation.

It appears probable that, for some unexplained reason, the warmouth sunfish was absent in southern Ohio before 1925; or if present, then only as strays. None was recorded until several years after the State Conservation Department had begun the planting of small numbers of warmouths in with the larger plantings of other sunfish species. The first specimens in southern Ohio were taken July 27, 1929, when two (OSUM: 459) were captured from an overflow pond at the junction of the Scioto and Ohio rivers. By 1939, the species was locally established in the oxbows and overflow ponds adjacent to the Scioto and Ohio rivers. Recent introductions in the many recently constructed impoundments should result in the establishment of small, local populations throughout Ohio.

Habitat—The warmouth sunfish was most numerous in lakes, ponds, oxbows, marshes, and streams of base or very low gradients which had silt-free water, an abundance of aquatic vegetation, and a mucky bottom which was often covered with organic debris. The species was present only in small numbers in weedless oxbows and ponds which had a

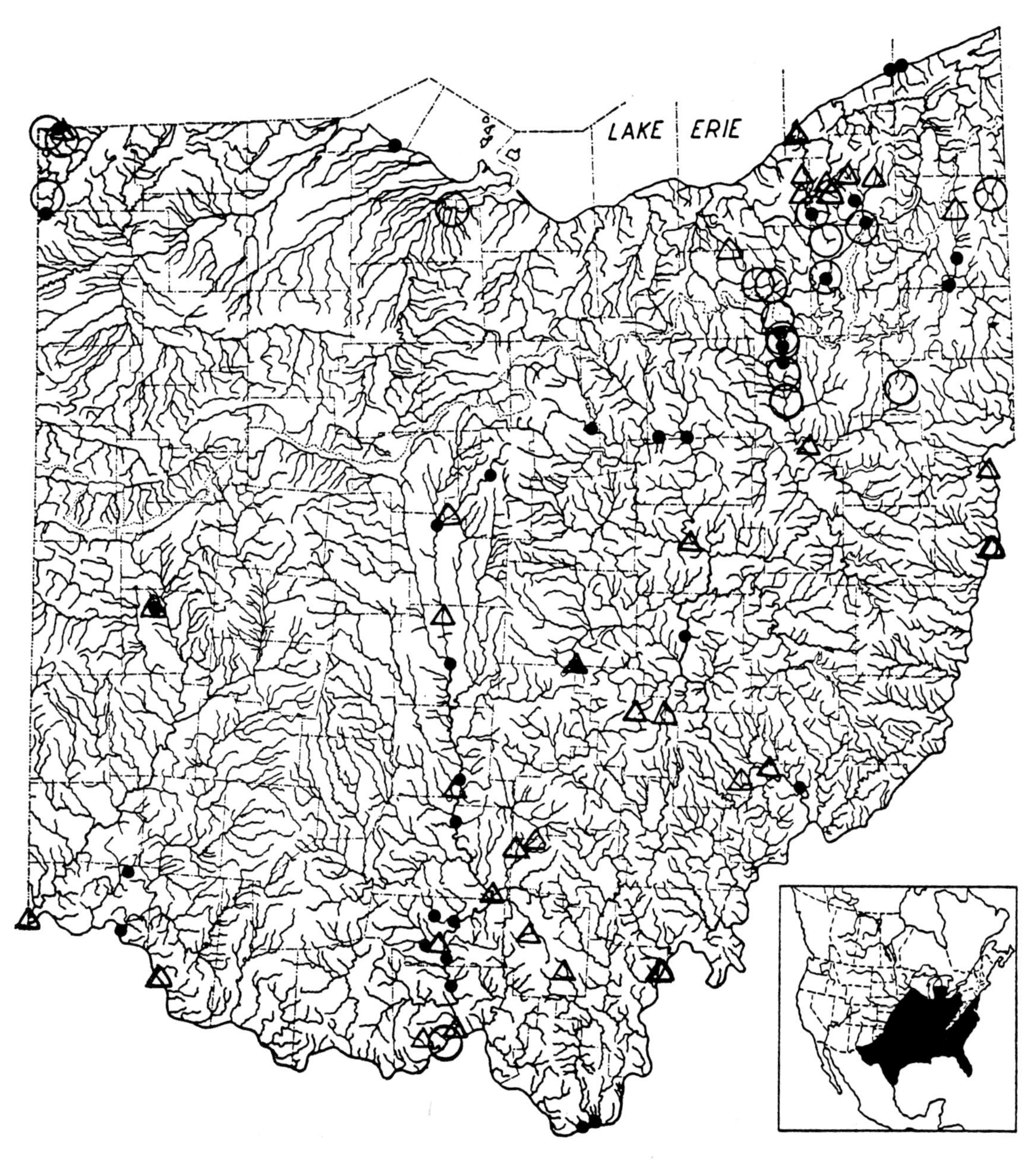

MAP 138. Warmouth sunfish

Locality records. ○ Before 1939. ● 1939–50. △ 1955–80.
Insert: Successfully introduced within and outside of this range; western limits of range indefinite.

yellow-silt bottom, and although its colloquial name was "Mud Bass" it seemed to be less tolerant to turbidity and siltation than was the green sunfish.

Years 1955–80—The general distributional pattern and the numerical abundance of the warmouth have remained relatively unchanged during this period, although additional populations were found in Lake, Jefferson, Tuscarawas, Coshocton, Perry, Vinton and Jackson counties. As was previously the case, the species usually remained uncommon wherever found. Comparatively few of the many recent introductions of the warmouth into the streams, ponds and reservoirs of Ohio have resulted in the possible establishment of permanent populations.

H. Ronald Preston reported that during the

1968–70 surveys of the Ohio River the species was recorded for only one of the nine collecting stations which border the state of Ohio. The station is opposite northern Jefferson County. Between 1976 and 1978 Margulies and Burch collected the warmouth at four localities in the Ohio River from Jefferson County downstream to Hamilton County, Ohio.

Many ichthyologists have long believed that the genus *Chaenobryttus* should be synonomized with the genus *Lepomis*. This was done in 1970 (Bailey et al.:36).

GREEN SUNFISH

Lepomis cyanellus Rafinesque

Fig. 139

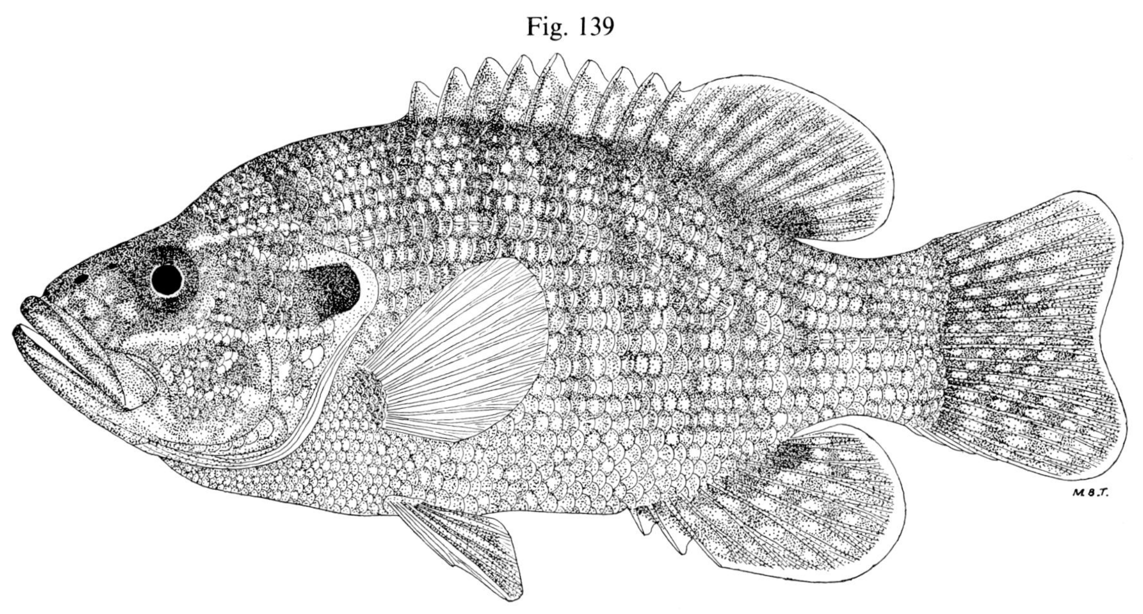

St. Joseph River, Williams County, O.

Oct. 24, 1939.
Adult female.

115 mm SL, 5.7" (14 cm) TL.
OSUM 1230.

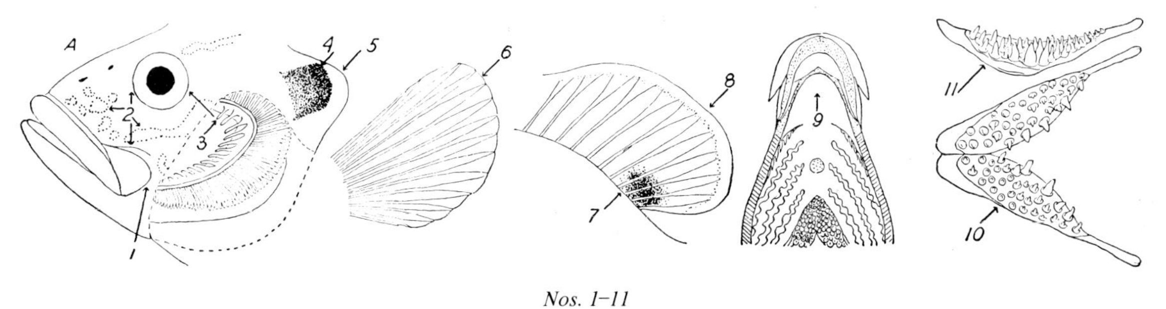

Nos. 1–11

(1) mouth large; posterior end of upper jaw extending well beyond anterior edge of eye; in large adults upper jaw may extend to posterior edge of eye or slightly beyond.

(2) sides of head with emerald mottlings; in some individuals 3–5 wavy, broken, emerald bars radiate backward from snout and eye.

(3) dotted portion of gill cover removed, exposing first gill arch; note long, thin gill rakers, the longest longer than pupil diameter.

(4) black opercle flap stiff, the inflexible opercle bone extending posteriorly to opercle membrane.

(5) opercle membrane slaty-white, without red but occasionally yellow-tinged; light color continues forward forming a light dorsal and ventral border for the black opercle spot.

(6) pectoral fin short; its tip broadly rounded.

(7) usually a dusky spot at bases of last three rays of dorsal and anal fins; absent in smallest young, occasionally absent in larger fishes.

(8) dorsal, caudal and anal fins usually margined with whitish, pale yellow or pale orange in large young and adults; very conspicuous in breeding adults.

(9) normally no teeth on tongue.

(10) dorsal view of lower pharyngeal arches, showing their narrow width and bluntly conical teeth.

(11) lateral view of a pharyngeal arch; note heavy basal portion of the conical teeth.

Identification

Characters: *General—See* white crappie. *Generic*—No teeth on tongue, *see* No. 9, except in warmouth and rarely a few teeth in a green sunfish; supra-maxilla very small or lacking. Anal spines 3. Dorsal spines 10, rarely 9 or 11. Fewer than 55 scales in the complete lateral line, rarely more than 50. Body depth usually contained 1.9–2.8, extremes 1.7 (largest adults)—3.0 (smallest young) times in standard length. *Specific—See* Nos. 1–11 for these characters.

Differs: Warmouth sunfish has teeth on tongue. Bluegill, orangespotted and longear have flexible opercle flaps. Redear and pumpkinseed have pointed pectoral fins. Blackbasses have more lateral line scales. White bass, walleye, sauger, perch, drum have separated dorsal fins.

Most like: Warmouth and longear sunfishes.

Coloration: *Adults*—Dorsally olive-green or olive-slate, with many emerald-green reflections. Sides lighter; often a light yellow-green. Ventrally white, yellow-white, or pale yellow. Sides usually contain 7–12 dark, vertical bars; these bars absent on smallest young and on fishes taken from turbid waters. Sides of head with mottlings of light emerald, some of which tend to form wavy backward-radiating lines. No red on opercle flap. A dusky spot usually present at bases of last three soft dorsal and anal rays. Dorsal, caudal, and anal fins olive, sometimes plain, sometimes with light spots and their margins whitish, yellowish, or orangish. Paired fins olive or transparent; in breeding males pelvics are dusky. *Young*—Similar to adults, except that very small young lack bands on sides and spots on cheeks and fins.

Lengths and **weights:** *Young* of year in Oct., 0.8″–2.5″ (2.0–6.4 cm) long. Around **1** year, 1.0″—3.2″ (2.5–8.1 cm). *Adults*, usually 2.5″ (6.4 cm) (dwarfed individuals)—7.5″ (19 cm) long, weight 0.3 oz–10 oz (8.5–283g). Largest specimen, 10.8″ (27.4 cm) long, weight 14.5 oz (411 g).

Hybridizes: naturally in Ohio with bluegill, orangespotted, longear, redear and pumpkinseed sunfishes; artificially produced elsewhere with largemouth blackbass and warmouth (Hester, 1970: 102; Childers, 1967:167).

Distribution and Habitat

Ohio Distribution—Preserved specimens from many localities and reliable literature references indicate that previous to 1900 the green sunfish was present in all sections of Ohio, and that it was generally distributed and abundant west of a line from Cleveland south to Marietta. East of that line it was less generally distributed and only locally abundant. Since 1920, the species was most abundant in the streams and overflow ponds of the Maumee and Scioto river drainages, and it was evident that small populations or strays occurred in every large and most of the small stream systems in the state. The population density of the green sunfish was comparatively low in northeastern Ohio, was very low in the high-gradient streams east of the Flushing Escarpment, and for no apparent reason was surprisingly uncommon in the mature streams of the northern half of the unglaciated section of Ohio. In all three sections it usually became well established or abundant in impoundments and natural lakes of all sizes after it had invaded these or was introduced. It increased noticeably in abundance in several streams during the 1920–50 period, and especially in those where the decrease in abundance of the longear sunfish was most marked. Most of the Ohio River records were for the colder months, and only isolated populations or strays were present in Lake Erie.

Habitat—The green sunfish was essentially an inhabitant of brooks and small creeks, especially during the breeding season, and of oxbows, overflow ponds, impoundments and natural lakes. It was present in fewer numbers in the large streams, and in these large streams the populations were larger in winter than in the breeding season. The species principally inhabited low- or moderate-gradient streams, and displayed no particular preference for any type of bottom. It frequented sunken brush heaps, and was present in moderate numbers in beds of aquatic vegetation. This was especially true of the young. The largest populations occurred in those favored habitats where there was little competition from other sunfish species, and since the green sunfish was more tolerant to turbidity and siltation than were the other sunfishes except for the orangespotted, it occurred, sometimes in large populations, in waters too silty for these others.

Years 1955–80—All except a few of the routine collections of fishes made in inland Ohio waters during this period produced at least one green sunfish. From these investigations it was apparent that the species remained one of the most widely distributed and numerically abundant of any fish species found in the state. It was present and usually

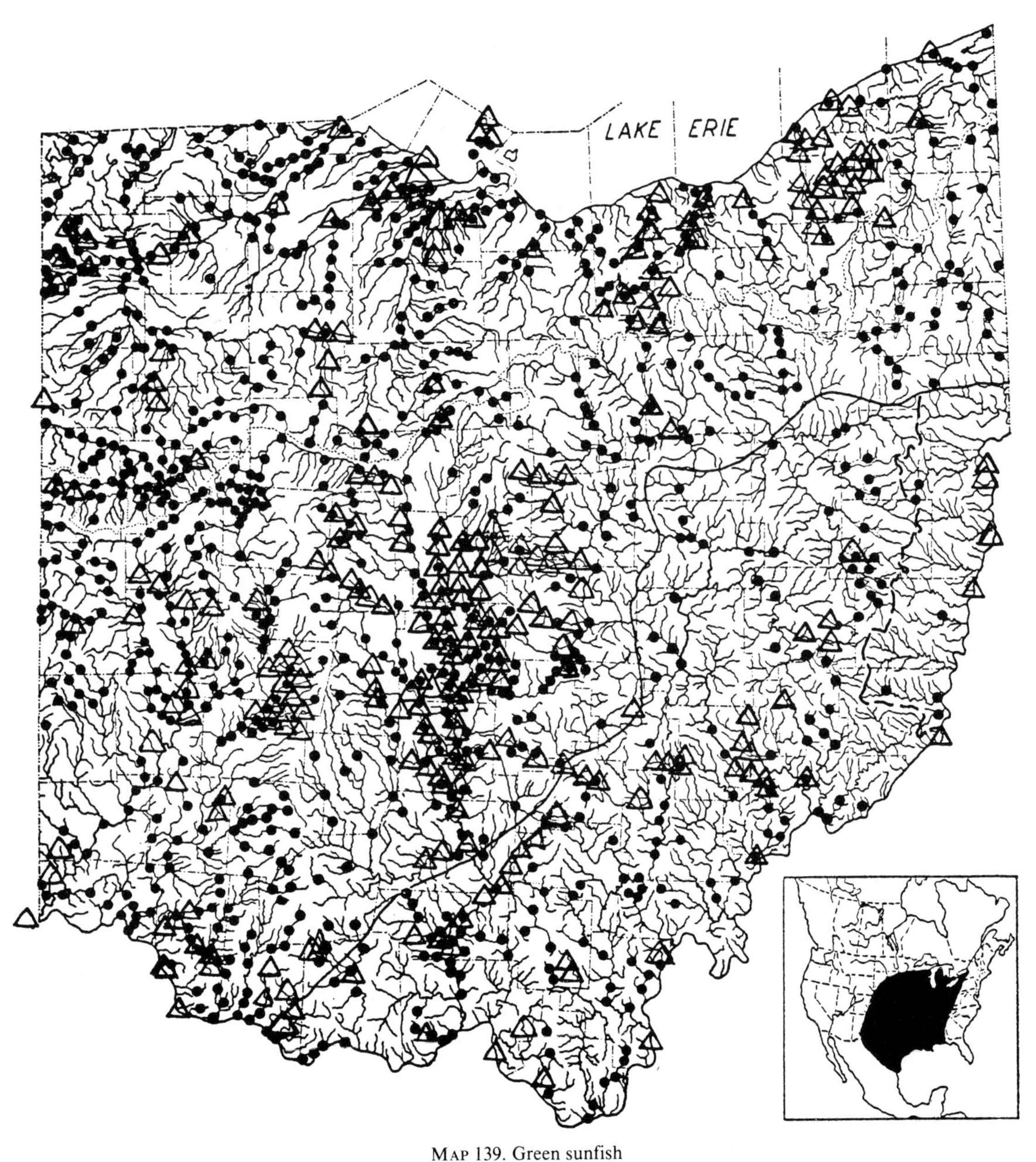

MAP 139. Green sunfish
Locality records. ● Before 1955. △ 1955–80. ~ Glacial boundary. /\ Flushing Escarpment.
Insert: Introduced, often inadvertently, into eastern and western states outside of this range.

numerous in those farm ponds where it had advertently or inadvertently been introduced. The species hybridizes freely with other sunfish species (*see* Hybridizes above), and in some ponds the parent species were outnumbered by their hybrid young. Together with the orangespotted sunfish, the green appeared to be more tolerant of disturbed conditions and many pollutants, including silt, than were the other sunfish species.

The green sunfish occurred in the wave-protected and/or vegetated shallows of western Lake Erie (Van Meter and Trautman, 1970:75). It was fairly numerous in the Ohio River as indicated by the 1957–59 investigation of that entire stream (ORSANCO, 1962:151). During the 1968–69 investigations of that river, H. Ronald Preston recorded it from six of the nine collecting stations that bordered the state of Ohio.

NORTHERN BLUEGILL SUNFISH

Lepomis macrochirus macrochirus Rafinesque

Fig. 140

Lake Erie, Ottawa County, O.

June 30, 1942.
Adult male.

107 mm SL, 5.3″ (13.5 cm) TL.
OSUM 5105.

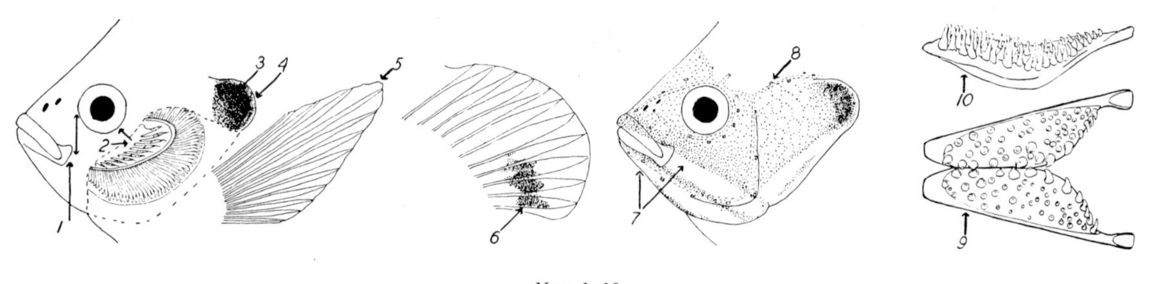

Nos. 1–10

(1) mouth very small, only in some individuals does the posterior end of upper jaw extend as far backward as the anterior edge of the eye.

(2) dotted portion of gill cover removed, exposing first gill arch; note thin, straight, long gill rakers; the longest may be longer than half the eye diameter; gill rakers shortest in smallest and largest fishes and those from polluted waters; in polluted waters gill rakers may be crooked and deformed.

(3) black opercle flap very flexible; the opercle bone so flexible that the flap may be doubled forward until its posterior margin nearly touches the cheek, without fracturing the bone.

(4) opercle membrane as black as opercle spot or slightly lighter; no red.

(5) pectoral fin long, its tip moderately or sharply pointed.

(6) normally a dusky spot on webbing near base of last three soft dorsal rays; usually a similar spot at base of last three soft anal rays; spots least conspicuous or absent in fishes from turbid waters and in the small young.

(7) large young and adults have two bluish bars extending backward from mouth and chin, and usually the lower edge of gill cover is likewise bluish, hence the name "Bluegill"; bars absent in small young.

(8) sensory openings small and circular, as they are in all species of sunfishes *except* the orangespotted.

(9) dorsal view of lower pharyngeal arches, showing their moderate width.

(10) lateral view of a pharyngeal arch; note thin and sharply-pointed teeth.

Identification

Characters: *General—See* white crappie. *Generic—See* green sunfish. *Specific—See* Nos. 1–10 for these characters.

Differs: Orangespotted, longear and green sunfishes have short, rounded pectoral fins and large mouths. *See* green sunfish.

Most like: Redear sunfish.

Coloration: *Adults*—Varies dorsally from light yellow-olive to dark olive-green, with an emerald-bluish luster. Sides more bluish with some silvery reflections. Breast and belly whitish, yellow, orange, or rusty-red. Body normally with 5–9 chainlike double-bars; these bars sometimes absent; almost invariably absent in fishes from turbid waters. Bluish bars extending backward from mouth and chin, see No. 7. Checks without emerald or reddish spots. Opercle flap black. Fins plain, except for dusky dorsal and anal spots, *see* No. 6. In some adults the small scales on nape and predorsal area of back have bold, black spots at their bases. *Breeding male*—Has bright orange or rust-red breast; a bluish sheen over the body; tail slaty-bluish; the two blue streaks very intense that go backward from mouth and cheek; pelvics dusky. *Young*—Similar to adults, except that they are more silvery; spots on posterior portions of the anal and dorsal fins not developed; neither are the blue streaks on lower sides of head. *Young—less than 2.0″ (5.1 cm) long*—Have a few large melanophores on breast, which greatly aid in separating the bluegill from young pumpkinseeds, which have no spots.

Lengths and **weights:** *Young* of year in Oct., 0.7″–3.2″ (1.8–8.1 cm) long. Around **1** year, 1.0″—–4.0″ (2.5–10 cm). *Adults*, usually 3.5″ (8.9 cm) (dwarfed breeding adults in small ponds)–10.0″ (25.4 cm). Fishes 4.0″–6.0″ (10–15 cm) long, usually weigh 0.5 oz–3.0 oz (14–85 g); fishes 6.0″–8.0″ (1.5–20 cm) usually weigh 2 oz–7 oz (57–198 g); fishes 8.0″–10.0″ (20–25.4 cm) usually weigh 6 oz–14 oz (170–397 g). Longest specimen, 11.8″ (29.9 cm), weight 1 lb, 3 oz (0.5 kg). Heaviest specimen, 10.5″ (26.7 cm) weight 1 lb, 12.5 oz (0.8 kg). For growth-weight relationship, see Beckman (1949:68).

Hybridizes: naturally in Ohio with warmouth, green, orangespotted, longear, redear and pumpkinseed sunfishes; naturally or artificially produced elsewhere with black crappie, rockbass and largemouth blackbass (Childers, 1967:161; Birdsong and Yerger, 1967:62–71; Hester, 1970:102).

Distribution and Habitat

Ohio Distribution—Preserved specimens from Hamilton, Franklin, Licking, Allen, Lucas, Ottawa, Sandusky, Erie, Lorain, Cuyahoga, Lake and Ashtabula counties, taken between 1853–99, indicate that the bluegill was present in both drainages since early times; from reliable literature references that it was abundant in the harbors, marshes and bays of western Lake Erie and the larger inland lakes. The species reached commercial importance about western Lake Erie before 1850, for this species and the pumpkinseed could be taken readily in the shallow waters with the small seines in use in those days. It continued to be of commercial importance from 1850–1900, especially in Sandusky Bay, West, Middle and East harbors, and the marshes from Port Clinton to Toledo (Smith and Snell, 1891:255–65); but how important it was commercially is not known, for the annual poundage of this species and several others was included together under the term "soft fish." Between 1840–1900 the bluegill was commercially important in such large impoundments as Buckeye, Indian and St. Marys lakes, where during the height of the seining and hoop-netting season, dozens of barrels of bluegills were shipped weekly from each lake.

Before 1900, the bluegill probably occurred in all or most of the pothole lakes of glaciated Ohio, and in many overflow ponds and oxbows. It quickly became abundant in the canals. Canals may have aided in distributing the species more generally.

Between 1920–50 the State Conservation Department planted millions of bluegills annually, the shipments going to every county. Introductions into impounded waters, including farm ponds, were frequently successful, and the species soon became

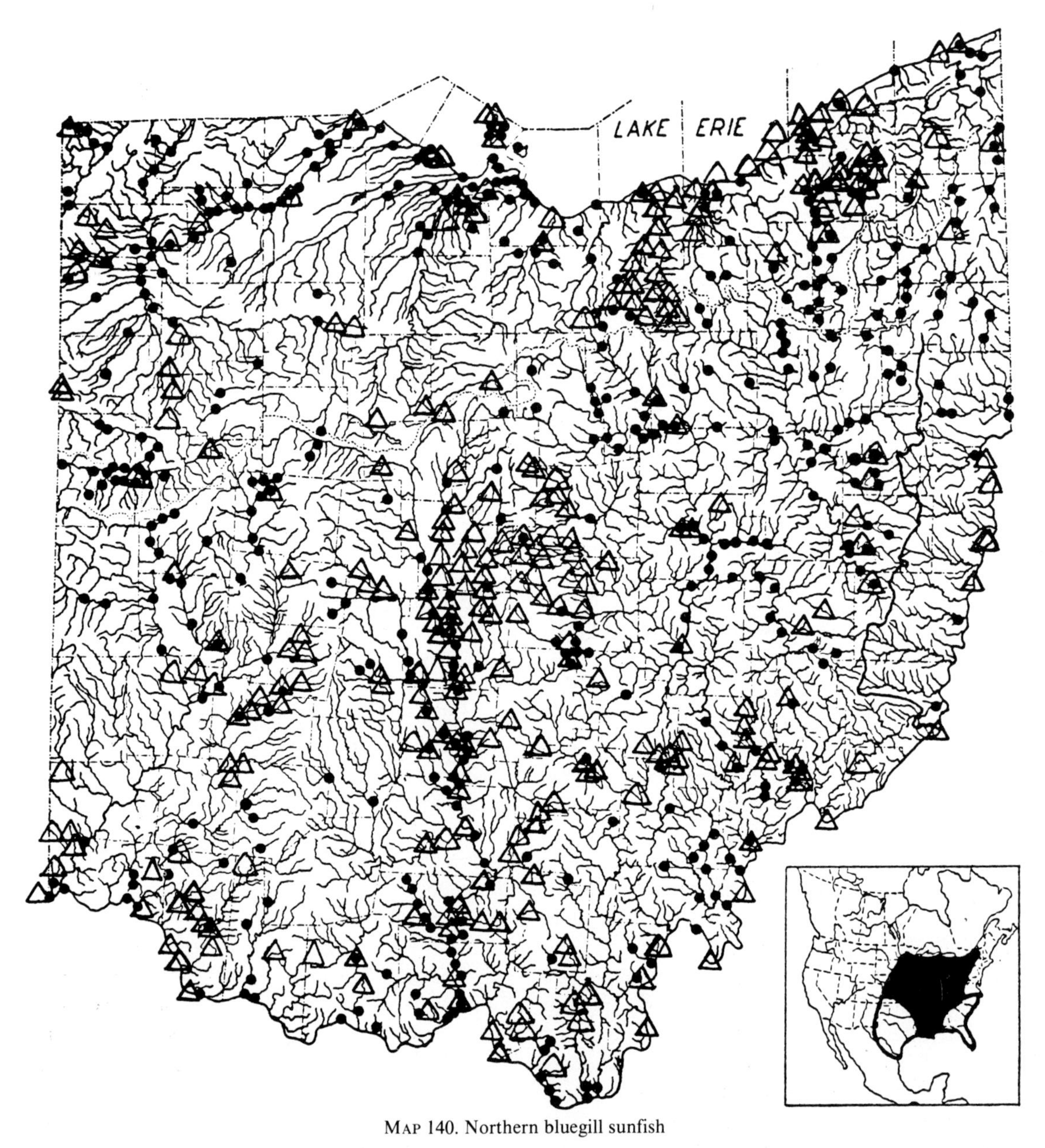

Map 140. Northern bluegill sunfish

Locality records. ● Before 1955. △ 1955–80. ~ Flushing Escarpment.
Insert: Black area represents original range of nominate form, outlined areas of nominal forms; widely and often successfully introduced outside of, and within, this range.

abundant in them; also fishes drifting from these impoundments into their outlets resulted in producing small populations in these streams. Introductions into flowing waters were not so successful, except when some of the fishes reached the oxbows and overflow ponds, or when they found a more or less permanent habitat in a large stream of base or low gradient containing clear water and some aquatic vegetation. The bluegill was rare or absent in high-gradient streams, such as those east of the Flushing Escarpment.

Despite repeated plantings of fry and adults, the bluegill decreased markedly in abundance during the 1920–50 period in the now turbid Maumee and Auglaize rivers, where in 1893 Kirsch (1895A:330) found it to be present at nearly every collecting station, but where since 1940 only strays were found. Likewise the huge bluegill population in Lake St. Marys, present before 1900, almost disappeared with the increased turbidity and virtual disappearance of stumps and aquatic vegetation, and by 1940 only a remnant population remained (Clark, 1942:134, and 1951:16–19).

Habitat—The largest bluegill populations were in non-flowing waters which were clear or contained little suspended clayey silts, which had bottoms consisting of sand, gravel, or muck containing organic debris, and usually where there were scattered beds of aquatic vegetation. If conditions in small impoundments or farm ponds were very favorable, this prolific species became too numerous, resulting in a drastic dwarfing of the many individuals. In extreme cases the dwarfed adults would be less than 4.0″ (10 cm) in total length.

Years 1955–80—Before 1900 the published annual commercial catches for Ohio of bluegills and pumpkinseeds, plus insignificant poundages of other sunfishes (excluding crappies, rockbass and blackbasses), were combined under the title "Sunfish." Although the annual commercial poundage of bluegills is unknown, it must have been large because of the large amount of prime habitat then available. This species was highly desirable for the market because in Ohio it averaged the largest of the Lepomine sunfishes and was of excellent flavor. Smith (1898B:510–11) states that 315,876 lbs (143,279 kg) of "Sunfish" were commercially taken in 1894 from four reservoirs: Grand (St. Marys), Lewiston (Indian), Loramie and Licking (Buckeye).

After 1900, if not before, the Ohio population of bluegills began to decrease in size, continuing until at least 1940, despite annual introductions of millions of adults and young that were produced in hatcheries and taken from Lake Erie or elsewhere. The decrease was coincidental with the decreasing size and eroding of its habitat, especially in the larger inland impoundments (Clark, 1942:16–19, 134, 151).

Before 1950 Ohio had relatively few farm ponds or similar impoundments. By 1970 the number had increased so rapidly that there were a reported 40,000 (White et al., 1975:106–7). Bluegills were well-suited to the average farm impoundment, especially when managed properly, and all except a small number of these ponds were stocked with this sunfish. In these impoundments the species was normally most successful in maintaining or increasing its numbers, particularly when their major controlling predator, the largemouth blackbass, was sharply reduced in numbers, such as through over-fishing. A greatly reduced bass population normally resulted in a population explosion of bluegills unless the latter were otherwise controlled.

By 1970 the bluegill was presumably caught by more men, women, boys and girls than was any other fish species in Ohio, although the blackbasses remained the most important game fishes to many sportsmen. The popularity of the bluegill was increased because of the ease with which it could be caught. Larger numbers of bluegills were taken between 1955–80 than were normally captured in the same length of time during the 1930–50 period.

Bluegill populations obviously increased greatly in size in the Ohio River during this period in part probably because of reinforcements from inland streams. The 1957–59 investigations of the entire river proved that this species was the most numerous sunfish in that stream (ORSANCO, 1962:151). H. Ronald Preston reported that during 1968–69 bluegills were taken in all nine collecting stations in that portion of the river bordering the state of Ohio, as did Margulies and Burch at their five collecting stations during the years 1976 to 1978.

For life history at Buckeye Lake, *see* Morgan (1951).

ORANGESPOTTED SUNFISH

Lepomis humilis (Girard)

Fig. 141

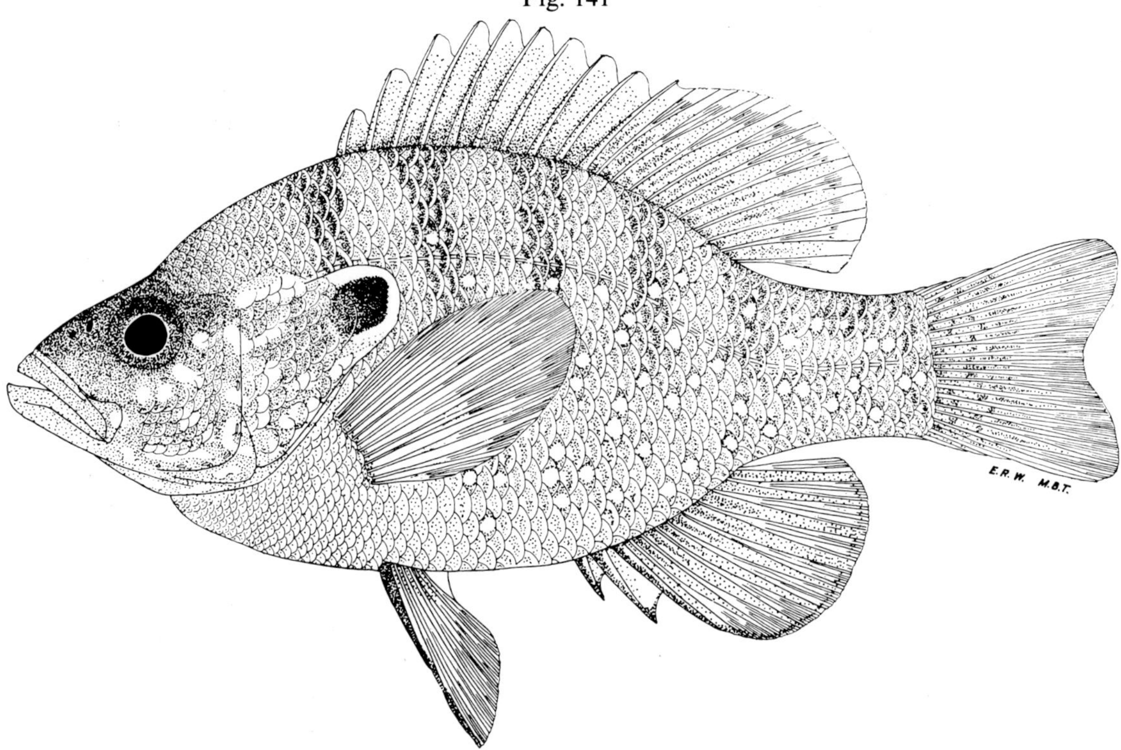

Auglaize River, Auglaize County, O.

July 9, 1941. 60 mm SL, 2.8″ (7.1 cm) TL.
Breeding male. OSUM 3358.

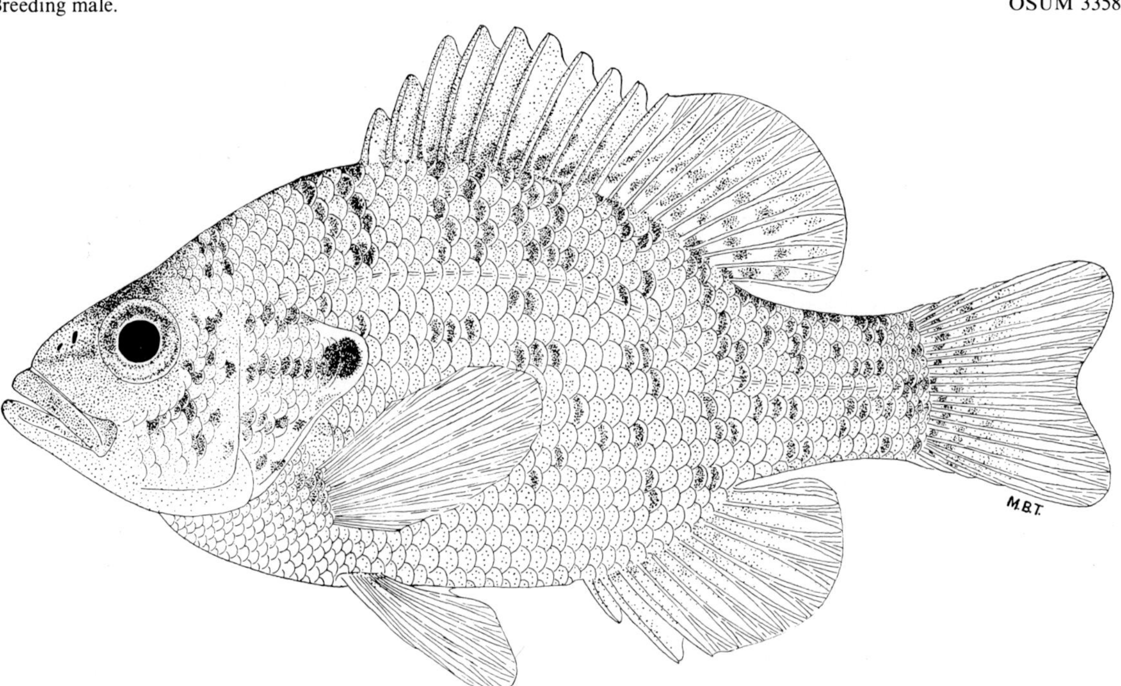

Same locality, date and OSUM number.

Breeding female. 51 mm SL, 2.5″ (6.4 cm) TL.

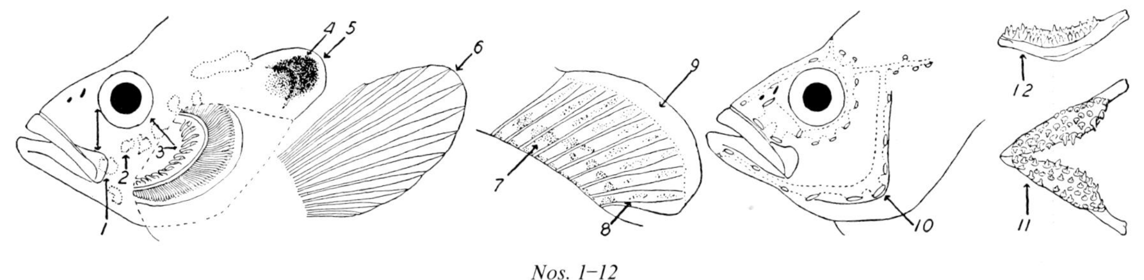

Nos. 1–12

(1) mouth moderately large; the posterior end of upper jaw extends beyond the anterior edge of eye.

(2) cheeks, also body, of males with orange spots; spots brownish in females and young.

(3) dotted portion of gill cover removed; exposing first gill arch; note moderately long gill rakers, about a third the diameter of eye.

(4) black opercle flap extremely flexible; the opercle bone so flexible that the flap may be doubled forward until its posterior margin touches the cheek without fracturing the bone.

(5) opercle membrane white; flushed with orange in breeding males.

(6) pectoral fin short and broadly rounded at the tip.

(7) soft dorsal spotted or mottled in females and young males; transparent without spots in smallest young.

(8) soft dorsal flushed with orange in breeding males.

(9) soft dorsal margined broadly with white or light orange in males.

(10) sensory openings along free edges of preopercle greatly lengthened; the longest usually longer than diameter of anterior nasal opening.

(11) dorsal view of the narrow, lower pharyngeal arches.

(12) lateral view of pharyngeal arch; note sharply pointed teeth.

Identification

Characters: *General—See* white crappie. *Generic—See* green sunfish. *Specific—See* Nos. 1–12 for these characters.

Differs: All other species of sunfishes have the sensory pores along the free edges of the preopercle circular and small, *see* central longear sunfish, No. 8.

Most like: Green, longear, and pumpkinseed sunfishes.

Coloration: *Adult male*—Dorsally olive-green with much blue and emerald luster. Sides lighter and more silvery. Ventrally milk-white or yellowish. Back and sides of some males contain 4–7 broad, dark olive bands. Sides of head and body with many orange spots; spots on head sometimes merging to form backward radiating, wavy bands. Black spot on opercle flap broadly margined with white dorsally, ventrally, and posteriorly; occasionally the white border is flushed with orange. Vertical fins usually unmottled, but flushed and broadly margined with reddish-orange; the tail with the least flush. Vertical fins transparent, or tinged with yellow or light orange. *Breeding male*—Intensely brilliant, the dark bands, orange spots and fins boldly conspicuous; iris of eyes orange-red; first ray of pelvics blackish; remaining rays bright orange. *Female* and *immature male*—Have the same general color-pattern, but lack the orange blush; spots on side of head and body brown; soft dorsal and sometimes the anal and caudal base mottled with dark spots; remaining fins silvery; entire fish with a silvery cast. *Small young*—Have all fins transparent; brown spots undeveloped; bands on the sides often very conspicuous.

Lengths and **weights:** *Young* of year in Oct., 0.8″–2.0″ (2.0–5.1 cm) long. Around **1** year, 1.0″—–2.5″ (2.5–6.4 cm). *Adults*, usually 1.5″–3.5″ (3.8–8.9 cm) long, weight 0.5 oz–1.0 oz (14–28 g).

Hybridizes: with green, bluegill, longear, redear and pumpkinseed sunfishes.

Distribution and Habitat

Ohio Distribution—In 1888, Jordan and Evermann, investigating the Wabash and other streams in Indiana, found the orangespotted sunfish only in its extreme southwestern portion (Jordan, 1890:163). Since 1888, agricultural practices similar to those conducted in Ohio have converted the once clear prairie-type streams of Indiana into turbid plains-type streams, thereby establishing orange-spotted habitat throughout the Wabash River drainage (Gerking, 1945:101). As a result, this species moved across Indiana and into western Ohio, where in 1920 it was found to be rather common in the Wabash system of Mercer County, Ohio. At that time it apparently had not become established in Lake St. Marys; however, by 1929 it

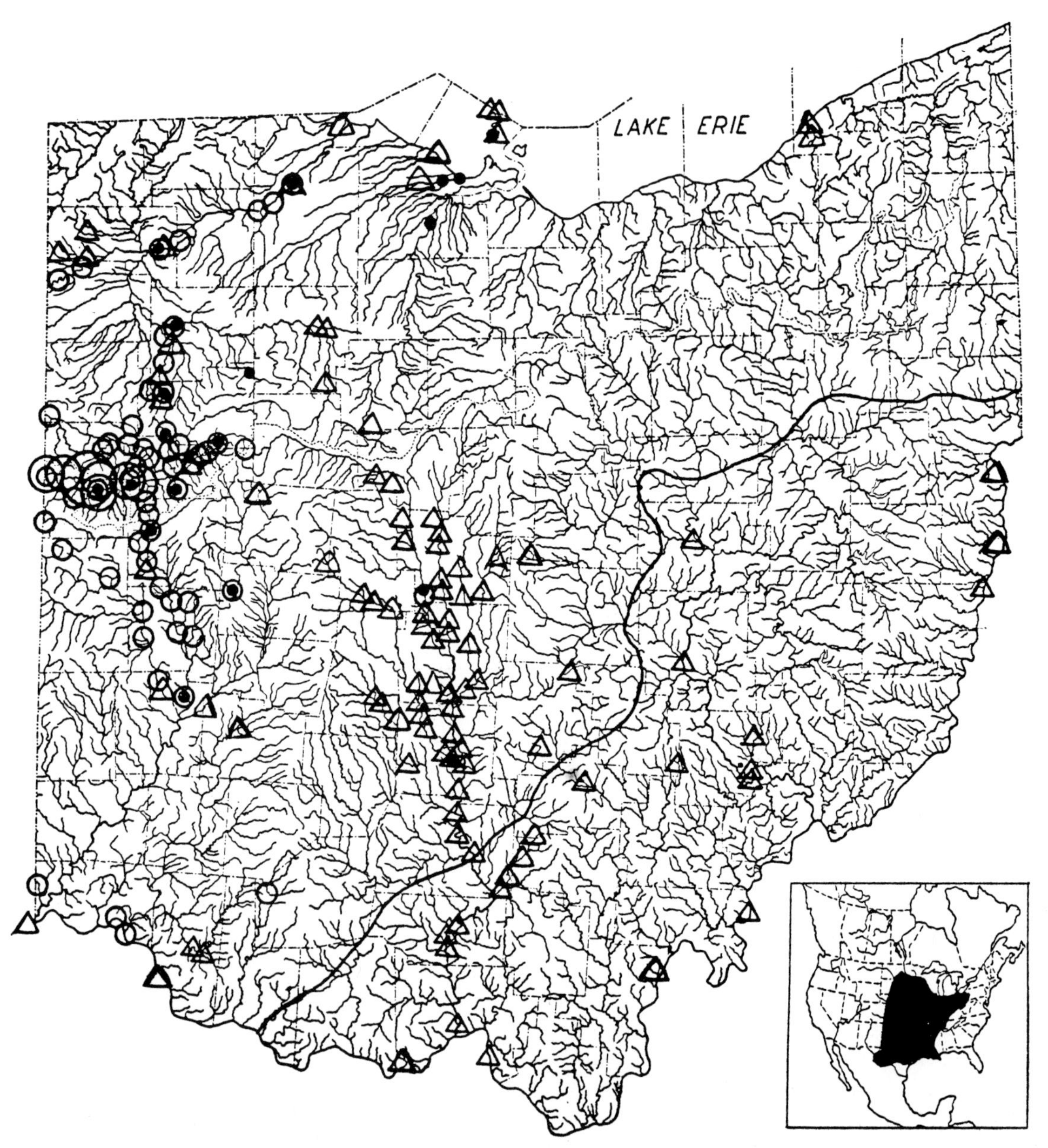

MAP 141. Orangespotted sunfish

Locality records. ○ 1920. ○ 1921–29. ○ 1930–54. △ 1955–80. ~ Glacial boundary.
Insert: Recent northeastward extension of range; western limits of range and those in Mexico indefinite.

had surmounted the spillway (presumably by the aid of man) which connected Lake St. Marys and the Wabash river systems, and in that year was found in the eastern outlet of Lake St. Marys, which is part of the Lake Erie drainage. After that it rapidly invaded the Maumee River system, so that by 1941 fair-sized populations occurred as far downstream as Waterville, Wood County. It was first taken in the Portage River and tributaries of Sandusky Bay in 1948, and in 1952 the first specimens were taken in Terwilliger's Pond of South Bass Island. In the Ohio drainage it had become established in the upper Great Miami River drainage by 1930, by 1942 it was in the Little Miami, and by 1945 in the Scioto River system of central Ohio.

Throughout the 1925–50 period I attempted to follow the orangespotteds as they moved eastward across western Ohio. Almost invariably the first specimens collected along the then-existing eastern frontier were hybrids between the orangespotted and some other sunfish species. Collecting in the same locality a few years later, after the eastern frontier had moved beyond this locality, pure orangespotted sunfishes were taken, and usually they far outnumbered the hybrids. Apparently the first orangespotteds, arriving in a new locality, were too few in number to be taken with seines and possibly were so few that they were forced to spawn with other sunfish species. It was therefore not until their numerous hybrid young appeared that we were able to take a few of them in seines. After a few years had elapsed however, orangespotteds had invaded this locality in sufficient numbers so that they bred together, and after that the orangespotted sunfishes outnumbered their infertile hybrids.

In 1888 Henshall (1888:79) wrote that this species was "Common in Ross Lake and Clough Creek" of Hamilton County, Ohio. Examining Henshall's sunfish collections in the Cincinnati Society of Natural History, which had been labelled by Henshall as *Lepomis humilis*, I found that these sunfishes were not *humilis* but one of three other species. Therefore Henshall's records are not accepted.

Habitat—This species was present in the largest numbers in streams of low gradients, in lakes and other impoundments, where the waters were usually turbid and the bottoms covered with silt. In such situations it apparently was a competitor of the green sunfish, with which it freely hybridized. The orangespotted appeared to avoid high gradients, and in clear waters was usually far outnumbered by longears, pumpkinseeds and bluegills. It seems logical to assume that in the future the largest populations will occur in the sluggish, turbid waters of glaciated Ohio, with possible isolated populations in impoundments and low-gradient streams of the glaciated section.

Years 1955–80—The orangespotted sunfish continued its eastward invasion of Ohio (White et al., 1975:108), making its greatest gains in glaciated territory, as suggested above in the last sentence under Habitat. This was especially true of the glaciated Scioto River system. In the lower section of the Scioto south of the Allegheny Front Escarpment, its numerical increase was not so pronounced, although this glacial outwash section is in fact glaciated. Whether through introductions by man or by invasion, the species likewise appeared in the upper Hocking and Muskingum rivers to the east of the Escarpment, where these also are primarily glacial outwash streams.

The distribution of the species in Ohio waters appears to have been aided, inadvertently or otherwise, by introductions into farm ponds and larger reservoirs, such as Burr Oak Reservoir in unglaciated Athens and Morgan counties. During the 20-year period the older populations of the Maumee and upper Miami drainages definitely appear to have become stabilized as regards numbers, and the numbers of hybrids observed have decreased sharply. In the Scioto and to a lesser extent in the Muskingum River, their numbers seem to have been increasing rapidly as was hybridization with other sunfish species.

CENTRAL LONGEAR SUNFISH

Lepomis megalotis megalotis (Rafinesque)

Fig. 142

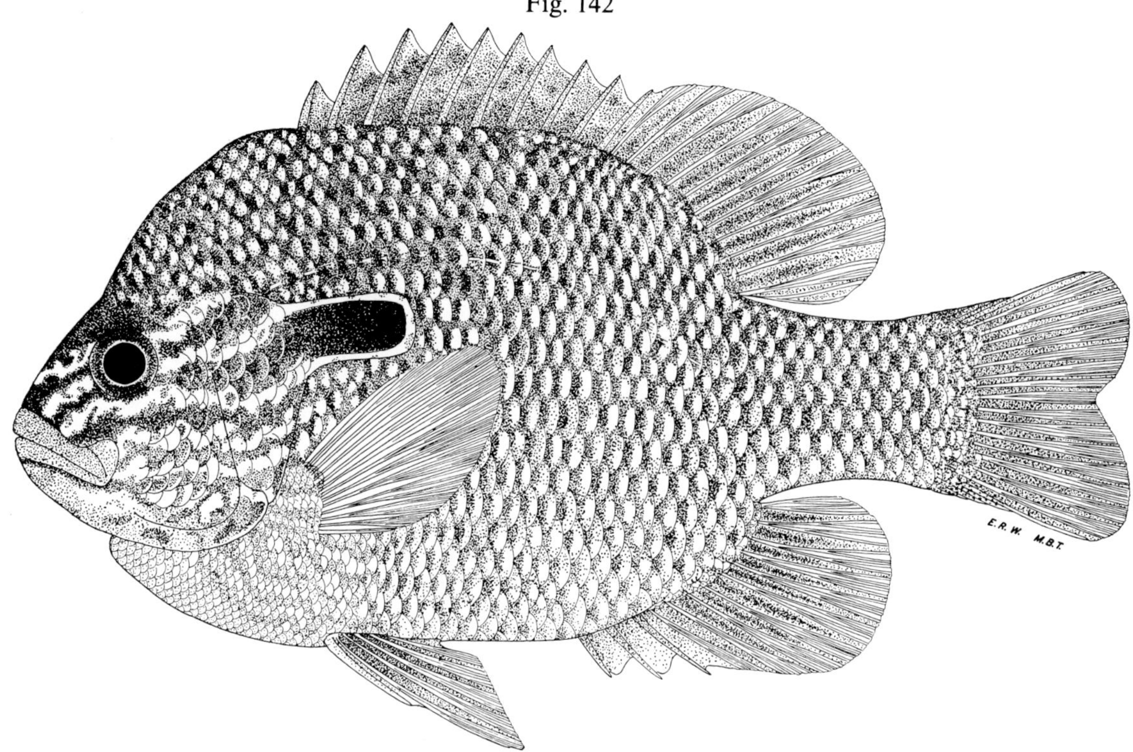

Hocking River, Athens County, O.

May 25, 1939.
Breeding male.

112 mm SL, 5.3″ (13 cm) TL.
OSUM 262.

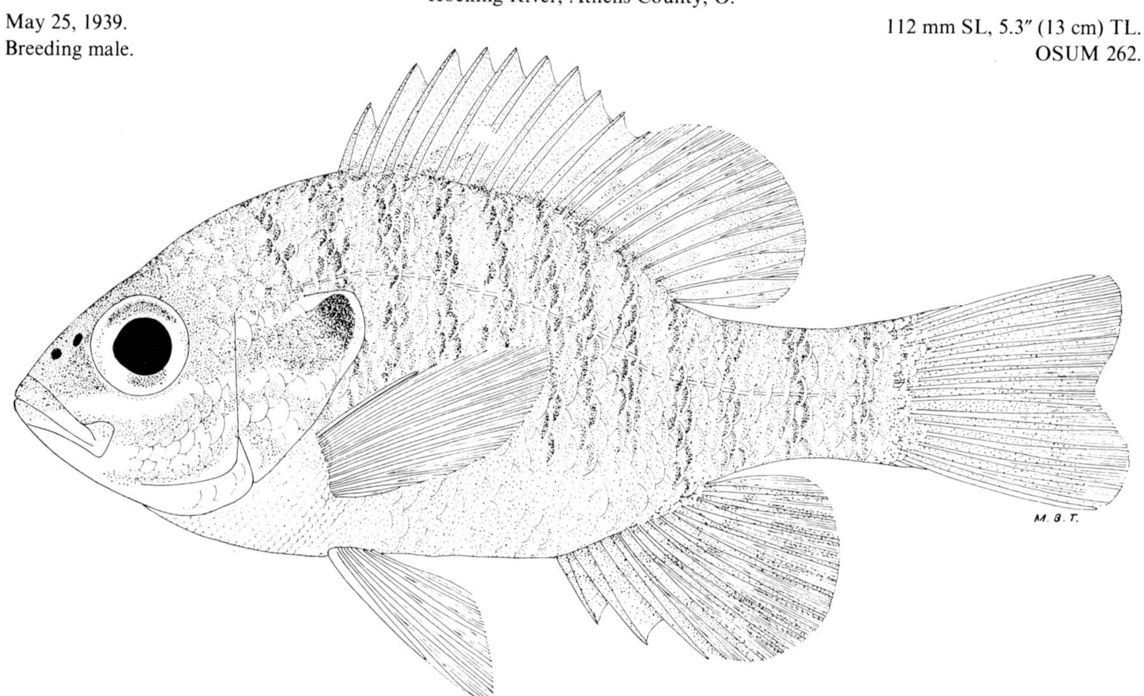

North Fork, Paint Creek, Fayette County, O.

Nov. 21, 1939.
Immature.

48 mm SL, 2.4″ (6.1 cm) TL.
OSUM 1342.

Compare with adult; note in young the larger eye; higher fins; angle and smaller size of opercle flap; distinct banding on sides; more slender body; other differences in body proportions.

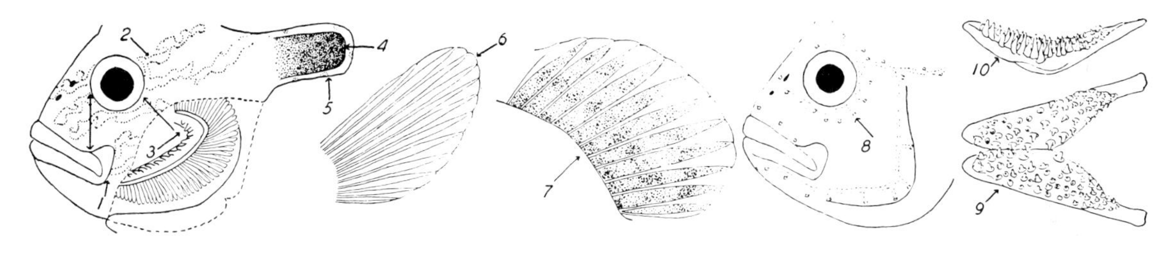

Nos. 1–10

(1) mouth moderately large; the posterior end of upper jaw extends to about the center of eye; always beyond anterior edge of eye.

(2) sides of head in large young and adults with several wavy, emerald-blue or light bars which radiate backward from mouth and eye.

(3) dotted portion of gill cover removed, exposing first gill arch; note very short gill rakers, almost as wide as long.

(4) black opercle flap greatly elongated horizontally in adult males; less so in females and young; opercle bone so flexible that the flap may be doubled forward until its posterior margin touches the cheek, without fracturing the bone.

(5) black opercle spot bordered with whitish, dorsally, ventrally and posteriorly in large young and adults; sometimes flushed with an orange blush, or with 1–9 small orange-red spots.

(6) pectoral fin moderately short and rounded at its tip.

(7) soft dorsal without distinct spots in large young and adults; flushed with rusty-orange in males; some spotting of fins in small young.

(8) sensory openings, especially on free edges of preopercle, circular and very small, as it is in all sunfishes except the orangespotted (*see* No. 10 of that species).

(9) dorsal view of the narrow, lower pharyngeal arches.

(10) lateral view of pharyngeal arch, note sharply pointed teeth.

Identification

Characters: *General—See* white crappie. *Generic—See* green sunfish. *Specific* and *subspecific—See* Nos. 1–10 for these characters. Size larger; 4.5″–9.0″ (11–23 cm) long. Adult males, some adult females, and large young have opercle flap greatly elongated horizontally; flap with whitish margins that may be flushed with orange-red, or which may contain several, small reddish spots. Bars on sides of body, and spotting on soft dorsal and anal fins usually faint or absent in large young and adults. Small young of this subspecies and the northern longear have characters so similar as to be impossible to separate with certainty.

Differs: *Adults*—Northern longear has a smaller maximum size; less produced and less horizontal opercle flap; male has only a single large red spot on opercle margin. Bluegill, redear, and pumpkinseed have pointed pectoral fins. Warmouth has teeth on tongue. Green and orange-spotted sunfishes have long gill rakers. *See* green sunfish.

Most like: Northern longear. **Superficially like:** Bluegill and redear.

Coloration: *Adult male*—Dorsally olive-green with specks of yellow, emerald, and green. Sides the same, except that the olive-green is lighter and the emerald and blue spotting more intense. Breast and belly yellow, yellow-orange, or orange-red. The 8–12 bands on sides of body usually very faint, often absent. Sides of head with much emerald-blue and several wavy, light orange, or orange-red bars radiating backward from mouth and eye. Black opercle spot bordered with white; the white may be flushed with red or may contain 1–9 small orange or red spots. Dorsal and anal fins olive-slate, often flushed with rusty-orange. Caudal more olive-bluish. *Breeding male*—Intensely brilliant, the wavy bands across cheeks often continuing across the reddish breast; the vertical fins a deep rusty-orange; the pelvics blue-black. *Adult female*—Like adult male except less brilliantly colored; reds are replaced by orange or pale whitish; blues are paler; orange or red is usually lacking on the shorter opercle flap; there is a greater tendency towards bands on sides of body and spots on fins. *Young*—Like female, except that oranges and blues are lacking; bands on sides usually prominent, as are the spots on the fins of the larger young; smallest young have transparent fins.

Lengths and **weights:** *Young* of year in Oct., 0.8″–2.2″ (2.0–5.6 cm) long. Around **1** year, 1.2″–2.8″ (3.0–7.1 cm). *Adults* usually 2.5″ (6.4 cm) (dwarf or intergrading populations)–7.0″ (18 cm) long, weight 0.3 oz–7.0 oz (9–198 g). Largest specimen, 9.3″ (24 cm) long, weight 10 oz (283 g).

Hybridizes: with green, orangespotted, bluegill, redear and pumpkinseed sunfishes.

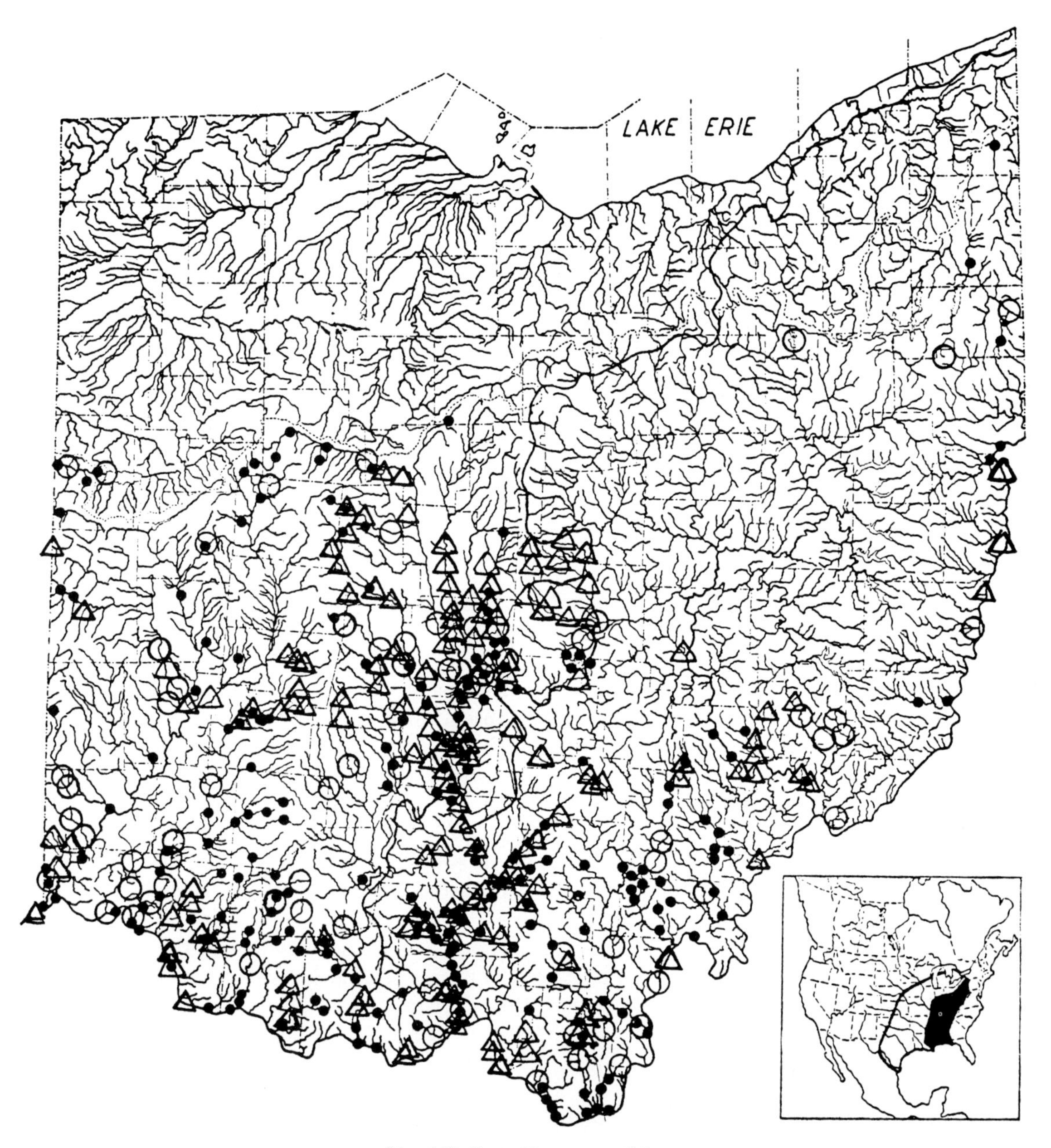

MAP 142. Central longear sunfish

Locality records. ○ Before 1926. ● 1926–54. △ 1955–80. ~ Allegheny Front Escarpment.
Insert: Black area represents range of this subspecies, outlined area of other subspecies.

Distribution and Habitat

Ohio Distribution—It is apparent from the many preserved specimens and fewer reliable published references, that before 1900 the longear sunfish as a species occurred in all sections of inland Ohio. These data likewise indicate that the species occurred abundantly only along the two southernmost tiers of counties and to the west of the Allegheny Front Escarpment. In this western half of Ohio Kirsch (1895A:331) found the longear to be present in almost every locality in the Maumee drainage which he investigated in 1893 and Osburn (1901:81) found it to be "apparently the most common sunfish" in the Scioto River drainage of central Ohio. In the north-central section of the state its scarcity was commented upon by McCormick (1892:27), for he found only two specimens during his investigations of Lorain County, and in northeastern Ohio Kirtland (1850M:117, as *Pomotis nitida*) found the longear only in the Mahoning River. Further proof of abundance to the westward are the many locality records for Indiana (Gerking, 1945: 102, map 106); the few locality records for Pennsylvania (Raney, 1938:75) are proof of its scarcity eastward.

Between 1920–50 the central longear remained abundant only in the low gradient, clearer sections of streams in the two southernmost tiers of counties, from the Indiana state line to Washington County. It was far less numerous throughout the middle third of the state, and was much less abundant in Franklin County after 1920 than it had been before 1900; in fact, there was a continuous decrease in abundance in central Ohio throughout the 1920–50 period. This decrease was undoubtedly correlated with increased turbidity and siltation. Only strays occurred in the Ohio River, and strays or isolated populations in the headwater streams along the Lake Erie-Ohio River watershed.

Habitat—The central longear remained abundant only in the long, sluggish pools in the clearer sections of low-gradient streams of southern Ohio, where the sand and gravel bottoms remained relatively free of yellow silts. It occurred frequently among beds of aquatic vegetation. In southern Ohio, it was a close associate of the spotted blackbass, largely avoiding the higher gradients occupied by the smallmouth blackbass, and the overflow ponds and oxbows inhabited by the largemouth blackbass. This longear avoided turbid impoundments, and occurred in small populations only in a few of the clear-water lakes. It frequented all sizes of streams, but most often occurred in the largest populations in small headwaters and in moderate-sized streams. Its occurrence in few numbers in the large streams may have been due to its intolerance to turbid conditions, for many old fishermen have told me of the great abundance of the longear in the larger streams of southern Ohio about 1900, and in the Ohio River before it was ponded. In several streams during the 1920–50 period I noted that as the longear decreased in abundance, the turbid-tolerant green sunfish increased. Contrary to Osburn's statements for 1898, after 1920 the green sunfish was far more abundant and generally distributed in Franklin County streams than was the longear.

Intergradation between the two subspecies of longear sunfish. The central longear, subspecies *megalotis*, differs from the northern longear, subspecies *peltastes*, in its much greater maximum length, larger and differently colored earflap, and habitat preferences. The differences are so great as to suggest that these two may be incipient species rather than well-marked subspecies. The area of intergradation, if any, lies in a narrow band along the Lake Erie-Ohio River divide. Intergradation was not thoroughly studied or proven because of the few adult specimens obtainable in the headwaters of the Ohio drainage along the divide, and because of the pronounced tendency of the larger central longear, like many species of sunfishes, to dwarf in size when confined in small, headwater streams.

Years 1955–80—Since 1950 the general distributional pattern of this subspecies of the longear sunfish has remained the same in the Ohio River drainage, and its center of population abundance remained virtually unchanged. The greatest numerical densities continued to exist in the southernmost two tiers of counties. This subspecies occurred rarely in the upper two-thirds of the Muskingum River system, and none were collected despite considerable effort to capture them.

Although a numerous and often abundant sunfish in the streams of southern Ohio, it was taken in fewer numbers in the Ohio River than were the bluegill and white crappie during the 1957–59 investigations (ORSANCO, 1962:151); during the 1968–69 investigations H. Ronald Preston recorded the longear in only three of the nine Ohio River collecting stations adjacent to the state of Ohio, Margulies and Burch at all of their collecting stations between 1976 and 1978.

NORTHERN LONGEAR SUNFISH

Lepomis megalotis peltastes Cope

Fig. 143

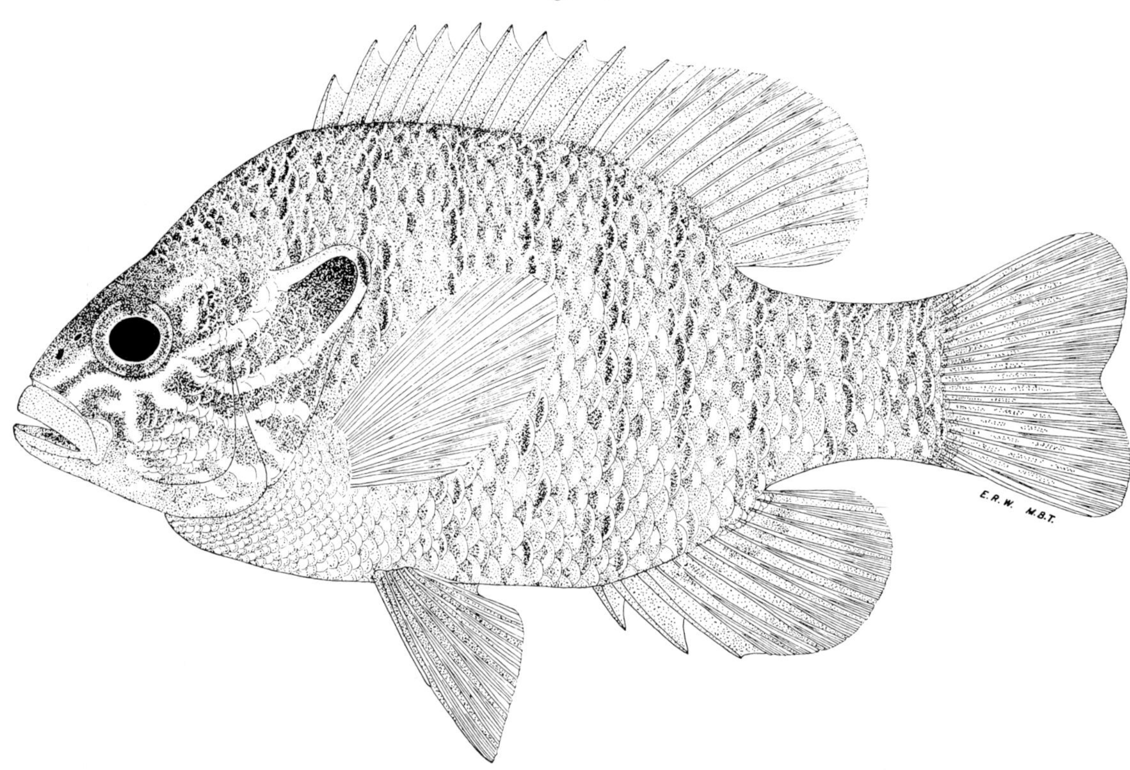

Beaver Creek, Williams County, O.

June 30, 1939.
Breeding male.

68 mm SL, 3.5″ (8.9 cm) TL.
OSUM 1629.

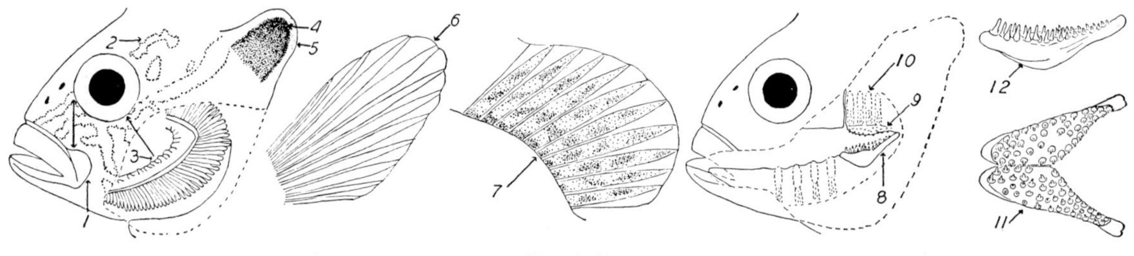

Nos. 1–12

(1) mouth moderately large; the posterior end of upper jaw extends to about the center of eye, always beyond anterior edge of eye.

(2) sides of head in large young and adults with several wavy, emerald-blue or light bars radiating backward from mouth and eye.

(3) dotted portion of gill cover removed, exposing first gill arch; note very short gill rakers, almost as wide as long.

(4) black-centered opercle flap pointing obliquely upward, as in young of central longear and usually not elongated horizontally as in the male of that species; opercle bone so flexible that flap may be doubled forward until its posterior margin touches cheek, without fracturing the bone.

(5) black opercle spot in large young and adults bordered with whitish dorsally and ventrally, and posteriorly with a large orange or red spot.

(6) pectoral fin moderately short and rounded at its tip.

(7) soft dorsal vaguely spotted in some adults; usually spotted in young; usually flushed with rusty-red in males.

(8) lateral view of head with that portion removed which is encircled by dotted lines in order to expose the lower left pharyngeal arch, showing its position in the throat; position similar in all species of sunfishes.

(9) as in 8, except that upper pharyngeal arch is exposed.

(10) as in 8, except that the dotted rectangles show where the 4 gill arches were cut so that they could be removed.

(11) dorsal view of the rather narrow lower pharyngeal arches.

(12) lateral view of pharyngeal arch; note pointed teeth.

Identification

Characters: *General—See* white crappie. *Generic—See* green sunfish. *Specific* and *subspecific—See* Nos. 1–12 for these characters. Size smaller; largest adults usually less than 5.0″ (13 cm) long. Adult males of this subspecies appear to have retained more of their juvenile characters than have males of the central longear; such as the opercle flap of moderate length which generally extends upward at a 45° angle, instead of being greatly lengthened and lying horizontally as in the central longear; in retention in adults of the bands on the sides of the body, and spotting on dorsal and anal fins. Adult males of the two subspecies differ considerably in the smaller maximum size of the northern longear; and in presence of a single large, red or orange spot on the extreme posterior end of the opercle margin, instead of having the entire margin whitish or with 2–9 small, reddish spots scattered along the margin as in the central longear.

Differs: Central longear attains a larger maximum size; male has an elongated, horizontal opercle flap, which usually has more than one small reddish spot on opercle margin. *See* central longear.

Most like: Central longear. **Superficially like:** Bluegill and redear.

Coloration: *Adult male*—Like central longear male, except that the 8–12 bands on sides of body are usually present and often sharply outlined; soft dorsal and anal fins generally vaguely spotted; margin of opercle flap with only one, rather large red spot. *Adult female*—Like central longear female, except that banding on sides and spotting on fins generally are more prominent. *Young*—Like young of central longear.

Lengths and **weights:** *Young* of year in Oct., 0.8″–1.8″ (2.0–4.6 cm) long. Around **1** year, 1.1″–2.2″ (2.8–5.6 cm). *Adults,* usually 2.1″–4.0″ (5.3–10 cm). Largest specimen, 4.8″ (12 cm) long, weight 2 oz (57 g).

Hybridizes: with green and orangespotted sunfishes.

Distribution and Habitat

Ohio Distribution—Between the years 1920–50 the northern longear was numerous in Ohio only in the clearer streams and a few clear lakes of the Maumee River drainage in the three counties bordering Michigan, and in Auglaize County. Since 1927, it has become rare or absent in the turbid St. Marys and Maumee rivers, where before 1900 Meek (1889:439) and Kirsch (1895A:331) found it at almost every locality investigated, and where between 1920–26 small populations still occurred. After 1930 the species was apparently absent from Sandusky Bay where between 1900–26 Osburn and others collected many specimens.

Habitat—The dwarfish northern longear remained numerous in northern Ohio only in the larger pools of the low-gradient, clearer streams and in lakes, where the sand, gravel and blackish muck were comparatively free from yellow silts, and where some aquatic vegetation was present. During the 1920–50 period it was more often found in smaller, rather than larger, streams which was in direct contrast to Kirsch's (1895A:331) statement of "Found in all the larger streams" of the Maumee River system. This difference was undoubtedly caused by its extirpation from the larger streams because of the recent turbid and silted conditions. It differed ecologically from the central longear in showing much more of a preference for lakes and ponds, and for submerged aquatic vegetation. Like the central longear, the northern longear did much of its feeding near the surface of the water when it was clear, and rose readily to an artificial fly.

Years 1955–80—The numerical abundance of this apparently silt-intolerant subspecies inhabiting the tributaries and shallows of Lake Erie, continued to decrease as it had done since before 1900. Several streams were thoroughly investigated, where formerly this sunfish was numerous, without producing a specimen. In the Auglaize and Blanchard river systems populations were found only in those stream sections where little siltation was evident

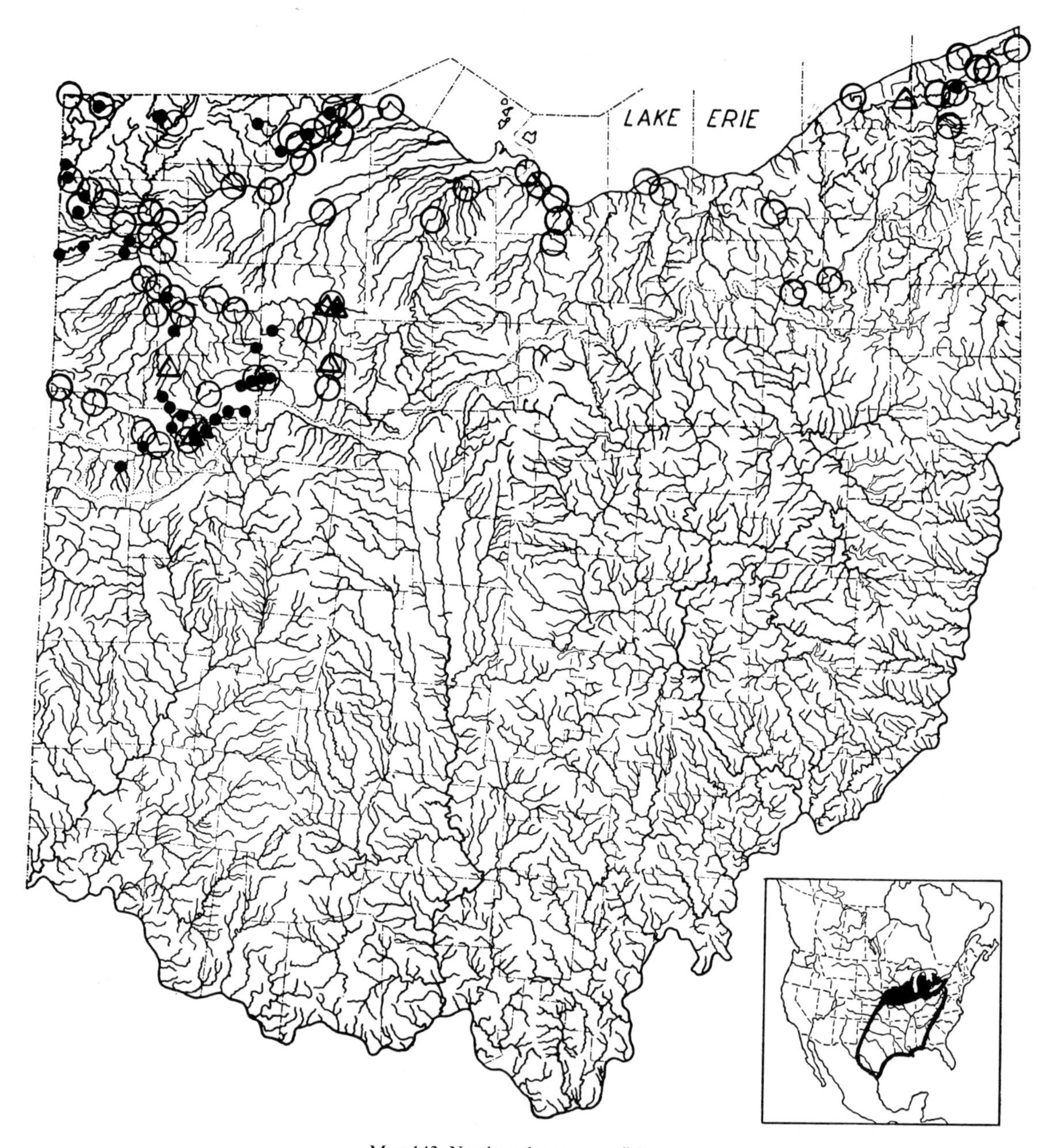

MAP 143. Northern longear sunfish

Locality records. ○ Before 1926. ● 1926–54. △ 1955–80.
Insert: Black area represents range of this subspecies, outlined area of other subspecies.

and/or there remained some aquatic vegetation. None was found in the shallows and marshes of Lake Erie or in Sandusky Bay where Osburn (1901:81) obtained numerous specimens.

In several localities in northwestern Ohio, previous to 1934, I had abundant opportunity to investigate the habitats, habits, coloration and morphology of this distinctive sunfish, as I had between the years 1934 and 1938 in the Huron River system in southern Michigan, where at that time it was quite numerous. These investigations have convinced me that this taxa is probably distinct specifically from the typical subspecies *Lepomis m. megalotis*; if not, then it is an incipient species.

REDEAR SUNFISH

Lepomis microlophus (Günther)

Fig. 144

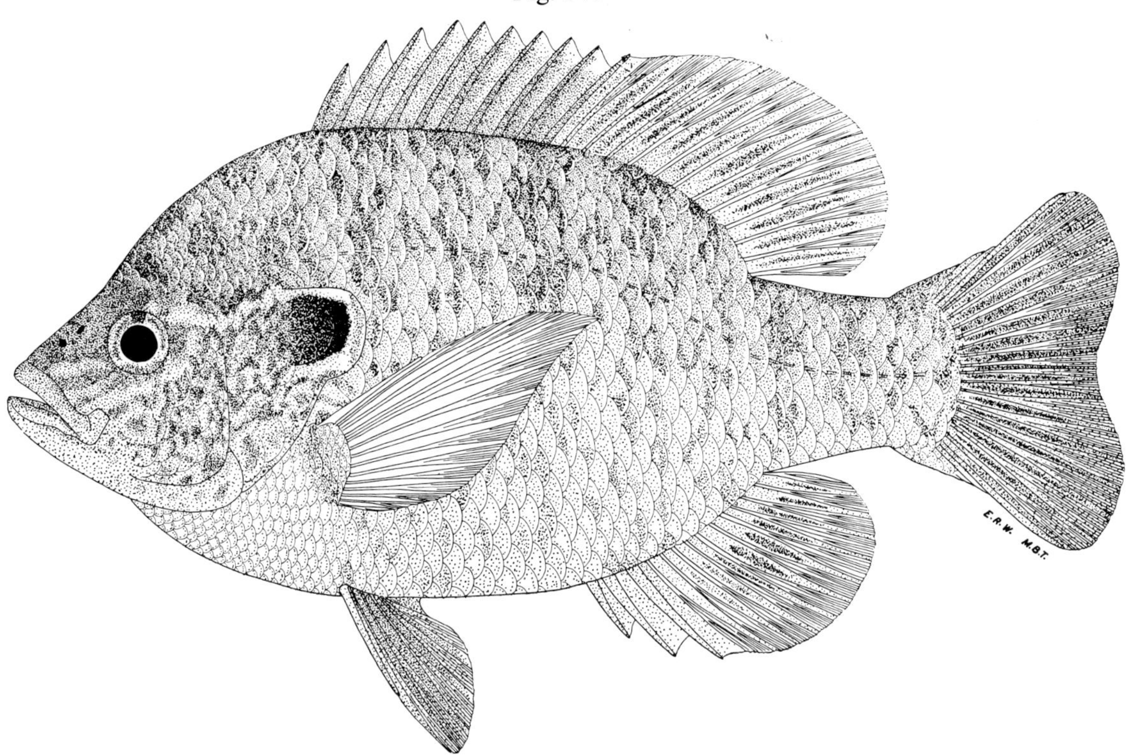

Buckeye Lake, Licking County, O.

June 22, 1939.
Breeding male.

132 mm SL, 6.5″ (17 cm) TL.
OSUM 1441.

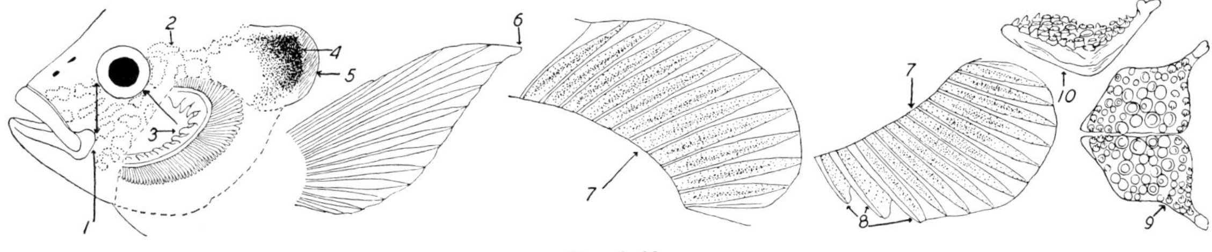

Nos. 1–10

(1) mouth moderately small; the posterior end of upper jaw extends almost to, or to, anterior edge of eye.

(2) sides of head of adults spotted with dark olive; wavy bands not well developed.

(3) dotted portion of gill cover removed, exposing first gill arch; note short, blunt, and often crooked gill rakers.

(4) black-centered opercle flap moderately flexible; may be bent forward until the flap is at right angles to the body, without fracturing opercle bone, except in some old fishes.

(5) black opercle spot bordered with white or light slate in adults and large young; males with a conspicuous red spot on posterior border of flap; in females this spot is orange.

(6) pectoral fin very long and pointed at its tip.

(7) soft dorsal and anal fins without definite spots.

(8) this species has only three anal spines as do all other species of sunfishes in the genera *Micropterus* and *Lepomis*.

(9) dorsal view of the very wide lower pharyngeal arches.

(10) lateral view of pharyngeal arch; note bluntly pointed or rounded molar-like teeth, and heavy bone of the arch itself.

Identification

Characters: *General—See* white crappie. *Generic—See* green sunfish. *Specific—See* Nos. 1–10 for these characters.

Differs: Warmouth, green, orangespotted, and longear sunfishes have short and rounded pectoral fins. Bluegill has an all black, flexible opercle flap. Pumpkinseed has a stiff opercle flap; spotted dorsal fin; adults have wavy light or bluish bands crossing cheeks. *See* green sunfish.

Most like: Pumpkinseed. **Superficially like:** Longear and bluegill.

Coloration: *Adult male*—Dorsally olive-green speckled with darker. Sides lighter with some brassy reflections. Breast and belly brassy. Some males have 5–10 faint vertical bands of dark dusky-olive on sides of body. Sides of head a light bluish-brown or slaty-olive, spotted and mottled with darker. Black-centered opercle spot broadly bordered dorsally and ventrally with lighter, usually some shade of slaty-white, posteriorly with a large blood red blotch; this blotch orange-red in younger males. Fins light greenish-olive without definite spots; basal half of caudal olive-slate. *Breeding male*—Like adult male, except that pelvics are dusky; breast brighter; entire body more brassy, with the green and blue reflections more intense. *Adult* and *breeding females*—Like males, except less brilliantly colored; breast paler and more yellowish; opercle spot orange-red or orange; the 5–10 bands on sides of body more distinct. *Young*—Like adult female, except that breast is whitish or white-yellow; opercle spot light gray or pale yellow; the 5–10 bands on sides of body more distinct; body more silvery.

Lengths and **weights:** *Young* of year in Oct., 0.8″–3.0″ (2.0–7.6 cm) long. Around **1** year, 1.0″—–4.0″ (2.5–10 cm). *Adults*, usually 4.5″–9.0″ (11–23 cm) long, weight 1 oz–11 oz (28–312 g). Largest specimen, 10.4″ (16.7 cm) long.

Hybridizes: naturally in Ohio with green, bluegill and pumpkinseed sunfishes; artificially produced elsewhere with warmouth, green and bluegill sunfishes (Childers, 1967:161).

Distribution and Habitat

Ohio Distribution—During the late winter or early spring of 1931 A. F. Greter received 14 redears which had been taken from a northern Indiana lake. These fishes were liberated in Greter's Lake, in Springfield Township, Richland County. Three years later the offspring of these 14 redears appeared to be as numerous in the lake as were the offspring of the long-established bluegills and pumpkinseeds.

On May 23, 1935 the Ohio Conservation Department obtained 53 breeder redears from the Indiana Conservation Department. These breeders were liberated in the Honey Creek sanctuary at the eastern end of Buckeye Lake. In 1939 the young produced by these 53 breeders were taken in the experimental nets in larger numbers than were those of the long-established pumpkinseeds and in somewhat smaller numbers than were the bluegills (Trautman, 1939B:5 and 31).

In 1939, 142 adult redears, captured at Buckeye Lake, were liberated in Pippen Lake of Portage County, and in 1944 their young were trapped and shipped to various Ohio localities for stocking purposes. Several of these recent stockings appear to have been successful.

In 1889 Henshall (1889:125, as *Lepomis notatus*) recorded the redear, without comment, from the Little Miami River of southwestern Ohio, a record which has become established in the literature (Osburn, 1901:83; Forbes and Richardson, 1920:260). In 1931 I examined the sunfish material which was in the Cincinnati Society of Natural History collections and which had been identified by Henshall. I found that the majority of the sunfishes had been misidentified and that the only supposed redear in the collection, listed without a locality and its identification questioned in the catalogue by some later examiner, was a pumpkinseed. Because of the misidentification of much of the sunfish material, including the supposed redear, the Henshall record for the Little Miami River is considered invalid. As a further argument it now appears that the original range of this southern sunfish did not extend up the Ohio River beyond the mouth of the Wabash River (Gerking, 1945:100).

Habitat—Wherever the redear has been introduced into waters which are north of its original range, the species has essentially inhabited nonflowing waters which were relatively clear and which contained some aquatic vegetation. At Buckeye Lake the redear seemed to require less vegetation than did the pumpkinseed, and as much as, or more, than did the bluegill. Like the latter, the redear frequented open waters. It congregated about brush, stumps and logs more than did the bluegill and pumpkinseed, hence its colloquial name of "Stumpknocker."

Years 1955–80—This much-desired sport and

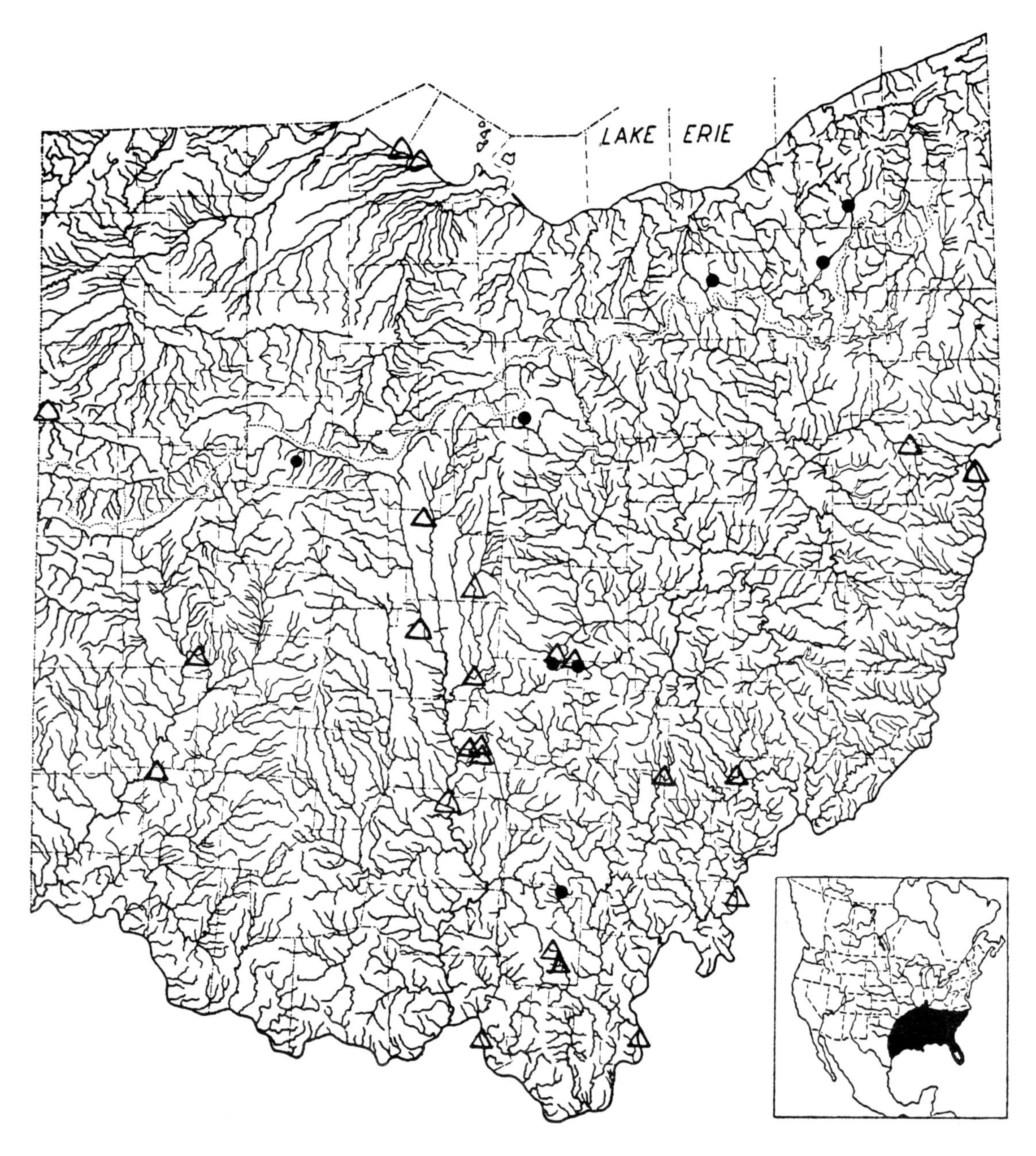

Map 144. Redear sunfish

Locality records. ● Before 1955. △ 1955–80.
Insert: Widely and often successfully introduced outside of this range.

pan fish increased steadily in popularity throughout the 25-year period. It was first introduced, or reintroduced, into many farm ponds and especially in the southern half of the state. The species had difficulty in maintaining its numbers in many waters; it was subject to winter killing, apparently had difficulty in competing for living space and hybridized readily with the green, bluegill and pumpkinseed. Hybrid vigor, as indicated by increased growth and voracious feeding, between the redear and other sunfish species was repeatedly demonstrated, and some of the longest and/or heaviest sunfishes I have seen in Ohio were hybrids between the redear and the bluegill.

The redear was seldom found in flowing waters, even those adjacent to ponds containing this species, giving the impression that this species "escaped" less frequently from impoundments than did the bluegill. H. Ronald Preston reported that the 1968–69 investigations of the Ohio River produced specimens adjacent to Scioto, Gallia and Meigs counties.

PUMPKINSEED SUNFISH

Lepomis gibbosus (Linnaeus)

Fig. 145

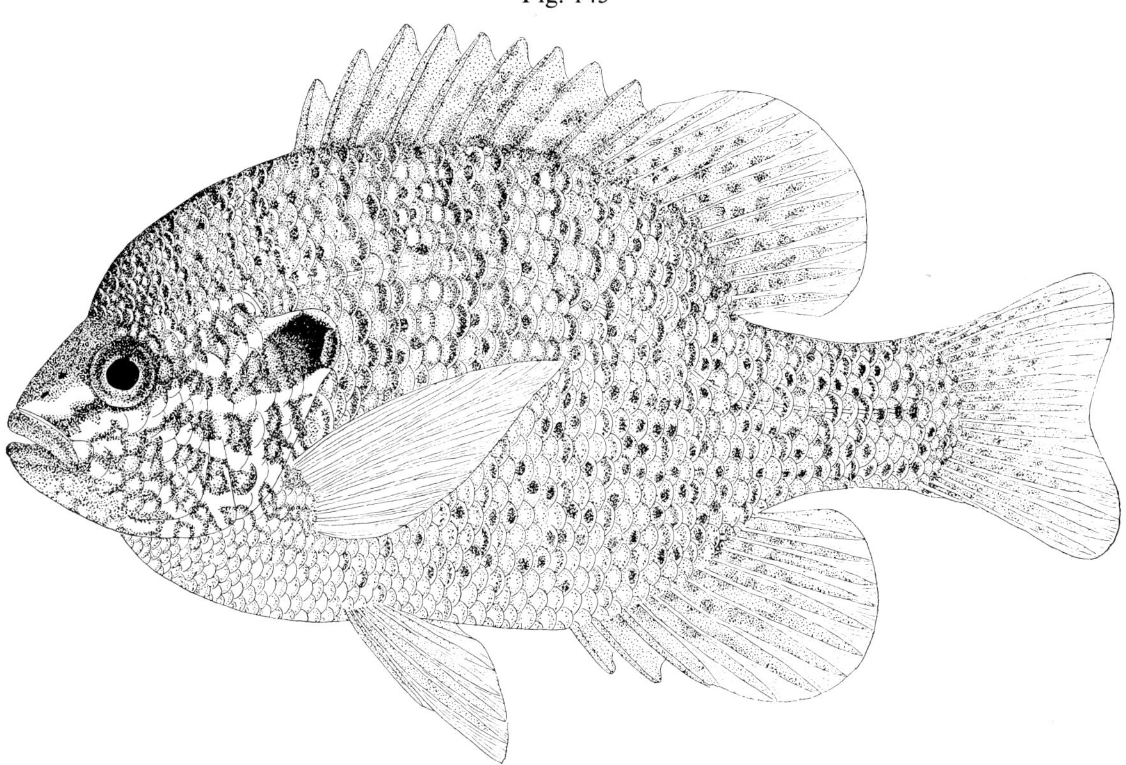

Lake Erie, Ottawa County, O.

June 30, 1942.
Breeding male.

110 mm SL, 5.4" (8.7 cm) TL.
OSUM 5106.

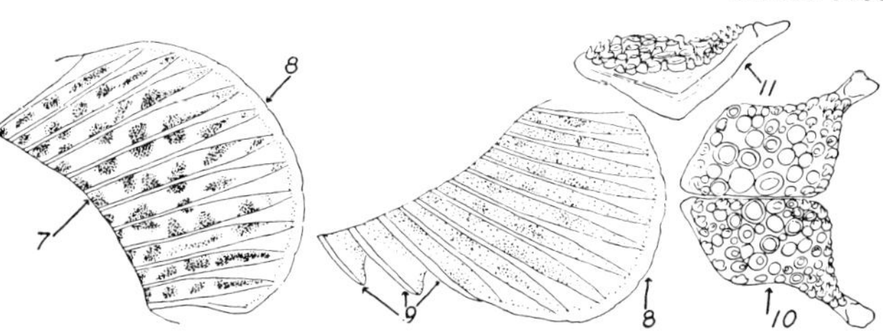

Nos. 1–11

(1) mouth small; the posterior end of upper jaw extends backward almost to, or to, anterior edge of eye.

(2) sides of head of adults with wavy emerald-blue bands radiating backward from snout and eye; these bands pale blue and poorly developed in large young; bands absent in small young.

(3) dotted portion of gill cover removed, exposing first gill arch; note short, blunt, and often crooked gill rakers.

(4) black-centered opercle flap very stiff; cannot be bent forward without fracturing the heavy opercle bone.

(5) black opercle spot bordered dorsally and ventrally with slaty-white, posteriorly with a large spot which is turkey-red in males, orange in females, yellowish or pale whitish in young.

(6) pectoral fin long and sharply pointed.

(7) soft dorsal heavily and distinctly spotted.

(8) margins of soft dorsal and anal fins light colored, either whitish or yellowish.

(9) anal spines three, as in all species in genera *Micropterus* and *Lepomis*.

(10) dorsal view of the very broad and heavy, lower pharyngeal arches.

(11) lateral view of pharyngeal arch showing the large, round, molar-like teeth.

Identification

Characters: *General—See* white crappie. *Generic—See* green sunfish. *Specific—See* Nos. 1–11 for these characters.

Differs: Warmouth, green, orangespotted, and longear sunfishes have short, rounded pectorals. Bluegill has a flexible, all black earflap. Redear lacks spotting on soft dorsal and wavy distinct bands on cheeks. *See* green sunfish.

Most like: Redear. **Superficially like:** Longear and orangespotted.

Coloration: *Adult male*—Dorsally golden- or olive-green, much mottled and spotted with rusty and orange. Sides lighter, with many spots and reflections of orange, yellow, blue, emerald, and green. Breast and belly orange-yellow. Sides of body with 7–10 vertical, dark, chainlike, double-bands which may be faint or absent in some males. Sides of head with several wavy, emerald or blue bands radiating backward from snout and eye. Black-centered opercle spot broadly bordered dorsally and ventrally with light slate or blue, posteriorly with a turkey- or orange-red spot. Soft dorsal and anal spotted with orange, yellow, red (usually fishes from bog waters) or dark brown; spots on dorsal tending to form 5–7 oblique rows. Caudal brownish-olive basally, bluish-yellow distally. Pelvics olive-yellow without any dusky. *Breeding male*—Like adult female, except that it is very brilliantly colored; wavy bands on sides of head intensely emerald; breast bright orange; pelvics bright yellow-orange; the bluish distal third of caudal fin most conspicuous as the fish swims about. *Adult* and *breeding females*—Like adult male, except less brilliantly colored. *Young*—Like adult female, except that much of the bright coloring is replaced with silvery and light olive.

Lengths and **weights:** *Young* of year in Oct., 0.8″–3.2″ (2.0–8.1 cm) long. Around **1** year, 1.2″—–3.5″ (3.0–8.9 cm). *Adults*, usually 2.5″ (6.4 cm) (dwarf populations)–8.0″ (20 cm) long, weight 0.3 oz–9.0 oz (8.5–255 g). Largest specimen, 8.8″ (22 cm) long, weight, 11 oz (312 g).

Hybridizes: naturally in Ohio with green, bluegill, orangespotted, redear, and longear sunfishes; artificially produced elsewhere with warmouth sunfish (Childers, 1967:161).

Distribution and Habitat

Ohio Distribution—Preserved specimens, taken in the northwestern and northern counties from Defiance to Ashtabula, and early literature references demonstrate that the pumpkinseed was well distributed in northern Ohio previous to 1900. It was particularly abundant in the marshes, harbors and bays of western Lake Erie where, with the bluegill (*see* that species), it was of commercial importance before 1850, to remain so until about 1900 when its sale in Ohio was prohibited. In 1899, Osburn (1901:84) considered it to be the most abundant sunfish in Sandusky Bay, and to be locally abundant in northeastern Ohio. Kirsch (1895A: 331) in 1893 found the pumpkinseed to be abundant in all lakes, and common in the streams, of the Maumee River system except the Auglaize River and its tributaries (in which the gradients were too high).

The pumpkinseed occurred only in a few localities in the Ohio River drainage of central and southern Ohio, and large populations were present only in St. Marys, Indian and Buckeye lakes. In these old canal impoundments, the pumpkinseed was of considerable commercial importance between the years 1840–1900 (*see* bluegill); Osburn (1901:84) considered it to be the most abundant sunfish in Buckeye Lake (then called Licking Reservoir) as late as 1899, a statement with which the older fishermen are in agreement. Before 1900 the pumpkinseed was present in the canals in central Ohio (Williamson and Osburn, 1898:44), and Henshall (1889:125) found it in Ross Lake, which was a basin of the canal in Hamilton County at Cincinnati. Strays or small populations appeared to have been present in the west-central prairie areas, and especially in the vicinity of the canals.

The scarcity of records previous to 1900 suggests that the pumpkinseed was probably absent throughout the southern half of Ohio before the building of the canals. This presumption is substantiated by the absence of records for the southern half of Indiana (Gerking, 1945:99–100, and map 102), and from southwestern Pennsylvania (Raney, 1938:76).

Between 1920–50 the State Conservation Department poured hundreds of thousands of fry, young and adults, into Ohio waters. Those placed in the flowing waters of southern Ohio failed to establish themselves, but those introduced into weedy impoundments, flourished, even in the impoundments of unglaciated Ohio. Apparently the species was absent from most of the unglaciated portion of the state prior to 1929.

Despite the many introductions, the

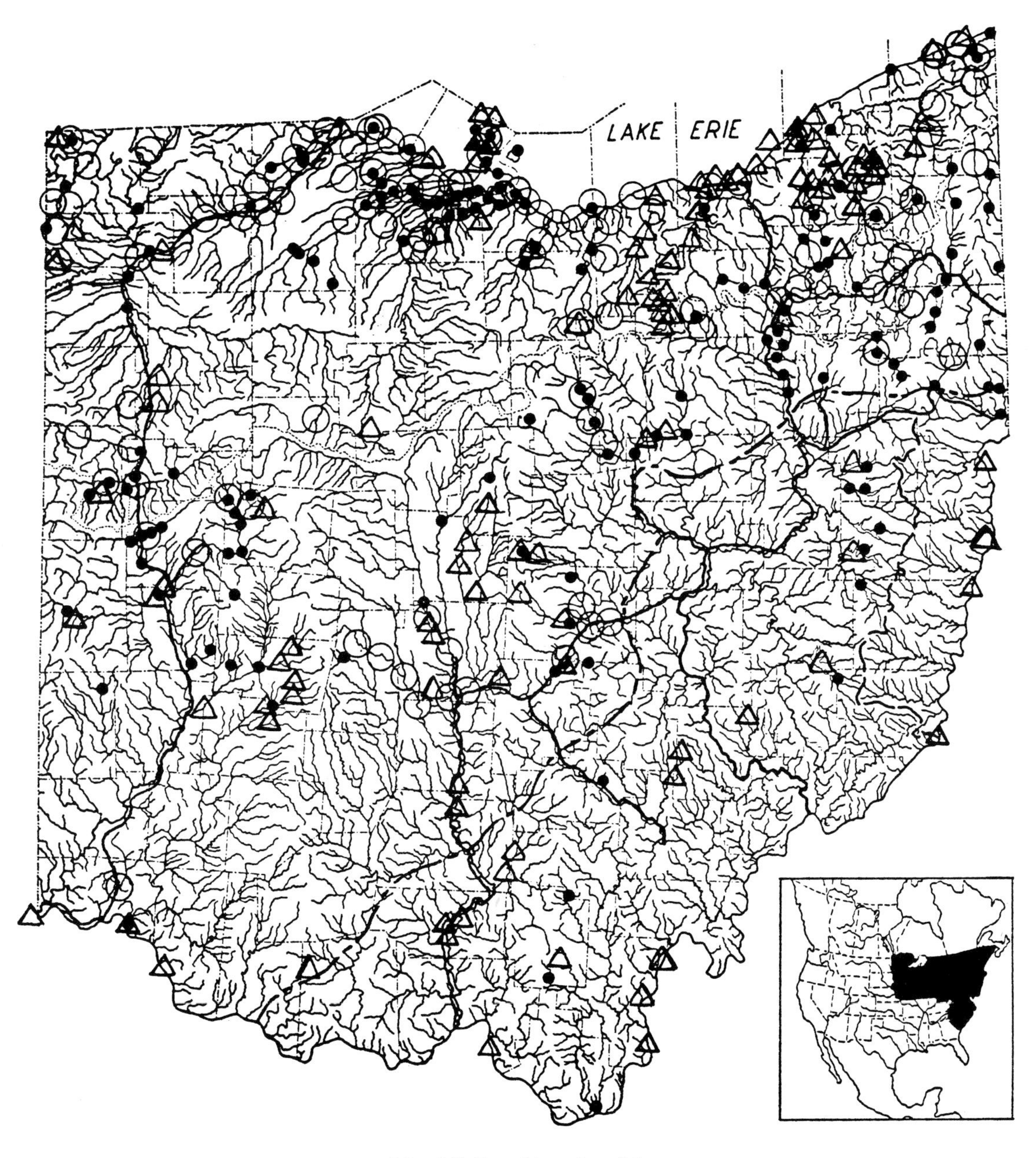

MAP 145. Pumpkinseed sunfish

Locality records. ○ Before 1926. ● 1926–54. △ 1955–80. ~ Canals. – — Glacial boundary. /\ Flushing Escarpment.
Insert: Widely introduced, often successfully, outside of this range; western limits of range indefinite.

pumpkinseed showed notable decreases in abundance in several northern streams and lakes during the 1900–50 period. As a small boy, between 1906–10, I well remember the many "Punkies" which my father and I caught "pole fishing" the scattered, small, open holes in the dense beds of submerged aquatics (principally hornwort) which then covered the "New Reservoir" at Buckeye Lake (Trautman, 1940:33–38 and 42). Also, we caught large numbers in Long Pond in the northeast corner of Indian Lake, where in August the submerged aquatics were so dense that a rowboat could be poled through them only with difficulty, and where we were forced to rake out a hole in the vegetative cover before we could fish and observe the "Punkies" in the clear water as they bit. Since 1930 the vast mass of submerged aquatics at the western end of Buckeye Lake has largely disappeared, as had 80% of it in Long Pond; likewise gone are the many logs, stumps and clear waters. With these changes all except a remnant of the pumpkinseed population has likewise vanished.

Habitat—The largest populations of pumpkinseeds were in nonflowing waters which were clear, or at least did not contain much clayey silt in suspension, where the bottoms were of muck or sand partly covered with organic debris, and there were dense beds of aquatic vegetation of the submerged type. The pumpkinseed required larger and denser masses of aquatic vegetation than did the bluegill, and was less numerous in open, deeper waters than was that species. It did not so readily overpopulate a lake as did the bluegill, probably because it fed more upon small fishes, including its own young.

Years 1955–80—The populations of pumpkinseeds present in the shallows of Lake Erie including its adjacent marshes and Sandusky Bay continued to lessen in size during this period as the vegetated, clear-water habitat was reduced in amount and quality. In the streams of the northern half of Ohio, many formerly isolated or continuous populations were greatly reduced in size or were extirpated, and it was only in the densely vegetated natural ponds and impoundments that the species retained its numerical abundance (White et al., 1975:110–11).

Prior to 1950 many millions of pumpkinseeds were liberated into the streams and impoundments of the southern half of Ohio. Such introductions were continued during the present period but to a lesser degree. Liberations were made in the newer and usually larger impoundments, a few resulting in presumably permanent populations such as are present in Burr Oak Lake. With few exceptions the species was of rare or accidental occurrence in the flowing waters of southern Ohio.

During the 1968–69 investigations of the Ohio River, H. Ronald Preston reported capturing the species as far downstream as Meigs County and from four of the nine collecting stations which were adjacent to the state of Ohio. Between 1976 and 1978 Margulies and Burch recorded the species from all five of their Ohio River stations that were adjacent to Ohio.

SAUGER

Stizostedion canadense (Smith)

Fig. 146

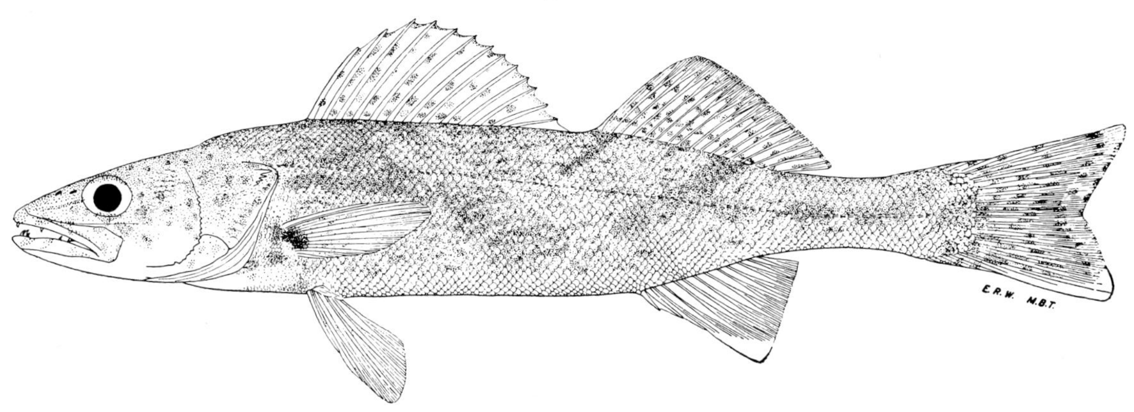

Lake Erie, Ottawa County, O.

Late Sept., 1939. 255 mm SL, 12.2″ (30.9 cm) TL.
Male. OSUM 1140

Identification

Characters: *General*—Tail moderately or deeply forked. Two anal spines. Free edges of preopercle strongly serrated (toothed), *see* blue pike, fig. C. The two dorsal fins well separated; the first usually of 12–13 spines, extremes 10–14. Opercle with a sharp spine. Scales strongly ctenoid. Belly scaled normally, the scales small. *Generic*—Sharp, large canine teeth on jaws and palatines. Pelvic fins widely separated; distance between pelvic fin insertions almost equaling width of base of either fin. Anal rays 12–13. Body slender; body depth almost invariably less than head length, rarely equal. *Specific*—Spinous dorsal with round, dusky spots which form oblique rows; spots faint or absent in smallest young. No black basal blotch on webbing between three posteriormost spines. Cheeks partly scaled. Rays of soft dorsal 17–21, usually fewer than 20. Back crossed with 3–4 saddle-bands; these usually expanded laterally along the side forming three oblong, horizontal blotches, one beneath each dorsal fin and one on the caudal peduncle; these bands often faint or absent in fishes from the Lake Erie drainage; bands very distinct in fishes from the Ohio River drainage. Pyloric caeca 5–8, each shorter than length of stomach.

Differs: Blue pike and walleye have a dusky blotch at base of last three dorsal spines; has an average higher number of soft dorsal rays. Perch, all species of darters, and drum lack canine teeth in jaws. White bass lacks spotting or blotching on spinous dorsal; has fewer dorsal spines. All sunfishes have three or more anal spines.

Most like: Blue pike and walleye. Several species of darters are sometimes mistaken for young saugers.

Coloration: Dorsally slaty-yellow-, or golden-olive; more silvery and lighter on sides. Ventrally silvery with a milk-white breast and belly. *See* under Characters and text for description of the three oblong blotches, and intensity of color pattern. Dorsal fins transparent in young; tinged with golden-yellow in adults, and with the dark spots tending to form 2–4 oblique rows. Caudal fin spotted vaguely or barred. Tip of lower lobes of caudal and anterior tips of anal fins milk-white. A black spot at base of pectoral fin.

Lengths and **weights:** *Young* of year in Oct., 3.0″–6.0″ (7.6–15 cm) long. Around **1** year, 5.0″–8.0″ (13–20 cm). *Adults*, usually 9.0″ (23 cm) (males) and 10.0″ (25.4 cm) (females)–16.0″ (40.6 cm). Fishes 5.0″ (13 cm) long usually weigh 0.5 oz–1 oz (14–28 g); 10.0″ (25.4 cm) long, 5 oz–9 oz (142–255 g); 12.0″ (30.5 cm) long, 8 oz–12 oz (227–340 g); 16.0″ (40.6 cm) long, 1 lb–1 lb, 8 oz (0.5–0.7 kg); 18.0″ (45.7 cm)

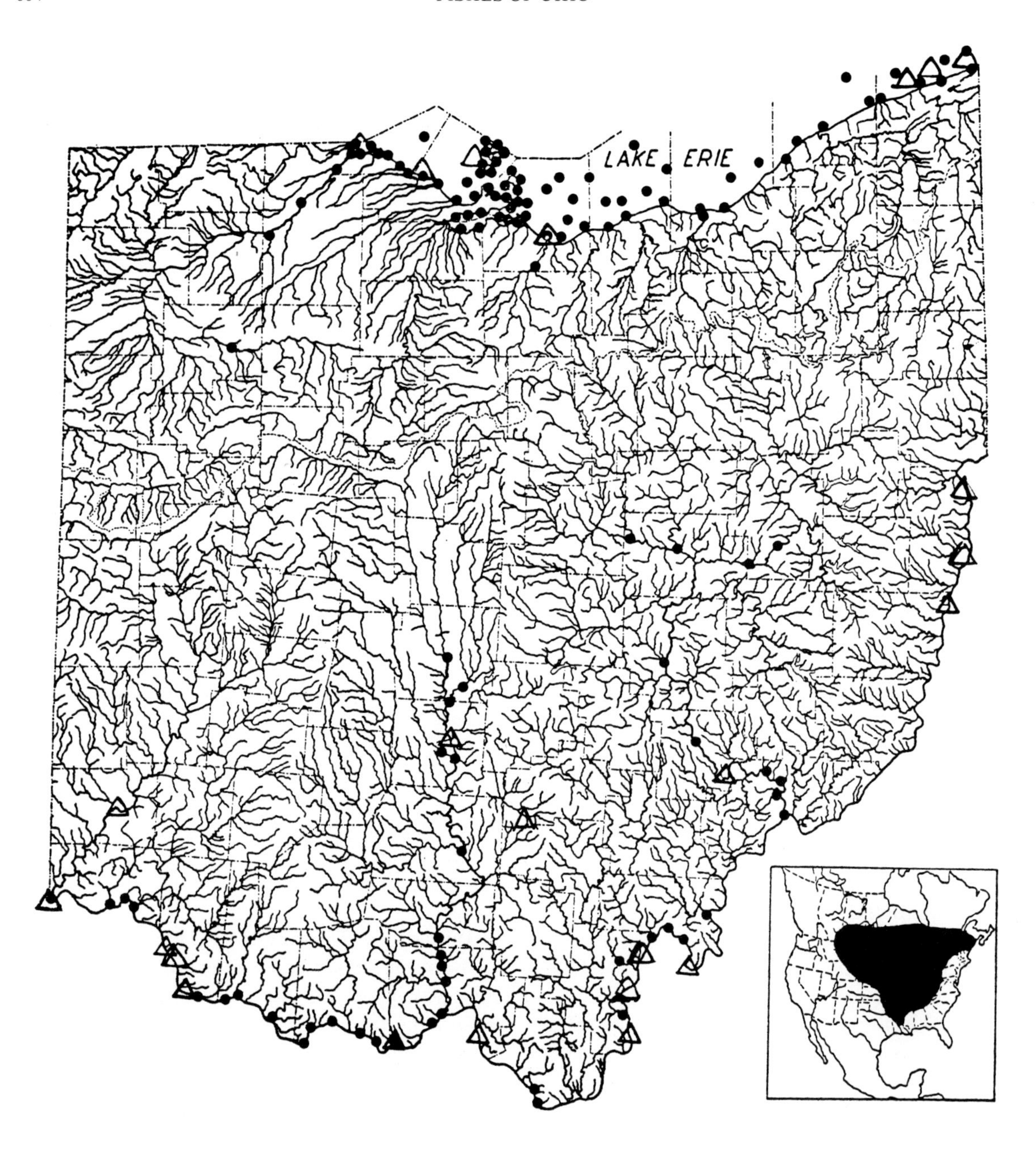

MAP 146. Sauger

Locality records. ● Before 1955. △ 1955–80.
Insert: Large unoccupied areas occur within this range; northern limits of range indefinite.

long, 1 lb, 14 oz–2 lbs, 5 oz (0.9–1.0 kg). Largest specimen, 19.3″ (49.0 cm) long, weight 3 lb, 3 oz (1.4 kg) (gravid female). Lake Erie fishermen give maximum weights of 3–4 lbs (1.4–1.8 kg); possibly some of the larger supposed saugers are hybrids between sauger and blue pike or walleye. For growth and weight, *see* Roach (1949:13).

Hybridizes: rarely in Ohio with walleyes and blue pike.

Distribution and Habitat

Ohio Distribution—Few reliable data exist concerning the distribution and abundance of the sauger in Ohio before 1850, principally because early writers considered the sauger and the walleye to be a single species, usually considering the more distinctly marked saugers to be the male ("Jack-Salmon" of the Ohio River fishermen) and the more plainly colored walleyes to be the female ("White-Salmon" of Ohio River fishermen). Rafinesque's (1820: 65–66) description of *Perca salmonea* from the Ohio River is plainly a combination of the sauger and the walleye, and his statements that "all fins are spotted" and the "back and sides are gilt by patches and head variegated with small gilt spots" are referable to the sauger. It is surprising that Klippart (1878:87), and Jordan (1882:962) believed statements made by fishermen that the sauger was introduced after 1825 into the Ohio River by its migrating through the newly constructed canals. As outlined by Hubbs and Lagler (1947:86) and Rostlund (1952:283), the sauger has been present in the Ohio drainage throughout Columbian time, including the Tennessee River system. Beyond all doubt the sauger was present in both the Ohio River and Lake Erie drainages at the time the white man came into the Ohio Country, and as Greene (1935:164) suggested, it probably used the outlets of glacial lakes Agassiz, Chicago and Maumee as invasion routes to enter the Lake Erie drainage.

By 1885 saugers and walleyes were considered as distinct species, and for that year Smith and Snell (1891:241) reported a total catch of 5,272,300 lbs (2,391,475 kg) of saugers for the Ohio ports of Lake Erie. In 1888, Henshall (1888:80) considered the sauger to be "Common in Ohio River."

Between 1925–50 the commercial catch for the Ohio ports of Lake Erie averaged 656,972 lbs (284,-390 kg) (Cummins, 1952:1) annually. During that period it occurred sparingly in the lower portions of the Lake Erie tributaries. In the Ohio drainage it was fairly numerous in the Ohio River from the Indiana state line to Washington County, and in the Scioto River upstream as far as Pickaway County. It was much less abundant in the Muskingum River, occurring in the largest numbers in that stream from its mouth upstream to Coshocton County.

A careful study of the Ohio drainage and Lake Erie populations of saugers should prove that the two populations are subspecifically distinct. The Ohio River drainage population inhabits waters that normally are much more turbid than are Lake Erie waters; therefore, the Ohio River population should have a more faded color pattern. Yet the reverse is true. The Ohio drainage saugers, even those taken from the most turbid waters, have the basic color of the dorsal half of head and body a brilliant golden-bronze and the black saddle and head markings are in sharp contrast to this basic color; in many specimens there are solid black areas of considerable size. The Lake Erie saugers have a pale yellow or bluish-yellow basic color, and the saddle marks are often very faint or absent, never with solid black areas. The Ohio drainage saugers usually have 13 dorsal spines; the distance between the two dorsal fins averages less than the horizontal diameter of the eye, and the two fins are sometimes contiguous; the Lake Erie saugers usually have 12 dorsal spines, and the distance between the two dorsal fins usually equals the diameter of the eye. The head of Ohio drainage saugers appears blunter and more triangulate, and the body is wider and more robust.

Habitat—In the Lake Erie drainage the largest populations of saugers occurred in the shallower and more turbid waters of Lake Erie, and since such conditions were more prevalent at the western end the species was most abundant there (Smith and Snell, 1891:248); in 1951 there were 235,499 lbs (106,821 kg) of sauger taken in Ohio District 1, which extended from the port of Huron westward, whereas only 152,534 lbs (69,188.3 kg) were taken in Districts 2 and 3 which extended from Huron eastward (Cummins, 1952:1–3). In Lake Erie the sauger was more tolerant of a silted bottom and turbid waters than was the walleye (Doan, 1941:449–52).

The largest Ohio drainage populations occurred in the larger, deeper waters of rather low gradients, and the species was notably more tolerant of turbid waters and silted bottoms than was the walleye. This difference in tolerance to turbid waters was graphically illustrated when recent trap and sport fishermen's catches from the Ohio and Scioto rivers were compared with catches from the Muskingum River. In the turbid Ohio and Scioto rivers, the sauger was far more abundant than was the walleye, whereas in the less turbid Muskingum the walleye greatly outnumbered the sauger.

Years 1955–80—Smith and Snell (1891:241) reported that in 1885 the commercial catch of saugers from Lake Erie that was brought into Ohio ports totaled 5,272,300 lbs (2,391,475 kg). Applegate and Van Meter (1970:14) demonstrated a fluctuating decrease in abundance from 1915 to

1968. In 1968 the commercial catch for all Ohio ports of Lake Erie was 62 lbs (28 kg). We (White et al., 1975:112) considered it to be "probably extirpated" from the Cleveland Metropolitan area by 1975. The above illustrates the change in numerical status of a formerly commercially important, renewable natural resource to commercial extinction during the space of 100 years. Since 1968 commercial fishing for saugers has not been permitted in the Ohio waters of Lake Erie. Recently the Ohio Division of Wildlife began propagating and releasing saugers in Ohio waters and especially in Lake Erie. Successful spawning of adults is now occurring, especially in the Sandusky and Maumee rivers (Anonymous D, 1979:19).

The inland Ohio populations were never large, especially since 1900, and were principally confined to the larger streams. Apparently there has been little change in overall abundance since 1950. During 1968 and 1969 H. Ronald Preston reports taking the species from five of the nine collecting stations situated in the Ohio River adjacent to the state of Ohio. During their thorough investigations of the Ohio River between the years 1976 and 1978, Margulies and Burch collected saugers in all localities adjacent to Ohio.

WALLEYE

Stizostedion vitreum vitreum (Mitchill)

Fig. 147

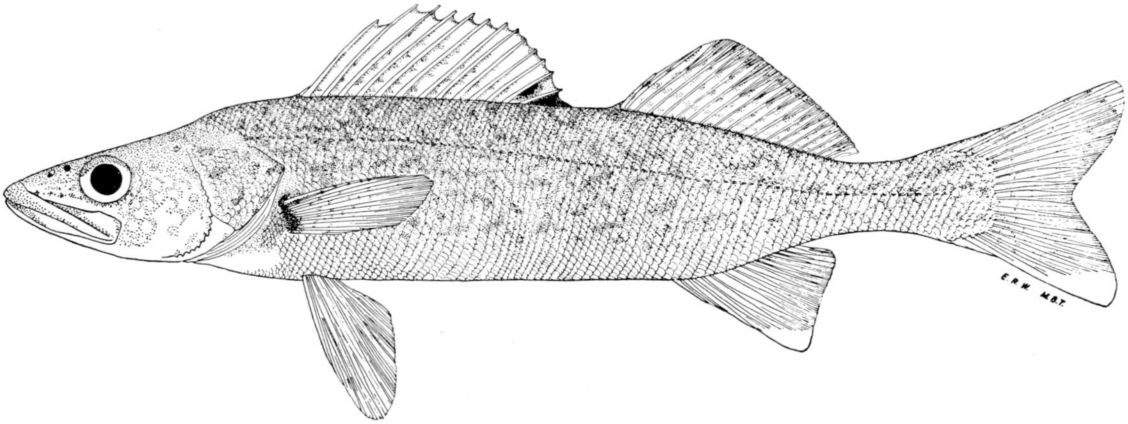

Lake Erie, Ottawa County, O.

Feb. 25, 1941.
Immature male.
Note: See fig. 148 for eye characters.

213 mm SL, 10.0″ (25.4 cm) TL.
OSUM 4235.

Identification

Characters: *General* and *generic—See* sauger. *Specific*—Spinous dorsal with a large dusky blotch on webbing between the last three dorsal spines; blotch absent only in smallest young. Remainder of spinous dorsal vaguely marked, but lacks definite spots or rows of spots. Cheeks usually without scales or with only a few. Rays of soft dorsal 19–23, usually more than 19. Between 4–14 dark, vertical bands cross back and sides of some fishes; bands most prominent in fishes 4.0″–12.0″ (10–30.5 cm) long which are taken from clear waters. Pyloric caeca three, each about as long as stomach. *Subspecific*—Pelvic fins yellowish; body of living fishes has mottlings and reflections of brassy-yellow; this coloration most intense in fishes from clear waters; fishes from turbid waters of Ohio River drainage may lack yellow coloration. Eyes smaller and farther apart; the bony interorbital width usually contained 1.1–1.4 times in length of eye in fishes less than 10.0″ (25.4 cm) long, and 0.8–1.2 times in eye length of adults; fig. 148 B, note arrows pointing to interorbital and eye widths. Formerly walleye and blue pike intergraded or hybridized commonly in Lake Erie. Maximum length longer than 20.0″ (50.8 cm).

Differs: Sauger has oblique rows of spots on spinous dorsal; fewer soft dorsal rays. Blue pike has larger eyes; lacks yellow on pelvic fins and body. *See* sauger.

Most like: Blue pike and sauger. Several species of darters have been mistaken for young walleyes.

Coloration: Dorsally yellow-olive or yellow-slate with a blue and brassy overcast. Sides lighter, more brassy and more yellow with a pronounced silvery cast. Ventrally milk-white. For description of bands on back and sides *see* under Characters. Pelvic fins usually with a pronounced yellow cast. Spinous dorsal with a dusky blotch on webbing between last three spines. Tip of lower lobe of caudal, and anterior tip of anal fins milk-white. A dusky blotch at base of pectoral fin; this blotch may be absent in fishes from turbid waters. Color variations not uncommon; Ohio drainage walleyes often lack the brassy-yellow tinge and are as bluish as Lake Erie blue pikes; others lack the blue-slate and all dark pigments and consequently are a spectacular golden-yellow (OSUM 5032); this golden-yellow variation noted in 1850 by Kirtland (1850P:61).

Lengths and **weights:** *Young* of year in Oct., 3.5″–8.0″ (8.9–20 cm) long. Around **1** year 4.0″—10.5″ (10–26.7 cm). *Adults*, usually 11.0″ (27.9 cm) (males)–12.0″ (30.5 cm) (females)–30″ (76 cm) long.

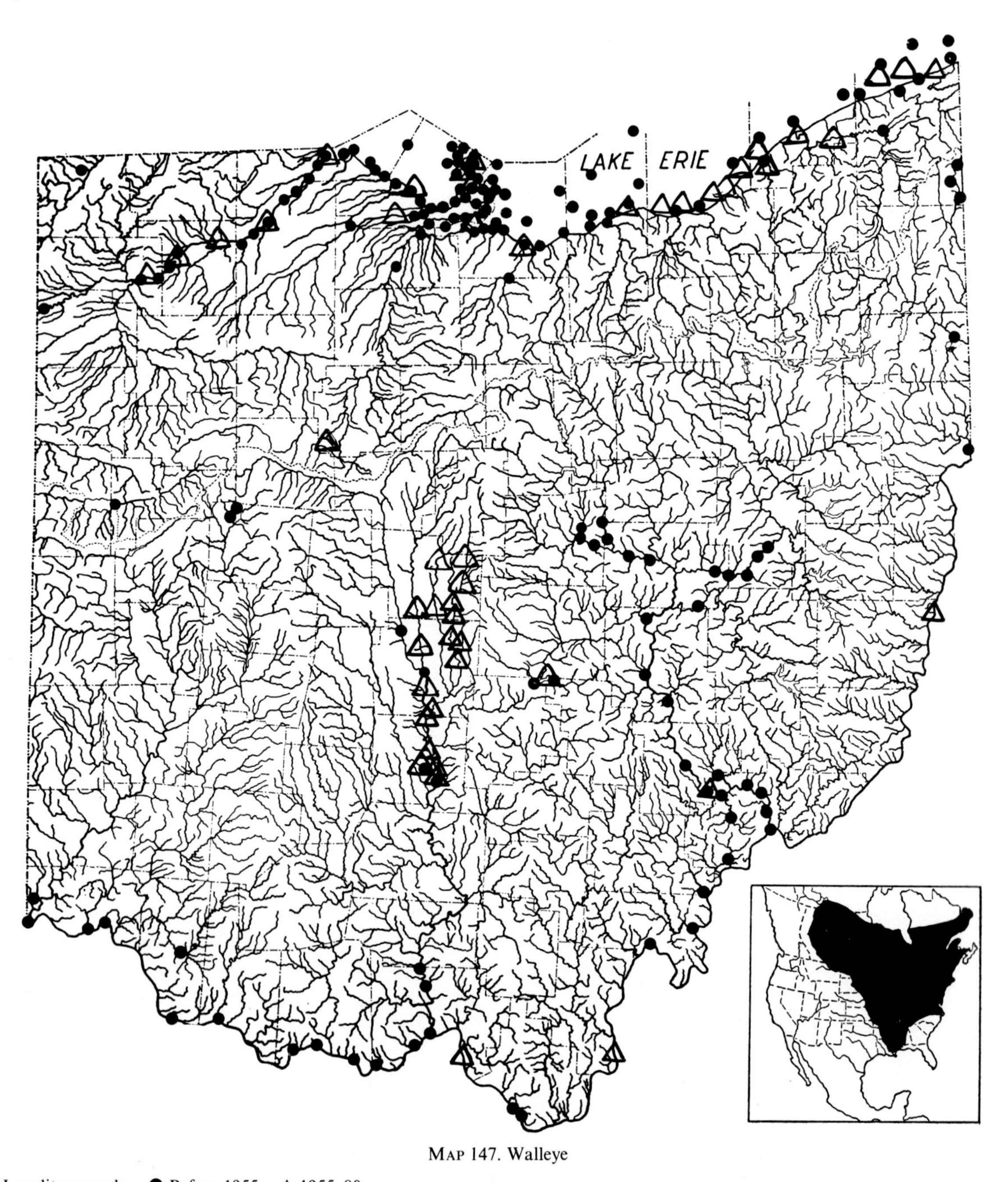

MAP 147. Walleye

Locality records. ● Before 1955. △ 1955–80.
Insert: Widely introduced within and outside of this range, western and northern limits of range indefinite.

Fishes 8.0″–10.0″ (20–25.4 cm) long usually weigh 3.5 oz–7.5 oz (99–213 g); 12.0″ (30.5 cm) long, usually 8 oz–1 lb (227g–0.5 kg); 14.0″ (35.6 cm), usually weigh 13 oz–1 lb, 4 oz (369 g–0.6 kg); 16.0″ (40.6 cm), usually 1 lb–1 lb, 13 oz (0.5–0.8 kg); 18.0″ (45.7 cm), usually 1 lb, 8 oz–2 lbs, 5 oz (0.7–1.0 kg); 20.0″ (50.8 cm), usually 2 lbs, 4 oz–3 lbs, 4 oz (1.0–1.5 kg); 24.0″ (60.9 cm), usually 3 lbs, 4 oz–5 lbs, 8 oz (1.5–2.5 kg); 26.0″ (66.0 cm), usually 5 lbs, 3 oz–8 lbs, 3 oz (2.4–3.7 kg). Largest specimen (Lake Erie) 31.0″ (78.7 cm) long, weight 11 lbs, 14 oz (5.4 kg). Nathan Ladd, of South Bass Island, caught one in the 1920's that weighed 16 lbs (7.2 kg). About 1900 J. B. Ward, then of South Bass Island, caught two while fishing through the ice that weighed 17 and 18 lbs (7.7 and 8.2 kg). Other fishermen give maximum weights of 12–18 lbs (5.4–8.2 kg). For age and growth, *see* Deason (1933:348–60).

Hybridizes: rarely in Ohio with sauger.

Distribution and Habitat

Ohio Distribution—The walleye was abundant and was widely distributed throughout inland Ohio and in Lake Erie when the first white men entered the present state of Ohio. It was an important source of food for the early pioneers, as it had been previously for the Indians. The literature contains many references to its presence and abundance. Zeisberger (Hulbert and Schwarze, 1910:240) notes the presence of the "yellow perch"* in the Ohio drainage prior to 1780.

Hildredth (1848:498) records the presence of the "salmon" in the Muskingum and Ohio rivers between 1778–1800. Scott (Journal 1893–94) records the "salmon" in the St. Marys, Auglaize and Maumee rivers during August, 1794. Smith and Snell (1891:235 and 248) mention a seine fishery which was begun in the Maumee River "about seventy years ago [about 1815]," in which the walleye was one of the principal species taken. In 1844 Kirtland† (1844:238; as *Lucio-perca americana*) mentions the great quantities which were captured in the Maumee River; he also mentions (1850P:61) that "All the larger streams of our state contain it; but Lake Erie seems to be its favorite residence."

Kirtland (1850A:1; as "Pickerel") as early as 1850 noted a decrease in its abundance in the Mahoning River, after that stream had been obstructed by dams. In 1874 Klippart (1874:6) likewise recorded its early decrease in abundance, stating that "In the Scioto River in days gone by" the "jack salmon" (obviously both saugers and the "other salmon" which was the walleye because it reached a weight of 18–20 lbs [8.2–9.1 kg]) was very abundant, but after the building of the State dam across the Scioto River a few miles below Chillicothe it became "comparatively rare" and has "nearly disappeared." The late Howard Jones, of Circleville, told me that when he was a boy, about 1860, he and his family fished in the Scioto River at the Circleville dam each Thanksgiving Day, catching five or more walleyes which weighed from 2.0 to 10.0 lbs (0.9–4.5 kg) each. After 1940 walleyes occurred in the Scioto River only as strays. In 1887 (1888:80) Henshall considered the "Ohio Salmon" to be abundant in the Ohio River. Old rivermen told me that the "white salmon" was numerous in the Ohio River prior to 1911, before the construction of many dams but that shortly after these dams were in operation the species decreased sharply in numbers.

During the years from 1925 to 1950 the walleye was abundant only in Lake Erie, and especially in its western third. In inland Ohio it was most numerous in the Muskingum River and Pymatuning Reservoir, and was uncommon, rare or absent in the Ohio River and its other tributaries. A small spring migration still occurred in the Maumee River as far upstream as Grand Rapids.

For many years the Conservation Department has introduced walleyes into inland streams and impoundments. The many plantings, made during the past 50 years in Buckeye, St. Marys and many other lakes and impoundments have met with little or no success. An exception was Pymatuning Reservoir, where the introduction has been markedly successful.

Habitat—In Lake Erie waters the walleye was an inhabitant of the shallow waters, occurring in the greatest abundance over gravel, bedrock, and other types of firm bottoms, where the turbidity of the waters was the least. The rock-bottomed reefs about the island region were concentration areas. The species was present in the fewest numbers in the most turbid portions of Sandusky and Maumee bays, and it was uncommon or absent in the vegetated portions of marshes, even if the waters were comparatively free from turbidity. In inland Ohio it was present in the largest numbers in those portions of the larger and deeper streams where the waters were the clearest, and it avoided the more turbid rivers such as the Scioto. Its decrease in numbers prior to 1900 appears to have been caused chiefly by dams which prevented this highly-migratory species from ascending the streams. Since 1900 dams, the ever increasing turbidity of the waters, and the silting over of the hard bottoms with soft clayey silts, have been major factors.

* The "yellow perch" of Zeisberger was the walleye and sauger, for Zeisberger mentions its "sharp teeth like those of a pike"; it was not the yellow perch (*Perca flavescens*) as suggested by Mahr (1949:66), for the latter species lacks sharp teeth and it presumably was absent from much of the Ohio drainage; *see* yellow perch, under Distribution.

† Kirtland believed that the brighter colored saugers were the males, and the paler walleyes were the females, of a single species.

Years 1955–80—Smith and Snell (1891:241) reported that the 1885 commercial catch of the walleye (then known as "Wall-eyed Pike") for the ports of Lake Erie (exclusive of Canada) was 1,401,200 lbs (635,573.6 kg) Applegate and Van Meter (1970:15, fig. 5) gave the catch totals between the years 1950–55 for both U.S. and Canadian ports as fluctuating from less than 1,000,000 (453,592.4 kg) to more than 10,000,000 lbs (4,535,924 kg) annually. In 1956 the annual poundage peaked at more than 15,000,000 (6,803,886 kg), after which it dramatically declined to 162,820 lbs (73,853.9 kg) in 1966 and 303,875 lbs (137,835 kg) in 1968 (Anonymous B, 1968:8). In the recent intensive investigations of the Cleveland Metropolitan area we (White et al., 1975:12–13) found the species to be "rare." Commercial fishing for walleyes has not been permitted in the Ohio waters of Lake Erie since 1970. Annual spawning success continues to fluctuate greatly. As examples, it was extremely poor in 1971, excellent in 1974, fair in 1975, extremely high in 1977 and poor in 1978. At present there is some indication that without the drain imposed by commercial fishing, the Lake Erie population may be increasing (Anonymous D, 1979:7).

Prior to 1900 the walleye populations were rather high in the larger, less turbid inland streams, after which a decrease became apparent. The species was planted in many inland waters prior to 1950, this activity increasing greatly thereafter. Clarence F. Clark informed me that the walleye has been stocked in more than 50 Ohio counties, and that there were one or more introductions into 99 impoundments and 51 streams. Some introductions were markedly successful, and natural reproduction has taken place in a few impoundments, such as at Hoover Reservoir, Delaware County. Possibly because of "escapes" from Hoover and other reservoirs into the upper Scioto River system, there has been fair walleye fishing recently in the Scioto River from the confluence of Little Walnut Creek downstream to Circleville. In late July and early August of 1962 and late August of 1963, the Ohio Division of Wildlife set experimental fyke nets in the Scioto River near the mouth of Big Darby Creek, catching several walleyes daily which were 13–25″ (33–64 cm) TL. Comparing lengths and weights with the average of 300 walleyes taken from Lake Erie before 1940, it was noted that Scioto River walleyes weighed significantly heavier than did those from Lake Erie.

The 1957–59 investigations of the entire Ohio River gave a total of only 29 walleyes (ORSANCO, 1962:152). H. Ronald Preston informed me that during the 1968–69 investigations the species was reported from only three of those nine collecting stations in the river that were opposite the state of Ohio.

BLUE PIKE

Stizostedion vitreum glaucum Hubbs

Fig. 148

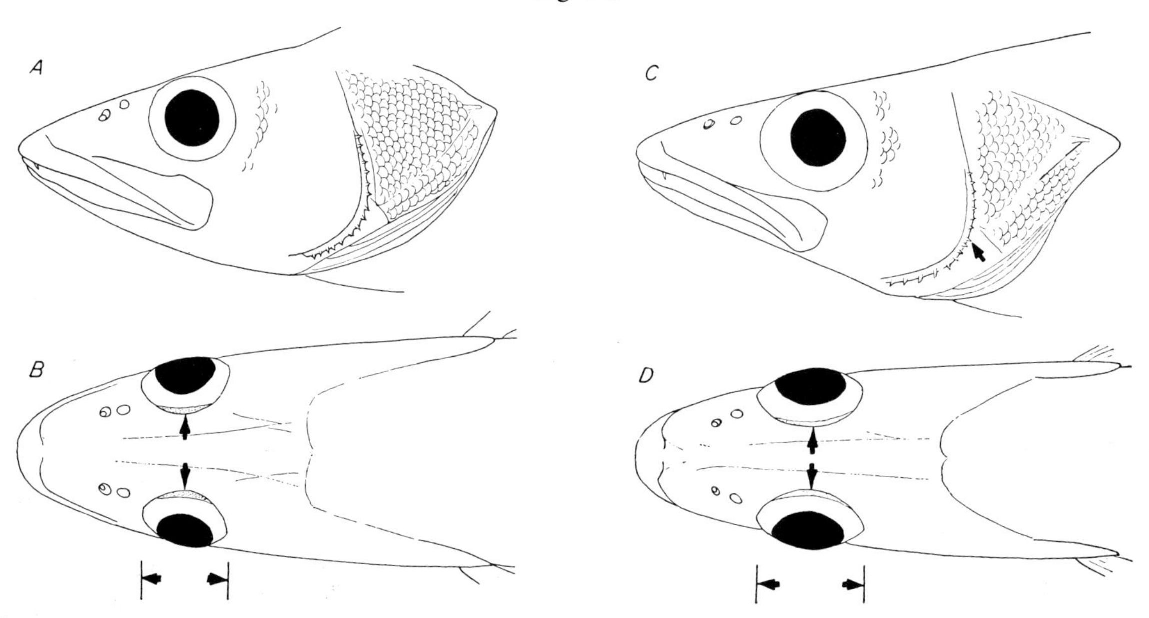

Note: No figure of this subspecies was drawn because all proportions are similar to those of the walleye except the eye size, *see* under walleye.

Figs. A and B: Head of the specimen shown in fig. 147.

Fig. A: lateral view of head of walleye; compare eye size with that of blue pike, fig. C.

Fig. B: dorsal view of head of walleye; compare wide bony interorbital space (arrows between eyes) with diameter of the small eye (diameter indicated with arrows beneath eye).

Figs. C and D: Lake Erie, Ashtabula County, O.

Sept. 18, 1940. 212 mm SL, 10.0″ (25.4 cm) TL.
Male. OSUM 1868.

Fig. C: lateral view of head of blue pike; compare eye size with that of walleye, fig. A.

Fig. D: dorsal view of head of blue pike; compare narrow bony interorbital space (arrows between eyes) with diameter of the large eye (diameter indicated with arrows beneath eye).

Identification

Characters: *General* and *generic—See* sauger. *Specific—See* walleye. *Subspecific*—Pelvic fins whitish-blue and body grayish-blue without brassy or yellow mottlings or overcast. Eyes larger and closer together; the bony interorbital width usually contained 1.4–2.0 times in eye length, fig. D; note arrows pointing to interorbital and eye widths. Maximum length less than 20.0″ (50.8cm).

Differs: Walleye has yellowish cast and smaller eyes. *See* walleye and sauger.

Most like: Walleye and sauger.

Coloration: Dorsally pale bluish-gray. Sides lighter and more silvery-bluish. Ventrally silvery and milk-white. No prominent brassy reflections on body. Between 4–14 dark bands cross back and sides of many individuals; these are most pronounced on small individuals. Pelvic fins transparent in small young; silvery-blue in larger fishes. Dusky blotch distinct on webbing between three last dorsal spines. Dusky spot usually prominent at base of pectoral fin. Tip of lower lobe of caudal and anterior tip of anal fins milk-white.

Lengths and **weights:** *Young* of year in Oct., 2.0″–5.0″ (5.1–13 cm) long. Around **1** year, 3.5″–6.5″

MAP 148. Blue pike

○ Type locality, Lake Erie, Ashtabula County "off Ashtabula." ● Locality records.

Insert: Isolated populations which are superficially similar to this subspecies occur in bays or other parts of Lakes Winnipeg, Ontario and Huron. For range of the species *see* **Insert**, Map 147.

(8.9–17 cm). *Adults*, usually 9.0″–16.0″ (23–40.6 cm) long, weight 5 oz–1 lb, 8 oz (142 g–0.7 kg). Maximum size unknown because of intergradation with walleye. Many fishermen consider the "Gray Pike" to be the large adult of the blue pike; these "Grays" have a maximum weight of 7 lbs (3.2 kg) and presumably are intergrades because of their intermediate characters. For rate of growth, *see* Adamstone (1922:80–83).

Hybridizes: with sauger (rarely); intergrades formerly abundant between this subspecies and the walleye.

Distribution and Habitat

Ohio Distribution—Until the mid-1950's the blue pike was abundant in Lake Erie and presumably has been present since early postglacial times. Since about 1850 it has been of great economic importance to the fishermen of the eastern two-thirds of the lake. The subspecies appears to be confined to the lake itself, and no typical specimens have been found in the tributaries.

Until recently much confusion existed as to whether the blue pike was a distinct species, or whether it was only a subspecies or variation of the walleye. For several generations commercial fishermen, impressed with its small size, soft flesh and bluish coloration, have considered the blue pike to be a distinct species and the intergrades between the blue pike and walleye, known as the "Gray Pike," were considered to be "jumbos" or "mules" (hybrids) between the two species. Until 1926 many ichthyologists believed that the blue pike of Lake Erie was the same species as the "White Salmon" of the Ohio drainage, and applied Rafinesque's (1818:354 and 1820:65–66; as *Perca salmonea*) name of *salmonea* to the Ohio drainage and Lake Erie populations. In 1882 Jordan (1882:962–64) considered the blue pike to be only a "variation" (subspecies) of the walleye but made the mistake of assuming that the Lake Erie and Ohio drainage populations were the same, obviously because in the color-unstable walleye decidedly bluish Ohio drainage specimens are not uncommon. Jordan was further misled by Klippart's (1877B: 65–71) incorrect statement, which Jordan quoted, that the blue pike frequented shallow "bayous" whereas the walleye inhabited deeper waters. In 1926 Hubbs (1926: 58 59), realizing that the blue pike had not been described technically, named it *Stizostedion glaucum* (type locality, Lake Erie off Ashtabula, Ohio). Later the blue pike was reduced from specific to subspecific rank because of the number of intergrades.

Intergrades—Known as "Gray Pike," these occur occasionally in the commercial catch in rather large numbers. At infrequent intervals intergrades are sometimes numerous in the ice-fishery catch about the Bass Islands.

It is difficult or impossible to identify satisfactorily many individuals believed to be intergrades (presumably back crosses) because of their close similarity to either the walleye or blue pike, and because coloration is particularly unstable in the walleye. As related elsewhere, golden individuals lacking bluish chromatophores are not rare, and Ohio drainage specimens lacking yellow are rather common. In addition, occasional bluish Lake Erie specimens of 8 (3.6 kg) or more pounds in weight are noted, which lack yellow and which are not blue pikes because of their large size, eye size, and firm flesh.

Habitat—The blue pike inhabited the deeper and clearer waters of Lake Erie, the largest populations occurring in the eastern two-thirds of the lake. It was less numerous in the shallow waters of the western third, especially during the summer, and was absent or present only as strays in the shallow, turbid waters. There appeared to be an annual movement into the shallower inshore waters and into the western third of the lake, during fall and early winter.

Years 1955–80—This subspecies of walleye has been until recently of great commercial importance since early in the history of the Lake Erie fisheries. For many years prior to its scientific description in 1926, the commercial fishermen had recognized the blue pike as distinct from the walleye.

Smith and Snell (1891:241) reported that the commercial catch for 1885, brought into the ports of Michigan, Ohio, Pennsylvania and New York (exclusive of Canadian ports) was 3,152,400 lbs (1,429,905 kg). Applegate and Van Meter (1970:14, fig. 4) gave the blue pike annual poundages for all U.S. and Canadian ports between the years 1950 and 1957 as fluctuating between 2,000,000 and 26,000,000 lbs (907,184 and 11,793,402 kg).

In 1959 the fishery collapsed with a total of only 79,000 lbs (35,834 kg). In 1964 wholesale fish dealers reportedly sold less than a total of 200 lbs (90.7 kg). At present some believe that this subspecies may be extinct. None was collected during the intensive investigations in the Cleveland Metropolitan area between 1971 and 1975, and we consider the species to be "probably extirpated" (White et al., 1975:115). For summary of blue pike 1951–60 *see* Woner, 1961.

In 1970 the committee revising *A List of Common and Scientific Names of Fishes* (Bailey et al.: 39) accepted the name of blue pike, a name long used by commercial fishermen, for this subspecies. The name blue pike gives two specific names to one species (walleye and blue pike), which is undesirable because it gives no indication of their conspecific relationship and leaves the impression with some that the blue pike is a true pike (*Esox*).

YELLOW PERCH

Perca flavescens (Mitchill)

Fig. 149

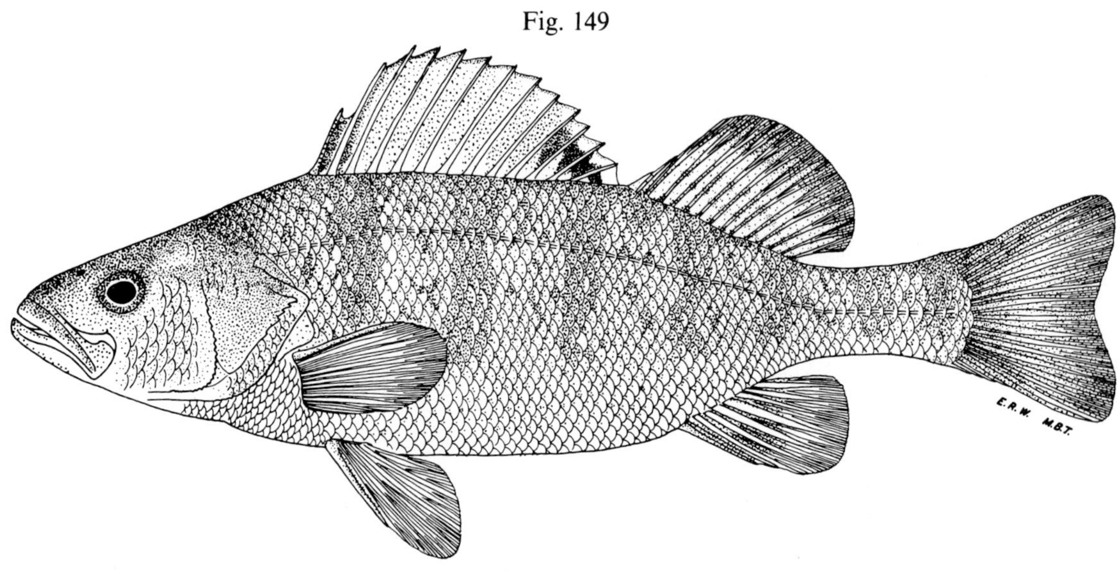

Lake Erie, Ottawa County, O.

Dec. 6, 1939.
Adult female.

205 mm SL, 9.5″ (24 cm) TL.
OSUM 1122.

Identification

Characters: *General—See* sauger. *Generic* and *specific*—No canine teeth on jaws. Free edges of preopercle strongly serrated (toothed); *see* blue pike, fig. 148 C. Pelvic fins inserted closely together; distance between insertions of the last ray of each pelvic fin less than diameter of pupil of eye. Anal spines two; rays 6–8. Body the deepest of all percid fishes (walleyes; darters) found in Ohio; body depth contained 3.5–4.2 times in standard length in small young, 3.0–3.8 times in adults; 6–9 blackish vertical bands crossing back and sides; bands most prominent in young and small adults which live in clear, vegetated waters; faint or absent in old fishes living in turbid waters. A dusky blotch on webbing between four last dorsal spines; blotch may be undeveloped in small young.

Differs: Sauger and walleye have canine teeth on jaws. Darters lack the large serrations on preopercle. White bass lacks vertical bands on sides; dusky blotch on spinous dorsal; has more than eight anal rays. Drum has first anal spine very short; second very long and stout. Blackbasses and other sunfishes have three or more anal spines. Trout-perch and pirate perch have fewer dorsal spines.

Most like: Dusky darter. **Superficially like:** Small blackbasses and small white bass.

Coloration: Dorsally brassy-green, greenish-olive or golden-yellow (usually turbid waters). Sides lighter; more greenish-yellow or golden-yellow. Ventral surface white, whitish-yellow, or yellow; 6–9 blackish or dark olive bands across the back and sides. Vertical fins olive-yellow; spinous dorsal has a dusky blotch on webbing between last four spines; often a dusky blotch between the first two spines. Pelvics silvery or yellow. *Breeding adult*—Very brassy and golden; pelvic and anal fins a rich orange-yellow; bands usually blackish. *Small young*—Usually have the yellows replaced with transparent or silvery.

Lengths and **weights:** *Young* of year in Oct., 1.8″–4.0″ (4.6–10 cm) long. Around **1** year, 2.0″–4.5″ (5.1–11 cm). *Adults*, usually 4.5″–12.0″ (11–30.5 cm); dwarfed breeding adults in small impoundments only 3.0″ (7.6 cm) long. Fishes 5.0″ (13

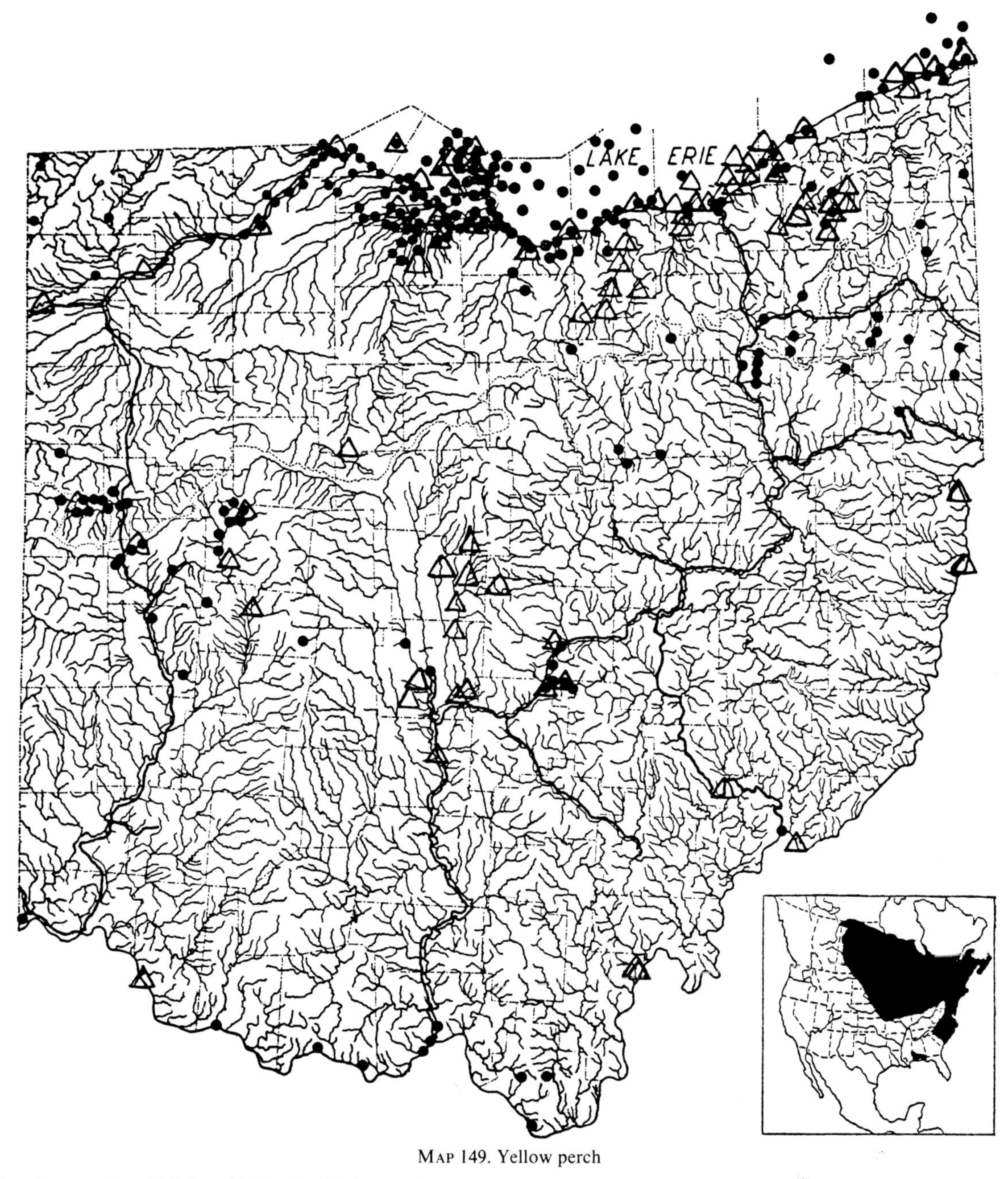

MAP 149. Yellow perch

Locality records. ● Before 1955. △ 1955–80. ~ Canals.

Insert: Widely introduced throughout the United States and some portions of Canada, both inside and outside of this range; northern and western limits of range indefinite.

cm) long, usually weigh 0.8 oz–1.3 oz (23–37 g); 6.0″ (15 cm) long, usually 1.1 oz–3.0 oz (31–85 g); 7.0″ (18 cm) long, usually 2.5 oz–4.2 oz (71–119 g); 8.0″ (20 cm) long, usually 3.0 oz–5.2 oz (85–147 g); 9.0″ (23 cm) long, usually 4.5 oz–7.0 oz (128–198 g); 10.0″ (25.4 cm) long, usually 7.5 oz–10.3 oz (213–292 g); 11.0″ (27.9 cm) long, usually 9.0 oz–14.5 oz (255–411 g); 12.0″ (30.5 cm) long, usually 12 oz–1 lb, 2 oz (340 g–0.5 kg); 13.0″ (33 cm) long, usually 14 oz–1 lb, 4 oz (397 g–0.6 kg). Largest specimen, 15.7″ (39.9 cm) long, weight 2 lbs, 8 oz (1.1 kg). For growth-weight relationship for Lake Erie fishes, *see* Jobes (1952) and Harkness (1922).

Distribution and Habitat

Ohio Distribution—The invasion of yellow perch into Lake Erie waters probably occurred in early post-glacial times, possibly through the Maumee Outlet (Greene 1935:162). The earliest literature references indicate its early abundance in Lake Erie, and since it was a shallow water species, it was abundant in Sandusky and Maumee bays, and about the Bass Islands and the three harbors on the Catawba peninsula. Because of its presence in the shallow waters it was among the earliest of the fish species to be captured with shore seines and to attain commercial importance (Smith and Snell, 1891:236;264;268). Before 1877, however, the flesh of the perch was not held in the universally high esteem as food as it has been since 1900 after the more desirable species such as the muskellunge had become scarce. Klippart (1877B:66–67) voices this early disapproval of the perch by stating that its "flesh is soft, rather coarse, and insipid; and at best it is nothing more than a third-rate pan fish" and "The writer's opinion is that perch make better glue than food." E. Sterling, of Cleveland, wrote to Klippart that on the whole Lake Erie perch were "a most worthless animal . . . you can have all you want [from fish dealers] for the trouble of carrying them away" and "I once saw three tons sold for manure at Kelley's Island for as many dollars." Despite this early dislike Smith and Snell (1891:241) report that in 1885 the total catch of perch brought into the Ohio ports of Lake Erie was 1,265,500 lbs (574,021 kg). The 1942–51 yearly average was 2,-030,669 lbs (921,095.9 kg) (Cummins, 1952:1).

Kirtland (1838:190; 1847:338; 1850:13) appears to have been the first to record that, except for some headwater streams, the perch was absent from the Ohio River drainage before the building of the Ohio canals, stating that "The yellow perch is found in Lake Erie and most of the small lakes in the northern parts of the State, but did not exist in the waters of the Ohio until it found its way into them through the medium of the Ohio canal after 1825." Later Jordan (1882:959) wrote that "west of the Alleghenies it does not occur, except in the lake region and in the upper waters of such streams as the Scioto, Wabash, Illinois, Rock, etc., rising in the same water shed with streams flowing into the great lakes." Recently Rostlund (1952:282), outlining its range, shows the early absence of perch from the Ohio drainage.

Kirtland (1850 O:13) in 1850 believed that the perch was moving southward through the newly constructed canals and would eventually extend "its migrations into the Ohio river"; but later events proved that this southward extension progressed no farther than central Ohio (Buckeye Lake).

After 1900 many thousands of perches were taken from Lake Erie and liberated in the streams of southern Ohio, but these introductions failed except for occasional strays in streams or establishment, often of dwarfed individuals, in some impoundments. During the 1920–32 fish surveys a total of fewer than 20 perches were captured in southern Ohio streams despite the fact that I made a practice of seining in the vicinity of recent plantings (Trautman, 1932:259).

Habitat—Yellow perches occurred in greatest numbers in clear waters of base and very low gradients where there was an abundance of rooted aquatics and the bottoms were of muck, organic debris, sand or gravel. Its numbers decreased drastically with increased turbidity and siltation, and the subsequent reduction and final disappearance of rooted aquatics. Excellent examples of decreases in abundance were observed in Lake St. Marys, and in Middle and West harbors, where the species was relatively quite abundant as late as 1925 but where it had become uncommon or almost absent by 1950. The species became dwarfed, with the adults breeding when less than 4.0″ (10 cm) in total length, in areas where there were excellent spawning conditions for adults and growth conditions for young and/or a comparative lack of predators, and/or too much parasitism.

For food of young, *see* Turner (1920:137–52).

Years 1955–80—Before 1900 the yellow perch was considered to be of only secondary importance, commercially, to the Lake Erie fishery at the time when large poundages of the more desirable whitefish, cisco and walleye were available. Later, when the annual poundages of some of these more desirable species had decreased, the perch became increasingly important commercially until in the late 1960's it had become of major importance.

The poundages brought annually into Lake Erie ports have varied greatly, caused in part by demands of the market and survival success of the various year-classes. Applegate and Van Meter (1970:16, fig. 6) demonstrated annual fluctuations between the years 1950–56, the annual total poundages for Lake Erie ranging from a low of 2 million lbs (907,184 kg) in 1917, rising irregularly to 20 million lbs (9,071,847 kg) in 1934, followed by 2 million lbs (907,184 kg) in 1943, and highs of 29 million lbs (13,154,178 kg) in 1959 and 28 million lbs (12,700,586 kg) in 1962.

Some commercial fishermen and fishery workers believed that an inverse relationship existed between the perch and walleye in Lake Erie, that whenever the population of one species was large, the other was small, and *vice versa*. A comparison of Applegate and Van Meter's figures (1970:15, fig. 5; 16, fig. 6) does show an inverse relationship between the years 1915 and 1953. However, in 1956, the walleye poundage for Lake Erie was approximately 16 million lbs (7,257,477.9 kg), after which it collapsed to less than 3 million lbs (1,360,777 kg) annually. By 1956 the perch populations had also greatly increased to approximately 20 million lbs (9,071,847 kg), but unlike the walleye, poundages remained very high, with from 9 million to 29 million lbs (4,082,331–13,154,178 kg) taken annually. The intensive investigations of Lake Erie in the Cleveland Metropolitan area after 1970 indicate that it was the most important and abundant food species in the area, but that recently it appears to be declining in numbers (White et al., 1975:116).

After 1950 the capturing of perch from Lake Erie and releasing of them into inland waters decreased sharply. These introductions were largely confined to natural lakes, reservoirs and other impoundments. Small populations infrequently became established below the larger impoundments, especially in the northern half of the state; but on the whole, few were taken from flowing waters. Few perch were stocked in farm ponds.

As prior to 1950, the yellow perch remained of rare or accidental occurrence in the Ohio River, in 1968 and 1969 H. Ronald Preston reporting the species from only one of nine stations bordering the state of Ohio.

In fishes (and birds) melanophores and black pigments appear to be more unstable than are other colors, and individuals of many species are rather frequently seen in which black is muted or wholly absent. The genera *Stizostedion* and *Perca* appear to be unusually prone to a muting or lack of melanophores or black pigment. Above, under Coloration of *Stizostedion vitreum vitreum*, I mention individuals of this species which lack blue, slate and other dark pigments, and as a consequence are a golden yellow. I have seen more than a dozen *Perca flavescens* from Lake Erie which lacked, or mostly lacked, black coloration, some of which were similar to the yellow perch color mutant from Lake Erie described by E. J. Crossman (1962: 224–25).

On the other hand, and contrary to the more usual conditions, xanthophores and yellow pigments appear to be more unstable than is black in the genera *Stizostedion* and *Perca*. In the inland streams of the Ohio River drainage of Ohio, the yellowish cast of the walleye (*S. v. vitreum*) is muted or lacking, giving rise to the former name of white salmon (*see* above). Some lakes draining into the St. Lawrence River, upper Great Lakes and Lake Winnipeg have populations which lack yellow but otherwise agree morphologically with the walleye. The blue pike (*S. v. glaucum*) appears to have lost permanently the yellow coloration. "Blue Perch" in which yellow is lacking appear to be as frequently observed as are the "Golden Perch."

NORTHERN DUSKY DARTER

*Percina sciera sciera** (Swain)

Fig. 150

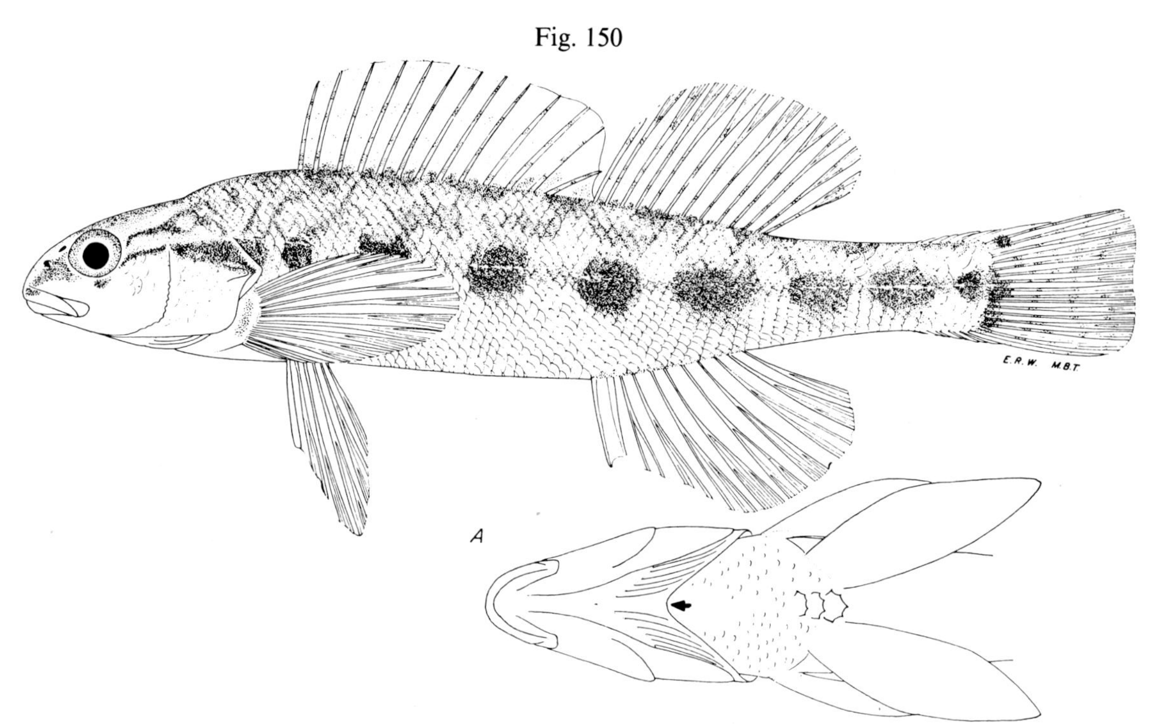

Upper fig.: Big Darby Creek, Pickaway County, O.

April 21, 1942. 80 mm SL, 3.6″ (9.1 cm) TL.
Male. OSUM 4859.

Fig. A: ventral view of head and breast; arrow points to the broad angle (the isthmus), formed by membrane which lies across the upper breast and which connects one gill cover with the other, note also the large specialized scales between pelvic insertions.

Identification

Characters: *General*—Posterior edges of tail rounded, straight, or slightly emarginated; definitely forked only in the crystal darter. Anal spines *two*; except in johnny, crystal, and sand darters which have only *one*. Anal rays 5–13. First dorsal of 5–16 spines; the two dorsals well separated or slightly conjoined at their bases. No canine teeth on jaws. Free edges of preopercle smooth; weakly roughened only in some specimens of dusky darters. Body slender; never sunfish-like in shape; body depth almost invariably less than head length, rarely equal. Scales strongly ctenoid. Opercles with a spine. Branchiostegals on each side 6, rarely 5. Genital papilla large. Usual maximum length of most species less than 4.5″ (11 cm) long; more than 5.0″ (13 cm) long only in logperch and crystal darters. *Generic*—Breast and midline of belly either partly or entirely naked; *or* naked with one or more large specialized, caducous scales along the midline, *see* blackside darter, fig. B; *or* naked except for a narrow bridge of scales crossing belly immediately before anus (river darter only, fig. A). Pelvic fins inserted closely together; distance between insertions of last rays of each pelvic fin almost equal to, or equal to, the basal width of either pelvic fin. Complete lateral line. Mouth terminal, or snout projects beyond the inferior mouth. *Specific*—No groove separating tip of upper jaw and snout; fig. 151, A. Gill covers rather broadly connected across isthmus of breast with membrane; *see* arrow, fig. A; membrane forming a wide angle; distance from

* Formerly in the genus *Hadropterus*; *see* Bailey and Gosline, 1955:10.

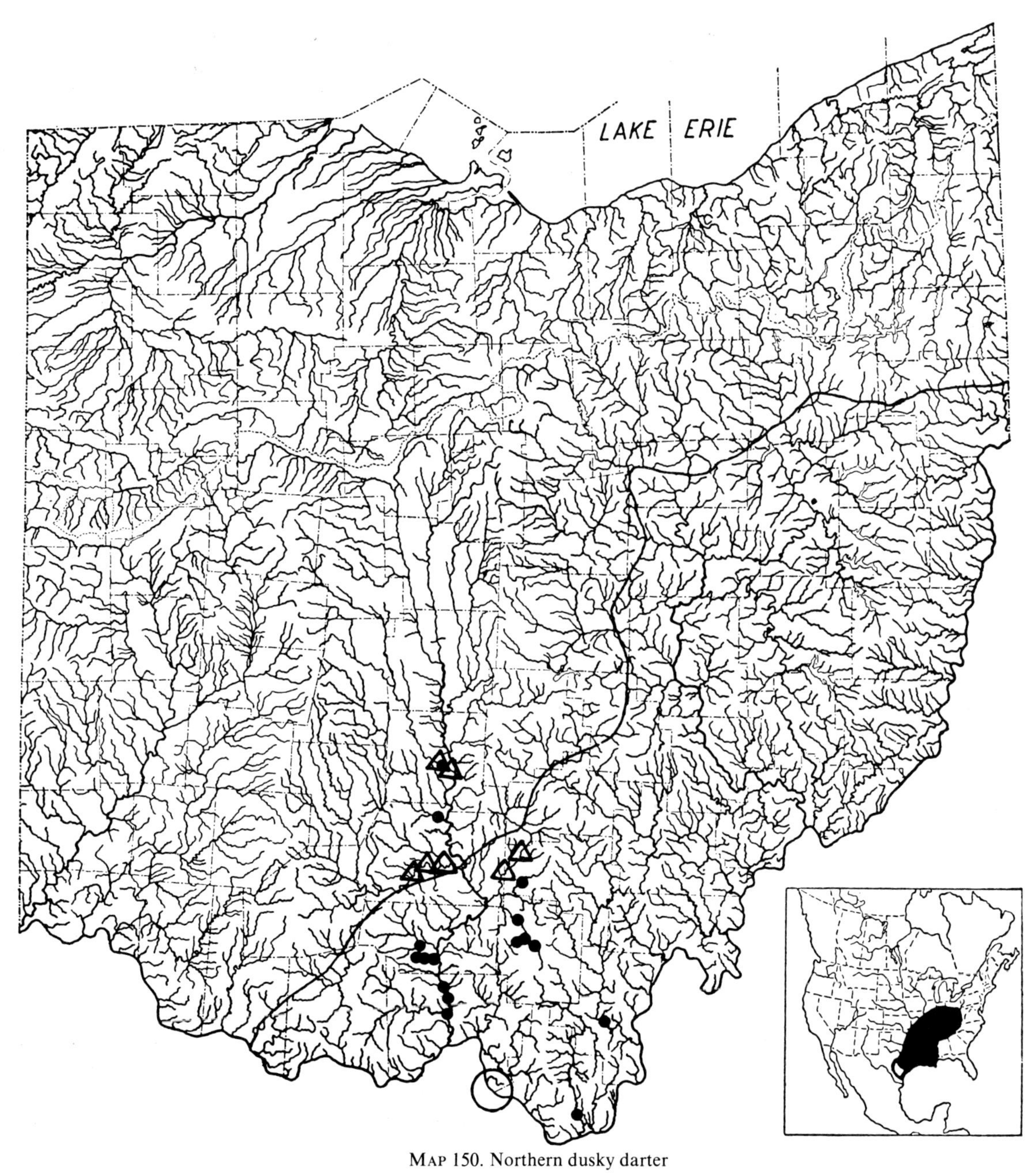

MAP 150. Northern dusky darter

○ Literature record, no specimen. ● Locality records after 1924, with specimens. △ 1955–80. ~ Glacial boundary.
Insert: Black area represents range of this subspecies, outlined area another.

apex of this membranous angle to tip of upper jaw, when measured carefully with dividers, extends from tip of jaw to center of cheek; is contained 1.3–1.8 times in length of head. Cheeks partly or entirely scaled. Three dark blotches form a broken vertical bar at caudal base; the two lower blotches often confluent. Free edge of preopercle sometimes roughened. Dorsal spines usually 12–13. Lateral line scales usually 61–65, extremes 60–70. Teardrop much reduced or absent.

Differs: Blackside darter has gill covers not connected with membrane. Longhead, slenderhead, channel, and gilt darters normally have scaleless cheeks; slenderhead and longhead have attenuated snouts; river darter has a narrow bridge of scales crossing belly before anus. Logperch has an overhanging snout. Crystal, sand, and johnny darters have one anal spine. All other darters have a scaled belly; *or* if belly is partly or entirely scaleless, then lateral line is incomplete or absent. Walleye, sauger and perch have forked tails. Young blackbasses have three anal spines.

Most like: Blackside, river, channel, and gilt darters.

Coloration: *Non-breeding adults*—Dorsally olive-green or dark olive with darker mottlings. Sides olive-yellow; a dark band extending across snout and opercles, forming a broken lateral band of 8–12 dark olive-green or dusky, oblong blotches. Ventrally white or yellowish with some silvery overcast. Back with short, quadrate saddle-bands; usually four on back beneath spinous dorsal, two under soft dorsal, and two on caudal peduncle. Usually three spots form a broken, vertical bar at base of caudal; often the lower two spots are confluent. Tear-drop short or absent. Fins transparent-olive and usually unmarked, except for a dusky blotch on webbing between the last four rays of spinous dorsal. *Breeding male*—Has much of the yellow replaced with a suffusion of dusky, hence the name dusky darter. *Breeding female*—Chiefly yellow, heavily mottled with dark olive or black. *Young*—Similar to adults except that yellow is mostly replaced with silvery and fins are colorless.

Lengths: *Young* of year in Oct., 1.2″–2.0″ (3.0–5.1 cm) long. *Adults*, usually 2.0″–4.0″ (5.1–10 cm). Largest specimen, 4.5″ (11 cm) long; male appears to grow larger than does the female.

Hybridizes: with logperch, greenside, rainbow and orangethroat darters. No Ohio specimens.

Distribution and Habitat

Ohio Distribution—The dusky darter, collected in the Ohio River near the mouth of Little Sandy River in August, 1888, by Charles H. Gilbert and James A. Henshall (Henshall 1889:126) was originally preserved in the Cincinnati Society of Natural History. Although the specimen could not be found in 1931, there is no reason to doubt the validity of the record, for it was taken within the range of the species during the 1925–50 period, and in a locality where a suitable habitat obviously existed before impoundment of the Ohio River.

Between 1925–50 the dusky darter was restricted to the broad, partly-filled valleys of the old Teays Stage drainage (map II) which were later flooded by proglacial Lake Tight (Wolfe 1942: fig. 1), and which today are below 650′ (198 m) in elevation. These old valleys are more obvious, and the habitat for the darter more extensive, in the unglaciated portion of Ohio.

Habitat—The dusky darter was most frequently taken in moderate- or large-sized streams of low or moderate gradients. Most of the spring specimens, especially breeding adults, were found on gravelly riffles, particularly among beds of such aquatics as pondweeds or along the cutting edge of riffles where the water was more than a foot (0.3m) in depth and there was an accumulation of brush, timber or roots of trees. In summer and fall the majority were taken about debris in deeper water of less current, or in the denser patches of aquatic vegetation. During high floods in spring, occasional specimens were taken at the water's edge in flooded fields and meadows far from the river's channel, and it is assumed that these fishes were taken in the act of migrating upstream.

Years 1955–80—During this period dusky darters were found in several sections of Paint Creek in Ross County. Specimens were occasionally taken in Big Darby Creek, Pickaway County, in the vicinity of Fox, Ohio, and in Salt Creek in Vinton County. No other individuals were found despite considerable efforts to locate them elsewhere in the Ohio River drainage.

During the past decade I observed dusky darters feeding about the sand, gravel and boulders in exposed riffles. When frightened while feeding, they darted beneath an undercut bank where rootlets of willows, soft maples and other plant species extended into the water affording excellent escape cover. If rootlets, debris or other types of vegetation were absent, they secreted themselves among the water willow. They rarely scurried beneath stones or hid in *Cladophora* or other algae as did such species as the variegate, banded and greenside darters.

For life history of the dusky darter *see* Page and Smith (1970).

BLACKSIDE DARTER

*Percina maculata** (Girard)

Fig. 151

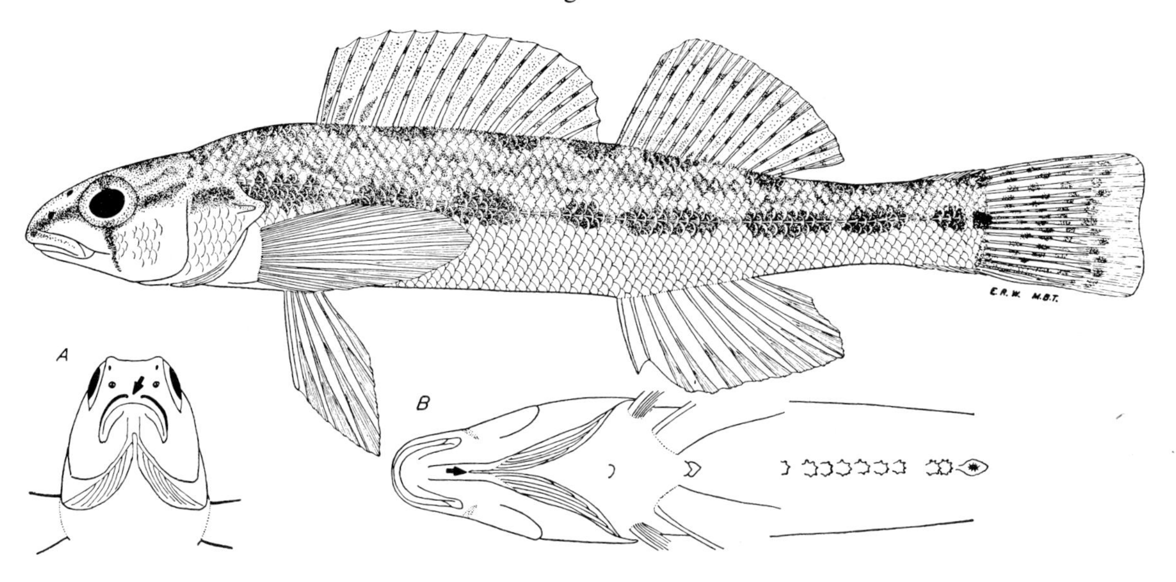

Upper fig.: Fish Creek, Williams County, O.

July 13, 1939. 70 mm SL, 3.3″ (8.4 cm) TL.
Female. OSUM 363.

Fig. A: frontal view of head; arrow points to the frenum which binds the tip of the upper lip to the tip of the snout, thereby resulting in inability of the non-protractile upper jaw to be projected outward (*see* fig. B of the two forms of johnny darters for a protractile upper jaw).

Fig. B: ventral surface of head and body; arrow points to the junction of the gill covers; note that this junction is far forward because there is no membrane joining the gill covers together across the upper breast as there is in fig. A of the dusky darter; note also the large, specialized scales along midline of belly.

Identification

Characters: *General* and *generic—See* dusky darter. *Specific*—No groove separating tip of upper jaw and snout, *see* arrow, fig. A. Gill covers not connected across isthmus with membrane; distance from junction of gill covers, *see* arrow, fig. B, to tip of upper jaw when carefully measured with dividers, extends from tip of upper jaw almost to, or to, posterior edge of eye; this distance contained 2.2–2.8 times in head length. Cheeks usually partly scaled. Longitudinal blotches along lateral line usually more or less confluent; never forming vertical bars. Dorsal spines normally 13–15, extremes 12–16. Lateral line scales usually 65–75, extremes 63–81. Tear-drop faint or absent in small young.

* Formerly in the genus *Hadropterus*; *see* Bailey and Gosline, 1955:10.

Differs: Dusky darter has gill covers connected rather broadly with membrane. *See* dusky darter.

Most like: Dusky, slenderhead, river, and gilt darters.

Coloration: *Non-breeding adults*—Dorsally bright olive-yellow, vermiculated with dusky-olive. Usually 6–11 dark, quadrate saddle-bands are present. Sides yellow; a dark brown or blackish band crosses snout, opercles, and continues across body as a broken lateral band in a series of 7–9 more or less confluent, longitudinal blotches. Caudal spot prominent. Ventrally white, or whitish-yellow; ventral surface of head unspotted, except for the presence of the ventral ends of the two tear-drops. Dorsals and caudal fins transparent-olive; in fishes more than 2.0″ (5.1 cm) long these fins have one or more brownish dots on each spine or ray, resulting in 2–3 rows of spots on the dorsals and 2–6 wavy

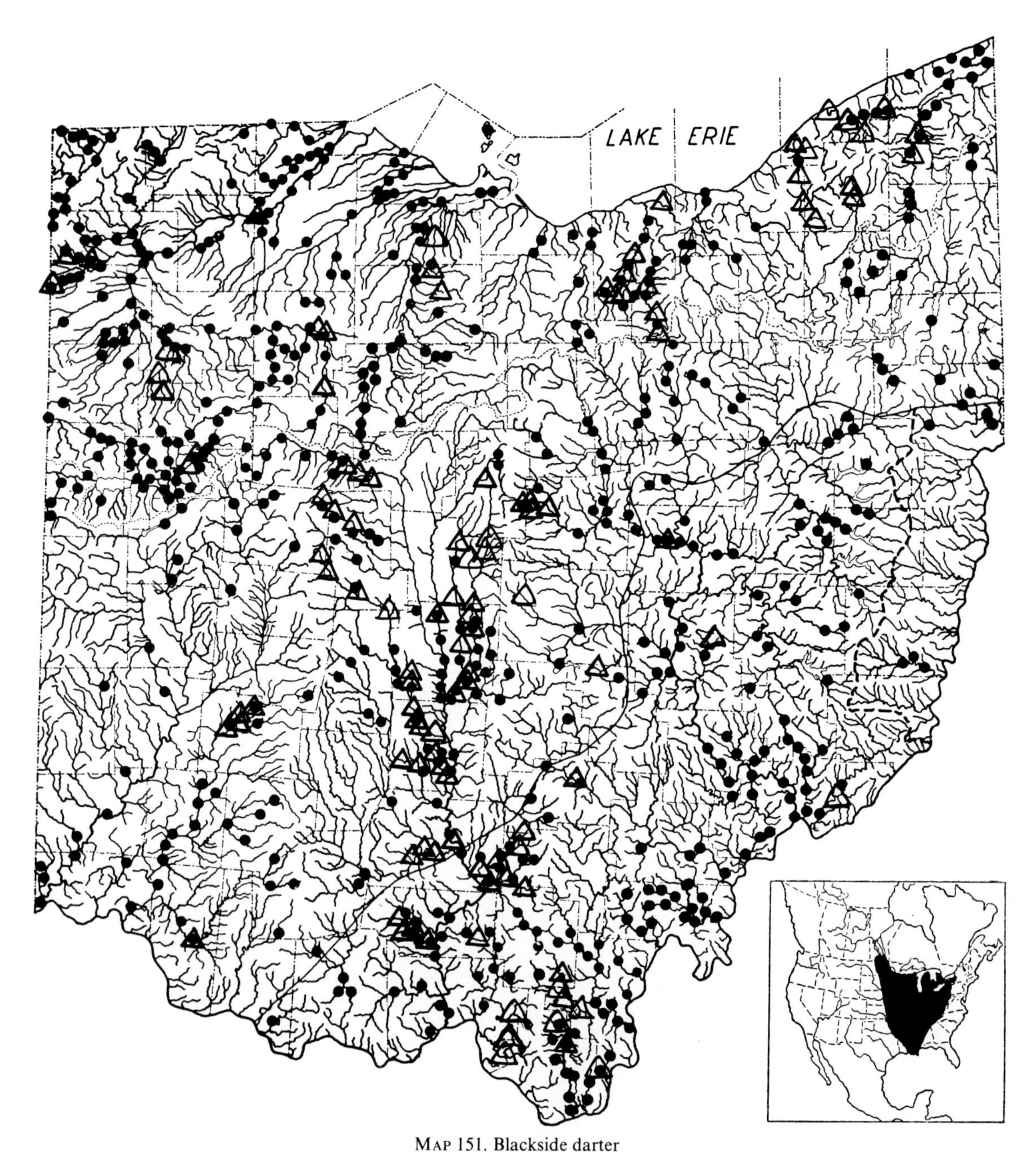

MAP 151. Blackside darter

Locality records. ● Before 1955. △ 1955–80. ~ Glacial boundary. /\ Flushing Escarpment.
Insert: Western limits of range indefinite.

bands on the caudal. Anal and paired fins transparent or yellowish. *Breeding male*—Intensely yellow and black; blotches along sides often coalesced, forming a solid lateral band. *Breeding female*—Similar to male, but colors less intense. *Young*—Similar to the female, except that yellows are pale or are replaced by white or silvery, especially on ventral half of body; tear-drop small or undeveloped.

Lengths: *Young* of year in Oct., 1.2″–2.2″ (3.0–5.6 cm) long. *Adults*, usually 2.0″–4.0″ (5.1–10 cm). Largest specimen, 4.2″ (11 cm). *Male*—usually larger than female.

Hybridizes: With logperch; specimens taken in Ohio, *see* Trautman (1948:173); natural hybridization elsewhere with slenderhead darter (OSUM 14019, from Illinois).

Distribution and Habitat

Ohio Distribution—Literature references, the preserved material captured before 1900, and the unpublished Muskingum River notes of Henshall depict the early presence of the blackside darter in widely separated Ohio localities in both drainages. That it was quite abundant in several localities is indicated in such statements as those of Kirsch (1895A:331) for the Maumee River system "abundantly distributed in all streams examined" and Williamson and Osburn (1898:49) for Franklin County "A common darter of general distribution."

During the 1925–50 surveys blackside darters were recorded at the majority of the seining stations; however, usually fewer than three individuals were taken at a locality. It was common in the upper Auglaize River system, but in the Maumee River where Kirsch in 1893 found the species abundant, it was usually absent from collections after 1940, as it was in Franklin County where in 1897 the species was common and of general distribution. This darter was fairly numerous in the Ohio River counties from Scioto to Washington, but was absent or rare in the high-gradient streams east of the Flushing Escarpment, and in the Chagrin and Mad River drainages. For some unknown reason it was almost absent from the prairie areas in Darke, Fayette and Marion counties. There is one Ohio River record, in Belmont County (Osburn, 1901:92), and it is possible that the species was present in the Ohio River before impoundment and increased siltation. I have seen one Lake Erie specimen, and during a drought and period of low stream levels in August, 1939, two were taken along the shores of Sandusky Bay.

Habitat—The blackside darter chiefly inhabited moderate-sized streams of moderate gradients, fairly clear water, and bottoms of sand and gravel, occurring most often in pools with current, in small pools in riffles and in the more sluggish portions of the riffles, and particularly in areas of quickening and lessening currents connecting the pools and riffles. The blackside also frequented, especially as emergency cover, brush heaps and tree roots under cut-banks. The species and its young especially, were not adverse to moderate stands of water-willow and pondweeds. It appeared to be highly intolerant to certain organic pollutants, such as mine wastes.

The blackside was essentially a midwater swimmer, which frequently rose to the water's surface to obtain food. I have often seen it rising for ovipositing Diptera, or jumping into the air for flying insects. It appeared to be highly migratory, and strays could be expected almost anywhere after the spawning season. Small spawning populations were also found in decidedly submarginal habitats (Trautman, 1948:173).

Years 1955–80—The blackside darter was captured in more than 100 seining localities in inland Ohio during the 1955–80 period; however, seldom more than one or a few were captured at any seining station. The species was generally distributed in those stream systems or localities where the gradient was moderate and the water normally clear (White et al., 1975: 116–17). It was usually conspicuously absent in the turbid Maumee River and lower portions of its larger tributaries.

The blackside was reported as straying rarely into Lake Erie (Van Meter and Trautman, 1970:76) and to my knowledge was not observed in that portion of the Ohio River bordering the state of Ohio.

LONGHEAD DARTER

*Percina macrocephala** (Cope)

Fig. 152

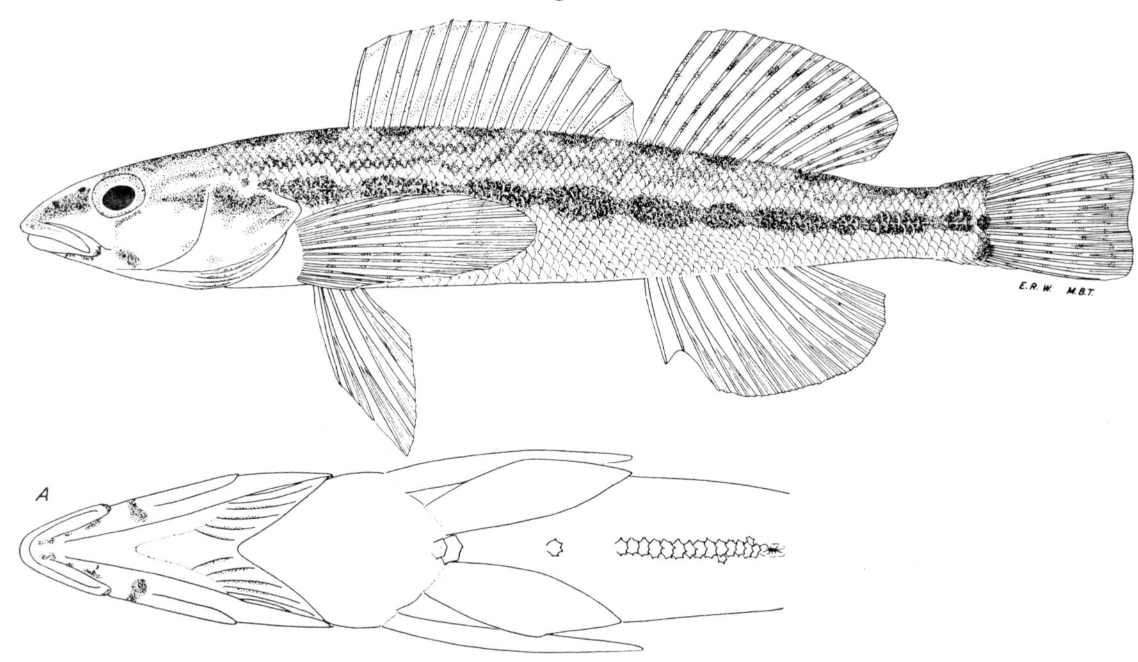

Upper fig.: Walhonding River, Coshocton County, O.

Oct. 20, 1939. 88 mm SL, 3.9″ (9.9 cm) TL.
Male. OSUM 977.

Fig. A: ventral surface of head, breast and belly; note attenuated snout; amount of membrane connecting gill covers and the resulting angle between gill covers; specialized scales along midline of belly; and pairs of dusky spots on ventral surface of head.

Identification

Characters: *General* and *generic—See* dusky darter. *Specific*—No groove separating tip of upper jaw and snout, fig. 151 A. Gill covers moderately connected across isthmus with membrane; distance from gill cover angle to tip of upper jaw usually contained 1.7–2.0 times in head length. Snout long and head narrow. Body depth of adults usually contained considerably more than 1.5 times in head length. Cheeks and opercles scaleless; 1–3 dusky spots on each side of ventral surface of head, the most posterior of which is directly below eye; in adults this posterior spot connects with the suborbital bar to form a sickle-shaped tear-drop, fig. A; these spots undeveloped in smallest young. Blotches along lateral band confluent. Dorsal spines 13–14, extremes 13–16. Lateral line scales usually 74–80.

Differs: Slenderhead darter lacks spots on ventral surface of head; has fewer lateral scales. Channel, gilt and dusky darters have blunt snouts; fewer lateral scales. *See* dusky darter.

Most like: Slenderhead and blackside darters.

Coloration: *Adult male*—Dorsally bright olive-yellow; mid-line of back usually crossed with more than 12 brownish, quadrate saddle-bands; many short, longitudinal vermiculations on upper back, a light yellow streak separating these vermiculations from the lateral band. The brownish-black band crosses snout, opercles, and continues as a series of confluent blotches along lateral line. A small caudal spot; beneath it is a short, dusky, vertical bar. Ventrally yellowish-white and unspotted, except for the 1–3 dusky spots on each side of the ventral surface

* Formerly in the genus *Hadropterus*; *see* Bailey and Gosline, 1955:10.

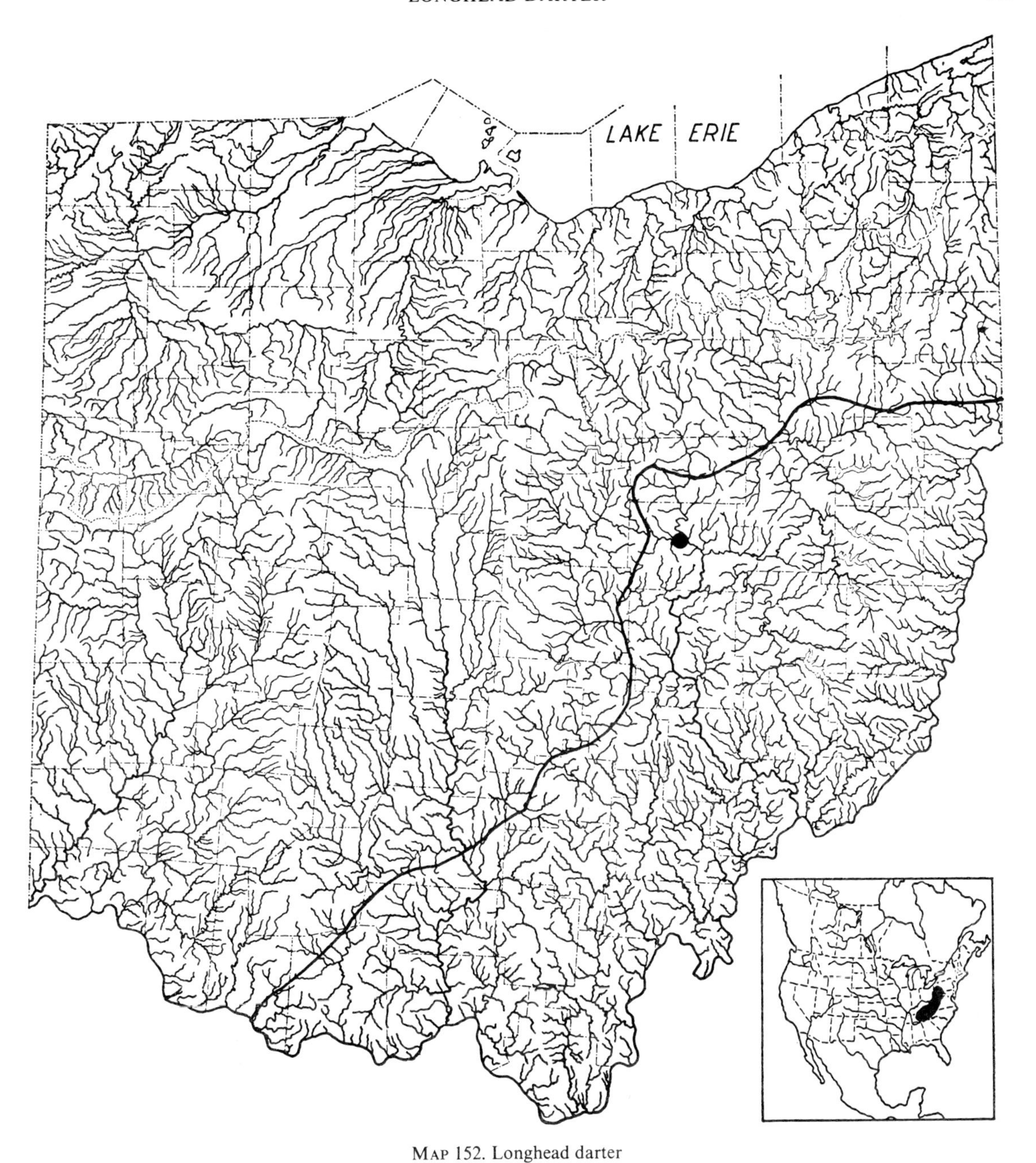

Map 152. Longhead darter

Locality record. ● After 1938. ~ Glacial boundary.
Insert: Large unoccupied areas apparently occur within this range.

of head, *see* descriptions under Characters. Fins transparent-yellow; dorsals and caudal with faint spots and vague blotches. *Adult female*—Like male but color less intense. *Young*—Like female except that the yellows of the ventral half of body are replaced with silvery-white; all fins are transparent and spotless; the lateral band is usually broken up into a series of oblong blotches.

Lengths: 7 Ohio specimens, 2.8″–4.0″ (7.1–10 cm) long.

Distribution and Habitat

Ohio Distribution—On September 30, 1939, Edward L. Wickliff and I collected two specimens of the longhead darter in the Walhonding River, in the deep riffles immediately below Six Mile Dam, Bethlehem Township, Coshocton County. On October 20, 1939, we took five additional specimens, 1 to 2 miles upstream from the dam, in Bethlehem and Jefferson townships. Since then infrequent attempts to collect the species have failed, as have attempts in presumably favorable localities elsewhere in Ohio.

The nearest localities to the Ohio one where the species has been taken are in western Pennsylvania, in the French River at the Pennsylvania-New York state line (Raney, 1938:63), Allegheny River at Foxburg (Bean, 1884:466), and the Youghiogheny River (Cope, 1869:400–01). The nearest of these localities is more than 100 air miles (161 km) distant.

Habitat—The longhead darter inhabits the deeper riffles and pools with current, of the mountain-torrent type of moderate- or large-sized Appalachian streams which normally have fairly clear waters and clean bottoms of gravel and boulders. The isolated Ohio locality met these requirements, for in this stream section of 2.4 miles (3.9 km) in length, the gradient is 8.3′/mile (1.6 m/km), and the average monthly discharge in second-feet is 1,468 (Owens, 1939:56). So high a gradient for so large a stream is unusual in Ohio, for with few exceptions such large streams have a fall of less than 5.0′/mile (0.95 m/km); immediately above and below the above-mentioned section of the Walhonding River, the stream gradients are below 3′/mile (0.57m/km) Trautman, 1942:218–19, fig. 4).

Years 1955–80—During this period many attempts were made by R. F. Jezerinac, K. R. Troutman, me, and others to discover the present range of the longhead darter in the Walhonding River, Coshocton County, and to find out if a population existed in the adjacent portion of the Muskingum River. All recognizable habitats were investigated but no longheads were found. It is possible that the relict population of this essentially mountain-torrent species has become extirpated. For recent distribution *see* Page (1978).

SLENDERHEAD DARTER

*Percina phoxocephala** (Nelson)

Fig. 153

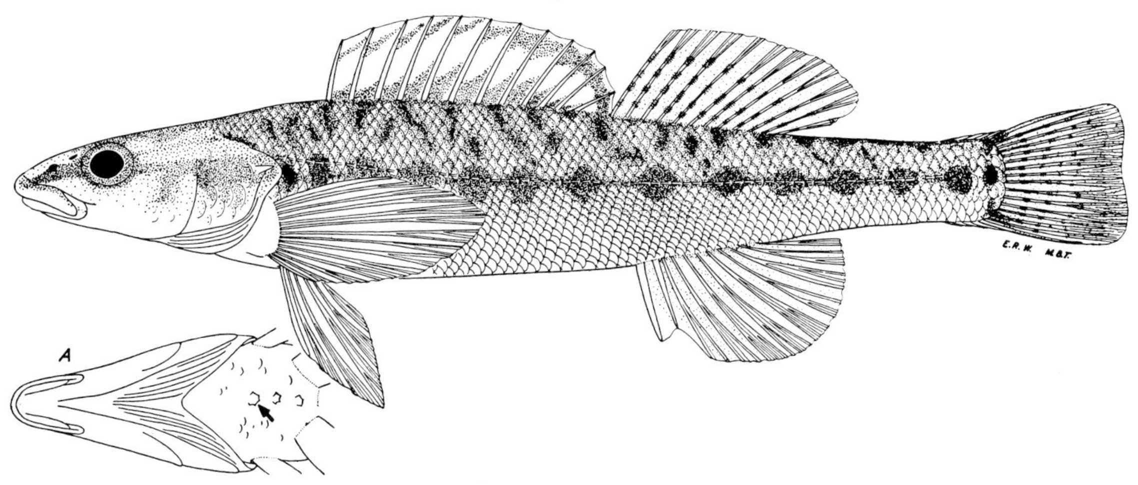

Upper fig.: Muskingum River, Washington County, O.
Oct. 18, 1939. 76 mm SL, 3.8″ (9.7 cm) TL.
Female. OSUM 870.

Fig. A: ventral surface of head and breast; note attenuated snout; amount of membrane connecting gill covers and the resulting angle; arrow points to specialized scale on breast; no spots on ventral surface of head.

Identification

Characters: *General* and *generic*—*See* dusky darter. *Specific*—No groove separating tip of upper jaw and snout, *see* fig. 151 A. Gill covers rather broadly connected by membrane; distance from gill cover angle to tip of upper jaw usually contained 1.4–1.7 times in head length. Snout long; head narrow. Body depth of adults usually contained more than 1.5 times in head length. Opercles with scales; cheeks sometimes contain a few, deeply-embedded scales. Ventral surfaces of head unspotted, fig. A. The 10–16 short, vertical bars along the lateral series of scales only slightly confluent at most. No teardrop. Dorsal spines usually 12–13, extremes 11–13. Lateral line scales usually 60–70, extremes 59–72.

Differs: Longhead darter has spots on ventral surface of head; has more lateral line scales. Channel, gilt and dusky darters have blunter snouts. *See* dusky darter.

Most like: Longhead and blackside darters.

Coloration: *Adult male*—Dorsally tan-brown, much vermiculated and spotted with darker brown; 14–22 squarish spots usually present along the dorsal ridge. Sides yellower; a dark brown band crosses snout, is faint on cheeks and opercles, continues along lateral line as a series of 11–16 blotches, these blotches generally higher than long and connected with each other by a very narrow band. A very small, blackish caudal spot. Ventrally pale straw-color and whitish. Spinous dorsal largely transparent, except for an orange-brown basal band; an orange-yellow subdistal band; a distal band of light bluish. Soft dorsal and caudal fins have brownish spots on their rays which tend to form rows and bands. Lower fins transparent, with a pale olive cast. *Adult female*—Like male, except that colors are less intense; the orange bands on dorsal are absent. *Young*—Like female but yellows partly replaced by white or silvery; fins transparent and sometimes unspotted.

Lengths: *Young* of year in Oct., 1.1″–2.0″ (2.8–5.1 cm) long. *Adults*, usually 2.2″–3.8″ (5.6–9.7 cm). Largest specimen, 4.0″ (10 cm) long.

---

* Formerly in the genus *Hadropterus*; *see* Bailey and Gosline, 1955:10.

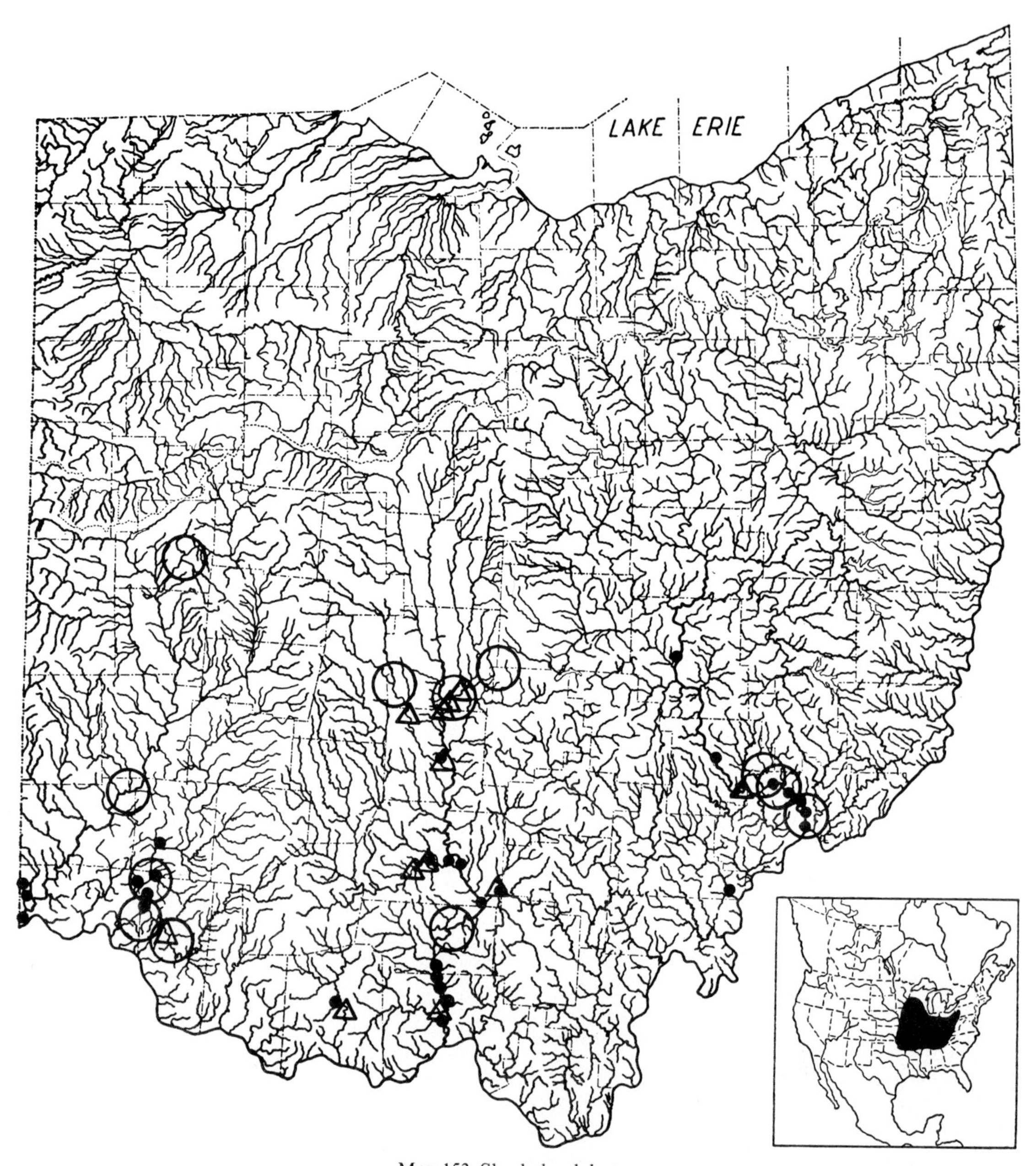

MAP 153. Slenderhead darter

Locality records. ○ Before 1920. ● 1920–54. △ 1955–80.
Insert: Unoccupied areas within this range increasing in size and number, especially northward.

Hybridizes: Natural hybridization with logperch darter in Ohio and elsewhere with blackside darter (OSUM 14019, from Illinois).

Distribution and Habitat

Ohio Distribution—Early literature records, further substantiated by a few preserved specimens and an identifiable photograph of a specimen collected in central Ohio, June 24, 1897, by Williamson and Osburn (1898:pl. 36), prove that before 1900 the slenderhead darter was present in the upper Ohio River drainage of Ohio and Pennsylvania (Evermann and Bollman, 1886:339); also that it probably was more widely distributed than it has been since.

Between 1925–50 no specimens were taken in the Great Miami River except near its mouth, or from the Scioto River system in Franklin County where extensive seining was done; also that the only central Ohio specimens taken were a stray in Big Darby Creek near Fox, Pickaway County, and one in the Muskingum River near Zanesville, Muskingum County. Since 1925, the species appears to have decreased greatly in abundance in the lower Scioto and Muskingum rivers, where before 1935 series of 10–20 individuals could be taken in a short period, whereas since then only 1–6 specimens were taken during prolonged and intensive seinings. This decrease in abundance is correlated with a decrease in the amount of habitat.

McCormick (1892:30) states that there was "One specimen taken near Lorain." In McCormick's unpublished notes, preserved in the Oberlin College Museum library, he wrote concerning this species "Lake Erie—not common" and "Yellowish-brown with x-shaped side markings." It is difficult to believe that this species of southern rivers was present in Lake Erie; also the species shows no "x-shaped side markings." Furthermore, in McCormick's unsatisfactory treatment of the Hadropterine group of darters, he lists *Etheostoma peltatum* (*Percina peltata*) which is a species confined to the Atlantic coast drainage (Osburn, 1901:92). Because of doubts concerning McCormick's statements, and lack of specimens, the Lake Erie record of the slenderhead darter is considered invalid (Osburn, 1901:92).

Habitat—The largest numbers of slenderhead darters were taken on the extensive bars and riffles which contained clean sand and small gravel that was almost free of silt. Such sand and gravel areas occurred only where the current was not sufficiently strong to wash this material away, and these areas could remain free of silt only when the stream had fairly clear waters. With greatly increased turbidity, especially during periods of low stream levels, the current was not strong enough to keep the suspended silt from settling over the sand and gravel, thereby smothering out the habitat of this darter and its food supply. I believe that it was this silting of the sand and gravel areas of the larger southern Ohio streams that has caused the recent decreases in abundance of this darter, as it has with other gravel species such as the gravel and speckled chubs.

Years 1955–80—During this period no distinct portions of any stream system were discovered in which the slenderhead darter had not been recorded previously, and this despite numerous attempts to discover such localities in the Ohio River drainage. However, a moderate-sized population was found in the lower portion of Big Walnut Creek, Franklin County, where between 1897 and 1950 only three individuals had been taken and these at widely separated intervals. It is not known why the slenderhead population increased in lower Big Walnut Creek since 1950. It is possible that the establishment of the Hoover Reservoir acted sufficiently as a silt-settling basin, thereby creating a more favorable slenderhead environment downstream.

RIVER DARTER

*Percina shumardi** (Girard)

Fig. 154

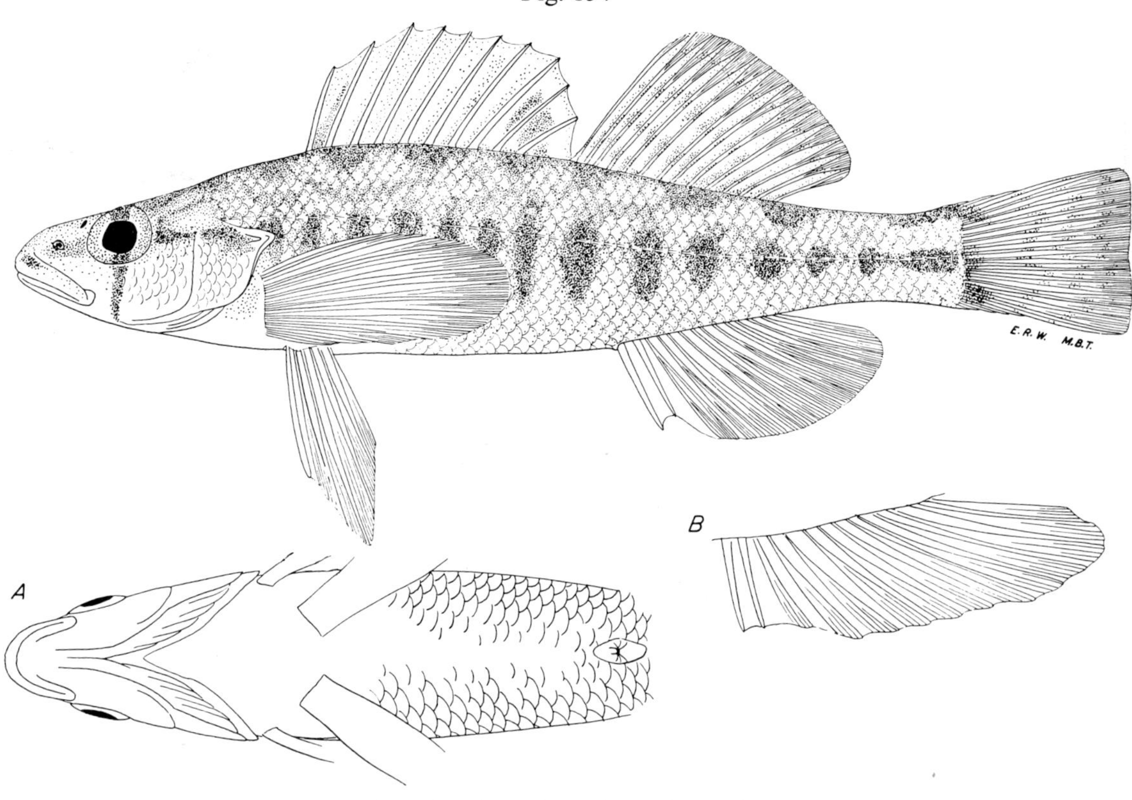

Upper fig.: Scioto River, Scioto County, O.

Nov. 16, 1939. 65 mm SL, 3.0″ (7.6 cm) TL.
Female. OSUM 40.

Fig. A: ventral surface of head and body; note short, rounded snout; sharp angle formed by junction of gill covers; scaleless midline of belly, except for a narrow bridge of scales before the anus; lack of large, specialized scales on the midline which are never present in this species.

Fig. B: Same locality, date and OSUM number.

Adult male. 52 mm SL, 2.4″ (6.1 cm) TL.
Anal fin of male; note the greatly elongated posterior rays.

Identification

Characters: *General* and *generic—See* dusky darter. *Specific*—Midline of belly naked, except that usually a narrow bridge of scales crosses the midline directly in front of anus, fig. A; the large, specialized scales along the midline, present in several other species of *Percina*, always *absent* in this one. Usually the tip of upper jaw is bound to snout by a narrow frenum whose width is less than 1.5 times the diameter of the anterior nostril; occasional individuals have a very shallow groove separating tip of upper jaw and snout. Gill covers rather narrowly joined by membrane across isthmus, the angle rather acute, fig. A; distance from apex of gill-cover angle to tip of upper jaw usually contained more than 1.7 times in head length. Snout short and

* Formerly in the genus *Hadropterus*; *see* Bailey and Gosline, 1955:10.

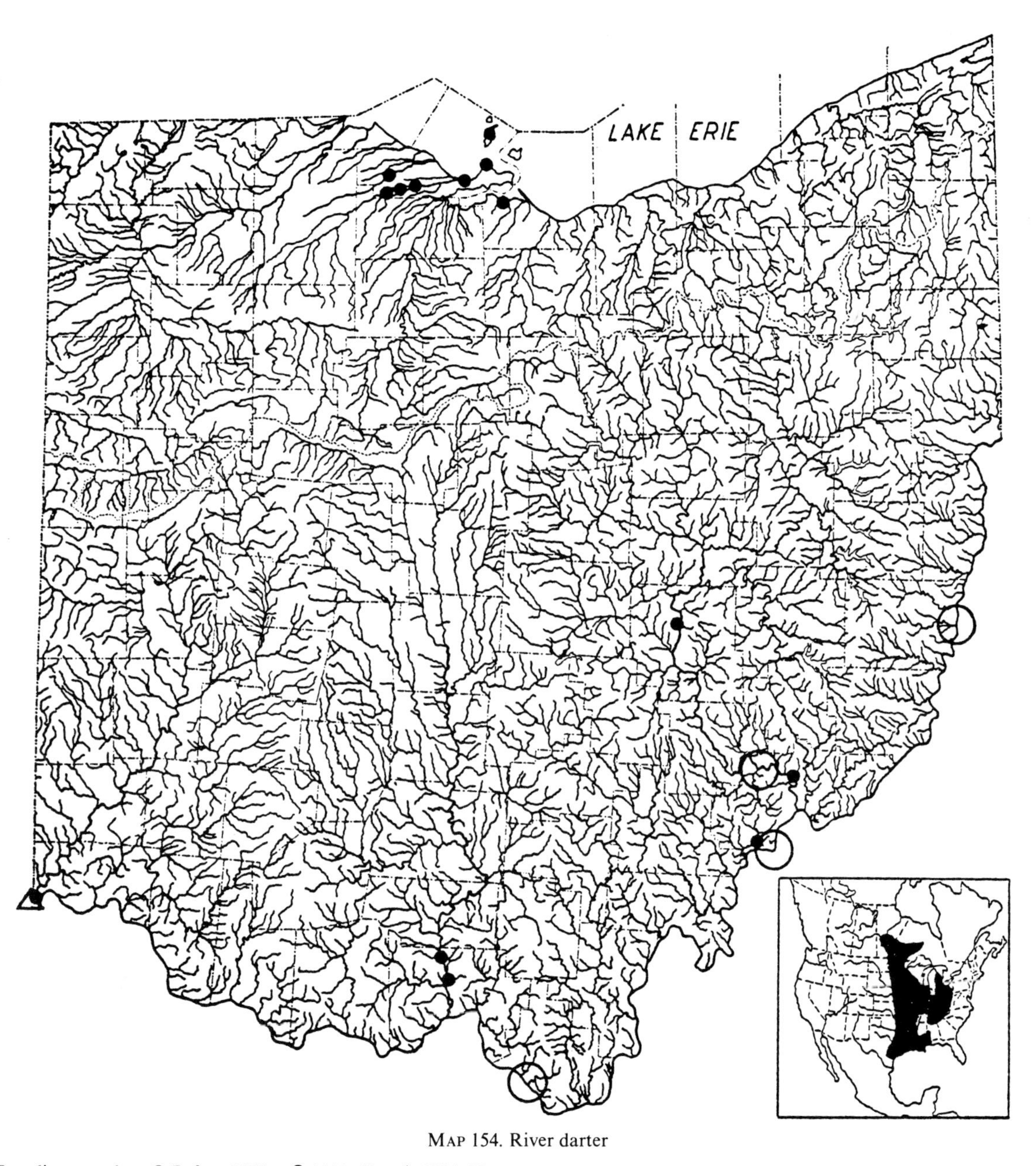

MAP 154. River darter

Locality records. ○ Before 1900. ● 1920–50. △ 1951–80.
Insert: Large unoccupied areas within this range; western limits of range indefinite.

rounded. Body rather deep for a darter; body depth contained fewer than 1.5 times in head length. Cheeks and opercles scaled. The 10–15 short, vertical bars along the lateral line are usually more or less confluent only on the caudal peduncle. Adults normally have a dusky, basal blotch on webbing between the first two dorsal spines; other blotches between the last three spines. Dorsal spines usually 9–11, rarely 12. Lateral line scales usually 52–58, extremes 48–60. Tear-drop well developed, except in small young. Anal fin of male enlarged, fig. B.

Differs: Dusky, blackside, longhead, slenderhead, and gilt darters have wide frenums. Channel darter has scaleless cheeks. Logperch has overhanging snout. Crystal, johnny and sand darters have only one anal spine. Other darters have scaled bellies; or if midline is partly or entirely naked, then lateral line is incomplete.

Most like: Dusky and gilt darters. **Superficially like:** Iowa and blackside darters.

Coloration: *Adult male*—Dorsally brownish- or yellowish-olive; sides light tan-yellow; ventrally pale yellow and whitish. Between 6–12 squarish, dark brown, saddle-bands cross dorsal ridge; these usually prominent. A longitudinal band crosses snout, cheeks and opercles; along the lateral line this band consists of 10–15 short, vertical bars which become somewhat confluent on the caudal peduncle. Tear-drop well developed. All fins transparent-olive. A dusky basal spot on webbing between first two dorsal spines; dusky blotches on webbing between last three dorsal spines. Soft dorsal and caudal fins vaguely spotted. *Adult female*—Like male, except colors and color pattern more subdued. *Young*—Like female, except that much of the yellow is replaced by whitish; fins transparent and lack dusky markings; teardrop undeveloped.

Lengths: *Young* of year in Oct., 1.1″–2.0″ (2.8–5.1 cm) long. *Adults*, usually 1.8″–2.8″ (4.6–7.1 cm) long. Largest specimen, 3.2″ (8.1 cm) long.

Hybridizes: With logperch darter.

Distribution and Habitat

Ohio Distribution—Between 1925–50, the river darter was found singly or in small numbers in the low-gradient and turbid streams entering western Lake Erie, in Lake Erie and Sandusky Bay (late summer only), and in the larger streams of southern Ohio. It was present in the Ohio River drainage before 1900. Definite proof of its presence in Lake Erie prior to 1928 is lacking, but there is no reason to believe that it did not enter Lake Erie early by way of the Maumee outlet.

In 1892, McCormick (1892:30) described *Etheostoma wrighti*, which later Jordan and Evermann (1896:Pt. 1, 1047) and Osburn (1901:93) considered to be *shumardi*. The river darter surely could have been present in the Vermilion River in 1892 as it has been known to be in the tributaries of southwestern Lake Erie since 1928. However, McCormick's statements concerning the non-protractile premaxillaries and confluent blotches along the sides, suggest a hybrid between *shumardi* and *Percina caprodes* or *Percina maculata*, or between *caprodes* and *maculata*. Several *caprodes* x *maculata* hybrids have been taken in the low-gradient tributaries of southwestern Lake Erie (Trautman, 1948:173). I failed to find the type of *wrighti* at the Oberlin College Museum.

Habitat—Small numbers of river darters were found most frequently in waters deeper than 3′ (0.9 m), in the well-defined, rather swift chutes and riffles of moderate- or large-sized streams having a gravel or rocky bottom. A few were also taken in shallower water, in the sluggish, sandy-silt riffles of the habitually turbid tributaries of southwestern Lake Erie. Apparently the river darter inhabited the deeper waters when they were clear, for we failed to find them in shallow, clear water even at night; it was only in turbid water that the species was taken in the shallows. Upon each of two occasions in spring when the Scioto River was in flood, an individual was caught at the water's edge in a field far from the river channel, suggesting that these fishes were caught in the act of migrating upstream.

Years 1955–80—To my knowledge no river darter has been collected in inland Ohio during this period, although areas where it formerly existed have been investigated. It may still be present in these localities because this species normally inhabits deep waters where collecting them is very difficult.

A specimen collected August 1, 1946, at South Bass Island and preserved as OSUM 6735 was not recorded in the first edition of this report (Van Meter and Trautman, 1970:76).

CHANNEL DARTER

*Percina copelandi** (Jordan)

Fig. 155

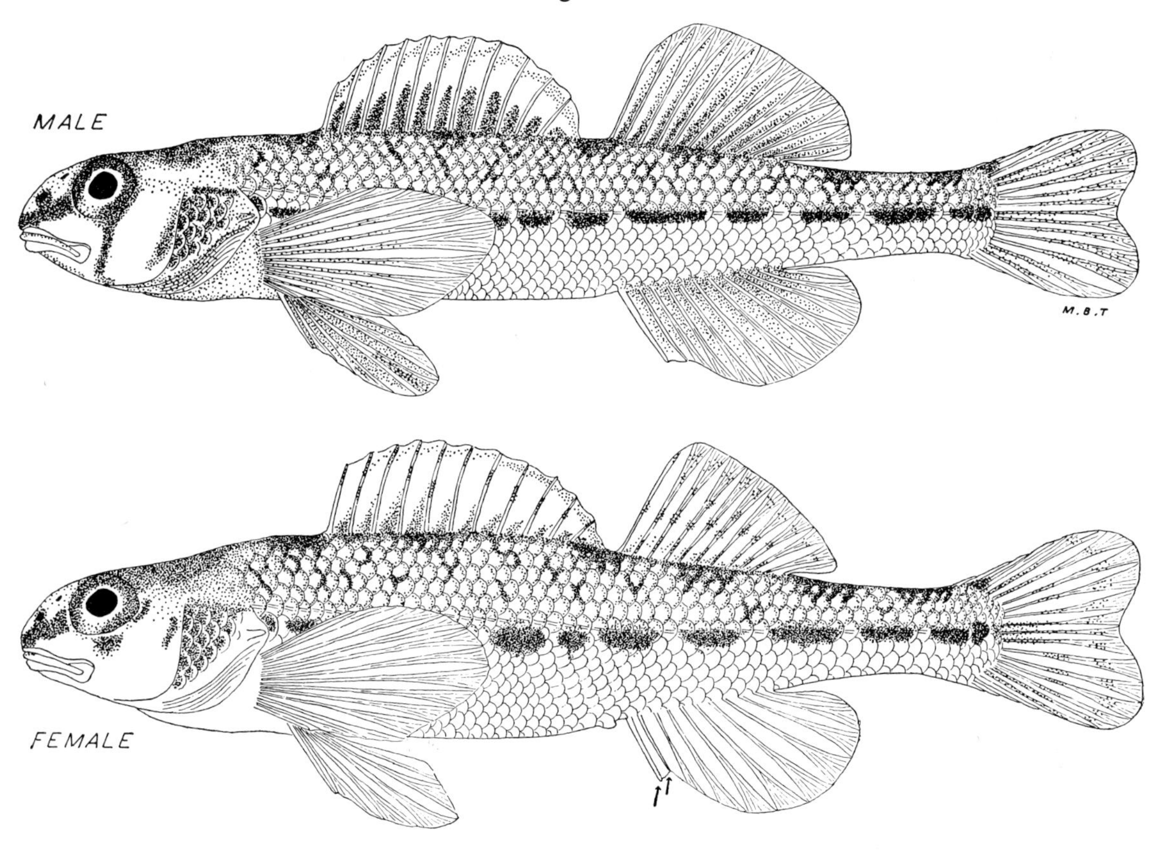

Upper fig.: Lake Erie, Ottawa County, O.

June 24, 1947. 43 mm SL, 2.0″ (5.1 cm) TL.
Breeding male. OSUM 7137.

Lower fig.: Same locality, date and OSUM number.

Breeding female. 42 mm SL, 1.9″ (4.8 cm) TL.
Note arrows pointing to the two anal spines; not one spine as in the johnny darter.

Identification

Characters: *General* and *generic—See* dusky darter. *Specific*—Usually a definite groove separates tip of upper jaw and snout; in occasional individuals this groove is very shallow. Midline of belly naked or with specialized scales, fig. 152 A. Gill covers narrowly joined across isthmus with membrane, the angle acute; distance from apex of this angle to tip of upper jaw usually extends from tip of upper jaw to posterior edge of eye; distance is contained more than 1.8 times in head length. Snout blunt, rounded, and in some individuals slightly overhangs upper lip. Cheeks scaleless; opercle scaled. Body and head shape very similar to that of johnny darter (which species has only *one* anal spine). Body depth usually contained 5.0 or more times in standard length. Tear-drop long in males, upper fig.; a dusky circle in females, lower fig; poorly developed or absent in young. Dorsal spines usually 10–11, extremes 9–12.

* Formerly in the genus *Hadropterus*; *see* Bailey and Gosline, 1955:10.

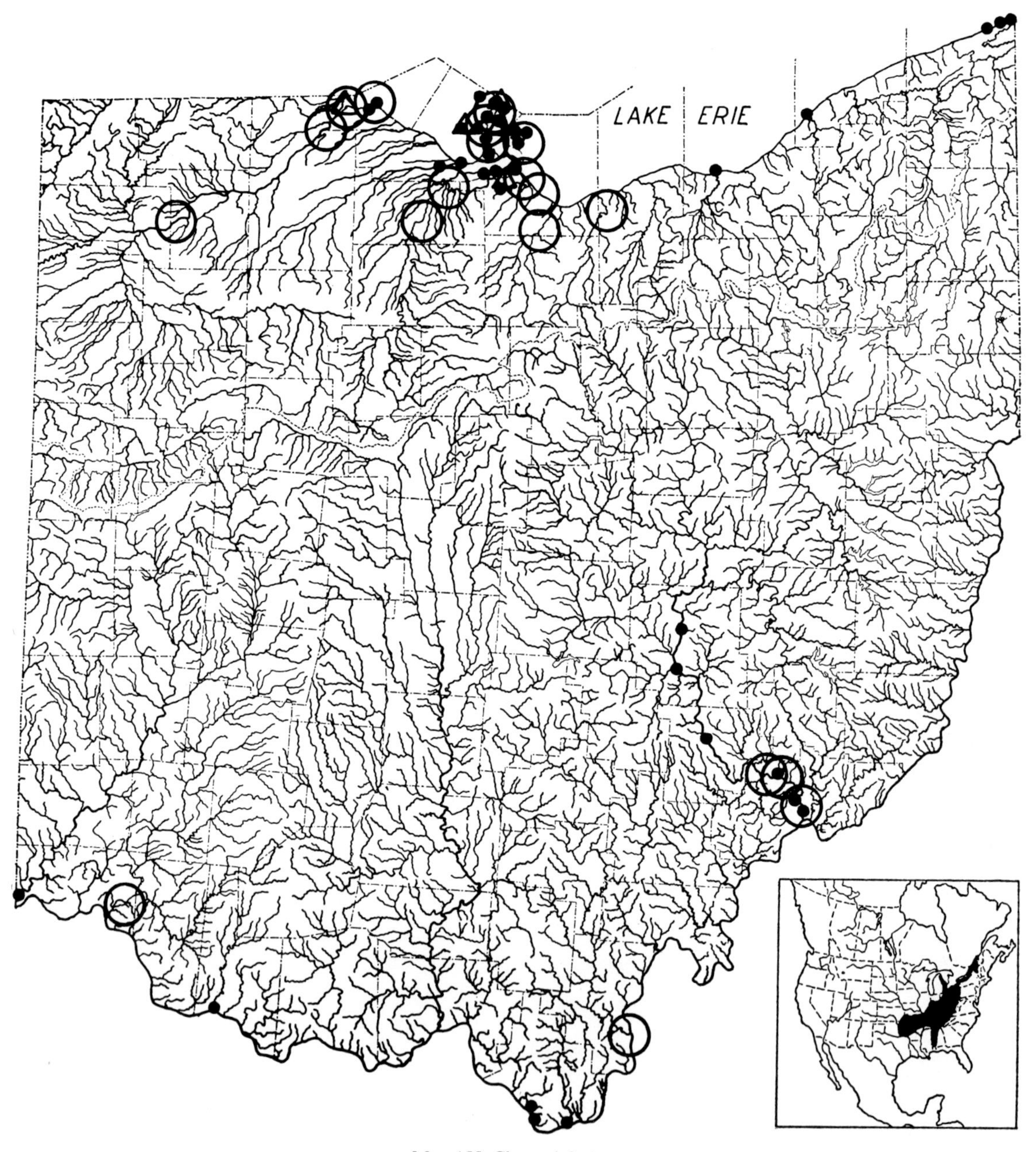

MAP 155. Channel darter

Locality records. ○ Before 1924. ● 1924–54. △ 1955–80.
Insert: Large unoccupied areas within this range, especially northeastward.

Anal spines *two, see* arrows, lower fig.; anal rays 7–9. Lateral line scales usually 48–52, extremes 44–56.

Differs: Johnny darter has only one anal spine. River darter has scaled cheeks. Greenside darter has more scales in the lateral line. All other species of darters in Ohio have a well-developed frenum.

Most like: Johnny darter. **Superficially like:** Iowa and young of greenside darter.

Coloration: *Adult male*—Dorsally yellow-olive, the scales outlined with brown. Sides a lighter yellow-olive; ventrally whitish-yellow or whitish. Between 6–9 squarish, dark brown saddle-bands cross dorsal ridge. Usually W- and V-shaped markings scattered over back. A dark band across snout. A lateral band along the sides, containing 12–18 oblong blotches. A long tear-drop. Webbing of basal half of spinous dorsal dark or dusky; margin narrowly bordered with dusky; remainder of fin transparent. Remaining fins transparent or flushed with olive-yellow. *Breeding male*—Body suffused with dusky; breast, pelvic fins, and ventral surface of head blackish; a more or less solid lateral band. *Adult female*—Like adult male but coloration lighter; tear-drop smaller; spinous dorsal coloration more restricted and less intense; some yellow coloration replaced with white or silvery. *Young*—Like female but more silvery; fins more transparent; tear-drop undeveloped.

Lengths: *Young* of year in Oct., 0.8″–1.5″ (2.0–3.8 cm) long. *Adults*, usually 1.5″–2.2″ (3.8–5.6 cm). Largest specimen, 2.5″ (6.4 cm) long.

Hybridizes: With logperch darter.

Distribution and Habitat

Ohio Distribution—The channel darter was present in both drainages before 1900, as indicated by literature references and preserved specimens. Strays were recorded in the Maumee River until 1922. A large population should have been present on the extensive sand and gravel bars of the Ohio River before impoundment and great siltation, and at least a small population was present on the former sand and gravel bars of the now turbid and silted lower Scioto River.

During the 1925–50 period there were large populations on the cleanest of the coarse sand and fine gravel beaches of Lake Erie, especially about the Bass Islands; smaller populations were present in the Muskingum River, particularly on the sand and gravel bars below the dams; and a small population, probably a decreasing one, in the Ohio River from the Indiana state line to the mouth of the Muskingum River.

Habitat—The channel darter was present in goodly numbers on the coarse-sand, fine-gravel beaches of Lake Erie and the largest rivers, usually remaining throughout the day in waters more than 3′ (0.9 m) in depth whenever they were clear, and invading the very shallow waters at night. It was essentially an inhabitant of extensive beaches and bars where currents were sluggish, and was seldom found in moderate- or fast-flowing riffles.

Years 1955–80—The present numerical status of the channel darter in Lake Erie east of Sandusky is unknown because of the lack of adequate collections, except for the intensive investigations in the vicinity of Cleveland, after 1970, which failed to capture the species (White et al., 1975:117). Its numerical status west of Sandusky is better known because of annual investigations during this period. The species was taken annually and in considerable numbers about the Bass Islands until 1954. Since then, one was captured in 1969 by a class of students and me at North Bass Island, and in 1972 Ted Cavender and his students took two (Van Meter and Trautman, 1962:76).

In the Ohio River the channel darter was taken adjacent to Jefferson County, Ohio, and then only during 1978.

GILT DARTER

*Percina evides** (Jordan and Copeland)

Fig. 156

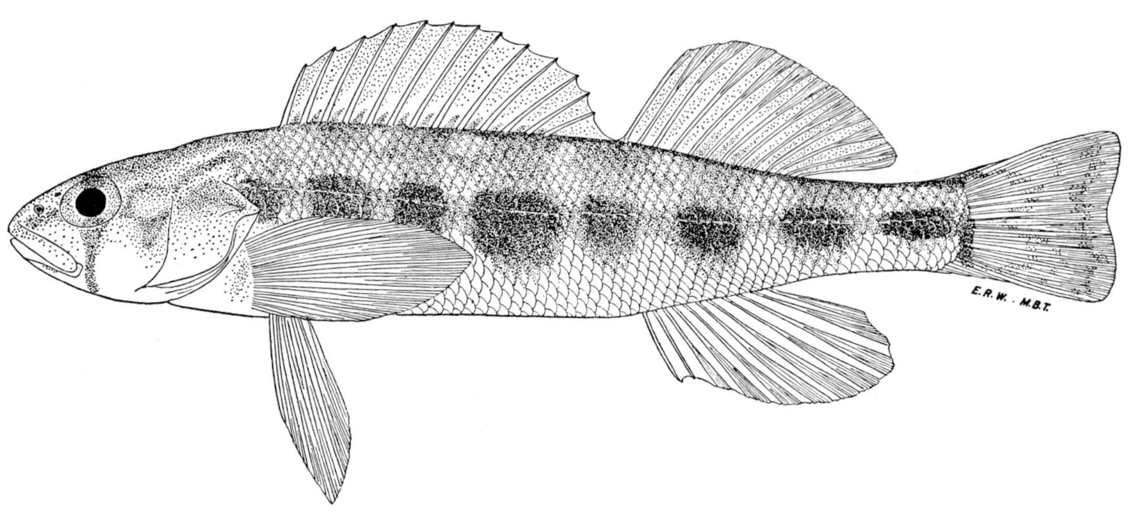

Tippecanoe River, Indiana.

July, 1925.
Male.

54 mm SL, 2.5″ (6.4 cm) TL.
OSUM 9270.

Identification

Characters: *General* and *generic—See* dusky darter. *Specific*—No groove separating tip of upper lip and snout. Midline of belly naked or with one or more large, specialized scales, fig. 152 A. Gill covers slightly connected across isthmus with membrane; distance from apex of gill-cover angle to tip of upper lip usually contained 2.0 or more times, extremes 1.9–2.2, in head length; this distance extending from tip of upper lip to immediately beyond posterior edge of eye. Mouth terminal. Snout short. Cheeks and opercles usually scaleless. Dorsal spines usually 11–13, rarely 10. Lateral line scales usually 55–60, extremes 52–67. Only species of *Percina*, recorded for Ohio, in which the male has reds and blues. Body shape similar to that of the bluebreast and river darters. (Lacking Ohio specimens, all counts, measurements, and color descriptions were taken from Indiana specimens.)

Differs: No other species of *Percina* in Ohio has reds and blues. Johnny darter has no frenum. All species of *Etheostoma* in Ohio either have the belly normally scaled, *or* the midline of belly is partly or entirely naked and the lateral line incomplete.

Most like: River and dusky darters.

Coloration: *Adult male*—Dorsally dark olive, tasselated with darker. Between 5–8 dark, squarish saddle-bands cross the dorsal ridge; each saddle-band is directly above a squarish, blue-green blotch which is situated on the lateral line; interspaces between the lateral blotches yellow, or copper-red in some individuals. Two roundish spots at base of caudal fin. Ventrally yellowish or whitish. Fins vaguely marked. *Breeding male*—Each saddle-band extends downward to unite with the lateral blotch beneath it, thereby forming 5–8 broad, blue-green vertical bands; spaces along the lateral line and between the vertical bands copper-red; two round spots at base of caudal fin are brilliant orange; ventral half of head and breast orange-red; dorsal fins mostly orange; anal and pelvics blue-black. *Adult female*—Pattern similar to that of male except that all colors are subdued or lacking; vertical bands yellowish; fins more transparent.

Length: Maximum for Indiana specimens, 3.0″ (7.6 cm) long.

* Formerly in the genus *Hadropterus*; *see* Bailey and Gosline, 1955:10.

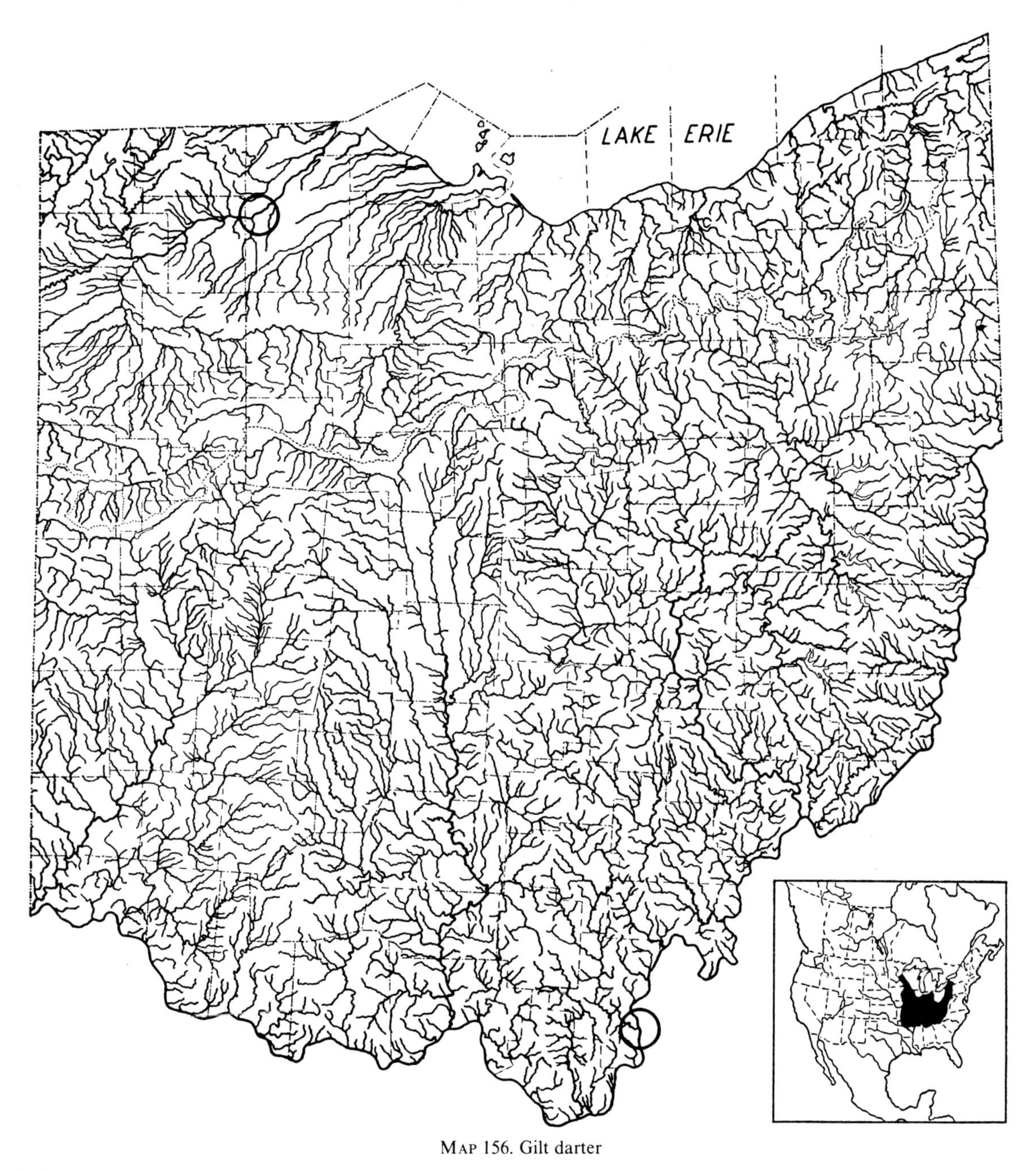

MAP 156. Gilt darter

○ Literature records before 1900.

Insert: Limits of original range appear to be shrinking in size, and unoccupied areas within this range increasing in size and number.

Distribution and Habitat

Ohio Distribution—On September 5, 1888, Charles H. Gilbert and James A. Henshall collected a gilt darter in the Ohio River near Raccoon Island, Gallia County (Henshall, 1889:126, and from Henshall's unpublished list of fishes collected near Raccoon Island). The specimen, identified by both men, was catalogued in the collections of the CSNH. I failed to find the specimen in the Cincinnati Society collections in 1931. Rivermen have told me that before impoundment large rock and stone riffles existed in the Ohio River at Raccoon Island, an excellent habitat for this species.

On August 21 or 22, 1893 Kirsch (1895A:331) took "one specimen below the dam in the Maumee

River at Grand Rapids, Ohio." He also took five specimens near Ft. Wayne, Ind., where the St. Marys and St. Joseph rivers converge to form the Maumee River. Attempts to locate Kirsch's specimen have also failed. The Maumee River at Grand Rapids, before the greatly increased turbidity of its waters, should have contained an excellent habitat for this darter.

The above records, although not substantiated by specimens, are accepted as valid, for this species has been taken near Ohio's borders in Pennsylvania (Raney, 1938:62), Kentucky (Clark, 1937:8; Big and Little Sandy rivers) and Indiana. Gerking (1945:87) suggests that in recent years this darter has decreased greatly in abundance in Indiana, and between 1940–43 he failed to capture it at five localities where it was formerly known to occur.

Habitat—The gilt darter inhabited fast, deep, rocky, and strong riffles of clear-water streams, such as were the Maumee, Mahoning, Scioto, Muskingum, Great Miami and Ohio rivers before they were impounded, and polluted with soil and industrial pollutants.

Years 1955–80—Investigations during this period indicated that the gilt darter no longer existed in Ohio waters.

OHIO LOGPERCH DARTER

Percina caprodes caprodes (Rafinesque)

Fig. 157

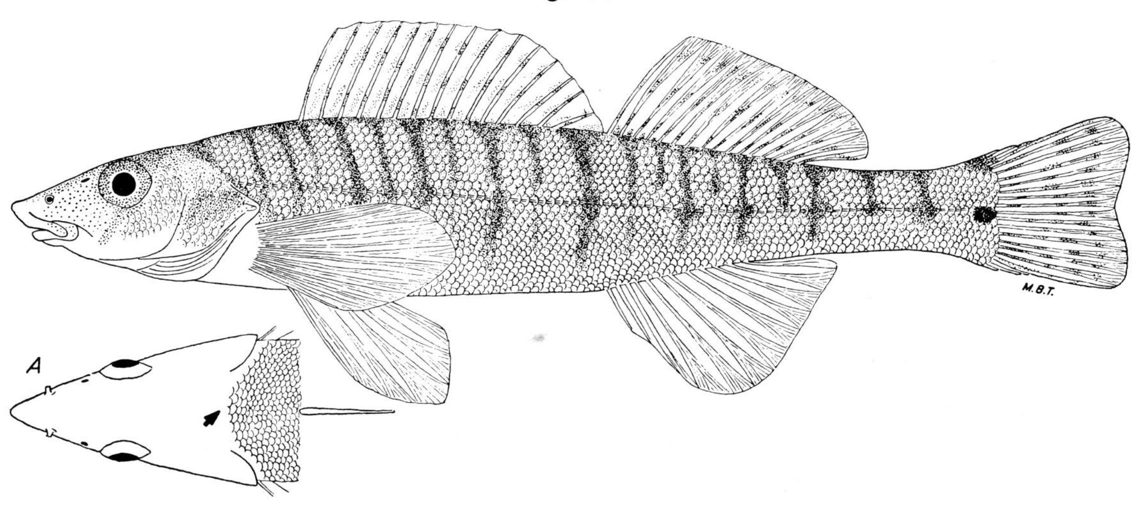

Upper fig.: Scioto River, Scioto County, O.

April 24, 1940.
Male.

85 mm SL, 3.9″ (9.9 cm) TL.
OSUM 2504.

Fig. A: dorsal view of head and nape; arrow points to the scaled nape; note also the very conical and pointed snout.

Identification

Characters: *General* and *generic*—*See* dusky darter. *Specific*—The pointed, conical snout projects considerably beyond the wholly inferior mouth and the very wide frenum makes the logperch the most readily recognizable of the darters. Midline of belly partly or entirely scaleless, with some individuals having large, specialized scales, *see* fig. 153 A. Interorbital space somewhat depressed as in the hog sucker, *see* fig. 101 A. Interpelvic space equal in length to the length of the base of either pelvic fin. Gill covers narrowly connected across isthmus with membrane; distance from apex of gill-cover angle to tip of snout extends from snout tip to posterior half of eye. Complete lateral line has more than 78 scales. Anal spines two. Dorsal spines 13–16, rarely 12. Between 15–25 narrow bands cross back and extend downward on sides; these bands alternate in length, with every other one extending downward past the lateral line, whereas its neighbor on each side only extends to the lateral line. *Subspecific*—Nape closely and evenly scaled in typical examples, fig. A. Bands on sides of body more regular in width and outline, having little tendency to become confluent along the lateral line.

Differs: Northern logperch has a scaleless or almost scaleless nape. Other species of darters in Ohio lack the long, pointed, conical snout. Small saugers and walleyes have large mouths and canine teeth. Small yellow perch has fewer vertical bands on the sides; tail definitely forked.

Most like: Northern logperch. **Superficially like:** River darter; small saugers, walleyes and perch are sometimes mistaken by fishermen for the logperch or *vice versa*.

Coloration: *Adults*—Basic color of dorsal surface ranging from a dark olive- to pale straw-yellow; sides lighter; ventrally white or tinged with pale yellow. The 15–25 vertical bands on the sides as described under Characters. Usually a black caudal spot about the size of pupil of eye. A faint, oblique tear-drop present in some individuals. Dorsal and caudal fins transparent-olive and having small, brownish spots which tend to form wavy rows or bands. Lower fins transparent. *Breeding male*—Like adult except that colors are more intense. *Young*—Like adult except that the basic coloration

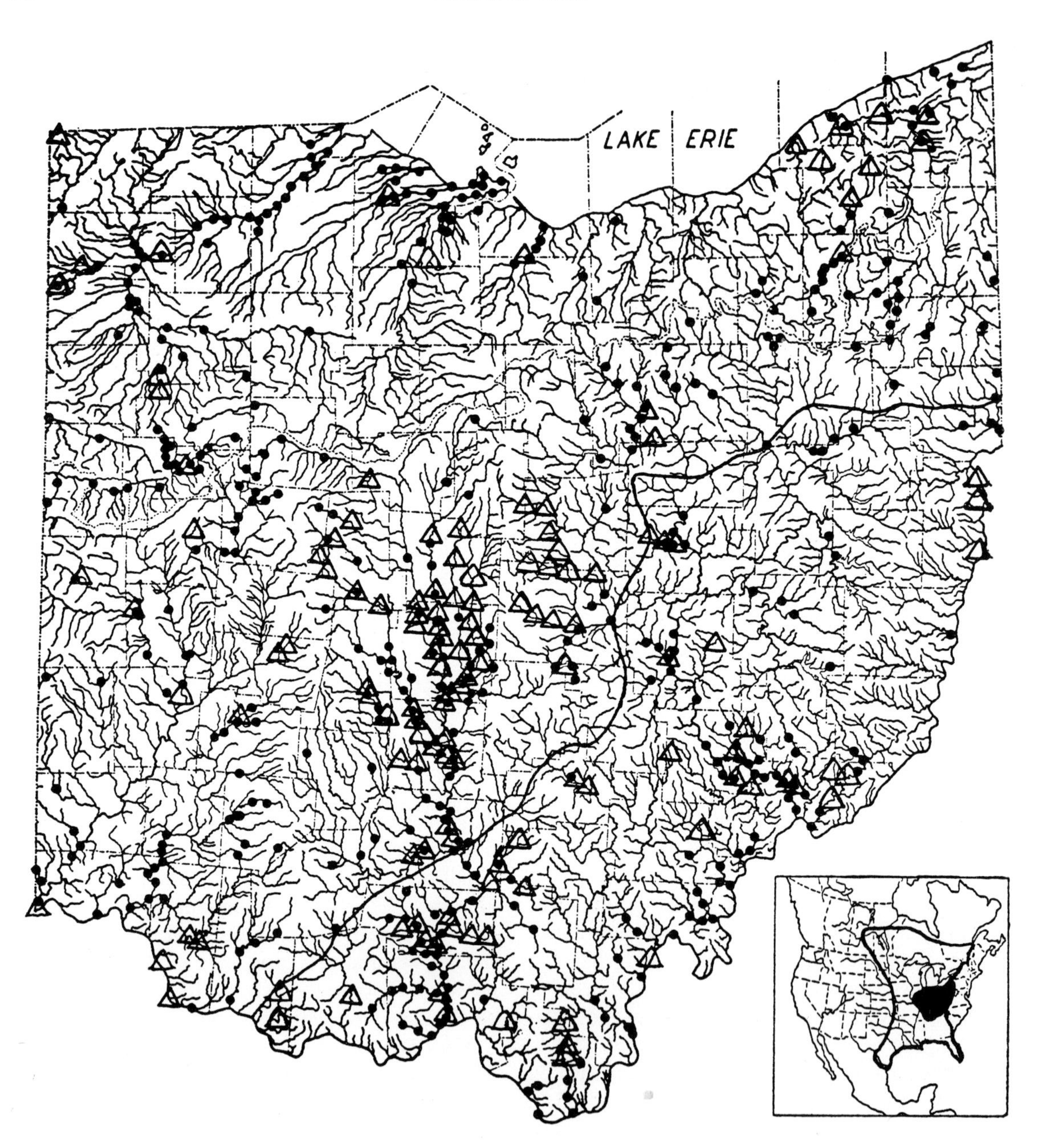

MAP 157. Ohio logperch darter

Locality records. ● Before 1955. △ 1955–80. ~ Glacial boundary.
Insert: Black area represents range of this subspecies, outlined area of other subspecies.

is more whitish and silvery; the bands more prominent; fins transparent and with few or no spots.

Lengths: *Young* of year in Oct., 1.1″–2.8″ (2.8–7.1 cm) long. Around **1** year, 2.0″–3.5″ (5.1–8.9 cm) long. *Adults*, usually 2.8″–6.0″ (7.1–15 cm). Largest specimen, 7.1″ (18 cm) long (Buckeye Lake).

Hybridizes: with dusky, blackside, slenderhead, river, channel, greenside and orangethroat darters.

Distribution and Habitat

Ohio Distribution—After reviewing the literature, Osburn (1901:90) concluded that the logperch as a species was widely distributed and abundant in Ohio prior to 1900. After 1920 the species was numerous only in Lake Erie and in some of the larger inland streams, and it was apparent that a decrease in abundance had occurred since 1900.

This decrease was most apparent in the heavily silted section of the Maumee River in Paulding, Defiance, and Henry counties, where in 1893 Kirsch (1895A:331) found the species to be "rather common." It was absent from the most polluted sections of the Great Miami and Mahoning rivers, where the sandy gravel bars and riffles were covered with industrial wastes. The species was likewise absent or rare in most of the Ohio River, where in 1820 Rafinesque (1820:88) considered it to be a "most common species" and where, according to the few old rivermen who knew the "Hogfish," it had been abundant on the extensive sandy gravel riffles and bars before impoundment of the river and subsequent silting.

The populations of logperches in unglaciated Ohio seemed to be composed entirely of this subspecies. Specimens considered to be fairly typical of the subspecies were found elsewhere in both drainages, but were represented only by an occasional individual in Sandusky Bay and the shores of Lake Erie where the northern logperch and intergrades were numerous. In Lake Erie no typical Ohio logperch was found during the spawning season.

Habitat—The largest inland Ohio populations were present on the riffles and bars of moderate- or large-sized streams of moderate gradients, where the bottom was of clean sand and gravel. The logperch retired to deeper water during sunny days when the water was clear, where it hid under rocks, or buried itself in sand until only the eyes were exposed, returning to shallower water or coming out of hiding at the approach of darkness.

Intergradation—So-called "typical" Ohio logperches, subspecies *caprodes*, differ from the northern logperches, subspecies *semifasciata*, in having the nape closely scaled instead of naked (occasional specimens have a small scaleless semicircular area adjacent to the occiput), and in having both the long and short alternating bars of equal width throughout instead of having these bars expanded along the lateral line to form a more or less confluent band. All combinations of intergradation occurred. Specimens were considered typical of a subspecies when both of the above-mentioned characters were plainly evident; they were considered intergrades when one or both characters were more or less intermediate.

Years 1955–80—This subspecies of the logperch darter (*Percina caprodes caprodes*) was generally distributed in the larger streams of the Ohio River drainage and especially in those portions where there was considerable sandy gravel. Usually fewer than 15 were taken at a seining locality, and rarely were more than 15 captured or observed.

Typical-appearing *Percina c. caprodes* and intergrades were present in the larger streams tributary to Lake Erie but in smaller populations. It was apparent that in many localities in this watershed where formerly large populations were present these had diminished in size since 1950, especially the Maumee River population. During the intensive investigations of Lake Erie and its tributaries in the vicinity of Cleveland after 1970, only one typical *P. c. caprodes* was taken in the Cuyahoga River and only one in the lower Chagrin River. The latter we believe to be an intergrade between this subspecies and *P. c. semifasciata* (White et al., 1975:118–19).

The Muskingum River populations in unglaciated territory were typically *Percina c. caprodes*, and rarely indicated any intergradation as did some specimens in the glaciated upper Scioto River system. It continued to be rather rare in the Ohio River (ORSANCO, 1962:152). However, during their thorough researches of the Ohio River between the years 1976 to 1978 Margulies and Burch obtained the species from all of their five collecting stations adjacent to Ohio in at least one year.

NORTHERN LOGPERCH DARTER

Percina caprodes semifasciata (DeKay)

Fig. 158

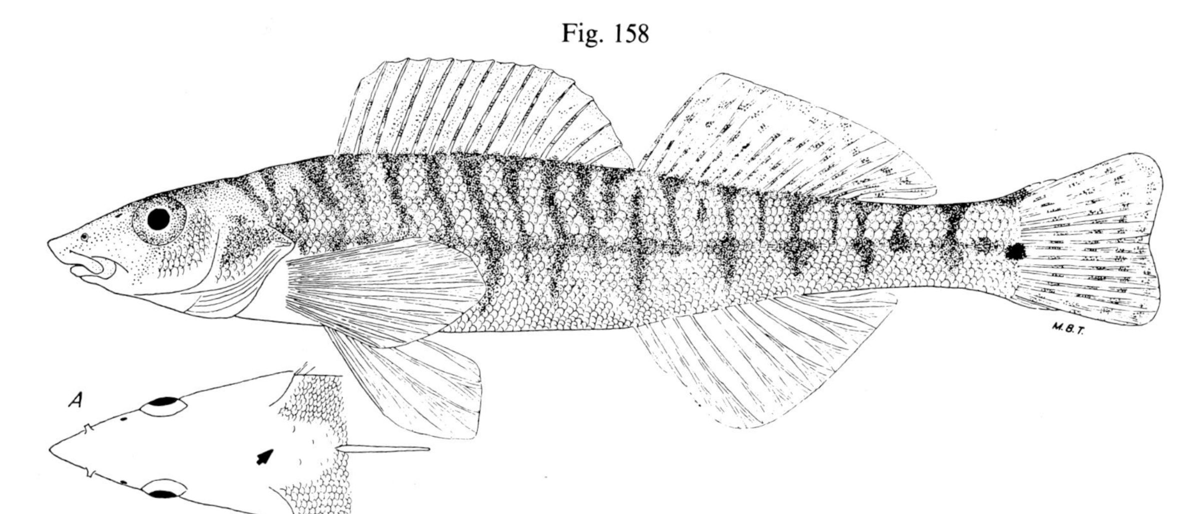

Upper fig.: Lake Erie, Ottawa County, O.

Sept. 18, 1946. 92 mm SL, 4.3″ (11 cm) TL.
Male. OSUM 6786.

Fig. A: dorsal view of head and nape; arrow points to the almost scaleless nape.

Identification

Characters: *General* and *generic—See* dusky darter. *Specific—See* Ohio logperch. *Subspecific—* Nape with a completely scaleless triangular area, or area contains a few, scattered scales, *see* arrow, fig. A. Bands on sides of body more irregular in length, width, shape and outline, having a tendency to widen laterally along the lateral line and to become more or less confluent along that line.

Differs: Ohio logperch has scaled nape. *See* Ohio logperch.

Most like: Ohio logperch. **Superficially like:** River darter; small saugers, walleyes, and perch are sometimes mistaken by fishermen for the logperch or *vice versa.*

Coloration: Similar to Ohio logperch. *See* Ohio logperch.

Lengths: *Young* of year in Oct., 1.2″–2.5″ (3.0–6.4 cm) long. Around **1** year, 2.0″–3.5″ (5.1–8.9 cm). *Adults*, usually 2.8″–5.8″ (7.1–15 cm). Largest specimen, 6.5″ (17 cm) long.

Hybridizes: with blackside darter. *See* Trautman (1948:173).

Distribution and Habitat

Ohio Distribution—Ohio populations of northern logperches containing little or no evidence of intergradation were present only about the islands of western Lake Erie. Evidence of intergradation was more evident along the south shore of Lake Erie, where the ratio of northern logperches to intergrades was about 5 to 1; in Sandusky Bay the ratio was almost equal. In the western Lake Erie tributary streams, from their mouths upstream to where they crossed the Warren Beach Ridge, the intergrades outnumbered the northern logperches. In the Lake Erie streams above the Warren Beach Ridge typical northern logperches were virtually absent; Ohio logperches were more numerous in them than were the intergrades and particularly as the crest of the watershed was approached. However, indications of northern logperch influence were present throughout the entire Lake Erie drainage, especially in the Auglaize River system, spilling over into the headwaters of the Ohio River drainage to disappear finally near the line of glaciation. Raney (1938:65) noticed a tendency toward possible inter-

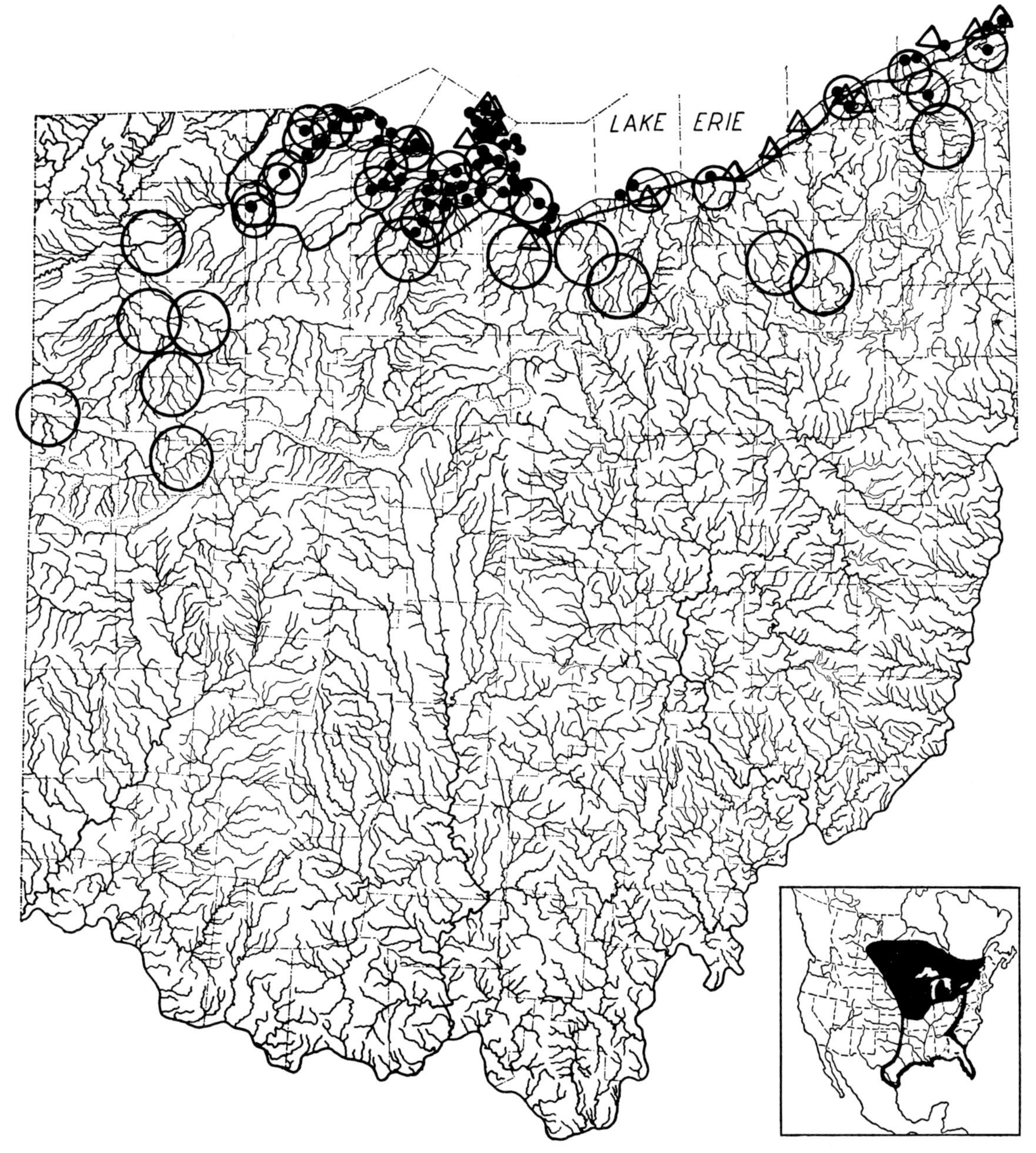

MAP 158. Northern logperch darter

○ Northern logperch influence evident. ○ Intergrades between northern and Ohio logperches.
● Northern logperch locality records. △ 1955–80. ~ Warren Beach Ridge.
Insert: Black area represents range of this subspecies, outlined area of other subspecies; northern limits of range indefinite.

gradation in the unglaciated portion of the upper Ohio drainage in western Pennsylvania. Specimens collected before 1900 indicate that the northern logperch occurred in typical form farther inland before 1900 than it did later, and that recent changes in stream conditions appear to favor the Ohio logperch.

The northern logperch could have invaded the Lake Erie drainage from the west through one of the post-glacial outlets, or from the east through an

Atlantic coast drainage outlet (Radforth, 1944:13–24), or through both. The Ohio logperch presumably was present in part of the Ohio drainage during late glacial time, and invaded glaciated Ohio after the glacial retreat. As a result of these invasions, a merging occurred which has produced a broad band of intergradation between the two throughout most of the glaciated portion of Ohio.

Habitat—An inhabitant of the sandy and/or fine gravel beaches, bars, and bottoms of Lake Erie, and of the more slugggish sand and/or gravel riffles of its larger tributaries below the Warren Beach Ridge. This form is frequently found about moderately dense beds of submerged aquatic vegetation, especially the young. Specimens were taken with trawls in western Lake Erie to a depth of 30′ (9.1 m); in eastern Lake Erie (Fish, 1929:176) to a depth of 40 meters or 131 feet.

Years 1955–80—The northern logperch, subspecies *Percina caprodes semifasciata*, continued to contain little evidence of intergradation with *P. c. caprodes* in the island region of western Lake Erie. During this period the population about the islands showed a marked decrease. Before 1950 fifty or more individuals could be taken in a locality within a short period of time; since then fewer than three were generally captured. The tributary populations of *P. c. semifasciata* and intergrades were almost without exception smaller than before 1950, and none were found in many areas formerly containing sizeable populations (Van Meter and Trautman, 1970:76). The investigations in Lake Erie in the vicinity of Cleveland since 1970 indicate that this subspecies is decreasing in numbers (White et al., 1975:119).

CRYSTAL DARTER

Ammocrypta asprella* (Jordan)

Fig. 159

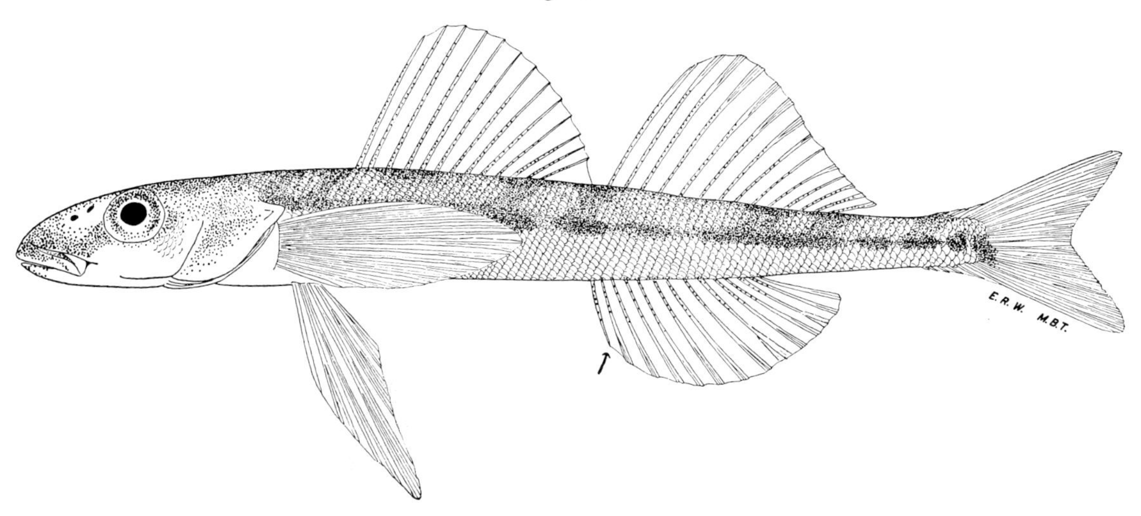

Ohio River, Lawrence County, O.

May 31, 1899.
Male.

51 mm SL, 2.4″ (6.1 cm) TL.
OSUM 8.

Arrow points to the single anal spine.

Identification

Characters: *General—See* dusky darter. *Generic*—Only *one* anal spine; spine rather long, thin, flexible, *see* arrow. Midline of belly scaleless. No groove separates tip of upper lip and snout; frenum quite wide. More than 65 scales in complete lateral line. *Specific*—Tail forked. Body extremely elongate; body depth contained 7.1–10.0 times in standard length. More than 80 scales in complete lateral line; four or more pored scales on caudal base. Dorsal spines and anal rays 12–14. Between 3–7 wide, oblique saddle-bands across back.

Differs: Sand and johnny darters have fewer lateral line scales and fewer anal rays. All other species of darters in Ohio have two anal spines.

Most like: Sand darter.

Coloration: Entire fish opaque; dorsal half of head and body pale straw-color with a silvery luster; ventral half more silvery. Between 3–7 golden-brown, wide saddle-bands cross back, extending obliquely forward and downward to lateral line. Between 8–12 oblong, golden-brown blotches along the lateral line which tend to be confluent. Fins transparent. *Note:* Color description taken from specimens collected in upper Mississippi River.

Lengths: Of the two Ohio specimens 2.4″ (6.1 cm) and 3.5″ (8.9 cm).

Distribution and Habitat

Ohio Distribution—In the Museum of Stanford University, in June, 1950, I found a crystal darter labelled "200, *Crystallaria asprella*, 2631: Beverly, Ohio, Gilbert and Henshall" (No. 8996 in OSUM collection). This specimen is presumably the one collected by Charles H. Gilbert and James A. Henshall in the Muskingum River (Henshall, 1889:125) on August 31, 1888.

On May 31, 1899, Osburn (1901:96) took a specimen (OSUM:8) "on the sandy bottom in the Ohio River at Ironton." No other specimens have been reported.

It is quite probable that before 1900 the crystal darter was rather well distributed in the lower

* Until recently in the genus *Crystallaria*; *see* Bailey and Gosline, 1955:9–10.

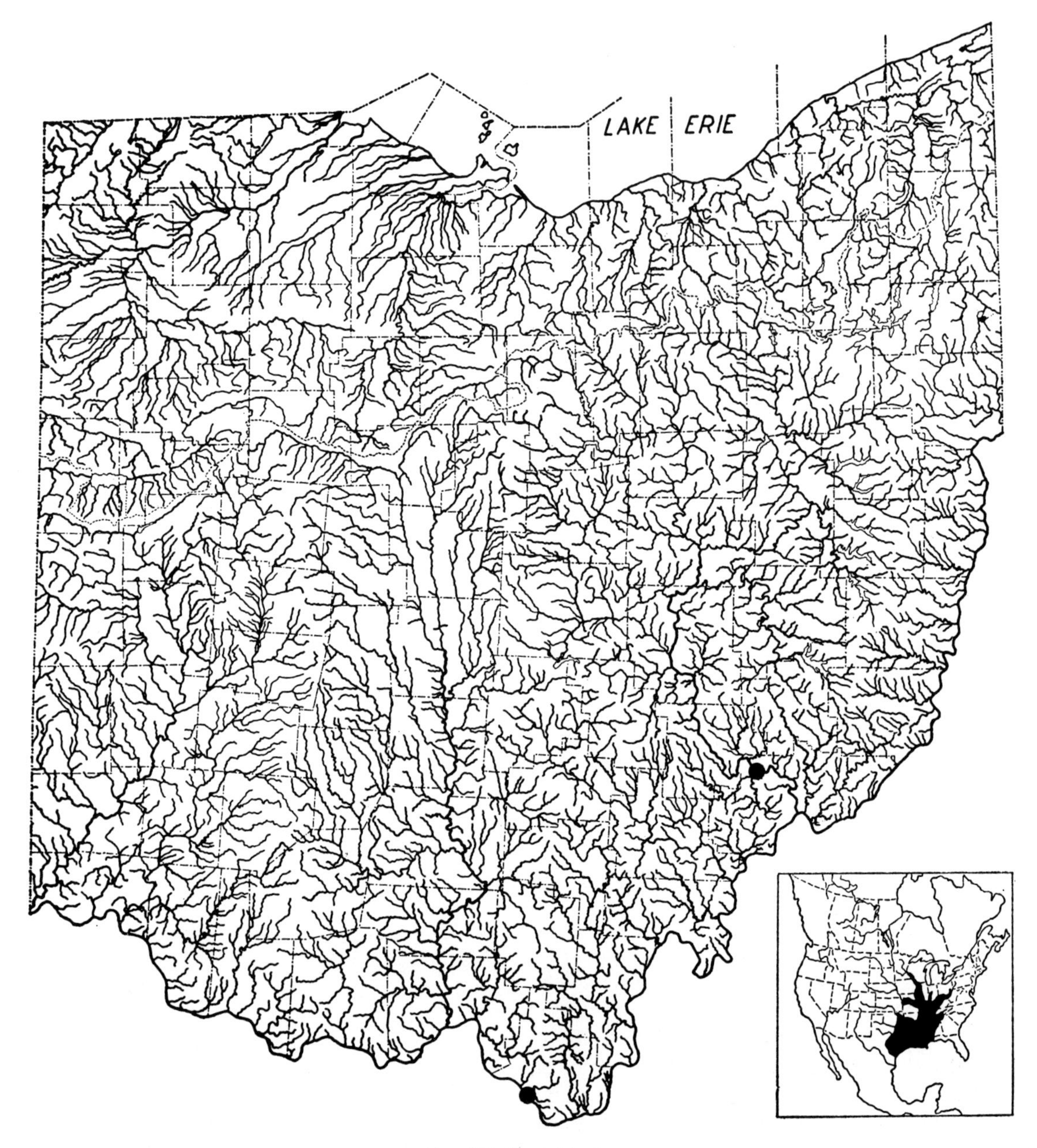

MAP 159. Crystal darter

Locality record. ● Before 1900.

Insert: Large unoccupied areas within this range increasing in size and number.

courses of the southern Ohio tributaries and in the Ohio River, before the greatly increased silt load and subsequent silting smothered the sandy riffles, extensive sand bars, and sand-bottomed pools. The habitat of few other Ohio fishes seemed to be so vulnerable to annihilation, and I very much doubt whether the species occurred in Ohio after 1925.

Habitat—An inhabitant of extensive sandy riffles, bars, and pool bottoms, as a rule occupying larger rivers and/or deeper waters than does the sand darter.

Years 1955–80—Investigations during this period indicated that the species no longer existed in Ohio waters.

EASTERN SAND DARTER

Ammocrypta pellucida (Putnam)

Fig. 160

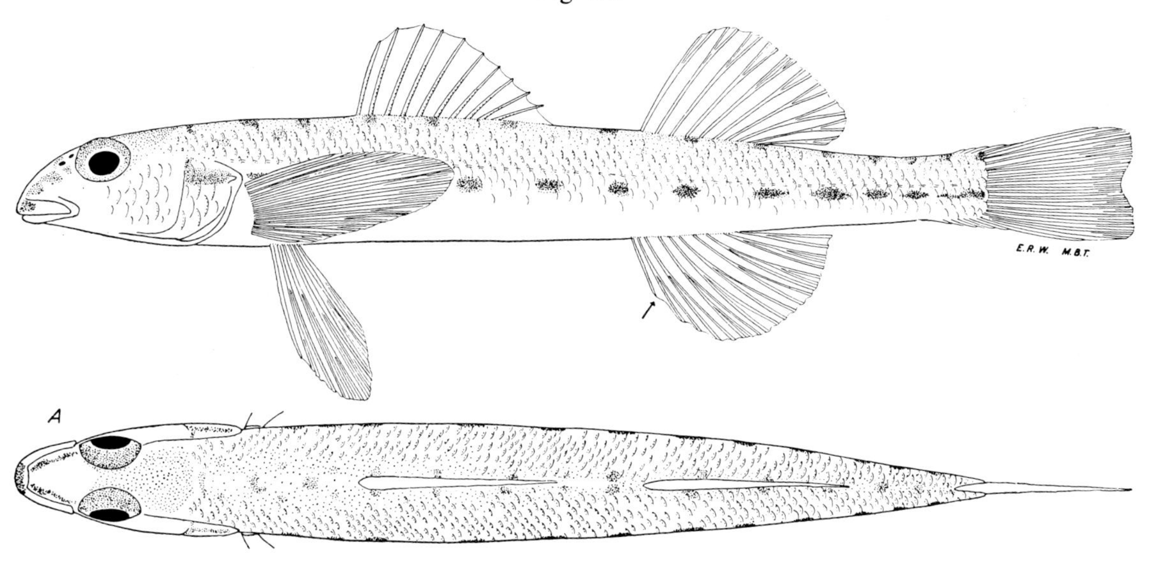

Upper fig.: Big Darby Creek, Pickaway County, O.

June 9, 1942.
Breeding female.

51 mm SL, 2.3″ (5.8 cm) TL.
OSUM 5056.

Arrow points to the single anal spine.

Fig. A: dorsal view; note color pattern.

Identification

Characters: *General—See* dusky darter. *Generic—See* crystal darter. *Specific*—Only one anal spine which is very slim and weak, *see* arrow, upper fig. Tail slightly emarginated or straight distally. Ventral half of body almost scaleless, except for 1–3 rows immediately beneath the lateral line scales. No groove separates tip of upper lip and snout; frenum rather wide. Body extremely elongate; body depth contained 7.0–11.2 times in standard length. Complete lateral line with 65–78 scales. Dorsal spines 9–11. Anal rays 8–10. A series of 12–19 spots along lateral line; another series along dorsal ridge, lower fig.

Differs: Crystal darter has more dorsal spines and anal rays. Johnny darter has a groove between upper lip and snout. All other species of darters in Ohio have two anal spines.

Most like: Crystal darter. **Superficially like:** Johnny and channel darters.

Coloration: *Adults*—Dorsal half of head and body a pellucid-white with a yellowish cast. Ventral half of head and body white or silvery. A series of 12–16 small, olive spots along dorsal ridge; these become rows of paired spots along thc base of the dorsal fins, with one row on each side of fin. Series of 9–14 oblong, dusky-olive spots along lateral line; these posteriorly tending to become confluent; in some specimens a suffused band of yellow is present along the lateral line. Webbing of fins transparent; some have a yellowish tinge. *Breeding male*—Like adults, except that the body is flushed with yellowish. *Young*—Like adults, except more silvery, and with little or no yellow.

Lengths: *Young* of year in Oct., 1.1″–2.1″ (2.8–5.3 cm) long. *Adults*, usually 1.8″–2.8″ (4.6–7.1 cm) long. Largest specimen, 3.2″ (8.1 cm) long.

Distribution and Habitat

Ohio Distribution—Literature references and the many preserved specimens from every section testify to the general distribution of the sand darter

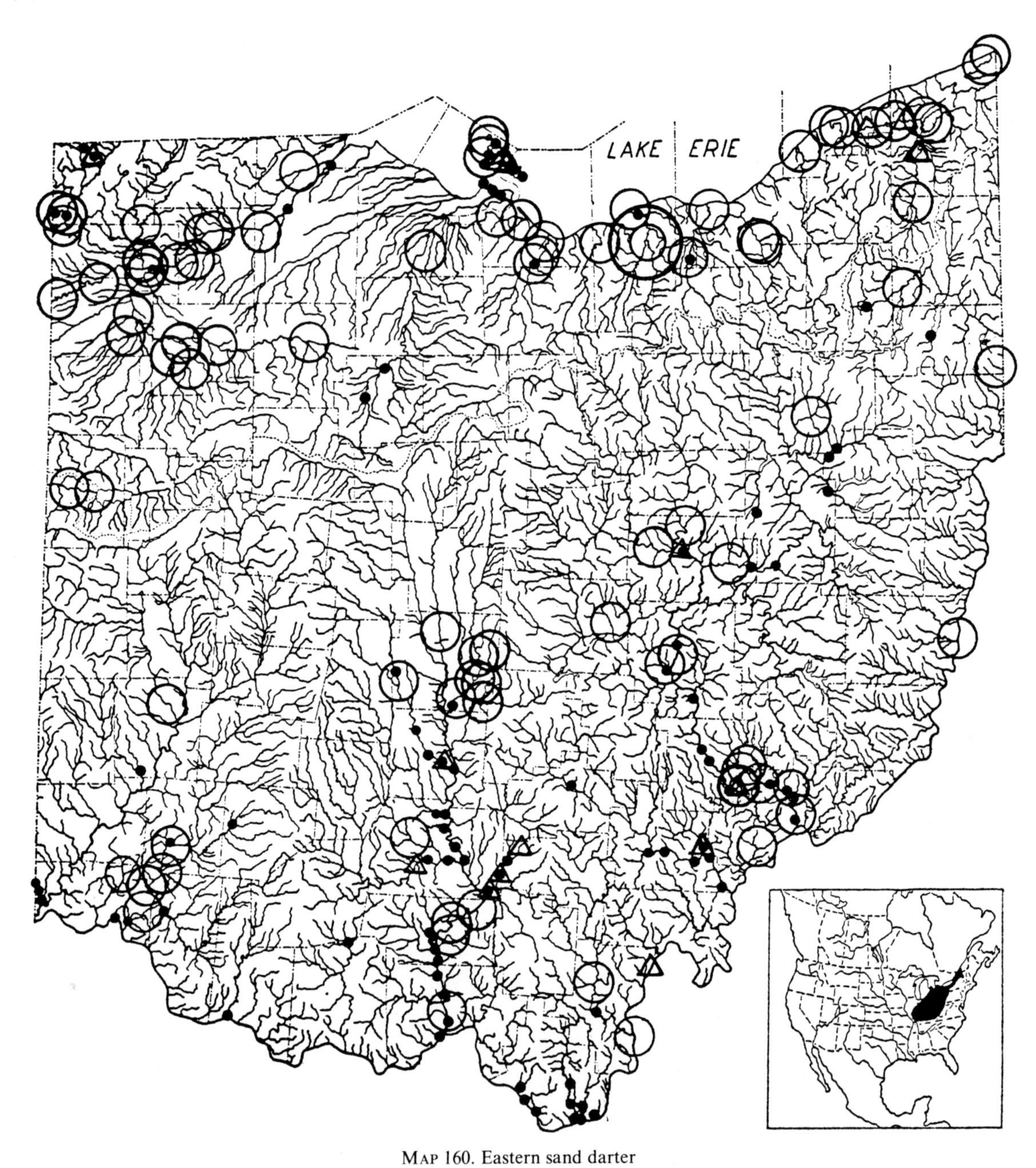

MAP 160. Eastern sand darter

○ Type locality, Black River, Lorain County. Locality records. ○ Before 1924. ● 1924–54. △ 1955–80.
Insert: Unoccupied areas within this range increasing in size and number.

throughout Ohio prior to 1900. Several references note its early abundance: Kirsch (1895A:331) in 1893 stated that it was "Common everywhere on sandy bottom in the Maumee River and in lower courses of the larger tributaries;" Osburn (1901:96) that it was "common locally in larger streams on sandy bottoms." This darter presumably was abundant until about 1900 in the inland streams, in the then-unponded Ohio River, and on the wave-protected sand beaches of Lake Erie.

In 1853 S. F. Baird collected the types of this species in the Black River below the falls, near Elyria, Lorain County, Ohio, but it was not described until 1863 (as *Pleurolepis pellucidus*, in Putnam, 1863:5). The types are number 1311 in the USNM.

Throughout the 1925–50 surveys the sand darter displayed a continuous decrease in abundance which was correlated with a decrease in amount of suitable habitat, and especially in the Scioto and Maumee drainages.

Habitat—The sand darter inhabited the sandy areas of streams ranging in size from small creeks to large rivers. However, the species was most abundant in the larger, sandy areas of those sections of moderate- or large-sized streams where the silting-over of the sand was at a minimum and the current was not strong enough to wash away the sand. The sandy habitats of the sand darters were readily recognized, and during periods of low, clear water one could walk along a stream and readily count them. During the 1925–50 period there was a continuous decrease in the size and numbers of sandy areas in some streams. As an example: before 1930 I found 1–10 sandy areas, containing sand darters, in every pool in a mile-long stretch of Big Darby Creek in Pickaway County. Between 1930–45 many of these areas became silt-smothered and after 1945 only 8 sandy areas remained in the mile stretch of stream, of which only 4 areas contained sand darters. In the clean, sandy areas the sand darter buried itself in the sand, with only its eyes exposed. From such a retreat it would dash out to catch passing prey, after which it rapidly reburied itself, tail first, in the sand.

Years 1955–80—During this period the sand darter was captured in few inland localities in Ohio and these only where previously moderately large populations had existed. In many other localities where the species had been present previously, none were found despite thorough investigations (White et al., 1975:119). Only one specimen was taken about the Bass Islands of Lake Erie (Van Meter and Trautman, 1970:77–78).

None were taken during the investigations of the Ohio River by personnel from the University of Louisville between 1957–59 (ORSANCO, 1962) or by H. Ronald Preston in his 1968–69 surveys. In 1977 Margulies and Burch obtained a specimen from the Ohio River adjacent to Gallia County, Ohio.

In the first edition of this report Baird was considered to be the describer of *Ammocrypta pellucida*. Recently, through a technicality, F. W. Putnam is given credit for the description of the species, which now is *Ammocrypta pellucida* (Putnam); *see* Bailey et al., 1970:75.

CENTRAL JOHNNY DARTER

Etheostoma nigrum nigrum Rafinesque

Fig. 161

Upper fig.: Deer Run, Stark County, O.

May 28–29, 1941. 40 mm SL, 1.8″ (4.6 cm) TL.
Breeding male. OSUM 3240.

Lower fig.: Same locality, date and OSUM number.

Breeding female. 40 mm SL, 1.8″ (4.6 cm) TL.
Note: Arrow points to the single anal spine.

Identification

Characters: *General—See* dusky darter. *Generic*—The genus *Etheostoma* contains a large number of species with rather negative characters. The anal fin is smaller than is the soft dorsal. Species found in Ohio have two anal spines, except the johnny darter which has only one. The belly is scaled in most species; in those species in which the midline is partly or entirely naked the lateral line is either incomplete (lateral line complete in genera, *Percina* and *Ammocrypta*), *or* there is only one anal spine (*Percina* has two anal spines). There are no specialized scales on the midline. A frenum is present in Ohio species, except in the johnny and greenside darters. Scales on lateral series range from 30–70. *Specific—One* long, flexible anal spine, *see* arrow, lower fig. A deep groove separates tip of upper lip and snout, *see* head of johnny darter, fig. 162 B. Midline of belly naked or scaled. Gill covers

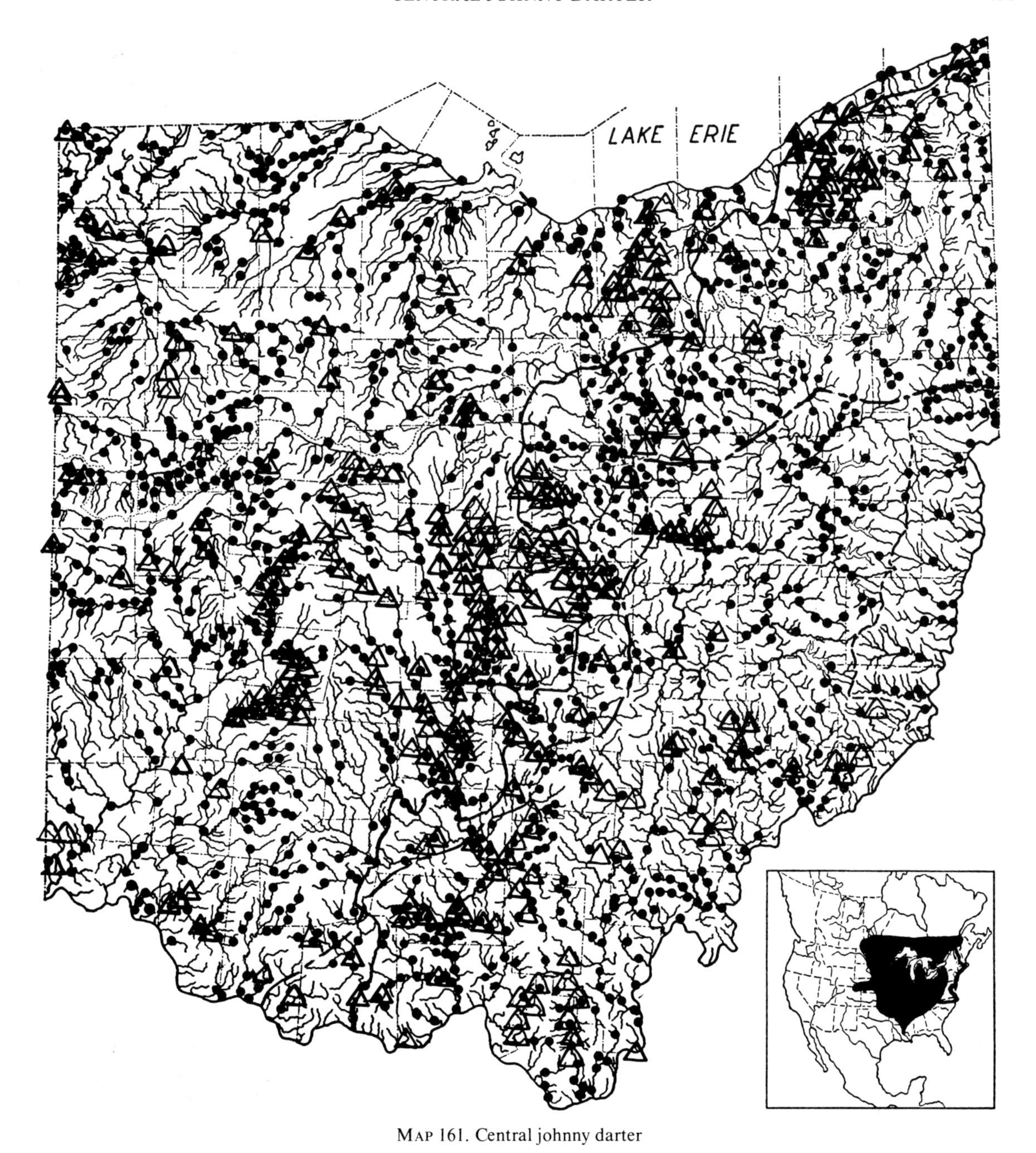

MAP 161. Central johnny darter

Locality records. ● Before 1955. △ 1955–80. ~ Allegheny Front Escarpment. /\ Glacial boundary. --- Flushing Escarpment.

Insert: Black area represents range of this subspecies, outlined area of other subspecies; western limits of range indefinite.

narrowly joined across isthmus with membrane; distance from apex of gill-cover angle to tip of upper lip usually contained more than 1.8 times in head length; this distance extends from tip of upper lip to posterior edge of eye. Snout rounded and blunt. Dorsal rays 7–10. Anal rays 6–9. Complete lateral line with 39–55 scales. Body shape similar to that of the channel darter. *Subspecific*—Nape, cheeks, and breast scaleless, *see* head of johnny darter, fig. 162 A. Few or no scales on belly immediately behind pelvic fins; in some males in spring, midline is scaleless from pelvics to anal fin origin, except for a narrow bridge of scales before anus, as it is in the river darter.

Differs: Scaly johnny darter has scaled nape, cheeks, breast and belly. Sand and crystal darters have more lateral line scales. Channel and river darters, and remaining species of darters in Ohio have two anal spines.

Most like: Scaly johnny. **Superficially like:** Channel and river darters.

Coloration: *Non-breeding adults*—Dorsally olive- or straw-yellow; sides a paler yellow; ventrally pale yellow becoming milk-white on belly. Usually 6, extremes 4–7, quadrate, dark brown saddle-bands cross dorsal ridge; normally the first is before the first dorsal origin; second and third, beneath the spinous dorsal; fourth, beneath soft dorsal; fifth and sixth, on caudal peduncle. W- and S-shaped brownish markings scattered over back and sides; some of these markings line up along the lateral line suggesting a broken lateral band. Tear-drop small or absent. No caudal spot. Dorsals and caudal fins with spotting; these spots tend to form oblique rows on dorsals, and wavy, vertical bands on the caudal. Lower fins usually transparent and unspotted. *Breeding male*—As breeding season approaches the entire head, breast, pelvics, anal and spinous dorsal fins become blackish; remainder of body and remaining fins become dusky; between 4–8 vague, blackish, vertical bands cross the body and these are most distinct posteriorly; the first six dorsal spines develop soft, whitish pads on their tips, as do the ends of the pelvic spine and pelvic rays and anal spine and first anal ray. *Breeding female*—Color pattern as in non-breeding adult but dark browns and yellows intensified. *Young*—Like adults, except that the yellow is more or less replaced with silvery; body and fin markings less developed.

Lengths: *Young* of year in Oct., 1.1″–2.1″ (2.8–5.3 cm) long. Around **1** year, 1.3″–2.5″ (3.3–6.4 cm). *Adults*, usually 1.5″–2.8″ (3.8–7.1 cm) long. Largest specimen, 3.0″ (7.6 cm) long.

Distribution and Habitat

Ohio Distribution—The johnny darter as a species before 1900 was considered to be abundant and to be found "all over Ohio" (Osburn, 1901:95); since 1925 it has been the most universally distributed species of darter inhabiting Ohio waters. The johnny was present in all waters capable of sustaining fish life, from the deep waters of western Lake Erie to the shallowest of small brooks, excepting some farm and bog ponds without accessible stream connections. It was very tolerant, for a darter, of many organic and inorganic pollutants, and was able to inhabit waters more silty than those tolerated by many of the other fish species. Its eggs developed successfully in silty waters because they were deposited on the underside of stones, tin cans, and other objects, where the settling silt had little effect upon them and where the guarding male could keep the eggs clean by brushing them with his spongy nape or by turning upside down and cleaning them with the soft pads on the tips of his pelvic spines, and rays. In waters habitually low in oxygen, the johnny darter deserted its usual pool and sluggish riffle habitats and occupied the faster riffles, which in streams having a greater amount of oxygen, were inhabited by other darter species. In high-gradient streams the johnny was restricted largely to the quiet waters in the pools. It inhabited all types of bottom, and could be found among aquatic vegetation. No other darter species appeared to be so tolerant of so many diverse conditions.

During the 1925–50 surveys the central johnny darter was most numerous along and between the glacial boundary and the Allegheny Front Escarpment. It was least numerous in the Ohio River, in small impounded waters, in very turbid or otherwise polluted waters, and along the shores of Lake Erie where this subspecies usually could be found only in late summer and fall after it had been forced into the lake by the drying up of the small tributaries. Those specimens taken east of the Allegheny Front Escarpment showed no scaly johnny darter influence; those found in the remaining portion of Ohio are discussed under that subspecies.

Habitat—The central johnny darter principally inhabited flowing waters, with the greatest population densities in small- and moderate-sized streams of moderate gradients having a sandy gravel bottom. This subspecies was less tolerant of submerged aquatic vegetation, and more tolerant to flowing waters, than was the scaly johnny darter.

Years 1955–80—During this period the subspecies *Etheostoma nigrum nigrum* continued to be present in every Ohio stream capable of sustaining fish life and was as universally distributed throughout a stream as was the bluntnose minnow. The largest populations occurred in moderate- and small-sized streams, were least numerous in large rivers and the Ohio River, and were usually absent from farm ponds and other small static waters.

SCALY JOHNNY DARTER

Etheostoma nigrum eulepis (Hubbs and Greene)

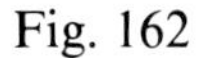

Fig. 162

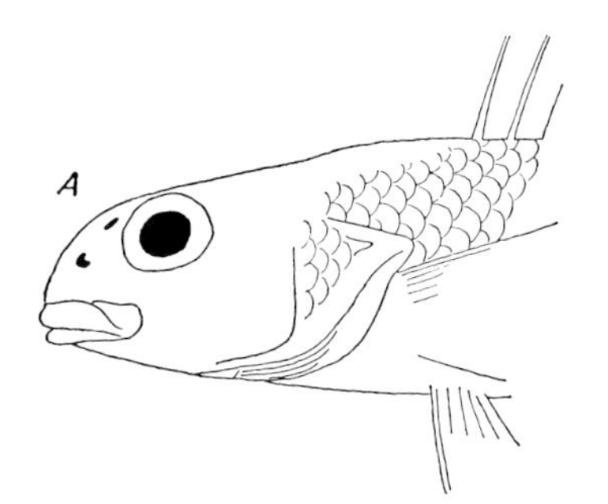

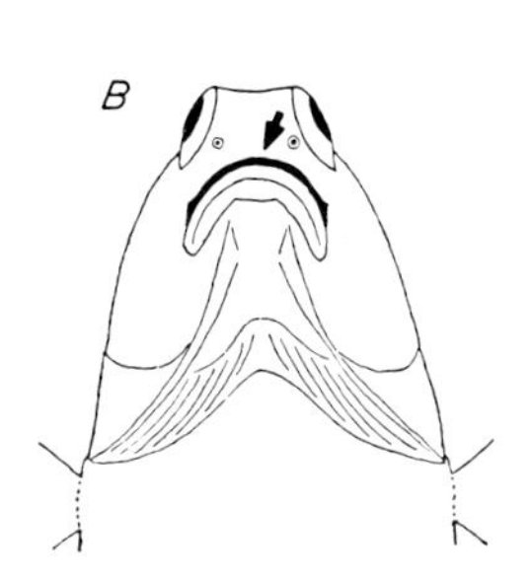

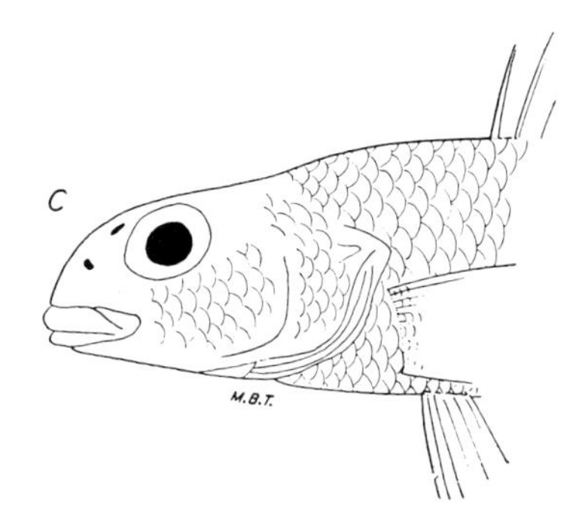

Note: No figure of this subspecies was drawn because all proportions are similar to those of the central johnny darter, *see* fig. 161.

Fig. A: lateral view of head of central johnny darter; note absence of scales on nape, cheek and breast.

Fig. B: frontal view; note groove which separates tip of upper jaw or lip from remainder of snout.

Fig. C: lateral view of head of the scaly johnny darter; note scales on nape, cheek and breast.

Identification

Characters: *General—See* dusky darter. *Generic* and *specific—See* central johnny darter. *Subspecific*—Nape, cheeks and breast scaled, *see* head of johnny darters, fig. C. Belly normally scaled.

Differs: Central johnny darter has nape, cheeks, and breast scaleless. *See* central johnny darter.

Most like: Central johnny and channel darters.

Coloration: Similar to central johnny darter at all ages and while breeding, except that the yellows are darker, more clouded, and browner; the browns contain more chocolate; the young are less silvery.

Lengths: *Young* of year in Oct., 1.1″–2.1″ (2.8–5.3 cm) long. Around **1** year, 1.3″–2.5″ (3.3–6.4 cm). *Adults*, usually 1.5″–2.8″ (3.8–7.1 cm). Largest specimen, 3.0″ (7.6 cm) long.

Distribution and Habitat

Ohio Distribution—In 1935 Hubbs and Greene (1935:12) described this distinctive subspecies. Since becoming aware of its presence in Ohio, about 1934, all except a few individuals have been found to be typical scaly johnny darters (solid circles) which were taken during the spawning season in the waters about the islands of western Lake Erie, along the shores of the lake from the Michigan state line to the city of Sandusky, in Sandusky Bay, and in the tributary streams of western Lake Erie upstream to the first, well defined riffles.

The first indication of intergrading tendencies in breeding adults was observed at the first well defined riffles above the mouths of the Lake Erie tributaries. From these riffles upstream to where the gradient rose above 2.0′/mile (0.38 m/km), or where the stream crossed the Lake Warren Beach Ridge, there were a few typical-appearing scaly johnnies and some obvious intergrades in the pools, and some typical-appearing central johnnies in the faster riffles. Only central johnnies and intergrades occurred in those sections of the tributaries where the gradient was higher than 2.0′ (0.6 m), or in those sections above the Beach Ridge. The Maumee River population from the ridge upstream to the Indiana state line consisted of individuals which were more or less intermediate in characters, as was the population in the St. Marys River in Paulding, Mercer and Auglaize counties (medium-sized hollow circles). Elsewhere in Ohio, scaly johnny darter influence was observed only in occasional individuals which had a partly-scaled nape (large-sized hollow circles). East of Sandusky such partly scaled individuals were found during the breeding season in Lake Erie, in the short distance in the tributaries between the lake and the ridge, and in the marshy headwaters of such rivers as the Vermilion in Ashland County. Occasional partly scaled individuals

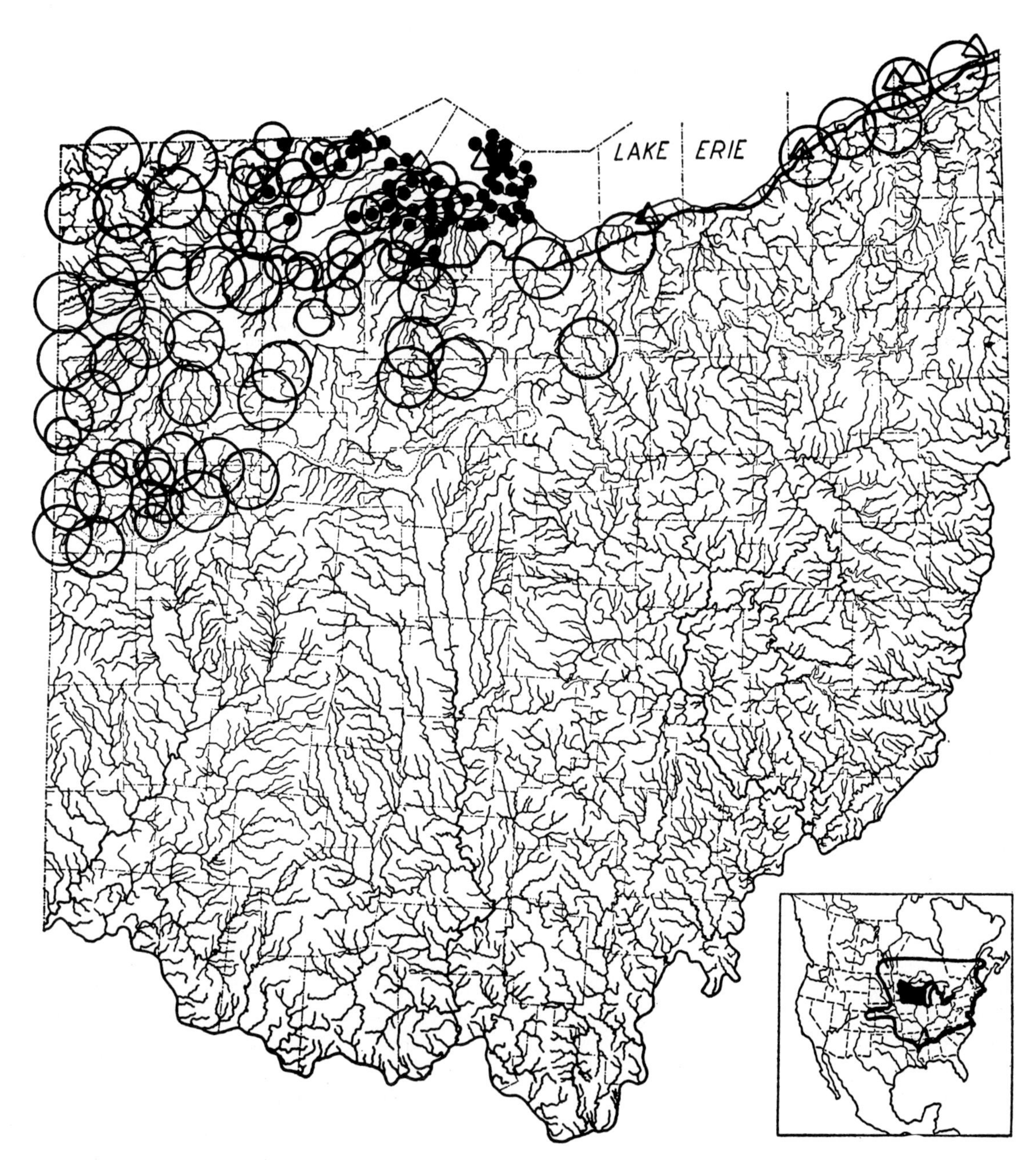

MAP 162. Scaly johnny darter

○ Specimens with partly scaled nape, last evidence of scaly johnny darters.
○ Intergrades between scaly and central johnny darters.
Scaly johnny darter locality records. ● Before 1955. △ 1955–80.
~ Warren Beach Ridge.
Insert: Range (black area) of this distinctive subspecies lies wholly within the range (outlined area) of the nominate form.

were found in the Wabash River tributaries of the Ohio River drainage in Mercer County, but were lacking elsewhere in that drainage except for a very occasional individual in the lowest-gradient portions of the prairie streams of west-central Ohio. During late summer, fall, and winter more or less typical central johnnies and intergrades were found in the estuaries of the tributary streams in Sandusky Bay, and along the shores of Lake Erie, and especially during drought years when the drying-up of the headwaters forced the darters into the estuaries and the lake.

Presumably the scaly johnny utilized the Maumee Outlet (Greene, 1935:12) to invade the Lake Erie drainage. It is quite probable that before 1850 populations of more or less typical scaly johnnies were present in the low-gradient streams of northwestern Ohio, and especially of the St. Marys drainage. Later, intensive ditching and dredging increased the amount of habitat of the central johnny, thereby enabling it to "swamp out" the typical scaly johnny populations, leaving only central johnnies and intergrades.

Habitat—The scaly johnny chiefly inhabited non-flowing waters, with its greatest population densities about the sandy gravel and somewhat silty bottoms of the bays and protected beaches about the western Lake Erie islands, and in the estuaries and quiet pools of the streams and southwestern shore of Lake Erie, including Sandusky Bay. It has been taken in trawls about the Bass Islands at a depth of 30′ (9.1 m). This form was quite tolerant to a silty bottom and moderate amount of submerged aquatic vegetation, but avoided currents on the Lake Erie bars and reefs, and in the streams.

Years 1955–80—Before 1950 the scaly johnny darter, *Etheostoma nigrum eulepis*, was locally abundant about the islands of western Lake Erie and in the static portions of the tributary streams and estuaries. At that time the species appeared to be as tolerant of pollutants as was the inland Ohio subspecies, *Etheostoma nigrum nigrum*.

As indications of its former abundance in western Lake Erie: on March 25, 1947 three of us captured 161 *E. n. eulepis* along a small portion of the shore of South Bass Island, and on April 9, 1949 Karl F. Lagler and Reeve M. Bailey captured a large number of *eulepis* in the same locality, retaining 60 for experimental purposes.

Shortly after 1950 a decrease in the population was observed. Between 1960 and 1965 no more than eight specimens were taken by an ichthyology class of the Franz Theodore Stone Laboratory during any daylong seining expedition and none appears to have been captured since 1966 (Van Meter and Trautman, 1970:77). The inland population, *E. n. nigrum*, apparently has maintained its numerical abundance throughout Ohio during the past 20-year period. During the intensive collections in Lake Erie in the vicinity of Cleveland after 1970 only one typical *E. n. eulepis* was collected (White et al., 1975:120).

For the genetic fixity of subspecies of *Etheostoma nigrum*, *see* Lagler and Bailey (1947:50–59).

EASTERN GREENSIDE DARTER

Etheostoma blennioides blennioides Rafinesque

Fig. 163

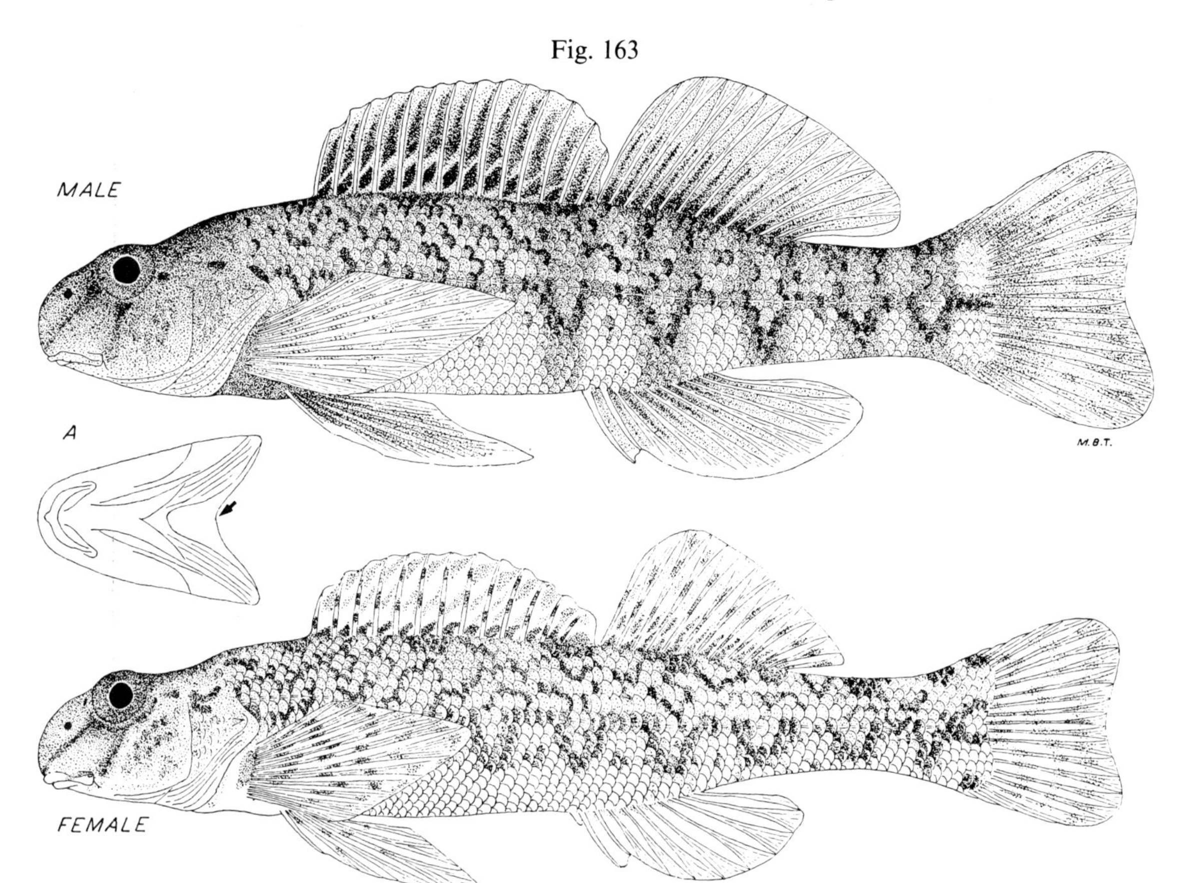

Upper fig.: Big Darby Creek, Pickaway County, O.

Oct. 10, 1947. 74 mm SL, 3.5″ (8.9 cm) TL.
Adult male. OSUM 7419.

Lower fig.: Same locality, date and OSUM number.

Adult female. 67 mm SL, 3.2″ (8.1 cm) TL.

Fig. A: ventral view of head; note arrow pointing to shallow angle of the broadly joined gill covers; also note the knob on the tip of the upper lip.

Identification

Characters: *General—See* dusky darter. *Generic—See* central johnny darter. *Specific*—Two anal spines. Belly scaled. A deep *groove* separates tip of upper jaw and snout. Groove on side of snout, which separates the maxillary from skin of the preorbital portion of snout, is extremely short and is restricted to the posterior third of the maxillary bone (in all other species of *Etheostoma* this groove extends along the entire length of the maxillary). Snout blunt, rounded, and usually slightly overhangs tip of upper lip; tip of upper lip often contains a knob, fig. A. Ventral surface of head flattened, giving head a triangular appearance in cross section. Gill covers broadly joined with membrane across isthmus, fig. A, arrow. Cheeks usually have some deeply embedded scales; opercles conspicuously scaled. Dorsal spines 12–14, extremes 11–16. Complete lateral line usually with 60–67 scales, extremes 57–70. Color predominantly green or olive-green.

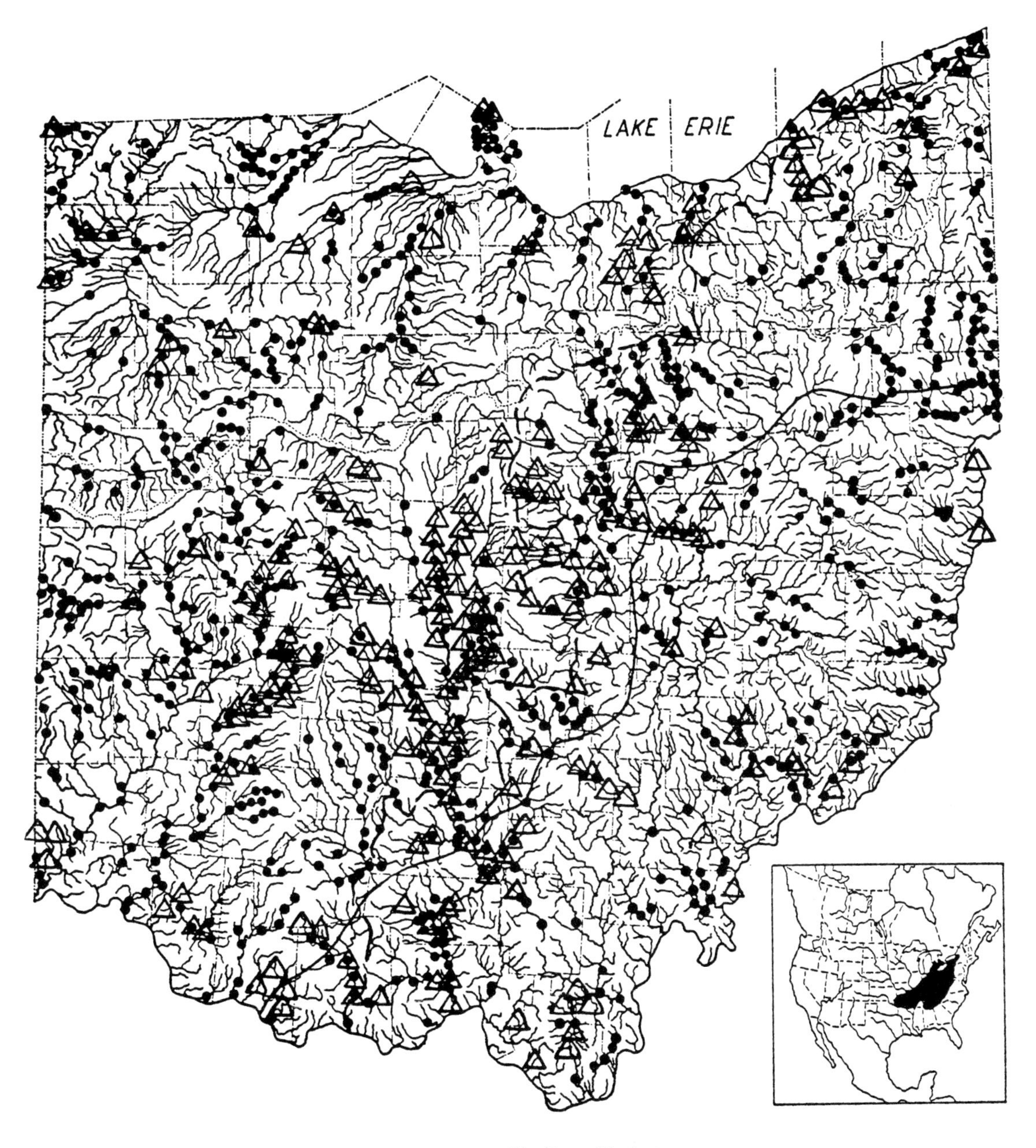

MAP 163a and 163b. Greenside darter

Locality records. ● Before 1955. △ 1955–80. /\ Allegheny Front Escarpment. ~ Glacial boundary.
Insert: Rather uniformly distributed throughout most of its range.

Differs: No other species of darter in Ohio has combination of: two anal spines; deep groove separating upper lip and snout; gill covers broadly connected with membrane; normally scaled belly; complete lateral line with 57–70 scales, color predominantly green or olive-green.

Most like: Banded darter; small young like johnny darter. **Superficially like:** River and variegate darters.

Coloration: *Non-breeding adults*—Dorsally olive-green or green; sides greenish-yellow; ventrally whitish-green or whitish. Usually 6 (extremes 4–7) quadrate, dark, brownish-green saddle-bands cross dorsal ridge. Back contains many specialized scales scattered over it; these scales are olive or brown in some individuals, brick-red in others, *see* Habitat. Sides have 4–10 V- or W-shaped marks, the upper ends of which tend to be confluent along the lateral line; these marks most distinct posteriorly. In most specimens both dorsals have a basal bar of rusty-red; remaining portions of fins have greenish webbing and olive-green spines and rays and vague spottings on both. Caudal olive-green and lightly spotted; spots forming wavy bands. Pelvics and pectorals often spotted. *Breeding male*—Body brilliantly dark green with 4–7 dark green vertical bands crossing posterior half of body. Ventral surface of head and breast dark bluish-green. Spinous dorsal has the basal bar intensely brick-red; remaining portions dark green. Caudal, anal, and pelvics green. Tip of pelvic spine and rays thickened and whitish. *Breeding female*—Similar to non-breeding adults, except that colors are intensified; breast bluish-white; spotting on dorsals and caudal prominent; flush of green present on lower fins; tear-drop usually evident. *Young*—Like adult, but with more W- than V-shaped markings on sides; colors less intense, greens replaced by greenish-browns, yellows by white; fins more transparent and less spotted. *Small young*—Often mistaken for johnny darters and have a rather similar coloration.

Lengths: *Young* of year in Oct., 1.2″–2.0″ (3.0–5.1 cm) long. Around **1** year, 1.5″–2.5″ (3.8–6.4 cm). *Adults*, usually 2.4″–3.5″ (6.1–8.9 cm). Largest specimen, 4.3″ (11 cm) long.

Hybridizes: with dusky and logperch darters.

Distribution and Habitat

Ohio Distribution—Rafinesque (1820:88–89) and other early writers indicated that prior to 1900 the greenside darter was abundant and widely distributed throughout Ohio. During the 1920–50 surveys it was presumably present in every stream system, at least as strays. The largest populations occurred in the glacially-rejuvenated streams along and between the glacial boundary and the Allegheny Front Escarpment, in the streams of moderate and high gradients of southwestern and westcentral Ohio, and about the Lake Erie islands. The numbers present in the Maumee system between 1925–30 were rather small, and since 1930 a decrease in abundance has been noted in several localities in the Maumee and lower Auglaize rivers where turbidity increase was greatest. The greenside was absent, except for isolated populations or strays, in the mature, base- or low-gradient, unglaciated streams of southeastern Ohio between the line of glaciation and the short, glacially-rejuvenated streams along the Ohio River. It was rare during the surveys in the Ohio River, except as wintering strays, but it must have been present as a permanent resident in the river before impoundment which resulted in the elimination of riffles and increase of siltation.

Habitat—The largest breeding populations of greensides occurred in those riffles of moderate- and high-gradient streams in which the rocks and stones larger than 2″ (5.1 cm) in diameter contained growths of algae and aquatic mosses, the riffles were between 30′–80′ (9.1–24 m) in width, and moderately clear, unpolluted waters prevailed during most of the spring. In such streams the species was most numerous near the head of riffles in depths of 1′–4′ (0.3–1.2 m), where the current was increasing in speed. Adults have been found spawning among the strands of algae (usually *Cladophora* spp.) and aquatic mosses (Clarence E. Taft identified some of this moss as *Drepanocladus exannulatus* [Bry. Eur.] Warnst); males were observed guarding territory, and eggs were found among the strands of vegetation. Smaller numbers of breeding adults were found in riffles of small streams having an average width of less than 20′ (6.1 m), and in the riffles of the largest inland streams. After the spawning season most of the adults and many of the young began to drift downstream, to winter in deeper riffles and less swift waters. Those individuals, trapped in pools after the connecting riffles had ceased to flow, were found either near a spring or in the deepest part of the pool.

Two color types of greenside darters inhabited Ohio; one may be designated as the "prairie type," the other as the "Allegheny type." The prairie type occupied the clearer streams of low or moderate gradients of the prairie areas of westcentral and northwestern Ohio, especially in the Wabash and Maumee river drainages, extending as far eastward as the Sandusky Bay tributaries and southward into Darke County. This type was chiefly associated with such rooted aquatics as lizard's-tail, turtlehead, pondweeds and smartweed, as well as with the algae and mosses. The males of the prairie type had many scales of a brilliant, rusty-red scattered over their backs and sides, and the symphysis of the upper jaw in both sexes was normally without a tip.

The Allegheny type inhabited the remainder of Ohio and was best represented in the unglaciated southeastern section. This type was not associated with vegetation, except algae and aquatic mosses; the males had few, usually no, rusty scales on back and sides, or these scales were brownish as on the females; in both males and females (but best developed in males) the symphysis of the upper jaw often contained a tip, *see* fig. A. It is suggested that after the retreat of the last glaciers, the prairie type invaded Ohio from the west following the prairie eastward; that the Allegheny type inhabited the Allegheny plateau during glacial times and after the glacial retreat invaded portions of glaciated Ohio.

The habitat niches of the greenside and banded darters were similar in many respects; yet competition between the two was apparently not great. The greenside spawned chiefly in April when water temperatures were usually below 65°F (18°C), utilizing the deeper, swifter portions of riffles and the more luxuriant clumps of algae and mosses. The banded spawned principally in May, in water temperatures usually above 65°F (18°C), utilizing the shallower, more sluggish portions of riffles and the shorter, sparser clumps of algae and mosses. Competition appeared to be much greater among their young in the "nursery" sections of the riffles and among the feeding adults, than it was among the territory-holding males.

Years 1955–80—The "Allegheny type" of the greenside darter, referred to above, maintained its general distributional range and abundance in inland Ohio throughout this period. As previously, this subspecies was numerous in the Ohio River drainage and especially so in those glacially rejuvenated streams immediately west of the glacial boundary.

A spectacular "population explosion" of this subspecies occurred in the lower portion of Big Darby Creek in 1962. In late April the *small* population of adults was ready to spawn. Conditions were particularly favorable because the unusually clear waters, suitable water temperature, and considerable sunshine had produced a profuse growth of algae, principally *Drepanoclatus exannulatus* and *Cladophora glomerata*, in and among which this form of the greenside could deposit its eggs. By May 14 egg-laying and hatching of fry had been largely completed, resulting in this small breeding population producing the largest hatch of greenside fry that I have observed between the years 1925 and 1962.

Unlike the small numbers of greenside breeders, the banded darter (*E. zonale*) population was the largest I have ever seen. This latter species, however, did not begin to spawn abundantly until about May 14, at which time there began an unprecedented 5-day heat wave. Banded darters also utilize algae for spawning; however, water temperatures rose so high that the algae died and rapidly disintegrated, leaving the bandeds largely without a suitable spawning habitat, presumably resulting in an almost total spawning failure.

Apparently sufficient nutrients were released from the decomposing algae to produce a huge pulse of zooplankton at that period when the greenside young could utilize this food. The result was a huge population of greenside young which dominated not only their nursery and riffle habitats but apparently every other stream habitat, even in the deepest pools. During late spring we observed groups of young moving downstream from Big Darby into the Scioto River. Even with this loss of young, the following winter found greensides abundant in the many habitat types of Big Darby Creek. This species occurs sparingly in the Ohio River (ORSANCO, 1962:151). Margulies and Burch obtained the greenside darter in the Ohio River adjacent to Jefferson County, Ohio.

For the life history of this species *see* Fahy, 1954:139–205.

CENTRAL GREENSIDE DARTER

Etheostoma blennioides pholidotum Miller

"Prairie" type

Note: No figure of this subspecies was drawn because it and the typical subspecies are superficially very similar. For a figure of the greenside darter *see* fig. 163.

Identification

Characters: *General—See* dusky darter. *Generic—See* central johnny darter. *Specific—See* greenside darter ("Allegheny" type). *Subspecific—* Complete lateral line in Lake Erie specimens of this subspecies average 59 scales, fewer than in *E. b. blennioides*, which average 65 scales (highest in specimens from southern and southeastern Ohio); transverse scale rows from below central portion of dorsal fin to the lateral line are 6–8, average 7 in *E. b. pholidotum*; 7–10, average 8.5, in *E. b. blennioides*; symphysis of upper jaw normally without a tip.

Differs: *E. b. blennioides* has more scales in the lateral line, more transverse scale rows above the lateral line; symphysis of upper jaw more often has a tip; scattered rusty-red scales on back and sides of adult males poorly developed or absent.

Most like: Banded darter; small young like johnny darter. **Superficially like:** River and variegate darters.

Coloration: Females and small adults similar to *E. b. blennioides*; in adult males there are usually some brilliantly rusty-red scales scattered over the back and sides.

Distribution and Habitat

In 1968 Robert V. Miller (Copeia:26–36) described the "Prairie" type of the greenside darter as *Etheostoma blennioides pholidotum*. In Ohio this subspecies occurs in Lake Erie and, in more or less typical form, in streams tributary to Lake Erie, and especially those lower portions having basic or moderate gradients. *E. b. pholidotum* is largely absent from headwater streams having higher gradients and chiefly those adjacent to the Lake Erie–Ohio River watershed; these sections are inhabited primarily by typical *E. b. blennioides* or intergrades.

In northeastern Ohio *E. b. blennioides* usually extends farther downstream in the higher-gradient tributaries, such as the Chagrin River, than it does in the lower-gradient headwaters in the northwestern portion of the state. For additional information concerning habitat, *see* under *E. b. blennioides.*

Years 1955–80—Before 1950 *E. b. pholidotum* was abundant about the islands of western Lake Erie, where sometimes dozens could be taken with a small seine in an hour (OSUM 4270:55 specimens). Since 1965 collecting has resulted in capturing fewer than five in the same period of time it took to capture dozens formerly (Van Meter and Trautman, 1970:77).

EASTERN BANDED DARTER

Etheostoma zonale zonale (Cope)

Fig. 164

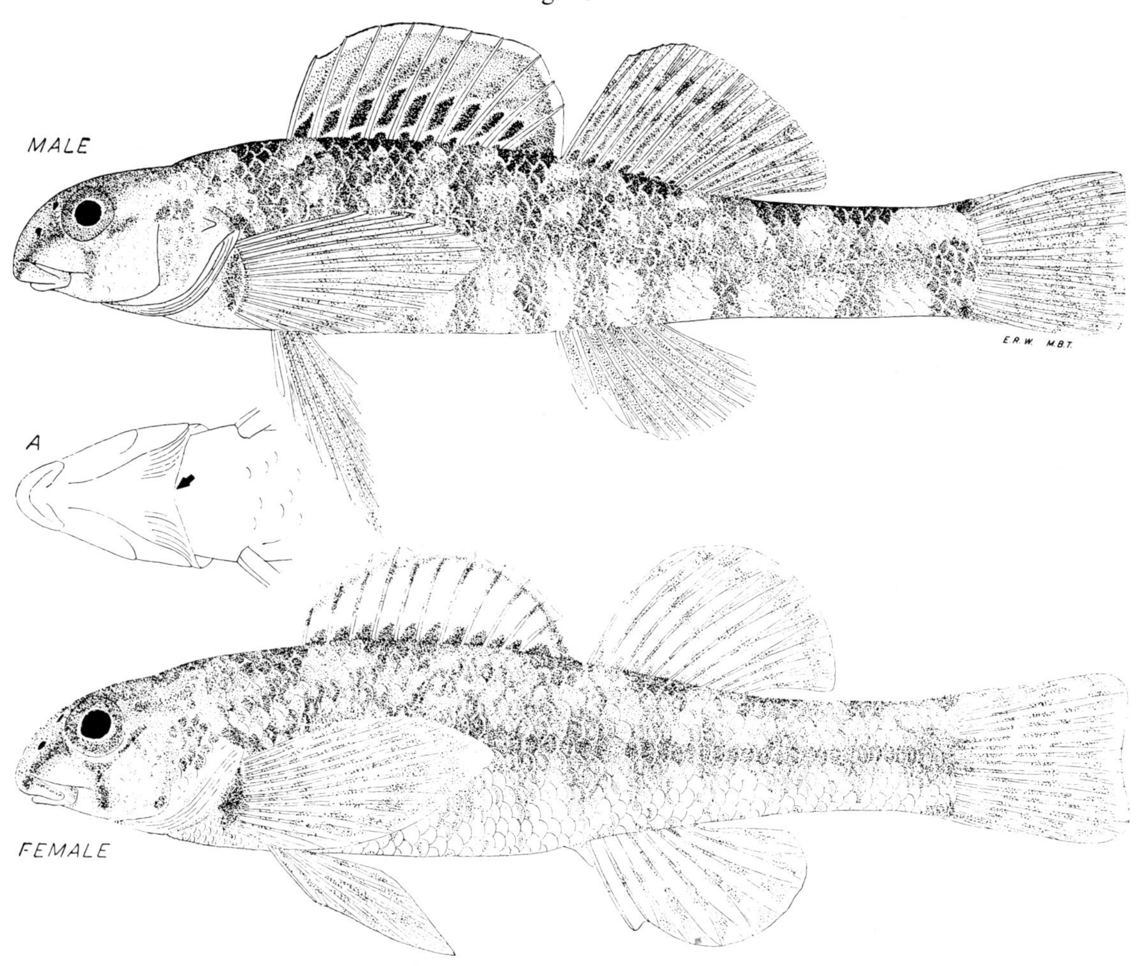

Upper fig.: Mahoning River, Portage County, O.

May 10, 1941. 46 mm SL, 2.2″ (5.6 cm) TL.
Breeding male. OSUM 3147.

Lower fig.: Same locality, date and OSUM number.

Breeding female. 38 mm SL, 1.8″ (4.6 cm) TL.

Fig. A: ventral view of head; arrow points to very broadly joined gill covers.

Identification

Characters: *General—See* dusky darter. *Generic—See* central johnny darter. *Specific*—Two anal spines. Belly scaled. *No* groove separating tip of upper jaw from snout. Snout blunt and rounded; mouth horizontal. Ventral surface of head flattened, giving head a triangular appearance in cross section. Gill covers very broadly joined with membrane across isthmus, fig. A., arrow. Cheeks and opercles scaled. Dorsal spines 10–11, extremes 9–12. Complete lateral line usually with 43–47 scales, extremes 42–53. Color predominantly yellow-green or green.

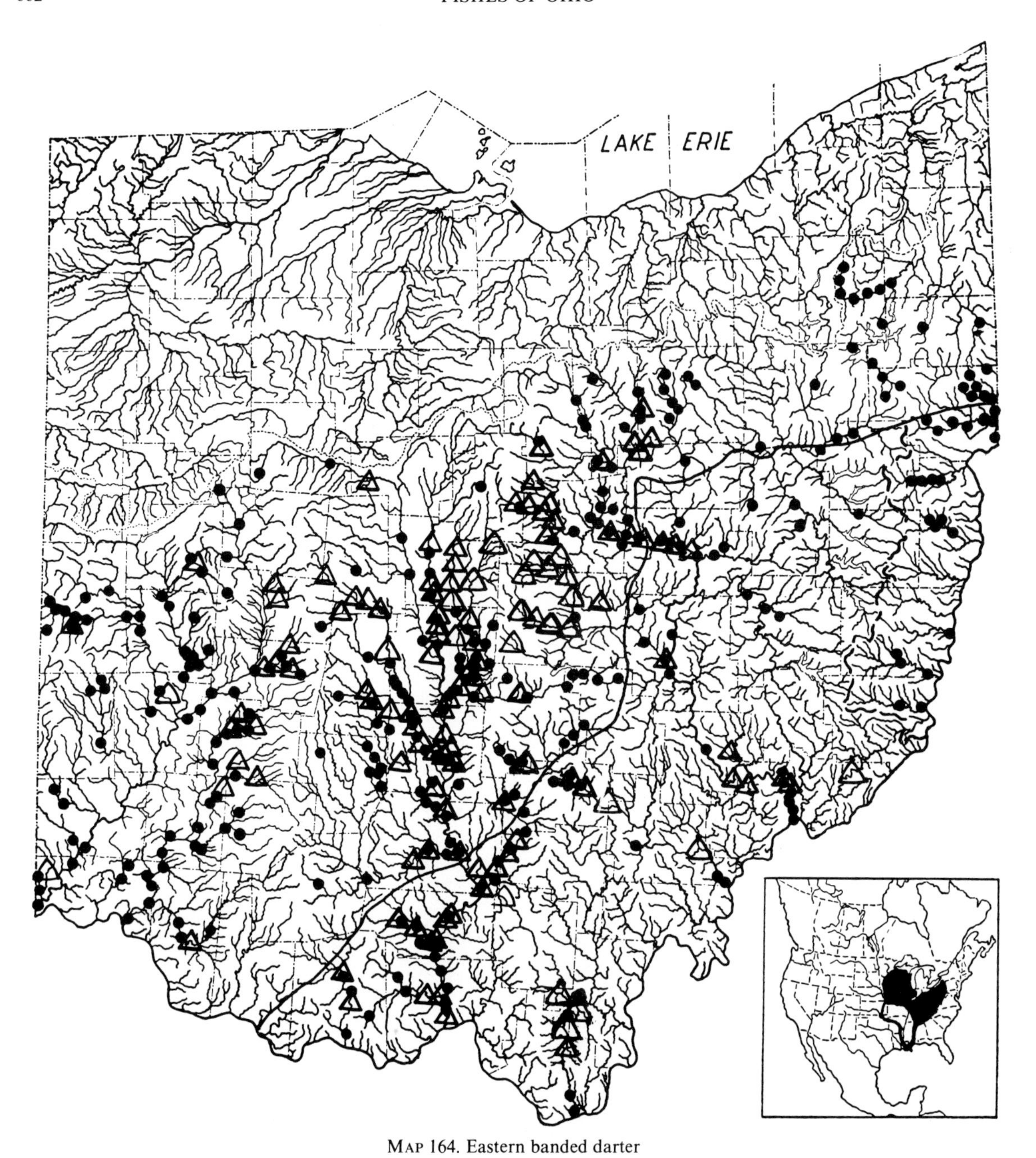

MAP 164. Eastern banded darter

Locality records. ● Before 1955. △ 1955–80. ~ Glacial boundary. /\ Flushing Escarpment.
Insert: Black area represents range of this subspecies, outlined area of other nominal subspecies.

Differs: No other species of darter in Ohio has combination of: two anal spines; no groove separating tip of upper jaw and snout; gill covers very broadly connected with membrane; normally scaled belly; complete lateral line with 42–53 scales; color yellow-green or green.

Most like: Greenside darter. **Superficially like:** Johnny and channel darters.

Coloration: *Non-breeding adults*—Dorsally green, olive- or yellow-green; sides greenish-yellow; ventrally whitish-green, yellowish-white or silvery-white. Usually 6, extremes 4–7, quadrate, dark brownish-green saddle-bands cross dorsal ridge. Usually 8–13 horizontal markings on sides of body which tend to form, or do form, vertical bands. Tear-drop usually prominent. In many specimens, both dorsals have a basal bar of rusty-red; remaining portions of fins transparent, with olive-green on webbing and rays and/or with spottings on soft dorsal which tend to form oblique rows. Caudal olive with dark spots which form wavy vertical bands. Lower fins transparent, sometimes vaguely spotted. Pectorals with wavy bands. *Breeding male*—8–13 dark green bands encircle body. Entire head and breast intensely green. A broad basal bar of brick-red on dorsals; remainder of dorsals green, except for a pale border on spinous dorsal. Caudal olive-green. Anal and pelvics emerald-green, with tips of spines and anteriormost rays whitish and swollen. *Breeding female*—Less intensely colored than male; greens are largely replaced with olives on back and sides, and whitish on breast; bands faint and/or reduced to vertical, lateral blotches; fins contain little green, are more transparent and spotted. *Young*—Similar to adults, but bands are absent; lateral blotches more triangular and they tend to form a broken lateral band; basic colors more olive and yellow; belly silvery-white; fins transparent with or without spots; tear-drop and saddle-bands very distinct.

Lengths: *Young* of year in Oct., 0.8″–1.7″ (2.0–4.3 cm) long. *Adults*, usually 1.4″–2.5″ (3.6–6.4 cm). Largest specimen, 3.2″ (8.1 cm) long.

Hybridizes: with rainbow darter.

Distribution and Habitat

Ohio Distribution—Preserved specimens of banded darters, collected from streams in Mahoning, Franklin and Hamilton counties, and accepted literature records for Knox (Parker, Williamson, Osburn, 1899:30), Montgomery, Licking and Belmont (Osburn, 1901:98) counties present the same general distributional pattern for the banded darter in the Ohio drainage prior to 1900 as has existed since 1925. During the 1925–50 period the largest concentrations of banded darters occurred in a broad band along the glacial boundary and in the two Miami River systems of southwestern Ohio. In unglaciated territory moderately-sized populations were present in such glacial outlet streams as the Scioto and Muskingum rivers. Small populations were present in the streams east of the Flushing Escarpment. It was almost absent from the mature, unglaciated streams of southeastern Ohio which had been relatively undisturbed by glacial action.

Osburn (1901:98) reported that in 1898 he captured banded darters in the Huron River, Erie County, at Milan. This is the only record for any Lake Erie tributary (Hubbs and Lagler, 1947:88). As discussed in the bluebreast darter under Distribution, Osburn considered that the records of the banded and bluebreast darters for the Huron River are of doubtful validity. Since 1920 all of the many attempts to collect the banded darter in the Huron River have failed. Because of lack of specimens and factors mentioned above, this Lake Erie drainage record is considered to be invalid.

Habitat—In the Ohio drainage banded darters were found on the riffles of streams of all sizes, from the smaller brooks to the largest inland rivers. In the spawning season the largest concentrations occurred in riffles of streams which had moderate and high gradients, where the riffle width was less than 50′ (15 m), its depth less than 2′ (0.6 m), and where many of the stones, boulders and bedrock contained an abundance of algae and aquatic mosses in which the females deposited their eggs. Such aquatic growths were often profuse in those glacially rejuvenated stream sections which contained large quantities of limestone and igneous gravel and boulders, but were sparse in sections where the gravel and boulders consisted of a soft, crumbling type of sandstone. During the 1925–50 surveys a direct correlation existed between the yearly numbers of breeding and wintering banded darters and the amount of algae and aquatic mosses present on the riffles of Big Darby Creek in Pickaway County. After the spawning season was over most of the darters in the smaller waters drifted downstream to winter in deeper waters.

Years 1955–80—The sizes of the populations and distributional pattern of the banded darter in the Ohio drainage remained quite constant during this period. Their numbers in a given locality were usually determined by the amount of algae present, and this was especially true throughout the spring and summer.

The banded darter appeared to be rather tolerant of several types of organic pollutants. It was usually present in larger numbers farther upstream and in smaller tributaries than was its larger associate, the greenside darter (*see* under *E. blennioides*).

For systematics and distribution of the banded darter, *see* Tsai and Raney, 1974.

VARIEGATE DARTER

Etheostoma variatum Kirtland

Fig. 165

Upper fig.: Big Darby Creek, Pickaway County, O.

April 21, 1942. 71 mm SL, 3.5″ (8.9 cm) TL.
Breeding male. OSUM 4862.

Lower fig.: Same locality, date and OSUM number.

Breeding female. 50 mm SL, 2.4″ (6.1 cm) TL.

Fig. A: ventral view of head; arrow points to the rather acute angle of the moderately joined gill covers.

Identification

Characters: *General—See* dusky darter. *Generic —See* central johnny darter. *Specific*—Two anal spines. Belly normally scaled. No groove separating tip of upper jaw and snout. Dorsal profile an even curve from upper jaw tip to dorsal origin. Mouth almost horizontal. Ventral surface of head flattened, giving head a triangular appearance in cross section. Gill covers moderately or rather broadly joined with membrane across isthmus, fig. A, arrow. Cheeks and opercles scaleless. Dorsal spines usually 12–13, extremes 11–14. Complete lateral line usually with 50–57 scales, extremes

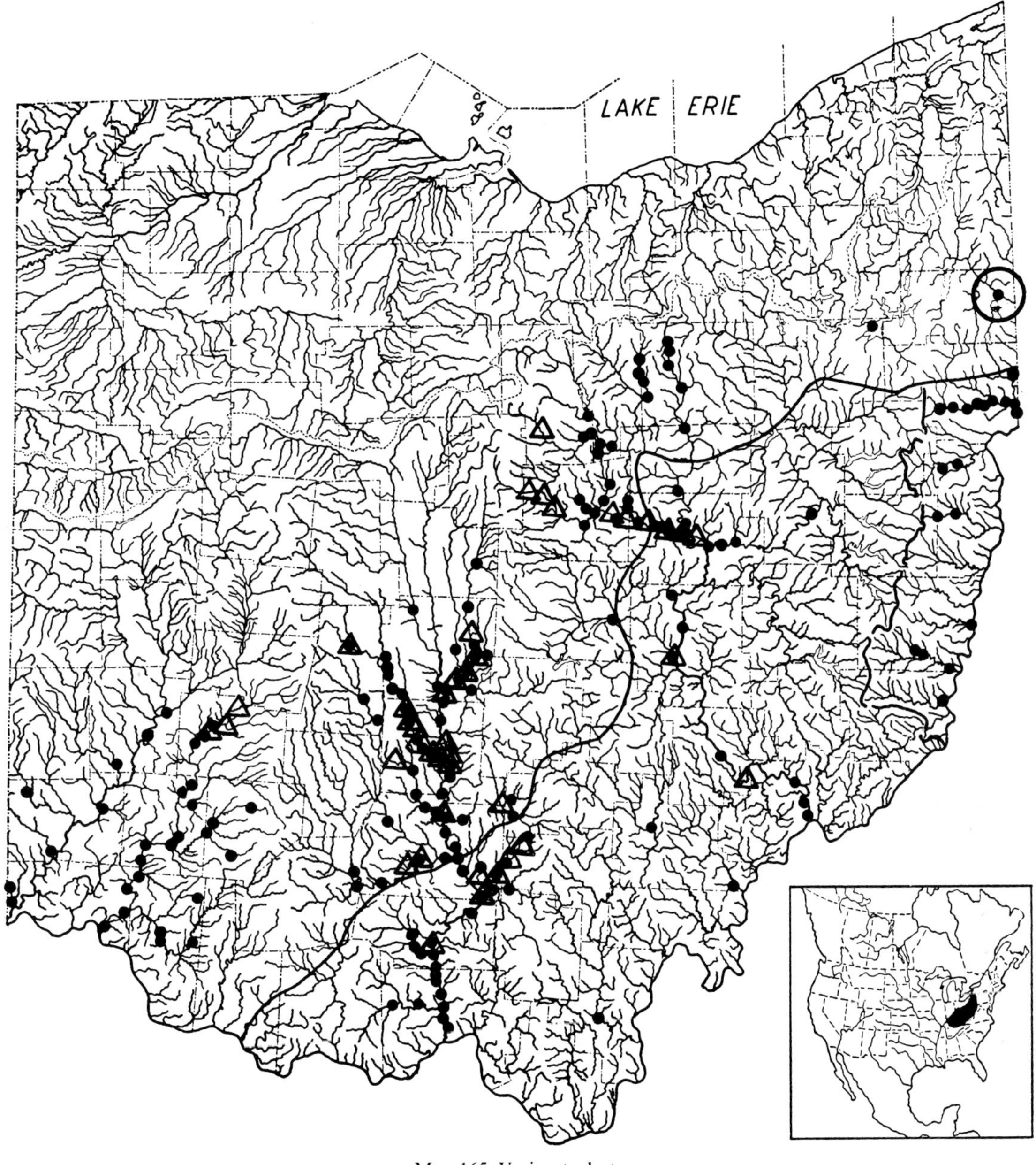

MAP 165. Variegate darter

○ Type locality, Mahoning River, Mahoning County.
Locality records. ● Before 1955. △ 1955–80. ~ Glacial boundary. /\ Flushing Escarpment.
Insert: Unoccupied areas occur within this range.

48–60. Large immature and adult males have pairs of orange spots on sides between the darker bars.

Differs: No other species of darter in Ohio has combination of: two anal spines; no groove separating tip of upper jaw and snout; gill covers rather broadly connected; belly normally scaled; complete lateral line with 48–60 scales; males with yellow and orange-red spots.

Superficially like: Young greenside darter.

Coloration: *Non-breeding adult male*—Dorsally dark greenish-olive; ventrally light green with breast bluish-white or blue; 4–6 saddle-bands cross back; the first begins at pectoral base, extending obliquely upward to before spinous dorsal origin; second, often obsolete, situated under middle of spinous dorsal; third, usually bold, under posterior end of spinous dorsal; fourth, often obsolete, under middle of soft dorsal; fifth, usually bold, under posterior end of soft dorsal; sixth, usually bold, on caudal peduncle; 4–6 blue-green bands encircle posterior half of body; interspaces between these bands golden-yellow with scattered orange-red spots. A bright orange-red bar along lower sides extends from behind pectoral fins to above anus. Spinous dorsal with six horizontal bars; colors of these bars from base to fin margin are: red-brown; transparent; blue-black; transparent; orange-red; transparent. Soft dorsal a dusky olive-green speckled with orange-red dots which tend to form rows; a dusky-blue margin. Base and posterior edge of caudal fin dusky blue-black; central portion orange-green with red spots. Anal and pelvics dusky-green; tips of spines and anterior rays whitish. Pectoral fin olive with rows of reddish or brown spots. *Breeding male*—Exceedingly brilliant; ventral surface of head and breast a deep blue-black; horizontal bar on lower sides scarlet centrally, bordered dorsally and ventrally with yellow-orange; vertical bars on sides very green-blue; interspaces bright golden-yellow, the reddish spots very conspicuous. *Adult female*—Duller colored than male; the orange, blue, green, and red largely replaced with olives and greenish-browns; bands on sides ill-defined or mere blotches; spinous dorsal with little color; spotting of dorsal and caudal more distinct; lower fins more transparent; breast and ventral surface of head blue-white; brown saddle-bands more prominent. *Young*—Similar to female but duller; fins more transparent; body more olive and less green; no orange or red; belly and breast silver-white; saddle-bands dark and very conspicuous.

Lengths: *Young* of year in Oct., 1.2″–2.3″ (3.0–5.8 cm) long. Around **1** year, 1.4″–2.7″ (3.6–6.9 cm). *Adults*, usually 2.0″–3.5″ (5.1–8.9 cm). Largest specimen, 4.3″ (11 cm) long. For growth-age relationship, *see* Lachner, Westlake, and Handwerk (1950:92–111).

Distribution and Habitat

Ohio Distribution—Kirtland (1841A:274–75) described this species from specimens captured before 1838 in the Mahoning River, Ohio, presumably at Loveland's Ripple near Youngstown (Kirtland, 1850L:21). In the latter publication Kirtland also mentioned obtaining specimens from the Scioto River, (from his friend, Joseph Sullivant, Franklin County [?]). Kirtland's records, Osburn's (1901:98) for the Ohio River, Belmont County, and preserved specimens from Knox (CHNM:1678), Franklin (CAS:8863), and Hamilton (SMHN:3670) counties, indicate roughly the areas in which since 1920 the variegate darter was present. Between 1920–50 the species was numerous in the Little Miami River system and along the glacial boundary; small populations occurred east of the Flushing Escarpment, and in the glacial outlets flowing through unglaciated territory, such as the Hocking and lower Scioto and Muskingum rivers. None was found in the Ohio River since 1920, but it surely was present in that stream before ponding and silting, especially in sections of higher gradients above Portsmouth.

Habitat—The variegate darter inhabited in greatest abundance those sections of moderate- or large-sized streams where the riffle current was rapid, riffle width 25′–75′ (7.6–23 m) and depth 1′–5′ (0.3–1.5 m), and the bottom of clean, glacial rubble and boulders. Few or no variegate darters were present on riffles whose boulders contained much silt, or were covered with coal mine wastes or other pollutants. Most of the larger darters moved downstream in late summer and fall, to winter in deeper riffles which had a slower current. Upon a few occasions in early spring, when the rivers were in high flood, I fin-clipped individuals which I caught along the water's edge in flooded fields far from permanent riffles; shortly thereafter I recaptured them several hundred feet upstream from where I had originally caught them. Presumably, these individuals, when taken on each occasion, were migrating upstream.

Years 1955–80—During their 1897 survey of the Scioto River system in Franklin County, Osburn and Williamson (1898:16 and 20) found the variegate darter to be rare. These men collected the species only in Big Walnut, Blacklick and Little Walnut creeks. They told me that they had seldom captured more than five individuals at any seining station.

Between 1925–30 it was possible to capture a dozen while seining favored riffles, and by 1940 several dozens could be taken from the same riffles. By 1960 the species had become so numerous that occasionally hundreds could be caught on one riffle during an hour's seining.

Through monies supplied by the Ohio Division of Wildlife and The Ohio State University, I conducted a central Ohio stream survey between the years 1960–65, aided by several enthusiastic students. We found no extension of the range of the variegate in Ohio. Neither did Bruce May for central Ohio while collecting data for his master's thesis (1966:9). For reproductive biology, larval development and migration *see* May (1969:85–92).

There appear to be only two possible locality records for the variegate darter in the Scioto system from Columbus upstream. Before 1950 Kirtland (1850L:21) had "received specimens from the Scioto," presumably from the vicinity of Columbus. In 1928 a fisherman showed me one which he had just captured in a riffle on the Scioto River near its junction with Indian Run, northwestern Franklin County. Despite an immediate and thorough seining of this riffle, and several times since, no more have been taken. It is believed that the specimen had escaped, or was discarded, from a minnow bucket and had been recaptured. Kirtland (1850L:21) stated that this species was "preferred to common minnows" for fish bait.

No variegate darter has ever been recorded from the Olentangy River. The riffles of the lower half of this stream appear to be similar to those of Big Walnut and Big Darby creeks, and the fish associates, present on riffles in these creeks, are similar to those in the Olentangy. We once believed that some pollutant had eliminated the variegate from the Olentangy, and to prove this assumption Raymond F. Jezerinac, Donald Veth, and I, on April 18, 1963 captured 239 adults about to spawn in Big Darby Creek, releasing them in the Olentangy one mile (1.6 km) northwest of Worthington, Franklin County. They were released on a riffle that was remarkably similar to the one from which they came and where the spotted darter occurred. Investigations disclosed that no variegates could be taken a month after liberating the spawning adults, nor were young ever found. Such introduction experiments should be repeated; and if they are unsuccessful, attempts should be made to learn why.

SPOTTED DARTER

Etheostoma maculatum Kirtland

Fig. 166

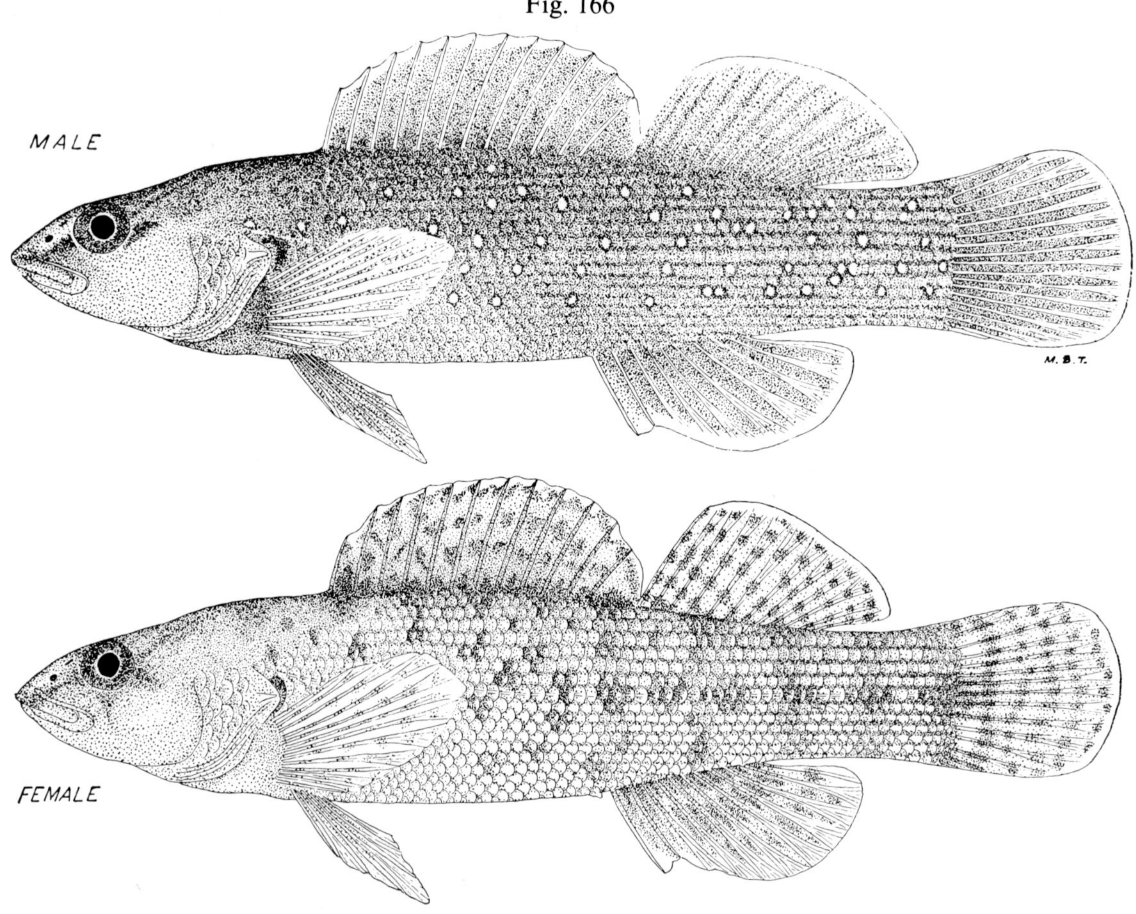

Upper fig.: Big Darby Creek, Pickaway County, O.

Oct. 22, 1943. 62 mm SL, 2.8″ (7.1 cm) TL.
Adult male. OSUM 5880.

Lower fig.: Same locality as above.

Nov. 4, 1943. 50 mm SL, 2.3″ (5.8 cm) TL.
Adult female. OSUM 5917.

Identification

Characters: *General—See* dusky darter. *Generic—See* central johnny darter. *Specific*—Two anal spines. Belly normally scaled. No groove separating tip of upper lip and snout. Snout sharp; dorsal profile from tip of upper jaw to eye almost straight. Snout length about as long as diameter of eye in fishes less than 2.0″ (5.1 cm) long; snout longer than eye diameter in larger fishes. Head compressed laterally, the ventral surface evenly rounded. Gill covers not connected across isthmus with membrane. Cheeks scaleless; opercles scaled. Tail rounded. Dorsal spines usually 12, extremes 11–13. Incomplete lateral line usually with 57–62 scales, extremes 54–64; 1–10 pores missing posteriorly. Whitish margin present on soft dorsal, caudal and anal of large young and adults.

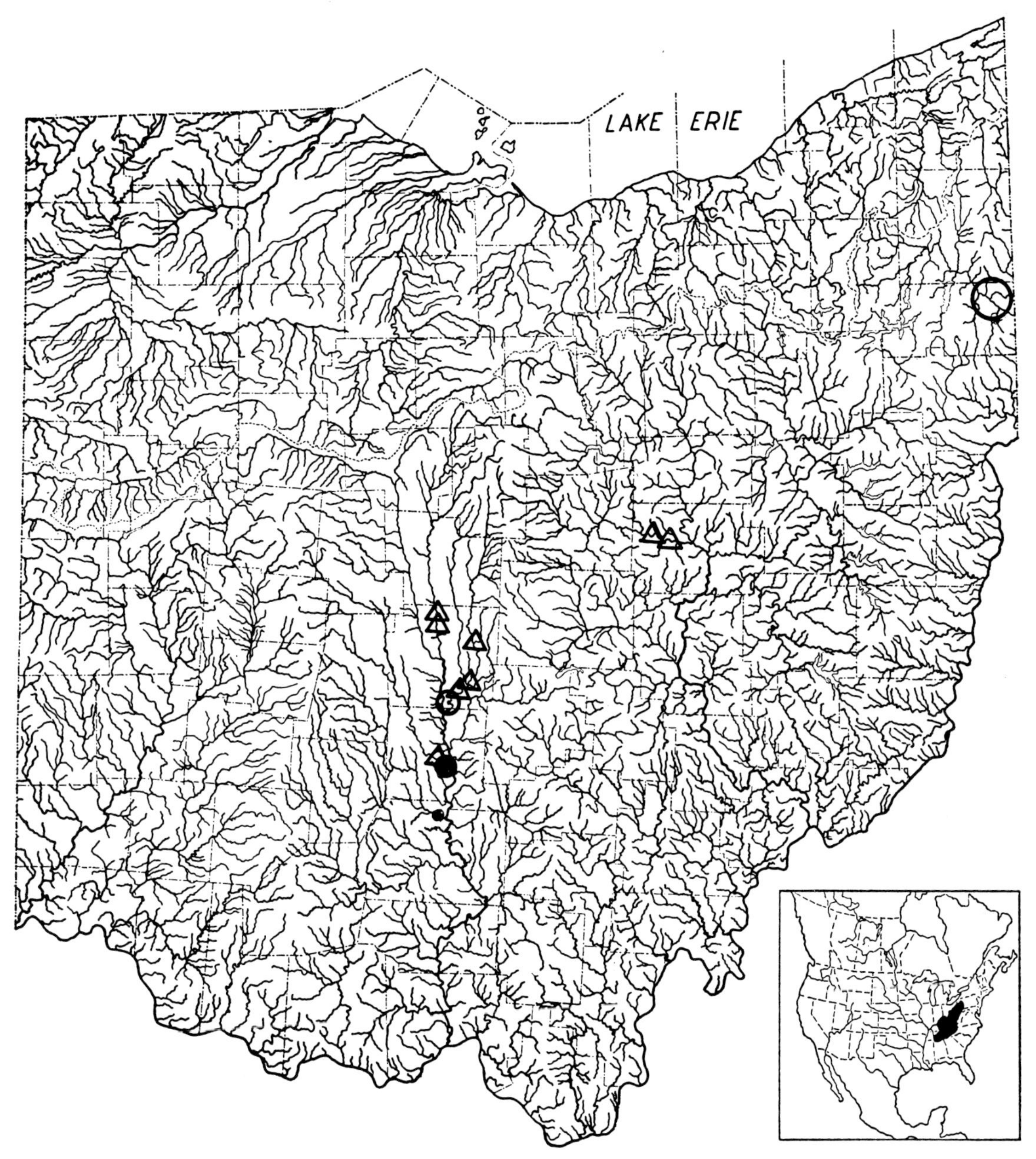

MAP 166. Spotted darter

○ Type locality, Mahoning River, Mahoning County, specimens collected in 1838 or before.
Locality records. ○ 1897. ● 1930–48. ● 1956. △1955–80.
Insert: Unoccupied areas occur within this range.

Differs: Bluebreast darter averages fewer lateral line scales; tail straight or emarginated posteriorly; snout blunt and rounded; eye larger; soft dorsal, caudal, and anal fins have dusky margins. Tippecanoe darter has fewer lateral line scales; midline of belly partly or entirely scaleless between pelvics and anus. No other species of darter in Ohio has combination of: two anal spines; no groove separating tip of upper jaw and snout; gill covers not connected with membrane; belly normally scaled; incomplete lateral line; sharp snout; whitish border to ventral fins.

Most like: Bluebreast and Tippecanoe darters.

Coloration: *Non-breeding adult male*—Dorsally

dark olive-green; sides lighter; ventrally pale olive-green with breast whitish or light greenish-slate. Scattered over each side of body are 30–80 carmine-red scales with dusky borders. Dorsal and ventral edges of body scales dusky, resulting in horizontal lines between the scale rows. Usually a narrow, light band extends from snout tip, along dorsal ridge to dorsal origin. Dorsal fins golden-olive, their margins lighter. Caudal olive-brown, bordered with lighter. Anal olive, marginated with lighter. Pelvics olive-green, margined with lighter. Pectorals pale olive. *Breeding male*—Like adult male, except that ventral surface of head and breast is blue-black; whole body has a dusky tinge; narrow line along dorsal ridge a conspicuous white, yellow, or blue-green. Fins brightly colored; the white margins very conspicuous. *Adult female*—Chiefly olive-green, the sides mottled and barred with darker, with traces of vertical bands on posterior half of body; no carmine spots; horizontal lines between scale rows often distinct; breast slaty-white; dorsals and caudal fins spotted, the spots forming horizontal rows or vertical bands. *Young*—Similar to adult female, except that the general color is a light brownish-olive; 4–7 saddle-bands often evident; spotting on sides rather faint; fins spotted or largely transparent.

Lengths: Five *young* of year in Oct., 1.2″–1.7″ (3.0–4.3 cm) long. *Adults*, usually 2.2″–3.0″ (5.6–7.6 cm). Largest adult, 3.1″ (7.9 cm) long.

Distribution and Habitat

Ohio Distribution—Kirtland (1841A:276) described this species from fishes he (1850K:29) collected in 1838 or before "at Loveland's Ripple, in the Mahoning River, near the village of Youngstown." The Museum of Comparative Zoology of Harvard has specimens collected by Kirtland and Baird at "Poland, Mahoning County." The UMMZ (86343) has a specimen collected by Baird in the Mahoning River tributary of Yellow Creek, which presumably was taken in 1853 when Baird and Kirtland collected fishes there. The USNM (1319) has twelve specimens from Yellow Creek. Recently Raney (1938:67–68) found this darter in northwestern Pennsylvania within 20 miles (32 km) of the above Ohio localities, and in the Beaver River system of which the Mahoning River is a tributary. That section of the Mahoning River where Kirtland captured his specimens is now almost devoid of fish life because of steel mill and other pollution.

On June 26, 1897, Williamson and Osburn (Osburn, 1901:99) collected a ripe female (OSUM 6) in Big Walnut Creek near Lockbourne, Franklin County, the first central Ohio record.

Despite the many annual collections made in central Ohio since 1920, no spotted darter was taken until October 27, 1930 when Edward L. Wickliff and I collected a small young (OSUM 613) in Big Darby Creek, Pickaway County, near Fox, Ohio. Although several collections were made at Fox each year after 1930, the next specimens (7) were not taken until 1943 (OSUM 5917 and 5880); then no more until 1947 when Mary A. Trautman and I took six (OSUM 7415). In 1948 we took seven more (OSUM 8411 and 8423); and on Nov. 30, 1956, Donald I. Mount took one. Attempts have failed to collect the species elsewhere in Big Darby Creek. On May 26, 1956 Mount collected a large male (OSUM 9501) in Deer Creek, Union Township, Ross County.

Habitat—Kirtland's (1850K:29) description of conditions in the Loveland Ripple of the Mahoning River is similar to conditions existing on the Big Darby riffle; both riffles had a rapid current flowing over a gravel- and boulder-strewn bottom. All specimens taken on the Big Darby riffle were captured in the central portion of the riffle, where the current was approaching, or had attained, maximum swiftness. Raney and Lachner (1939:157–65), in describing the spawning habits of this darter, also noted its presence in a similar type of habitat. Other species of fishes associated with the spotted darter, as given by Raney and Lachner, are the same as those found associating with this darter on the Big Darby riffle.

Years 1955–80—During this period spotted darters were captured in the Olentangy River and Big Walnut Creek, two tributaries of the Scioto River. None had been recorded from either stream during the present century; however, in 1897 Williamson and Osburn captured one in Big Walnut Creek. It was most unexpected to discover small populations in the Olentangy River, especially since its usually more numerous associate, the variegate darter, has not been taken in that stream.

Another spotted darter population was discovered in the lower section of the Walhonding River, a tributary of the Muskingum River.

No marked increase in population size was noted in any locality either in those discovered before 1950 or since, although there may be considerable variations in the numbers from one year to another.

BLUEBREAST DARTER

Etheostoma camurum (Cope)

Fig. 167

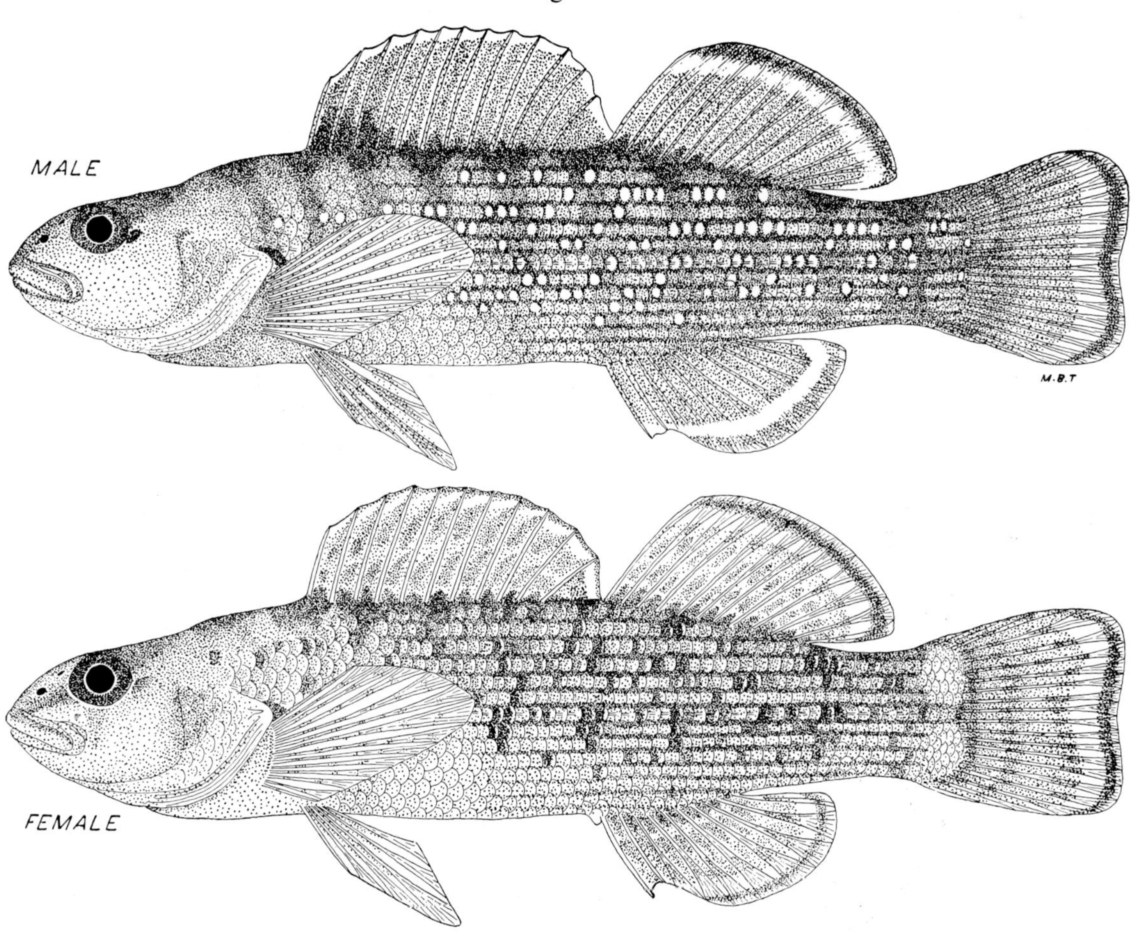

Upper fig.: Big Darby Creek, Pickaway County, O.

Sept. 26, 1939. 54 mm SL, 2.5″ (6.4 cm) TL.
Adult male. OSUM 552.

Lower fig.: Same locality, date and OSUM number.

Adult female. 42 mm SL, 2.1″ (5.3 cm) TL.

Identification

Characters: *General—See* dusky darter. *Generic —See* central johnny darter. *Specific*—Two anal spines. Belly normally scaled. No groove separating tip of upper jaw and snout. Snout short, rounded, and blunt. Profile from tip of upper jaw to above eye an even curve. Snout length shorter than eye diameter in young less than 1.8″ (4.6 cm) long; snout length in larger fishes about as long as eye diameter. Head moderately compressed laterally; ventral surface rounded; not flattened ventrally as in the variegate darter. Gill covers not connected across isthmus with membrane. Cheeks scaleless; opercles scaled. Posterior edge of tail straight in young, emarginated in adults. Dorsal spines usually 11–12, extremes 10–13. Scales in lateral line usually 53–58, extremes 50–60, with 1–10 pores missing posteriorly. Dorsal, caudal, and anal fins with a margin or submargin of dusky.

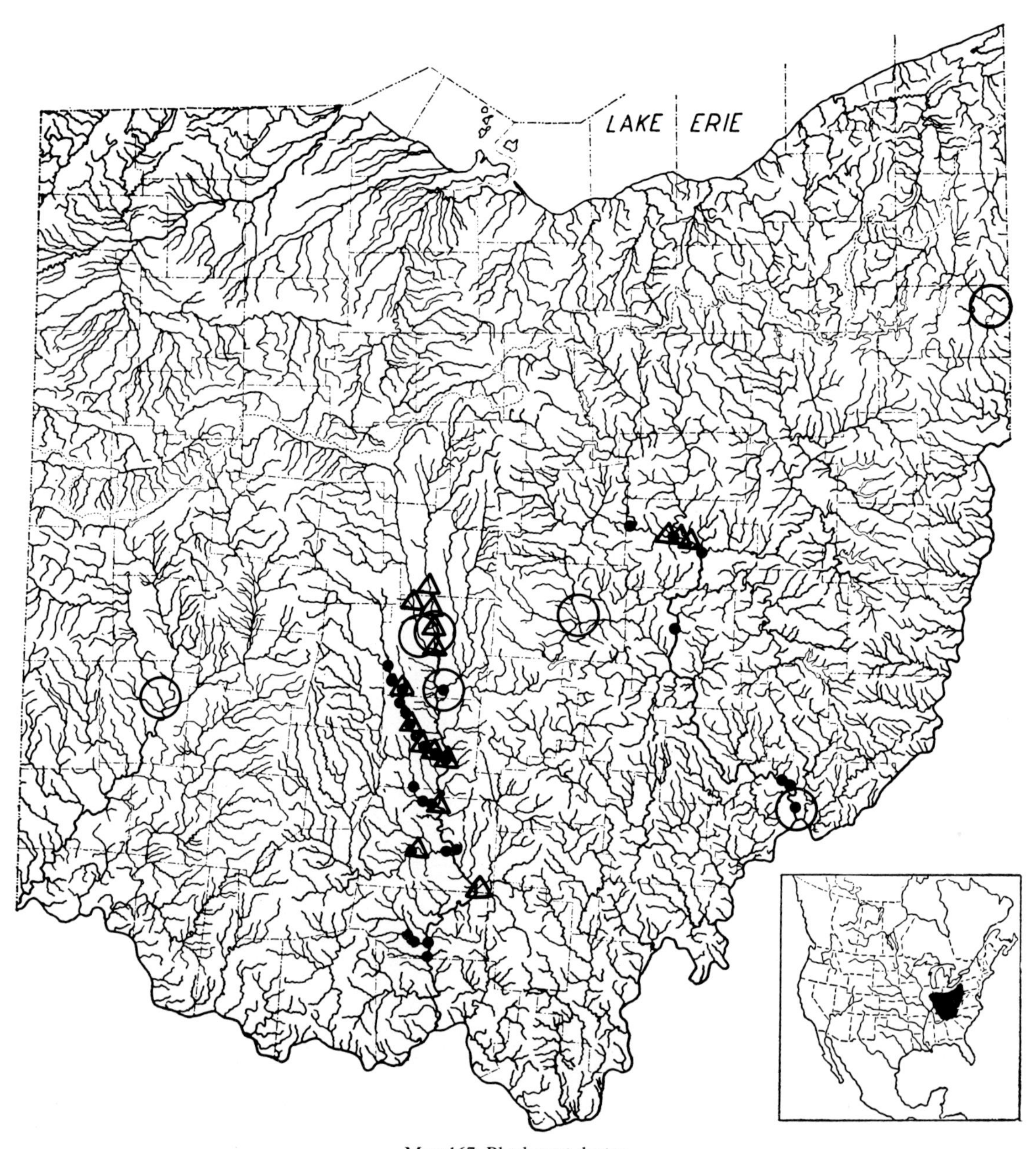

MAP 167. Bluebreast darter

Locality records. ○ Before 1900. ● 1920–50. △ 1955–80.
Insert: Unoccupied areas within this range appear to be increasing in size and number.

Differs: Spotted darter has a greater average number of scales in the lateral series; a rounded caudal; more pointed snout; smaller eye; vertical fins margined with lighter. Tippecanoe darter has midline of belly partly or entirely scaleless. No other species of darter in Ohio has combination of: two anal spines; belly normally scaled; incomplete lateral line with 50–60 scales; rounded snout; dusky margins on the vertical fins.

Most like: Spotted and Tippecanoe darters.

Coloration: *Non-breeding adult male*—Dorsally dusky olive-green; sides light olive-green; ventrally pale olive; the breast pale blue, green, or white. Between 7–10 faint saddle-bands of dark olive-brown astride dorsal ridge. Between 8–12 vague vertical bands on sides; most distinct posteriorly. Between 70–300 carmine scales scattered over each side of body; these scales occurring singly or in horizontal rows of 2–6. Two light areas on caudal base, one above, the other below the lateral line. Dorsal and ventral edges of scales on sides of body dusky, resulting in horizontal, dusky lines between scale rows. A golden-orange bar extends from pectoral base to above anus. Webbing of dorsal fins dusky-olive or reddish-brown; margin of spinous dorsal whitish or orange-red; soft dorsal, caudal, and anal fins margined with dusky, below which is a whitish submarginal band. Pelvics and pectorals olive. *Breeding male*—Has belly and ventral surface of head bluish; breast blue-black; carmine spots very brilliant; 8–12 bands on sides often conspicuous. *Adult female*—Like adult male, except that colors are subdued; blues and greens duller or replaced by olives and browns; carmine spots replaced by brown; no orange-red bar on lower sides, groups of dark scales on sides; vertical bands, horizontal rows, and two light spots on caudal base often very prominent; fins vaguely spotted, the spots tending to form rows and wavy bars; vertical fins margined with dusky, or faintly margined with whitish and having submargin of dusky. *Young*—Like adult female, except that color pattern is not so definite; colors are mostly grays, browns, olives; some silvery overcast especially ventrally; fins more transparent; spots and dusky margins on fins present or absent.

Lengths: *Young* of year in Oct., 1.2″–1.7″ (3.0–4.3 cm) long. *Adults*, usually 1.5″–2.8″ (3.8–7.1 cm). Largest specimen, 3.2″ (8.1 cm) long.

Hybridizes: with Tippecanoe darter; one specimen.

Distribution and Habitat

Ohio Distribution—There are preserved specimens in the U. S. National Museum (1320) which were taken by S. F. Baird in Yellow Creek of the Mahoning River system near Poland, Ohio in August, 1853.* Recently Raney (1938:68) found the bluebreast in the Beaver River system of Pennsylvania of which the Mahoning River is a tributary. No specimens have been taken in the Mahoning River system in Ohio in recent years presumably because pollution had destroyed the former habitat of the species. On August 31, 1888 Charles H. Gilbert and James A. Henshall (unpublished notes) collected the bluebreast in the Muskingum River, Washington County, at Lock No. 2, a locality where it has been repeatedly taken since 1929. In 1898 Osburn (1901:99) captured it in three streams of central Ohio, and in 1899 in the North Fork of the Licking River at Newark and the Stillwater River near Dayton. There is no reason to doubt the Stillwater River record despite our failure to collect it in the Great Miami River system during the 1925–50 surveys. The bluebreast probably was present in the Ohio River, and was more abundant in such large, inland streams as the Muskingum, before these streams were dammed and the fast flowing riffles eliminated.

Osburn (1901:99) recorded in 1899 the bluebreast as present in the Huron River at Milan, in Erie County. This is the only record for the Great Lakes drainage and as such has been published repeatedly (Hubbs and Lagler, 1947:88). I consider this record to be invalid because in 1930 R. C. Osburn informed me that he considered it to be of doubtful validity because of some clerical errors which occurred during the publication of his "Fishes of Ohio." Repeated collections in the Huron River have failed to produce this species or several of its closest associates.

During the 1925–50 surveys the bluebreast was captured repeatedly in the lower Muskingum River. It was moderately numerous in the Walhonding River before 1930, but since 1940 only one has been taken in that stream despite repeated collecting. The annual numbers of bluebreasts fluctuated greatly in the Scioto drainage during the 1925–50 period. In Big Darby Creek, Pickaway County, their numbers varied from uncommon during such a year as 1944, to their being the most abundant species of darter on the riffles in such a year as 1948.

Habitat—The bluebreast was numerous only in those sections of moderate- or large-sized streams where turbidity was unusually low. In such streams it inhabited the faster flowing and deeper riffles whose bottoms contained many large stones, boulders, and some sandy gravel, behind, beside or under which the species remained. As a species the bluebreast occupied those portions of the riffle where the current was less swift than were those areas inhabited by the spotted darter, and where the current was more swift than those sections frequented by the variegate darter. Fluctuations in abundance seemed to be correlated with stream flow and turbidity, for the largest populations were present during or following those years when there were neither extended periods of low stream levels

* The bluebreast was not described until 1870, Cope (1870:265) describing it from specimens taken from the Cumberland River in Tennessee.

nor floods which caused the waters to remain turbid for long periods of time.

My studies of the fishes of Big Darby Creek between the years 1925 and 1950 abundantly demonstrate the migratory nature of the bluebreast and several other darter species. The bluebreast was usually present in all suitable riffles of Big Darby throughout the breeding season, from its confluence with the Scioto River upstream to its junction with Little Darby, a distance of approximately 32 miles (51 km). A diminution of numbers usually became apparent in the upper half of this 32-mile (51-km) section during September, and by December few individuals could be found in the upper half, and especially during drought years of unusually low water levels. Coincidental with this decrease in numbers in the upper half there was a marked increase in numbers in the lower half. This was particularly true in the stream section adjacent to the village of Fox, where during some years the late fall and winter concentrations were 5 to 15 times greater than were the summer numbers. Such shifts in abundance were well demonstrated in 1948: during spring and summer of that year the largest number of bluebreasts recorded in a day from the Fox riffles was only 16 (August 11), whereas on October 3 my wife and I captured 100 individuals, from which 70 adult males were preserved. These fishes were taken from a measured area in a riffle of only 900 square feet (83.6 m^2). Two days later we reseined the same area, capturing 129 individuals of which 82 were adult males. Apparently a downstream movement was in progress. A decrease in numbers in the lower half, and corresponding increase in the upper half, became evident during late winter and early spring, after water temperatures had risen above 45° F (7.2° C).

Upon several occasions I observed in spring what apparently were upstream migrations of the bluebreast and other darter species. On April 13, 1944, Big Darby was more than 8″ (2.4 m) above normal, and was widely flooding the fields in the vicinity of Fox. The scouring current was dislodging boulders. On that day my wife and I seined a shallow area in a flooded field, the area being several hundred feet from midstream where a favorite riffle was deeply submerged. In this seining area we captured eleven species of darters and a total of 78 individuals, the majority of which were immediately finclipped and released at the point of capture. Later in the day several finclipped individuals, including three bluebreasts, were recaptured 300′ to 500′ (91.4 to 152 m) upstream from the point of release, indicating an upstream movement. Presumably this movement took place in the relatively quiet shore waters of the flooded field. None was recaptured downstream.

In late September and October, when seining the deeper waters of some long, currentless pool, I have taken an occasional bluebreast or other species of riffle-inhabiting darter. I assume that these individuals were captured while they were in the process of moving downstream.

Years 1955–80—The fluctuations in numbers and general distributions of the bluebreast darter throughout the past 80 years in Ohio illustrates the fluctuations and distributions of other species of riffle darters.

During their summer survey of 1897 Osburn and Williamson (1898:17 and 30) found the bluebreast in only three streams in Franklin County: the Scioto and Olentangy rivers and Big Walnut Creek. During conversations with these men before their deaths, they stated that the bluebreast was very rare in Franklin County streams, and none had been recorded from that section of Big Darby Creek which transversed this county. By 1930 the bluebreast had become fairly numerous in some central Ohio streams and especially in Big Darby Creek, where by 1940 it was abundant.

Osburn and Williamson told me that they captured only one or two specimens in Big Walnut Creek in 1897. Edward L. Wickliff took one in the lower portion of this creek about 1926. None of these specimens appear to be extant. None have been taken in this stream since, although Big Walnut has been seined upon dozens of occasions.

On October 1, 1896 R. C. Osburn captured a specimen (OSUM 9857) in the Olentangy River. The species was not again taken in that stream until October 9, 1958 when D. I. Mount captured two specimens at the Dodridge Street dam, located immediately north of The Ohio State University campus. Since then as many as a dozen have been taken during a half hour in several localities along that stream.

In the Franklin County portion of Big Darby Creek, where Osburn and Williamson took none in 1897, the bluebreast has, since 1926, been numerous.

No reasons are apparent why one species of darter is present, or why it has decreased or increased in one tributary and not in another. The data do indicate that all riffle darter populations were low about 1900 and, in general, have become larger. We might theorize that populations have increased because of improved habitat conditions. Between 1818 and 1880 hundreds of small mill dams were erected, usually where gradients were high, thereby eliminating some of the faster-flowing riffles and in general preventing upstream migration. At first much of the humus from the cut-over forests washed into the streams smothering the substrate and at times removing oxygen. This, and later the introduction of untreated domestic and industrial wastes, apparently resulted in the low darter populations about 1900. Conditions in most stream sections have improved since. O'Shaughnessy, Hoover and other huge impoundments act largely as settling basins, and water quality below them has been improved. Even so there appear to be no plausible reasons why one species is present in one tributary and not in another, such as the absence of the bluebreast in Big Walnut and its presence in the Olentangy, and the absence of the variegate in the Olentangy and abundance in Big Darby.

For spawning behavior of the bluebreast *see* Mount (1959:240–43).

TIPPECANOE DARTER

Etheostoma tippecanoe Jordan and Evermann

Fig. 168

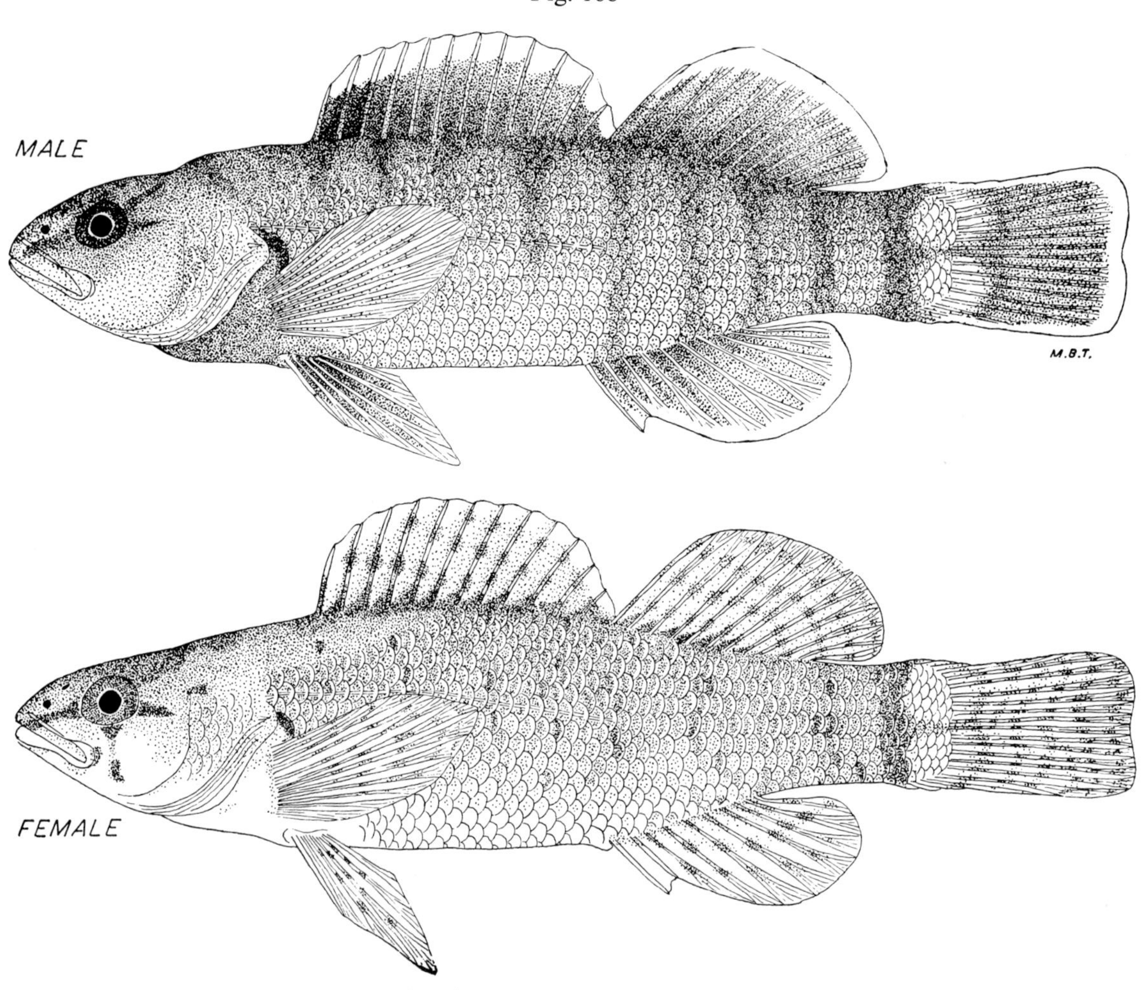

Upper fig.: Big Darby Creek, Pickaway County, O.

June 9, 1942. 29 mm SL, 1.4″ (3.6 cm) TL.
Breeding male. OSUM 5062.

Lower fig.: Same locality, date and OSUM number.

Breeding female. 20 mm SL, 0.9″ (2.3 cm) TL.

Identification

Characters: *General—See* dusky darter. *Generic—See* central johnny darter. *Specific—* Belly behind pelvics normally scaleless, the naked strip extending part or all the way to the anus. Two anal spines. No groove between tip of upper jaw and snout. Snout rather pointed. Eye small. Snout length equal to, or longer than, diameter of eye. Head moderately compressed laterally and rounded ventrally. Gill covers slightly connected across isthmus with membrane. Cheeks scaleless. Opercles scaled. Posterior edge of tail straight or emarginated. Dorsal spines usually 12–13, extremes 11–14. Pored scales in the lateral series ending

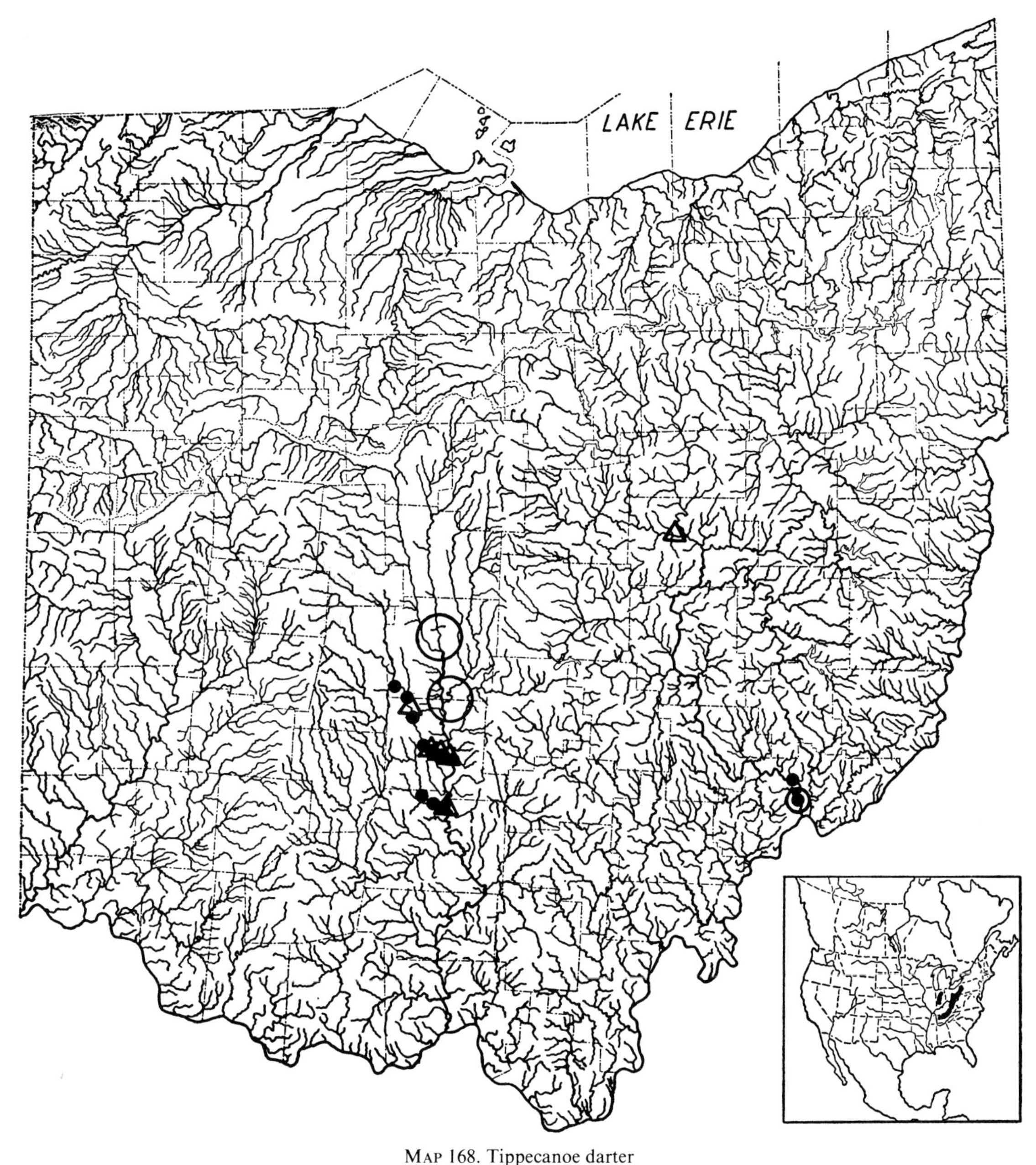

Map 168. Tippecanoe darter

Locality records. ○ Before 1900. ○ 1929. ● 1938–50. △ 1955–80.
Insert: Only in widely separated colonies within its range.

beneath soft dorsal; usually 44–49 scales in lateral series, rarely as many as 52. Two light or orange spots on caudal base encircled with dusky; these spots often very prominent. The golden-colored male is most distinctive.

Differs: Spotted darter has more lateral line scales. Bluebreast darter has larger eye; averages more scales in the lateral series; has dark horizontal streaks between scale rows; young are heavily spotted. The above two species and all other darters occurring in Ohio do not have the following combination: two anal spines; incomplete lateral line; midline of belly partly or entirely naked; 44–52 lateral scales; 11–14 dorsal spines.

Superficially like: Spotted and bluebreast darters.

Coloration: *Non-breeding adult male*—Dorsally dark golden; sides orange or yellow-golden; belly light golden; breast pale blue. Between 4–11 dark, saddle-bands cross back, these often faint or absent. Between 1–10 pale vertical emerald-blue bars on sides; anterior ones confined to upper sides; posterior ones encircle body; these bars may be entirely absent on some individuals. All specimens show two light golden or yellow spots on caudal base, one above the lateral line, the other below; these spots encircled with blue or dusky. Sides and ventral surface of head golden-orange. Basal two-thirds of spinous dorsal a rich golden-olive; marginal third a light golden; basal halves of webbing between the first three spines blackish. Soft dorsal, anal and caudal olive-golden basally, margined with light golden. Central portions of pelvics dusky-olive, remainder golden. *Breeding male*—Like adult male, except that colors are very brilliant; bright yellow-golden sides; intensely blue bands (when present); a blue-black breast. *Adult female*—Dorsally olive-brown; sides brownish-yellow; ventrally very pale slate; entire body with a faint golden overcast. Sides faintly mottled but lack bands of male. The two light areas on caudal base are very prominent; sometimes with a vertical, dusky band preceding them. A poorly or well developed tear-drop; band across snout. All fins vaguely spotted; spots on dorsals tending to form rows; those on caudal form wavy bands. *Young*—Like adult female, except that markings are not as distinct; general coloration is more grayish-brown, overlaid with silvery; fins mostly transparent.

Lengths: *Young* of year in Oct., 0.8″–1.1″ (2.0–2.8 cm) long. *Adults*, usually 1.0″–1.6″ (2.5–4.1 cm). Largest specimen, 1.8″ (4.6 cm) long.

Hybridizes: with bluebreast darter.

Distribution and Habitat

Ohio Distribution—In September of 1896 R. C. Osburn and E. B. Williamson collected one of these darters in the Olentangy River at Columbus, and in June, 1897, they caught three more in Big Walnut Creek near Lockbourne (*see* large hollow circles). These specimens became the types of *Etheostoma sciotense* (Williamson and Osburn, 1898:17–18); later the name *sciotense* was synonymized (Osburn and Williamson, 1899:33) with the older name of *Etheostoma tippecanoe* (Jordan and Evermann, 1891:3–4). On September 29, 1929 Robert B. Foster and I collected two immatures below Dam 2 in the Muskingum River, Washington County (OSUM 472; UMMZ 104321; *see* medium-sized circle).

Between the years 1920–39 Edward L. Wickliff and I made many unsuccessful attempts to capture the Tippecanoe darter in central Ohio, and between 1925–38 I yearly seined Big Darby Creek near Fox, Pickaway County, without taking it. On August 25, 1939, while making yearly routine collections in Big Darby at Fox, John Addair, William McLane and I collected 17 specimens. Since 1939 the species has been taken annually; however its yearly numbers fluctuated greatly. Only a few individuals were taken in those years following an October in which few or no young could be captured, such as in 1941, 1945, and 1950. But 50–100 could be taken in those years following an October in which dozens of young were captured such as in 1943 and 1948. Deer Creek in Ross County also has been seined yearly since 1925, but it was not until 1946 that the first specimens were taken; since 1946 the Deer Creek population has fluctuated in abundance in the same manner as the Big Darby population.

Habitat—The largest numbers of Tippecanoe darters were found on those portions of riffles having a rather slow or moderate current and a bottom of clean gravel and sand. Throughout the spawning season groups of 2–18 males were observed guarding their territories and eggs, in water between 3″–18″ (7.6–46 cm) in depth; these territories were usually at the heads or tails of riffles or along their edges where the current was gentle but sufficiently strong to keep the sandy gravel free from clayey silt. The males usually deserted their territories whenever storms caused the water to become turbid, and siltation over the sandy gravel was rapid. It was interesting to note that some turbid waters which caused Tippecanoe darters to desert their territories were not sufficient to cause territory desertion by males of other darter species which inhabited waters that flowed too rapidly to allow silt to accumulate on their territories. It is possible that desertion of territory and egg guarding by the Tippecanoe darter was a major factor in producing the great fluctuations in annual numbers. In winter the species usually retired into waters where the current was very sluggish and the depth between 2′–5′ (0.6–1.5 m).

Years 1955–80—On October 28, 1962 K. R.

Troutman and Steve Rogers collected a Tippecanoe darter in the Walhonding River, Coshocton County, one mile (1.6 km) above the Six Mile Dam. This is the only extension of its Ohio range made during the 1955–80 period.

The many investigations conducted in central Ohio since 1951 resulted in finding no populations not known previously. The riffle habitat of this darter was especially explored in Big Walnut Creek and the Olentangy River.

The numbers which could be captured on a given riffle fluctuated moderately from one year to the next; however, a few were taken on the better riffles in Big Darby Creek annually. There were no years such as 1943 and 1948, when the population was so great that 50 to 100 could be taken in a few hours.

IOWA DARTER

Etheostoma exile (Girard)

Fig. 169

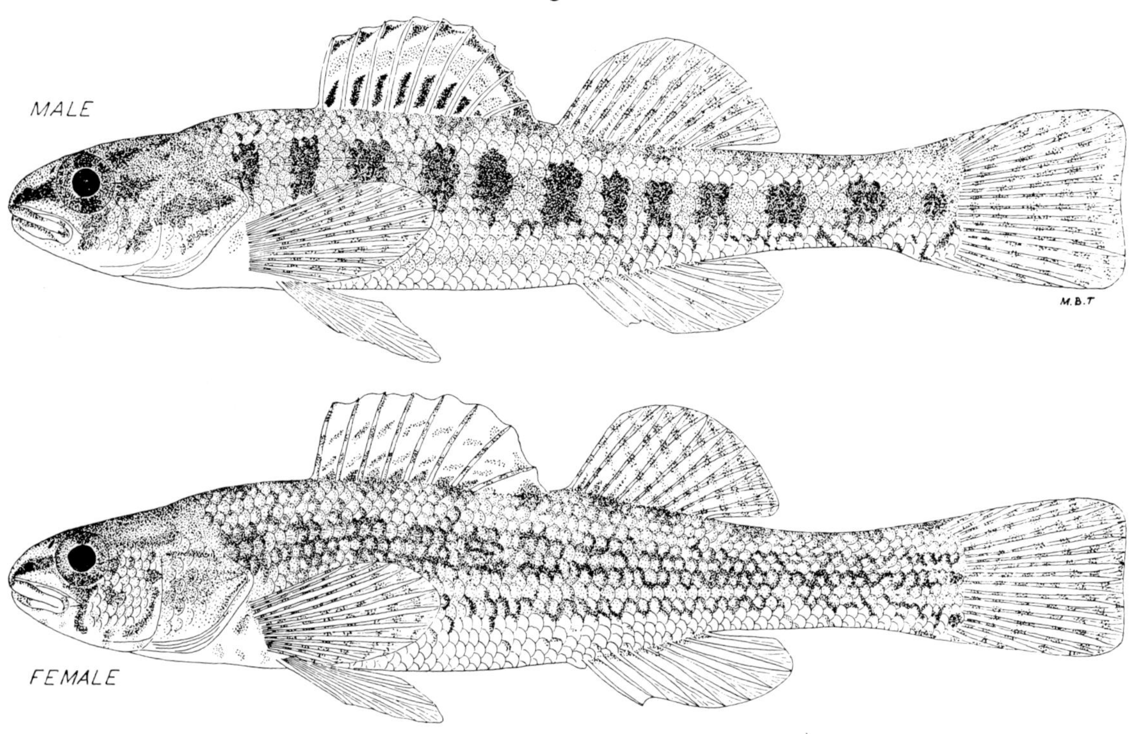

Upper fig.: Buckeye Lake, Licking County, O.

July 10, 1928. 44 mm SL, 2.2″ (5.6 cm) TL.
Adult male. OSUM 1779.

Lower fig.: Buckeye Lake, Perry County, O.

Oct. 21, 1943. 45 mm SL, 2.3″ (5.8 cm) TL.
Adult female. OSUM 5868.

Identification

Characters: *General—See* dusky darter. *Generic—See* central johnny darter. *Specific*—Two anal spines. No groove separating tip of upper lip and snout. Snout rounded, short and blunt; its length always shorter than diameter of eye. Gill covers slightly connected with membrane across isthmus. Cheeks and opercles scaled. Dorsal spines usually 8–9, extremes 7–11. Lateral line very short, ending beneath spinous dorsal. Scales in lateral series usually 55–60, extremes 53–62. Body slender; its depth contained 5.4–6.8 times in standard length.

Differs: No other species of darter in Ohio has combination of: very short lateral line; 53–62 lateral scales; fully scaled cheeks; fully scaled belly; low dorsal spine number.

Superficially like: Johnny, channel, least, orange-throat darters.

Coloration: *Breeding male*—Dorsally olive-green with 7–10 dusky saddle-bands crossing back. Sides with 9–12 dark blue, quadrate blotches, the interspaces brick-red. Lower sides from pectoral base to above anal fin orange-yellow, with an orange-red blush. Ventral surface of head, breast and belly whitish; sides of head mottled. Tear-drop prominent. Basal half of spinous dorsal with slaty blue spots between the spines; each spot surrounded by transparent webbing; above this a broad orange band; margin of fin blue with a transparent band

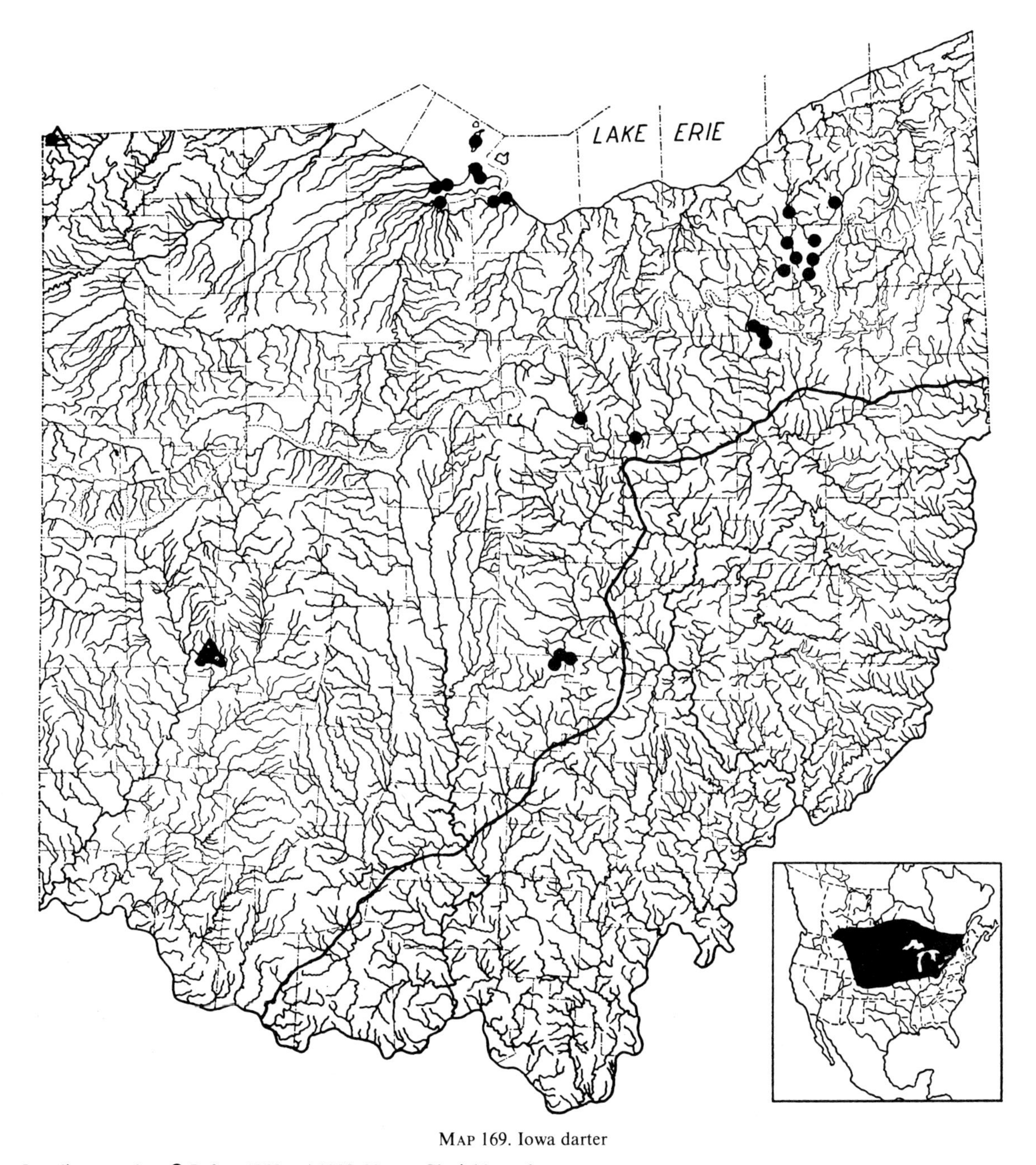

MAP 169. Iowa darter

Locality records. ● Before 1955. △1955–80. ~ Glacial boundary.
Insert: Large unoccupied areas occur within this range, especially along the southern border; northern limits of range indefinite.

between margin and the orange band. Second dorsal and caudal fins transparent, except for brownish spots which tend to form rows and wavy bands. Pectorals with spots; lower fins transparent. *Nonbreeding adult male*—Much less brilliant than is breeding male; the reds subdued or absent; blues replaced by olive-browns. *Adult female*—Dorsally olive-brown with 7–10 dark saddle-bands crossing back. Sides without reds and blues of the breeding male; when quadrate blotches are present they are less distinct and of a greenish-olive and yellow-brown; when quadrate blotches are absent the sides are vaguely mottled with greenish-brown. Ventrally silvery with a pale slaty tinge. Dorsal and caudal fins spotted, with much of the fins transparent or tinged with olive. Tear-drop well developed. *Young*—Like adult female, except that body is mottled with grays and browns and overcast with silvery; tear-drop less developed or absent; fins slightly spotted or unspotted.

Lengths: *Young* of year in Oct., 1.0″–1.8″ (2.5–4.6 cm) long. *Adults*, usually 1.8″–2.5″ (4.6–6.4 cm). Largest specimen, 2.7″ (6.9 cm) long.

Distribution and Habitat

Ohio Distribution—The Iowa darter is obviously a glacial relict species, whose distribution throughout glaciated Ohio presumably was more extensive and general before 1900 than it has been since. The clear-water, clean-bottomed, vegetated pothole ponds, bogs, marshes and streams, which formerly were very numerous in the northern half of the state and which comprised the habitat of this darter, were readily destroyed. During the 1925–50 survey several populations were observed to become greatly reduced in numbers or to disappear following a destruction of their habitat. As an example: before 1930 the Iowa darter was a numerous species in the vegetated shallows of Buckeye Lake, and as many as 50 individuals could be collected in an hour. After 1945 the population had so decreased in size that only an occasional specimen was taken during several hours of intensive seining, and then only at a few localities. No Iowa darters have been captured in Indian and St. Marys lakes, although the glacial pothole lakes which were flooded by these reservoirs undoubtedly must have contained this species.

Habitat—The Iowa darter was confined to those inland glacial lakes, springs and small streams, and the marshes, bays and harbors of Lake Erie, where the waters were habitually clear, the bottoms were of sand, peat, muck or organic debris. It is a northern species that displayed a preference for the colder waters. It was intolerant to turbid waters which deposited silts on the bottoms of the ponds and streams.

Years 1955–80—Investigations during this period disclosed that, with few exceptions, populations that had existed before 1952 have been extirpated. Drastic "improvements" which eliminated its habitat have been largely responsible for the disappearance of the species.

Before 1930 dozens of Iowa darters could be captured in an hour's seining at Buckeye Lake, but after 1945 only an occasional specimen was taken; the last one was captured on June 3, 1948. It is possible that a few may still exist in this body of water.

Before 1930 large populations were present in the weedy shallows about South Bass Island and East Harbor, Ottawa County. Annual investigations by classes of students, others and me have resulted in the taking of only one specimen at South Bass Island in 1947 and another at East Harbor in 1949. During investigations in the vicinity of Cleveland after 1970, Andrew M. White, others and I seined in all localities in the Cuyahoga and Chagrin drainages, including pothole lakes, where before 1955 others and I collected this species, sometimes in fair numbers. We also investigated other localities which appeared promising. We did not take a specimen (White et al., 1975:121).

During the 1955–80 period the species was captured only four times to my knowledge. On May 14, 1963, R. A. Zura took three specimens at Nettle Lake; on April 22 and 27, 1971 S. L. Phillips caught several at Nettle Lake, Williams County; on August 1, 1972 David F. Ross, Charles F. Willis, my wife and I captured several at Silver Lake, Miami County, where previously on July 2, 1948 my wife and I took a series of 68. In the first edition of this report the distributional dot representing the collection of the species in Miami County was erroneously placed in southeast Clark County.

RAINBOW DARTER

Etheostoma caeruleum Storer

Fig. 170

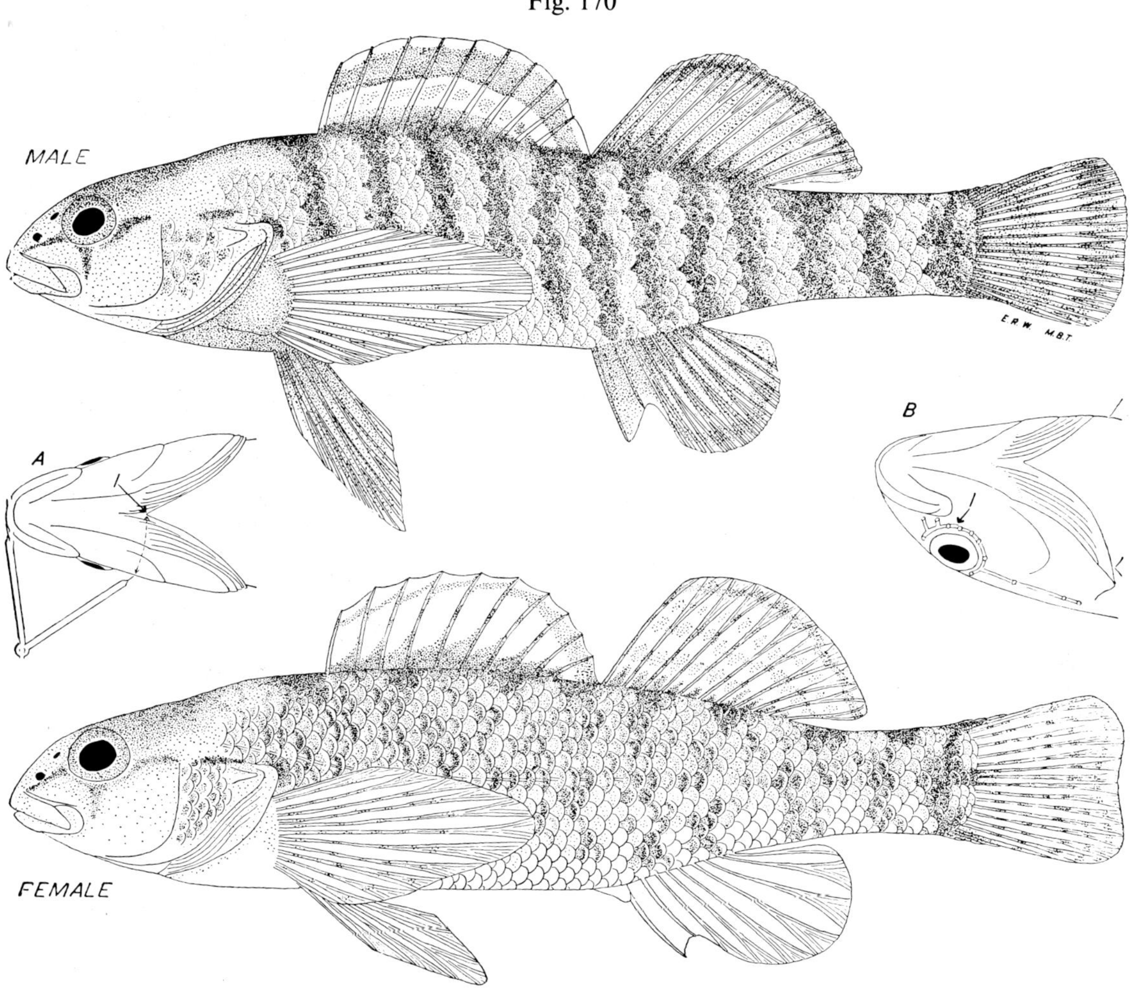

Upper fig.: Olentangy River, Delaware County, O.

Dec. 13, 1940. 47 mm SL, 2.3″ (5.8 cm) TL.
Adult male. OSUM 2391.

Lower fig.: Bend Fork, Captina Creek Drainage, Belmont County, O.

April 19, 1946. 42 mm SL, 2.0″ (5.1 cm) TL.
Breeding female. OSUM 6862.

Fig. A: ventral view of head, distance from apex of the gill-cover angle (at arrow) to tip of upper jaw, when projected alongside of head with dividers, extends from top of upper jaw to a point *beyond* the eye.

Fig. B: ventro-lateral view of head; arrow points to the unbroken infraorbital canal.

Identification

Characters: *General—See* dusky darter. *Generic—See* central johnny darter. *Specific*—Two anal spines. No groove separates tip of upper jaw and snout. Gill covers slightly joined with membrane across isthmus; distance from gill-cover angle (*see* fig. A, arrow), to tip of upper jaw, when measured carefully with dividers as shown, extends from upper jaw tip to well past posterior edge of eye.

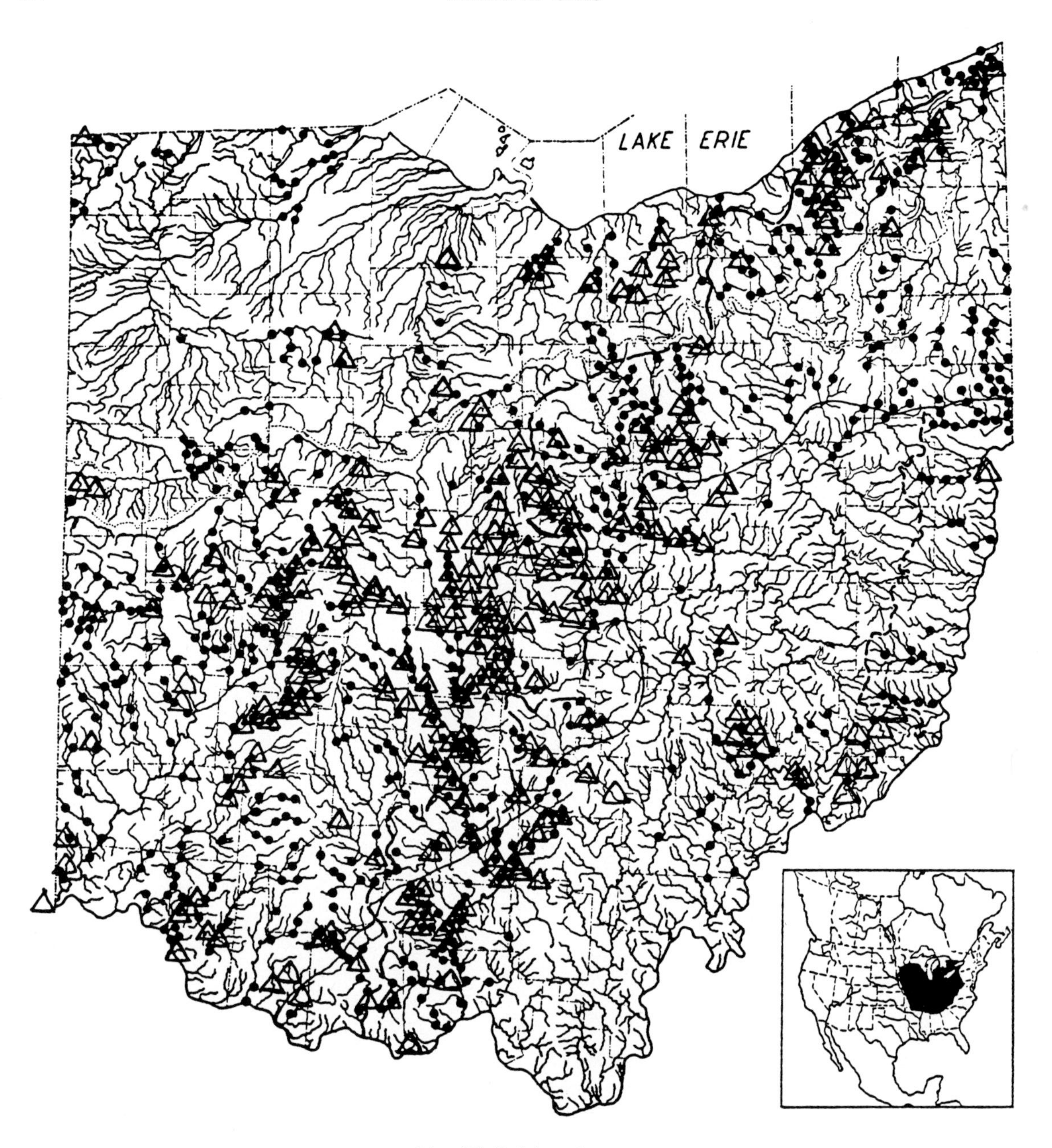

MAP 170. Rainbow darter

Locality records. ● Before 1955. △ 1955–80. ~ Glacial boundary. --- Flushing Escarpment. / \ Allegheny Front Escarpment.

Insert: Rather uniformly distributed over most of this range.

Cheeks naked; opercles scaled. Dorsal spines usually 10–11, extremes 9–12. Incomplete lateral line with 18–30 pored scales; lateral line ending beneath soft dorsal. Scales in lateral series 39–50. Belly normally scaled. Body deepest beneath middle portion of spinous dorsal. Usually 13 pectoral rays. Infraorbital canal complete, fig. B, arrow. Adult male has vertical bands completely encircling body posteriorly; brick-red on anal fin.

Differs: Orangethroat darter has the distance shorter from gill-cover angle to tip of upper jaw; usually fewer than 13 pectoral rays; greatest body depth before dorsal origin; interrupted infraorbital canal; vertical bands in male not completely encircling posterior half of body. No other species of darter in Ohio has combination of: short lateral line; belly normally scaled; 39–50 scales in lateral series; usually 10–11 dorsal spines; gill covers slightly joined across isthmus; straight or emarginated tail.

Most like: Orangethroat darter.

Coloration: *Breeding male*—Dorsally brownish-olive with 3–11 saddle-bands; the saddle-band before, the one between, and the one after the dorsal fins are the most prominent. Body with 8–13 dark, blue-green bands; those posterior to soft dorsal origin encircle body; interspaces between bands brick-red and orange. Lower sides of body and belly greenish; breast blue-green; ventral surface of head orange. Spinous dorsal usually with four horizontal bands; the basal-most brick-red; above this a band of reddish spots which are often bordered with transparent webbing (this band sometimes eliminated by the band above it, except for the posterior-most 2 spots); above this band a broad blue one; lastly a margin of pale greenish-blue. Soft dorsal greenish basally; remainder of fin orange, greenish, and red, except for a dusky blue-green margin. Caudal largely olive-green with a reddish center. Anal fin deep green, except for the middle portion which is brick-red. Pelvics blue-green; anterior ray tips whitish. *Non-breeding adult male*—Similar to breeding male, except that colors are subdued; breast light slate; fins more transparent. *Adult female*—Dorsally olive-green; sides lighter; breast and belly slaty-white. Saddle-bands similar to those of adult male, but often more distinct. Body mottled with olives and dark greens, little or no red or yellow; vertical bands ill-defined. Spinous dorsal with central portion largely transparent. Soft dorsal spotted; the spots tending to form rows; margin of fin bordered with dark brown. Caudal with wavy bars; remaining fins largely transparent. *Young*—Similar to adult female, but lacks greens, yellows or reds; vertical bands absent or poorly defined; fins largely transparent.

Lengths: *Young* of year in Oct., 1.1″– 1.8″ (2.8–4.6 cm) long. *Adults*, usually 1.4″–2.5″ (3.6–6.4 cm). Largest specimen, 3.0″ (7.6 cm) long.

Hybridizes: with dusky, banded and orangethroat darters.

Distribution and Habitat

Ohio Distribution—Between 1920–50 the rainbow darter was particularly abundant along the glacial boundary, along the Allegheny Front Escarpment, and in the two Miami River systems. Small populations were present in the least polluted of the high-gradient streams east of the Flushing Escarpment, and along such glacial outlets as the Muskingum River. The species was absent, except for strays, in mature, unglaciated streams, and in the former Black Swamp of northwestern Ohio. In the latter section it was present in moderate populations only in the streams of higher gradients, where there was much sand, as in Lucas County, or much gravel, as in Auglaize and Williams counties. Since 1920, the population in the Blanchard River, Hancock County, has decreased greatly in abundance (*see* orangethroat darter).

Most of the literature records, published before 1900, are unreliable because of inability of early workers to separate the rainbow and orangethroat darters; however, preserved specimens from Auglaize, Putnam, Huron, Lorain, Cuyahoga, Lake, Geauga, Ashtabula, Mahoning, Franklin, Washington and Hamilton counties show that the rainbow darter was widely distributed in Ohio between 1850–1900.

Habitat—Large populations occurred in moderate-sized streams of moderate or high gradients wherever the riffles were 15′–70′ (4.6–21 m) in width, the average depth was one foot (0.3 m), and the bottoms were of sand, gravel and boulders. Marked interspecific competition was apparent on many riffles, especially in the feeding and nursery areas. This was particularly true with the orangethroat darter on the smaller and/or more sluggish riffles, with banded and variegate darters on the larger and/or swifter riffles, and with the greenside darter on all types of riffles. Whenever interspecific competition from one or more of these species was lessened or absent, the rainbow darter numbers showed a definite increase, and in the few localities where such competition was absent, the numbers of rainbow darters were phenomenally great. This darter appeared to be less tolerant to most pollutants than was the johnny darter, and more tolerant than were the variegate or bluebreast darters.

Years 1955–80—Comparing those collections of the rainbow darter that were taken since 1955 (hollow triangles) and those taken previously (black circles), it can be noted that little change in the Ohio distribution of this species has occurred during the past 25 years. Large populations continued to be present along, and principally west of, the Allegheny Front Escarpment and glacial boundary, and in the higher-gradient streams of southwestern and central western Ohio (White et al., 1975:121–22). Investigations in the Maumee drainage disclosed only three small relict populations, one at Nettle Lake, one in the upper Blanchard River, and the other in a tributary (*see* map). Likewise, few were located in the eastern tributaries of the Muskingum River drainage. Marked decreases in numbers occurred wherever there was a considerable increase in siltation or in some types of pollutants. There were increases in numbers below the dams of some large impoundments, such as the Hoover Dam on Big Walnut Creek.

J. P. Kirtland (1854:4–5) described this species as *Poecilosoma erythrogastrum*; type locality, Rocky River, 7 miles (11 km) west of Cleveland.

For description of eggs and larvae, *see* Cooper, 1979:46–56.

NORTHERN ORANGETHROAT DARTER

Etheostoma spectabile spectabile (Agassiz)

Fig. 171

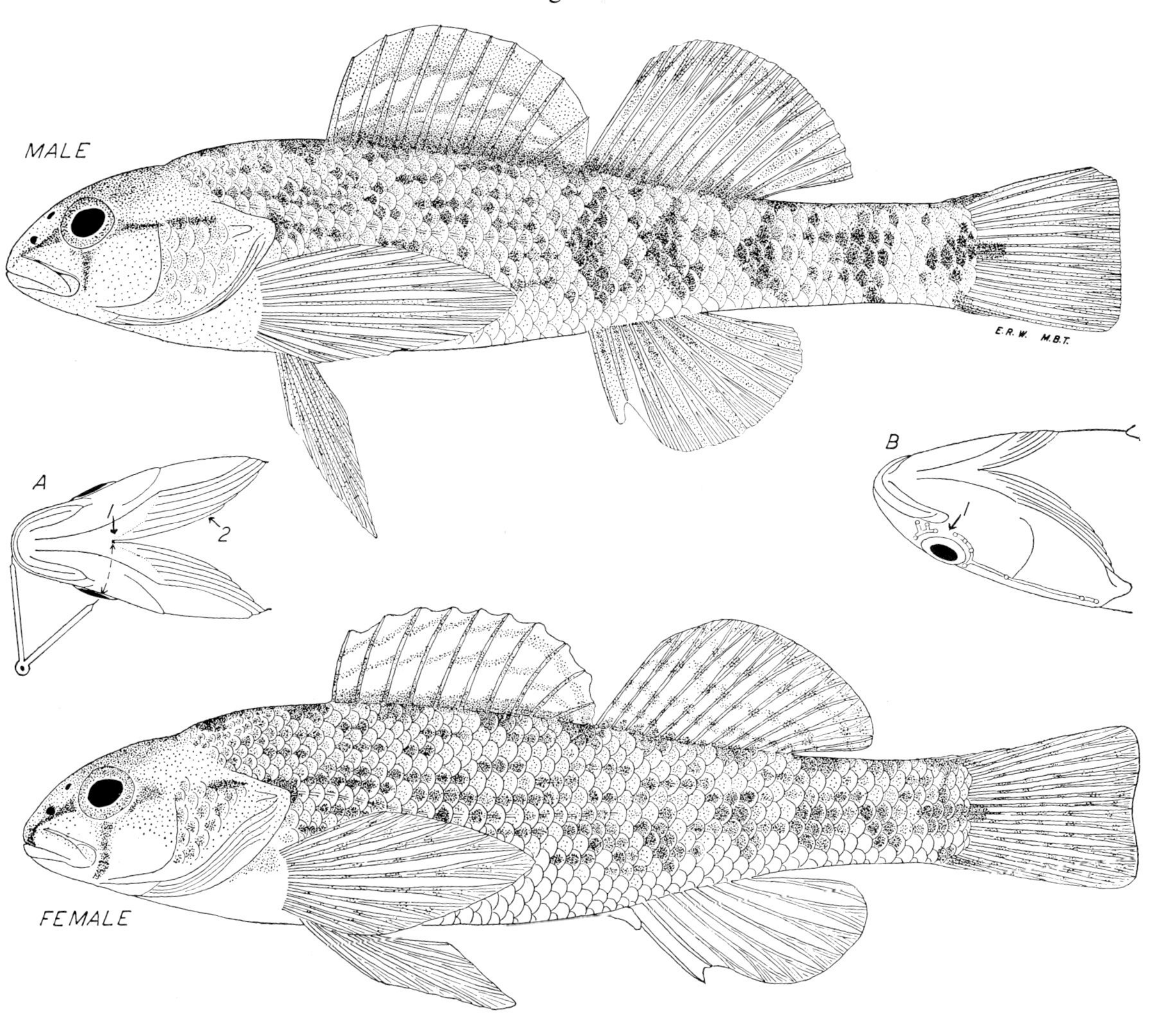

Upper fig.: Six Mile Creek, Auglaize County, O.

July 8, 1941. 45 mm SL, 2.2″ (5.6 cm) TL.
Adult male. OSUM 3273.

Lower fig.: Shepherd's Run, Paint Creek Drainage, Fayette County, O.

April 3, 1939. 33 mm SL, 1.6″ (4.1 cm) TL.
Breeding female. OSUM 1615.

Fig. A: ventral view of head; distance from apex of the gill-cover angle (at arrow) to tip of upper jaw, when projected along side of head with dividers, extends from tip of upper jaw to a point within the eye, rarely beyond. No. 2 points to the six branchiostegal rays, a characteristic of darters; the sauger, walleye and perch have seven.

Fig. B: ventro-lateral view of head; arrow points to the interrupted infraorbital canal, with a section of canal missing between two sensory openings or pores.

Identification

Characters: *General—See* dusky darter. *Generic—See* central johnny darter. *Specific*—Two anal spines. No groove separates tip of upper jaw and snout. Gill covers without membrane across isthmus; distance from gill-cover angle (*see* fig. A, arrow) to tip of upper jaw, when carefully measured with dividers as shown extends from upper jaw tip to posterior half of eye, seldom beyond. Cheeks naked; opercles scaled. Dorsal spines usually 10–11, extremes 9–12. Incomplete lateral line with 18–25 pored scales, lateral line ending beneath spinous or soft dorsals; scales in lateral series 42–50. Belly normally scaled. Body deepest immediately before spinous dorsal. Usually 12 or fewer pectoral rays. Infraorbital canal disconnected beneath eye, leaving an isolated group of four pores before eye, fig. B, arrow. Males with 5–7 triangular blotches posteriorly which never form bands to completely encircle body; no brick-red on anal fin.

Differs: Rainbow darter has the distance longer from gill-cover angle to tip of upper jaw; usually 13 pectoral rays; greatest body depth under spinous dorsal; a complete interorbital canal; vertical bands in males encircling body posteriorly. No other species of darter in Ohio has combination of: a short lateral line; belly normally scaled; 42–50 scales in lateral series; 10–11 dorsal spines; gill covers not connected across isthmus ith membrane; a straight or emarginated tail.

Most like: Rainbow darter.

Coloration: *Breeding male*—Dorsally yellow-olive with 3–11 saddle-bands; the saddle-band before, the one between, and the one after the dorsal fins the most prominent. Posterior sides of body contain 4–6 olive-green blotches which do not usually encircle body as bands but are somewhat triangular in shape (note upper fig.); between these blotches is much yellow, orange, and brick-red. Anterior sides of body have short, horizontal bars covering 2–12 dark scales; lower sides of body and belly yellowish. Breast and ventral surface of head deep orange. The most highly-colored males have base of spinous dorsal brick-red; above this a wider, bluish band; above this a brick-red band which fades out posteriorly; then a whitish bar; marginal fourth or third of fin a bluish-green. Soft dorsal has a bluish-slate basal bar; a distal margin of slate; between mostly orange and olive. Caudal largely pale green. Basal two-thirds of anal green; distal third whitish-green; no brick-red as in rainbow darter. Pelvics greenish; anterior rays tipped with whitish. *Non-breeding male*—Colors more subdued than in breeding male; breast whitish. *Adult female*—Dorsally yellow-olive mottled with darker; saddle-bands and short horizontal bars on sides usually more conspicuous than in adult male. Posterior blotches of sides poorly defined. Fins chiefly transparent; reds, blues, and greens largely absent. Soft dorsal and caudal with spots which form rows or wavy bars. *Young*—Like adult female except that body is less boldly marked; fins more transparent.

Lengths: *Young* of year in Oct., 1.1″–1.8″ (2.8–4.6 cm) long. *Adults*, usually 1.3″–2.5″ (3.3–6.4 cm). Largest specimen, 2.8″ (7.1 cm) long.

Hybridizes: with dusky, logperch and rainbow darters.

Distribution and Habitat

Ohio Distribution—Throughout the years 1920–50 the orangethroat was abundant only in westcentral Ohio. It was uncommon or absent in much of the Maumee River drainage and was usually uncommon, when present, in the unglaciated streams of southern Ohio. The species was present in some of the prairie-type headwaters of the Hocking River drainage in Fairfield County. Apparently it had long ago invaded this drainage when marshes connected the headwaters of the Hocking and Scioto river drainages. An example is found in the headwaters of Toby Creek (Hocking River drainage). Formerly about a mile (1.6 km) of wet prairie separated the Scioto and Hocking river drainages, and during high waters these two systems were presumably connected. This old, wet prairie was formerly called the "Walnut Plains" or "Walnut Prairies" (Jones, 1774:86).

Preserved specimens collected prior to 1900 indicate its early presence in the Lake Erie and Ohio River drainages in western and central Ohio.

Habitat—The orangethroat was present in largest numbers in small- and moderate-sized streams of low- and moderate-gradients where the riffles were 3′–20′ (0.9–6.1 m) in width and had an average depth of 5″ (13 cm), and where the bottoms were of sand and gravel with or without a slight covering of silt. The numbers of orangethroats in such streams apparently were influenced by the numbers of rainbow darters also present, for whenever one species was

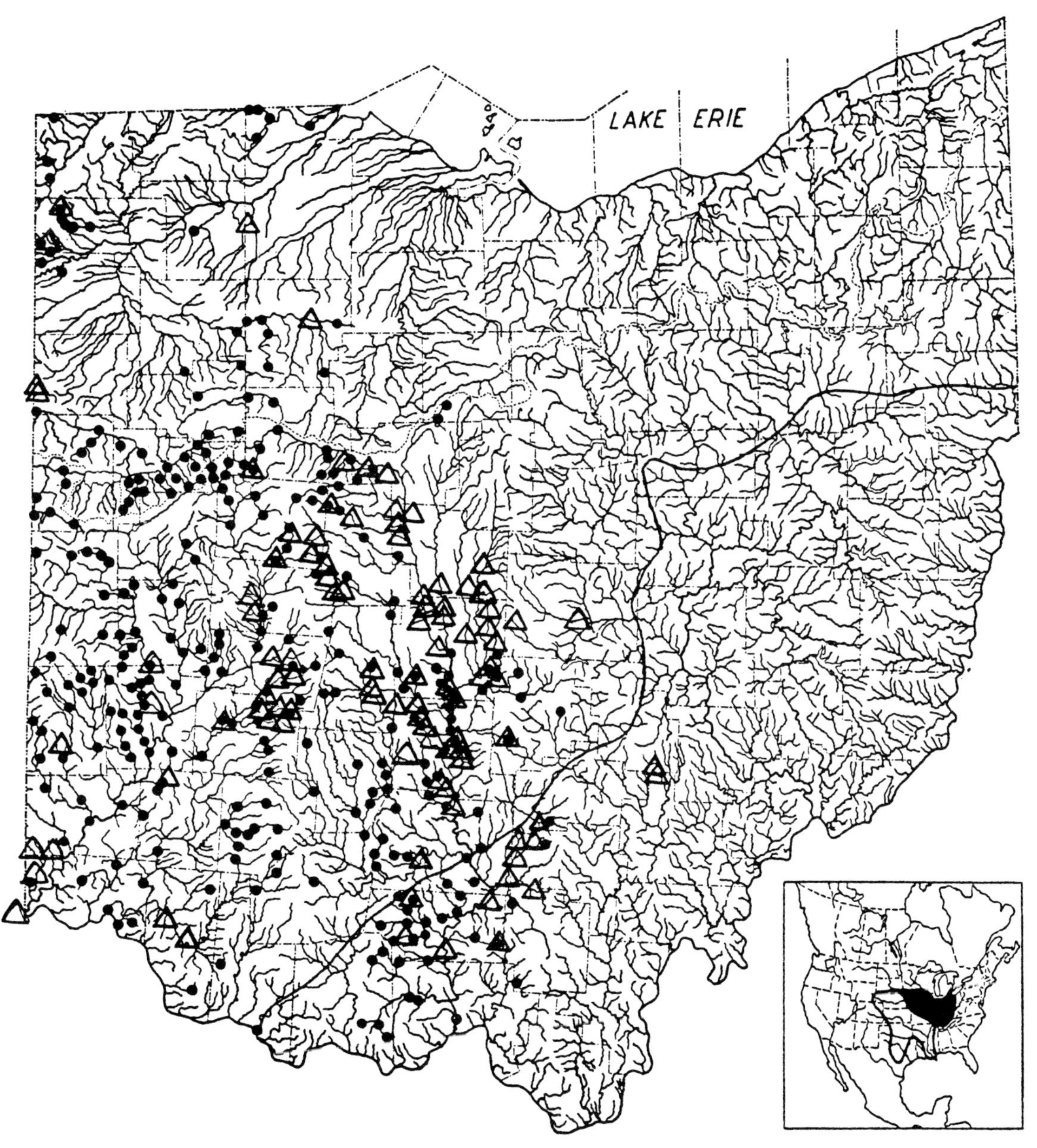

MAP 171. Northern orangethroat darter

Locality records. ● Before 1955. △ 1955–80. ~ Glacial boundary.
Insert: Black area represents range of this subspecies, outlined area of other subspecies.

abundant the other usually was rare or absent; presumably there was considerable interspecific competition between these two. The orangethroat was more tolerant to turbid waters and silted bottoms, and inhabited smaller waters than did its rival; consequently, although both species were rather abundant in the Auglaize River system of Auglaize County, the orangethroat was the more numerous only on the smaller, more sluggish and more silted riffles, whereas the rainbow was more numerous on the larger, faster riffles having clean gravel or sand bottoms. In the St. Marys and Wabash river systems in Auglaize and Mercer counties the streams were sluggish, the waters usually turbid, the bottoms silted; in these streams the orangethroat was numerous whereas the rainbow was uncommon. Effects of stream size, depth and speed of current upon the two species was aptly demonstrated in Toby Creek, Fairfield County. In the prairie-type headwaters there existed a thriving population of orangethroats, but farther downstream in deeper, faster waters and where the stream width averaged more than 10′ (3.0 m), the rainbow supplanted the orangethroat.

During the 1920–50 surveys I observed changes in abundance between the orangethroat and rainbow when a stream was dredged. Before dredging the orangethroat was the more numerous; after dredging had eliminated the pools and increased the flow over the deeper riffles, the rainbow became the more numerous. Decrease in abundance of orangethroats was noted in some lowland streams of the Black Swamp section of the Maumee River drainage after these streams had become exceedingly turbid and siltation had greatly increased.

Years 1955–80—With few exceptions the distribution of the orangethroat darter in Ohio has remained unchanged during the past 25 years. The exceptions are Indian Run, a tributary of Sunday Creek, southwestern Monroe Township, Perry County, and Johnson Run, a tributary of Sunday Creek, northern Trimble Township, Athens County, Raccoon Creek, Granville Township and upper South Fork of the Licking, Lima Township, both Licking County.

On August 14, 1965, James Norrorky and James Matthews captured 7 of these darters in Indian Run; and on January 1, 1966, Norrorky and Raymond Jezerinac collected 13 there. On March 14, 1971, Jezerinac revisited this creek, capturing 17; and on June 3, 1971, Ted M. Cavender, Richard Moerchen, David Ross and I collected 12 at Indian Run and 2 specimens in Johnson Run. It is surprising to find this species in creeks several miles east of the Appalachian Front Escarpment and the glacial boundary. These streams rise in hills surrounded by strip mines, have gradients of approximately 40′/mile (7.6 m/km) and flow eastward a few miles into Sunday Creek, which is heavily polluted with mine wastes.

It is interesting to contemplate when and how the orangethroat invaded these hilly, unglaciated streams, which appear unsuited for the low-gradient, prairie-type orangethroat. If the species invaded from the west, why is it not present in many other unglaciated hill streams of similar types? It is possible that the species was inadvertently introduced within the last 100 years and found local conditions suitable for survival.

Apparent reduction in population size of orangethroats was observed wherever silting or other pollutants became excessive or where the rainbow darter population had greatly increased recently.

BARRED FANTAIL DARTER

Etheostoma flabellare flabellare Rafinesque

Fig. 172

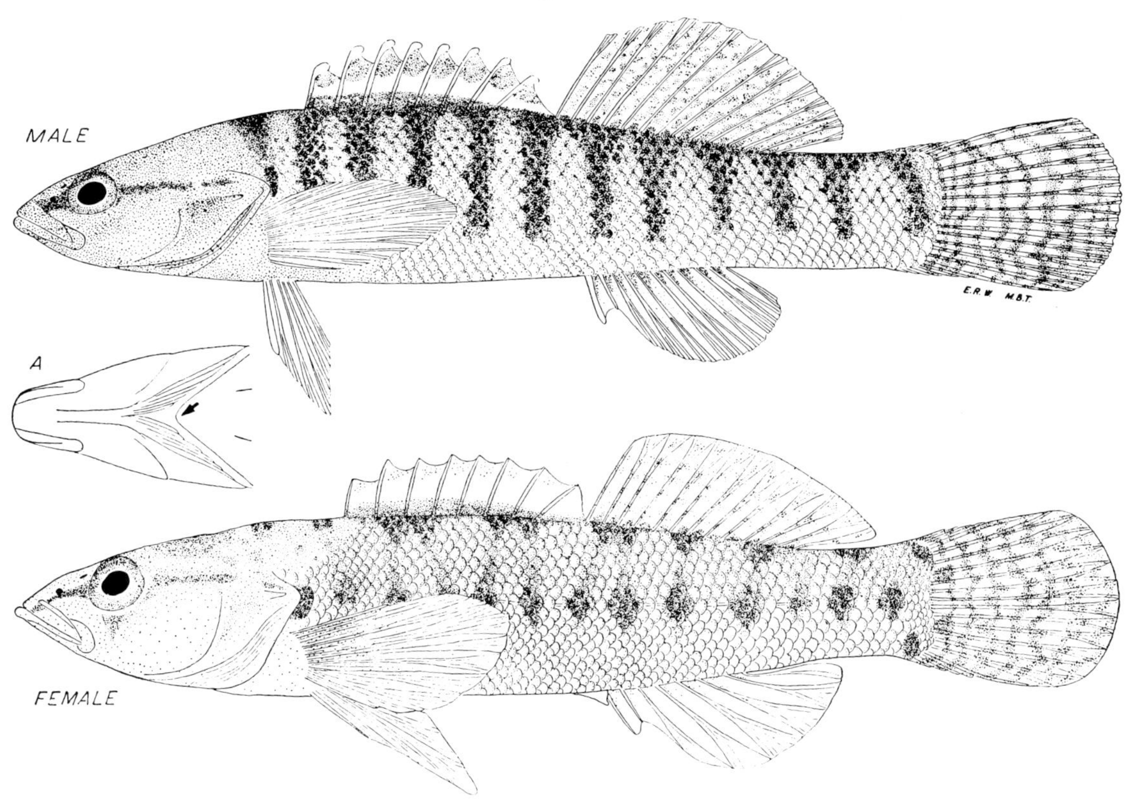

Upper fig.: Deer Creek, Stark County, O.

May 9, 1942. 49 mm SL, 2.3″ (5.8 cm) TL.
Breeding male. OSUM 4899.

Lower fig.: Same locality, date and OSUM number.

Breeding female. 35 mm SL, 1.7″ (4.3 cm) TL.

Fig. A: ventral view of head; arrow points to the rather acute angle of the moderately joined gill covers.

Identification

Characters: *General—See* dusky darter. *Generic—See* central johnny darter. *Specific*—The first of the two anal spines stout; the second spine frail. No groove between tip of upper lip and snout. Mouth terminal; large young and adults may have the lower jaw the longer. Gill covers broadly connected across isthmus with membrane; distance from gill-cover angle (fig. A, arrow) to tip of upper jaw extending from tip of upper jaw to opercle in adults; to posterior edge of cheek in young. Cheeks and opercles scaleless. Dorsal spines usually 7–8, extremes 6–9. Incomplete lateral line usually ending beneath soft dorsal fin, of 20–35 pored scales; scales in lateral series 46–55. Tail rounded. Adults brownish with 10–15 dusky bands; no reds, blues, or greens.

Differs: No other species of darter in Ohio has combination of: terminal mouth; gill covers broadly connected; dorsal spines nine or fewer; belly normally scaled; rounded tail; incomplete lateral line; ventral surface of head speckled with black; brown body banded and suffused with dusky in adults.

Superficially like: Iowa, johnny and river darters.

Coloration: *Breeding male*—Entire head except

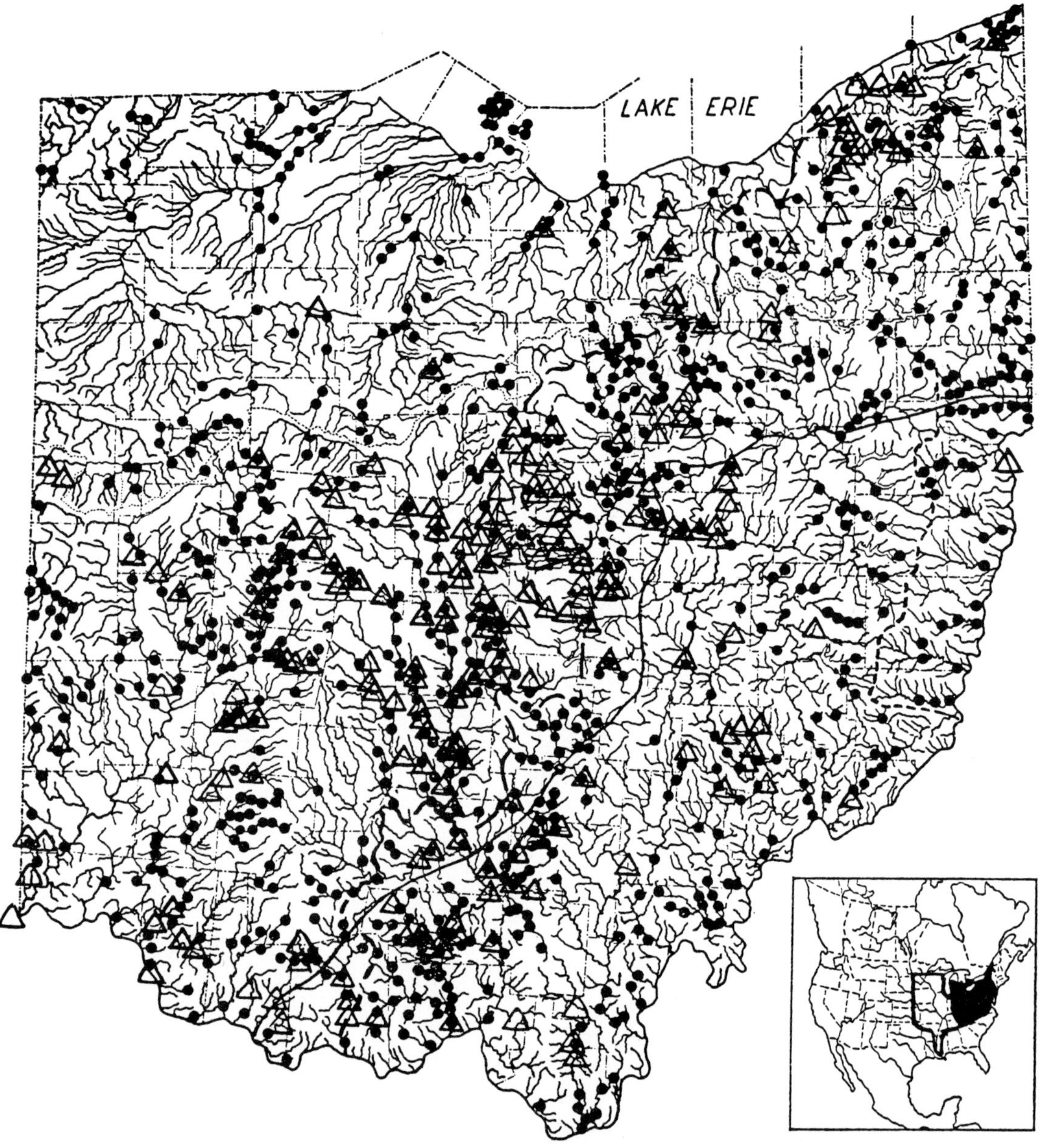

MAP 172. Barred fantail darter

Locality records. ● Before 1955. △ 1955–80. /\ Allegheny Front Escarpment. ~ Glacial boundary. --- Flushing Escarpment.

Insert: Black area represents range of this subspecies, outlined area of other subspecies.

cheeks, dusky or black; cheeks olive-green. Back dark olive-green, lightening to tan or light brown on sides, to olive-yellow on belly. Breast dusky. Back crossed with 10–15 chocolate or black bands which extend downward to lower sides. Spinous dorsal with orange bands basally and distally; spines tipped with orange-white pads (used by the male to clean the eggs). Soft dorsal light olive, speckled with darker. Caudal olive with 8–11 wavy vertical bands.

Non-breeding adult male—Similar to above, except that it lacks dusky suffusion on head and pads on spines. *Adult female*—Lighter colored than adult male and color pattern less contrasting; bands on body imperfectly formed, resulting in saddle-bands being more conspicuous and a tendency towards a series of blotches along the lateral line; faint tear-drop usually present; fins more transparent. *Young*—Similar to adult female, except that color pattern is less contrasting, and vertical bands often absent entirely.

Lengths: *Young* of year in Oct., 1.1"–1.9" (2.8–4.8 cm) long. *Adults*, usually 1.2"–3.0" (3.0–7.6 cm). Largest specimen, 3.4" (8.6 cm) long.

Distribution and Habitat

Ohio Distribution—Between 1920–50 the fantail darter was well distributed throughout much of Ohio. The species was present in largest numbers in the shoal waters about the Lake Erie islands, in the rejuvenated streams along and between the glacial boundary and the Allegheny Front Escarpment, in the Mad River system, and in the clear-water prairie-streams of westcentral Ohio. It was present only in small numbers in those mature streams not affected by glaciation which lie between the line of glaciation and the Flushing Escarpment, and it was casual in the Ohio River. The fantail was likewise uncommon or absent in most of the Maumee River drainage, particularly in the Black Swamp streams of low gradient and great turbidity, and in this drainage was absent in what apparently were favorable habitats such as were present in the headwaters of such streams as Lost Creek, Defiance County. Its absence in these isolated waters may have been caused by inability to penetrate the unfavorable lowland streams of the Black Swamp, and "thus reach the upland tributaries with their apparently suitable habitats." (Trautman, 1939:281.)

Early references in the literature (Williamson and Osburn, 1898:55; Osburn, 1901:101–102) and preserved specimens indicate that before 1900 the fantail was an abundant darter of rather general distribution throughout the Ohio River drainage, and in the Lake Erie drainage of northeastern Ohio. It apparently was less generally distributed and less abundant in northcentral and northwestern Ohio (McCormick, 1892:31; Kirsch, 1895A:331).

Many specimens from the prairie streams of western Ohio had rather well-developed, lengthwise rows of dusky dots along their sides, suggesting possible striped fantail darter (*E. flabellare lineolatum*) influence in the prairie areas of Ohio. Few specimens from the high-gradient streams east of the Flushing Escarpment contained spots, and the longitudinal rows of spots were entirely absent.

Habitat—Throughout the breeding season the fantail was most numerous in streams which had an average width of less than 30′ (9.1 m). In such streams the species chiefly inhabited the more sluggish riffles and those sections of pools where there was a fair current, the water was less than 18" (46 cm) deep, and the bottoms were of gravel, flat stones and boulders. The species was rather tolerant (for a darter) to most pollutants, including silty clays. When silt became so deep that it covered the nesting sites in the pools and sluggish riffles, the fantail would deposit its eggs on the underside of flat stones and other objects in the faster riffles. When clayey silt or other pollutants caused the partial or complete desertion of the riffles by those darter species which normally inhabited the faster portions of these riffles, the fantails would occupy their former nesting territories. During midsummer, late summer and fall the majority of the fantails moved downstream into larger and deeper waters where they wintered. The species avoided riffles of bare sand or bedrock, but inhabited these if there were present some water-willow, pondweeds, cladophora or other vegetation in which these darters could hide.

Years 1955–80—During this period the fantail darter continued to be one of the very numerous darter species in most of the streams of Ohio. It continued to be numerous in those streams adjacent to the glacial boundary and the Allegheny Front Escarpment. In inland Ohio the species appeared to be decreasing numerically only in the most heavily silted waters in northwestern Ohio or in recently heavily polluted sections throughout the state (White et al., 1975:122).

Prior to 1950 the fantail was very numerous in the shoal waters about the Lake Erie islands, and dozens could be captured daily with a small seine. As an illustration: as late as March 25, 1947 more than 200 specimens (of which 101 were preserved) were captured during a 3-hour period during a strong westerly wind which lowered the lake level approximately 3′ (0.9 m) and which also stranded several other fish species that normally hide beside or under stones and boulders. Since 1960 none has been found in the shoal areas about the islands despite determined efforts (Van Meter and Trautman, 1970:77). It is unknown what caused this marked decrease or possible extirpation, and it is rather surprising because in inland Ohio streams the fantail seems to be more tolerant to many pollutants than do most other darter species.

For description of eggs and larvae, *see* Cooper, 1979:46–56.

LEAST DARTER

Etheostoma microperca Jordan and Gilbert

Fig. 173

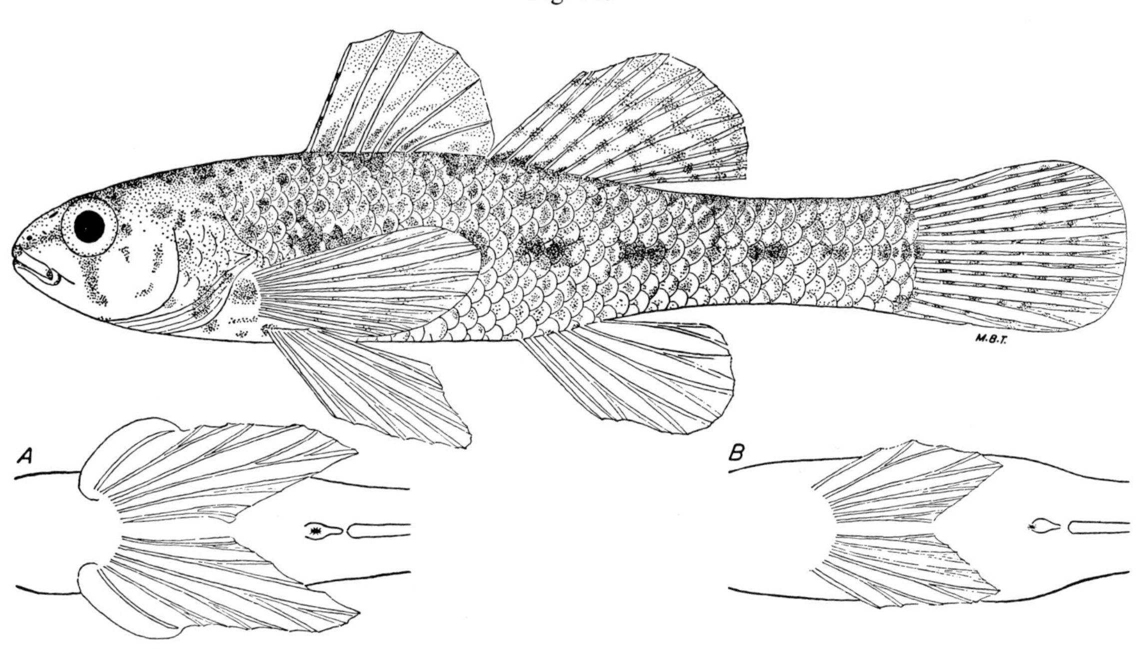

Upper fig.: Greenville Creek, Darke County, O.

Nov. 7, 1940. Adult female. 26 mm SL, 1.3" (3.3 cm) TL. OSUM 2280.

Fig. A: Twin Creek, Preble County, O.

April 20, 1942. Breeding male. 27 mm SL, 1.4" (3.6 cm) TL. OSUM 4842.

Ventral view of body showing greatly enlarged pelvic fins of the male.

Fig. B: ventral view of body of the upper fig.; showing the smaller pelvic fins of the female.

Identification

Characters: *General—See* dusky darter. *Generic* and *specific*—Lateral line absent or with fewer than eight pored scales. Scales in lateral series 32–37, extremes 30–39. Dorsal spines usually six, extremes 5–8. Anterior half of belly scaleless as is the breast. Cheeks scaleless; opercles partly scaled. Gill covers rather broadly connected across isthmus with membrane. Anal spines two. No groove between tip of upper lip and snout. Tear-drop conspicuous, except in small young. Pelvic fins of breeding male large, fig. A; those of adult female small, fig. B.

Differs: No other species of Ohio darter has combination of: 0–8 pored scales in lateral line; 30–39 scales in lateral series; 5–8 dorsal spines; cheeks naked; two anal spines; belly partly scaleless.

Superficially like: Young of Iowa and orangethroat darters.

Coloration: *Breeding male*—Dorsally olive-brown; sides olive-yellow; belly yellowish-white. Body mottled and speckled with browns; 3–11 vague saddle-bands on some specimens; usually 7–15 blotches along the lateral series of scales. Tear-drop large and conspicuous. Spinous dorsal transparent, and/or blotched with olive-brown, and with a basal row of reddish spots. Soft dorsal and caudal pale olive, barred with brown. Anal and

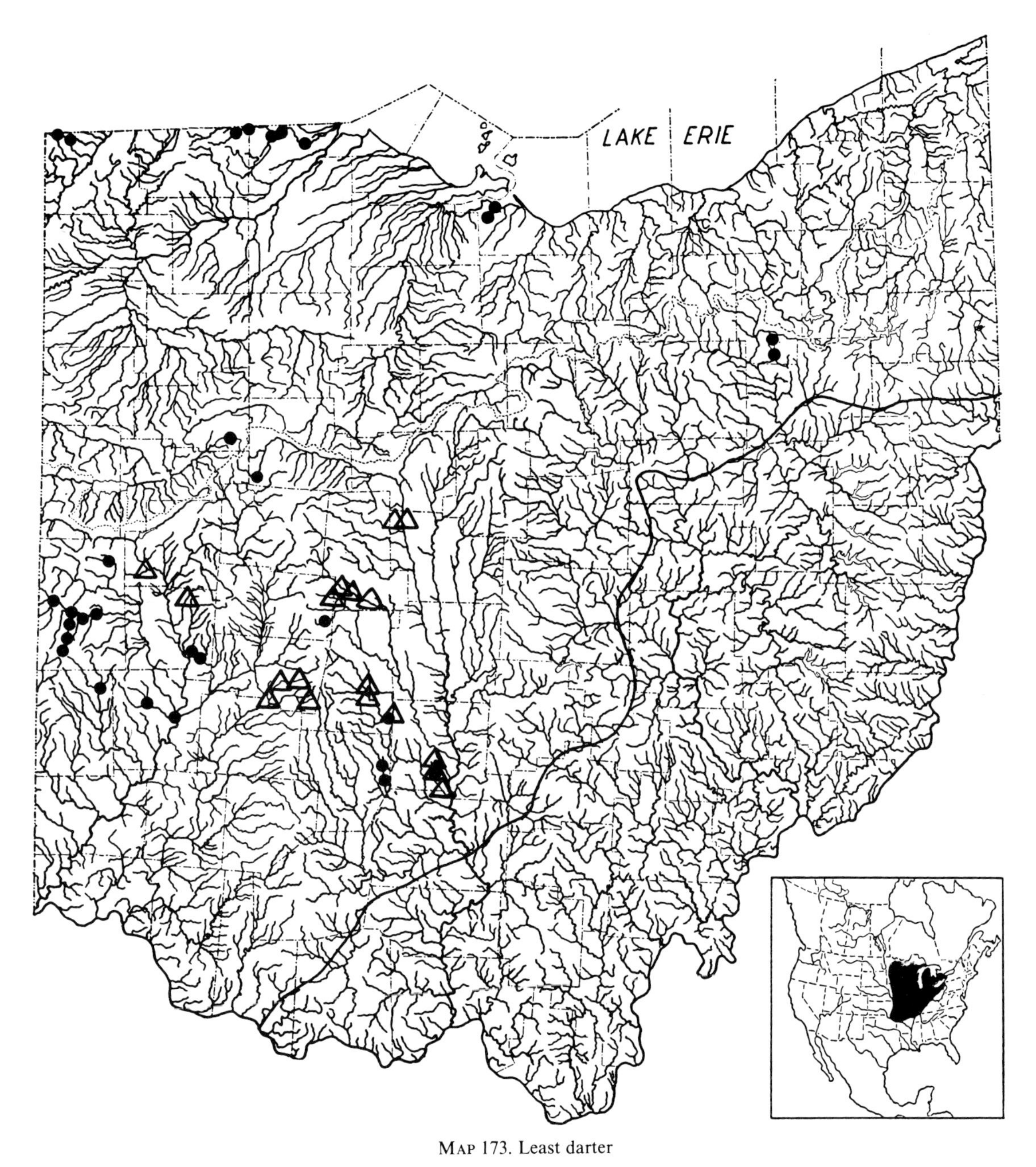

MAP 173. Least darter

Locality records. ● Before 1955. △ 1955–80. ~ Glacial boundary.
Insert: Large unoccupied areas occur within this range.

pelvic fins reddish-orange. *Adult female* and *young*—Similar to male except that spinous dorsals lack reddish spots; anal and pelvic fins transparent; body pattern less contrasting.

Lengths: *Young* of year in Oct., 0.8″–1.2″ (2.0–3.0 cm) long. *Adults*, usually 1.0″–1.5″ (2.5–3.8 cm) long. Largest specimen, 1.8″ (4.6 cm) long. For breeding habits, *see* Petravicz (1936:77–82).

Distribution and Habitat

Ohio Distribution—Although not recorded until 1922 for Ohio waters (Osburn, Wickliff, Trautman, 1930:175), the least darter unquestionably invaded both drainages long before, probably utilizing the Maumee outlet to invade the lake drainage, and entering the glaciated portion of the Ohio River drainage before or during establishment of the prairies.

Since 1925, the least darter occurred in goodly numbers only in Cold Creek, Erie County, and in the prairie streams of Darke County. The populations in the Portage (Summit County) and Nettle (Williams County) lakes may have decreased during the 1930–50 period, and because of destruction of habitat through dredging, the species was obviously less numerous in Ten Mile Creek after 1940 then it was before.

This small, secretive species was difficult to capture, and in several localities where its presence was suspected, it was taken only after prolonged or repeated attempts to capture it were made. Because of this difficulty in capturing it, the species was probably present during the 1920–50 period in many prairie areas from which we have no records. It is also assumed that because of the extensive ditching, dredging, draining and polluting of prairie streams, the least darter was far more abundant and widely distributed before 1900 than it has been since. Then it should have been present in the extensive prairies centering in Marion County, and was possibly present in the St. Marys and Auglaize river systems, and streams entering Sandusky Bay.

Habitat—An inhabitant of the dense beds of filamentous algae and higher aquatic vegetation of the clearest, pothole lakes and of the clearest prairie streams of base or low gradients, where the bottom was of soft muck, debris, sand or gravel, but not of yellow clayey silts.

Years 1955–80—Despite the continued ditching and dredging operations in prairie-type streams of west central Ohio, at least seventeen distinct populations of least darters were located which had not been discovered prior to 1950. Some of these populations were quite large, and more than a hundred individuals could be captured in an hour with the aid of a small seine. Obviously, there were many other localities in this section of the state containing populations which were not located. The species undoubtedly was more widespread in its Ohio distribution, and the total population much larger, before 1950 than it has been since.

FRESHWATER DRUM

Aplodinotus grunniens Rafinesque

Fig. 174

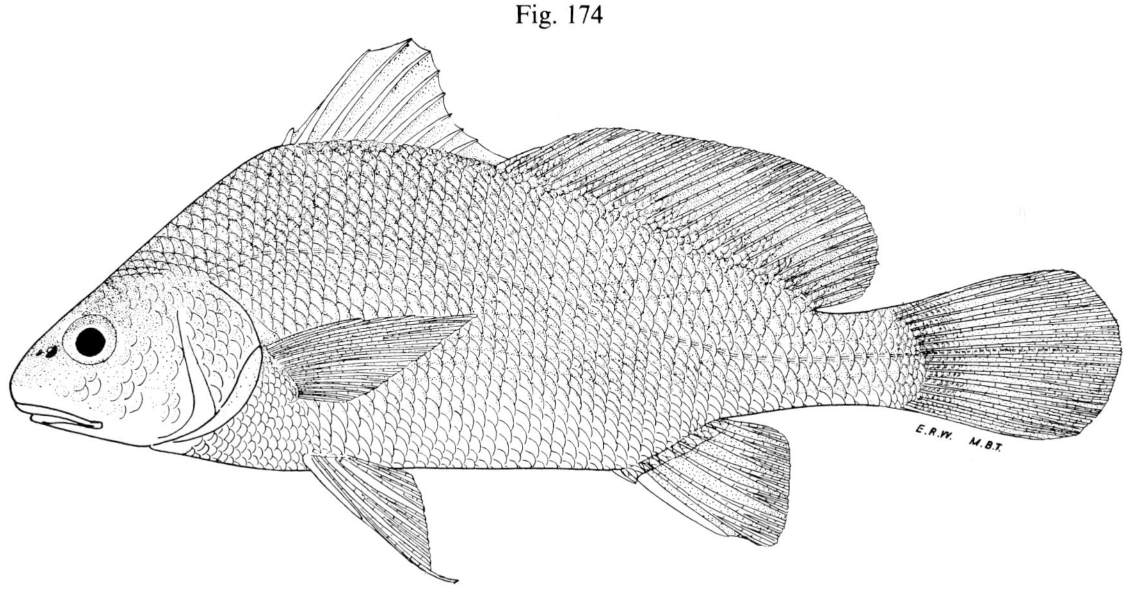

Scioto River, Scioto County, O.

May 19, 1940.
Young male.

218 mm SL, 11.0″ (27.9 cm) TL.
OSUM 2981.

Identification

Characters: Two anal spines, the first shorter than diameter of eye, the second longer than snout length and very stout. Dorsal spines 8–9. Soft dorsal very long, of 24–32 rays. Mouth subterminal; jaws have small comb-like teeth but lack canines. Top of head, cheeks and opercles scaled. Tail rounded; lateral line pores extend across tail (pores on tail often difficult to see). Coloration silvery. Otoliths (ear bones) very large, circular and with an "L" on one of the two flat surfaces; these are the "lucky stones" which are prized by children.

Differs: White bass has forked tail, terminal mouth; fewer than 15 soft dorsal rays. Blackbasses have lower jaw longer than upper. Sauger and walleye have forked tails; canine teeth in jaws. Carpsuckers lack dorsal spines. Carp and goldfish have forked tails; only one dorsal spine.

Superficially like: White bass and carpsuckers.

Coloration: Dorsally pearl-gray with bronze, blue and silver reflections; sides lighter and more silvery. Belly, breast, and ventral surface of head milk-white and silvery. Fins of same colors as adjacent parts. Fishes from clearest waters are bronzy; those from turbid waters yellowish-white.

Lengths and weights: *Young* of year in Oct., 4.0″–7.0″ (10–18 cm) long. Around **1** year, 5.0″–9.0″ (13–23 cm); weight 0.7 oz–8 oz (20–227 g). *Adults*, usually 12.0″–30.0″ (30–76.2 cm) long; weight 1 lb–17 lbs (0.5–7.7 kg). Specimens 8.0″ (20 cm) long usually weigh 3.0 oz–5.0 oz (85–142 g); 10.0″ (25.4 cm) usually weigh 5.0 oz–10.0 oz (142–283 g); 12.0″ (30.5 cm) usually 10 oz–1 lb (283g–0.5 kg); 14.0″ (35.6 cm) usually 1 lb–1 lb, 8 oz (0.5–0.7 kg); 16.0″ (40.6 cm), usually 1 lb, 8 oz–2 lbs, 8 oz (0.7–1.1 kg); 24.0″ (60.9 cm), usually 5 lbs–7 lbs, 8 oz (2.3–3.4 kg). Largest specimen (from Ohio River), between 35.0″–39″ (88.9–99.1 cm) long; weight 36 lbs (16 kg). Lake Erie drums seldom are heavier than 10 lbs (4.5 kg); Ohio River drums seldom more than 25 lbs (11 kg). *See* Van Oosten (1938) for growth and weight of drums in Lake Erie.

MAP 174. Freshwater drum

Locality records. ● Before 1955. △ 1955–80.
Insert: Rare or absent from large upland areas within this range; Mexican limits of range indefinite.

Distribution and Habitat

Ohio Distribution—As early as 1780, Zeisberger (Hulbert and Schwarze, 1910:73 as buffalofish, 74 as white perch) described the drum and its habits in Ohio waters, and explained that to capture it the Indians would "commonly peirce this fish with an iron prong [spear];" Rafinesque (1820:70–72) in 1820 stated that among Ohio drainage fishes it was "one of the most common, being found all over the Ohio, and even the Monongehela and Allegheny"; and Kirtland (1850–I:133 as *Corvina oscula*) wrote in 1850 that "In early days large numbers ascended the Mahoning and other streams from the Ohio river, . . . Mill-dams and other causes have long since exterminated them from the tributaries of the

Ohio. In the latter stream they are still taken. Lake Erie abounds with them." Elsewhere Kirtland (1850A:1) mentioned that "large numbers" migrated into the upper waters of the Mahoning River in the early days. From the above, and many other literature references, it is obvious that the drum was an abundant species in the larger waters of both drainages before 1860.

Between 1920–50 the freshwater drum was abundant in Lake Erie, and during the 1939–49 decade an annual average of 3,543,025 lbs (1,607,171 k) (Shafer, 1950:1) was brought into Ohio ports. However, the poundage captured was far greater, and the excess, which could not be marketed, was returned to the water. There was a small population in the Maumee River. Unlike the Lake Erie population, that of the Ohio River has greatly decreased in abundance since 1860, and after 1925, the species was moderately common in that drainage only in the Ohio River as far upstream as Marietta, the Scioto River upstream to Waverly, and the Muskingum River to Zanesville. Elsewhere in the Ohio drainage the drum was rare, and apparently it was entirely absent from the Mahoning River and other upper Ohio River tributaries where Kirtland formerly noted its great abundance. The drum likewise was absent during recent years from the Ohio waters of western Pennsylvania (Raney, 1938:78).

In the past 100 years much has been written concerning the marked differences in edible qualities between drums from Lake Erie and those from the Ohio River. More than a hundred years ago Kirtland (1841B:352) voiced the general opinion that drums from Lake Erie were "hardly eatable" whereas those from the Ohio drainage were "always fat, tender and delicious"; in 1858 he (1858:281) suggested that drums from the Ohio River be introduced into Lake Erie in the hope that "the rich and delicate White Perch or Sheepshead" of the Ohio River might supplant the "worthless" sheepshead of Lake Erie. There was also a great difference in the maximum weights attained in the two drainages. Older commercial fishermen have told me that in Lake Erie before 1910 they occasionally took a drum which weighed as much as 18 lbs (8.2 kg), but since then, 13 lbs (5.9 kg) was the maximum. In 1927–28 Van Oosten (1938:653) examined 2,183 fishes from Lake Erie waters of which the largest weighed only 7 lbs, 14 oz (3.6 kg). But in the Ohio drainage, individuals weighing between 15–20 lbs (6.8–9.1 kg) were still frequently taken, and I took one in 1930 from a hoop net in the Ohio River that weighed 36 lbs (16.3 kg).

Before 1932, the much prized drum of southern Ohio was known by several names, including grunter, bubbler, gray or white bass, white or gray perch, and buffalofish. This latter name was used by the Indians because the fish grunts exactly like a buffalo. Since 1932, white perch was the name almost universally used in the Ohio River drainage, and sheepshead about Lake Erie.

Habitat—The drum normally frequented the deeper pools of rivers, and in Lake Erie, waters between 5′–60′ (1.5–18 m) in depth. At twilight during the warmer months, these fishes came into waters less than 5′ (1.5 m) in depth where they fed by moving rocks and stones with their snouts, capturing the crayfishes, aquatic insects, darters, and other fishes thus disturbed. Many of the older references (Hulbert and Schwarze, 1910:73; Kirtland, (1850I:133) indicate that before 1910 snails and "clams" constituted a principal food of the drum. This appears plausible, for in early days most Ohio streams contained huge snail and mollusk populations, which were destroyed later by silting and pollution. With only remnant populations of mollusks remaining in most waters, there apparently has been a food shift from snails and mollusks to crayfishes, fishes, and insects (Daiber, 1952:45). Although the drum seemingly can tolerate turbid water, its preference for clear water and clean bottoms is very evident; as an example, there is a considerably larger population of drums in the less turbid Muskingum River than in the more turbid Scioto River.

Years 1955–80—Annual Lake Erie populations of freshwater drums continued to be large as evidenced by the commercial landings, which averaged more than 3 million lbs (1,360,777 kg) yearly, as they had since 1900 (Applegate and Van Meter, 1970:19). As was apparent before 1950, unknown poundages of trap-netted or seined drums were released yearly by commercial fishermen because of a lack of market.

Increased efforts are being made to interest the public to consume more drums, with comparatively little success to date. Apparently this dislike for Lake Erie drums as food was not confined to the white man. Keeler (1904:223) quotes the Rev. James Finley, an early and outstanding M.E. minister of northern Ohio, "I left my horse at Fort Ball and hired two young Indians to take me to Portland [the present Sandusky on the bay] in a bark canoe. We started about noon and the Sandusky River being very full, our bark canoe went over the rapids almost with the swiftness of a bird. But when we got to eddy water which we reached a short distance below Lower Sandusky [Fremont], we met schools of fish called sheep-head; and they much annoyed us by sticking fast to the bottom of our canoe. Once in a while one of the Indians who steered for us would take his butcher knife out of his belt and slip down his arm into the water and stab one of them and it would almost jump on board. But they not being good to eat, we cared not to take them."

The above is the only reference I have noted concerning drums gathering beneath a moving boat or canoe. I have seen this phenomenon upon a few occasions in Lake Erie and the Ohio River. It was always on still evenings in quiet waters. The fishes did not "stick fast" to the bottom of the boat, but some did press their backs against the boat bottom. Upon one occasion I was in a sheet iron boat when several drums not only followed closely beneath the boat but also began to grunt. When one of them pressed its back against the boat's bottom and grunted, the steel boat acted as a sounding board, and the grunting notes were magnified.

In his account of the sheepshead or drum Clinton (1815) gives another early reference to the undesirability of drums for human consumption, stating that "a very ill-tasted fish in Erie, is called the sheep's head, on account of a supposed resemblance to its salt water namesake."

I am impressed with the great similarity between the grunt of a drum and that of one of the grunts of the American buffalo or bison. It is my belief, based on circumstantial evidence, that the fish species which the Indian called the buffalofish was the drum and not a species of *Ictiobus*.

Van Meter and Trautman (1970:77) believe that the species "apparently has increased in Lake Erie waters in recent years, possible because of the decline in numbers of other species" and/or the result of less competition for food and living space. For some aspects of the spawning population in western Lake Erie, *see* Daiber (1953:159–71).

Populations continued to be small in the larger inland streams. The Ohio River populations, however, were rather large and may have increased in size since 1950. In their Ohio River investigations, in 1957–59, William M. Clay and associates found that the drum was one of the ten most abundant species in population samples at those collecting stations in, and adjacent to, the state of Ohio (ORSANCO, 1962:126, 129). In 1968 H. Ronald Preston captured the species in eight of nine collecting stations on the River adjacent to the state of Ohio. Margulies and Burch obtained specimens from all five of their collecting stations adjacent to Ohio.

SPOONHEAD SCULPIN

Cottus ricei (Nelson)

Fig. 175

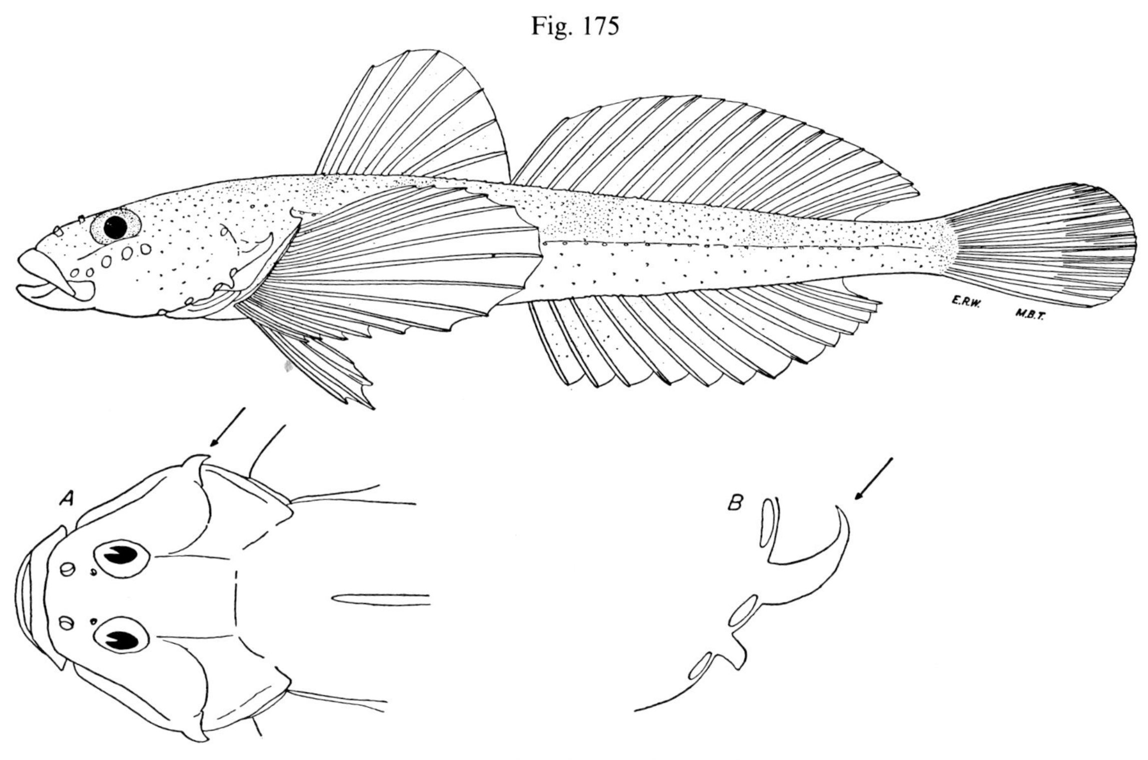

Upper fig.: Lake Erie, Erie County, O.

Nov. 7, 1928. 32 mm SL, 1.4″ (3.6 cm) TL.
Female. OSUM 205.

Fig. A: dorsal view of head; note triangulate shape of head; arrow points to position of the preopercle spine on the right side of the head.

Fig. B: lateral view of the preopercle spine with skin removed, showing its length and curved shape; note the three epeopercle pores.

Identification

Characters: *General* and *generic*—No scales; body sometimes partly or entirely covered with prickles; body catfish-like in shape. Two dorsals, the first of 6–9 weak, flexible spines; these spines bend as readily as, and look like, unbranched rays. Soft dorsal of 15–19 rays. Length of base of soft dorsal as long as or longer than, anal fin base. Anal fin without spines. Pelvic fins each with a hidden, very small spine and four rays. Gill covers attached directly to isthmus; not directly to each other. Tail rounded. *Specific*—Lateral line complete. Preopercle spine (arrow, fig. A) long and usually curved into a half circle, fig. B, arrow. When viewed from above, head outline appears to be somewhat triangular and snout rather pointed, fig. A. Saddle-bands faint or absent. Skin often prickly.

Differs: Mottled sculpins have incomplete lateral line; shorter, less curved preopercle spines. Small burbot has a chin barbel; more rays in soft dorsal. Catfishes have adipose fin. No other species of fish in Ohio has combination of: scaleless body; flexible dorsal spines; no stiff anal or pectoral spines; no barbels about mouth.

Most like: Mottled sculpin. **Superficially like:** Young burbot and catfishes.

Coloration: Dorsally olive-yellow, speckled and

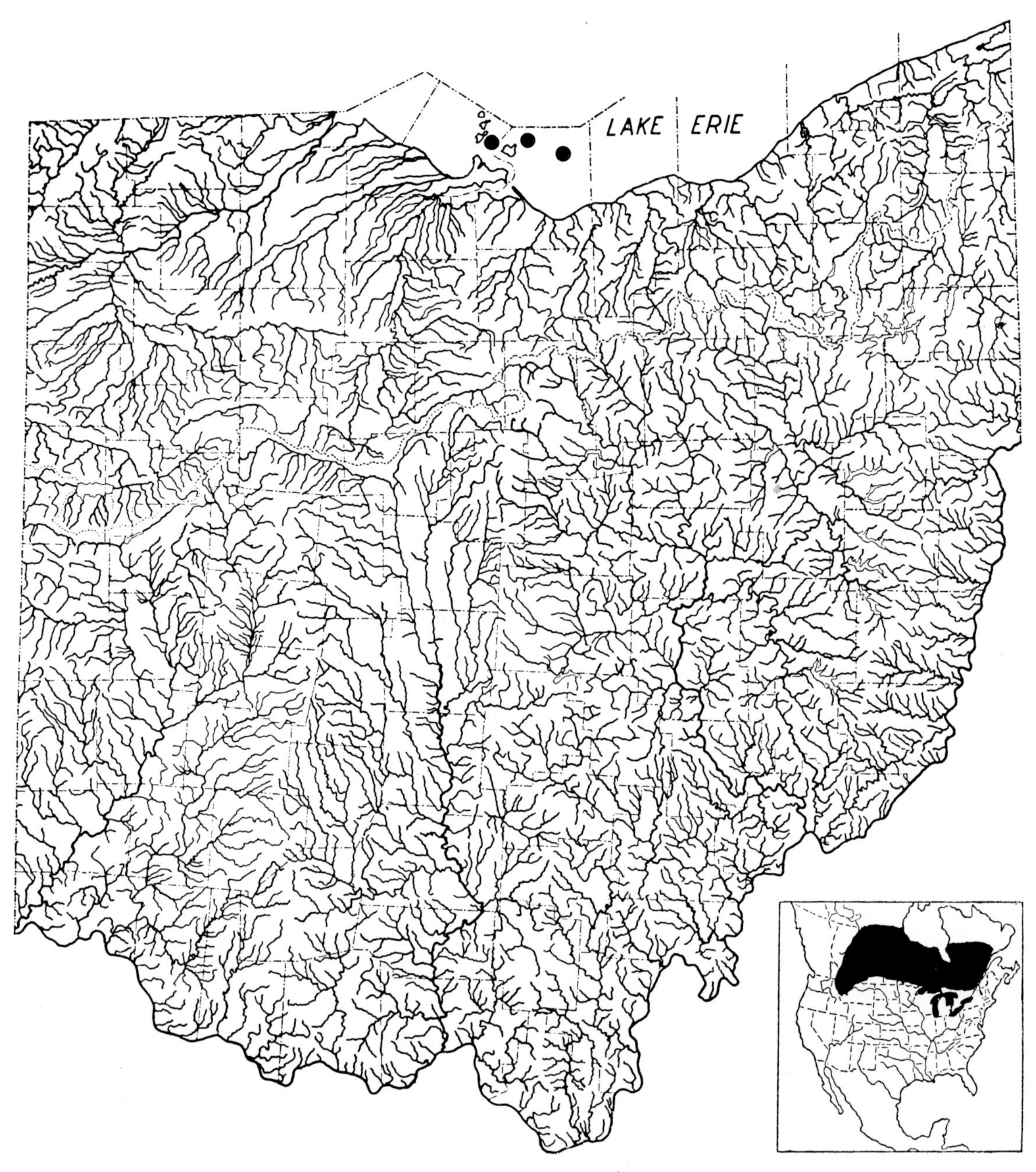

MAP 175. Spoonhead sculpin

● Locality records.
Insert: Apparently large unoccupied areas occur within this range; Canadian range limits indefinite.

mottled with olive-brown. Sides paler with some light yellow. Belly white. Usually 4–8 vague, ill-defined saddle-bands or blotches crossing back. An ill-defined dark, vertical caudal bar. Color of fins similar to adjacent body parts; dorsals and caudal faintly barred in the larger fishes.

Lengths: of the three Ohio specimens—1.5″ (3.8 cm); 2.4″ (6.1 cm); and 2.4″ (6.1 cm).

Distribution and Habitat

Ohio Distribution—The three records for the Ohio waters of Lake Erie for this deep-water sculpin are: one taken November 7, 1928, by Edward L. Wickliff and Wilbur M. Tidd, while they were trawling on the bottom 6 miles (9.7 km) northeast of Kelleys Island, in water about 40′ (12 m) deep (OSUM:205); one taken June 26, 1950, by Franklin

C. Daiber, while he was trawling on the bottom between South Bass and Kelleys islands in water about 25′ (7.6 m) deep (OSUM:9277); one taken June 21, 1950, by Edward C. Kinney and Daiber, while they were trawling on the bottom 13 miles (21 km) east of Kelleys and 13 miles (21 km) northwest of Vermilion in water about 50′ (15 m) deep (OSUM:9248).

The first Lake Erie record for this species was taken to the east of Ohio on August 25, 1928, in a Helgoland trawl, at a depth of 22 m (132′); later in the same year two large specimens were taken from the stomach of a burbot which had been captured near Dunkirk, N.Y. (Fish, 1932:391).

Habitat—From the above records it is assumed that the spoonhead sculpin occurred in at least small numbers in the deeper waters of Lake Erie, presumably over a sand, gravel, or bedrock bottom.

Years 1955–80—A moderate amount of trawling in waters of more than 25′ (7.6 m) in depth, in the Ohio portion of Lake Erie have, to my knowledge, produced no spoonhead sculpins since 1950.

CENTRAL MOTTLED SCULPIN*

Cottus bairdi bairdi Girard

Fig. 176

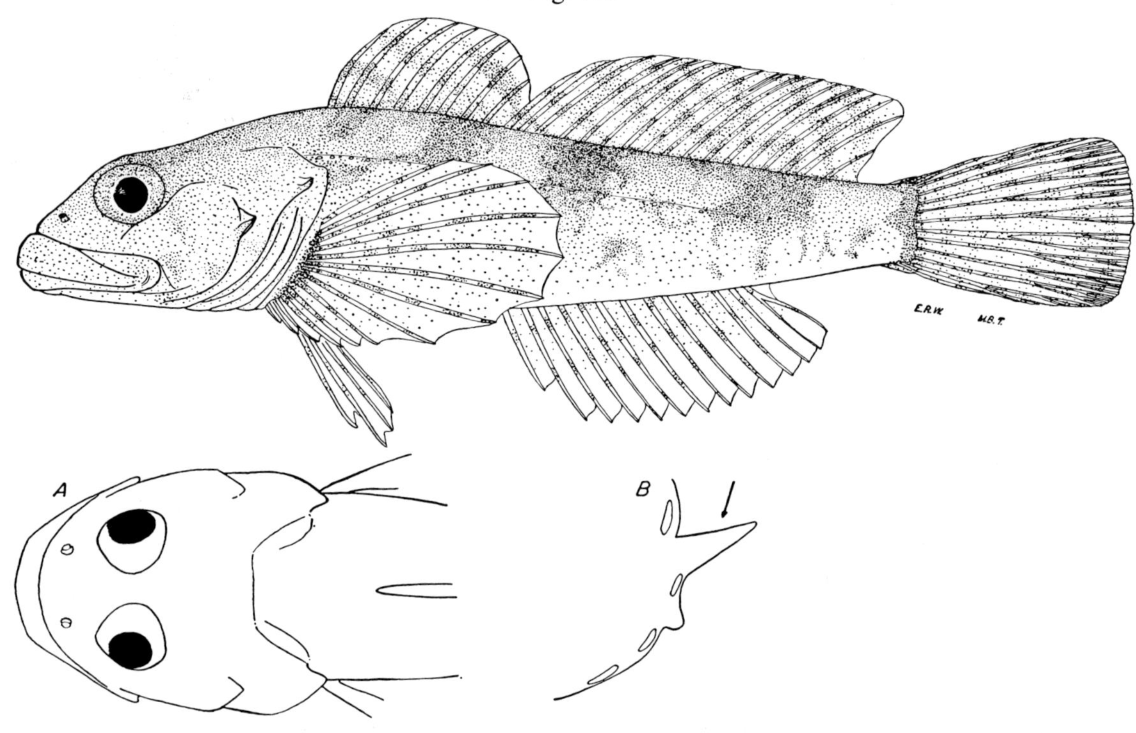

Upper fig.: Mad River, Logan County, O.

April 6, 1942. 54 mm SL, 2.6″ (6.6 cm) TL.
Breeding male. OSUM 4764.

Fig. A: dorsal view of head; note broadly rounded snout and the quadrate-shaped head.

Fig. B.: lateral view of the preopercle spine with skin removed, showing its stout appearance and almost straight shape; note the four preopercle pores.

Identification

Characters: *General* and *generic–See* spoonhead sculpin. *Specific*—Lateral line incomplete, ending beneath soft dorsal. Preopercle spine (for position on head, *see* spoonhead sculpin, fig. A, arrow) short, stout and only slightly curved, fig. B. Saddle-bands more or less prominent. Skin seldom prickly. *Subspecific*—A deeper more robust and more tadpole-shaped body; when viewed from above the head is squarish and the snout broadly rounded, fig. A; saddle-band or bands under first dorsal usually conspicuous and well developed; those under second dorsal almost invariably bold and clear cut (except in smallest young), as is the vertical or triangulate caudal bar; incomplete lateral line averaging longer, usually contains 20–24 pores, rarely fewer; anal papilla of adult male moderate in size and never as long as first anal ray.

Differs: Spoonhead sculpin has complete lateral line. Northern mottled sculpin has a more triangular-shaped head; a less contrasting color pattern, and more lateral line pores.

Most like: Northern mottled and spoonhead sculpins.

Superficially like: Young burbot and catfishes.

Coloration: Dorsally slaty- or olive-brown, heavily mottled and speckled with chocolate-browns and

* Formerly redfin sculpin.

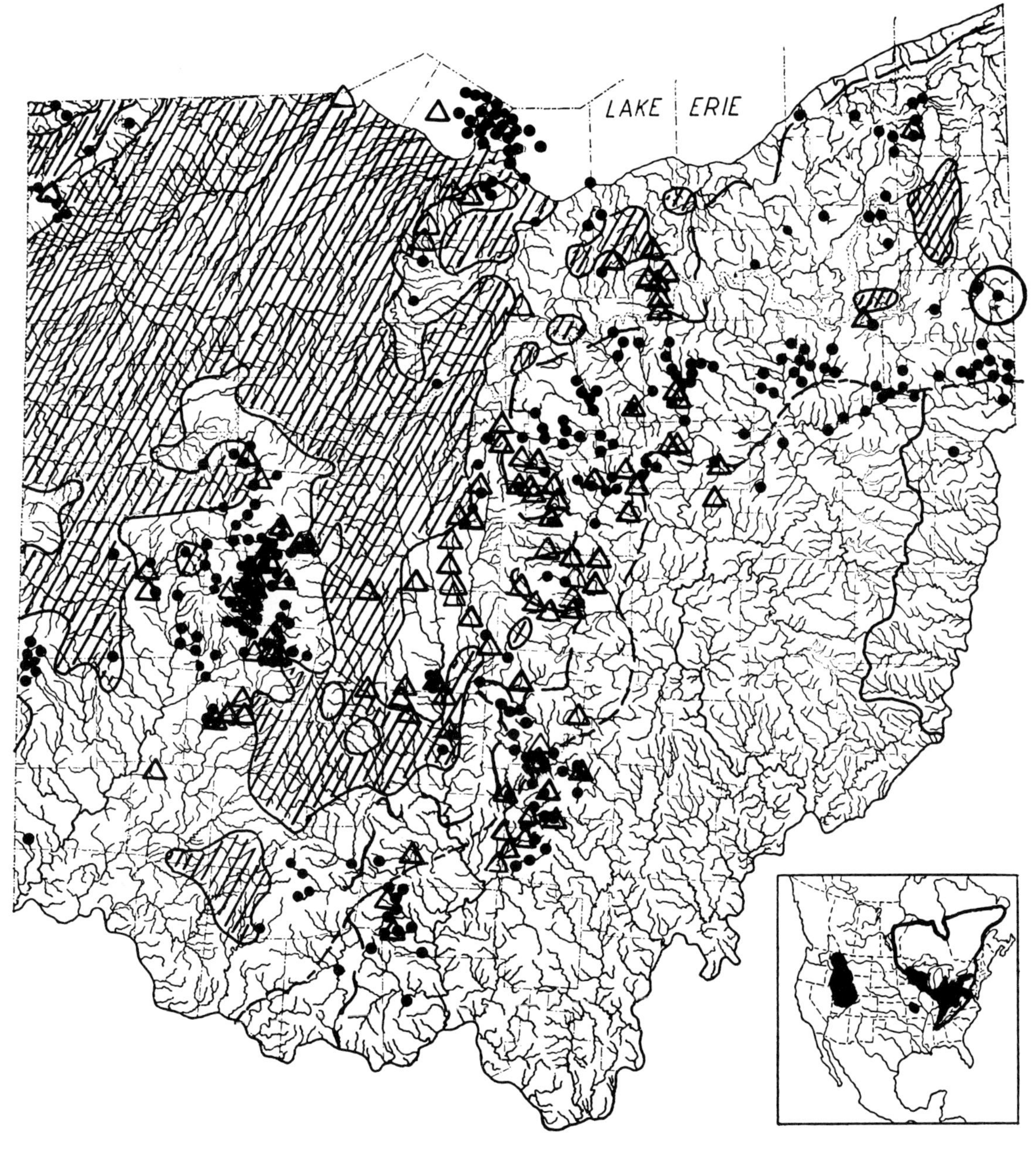

MAP 176. Central mottled sculpin

○ Type locality, Mahoning River at Poland, Mahoning County, O. Locality records. ● Before 1955. △ 1955–80.
/\ Allegheny Front Escarpment. ----- Glacial boundary. ~ Flushing Escarpment.
Shaded areas have maximum relief of less than 100 feet (30.5 m).

Insert: Largest black area represents range of this subspecies, two smaller black areas appear to be relict populations of this or another form; outlined areas represent ranges of other subspecies.

blacks. Four saddle-bands; the two beneath the first dorsal are squarish, short and often inconspicuous; the two beneath the second dorsal are long and extend obliquely across sides to form conspicuous dark bands. Sides paler. A very black, vertical caudal bar which often is triangular in outline. Ventral surface of head and body whitish, ventral head surface generally speckled with dusky. All fins dark olive or brown basally; dorsals heavily spotted, the spots tending to form oblique bands; caudal with wavy bars; anal rather transparent distally; pelvics pale and with few or no spots; spots on pectorals forming wavy bands. *Breeding male*—Contains much dusky; spotting chocolate-black; belly bluish-white. *Young*—Like adults, but fins more transparent.

Lengths: *Young* of year in Oct., 1.2″–2.1″ (3.0–5.3 cm) long. *Adults*, usually 1.8″–4.0″ (4.6–10 cm). Largest specimen, 4.5″ (11 cm) long.

Distribution and Habitat

Ohio Distribution—Except for Lake Erie, the distribution of the central mottled sculpin in glaciated Ohio since 1900 has been correlated with the presence of glacial moraines (*see* glacial moraines, map VII) and a relative relief of more than 100′ (30.5 m) (*see* relative relief, map VI). The largest and densest populations in the glaciated portions were centered about the Mad River drainage. A large but less dense population extended from Morrow eastward to Columbiana County. A minor population was in the morainal country that is drained by the headwaters of the Whitewater River in Darke and Preble counties; a scattered population was in Geauga and southwestern Ashtabula counties and was centered about the Cleveland moraine; and another scattered population was present about the Fort Wayne and Wabash moraines in Williams and Defiance counties. The largest populations of this subspecies in Lake Erie were centered about the rocky islands of western Lake Erie (*see* northern mottled sculpin).

In unglaciated Ohio there were two distributional centers; one was in western Pike County and was confined primarily to the clear, upland, limestone streams; the other was in Hocking County in the headwater brooks which were cutting into the acid sandstones and soils of the hills. The absence of this sculpin from the high-gradient gravel and bedrock brooks east of the Flushing Escarpment is puzzling, and may be largely the result of too much pollution and presence of flat shales instead of rounded gravel under which this species hides. Its absence from the mature streams in the area from Jackson northeast to Coshocton County is understandable, since only isolated and usually widely separated habitats occurred there.

It is apparent from the early literature relative to conditions in Ohio, that before 1850 the habitat for this small-stream, clear-brook species was far more widespread than it has been since. Then there were innumerable small, permanent, spring-fed brooks which since have been completely eliminated, as were the many brooks in areas now occupied by large cities; or which have been rendered uninhabitable because of habitat destruction through dredging, ditching, polluting or silting, or from a retreating ground-water table which caused the brooks to cease flowing. Between 1925–30 I observed more than a score of flourishing populations, in as many small brooks, that since then have disappeared because of destruction of their habitat.

The distributional patterns of Ohio, western Pennsylvania (Raney, 1939A: map 51) and Indiana (Gerking, 1945: map 113) form a closely knit unit.

Habitat—The densest populations occurred in brooks or streams of clear water, having a rocky or sandy gravel bottom, with or without aquatic vegetation. The majority of the more flourishing populations were in brooks of high gradients; but occasionally large populations were present in low-gradient streams of unusually clear water, having bottoms relatively free from clayey silts or other pollutants, and a permanent flow. Permanency of stream flow was a prime requisite; for this sculpin was essentially a headwater fish which did not normally migrate downstream in late summer, but remained in the headwaters until isolated in the non-flowing pools. The Lake Erie population lived under and among the boulders and gravel or in the bedrock crevices of the wave-washed beaches, and in the deeper waters.

The clear-cut Ohio distributional pattern of this sculpin strikingly demonstrates the absolute need of a suitable habitat to a species of animal. Although this sculpin was present in all major stream systems in numbers sufficient to "stock" these entire systems, it was confined to only those portions of these systems where its habitat niche occurred. A few isolated habitats contained no sculpins, but obviously these had once contained populations which were subsequently destroyed by pollutants, or these localities were so isolated from existing populations that they had as yet not received any strays of this sedentary form to "stock" them.

Years 1955–80—Before 1950 this subspecies of the mottled sculpin was numerous to abundant about the rocky shores of the islands of western Lake Erie. It was also apparently numerous in the deeper waters of that region. As an example of its abundance:

About the Bass Islands on March 25, 1947, a high southwest wind lowered the lake level approximately 3′ (0.9 m). In a small area at Peach Point, South Bass Island, three of us captured 282 sculpins in an hour, securing them with our hands from beneath boulders where they had been trapped by the lowering water level. It was apparent at that time that the population present about South Bass Island numbered in the many thousands.

During the summers of 1967–69, ichthyology classes and I collected about the shores of the Bass Islands without taking a specimen. The last individual obtained from the Island region was captured in 1953; *see* distribution map.

During the 1955–80 period individuals of this subspecies were taken in Ohio streams upon more than sixty occasions. Flourishing populations continue to exist in the clearest, cleanest brooks of higher gradients; but there was a reduction in, or extirpation of, populations in those streams which had increased silts or other pollutants, were dredged or otherwise modified. It is interesting to note on the distribution map that the collections made since 1950 fall within the pre-1950 distributional pattern.

I am following the American Fisheries Society (Bailey et al., 1970:58) in changing the common name of this species from the redfin sculpin, used in the 1957 edition of the *Fishes of Ohio*, to mottled sculpin. For food of sculpins *see* Turner, 1922:195–96; and Koster, 1937:374–82.

NORTHERN MOTTLED SCULPIN*

Cottus bairdi kumleini (Hoy)

Fig. 177

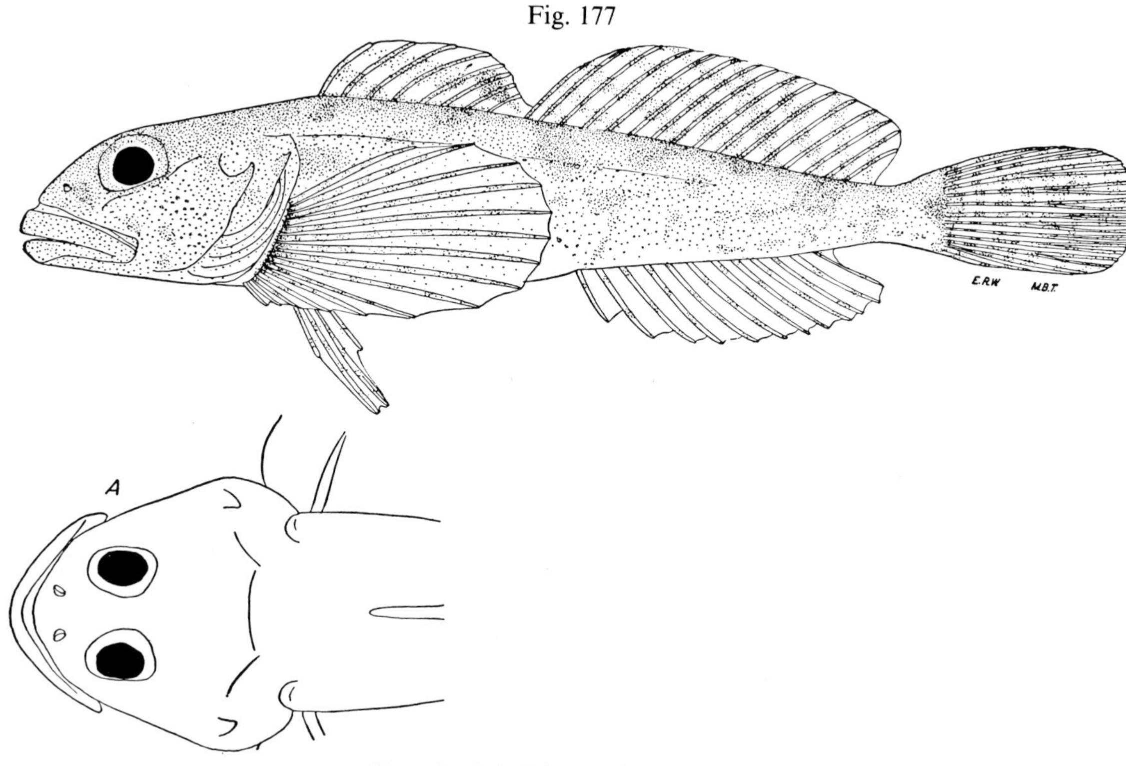

Upper fig.: Lake Erie, Ashtabula County, O. Nov. 11, 1943. Male. 53 mm SL, 2.5″ (6.4 cm) TL. OSUM 5962.

Fig. A: dorsal view of head; note the triangular-quadrate shape of head.

Identification

Characters: *General* and *generic—See* spoonhead sculpin. *Specific—See* central mottled sculpin. *Subspecific*—Body averaging more slender; when viewed from above the head is more triangular than square in outline, the snout more pointed; saddle-band or bands under first dorsal usually absent or if present are vague blotches which are never distinct; caudal bar triangular and faint when present; incomplete lateral line averaging shorter, usually contains fewer than 19 pores; anal papilla of adult male long, sometimes exceeding length of first anal ray.

Differs: *See* central mottled and spoonhead sculpins.

Most like: Central mottled and spoonhead sculpins. **Superficially like:** Young burbot and catfishes.

Coloration: Similar to central mottled sculpin, except that the colors average lighter, there are more yellows and light browns, mottlings are less bold, saddle-bands less sharply defined.

Lengths: *Young* of year in Nov., 1.5″–2.0″ (3.8–5.1 cm) long (only 12 specimens). *Adults*, 1.8″–2.5″ (4.6–6.4 cm) (only five specimens, the largest and smallest with well-developed eggs).

Distribution and Habitat

Ohio Distribution—Typical examples of this distinctive subspecies have been taken in the Ohio

* Formerly redfin sculpin.

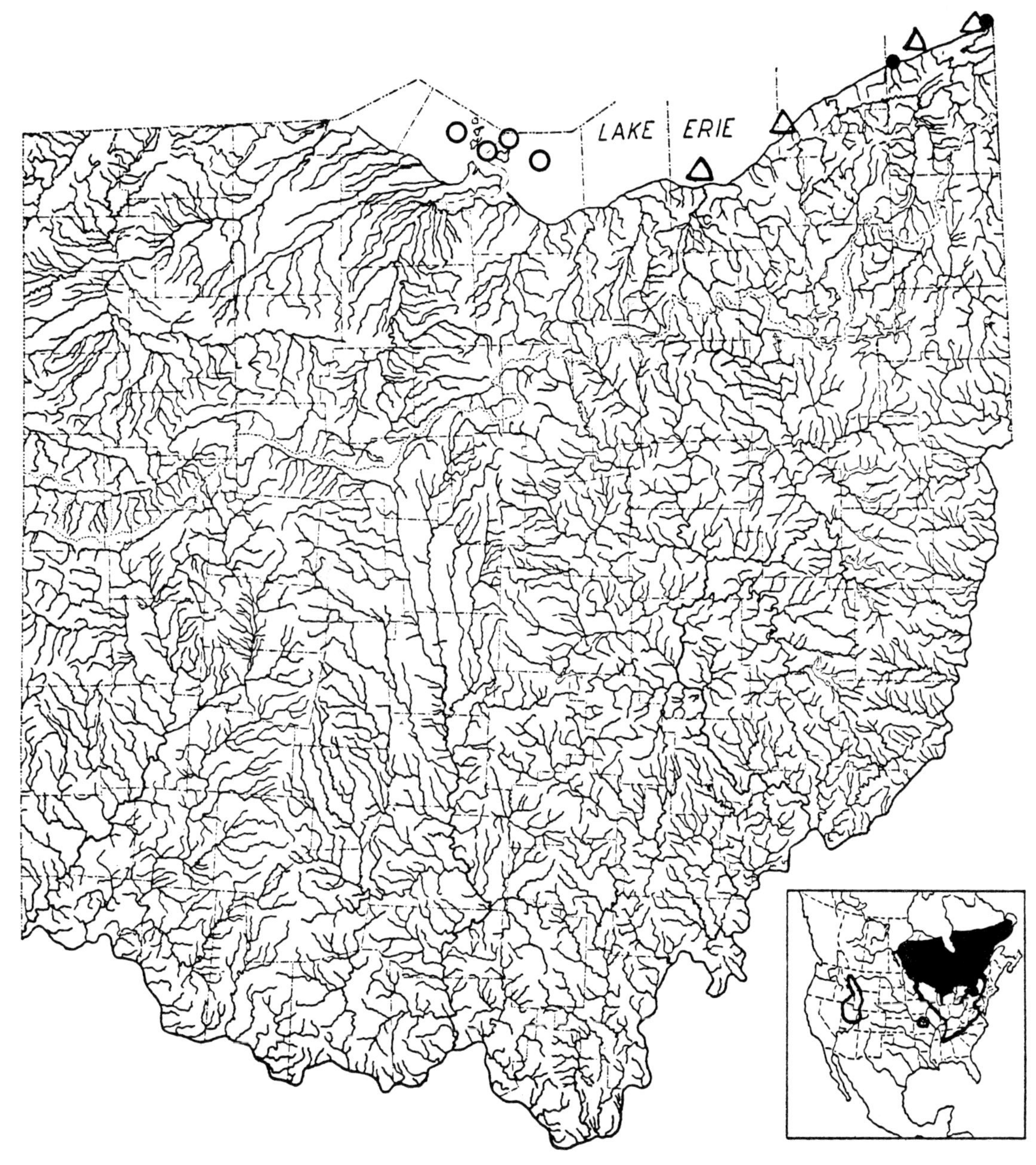

MAP 177. Northern mottled sculpin

○ Possible intergrades between northern and central mottled sculpins.
Locality records of northern mottled sculpins. ● Before 1955. △ 1955–80.
Insert: Black area represents range of this subspecies, outlined areas represent ranges of other subspecies; northern limits of range indefinite.

waters of Lake Erie only in Ashtabula County, and then only during a few days or weeks previous to shore ice formation in fall (November 11, 1954: OSUM:5962) and immediately after the shore ice break-up in spring (March 25–26, 1946: OSUM: 6808 and 6810). Collecting along the Lake Erie shore in Ashtabula County during the warmer months failed to produce either northern or central mottled sculpins. It is assumed therefore that although the northern mottled sculpin may have been numerous in the deep waters of the eastern end of Lake Erie (Fish, 1932:387–89), it inhabited the shallow waters of the beaches in that section only during the colder months.

There were large populations of mottled sculpins about the Bass Islands of western Lake Erie. A comparatively few specimens, taken about these islands in trawls in water deeper than 25′ (7.6 m), have somewhat intermediate characters between the northern and central mottled sculpins; but the hundreds of specimens taken in the shallow waters along the gravel, rock-strewn island beaches appear to be typically central mottled sculpins. This appearance of intermediacy in specimens from the deeper waters may be caused by actual intergradation between the two forms, or may be an environmental response in the central mottled to deep water. In inland Ohio, where only the central mottled occurred, this subspecies showed marked differences in body proportions, coloration, and number of pectoral rays in response to different environments. As an example: the average specimen from the open riffles of the Mad River had body proportions, coloration, and pectoral ray counts which showed an approach to the characters of the Northern mottled sculpins, whereas the average specimen from an adjacent, small tributary, which was sluggish and filled with watercress, was an extreme type of the central mottled sculpin.

Habitat—The northern mottled sculpin was presumably an inhabitant of the deeper waters of eastern Lake Erie throughout the year, invading the shallow waters of the beaches of that section of the lake in small numbers during the colder months. When inshore, they hid under stones, logs, or similar objects, and were then usually associated with the longnose dace.

Years 1955–80—During this period this subspecies was captured in Lake Erie along the shores from Cuyahoga, Lake and Ashtabula counties. On May 22, 1974 one was taken in Lake Erie from the city of Cleveland's water intake crib, situated approximately 2 miles (3.2 km) north of the Cleveland shoreline and in water about 35′ (11 m) deep. Since 1975 Andrew White (et al., 1975:124–25) and his associates have, with improved collecting equipment and techniques, found this subspecies to be fairly numerous in Lake Erie, especially in Ashtabula County, and have captured about 75 specimens including both adults and small young.

Literature Cited*

ABBOTT, CHARLES C.

1860. Descriptions of new species of American fresh-water fishes. Proc. Acad. Nat. Sci., Phil.: 327–28.

ADAMS, CHARLES C., AND THOMAS L. HANKINSON

1928. The ecology and economics of Oneida Lake fish. Bull. New York St. Coll. of Forestry, Syracuse Univ. 1. No. 4a (Roosevelt Wild Life Annals, 1, Nos. 3 and 4): 235–548, pls. 4, figs. 175–244, map 16.

ADAMSTONE, F. B.

1922. Rates of growth of the blue and yellow pike perch. Univ. Toronto Studies, Biol. Ser. No. 5 (Publ. Ont. Fish. Res. Lab., No. 37): 77–86.

ALBAUGH, DOUGLAS WILLIAM

1966. A study of the distribution of the fishes of the Meigs Creek, Ohio watershed. Unpublished, Ohio Univ.: 1–78, 41 figs. (mostly distribution).

AMERICAN ORNITHOLOGISTS' UNION

1957. Check-list of North American birds. 5th ed. Baltimore, Md.: 1–691.

Anonymous

1940. The Ohio guide. Oxford Univ. Press: New York (Compiled by workers of the Writers' Program of the Works Projects Admin. in State of Ohio): 634p.

Anonymous A

1960. Fish populations recovery in the Scioto River. Clean waters, Ohio Water Poll. Control Board 9 (3): 8–11.

Anonymous B

1968. Commercial fish landings for Lake Erie, Ohio, 1968. Pub. 200, Ohio Dept. of Nat'l. Res., Div. of Wildlife: 1–9.

Anonymous C

Where the sportsman's dollar goes. Financial Report of the Division of Conservation for 1932, 1933 and Accomplishments to May 31st, 1934. [Ohio] Div. of Conservation: unnumbered.

Anonymous D

1979. Status of Ohio's Lake Erie fisheries. Ohio Dept. of Natur. Resources. Div. of Wildlife. Sandusky, Ohio: 1–19, 10 tabs.

APPLEGATE, VERNON C.

1943. Partial analysis of growth in a population of mudminnows, *Umbra limi* (Kirtland). Copeia (2): 92–96.

1947. The menace of the sea lamprey. Michigan Dept. of Cons. 16, (4): 6–7 and 10, pls. 4.

1950. Natural history of the sea lamprey, *Petromyzon marinus*, in Michigan. U. S. Dept. of Interior, Fish and Wildlife Serv., Spec. Sci. Rept.: Fish. 55: 1–235, figs. 65.

APPLEGATE. VERNON C., AND HARRY D. VAN METER

1970. A brief history of commercial fishing in Lake Erie. Fishery Leaflet 630, U. S. Fish and Wildlife Serv., Bur. of Commercial Fisheries: 1–28, 16 photos, 8 figs., 1 tab.

BAILEY, REEVE M.

1938. The fishes of the Merrimack watershed. *In*: Biological survey of the Merrimack watershed. New Hampshire Fish and Game Comm.: 129–85.

1951. A check-list of the fishes of Iowa, with keys for identification. *In*: Iowa fish and fishing, by James R. Harlan and Everett B. Speaker. State of Iowa; 186–257, figs. 1–9.

1959. Distribution of the American cyprinid fish *Notropis anogenus*. Copeia (2): 119–23, 1 fig., 1 tab.

BAILEY, REEVE M., AND MARVIN O. ALLUM

1962. Fishes of South Dakota. Misc. Publ. Mus. Zool. Univ. Mich. 119: 131 p., 7 figs., 9 tabs., 1 pl.

BAILEY, REEVE M., AND FRANK B. CROSS

1954. Fishes from the Escambia River, Alabama and Florida, with ecologic and taxonomic notes. Proc. Acad. Mich. Acad. Sci., Arts and Letters 39: 169–208, 10 figs., 4 tabs.

* Several of the rarer publications and unpublished manuscripts are in the Ohio State Museum library.

BAILEY, REEVE M., JOHN E. FITCH, EARL S. HERALD, ERNEST A. LACHNER, C. C. LINDSEY, C. RICHARD ROBINS, AND W. B. SCOTT

1970. A list of common and scientific names of fishes from the United States and Canada, 3d ed. Amer. Fish. Soc. Spec. Publ. 6: 150 p.

BAILEY. REEVE M., AND CARTER R. GILBERT

1960. The American cyprinid fish *Notropis kanawha* identified as an interspecific hybrid. Copeia (4): 354–57.

BAILEY, REEVE M., AND WILLIAM A. GOSLINE

1955. Variation and systematic significance of vertebral counts in the American fishes of the family Percidae. Misc. Publ. Mus. Zool. Univ. Mich. 93: 1–44, figs. 1–4.

BAILEY, REEVE M., ERNEST A. LACHNER, C. C. LINDSEY, C. RICHARD ROBINS, PHIL M. ROEDEL, W. B. SCOTT, AND LOREN P. WOODS

1960. A list of common and scientific names of fishes from the United States and Canada, 2d ed. Amer. Fish. Soc. Spec. Publ. 2: 1–102.

BAILEY, REEVE M. AND KARL F. LAGLER

1938. An analysis of hybridization in a population of stunted sunfishes in New York. Pap. Mich. Acad. Sci., Arts and Letters 23 (1937): 577–606.

BAILEY, REEVE M., AND WILLIAM RALPH TAYLOR

1950. *Schilbeodes hildebrandi*, a new Ameiurid catfish from Mississippi. Copeia (1): 31–38, 2 pls.

BAILEY, REEVE M., HOWARD ELLIOTT WINN, AND C. LAVETT SMITH

1954. Fishes from the Escambia River, Alabama and Florida, with ecologic and taxonomic notes. Proc. Acad. Nat. Sci. Phila. 106: 109–64.

BANARESCU, PETRU

1964. Fauna Republicii Populare Romine (Pisces-Osteichthyes). Bucuresti, 13: 1–959.

BARBOUR, THOMAS

1911. The smallest *Polyodon*. Biol. Bull., 21, (4): 207–14, pls. 3.

BARNES, M. D., AND R. F. CARLINE

1977. A survey of the eastern sand darter, *Ammocrypta pellucida* (Putnam), in streams of the Wayne National Forest, Ohio. Final Project Rpt., National Forest Serv., Bedford, IN.: 1–40, 2 figs., 4 tabs.

1978. A survey of the eastern sand darter, *Ammocrypta pellucida* (Putnam), and other fishes in Symmes Creek and Pine Creek, Wayne National Forest, Ohio. Cooperative Project of the Ohio Cooperative Research Unit and the U. S. National Forest Service, Eastern Region: 1–24, 2 figs., 4 tabs.

BATTLE, HELEN I., AND WILLIAM M. SPRULES

1960. A description of the semi-buoyant eggs and early developmental stages of the goldeye, *Hiodon alosoides* (Rafinesque). J. Fish. Res. Bd. Can. 17 (2): 245–66, 11 figs., 2 tabs.

BAUER, BRUCE H., BRANLEY A. BRANSON, AND STRANT T. COLWELL

1978. Fishes of Paddy's Run Creek and the Dry Fork of the Whitewater River, southwestern Ohio. Ohio J. Sci. 78 (3), May: 144–51.

BEAN, TARLETON H.

1884. Catalogue of the collections of fishes exhibited by the United States National Museum. Bull. U. S. Nat. Mus. 27 (1883): 387–510.

1892. The fishes of Pennsylvania, with descriptions of the species and notes on their common names, distribution, habits, reproduction, rate of growth and mode of capture. Harrisburg. 149p. pls. 16–35.

BECKER, HERBERT RAY

1923. The habitat of *Aphredoderus sayanus* in Kalamazoo County, Michigan. Occ. Pap. Mus. Zool. Univ. Mich. 138: 1–4.

BECKMAN, WILLIAM C.

1949. The rate of growth and sex ratio for seven Michigan fishes. Trans. Amer. Fish. Soc. Ann Arbor 76: 63–81, figs. 7, tabs. 7.

BEETON, ALFRED M.

1961. Environmental changes in Lake Erie. Trans. Am. Fish. Soc., 90: 153–59.

1965. Eutrophication of the St. Lawrence Great Lakes. Limnol. and Oceanog. Lawrence, Kans., 10 (2): 240–54.

1969. Eutrophication: causes, consequences, correctives. Nat. Acad. Sci.: 150–87.

BERG, LEO S.

1931. A review of the lampreys of the Northern Hemisphere. Ann. Mus. Zool. Acad. Sci. U.R.S.S. 32: 87–166, pls. 8.

1940. Classification of fishes, both recent and fossil. Trav. Inst. Zool. Acad. Sci. U.R.S.S. Tome 5, livr. 2: 87–517, figs. 190 [Both Russian and English text.]

BIGELOW, HENRY B., AND WILLIAM C. SCHROEDER

1948. Fishes of the Western North Atlantic. Part 1, Lancelets, Cyclostomes, Sharks. Yale Univ.: 576 p.

BIRDSONG, RAY S., AND RALPH W. YERGER

1967. A natural population of hybrid sunfishes: *Lepomis macrochirus* × *Chaenobryttus gulosus*. Copeia (1): 62–71.

BLACK, JOHN D.

1945. Natural history of the northern mimic shiner *Notropis volucellus volucellus* Cope. Invest. Ind. Lakes and Streams. Ind. Dept. of Cons. and Ind. Univ., Indianapolis 2: 449–69.

BLACK, JOHN D., AND LYMAN O. WILLIAMSON

1947. Artificial hybrids between muskellunge and northern pike. Trans. Wisc. Acad. Sci. and Letters 38: 299–314, figs. 1–9.

BORROR, DONALD J.

1950. A check list of the birds of Ohio. Ohio J. Sci. 50, (1): 1–32.

BOWMAN, MILTON L.

1970. Life history of the black redhorse, *Moxostoma duguesnei* (Lesueur), in Missouri. Trans. Amer. Fish. Soc. 99 (3): 546–59.

BOWNOCKER, JOHN A.

1920. Geologic map of Ohio. Ohio Geol. Surv.

BRAUN, E. LUCY

1934. A history of Ohio's vegetation. Ohio J. Sci., 34, (4): 247–57.

1951. Plant distribution in relation to the glacial boundary. Ohio J. Sci. 51, (3): 139–46.

BREEDER, C(HARLES) M., JR., AND D. R. CRAWFORD

1922. The food of certain minnows. Zoologica. Sci. Contrib. of New York Zool. Soc. 2 (14): 287–327.

BRITT, N. WILSON

1955. Stratification in western Lake Erie in summer of 1953: Effects on the Hexagenia (Ephemeroptera) population. Ecology 36: 239–44.

1966. Benthic changes in the island area of western Lake Erie during the past 15 years as indicated by 1959–1965 bottom fauna collections. Wheaton Club Bull., Columbus, Ohio, 11: 14–15.

BROWN, C. J. D.

1966. Natural hybrids of *Salmo trutta* × *Salvelinus frontalis*. Copeia (3): 600–601.

BROWN, EDWARD J., JR.

1960. Little Miami River headwater-stream investigations. Ohio Dept. of Nat'l. Res., Div. of Wildlife: 143 p.

BROWN, ROLAND WILBUR

1956. Composition of scientific words. Published by author. Reese Press, Baltimore: 882 p.

BROWN, SAMUEL R.

1815. Views of the campaigns of the north-western Army, etc.—comprising sketches of the campaigns of generals Hull and Harrison, a minute and interesting account of the naval conflict on Lake Erie. William G. Murphey, Printers, Phila: 1–156 (includes, "View of the Lake Coast from Sandusky to Detroit": 127–56).

1817. The Western Gazetter or emigrants directory—containing a geographical description of the western states and territories, viz. the states of Kentucky, Indiana, Louisiana, Ohio, Tennessee, Mississippi and the territories of Illinois, Missouri, Alabama, Michigan and North Western. Printed by H. C. Southwick, Auburn, N. Y.: 1–352.

BURR, BROOKS M., AND MICHAEL A. MORRIS

1977. Spawning behavior of the Shorthead Redhorse, *Moxostoma macrolepidotum*, in Big Rock Creek, Illinois. Trans. Amer. Fish. Soc. 106 (1):80–82.

BURT, WILLIAM HENRY, AND RICHARD PHILIP GROSSENHEIDER

1952. A field guide to the mammals. Houghton Mifflin Co., Boston: 1–200.

BURTON, GEORGE W., AND EUGENE O. ODUM
1945. The distribution of stream fish in the vicinity of Mountain Lake, Virginia. Ecology 26, (2): 182–94, figs. 2, tabs. 6.

BUSCH, WOLF-DIETER N., DAVID H. DAVIES AND STEPHEN J. NEPSZY
1977. Establishment of white perch, *Morone americana*, in Lake Erie, J. Fish. Res. B. of Can. 34(7): 1039–41.

CAHN, ALVIN R.
1936. Observations on the breeding of the lawyer, *Lota maculosa*. Copeia (3): 163–65.

CARMAN, J. ERNEST
1946. The geologic interpretation of scenic features in Ohio. Ohio J. Sci. 46, (5): 241–83.

CAROOTS, MARK S.
1976. A study of the gizzard shad *Dorosoma cepedianum* (LeSueur) from Lake Erie near Cleveland, Ohio. John Carroll University M.S. Thesis: 1–128.

CAVENDER, TED M.
1981. Introduction and spread of native and non-native fishes into Ohio waters. O. J. Sci. 81(Apr.):18.

CAVENDER, TED M., AND RON L. CRUNKILTON
[1976.] Impact of a mainstream impoundment on the fish fauna of Big Walnut Creek a Scioto River tributary in central Ohio. Report 449. Water Resources Center Engineering Experiment Station. The Ohio State Univ. Supported in part by Office of Water Res. and Technol. U. S. Dept. of the Interior Proj. A-037-Ohio: 191 p., 43 figs., 58 graphs.

CAVENDER, TED M., AND CHRIS YODER
1973. The fishes of Cedar Bog - past and present. Cedar Bog Symposium. Ohio Biol. Surv. Circ. 4: 24–29.

CHILDERS, WILLIAM F.
1967. Hybridization of four species of sunfishes (Centrarchidae). Ill. Nat. Survey Bull. 29 (3): 159–214.

CHILDERS, WILLIAM F., AND GEORGE W. BENNETT
1961. Hybridization between three species of sunfishes (*Lepomis*). Ill. Nat. Hist. Survey Div., Biol. Notes 46: 1–15, 6 figs.

CLARK, CLARENCE F.
1942. The fishes of Auglaize County. Unpublished Master's Thesis, Ohio State Univ.: 165 p.
1949. New records of the pirate perch, *Aphredoderus sayanus* (Gilliams) in central-western Ohio. Copeia, (3): 219–20.
1951. The "Lake St. Marys" or "Grand Lake" story. Ohio Cons. Bull., Dept. Nat'l. Res., Columbus: 16–19.

CLARK, CLARENCE F., AND DARRELL ALLISON
1966. Fish population trends in the Maumee and Auglaize rivers. Ohio Dept. of Nat'l. Res. Div. of Wildlife: 1–51.

CLARK, FRANK N.
1887. 116. Report on distribution of fish and eggs from Northville and Alpena stations for season of 1885–86. Bull. U. S. Fish Comm. 6: 395–99.

CLARK, H. WALTON, AND CHARLES B. WILSON
1912. The mussel fauna of the Maumee River. Dept. of Commerce and Labor. Bur. of Fish. Doc. 757, Washington: 72 p. pls. 2.

CLARK, MINOR E.
[1937]. A list of the fishes of northeastern Kentucky. [Kentucky Game and Fish Comm.]: 11 p. unnumbered, Mimeo.

CLAY, WILLIAM M.
1975. The fishes of Kentucky. Kentucky Dept. of Fish and Wildlife Res. Frankfort, Ky.: 416 p., many photos.

CLEMENS, HOWARD PAUL
1951. The growth of the burbot *Lota lota maculosa* (Lesueur) in Lake Erie. Trans. Amer. Fish. Soc. 80 (1950): 163—73, 3 figs., 6 tabs.

CLEMENS, JAMES W.
1827. Art. 1. Notice on the spoonbill sturgeon, or paddle fish, of the Ohio, (*Polyodon feuille* of Lacépède). Amer. J. Sci. and Arts. 12: 201–05, pl. 1.

CLINTON, DE. W.
1815. Some remarks on the fishes of the western waters of the State of New York, in a letter to S. L. Mitchell. Trans. Lit. and Philos. Soc. N.Y. Vol. I: 493–501.

COBURN, MILES M.

1975. Evidence for a subspecific relationship between the common and striped shiners in the Chagrin River, northeastern Ohio. John Carroll University M.S. Thesis: 137 p.

COCKERELL, T. D. A., AND E. M. ALLISON

1909. The scales of some American Cyprinidae. Proc. Biol. Soc. Washington 22: 157–63.

COKER, ROBERT E.

1930. Studies of common fishes of the Mississippi River at Keokuk. Bull. U. S. Bur. Fish. 45(1929): 141–225, figs. 30.

COLE, LEON J.

1905. The German carp in the United States. Rept. U. S. Bur. of Fish. (1904): 523–641, pls. 1–3.

CONANT, ROGER, EDWARD S. THOMAS, AND ROBERT L. RAUSCH

1945. The plains garter snake, *Thamnophis radix*, in Ohio. Copeia (2): 61–68.

COOK, FANNYE A.

1959. Freshwater fishes in Mississippi. Mississippi Game and Fish Comm., Hederman Brothers, Jackson: 239 p., 2 maps, 4 tabs., 40 figs.

COOPER, JOHN E.

1979. Description of eggs and larvae of fantail (*Etheostoma flabellare*) and rainbow (*E. caeruleum*) darters from Lake Erie tributaries. Trans. of Amer. Fish. Soc. 108 (1): 46–56.

COPE, EDWARD D.

1865. Partial catalogue of the cold-blooded vertebrata of Michigan. Pt. 2, Proc. Acad. Nat. Sci., Phila. 17: 77–88.

1869. Synopsis of the Cyprinidae of Pennsylvania. Trans. Amer. Philos. Soc. New Series 13: 351–410, pls. 10–13, figs. 2.

1870. On some Etheostomine perch from Tennessee and North Carolina. Proc. Amer. Philos. Soc. 11: 261–70.

1871. A partial synopsis of the fishes of the fresh waters of North Carolina. Proc. Amer. Philos. Soc. 11(1869–70): 448–95, figs. 2.

1881. The fishes of Pennsylvania. Rept. [Pa.] State Comm. Fisheries (1879–80): 59–145, pls. 1–26, figs. 1–44.

1883. The fisheries of Pennsylvania. Rept. Fish. Comm. Pa. 1881–82: 103–94, figs. 26–44.

[CRAMER, ZADOK]

1818. The navigator, containing directions for navigating the Monongahela, Allegheny, Ohio and Mississippi rivers; with an ample account of these much admired waters, from the head of the former to the mouth of the latter; and a concise description of their towns, villages, harbors, settlements, etc. with maps of the Ohio and Mississippi to which is added an appendix containing an account of Louisiana, and of the Missouri and Columbia rivers, as discovered by the voyage under Capts. Lewis and Clark. 10th ed., Pittsburgh, Cramer and Spear: 1–304.

CREASER, CHARLES W.

1926. The structure and growth of the scales of fishes in relation to the interpretation of their life history, with special reference to the sunfish, *Eupomotis gibbosus*. Misc. Publ. Mus. Zool. Univ. Mich. No. 17: 82 p., pl. 1, figs. 12.

CREASER, CHARLES W., AND CLARE S. HANN

1929. The food of larval lampreys. Pap. Mich. Acad. Sci., Arts and Letters, 10 (1928): 433–37.

CREASER, CHARLES W., AND CARL L. HUBBS

1922. A revision of the holartic lampreys. Occ. Pap. Mus. Zool. Univ. Mich. No. 120: 1–14, pl. 1.

CROSS, FRANK B.

1967. Handbook of fishes of Kansas. Mus. of Nat. Hist. Univ. of Kansas, Robert R. Sanders, State Printer, Topeka, Kansas: 1–334, many unnumbered figs. and maps.

CROSS, FRANK B., AND W. L. MINCKLEY

1960. Five natural hybrid combinations in minnows, (Cyprinidae). Kan. Univ. Mus. Nat. Hist. Publ. 13(1): 1–18.

CROSSMAN, E. J.

1962. A colour mutant of the yellow perch from Lake Erie. Canadian Field-Naturalist 76 (4): 224–25.

CROSSMAN, E. J., AND H. D. VAN METER

1979. Annotated list of the fishes of the Lake Ontario watershed. Tech. Rep. 36. Great Lakes Fishery Comm. Ann Arbor, June: 1–25.

CUMMINS, ROBERT

1952. General catch reports of Ohio Lake Erie commercial fisheries for 1951. Ohio Dept. Nat. Res., Div. of Wildlife, Sec. of Fish Management: 1–10, Mimeo.

CUVIER, GEORGES L. C. F. D., ET ACHILLE VALENCIENNES

1844. Histoire naturelles des Poissons, 17: 1–497, pls. 487–519. (This edition prepared by Valenciennes alone.)

DAIBER, FRANKLIN C.

1952. The food and feeding relationships of the freshwater drum, *Aplodinotus grunniens* Rafinesque in western Lake Erie. Ohio J. Sci. 52, (1): 35–46, fig. 1.

1953. Notes on the spawning population of the freshwater drum, *Aplodinotus grunniens* (Rafinesque) in western Lake Erie. Amer. Midland Natur. 50 (1) : 159–71, 4 figs., 5 tabs.

DALL, WILLIAM HEALEY

1915. Spencer Fullerton Baird a biography. J. B. Lippincott Co., Phila. and London: 462 p.

DAVIS, B. J., AND R. J. MILLER

1967. Brain patterns in minnows of the genus *Hybopsis* in relation to feeding habits and habitat. Copeia (1): 1–39.

DEASON, HILARY J.

1933. Preliminary report on the growth rate, dominance, and maturity of the pike-perches (*Stizostedion*) of Lake Erie. Trans. Amer. Fish. Soc. 63: 348–60.

DEEVEY, E. S., JR.

1949. Biogeography of the Pleistocene. Bull. Geol. Soc. Amer. 60: 1315–416.

DISTLER, DONALD A.

1968. Distribution and variation of *Etheostoma spectabile* (Agassiz) (Percidae, Teleostei). Univ. of Kans. Sci. Bull., 48 (5) : 143–208, 1 fig., 16 tabs., 1 map.

DOAN, KENNETH H.

1941. Relation of sauger catch to turbidity in Lake Erie. Ohio J. Sci., 41, (6): 449–52.

1944. The winter fishery in western Lake Erie, with a census of the 1942 catch. *Ibid.* 44, (2): 69–74.

DRAKE, DANIEL

1850. Systematic treatise, historical, etiological, and practical of the principal diseases of the interior valley of North America, etc. Winthrop B. Smith & Co., Publ., Cincinnati: 878 p.

DUN, WALTER A.

1884. A brief sketch of the floods in the Ohio River. J. Cincinnati Soc. Nat. Hist. 7 (3): 104–24.

DYMOND, JOHN R.

1922. A provisional list of the fishes of Lake Erie. Univ. Toronto Studies, Biol. Ser. 20. Publ. Ont. Fish. Res. Lab. (4): 55–74.

1932. Records of the alewife and steelhead (rainbow) trout from Lake Erie. Copeia (1): 32–3.

EDDY, SAMUEL

1941. Muskellunge and musky hybrids. Conservation Volunteer: 41–44.

1944A. Hybridization between northern pike (*Esox lucius*) and muskellunge (*Esox masquinongy*). Proc. Minnesota Acad. Sci. 12: 38–43.

1944B. Hybridization between northern pike (*Esox lucius*) and muskellunge (*Esox masquinongy*). Reprint of Proc. Minn. Acad. Sci. 12: 38–43.

EDDY, SAMUEL, AND PARKE H. SIMER

1929. Notes on the food of the paddlefish and the plankton of its habitat. Trans. Ill. State Acad. Sci. 21: 59–63.

EDDY, SAMUEL, AND THADDEUS SURBER

1943. Northern fishes with special reference to the Upper Mississippi Valley. Univ. Minn. Press., Minneapolis: 1–252, pls. 57.

EDMISTER, J. O.

1940. Report of Ohio Lake Erie commercial fisheries 1939 season. Ohio Div. of Cons. and Nat'l. Res., Sec. of Fish Man. and Prop., Put-in-Bay, Ohio: 1–11, Mimeo.

EIGENMAN, CARL H.
1909. Cave vertebrates of America. Carnegie Inst. of Wash., Washington, 104: 241 p., pls. 1–29.

EMBODY, G. C.
1918. Artificial hybrids between pike and pickerel. J. of Hered. 9: 253–56, figs. 4–5.

ESCHMEYER, PAUL H.
1957. The lake trout (*Salvelinus namaycush*). U.S. Dept. of Interior, Fish and Wildlife Serv., Leaflet 441: 1–11.

EVERMANN, BARTON WARREN
1902. Description of a new species of shad (*Alosa ohiensis*), with notes on other food-fishes of the Ohio River. Rept. U. S. Comm. Fish and Fisheries 27 (1901): 273–88, pls. 6.
1918. The fishes of Kentucky and Tennessee: a distributional catalogue of the known species. Bull. U. S. Bur. Fish. 25, Doc. No. 858 (1915–16): 295–368.

EVERMANN, BARTON W., AND CHARLES H. BOLLMAN
1886. XIX.-Notes on a collection of fishes from the Monongahela River. Annals New York Acad. Sci. 3 (1885): 335–40.

EVERMANN, BARTON WARREN, AND HOWARD WALTON CLARK
1920. Lake Maxinkuckee—A physical and biological survey. Dept. Cons., State of Indiana 1: 1–660.

EVERMANN, BARTON W., AND OLIVER P. JENKINS
1888. Notes on Indiana fishes. Proc. U. S. Nat. Mus. 11: 43–57.

FAHY, WILLIAM E.
1954. The life history of the northern greenside darter, *Etheostoma blennioides blennioides* Rafinesque. J. Elisha Mitchell Sci. Soc. 70(2): 205 p.

FENNEMAN, NEVIN M.
1938. Physiography of eastern United States. New York and London, McGraw-Hill Book Co.: 714 p.; pls 6, figs. 197.

FERNALD, MERRITT LYNDON
1950. Gray's manual of botany. Amer. Book Co.: 1632 p.

FINGERMAN, SUE WHITSELL, AND ROYAL D. SUTTKUS
1961. Comparison of *Hybognathus hayi* Jordan and *Hybognathus nuchalis* Agassiz. Copeia (4): 462–67.

FISH, MARIE POLAND
1927. Contribution to the embryology of the American eel (*Anguilla rostrata* LeSueur). Zoologica 8, (5): 289–324, figs. 103–16.
1929. Contributions to the early life histories of Lake Erie fishes. Bull. Buffalo Soc. Natur. Sci. Buffalo 14, (3): 136–87, figs. 20–72.
1930. Contributions to the natural history of the burbot *Lota maculosa* (LeSueur). *Ibid.* 15, (1): 5–20.
1932. Contributions to the early life histories of sixty-two species of fishes from Lake Erie and its tributary waters. Bull. U. S. Bur. Fish. 47, (1931–33): 293–398, figs. 44.

FISHER, JOHN C., JOHN H. KLIPPART, AND ROBERT CUMMINGS
1878. Annual report. Second annual report of the Ohio State Fish Comm. made to Gov. of State of Ohio, 1877. Nevins and Myers, Columbus: 1–69.

FORBES, STEPHEN ALFRED, AND ROBERT EARL RICHARDSON
1920. The fishes of Illinois. Nat. Hist. Surv. Ill. 3: 1–358, pls. 55, figs. 1–76; Atlas, maps 103, 2d ed.

FOWLER, HENRY W.
1902. Types of fishes (preserved in the Phila. Acad. Nat. Sci.). Proc. Acad. Nat. Sci. Phila., 53 (1901): 327–41, pls. 12–15.
1907. Notes on lancelets and lampreys. *Ibid.*, 1907–08, 59: 461–66, figs. 2.
1919. List of the fishes of Pennsylvania. Proc. Biol. Soc. Wash. 32: 49–74.

GAGE, SIMON HENRY
1928. The lampreys of New York State—life history and economics. *In:* A biological survey of the Oswego River system. Suppl. 17th Annual Rept. New York St. Cons. Dept. (1927): 159–91, figs. 7.

GALE, WILLIAM F., AND CYNTHIA A. GALE
1977. Spawning habits of spotfin shiner (*Notropis spilopterus*) - A fractional, crevice spawner. Trans. Amer. Fish. Soc. 106(2): 170–77.

GARLICK, THEODATUS

1857. A treatise of the artificial propagation of certain kinds of fish, with the description and habits of such kinds as are the most suitable for pisciculture. Tho. Brown, Publ., Ohio Farmer Office, Cleveland: 142 p.

GERKING, SHELBY D.

1945. 1. The distribution of the fishes of Indiana. Invest. Ind. Lakes and Streams. Ind. Dept. Cons. and Ind. Univ. 3 (1): 1–137, maps 113.

GILBERT, CARTER ROWELL

1960. A systematic and geographical study of the subgenus *Luxilus* (family Cyprinidae). Ph.D. diss., Univ. of Mich.: 386 p.

1961. Hybridization versus intergradation: An inquiry into the relationship of two cyprinid fishes. Copeia (2): 181–92, 6 figs., 2 tabs.

1969. Systematics and distribution of the American cyprinid fishes *Notropis ariommus* and *Notropis telescopus*. Copeia (3): 474–92, 4 figs., 4 tabs.

GILBERT, CARTER R., AND REEVE M. BAILEY

1972. Systematics and zoogeography of the American cyprinid fish *Notropis* [*Opsopoeodus*] *emiliae*. Occ. Pap. Mus. Zool. Univ. Mich. No. 664, March 9: 1–35, pl. 2.

GIRARD, C.

1856. Researches upon the cyprinoid fishes inhabiting the fresh waters of the United States of America, west of the Mississippi Valley, from specimens in the Museum of the Smithsonian Institution. Proc. Acad. Nat. Sci. Phila. 8: 165–213.

GIUDICE, JOHN J.

1966. Growth of a blue × channel catfish hybrid as compared to its parent species. Prog. Fish-Culturist 28: 142–45.

GOLDTHWAIT, RICHARD P.

1953. What the glaciers did to Ohio. Ohio Div. Geol. Surv. and Ohio St. Univ. Geol. Mus., Educational Leaflet Series No. 3: 1–17, pls., figs. and maps 5.

GOSLIN, ROBERT M.

1943. Animal remains. The Fairport Harbor village site. Ohio State Arch. and Hist. Quarterly 52, (1): 45–51.

GRAY, JOHN W.

1942. Report of Ohio Lake Erie commercial fisheries 1941 season. Ohio Div. Conserv. and Natur. Res., Put-in-Bay, Ohio: 1–14. Mimeo.

1946. 8th annual report of Ohio Lake Erie commercial fisheries 1945 season. Ohio Div. Conserv. and Natur. Res. and Franz Theo. Stone Lab., Put-in-Bay, Ohio: 1–11. Mimeo.

1947. 9th annual report of Ohio Lake Erie commercial fisheries 1946 season. *Ibid.*: 1–11. Mimeo.

GREELEY, JOHN R.

1927. Fishes of the Genesee region with annotated list. *In:* A biological survey of the Genesee River system. Suppl. 16th Ann. Rept. New York St. Cons. Dept. (1926): 47–66, pls. 8.

1928. Fishes of the Oswego watershed [with annotated list]. *In:* A biological survey of the Oswego River system. Suppl. 17th Ann. Rept., *Ibid.* (1927): 84–107, col. pls. 12.

1929. Fishes of the Erie-Niagara watershed [with annotated list]. *In:* A biological survey of the Erie-Niagara system. Suppl. 18th Ann. Rept., *Ibid.* (1928): 150–79, col. pls. 8.

1930. Fishes of the Lake Champlain watershed [with annotated list]. *In:* A biological survey of the Champlain watershed. Suppl. 19th Ann. Rept. *Ibid.* (1929): 44—87, col. pls. 16.

1934. Fishes of the Raquette watershed with annotated list. *In:* A biological survey of the Raquette watershed. Suppl. 23rd. Ann. Rept. *Ibid.* (1933): 53–108, figs. 4, illust. 18, col. pls. 12.

1935. Fishes of the watershed with annotated list. *In:* A biological survey of the Mohawk-Hudson watershed. Suppl. 24th Ann. Rept., *Ibid.* (1934): 63–101, illust. 5, col. pls. 4.

1936. Fishes of the area with annotated list. *In:* A biological survey of the Delaware and Susquehanna watersheds. Suppl. 25th Ann. Rept., *Ibid.* (1935): 45–88. col. pls. 4.

1937. Fishes of the area with annotated list. *In:* A biological survey of the lower Hudson watershed. Suppl. 26th Ann. Rept., *Ibid.* (1936): 45–103, figs. 3, illust. 10, col. pls. 4.

1938. Fishes of the area with annotated list. *In:* A biological survey of the Allegheny and Chemung watersheds. Suppl. 27th Ann. Rept. *Ibid.* (1937): 48–73, illust. 3, col. pls. 2.

1939. The freshwater fishes of Long Island and Staten Island with annotated list. *In:* A biological survey of the fresh waters of Long Island. Suppl. 28th Ann. Rept., *Ibid.* (1938): 29–44.

1940. Fishes of the watershed with annotated list. *In:* A biological survey of the Lake Ontario watershed. Suppl. 29th Ann. Rept., *Ibid.* (1939): 42–81, illust. 2, col. pls. 4.

GREELEY, JOHN R., AND SHERMAN C. BISHOP

1932. Fishes of the area with annotated list. *In:* A biological survey of the Oswegatchie and Black River systems. Suppl. 21st. Ann. Rept., *Ibid.* (1931): 54–92, figs. 3, col. pls. 12.

1933. Fishes of the upper Hudson watershed with annotated list. *In:* A biological survey of the upper Hudson watershed. Suppl. 22nd. Ann. Rept., *Ibid.* (1932): 64–101, figs. 9, col. pls. 12.

GREELEY, JOHN R., AND C. WILLARD GREENE

1931. Fishes of the area with annotated list. *In:* A biological survey of the St. Lawrence watershed (including the Grass, St. Regis, Salmon, Chateaugay systems and the St. Lawrence between Ogdensburg and the International Boundary). Suppl. 20th Ann. Rept., *Ibid.* (1930): 44–94, col. pls. 12.

GREENE, C. WILLARD

1935. The distribution of Wisconsin fishes. St. Wis. Cons. Comm., Madison, Wis.: 1–235, maps 96.

GREENFIELD, DAVID W., FATHI ABDEL-HAMEED, GARY D. DECKERT, AND RONALD R. FLINN

1973. Hybridization between *Chrosomus erythrogaster* and *Notropis cornutus* (Pisces: Cyprinidae). Copeia (1): 54–60, 7 figs. 2 tabs.

GREGORY, WILLIAM K.

1933. Fish skulls: a study of the evolution of natural mechanisms. Trans. Amer. Phil. Soc., Phila. 23 (n.s.) Pt. 2: 75–481.

GRIBBLE, LLOYD R.

1939. An ecological note on the brook lamprey. Ecology 20: 107.

GRUCHY, C. G., R. H. BOWEN, AND I. M. GRUCHY

1973A. First records of the stoneroller (*Campostoma anomalum*) and the blackstripe topminnow (*Fundulus notatus*) from Canada. J. Fish. Res. Board Can. 30 (5): 683–84.

1973B. First records of the silver shiner, *Notropis photogenis*, from Canada. J. Fish. Res. Board, Can. 30 (9): 1379–82.

GUDGER, E. W.

1930. *Ichthyomyzon concolor* in the Coasa River at Rome, Ga., with notes on other lampreys in our South Atlantic and Gulf drainages. Copeia, (4): 145–47.

1942. Giant fishes of North America. Nat'l. Hist. Mag. 49, (2): 115–21.

GÜNTHER, ALBERT

1868. Catalogue of the Physostomi, containing the families Heteropygii, Cyprinidae, Gonorhynchidae, Hyodontidae, Osteoglossidae, Clupeidae, Chirocentridae, Alepocephalidae, Notopteridae, Halosauridae, in the collection of the British Museum. Cat. Fishes Brit. Mus., 7: 1–512.

HANKINSON, THOMAS L.

1930. Breeding behavior of the silverfin minnow, *Notropis whipplii spilopterus* (Cope) Copeia (3): 73–4.

1932. Observations on the breeding behavior and habitats of fishes in southern Michigan. Pap. Mich. Acad. Sci., Art and Letters 15 (1931): 411—25, pls. 33–4.

HARKNESS, W. J. K.

1922. The rate of growth of the yellow perch (*Perca flavescens*) in Lake Erie. Univ. Toronto Studies, Biol. Ser., No. 6 (Publ. Ont. Fish. Res. Lab., No. 20): 89–95.

1924. The rate of growth and the food of the lake sturgeon (*Acipenser rubicundus* LeSueur). *Ibid.* No. 24 *Ibid.* No 18: 13–42.

HARKNESS, W. J. K., AND J. R. DYMOND

1961. The lake sturgeon: the history of its fishery and problems of conservation. Ont. Dept. of Lands and Forests: 121 p., 9 tabs.

HARLAN, JAMES R., AND EVERETT B. SPEAKER

1951. Iowa fish and fishing. State of Iowa: 139 p.

[HARRIS, L. A., C. W. BOND, AND H. C. POST]

1884. Annual report. 8th ann. rep. Ohio Fish Comm. (1883): Columbus: 55 p.

HARTMAN, WILBUR L.
1970. Resource crises in Lake Erie. Explorer 12 (1): unnumbered [6 p.].
1972. Lake Erie: effects of exploitation, environmental changes and new species on the fishery resources. J. Fish. Res. Board, Can. 29: 899–912.

HAY, OLIVER P.
1880. On a collection of fishes from eastern Mississippi. Proc. U. S. Nat. Mus., 3: 488–515.
1894. The lampreys and fishes of Indiana. *In:* 19th Ann. Rept. Indiana Dept. Geol. and Nat. Res.: 146–296.

HECKEWELDER, REV. JOHN
1876. An account of the history, manners, and customs of the Indian nations who once inhabited Pennsylvania and the neighboring states. Mem. of Hist. Soc. of Penna., Phila. Publ. by Abraham Small (rev. ed.) (1819): 47–465.

HELD, JOHN W.
1969. Some early summer foods of the shovelnose sturgeon in the Missouri River. Trans. Am. Fish. Soc. 98 (3): 514—17.

HENDERSON, I. F., AND W. D. HENDERSON [JOHN H. KENNETH, 5th Ed.]
1953. A dictionary of scientific terms. D. Van Nostrand Co., N. Y.: 506 p.

HENSHALL, JAMES A.
1888. Contributions to the ichthyology of Ohio No. 1. Journ. Cincinnati Soc. Nat. Hist. 11: 76—80.
1889. Contribution to the ichthyology of Ohio. No. 2. *Ibid:* 122–26.
1917. Book of the black bass. Stewart and Kidd Co., Cincinnati: 1–452 (other editions between 1904 and 1915 have same pagination).
1919. Bass, pike, perch and other game fishes of America. *Ibid:* 410 p., pls. and figs. 20.

HESTER, F. EUGENE
1970. Phylogenetic relationships of sunfishes as demonstrated by hybridization. Trans. Am. Fish. Soc. 99 (1): 100–4, 3 figs., 1 tab.

HIGGINS, FRANK
1858. A catalogue of the shell-bearing species, inhabiting the vicinity of Columbus, Ohio, with some remarks thereon. 12th Ann. Rept. Ohio State Bd. of Agr. for 1857: 548–55.

HILDRETH, S. P.
1848. Pioneer history: being an account of the first examinations of the Ohio Valley, and the early settlement of the Northwest Territory. Hist. Soc. of Cincinnati. H. W. Derby and Co., Cincinnati: 525 p.

HILE, RALPH
1937. Morphometry of the cisco, *Leucichthys artedi* (LeSueur) in the lakes of Northeastern Highlands, Wisconsin. Internationale Revue der gesamten Hydrobiologie und Hydrographie. Leipzig, 37: 57–130, pls. 3, figs. 5, tabs. 61.

HOUGH, JOHN N.
1953. Scientific terminology. Rinehart and Company, Inc., N. Y.: 231 p.

HOWE, HENRY
1900. Historical collections of Ohio. C. J. Krehbiel and Co., Cincinnati, 1: 1–992 and 2: 1–911.

[HOWELL, D. Y.]
1882. Report of the superintendent of Ohio fish hatcheries. Ohio Fish Comm. 6th Ann. Rept. (1881): 9–14.
1883. Report of the superintendent of Ohio Fish hatcheries. *Ibid.* 7th Ann. Rept. (1882): 4–12.

HOWLAND, JOE H.
1931. Studies on the Kentucky black bass (*Micropterus pseudaplites* Hubbs). Trans. Amer. Fish. Soc. 61: 89–94.

HUBBS, CARL L.
1920. Notes on hybrid sunfishes. Aquatic Life 5(9): 101–3.
1921. An ecological study of the life-history of the fresh-water atherine fish *Labidesthes sicculus*. Ecol. 2, (4): 262–76, figs. 1–4.
1925. The life-cycle and growth of lampreys. Pap. Mich. Acad. Sci., Arts and Letters. (1924): 587–603, figs. 16–22.
1926. A check list of the fishes of the Great Lakes and tributary waters, with nomenclatorial notes and analytical keys. Misc. Publ. Mus. Zool. Univ. Mich. No. 15: 1–77, pls. 1–4.

1927. *Micropterus pseuadaplites*, a new species of black bass. Occ. Pap. Mus. Zool. Univ. Mich. No. 184: 1–15, pls. 2.

1930A. Materials for a revision of the catostomid fishes of eastern North America. Misc. Publ. Mus. Zool. Univ. Mich. No. 20: 1–47.

1930B. Further additions and corrections to the list of fishes of the Great Lakes and tributary waters. Pap. Mich. Acad. Sci., Arts and Letters, No. 11, 1929: 425—36.

1931. *Parexoglossum laurae*, a new cyprinid fish from the Upper Kanawha River System. Occ. Pap. Mus. Zool. Univ. Mich. No. 234: pp. 1–2, pls. 2.

1940A. The cranium of a fresh-water sheepshead from postglacial marl in Cheboygan County, Michigan. Pap. Mich. Acad. Sci., Arts and Letters 25 (1939): 293–97, pl. 1.

1940B. Speciation of fishes. Amer. Nat. 74: 198–211.

1955. Hybridization between fish species in nature. Systematic Zoology 4(1): 1–20.

HUBBS, CARL L., AND REEVE M. BAILEY

1938. The small-mouth bass. Bull. Cranbrook Inst. Sci. (10): 1–89, front., pls. 9, figs. 5.

1940. A revision of the black basses (*Micropterus* and *Huro*) with descriptions of four new forms. Misc. Publ. Mus. Zool. Univ. Mich. (48): 1–51, pls. 6, fig. 1, maps 2.

1952. Identification of *Oxygeneum pulverulentum* Forbes, from Illinois, as a hybrid cyprinid fish. Pap. Mich. Acad. Sci., Arts and Letters 37 (1951): 143–52.

HUBBS, CARL L., AND JOHN D. BLACK

1947. Revision of *Ceratichthys*, a genus of American cyprinid fishes. Misc. Publ. Mus. Zool. Univ. Mich. (66): 1–56, pls. 2, fig. 1, maps. 2.

HUBBS, CARL L., AND DUGALD E. S. BROWN

1929. Materials for a distributional study of Ontario fishes. Trans. Roy. Can. Inst., Pt. 1, 17: 1–56.

HUBBS, CARL L., AND GERALD P. COOPER

1936. Minnows of Michigan. Bull. Cranbrook Inst. Sci. (8): 1–95, pls. 10, figs. 2.

HUBBS, CARL L., AND WALTER R. CROWE

1956. Preliminary analysis of the American cyprinid fishes, seven new, referred to the genus *Hybopsis*, subgenus *Erimystax*. Occ. Pap. Mus. Zool. Univ. Mich. No. 578: 1–8.

HUBBS, CARL L., AND C. WILLARD GREENE

1928. Further notes on the fishes of the Great Lakes and tributary waters. Pap. Mich. Acad. Sci., Arts and Letters 8 (1927): 371–92.

1935. Two new subspecies of fishes from Wisconsin. Trans. Wis. Acad. Sci., Arts and Letters 29: 89–101, pls. 2–3.

HUBBS, CARL L., AND LAURA C. HUBBS

1931. Increased growth in hybrid sunfishes. Pap. Mich. Acad. Sci., Arts and Letters 13 (1930): 291–301, figs. 45–46.

1932A. Experimental verification of natural hybridization between distinct genera of sunfishes. Pap. Mich. Acad. Sci., Arts and Letters 15 (1931): 427–37.

1932B. Hybridization between fish species in nature. Proc. 6th Intern. Cong. Geneticists 2: 245–46.

1933. The increased growth, predominant maleness, and apparent infertility of hybrid sunfishes. Pap. Mich. Acad. Sci., Arts and Letters 17 (1932): 613–41.

1947. Natural hybrids between two species of catostomid fishes. *Ibid.* 31 (1945): 147–67.

HUBBS, CARL L., LAURA C. HUBBS, AND RAYMOND E. JOHNSON

1943. Hybridization in nature between species of catostomid fishes. Contrib. from Lab. Vert. Biol. Univ. Mich., Ann Arbor No. 22: 1–76, pls. 1–7.

HUBBS, CARL L., AND KATSUZO KURONUMA

1942. Analysis of hybridization in nature between two species of Japanese flounders. Pap. Mich. Acad. Sci., Arts and Letters 27 (1941): 267–306.

HUBBS, CARL L., AND KARL F. LAGLER

1941. Guide to the fishes of the Great Lakes and tributary waters. Cranbrook Inst. of Sci., Bull. 18: 1–99, figs. 118.

1947. Fishes of the Great Lakes region. *Ibid.* Bull. 26: 1–186, pls. 26, figs. 251.

Hubbs, Carl L., and Robert R. Miller

1943. Mass hybridization between two genera of cyprinid fishes in the Mohave Desert, California. Pap. Mich. Acad. Sci., Arts and Letters 28 (1942): 343–78, pls. 4.

Hubbs, Carl L., and Robert Rush Miller

1953. Hybridization in nature between the fish genera *Catostomus* and *Zyrauchen*. Mich. Acad. Sci., Arts and Letters 38: 207–33, 4 pls.

Hubbs, Carl L., and T. E. B. Pope

1937. The spread of the sea lamprey through the Great Lakes. Michigan Conservation, July (1937): 1–7, figs. 4.

Hubbs, Carl L., and I. C. Potter

1971. The biology of lampreys: distribution, phylogeny and taxonomy, Part 1. Academic Press, London: 1–65.

Hubbs, Carl L., and Edward C. Raney

1944. Systematic notes of North American siluroid fishes of the genus *Schilbeodes*. Occ. Pap. Mus. Zool. Univ. Mich. No. 487: 1–36, pl. 1, map 1.

Hubbs, Carl L., and Milton B. Trautman

1937. A revision of the lamprey genus *Ichthyomyzon*. Misc. Publ. Mus. Zool. Univ. Mich. No. 35: 1–109, pls. 2, figs. 5, map 1, tabs. 22.

Hubbs, Carl L., Boyd W. Walker, and Raymond E. Johnson

1943. Hybridization in nature between species of American cyprinodont fishes. Contrib. from Lab. Vert. Biol. Univ. Mich., Ann Arbor No. 23: 1–21, pls. 1–6.

Hubbs, Clark

1959. Laboratory hybrid combinations among etheostomatine fishes. Tex. J. Sci. 11 (1): 49–56.

Hubbs, Clark, and Charles Michael Laritz

1961A. Occurrence of a natural intergeneric etheostomatine fish hybrid. Copeia (2): 231–32.

1961B. Natural hybridization between *Hapropterus scierus* and *Percina caprodes*. Southwestern Naturalist (6): 188–92.

Hubbs, Clark, and Kirk Strawn

1957A. Survival of F_1 hybrids between fishes of the subfamily Etheostominae. J. of Exper. Zool. 134 (1): 33–62.

1957B. Relative variability of hybrids between the darters *Etheostoma spectabile* and *Percina caprodes*. Evolution 11 (1): 1–10.

Hulbert, Archer Butler

1903. Waterways of westward expansion. The Ohio River and its tributaries. Historic waterways of America. A. H. Clark Co., Cleveland, 9: 1–220.

Hulbert, Archer, and William Nathaniel Schwarze [editors]

1910. David Zeisberger's history of the Northern American Indians. Ohio Archaeol. and Hist. Publ., Columbus, 19: 1–189.

Hunter, John R., and Warren J. Wisby

1961. Utilization of the nests of green sunfish (*Lepomis cyanellus*) by the redfin shiner (*Notropis umbratilis cyanocephalus*). Copeia (1): 113–15.

Hussakof, L.

1912. The spawning habits of the sea lamprey, *Petromyzon marinus*. Amer. Naturalist 46: 729–40, figs. 5.

Hutchins, Ed

1969. Hybrid bass has great promise. Columbus Dispatch, June 8: 13B.

Imlay, G[ilbert]

1793. A topographical description of the Western Territory of North America containing a succinct account of its climate and history. London, 2d. ed.: 433 p. plus index.

Irwin, William Henry

1945. Methods of precipitating colloidal soil particles from impounded waters of central Oklahoma. Bull. Okla. Agric. and Mech. Coll., Stillwater 42, (11): 3–16.

Irwin, W. H., and James H. Stevenson

1951. Physiochemical nature of clay turbidity with special reference to clarification and productivity of impounded waters. *Ibid.* 48, Biol. Ser. 4: 1–54.

JENKINS, ROBERT ELLIS
1970. Systematic studies of the catostomid fish tribe Moxostomatini. Univ. Microfilms, Ann Arbor, Mich.: Pt.1: 1–399; Pt.2: 400–799. A Thesis presented to the Faculty of the Graduate School of Cornell Univ.
1971. Systematic studies of the catostomid fish tribe Moxostomatini. Diss. Abst. Intern. 31 (12): 818.

JENKINS, ROBERT E., AND ERNEST A. LACHNER
1971. Criteria for analysis and interpretation of the American fish genera *Nocomis* Girard and *Hybopsis* Agassiz. Smithsonian Contrib. to Zool. Smithsonian Inst. Press (90): 1–15.

JOBES, FRANK W.
1952. Age, growth, and production of yellow perch in Lake Erie. Fish. Bull. Fish and Wildlife Ser. 52, (70): 204–66.

JONES, DAVID [HORATIO GATER JONES, JR., ed.]
1774. A journal of two visits made to some nations of Indians on the west side of the River Ohio, in the years 1772 and 1773. Printed and sold by Isaac Collins: 11–127. (Reprint: New York, reprinted for Joseph Sabin: with a bibliographical note by the editor.)

JORDAN, DAVID STARR
1877A. On the fishes of northern Indiana. Proc. Acad. Nat. Sci. Phil. 24: 42–104.
1877B. Contributions to North American ichthyology; based primarily on the collections of the United States National Museum, No. 1. Review of Rafinesque's memoirs on North American fishes. Bull. U. S. Nat. Mus. 9: 1–53.
1878. Contributions to North American ichthyology, based primarily on the collections of the United States National Museum, 3B. A synopsis of the family Catostomidae. *Ibid.* 12: 97–237.
1880. Description of new species of North American fishes. Proc. U. S. Nat. Mus. 2 (1879): 235–41.
1882. Sec. IV. Report on the fishes of Ohio. Geol. Surv. Ohio, 4: 738–1002.
1886. Note on the scientific name of the yellow perch, the striped bass and other North American fishes. Proc. U. S. Nat. Mus. 8 (1885): 72–3.
1890. 2. Report of explorations made during the summer and autumn of 1888, in the Allegheny region of Virginia, North Carolina and Tennessee, and in western Indiana, with an account of the fishes found in each of the river basins of those regions. Bull. U. S. Fish. Comm. 8 (1888): 97–192, pls. 15.
1924. Concerning the genus *Hybopsis* of Agassiz. Copeia 126: 51–52.
1929. Manual of the vertebrate animals of the northeastern United States inclusive of marine species. World Book Co., Yonkers on Hudson, New York: 446 p.
1929. Manual of the vertebrate animals of the northeastern United States. 13th ed. World Book Co.: 446 p.

JORDAN, DAVID STARR, AND BARTON WARREN EVERMANN
1891. Description of a new darter (*Etheostoma tippecanoe*) from the Tippecanoe River, Indiana. Proc. U. S. Nat. Mus. 13 (1890): 3–4.
1896–1900. The fishes of North and Middle America. Bull. U. S. Nat. Mus. 47; 1896, Pt. 1: 1–1240; *Ibid.* 1898, Pt. 2: 1241–2183; *Ibid.* 1898, Pt. 3: 2183a–3136; *Ibid.* 1900, Pt. 4: 3137–313, pls. 392.
1902. American food and game fishes. A popular account of all the species found in America north of the Equator, with keys for ready identification, life histories and methods of capture. Doubleday, Page and Co., New York: 1–1573, Col. front., col. pls. 9, pls. 66, figs. 221.

JORDAN, DAVID STARR, BARTON WARREN EVERMANN, AND HOWARD WALTON CLARK
1930. Check list of the fishes and fishlike vertebrates of North and Middle America north of the northern boundary of Venezuela and Colombia. Rept. U. S. Comm. Fish, Pt. 2: 1–670.

JUDD, JOHN BAYLESS
1971. The effects of ecological changes on the largemouth bass of Buckeye Lake, Ohio. Master's thesis, Ohio State Univ.: 1–86, 5 figs., 16 tabs.

KEELER, LUCY ELLIOT
1904. The Sandusky River. Ohio Archaeol. and Hist. Pub. Columbus (13): 190–247, 1 map.

KENNEDY, C. H.
1931. James Stewart Hine. Ohio J. Sci. 31, (6): 510–11.

KENWORTHY, HARRY F.
1931. A review of the literature on Ohio fishes. Unpub. M. Sci. Thesis, Ohio State Univ.: 190 p.

KERBY, JEROME HOWARD

1979. Meristic characters of two *Morone* hybrids. Copeia (3): 513–18.

[KIDDNEY, MR. AND MRS. HAROLD]

1951. Giant shovelhead. Ohio Cons. Bull., Sept.: 27.

KINNEY, E. C.

1950. The life history of the trout-perch, *Percopsis omiscomaycus* (Walbaum), in western Lake Erie. M. Sc. Thesis, Ohio State Univ. 75 p.

KIRSCH, PHILIP H.

1895A. 20. A report upon investigations in the Maumee River basin during the summer of 1893. Bull. U. S. Fish Comm. 14 (1894): 315–37, tab. 1.

1895B. Report upon explorations made in Eel River basin in the northeastern part of Indiana in the summer of 1892. *Ibid.* 14, (1894): 31–41.

KIRTLAND, JARED P.

1838. Report on the zoology of Ohio. Ann. Rep. Geol. Surv. St. of Ohio 2: 157–97.

1841A. Art. 3. Descriptions of four new species of fishes. Boston J. Nat. Hist. 3 (1840): 273–77. pls. 2.

1841B. Art. 10. Descriptions of the fishes of the Ohio River and its tributaries. *Ibid.* 3 (1840): 338–52, pls. 4–6.
Art. 17. Description of the fishes of the Ohio River and its tributaries. *Ibid.* 3 (1840): 469–82, pls. 27–29.

1844. Art. 2. Descriptions of the fishes of the Ohio River and its tributaries. *Ibid.* 4 (1842): 16–26, pls. 1–4.
Art. 7. Descriptions of the fishes of Lake Erie, the Ohio River and their tributaries. *Ibid.* 4 (1842): 231–40, pls. 9–11.
Art. 25. Descriptions of the fishes of the Ohio River and its tributaries. *Ibid.* 4 (1843): 303–08, pls. 14–15.

1847. Art. 2. Descriptions of the fishes of the Ohio River and its tributaries. *Ibid.* 5 (1845): 21–32, pls. 7–9.
Art. 16. Descriptions of the fishes of Lake Erie, the Ohio River and their tributaries. *Ibid.* 5 (1845): 265–76, pls. 19–22.
Art. 24. Descriptions of the fishes of Lake Erie, the Ohio River and their tributaries. [Kirtland, J. P.] *Ibid.* 5 (1846): 330–44, pls. 26–9.

1850A. Fragments of natural history. The Family Visitor. Cleveland, Ohio, 1, No. 1, Jan. 3: 1.

1850B. *Leusiscus Erythrogaster.* Raf.—Red-bellied minnow. *Ibid.* Cleveland and Hudson, Ohio, 1, No. 27, Oct. 3: 213.

1850C. *Chatoessus ellipticus*; or, gizzard shad. *Ibid.* 1, No. 1, Jan. 3: 1–2.

1850D. Chinese gold-fish. *Ibid.* 1, No. 4, Jan. 24: 29.

1850E. *Leuciscus Atromaculatus.* Mitchell—chub—dace. *Ibid.* 1, No. 27, Oct. 3: 213.

1850F. *Leuciscus Plagyrus.* Raf.—common brook minnow. *Ibid.* 1, No. 30, Nov. 14: 237.

1850G. *Pimelodus Catus* Linn.—bull-head—bull-pout. *Ibid.* 1, No. 18, May 30: 141.

1850H. *Pimelodus Coerulescens* Raf.—blue catfish, black catfish and silvery catfish. *Ibid.* 1, No. 22, July 25: 173.

1850I. *Corvina Oscula* LeSueur—sheepshead of Lake Erie. White perch of the Ohio River. *Ibid.* 1, No. 17, May 16: 133.

1850J. *Labrax multilineatus.*—white bass, striped bass, or white perch of Lake Erie. *Ibid.* 1, No. 7, Feb. 14: 53.

1850K. *Etheostoma maculata.* Kirtland—black-darter;—spotted hog-fish. *Ibid.* 1, No. 4, Jan. 24: 29.

1850L. *Etheostoma variata.* Kirtland—variegated rock fish. *Ibid.* 1, No. 3, Jan. 17: 21.

1850M. *Pomotis Nitida* Kirtland—brilliant sun fish—red-eyed sun fish. *Ibid.* 1, No. 15, April 18: 117.

1850N. *Centrarchus hexacanthus.* Val.—grass bass, bank lick bass, roach. *Ibid.* 1, No. 9, Feb. 28: 69.

1850O. *Perca flavescens.* Mitchell—yellow perch. *Ibid.* 1, No. 2, Jan. 10: 13.

1850P. *Lucio-perca Americana.* Cuvier—pike and pickerel of Lake Erie; salmon of the Ohio River; sandre of the Canadians. *Ibid.* 1, No. 8, Feb. 21: 61.

1850Q. *Leuciscus Dissimilis.* Kirtland—spotted minnow. *Ibid.* 1, No. 24, Aug. 22: 189.

1850R. *Leuciscus Kentuckiensis.* Raf.—white and yellow-winged minnow. *Ibid.* 1, No. 31, Nov. 28: 245.

1850S. *Leuciscus Storerianus* Kirtland—storeis minnows. *Ibid.* 1, No. 32, Dec. 12: 256.

1850T. *Leuciscus Elongatus* Kirtland.—slim minnow—red-sides. *Ibid.* 1, No. 23, Aug. 8: 181.

1850U. *Leuciscus longirostris.* Kirtland.—long-nosed minnow. *Ibid.* 1, No. 11, March 14: 85.

1850V. *Leuciscus Americanus.* Lacépède.—gold-shiner. *Ibid.* 1, No. 28, Oct. 17: 221.

1850W. *Leuciscus Compressus.* Raf.—flat-shiner. *Ibid.* 1, No. 29, Oct. 31: 229.

1850X. *Pimelodus Cupreus* Raf.—yellow catfish. *Ibid.* 1, No. 20, June 27: 157.

1850Y. *Pimelodus Limosus* Raf. *Ibid.* 1, No. 21, July 11: 165.

1850Z. *Etheostoma caprodes.* Raf. *Ibid.* 1, No. 5, Jan. 31: 37.

1850AA. *Etheostoma blennoides*. Rafinesque.—blenny-like hog-fish. *Ibid.* 1, No. 6, Feb. 7: 45.
1850BB. *Centrarchus aenus.*—rock bass, goggle-eyed bass, black sun-fish. *Ibid.* 1, No. 10, March 7: 77.
1850CC. *Centrarchus fasciatus.*—black bass. *Ibid.* 1, No. 12, March 21: 93.
1850DD. *Pomotis machrochira.* Raf.—gilded sunfish. *Ibid.* 1, No. 13, March 28: 101.
1850EE. *Pomotis vulgaris.* Cuv.—roach, sunfish. *Ibid.* 1, No. 14, April 4: 109.
1850FF. *Gasterosteus Inconstans* Kirtland—stickle back. *Ibid.* 1, No. 16, May 2: 125.
1850GG. *Leuciscus Biguttatus.* Kirtland—two spotted minnow or jerker. *Ibid.* 1, No. 25, Sept. 5: 197.
1851A. *Esox Estor.* Les.—muskallonge. *Ibid.* 2, No. 8, whole No. 60. July 1: 61.
1851B. Piscatoriana. No. 5. *Ibid.* 2, No. 11, whole No. 63, July 22: 87.
1851C. *Catostomus Communis* LeSueur. *Ibid.* 1, No. 40, Feb. 13: 317.
1851D. *Alosa Chrysochloris.* Raf.—skip-jack of fishermen. gold-herring. *Ibid.* 2, No. 15., Aug. 19: 117.
1851E. *Catostomus Elongatus.* LeSueur.—long sucker, black horse, and Missouri sucker of fishermen. *Ibid.* 1, No. 44, March 13: 349.
1851F. *Catostomus Melanops.* Raf.—spotted sucker. *Ibid.*, 1, No. 52, May 6: 413.
1851G. *Catostomus Aureolus.* Les.—mullet of Lake Erie. *Ibid.* 1, No. 39, Feb. 6: 309.
1851H. *Accipenser Platorynchus.* Raf.—shovel-nosed sturgeon. *Ibid.* 2, No. 30, Dec. 2: 233.
1851I. *Accipenser rubicundus.* Les.—common sturgeon. *Ibid.* 2, No. 29, Nov. 25: 229.
1851J. *Lepisosteus ferox.* Raf.—alligator gar-fish. *Ibid.* 2, No. 19, Sept. 16: 149.
1851K. *Lepisosteus platostomus.* Raf.—duck-bill gar-fish. *Ibid.* 2, No. 20, Sept. 23: 157.
1851L. *Lepisosteus Osseus.* Lin.—common gar-fish. *Ibid.* 2, No. 18, Sept. 9: 141.
1851M. *Amia Calva.* Lin.—dog-fish. Lake lawyer. *Ibid.* 2, No. 11, Aug. 12: 109.
1851N. *Lota Maculosa.* Les.—eel-pout. *Ibid.* 2, No. 23, Oct. 14: 181.
1851O. *Anguilla lutea.* Raf.—eel. *Ibid.* 2, No. 24, Oct. 21: 189.
1851P. *Salmo Amethystus* Mitchell—Mackinaw trout. *Ibid.* 2, No. 13, Aug. 5: 101.
1851Q. *Petromyzon Argenteus.* Kinland [sic]—small lamprey. *Ibid.* 2, No. 26, Nov. 4: 205.
1851R. *Ammocoetes Concolor* Kirtland.—mud eel, blind-eel. *Ibid.* 2, No. 27, Nov. 11: 213.
1851S. *Scleroghathus Cyprinus.* LeSueur.—*Catastomus Cyprinus* of Les.—carp of the Ohio River; shad of Lake Erie. *Ibid.* 1, No. 47, Apr. 3: 373.
1851T. *Salmo fontinalis.* Mitchell.—brook trout. *Ibid.* 2, No. 9, July 8: 69.
1851U. *Polyodon folium.* Lacépède.—spoon-bill paddlefish. *Ibid.* 2, No. 32, Dec. 16: 249.
1851V. *Hyodon tergisus.* Les.—toothed herring, moon-eyes. *Ibid.* 2, No. 17, Sept. 2: 133.
1851W. *Catostomus bubalus.* Raf.—buffalo sucker. *Ibid.* 1, No. 43, March 6: 341.
1851X. *Noturus Flavus.* Raf.—yellow back-tail of vulgar language. *Ibid.* 1, No. 35, Jan. 9: 277.
1851Y. *Catostomus Nigricans.* LeSueur.—mud-sucker. *Ibid.* 1, No. 50, April 24: 397.
1851Z. *Catostomus Duquesnie.*—red horse.—Pittsburg sucker of fishermen. *Ibid.* 1, No. 46, March 27: 365.
1851AA. *Catostomus Gibbosus.* LeSueur.—horned sucker, chub-sucker. *Ibid.* 1, No. 42, Feb. 27: 333.
1851BB. *Catostomus Anisurus.* Raf.—white sucker., *Ibid.* No. 49, April 17: 389.
1851CC. *Leuciscus Cornutus.* Mitchell.—red sided chub. *Ibid.* 1, No. 34, Jan. 2: 269.
1851DD. *Pimephales Promelas.* Raf. *Ibid.* 1, No. 36, Jan. 16: 285.
1851EE. *Esox reticulatus.* Les.—pike, pickerel. *Ibid.* 2, No. 6, June 17: 45.
1851FF. *Hydrargyra limi* Kirtland.—mud-fish. mud-minnow. *Ibid.* 2, No. 5, June 10: 37.
1854A. Revision of the species belonging to the genus *Esox*, inhabiting Lake Erie and the River Ohio. Anals of Sci., Cleveland, Ohio, 2, No. 3: 78–79.
1854B. *Alburnus nitidus.* (silver minnow.) *Ibid.* 2, No. 2: 44–45.
1854C. *Poecilosoma erythrogastrum. Ibid.* 2, No. 1: 4–5.
1858. Fish culture. Small lakes. The Ohio Farmer, Cleveland, Ohio. 7, No. 36, Sept. 4: 281.

KLIPPART, J. H.
1877A. History of Toledo and Sandusky fisheries. 1st Ann. Rept. Ohio State Fish. Comm., years 1875–1876. Nivins and Myers, Columbus: 31–42.
1877B. Catalogue of fishes of Ohio (includes "Scientific Description of Ohio Fishes"). *Ibid.*: 43–88.
1878. Descriptions of Ohio fishes, arranged from manuscript notes of Professor D. S. Jordan, by his assistant, Ernest R. Copeland. 2d Ann. Rept. *Ibid* (1877): 83–116.

KLIPPART, JOHN H.
1878A. Annual Report. 2d Ann. Rept. of the Ohio State Fish Comm. for the Year 1877. Nevins & Myers, Columbus: 5–58.

KLIPPART, JOHN H., AND JOHN HUSSEY

1874. Report of the commissioners of fisheries of the State of Ohio for the year ending December, 1873. Nevins and Myers, Columbus: 1–32 (Also an appendix: 33–40).

KOELZ, WALTER

1926. Fishing industry of the Great Lakes. Appendix XI to: Rept. of U. S. Comm. of Fish. Bur. Fish. Doc. No. 1001 (1925): 553–617.

1929. Coregonid fishes of the Great Lakes. Bull. U. S. Bur. Fish. 43, Pt. 2 (1927): 297–643, figs. 31.

1931. The coregonid fishes of northeastern North America. Pap. Mich. Acad. Sci. Arts and Letters 13 (1930): 303–482, 1 pl.

KOSTER, WILLIAM J.

1937. The food of sculpins (Cottidae) in central New York. Trans. Amer. Fish. Soc. 66 (1936): 374–82, 3 tabs.

KRAATZ, WALTER C.

1924. The intestine of the minnow *Campostoma anomalum* (Rafinesque), with special reference to the development of its coiling. Ohio J. Sci. 24, (6): 265–98.

1928. Study of the food of the blunt-nosed minnow, *Pimephales notatus. Ibid.* 28, No. 2: 86–98.

KRECKER, FREDERICK H.

1928. Periodic oscillations in Lake Erie. Franz Theodore Stone Lab., Ohio State Univ. Press, Columbus, No. 1: 1–22, figs. 6.

KRUMHOLZ, LOUIS A.

1949. Length-weight relationship of the muskellunge, *Esox m. masquinongy*, in Lake St. Clair. Trans. Amer. Fish. Soc. 77 (1947): 42—48.

1950. Further observations on the use of hybrid sunfish in stocking small ponds. Trans. Amer. Fish. Soc. 79 (1949): 112–24.

KUHNE, EUGENE R.

1939. A guide to the fishes of Tennessee and the Mid-South. Div. of Game and Fish, Tenn. Dept. of Cons., March: 124 p., figs, 81.

LACÉPÈDE (LACÉPÈDE)

1798. Histoire naturelle des poissons, 4: 728 p. pls. 1–16.

LACHNER, ERNEST ALBERT

1946. Studies of the biology of the chubs (Genus *Nocomis*, family Cyprinidae) of northeastern United States. Reprint from Cornell Univ. Abstracts of Theses. Ithaca, New York: 207—10.

1950. Ichthyology–The comparative food habits of the cyprinid fishes *Nocomis biguttatus* and *Nocomis micropogon* in western New York. J. Wash. Acad. Sci. 40, (7), July 15: 229–36.

1952. Studies of the biology of the cyprinid fishes of the chub genus *Nocomis* of northeastern United States. Amer. Midl. Nat. 48, (2): 433–66.

LACHNER, ERNEST A., AND EARL E. DEUBLER, JR.

1960. *Clinostomus funduloides* Girard to replace *Clinostomus vandoisulus* (Valenciennes) as the name of the rosyside dace of eastern North America. Copeia (4): 358–60.

LACHNER, ERNEST A., AND ROBERT E. JENKINS

1971. Systematics, distribution and evolution of the chub genus *Nocomis* Girard (Pisces, Cyprinidae) of eastern United States, with descriptions of new species. Smithsonian Contrib. to Zool., Smithsonian Inst. Press (85): 97 p., 30 figs., 26 tabs.

LACHNER, ERNEST A., EDWARD F. WESTLAKE, AND PAUL S. HANDWERK

1950. Studies on the biology of some percid fishes from western Pennsylvania. Amer. Midl. Nat. 43, (1): 92–111.

LAGLER, KARL F., AND REEVE M. BAILEY

1947. The genetic fixity of differential characters in subspecies of the percid fish, *Boleosoma nigrum*. Copeia (1): 50–9, tabs. 7.

LAGLER, KARL F., CARL B. OBRECHT, AND GEORGE V. HARRY

1943. The food and habits of gars (*Lepisosteus* spp.) considered in relation to fish management. Invest. Ind. Lakes and Streams. Ind. Dept. Cons. and Ind. Univ. 2 (1942): 117—35.

LAGLER, KARL F., AND CHARLES STEINMETZ, JR.

1957. Characteristics and fertility of experimentally produced sunfish hybrids, *Lepomis gibbosus* and *L. macrochirus*. Copeia (4): 290–92.

LANGLOIS, THOMAS H.

1946. The herring fishery of Lake Erie. Inland Seas. Quarterly Bull. Great Lakes Hist. Soc. Sponsored by Cleveland Public Library 2, (2): 101–04.

1949. The biological station of The Ohio State University. Contrib. 11, Franz Theodore Stone Lab., Ohio State Univ.: 1–57.

LANGLOIS, T. H., AND JAMES EDMISTER

1939. First annual statistical report of the Ohio commercial fisheries in Lake Erie. Ohio State Univ. and Ohio Div. Cons., Put-in-Bay, Ohio: 1—12. Mimeo.

LARSEN, ALFRED

1954. First record of the white perch (*Morone americana*) in Lake Erie. Copeia (2): 154.

LEACH, W. JAMES

1939. The endostyle and thyroid gland of the brook lamprey, *Ichthyomyzon fossor*. J. Morph. 65, (3): 549–605, figs. 41, pls. 6.

1940. Occurrence and life history of the northern brook lamprey, *Ichthyomyzon fossor*, in Indiana. Copeia (1): 21–34, pls. 2.

LEE, ALFRED E.

1892. History of city of Columbus, capital of Ohio. Munsell and Co.: New York and Chicago, 921 p.

LEGENDRE, VIANNEY

1942. Redécouverte après un siècle et reclassification d'une espèce de Catostomidé. Le Naturaliste Canadien 69, Nos. 10—11: 228–33.

1952. Les poissons d'eau duce. Clef des poissons de pêche sportive et commerciale de la Province de Quebec. La Société Canadienne d'Ecologie, Montreal, Canada. 1: 1–84, figs. 1—80.

1954. The Freshwater fishes—key to the game and commercial fishes of the province of Quebec. *Ibid.* 1: 1—180, figs. 1—80.

LESUEUR, C. A.

1827. American ichthyology or, natural history of the fishes of North America: with colored figures from drawings executed from nature. New Harmony, Indiana: 44 p. unnumbered, pls. 9.

LEVERETT, FRANK

1902. Glacial formations and drainage features of the Erie and Ohio basins. U. S. Geol. Surv.: 1–802, pls. 26.

LONG, WILBUR L. AND WILLIAM W. BALLARD

1976. Normal embryonic stages of the white sucker, *Catostomus commersoni*. Copeia (2): 342–51.

LUCE, WILBUR M.

1937. Hybrid crosses in sunfishes. Trans. Ill. Acad. Sci. 30 (2): 309–10.

LUTTERBIE, GARY W.

1975. Illustrated key to the Percidae of Wisconsin (walleye, sauger, perch and darters). Reports of the fauna and flora of Wisconsin. The Mus. of Natur. Hist. Univ. of Wisc., Stevens Point: 17–43.

MACKAY, H. H.

1934. Record of the alewife from Lake Huron. Copeia (2): 97.

MAGNUSON, JOHN L., AND LLOYD L. SMITH

1963. Some phases of the life history of the trout-perch. Ecology 44 (1): 83–95, 3 figs., 12 tabs.

MAHR, AUGUST. C.

1949. A chapter of early Ohio natural history. Ohio J. Sci. 49: 45–69.

MANION, PATRICK J., AND A. L. MCLAIN

1971. Biology of larval sea lampreys (*Petromyzon marinus*) of the 1960 year class, isolated in the Big Garlic River, Michigan, 1960–65. Great Lakes Fish. Comm., Ann Arbor, Mich., Tech. Rep. No. 16: 1–35.

MARKUS, HENRY C.

1934. Life history of the blackhead minnow (*Pimephales promelas*). Copeia, (3): 116–22.

MARSHALL, NELSON

1939. Annulus formation in scales of the common shiner, *Notropis cornutus chrysocephalus* (Rafinesque). Copeia, (3): 148–54.

MARTIN, MAYO

1952. Age and growth of the goldeye *Hiodon alosoides* (Rafinesque) of Lake Texoma, Oklahoma. Proc. Okla. Acad. Sci. 33: 37–49, 3 figs., 10 tabs.

MATSUI, Y.

1931. (19) Genetical studies of fresh water fish. 2. On the hybrid of *Cyprinus carpio* L. and *Carassius carassius* (L.) *auratus* (L.) (Abstract). J. of the Imperial Fisheries Experimental Sta., Tokyo, Japan, No. 2 (Papers No. 13–23) Sept.: 135—37, pls. 15–16.

MAY, BRUCE

1966. A study of some phases of the life history of the variegated darter, *Etheostoma variatum* (Kirtland). Master's thesis, Ohio State Univ., Columbus: 63 p., 9 figs., 8 tabs.

1969. Observations on the biology of the variegated darter, *Etheostoma variatum* (Kirtland). Ohio Sci. 69 (2): 85–92, 4 figs.

MAYHEW, J. K.

1969. Age and growth of flathead catfish in the Des Moines River, Iowa. Trans. Amer. Fish. Soc., 98 (1): 118–21.

MAYR, ERNST

1947. Systematics and the origin of species from the viewpoint of a zoologist. Columbia Univ. Press, New York: 334 p.

MCALLISTER, D. E.

1963. A revision of the smelt family, Osmeridae. Nat. Mus. Can. Bull. 191, Ser. 71: 53 p.

MCCORMICK, LEWIS M.

1890. List of fishes of Lorain County, Ohio. J. Cincinnati Soc. Nat. Hist. 12: 126–28.

1892. Descriptive list of the fishes of Lorain County, Ohio. Bull., Publ. by Oberlin College, No. 2: 1–33, pls. 1–14.

MCDONALD, MARSHALL

1887. 115. Report on distribution of fish and eggs by the U.S. Fish Commission for the season of 1885–86. Bull. U. S. Fish Comm. 6 (1886): 385–94.

MCKAY, CHARLES L.

1881. A review of the genera and species of the family Centrarchidae, with a description of one new species. Proc. U. S. Nat. Mus. 4: 87–93.

MEEK, SETH E.

1889. Notes on a collection of fishes from the Maumee Valley, Ohio. Proc. U. S. Nat. Mus. 2 (1888): 435–40.

METCALF, ARTIE L.

1966. Fishes of the Kansas River system in relation to zoogeography of the Great Plains. Mus. of Natur. Hist., Univer. of Kansas Publ. Lawrence 17 (3): 23–189, 4 figs., 51 maps.

MEYERS, DEXTER

1947. Norwalk. Ohio Cons. Bull., Columbus 11, No. 7: 30.

MILLER, ROBERT RUSH

1950. A review of the American clupeid fishes of the genus *Dorosoma*. Proc. U. S. Nat. Mus. 100, No. 3267: 387–410.

1952. Bait fishes of the lower Colorado River from Lake Mead, Nevada, to Yuma, Arizona, with a key for their identification. Calif. Fish and Game 38, No. 1: 7–42 (reprint).

1960. Systematics and biology of the gizzard shad (*Dorosoma cepedianum*) and related fishes. U. S. Fish and Wildlife Serv. Fish. Bull. 173, 60: 371–92, 4 figs. 4 tabs.

MILLER, ROBERT VICTOR

1968. A systematic study of the greenside darter, *Etheostoma blennioides* Rafinesque (Pisces: Percidae). Copeia (1): 1–40, 5 figs., 9 tabs.

MILLER, RUDOLPH J.

1968. Speciation in the common shiner: an alternate view. Copeia (3): 640–47.

MILLER, RUDOLPH J., AND HENRY W. ROBISON

1973. The fishes of Oklahoma. Okla. State Univ. Press: 1–246, 9 figs., plus many unnumbered figs. and distribution maps.

MILNER, JAMES W.

1874. Report on the fisheries of the Great Lakes; the result of inquiries prosecuted in 1871 and 1872. Rept. U. S. Comm. Fish and Fish. 73 (1872): 1–78.

MINCKLEY, W. L., AND LOUIS A. KRUMHOLZ

1960. Natural hybridization between the clupeid genera *Dorosoma* and *Signalosa*, with a report on the distribution of *S. petenensis*. Zoologica 45 (4): 171–80, 2 figs., 2 pl.

MOENKHAUS, W. J.

1911. Cross fertilization among fishes. Proc. Ind. Acad. Sci. for 1910: 353–93.

MOORE, GEORGE A.

1968. Fishes. *In* Vertebrates of the United States, W. Frank Blair et al., Part 2. McGraw-Hill Co.: 22–165.

MOORE, GEORGE A., MILTON B. TRAUTMAN, AND MILTON R. CURD

1973. A description of postlarval gar (*Lepisosteus spatula* Lacépède, Lepisosteidae), with a list of associated species from the Red River, Choctaw County, Oklahoma. Southwestern Naturalist 18 (3): 343–44.

MORGAN, GEORGE D.

1951. The life history of the bluegill sunfish, *Lepomis macrochirus*, of Buckeye Lake, Ohio. J. Scientific Laboratories, Denison Univ. 42 (4): 21–59, 7 figs., 15 tabs., 9 graphs, 2 pls.

1954. The life history of the white crappie (*Poxomis annularis*) of Buckeye Lake, Ohio. J. Scientific Laboratories, Denison Univ. 43: Arts. 6, 7, 8, 12 figs., 7 tabs.

MORMAN, ROBERT H.

1979. Distribution and ecology of lampreys in the Lower Peninsula of Michigan, 1957–75. Tech. Rep. 33 Great Lakes Fish. Comm. Ann Arbor, Mich. April: 1–59.

MORSE, H. HOWE

1939. Erosion and related land use conditions on the Muskingum River watershed. U. S. Dept. Agri., Washington D.C.: 1–36.

MORSE, MAX

1903. News and Notes. Ohio Nat., Columbus, 4, No. 1: 24.

MOSELEY, E. L.

1899. Sandusky flora. A catalogue of the flowering plants and ferns. Ohio Acad. Sci., Spec. Pap. No. 1, Clapper Printing Co., Wooster: 1–167, map. 1.

1939. Long time forecasts of Ohio River floods. Ohio J. Sci. 39, (4): 220–31.

MOUNT, DONALD I.

1959. Spawning behavior of the bluebreast darter, *Etheostoma camurum* (Cope). Copeia (3): 240–43, 1 fig.

NATURAL HISTORY SURVEY

1968. Hybrid Bass. Ill. Nat'l. Hist. Survey, Urbana, Ill. Dec. No. 74.

NELSON E. W.

1876. A partial catalogue of the fishes of Illinois. Bull. Ill. Mus. Nat. Hist., Bloomington. 1: 1–52.

NELSON, GARETH J.

1972. Cephalic sensory canals, pitlines, and the classification of esocoid fishes, with notes on galaxiids and other teleosts. Amer. Mus. Novitates. Amer. Mus. of Nat'l. Hist. 2492: 1–49.

NELSON, GARETH, AND N. NORMA ROTHMAN

1973. The species of gizzard shads (Dorosomatinae) with particular reference to the Indo-Pacific region. Bull. Amer. Mus. of Nat'l. Hist. 150 (2): 135–204.

NELSON, JOSEPH S.

1968. Hybridization and isolating mechanisms between *Catostomus commersonii* and *C. marcocheilus* (Pisces: Catostomidae). J. Fish. Res. Bd. Can. 25 (1): 101–50.

1969. Geographic variation in the brook stickleback, *Culaea inconstans*, and notes on nomenclature and distribution. J. Fish. Res. Bd. Can.: 2431–2447.

NELSON JOSEPH S., AND SHELBY D. GERKING

1968. Annotated key to the fishes of Indiana. Dept. of Zoology, Indiana Univ., Bloomington, Ind.: 83 p.

NETSCH, LT. NORVAL F., AND ARTHUR WITT, JR.

1962. Contributions to the life history of the longnose gar (*Lepisosteus osseus*) in Missouri. Trans. Amer. Fish. Soc. 91 (3): 251–62.

OKKLEBERG, PETER

1922. Notes on the life-history of the brook lamprey, *Ichthyomyzon unicolor*. Occ. Pap. Mus. Zool. Univ. Mich. No. 125: 1–14, figs. 4.

ORR, LOWELL P., AND RUSSELL G. RHODES
1967. The algae and fishes of the Upper Cuyahoga River. Dept. of Biol. Sci., Kent State Univ., unpubl.: 1–58.
ORSANCO
1962. Aquatic life resources of the Ohio River. Ohio River Valley Water Sanitation Commission (ORSANCO). Cincinnati: 218 p.
OSBURN RAYMOND C.
1901. The fishes of Ohio. Ohio St. Acad. Sci., Spec. Paper, 4: 1—105.
OSBURN, RAYMOND C., EDWARD L. WICKLIFF, AND MILTON B. TRAUTMAN
1930. A revised list of the fishes of Ohio. Ohio J. Sci. 30 (3): 169–76.
OSBURN, R. C., AND E. B. WILLIAMSON
1898. A list of the fishes of Franklin County, Ohio, with a description of a new species of *Etheostoma*. 6th Ann. Rept. Ohio St. Acad. Sci.: 11—20.
1899. Additional notes on the fishes of Franklin, County Ohio. 7th Ann. Rept. Ohio St. Acad. Sci. 7: 33–4.
[OWENS, W. E.]
1939. Compilation of stream flow of Ohio. Ohio Dept. Agri. Div. Cons. and Nat. Res. Bull. 200: 1–129.
PAGE, LAWRENCE M.
1978. Redescription, distribution, variation and life history notes on *Percina macrocephala* (Percidae). Copeia (4): 655–64, 3 figs., 6 tabs.
PAGE, LAWRENCE M., AND PHILIP W. SMITH
1970. The life history of the dusky darter, *Percina sciera*, in the Embarras River, Illinois. Ill. Nat'l. Hist. Survey, Biol. Notes, No. 69: 15 p., 11 figs., 2 tabs.
PARKER, J. B., E. B. WILLIAMSON, AND R. C. OSBURN
1899. A descriptive list of the fishes of Big Jelloway Creek and tributaries, Knox County, Ohio. 7th Ann. Rept. Ohio St. Acad. Sci. 7: 18–32.
PARSONS, JOHN W.
1973. History of the salmon in the Great Lakes, 1850–1970. Technical Papers, Bur. of Sport Fish. and Wildlife, Washington: 1–76.
1974. Striped bass (*Morone saxatilis*): distribution, abundance, biology, and propagation in North America and suitability for introduction into the Great Lakes. Bur. of Sport Fish. and Wildlife, Ann Arbor: 1–64.
PARSONS, JOHN W., AND J. BRUCE KIMSEY
1954. A report on the Mississippi threadfin shad. U. S. Fish and Wildlife Serv. Progressive Fish-Culturist, 16 (4): 179–81, 1 fig.
PEATTIE, RODERICK
1923. Geography of Ohio. Geol. Surv. of Ohio. 4th Ser., Bull. 27, Columbus: 1–137.
PETRAVICZ, JOSEPH J.
1936. The breeding habits of the least darter, *Microperca punctulata* Putnam. Copeia (2): 77–82.
PFLIEGER, WILLIAM L.
1965. Reproductive behavior of the minnows *Notropis spilopterus* and *Notropis whipplii*. Copeia (1): 1–8.
1966. Young of the orangethroat darter (*Etheostoma spectabile*) in nests of the smallmouth bass (*Micropterus dolomieui*). Copeia (1): 139–40.
1971. A distributional study of Missouri fishes. Univ. of Kansas Publ. 20 (3) 225–570, 15 figs., 193 maps.
1975. The fishes of Missouri. Missouri Dept. of Cons.: 343 p., many distribution maps.
PIETERS, A. J.
1902. The plants of western Lake Erie, with observations on their distribution. Bull. U. S. Fish. Comm., Washington, 21 (1901): 57–79, pls. 11—20.
PIRNIE, MILES DAVID
1935. Michigan waterfowl management. Mich. Dept. of Cons., Game Div., Lansing: 328 p., 212 figs., 14 tabs.
POLLARD, JAMES E.
1935. The journal of Jay Cooke or the Gibraltar records 1865–1905. Ohio State Univ. Press, Columbus: 1–359.
POTTER, EMERY D.
1877. Report of superintendent of hatchery. First Ann. Rept. Ohio State Fish Comm., Nevins and Myers, Columbus: 27–30.

1878. Report of the superintendent. 2d Ann. Rept. of the Ohio State Fish Comm., for the year 1877. Nevins & Myers, Columbus: 70–76.

1879. Appendix to annual report of superintendent of hatcheries. 3d Ann. Rept. *Ibid.* (1878): 13–20.

1881. Report of the superintendent of fish hatcheries, for the State of Ohio, for the year 1880, and the first month of 1881. 5th Ann. Rept. *Ibid.* (1880): 13–21.

PUTNAM, F. W.

1863. No. 1. List of the fishes sent by the Museum to different institutions, in exchange for other specimens, with annotations. Bull. Mus. Comp. Zool. Harvard College 1: 1–16.

RADFORTH, ISOBEL

1944. Some considerations on the distribution of fishes in Ontario. Contrib. Roy. Ont. Mus. Zool. 25: 1–116, figs. 32.

RAFINESQUE, C. S.

1818. Discoveries in natural history, made during a journey through the Western Region of the United States. Amer. Monthly Mag. and Journ., 3 (5): 354–56.

1819. Prodrome de 70 nauveaux generes d'animaux decouvertes dans l'interieur des Etats-Unis d'Amerique durant l'annee 1818. J. de physique, de chimie, d'histoire naturelle et des arts, 88: 417–29.

1820. Icthyologia Ohiensis, or natural history of the fishes inhabiting the river and its tributary streams, preceded by a physical description of the Ohio and its branches. Lexington. W. G. Hunt: 1–175 (Call's 1899 reprint and pagination used.)

RANEY, EDWARD C.

1938. The distribution of the fishes of the Ohio drainage basin of western Pennsylvania. Unpub. Ph.diss., Cornell Univ.: 1—102.

1939A. The distribution of the fishes of the Ohio drainage basin of western Pennsylvania. Cornell Univ., Abstracts of theses (1938): 273–77, maps 1–122.

1939B. The breeding habits of *Ichthyomyzon greeleyi* Hubbs and Trautman. Copeia. (2): 111–12.

1939C. The breeding habits of the silvery minnow, *Hyboganthus regius* Girard. Amer. Midl. Nat., 21 (3): 674–80.

1940A. *Rhinichthys bowersi* from West Virginia a hybrid, *Rhinichthys cataractae* X *Nocomis micropogon*. Copeia (4): 270–71.

1940B. The breeding behavior of the common shiner *Notropis cornutus* (Mitchill). Zoologica. New York Zool. Soc., 25, Pt. 1–14, pls. 1–4.

1940C. Reproductive activities of a hybrid minnow. *Notropis cornutus* × *Notropis rubellus*. Zoologica 25 (3): 361–67.

1941A. Records of the brook lamprey, *Lampetra aepyptera* (Abbott), from the Atlantic drainage of North Carolina and Virginia. J. Elisha Mitchell Sci. Soc., 57 (2): 318–20.

1941B. Range extensions and remarks on the distribution of *Parexoglossum laurae* Hubbs. Copeia (4): 272.

1942. Alligator gar feeds upon birds in Texas. Copeia (1): 50.

1947. A tentative list of the fishes of West Virginia. Cons. Comm. of West Va.: 1–21, Mimeo.

1957. Natural hybrids between two species of pickerel (*Esox*) in Stearns Pond, Massachusetts. Fish. Rept. for some central, eastern, and western Massachusetts lakes, ponds, and reservoirs, 1951–52. Commonw. of Mass. (1955): 1–15.

RANEY, EDWARD C., AND ERNEST A. LACHNER

1939. Observations on the life history of the spotted darter, *Poecilichthys maculatus* (Kirtland). Copeia (3): 157–65, figs. 2.

RANNEY, LEO

1940. Ohio's falling water tables—suggestions for recharging. Ohio Chamber of Commerce: 1–8, unnumbered.

REEVES, CORA D.

1907. The breeding habits of the rainbow darter (*Etheostoma coeruleum* Storer), a study in sexual selection. Biol. Bull., 14: 35–59, figs. 3.

REIGHARD, JACOB

1903. The natural history of *Amia calva* Linnaeus. *In:* Mark anniversary volume, Henry Holt and Co., New York, No. 4: 57–109, pls. 7.

1910. Methods of studying the habits of fishes with an account of the breeding habits of the horned dace. Bull. U. S. Bur. Fish., 28. (2) (1908): 1113–36, pls. 114–20, figs. 5.

1913. The breeding habits of the log-perch (*Percina caprodes*). 15th Ann. Rept. Mich. Acad. Sci.: 104–05.
1920. The breeding habits of the suckers and minnows. Biol. Bull., 38: 1–32.
1943. The breeding habits of the river chub, *Nocomis micropogon* (Cope). Pap. Mich. Acad. Sci., Arts and Letters, 28 (1942): 397–423.

REIGHARD, JACOB, AND HAROLD CUMMINS
1916. Description of a new species of lamprey of the genus *Ichthyomyzon*. Occ. Pap. Mus. Zool. Univ. Mich., No. 31:1–12, pls. 2.

RENO, H. W.
1969. Cephalic lateral-line systems of the cyprinid genus *Hybopsis*. Copeia (4): 736–73.

RICHARDSON, JOHN
1836. Fauna Boreali-Americana: on the zoology of the northern parts of British America, containing descriptions of the objects of natural history collected on the Late Northern Land Expeditions, under the Command of Sir John Franklin, R. N., Pt. 3, The Fish. London and Norwich: 1–327, pls. 74–97.

RICKER, WILLIAM E.
1948. Hybrid sunfish for stocking small ponds. Trans. Amer. Fish. Soc., 75 (1945): 84–96.

RIDDELL, JOHN L.
1837. Report of John L. Riddell, M.D., etc.; Rept. No. 60 to Gov. and Legisl. of Ohio, reports, etc., made to the Thirty-fifth General Assembly of the State of Ohio. James B. Gardiner, Columbus (1836): 1–34.

RIGGS, CARL D.
1955. Reproduction of the white bass, *Morone chrysops*. Invest. Indiana Lakes and Streams 4 (3): 87–110, 2 figs., 4 tabs.

ROACH, LEE S.
1943. Buckeye Lake white bass. Ohio J. Sci., 43 (6): 263–69.
1949. Sauger: *Stizostedion canadense*. Ohio Cons. Bull., No. 2: 12–13.

ROBINS, CHARLES RICHARD
1954. A taxonomic revision of the *Cottus bairdi* and *Cottus carolinae* species groups in eastern North America (Pisces, Cottidae). Unpublished Ph.D. Thesis, Cornell Univ.: 1–248.

ROBINS, C. RICHARD, REEVE M. BAILEY, CARL E. BOND, JAMES R. BOOKER, ERNEST A. LACHNER, ROBERT H. LEA, AND W. B. SCOTT
1980. A list of common and scientific names of fishes from the United States and Canada. Fourth Edition, Ann. Fish. Soc. Spec. Publ. 12:1–174.

ROBINS, C. RICHARD, AND EARL E. DEUBLER, JR.
1955. The life history and systematic status of the burbot, *Lota lota lacustris* (Walbaum), in the Susquehanna River system. N. Y. State Mus. and Sci. Serv., Circ. 39: 1–49, 3 figs., 5 tabs., 2 pls.

ROSS, D. F.
1972. Aspects of the life history and ecology of the silverjaw minnow, *Ericymba buccata*, in Yellowbud Creek, Ohio. Master's thesis, Ohio State Univ.: 78 p.

ROSS, M. R.
1976. Nest-entry behavior of female creek chubs, *Semotilus atromaculatus*, in different habitats. Copeia (2): 378–80.
1977. Function of creek chub (*Semotilus atromaculatus*) nest-building. Ohio J. Sci. 77(1): 36–37.

ROSS, M. R., AND TED M. CAVENDER
1977. First report of the natural cyprinid hybrid, *Notropis cornutus* × *Rhinichthys* cataractae, from Ohio. Copeia (4): 777–78.

ROSTLUND, ERHARD
1951. Three early historical reports of North American freshwater fishes. Copeia (4): 295–96.
1952. Freshwater fish and fishing in native North America. Univ. of California Publ. in Geography, Univ. Calif. Press, 9: 1–313, maps 1–47.

SALDANA, LUPI
1971. Bass, bluegill mate in Hawaii reservoir. Los Angeles Times, August 6: 11.

SCHAFFNER, JOHN H.
1928. Field manual of the flora of Ohio and adjacent territory. Adams and Company, Columbus: 1–638.

SCHLOEMER, CLARENCE L.

1936. The growth of the muskellunge, *Esox masquinongy immaculatus* (Garrard), in various lakes and drainage areas of northern Wisconsin. Copeia (4): 185–93.

SCHMIDT, KARL P.

1938. Herpetological evidence for the postglacial eastward extension of the steppe in North America. Ecol., 19, (3): 396–407.

SCHNEBERGER, EDWARD, AND LOWELL A. WOODBURY

1944. The lake sturgeon, *Acipenser flavescens* Rafinesque, in Lake Winnebago, Wisconsin. Trans. Wisc. Acad., Sci., Arts and Letters, 36: 131–40.

SCHOFFMAN, ROBERT J.

1944. Age and growth of the smallmouth buffalo in Reelfoot Lake J. Tenn. Acad. Sci., 29 (1): 3–9.

SCHRENKELSEN, RAY

1938. Field book of fresh-water fishes of North America north of Mexico. G. P. Putnam's Sons, New York: 312 p.

Scott Journal [by CHARLES SCOTT]

1793–94. Journals of the campaigns of 1793 and 1794 under the command of Major General Charles Scott. Unpublished manuscript. Original in Filson Manuscript Collection of Louisville, Ky., a transcript in the Ohio State Museum Library (which granted permission to cite): pp. unnumbered.

SCOTT, W. B.

1951. Fluctuations in abundance of the Lake Erie cisco (*Leucichthys artedi*) population. Contrib. Roy. Ont. Mus. Zool. 32: 1–41.

SCOTT, W. B. AND E. J. CROSSMAN

1973. Freshwater fishes of Canada. Fish. Res. Board of Can. Ottawa: 966 p.

SEARS, PAUL B.

1941. Postglacial vegetation in the Erie - Ohio area. Ohio J. Sci., 41, (3): 225–34.

SERNS, S. L., AND TERRENCE C. MCKNIGHT

1977. The occurrence of northern pike × grass pickerel hybrids and an exceptionally large grass pickerel in a northern Wisconsin stream. Copeia (4): 780–81.

SHAFER, PAUL V.

1950. General catch reports of Ohio Lake Erie commercial fisheries for 1949. Dept. Nat'l. Res., Div. of Wildlife, Sec. of Fish Man., Bull. 251: 1–10, Mimeo.

SHERMAN, C. E.

1933. Ohio cooperative topographic survey. Miscellaneous data. Final Report, Ohio State Univ. Press, 4: 1–327.

SHETTER, DAVID S.

1949. A brief history of the sea lamprey problem in Michigan waters. Trans. Amer. Fish. Soc. 76 (1946): 160–76.

SLASTENENKO, E. P.

1954. The relative growth of hybrid char (*Salvelinus fontinalis* × *Cristivomer namaycush*). Jour. Fish. Res. Bd. Can. 11 (5): 652—59, 3 figs., 1 tab.

SMILEY, CHARLES W.

1884. 39. Report on the distribution of carp to July 1, 1881, from young reared in 1879 and 1880. Rept. U. S. Comm. Fish and Fish., 10 (1882): 943–1008.

1889. 11. Loch Leven trout introduced in the United States. Bull. U. S. Fish Comm. 7 (1887): 28–32.

SMITH, BERTRAM G.

1908. The spawning habits of *Chrosomus erythrogaster*. Biol. Bull., 15: 9–18.

1922. Notes on the nesting habits of *Cottus*. Pap. Mich. Acad. Sci., Arts and Letters 2: 221–24, pls. 2.

SMITH, GUY-HAROLD

1935. The relative relief of Ohio. Geographical Review 25: 272–84, fig. 1.

SMITH, HUGH M.

1898A. Report of the Division of Statistics and methods of the fisheries. U.S. Comm. of Fish and Fisheries, Rept. of the Commissioner, Part 22 (1896): 119–45.

1898B. Statistics of the fisheries of the interior waters of the United States. U.S. Comm. of Fish and Fisheries, Rept. of the Commissioner, Part 22 (1896): 489–574.

SMITH, HUGH M., AND L. G. HARRON

1904. Breeding habits of the yellow catfish. Bull. U. S. Fish Comm. 22 (1902): 149–54.

SMITH, HUGH M., AND MERWIN-MARIE SNELL

1891. Review of the fisheries of the Great Lakes in 1885, compiled by Hugh M. Smith and Mervin-Marie Snell, with introduction and description of fishing vessels and boats by J. W. Collins. U. S. Comm. Fish and Fish., Misc. Doc. 133 (1887): 1–333.

SMITH, O. W.

1922. The book of the pike. Stewart Kidd Co., Cincinnati: 197 p., pls. 13.

SMITH, S. B.

1885. 150. Efforts to raise trout. Bull. U. S. Fish Comm. 5: 435–36.

SNELSON, FRANKLIN F., AND WILLIAM L. PFLIEGER

1975. Redescription of the redfin shiner, *Notropis umbratilis*, and its subspecies in the Central Mississippi River Basin. Copeia (2): 231–49, 6 figs., 6 tabs.

SPIERS J. MURRAY

1952. Nomenclature of the channel catfish and the burbot of North America. Copeia (2): 99–103.

STANSBERY, DAVID H.

1961. A century of change in the naiad population of the Scioto River System in central Ohio. Ann. Rept. for 1961 of the American Malacological Union: 20–22.

STEWART, NORMAN H.

1926. Development, growth and food habits of the white sucker, *Catostomus commersonii* LeSueur. Bull. U. S. Bur. Fish. (42): 147–84, figs. 55.

STOUT, WILBER, AND G. F. LAMB

1938. Physiographic features of southeastern Ohio. Ohio J. Sci. 38, (2): 49–83.

STOUT, WILBER, KARL VER STEEG, AND G. F. LAMB

1943. Geology of water in Ohio. Geol. Surv. of Ohio. Bull. 44, Ser. 4: 1—694, maps 8.

SULLIVANT, JOSEPH

1874. A genealogy and family memorial. Ohio State Journal and Job Rooms: 1–372.

SUMMERFELT, ROBERT C., AND CHARLES O. MINCKLEY

1969. Aspects of the life history of the sand shiner, *Notropis stramineus* (Cope), in the Smoky Hill River, Kansas. Trans. Amer. Fish. Soc. 98 (3): 444–53, 4 figs., 1 tab.

SUTTKUS, ROYAL D.

1958. Status of the nominal cyprinid species *Moniani deliciosa* Girard and *Cyprinella texana* Girard. Copeia (4): 307–18, 8 figs., 5 tabs.

1963. Order Lepisostei. *In* Fishes of the Western North Atlantic. Memoir Sears Found. for Marine Res., 1 (3): 61–88.

TARTER, DONALD C.

1969. Some aspects of reproduction in the western blacknose dace, *Rhinichthys atratulus meleagris* Agassiz, in Doe Run, Meade County, Kentucky. Trans. Amer. Fish. Soc. 98 (3): 454–59.

TAUB, STEPHEN H.

1969. Fecundity of the white perch. Progressive Fish-Culturists 31 (3): 166–68.

TAYLOR, WILLIAM RALPH

1954. Records of fishes in the John N. Lowe Collection from the Upper Peninsula of Michigan. Misc. Publ. Mus. Zool. Univ. Mich., (87): 1–50, map 1.

1955. A revision of the genus *Noturus* Rafinesque with a contribution to the classification of the North American catfishes. Unpubl. Ph.D. Thesis, Univ. of Mich.: 1–583.

1957. A revision of the genus *Noturus* Rafinesque with a contribution to the classification of the North American catfishes (Abst.). Diss. Abst., 17: 192. Reprinted (1957), Sport Fish. Abstr., 2: 151–52.

1969. A revision of catfish genus *Noturus* Rafinesque with an analysis of higher groups in the Ictaluridae. U. S. Nat. Mus., Bull. 282: 315 p., 28 tabs., 21 pls.

TESTER, ALBERT L.

1931. Spawning habits of the small-mouthed black bass in Ontario waters. Trans. Amer. Fish. Soc., 60 (1930): 53–61.

1932A. Foods of the small-mouthed black bass (*Micropterus dolomieu*) in some Ontario waters. Univ. Toronto Studies, Biol. Ser., (36) (Publ. Ont. Fish. Res. Lab., No. 46): 169–203.

1932B. Rate of growth of the small-mouth black bass (*Micropterus dolomieu*) in some Ontario waters. *Ibid.* (*Ibid.* No. 47): 205–21.

THOMAS, EDWARD S.

1951. Distribution of Ohio animals. Ohio J. Sci. 51, (4): 153–67.

THOMPSON, DAVID H.

1935. Hybridization and racial differentiation among Illinois fishes. Ill. Nat. Hist. Surv. Bull., 20 (5): 492–94. (Appendix to Annotated List of the Fishes of Illinois, by D. John O'Donnell.)

TRANSEAU, EDGAR NELSON

1935. The Prairie Peninsula. Ecol., 16, (3): 423–37.

1941. Prehistoric factors in the development of the vegetation of Ohio. Ohio J. Sci., 41 (3): 207–11.

TRAUTMAN, MILTON B.

1928. Ducks feeding upon gizzard shad. Ohio State Mus. Sci. Bull., No. 1: 78.

1930. The specific distinctness of *Poecilichthys coeruleus* (Storer) and *Poecilichthys spectabilis* Agassiz. Copeia (1): 12–13.

1931A. *Parexoglossum hubbsi*, a new cyprinid fish from western Ohio. Occ. Pap. Mus. Zool. Univ. Mich., No. 235, Dec. 1: 1–11, pl. 1.

1931B. *Notropis volucellus wickliffi*, a new subspecies of cyprinid fish from the Ohio and Upper Mississippi rivers. Ohio J. Sci., 31 (6): 468–74, fig. 1.

1932. The practical value of scientific data in a stocking policy for Ohio. Trans. Amer. Fish. Soc. Vol. 62: 258–60.

1933. The general effects of pollution on Ohio fish life. Trans. Amer. Fish. Soc., 63: 69–72.

1939A. The effects of man-made modifications on the fish fauna in Lost and Gordon creeks, Ohio, between 1887–1938. Ohio J. Sci., 39 (5): 275–88, tab. 1.

1939B. A new sunfish for Ohio fishermen. Ohio Cons. Bull. Div. Cons. and Nat. Res., Columbus: 5 and 31.

1939C. The numerical status of some mammals throughout historic time in the vicinity of Buckeye Lake, Ohio. Ohio J. Sci., 39 (3): 133–43.

1940A. The birds of Buckeye Lake, Ohio. Misc. Publ. Mus. Zool. Univ. Mich., No. 44: 1–466, pls. 15 and front., maps 2.

1940B. Artificial keys for the identification of the fishes of the State of Ohio. Franz Theo. Stone Lab., Put-in-Bay, Ohio, March: 1–38. Mimeo.

1941. Fluctuations in lengths and numbers of certain species of fishes over a five-year period in Whitmore Lake, Michigan. Trans. Amer. Fish. Soc., 70 (1940): 193–208, figs. 3, tabs. 3.

1942. Fish distribution and abundance correlated with stream gradient as a consideration in stocking programs. Trans. 7th North Amer. Wildlife Conf., Washington: 211–23, figs. 5, tabs. 14.

1946. Artificial keys for the identification of the fishes of the State of Ohio. The Franz Theodore Stone Institute of Hydrobiology and Ohio Div. of Wildlife. Columbus, Ohio. Aug. 1: 1–52, Mimeo.

1948. A natural hybrid catfish, *Schilbeodes miurus* x *Schilbeodes mollis*. Copeia (3): 166–74, pl. 1.

1949. The invasion, present status, and life history of the sea lamprey in the waters of the Great Lakes, especially the Ohio waters of the Lake. Contrib. of and Mimeographed by the Franz Theodore Stone Laboratory, Put-in-Bay, Ohio, Dec. 1: 1–9.

1950. Artificial keys for the identification of the fishes of the State of Ohio. The Franz Theodore Stone Institute of Hydrobiology and Ohio Div. of Wildlife. Columbus, Ohio. June 1: 1–52, Mimeo. (Similar to the 1946 revision).

1956. *Carpiodes cyprinus hinei*, a new subspecies of carpsucker from the Ohio and Upper Mississippi River systems. Ohio J. Sci., 56 (1): 33–40, pl. 1, map 1.

1957. The fishes of Ohio with illustrated keys. The Ohio State University Press in collaboration with the Ohio Div. of Wildlife and the Ohio State Univ. Dev. Fund: 1–683, 7 color pls., 202 figs., 183 maps.

1960. Diversions, canals and conduits—their role in introducing aquatic organisms into drainage basins where they did not reside formerly. International Union for Cons. of Nature and Nat'l Res. 7th Tech. Meeting, Athens (1958). Soil and Water Conser. Theme 1; Nat'l Aquatic Res. 4, Brussels: 175–84.

1973. A guide to the collections and identification of presmolt Pacific salmon in Alaska with an illustrated key. NOAA Tech. Mem. NMFS ABFL-2. U.S. Dept. of Commerce, Seattle: 1–111, 1–20, 7 figs.

1976. The fishes of the Sandusky river system, Ohio. p. 233–41. *In* David B. Baker, William B. Jackson, Bayliss L. Prater, eds. Sandusky River Basin Symposium, May 2–3, 1975. Tiffin, Ohio. Intern. Ref. Group on Great Lakes Pollution from Land Use Activities. Inter. Joint Commissioner, U.S. Environmental Protection Agency. Chicago: 475 p.

1977. The Ohio Country from 1750 to 1977 - a naturalist's view. Ohio Biol. Surv. Bio. Notes No. 10. The Ohio State Univ.: 1—25.

1981. Some vertebrates of the Prairie Peninsula. *In* Ronald L. Stukey and Karen J. Reese, eds. The Prairie Peninsula in the Shadow of Transeau. Proc. 6th N. Amer. Prairie Conf., The Ohio State Univ., Columbus, Ohio, Aug. 12–17, 1978. Ohio Biol. Surv. Biol. Notes 15.

TRAUTMAN, MILTON B., AND DONALD K. GARTMAN

1974. Re-evaluation of the effects of man-made modifications on Gordon Creek between 1887 and 1973 and especially as regards its fish fauna. Ohio J. Sci. 74 (3): 162–73, 3 figs., 1 tab.

TRAUTMAN, MILTON B., AND ROBERT G. MARTIN

1951. *Moxostoma aureolum pisolabrum*, a new subspecies of sucker from the Ozarkian streams of the Mississippi River System. Occ. Pap. Mus. Zool. Univ. Mich. No. 534: 1–10, pl. 1, map 1.

TRAVER, JAY B.

1929. The habits of the black-nosed dace, *Rhinichthys atronasus* (Mitchill). J. Elisha Mitchell Sci. Soc. 45 (1): 101–20.

TSAI, CHU-FA, AND EDWARD C. RANEY

1974. Systematics of the banded darter, *Etheostoma zonale* (Pisces: Percidae). Copeia (1): 1–24.

TURNER, CLARENCE L.

1920. Distribution, food and fish associates of young perch in the Bass Island region of Lake Erie. Ohio J. Sci., 20 (5): 137–52.

TURNER, C. L.

1922. Notes on the food habits of young of *Cottus ictalops* (Millers Thumb.). Ohio J. Sci. 22 (3): 95–96.

UNDERHILL, A. HEATON

1939. Cross between *Esox niger* and *E. Lucius*. Copeia (4): 237.

U. S. Geological Survey.

1915. Ohio East Liberty quadrangle. Scale 1/62500.

[United States Army, compiled under directions of Chief of Engineers.]

1935. The Ohio River. U. S. Gov't. Printing Office. Washington: 1–438.

VAN CLEAVE, HARLEY J., AND HENRY C. MARKUS

1929. Studies on the life history of the blunt-nosed minnow. Amer. Nat. 63: 530–39.

VAN DENBURGH, JOHN

1920. A further study of variation in the gopher snakes of western North America. Proc. Calif. Acad. Sci. 4th Ser., 10: 1–27, pls. 1–2, figs. 1–7.

VAN METER, HARRY D., AND MILTON B. TRAUTMAN

1970. An annotated list of the fishes of Lake Erie and its tributary waters exclusive of the Detroit River. Ohio J. Sci. 70 (2): 65–78.

VAN OOSTEN, JOHN

1930. The disappearance of the Lake Erie cisco.–A preliminary report. Trans. Amer. Fish Soc. 60: 204–14.

1937A. The dispersal of smelt, *Osmerus mordax* (Mitchill), in the Great Lakes region. *Ibid.* 66: 160–71, map 1.

1937B. First records of the smelt, *Osmerus mordax*, in Lake Erie. Copeia (1): 64–65.

1938. The age and growth of the Lake Erie sheepshead *Aplodinotus grunniens* Rafinesque. Pap. Mich. Acad. Sci., Arts, and Letters 23 (1937): 651–68., tabs. 1–8.

1942. The age and growth of the Lake Erie white bass, *Lepibema Chrysops* (Rafinesque). Pap. Mich. Acad. Sci., Arts and Letters 27 (1941): 307–34.

VAN OOSTEN, JOHN, AND RALPH HILE

1949. Age and growth of the lake whitefish, *Coregonus clupeaformis* (Mitchill), in Lake Erie. Trans. Amer. Fish. Soc. 77 (1947): 178–249.

VLADYKOV, VADIM D.
1949. Quebec lampreys. (*Petromyzonidae*) 1. List of species and the economical importance. Dept. of Fisheries, Quebec, Contrib. 26: 1–67, figs. 21.

VLADYKOV, VADIM D., AND BEAULIEU GERARD
1946. Etudes sur l'Esturgeon (Acipenser) de la Province de Quebec. Cont. du Dept. des Pecheries, Quebec (18): 1–62, pls. 20, figs. 23, tabs. 18 (reprinted from: Naturaliste Canadien, Quebec [73]: 143–204).

WALLEN, I. EUGENE
1951. The direct effect of turbidity on fishes. Bull. Okla. Agric. & Mech. College 48 (2): 1–27.

WARD, HENRY B.
1919. Notes on North American Myxosporidia. J. of Parasitology 6 (2): 49–64, pls. 1–5.

WASCKO, HAROLD
1950. Gambusia in northwestern Ohio. Ohio Dept. of Nat'l Res., Div. of Wildlife. Columbus. Bull. 240, Jan.: 1–4, Mimeo.

WASCKO, HAROLD, AND CLARENCE F. CLARK
1948. Pond propagation of bluntnose and blackhead minnows. Wildlife Cons. Bull., Ohio Div. of Cons. and Nat'l Res., Bull. 4: 1–16.

WEED, ALFRED C.
1927. Pike, pickerel and muskalonge. Field Mus. Nat. Hist., Zool. Leaflet No. 9: 1–52, pls. 8, figs. 4.

WEST, JERRY L., AND F. EUGENE HESTER
1966. Intergeneric hybridization of centrarchids. Trans. Amer. Fish Soc. 95 (3): 280–88, 6 figs., 2 tabs.

[Western Intelligencer, The]
1811. Historical and geographical sketches of the State of Ohio. Of the Scioto Country. Western Intelligencer. Worthington, 1, No. 5, Sept.: 2 p. unnumbered.

WHITE, ANDREW M., MILTON B. TRAUTMAN, ERIC J. FOELL, MICHAEL P. KELTY, AND RONALD GABY
1975. Water quality baseline assessment for the Cleveland area–Lake Erie. Vol. 2. The Fishes of the Cleveland Metropolitan Area including the Lake Erie shoreline. U. S. Environmental Protection Agency, Region 5, Chicago: 181 p., 43 figs., 12 tabs.

WHITEHEAD, P. J. P., AND A. C. WHEELER
1966. Am. Mus. Civ. Stor. Nat. Genova, 76: 23.

WHITLEY, G. P.
1950. New fish names. Proc. Royal Zool. Soc. New So. Wales, 1948–49: 44.

WICKLIFF, E. L.
1933. Are newly impounded waters in Ohio suitable for fish life? Trans. Amer. Fish. Soc. 62 (1932): 275–77.

WICKLIFF, EDWARD L. AND MILTON B. TRAUTMAN
1934. List of the fishes of Ohio. Ohio Dept. Agri., Div. of Cons., Bur. Sci. Research. Bull. 1, Jan.: 1–3 Mimeo.
1947. Some food and game fishes of Ohio. Bull. Ohio Dept. Agri., Div. of Cons. and Nat'l Res., Columbus, May: 1–38, figs. 15, maps 15 (First printing, May, 1931; reprinted on Jan. 1932, Jan. 1935, June 1941, May 1947).

WILLIAMS, JOHN E.
1948. The muskellunge in Michigan. Mich. Conserv. 17, No. 10: 10–11 and 15.

WILLIAMSON, E. B.
1901. Fishes taken near Salem, Ohio. Ohio Naturalist. Columbus, 2, No. 2, Dec.: 165–66.

WILLIAMSON, E. B., AND R. C. OSBURN
1898. A descriptive catalogue of the fishes of Franklin County, Ohio. Unpubl. Joint Bachelor's Thesis, Ohio State Univ., June: 1–56 p., pls. 1–26.

WILSON, FRAZER E. [editor]
1935. Journal of Capt. Daniel Bradley. (An epic of the Ohio frontier, with copious comment by Frazer E. Wilson.) Frank H. Jobes and Son, Greenville, Ohio: 1–76.

WOLFE, JOHN N.
1942. Species isolation and a proglacial lake in southern Ohio. Ohio J. Sci. 42 (1): 2–12.
1951. The possible role of microclimate. *Ibid.* 51, (3): 134–38.

WONER, PAUL C.

1961. Commercial fisheries catch summary for Lake Erie, Ohio in 1960. Ohio Dept. of Nat'l Res. Div. of Wildlife Publ. W-200, 1–7 mimeo.

WOODS, ROBERT S.

1944. The naturalist's lexicon. Abbey Garden Press, Pasadena: 1–282 p.

1947. Addenda to the naturalist's lexicon. *Ibid.*: 1–47.

ZAHURANIC, BERNARD JOHN

1961. A study of the distribution of the fishes of the Black River, Ohio, watershed. Unpubl., Ohio Univ.: 131., 57 figs. (mostly distribution).

1962. Range extensions of some cyprinid fishes in southeastern Ohio. Copeia (4): 842–43.

ZURA, RICHARD A., AND LINDA B. SCOTHORN

1966. A new locality record for the American brook lamprey, *Lampetra lamottei* (Lesueur) in Ohio. Ohio J. Sci. 66 (6): 581.

PLATES

PLATE I

Fig. 1

NORTHERN SPOTTED BLACKBASS

Adult and immature. From specimens taken from Campaign Creek, Gallia County, O., on August 13, 1930 (*see* p. 565).

Fig. 2

WHITE BASS

From specimens taken from Buckeye Lake, Fairfield County, O., during summer of 1930 (*see* p. 542).

Fig. 3

COMMON SHINER

Structural features from specimens taken from Little Jelloway Creek, Knox County, O., on September 22, 1928; color characters from specimens taken from Big Darby Creek, Pickaway County, O., during summer of 1930 (*see* p. 340).

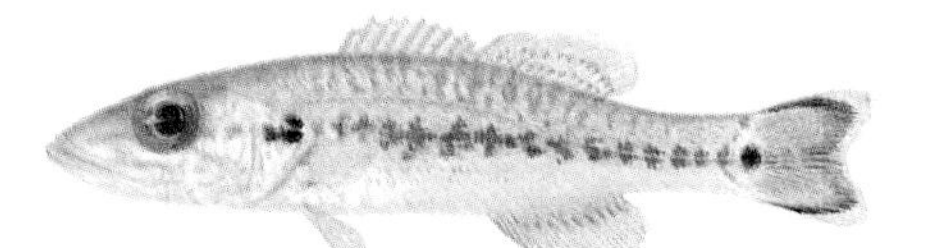

PLATE II

Fig. 4

CENTRAL MOTTLED SCULPIN

Structural features from specimen taken from Mad River, Logan County, O., on July 21, 1928; color characters from specimens taken from Mad River, Champaign County, O., on November 18, 1930 (*see* p. 704).

Fig. 5

NORTHERN ORANGETHROAT DARTER

Structural features fom specimen taken from Sugar Run, Union County, O., on October 16, 1928; color characters from specimens taken from same locality on September 1, 1930 (*see* p. 687).

Fig. 6

REDSIDE DACE

From specimens taken from Pine Run, Ashland County, O., on November 12, 1930 (*see* p. 309).

Fig. 7

BLUNTNOSE MINNOW

From specimens taken from Pine Run, Ashland County, O., on November 12, 1930 (*see* p. 394).

PLATE III

Fig. 8

WESTERN TONGUETIED MINNOW

Structural features from specimen taken from the Mad River, Green County, O., on July 20, 1929; color characters from specimens taken from the Mad River, Champaign County, O., on November 18, 1930 (*see* p. 300).

Fig. 9

GREATER REDHORSE

From specimen taken from Gordon Greek, Paulding County, O., on August 31, 1930 (*see* p. 444).

Fig. 10

OHIO SHORTHEAD REDHORSE

Structural features from specimens taken from Deer Creek, Ross County, O., on April 19, 1929 and from the Kokosing River, Knox County, O., on April 28, 1929; color characters from specimens taken from the Scioto River, Pike County, O., on September 14, 1930 (*see* p. 441).

PLATE IV

Fig. 11

NORTHERN CREEK CHUB

Structural features from specimen taken from Honey Creek, Perry County, O., on January 1, 1929; color characters from specimens taken from same locality during summer of 1930 (*see* p. 297).

Fig. 12

SILVER REDHORSE

From specimens taken from the Scioto River, Pike County, O., on September 14, 1930 (*see* p. 428).

Fig. 13

GOLDEN REDHORSE

Structural features from specimen taken from the Scioto River, Pike County, O., on May 3, 1930; color characters from specimens taken from the above locality on September 14, 1930 (*see* p. 434).

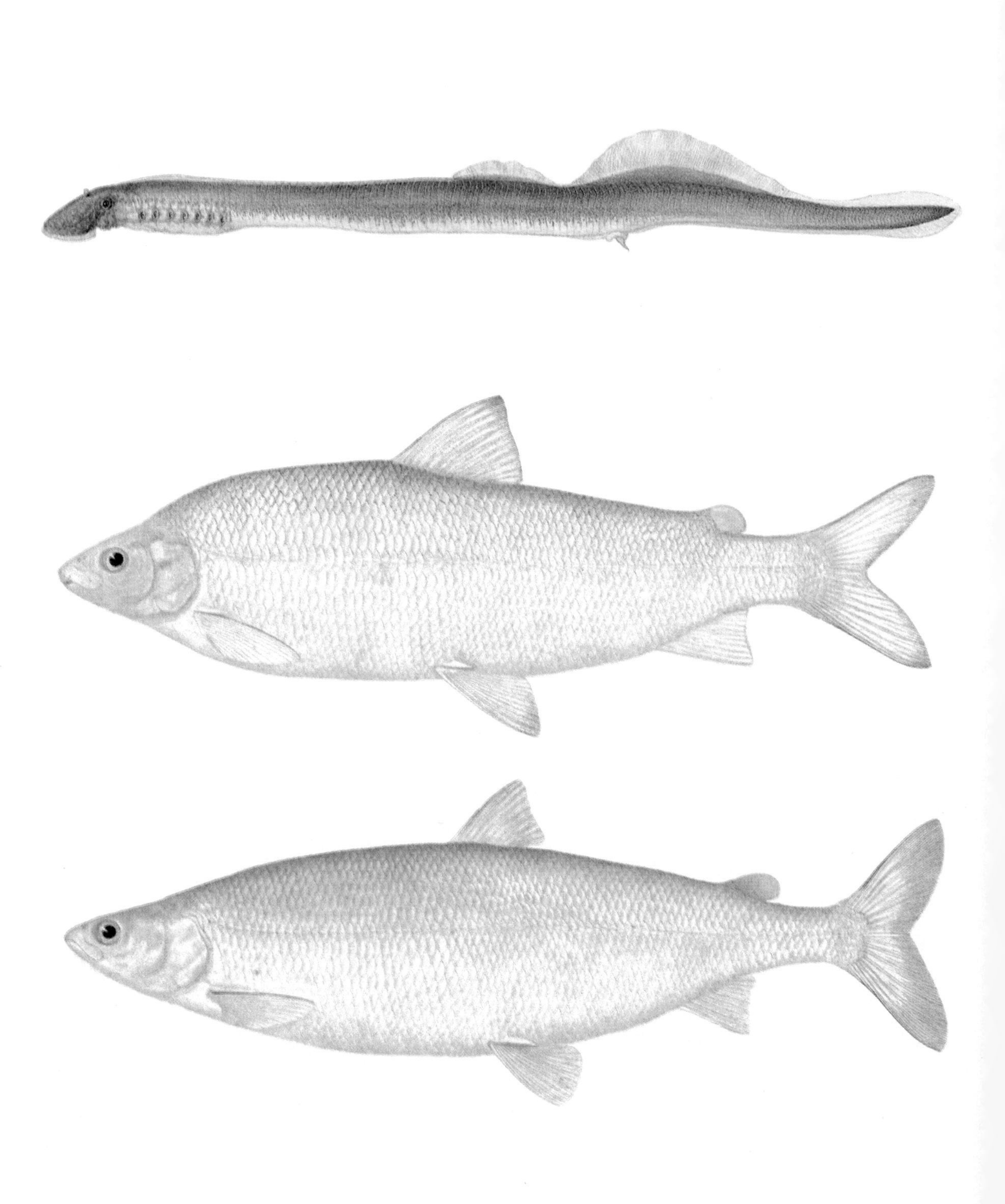

PLATE V

Fig. 14

LEAST LAMPREY

From specimen taken from a tributary of Clear Creek, Hocking County, O., on April 16, 1930 (*see* p. 160).

Fig. 15

LAKE WHITEFISH

From specimen taken near Kelleys Island, Erie County, O., on October 30, 1930 (*see* p. 236).

Fig. 16

LAKE ERIE CISCO

From specimen from same locality and date as above (*see* p. 232).

General Index

Pagination in boldface type indicates an important reference

Index to Common Fish Names

N. B. Pagination in bold type indicates an important reference.

Index to Scientific Fish Names

N. B. Pagination in bold type indicates an important reference.